P9-CCU-926

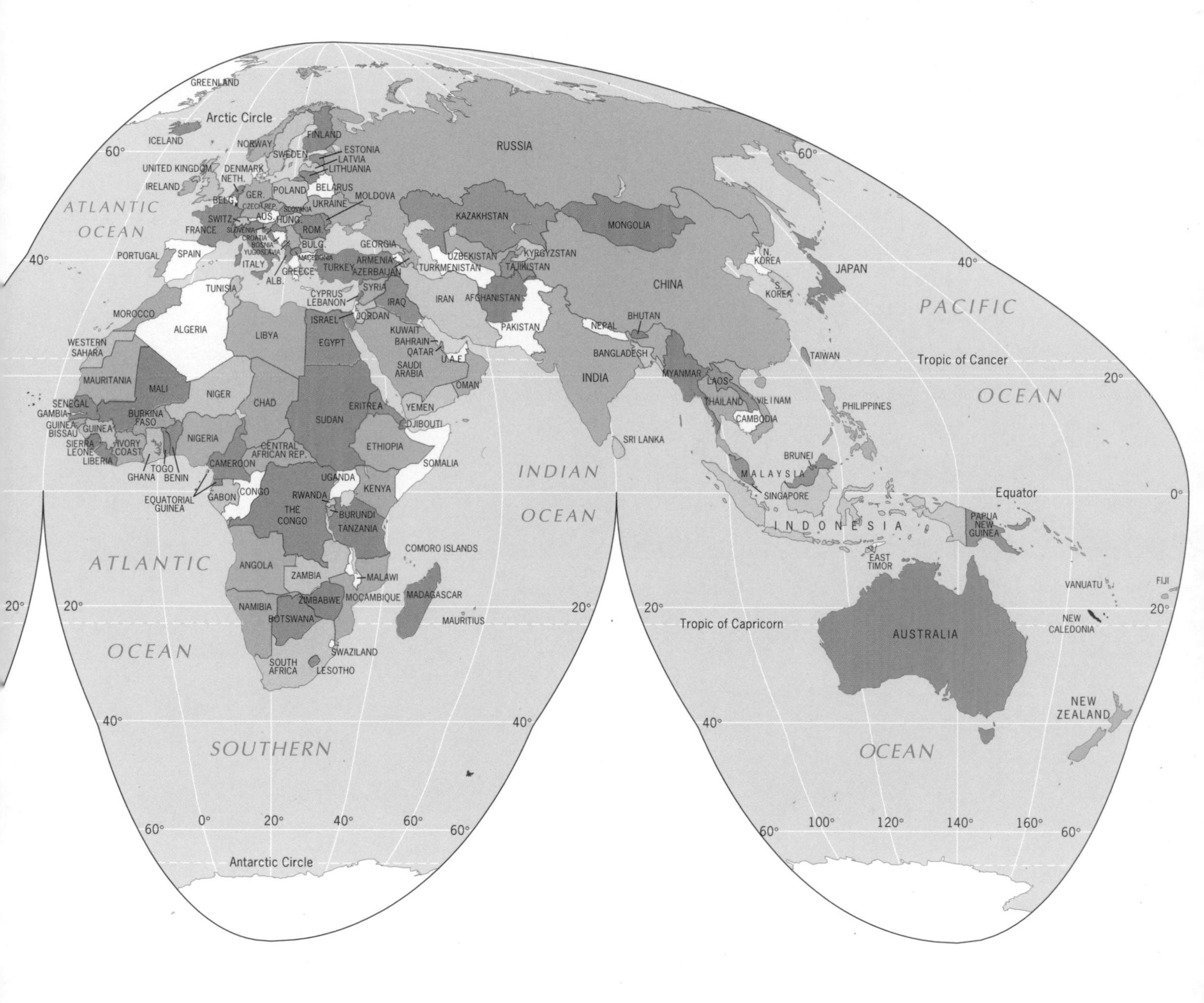
GREENLAND
Arctic Circle
ICELAND
NORWAY
FINLAND
SWEDEN
ESTONIA
LATVIA
LITHUANIA
RUSSIA
UNITED KINGDOM
DENMARK
NETH.
IRELAND
BELARUS
POLAND
GER.
BELG.
CZECH REP.
SLOVAKIA
UKRAINE
MOLDOVA
ATLANTIC
OCEAN
SWITZ.
AUS.
HUNG.
FRANCE
SLOVENIA
CROATIA
BOSNIA
YUGOSLAVIA
ROM.
BULG.
MACEDONIA
PORTUGAL
SPAIN
ITALY
ALB.
GREECE
TURKEY
GEORGIA
ARMENIA
AZERBAIJAN
KAZAKHSTAN
MONGOLIA
UZBEKISTAN
KYRGYZSTAN
TURKMENISTAN
TAJIKISTAN
N. KOREA
S. KOREA
JAPAN
CHINA
TUNISIA
CYPRUS
LEBANON
SYRIA
IRAQ
IRAN
AFGHANISTAN
MOROCCO
ALGERIA
LIBYA
EGYPT
ISRAEL
JORDAN
KUWAIT
BAHRAIN
QATAR
U.A.E.
SAUDI ARABIA
OMAN
PAKISTAN
NEPAL
BHUTAN
BANGLADESH
INDIA
MYANMAR
LAOS
THAILAND
VIETNAM
CAMBODIA
PHILIPPINES
TAIWAN
PACIFIC
OCEAN
Tropic of Cancer
WESTERN SAHARA
MAURITANIA
MALI
NIGER
CHAD
SUDAN
ERITREA
YEMEN
DJIBOUTI
SENEGAL
GAMBIA
GUINEA-BISSAU
GUINEA
BURKINA FASO
NIGERIA
SIERRA LEONE
LIBERIA
IVORY COAST
GHANA
TOGO
BENIN
CENTRAL AFRICAN REP.
CAMEROON
ETHIOPIA
SOMALIA
SRI LANKA
UGANDA
KENYA
EQUATORIAL GUINEA
GABON
CONGO
RWANDA
BURUNDI
THE CONGO
TANZANIA
INDIAN
OCEAN
BRUNEI
MALAYSIA
SINGAPORE
INDONESIA
PAPUA NEW GUINEA
Equator
COMORO ISLANDS
EAST TIMOR
ATLANTIC
ANGOLA
ZAMBIA
MALAWI
MADAGASCAR
MOÇAMBIQUE
NAMIBIA
ZIMBABWE
BOTSWANA
MAURITIUS
VANUATU
FIJI
Tropic of Capricorn
AUSTRALIA
NEW CALEDONIA
OCEAN
SOUTH AFRICA
SWAZILAND
LESOTHO
NEW ZEALAND
SOUTHERN
OCEAN
Antarctic Circle
60°
40°
20°
0°
20°
40°
60°
100°
120°
140°
160°

GEOGRAPHY

Realms, Regions, and Concepts

Tenth Edition

GEOGRAPHY

Realms, Regions, and Concepts

Tenth Edition

H. J. de Blij
Michigan State University

Peter O. Muller
University of Miami

The first map figure in each chapter that appears in this text comes from *Goode's World Atlas*, Rand McNally, R.L. 93-S-115, and is used with permission.

JOHN WILEY & SONS, INC.

New York ■ Chichester ■ Weinheim ■ Brisbane ■ Toronto ■ Singapore

ACQUISITIONS EDITOR: Ryan Flahive
MARKETING MANAGER: Clay Stone
SENIOR PRODUCTION EDITOR: Kelly Tavares
SENIOR DESIGNER: Karin Kincheloe
PHOTO EDITOR: Jennifer MacMillan
PHOTO RESEARCHER: Alexandra Truitt & Jerry Marshall
ILLUSTRATION EDITOR: Sigmund Malinowski
TEXT DESIGN: Lee Goldstein
COVER DESIGN: David Levy
FRONT/BACK COVER PHOTO: Keren Su/Stone
ELECTRONIC ILLUSTRATIONS: Mapping Specialists

This book was set in Times Roman by UG / GGS Information Services, Inc. and printed and bound by Von Hoffman Press.

Copyright © 2002 by John Wiley & Sons, Inc.

No part of this publication may be reproduced, stored in a retrieval system or transmitted in any form or by any means, electronic, mechanical, photocopying, recording, scanning or otherwise, except as permitted under Sections 107 or 108 of the 1976 United States Copyright Act, without either the prior written permission of the Publisher, or authorization through payment of the appropriate per-copy fee to the Copyright Clearance Center, 222 Rosewood Drive, Danvers, MA 01923, (508)750-8400, fax (508)750-4470. Requests to the Publisher for permission should be addressed to the Permissions Department, John Wiley & Sons, Inc., 605 Third Avenue, New York, NY 10158-0012, (212)850-6011, fax (212)850-6008, E-Mail:PERMREQ@WILEY.COM. To order books or for customer service please call 1(800)225-5945.

ISBN 0-471-40775-5

Printed in the United States of America

10 9 8 7 6 5 4 3 2 1

To

Arend D. Lubbers
President of Grand Valley State University
1969–2001 and
Advocate of Geography

Preface

For more than three decades, *Geography: Realms, Regions, and Concepts* has reported (and sometimes anticipated) trends in the discipline of Geography and developments in the world at large. In nine preceding editions, *Regions*, as the book is generally called, has explained the modern world's great geographic realms and their physical and human contents, and has introduced geography itself, the discipline that links the study of human societies and natural environments through a fascinating, spatial approach. From old ideas to new, from environmental determinism to expansion diffusion from decolonization to devolution, *Regions* has provided geographic perspective on our transforming world.

The book before you, therefore, is an information highway to geographic literacy. The first edition appeared in 1971, at a time when school geography in the United States (though not in Canada) was a subject in decline. It was a precursor of a dangerous isolationism in America, and geographers foresaw the looming cost of geographic illiteracy. Sure enough, the media during the 1980s began to report that polls, public surveys, tests, and other instruments were recording a lack of geographic knowledge at a time when our world was changing ever faster and becoming more competitive by the day. Various institutions, including the National Geographic Society, banks, airline companies, and a consortium of scholarly organizations mobilized to confront an educational dilemma that had resulted substantially from a neglect of the very topics this book is about.

Before we can usefully discuss such commonplace topics as our "shrinking world," our "global village," and our "distant linkages," we should know what the parts are, the components that do the shrinking and linking. This is not just an academic exercise. You will find that much of what you encounter in this book is of immediate, practical value to you—as a citizen, a consumer, a traveller, a voter, a jobseeker. North America is a geographic realm with intensifying global interests and involvements. Those interests and involvements require countless, often instantaneous decisions. Such decisions must be based on the best possible knowledge of the world beyond our continent. That knowledge can be gained by studying the layout of our world, its environments, societies, resources, policies, traditions, and other properties—in short, its regional geography.

Realms and Concepts

This book is organized into thirteen chapters. The Introduction discusses the world as a whole, outlining the physical stage on which the human drama is being played out, providing environmental information, demographic data, political background, and economic geographical context. Each of the remaining twelve chapters focuses on one of the world's major geographic realms.

Geographic concepts and ideas are placed in their regional settings in all 13 chapters. Most of these approximately 150 concepts are primarily geographical, but others are ideas about which, we believe, students of geography should have some knowledge. Although such concepts are listed on the opening page of every chapter, we have not, of course, enumerated every geographic notion used in that chapter. Many colleagues, we suspect, will want to make their own realm-concept associations, and as readers will readily perceive, the book's organization is quite flexible. It is possible, for example, to focus almost exclusively on substantive regional material, or, alternatively, to concentrate mainly on conceptual issues.

The Tenth Edition

Ever since this book was conceived, it has been a challenge to keep abreast of the rapid changes affecting the world. The decolonization of Africa and other areas, the growing economic power of the Pacific Rim, the devolution of the Soviet Union, the transition in South Africa, the collapse of Yugoslavia—these are just a few of the more dramatic changes that have occurred over the period of the life of this book. The unification effort in Europe, the resurrection of Christian churches in Russia and of Islamic forces in Central Asia, the pandemic of AIDS, and the gradual weakening of many nation-states are among other processes transforming the world's geography.

As readers of successive editions of this book know, we introduce significant conceptual as well as substantive changes with every edition, many of them based on your suggestions and recommendations. In the Ninth Edition, for example, to mark the arrival of our new century and millennium, we included a foldout map of the world as it was in 1900 together with a narrative that described the momentous geographic changes of the twentieth century just past. In the Eighth Edition we reformulated the developed-underdeveloped dichotomy in our economic-geographic framework and replaced it with a core-periphery model. In the Seventh Edition, we took note of the emerging Western Pacific Rim region and focused on its expanding internal and external linkages. In the Sixth Edition, we took the lead in reintroducing the regional term *Turkestan* to conceptualize the cultural reorientation of the former Soviet Central Asian republics.

This book always has been known for its current and accurate data, now including those for the year 2002 as well as projections beyond the beginning of the twenty-first century. Our population data table is derived from as many as a dozen sources, including our own observations in the field (see "Data Sources" below).

As with its predecessors, the Tenth Edition of *Regions* contains a large number of significant changes in the text, cartography, and other contents. In the **Introduction**, a new discussion in economic geography focuses on national debt, accompanied by new Table I-1. Another addition is a consideration of the various processes of globalization and their impacts on countries and peoples. The map of the world's countries (entitled States of the World, 2002) has been shifted forward to the front endpapers to maximize its accessibility. Our unmatched table of Area and Demographic Data for the World's States, now numbered Table I-2 and located on pp. 35–39 between the Introduction and Chapter 1, has been updated to 2002 and reflects social conditions and changes in all of the countries for which data are available (except microstates and dependencies); for the United States, this includes the first information from the 2000 Census.

Chapter 1 (Europe) begins with a new introduction that supports the notion of beginning a world regional geographic survey with this particular realm. One of the prominent geographic issues here is the expansion of the European Union, and major new text addresses this topic. Also noteworthy is the growing role of unified Germany in Europe, requiring a thorough revision of the text on its political organization and economic system. New text also focuses on Ireland, the rising economic *Celtic Tiger*. Devolution remains a salient geographic topic here, and new inset maps of the Iberian Peninsula's Basque region and Gibraltar support the discussion. Changes in the political geography of Eastern Europe, notably the former Yugoslavia, also are covered in text and cartography.

Chapter 2 (Russia) contains a new map and text on the Russian Federation's newly established functional regions, heralding a new phase in the relations between the regions and the center. The social geography of Russia is also changing dramatically, with a health crisis and economic hardships producing an actual decline in population, the topic of new text in this chapter. Additional text also concentrates on the problematic relationships between Russia and its neighbor Georgia, as well as the secessionist campaign in Chechnya.

Chapter 3 (North America) now includes a new map on the realm's indigenous peoples and accompanying text. New material deals with regional developments in the "New Economy" including the technopole phenomenon. A new box focuses on the burgeoning soybean industry, which is changing the agricultural landscape of the Midwest. The 2000 parliamentary election in Canada afforded us the opportunity to reconsider the prospect of Canadian devolution. The changing economic geography of Canada's Atlantic Provinces is discussed through an account of new offshore oil finds near Newfoundland and natural gas near Nova Scotia.

Chapter 4 (Middle America) has new text on Mexico, where another election in 2000 provided insights into the changing regional geography of this important country. One of Mexico's regional-geographic challenges, the Chiapas problem, is discussed in some detail. In the Central America segment, new text addresses current notions of "dry-canal" links between Pacific and Atlantic, the growth of Costa Rica's semiconductor industry and its geographic implications, developments in Panama's canal corridor following the takeover of the waterway, the aftermath of environmental crises, and the altitudinal zonation of agriculture. In the Caribbean region, new material covers important social and economic developments in the Dominican Republic and Trinidad and Tobago.

Chapter 5 (South America) contains revised text on the pressures disrupting life in Colombia, including the insurgent-state phenomenon and the impact of the cocaine industry not only on Colombia but also on its neighbors. Even as centrifugal forces are tearing at Colombia, the country's energy industry experiences a boom based on new discoveries, the topic of new text. Revised material also addresses the phenomenon of indigenous peoples' uprisings in the Andean region, a development with major implications for the future. Brazil, South America's giant, receives more attention on its booming interior *cerrado* and Amazonian growth poles. In the country's core area, we update its technopoles such as São Paulo's *Silicon Village*, and gauge the impact of Mercosur/Mercosul and other supranational organizations.

Chapter 6 (North Africa/Southwest Asia) includes a new map on Egypt and Sudan, with new text on Egypt's changing social geography, the Nile River, Aswan High Dam, and alternate farming areas to be opened by the planned Nile diversion. The regional fragmentation of Sudan is given better perspective. The text on the Middle East also is substantially revised, with a new inset map on the Golan Heights and other cartographic improvements. New text also addresses the regional geography of divided Kazakhstan.

Chapter 7 (Subsaharan Africa) includes new text on the powerful centrifugal forces in West Africa's cornerstone state affected by Islamic revivalism in its north. The continuing disintegration of The Congo also required extensive revision of the text, as did developments in other Equatorial African states. The recent (and current) boundary problems in the Horn of Africa are discussed as well. New text on South Africa focuses on the geographic dimensions of the country's constitution, the post-Mandela leadership prospects, and the impact of AIDS in geographic context.

Chapter 8 (South Asia) required extensive revision of text for Pakistan as well as India. New and revised maps support the new text on Pakistan; in the case of India, the proclamation of three new States required the revision of maps, text, and population tables. We had foreshadowed the prospect of a new Jharkhand State in our Ninth Edition, but the authorization of two additional States was unexpected and did not come to world attention until after we had been alerted to it by colleagues in the field.

Chapter 9 (East Asia) includes a new discussion of China's physical geography, clarifying a complex mosaic. Revised and shortened text focuses on key events in the Great Leap Forward, the formation of the commune system, and the ongoing ideological struggle as it relates to the present-day map. A new map covers the Xinjiang-Uyghur Autonomous Region and is accompanied by new text on this frontier's integration into China as well as its Muslim challenge. Our new material focuses on Mongolia and its environmental and social crises. We take note of China's takeover of Portuguese Macao (now the Macau SAR).

Chapter 10 (Southeast Asia) includes a new map of Malaysia showing the geographic dimensions of that country's Islamic revival; new text focuses on Malaysian modernization. During the years since the appearance of the Ninth Edition, Indonesia has undergone significant change, and this transition is covered in new material. This edition also focuses in greater detail on the troubled province of Irian Jaya in the context of Indonesia's struggle with devolutionary forces. A new inset map of East Timor is part of the revised map of Indonesia, and new text covers the emergence of this newest state in the realm.

Chapter 11 (Austral Realm) has a new introduction on Australia's place in the world (and on the Pacific Rim). Revised text deals with several of Australia's internal problems, notably the persistent Aboriginal disadvantage reflected by social indicators. The immigration question is discussed from a new perspective. New discussion focuses on Australia's relationships with nearby Indonesia and East Timor.

Chapter 12 (Pacific Realm) now includes a discussion of the regional problems affecting the Solomon Islands, a lingering aftermath of World War II; new text also addresses the geographic implications of the political-ethnic crisis in Fiji. Since this chapter deals in part with maritime boundaries, the maritime consequences of East Timor's independence and a new EEZ delimitation are mentioned even though the final UN decision on this was still pending at press time.

This summary can only cover the highlights of the revisions that went into the Tenth Edition. No page, no paragraph, and no map escaped our scrutiny.

Data Sources

For all matters geographical, of course, we consult *The Annals of the Association of American Geographers, The Professional Geographer, The Geographical Review, The Journal of Geography,* and many other academic journals published regularly in North America—plus an array of similar periodicals published in English-speaking countries from Scotland to New Zealand.

As with every new edition of this book, all quantitative information was updated to the year of publication and checked rigorously. In addition to the major revisions described above, hundreds of other modifications were made, many in response to readers' and reviewers' comments. Some readers found our habit of reporting urban population data within the text disruptive, so we continue to tabulate these at the beginning of the "Regions of the Realm" section of each chapter. The stream of new spellings of geographic names continues, and we pride ourselves in being a reliable source for current and correct usage.

The statistical data that constitute Table I-2 (pp. 35–39) are derived from numerous sources. As users of such data are aware, considerable inconsistency marks the reportage by various agencies, and it is often necessary to make informed decisions on contradictory information. For example, some sources still do not reflect the rapidly declining rates of population increase or life expectancies in AIDS-stricken African countries. Others list demographic averages without accounting for differences between males and females in this regard.

In formulating Table I-2 we have used among our sources the United Nations, the Population Reference Bureau, the World Bank, the Encyclopaedia Britannica *Books of the Year*, the *Economist* Intelligence Unit, the *Statesman's Year-Book*, and the *The New York Times Almanac*.

The urban population figures—which also entail major problems of reliability and comparability—are mainly drawn from the most recent database published by the United Nations' Population Division. For cities of less than 750,000, we developed our own estimates from a variety of other sources. At any rate, the urban population figures used here are estimates for 2002 and they represent *metropolitan-area totals* unless otherwise specified.

Cartography

This newest version of *Regions* continues the innovation begun in the Seventh Edition, when atlas-style maps from the most recently available edition of *Goode's World Atlas* (currently the 20th, published in 2000) were first used as opening maps for each chapter. In the Eighth Edition, two maps were specifically drawn in the Rand McNally style to serve as matching openers: those of North Africa/Southwest Asia and the Pacific Realm. The South Asia map was substantially expanded from its *Goode's* base.

Readers of earlier editions of this book will note that our tradition of updating, enhancing, and improving our own thematic cartography continues. In this Tenth Edition of *Regions*, the following map figures are new:

- 2–8 Russia: New Federal Districts . . . 2000
- 3–6 North America: Indigenous Domains
- 6–10 Egypt and Sudan
- 7–11 The States of Federal Nigeria
- 9–16 Xinjiang-Uyghur Autonomous Region (China)
- 10–10 States of West Malaysia

The following maps were significantly redrawn:

- I–10 Cartogram of the World's National Populations, 2002
- I–11 States and Economies of the World
- 1–10 Europe: Foci of Devolutionary Pressures, 2002
- 1–11 European Supranationalism
- 1–21 Autonomous Communities of Spain
- 1–25 Yugoslavia and Its Neighbors
- 2–12 Oil and Gas Regions of Russia
- 4–7 Cuba
- 4–11 Maquiladora Employment in the U.S.-Mexico Border Zone, 2000
- 4–13 Altitudinal Zonation
- 5–10 Colombia
- 6–8 North Africa/Southwest Asia: Oil and Natural Gas
- 6–14 Israel in the Middle East
- 6–15 The West Bank
- 8–7 Pakistan
- 8–8 Partition of Jammu and Kashmir
- 8–9 States of Modern India
- 9–15 China's Western Flank
- 10–12 Indonesia

Users of this book should note that the spelling of some names on these thematic maps does not always match that on the *Goode's World Atlas* maps. This is not unusual; you will even find inconsistencies among various atlases. Almost invariably, we have followed the very latest standards set by the United States Board of Geographic Names.

Photography

The map undoubtedly is geography's closest ally, but there are times when photography is not far behind. Whether from space, from an aircraft, from the tallest building in town, or on the ground, a photograph can, indeed, be worth a thousand words. When geographers perform field research in some area of the world, they are likely to maintain a written record that correlates with the photographic one.

This Tenth Edition revision is not confined to the text and maps. As readers of *Regions* will note, the illustrations program was thoroughly overhauled with numerous new photographs, many by the authors, new accompanying "Field Notes" where appropriate, and new, detailed captions.

Pedagogy

We continue to devise ways to help students learn important geographic concepts and ideas, and to make sense of our complex and rapidly changing world. Continuing special features include the following:

Atlas Maps. As in previous editions, a comprehensive map of the region opens each chapter. The maps are reproduced from the 20th revised edition (2000) of *Goode's World Atlas* (the maps for Chapters 6 and 12 have been created in the Atlas style). In the Tenth Edition, each of these maps is assigned the first figure number in each chapter, which better facilitates the integration of this cartographic material into the text.

Concepts, Ideas, and Terms. Each chapter begins with a boxed sequential listing of the key geographic concepts, ideas, and terms that appear in the pages that follow. These are noted by numbers in the margins that correspond to the introduction of each item in the text.

Two-Part Chapter Organization. To help the reader to logically organize the material within chapters, we have broken the regional chapters into two distinct parts: first, "Defining the Realm" includes the general physiographic, historical, and human-geographic background common to the realm, and the second section, "Regions of the Realm," presents each of the distinctive regions within the realm (denoted by the symbol).

List of Regions. Also on the chapter-opening page, a list of the regions within the particular realm provides a preview and helps to organize the chapter. For ease of identification, the triangular symbol (shown 4 lines above) that denotes the regions list here also appears beside each region heading in the chapter.

Major Geographic Qualities. Near the beginning of each realm chapter, we list, in boxed format, the major geographic qualities that best summarize that portion of the Earth's surface.

Sidebar Boxes. Special topical and sometimes controversial issues are highlighted in boxed sections. These boxes allow us to include interesting and current topics without interrupting the flow of material within chapters.

Among the Realm's Great Cities (Boxes). This feature reflects the growing process and influence of urbanization worldwide. More than thirty profiles of the world's leading cities are presented, each accompanied by specially drawn maps.

Major Cities of the Realm Population Tables. Near the beginning of every "Regions" section of each chapter, we have included a table reporting the most up-to-date urban population data (based on 2002 estimates drawn from the sources listed above). Readers should find this format less disruptive than citation of the population when the city is mentioned in the text.

From the Field Notes. In the Eighth Edition we introduced a new feature that has proven effective in some of our other textbooks. Many of the photographs in this book were taken by the senior author while doing fieldwork. The more extensive captions, *From the Field Notes*, provide valuable insights into how a geographer observes and interprets information in the field.

Appendices, References, and Glossary. At the end of the book, the reader will find five sections that enrich and/or supplement the main text: (1) *Appendix A*, a guide to Map Reading and Interpretation; (2) *Appendix B*, an overview of Career Opportunities in Geography; (3) *Appendix C*, a Pronunciation Guide; (4) a detailed Bibliography (*References and Futher Readings*) that introduces the wide-ranging literature of the discipline and World Regional Geography; and (5) an extensive *Glossary*. The general index follows. A geographical index or *gazetteer* of the place names contained in our maps now appears in the book's Web site.

Ancillaries

A broad spectrum of print and electronic ancillaries are available to accompany the Tenth Edition of *Regions*. Additional information, including prices and ISBNs for ordering, can be obtained by contacting John Wiley & Sons.

FOR SALE TO THE STUDENT

Student Study Guide. Text co-author Peter O. Muller and his geographer daughter, Elizabeth Muller Hames, have written a popular Study Guide to accompany the book that is packed with useful study and review tools. For each chapter in the textbook, the Study Guide gives students and faculty access to chapter objectives, content questions-and-answers, outline maps of each realm, sample tests, and more.

***Goode's Atlas* from Rand McNally.** With the Tenth Edition of *Regions*, we are delighted to be able to continue offering the *Goode's Atlas* at a deeply-discounted price when shrink-wrapped with the text. Economies of scale allow us to provide this at a net price that is close to our cost. Our partnership with Rand McNally and the widely-popular *Goode's Atlas* is an arrangement that is exclusive to John Wiley & Sons.

Microsoft® Encarta® Interactive Atlas CD-ROM. This award-winning atlas CD-ROM will captivate the imaginations of your students and engage them in a spatial adventure, all the while exposing them to an abundance of resources appropriate for university-level geography. Our arrangement with Microsoft enables us to offer the Encarta® Interactive Atlas at a cost that is less than one-third the suggested retail price when shrink-wrapped with our text.

Take Note!. For a nominal cost, the *Take Note!* can be shrink-wrapped with the textbook. This useful tool includes a black-and-white version of the illustrations from the text. Students can take notes directly on the figure in the *Take Note!* ancillary while the instructor discusses the projected map or illustration, or as they study for the course.

Annenberg/CPB. Power of Place: World Regional Geography **Study Guide, Third Edition**. The Third Edition of the *Power of Place: World Regional Geography Study Guide* updates content and references so that the Annenberg/Power of Place Telecourse and Video Series connects to the Tenth Edition of de Blij and Muller's *Regions*. It was written by Gil Latz, Portland State University.

THE *REGIONS* WEB SITE: ON-LINE RESOURCES FOR INSTRUCTORS AND STUDENTS

PowerPoint Slides. Available for the Tenth Edition, these electronic files outline the main concepts of each chapter in *Regions* in a highly visual manner. These presentations are available on the Instructor's Web Site and the Resource CD-ROM, and can be uploaded to presentation programs such as PowerPoint, or to any popular word processing program.

Instructor's Manual. Distributed on-line to instructors via a secure, password-protected Instructor's Web Site, the *Instructor's Manual* by Wendy Shaw, Southern Illinois University, Edwardsville, provides outlines, descriptions, and key terms to help professors organize the concepts in the book for classroom use.

Test Bank. Prepared by long-term Test Bank author Ira Sheskin, University of Miami, the *Test Bank* for the Tenth

Edition of *Regions* contains over 3,000 test items including multiple-choice, fill-in, matching, and essay questions. It is distributed via the secure Instructor's Web Site as electronic files, which can be saved into all major word processing programs.

Computerized Test Bank. An easy to use program that can be used to create and customize exams.

Student Web Site. This comprehensive on-line resource will contain chapter-based self-quizzes and extensive links to Web material providing real-world examples and additional research tools.

Course Management. On-line course management assets are available to accompany the Tenth Edition of *Regions.*

OTHER RESOURCES FOR THE CLASSROOM

Overhead Transparencies and Slides. The book's maps and diagrams are available in their entirety for either transparency or slide projection in beautifully rendered, 4-color format.

***Regions* Resource CD-ROM.** This rich resource contains animations, videos, PowerPoint presentations, the Test Bank, and the Instructor's Manual. Organized by chapter, the *Regions* Resource CD-ROM has a tested, intuitive interface that allows for easy file management and presentation building. If the instructor prefers to use programs such as PowerPoint in the classroom, the text, map, and photo files can be uploaded easily from the Resource CD-ROM into your presentation program.

Regions is featured as the companion text to *The Power of Place—World Regional Geography*, a PBS television course and video resource produced in collaboration with the Annenberg/CPB Project. *The Power of Place* is a series of twenty-six half-hour video programs organized around the *Regions* text. Each program contains documentary-style case studies that focus on one of eleven geographic realms, and features on-screen commentary by H. J. de Blij. Videocassettes can be purchased individually or as a thirteen-tape set. A *Study Guide* and *Faculty Guide* are also available to supplement the programs. For information regarding the use of *The Power of Place* as a television course, contact the PBS Adult Learning Service at 1-800-257-2578. To purchase videocassettes for institutional or classroom use, contact The Annenberg/CPB Multimedia Collection at 1-800-LEARNER.

ACKNOWLEDGMENTS

Over the more than three decades since the publication of the First Edition of *Geography: Realms, Regions, and Concepts*, we have been fortunate to receive advice and assistance from literally hundreds of people. One of the rewards associated with the publication of a book of this kind is the steady stream of correspondence and other feedback it generates. Geographers, economists, political scientists, education specialists, and others have written us, often with fascinating enclosures. We make it a point to respond personally to every such letter, and our editors have communicated with many of our correspondents as well. Moreover, we have considered every suggestion made—and many who wrote or transmitted their reactions through other channels will see their recommendations in print in this edition.

STUDENT RESPONSE

A good part of the correspondence we receive comes from student readers. On this occasion, we would like to extend our deep appreciation to the several million students around the world who have studied from the first nine editions of our text. In particular, we thank the students from more than 100 different colleges across the United States who took the time to send us their opinions.

Students told us they found the maps and graphics attractive and functional. We have not only enhanced the map program with exhaustive updating, but have added 6 new maps to this Tenth Edition as well as making major changes in an additional 19 maps.

Several students also commented that the chapters were long and sometimes wordy. We now have a shorter Introduction, and the ensuing chapters on the world's 12 geographic realms are divided into two parts: the first part covers the realm's general physiography, history, and human geography; the second part is devoted to the various regions within the realm. To add interest for today's student, we have refined the relatively new feature that highlights the Great Cities of each realm from the point of view of a contemporary visitor. We have also enlivened the overall layout with a fresh new design.

Generally, students have told us that they found the pedagogical devices quite useful. We have kept the study aids the students cited as effective: a boxed list of each chapter's key concepts, ideas, and terms (now numbered for quick reference in both the box and text margins); a box summarizing each realm's major geographic qualities; a pronunciation guide in Appendix C; and an extensive Glossary.

FACULTY FEEDBACK

Faculty members from a large number of North American colleges and universities continue to supply us with vital feedback and much-appreciated advice. Our publishers commissioned a number of reviews, and we are most grateful to the following professors for showing us where the written text could be strengthened and made more precise:

RANDY BERTOLAS, Wayne State College
JONATHAN C. COMER, Oklahoma State University
MICHAEL CORNEBISE, University of Tennessee
FIONA M. DAVIDSON, University of Arkansas
MEL DROUBAY, University of West Florida
DAVID J. KEELING, Western Kentucky University
MOHAMEDEN OULD-MEY, Indiana State University
THOMAS W. PARADIS, Northern Arizona University
JAMES W. PENN, JR., University of Florida
JOHN D. REILLY, University of Florida
THOMAS C. SCHAFER, Fort Hays State University

In addition, several faculty colleagues from around the world assisted us with earlier editions, and their contributions continue to grace the pages of this book. Among the them are:

JAMES P. ALLEN, California State University, Northridge
STEPHEN S. BIRDSALL, University of North Carolina
J. DOUGLAS EYRE, University of North Carolina
FANG YONG-MING, Shanghai, China
EDWARD J. FERNALD, Florida State University
RAY HENKEL, Arizona State University
RICHARD C. JONES, University of Texas at San Antonio
GIL LATZ, Portland State University (Oregon)
IAN MACLACHLAN, University of Lethbridge (Alberta)
MELINDA S. MEADE, University of North Carolina
HENRY N. MICHAEL, Temple University (Pennsylvania)
CLIFTON W. PANNELL, University of Georgia
J. R. VICTOR PRESCOTT, University of Melbourne (Victoria)
JOHN D. STEPHENS, University of Washington
CANUTE VANDER MEER, University of Vermont

We also received input from a much wider circle of academic geographers. The list that follows is merely representative of a group of colleagues across North America to whom we are grateful for taking the time to share their thoughts and opinions with us:

MEL AAMODT, California State University-Stanislaus
R. GABRYS ALEXSON, University of Wisconsin-Superior
NIGEL ALLAN, University of California-Davis
JAMES P. ALLEN, California State University, Northridge
JOHN L. ALLEN, University of Connecticut
JERRY R. ASCHERMANN, Missouri Western State College
JOSEPH M. ASHLEY, Montana State University
THEODORE P. AUFDEMBERGE, Concordia College (Michigan)
EDWARD BABIN, University of South Carolina-Spartanburg
MARVIN W. BAKER, University of Oklahoma
THOMAS F. BAUCOM, Jacksonville State University (Alabama)
GOURI BANERJEE, Boston University (Massachusetts)
J. HENRY BARTON, Thiel College (Pennsylvania)
STEVEN BASS, Paradise Valley Community College (Arizona)
KLAUS J. BAYR, University of New Hampshire-Manchester
JAMES BELL, Linn Benton Community College (Oregon)
WILLIAM H. BERENTSEN, University of Connecticut
ROYAL BERGLEE, Indiana State University
RIVA BERLEANT-SCHILLER, University of Connecticut
THOMAS BITNER, University of Wisconsin
WARREN BLAND, California State University-Northridge
DAVIS BLEVINS, Huntington College (Alabama)
S. BO JUNG, Bellevue College (Nebraska)
MARTHA BONTE, Clinton Community College (Idaho)
GEORGE R. BOTJER, University of Tampa (Florida)
R. LYNN BRADLEY, Belleville Area College (Illinois)
KEN BREHOB, Elmhurst, Illinois
JAMES A. BREY, University of Wisconsin-Fox Valley
ROBERT BRINSON, Santa Fe Community College (Florida)
REUBEN H. BROOKS, Tennessee State University
LARRY BROWN, Ohio State University
LAWRENCE A. BROWN, Troy State-Dothan (Alabama)
ROBERT N. BROWN, Delta State University (Mississippi)
STANLEY D. BRUNN, University of Kentucky
RANDALL L. BUCHMAN, Defiance College (Ohio)
DIANN CASTEEL, Tusculum College (Tennessee)

JOHN E. COFFMAN, University of Houston (Texas)
DAWYNE COLE, Grand Rapids Baptist College (Michigan)
JONATHAN C. COMER, Oklahoma State University
BARBARA CONNELLY, Westchester Community College (New York)
WILLIS M. CONOVER, University of Scranton (Pennsylvania)
OMAR CONRAD, Maple Woods Community College (Missouri)
BARBARA CRAGG, Aquinas College (Michigan)
GEORGES G. CRAVINS, University of Wisconsin
ELLEN K. CROMLEY, University of Connecticut
JOHN A. CROSS, University of Wisconsin-Oshkosh
WILLIAM CURRAN, South Suburban (Illinois)
ARMANDO DA SILVA, Towson State University (Maryland)
DAVID D. DANIELS, Central Missouri State University
RUDOLPH L. DANIELS, Morningside College (Iowa)
SATISH K. DAVGUN, Bemidji State University (Minnesota)
JAMES DAVIS, Illinois College
JAMES L. DAVIS, Western Kentucky University
KEITH DEBBAGE, University of North Carolina-Greensboro
MOLLY DEBYSINGH, California State University, Long Beach
DENNIS K. DEDRICK, Georgetown College (Kentucky)
STANFORD DEMARS, Rhode Island College
THOMAS DIMICELLI, William Paterson College (New Jersey)
D.F. DOEPPERS, University of Wisconsin-Madison
ANN DOOLEN, Lincoln College (Illinois)
STEVEN DRIEVER, University of Missouri-Kansas City
WILLIAM ROBERT DRUEN, Western Kentucky University
ALASDAIR DRYSDALE, University of New Hampshire
KEITH A. DUCOTE, Cabrillo Community College (California)
WALTER N. DUFFET, University of Arizona
CHRISTINA DUNPHY, Champlain College (Vermont)
ANTHONY DZIK, Shawnee State University (Kansas)
DENNIS EDGELL, Firelands BGSU (Ohio)
JAMES H. EDMONSON, Union University (Tennessee)
M.H. EDNEY, State University of New York-Binghamton
HAROLD M. ELLIOTT, Weber State University (Utah)
JAMES ELSNES, Western State College
DINO FIABANE, Community College of Philadelphia (Pennsylvania)
G.A. FINCHUM, Milligan College (Tennessee)
IRA FOGEL, Foothill College (California)
ROBERT G. FOOTE, Wayne State College (Nebraska)
G.S. FREEDOM, McNeese State University (Louisiana)
RONALD FORESTA, University of Tennessee
OWEN FURUSETH, University of North Carolina-Charlotte
RICHARD FUSCH, Ohio Wesleyan University
GARY GAILE, University of Colorado-Boulder
EVELYN GALLEGOS, Eastern Michigan University & Schoolcraft College
JERRY GERLACH, Winona State University (Minnesota)
LORNE E. GLAIM, Pacific Union College (California)
SHARLEEN GONZALEZ, Baker College (Michigan)
DANIEL B. GOOD, Georgia Southern University
GARY C. GOODWIN, Suffolk Community College (New York)
S. GOPAL, Boston University (Massachusetts)
ROBERT GOULD, Morehead State University (Kentucky)
GORDON GRANT, Texas A&M University
DONALD GREEN, Baylor University (Texas)
GARY M. GREEN, University of North Alabama
MARK GREER, Laramie County Community College (Wyoming)
STANLEY C. GREEN, Laredo State University (Texas)
W. GREGORY HAGER, Northwestern Connecticut Community College
RUTH F. HALE, University of Wisconsin-River Falls
JOHN W. HALL, Louisiana State University-Shreveport
PETER L. HALVORSON, University of Connecticut
MERVIN HANSON, Willmar Community College (Minnesota)
ROBERT J. HARTIG, Fort Valley State College (Georgia)
JAMES G. HEIDT, University of Wisconsin Center-Sheboygan
CATHERINE HELGELAND, University of Wisconsin-Manitowoc
NORMA HENDRIX, East Arkansas Community College
JAMES HERTZLER, Goshen College (Indiana)
JOHN HICKEY, Inver Hills Community College (Minnesota)
THOMAS HIGGINS, San Jacinto College (Texas)
EUGENE HILL, Westminster College (Missouri)
LOUISE HILL, University of South Carolina-Spartanburg
MIRIAM HELEN HILL, Indiana University Southeast
SUZY HILL, University of South Carolina-Spartanburg
ROBERT HILT, Pittsburg State University (Kansas)
SOPHIA HINSHALWOOD, Montclair State University (New Jersey)

PRISCILLA HOLLAND, University of North Alabama
ROBERT K. HOLZ, University of Texas-Austin
R. HOSTETLER, Fresno City College (California)
LLOYD E. HUDMAN, Brigham Young University (Utah)
JANIS W. HUMBLE, University of Kentucky
WILLIAM IMPERATORE, Appalachian State University (North Carolina)
RICHARD JACKSON, Brigham Young University (Utah)
MARY JACOB, Mount Holyoke College (Massachusetts)
GREGORY JEANE, Samford University (Alabama)
SCOTT JEFFREY, Catonsville Community College (Maryland)
JERZY JEMIOLO, Ball State University (Indiana)
SHARON JOHNSON, Marymount College (New York)
SARA MAYFIELD, San Jacinto College, Central (California)
DAVID JOHNSON, University of Southwestern Louisiana
JEFFREY JONES, University of Kentucky
MARCUS E. JONES, Claflin College (South Carolina)
MOHAMMAD S. KAMIAR, Florida Community College, Jacksonville
MATTI E. KAUPS, University of Minnesota-Duluth
COLLEEN KEEN, Gustavus Adolphus College (Minnesota)
GORDON F. KELLS, Mott Community College
SUSANNE KIBLER-HACKER, Unity College (Maine)
JAMES W. KING, University of Utah
JOHN C. KINWORTHY, Concordia College (Nebraska)
ALBERT KITCHEN, Paine College
TED KLIMASEWSKI, Jacksonville State University (Alabama)
ROBERT D. KLINGENSMITH, Ohio State University-Newark
LAWRENCE M. KNOPP, JR., University of Minnesota-Duluth
TERRILL J. KRAMER, University of Nevada
ARTHUR J. KRIM, Cambridge, Massachusetts
ELROY LANG, El Camino Community College (California)
CHRISTOPHER LANT, Southern Illinois University-Carbondale
A.J. LARSON, University of Illinois-Chicago
LARRY LEAGUE, Dickinson State University (North Dakota)
DAVID R. LEE, Florida Atlantic University
JOE LEEPER, Humboldt State University (California)
YECHIEL M. LEHAVY, Atlantic Community College (New Jersey)
JOHN C. LEWIS, Northeast Louisiana University
CAEDMON S. LIBURD, University of Alaska-Anchorage
T. LIGIBEL, Eastern Michigan University
Z.L. LIPCHINSKY, Berea College (Kentucky)
ALLAN L. LIPPERT, Manatee Community College (Florida)
JOHN H. LITCHER, Wake Forest University (North Carolina)
LI LIU, Stephen F. Austin State University (Texas)
WILLIAM R. LIVINGSTON, Baker College (Michigan)
CYNTHIA LONGSTREET, Ohio State University
TOM LOVE, Linfield College (Oregon)
K.J. LOWREY, Miami University (Ohio)
ROBIN R. LYONS, University of Hawai'i-Leeward Community College
SUSAN M. MACEY, Southwest Texas State University
CHRISTIANE MAINZER, Oxnard College (California)
HARLEY I. MANNER, University of Guam
JAMES T. MARKLEY, Lord Fairfax Community College (Virginia)
SISTER MAY LENORE MARTIN, Saint Mary College (Kansas)
GARY MANSON, Michigan State University
KENT MATHEWSON, Louisiana State University
DICK MAYER, Maui Community College (Hawai'i)
DEAN R. MAYHEW, Maine Maritime Academy
J.P. MCFADDEN, Orange Coast College (California)
BERNARD MCGONIGLE, Community College of Philadelphia (Pennsylvania)
PAUL D. MEARTZ, Mayville State University (North Dakota)
DALTON W. MILLER, JR., Mississippi State University
RAOUL MILLER, University of Minnesota, Duluth
INES MIYARES, Hunter College, CUNY (New York)
BOB MONAHAN, Western Carolina University
KEITH MONTGOMERY, University of Wisconsin-Marathon
JOHN MORTON, Benedict College (South Carolina)
ANNE MOSHER, Syracuse University (New York)
BARRY MOWELL, Broward Community College (Florida)
ROBERT R. MYERS, West Georgia College
YASER M. NAJJAR, Framingham State College (Massachusetts)
JEFFREY W. NEFF, Western Carolina University
DAVID NEMETH, University of Toledo (Ohio)
RAYMOND O'BRIEN, Bucks County Community College (Pennsylvania)

JOHN ODLAND, Indiana University
JOSEPH R. OPPONG, University of North Texas
RICHARD OUTWATER, California State University, Long Beach
PATRICK O'SULLIVAN, Florida State University
BIMAL K. PAUL, Kansas State University
JAMES PENN, Southeastern Louisiana University
PAUL PHILLIPS, Fort Hays State University (Kansas)
MICHAEL PHOENIX, ESRI (California)
JERRY PITZL, Macalester College (Minnesota)
BILLIE E. POOL, Holmes Community College (Mississippi)
VINTON M. PRINCE, Wilmington College (North Carolina)
RHONDA REAGAN, Blinn College (Texas)
DANNY I. REAMS, Southeast Community College (Nebraska)
JIM RECK, Golden West College (California)
ROGER REEDE, Southwest State University (Minnesota)
JOHN RESSLER, Central Washington University
JOHN B. RICHARDS, Southern Oregon State College
DAVID C. RICHARDSON, Evangel College (Missouri)
SUSAN ROBERTS, University of Kentucky
WOLF RODER, University of Cincinnati
JAMES ROGERS, University of Central Oklahoma
PAUL A. ROLLINSON, AICP, Southwest Missouri State University
JAMES C. ROSE, Tompkins/Cortland Community College (New York)
THOMAS E. ROSS, Pembroke State University (North Carolina)
THOMAS A. RUMNEY, State University of New York-Plattsburgh
GEORGE H. RUSSELL, University of Connecticut
RAJAGOPAL RYALI, Auburn University at Montgomery (Alabama)
PERRY RYAN, Mott Community College
ADENA SCHUTZBERG, Middlesex Community College (Massachusetts)
SIDNEY R. SHERTER, Long Island University (New York)
NANDA SHRESTHA, Florida A&M University
WILLIAM R. SIDDALL, Kansas State University
DAVID SILVA, Bee County College (Texas)
DEBRA STRAUSSFOGEL, University of New Hampshire
MORRIS SIMON, Stillman College (Alabama)
KENN E. SINCLAIR, Holyoke Community College (Massachusetts)
ROBERT SINCLAIR, Wayne State University (Michigan)
EVERETT G. SMITH, JR., University of Oregon
RICHARD V. SMITH, Miami University (Ohio)
CAROLYN D. SPATTA, California State University-Hayward
M.R. SPONBERG, Laredo Junior College (Texas)
DONALD L. STAHL, Towson State University (Maryland)
ELAINE STEINBERG, Central Florida Community College
D.J. STEPHENSON, Ohio University Eastern
HERSCHEL STERN, Mira Costa College (California)
REED F. STEWART, Bridgewater State College (Massachusetts)
NOEL L. STIRRAT, College of Lake County (Illinois)
GEORGE STOOPS, Mankato State University (Minnesota)
JOSEPH P. STOLTMAN, Western Michigan University
PHILIP SUCKLING, University of Northern Iowa
CHRISTOPHER SUTTON, Western Illinois University
T. L. TARLOS, Orange Coast College (California)
MICHAEL THEDE, North Iowa Area Community College
DERRICK J. THOM, Utah State University
CURTIS THOMSON, University of Idaho
S. TOOPS, Miami University (Ohio)
ROGER T. TRINDELL, Mansfield University of Pennsylvania
DAN TURBEVILLE, East Oregon State College
NORMAN TYLER, Eastern Michigan University
GEORGE VAN OTTEN, Northern Arizona University
C.S. VERMA, Weber State College (Utah)
GRAHAM T. WALKER, Metropolitan State College of Denver
DEBORAH WALLIN, Skagit Valley College (Washington)
MIKE WALTERS, Henderson Community College (Kentucky)
J.L. WATKINS, Midwestern State University (Texas)
P. GARY WHITE, Western Carolina University (North Carolina)
W.R. WHITE, Western Oregon University
GARY WHITTON, Fairbanks, Alaska
GENE C. WILKEN, Colorado State University
STEPHEN A. WILLIAMS, Methodist College
P. WILLIAMS, Baldwin-Wallace College
MORTON D. WINSBERG, Florida State University
ROGER WINSOR, Appalachian State University (North Carolina)
WILLIAM A. WITHINGTON, University of Kentucky
A. WOLF, Appalachian State University, N.C.)
JOSEPH WOOD, University of Southern Maine
RICHARD WOOD, Seminole Junior College (Florida)
GEORGE I. WOODALL, Winthrop College (North Carolina)

STEPHEN E. WRIGHT, James Madison University (Virginia)

LEON YACHER, Southern Connecticut State University

DONALD J. ZEIGLER, Old Dominion University (Virginia)

In assembling the Tenth Edition, we are indebted to the following people for advising us on a number of matters:

THOMAS L. BELL, University of Tennessee

KATHLEEN BRADEN, Seattle Pacific University

JESUS CAÑAS, Research Deparment, Federal Reserve Bank of Dallas, El Paso Branch

STUART E. CORBRIDGE, University of Miami

WILLIAM V. DAVIDSON, Louisiana State University

JAMES D. FITZSIMMONS, U.S. Bureau of the Census (D.C.)

GARY A. FULLER, University of Hawai'i

RICHARD J. GRANT, University of Miami

MARGARET M. GRIPSHOVER, Marshall University (West Virginia)

TRUMAN A. HARTSHORN, Georgia State University

PHILIP L. KEATING, Indiana University

DAVID LEY, University of British Columbia

RICHARD LISICHENKO, Fort Hays State University (Kansas)

GLEN M. MACDONALD, University of California, Los Angeles

IAN MACLACHLAN, University of Lethbridge (Alberta)

DALTON MILLER, Mississippi State University

ANNE MOSHER, Syracuse University

VALIANT C. NORMAN, Lexington Community College (Kentucky)

PAI YUNG-FENG, New York City

EUGENE J. PALKA, U.S. Military Academy (New York)

JOSEPH L. SCARPACI, JR., Virginia Tech

ROLF STERNBERG, Montclair State University (New Jersey)

COLLEEN J. WATKINS, Linfield College (Oregon)

BARBARA A. WEIGHTMAN, California State University, Fullerton

KRISTOPHER D. WHITE, University of Connecticut

For assistance with the new map of North American indigenous peoples, we are greatly indebted to Jack Weatherford, Professor of Anthropology at Macalester College (Minnesota); Henry T. Wright, Professor and Curator of Anthropology at the University of Michigan; and George E. Stuart, President of the Center for Maya Research (North Carolina), who also assisted with the Altun Ha site in Belize. The new map of Russia's federal regions could not have been compiled without the invaluable help of David B. Miller, Senior Edit Cartographer at the National Geographic Society, and Leo Dillon of the Russia Desk of the U.S. Department of State. And special thanks, too, go to Charles Pirtle, Professor of Geography at Georgetown University's School of Foreign Service for his advice on Chapter 4, and to Charles Fahrer of the Department of Geography at the University of South Carolina for his suggestions on Chapter 6.

We also record our appreciation to those geographers who ensured the quality of this book's ancillary products: Ira M. Sheskin (University of Miami) prepared the *Test Bank* and manipulated a large body of demographic data to derive the tabular display in Table I-2; Eugene J. Palka (U.S. Military Academy) prepared the *Instructor's Manual*; and Elizabeth Muller Hames (M.A. in Geography, University of Miami) co-authored the *Study Guide* and prepared the Geographical Index found on the Web site. At the University of Miami's Department of Geography and Regional Studies, Peter Muller is most grateful for the advice and support he continues to receive from all his faculty colleagues: Tom Boswell, Stuart Corbridge, Richard Grant, Jennifer Mandel, Jan Nijman, Ira Sheskin, and Michael Shin plus GIS Lab Manager Chris Hanson. Moreover, his departmental office staff tirelessly performs an array of critical supporting tasks; in addition to the excellent supervision of Assistant to the Chair/Office Manager Julia Lemus-Rodriguez, he wishes to thank Melissa Blankson, Felicia Price, and Scarleth Padilla.

PERSONAL APPRECIATION

We are privileged to work with a team of professionals at John Wiley & Sons that is unsurpassed in the college textbook publishing industry. As authors we are acutely aware of these talents on a daily basis during the crucial production stage, especially the outstanding coordination and leadership skills of Senior Production Editor Kelly Tavares, Illustration Editor Sigmund Malinowski, Photo Coordinator Jennifer MacMillan, and Production Assistants Rebecca Rothaug and Carmen Hernandez. Others who played a leading role in this process were Senior Designer Karin Kincheloe, Copy Editor Betty Pessagno, Photo Director Marge Graham, Photo Manager Hilary Newman, Photo Researchers Alexandra Truitt and Jerry Marshall, and Don Larson of Mapping Specialists, Ltd. in Madison, Wisconsin. Geography Editor Ryan Flahive was the prime mover in introducing a number of major innovations for the Tenth Edition, and together with his assistants, Karen Ayoub and Denise Powell, smoothly propelled and guided this latest revision. (We also appreciated the many interim tasks undertaken by Tom Kulesa before Ryan's arrival just after the project had begun.) Our College Marketing Manager, Clay Stone, gave us a great deal of attention throughout the revision process, and his advice and tremendous enthusiasm were always most welcome. Beyond this immediate circle, we acknowledge the support and encouragement we have received over the years from others at Wiley including Publisher Anne Smith, Vice-President for Production Ann Berlin, Executive Editor Kaye Pace, and Ancillaries Editor Mark Gerber.

Finally, and most of all, we thank our wives, Bonnie and Nancy, for yet again seeing us through the challenging schedule of our seventh collaboration on this volume in the past 17 years.

H. J. de Blij
Boca Grande, Florida

Peter O. Muller
Coral Gables, Florida

March 13, 2001

Brief Contents

Contents

CONCEPTS, IDEAS, AND TERMS

1. Geographic realm
2. Spatial perspective
3. Taxonomy
4. Transition zone
5. Geographic change
6. New World Order
7. Regional concept
8. Regional boundaries
9. Location
10. Absolute location
11. Relative location
12. Formal region
13. Spatial system
14. Hinterland
15. Functional region
16. Scale
17. Natural landscape
18. Physical geography
19. Continental drift
20. Tectonic plate
21. Subduction
22. Pacific Ring of Fire
23. Climate
24. Desertification
25. Glaciation
26. Ice age
27. Interglaciation
28. Hydrologic cycle
29. Climatic region
30. Physiography
31. Culture
32. Regional character
33. Cultural landscape
34. Sequent occupance
35. Central business district (CBD)
36. Ethnicity
37. Population distribution
38. Population density
39. Megalopolis
40. Cartogram
41. Urbanization
42. State

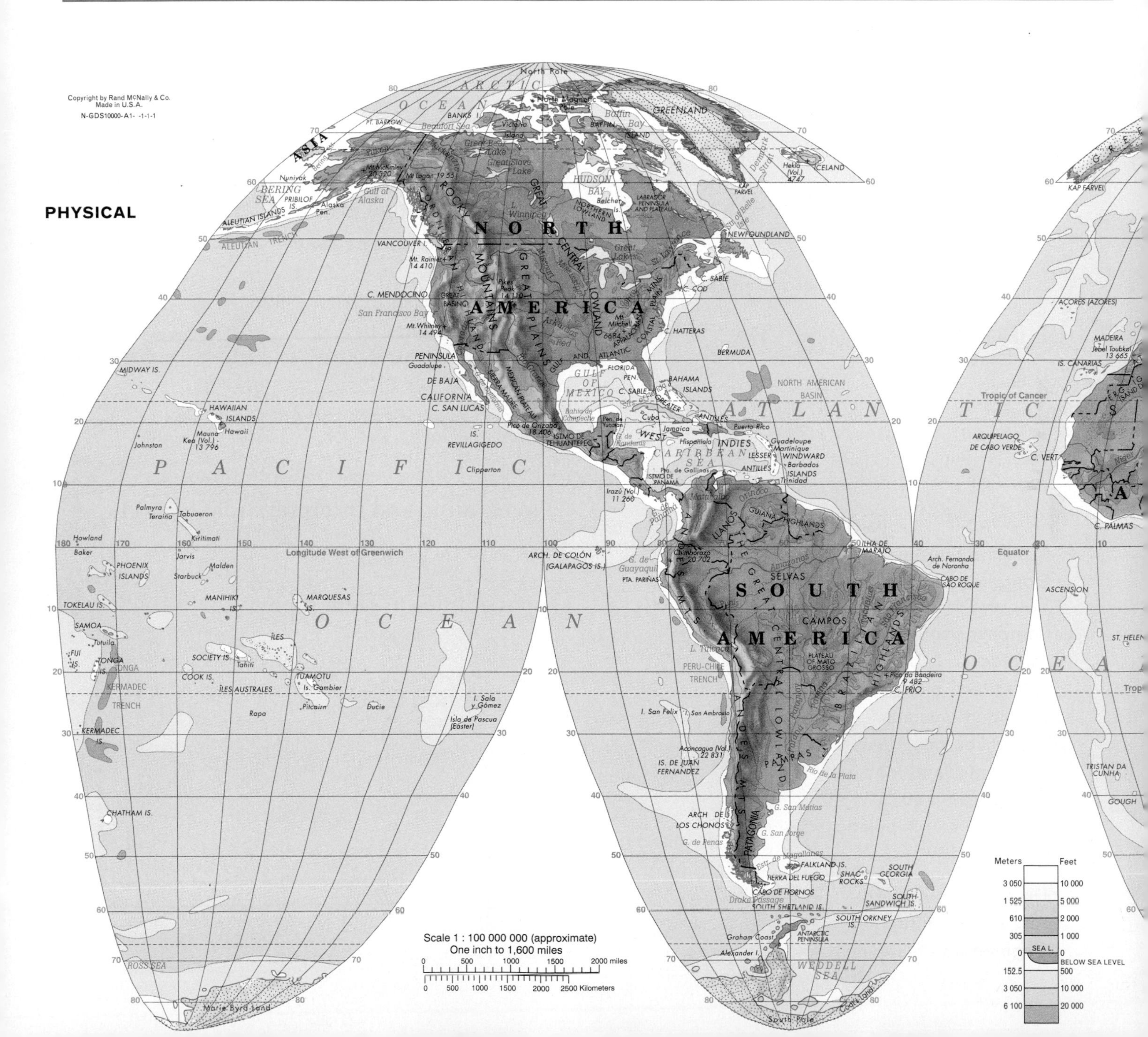

introduction

World Regional Geography

43	European state model	46	Core	49	Regional disparity	52	Globalization
44	Development	47	Periphery	50	Advantage	53	Regional geography
45	Economic geography	48	Core area	51	Neocolonialism	54	Systematic geography

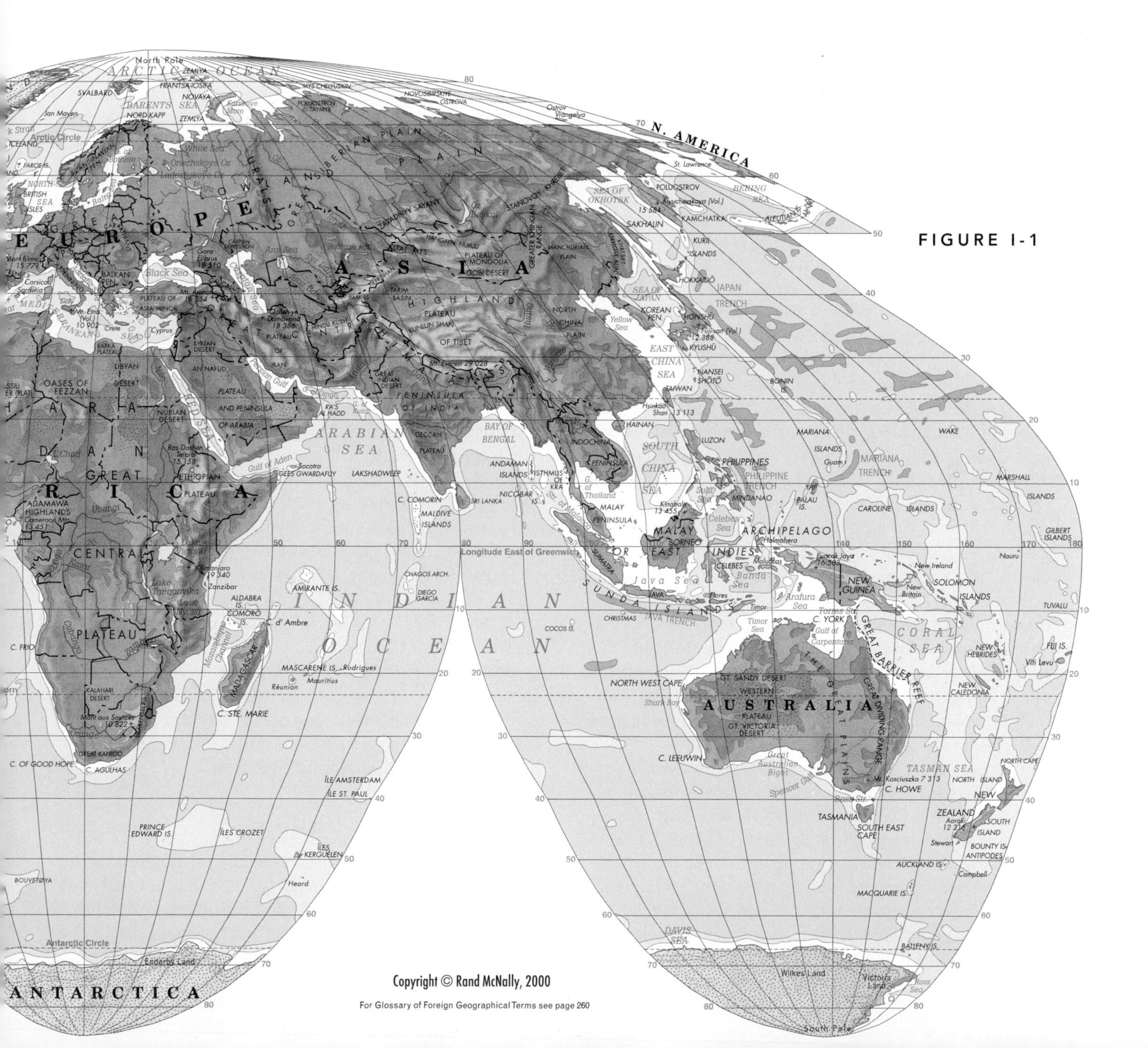

FIGURE I-1

What a time this is to be studying geography! The world is undergoing a historic transformation, one of those momentous political, economic, and social upheavals about which we have read in history books. It is happening today, and as in the past, new alliances are forming, old unions are fracturing, novel ideas are spreading, older notions are fading. We hear from our political leaders talk about the emergence of a "New World Order," but such pronouncements are premature. A future world order is indeed in the offing, but it is too early to tell what it will be like. Of this we can be sure: the new world map will look quite different from the old. Our task is to understand the ongoing changes, to make sense of the new directions our world is taking. Geography is our most powerful ally in this mission.

GEOGRAPHIC PERSPECTIVES

In this book we take a penetrating look at the geographic framework of the contemporary world, the grand design that is the product of thousands of years of human achievement and failure, movement and stagnation, revolution and stability, interaction and isolation. Ours is an interconnected world of travel and trade, tourism and television, a global village—but the village still has neighborhoods. Their names are Europe, South America, Southeast Asia, and others familiar to all of us. We call such global neighborhoods **geographic realms**, and when we subject these realms to geographic scrutiny, we find that each has its own identity and distinctiveness. 1

Geographers study the location and distribution of features on the Earth's surface. These features may be the landmarks of human occupation, the properties of the natural environment, or both: one of the most interesting themes in geography is the relationship between natural environments and human societies. Geographers investigate the reasons for these distributions. Their approach is guided by a **spatial perspective**. Just as historians focus on chronology, geographers concentrate on space and place. The spatial structure of cities, the layout of farms and fields, the networks of transportation, the system of rivers, the pattern of climate—all these and more go into the study of a geographic realm. As you will discover, geography is full of spatial terms: area, distance, direction, clustering, proximity, accessibility, isolation, and many others which we will encounter in the pages that follow. 2

In this book we use the geographic perspective and geography's spatial terminology to investigate the world's great geographic realms. We will find that each of these realms possesses a special combination of cultural, organizational, and environmental properties. These characteristic qualities are imprinted on the landscape, giving each realm its own traditional attributes and social settings. As we come to understand the human and natural makeup of those geographic realms, we learn not only *where* they are located (always a crucial question in geography, and often the answer is not as simple as it may appear), but also *why they are located where they are*, how they are constituted, and what their future is likely to be in our changing world.

It is not enough, however, to study the world from a geographic viewpoint without learning something about the discipline of geography itself. Beginning in this introductory chapter, we introduce the fundamental ideas and concepts that make modern geography what it is. Not only will these geographic ideas enhance your awareness of the many dimensions of our complex, multicultural, interconnected world, but also many of them will remain useful to you long after you have closed this book. Welcome to geography . . . realms, regions, *and* concepts.

REALMS AND REGIONS

Geographers, like other scholars, seek to establish order from the countless data that confront them. Biologists have established a system of classification, or **taxonomy**, to categorize the many millions of plants and animals into a hierarchical system of seven ranks. In descending order, we humans belong to the animal *kingdom*, the *phylum* (division) named chordata, the *class* of mammals, the *order* of primates, the *family* of hominids, the *genus* designated *Homo*, and the *species* known as *Homo sapiens*. Geologists classify the Earth's rocks into three major (and many subsidiary) categories and then fit these categories into a complicated geologic time scale that spans hundreds of millions of years. Historians define eras, ages, and periods to conceptualize the sequence of the events they study. 3

Geography, too, employs systems of classification. When geographers deal with urban problems, for instance, they use a classification scheme based on the sizes and functions of the places involved. Some of the terms in this classification are part of our everyday language: megalopolis, metropolis, city, town, village, hamlet.

In regional geography, which is the focus of this book, our challenge is different. We, too, need a hierarchical framework for the areas of the world we study, from the largest to the smallest. But our classification scheme is horizontal, not vertical. It is *spatial*. Our equivalent of the biologists' overarching kingdoms (of plants and animals) is the Earth's natural partitioning into landmasses and oceans. The next level is the division of the inhabited landmasses into geographic realms based on human as well as natural properties.

The Criteria for Realms

In any classification system, criteria are the key. Not all animals are mammals; the criteria for inclusion in that biological class are more specific and restrictive. A dolphin may look and act like a fish, but both anatomically and functionally dolphins belong to the class of mammals.

FROM THE FIELD NOTES

"The Canadian city of Victoria, capital of the province of British Columbia, is located on the southeastern tip of Vancouver Island and is a major Pacific Rim port. Having just seen the social transformation going on in the nearby city of Vancouver, I was interested to learn to what extent Victoria, too, is affected by ethnic change. I asked the way to Chinatown, a short walk from the inner harbor, and found this imposing gate, flanked by Chinese shops and restaurants. 'I expected your Chinatown to be larger,' I said to the owner of a shop, 'given what I saw in Vancouver.' He shook his head. 'Our Chinatown is getting smaller every year,' he said. 'Chinese are coming to Vancouver from Hong Kong and other places in Asia, but the same attraction is pulling Chinese from here to Vancouver too. Now, there are shops on this street owned by non-Chinese, sold by Chinese families who did business here for a very long time.' I asked about this in a government office. 'It's jobs and money,' said a planner there. 'The young Chinese hear about opportunities and higher wages, and the oldsters die off, and the community shrinks. So the burgeoning of the Pacific Rim in Asia is changing the demographic balance among cities in Canada.'"

Geographic realms are based on sets of spatial criteria. First, they are the largest units into which the inhabited world can be divided. The criteria on which such a broad regionalization is based include both physical (that is, natural) and human (or social) yardsticks. South America is a geographic realm, for example, because physically it is a continent and culturally it is dominated by a set of social norms. The realm called South Asia, on the other hand, lies on a Eurasian landmass shared by several other geographic realms; high mountains, wide deserts, and dense forests combine with a distinctive social fabric to create this well-defined realm centered on India.

Second, geographic realms are the result of the interaction of human societies and natural environments, a *functional* interaction revealed by farms, mines, fishing ports, transport routes, dams, bridges, villages, and countless other features that mark the landscape. According to this criterion, Antarctica is a continent but not a geographic realm.

Third, geographic realms must represent the most comprehensive and encompassing definition of the great clusters of humankind in the world today. China lies at the heart of such a cluster, as does India. Africa constitutes a geographic realm from the southern margin of the Sahara (an Arabic word for "desert") to the Cape of Good Hope and from its Atlantic to its Indian Ocean shores.

Figure I-2 displays the 12 world geographic realms based on the these criteria. As we will show in more detail later, waters, deserts, and mountains as well as cultural and political shifts mark the borders of these realms. We will discuss the position of these boundaries as we examine each realm. For the moment, keep in mind the following:

- *Where geographic realms meet,* **transition zones**, *not sharp boundaries, mark their contacts.* 4

We need only remind ourselves of the border zone between the geographic realm in which most of us live, North America, and the adjacent realm of Middle America. The line on Figure I-2 coincides with the boundary between Mexico and the United States, crosses the Gulf of Mexico, and then separates Florida from Cuba and the Bahamas. But Hispanic influences are strong in North America north of this boundary, and the U.S. economic influence is strong south of it. The line, therefore, represents an ever-changing zone of regional interaction. Again, there are many ties between South Florida and the Bahamas, but the Bahamas resemble a Caribbean more than a North American society.

In Africa, the transition zone from Subsaharan to North Africa is so wide and well defined that we have put it on the world map; elsewhere, transition zones tend to be narrower and less easily represented. In these early years of the twenty-first century, such countries as Belarus (between Europe and Russia) and Kazakhstan (between Russia and Muslim Southwest Asia) lie in inter-realm transition zones. Remember, over much (though not all) of their length, borders between realms are zones of regional change.

- *Geographic realms* **change** *over time.* 5

Had we drawn Figure I-2 before Columbus made his voyages (1492–1504), the map would have looked different: Amerindian states and peoples would have determined the boundaries in the Americas; Australia and New Guinea would have constituted one realm, and New Zealand would

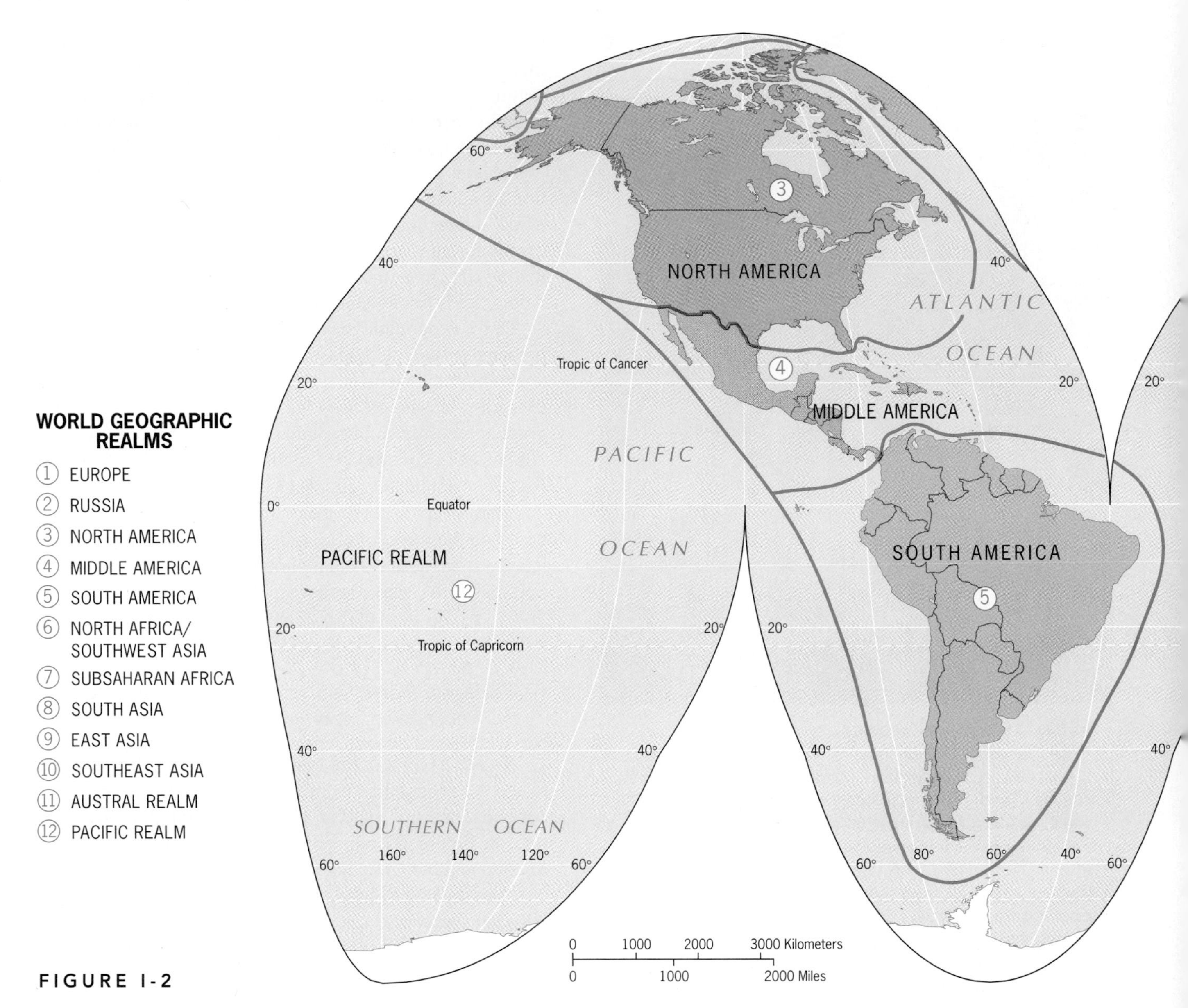

FIGURE I-2

have been part of the Pacific Realm. The colonization, Europeanization, and Westernization of the world changed that map dramatically. During the four decades after World War II relatively little change took place, but since 1985 far-reaching realignments have again been occurring.

6 As we try to envisage what a **New World Order** will look like on the map, note that the 12 geographic realms can be divided into two groups: (1) those dominated by one major political entity, in terms of territory or population or both (North America/United States, Middle America/Mexico, South America/Brazil, South Asia/India, East Asia/China, Southeast Asia/Indonesia as well as Russia and Australia), and (2) those that contain many countries but no dominant state (Europe, North Africa/Southwest Asia, Subsaharan Africa, and the Pacific Realm). For several decades two major powers, the United States and the former Soviet Union, dominated the world. Will a multipolar world arise from the rubble of that bipolar world? We will address this question in the pages that follow.

The Criteria for Regions

The spatial division of the world into geographic realms establishes a broad global framework, but for our purposes a more refined level of spatial classification is needed. This brings us to an important organizing concept in geography: the **regional concept**. To continue the analogy with biological taxonomy, we now go from phylum to order. To establish regions within geographic realms, we need more specific criteria. 7

Let us use the North American realm to demonstrate the regional idea. When we refer to a part of the United States or Canada (e.g., the South, the Midwest, or the Prairie Provinces), we employ a regional concept—not scientifi-

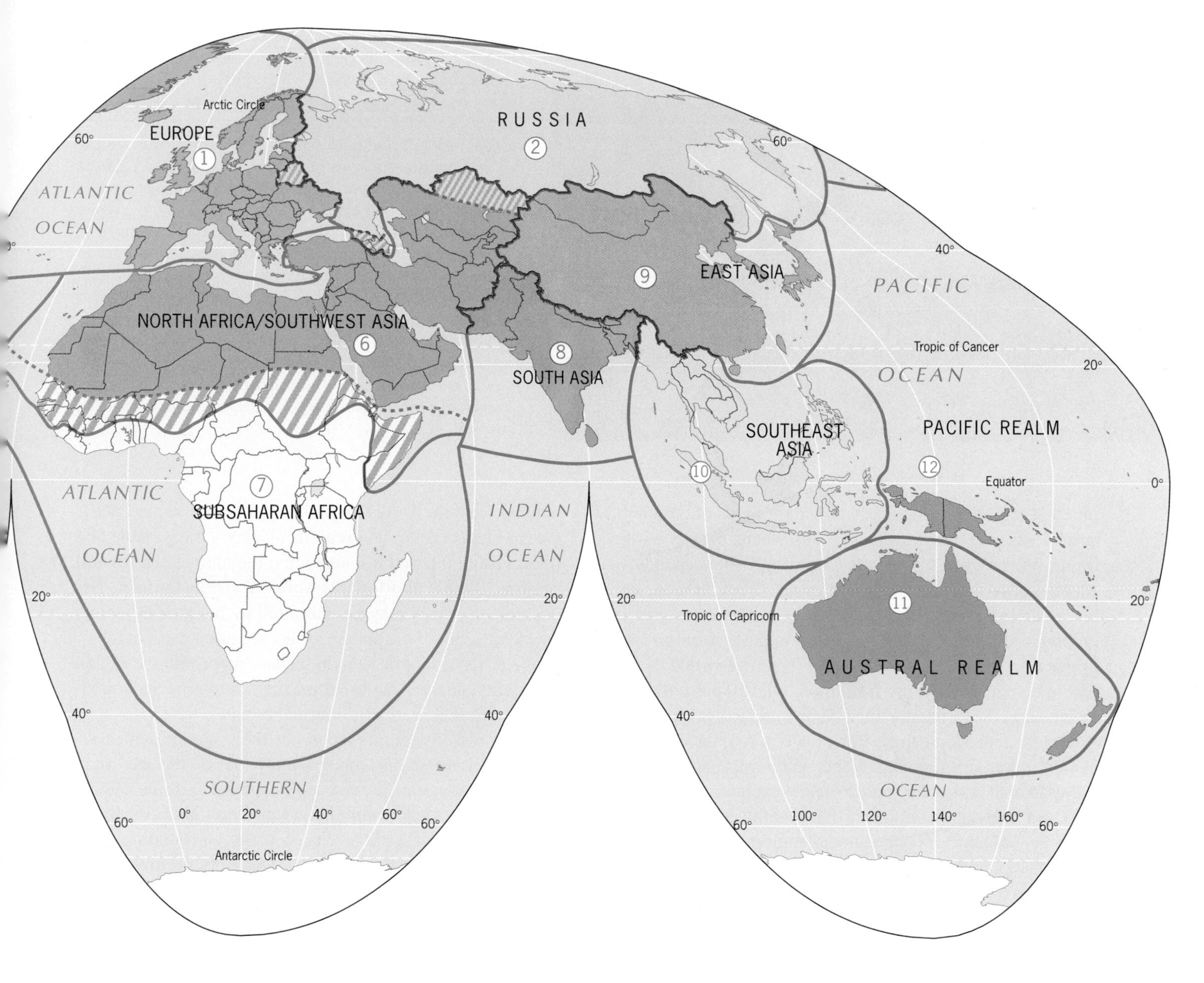

cally but as part of everyday communication. We reveal our *perception* of local or distant space as well as our mental image of the region we are describing.

But what exactly is the Midwest? How would you draw this region on the North American map? Regions are easy to imagine and describe, but they can be difficult to outline on a map. One way to define the Midwest is to use the borders of States: certain States are part of this region, others are not. You could also use agriculture as the chief criterion: the Midwest is where corn and/or soybeans occupy a certain percentage of the farmland. Each method results in a different delimitation; a Midwest based on States is different from a Midwest based on farm production. Therein lies an important principle: regions are scientific devices that allow us to make spatial generalizations, and they are based on artificial criteria we establish to help us construct them. If you were studying the geography behind politics, then a Midwest region defined by State boundaries would make sense. If you were studying agricultural distributions, you would need a different definition.

Given these different dimensions of the same region, we can identify properties that all regions have in common. To begin with, all regions have *area*. This observation would seem obvious, but there is more to this idea than meets the eye. Regions may be intellectual constructs, but they are not abstractions: they exist in the real world, and they occupy space on the Earth's surface.

It follows that regions have *boundaries*. Occasionally, nature itself draws sharp dividing lines, for example, along the crest of a mountain range or the margin of a forest. More often, **regional boundaries** are not self-evident, and we must determine them using criteria that we establish for that purpose. Take, for instance, the notion of a Corn Belt in the agricultural heartland of the United States. When you travel

8

Interconnections in the spatial system of the Corn Belt: Garden City, Iowa. Roads and grain elevators underscore the linkages between the town and its surroundings. The spatial homogeneity of the region is quite evident from the agricultural landscape beyond (evenly dispersed farmsteads, repetition of field size and land uses). Also visible is the signature geometry of the Township-and-Range system of land division (see p. 22) which, beginning in 1785, superimposed a rectangular grid of 6-mile by 6-mile townships on all but the 13 original U.S. States. Within that geographical matrix, townships were subdivided into 36 square-mile (640-acre) properties that were further split into 160-acre (and sometimes smaller) farms. As the photo shows, this land division scheme dominates every aspect of the cultural landscape: not only the rectangular fields and the north-south/east-west layout of roads and railways in the countryside, but also the configuration and spacing of streets, buildings, and properties in Garden City itself.

into that region, you see more and more fields of corn until corn dominates the farmlands. But where exactly is the limit of this region we call the Corn Belt? That depends on specific criteria. First we would establish a unit area, say, one-quarter square mile. Next we would create a grid of quarter-square-mile units and superimpose it on an area larger than the corn-growing zone. Then we would decide that to be part of the Corn Belt, more than 50 percent of the farmland in any unit area must be devoted to growing corn. We might also decide that a unit area is part of the Corn Belt even if it contains less than 50 percent corn, as long as spatial units that do qualify completely surround it. The result is a Corn Belt region boundary that is both visible and quantitatively defined. Of course, not all regional boundaries can be so specifically delimited, and neighboring regions, like adjacent realms, can display transitional borderlands.

[9] All regions also have **location**. Often the name of a region contains a locational clue, as in Amazon Basin or Indochina (a region of Southeast Asia lying between India and China). [10] Geographers refer to the **absolute location** of a place or region by providing the latitudinal and longitudinal extent of the region with respect to the Earth's grid coordinates. [11] A far more practical measure is a region's **relative location**, that is, its location with reference to other regions. Again, the names of some regions reveal aspects of their relative locations, as in *Eastern* Europe and *Equatorial* Africa.

Many regions are marked by a certain *homogeneity* or sameness. Homogeneity may lie in a region's human (cultural) properties, its physical (natural) characteristics, or both. Siberia, a vast region of northeastern Russia, is marked by a sparse human population that resides in widely scattered, small settlements of similar form; frigid climates; extensive areas of permafrost (permanently frozen subsoil); and cold-adapted vegetation. This dominant uniformity makes it one of Russia's natural and cultural regions, extending from the Ural Mountains in the west to the Pacific Ocean in the east. When regions display a measurable and often visible internal homogeneity, they are called **formal regions**. [12] But not all formal regions are visibly uniform. For example, a region may be delimited by the area in which, say, 90 percent of the people or more speak a particular language. This cannot be seen in the landscape, but the region is a reality, and we can use this criterion to draw its boundaries accurately. It, too, is a formal region.

Other regions are marked *not* by their internal sameness, but by their functional integration—that is, the way they work. These regions are defined as **spatial systems** [13] and are formed by the areal extent of the activities that define them. Take the case of a large city with its surrounding zone of suburbs, urban-fringe countryside, satellite towns, and farms. The city supplies goods and services to this encircling zone, and it buys farm products and other commodities from it. The city is the heart, the *core* of this region, and we call the surrounding zone of interaction the city's **hinterland**. [14] But the city's influence wanes on the outer periphery of that hinterland, and there lies the boundary of the functional region of which the city is the focus. A **functional region**, [15] therefore, is forged by a structured, urban-centered system of interaction. It has a core and a periphery. As we shall see, core-periphery contrasts in some parts of the world are becoming strong enough to endanger the stability of countries.

All human-geographic regions are *interconnected*, being linked to other regions. We know that the borders of geographic realms sometimes take on the character of transition zones, and so do neighboring regions. Trade, migration, education, television, computer linkages, and other interactions blur regional boundaries. These are just some of the links in the fast-growing interdependence among the world's peoples, and they reduce the differences that still divide us. Understanding these differences will lessen them further.

REGIONS AT SCALE

In this book we examine the geography of the world at the level of the region, using regional concepts as we proceed. But all regions are parts of realms, which is why this book is called *Geography: Realms, Regions, and Concepts*.

As we have just noted, regions come in many sizes. Some are huge, for example, the Russian formal region of Siberia. Others are comparatively small, for example, the functional region constituted by Boston and its hinterland. Some geographers have tried to distinguish between larger and smaller regions; occasionally, you will see a reference to a subregion within a region. But no generally accepted nomenclature has ever emerged, so we use the term *region* for bounded spaces large and small. At the level of detail in our maps, we must focus on the larger regions, referring occasionally to smaller regions "nested" within the larger ones. And this brings us to the geographic concept of scale.

The map is the geographer's strongest ally. It does for geography what taxonomic (and other) classification systems do for the other sciences. Maps display enormous quantities of information; they suggest relationships; they answer questions; they lay out spatial problems for researchers to investigate (see Appendix A, Map Reading and Interpretation). Many maps, moreover, are fascinating. No personal library is complete without a good atlas.

Maps represent the surface of the Earth (and other features of the planet, present and past) at various levels of generalization. We can gauge these levels from the map's **scale**, that is, the ratio of the distance between two places on a map and the actual distance between those two places on the Earth's surface.

16

Consider the four maps in Figure I-3. The first map (upper left) shows most of the North American realm but gives little spatial information, although it does show the political boundary between Canada and the United States.

EFFECT OF SCALE

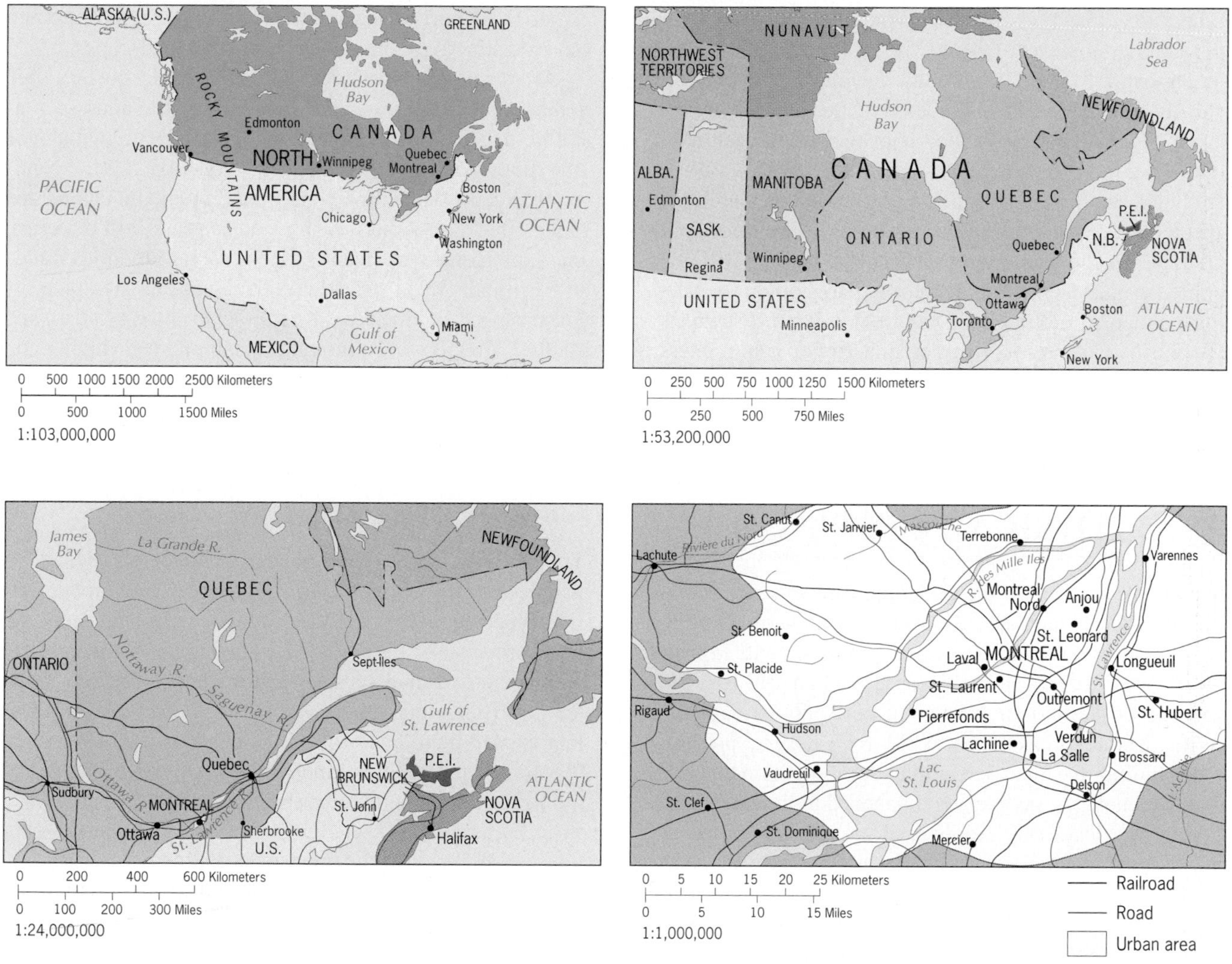

FIGURE I-3

The second map (upper right) depicts eastern and central Canada in sufficient detail to display the provinces, several cities, and some physical features (Manitoba's major lakes) not shown on the first map. The third map (lower left) shows the main surface communications of the province of Quebec and immediate surroundings, the relative location of Montreal, and the St. Lawrence and Hudson/James Bay drainage systems. The fourth map (lower right) reveals the metropolitan layout of Montreal and its adjacent hinterland in considerable detail.

Each of the four maps has a scale designation, which can be shown as a bar graph (in miles and kilometers [km] in this case) and as a fraction—1:103,000,000 on the first map. The fraction is a ratio indicating that one unit of distance on the map (one inch or one centimeter) represents 103 million of the same units on the ground. The smaller the fraction (in other words, the larger the number in the denominator), the smaller the scale of the map. Clearly, this representative fraction (RF) on the first map (1:103,000,000) is the smallest, and that of the fourth map (1:1,000,000) is the largest. Comparing maps 1 and 3, we find that on the linear scale, map 3 has a representative fraction that is more than four times larger than that of map 1. When it comes to areal representation, however, 1:24,000,000 is more than 16 times larger than 1:103,000,000 because the linear difference prevails in both the length and breadth of the map.

In a book that surveys the major regions of each world realm, we have to operate at relatively small scales. When studying smaller regions in greater detail, our ability to specify criteria and to "filter" the factors we employ increases as we work at larger scales. That method will often come into play when urban areas are the topic of concern (as the map of metropolitan Montreal in Fig. I-3 suggests). But mostly our view will be more macroscopic and general: a small-scale perspective on the world's geographic realms and regions.

THE PHYSICAL SETTING

This book focuses on the geographic realms and regions produced by human activity over thousands of years. But we should never forget the natural environments in which all this activity took place because we can still recognize the role of these environments in how people make their living. Certain areas of the world, for example, presented opportunities for plant and animal domestication that other areas did not. The people who happened to live in those favored areas learned to grow wheat, rice, or root crops, and to domesticate oxen, goats, or llamas. We can still discern those early "patterns of opportunity" on the map in the twenty-first century. From such opportunities came adaptation and invention, and thus arose villages, towns, cities, and states. But people living in more difficult environments found it much harder to achieve this organization. Take tropical Africa. There, the human communities could not domesticate any of the many species of wildlife, from gazelles and zebras to giraffes and buffalo. Wild animals were a threat, not an opportunity. Eventually humans domesticated only one African animal, the guinea fowl. Early African peoples faced environmental disadvantages that persist today. The modern map carries many imprints of the past.

Natural Landscapes

17

Planet Earth presents a jumble of **natural landscapes** ranging from rugged mountain chains to smooth coastal plains (Fig. I-1). As we investigate each geographic realm, we will note the natural landscapes that form its physical base because these landscapes continue to play a role in their formation. Mountain ranges form barriers to movement, but they also protect peoples against enemies. River basins in Asia still contain some of the world's largest population concentrations because they are well-watered and have fertile soils, but river basins in tropical Africa and South America support no such numbers. Every geographic realm has its distinctive combination of natural landscapes.

All the continents contain old geologic cores called *shields* that create vast plains (low-lying flatlands such as northern Canada) or plateaus (higher-elevation flatlands like those that prevail over much of Africa). All the continents except Africa have long, linear mountain chains: the Andes in South America, the Rockies in North America, the Alps-Himalayas in Eurasia, the Great Dividing Range in Australia. Only Africa lacks such a mountain spine, its linear ranges confined to the extreme northwest (Atlas) and south (Cape). Between mountains and plains lie higher and lower plateaus, and higher and lower hills. There is no generalizing about the Earth's natural landscapes, and we will examine them one realm at a time.

Continents and Plates

18

The study of natural landscapes is part of **physical geography**. Nearly a century ago a physical geographer, Alfred Wegener, used spatial analysis to explain the details he saw on maps like Figure I-1. He studied the outlines of the continents and, marshalling vast evidence, theorized that the Earth's landmasses are pieces of a giant supercontinent that existed hundreds of millions of years ago. He named that postulated supercontinent *Pangaea*, and Africa lay at the heart of it. When North and South America split away from Eurasia and Africa, the Atlantic Ocean opened—but you can still see how the opposite coastlines fit together.

19

Wegener's hypothesis of **continental drift** seemed to explain much of what Figure I-1 shows, including the global distribution of major mountain ranges. As South America "drifted" westward, away from Africa, the rocks

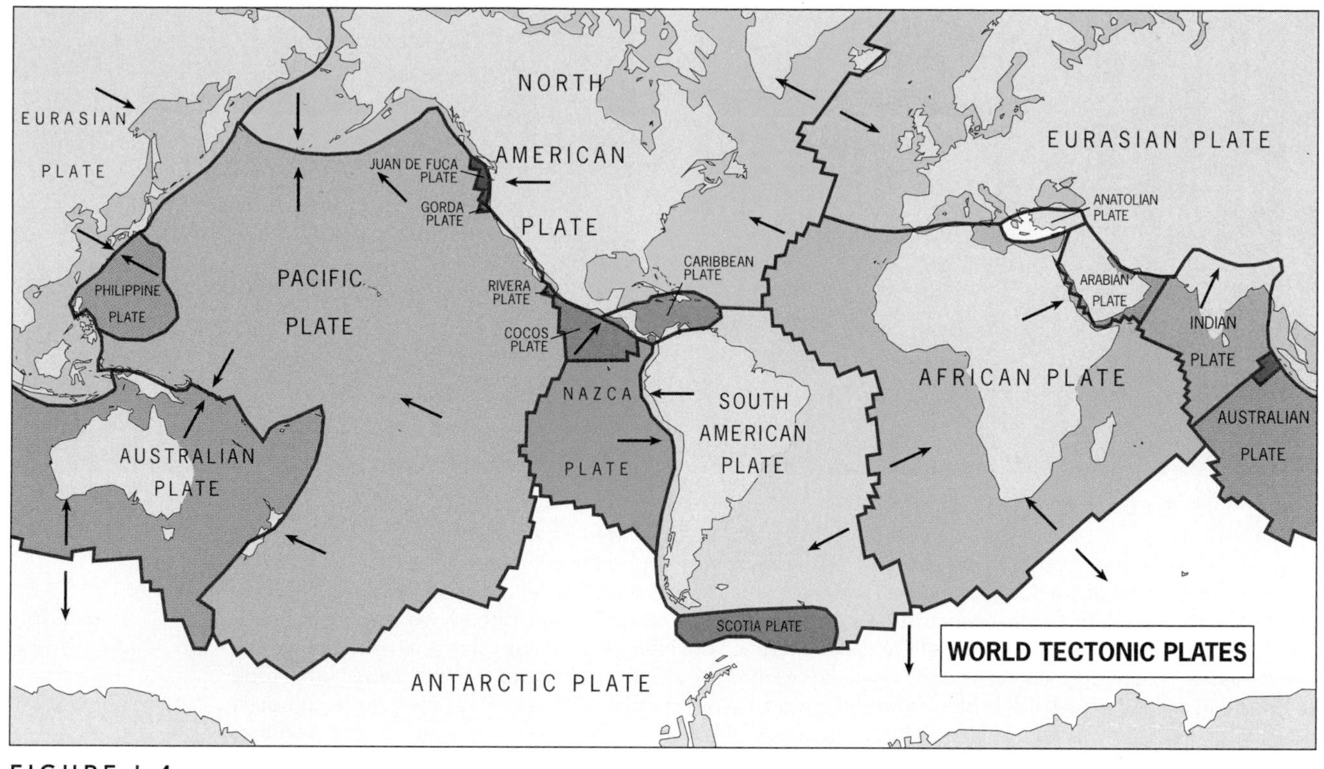

FIGURE I-4

of its leading edge crumpled like folds of an accordion, creating the Andes Mountains. On the opposite side of Africa, Australia moved eastward, so that its major mountain chain now lies near its eastern margin. Africa itself moved little, which is why no major linear mountain range formed there.

Wegener did not live to see his hypothesis accepted; in his day, many prominent geologists derided it as nonsense. But as is so often true, a geographer's spatial theory pointed the way for the research of others. Eventually, geologists discovered that the crust of the Earth consists of a set of **tectonic plates**, great slabs of solid rock that form the ocean floor and carry the continents (Fig. I-4). To no geographer's surprise, those plates are indeed mobile. They move slowly, a few centimeters (an inch or so) a year, propelled by heat-driven convection cells in the molten rock below. An inch per year may not be much, but the Earth's history is measured in millions of years. In 100 million years, a landmass would move nearly 1600 miles (2550 km)!

20

Since the Earth's tectonic plates move, there are places where they collide. When this happens, the plate that consists of lighter (less heavy) rock rides up over the weightier plate, which is pushed downward in a process called **subduction**. Such gigantic collisions create the mountain chains that girdle the Earth, causing earthquakes and volcanic eruptions in the process. In Figure I-4, note that the North American and South American Plates are moving toward the Pacific from the east, while the Eurasian Plate is coming in from the west. Thus the Pacific Ocean is surrounded by a nearly continuous plate-collision zone called the **Pacific Ring of Fire**. Several of the realms and regions we will investigate are strongly affected by this dangerous crustal instability. Disastrous earthquakes, volcanic eruptions, landslides, mudslides, and other threats are facts of daily life there (Fig. I-5).

21

22

The complex and varied natural landscapes of the continents result not only from such geologic forces but also from the processes of *erosion*. Rivers and smaller streams, glaciers, and even the wind all modify the surface. The mosaic of natural landscapes depicted on Figure I-1 is the product of forces and processes that continue to transform the landmasses on which we live.

Climate

It would be impossible to discuss the world's geographic realms and regions without reference to their **climates**. Many of humanity's inventions were motivated by the need to overcome problems posed by climate: clothing against cold, irrigation against drought, air conditioning against heat. When we think of some distant region, the first question we are likely to ask is, "What is its climate?"

23

A more appropriate question would be "What is its climate *today*?" Global climates change, sometimes rapidly. Archeological evidence tells us that farms, villages, and

FROM THE FIELD NOTES

"The land of New Zealand's South Island dropped off precipitously under the waters of the Tasman Sea here, along the southwestern coast. As we approached the shore we crossed the contact zone between the Pacific Plate, on which most of the South Island lies, and the Australian Plate (see Fig. I-4). Subduction along this zone creates some of the world's most spectacular scenery as the crust is pushed precipitously downward, to be recycled in the mantle below . . . diagonally across the Pacific from New Zealand lies volcanically and seismically active Alaska. Mount Edgecumbe, about 16 miles (26 km) west of Sitka, dominates the landscape here. I took the day and walked across the bridge toward the old volcano, chatting with locals as I went. Edgecumbe is listed as an extinct volcano, but Sitkans have their doubts. Tlingit oral history tells of a violent eruption that buried the whole countryside under a deep blanket of ash. 'The earth tremors around here tell me that it's a little early to call it extinct,' said one resident. 'A few centuries is nothing when it comes to geologic activity. I'd say the old lady is dormant.' The scars of erosion, outlined by the summer snow in the picture, suggest a lengthy dormancy, but the crater rim still rises nearly 3300 feet (1000 m) above the water."

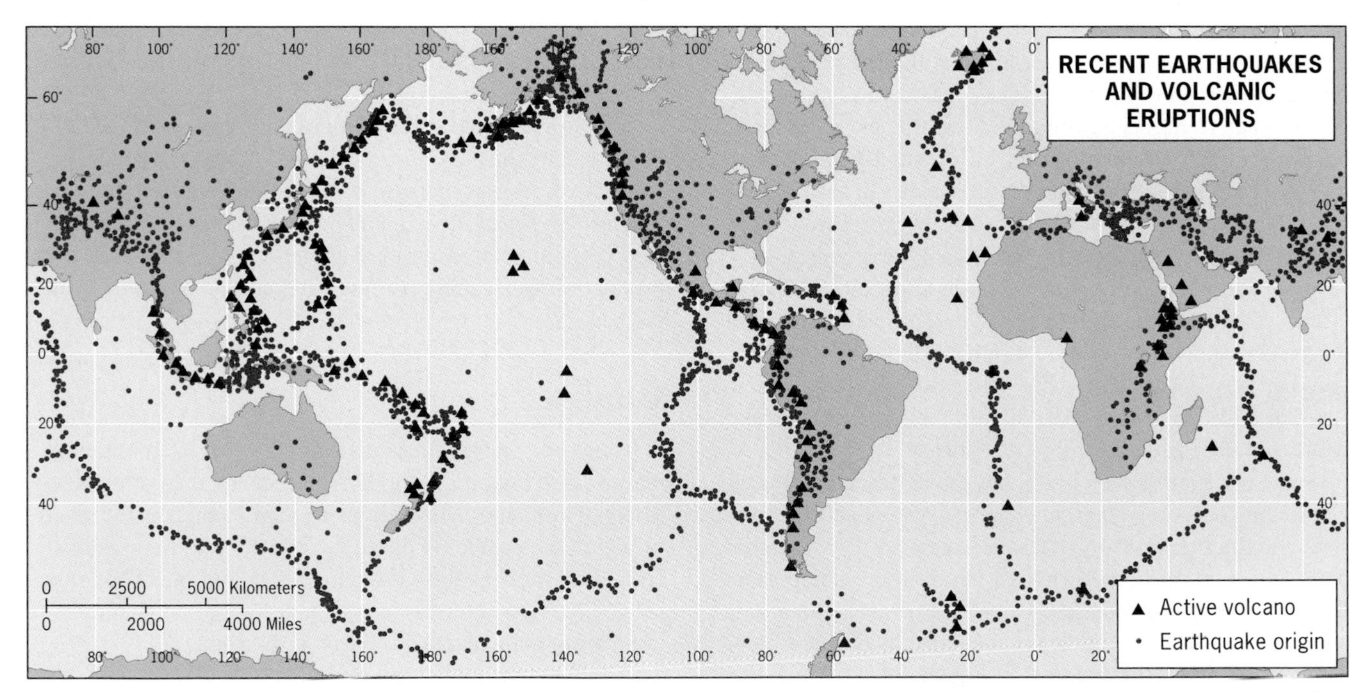

FIGURE I-5

towns that sprang up in well-watered, fertile areas of Southwest Asia were overtaken by **desertification**. Dependable rainfall cycles gave way to intermittent droughts, river levels dropped, wells failed. Eventually the expanding desert claimed what had once been thriving farmlands and bustling settlements. Some settlements were buried under advancing dunes.

Over the longer term, too, climate is subject to wide swings. Little more than 12,000 years ago, an eyeblink in geologic history, giant icesheets covered virtually all of Canada and much of the U.S. Midwest (Fig. I-6) as far south as the Ohio River. Most of Europe was covered from the Alps to Scandinavia. This was only the latest phase of **glaciation**, a period of lowered temperatures, ice surges, and dropping sea levels that has prevailed over the past 12 million years. It all started with a cooling trend, perhaps as long as 20 million years ago, the beginning of a global **ice age**. Eventually this gradual cooling gave way to massive advances of icesheets and valley glaciers, marking the beginning of what geologists call the Pleistocene epoch less than 3 million years ago. Ever since, frigid glaciations have been followed by relatively warm spells called **interglaciations**. During the current interglaciation, named the Holocene and already more than 10,000 years old, human civilizations have evolved from the smallest and simplest communities to the megacities of today.

Given this long-term as well as short-term variability, we should not be surprised that climates continue to change. We have learned to record and interpret climatic data, and we recognize that what happened to those ancient farms and towns could also happen to us: croplands could be threatened, sea levels could rise, coasts could be inundated. That is why global warming has become an important issue in the international arena. In the late twentieth century, evidence of a global warming cycle mounted. (From approximately 1940 to 1970, there had been a marked cooling cycle.) Although a majority of scientists and politicians attributed this warming cycle to human activity, notably the production of so-called greenhouse gases, evidence suggests that the present warming is the result mainly of natural causes. This makes the problem more serious because even significantly reducing greenhouse-gas emissions may not materially slow the warming. About 1000 years ago, the Earth also experienced a strong warming cycle. Agricultural frontiers moved to ever-higher latitudes, settlements thrived in Iceland and

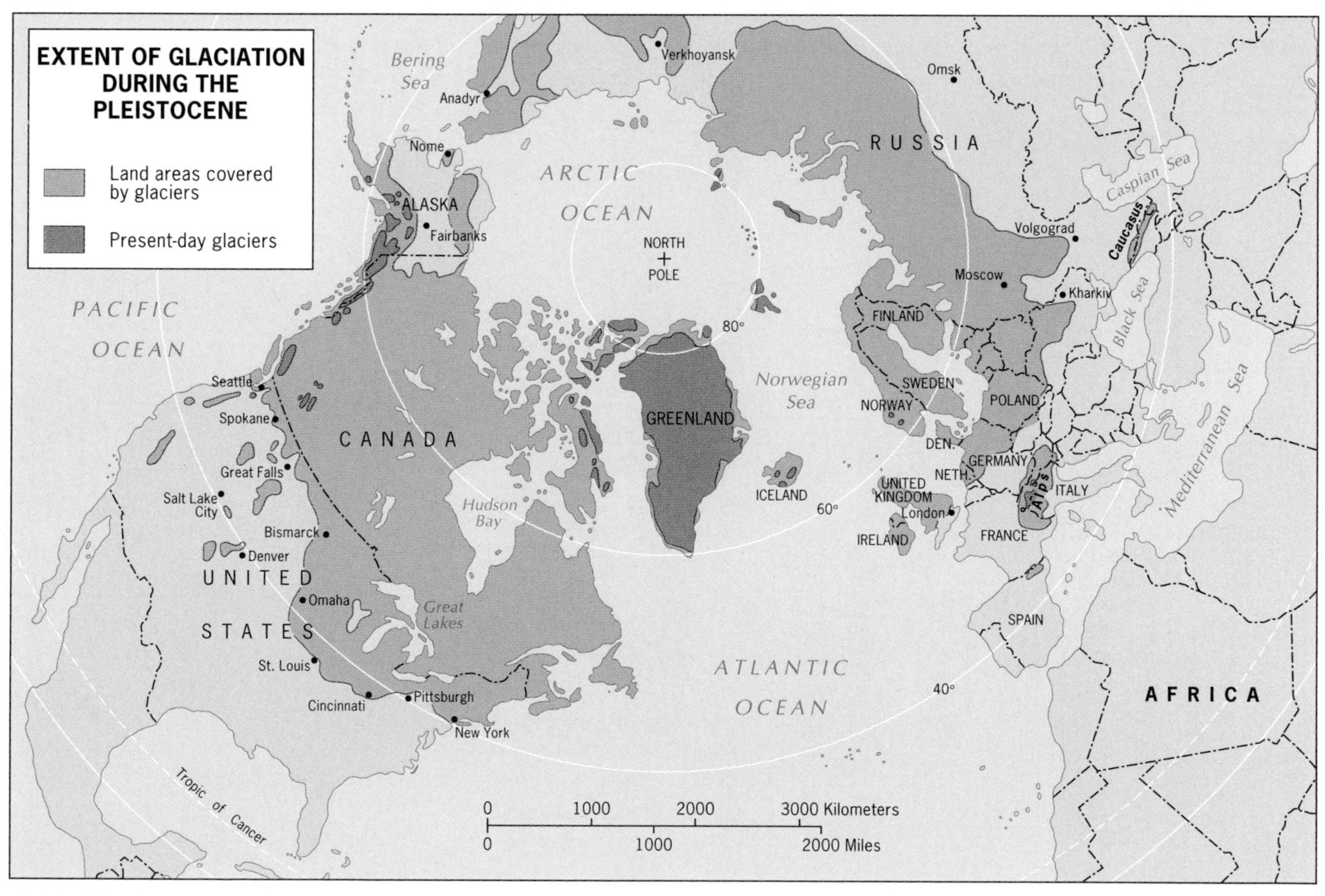

FIGURE I-6

Greenland, the Maori reached a warming New Zealand in their canoes. Anthropogenic (human-created) greenhouse gases had nothing to do with this warming cycle. But then, suddenly, the good times ended. In her book, *The Little Ice Age*, Jean Grove describes what happened when rapid cooling replaced global warming:

> *Grain would no longer ripen in Iceland . . . in Europe, disastrous harvests were experienced in the latter part of the thirteenth and in the early fourteenth century . . . extremes of weather were greater, with severe winters and unusually hot or wet summers.*

Thus the most recent warming cycle came to a turbulent end, at least in those areas for which we have a record. We know much less about what may have happened to Chinese civilization, which also lay in the path of sudden climate change, or of hurricanes in the Atlantic or El Niños in the Pacific. But the evidence indicates that a warming cycle of the kind the world is now experiencing will be followed by a reversal accompanied by environmental extremes and, unless we prepare for it, social and political dislocation.

Climatic Regions

We must therefore understand the climatic environments of the world's geographic realms and regions. Our technological advances notwithstanding, climate plays a key role in the lives and livelihoods of billions of people. A perturbation in the system—such as an El Niño event or a drought—can have widespread repercussions on economies and even politics.

It is no easier to generalize about climates than it is about natural landscapes. A regional climate is the average of countless weather observations among which measures of precipitation and temperature are key. Also taken into account are seasonal variations and extremes. In terms of precipitation, the Earth displays enormous spatial variation (Fig. I-7). Equatorial and tropical areas tend to be well-watered because the **hydrologic cycle**—the system that carries evaporated ocean water (leaving the salt behind) over the landmasses where it falls as rain or snow—is most efficient in the high heat and humidity of low latitudes. But in Africa this equatorial bounty soon gives way to subtropical aridity. From over 80 inches (200 cm) of rain in The Congo near the equator, the annual total drops to 4 inches (10 cm) or less in the Sahara to the north and the Kalahari-Namib to the south. In higher latitudes, only certain favored areas (notably Western Europe) receive ample precipitation.

28

Average temperatures decline from equatorial toward polar latitudes, but the gradation is irregular, affected by elevation (mountains have a cooling effect) and location relative to coastlines (interiors of continents are hotter in summer, cooler in winter). The size, shape, and terrain of the landmasses complicate the map of average temperatures. Ocean currents also influence the pattern. Warm waters offshore,

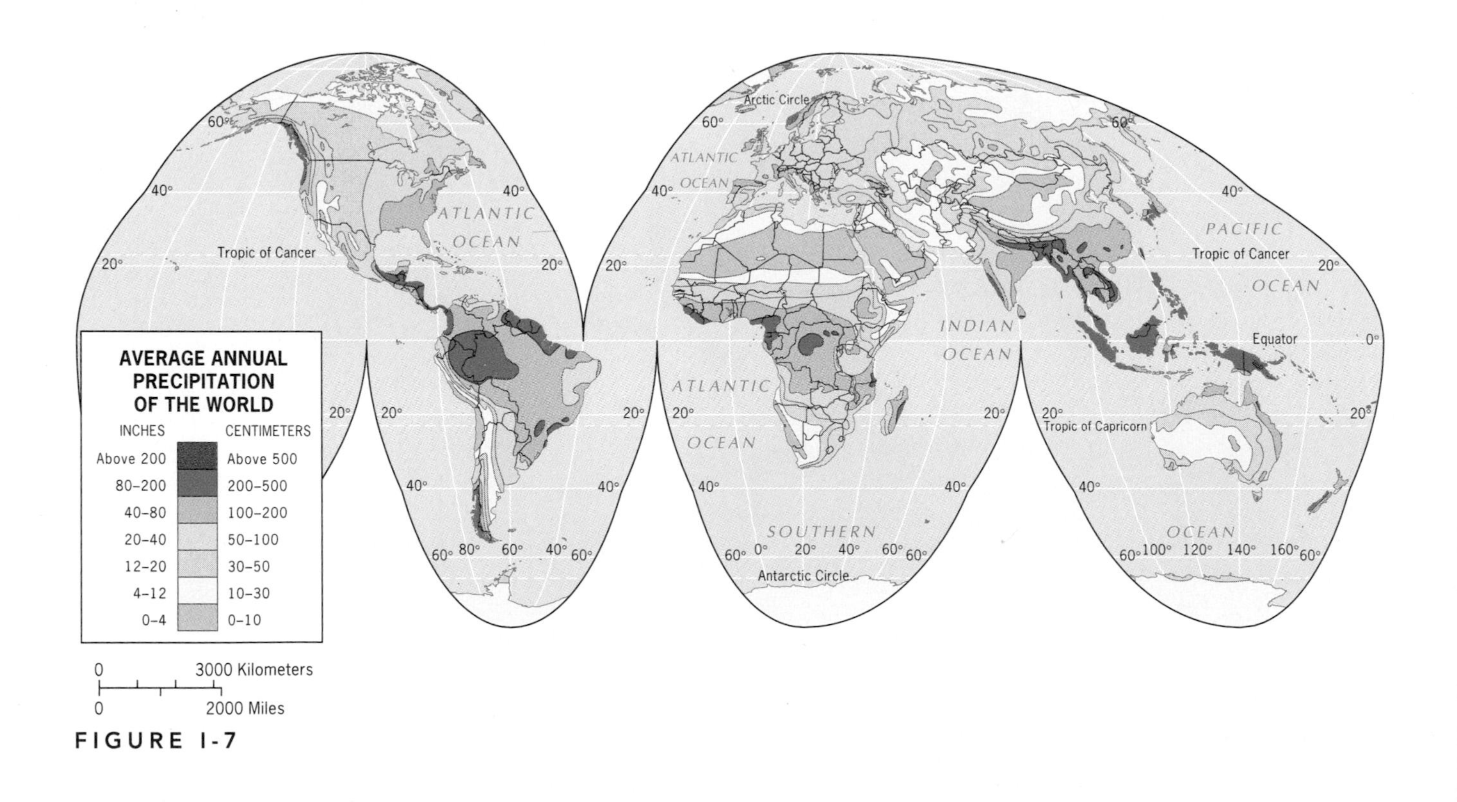

FIGURE I-7

FROM THE FIELD NOTES

"Walking the hilly countryside anywhere in Indonesia leaves you in no doubt about the properties of *Af* climate. The sweltering sun, the hot, humid air, the daily afternoon rains, and the lack of relief even when the sun goes down—the still atmosphere lies like a heavy blanket on the countryside to make this a challenging field trip. Deep, fertile soils form rapidly here on the volcanic rocks, and the entire landscape is green, most of it now draped by rice fields and terraced paddies. The farmer whose paddies I photographed here told me that his land produces three crops per year. Consider this: the island of Jawa, about the size of Louisiana, has a population of more than 125 million—its growth made possible by this combination of equatorial circumstances . . . Alaska, almost a dozen times as large as Jawa, has a population under three-quarters of a million. Here climates range from *Cfc* to *E*, soils are thin and take thousands of years to develop, and the air is arctic. We sailed slowly into Glacier Bay, in awe of the spectacular, unspoiled scenery, and turned into a bay filled with ice floes, some of them serving as rafts for sleeping seals. Calving (breaking up) into the bay was the Grand Pacific Glacier, with evidence of its recent recession all around. Less than 300 years ago, all of Glacier Bay was filled with ice; today you can sail miles to the Johns Hopkins (shown here) and Margerie tidewater glaciers' current outer edges. Global warming in action—and a reminder that any map of climate is a still picture of a changing world."

combined with onshore winds, give most of Europe a warmth that is not experienced at similar latitudes elsewhere.

How can we forge regional coherence out of such a jumble of data? The effort to create a relatively simple, small-scale world map of climates has continued for nearly a century. Figure I-8 is based on the system Wladimir Köppen devised and Rudolf Geiger then modified. This system, which has the advantage of simplicity, is represented by a set of letter symbols. The first (capital) letter is the critical one: the *A* climates are humid and tropical, the *B* climates are arid, the *C* climates are mild and humid, the *D* climates show increasing extremes of seasonal heat and cold, and the *E* climates reflect the frigid conditions at and near the poles.

Figure I-8 merits your attention because familiarity with it will help you understand much of what follows in this book. The map has practical utility, too. Although it 29 depicts **climatic regions**, daily weather in each color-coded region is relatively standard. If, for example, you are familiar with the weather in the large area mapped as *Cfa* in the southeastern United States, you will feel at home in Uruguay (South America), Kwazulu-Natal (South Africa), New South Wales (Australia), and Fujian Province (China). Let us look at the world's climatic regions in some detail.

Humid Equatorial (*A*) Climates

The humid equatorial, or tropical, climates are characterized by high temperatures all year and by heavy precipitation. In the *Af* subtype, the rainfall arrives in substantial amounts every month; but in the *Am* areas, the arrival of the annual wet *monsoon* (the Arabic word for "season" [see p. 387]) marks a sudden enormous increase in precipitation. The *Af* subtype is named after the vegetation that develops there—the tropical rainforest. The *Am* subtype, prevailing in part of peninsular India, in a coastal area of West Africa, and in sections of Southeast Asia, is appropriately referred to as the monsoon climate. A third tropical climate, the savanna (*Aw*), has a wider daily and annual temperature range

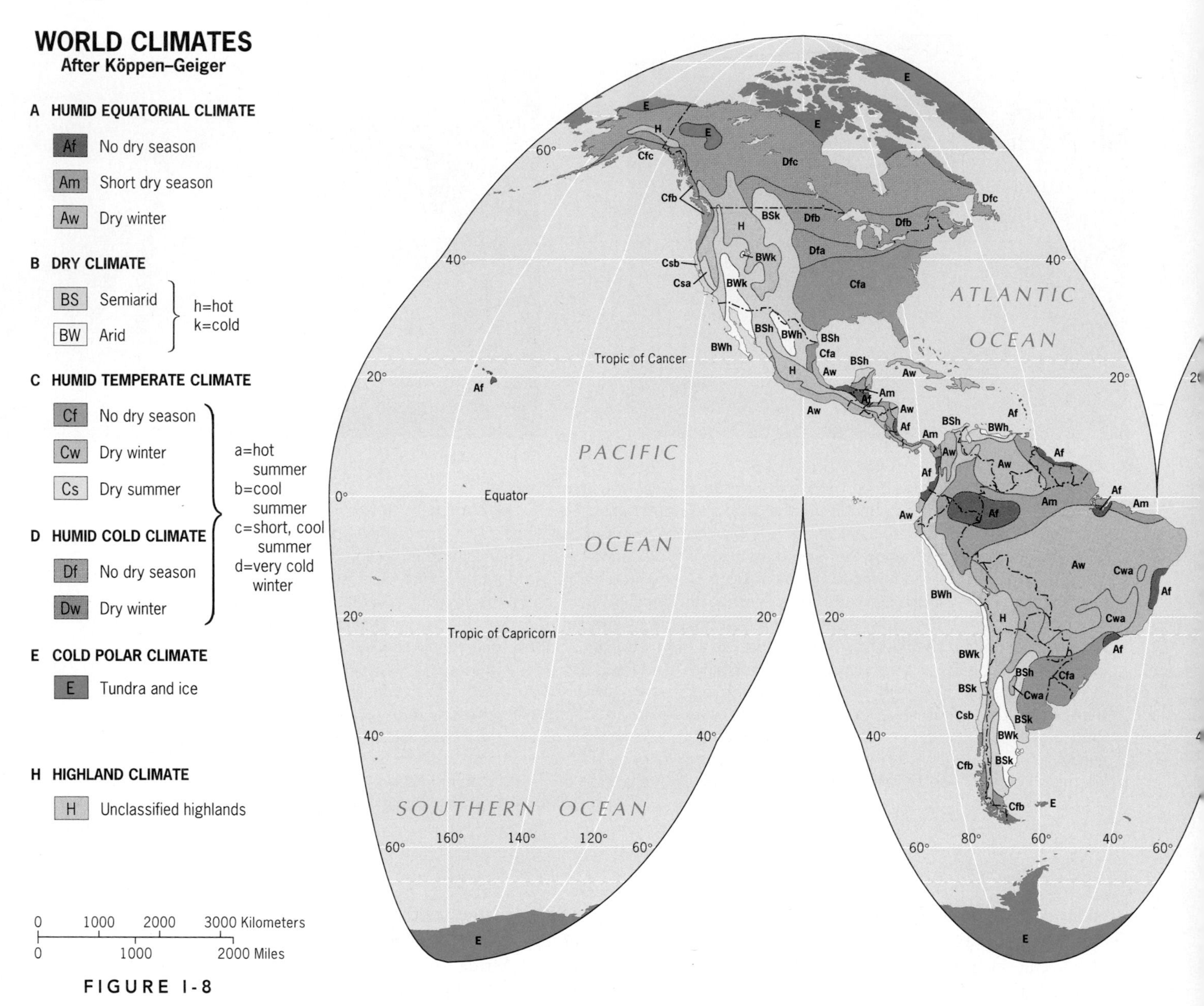

FIGURE I-8

and a more strongly seasonal distribution of rainfall. As Figure I-7 indicates, savanna rainfall totals tend to be lower than those in the rainforest zone, and savanna seasonality is often expressed in a "double maximum." Each year produces two periods of increased rainfall separated by pronounced dry spells. In many savanna zones, inhabitants refer to the "long rains" and the "short rains" to identify those seasons; a persistent problem is the unpredictability of the rain's arrival. Savanna soils are not among the most fertile, and when the rains fail hunger looms. Savanna regions are far more densely peopled than rainforest areas, and millions of residents of the savanna subsist on what they cultivate. Rainfall variability is their principal environmental problem.

Dry (*B*) Climates

Dry climates occur in both lower and higher latitudes. The difference between the *BW* (true desert) and the moister *BS* (semiarid steppe) varies but may be taken to lie at about 10 inches (25 cm) of annual precipitation. Parts of the central Sahara in North Africa receive less than 4 inches (10 cm) of rainfall. Most of the world's arid areas have an enormous daily temperature range, especially in subtropical deserts. In the Sahara, there are recorded instances of a maximum daytime shade temperature of over 120°F (49°C) followed by a nighttime low of 48°F (9°C). Soils in these arid areas tend to be thin and poorly developed; soil scientists have an appropriate name for them—aridisols.

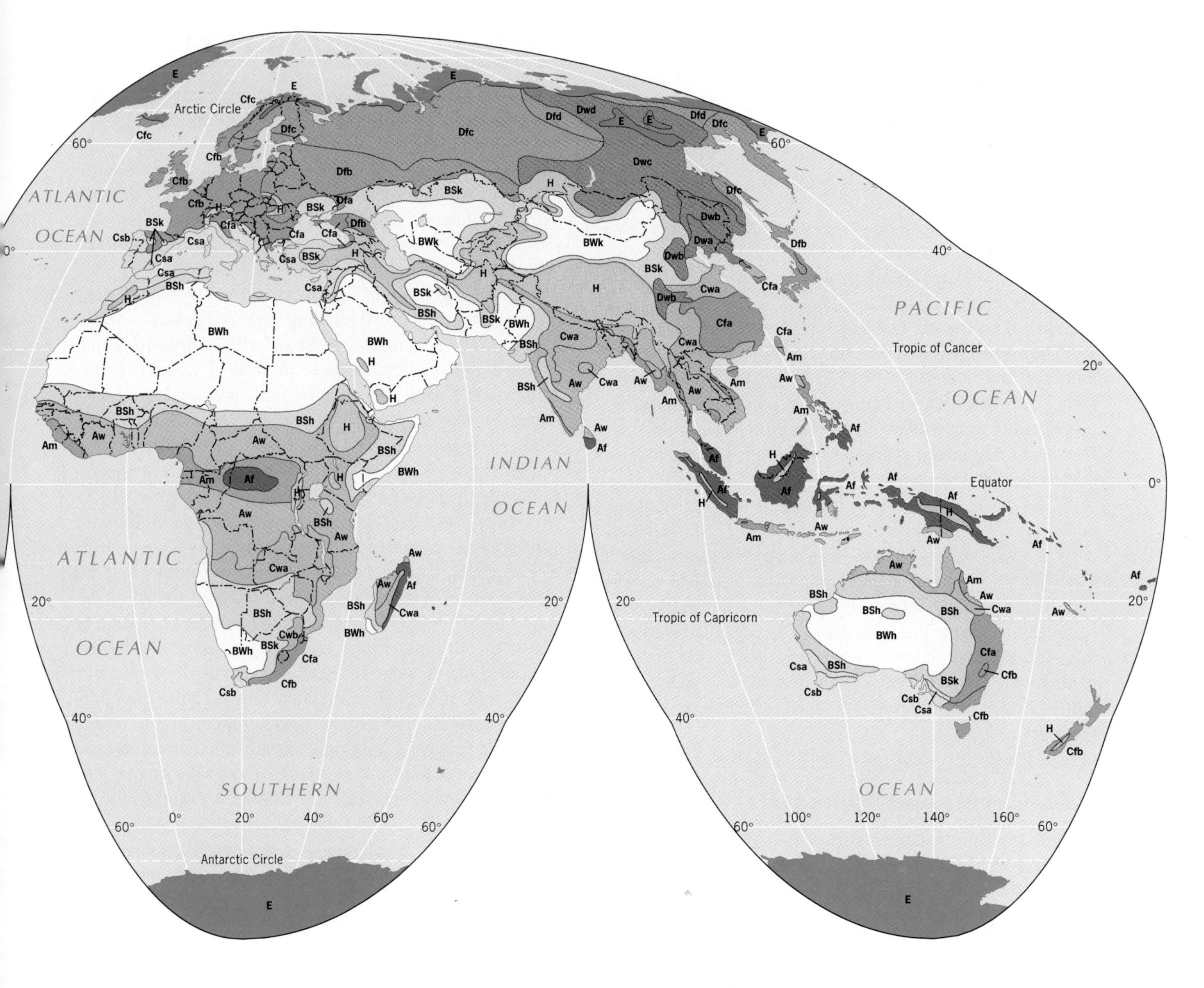

Humid Temperate (*C*) Climates

As the map shows, almost all these mid-latitude climate areas lie just beyond the Tropics of Cancer and Capricorn (23½ ° North and South latitude, respectively). This is the prevailing climate in the southeastern United States from Kentucky to central Florida, on North America's west coast, in most of Europe and the Mediterranean, in southern Brazil and northern Argentina, in coastal South Africa, in eastern Australia, and in eastern China and southern Japan. None of these areas suffers climatic extremes or severity, but the winters can be cold, especially away from water bodies that moderate temperatures. These areas lie midway between the winterless equatorial climates and the summerless polar zones. Fertile and productive soils have developed under this regime, as we will note in our discussion of the North American and European realms.

The humid temperate climates range from moist, as along the densely forested coasts of Oregon, Washington, and British Columbia, to relatively dry, as in the so-called Mediterranean (dry-summer) areas that include not only coastal southern Europe and northwestern Africa but also the southwestern tips of Australia and Africa, central Chile, and Southern California. In these Mediterranean environments, the scrubby, moisture-preserving vegetation creates a natural landscape different from that of richly green Western Europe.

Humid Cold (*D*) Climates

The humid cold (or "snow") climates may be called the continental climates, for they seem to develop in the interior of large landmasses, as in the heart of Eurasia or North America. No equivalent land areas at similar latitudes exist in the Southern Hemisphere; consequently, no *D* climates occur there.

Great annual temperature ranges mark these humid continental climates, and cold winters and relatively cool summers are the rule. In a *Dfa* climate, for instance, the warmest summer month (July) may average as high as 70°F (21°C), but the coldest month (January) might average only 12°F (−11°C). Total precipitation, much of it snow, is not high, ranging from over 30 inches (75 cm) to a steppe-like 10 inches (25 cm). Compensating for this paucity of precipitation are cool temperatures that inhibit the loss of moisture from evaporation and evapotranspiration (moisture loss to the atmosphere from soils and plants).

Some of the world's most productive soils lie in areas under humid cold climates, including the U.S. Midwest, parts of southern Russia and Ukraine, and Northeast China. The winter dormancy (when all water is frozen) and the accumulation of plant debris during the fall balance the soil-forming and enriching processes. The soil differentiates into well-defined, nutrient-rich layers, and substantial organic humus accumulates. Even where the annual precipitation is light, this environment sustains extensive coniferous forests.

Cold Polar (*E*) and Highland (*H*) Climates

Cold polar (*E*) climates are differentiated into true icecap conditions, where permanent ice and snow keep vegetation from gaining a foothold, and the tundra, which may have average temperatures above freezing up to four months of the year. Like rainforest, savanna, and steppe, the term *tundra* is vegetative as well as climatic, and the boundary between the *D* and *E* climates in Figure I-8 corresponds closely to that between the northern coniferous forests and the tundra.

Finally, the *H* climates—unclassified highlands mapped in gray (Fig. I-8)—resemble the *E* climates. High elevations and the complex topography of major mountain systems often produce near-Arctic climates above the tree line, even in the lowest latitudes such as the equatorial section of the high Andes of South America.

Let us not forget an important qualification concerning Figure I-8: this is a still-picture of a changing scene, a single frame from an ongoing film. Climate is still changing, and less than a century from now climatologists are likely to be modifying the climate maps to reflect new data. Who knows: we may have to redraw even those familiar coastlines. Environmental change is a never-ending challenge.

REGIONS AND CULTURES

Whenever we explore a geographic realm or region, we should assess the physical stage that forms its base—its total physical geography or **physiography**. Still, the realms **30** and regions we discuss in the chapters that follow are defined primarily by human-geographic criteria, and one criterion, culture, is especially significant. Therefore, we should carefully consider the culture concept in a regional context.

When anthropologists define **culture**, they tend to con- **31** centrate on abstractions: learning, knowledge and its transmission, and behavior. More than a half century ago, Ralph Linton defined culture as "the sum total of the knowledge, attitudes, and habitual behavior patterns shared and transmitted by the members of a society." Marvin Harris, in 1971, wrote that culture is "the learned patterns of thought and behavior characteristic of a population or society." Hundreds of definitions of the culture concept exist. As with the regional concept in geography, such definitions are arbitrary and designed for particular purposes.

Geographers are most interested in how culture and the patterns of behavior associated with it are imprinted on the landscape. Thus we will examine how members of a society perceive and exploit their available resources, how they maximize the opportunities and adapt to the limitations of their natural environment, and how they organize their portion of the Earth. Some human works remain etched on the Earth's surface for a long time: the Egyptian pyramids, the Great Wall of China, and Roman roads and bridges are still in place millennia after they were built. Over time, regions take on dominant qualities that collectively constitute a **regional character**, a personality, a distinct atmosphere. **32** Regional character is a crucial criterion in ascertaining how we divide the human world into major geographic realms and regions.

Cultural Landscapes

As we noted, geographers are particularly concerned with the impress of culture on the Earth's physical surface. Culture gives visible character to a region in many ways. Often a single scene in a photograph or picture can reveal, in general terms, where the photo was taken. The architecture, forms of transportation, goods being carried, and clothing of the people (all these are part of culture) help us guess the region in the photo. We can do this because the people of any culture are active agents of change; they transform the land they occupy by building structures, creating lines of transport and communication, parceling out fields, and tilling the soil (among countless other activities).

The composite of human imprints on the Earth's surface is called the **cultural landscape**, a term that came into gen- **33**

FROM THE FIELD NOTES

"The Atlantic-coast city of Bergen, Norway displayed the Norse cultural landscape more comprehensively, it seemed, than any other Norwegian city, even Oslo. The high-relief site of Bergen creates great vistas, but also long shadows; windows are large to let in maximum light. Red-tiled roofs are pitched steeply to enhance runoff and inhibit snow accumulation; streets are narrow and houses clustered, conserving warmth . . . The coastal village of Mengkabong on the Borneo coast of the South China Sea represents a cultural landscape seen all along the island's shores, a stilt village of the Bajau, a fishing people. Houses and canoes are built of wood as they have been for centuries. But we could see some evidence of modernization: windows filling wall openings, water piped in from a nearby well."

eral use in geography during the 1920s. Carl Ortwin Sauer (a professor of geography at the University of California, Berkeley) developed a school of cultural geography that focused on the concept of cultural landscape. In a paper titled "Recent Developments in Cultural Geography" (1927), Sauer proposed his most straightforward definition of the cultural landscape: *the forms superimposed on the physical landscape by the activities of man*. Such forms result from cultural processes—causal forces that shape cultural patterns—that unfold over a long time and involve the cumulative influences of successive occupants.

Sometimes these successive groups are not of the same culture. Africans now occupy settlements European colonizers built over a century ago; minarets of Islam still rise above certain Eastern European cities, recalling dominance by the Muslim Ottoman Empire. In 1929, Derwent Whittlesey introduced the term **sequent occupance** to categorize these successive stages in the evolution of a region's cultural landscape.

34

The durability of the concept of cultural landscape is underscored by its redefinition in 1984 by John Brinckerhoff Jackson. His statement closely parallels Sauer's: *a composition of man-made or man-modified spaces to serve as infrastructure or background for our collective existence*. Thus the cultural landscape consists of buildings, roads, and fields—and more. But it also possesses an intangible quality, an atmosphere or flavor, a sense of place that is often easy to perceive yet difficult to define. The smells, sights, and sounds of a traditional African market are unmistakable, but try recording those qualities on maps or in another objective way for comparative study!

More concrete properties are easier to observe and record. Take, for instance, the urban "townscape"—a prominent element of the overall cultural landscape—and compare a major U.S. city with, say, one in Japan. Photographs of these two metropolitan scenes would reveal the differences quickly, of course, but so would maps. The American city with its square-grid layout of the **central business district (CBD)** and its widely dispersed, now heavily urbanized suburbs contrasts sharply with the clustered, space-conserving Japanese metropolis. In a rural example, the spatially lavish subdivision and ownership patterns of American farmland look unmistakably different from the traditional African countryside, with its irregular, often tiny patches of land surrounding a village. Still, the totality of a cultural landscape can never be captured by a

35

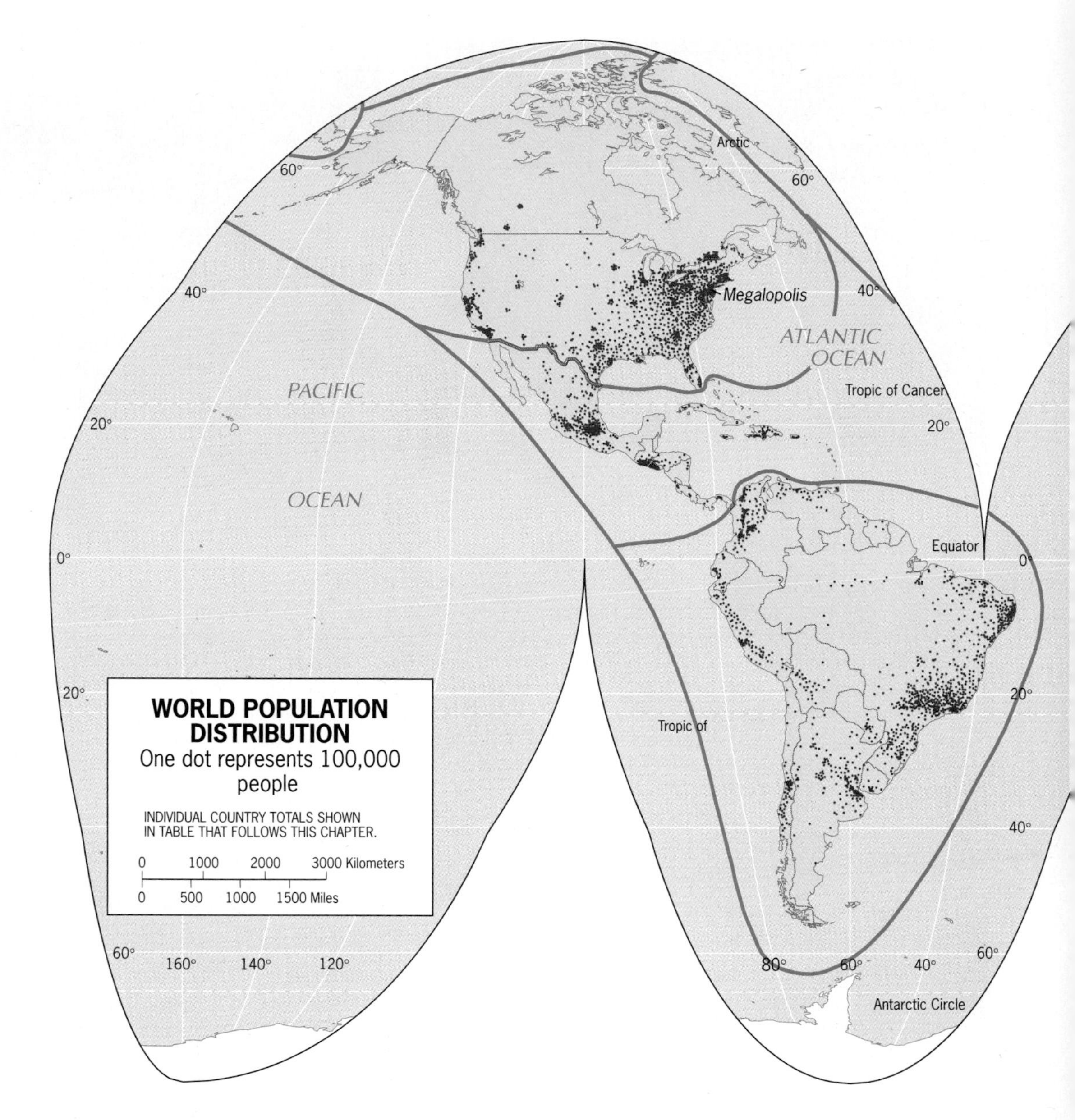

FIGURE 1-9

photograph or map because the personality of a region involves more than its prevailing spatial organization. One must also include the region's visual appearance, its noises and odors, the shared experiences of its inhabitants, and even their pace of life.

Culture and Ethnicity

Language, religion, and other cultural traditions often are durable and persistent. Culture is not necessarily based on 36 **ethnicity**, however, and as we study the world's human-geographic regions, we should be aware that peoples of different ethnic stocks can achieve a common cultural landscape, while people of the same ethnic background can be divided along cultural lines.

The recent events in the former Eastern European country of Yugoslavia are a case in point. As the old Yugoslavia broke up in the early 1990s, its component parts were first affected and then engulfed by what was often described as "ethnic" conflict. When the crisis reached Bosnia-Herzegovina in the heart of Yugoslavia, three groups fought a bitter civil war. These groups were identified as Bosnian Muslims, Serbs, and Croats—but in fact all were ethnic Slavs (Yugoslavia means "Land of the South Slavs"). What distanced them from one another, and what kept the conflict going, was cultural tradition, not ethnicity. The Bosnian

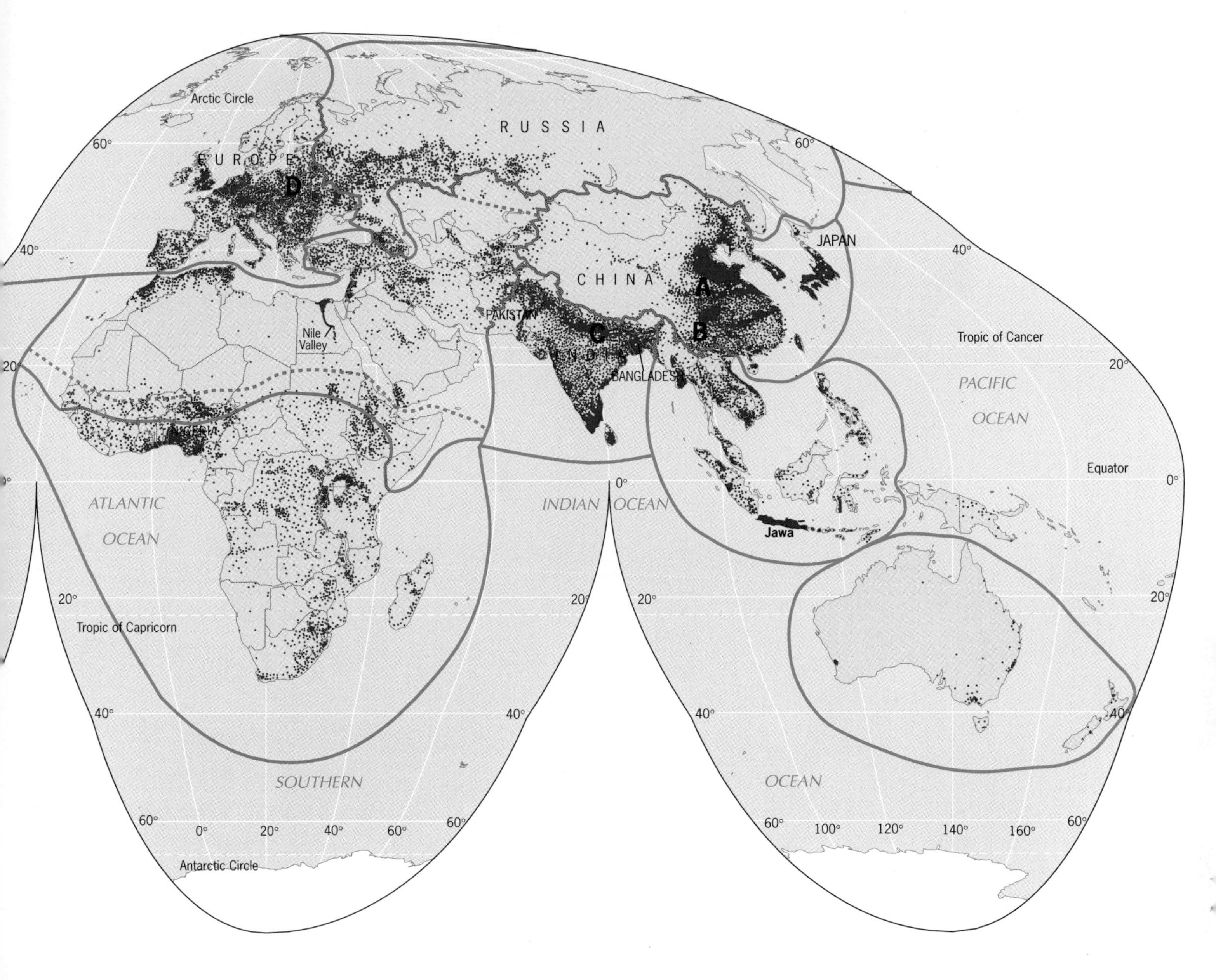

Muslims and the Serbs had developed different communities in different parts of the former Yugoslavia. The Muslims were descendants of Slavs who, a century ago or more, had been converted by the Turks (who once ruled here) from Christianity to Islam. Each of these groups feared domination by the others, and so South Slav turned against South Slav in what was, in truth, a culture-based conflict.

Even though the post–Cold War world is young, it has already witnessed numerous intraregional conflicts; in later chapters, we will explain some of these conflicts in geographic context. Not all of them have ethnicity at their roots. Culture is a great unifier; it can also be a powerful divider.

REALMS OF POPULATION

Earlier we noted that population numbers by themselves do not define geographic realms or regions. Population distributions, and the functioning society that gives them common ground, are more significant criteria. That is why we can identify one geographic realm (the Austral) with barely more than 20 million people and another (East Asia) with 1.5 billion inhabitants. Neither population numbers nor territorial size alone can delimit a geographic realm. Nevertheless, the map of world population distribution (Fig. I-9) suggests the relative location of several of the world's geographic realms, based on the strong clustering of population

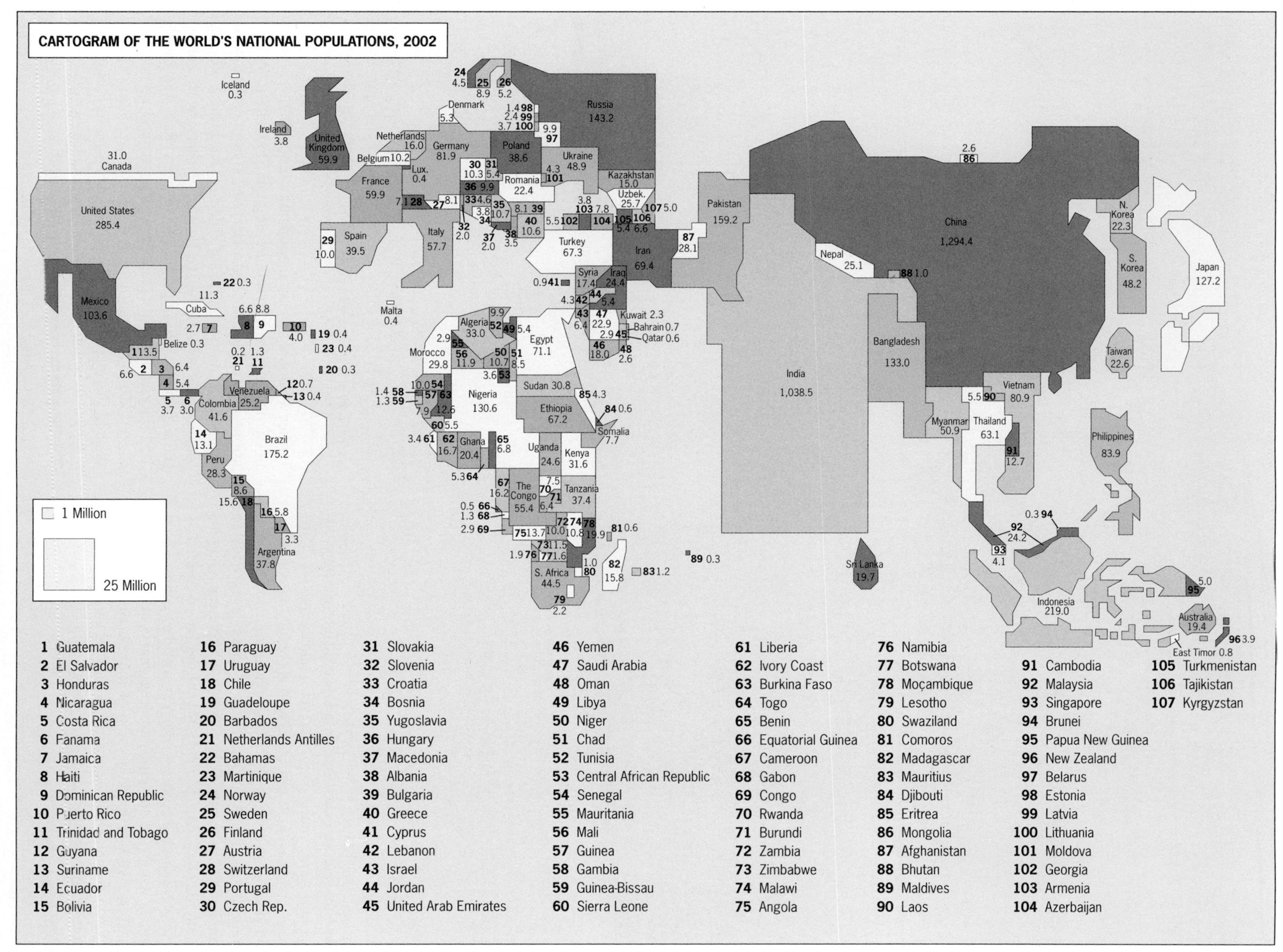

1 Guatemala	**16** Paraguay	**31** Slovakia	**46** Yemen	**61** Liberia	**76** Namibia	**91** Cambodia	**105** Turkmenistan
2 El Salvador	**17** Uruguay	**32** Slovenia	**47** Saudi Arabia	**62** Ivory Coast	**77** Botswana	**92** Malaysia	**106** Tajikistan
3 Honduras	**18** Chile	**33** Croatia	**48** Oman	**63** Burkina Faso	**78** Moçambique	**93** Singapore	**107** Kyrgyzstan
4 Nicaragua	**19** Guadeloupe	**34** Bosnia	**49** Libya	**64** Togo	**79** Lesotho	**94** Brunei	
5 Costa Rica	**20** Barbados	**35** Yugoslavia	**50** Niger	**65** Benin	**80** Swaziland	**95** Papua New Guinea	
6 Panama	**21** Netherlands Antilles	**36** Hungary	**51** Chad	**66** Equatorial Guinea	**81** Comoros	**96** New Zealand	
7 Jamaica	**22** Bahamas	**37** Macedonia	**52** Tunisia	**67** Cameroon	**82** Madagascar	**97** Belarus	
8 Haiti	**23** Martinique	**38** Albania	**53** Central African Republic	**68** Gabon	**83** Mauritius	**98** Estonia	
9 Dominican Republic	**24** Norway	**39** Bulgaria	**54** Senegal	**69** Congo	**84** Djibouti	**99** Latvia	
10 Puerto Rico	**25** Sweden	**40** Greece	**55** Mauritania	**70** Rwanda	**85** Eritrea	**100** Lithuania	
11 Trinidad and Tobago	**26** Finland	**41** Cyprus	**56** Mali	**71** Burundi	**86** Mongolia	**101** Moldova	
12 Guyana	**27** Austria	**42** Lebanon	**57** Guinea	**72** Zambia	**87** Afghanistan	**102** Georgia	
13 Suriname	**28** Switzerland	**43** Israel	**58** Gambia	**73** Zimbabwe	**88** Bhutan	**103** Armenia	
14 Ecuador	**29** Portugal	**44** Jordan	**59** Guinea-Bissau	**74** Malawi	**89** Maldives	**104** Azerbaijan	
15 Bolivia	**30** Czech Rep.	**45** United Arab Emirates	**60** Sierra Leone	**75** Angola	**90** Laos		

FIGURE I-10

in certain areas. Before we examine these clusters in some detail, remember that the Earth's human population now totals over 6 billion—6 thousand million people confined to the landmasses that constitute less than 30 percent of our planet's surface, much of which is arid desert, rugged mountain terrain, or frigid tundra. (Remember that Fig. I-9 is a turn-of-the-millennium still-picture of an ever-changing scene; the explosive growth of humankind continues.) After thousands of years of relatively slow growth, world population during the past two centuries has been expanding at an increasing rate. It took about 17 centuries after the birth of Christ for the world to add 250 million people; now we are adding that same number about every *four years*!

Demographic (population-related) issues will arise repeatedly in later chapters as we survey the world's most crowded realms. For the moment, we will confine ourselves to the overall global situation. For that purpose let us compare the map of world **population distribution** [37] to the demographic data shown in Table I-2 (pp. 35-39). This table provides information about the total populations in each geographic realm as well as their individual countries. Also, if you compare Figure I-9 to the maps of terrain (Fig. I-1) and climate (Fig. I-8), you will note that some of the largest population concentrations lie in the fertile basins of major rivers (China's Huang He and Chang Jiang, India's Ganges). We live in a modern world, but old ways of life still prescribe where hundreds of millions on this Earth live.

Major Population Clusters

The world's greatest population cluster, *East Asia*, lies centered on China and includes the Pacific-facing Asian coastal zone from the Korean Peninsula to Vietnam. The map indicates that the number of people per unit area—the **population density** [38]—tends to decline from the coastal zone toward the interior. Note, however, the ribbon-like extensions marked *A* and *B* (Fig. I-9). As the map of world natural landscapes (Fig. I-1) confirms, these are populations concentrated in the valleys of China's major rivers, the Huang He and the Chang Jiang, respectively. Here in East Asia, most people are farmers, not city dwellers. There are great cities in China (such as Beijing and Shanghai), but their total population still is far outnumbered by the farmers—those who live and work on the land and whose crops of rice and wheat feed not only themselves but also the people in those cities.

The *South Asia* population cluster lies centered on India and includes the populous neighboring countries of Bangladesh and Pakistan. This huge agglomeration of humanity focuses on the broad plain of the lower Ganges River (*C* in Fig. I-9). This cluster is nearly as large as that of East Asia and at present growth rates will overtake East Asia in less than 20 years. As in East Asia, most people are farmers, but in South Asia pressure on the land is even greater, and farming is less efficient. As we note in Chapter 8, the population issue looms large in this realm.

The third-ranking population cluster, *Europe*, also lies on the world's biggest landmass but at the opposite end from China. The European cluster, including western Russia, counts over 700 million inhabitants, which puts it in a class with the two larger Eurasian concentrations—but there the similarity ends. In Europe, the key to the linear, east-west orientation of the axis of population (*D* in Fig. I-9) is not a fertile river basin but a zone of raw materials for industry. Europe is among the world's most highly urbanized and industrialized realms, its human agglomeration sustained by forges and factories rather than paddies and pastures.

The three world population concentrations just discussed (East Asia, South Asia, and Europe) account for more than 3.6 billion of the world's 6.2 billion people. Nowhere else on the globe is there a population cluster with a total of even half of any of these. Look at the dimensions of the landmasses on Figure I-9: the populations of South America, Subsaharan Africa, and Australia together total less than the population of India alone. The next-ranking cluster, *Eastern North America*, comprising the east-central United States and southeastern Canada, is only about one-quarter the size of the smallest of the Eurasian concentrations. As Figure I-9 shows, this region does not possess the large, contiguous, high-density zones of Europe or East and South Asia. Like the European cluster, much of the population of this region is concentrated in several major metropolitan centers; the rural areas remain relatively sparsely settled. The heart of the North American cluster lies in the urban complex that lines the U.S. northeastern seaboard from Boston to Washington, D.C., and includes New York, Philadelphia, and Baltimore—the great multi-metropolitan agglomeration that urban geographers refer to as **Megalopolis**. [39]

The quantitative dominance of Eurasia's population clusters is revealed dramatically by a special map transformation called a **cartogram** [40] (Fig. I-10). Here the map of the world is redrawn, so that the area of each country reflects *not* territory but population numbers. China and India stand out because, unlike the European population cluster, their populations are not fragmented among many national political entities.

Neither Figure I-9 nor Figure I-10 can tell us how the populations they exhibit are changing. As we will see, varying rates of population growth mark the world's realms and regions; some countries' populations are growing rapidly, while populations in other countries are actually shrinking. Another key factor is **urbanization** [41], the percentage of the population living in cities and towns. Not only is there much geographic variation in the level of urbanization among realms and regions, but some realms are urbanizing

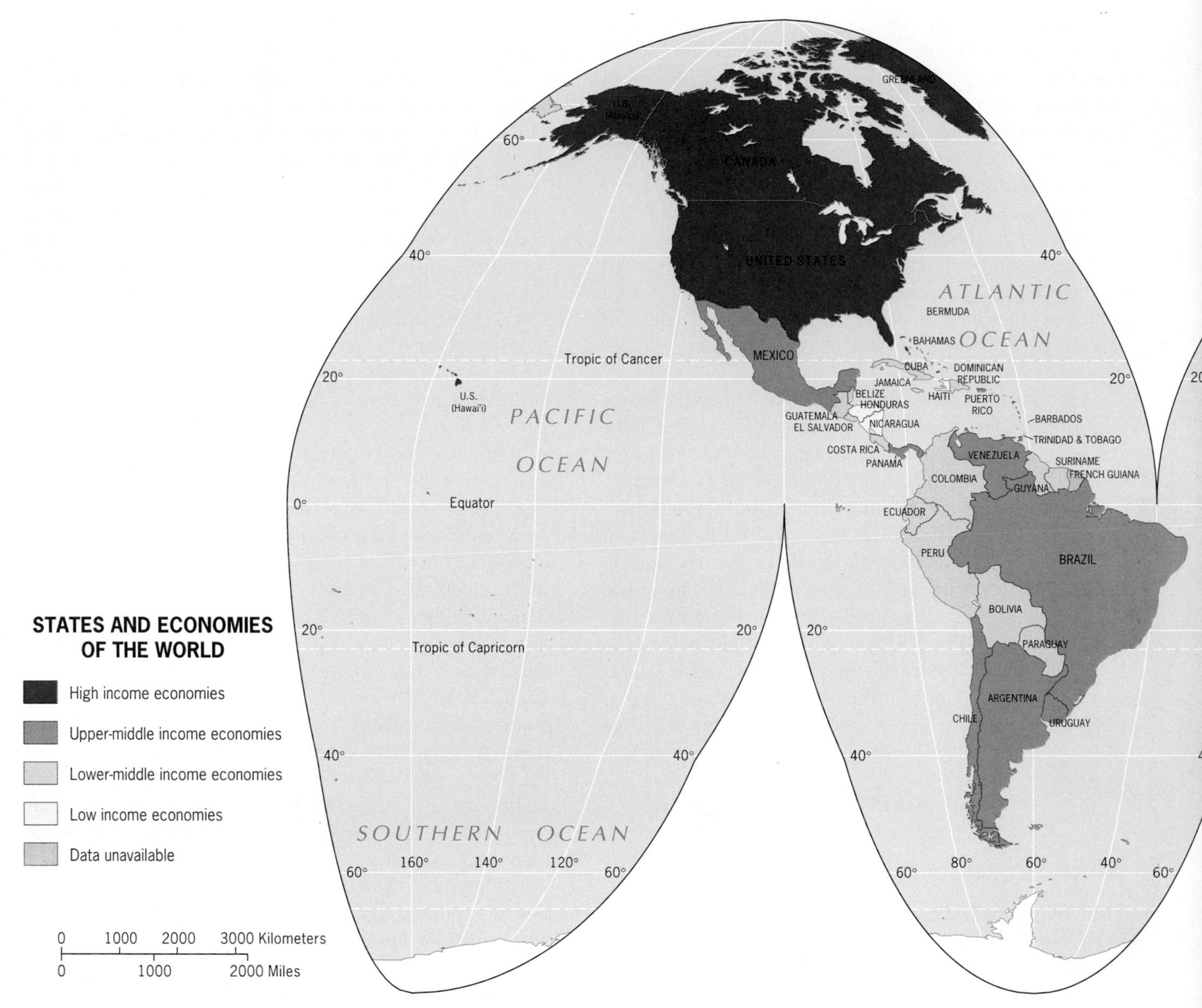

FIGURE I-11

much more rapidly than others. Urbanization is a defining property of geographic regions, and we will explain why they differ so markedly in this respect.

REALMS, REGIONS, AND STATES

Whenever geographers need to define (and justify) boundaries, they look for precedents—for example, an existing grid that can be put to use or data from a previous effort that may prove helpful. A case in point is the Township-and-Range system of land division in the United States west of the Appalachian Mountains. When you fly over this area, you can see a persistent rectangular pattern on the ground. This pattern is the product of the Ordinance of 1785, when it was decided to survey and delimit this then-remote area before it was opened to settlement. The country beyond the Appalachians was divided into squares of 6 by 6 miles, so that each square (or "township") contained 36 square miles. Next, each of these square miles (640 acres) was subdivided into four 160-acre squares, which was to be the smallest parcel of land a settler could purchase. Today, the Township-and-Range system is etched into the cultural landscape because it controlled the location of towns, roads, farms, and fields (see photo, p. 6). For

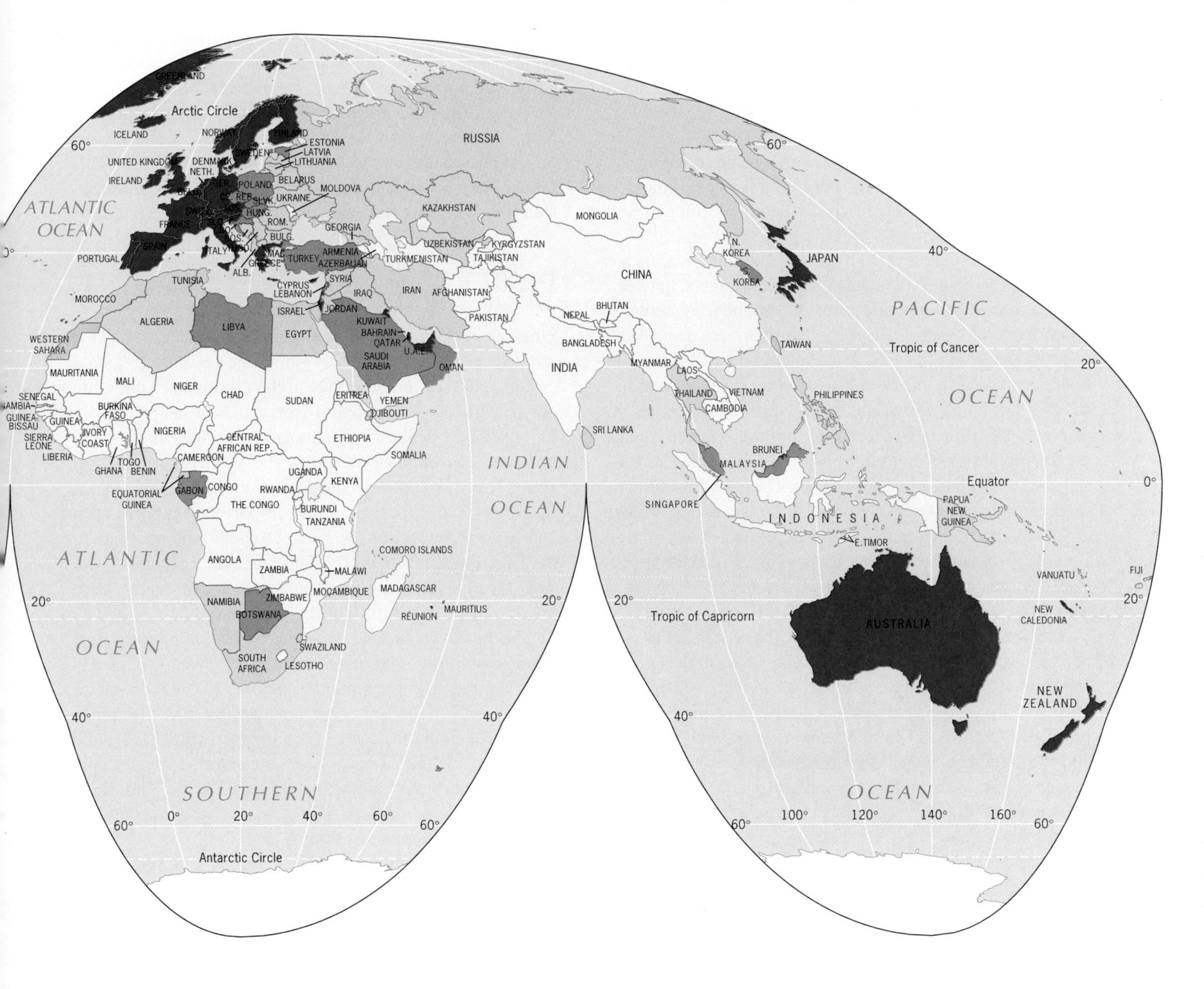

geographers, this ready-made grid is invaluable in delimiting a Corn Belt, a Wheat Belt, or some other economic region.

On a global scale we have at our disposal only one existing framework, and it is not a regular grid: the international boundary system that marks the territorial limits of the world's 180-plus countries (see Front Endpaper map; the world's countries are also shown in Fig. I-11). As the map shows, countries range in size from the largest, Russia (6.6 million square miles [17.1 million sq km]), to entities so tiny they cannot be shown. Irregular as this framework may be, it nonetheless helps us define regions within the world's great geographic realms. Although we will often refer to the political entities shown in Figure I-11 as *countries,* the appropriate geographic term for them is *states.**

The **state** has been developing for thousands of years, ever since agricultural surpluses made possible the growth of large and powerful cities that could command hinterlands and control peoples far beyond their walls. But the modern state is a relatively recent phenomenon. The

42

*This term may lead to some confusion because some states are administratively divided into units also called *states*—as in the United States of America. Our solution is to capitalize State when it is a subdivision of a country. Thus it is the State of California and the State of New South Wales (Australia), but the state of Japan and the state of Argentina.

boundary framework we see on the world political map today substantially came about during the nineteenth century, and the independence of dozens of former colonies (which made them states as well) occurred during the twentieth century. Today, the [43] **European state model**—a clearly and legally defined territory inhabited by a population governed from a capital city by a representative government—prevails in the aftermath of the collapse of colonial and communist empires.

As Figures I-2 and I-11 suggest, geographic realms are mostly assemblages of states, and the borders between realms frequently coincide with the boundaries between countries—for example, between North America and Middle America along the U.S.-Mexico boundary. But a realm boundary can also cut *across* a state, as does the one between Subsaharan Africa and the Muslim-dominated realm of North Africa/Southwest Asia. Here the boundary takes on the properties of a wide transition zone, but it still divides states such as Chad and Sudan. The transformation of the margins of the former Soviet Union is creating similar cross-country transitions. Newly independent states such as Belarus (between Europe and Russia) and Kazakhstan (between Russia and Muslim Southwest Asia) lie in zones of regional change.

Most often, however, geographic realms consist of groups of states whose boundaries also mark the limits of the realms. Look at Southeast Asia, for instance. Its northern border coincides with the political boundary that separates China (a realm practically unto itself) from Vietnam, Laos, and Myanmar (Burma). The boundary between Myanmar and Bangladesh (which is part of the South Asian realm) defines its western border. Here, the state boundary framework helps delimit geographic realms.

The global boundary framework is even more useful in delimiting regions *within* geographic realms. We shall discuss regional divisions every time we introduce a geographic realm, but an example is appropriate here. In the Middle American realm, we recognize four regions. Two of these lie on the mainland: Mexico, the giant of the realm, and Central America, which consists of the seven comparatively small countries located between Mexico and the Panama-Colombia border (which is the boundary with the South American realm). Central America often is misdefined in news reports; the correct regional definition is based on the politico-geographical framework.

To our earlier criteria of physiography, population distribution, and cultural geography, therefore, we now add political geography as a determinant of world-scale geographic regions. In doing so, we should be aware that the global boundary framework continues to change and that boundaries are created (as in 1993 between the Czech Republic and Slovakia) as well as eliminated (e.g., between former West and East Germany in 1990). But the overall system, much of it resulting from colonial and imperial expansionism, has endured, despite the predictions of some geographers that the "boundaries of imperialism" would be replaced by newly negotiated ones in the postcolonial period.

Toward the end of this book, we will discuss a recent, ominous development in boundary-making: the extension of boundaries onto and into the oceans and seas. This process has been consuming the last of the Earth's open frontiers, with uncertain consequences.

PATTERNS OF DEVELOPMENT

Finally, we turn to economic geography for criteria that allow us to place countries, regions, and realms in regional groupings based on their level of [44] **development.** [45] **Economic geography** focuses on spatial aspects of how people make a living; thus it deals with patterns of production, distribution, and consumption of goods and services. As with all else in this world, these patterns exhibit considerable variation. Individual states report their imports and exports, farm and factory production, and many other economic data to the United Nations and other international agencies. From such information we can determine the comparative economic well-being of the world's countries.

The World Bank, one of the agencies that monitor economic conditions, classifies countries into four groups: (1) high-income, (2) upper-middle-income, (3) lower-middle-income, and (4) low-income countries. These groupings display regional clustering when mapped (Fig. I-11). The higher-income economies are concentrated in Europe, North America, and along the western Pacific Rim, most notably in Japan and Australia. The low-income countries dominate in Africa and parts of Asia.

As Figure I-11 shows, several geographic-realm boundaries can be discerned on this economic map, including that between North America and Middle America, between Australia and Southeast Asia, and between Russia and East Asia. Within the great geographic realms, you can see some distinct regional boundaries: between Western and Eastern Europe, between Brazil and Andean South American (the region to the west), and between the oil-rich Arabian Peninsula and the countries of the Middle East to its northwest.

Figure I-11 reveals the level of economic development among the world's countries—based, we should remind ourselves, on the political framework as the reporting grid. Until recently, economists and economic geographers used this basis to divide the world into developed countries (DCs) and underdeveloped countries (UDCs). But that distinction has lost much of its relevance, for reasons Figure I-11 cannot show. Today, the world has [46] **core** areas where the richer countries are clustered. But many countries, whatever their level of income, display cores of development (where they often resemble the richest societies) as well as [47] **peripheries** of poverty and underdevelopment (see box titled "Core-Periphery Regional Relationships"). Even comparatively wealthy countries still have areas of severe underdevelopment. Other countries, mapped as low-income economies, have given rise to bustling, skyscrapered cities whose streets are jammed with luxury automobiles and shops selling opulent goods. As the rich cores get richer and the poor peripheries get poorer, the averages reported as "national" economic statistics lose much of their geographic meaning. Perhaps an index of **regional disparity** within each economy would give us a truer picture.

49

It is therefore no longer appropriate to divide the world geographically into developed and underdeveloped realms. East Asia is dominated by low-income economies, but it includes Japan, one of the world's richest states. Europe is one of the world's highest-income realms, but not in Albania, Moldova, or Ukraine.

Not long ago, countries also were categorized as being part of a First World (capitalist), Second World (socialist/communist), Third World (underdeveloped), Fourth World (severely underdeveloped), or even Fifth World (poorest). This classification, too, has lost most of its significance in the twenty-first century. The most meaningful distinction we can make is based on the notion of **advantage**: some countries have it and many others do not. Such advantage takes many forms: geographic location (to be on the coast is generally better than to be landlocked!), raw materials, government, political stability, productive skills, and much more. As the gap between the advantaged and the disadvantaged states widens (and it *is* widening), the stability of the political world is at risk, and with it the advantages of the "haves" over the "have nots."

50

As we proceed with our investigation of the world's geographic realms, the concept of development will be examined in a regional context and from several viewpoints. Figure I-11 reflects events that began long before the Industrial Revolution in the late eighteenth and nineteenth centuries: Europe, even by the middle of the eighteenth century, had laid the foundations for its colonial expansion. The Industrial Revolution then magnified Europe's demands for raw materials, and its manufactures increased its imperial control. Western countries thereby gained an enormous head start, while colonial dependencies continued to supply raw materials and to consume Western industrial products. Thus was born a system of international exchange and capital flow that changed little when the colonial period ended. Developing countries, well aware of their predicament, accused the developed world of perpetuating its long-term advantage through **neocolonialism**—the entrenchment of the old system under a new guise.

51

Symptoms of Underdevelopment

Although it is no longer appropriate to divide the world into developed and underdeveloped geographic realms, make no mistake: there still *are* regions within realms, and countries within regions, that suffer from underdevelopment. As we will discover during our global journey, underdevelopment takes many forms, displays many symptoms, and has many causes. To grasp some of the symptoms, see Table I-2 and compare, say, Bangladesh or Moçambique with Japan or Canada. Population growth rates in the periphery are higher, life expectancies shorter, urbanization rates lower, incomes smaller. As we will find, the poorer countries suffer from high infant and child mortality rates, poor overall health and sanitation, inefficient farming, insufficient diets, overcrowded cities, and many other ills. Many countries in the periphery are trapped in a global economic system in which the export of unfinished or partially finished raw materials is their only source of outside income.

The Specter of Debt

All of us are aware of the dangers of going too deeply into debt. Borrow too much money, and payments of interest and principal may leave too little for routine needs such as food and clothing, not to mention repairs and replacement of equipment. Individuals, families, villages, and cities must devise budgets and balance incomes against expenditures.

So it is with countries. National governments, like individuals and families, must sometimes borrow to make ends meet. If a drought curtails domestic food production, a government may borrow against future oil or ore exports to pay for urgently needed staples. Should domestic harvests fail in the following year as well, further borrowing will be necessary. This means that the government's future budgets will be burdened by ever-higher "debt retirement" payments, limiting its ability to spend on schools, roads, and clinics. Should the decision be made to raise taxes to help make the payments, the result may be a slowdown in the economy. The cycle of debt is hard to escape.

All too often it is not just a food emergency or some other unavoidable problem, but a government's mismanagement that leads to excessive debt. In the postcolonial period, many governments of former colonies sought shortcuts to boost their economies, building dams, factories, and ports that did not justify their huge investments. Of course the governments and corporations of the former colonial powers were all too willing to do the building, making large profits and even lending the new governments money to build still more.

Soon, many of the civilian governments that took over from the colonial powers, weakened by such errors and losing the confidence of their citizens, fell prey to military takeovers. And the military dictatorships that replaced them needed weapons to maintain their hold on power—weapons readily sold for credit by the arms manufacturers in the wealthy core countries. During the Cold War, dictatorial regimes in the periphery became close allies of both superpowers, and proxy wars were fought in Asia, Africa, and the Americas. By the time the Cold War ended and repre-

Core-Periphery Regional Relationships

Ours is a world of uneven development. The advantages lie with the developed countries (DCs), and economic indicators show that the gap between the DCs and many underdeveloped countries (UDCs) is growing. How has this situation become so extreme? And what perpetuates it?

Regional contrasts in development have been attributed to many factors. Climate has been blamed (as a conditioner of human capacity), as has cultural heritage (e.g., resistance to innovation), colonial exploitation, and, more recently, neocolonialism. Obviously, the distribution of accessible natural resources and the territorial location of countries play a major part. Certain areas have always had more opportunities for interaction and exchange than others. Isolation from the mainstreams of change has always been a disadvantage. These factors, to a greater or lesser degree depending on location, have been important.

48

One hallmark of national spatial organization is the evolution of **core areas**, foci of human activity that function as the leading regions of control and change. Today we tend to associate this regional concept with a country's heartland—its largest population cluster, most productive and influential region, the area possessing the greatest centrality and accessibility that usually contains a powerful capital city.

This is nothing new. Archeologists have unearthed evidence of the earliest states, proving that city-dominated cores formed their foci. These were small urban centers, but later they were recast as city-states, culture hearths, seats of empires, and sources of modern technological revolutions. But such cores would not have developed without contributions from their surrounding areas. One of the earliest developments in ancient cities was the creation of organized armed forces to help rulers secure taxes and tribute from the people living in the countryside. Thus core and periphery (margin) became functionally tied together: the core had requirements, and the periphery supplied them.

Have modern times ended this core-periphery relationship? The answer not only is no—the situation has worsened. You can see core (have) and periphery (have-not) relationships in every country in the world. In many African countries (Kenya, Senegal, and Zimbabwe, to name examples in three different regions), the former colonial capital now lies at the focus of the national core area. Its tall buildings and government centers symbolize the concentration of privilege, power, and control. Here are the trappings of development, the paved roads and streetlights, hospitals and schools, corporate headquarters and businesses, industries and markets. Farms producing high-priced goods, especially perishables, for the city dwellers who can afford to buy them surround the capital. But farther out into the countryside, life changes. In the periphery, the core area is distant, even ominous—always a concern. How much will markets in the core pay for crops? How much will goods made there, goods that the countryside needs, cost? How high will taxes be? This core-periphery system keeps regional development contrasts high. The core may appear to be a thriving metropolis and look like a city in a developed country. But in the periphery underdevelopment reigns.

Looking at the world from this perspective, we see that core-periphery systems encompass not just individual countries but the entire globe. Today the whole of Western Europe, North America, and Japan function as core areas. Therefore, what we said about that powerful city-region in an otherwise underdeveloped country also applies, at the macroscale, to these global core areas. Power, money, influence, and decision-making organizations are concentrated in Europe, North America, and Japan.

Imagine that you, as a geographer, are asked to locate a factory (say, a textile plant) to sell products in core-area markets. Because salaries and wages in the core area are high, you look for an alternative site in the periphery, where wages are much lower. Raw materials also can be bought locally. So you advise the corporation to establish its factory in a certain country in the periphery, thereby ini-

sentative government began to return to some (though not all) of these countries, they were so deeply in debt to the rich core countries that they had no prospect of recovery.

External debt is not confined to the countries in the periphery, and even the richer countries (including the United States) have national debts. What matters is a country's ability to service its debt and still have the capacity to pay for its other needs. In Table I-1, note, for example, that Poland, Brazil, and South Korea (with the highest per-capita debt in the table) have per-capita gross national products more than three times as high. (A country's *gross national product* or *GNP* is the total value of all goods and tiating a relationship between core-based business and periphery-based producers. Now, however, the corporate management is concerned about political stability in that periphery country. Soon you are talking to government representatives about guarantees, and the economic connection also becomes a political one.

Is all this necessarily disadvantageous to the countries in the periphery, the mainly underdeveloped countries, the countries of the disadvantaged world? After all, the core-area corporations and businesses make investments, stimulate economies, and put people to work in the countries of the periphery. However, a dependency relationship develops that soon diminishes any such advantage. The profits earned return mostly to the core area and benefit the periphery only minimally. When you see that ultramodern high-rise hotel towering above the sandy beaches of a Caribbean island and watch its staff at work, you observe both sides of it. An investment was made, and jobs were created. But the hotel's profits go back to the multinational corporation's bank account—in New York, Tokyo, or London.

Another problem concerns the sociopolitical effects. That textile factory we mentioned needs a manager and other administrative personnel. These people tend to be (or become) part of the elite, the "upper class" in that periphery country, a kind of cadre of the world core area in the disadvantaged realms. Because they represent core-area interests, these people have divided loyalties: what is best for core-area enterprises is not always best for their home country. Such examples of core-periphery interactions abound, and the entire world is now more interconnected than many of us realize. Superimposed on the traditional cultural landscapes of the disadvantaged countries is a growing global network of interaction that reaches into even the smallest village store in the remotest part of a UDC. The numerous geographic realms and regions we will come to recognize in this book are, above all, enmeshed in a worldwide economic-geographic system—a system tightly oriented to the world's core areas.

Thus the countries of the periphery find themselves locked within a global economic system over which they have no control. Even countries that possess commodities in high demand in the core (such as the oil of the OPEC states) have difficulty converting their temporary advantage into longer-term parity with core-area powers. Countries whose income depends heavily on the export of raw materials such as strategic minerals (or single crops like bananas or sugar) are at the mercy of those who do the buying. But there are other reasons for the depressing condition of many periphery countries. Like the contemporary city, the world core area constantly siphons off the skilled and professional people from the periphery. How many doctors, engineers, and teachers from India are working in England and the United States? And how badly India now needs such trained people! Every loss is magnified—but oppressive political circumstances in periphery countries contribute to this "brain drain."

The continuing underdevelopment of periphery countries also involves traditional cultural attitudes and values. The introduction of core-area tentacles into outlying regions where traditional culture is strong may cause a reaction, perhaps a resurgence of fundamentalism. This further restricts economic change. The invasions of modern ways can be unsettling. Do not forget that the long traditions of societies in the periphery have generated large bureaucracies that are threatened by the change development entails. Occasionally, the tangible presence of an external (core) installation such as an office building, airline agency, or retail establishment is violently attacked. Such incidents symbolize ideological opposition to the intrusion they represent.

The countries of the periphery, therefore, face severe problems. They are pawns in a global economic game whose rules they cannot change. Their internal problems are intensified by the aggressive involvement of core-area interests. The way they use resources (such as whether to produce food for local consumption or to produce export crops) is strongly affected by foreign interference. They suffer far more from environmental degradation, overpopulation, and mismanagement than core-area countries do. Their inherited disadvantages have grown, not lessened, over time. The widening gaps that result between enriching cores and persistently impoverished peripheries threaten the future of the world.

services produced by the citizens of a country, within or outside of its boundaries, during a calendar year.) Now compare these figures to those of Nicaragua, a Middle American Cold War victim, whose per-capita debt is nearly *four times* its per-capita GNP. Nicaragua is one of the three poorest countries in the Western Hemisphere, but similar debt situations abound in Africa. As a geographic realm, Subsaharan Africa today is the most debt-ridden on Earth, at a time when it faces medical, environmental, and political crises as well. But in the global periphery this realm is not alone in facing the poverty-deepening specter of debt.

Table I-1
EXTERNAL DEBT OF LOW- AND MIDDLE-INCOME COUNTRIES

GEOGRAPHIC REALM	COUNTRY	TOTAL EXTERNAL DEBT, 1998 ($ MILLION)	PER CAPITA DEBT, 1998 ($)	PER CAPITA GNP, 1998 ($)
EUROPE	Poland	47,708	1,236	3,910
	Romania	9,513	425	1,360
	Ukraine	12,718	252	980
RUSSIA	Russian Federation	183,601	1,252	2,260
	Georgia	1,647	305	970
MIDDLE AMERICA	Mexico	159,959	1,636	3,840
	Nicaragua	5,968	1,326	370
SOUTH AMERICA	Brazil	232,004	1,426	4,630
	Colombia	33,263	871	2,470
	Peru	32,397	1,301	2,440
NORTH AFRICA/SW ASIA	Egypt	31,964	483	1,290
	Algeria	30,665	1,005	1,550
	Jordan	8,484	1,885	1,150
SUBSAHARAN AFRICA	Nigeria	30,315	275	300
	Ghana	6,884	370	390
	Ivory Coast	14,852	964	700
	Congo	5,119	1,896	680
	Kenya	7,010	238	350
	South Africa	24,711	573	3,310
	Moçambique	8,208	434	210
SOUTH ASIA	India	98,232	99	440
	Pakistan	32,229	227	470
EAST ASIA	China	154,599	124	750
	South Korea	139,097	3,004	8,600
SOUTHEAST ASIA	Indonesia	150,875	726	640
	Vietnam	22,359	293	350
	Philippines	47,817	637	1,050
PACIFIC REALM	Papua New Guinea	2,692	598	890

Source: Data on external debt from The World Bank, *World Development Indicators, 2000* (Washington, DC: 2000), pp. 248–250.

Every year, the leaders of the seven richest economies plus Russia (the so-called G-8) meet to discuss world economic issues. In 1999, this group voted to forgive the world's poorest countries, mainly those in Subsaharan Africa, about $100 billion in debts. But those good intentions did not lead to action. By the time the G-8 met again in Okinawa, Japan in 2000, no debts had been canceled. One reason for this failure undoubtedly lay in the public reaction in the rich countries to the earlier decision. Newspaper and journal editorials asked whether such debt cancellation would simply lead to further mismanagement of budgets by rapacious military regimes, corrupt leaderships, greedy elites, and dishonest businesspeople with government ties. To what extent would the debt reduction mitigate ethnic and cultural tensions, misuse of aid, and entrenched traditionalisms? Here again was proof that the world economic system is controlled by the few rich countries of the core and works to the disadvantage of the many in the periphery.

Globalization

In August 1999 a Frenchman named José Bové, leader of the so-called Peasant Federation, became a national hero. He achieved his stardom because a group of his sympathizers destroyed a McDonald's restaurant being built in the town of Millau; when Mr. Bové came to trial ten months later, some 40,000 supporters rallied to his cause. To Mr. Bové, McDonald's is a symbol of **globalization**, a process 52 he views as the Americanization of France's traditions, and trashing it was a matter of cultural self-defense. In November 1999 he was in Seattle, working to disrupt a meeting of the World Trade Organization (WTO). As globalization proceeds, resistance to it mounts, not just in France but even in the United States itself.

Why does an apparently beneficial process arouse such heated passions? Globalization breaks down barriers to international trade, stimulates commerce, brings jobs to remote places, and promotes social, cultural, political, and other kinds of exchanges. High-tech workers in India are employed by computer firms based in California. Japanese cars are assembled in Thailand. American shoes are made in China. Thousands of McDonald's restaurants serve their familiar (and standard) menus to customers from Tokyo to Tel Aviv.

Opponents of globalization argue that the negatives far outweigh the positives. Those high-tech workers in India are paid a fraction of what their Californian counterparts earn, so that many of them will want to leave India for better pay abroad at a time when the country has great need of their skills. The wages and working conditions for the Chinese shoe-factory workers are far below those acceptable to American labor. And as for those McDonald's restaurants, they promote the consumption of junk food over better traditional fare. Opposition to globalization in the United States often centers on the loss of American jobs when corporations move their factories to foreign countries where labor is cheaper.

In a way, the current globalization process is a revolution—but it is not the first of its kind. The first "globalization revolution" occurred during the nineteenth and early twentieth centuries, when Europe's colonial expansion spread ideas, inventions, products, and habits around the world. Colonialism transformed the world as the European powers built cities, transport networks, dams, irrigation systems, power plants, and other facilities, often with devastating impact on local traditions, cultures, and economies. From goods to games (soap to soccer) people in much of the world started doing similar things. The largest of all colonial empires, that of Britain, made English a worldwide language, a key element in the current, second globalization process.

The present globalization is even more revolutionary than the colonial phase because it is driven by more modern, higher-speed communications. When the British colonists planned the construction of their ornate Victorian government and public buildings in (then) Bombay, the architectural drawings had to be prepared in London and sent by boat to India. When the Chinese government in the 1980s decided to create a Manhattan-like commercial district on the riverfront in Shanghai, the plans were drawn in the United States, Japan, and Western Europe and transmitted to Shanghai via the Internet. One container ship carrying products from China to the American market hauls more cargo than a hundred colonial-era boats.

And, as the pages that follow will show frequently, the world's national-political boundaries are becoming increasingly porous. Economic alliances enable manufacturers to send raw materials and finished products across borders that once inhibited such exchanges. Groups of countries forge unions whose acronyms (NAFTA, Mercosur) stand for freer trade. The ultimate goal of the World Trade Organization is to lower the remaining trade barriers the world over, boosting not just regional commerce but also global trade.

As with all revolutions, the overall consequences of the present globalization process are uncertain. Critics underscore that one of its outcomes is a growing gap between rich and poor, a polarization of wealth that will destabilize the world. Core-periphery contrasts are intensified, not lessened, by globalization as the poor in peripheral societies are exploited by core-based corporations. Proponents argue that, as with the Industrial Revolution, it will take time for the benefits to spread—but that globalization's ultimate effects will be advantageous to all.

Indeed, the world is functionally shrinking, and we will find evidence for this throughout this book. But the "global village" still retains its distinctive neighborhoods, and two revolutionary globalizations have failed to erase their particular properties. In the chapters that follow we use the vehicle of geography to visit and investigate them.

THE REGIONAL FRAMEWORK

At the beginning of this Introduction, we outlined a map of the great geographic realms of the world (Fig. I-2). We then addressed the task of dividing these realms into regions, and we used criteria ranging from physical geography to economic geography. The result is Figure I-12. Before we begin our survey, here is a summary of the 12 geographic realms and their regional components.

Europe (1)

Territorially small and politically fragmented, Europe remains disproportionately influential in global affairs. A core geographic realm, Europe has five regions: Western Europe, the British Isles, Northern (Nordic) Europe, Mediterranean Europe, and Eastern Europe.

Russia (2)

Territorially enormous and politically unified, Russia was the dominant force in the former Soviet Union that disbanded in 1991. Undergoing a difficult transition from dictatorship to democracy and from communism to capitalism, Russia is geographically complex and changing. We define four regions: the Russian Core, the Eastern Frontier, Siberia, and the Far East.

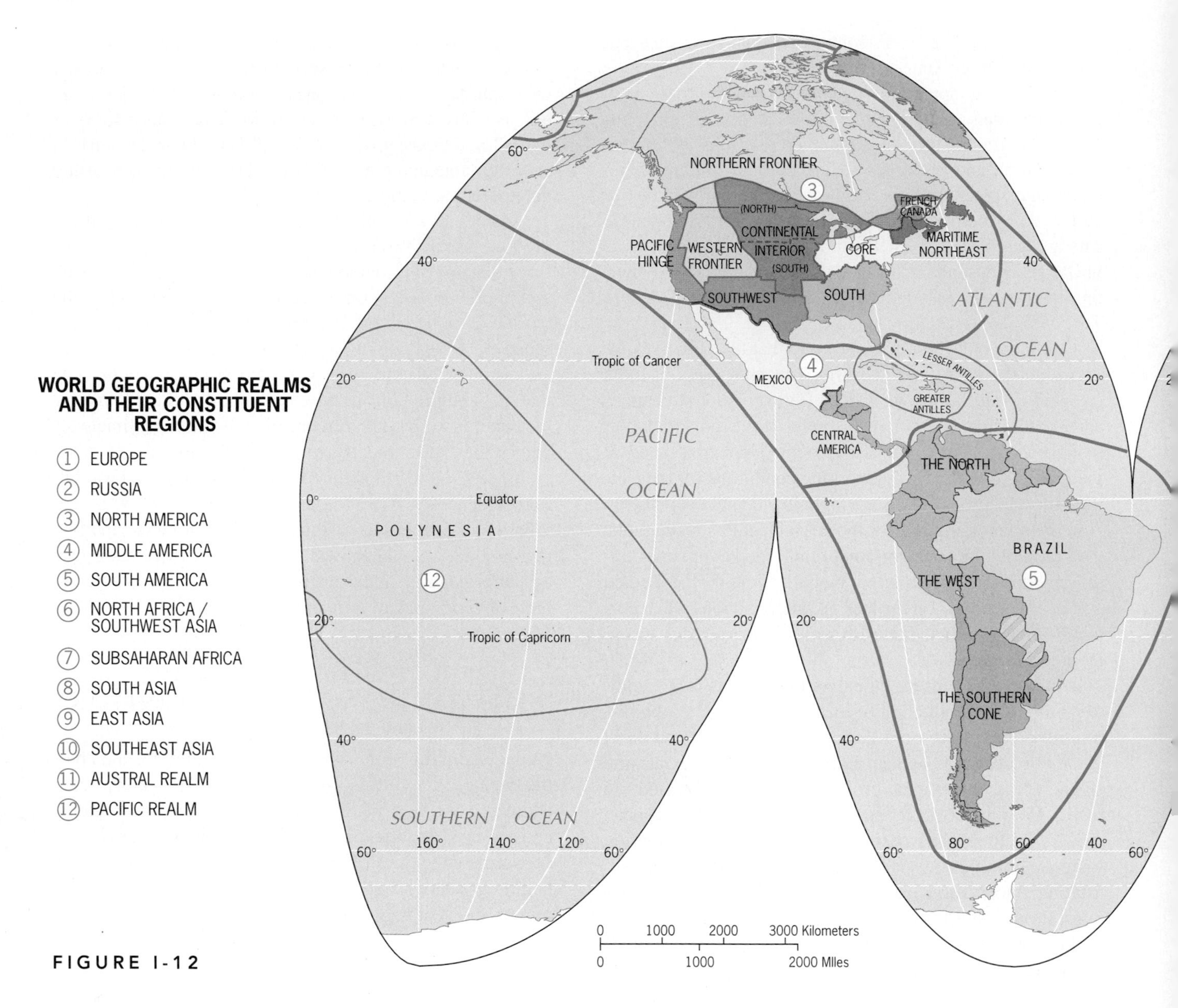

FIGURE I-12

North America (3)

Another realm in the global core, North America consists of the United States and Canada. We identify nine regions: the North American Core, the Maritime Northeast, French Canada, the Continental Interior, the South, the Southwest, the Western Frontier, the Northern Frontier, and the Pacific Hinge.

Middle America (4)

Nowhere in the world is the contrast between core and periphery as sharply demarcated as it is between North and Middle America. This small, fragmented realm clearly divides into four regions: Mexico, Central America, and the Greater Antilles and Lesser Antilles of the Caribbean.

South America (5)

The continent of South America also defines a geographic realm in which Iberian (Spanish and Portuguese) influences dominate the cultural geography but Amerindian imprints survive. We recognize four regions: Brazil, the realm's giant; the North, composed of Caribbean-facing states; the Andean West, with its strong Amerindian influences; and the Southern Cone.

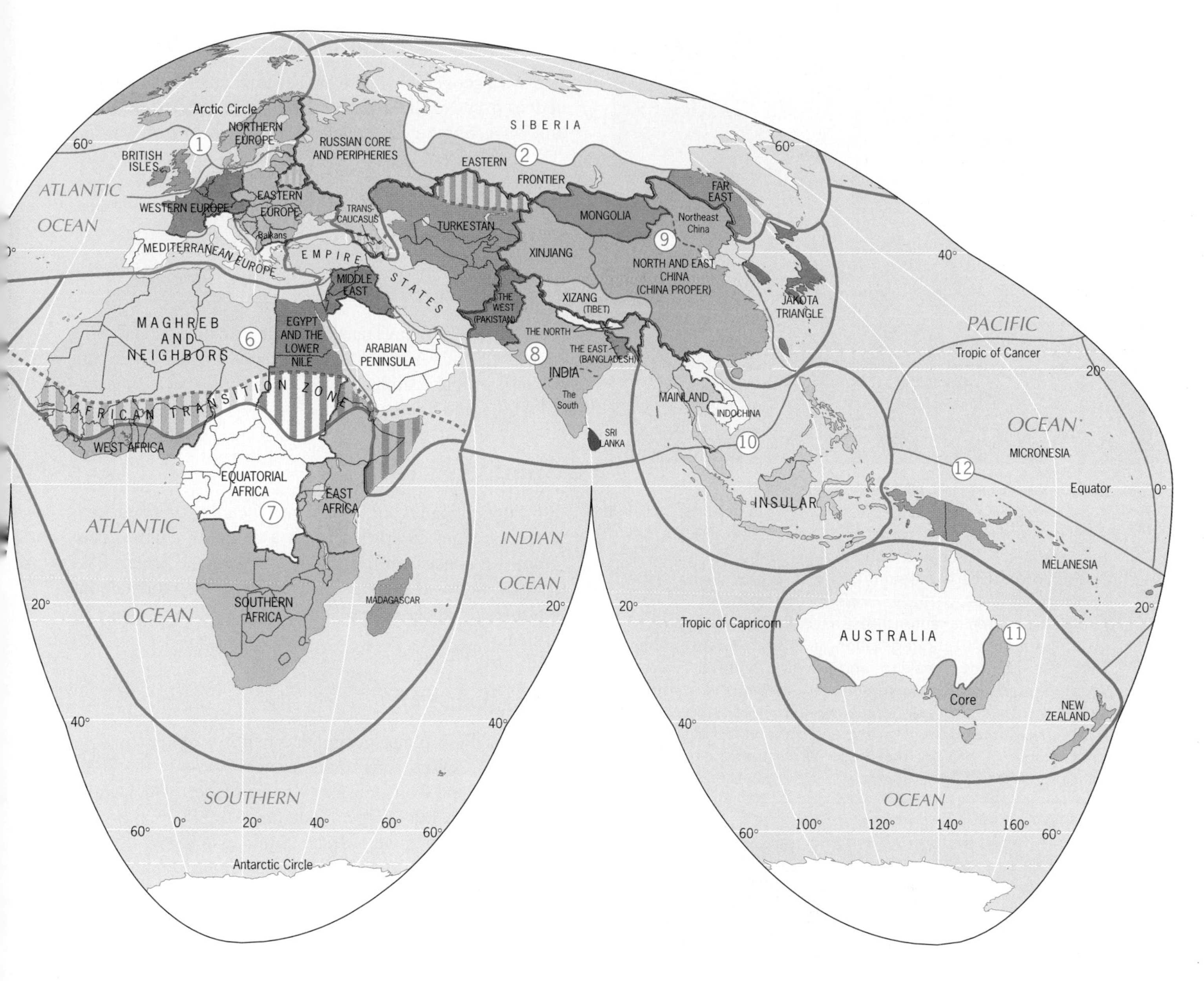

FROM THE FIELD NOTES

"It was late in the day in Morocco and I was hurrying back to the bus for the long ride from Marrakech to Casablanca when I saw this Berber tradesman, his wares arranged on a large blue carpet, in the middle of the square. Bicycle and pedestrian traffic streamed past him on all sides, but he was unperturbed. Minority traders have a difficult time almost everywhere, and Morocco's Berbers are no exception. Nearby, in Marrakech's *medina*, Arab traders sell almost everything you see here—spices, nuts, fruits, tea, coffee, beads, even teapots—from small but well-stocked stalls. But this Berber tradesman could not afford space in the *medina*, even if he could find someone to sell it to him. So he sits here out in the open, his prices just slightly better than those in the *medina*, working from dawn to dusk."

FROM THE FIELD NOTES
"Stuck in one of those timeless traffic jams in Chennai (India) on—what else—a hot day, I had the time to give the old British-built, Victorian-Moorish 'Agurchand Mansion' a closer look. Once an elegant residential building with an imposing entrance (partially visible in the lower left), it now houses commercial enterprises that reflect at least some of Chennai's priorities: refrigeration, finance, insurance, security. The arts, moviemaking, and publishing are also represented, all crowded into very limited, but also very visible, space. This scene is repeated countless times along the main arteries of India's cities. Starting and operating a business is not easy in a country that has yet to match its political achievements in the economic arena."

FROM THE FIELD NOTES
"In Indonesia, on my way from Semarang to Yogyakarta, I was struck by what seemed to be evidence of comparative prosperity even in Jawa's smaller villages—motor scooters, new bikes, luxury goods in store windows—compared to what I had seen in Sulawesi and the Malukus. These youngsters in the village of Ambarawa waited for a school bus."

North Africa/Southwest Asia (6)

This vast geographic realm has several names, extending as it does from North Africa into Southwest and, indeed, Central Asia. Some geographers call it *Naswasia* or *Afrasia*. There are seven regions: Egypt and the Lower Nile, the Maghreb, and the African Transition Zone in North Africa; the Middle East, the Arabian Peninsula, the Empire States, and Turkestan in Southwest Asia.

Subsaharan Africa (7)

Between the African Transition Zone and the southernmost Cape of South Africa lies Subsaharan Africa. The realm consists of four regions: West Africa, East Africa, Equatorial Africa, and Southern Africa.

South Asia (8)

Physiographically one of the most clearly defined geographic realms, South Asia has a complex cultural geography. It consists of five regions: India at the center; Pakistan to the west; Bangladesh to the east; the mountainous North; and the peninsular South, which includes island Sri Lanka.

East Asia (9)

The vast East Asian geographic realm extends from the deserts of Central Asia to the tropical coasts of the South China Sea and from Japan to Xizang (Tibet). We identify five regions: China Proper, including North Korea; Xizang (Tibet) in the southwest; desert Xinjiang in the west; Mongolia in the north; and the Jakota Triangle (Japan, South Korea, and Taiwan) in the east.

Southeast Asia (10)

Southeast Asia is a varied mosaic of natural landscapes, cultures, and economies. Influenced by India, China, Europe, and the United States, it includes dozens of religions and hundreds of languages plus economies representing both core and periphery. Physically, Southeast Asia consists of a peninsular mainland and an arc consisting of thousands of islands. The two regions (Mainland and Insular) are based on this distinction.

Austral Realm (11)

Australia and its neighbor New Zealand form the Austral geographic realm by virtue of continental dimensions, insular separation, and dominantly Western cultural heritage. The four regions are defined by physical as well as cultural geography: a highly urbanized, two-part core and a vast,

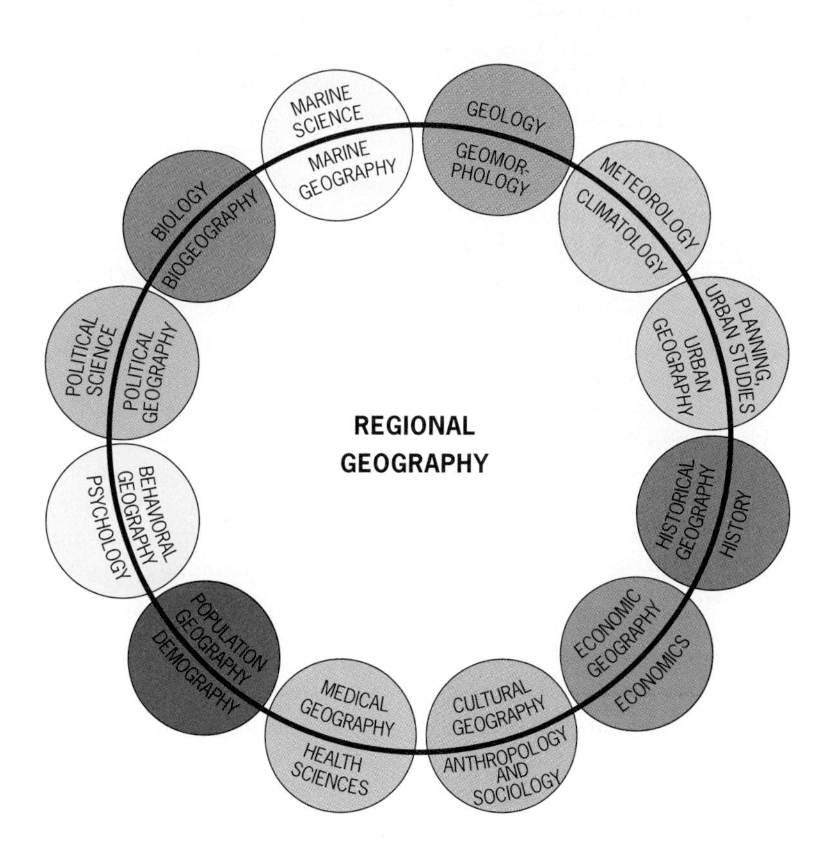

FIGURE I-13

desert-dominated interior in Australia; and two main islands in New Zealand that exhibit considerable geographic contrast.

Pacific Realm (12)

The vast Pacific Ocean, larger than all the landmasses combined, contains tens of thousands of islands large and small. Dominant cultural criteria warrant three regions: Melanesia, Micronesia, and Polynesia.

THE PERSPECTIVE OF GEOGRAPHY

As this introductory chapter demonstrates, our world regional survey is no mere description of places and areas. We have combined the study of realms and regions with a look at geography's ideas and concepts—the notions, generalizations, and basic theories that make the discipline what it is. We continue this method in the chapters ahead so that we will become better acquainted with the world and with geography. By now you are aware that geography is a wide-ranging, multifaceted discipline. It is often described as a social science, but that is only half the story: in fact, geography straddles the divide between the social and the physical (natural) sciences. Many of the ideas and concepts you will encounter have to do with the many interactions between human societies and natural environments.

Regional geography allows us to view the world in an all-encompassing way. As we have seen, regional geography borrows information from many sources to create an overall image of our divided world. Those sources are not random. They form topical or **systematic geography**. Research in the systematic fields of geography makes our world-scale generalizations possible. As Figure I-13 shows, these systematic fields relate closely to those of other disciplines. Cultural geography, for example, is allied with anthropology; it is the spatial perspective that distinguishes cultural geography. Economic geography focuses on the spatial dimensions of economic activity; political geography concentrates on the spatial imprints of political

53

54

What Do Geographers Do?

A systematic spatial perspective and an interest in regional study are the unifying themes and enthusiasms of geography. Geography's practitioners include physical geographers, whose principal interests are the study of geomorphology (land surfaces), research on climate and weather, vegetation and soils, and the management of water and other natural resources. There also are geographers whose research and teaching concentrate on the ecological interrelationships between the physical and human worlds. They study the impact of humankind on our globe's natural environments and the influences of the environment (including such artificial contents as air and water pollution) on human individuals and societies.

Other geographers are regional specialists, who often concentrate their work for governments, planning agencies, and multinational corporations on a particular region of the world. Still other geographers—who now constitute the largest group of practitioners—are devoted to topical or systematic subfields such as urban geography, economic geography, and cultural geography (see Fig. I-13). They perform numerous tasks associated with the identification and resolution (through policy-making and planning) of spatial problems in their specialized areas. And, as in the past, there remain many geographers who combine their fascination for spatial questions with technical know-how.

Computerized cartography, geographic information systems, remote sensing, and even environmental engineering are among the specializations listed by the 10,000-plus professional geographers of North America. In Appendix B, you will find considerable information on the discipline, how one trains to become a geographer, and the many exciting career options that are open to the young professional.

behavior. Other systematic fields include historical, medical, behavioral, environmental, agricultural, and coastal geography. We will also draw on information from biogeography, marine geography, population geography, and climatology (as we did earlier in this chapter).

These systematic fields of geography are so named because their approach is global, not regional. Take the geographic study of cities, urban geography. Urbanization is a worldwide process, and urban geographers can identify certain human activities that all cities in the world exhibit in one form or another. But cities also display regional properties. The model Japanese city is quite distinct from, say, the African city. Regional geography, therefore, borrows from the systematic field of urban geography, but it injects this regional perspective.

In the following chapters we call upon these systematic fields to give us a better understanding of the world's realms and regions. As a result, you will gain insights into the discipline of geography as well as the regions we investigate. This will prove that geography is a relevant and practical discipline when it comes to comprehending, and coping with, our fast-changing world (see box titled "What Do Geographers Do?" and Appendix B for a discussion of career opportunities).

Table I-2
AREA AND DEMOGRAPHIC DATA FOR THE WORLD'S STATES

	Land Area 1000 (sq mi)	Population 2002 (Millions)	Population 2010 (Millions)	Population Density Arithmetic	Population Density Physiologic	Birth Rate	Death Rate	Natural Increase	Doubling Time (years)	Infant Mortality per 1,000 (births)	Life Expectancy Males (years)	Life Expectancy Females (years)	Percent Urban Pop	Per Capita GNP ($US)
WORLD	**51510.8**	**6238.1**	**6866.8**	**117.0**		**22**	**9**	**1.4%**	**51**	**57**	**64**	**68**	**45**	**$4,890**
Europe	**2197.2**	**582.6**	**580.4**	**265.2**		**10**	**11**	**0.0%**		**7**	**72**	**79**	**73**	**$16,518**
Albania	10.6	3.5	3.9	329.1	1,567.4	18	5	1.3%	55	22	69	74	46	$810
Austria	31.9	8.1	8.1	253.9	1,493.6	10	10	0.0%		5	75	81	65	$26,830
Belarus	80.1	9.9	9.7	123.6	426.2	9	14	−0.5%		11	63	74	70	$2,180
Belgium	11.8	10.2	10.3	866.1	3,608.9	11	10	0.1%	770	6	75	81	97	$25,380
Bosnia	19.7	3.8	4.0	194.8	1,391.6	13	8	0.5%	141	12	71	76	40	
Bulgaria	42.7	8.1	7.5	189.7	512.8	8	14	−0.6%		14	67	74	68	$42,822
Croatia	21.6	4.6	4.5	212.5	1,012.1	11	12	−0.1%		8	69	76	54	$4,620
Cyprus	3.6	0.9	0.9	253.0	2,108.4	14	8	0.6%	124	8	74	79	64	$11,920
Czech Rep.	29.8	10.3	10.2	344.3	839.6	9	11	−0.2%		5	71	78	77	$5,150
Denmark	16.4	5.3	5.5	323.8	539.7	12	11	0.1%	472	5	74	79	85	$33,040
Estonia	16.3	1.4	1.4	85.0	386.5	8	13	−0.5%		9	64	75	69	$3,360
Finland	117.6	5.2	5.2	44.3	553.8	11	10	0.1%	433	4	74	81	60	$24,280
France	212.4	59.9	61.6	281.9	854.3	13	9	0.4%	204	5	75	82	74	$24,210
Germany	134.9	81.9	81.2	607.4	1,840.6	9	10	−0.1%		5	74	80	86	$26,570
Greece	49.8	10.6	10.5	212.9	1,120.3	10	10	0.0%		7	75	81	59	$11,740
Hungary	35.7	9.9	9.6	277.3	543.8	9	14	−0.5%		9	66	75	64	$4,510
Iceland	38.7	0.3	0.3	7.9		15	7	0.8%	81	3	77	82	92	$27,830
Ireland	26.6	3.8	4.1	144.6	1,112.1	15	9	0.6%	116	6	73	79	58	$27,135
Italy	113.5	57.7	55.6	508.2	1,639.5	9	10	−0.1%		6	75	81	90	$20,090
Latvia	24.0	2.4	2.3	98.8	365.9	8	14	−0.6%		11	64	76	69	$2,420
Liechtenstein	0.1	0.1	0.1	507.0		14	7	0.7%	105	18	67	78		
Lithuania	25.0	3.7	3.6	147.7	301.4	10	11	−0.1%		9	67	77	68	$2,540
Luxembourg	1.0	0.4	0.5	403.2	1,680.0	13	9	0.4%	198	5	74	80	88	$45,100
Macedonia	9.8	2.0	2.1	206.9	4,139.0	15	8	0.7%	112	16	70	75	59	$1,290
Malta	0.1	0.4	0.4	3,360.1		12	8	0.4%	182	5	74	80	89	$10,100
Moldova	12.7	4.3	4.4	338.6	638.8	11	11	0.0%		18	63	70	46	$380
Netherlands	13.1	16.0	16.5	1,223.5	4,531.4	13	9	0.4%	193	5	75	81	61	$24,780
Norway	118.5	4.5	4.7	38.2	1,273.4	13	10	0.3%	217	4	76	81	74	$34,310
Poland	117.5	38.6	38.6	328.5	699.0	10	10	0.0%		9	69	78	62	$3,910
Portugal	35.5	10.0	9.7	281.7	1,083.4	11	11	0.0%		5	72	79	48	$10,670
Romania	88.9	22.4	21.7	251.5	613.3	11	12	−0.1%		21	66	73	55	$1,360
Slovakia	18.6	5.4	5.4	290.9	938.4	11	10	0.1%	866	9	69	77	57	$3,700
Slovenia	7.8	2.0	2.0	255.9	2,132.5	9	10	−0.1%		5	71	79	50	$9,780
Spain	192.8	39.5	38.4	204.9	682.9	9	9	0.0%		6	74	82	64	$14,100
Sweden	158.9	8.9	9.0	55.9	798.5	10	11	−0.1%		4	77	82	84	$25,580
Switzerland	15.3	7.1	7.3	465.9	4,659.1	11	9	0.2%	315	5	77	83	68	$39,980
Ukraine	223.7	48.9	47.4	218.6	377.0	8	14	−0.6%		13	63	74	68	$980
United Kingdom	93.3	59.9	61.6	642.2	2,568.9	12	11	0.1%	546	6	74	80	89	$21,410
Yugoslavia	26.9	10.7	10.7	397.8	1,325.9	11	11	0.0%		10	70	75	52	

	Land Area 1000 (sq mi)	Population 2002 (Millions)	Population 2010 (Millions)	Population Density		Birth Rate	Death Rate	Natural Increase	Doubling Time (years)	Infant Mortality per 1,000 (births)	Life Expectancy		Percent Urban Pop	Per Capita GNP ($US)
				Arithmetic	Physiologic						Males (years)	Females (years)		
Russia	**6550.7**	**143.2**	**135.7**	**21.9**	**273.2**	**8**	**15**	**−0.7%**		**17**	**61**	**73**	**73**	**$2,260**
Armenia	10.9	3.8	3.9	351.4	2,067.2	10	6	0.4%	161	15	71	78	67	$460
Azerbaijan	33.4	7.8	8.6	234.7	1,303.9	15	6	0.9%	77	17	68	75	52	$480
Georgia	26.9	5.5	5.2	204.9	2,276.3	9	8	0.1%	462	15	69	76	56	$970
North America	**7567.5**	**316.4**	**342.6**	**41.8**		**14**	**9**	**0.5%**	**124**	**7**	**74**	**80**	**75**	**$28,230**
Canada	3849.7	31.0	33.0	8.1	174.4	11	7	0.4%	178	6	76	81	78	$19,170
United States	3717.8	285.4	309.6	76.8	415.1	15	9	0.6%	120	7	74	79	75	$29,240
Middle America	**1021.9**	**179.4**	**202.5**	**175.6**		**25.2**	**6**	**2.0%**	**35**	**37**	**72**	**73**	**66**	
Antigua and Barbuda	0.2	0.1	0.1	607.2	3,373.4	22	6	1.6%	45	17	69	74	37	$8,450
Bahamas	3.9	0.3	0.3	79.4	7,940.4	21	5	1.6%	45	18	70	77	84	
Barbados	0.2	0.3	0.3	1,782.4	4,817.3	14	9	0.5%	130	14	72	77	38	
Belize	8.8	0.3	0.3	36.0	1,797.8	32	5	2.7%	26	34	70	74	50	$2,660
Costa Rica	19.7	3.7	4.6	189.4	3,156.3	22	4	1.8%	39	13	75	79	45	$2,770
Cuba	42.8	11.3	11.4	265.5	1,106.1	14	7	0.7%	103	7	73	78	75	
Dominica	0.3	0.1	0.1	350.4	3,893.0	16	8	0.8%	83	15	75	80		$3,150
Dominican Rep.	18.7	8.8	10.1	469.2	2,234.2	28	6	2.2%	32	47	67	71	62	$1,770
El Salvador	8.0	6.6	7.9	824.1	3,052.4	30	7	2.3%	29	35	67	73	58	$1,850
Grenada	0.1	0.1	0.1	805.0	5,366.8	29	6	2.3%	30	14	68	73	34	$3,250
Guadeloupe	0.7	0.4	0.4	629.0	4,492.8	17	6	1.1%	61	10	73	80	48	
Guatemala	41.9	13.5	17.0	321.6	2,679.7	37	7	2.9%	24	45	61	67	39	$1,640
Haiti	10.6	6.6	7.8	624.5	3,122.4	33	16	1.7%	40	103	47	51	34	$410
Honduras	43.2	6.4	7.3	148.9	992.9	33	6	2.7%	25	42	66	71	45	$740
Jamaica	4.2	2.7	2.9	637.8	4,555.4	22	7	1.5%	45	24	70	73	50	$1,740
Martinique	0.4	0.4	0.4	993.2	12,415.6	15	6	0.9%	81	9	75	82	81	
Mexico	737.0	103.6	115.2	140.6	1,171.7	24	4	2.0%	36	32	69	75	74	$3,840
Netherlands Antilles	0.3	0.2	0.2	659.4	6,594.3	17	6	1.1%	62	14	72	78		
Nicaragua	46.9	5.4	6.7	115.4	1,281.8	36	6	3.0%	23	40	66	71	63	$370
Panama	28.7	3.0	3.3	104.5	1,493.0	22	5	1.7%	41	21	72	77	56	$2,990
Puerto Rico	3.4	4.0	4.1	1,167.8	29,195.0	17	8	0.9%	75	11	70	79	71	
Saint Lucia	0.2	0.2	0.3	855.1	10,689.3	19	6	1.3%	56	17	71	72	48	$3,660
St. Vincent and the Grenadines	0.2	0.1	0.1	682.8	1,796.7	19	7	1.2%	59	20	71	74	44	$2,560
Trinidad and Tobago	2.0	1.3	1.4	659.1	4,394.2	14	7	0.7%	103	16	68	73	72	$4,520
South America	**6763.3**	**356.8**	**400.1**	**52.8**		**23**	**6**	**1.7%**	**41**	**34**	**66**	**73**	**78**	**$4,270**
Argentina	1056.6	37.8	41.6	35.8	397.7	19	8	1.1%	62	19	70	77	90	$8,030
Bolivia	418.7	8.6	10.0	20.6	1,031.2	30	10	2.0%	34	67	59	62	62	$1,010
Brazil	3265.1	175.2	193.6	53.7	1,073.4	21	6	1.5%	45	38	64	71	78	$4,630
Chile	289.1	15.6	17.2	54.0	1,079.1	18	5	1.3%	54	11	72	78	85	$4,990
Colombia	401.0	41.6	48.3	103.8	2,594.5	26	6	2.0%	34	28	65	73	71	$2,470
Ecuador	106.9	13.1	15.0	122.9	2,047.8	27	6	2.1%	33	40	67	72	63	$1,520
French Guiana	34.0	0.2	0.3	6.2	616.8	27	3	2.4%	29	18	71	77	79	
Guyana	76.0	0.7	0.8	9.5	476.3	24	7	1.7%	40	63	63	69	36	$780

	Land Area 1000 (sq mi)	Population 2002 (Millions)	Population 2010 (Millions)	Population Density Arithmetic	Population Density Physiologic	Birth Rate	Death Rate	Natural Increase	Doubling Time (years)	Infant Mortality per 1,000 (births)	Life Expectancy Males (years)	Life Expectancy Females (years)	Percent Urban Pop	Per Capita GNP ($US)
Paraguay	153.4	5.8	7.2	37.7	629.0	32	6	2.6%	26	27	68	72	52	$1,760
Peru	494.2	28.3	32.6	57.2	1,905.4	27	6	2.1%	32	43	66	71	72	$2,440
Suriname	60.2	0.4	0.4	6.9	689.9	26	7	1.9%	37	29	68	73	69	$1,660
Uruguay	67.5	3.3	3.6	49.5	618.5	16	10	0.6%	107	15	70	78	92	$6,070
Venezuela	340.6	25.2	29.0	73.9	1,848.0	25	5	2.0%	34	21	70	76	86	$3,530
North Africa/ Southwest Asia	**7655.8**	**526.6**	**620.5**	**52.3**		**28**	**8**	**2.0%**	**35**	**52**	**63**	**69**	**54**	
Afghanistan	251.8	28.1	36.0	111.4	928.4	43	18	2.5%	28	150	46	45	20	
Algeria	919.6	33.0	38.4	35.8	1,194.9	29	6	2.3%	29	44	68	70	49	$1,550
Bahrain	0.3	0.7	1.1	2,692.0	269,204.7	22	3	1.9%	37	8	68	71	88	$7,640
Djibouti	9.0	0.6	0.8	69.8		39	16	2.3%	30	115	47	50	83	
Egypt	384.3	71.1	81.6	184.9	9,245.3	26	6	2.0%	35	52	64	67	44	$1,290
Eritrea	39.0	4.3	6.0	111.5	929.4	43	13	3.0%	23	82	52	57	16	$200
Iran	631.7	69.4	78.0	109.9	1,099.2	21	6	1.5%	48	31	68	71	63	$1,650
Iraq	168.9	24.4	31.0	144.5	1,204.4	38	10	2.8%	25	127	58	60	68	
Israel	8.0	6.4	7.2	800.0	4,705.9	22	6	1.6%	45	6	76	80	90	$16,180
Jordan	34.3	5.4	6.8	157.1	3,928.3	33	5	2.8%	24	34	68	70	78	$1,150
Kazakhstan	1031.2	15.0	14.9	14.6	121.4	14	10	0.4%	161	21	59	70	63	$1,650
Kuwait	6.9	2.3	2.9	333.0		24	2	2.2%	32	13	72	73	100	
Kyrgyzstan	74.1	5.0	5.3	68.1	973.2	22	7	1.5%	47	26	63	71	34	$380
Lebanon	4.0	4.3	4.8	1,083.9	5,161.3	23	7	1.6%	43	35	68	73	88	$3,560
Libya	679.4	5.4	6.5	7.9	788.7	28	3	2.5%	28	33	73	77	86	
Morocco	172.3	29.8	33.6	172.9	823.2	23	6	1.7%	41	37	67	71	54	$1,240
Oman	82.0	2.6	3.6	31.5		43	5	3.8%	18	25	69	73	72	
Palestinian Territ. (West Bank/Gaza)	2.4	3.3	5.0	1.4		41	5	3.6%	19	27	70	73		$1,560
Qatar	4.3	0.6	0.7	144.6	14,460.3	20	2	1.8%	38	20	70	75	91	
Saudi Arabia	830.0	22.9	29.7	27.6	1,380.4	35	5	3.0%	23	46	68	71	83	$6,910
Somalia	242.2	7.7	10.6	31.9	1,595.7	47	18	2.9%	24	126	45	48	24	
Sudan	917.4	30.8	37.0	33.5	670.4	33	12	2.1%	32	70	50	52	27	$290
Syria	71.0	17.4	21.2	245.1	875.4	33	6	2.7%	25	35	67	68	51	$1,020
Tajikistan	54.3	6.6	7.3	121.7	2,027.8	21	5	1.6%	43	28	66	71	27	$370
Tunisia	60.0	9.9	11.1	164.8	867.6	22	7	1.5%	44	35	67	70	61	$2,060
Turkey	297.2	67.3	75.6	226.4	707.4	22	7	1.5%	46	38	67	71	66	$3,160
Turkmenistan	181.4	5.4	5.9	29.5	984.4	21	6	1.5%	48	33	62	69	44	
United Arab Emirates	32.3	2.9	3.3	90.5		24	2	2.2%	32	16	73	76	84	$17,870
Uzbekistan	159.9	25.7	28.0	160.4	1,782.4	23	6	1.7%	40	22	66	72	38	$950
Western Sahara	102.7	0.3	0.4	3.1		46	18	2.8%	24	150	46	48		
Yemen	203.9	18.0	26.2	88.1	2,937.0	39	11	2.8%	25	75	58	61	26	$280
Subsaharan Africa	**7916.3**	**646.8**	**781.0**	**81.7**		**41**	**16**	**2.6%**	**27**	**93**	**48**	**50**	**29**	**$522**
Angola	481.4	13.7	18.2	28.4	1,418.7	48	19	2.9%	23	125	45	48	32	$380
Benin	42.7	6.8	8.7	158.4	1,218.4	45	17	2.8%	24	94	49	51	38	$380

	Land Area 1000 (sq mi)	Population 2002 (Millions)	Population 2010 (Millions)	Population Density		Birth Rate	Death Rate	Natural Increase	Doubling Time (years)	Infant Mortality per 1,000 (births)	Life Expectancy		Percent Urban Pop	Per Capita GNP ($US)
				Arithmetic	Physiologic						Males (years)	Females (years)		
Botswana	218.8	1.6	1.5	7.5	753.4	32	17	1.5%	45	57	38	40	49	$3,070
Burkina Faso	105.6	12.6	16.2	119.3	917.8	47	18	2.9%	24	105	47	47	15	$240
Burundi	9.9	6.4	8.0	646.0	1,468.3	42	17	2.5%	28	75	46	47	8	$140
Cameroon	179.7	16.2	19.6	90.0	692.6	37	12	2.5%	27	77	55	56	44	$610
Cape Verde Is.	1.6	0.4	0.5	264.2	2,401.8	37	9	2.8%	25	77	65	72	44	$1,200
Central African Republic	240.5	3.6	4.1	15.1	504.7	38	18	2.0%	34	97	43	46	39	$300
Chad	486.1	8.5	12.0	17.6	585.4	50	17	3.3%	21	110	46	51	22	$230
Comoros Is.	0.9	0.6	0.8	737.3	2,106.5	38	10	2.8%	25	77	57	62	29	$370
Congo	131.9	2.9	3.6	22.3	1,113.0	40	16	2.4%	29	109	45	50	41	$680
Congo, The	875.3	55.4	75.3	63.3	2,109.0	48	16	3.2%	22	109	47	50	29	$110
Equatorial Guinea	10.8	0.5	0.6	48.6	972.8	41	16	2.5%	28	108	48	52	37	$1,110
Ethiopia	386.1	67.2	86.3	174.1	1,450.7	45	21	2.4%	29	116	45	47	15	$100
Gabon	99.5	1.3	1.6	12.6	1,259.7	38	16	2.2%	32	87	51	54	73	$4,170
Gambia	3.9	1.4	1.7	349.5	1,941.8	43	19	2.4%	29	130	43	47	37	$340
Ghana	87.9	20.4	22.9	232.6	1,938.5	34	10	2.4%	29	56	56	59	37	$390
Guinea	94.9	7.9	9.8	82.9	4,143.5	42	18	2.4%	29	98	43	47	26	$530
Guinea-Bissau	10.9	1.3	1.5	115.0	1,045.4	42	20	2.2%	31	130	47	44	22	$160
Ivory Coast	122.8	16.7	19.3	136.1	1,701.1	38	16	2.2%	32	112	45	48	46	$700
Kenya	219.8	31.6	32.7	143.7	2,052.9	35	14	2.1%	33	74	48	49	20	$350
Lesotho	11.7	2.2	2.3	186.7	1,697.6	33	13	2.0%	33	85	52	55	16	$570
Liberia	37.2	3.4	4.4	91.8	9,179.3	50	17	3.3%	21	139	49	52	45	
Madagascar	224.5	15.8	20.9	70.4	1,760.3	44	14	3.0%	24	96	51	53	22	$260
Malawi	36.3	10.8	11.5	297.5	1,652.7	41	22	1.9%	36	127	38	40	20	$210
Mali	471.1	11.9	15.7	25.3	1,263.5	47	16	3.1%	22	123	55	52	26	$250
Mauritania	395.8	2.9	3.6	7.2	720.9	41	13	2.8%	25	92	52	55	54	$410
Mauritius	0.8	1.2	1.3	1,569.4	3,202.8	17	7	1.0%	66	19	67	74	43	$3,730
Moçambique	302.7	19.9	20.2	65.9	1,647.6	41	19	2.2%	32	134	40	39	28	$210
Namibia	317.9	1.9	2.0	5.8	584.5	36	20	1.6%	42	68	47	45	27	$1,940
Niger	489.1	10.7	13.9	21.9	730.3	54	24	3.0%	23	123	41	41	17	$200
Nigeria	351.7	130.6	160.1	371.2	1,124.9	42	13	2.9%	24	77	52	53	36	$300
Réunion	1.0	0.7	0.8	743.5	4,373.3	20	5	1.5%	49	9	70	79	73	
Rwanda	9.5	7.5	7.7	793.2	2,266.2	43	20	2.3%	30	121	39	40	5	$230
São Tomé and Principe	0.3	0.2	0.2	737.3	36,867.4	43	9	3.4%	20	51	63	66	44	$270
Senegal	74.3	10.0	12.7	135.1	1,126.0	41	13	2.8%	25	68	51	54	41	$520
Seychelles	0.2	0.1	0.1	511.1	25,553.0	18	7	1.1%	65	9	67	73	59	$6,420
Sierra Leone	27.7	5.5	7.2	197.6	2,823.1	47	21	2.6%	26	157	42	47	37	$140
South Africa	471.4	44.5	40.8	94.5	944.8	25	12	1.3%	55	45	54	57	45	$3,310
Swaziland	6.6	1.0	1.3	157.3	1,430.2	41	22	1.9%	37	108	36	39	22	$1,400
Tanzania	341.1	37.4	46.3	109.6	3,652.6	42	13	2.9%	24	99	52	54	20	$220
Togo	21.0	5.3	6.2	253.1	666.0	42	11	3.1%	23	80	48	50	31	$330
Uganda	77.1	24.6	34.0	319.4	1,277.5	48	20	2.8%	24	81	42	43	15	$310
Zambia	287.0	10.0	11.7	34.7	496.2	42	23	1.9%	35	109	37	38	38	$330
Zimbabwe	149.4	11.5	10.7	77.2	1,102.2	30	20	1.0%	69	80	41	39	32	$620

	Land Area 1000 (sq mi)	Population 2002 (Millions)	Population 2010 (Millions)	Population Density Arithmetic	Physiologic	Birth Rate	Death Rate	Natural Increase	Doubling Time (years)	Infant Mortality per 1,000 (births)	Life Expectancy Males (years)	Females (years)	Percent Urban Pop	Per Capita GNP ($US)
South Asia	**1592.0**	**1376.7**	**1558.5**	**864.8**		**28**	**9**	**1.9%**	**36**	**74**	**60**	**61**	**61**	**$436**
Bangladesh	50.3	133.0	150.7	2,644.4	3,622.5	27	8	1.9%	38	82	59	58	20	$350
Bhutan	18.2	1.0	1.1	52.6	2,628.2	40	9	3.1%	22	71			15	$470
India	1148.0	1038.5	1168.3	904.6	1,615.4	27	9	1.8%	39	72	60	61	28	$440
Maldives	0.1	0.3	0.4	2,652.3	26,522.5	35	5	3.0%	23	27	71	72	25	$1,130
Nepal	52.8	25.1	30.3	475.6	2,797.5	36	11	2.5%	28	79	58	57	11	$210
Pakistan	297.6	159.2	186.3	534.8	1,980.7	39	11	2.8%	25	91	58	59	33	$470
Sri Lanka	25.0	19.7	21.4	786.5	5,618.2	18	6	1.2%	60	17	70	74	22	$810
East Asia	**4450.1**	**1517.2**	**1574.2**	**340.9**		**15**	**7**	**0.8%**	**87**	**29**	**70**	**74**	**38**	**$3,880**
China	3705.8	1294.4	1349.0	349.3	3,594.2	15	6	0.9%	79	31	69	73	31	$750
Japan	145.4	127.2	124.7	874.5	7,950.1	9	8	0.1%	462	4	77	84	78	$32,350
Korea, North	46.5	22.3	23.7	479.8	3,427.3	21	7	1.4%	48	26	67	73	59	
Korea, South	38.1	48.2	50.2	1,263.9	6,652.2	14	5	0.9%	82	11	71	78	79	$8,600
Mongolia	604.8	2.6	2.9	4.2	424.2	20	7	1.3%	50	34	60	66	52	$380
Taiwan	14.0	22.6	23.7	1,615.2	6,730.1	13	6	0.7%	97	7	72	78	77	
Southeast Asia	**1735.4**	**546.1**	**614.5**	**314.7**		**24**	**7**	**1.7%**	**41**	**46**	**63**	**67**	**36**	**$1,240**
Brunei	2.0	0.3	0.4	156.7	15,667.3	25	3	2.2%	32	24	70	73	67	
Cambodia	68.2	12.7	16.1	186.8	1,436.7	38	12	2.6%	27	80	54	58	16	$260
East Timor	5.7	0.8	1.0	145.4		34	16	1.8%	39	143	45	47		
Indonesia	705.2	219.0	240.8	310.6	3,106.1	24	8	1.6%	44	46	62	66	39	$640
Laos	89.1	5.5	6.6	61.4	2,047.9	41	15	2.6%	26	104	50	52	17	$320
Malaysia	126.9	24.2	29.3	191.0	6,367.6	25	5	2.0%	34	8	70	75	57	$3,670
Myanmar/Burma	253.9	50.9	57.8	200.4	1,335.8	30	10	2.0%	35	83	53	56	26	
Philippines	115.1	83.9	97.2	728.7	3,835.2	29	7	2.2%	31	35	66	69	47	$1,050
Singapore	0.2	4.1	5.6	16,934.4	846,720.0	13	5	0.8%	84	3	76	80	100	$30,170
Thailand	197.3	63.1	66.7	319.9	941.0	16	7	0.9%	70	22	70	75	31	$2,160
Vietnam	125.7	80.9	92.5	643.7	3,786.7	20	6	1.4%	48	37	63	69	24	$350
Austral Realm	**3067.9**	**23.3**	**24.9**	**7.6**		**13**	**7**	**0.6%**	**109**	**5**	**76**	**82**	**85**	**$19,639**
Australia	2966.2	19.4	20.8	6.6	109.7	13	7	0.6%	110	5	76	82	85	$20,640
New Zealand	103.5	3.9	4.1	37.3	414.5	15	7	0.8%	89	5	74	80	85	$14,600
Pacific Realm	**207.7**	**7.5**	**9.0**	**36.3**		**29**	**8**	**2.1%**	**34**	**6**	**54**	**56**	**21**	**$912**
Federated States of Micronesia	0.3	0.1	0.1	389.9		33	7	2.6%	27	46	65	67	27	$1,800
Fiji	7.1	0.8	0.9	116.1	1,160.8	22	7	1.5%	46	13	65	69	46	$2,210
French Polynesia	1.4	0.2	0.2	147.5	14,746.5	21	5	1.6%	44	10	69	74	54	
Guam	0.2	0.2	0.2	998.6	9,078.6	28	4	2.4%	29	9	72	77	38	
Marshall Islands	0.1	0.1	0.1	1,492.1		26	4	2.2%	31	31	63	67	65	$1,540
New Caledonia	7.1	0.2	0.2	29.1		21	5	1.6%	42	7	69	77	59	
Papua New Guinea	174.9	5.0	6.1	28.8		34	10	2.4%	29	77	56	57	15	$890
Samoa	1.1	0.2	0.2	191.0	1,005.4	31	6	2.5%	28	25	65	72	21	$1,070
Solomon Is.	10.8	0.4	0.6	39.4	3,936.9	37	6	3.1%	23	25	69	74	13	$760
Vanuatu	4.7	0.2	0.2	45.0	2,248.5	35	7	2.8%	25	39	64	67	18	$1,260

Continued on pages 230-231

Scale 1: 16 000 000; one inch to 250 miles. Conic Projection

Elevations and depressions are given in feet

chapter 1

Europe

CONCEPTS, IDEAS, AND TERMS

1. Land hemisphere
2. Infrastructure
3. Areal functional specialization
4. Model
5. Von Thünen's Isolated State
6. Industrial Revolution
7. Nation
8. Nation-state
9. Centripetal forces
10. Centrifugal forces
11. Indo-European languages
12. Complementarity
13. Transferability
14. Intervening opportunity
15. Primate city
16. Metropolis
17. Devolution
18. Four Motors of Europe
19. Regional state
20. Supranationalism
21. Site
22. Situation
23. Conurbation
24. Landlocked location
25. Break-of-bulk point
26. Entrepôt
27. Shatter belt
28. Balkanization
29. Exclave
30. Irredentism

REGIONS

- WESTERN EUROPE
- THE BRITISH ISLES
- NORTHERN (NORDIC) EUROPE
- MEDITERRANEAN EUROPE
- EASTERN EUROPE

Copyright © Rand McNally, 2000 FIGURE 1-1

It is appropriate to begin our investigation of the world's geographic realms in Europe because over the past five centuries Europe and Europeans have influenced and changed the rest of the world more than any other realm or people has done. European empires spanned the globe and transformed societies far and near. European colonialism propelled the first wave of globalization. Millions of Europeans migrated from their homelands to the Old World as well as the New, changing (and sometimes nearly obliterating) traditional communities and creating new societies from Australia to North America. Colonial power and economic incentive combined to impel the movement of millions of imperial subjects from their ancestral homes to distant lands: Africans to the Americas, Indians to Africa, Chinese to Southeast Asia, Malays to South Africa's Cape, Native Americans from east to west. In agriculture, industry, politics, and other spheres, Europe generated revolutions—and then exported those revolutions across the world, thereby consolidating the European advantage.

But throughout much of that 500-year period of European hegemony, Europe also was a cauldron of conflict. Religious, territorial, and political disputes precipitated bitter wars that even spilled over into the colonies. And during the twentieth century, Europe twice plunged the world into war. The terrible, unprecedented toll of World War I (1914–1918) was not enough to stave off World War II (1939–1945), which ended with the first-ever use of nuclear weapons in Japan. In the aftermath of that war, Europe's weakened powers lost most of their colonial possessions, and a new rivalry emerged: an ideological Cold War between the communist Soviet Union and the capitalist United States. This Cold War lowered an Iron Curtain across the heart of Europe, leaving most of the east under Soviet control and most of the west in the American camp. Western Europe proved resilient, overcoming the destruction of war and the loss of colonial power to regain economic strength. Meanwhile the Soviet communist experiment failed at home and abroad, and in 1990 the last vestiges of the Iron Curtain were lifted. Since then, a massive effort has been underway to reintegrate and reunify Europe from the Atlantic coast to the Russian border, the key geographic story of this chapter.

◆ Major Geographic Qualities of Europe

1. The European realm lies on the western extremity of the Eurasian landmass, a locale of maximum efficiency for contact with the rest of the world.
2. Europe's lingering and resurgent world influence results largely from advantages accrued over centuries of global political and economic domination.
3. The European natural environment displays a wide range of topographic, climatic, and soil conditions and is endowed with many industrial resources.
4. Europe is marked by strong internal regional differentiation (cultural as well as physical), exhibits a high degree of functional specialization, and provides multiple exchange opportunities.
5. European economies are dominated by manufacturing, and the level of productivity has been high; levels of development generally decline from west to east.
6. Europe's nation-states emerged from durable power cores that formed the headquarters of world colonial empires. A number of those states are now plagued by internal separatist movements.
7. Europe's rapidly aging population is generally well off, highly urbanized, well educated, enjoys long life expectancies, and constitutes one of the world's three largest population clusters.
8. Europe has made important progress toward international economic integration. The push toward still stronger and broader coordination continues.

DEFINING THE REALM

As Figure 1-1 shows, Europe is a realm of peninsulas and islands on the western margin of the world's largest landmass, Eurasia. It is a realm of 583 million people and 39 countries, but it is territorially quite small. Yet despite its modest proportions it has had—and continues to have—a major impact on world affairs. For many centuries Europe has been a hearth of achievement, innovation, and invention.

Europe's peoples have benefited from a large and varied store of raw materials. Whenever the opportunity or need arose, the realm proved to contain what was required. Early on, these requirements included cultivable soils, rich fishing waters, and wild animals that could be domesticated; in addition, extensive forests provided wood for houses and boats. Later, mineral fuels and ores propelled industrialization.

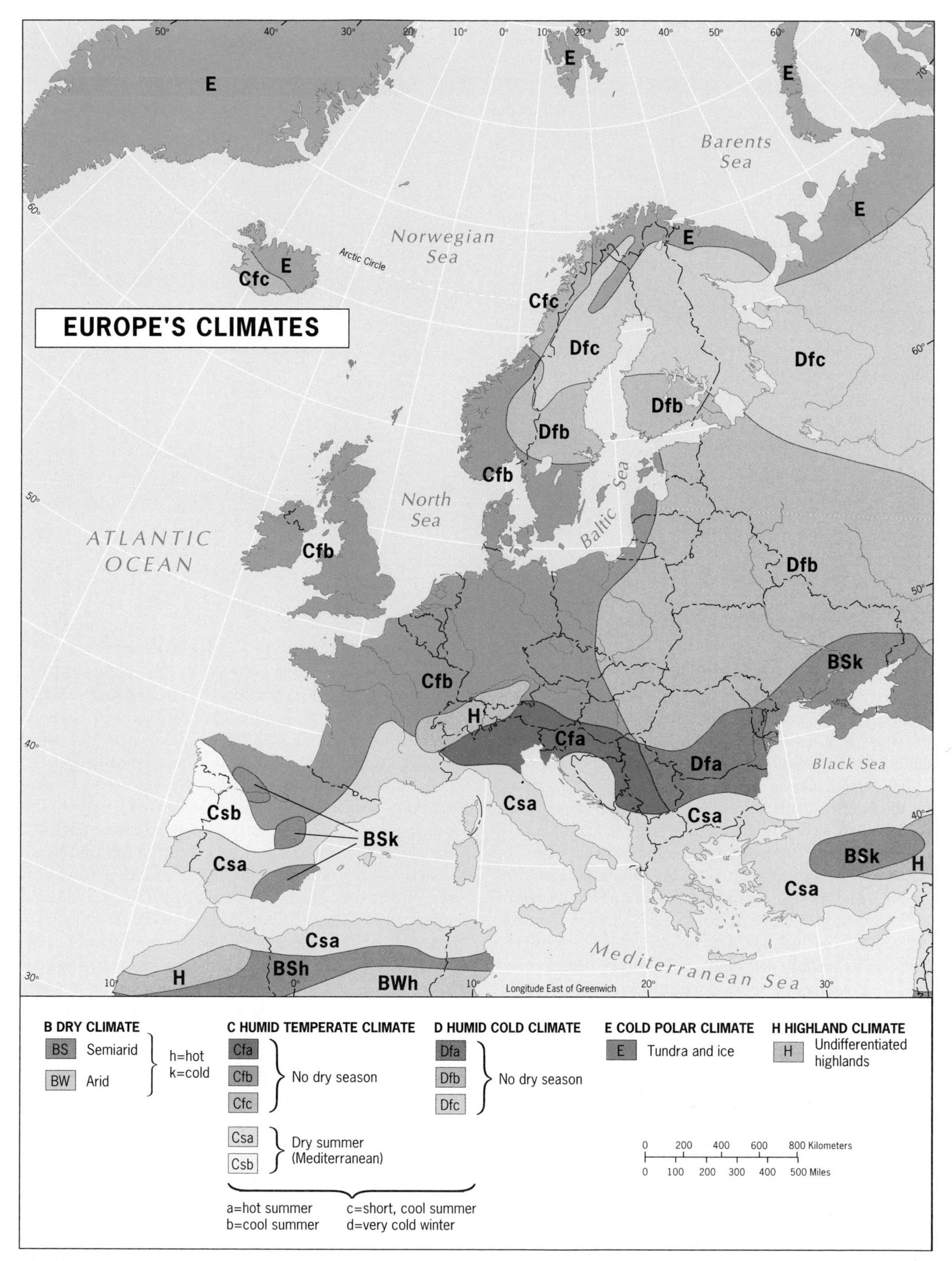
EUROPE'S CLIMATES
Norwegian Sea
Arctic Circle
Barents Sea
North Sea
Baltic Sea
ATLANTIC OCEAN
Black Sea
Mediterranean Sea
Longitude East of Greenwich
E
Cfc
Dfc
Dfb
Cfb
Cfa
Dfa
H
Csa
Csb
BSk
BSh
BWh
B DRY CLIMATE
BS Semiarid
BW Arid
h=hot
k=cold
C HUMID TEMPERATE CLIMATE
Cfa
Cfb
Cfc
No dry season
Csa
Csb
Dry summer (Mediterranean)
a=hot summer
b=cool summer
c=short, cool summer
d=very cold winter
D HUMID COLD CLIMATE
Dfa
Dfb
Dfc
No dry season
E COLD POLAR CLIMATE
E Tundra and ice
H HIGHLAND CLIMATE
H Undifferentiated highlands
0 200 400 600 800 Kilometers
0 100 200 300 400 500 Miles

FIGURE 1-2

RELATIVE LOCATION: EUROPE IN THE LAND HEMISPHERE

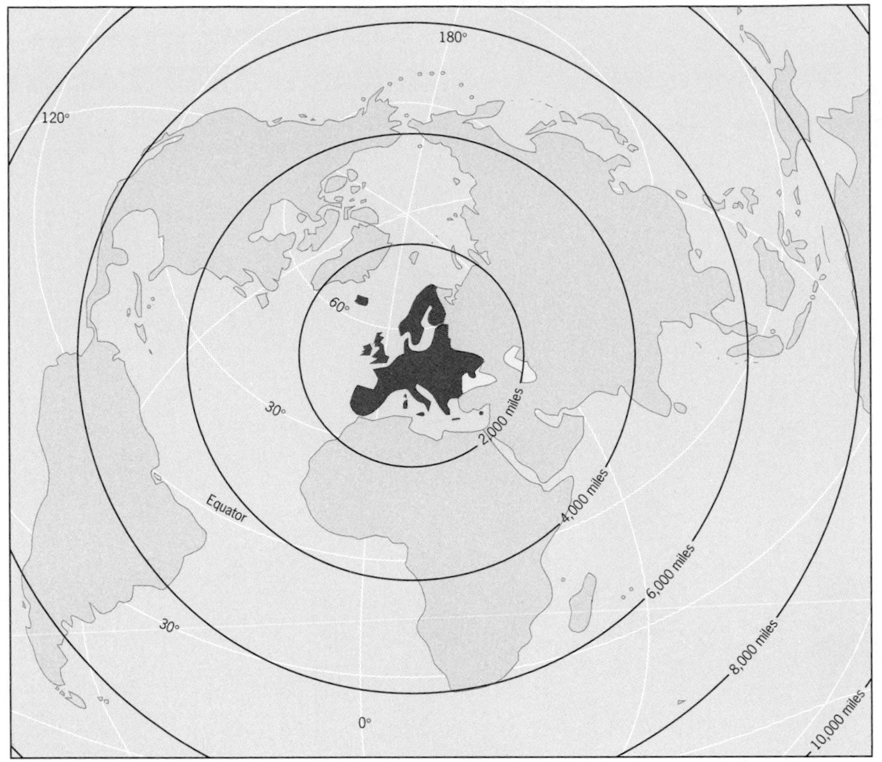

Azimuthal equidistant projection centered on Hamburg, Germany

FIGURE 1-3

From the balmy shores of the Mediterranean Sea to the icy peaks of the Alps, and from the moist woodlands and moors of the Atlantic fringe to the semiarid prairies north of the Black Sea, Europe presents an almost infinite range of natural environments (Fig. 1-2). The moderating influence of the Atlantic Ocean and its onshore windflow—warmed by the North Atlantic Drift current that originates as the Gulf Stream along the East Coast of the United States—reaches far to the north and east, pushing *Cfb* climatic conditions into Scandinavia as well as Poland and Hungary. On Figure I-8 (p. 14-15), compare northeastern North America to northwestern Europe at equivalent latitudes. Whereas Norwegians live in mild *Cfb* environments, Canadians confront Arctic cold. In the south, the warm dry summers and mild wet winters of the Mediterranean climatic regime dominate the peninsulas of Spain, Italy, and Greece. Full-fledged *continentality* (the more extreme temperatures and drier conditions associated with inland location remote from the sea) does not take over until the easternmost parts of the realm, and even there it occurs in its mildest form.

The European realm is home to peoples of numerous cultural-linguistic stocks, including not only Latins, Germanics, and Slavs but also many minorities such as Finns, Hungarians, Basques, and Celts. This diversity of ancestries continues to be an asset as well as a liability. It has generated interaction and exchange, but it has also caused conflict and war. Europeans have a history marked by efforts to unify and accommodate differences, punctuated by repeated failure and chaos. Sometimes one region of Europe made progress while another disintegrated. That remains the situation today.

Europe's cultural diversity and resource endowments are prime advantages, but this realm also has geographic assets in terms of location, scale, and proximities. Globally, Europe's *relative location*—at the heart of the **land hemisphere**—is one of maximum efficiency for contact with the rest of the world (Fig. 1-3). Regionally, Europe is also far more than a mere western extremity of the Eurasian landmass. Almost nowhere is Europe far from that essential ingredient of European development—the sea—and the water interdigitates with the land here as it does nowhere else on Earth. Southern and Western Europe consist almost entirely of peninsulas and islands, from Greece, Italy, and the Iberian Peninsula (Spain and Portugal) to the British Isles, Denmark, and the Scandinavian Peninsula (Norway and Sweden). Southern Europe faces the Mediterranean, and Western Europe virtually surrounds the North Sea as it looks out over the Atlantic Ocean. Beyond the Mediterranean lies Africa, and across the Atlantic are the Americas. Europe has long been a place of contact between peoples and cultures, of circulation of goods and ideas. The hun-

1

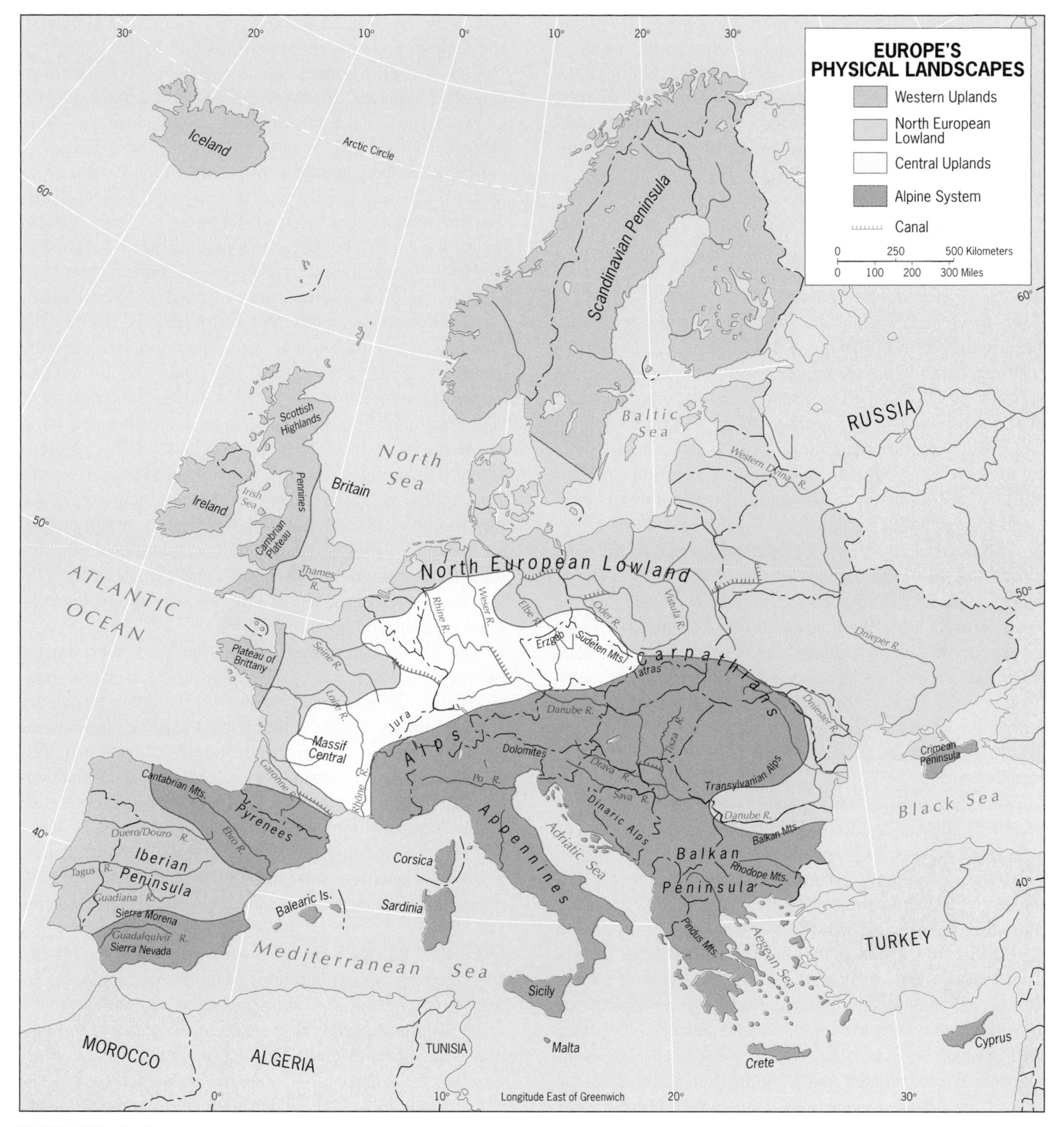

FIGURE 1-4

dreds of miles of navigable waterways; the easily traversed bays, straits, and channels between numerous islands and peninsulas and the mainland; and the highly accessible Mediterranean, North, and Baltic seas all provided the routeways for these exchanges. Later, even the oceans became avenues of long-distance spatial interaction.

This historic advantage of moderate distances applies on the mainland as well. Europe's Alps may form a transcontinental divider, but what they separate still lies in close juxtaposition (Alpine passes have for centuries provided several corridors for overland contact). Consider Rome and Paris: the distance between these long-time control points of Mediterranean and northwestern Europe is less than that between New York and Chicago. No place in Europe is very far from anyplace else on the continent, although the economy and outlook of nearby places often differ sharply. Short distances and large differences make for much interaction—and that has been the hallmark of Europe's geography for over a millennium.

LANDSCAPES AND OPPORTUNITIES

Europe's total area may be small, but its physical landscapes are varied and complex. It would be easy to identify a large number of physiographic regions, but in doing so we might lose sight of the broader regional pattern—a pattern that has much to do with how the European human drama unfolded. Accordingly, Europe's physical landscapes can be grouped regionally into four units: the Central Uplands, the Alpine Mountains in the south, the Western Uplands, and the great North European Lowland (Fig. 1-4).

The heart of Europe is occupied by an area of hills and small plateaus, with forest-clad slopes and fertile valleys. These *Central Uplands* also contain the majority of the realm's productive coalfields, part of a raw-material-laden backbone that extends across the middle of Europe from England eastward to Ukraine. When this region emerged from its long medieval quiescence and stirred with the stimuli of the Industrial Revolution, towns on the Uplands' flanks grew into cities, and farmlands gave way to mines and factories.

The Central Uplands are flanked on the south by the much higher Alpine Mountains and to the west, north, and east by the North European Lowland. The *Alpine Mountains* include not only the famous Alps themselves but also other ranges that belong to this great mountain system. The Pyrenees between Spain and France (one of Europe's few true physical barriers), Italy's Appennines, the Dinaric Ranges and the Carpathians of Eastern Europe are all part of this Alpine system, which extends even into North Africa (as the Atlas Mountains) and eastward into Turkey and beyond. Although the Alps are rugged and imposing, they have not prevented spatial interaction: traders have operated through their mountain passes for centuries.

Europe's western margins are also quite rugged, but the *Western Uplands* of Scandinavia, Scotland, Ireland, France's Brittany, Portugal, and Spain are not part of the Alpine system (maximum elevations are markedly lower than those in the Alps). This western arc of highlands represents older geologic mountain building, contrasting sharply with the relatively young, still active, earthquake-prone Alpine Mountains. Scandinavia's uplands form part of an ancient geologic shield underlain by old crystalline rocks now bearing the marks of the Pleistocene glaciation. Comparatively old rocks, now mostly worn down to a tableland, also support Spain's central plateau, the *Meseta*.

The last of Europe's landscape regions is its most densely populated. The *North European Lowland*—also known as the Great European Plain—extends in a gigantic arc from southern France through the Low Countries across northern Germany and then eastward through Poland, from where it fans out deeply into southwestern Russia. Note, too, that southeastern England, Denmark, and the southern tip of Sweden also belong to this region (Fig. 1-4). Most of the North European Lowland lies below 500 feet (150 m) in elevation, and local relief rarely exceeds 100 feet (30 m). Beyond this single topographic factor there is much to differentiate it internally. In France, this region includes the basins of three major rivers—the Garonne, Loire, and Seine. In the Netherlands, much of this region consists of land reclaimed from the sea, enclosed by dikes and lying below sea level. In southeastern England, the higher areas of the Netherlands, northern Germany and Denmark, southern Sweden, and farther eastward, it bears the marks of the Pleistocene glaciation that withdrew only a few thousand years ago (see Fig. I-6). Each of these particular areas affords its own opportunities as soils and climates vary, giving rise to some of the world's most productive (and prestigious) agricultural pursuits.

The North European Lowland has also been one of Europe's major avenues of human contact. Entire peoples have migrated across it, and armies have repeatedly marched through it. As settlement took place, agricultural diversity became the norm, and land use came to be dominated by intensive farming organized around myriad villages from which farmers commuted to their nearby fields. Today, centuries later, we still cannot speak of a "dairy belt" or "wheat belt" in Europe like those in North America; even where one particular crop dominates the farming scene, different crops are likely to grow just a few fields away.

Finally, Europe's great lowland possesses yet another crucial advantage: its multitude of navigable rivers, emerging from higher adjacent areas and wending their way to the sea. In addition to the three rivers of France already mentioned, the Rhine-Meuse (Maas) river system serves one of Europe's most productive industrial and agricultural areas, reaching the sea via the Netherlands; the Weser, Elbe, and Oder cross northern Germany; and the Vistula traverses Poland. In southeastern Europe, the Danube rivals the Rhine in regularity of flow and navigability. Thus, north of the Alpine Mountain system, Europe's major rivers create a radial pattern outward from the continent's interior highlands. In this way, the natural waterways as well as the land surface of the North European Lowland favor traffic and trade. Over many centuries, the Europeans have further improved the situation by connecting navigable stretches of rivers with artificial canals (Fig. 1-4). These waterways, as

Europe: The Eastern Boundary

The European realm is bounded on the west, north, and south by Atlantic, Arctic, and Mediterranean waters, respectively. Europe's eastern boundary, however, has always been debated. Some scholars place it at the Ural Mountains (deep inside Russia), thereby recognizing a "European" Russia and, presumably, an "Asian" one as well. Others argue that because there is a continuous transition from west to east (which continues into Russia), there is no point in trying to define any boundary.

Still, the boundary used in this chapter—marking Europe's eastern boundary as the border with Russia—has solid geographic justifications. Eastern Europe shares with Western Europe its fragmentation into several states that have distinct cultural geographies; this political condition sets it apart from the Russian giant to the east. Historical and cultural contrasts also mark large segments of this boundary (as can be observed in Fig. 1-8).

well as the roads and railroads that followed later, combined to bring tens of thousands of localities into contact with one another. Thus new techniques and innovations could spread rapidly, and trading connections and activity intensified continuously.

EUROPE'S PREMODERN HERITAGE

Modern Europe was peopled in the wake of Pleistocene's most recent glacial retreat—a gradual withdrawal that caused cold tundra to turn into deciduous forest and ice-filled valleys into grassy vales. On Mediterranean shores, Europe witnessed the rise of its first great civilizations—on the islands and peninsulas of Greece and later in Italy. Greece lay exposed to the influences radiating from the advanced civilizations of Mesopotamia and the Nile Valley (see map p. 284), and the intervening eastern Mediterranean was crisscrossed by maritime trade routes.

Ancient Greece

As the ancient Greeks forged their city-states and intercity leagues, they also made impressive intellectual achievements (which peaked during the fourth century BC). Their political science and philosophy have influenced politics and government ever since, and great accomplishments were also recorded in such fields as education, literature, architecture, and the arts. The fragmentation of Greece's habitat led to local experimentation and success, followed

FROM THE FIELD NOTES
"I got aboard a large ship in Dover, England and sailed up the Seine River to Rouen, France. What a lesson in regional geography: from the highest deck I had a panoramic view over the landscape of what I had learned to call the Paris Basin. Every square inch of the flat to undulating countryside of this part of the North European Lowland seemed to be in productive use in one of the most meticulously organized, parklike rural landscapes I had ever seen. Clusters of trees signaled villages; no road was without its line of poplars or oaks. The Seine River, narrow but surprisingly deep given the size of the ship, traverses its basin in wide meanders, which must have caused the land planners a headache. Still, the French pattern of 'long lots,' giving as many farmers as possible some frontage on a waterway and then extending the parcel of soil deep inland, had some regularity even here (you can also see it in Canada's Quebec). Two of those long lots, one fallow and the other under pasture, can be seen in the foreground."

Several cultures made lasting contributions to the architecture of Iberia: the Romans, the Arabs, and the Spanish and Portuguese themselves. The city of Segovia, capital of the Autonomous Community of Castile-Leon, was already more than six centuries old when it was taken by the Romans in about 80 BC; the Moors took it early in the eighth century AD, and the Spanish freed it in 1079. Today, the city's Old Town, encircled by ancient walls, lies centered on the *Plaza Mayor*; nearby lies the *Plaza Azoguejo*, another busy square. Crossing the *Plaza Azoguejo* is one of the best-preserved Roman aqueducts (water-carriers) still in existence, known locally as *El Puente*. As the background of this picture (behind the red-tiled roofs) shows, the occupiers of Segovia (and other towns), whether they were Romans, Arabs, or Spaniards, built their structures on the highest and most defensible hill in the area. This caused a water-supply problem, which the Romans solved by building aqueducts that brought water from far upstream to the center.

by active exchanges of ideas and innovations. But internal discord persisted as well; in the end this contributed to Greece's decline. By 147 BC, the Romans had defeated the last sovereign Greek intercity league. Nevertheless, what the ancient Greeks had accomplished was not undone: they had transformed the eastern Mediterranean into one of the cultural cores of the world, and Greek culture became a major component of Roman civilization.

The Roman Empire

The center of civilization and power now shifted to the Romans in present-day Italy. The Greeks never achieved politico-territorial organization on the scale that Imperial Rome would accomplish. At its greatest expansion (in the second century AD), the Roman Empire extended from Britain to the Persian Gulf and from the Black Sea to Egypt. The variety of cultures brought under Roman control and the resulting exchange of ideas and innovations yielded many opportunities for regional interaction—particularly in southern and western Europe. Areas that had hitherto supported only subsistence modes of life were drawn into the greater economic framework of the empire, and suddenly there were distant markets for products that had never found even local markets before. Foodstuffs and raw materials now flowed into Rome from most of the Mediterranean Basin. With a population that at its peak reached perhaps one million, the city itself was the greatest single marketplace of the empire and the first metropolitan-scale urban center in Europe.

That urban tradition came to characterize Roman culture throughout the empire, and many cities and towns founded by the Romans continue to prosper to this day. Roman urban centers were connected by an unparalleled network of highway and water routes, facilities that all formed part of an **infrastructure** needed to support economic growth and development. (Today, a modern state's infrastructure would include railroads, airports, energy-distribution systems, telecommunications networks, and the like.) More than anything else, however, the Roman Empire left Europe a legacy of ideas—concepts that long lay dormant but eventually played their part when Europe again discovered the path of progress. In political and military organization, effective administration, and long-term stability, the Empire was centuries ahead of its time. Moreover, never was a larger part of Europe unified by acquiescence than it was under the Romans, and at no time did Europe come closer to obtaining a *lingua franca* (common language) than during the age of Rome.

2

Finally, Europe's transformation under Roman rule heavily involved the geographic principle of **areal functional specialization**. Before the Romans brought order and connectivity to their vast domain, much of Europe was inhabited by tribal peoples who lived at a subsistence level. Many of these groups lived in virtual isolation, traded little, and fought over territory when outsiders encroached on it. Peoples under Rome's sway, however, were brought into its economic as well as political spheres, and farmlands, irrigation systems, mines, and workshops appeared. Thus Roman-dominated areas began to take on a characteristic that has marked Europe ever since: *particular peoples and particular places concentrated on the production of particular goods*. Parts of North Africa became granaries for urbanizing (European) Rome; Elba, a Mediterranean island, produced iron ore; the Cartagena area of southeastern Spain mined and exported silver and lead. Many other locales in

3

the Roman Empire specialized in the production of particular farm commodities, manufactured goods, or minerals. The Romans knew how to exploit their natural resources; at the same time, they also learned to use the diversified productive talents of their subjects.

Decline and Rebirth

Even the breakdown and collapse of the Empire in the fifth century AD could not undo what the Romans had forged by spreading their language, disseminating Christianity (in some ways the sole strand of permanence through the ensuing Dark Ages), and entrenching their systems of education, administration, and commerce. But ancient Rome's decline was attended by a momentous stirring of Europe's peoples as Germanic and Slavic populations moved to their present positions on the European stage. The Anglo-Saxons invaded Britain from Danish shores, the Franks moved into France, the Allemanni traversed the North European Lowland and settled in Germany. Capitalizing on the disintegration of Roman power, numerous kings, dukes, and lesser nobles established themselves as local rulers. Europe was in turmoil, and its weakness invited invasion from North Africa and Southwest Asia. In Iberia, the Arab-Berber Moors conquered a large area; in Eastern Europe, the Ottoman Turks extended their Islamic empire. The townscapes of southern Spain and the Balkans still carry the cultural imprints of these Muslim penetrations.

After nearly a thousand years of feudal fragmentation during the Dark and Middle Ages, modern Europe began its emergence in the second half of the fifteenth century. At home, monarchies strengthened at the expense of feudal lords and landed aristocracies and, in the process, forged the beginnings of nation-states. Abroad, Western Europe's developing states were on the threshold of discovery—the discovery of continents and riches across the oceans. Europe's emerging powers were fired by a new national consciousness and pride, and there was renewed interest in Greek and Roman achievements in science and government. Appropriately, this period is referred to as Europe's *Renaissance*.

The new age of progress and rising prosperity was centered in Western Europe, whose countries lay open to the new pathways to wealth—the oceans. Now the highly competitive monarchies of Western Europe engaged in economic nationalism that operated in the form of *mercantilism*. The objectives of this policy were to accumulate as large a quantity of gold and silver as possible and to use foreign trade and colonial acquisition to achieve that end. Mercantilism was promoted and sustained by the state; precious metals could be obtained either by the conquest of peoples in possession of them or indirectly by achieving a favorable balance of international trade. Thus there was a stimulus not only to seek new territories where such metals might lie, but also to produce goods at home that could be sold profitably abroad. The matrix of modern states in Western Europe was beginning to take shape, and the spiral had been entered that would lead to great colonial empires and a period of world domination.

THE REVOLUTIONS OF MODERNIZING EUROPE

Strife and dislocation punctuated Europe's march to world domination. Much of what was achieved during the Renaissance was destroyed again as powerful monarchies struggled for primacy, religious conflicts spread death and misery, and new tyrannies suppressed the beginnings of parliamentary government. Nonetheless, revolutions in several spheres were in the making. Economic developments in Western Europe ultimately undermined the absolute monarchs and their privileged, land-owning nobilities. The city-based merchant was gaining wealth and prestige, and the traditional measure of affluence—land—began to lose its status in these changing times. The merchants and businesspeople of Europe were soon able to demand political recognition on grounds the nobles could not match. Urban industries were thriving; Europe's population, more or less stable at about 100 million since the mid-sixteenth century, was on the rise.

The Agrarian Revolution

This transformation was heightened by an ongoing *agrarian revolution*—the significant metamorphosis of European farming that preceded the Industrial Revolution and helped make possible a sustained population increase during the seventeenth and eighteenth centuries. The Netherlands, Belgium, and northern Italy (soon joined by England and France) paved the way with their successes in commerce and manufacturing. The stimulus provided by expanding urbanization and markets led to the more efficient organization of land ownership and agriculture. At the same time, methods of soil preparation, crop rotation, cultivation, harvesting, and livestock-raising improved, and more effective farm equipment and storage and distribution systems developed. In the growing cities and towns, farm products fetched higher prices. New crops were introduced, especially from the Americas; the potato became a European staple. More and more of Europe's farmers were drawn from subsistence into profit-driven market economies. Later, the manufactured products of the Industrial Revolution further stimulated the transformation of the realm's agriculture.

As new forces and processes began to reshape Europe's economic geography, scholars tried to interpret the new spatial patterns they produced. In 1826, the economist Jo-

Models in Geography

A widely used approach to generalization in both human and physical regional geography is the development of conceptual *models*. Peter Haggett's seminal work, *Locational Analysis in Human Geography*, offers an especially good definition: *in model-building we create an idealized representation of reality in order to demonstrate its most important properties.*

Geographers need to use models because reality is so complex: to understand how things work, we must first filter out the main spatial processes and their responses from the myriad details within which they are embedded in a highly complicated world. Models therefore provide a simplified picture of reality, so that they can convey, if not the entire truth, then at least a useful and essential part of it. The theory-based derivation of the von Thünen model (Fig. 1-5) and its empirical, or real-world, application to contemporary Europe (Fig. 1-6) offer a classical demonstration of this geographic method.

hann Heinrich von Thünen (1783–1850) built one of the world's first geographical **models** (see box titled "Models in Geography"). For four decades von Thünen, who owned a large farming estate in northeastern Germany, studied the effects of distance and transportation costs on the location of productive activity. Eventually, he published a pioneering work entitled *The Isolated State*, and his methods in many ways constitute the foundations of modern location theory.

Von Thünen's Isolated State model was so named because he wanted to establish, for purposes of theoretical analysis, a self-contained country devoid of outside influences that would disturb the internal workings of the economy. He therefore created a sort of regional laboratory within which he could identify the factors that influence the spatial distribution of farms around a single urban center. To do this, he made a number of limiting assumptions. First, he stipulated that the soil and climate would be uniform throughout the region. Second, no river valleys or mountains would interrupt a completely flat land surface. Third, there would be a single centrally positioned city in the Isolated State, and the latter would be surrounded by an empty, unoccupied wilderness. Fourth, the farmers in the Isolated State would transport their own products to market by oxcart, directly overland and straight to the central city. This, of course, is the same as assuming a system of radially converging roads of equal and constant quality; within such a framework, transport costs would be directly proportional to distance.

Von Thünen integrated these assumptions with what he had learned from the actual data collected while running his estate. He now asked himself: What would be the ideal spatial arrangement of agricultural activities within the Isolated State? He concluded that farm products would be raised in a series of concentric zones outward from the central market city. Nearest to the city would be grown those crops that perished easily or yielded the highest returns (such as vegetables) because this readily accessible farmland was in great demand and therefore expensive; dairying would also be carried on in this innermost zone. Farther away would be potatoes and grains. Eventually, since transport costs to the city increased with distance, there would come a line beyond which it would be uneconomical to produce crops. There the wilderness would begin.

Von Thünen's model incorporated four zones or rings of agricultural land use surrounding the market center (Fig. 1-5). The first and innermost belt would be a zone of intensive farming and dairying. The second zone would be an

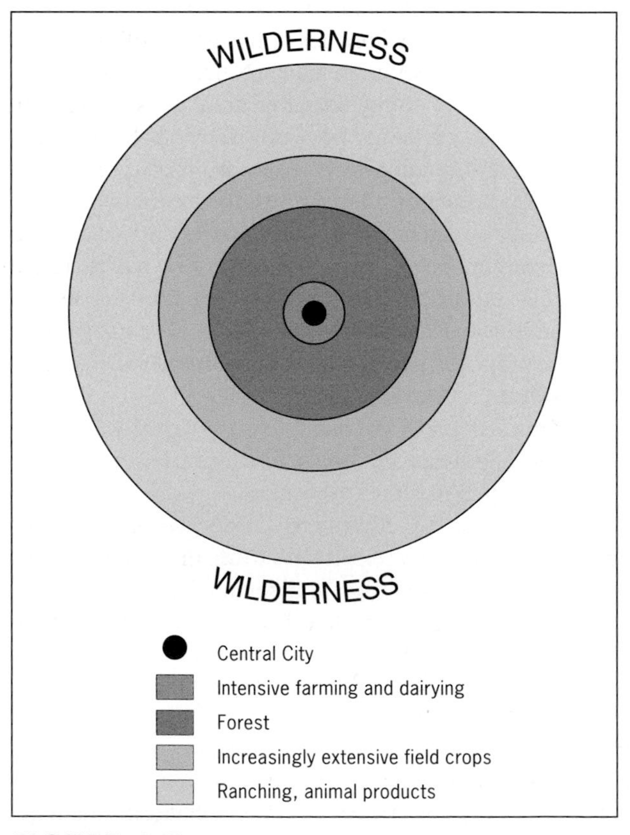

FIGURE 1-5

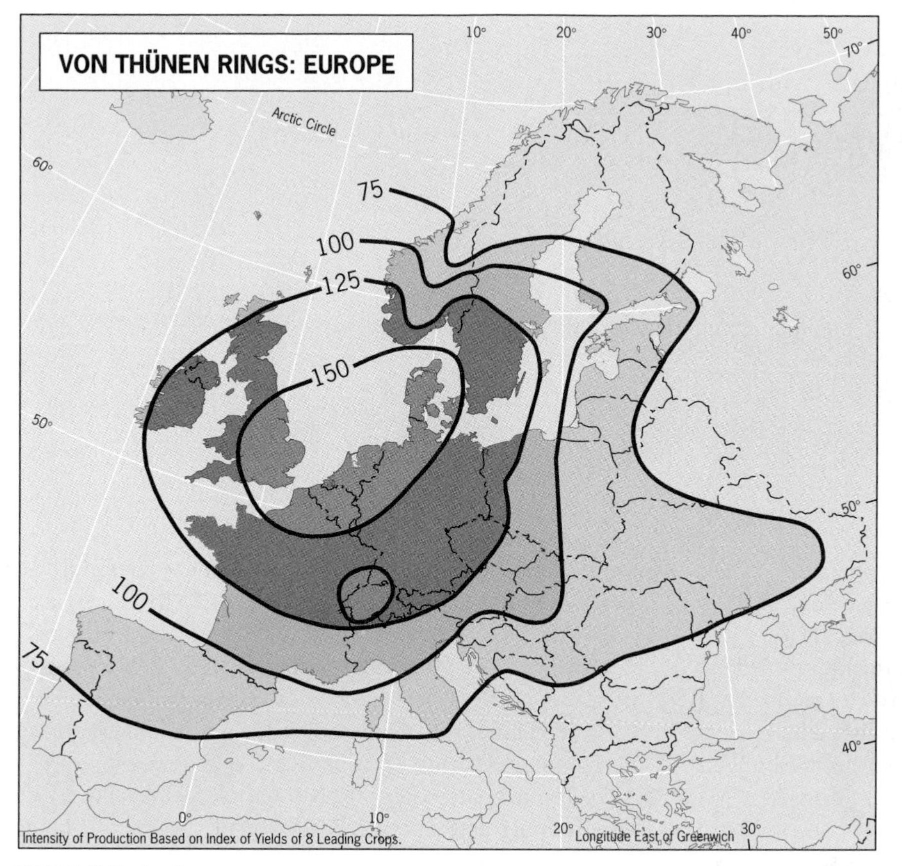

FIGURE 1-6

area of forest used for firewood and timber (still important as a building material in his time). Next, there would be a third ring of increasingly extensive field crops. The fourth and outermost zone would be occupied by ranching and animal products beyond which would begin the wilderness that isolated the region from the rest of the world.

Von Thünen knew, of course, that the real Europe (or world) did not present idealized situations exactly as he had postulated them. Transport routes serve certain sectors more efficiently than others. Physical barriers can impede even the most modern surface communications. External economic influences invade every area. But von Thünen wanted to eliminate these disruptive conditions in order to discern the fundamental processes that shaped the spatial layout of the agricultural economy. Later, the distorting factors could be introduced one by one and their influence examined. First, however, he developed his model in theoretical isolation, basing it on total regional uniformity.

It is a great tribute to von Thünen that his work still commands the attention of geographers. The economic-geographic landscape of Europe has changed enormously since his time, but geographers still compare present-day patterns of economic activity to the Thünian model. Such a comparison was made by Samuel van Valkenburg and Colbert Held, whose map of Europe's agricultural intensity reveals a striking ring-like concentricity (Fig. 1-6). The overriding spatial change since von Thünen's day is the improvement in transportation technology, which permitted the Isolated State to expand from *micro-* to *macroscale*. Thus the model is no longer centered on a single city but rather on the vast urbanized area lining the southern coasts of the North Sea—which now commands a continentwide Thünian agricultural system.

The Industrial Revolution

Alongside its advances in agriculture, Europe also had developed significant industries before the **Industrial Revolution** 6 began. In Flanders (western Belgium) and England, specialization had been achieved in the manufacturing of woolen and linen textiles; in eastern Germany's Saxony, iron ore was mined and smelted. European manufacturers produced a wide range of goods for local markets, but their quality was often surpassed by textiles and other wares from India and China. This gave European entrepreneurs a major incentive to refine and mass-produce their products. Because raw materials could be shipped home in virtually unlimited quantity, if they could find ways to mass-produce these commodities into finished goods, they could bury the Asian industries under growing volumes and declining prices.

Now the search for better machinery was on, especially improved spinning and weaving equipment. In the 1780s,

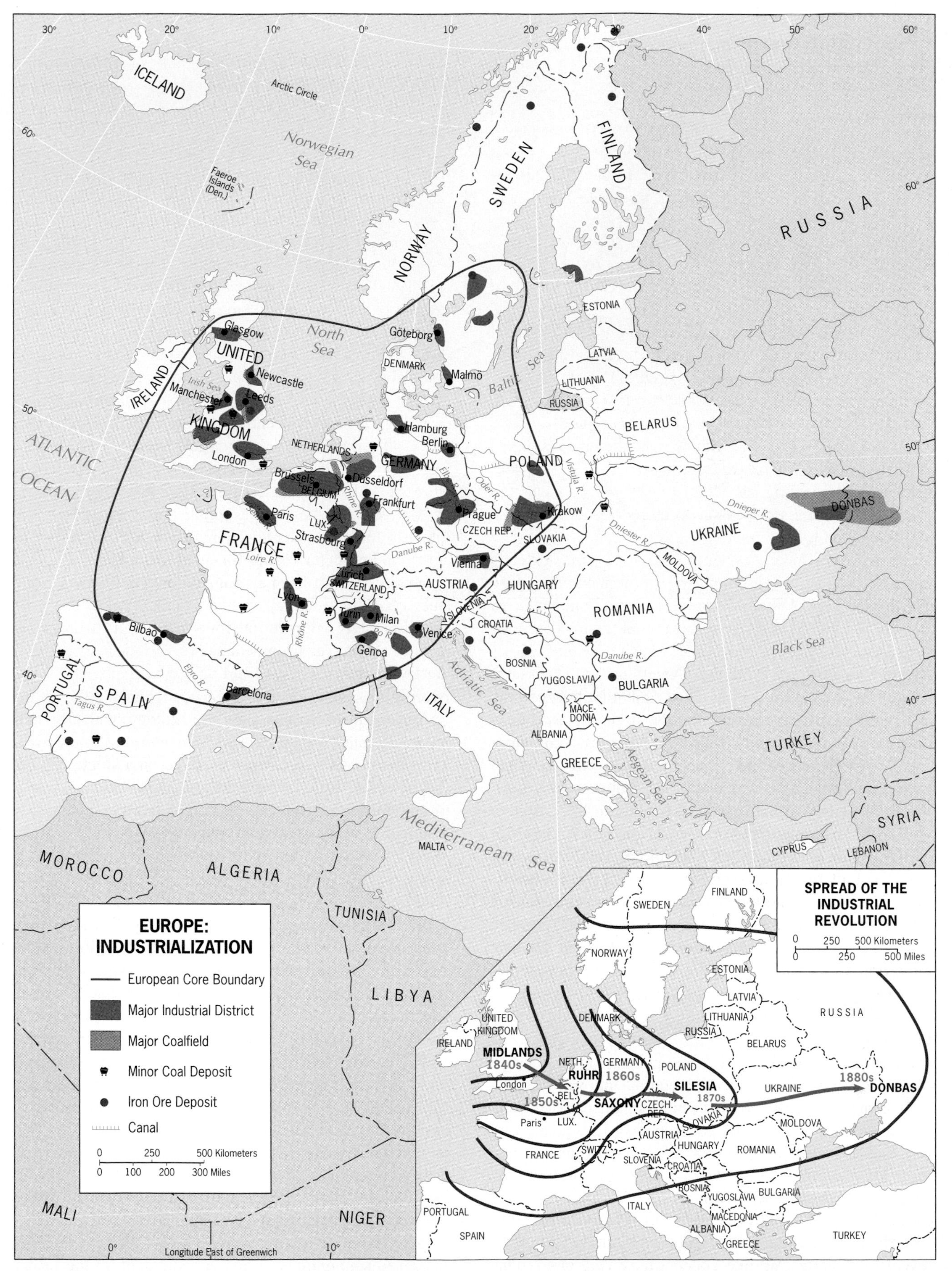

EUROPE: INDUSTRIALIZATION
European Core Boundary
Major Industrial District
Major Coalfield
Minor Coal Deposit
Iron Ore Deposit
Canal
0 250 500 Kilometers
0 100 200 300 Miles
SPREAD OF THE INDUSTRIAL REVOLUTION
0 250 500 Kilometers
0 250 500 Miles
MIDLANDS 1840s
1850s
RUHR 1860s
SAXONY
SILESIA 1870s
1880s
DONBAS
ICELAND
Arctic Circle
Norwegian Sea
Faeroe Islands (Den.)
NORWAY
SWEDEN
FINLAND
RUSSIA
ESTONIA
LATVIA
LITHUANIA
BELARUS
UKRAINE
POLAND
GERMANY
DENMARK
NETHERLANDS
BELGIUM
LUX.
FRANCE
SWITZERLAND
AUSTRIA
HUNGARY
SLOVAKIA
CZECH REP.
SLOVENIA
CROATIA
BOSNIA
YUGOSLAVIA
ROMANIA
MOLDOVA
BULGARIA
MACEDONIA
ALBANIA
GREECE
TURKEY
ITALY
SPAIN
PORTUGAL
UNITED KINGDOM
IRELAND
North Sea
Irish Sea
Baltic Sea
ATLANTIC OCEAN
Black Sea
Adriatic Sea
Aegean Sea
Mediterranean Sea
MOROCCO
ALGERIA
TUNISIA
LIBYA
MALI
NIGER
SYRIA
CYPRUS
LEBANON
MALTA
Glasgow
Newcastle
Leeds
Manchester
London
Göteborg
Malmö
Hamburg
Berlin
Brussels
Düsseldorf
Frankfurt
Paris
Strasbourg
Prague
Krakow
Vienna
Zurich
Lyon
Turin
Milan
Genoa
Venice
Bilbao
Barcelona
Loire R.
Rhône R.
Rhine R.
Elbe R.
Oder R.
Vistula R.
Danube R.
Dnieper R.
Dniester R.
Po R.
Ebro R.
Tagus R.
Longitude East of Greenwich

FIGURE 1-7

James Watt and others devised a steam-driven engine, which was soon adapted for various uses. About the same time, coal (converted into carbon-rich coke) was recognized as a vastly superior substitute for charcoal in smelting iron. These momentous innovations had a rapid effect. The power loom revolutionized the weaving industry. Iron smelters, long dependent on Europe's dwindling forests for fuel, could now be concentrated near coalfields. Engines could move locomotives as well as power looms. Ocean shipping entered a new age.

Britain had an enormous advantage, for the Industrial Revolution occurred when British influence reigned worldwide and the significant innovations were achieved in Britain itself. The British controlled the flow of raw materials, they held a monopoly over products that were in global demand, and they alone possessed the skills necessary to make the machines that manufactured the products. Soon the fruits of the Industrial Revolution were being exported, and the modern industrial spatial organization of Europe began to take shape. In Britain, manufacturing regions, densely populated and heavily urbanized, developed near coalfields in the English Midlands, at Newcastle to the northeast, in southern Wales, and along Scotland's Clyde River around Glasgow.

In mainland Europe, a belt of major coalfields extends from west to east, roughly along the southern margins of the North European Lowland, due eastward from southern England across northern France and Belgium, Germany (the Ruhr), western Bohemia in the Czech Republic, Silesia in southern Poland, and the Donets Basin in eastern Ukraine. Iron ore is found in a broadly similar belt, and the industrial map of Europe reflects the resulting concentrations of economic activity (Fig. 1-7). Another set of manufacturing regions emerged in and near the growing urban centers of Europe, as the same map demonstrates. London—already Europe's leading urban focus and Britain's richest domestic market—was typical of these developments. Many local industries were established here, taking advantage of the large supply of labor, the ready availability of capital, and the proximity of so great a number of potential buyers. Although the Industrial Revolution thrust other places into prominence, London did not lose its primacy: industries in and around the British capital multiplied.

Industrial and Urban Intensification

Not surprisingly, the industrialization of Europe—or, rather, its industrial *intensification* following the Industrial Revolution—also became a focus of geographic research. What influences shaped industrial location? How was Europe's industrialization channeled? Again, German scholars conducted the first important studies, mostly in the second half of the nineteenth century. Much of this work was incorporated in a volume by Alfred Weber (1868–1958), published in 1909 and titled *Concerning the Location of Industries*. Like von Thünen before him, Weber began with a set of limiting assumptions in order to minimize the complexities of the real Europe. But unlike von Thünen, Weber dealt with activities that take place at particular *points* rather than across large areas. Manufacturing plants, mines, and markets are located at specific places, and so Weber created a model region marked by sets of points where these activities would occur. He eliminated labor mobility and varying wage rates, and thereby could calculate the "pulls" exerted on each point in his theoretical region.

In the process, Weber discerned various factors that control industrial location. He recognized what he called "general" factors that would affect all industries—dominated by transport costs for raw materials and finished products—and "special" factors (such as perishability of foods). He also differentiated between "regional" factors (transport and labor costs) and "local" factors. The local factors, Weber argued, involve *agglomerative* (concentrating) and *deglomerative* (deconcentrating) forces. In the case of London, industries located there largely because of the advantages of locating together. The availability of specialized equipment, a technologically sophisticated labor force, and a large-scale market made London (as well as Paris and other big cities not positioned on rich deposits of natural resources) an attractive site for many manufacturing plants that could benefit from agglomeration. On the other hand, such concentration may eventually create strong disadvantages—chiefly competition for space, rising land prices, congestion, and environmental pollution. Ultimately, an industry might move away and deglomerative forces would set in.

Europe's industrialization also facilitated the growth of many of its cities and towns. In Britain in the year 1800, only about 9 percent of the population lived in urban areas, but by 1900, some 62 percent resided in cities and towns (today that proportion stands at 89 percent). All this was happening while the total population skyrocketed. As industrial modernization came to Belgium and Germany, to France and the Netherlands, and to other parts of Western Europe (see inset map, Fig. 1-7), the entire urban pattern changed. The nature of this process—the growth and strengthening of towns and cities—and related questions also became topics of geographic study, just as agriculture and industrialization had. (Urban geography became one of the discipline's leading subfields during the twentieth century; we discuss many of its principles in this chapter as well as in Chapters 3 and 5.)

Political Revolutions

Europe had had long experience with experiments in democratic government, but the *political revolution* that swept the realm after 1780 brought transformation on an unprecedented scale. Overshadowing these events was the French Revolution (1789–1795), but France's—and Europe's—political catharsis lasted into the twentieth century as the rising tide of nationalism eventually affected every monarchy on the continent.

In France, the popular revolution plunged the country into years of chaos and destruction. Only when Napoleon took control in 1799 was stability restored. Napoleon personified the new French republic, and he reorganized France so completely that he laid the foundations of the modern nation-state (a concept treated in the following section). He also built an empire that extended from German Prussia to Spain and from the Netherlands to Italy. Although his armies were ultimately repulsed, Napoleon had forever changed the political spatial structure of Europe. His French forces had been joined by nationalist revolutionaries from across the continent; one monarchy after another had been toppled. Even after Napoleon's final defeat at Waterloo in 1815, there were popular uprisings in Spain, Portugal, Italy, and Greece. By then Europe had had its first real taste of democracy and nationalist power—and it would not revert to its old ways.

The Rise of the Nation-State

As Europe went through its periods of rebirth and revolutionary change, the realm's politico-geographical map was transformed. Smaller entities were absorbed into larger units, conflicts resolved (by force as well as negotiation), boundaries defined, and internal divisions reorganized. European nation-states were in the making.

But what is a nation-state and what is not? The question centers in part on the definition of the term **nation**. That definition usually involves measures of homogeneity: a nation should comprise a group of tightly knit people who speak a single language, have a common history, share the same ethnic background, and are united by common political institutions. Accepted definitions of the term suggest that many states are not nation-states because their populations are divided in one or more important ways. But cultural homogeneity may not be as important as a more intangible "national spirit" or emotional commitment to the state and what it stands for. One of Europe's oldest states, Switzerland, has a population that is divided along linguistic, religious, and historical lines—but Switzerland has proved to be a most durable nation-state nonetheless, recently marking its seven hundredth anniversary.

A **nation-state**, therefore, may be defined as a political unit comprising a clearly delineated territory and inhabited by a substantial population, sufficiently well organized to possess a certain measure of power, the people considering themselves to be a nation, with certain emotional and other ties that are expressed in their most tangible form in the state's legal institutions, political system, and ideological strength. Supporting this structure, of course, is a govern-

Centripetal and Centrifugal Forces

Political geographers use the terms *centripetal* and *centrifugal* to identify forces within a state that tend, respectively, to bind that political system together and to pull it apart.

Centripetal forces tie the state together, unifying and strengthening it. A real or perceived external threat can be a powerful centripetal force. More important and lasting, however, is a sense of commitment to the governmental system, a recognition that it constitutes the best option. This commitment is sometimes focused on the strong charismatic qualities of a leader who personifies the state, who captures the population's imagination. Such charisma can submerge nearly everything else. Juan Peron's lasting popularity in Argentina is a case in point. Charles de Gaulle in France, Mao Zedong in China, and Jawaharlal Nehru in India all possessed similar charismatic qualities and played dominant roles (extending well beyond their lifetimes) in binding their states.

Centrifugal forces are disunifying or divisive. They can cause internal relationships to deteriorate. Religious conflict, racial strife, linguistic cleavages, and contrasting regional outlooks are among the major centrifugal forces. A third of a century ago, the Vietnam (Indochina) War became a significant centrifugal force in the United States, and we are still feeling its aftermath today. Newly independent countries often find tribalism a leading centrifugal force, sometimes strong enough to threaten the survival of the whole state system (as the Biafra conflict of the 1960s did in Nigeria).

The degree of strength and cohesion within the state depends on a surplus of centripetal forces over divisive centrifugal forces. Although it is difficult to measure such intangible qualities, some attempts have been made in this direction—for example, by surveying attitudes among minorities and by evaluating the strength of regionalism as expressed through voting in elections and referenda. When the centrifugal forces become too strong and cannot be checked, even by external forces, the state breaks up (as the Soviet Union and former Yugoslavia did in 1991, and Czechoslovakia in 1993). Or it undergoes revolutionary internal change that makes it, in effect, a new entity, as Iran became after the ouster of the shah and as Cuba became after the victory of Castro's forces.

ment that constantly works to ensure that the forces unifying the state prevail over those that would drive it apart (see box titled "Centripetal and Centrifugal Forces").

This definition of the nation-state essentially identifies the European model that emerged in the course of the realm's long period of evolution and revolutionary change. France is often cited as the best example among Europe's nation-states, but Poland, Hungary, Sweden, Spain, and reunified Germany are also among countries that satisfy the terms of the definition to a great extent. Belgium and Moldova are examples of European states that cannot at present be designated as nation-states. The former Yugoslavia and Czechoslovakia never were nation-states; their actual and latent centrifugal (disunifying) forces were too strong to allow long-term stability.

CONTEMPORARY EUROPE

The nation-states of Europe are among the world's oldest, and the colonial empires of European powers were among the most durable. Despite many disruptive wars and revolutions, European nations survived, and from their long-term stability they forged a confidence—based on strong individual identities—that is a hallmark of European culture. Although Europe constitutes a geographic realm, it exhibits little geographic homogeneity.

It is often postulated that Europe can be viewed as a regional unit because its peoples share **Indo-European languages** (Fig. 1-8), Christian religious traditions (see Fig. 6-2), and common European (Caucasian) racial ancestry. Yet these human cultural and physical traits extend well beyond Europe and, in any case, are not strong unifying elements within the European mosaic. As Figure 1-8 shows, Hungarians and Finns are among European groups that do not speak Indo-European languages; indeed, for so small a realm, Europe is a veritable Tower of Babel. As for common religious traditions, Europe has a history of intense and destructive conflict over issues involving religion; shared Christian principles, for instance, have done little to bring cohesion to Northern Ireland. And again, Europe's purported common racial ancestry masks strong differences in observable physical characteristics between Spaniard and Swede, Scot and Sicilian.

In terms of economic development, centuries of European exploitation of overseas domains led to the amassing of fortunes at home and established a global influence that survived two world wars and continued when the colonial era ended. Europe in 1945 entered a postwar era of reconstruction, realignment, and resurgence. Reconstruction was aided by the billions of dollars made available by the United States through the Marshall Plan (1948–1952). Realignment came in the form of the Iron Curtain, which between 1945 and 1989 separated the Soviet-dominated East from Western Europe; it also involved the emergence of several international "blocs" consisting of states seeking to promote multinational cooperation. The resurgence of Europe continued despite the loss of colonial empires, recurrent political crises, and other obstacles; once again, Europe's momentum had carried the day. Contemporary Europe emerged from the recently completed postwar era with many strengths. Among its most important geographic properties are those we discuss in the next three sections.

Intensifying Spatial Interaction

Greater international cooperation is a logical outcome of the peaceful postwar era because Europe's environments and resources have long presented outstanding opportunities for human contact and interaction. Conceptually, *spatial interaction* is effectively organized around a triad of principles developed by the American geographer Edward Ullman: (1) complementarity, (2) transferability, and (3) intervening opportunity.

Complementarity occurs when one area has a surplus of a commodity demanded by a second area. The mere existence of a resource in a locality is no guarantee that trade will develop; that resource must specifically be needed elsewhere. Thus complementarity arises from regional variations in both the supply and demand of human and natural resources. **Transferability** refers to the ease with which a commodity may be transported between two places. Sheer distance, in terms of both the cost and time of movement, may be the major obstacle to the transferability of a good. Therefore, even though complementarity may exist between a pair of areas, the problems of economically overcoming the distance separating them may be so great that trade cannot occur. The third interaction principle, **intervening opportunity**, holds that potential trade between two places, even if they satisfy the necessary conditions of complementarity and transferability, will develop only in the absence of a closer, intervening source of supply.

European economic geography has always been stimulated by internal complementarities. A specific example involves Italy, which ranks foremost among Mediterranean countries in economic development but lacks adequate coal supplies. For a long time its industries have depended on coal imports from the rich deposits of Western Europe. At the same time, Italian farmers grow crops that cannot be raised in the cooler climate north of the Alps. Citrus fruits, olives, grapes, and early vegetables are in high demand in Western Europe's markets. Hence, Italy imports northwestern Europe's coal, and northwestern Europe imports Italian fruits and wines. This case of double complementarity is not inhibited by transferability restrictions: the physical barrier of the Alps has long been breached by rail and highway routes, and the two-way trade flow is most attractive to shippers because freight-carrying vehicles will not have to return home empty. Moreover, because northwestern Europe and Italy are the closest sources of coal and fruits, re-

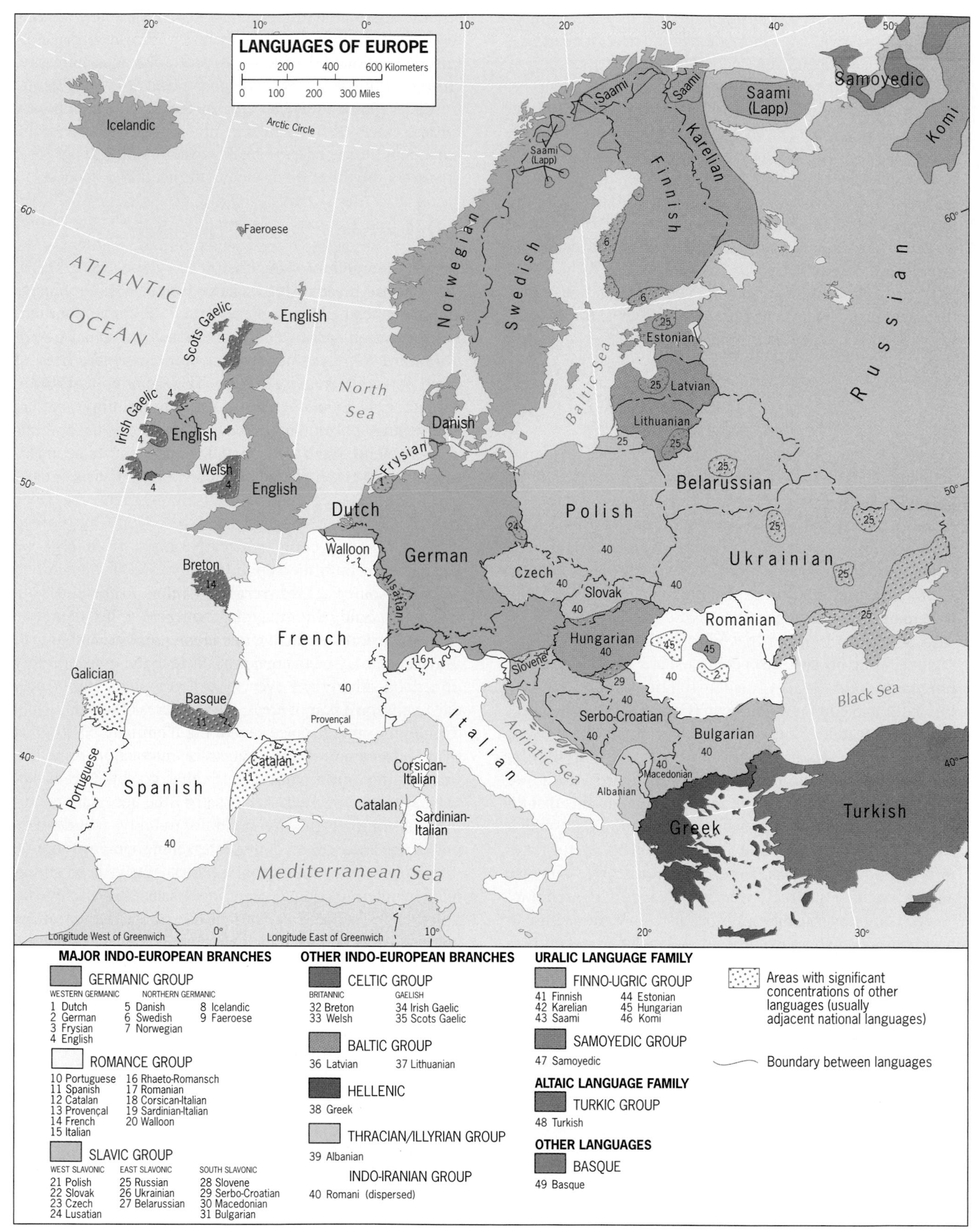
LANGUAGES OF EUROPE
0 200 400 600 Kilometers
0 100 200 300 Miles
Arctic Circle
ATLANTIC OCEAN
North Sea
Baltic Sea
Adriatic Sea
Black Sea
Mediterranean Sea
Icelandic
Faeroese
English
Scots Gaelic
Irish Gaelic
Welsh
Breton
Dutch
Frysian
Walloon
Alsatian
German
Danish
Norwegian
Swedish
Saami
Saami (Lapp)
Finnish
Karelian
Samoyedic
Komi
Russian
Estonian
Latvian
Lithuanian
Belarussian
Polish
Czech
Slovak
Ukrainian
Hungarian
Romanian
Slovene
Serbo-Croatian
Bulgarian
Macedonian
Albanian
Greek
Turkish
French
Provençal
Galician
Basque
Portuguese
Spanish
Catalan
Italian
Corsican-Italian
Sardinian-Italian
Longitude West of Greenwich
Longitude East of Greenwich
MAJOR INDO-EUROPEAN BRANCHES
GERMANIC GROUP
WESTERN GERMANIC
1 Dutch
2 German
3 Frysian
4 English
NORTHERN GERMANIC
5 Danish
6 Swedish
7 Norwegian
8 Icelandic
9 Faeroese
ROMANCE GROUP
10 Portuguese
11 Spanish
12 Catalan
13 Provençal
14 French
15 Italian
16 Rhaeto-Romansch
17 Romanian
18 Corsican-Italian
19 Sardinian-Italian
20 Walloon
SLAVIC GROUP
WEST SLAVONIC
21 Polish
22 Slovak
23 Czech
24 Lusatian
EAST SLAVONIC
25 Russian
26 Ukrainian
27 Belarussian
SOUTH SLAVONIC
28 Slovene
29 Serbo-Croatian
30 Macedonian
31 Bulgarian
OTHER INDO-EUROPEAN BRANCHES
CELTIC GROUP
BRITANNIC
32 Breton
33 Welsh
GAELISH
34 Irish Gaelic
35 Scots Gaelic
BALTIC GROUP
36 Latvian
37 Lithuanian
HELLENIC
38 Greek
THRACIAN/ILLYRIAN GROUP
39 Albanian
INDO-IRANIAN GROUP
40 Romani (dispersed)
URALIC LANGUAGE FAMILY
FINNO-UGRIC GROUP
41 Finnish
42 Karelian
43 Saami
44 Estonian
45 Hungarian
46 Komi
SAMOYEDIC GROUP
47 Samoyedic
ALTAIC LANGUAGE FAMILY
TURKIC GROUP
48 Turkish
OTHER LANGUAGES
BASQUE
49 Basque
Areas with significant concentrations of other languages (usually adjacent national languages)
Boundary between languages

FIGURE 1-8

A Eurostar train leaves Waterloo Station, London, for its three-hour run to Paris. The French and the Japanese have long been world leaders in fast-train technology, and it is now a European objective to link all parts of the European Union through a high-speed rail network. The completion of the tunnel under the English Channel was a major step in this process, but trains to and from the mainland are slowed by the still-inadequate tracks on the British side of the Channel. Nevertheless, these Eurostar trains symbolize an integrating, unifying Europe.

spectively, there are no intervening opportunities to disrupt spatial interaction, and a thriving mutual exchange of these commodities has been generated.

Historically, there have been countless examples of this kind throughout Europe, all dependent on efficient transportation linkages. The European realm has had a good circulation system since Roman times, and the steady improvement of transport technology spawned trading relationships involving ever more numerous and distant places. Today's excellent network of railroads, highways, and air routes is fully integrated because of the efforts made by the postwar planners of a united Europe. Moreover, several major projects have been launched to further upgrade northwestern Europe's already splendid passenger- and freight-rail operations: (1) the long-awaited tunnel beneath the English Channel, opened in 1994, now directly connects Britain to the continent; (2) in Western Europe a new high-speed rail network is being completed to link all of its major cities, modeled on France's TGV (*train à grande vitesse*) that connects Paris with most other French cities at average speeds of 185 miles (300 km) per hour; (3) in Northern Europe the Øresund Fixed Link, a bridge-tunnel opened in 2000 (see p. 85), now directly connects the Scandinavian Peninsula to the rest of Europe; (4) more than 100 miles (160 km) of new tunnels are planned beneath the Alps to expedite traffic flows and comply with Switzerland's mandate that after 2004 all through-freight move by rail; and (5) a new generation of magnetic levitation trains (capable of speeds exceeding 300 mph/480 kph) will be introduced in 2005 when the first line opens in Germany between Berlin and Hamburg. The outlook for road transportation, however, is less promising. As car ownership rates surge, traffic jams are becoming more frequent even on intercity expressways. The worst congestion by far occurs in Europe's myriad urban areas: highways end in the suburbs, and within the aging core city the automobile is poorly suited to navigate the maze of narrow, inefficiently connected streets that were laid out hundreds of years ago.

Urban Continuity and Change

Today, with 73 percent of its population residing in towns and cities, Europe ranks among the world's most highly urbanized realms. Internally, that urban proportion rises to 83 and 79 percent in the regions of Northern and Western Europe respectively. And in certain countries, it rises even higher: in Belgium, 97 percent; in Italy, 90 percent; in the United Kingdom, 89 percent; and in Germany, 86 percent. (One of the demographic by-products of the realm's long-standing urbanization trend is a very low birth rate, whose consequences we discuss in the box titled "Europe's Population Implosion.")

Large cities are the crucibles of their nations' cultures. In his 1939 study of the pivotal role of great cities in the development of national cultures, American geographer Mark Jefferson postulated the law of the **primate city**, which stated that "a country's leading city is always disproportionately large and exceptionally expressive of national capacity and feeling." Although imprecise, this "law" can readily be demonstrated using European examples. Certainly Paris personifies France in countless ways, and nothing in England rivals London. In both of these primate cities, the culture and history of a nation and empire are indelibly etched in the urban landscape. Similarly, Vienna is a microcosm of Austria, Warsaw is the heart of Poland, Stockholm typifies Sweden, and Athens is Greece. Today, each of these (together with the other primate cities of Europe) sits atop a hierarchy of urban centers that has captured the lion's share of its national population growth since World War II.

15

At the intraurban scale, the appearance of the European cityscape contrasts markedly with the urban scene in the United States. As already noted, European cities are much older, and the layout of their often haphazard street systems is far more cramped; available land is almost always scarce and extremely expensive. Moreover, in most countries, governments play a dominant role in shaping urban development by closely supervising the planning process, zealously implementing zoning codes, and imposing rent and real-estate price controls.

The urban pattern of the London region (Fig. 1-9) typifies the internal spatial structure of the European **metropolis** (i.e., the central city and its suburban ring). The metropolitan area remains focused on the large city at its center, especially the downtown *central business district* (*CBD*), which is the oldest part of the urban agglomeration and contains the region's largest concentration of business, government, and shopping facilities as well as its wealthiest and most prestigious residences. Wide residential sectors radiate outward from the CBD across the rest of the central city, each one home to a particular income group. Beyond the central city lies a sizeable suburban ring, but residential densities are much higher here than in the United States because the European tradition is one of setting aside recreational spaces (in "greenbelts") and living in apartments rather than in detached single-family houses. There is also a greater reliance on public transportation, which further concentrates the suburban development pattern. That has allowed many nonresidential activities to suburbanize as well, and today ultramodern outlying business centers increasingly compete with the CBD in many parts of urban Europe (see photo, p. 59).

16

Political and Economic Transformation

Today, a dozen years after the collapse of communism in both Eastern Europe and the former Soviet Union, the European realm has entered a new era. As the Soviet Empire unraveled, it fragmented into 15 newly independent states along the lines of the former republics of the USSR (see Chapter 2). Four of these fledgling countries—Latvia, Lithuania, Moldova, and Ukraine—are included in an expanded, redefined Eastern Europe that now stretches as far eastward as the Russian border; a fifth former Soviet republic, Estonia, is historically and culturally part of Northern Europe. Then there is Belarus, which has turned its back on Europe and whose future alignment is uncertain. Overall, this latest geographic reorganization has since 1991 enlarged the European realm's territorial dimensions by 21 percent, the number of its countries to 39, and its total population by 71 million (14 percent).

At the same time, a single new Europe has taken the place of the two Europes that until only a few years ago

Europe's Population Implosion

When a population urbanizes, average family size declines, and with it the overall rate of natural increase also declines. The data in Table I-2 (pp. 35–39) reveal the relationship between higher rates of urbanization and lower rates of population growth.

In the world as a whole, rates of population growth remain high, especially in the less urbanized periphery. But in the global core, and particularly in Europe, growth rates have decreased dramatically. To keep a population stable the (statistically) average woman must bear 2.1 children. For Europe as a whole, that figure in 2002 was 1.4. Not a single European country recorded more than 2.2 births per woman, and seven, including Italy and Spain, recorded below 1.3—the lowest ever in any human population. Italy in the late 1990s became the first society in world history with more people above the age of 60 than below the age of 20. Such negative population growth poses serious challenges for any nation.

Urbanization is the common denominator in all this. Confined living space and high costs of living impel urban families to have fewer children than rural ones, but increasing prosperity, later marriages, and family planning are also involved. More women prefer to achieve advanced education and pursue careers rather than raise children. More couples prefer to spend money on material things and on entertainment than on youngsters.

All this amounts to what population geographers call a *population implosion* (the opposite of the world's much-feared population explosion). Its impact on society is worrisome, for a shrinking number of younger working people will not be able to support the mushrooming elderly population who need pensions and health care. Tax increases to cover the deficit worsen the business climate, increasing unemployment and prompting the workers the country needs to emigrate. Europe's population implosion will be a formidable challenge as the twenty-first century unfolds.

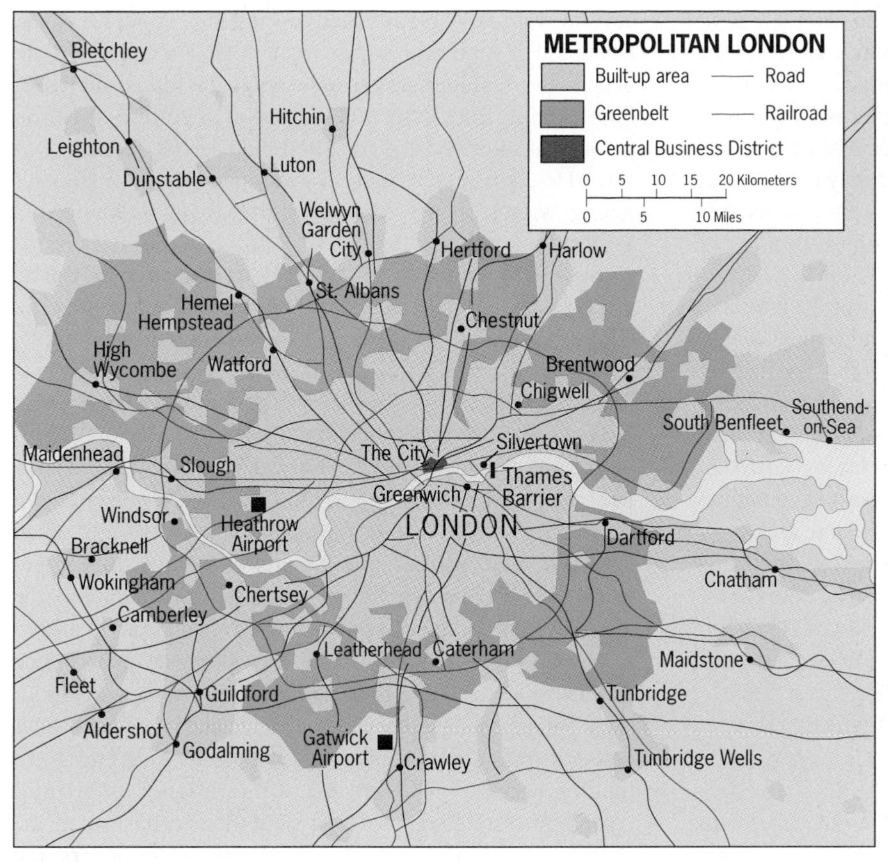

FIGURE 1-9

Booming *suburban downtowns* (see Chapter 3) are transforming the U.S. metropolitan landscape, and such "edge cities" are now multiplying in Western Europe as well. The biggest and most important of these complexes is *La Défense*, located just west of Paris, which has grown so robustly that it is now continental Europe's single largest business district, whose annual transactions exceed the GNP of the Netherlands! Besides its economic prowess, the cubic Grande Arche that forms the centerpiece of *La Défense* is an architectural jewel that is the western anchor of Paris's "Historic Axis" which includes the *Arc de Triomphe*, *Champs Elysées*, *Place de la Concorde*, and the *Louvre*.

threateningly faced each other across the Iron Curtain. While this political transformation has clearly lessened the danger of an East-West military confrontation, it has not materially reduced the realm's internal cultural divisions. Indeed, the swift revival of nationalisms and old hostilities in Eastern Europe—which almost immediately led to brutal civil war in disintegrating Yugoslavia—reminds us that many difficult challenges face a restructuring Europe in its quest for international cooperation.

These problems notwithstanding, the Europeans have long recognized the disadvantages of functioning within their highly fragmented mosaic and have already made considerable progress toward unification. For more than half a century, Western Europe has been forging the economic linkages that are the cornerstone of today's *European Union* (EU). This 15-member multinational organization, which incorporates almost all of the realm's core area (Fig. 1-7), offers the best hope for integrating the new Europe as it develops and implements plans to expand its operations. We now examine in some detail the interplay of forces involving the challenge of political disintegration and the opportunity for greater economic integration.

Devolution

In our earlier discussion of nation-states, we noted that centrifugal forces in former Yugoslavia and Czechoslovakia recently triggered the political disintegration of both countries. In other parts of the realm, too, similar tensions are now heightening and could lead to the breakup of other states. The term **devolution** has come into use to describe the powerful centrifugal process whereby regions or peoples within a state demand and gain political strength and sometimes autonomy at the expense of the center—through negotiation or active rebellion. Most states exhibit internal regionalism, but the process of devolution is set into motion when a key centripetal force—the nationally accepted idea of what a country stands for—erodes to the point that a regional secession movement is launched. This slide toward separatism is most likely to occur in countries whose governments have always had problems trying to forge a viable nation-state. Indeed, both Yugoslavia and Czechoslovakia were hastily patched together by powerful outsiders (the victorious Allies of World War I at the 1919 Versailles Peace Conference) from the Eastern European ethnic crazy-quilt left behind by the defeated Austro-Hungarian Empire.

17

The devolution concept itself was first used to summarize the course of political events in the United Kingdom, where the resurrection of regional separatism is something of a geographic irony. This state is dominated, in terms of population as well as political and economic power, by England, the historic core area of the British Isles. The country's three other politico-geographical entities—Scotland, Wales, and Northern Ireland—were acquired over several centuries and attached to England. Time, however, has failed to submerge regionalism in the United Kingdom, despite the development introduced by the Industrial Revolution and its aftermath, and the period of empire and comparative wealth. During the 1960s and 1970s, London was forced to confront a virtual civil war in Northern Ireland, together with a rising tide of separatism in Scotland and Wales.

Scottish nationalism proved to be an especially powerful force in British politics in part because Scotland was becoming the center of the country's vital new energy industries based on the discovery of major oil and gas deposits beneath the adjacent North Sea. Indeed, this became a key issue during the United Kingdom's 1997 election campaign, and the newly elected Labour Party swiftly gave the Scots (as well as the Welsh) the opportunity to vote—not for independence, but for the greater autonomy embodied in a new regional parliament that would have limited but significant powers over local affairs. In that referendum, the Scottish electorate voted overwhelmingly in favor of autonomy, and thus a major devolutionary step was taken in one of Europe's oldest and most stable states. To the surprise of the London government, however, instead of ending Scotland's flirtation with separatism, the referendum energized the Nationalists—who now are well-positioned as the principal opposition party in the new Scottish Parliament.

The emergence and strength of regionalism in the United Kingdom, one of Europe's most durable states, underscores the potential impact of devolution elsewhere on the continent. Another European state attempting to adjust to strong regional forces is Spain, which has signed autonomy agreements with the leaders of the northern Basque region and the northeastern province of Catalonia. These agreements have given both areas their own parliaments, recognized their languages as official and equal to Spanish, and transferred powers of local taxation and education. While these initiatives had good results in Catalonia, they were not enough to prevent extremist violence by Basque elements who demand nothing less than total independence from Spain. Many other indications of devolutionary tension on the map of Europe can be seen in Figure 1-10. France has had to contend with a drive for secession on its Mediterranean island of Corsica. A similar movement demands autonomy for the neighboring island of Sardinia, which is part of Italy; the Italians also face separatist drives in their far north in both South Tyrol and Lombardy (see the discussion of "Padania" on p. 90). Devolutionary pressures also affect Belgium—where a deep ethnic division persists between the country's Flemish and French speakers—as well as several of the newly independent states that emerged along the western rim of Russia following the dissolution of the Soviet Union in 1991. And, in addition to all this, there is still more to the geography of devolution in Europe: besides the already-collapsed states of Yugoslavia and Czechoslovakia, the map shows active movements in

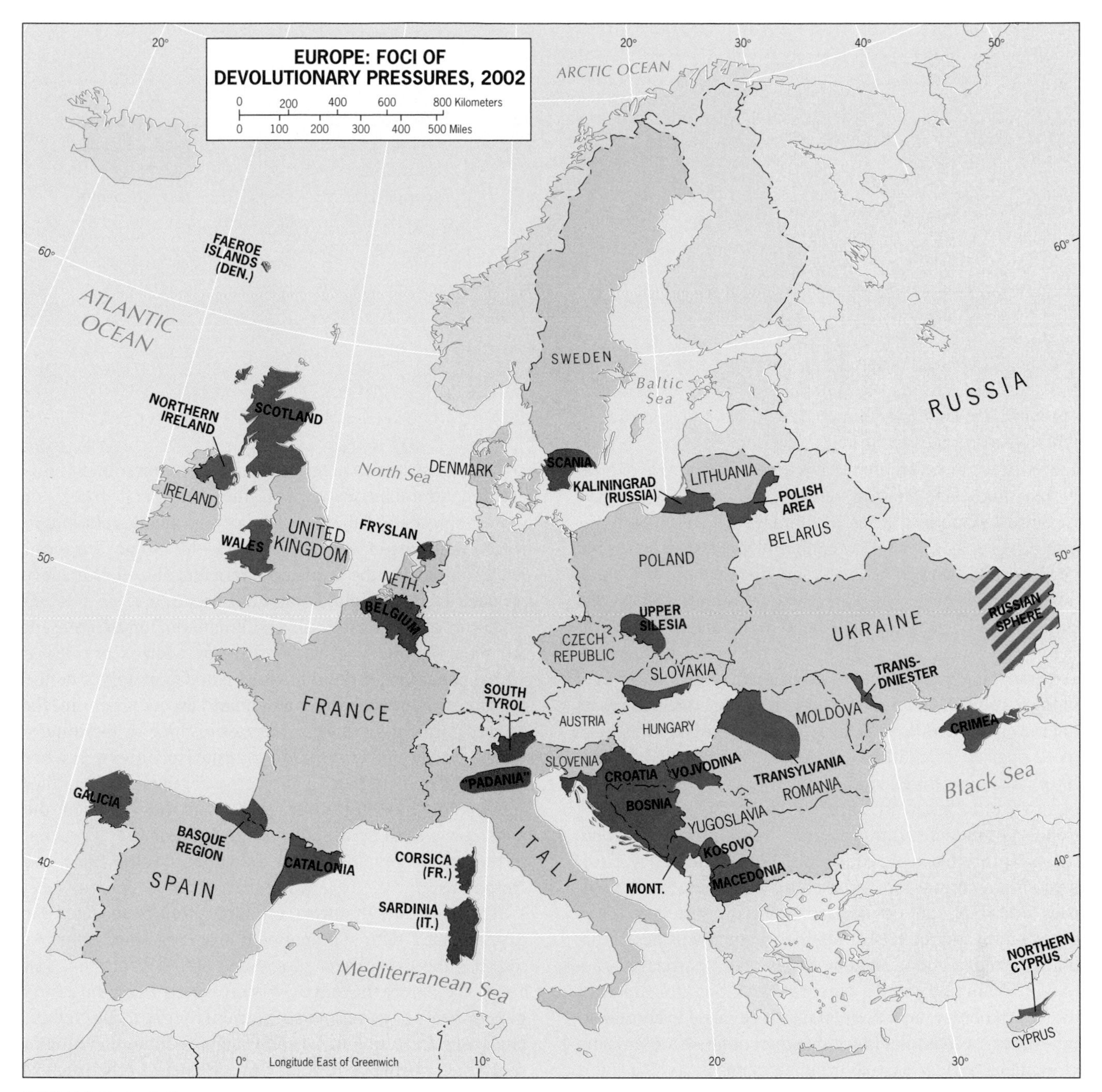

FIGURE 1-10

the Netherlands' Fryslan, Poland's Upper Silesia, and southern Sweden's Scania.

Regionalism in Europe today has been further heightened by the emergence of subnational regions as new hubs of economic power and influence. In the forefront of this development are the following city-regions, each anchored by a booming commercial center: (1) southeastern France's Rhône-Alpes region (anchored by Lyon); northern Italy's Lombardy (Milan); northeastern Spain's Catalonia (Barcelona); and southern Germany's Baden-Württemberg (Stuttgart). Most noteworthy is that these regions—known as the **Four Motors of Europe**—have developed direct linkages and relationships with one another, thereby bypassing the capital cities and central governments of their respective countries. Today, the Four Motors are moving to the next level by expanding their new business channels to forge direct economic ties with business centers all across the world. Within this broader international framework,

18

Some European Acronyms

EC European Commission. Sits in Brussels to consider major issues such as new membership in the European Union.

euro Europe's new currency.

EMU European Monetary Union. First stage began in 1999, when participating countries' banks started using the euro. Completed transition scheduled for 2002, when national currencies will disappear.

EU European Union. Formerly EC (European Community); previously EEC (European Economic Community).

ECB European Central Bank. The new umbrella international European bank that governs the transition to the euro.

each city-region is emerging as a textbook example of a *regional state*. This concept was recently introduced by the 19 Japanese economist Kenichi Ohmae, who defines a **regional state** as a "natural economic zone" that defies old borders and is shaped by the globalizing economy of which it is a part; its leaders deal directly with foreign partners and negotiate the best terms they can with the national governments under which they operate.

Interestingly, this new interregional (and increasingly global) web is being established with the encouragement of the European Union, with an eye not only toward forging a more open and flexible Europe but also toward shaping a future in which this expanding network of regions may supplant the old framework of nation-states. To accelerate that trend, more localized transnational regional structures are being nurtured as well, especially "Euroregions." A *Euroregion* is a formal territorial entity straddling one of Europe's international boundaries whose purpose is to foster cooperation and reduce inequalities on the two sides of that border. Poland and its neighbors have been particularly active in organizing about a dozen such regions along most of the Polish border, and progress has been made in coordinating strategies to attract investment, stimulate economic development, and resolve environmental problems. Since it is also important to get the inhabitants of a Euroregion to stop thinking of themselves as living in an area fragmented by a national boundary, encouraging the emergence of new cross-border institutions (such as joint sporting events and music festivals) is an integral part of this region-building process.

European Unification

Geographic contradictions abound in today's world. Even while centrifugal forces of devolution threaten the fabric of Europe's nation-states, those same states are trying to forge multinational associations that will strengthen the ties among them. This integration and disintegration is proceeding simultaneously, with the centripetal forces of union representing the new Europe and the devolutionary forces of division symbolizing the old. Some European leaders hope that the unification movement will ultimately produce a federal United States of Europe.

European efforts to form a multinational union represent the most significant manifestation of what is actually a global phenomenon: supranationalism. Geographers define **supranationalism** as the voluntary association in economic, political, or cultural spheres of three or more independent states willing to yield some measure of sovereignty for mutual benefit. In later chapters we will encounter other supranational organizations, but none has reached the plateau achieved by the *European Union* or *EU*. (For a list of current abbreviations, see box titled "Some European Acronyms.") 20

This is not the first time a group of European states has experimented with supranational programs, but never before has the experiment progressed as far. It all began in 1944, just before the end of World War II, when the exiled governments of three small countries (Belgium, the Netherlands, and Luxembourg) established an economic union to speed their postwar recovery. Called *Benelux*, this structure proved to be the vanguard of a coalition process that is still going forward today. It was a crucial example for a Europe devastated by conflict and in need of reorganization to facilitate reconstruction.

Soon after the war, U.S. President Harry S Truman's Secretary of State, George Marshall, proposed a massive infusion of American aid to help the struggling countries of postwar Western Europe. The Marshall Plan provided massive funding to boost European economies and to finance rebuilding. But it did much more than that: it reminded European leaders that they needed a multinational economic-administrative structure along the lines of Benelux. This was no time for divisive national boundaries to inhibit the flow of goods and raw materials or for troublesome tariffs to interfere with regional economic redevelopment. Soon, conferences were designing a new, cooperative Europe. One result was the formation of the Council of Europe, the beginnings of what might become a European Parliament, meeting in Strasbourg, France.

But the key initiatives occurred in the economic arena (see box titled "Supranationalism in Europe"). A series of agreements led in 1957 to the Treaty of Rome which established, as of 1958, the so-called Common Market. Also known as the European Economic Community (EEC), this organization consisted of the three Benelux countries as well as France, Italy, and West Germany. The EEC would become the core of expanding supranational Europe, and Belgium's capital, Brussels, its headquarters.

Not all eligible European countries joined the EEC, however. France initially vetoed British membership, and the British created a rival organization, the European Free Trade Association (EFTA), which included three Scandinavian states, landlocked Switzerland and Austria, and Portugal. But this discontiguous EFTA was no match for the powerful EEC and began to unravel as its members, including the United Kingdom itself, saw the advantages of EEC membership.

Now the EEC began to grow. Britain, Ireland, and Denmark joined in 1973, enlarging membership from six to nine countries. In 1981, when the organization had become known as simply the European Community (EC), Greece was admitted. And in 1986, with the admission of Spain and Portugal, membership reached 12. The organization

Supranationalism in Europe

1944	Benelux Agreement signed.
1947	Marshall Plan created (effective 1948–1952).
1948	Organization for European Economic Cooperation (OEEC) established.
1949	Council of Europe created.
1951	European Coal and Steel Community (ECSC) Agreement signed (effective 1952).
1957	Treaty of Rome signed, establishing European Economic Community (EEC) (effective 1958), also known as the Common Market and "The Six."
	European Atomic Energy Community (EURATOM) Treaty signed (effective 1958).
1959	European Free Trade Association (EFTA) Treaty signed (effective 1960).
1961	United Kingdom, Ireland, Denmark, and Norway apply for EEC membership.
1963	France vetoes UK EEC membership; Ireland, Denmark, and Norway withdraw applications.
1965	EEC-ECSC-EURATOM Merger Treaty signed (effective 1967).
1967	European Community (EC) inaugurated.
1968	All customs duties removed for intra-EC trade; common external tariff established.
1973	United Kingdom, Denmark, and Ireland admitted as members of EC, creating "The Nine."
	Norway rejects membership in the EC by referendum.
1979	First general elections for a European Parliament held; new 410-member legislature meets in Strasbourg.
	European Monetary System established.
1981	Greece admitted as member of EC, creating "The Ten."
1985	Greenland, acting independently of Denmark, withdraws from EC.
1986	Spain and Portugal admitted as members of EC, creating "The Twelve."
	Single European Act ratified, targeting a functioning European Union in the 1990s.
1987	Turkey and Morocco make first application to join EC. Morocco is rejected; Turkey is told that discussions will continue.
1990	Charter of Paris signed by 34 members of the Conference on Security and Cooperation in Europe (CSCE).
	Former East Germany, as part of newly reunified Germany, incorporated into EC.
1991	Maastricht meeting charts European Union (EU) course for the 1990s.
1993	Single European Market goes into effect.
	Modified European Union Treaty ratified, transforming EC into EU.
1995	Austria, Finland, and Sweden admitted into EU, creating "The Fifteen."
1999	European Monetary Union goes into effect; single currency (the euro) introduced in 11 of the EU's 15 countries.
	Helsinki summit discusses fast-track negotiations with six prospective members and applications from six others; prospects for Turkey considered in longer term.
2001	Greece becomes 12th EU country to adopt the euro.

FIGURE 1-11

changed its name again, to its current one (European Union or EU), and a flag was designed (see photo on p. 65). Expansion continued when Austria, Sweden, and Finland joined in 1995, and at the turn of the twenty-first century the European Union consisted of 15 members (Fig. 1-11).

But enlargement entailed problems. Not all member states of the EU are equally enthusiastic about—or committed to—the regulations of membership. As we noted earlier, supranationalism requires yielding some sovereignty, and EU participation compels member countries to adhere to a wide range of rules ranging from agricultural practices to financial procedures. This can lead to opposition from interest groups (farmers, bankers) with which governments must deal. Furthermore, the richer members of the EU are committed to subsidize the poorer ones, putting tax burdens on the former. All the member countries in 1986 signed a Single European Act that committed them to the shared pursuit of 279 specific goals. Some governments had difficulty persuading their voters that EU membership was worth the burden some of these goals imposed.

One of those goals involved a common currency. As anyone who has traveled in Europe knows, changing currencies at every international stop is a hassle. When it comes to common economic goals, such financial fragmentation is a huge disadvantage. (Imagine a United States in which every State had its own money!) But Europe's famous currencies long were part of their national cultures: France's franc, Germany's mark, Holland's guilder, Britain's pound sterling. During pivotal meetings in Maastricht, Amsterdam, and Brussels in the 1990s, EU members approved the principle of European Monetary Union (EMU). A single new currency was agreed upon (the *euro*), and the first phase of implementation took place in 1999. But participation was not compulsory. The United Kingdom was the principal defector, postponing the elimination of the pound even as Germany and France were set to abandon their mark and franc in favor of the euro. A few other countries also stayed out (Fig. 1-11): Denmark and Sweden in response to voter opinion and Greece because it could not meet the criteria for admission (which it finally did in 2001). For the others, however, the euro became a "parallel" currency on January 1, 1999, and for them national currencies are scheduled to disappear on July 1, 2002.

Another objective of the EU is further expansion. This has engendered a debate over whether it is better to have a "deeper" union of a limited number of states or a "shallower" union with more numerous members. To allow some of Europe's more peripheral, less prosperous countries to join, certain regulations must be relaxed—setting back the integration process. Even as this debate continued, the 15 EU members in 1999 held a summit in Helsinki, where the future of 13 potential members was discussed. Of these 13, six "first wave" countries already were in so-called full accession negotiations: the Greek part of the island of Cyprus, the Czech Republic, Estonia, Hungary, Poland, and Slovenia (Fig. 1-11). Six additional countries were invited to begin negotiations with the EU administration: Bulgaria, Latvia, Lithuania, Malta, Romania, and Slovakia. (Table I-2 contains data that will suggest why the inclusion of these countries will be a formidable undertaking: see the GNP figures, for example.)

But the candidate that dominated the discussion in Helsinki is not on either list: Turkey. For many years, European as well as Turkish leaders have harbored hopes that Turkey might some day join the European Union, a notion repeatedly torpedoed by the Turks' old adversaries, the Greeks, and indeed by anti-EU politicians in Turkey itself as well. The (still remote) possibility of Turkey, a country where most people are Muslims, joining a European organization sometimes called a "Christian Club" now became the topic of serious deliberation.

As so often happens in human history, an environmental event—or rather, a sequence of events—contributed to such political change. In August 1999 a devastating earthquake struck northwestern Turkey, killing almost 20,000 people and making many times that number homeless. Among the countries rushing in to help was Greece, and Greek aid workers distinguished themselves in their efforts to help the Turks. Then, less than a month later, a quake hit just north of Athens, and this time the Turks gave assistance to the Greeks. These efforts, much praised by officials and in the press alike, led to further exchange visits and a general conciliation. When the Helsinki summit took place in December 1999 these good feelings translated into constructive long-range planning for Turkey's EU candidacy, at least temporarily without Greek objection.

Why would the European Union want to include a Muslim (though officially secular) state, and one whose human-rights record and overall economy are well below EU standards? First, it would prove the EU's capacity to encompass cultures even more divergent than those of Europe itself.

FROM THE FIELD NOTES
"No place in Europe displays the flag of the European Union as liberally as does the Dutch city of Maastricht, where the European Union Treaty of 1991 was signed. The flag, seen here at City Hall, shows twelve yellow stars (representing the signatories to the Treaty) against a blue background. Although three additional states joined the EU in 1995, and others may join later, the flag will remain as is, and will not, as the U.S. flag does, change to reflect changing times."

Second, it could lead to the solution of one of Europe's most intractable problems, the political division of Cyprus (see pp. 93-94). Third, Turkey might become a bridge between Europe and energy-rich Central Asia, where the Turks have historic influence. Fourth, Turkey could become helpful in the long-range stabilization of the Balkans, where Muslim and Christian communities still coexist uneasily. And fifth, several million people of Turkish and Kurdish descent now live and work in Germany and other EU countries; with Turkish accession they would be in the European fold, not outsiders.

But the EU's rapid expansion plan also entails major risks. Greater breadth must mean less depth; when countries as diverse as Belgium and Bulgaria come under the same umbrella, some of the spokes must be bent. An ongoing dispute that will undoubtedly grow worse involves the so-called common agricultural policy (CAP). French farmers, long coddled by state subsidies of which farmers in Eastern Europe can only dream, are protected by their government, which argues that these farmers are an integral part of the French cultural fabric, so that protecting their farms and villages is a cultural as well as an economic priority. Imagine what will happen when the farm products of Eastern Europe reach Western European markets without the barriers that now keep them out (or raise their prices)! Another lingering problem is the EU's common defense policy. The United States, under NATO auspices, took the lead during the Yugoslavia-Kosovo campaign, and in truth Europe's combined forces could not have waged this war alone. European leaders want to improve the EU's military capacity but are not in agreement as to how.

Another growing concern has to do with the level of grass-roots support for the EU among ordinary people (not the bureaucrats) in the participating countries. Travel in Europe and ask taxi drivers, farmers, shopkeepers, and office workers and you will hear quite a different story from the one protrayed by the Helsinki negotiators. Grass-roots support for the EU has been declining in a number of member countries, even those that have been recipients of EU subsidies. This seems to be accompanied by a rise in support for far-right political parties opposed to immigration and asylum-granting. Austria saw the rise of such a party and its participation in the government, which led the other EU members to invoke sanctions—but Austria was not alone in experiencing the rise of xenophobic (foreigner-hating) movements. For voters supporting such parties, a prominent motive is their fear that their national culture (as they perceive it) will be submerged in a Greater Europe over which they have no control. As the EU matures and grows, it will continue to face an abundance of challenges.

And the momentum of European integration continues. The first general elections for the European Parliament were held in 1979; today this body, which is headquartered in Strasbourg, is a fixture in European affairs. The euro is replacing national currencies. A group of countries within the EU eliminated border-crossing formalities under the Schengen Agreement. All this has happened while the regional economy, overall, has done well (a recession in the 1990s did lead to some setbacks), and if EMU brings continued economic growth, the symbolism of a common currency may well translate into further political integration. A global economic downturn, on the other hand, would produce the first critical test for the new Europe. Will member countries reconsider the concessions they have made, or will they remain united in the face of economic adversity?

If the European Union succeeds, Europe will become an even more formidable presence in the economic and perhaps also the political geography of the world than it has ever been. As of 2001, the 15 member states contain nearly 380 million people who constitute one of the world's richest markets and who produce about 40 percent of all exports (Germany alone is the world's number two exporter). Functioning in unison, European countries can better promote their interests in the world than they can individually, and a united Europe would be a powerful force. But remember Europe's history of conflict, division and self-destruction, and its present-day struggle with devolution. Will this latest effort to defeat its centrifugal forces prove irreversible?

REGIONS OF THE REALM

So complex are Europe's physical and human geographies that it proves to be one of the most difficult geographic realms when it come to regionalization. Europe may not be large territorially, but it incorporates enormous environmental, cultural, political, and economic diversity. Europe extends from moist Atlantic shores to continental Russian borders and from Arctic cold to Mediterranean warmth. Small as Europe is, the realm contains nearly 40 countries.

To better comprehend Europe's geographic diversity, we group its countries into five regions based on their proximity, environmental similarities, historical associations, cultural congruities, social parallels, and economic linkages. On such bases, we map (1) Western Europe, (2) the British Isles, (3) Northern (Nordic) Europe, (4) Mediterranean Europe, and (5) Eastern Europe (Fig. 1-12).

In doing so we use the formal-region concept (see p. 6) to simplify a complex realm, but we can approach the problem in other ways. The first part of this chapter focused on Europe's ongoing integration, of which the European Union is only the most prominent manifestation. As the di-

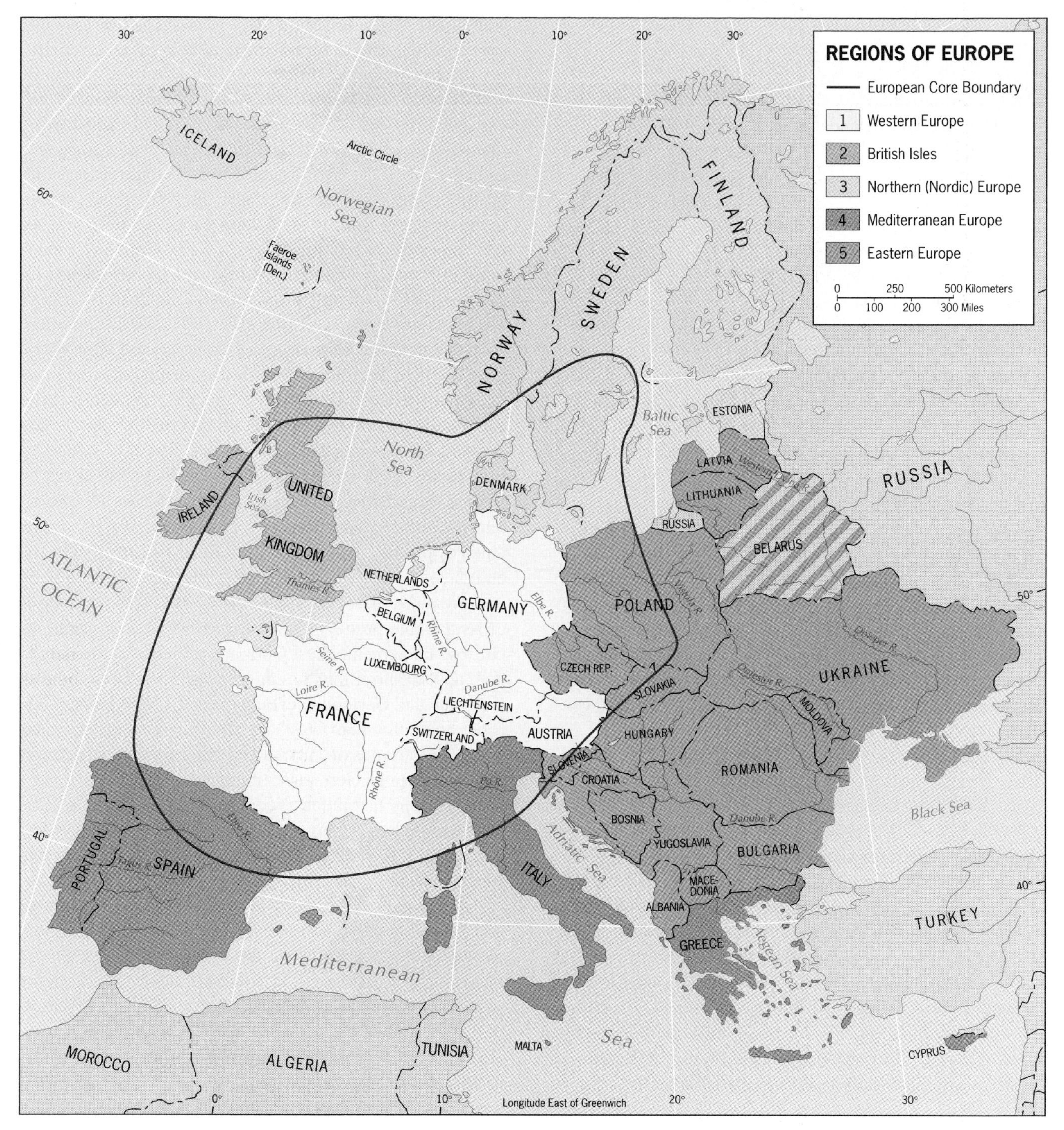

FIGURE 1-12

visive effects of European boundaries diminish and cross-border interconnections increase, a new regionalization of Europe is emerging. Using this functional-region concept, we would recognize a European core, consisting essentially of regions (1) through (4) on Figure 1-12, and a periphery mostly in region (5). Embedded in this functional framework are the "Four Motors," "Euroregions," and other spatial expressions of Europe's growing interconnections.

The framework in Figure 1-12 suits our purposes better because it provides insight into both Europe's enduring formal regions and its changing functional regions. For all its operational integration, Europe still is a realm whose differences in natural environments, economic opportunities, historical experiences, and relative locations have produced societies and countries with particular, often contrasting cultural traditions and outlooks. These contrasts make Eu-

MAJOR CITIES OF THE REALM

City	Population* (in millions)
Amsterdam, Netherlands	1.2
Athens, Greece	3.1
Barcelona, Spain	2.8
Berlin, Germany	3.3
Brussels, Belgium	1.1
Frankfurt, Germany	3.7
London, UK	7.6
Lyon, France	1.4
Madrid, Spain	4.1
Milan, Italy	4.3
Paris, France	9.6
Prague, Czech Rep.	1.2
Rome, Italy	2.7
Stuttgart, Germany	2.7
Vienna, Austria	2.1
Warsaw, Poland	2.3

*Based on 2002 estimates.

rope's political map what it is: one of the most tortuously fragmented in the world. And as we noted earlier, devolutionary forces are at work to fracture it even further. The outcome of Europe's contest between centripetal and centrifugal forces is far from certain.

WESTERN EUROPE

Western Europe is the heart of the realm, the hub of its economic power, the focus of its unifying drive. The region is dominated by two of Europe's giants, Germany and France, flanked by the three Benelux countries (Belgium, the Netherlands, and Luxembourg) in the northwest and the three Alpine states (Switzerland, Austria, and the microstate of Liechtenstein) in the east (Fig. 1-13). With a combined population of some 184 million representing eight of the world's richest economies, Western Europe is a powerful force not only in Europe but also on the international stage.

Dominant Germany

Germany—for nearly half a century a country divided into a larger, more prosperous, capitalist West and a smaller, poorer, communist East—was reunited in 1990 during the collapse of communism. Reunification erased the sealed border between West and East and toppled the infamous Berlin Wall in the divided capital. But after the celebrations were over, the Germans faced a formidable economic and social challenge. That so much progress has been made toward the reintegration of the impoverished former East into the economic sphere of the prosperous West in less than a decade is testimony to the strength of the national economy.

Before World War II, the young German Empire had expanded into what is today western Poland and incorporated a wide area along the Baltic Sea as far east as coastal Lithuania. At that time, Germany possessed three major industrial complexes: the Ruhr near the Netherlands border in the west, the Saxony area along the Czech border in the center, and Silesia in the east (Fig. 1-13). Germany's enormous industrial machine supported armed forces capable of plunging the world into war; domestic raw materials, from coal and fuels to iron and alloys, helped make this possible.

When defeated Germany was dismembered after World War II ended in 1945, and its eastern boundaries were redrawn, this economic geography changed drastically. Silesia became part of the newly delimited state of Poland. Saxony lay in communist East Germany. The new boundary framework left West Germany with the Ruhr, but also with a much-diminished resource base. Nevertheless, West Germany soon developed Europe's most successful economy, aided by the Marshall Plan, by political stability, and by the enlightened policies of the powers that had defeated it. In just 15 years, between 1949 and 1964, West Germany's gross national product (GNP) tripled while industrial output rose 60 percent. West Germany's market-driven economy absorbed millions of emigrants from communist East Germany and German-speaking refugees from Eastern Europe; unemployment was virtually nonexistent, and hundreds of thousands of Turkish and other foreign workers arrived to take jobs Germans could not fill or did not want.

Geography had much to do with West Germany's "economic miracle." As Figure 1-13 shows, Germany borders every one of the other Western European countries except Liechtenstein, and its surface transport systems (railroads, highways, navigable rivers, artificial waterways) are second to none in the region—indeed, in the realm. Seemingly somewhat landlocked by all these neighbors, Germany in fact benefits from the services and efficiency of one of the world's largest ports, Rotterdam (Netherlands). The industrial heartland of the former West Germany, which powered the country's economic recovery, was advantaged by its Rhine River link to this port. But West Germany never was a single-core country. In addition to the cluster of cities still marking the now-fading Ruhr, major urban centers anchor industrial and agricultural zones in several other parts of the country: Hamburg in the north (itself a major port), Frankfurt in the center, Stuttgart and Munich in the south.

The powerful West German economy stood among the world's largest producers of iron, steel, chemicals, motor vehicles, machinery, textiles, and a host of other goods in addition to a wide range of farm products. Unencumbered by major military expenditures, the West German economy consumed prodigious quantities of imported raw materials and sold its output in markets all over the world. As the

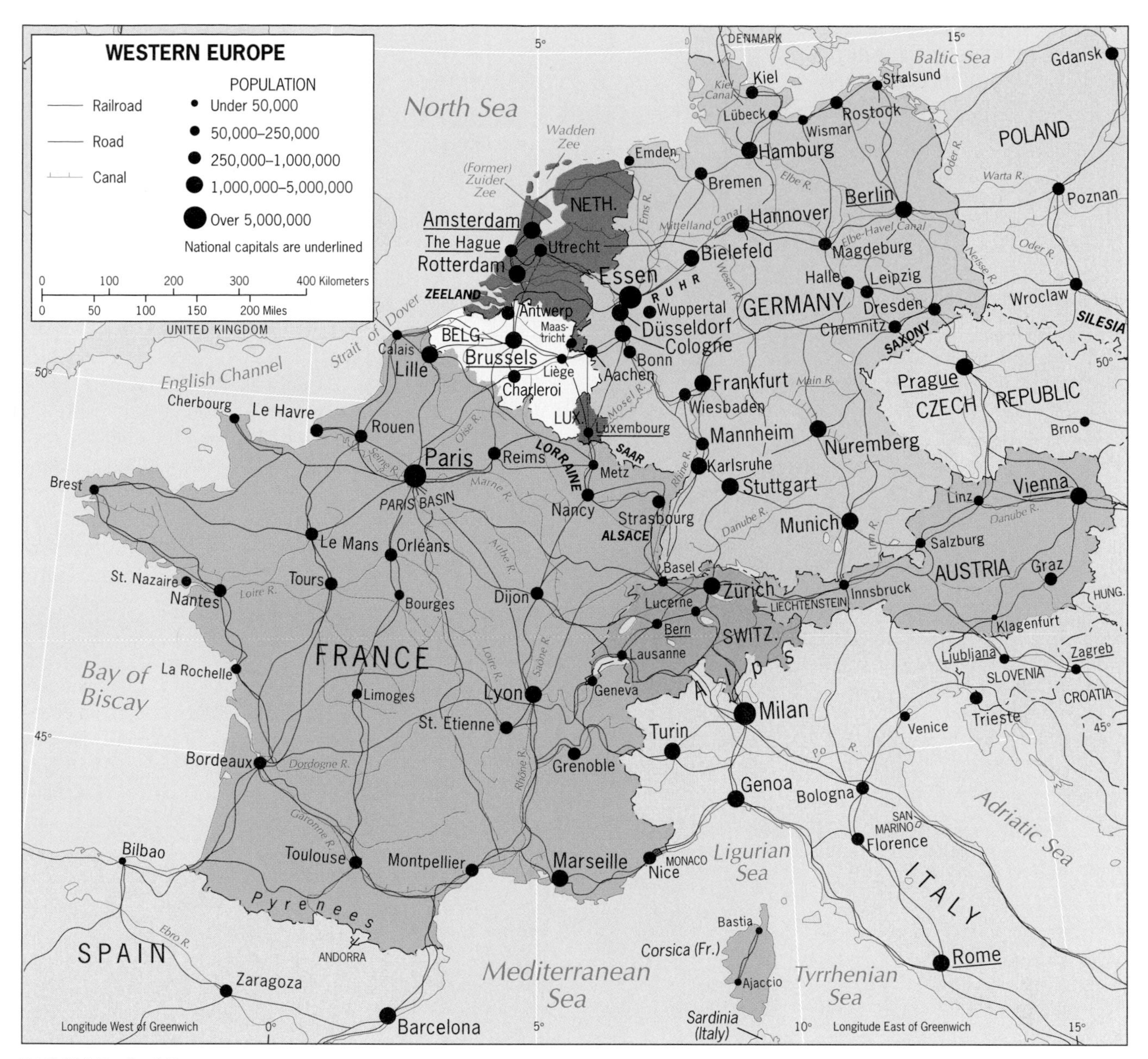

FIGURE 1-13

giant of the original Common Market and under the security umbrella of NATO, West Germany translated its geographic advantages into prosperity.

West Germany's boom could not continue indefinitely. When the economy slowed somewhat in the 1970s and the oil crisis raised fuel costs, the country's long-term problems came into focus: an aging population putting strain on the social services, a growing number of unemployed requiring help from the state, a loss of competitiveness on world markets as industrial modernization lagged, a rising xenophobia marked by murderous attacks on Turkish and African workers, and a revival of extremism in national politics.

Although economic conditions improved during the 1980s, this combination of problems continued to plague West Germany. While still Europe's largest economy and a leading force in European unification, West Germany was troubled by social and political tensions. Housing shortages, high taxation, and discord in government over financial policies roiled West German society. And in the midst of all this uncertainty, suddenly, came reunification with East Germany.

East Germany (except for the Western sectors of Berlin) had been under Soviet rule for 45 years, and as the Soviet Union disintegrated, the merger between capitalist West and communist East Germany became possible. This unification was accomplished in 1990, wiping "West" and "East" off the German map. Literally overnight, dominant West Germany absorbed a neighbor the size of Virginia

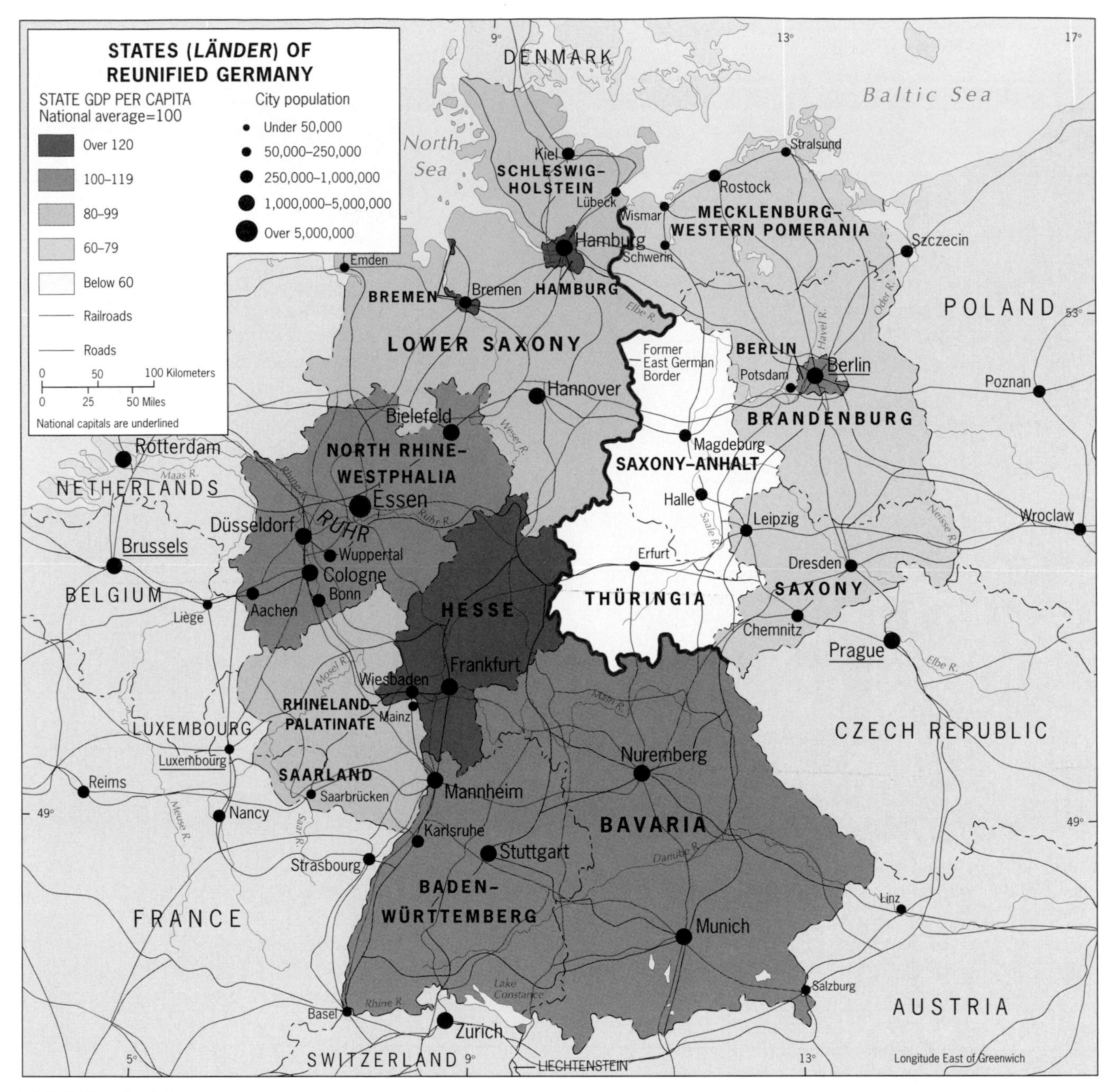

FIGURE 1-14

with a population of over 17 million, a collapsing economy with outdated factories and crumbling infrastructures, severely polluted environments, deteriorating cities, and collectivized, inefficient farming without property rights or adequate legal institutions. And although most Germans in the West as well as the East favored reunification, some did not. When a flood of *Ossies*, as East Germans were called, migrated westward in search of better lives but further complicated an already difficult employment situation, many *Wessies* had their doubts. When the German government imposed a higher sales tax and an income-tax surcharge on the former West to pay for the huge cost of reunification, the doubters multiplied. And when an economic recession hit Western Europe in the early 1990s, the very future of the new Germany seemed uncertain.

Conditions in the former East at first supported the skeptics' position. After nearly a decade of reunion, no amount of subsidies, capital investment, and tax breaks seemed enough to lift eastern Germany out of its economic misery. Unemployment, over 11 percent in Germany as a whole, was more than 18 percent in the former East. The German national debt was skyrocketing. In 1998, exports from the former East still contributed less than 5 percent of the German total.

But other developments were more favorable. German exports, ranging from luxury automobiles to chemicals, benefited from the weak euro (which lost over 20 percent of its value following its introduction) and production surged. The service industries boomed, unemployment started to decline, and even in the former East Germany economic conditions improved markedly—boosted further by the transfer of the national government from Bonn to Berlin, an East German city. After the inevitable decline of GNP per capita resulting from the West-East merger, Germany again surpassed France before the turn of the millennium. And the most powerful economy in Europe continued to contribute some 30 percent of the funds that make up the annual EU subsidy budget.

The Federal Republic

After World War II, Germany adopted a federal political framework, which was extended to East Germany following reunification. Germany is divided into 16 *Länder* (States), each *Land* representing a traditional cultural and/or political nucleus in the experience of the German nation (Fig. 1-14). Territorially the largest State is Bavaria in the south, but the most populous is North Rhine-Westphalia, where the once-mighty Ruhr industrial district provided the stimulus. Three of the *Länder* are actually cities: the capital, Berlin; the prosperous northern port city of Hamburg; and the dynamic industrial and port city of Bremen.

Six of Germany's *Länder* lie in the former East Germany. Here the leading State by many measures is Brandenburg, which surrounds Berlin. South of Brandenburg lies Saxony, an old German industrial complex with rich resources but outdated and inefficient factories from the communist period. The townscapes of the old cities of Leipzig and Dresden, the latter almost totally obliterated during World War II, still reflect the bleak urban culture of communism and the dislocation arising from the post-1990 transition to German-style capitalism.

As Figure 1-14 reveals, regional economic disparity is one of the new Germany's serious problems. In Table I-2 Germany's per-capita GNP is listed as one of the world's highest, but the GDP data in Figure 1-14 reflect the range of productivity from State to State. Hesse with its urban-industrial focus of Frankfurt, Germany's financial headquarters and air-transport hub, ranks as the leader. Adjacent lies North Rhine-Westphalia, in the hinterland of the great port of Rotterdam and containing six major cities in addition to the old federal capital of Bonn. And to the south of Hesse lie Baden-Württemberg, one of Europe's Four Motors and centered on Stuttgart, as well as Bavaria, anchored by Munich; here, in Germany's "Sunbelt," has emerged a booming high-technology heartland. Nothing comparable has yet emerged in the six *Länder* of the east, but national support for eastern Germany continues in the hope that the gap will narrow. Meanwhile, the federal map itself may undergo revision. Periodic proposals to combine certain *Länder* and to divide others have been raised, but no action has been taken.

With 81.9 million inhabitants, over 7 million foreigners and more than 4 million ethnic Germans born outside the country, Germany is Europe's most populous country by far. As the statistical setback of reunification (expressed in lower life expectancies, lowered health standards, higher crime rates, reduced urbanization) is overcome, Germany's formidable power will again loom over Europe. But it will be a different Europe from that of 1914 or 1939, and Germany is one of its chief architects. Still, some Europeans worry about a European Union dominated by a country that twice in the twentieth century plunged the world into war. Right-wing extremist activities and attacks on foreigners (especially the more than 2 million Turkish residents) seem to justify such fears, but the counterargument is that countries participating in multinational associations are less likely to take unilateral actions. On this point, proponents of the European Union agree.

France

One of those proponents is the other leading Western European country, France. The French and the Germans have been rivals in Europe for centuries. France is an old state, by most measures the oldest in Western Europe. Germany is a young country, created in 1871 after a loose association of German-speaking states had fought a successful war against . . . the French.

Territorially, France is much larger than Germany, and the map suggests that France has a superior relative location, with coastlines on the Mediterranean Sea, the Atlantic Ocean, and, at Calais, even a window on the North Sea. But France does not have any good natural harbors, and ocean-going ships cannot navigate its rivers and other waterways far inland. France has no equivalent to Rotterdam either internally or externally.

The map of Western Europe (Fig. 1-13) reveals a significant demographic contrast between France and Germany. France has one dominant city, Paris, at the heart of the Paris Basin, France's core area. No other city in France comes close to Paris in terms of population or centrality: Paris has 9.6 million residents, whereas its closest rival, Lyon, has only 1.4 million. Germany has no city to match Paris, but it does have a number of cities with populations between 1 and 5 million. And as Table I-2 shows, Germany is much more highly urbanized overall than France.

Why should Paris, without major raw materials nearby, have grown so large? Whenever geographers investigate the evolution of a city, they focus on two important locational qualities: its **site** (the physical attributes of the place it occupies) and its **situation** (its location relative to surrounding areas of productive capacity, other cities and

21 22

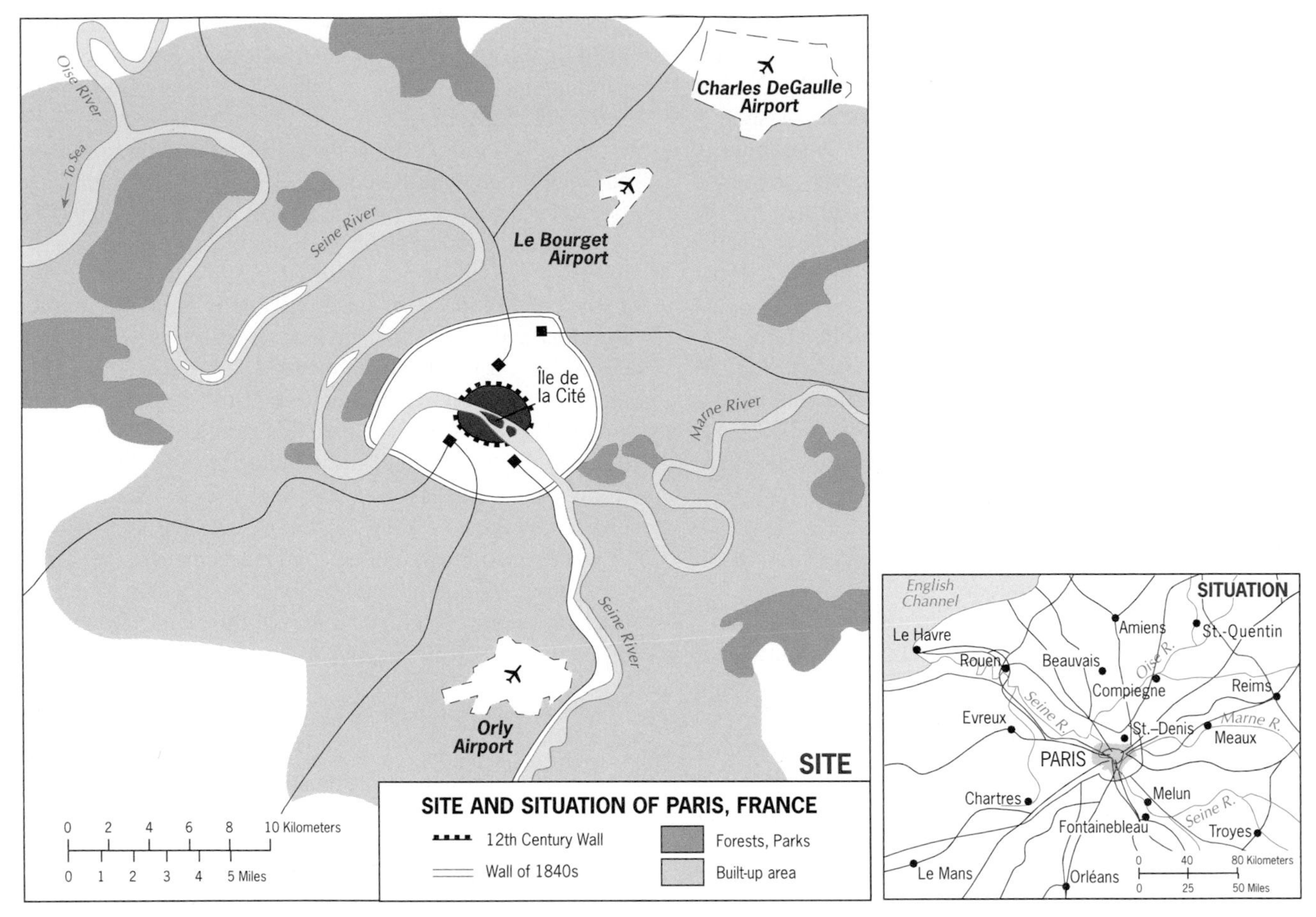

FIGURE 1-15

FROM THE FIELD NOTES

"Because the Seine has a meandering course in an often steep-sided valley, large ships can sail upstream only as far as Rouen, not to the French capital of Paris. At the river's mouth, the port of Le Havre is no great natural harbor, so that the lower Seine is lined with smaller riverside ports handling such commodities as oil, coal, and (as in the photo at the left) scrap iron . . . Beyond and to the southeast of Paris lies Bourgogne (Burgundy), famous for its vineyards and wines and known also for crops such as wheat, rye, mustard, and sunflowers. From the top of an old church I had a superb view of the rural landscape of Burgundy, with its nucleated settlements and its highly fragmented fields. The diversity of colors signifies the variety of crops even in this small area, but farming here is quite efficient. Farmers, who often own several separate parcels of land, use downsized mechanized equipment; in the Burgundy tradition they live in their clustered villages where they have vegetable gardens and keep their few livestock."

Among the Realm's Great Cities . . .

Paris

If the greatness of a city were to be measured solely by its number of inhabitants, Paris (9.6 million) would not even rank in the world's top 20. But if greatness is measured by a city's historic heritage, cultural content, and international influence, Paris has no peer. Old Paris, near the *Île de la Cité* that housed the original village where Paris began and carries the eight-century-old Notre Dame Cathedral, contains an unparalleled assemblage of architectural and artistic landmarks old and new. The *Arc de Triomphe*, erected by Napoleon in 1806 (though not completed until 1836), commemorates the emperor's victories and stands as a monument to French neoclassical architecture, overlooking one of the world's most famous streets, the *Champs Elysées*, which leads to the grandest of city squares, the *Place de la Concorde*, and on to the magnificent palace-turned-museum, the *Louvre*.

Even the Eiffel Tower, built for the 1889 International Exposition over the objections of Parisians who regarded it as ugly and unsafe, became a treasure. From its beautiful Seine river bridges to its palaces and parks, Paris embodies French culture and tradition. It is perhaps the ultimate primate city in the world.

As the capital of a globe-girdling empire, Paris was the hearth from which radiated the cultural forces of Francophone assimilation, transforming much of North, West, and Equatorial Africa, Madagascar, Indochina, and many smaller colonies into societies on the French model. Distant cities such as Dakar, Abidjan, Brazzaville, and Saigon acquired a Parisian atmosphere. France, meanwhile, spent heavily to keep Paris, especially Old Paris, well maintained—not just as a relic of history, but as a functioning, vibrant center, an example to which other cities can aspire.

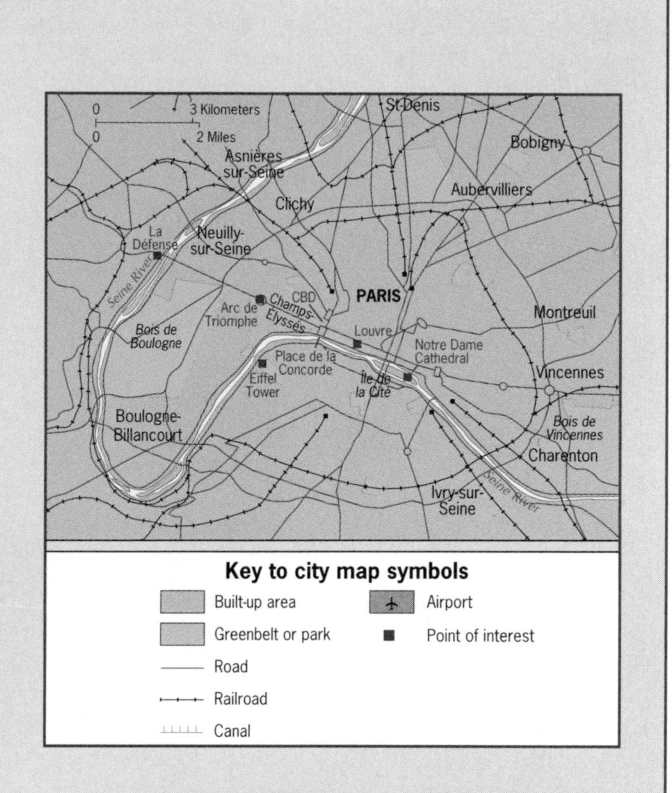

Today, Old Paris is ringed by a new and different Paris. Stand on top of the *Arc de Triomphe* and look away from the *Champs Elysées*, and the tree-lined avenue gives way to *La Défense*, an ultramodern high-rise complex that is one of Europe's leading business districts (see photo, p. 59). But from atop the Eiffel Tower you can see as far as 50 miles (80 km) and discern a Paris visitors rarely experience: grimy, aging industrial quarters, and poor, crowded neighborhoods where discontent and unemployment fester—and where Muslim immigrants cluster in a world apart from the splendor of the old city.

towns, barriers to access and movement, and other aspects of the greater regional framework in which it lies).

The site of the original settlement at Paris lay on an island in the Seine River, a defensible place where the river was often crossed. This island, the *Île de la Cité*, was a Roman outpost 2000 years ago; for centuries its security ensured continuity. Eventually the island became overcrowded, and the city expanded along the banks of the river (Fig. 1-15).

Soon the settlement's advantageous situation stimulated its growth and prosperity. Its fertile agricultural hinterland thrived, and, as an enlarging market, Paris's focality increased steadily. The Seine River is joined near Paris by several navigable tributaries (the Oise, Marne, and Yonne). When canals extended these waterways even farther, Paris was linked to the Loire Valley, the Rhône-Saône Basin, Lorraine (an industrial area), and the northern border with Belgium. When Napoleon reorganized France and built a radial system of roads—followed later by railroads—that focused on Paris from all parts of the country, the city's primacy was assured (Fig. 1-15, smaller map). The only disadvantage in Paris's situation lies in its seaward access: oceangoing ships can sail up the Seine River only as far as Rouen. Between the coastal port of Le Havre and Rouen, the meandering Seine is lined by rural villages interspersed with docking facilities (see left photo, p. 72).

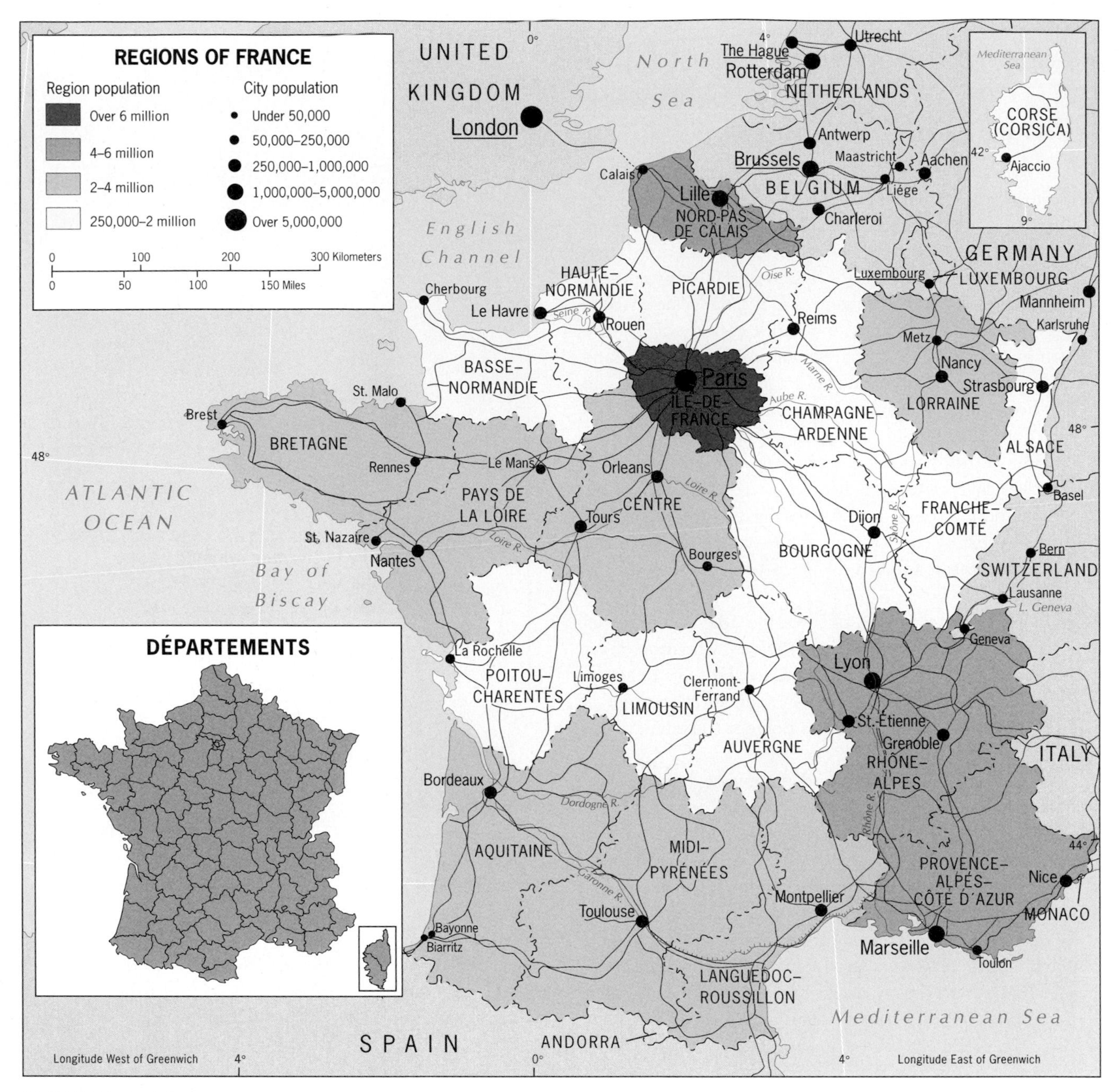

FIGURE 1-16

As Paris evolved into one of Europe's greatest cities, France's industrial development was less spectacular. Various explanations have been given for this lag, ranging from the lower availability of high-quality coal (and therefore its higher cost) and the lower efficiency of French transport systems to the notion that French manufacturers preferred to specialize in such products as high-quality textiles and precision equipment. Meanwhile, French agriculture remained Europe's most productive. From the Paris Basin to the Mediterranean coast and from the banks of the Rhine to the shores of the Bay of Biscay, France has a wide range of soils and climates, and its farms produce an almost limitless diversity of products. French wines, notably those of the Bordeaux, Burgundy, and Champagne districts, and French cheeses reach markets around the globe.

Today new high-tech industries, especially transportation and telecommunications, dominate France's economic geography. France is a leading producer of high-speed trains, aircraft, fiber-optic communications systems, and various space-related technologies. It also is the world leader in the development of a nuclear power grid, which today supplies more than 75 percent of the country's electricity and thereby reduces its dependence on imported oil.

When Napoleon reorganized France in the early 1800s, he broke up the country's large traditional subregions and established more than 80 small *départements* (additions and subdivisions later increased this number to 96). Each *département* had representation in Paris, but the power was concentrated in the capital, not in the individual *départements*. France became a highly centralized state and remained so for nearly two centuries (see inset map, Fig. 1-16). Only the island *département* of Corsica produced a rebel movement whose violent opposition to French rule continued for decades and even touched the mainland.

Today, France is decentralizing. A new subnational framework of 22 historically significant provinces, groupings of *départements* called *regions* (Fig. 1-16), has been established to accommodate the devolutionary forces felt throughout Europe and, indeed, throughout the world. These regions, though still represented in the Paris government, have substantial autonomy in such areas as taxation, borrowing, and development spending. The cities that anchor them benefit because they are the seats of governing regional councils that can attract investment, not only within France but also from abroad. Lyon, France's second city and headquarters of the region named Rhône-Alpes, has become a focus for growth industries and multinational firms. This region is evolving into a self-standing economic powerhouse that is becoming a driving force in the European economy; indeed, it is one of the Four Motors of Europe (see p. 61) with its own international business connections to countries as far away as China and Chile.

Until after 1950, France lay at the center of a global colonial empire that extended from Algeria to Madagascar and from West Africa to the South Pacific. That empire has been lost (except for a few remnants), and today France's cities are home to nearly 3 million immigrants from the former colonies (almost 5 percent of the total population of 59.9 million). Political and social crises in the former colonies tend to spill over into France; in the 1990s, a struggle between Islamic extremists and the government in Algeria led to acts of terrorism in Paris and other cities. The French have remained actively involved in their former empire, even providing armed support for threatened governments, and they show no signs of giving up the last of their holdings in South America (French Guiana), the Caribbean, the Atlantic, or the Pacific. France's South Pacific dependencies cost Paris billions of dollars annually, but despite opposition and even insurrection, the French Tricolor continues to fly over such far flung capitals as Papeete (Tahiti), Saint-Denis (Réunion), and Nouméa (New Caledonia).

This global girdle of colonial relics reflects France's ambivalence regarding its changing status in this changing world. Still harboring imperial (and superpower) notions, the French defy international opinion and explode nuclear bombs on an atoll in French Polynesia even as they take the lead in the international community's search for a solution to the conflict in former Yugoslavia. A French president visits (formerly French-colonial) Vietnam to secure closer ties even as Paris seeks to loosen its bonds with the states of its former Subsaharan African empire. French leaders want to make their country a cornerstone of the European Union, but French public opinion waxes and wanes on the matter. The still powerful French farmers besiege Paris when European Union policies hurt their interests.

France's regional reorganization, coupled with advances in internal and international surface and air transportation, reinforces one of the world's richest and most diversified economies. When the first-ever direct link between France and Britain, the Channel Tunnel, opened in 1994, another of France's regions—Nord-Pas de Calais—stood to gain, transforming its failing Rustbelt economy into one based on transport and trade. A pillar of Western Europe, France remains a powerful force in the European Union.

Benelux

Three political entities are crowded into the northwestern corner of Western Europe: Belgium, the Netherlands, and tiny Luxembourg, collectively referred to by their first syllables (*Be-Ne-Lux*). These states are also frequently called the Low Countries; this label is appropriate because most of the land is extremely flat and lies near (and in the Netherlands even below) sea level. Only toward the southeast, in Luxembourg and eastern Belgium's Ardennes, is there a hill-and-plateau landscape with elevations over 1000 feet (300 m). As is true of Germany and France, major differences exist between Belgium and the Netherlands—indeed so major that they find themselves in a position of economic complementarity.

Belgium is marked by two industrial corridors. One is the coal-based east-west axis through Charleroi and Liège, where there are heavy industries. The second corridor of lighter and more varied manufacturing extends north from Charleroi through Brussels to the major port of Antwerp. The diversified industrial products of these areas include metals, chemicals, furniture, and specialties such as pianos, soaps, and cutlery. In contrast, the Netherlands has a large agricultural base (along with its vitally important transport functions); it exports dairy products, meats, vegetables, and other foods. Hence the Benelux economic union of the 1940s facilitated the reciprocal flow of needed imports to both countries, and it doubled the domestic markets.

The Benelux countries are among the most densely populated on Earth. Space is truly at a premium: some 26.6 million people inhabit an area about the size of Maine (home to 1.3 million). For centuries the Dutch have been expanding their living space—not by warring with their neighbors, but by wresting land from the sea. The greatest project so far, the draining of almost the entire Zuider Zee (Sea), began in 1932 and continues. In the southwest the islands of Zeeland are being connected by dikes, and the

FROM THE FIELD NOTES
"Hundreds of bicycles are parked under the overhang of an office building in Utrecht, Netherlands. We watched the commuters arrive in the morning, filling the rows of bike stalls, then leave after work. From executives in business attire to office workers in jeans, all came to work by bicycle. Imagine the size of the parking lot required if these people had driven to work; imagine also the amount of gasoline saved, and pollution avoided. The continued use of the bicycle in the automobile era also has had an effect in limiting the sprawl of Dutch (and other European) cities. In the often cold and rainy weather of the Netherlands, the shorter the bike trip the better, so there is less attraction to living in a distant suburb."

water is being pumped out, creating additional *polders* (reclaimed lands).

Three cities (Amsterdam, Rotterdam, and The Hague) anchor the triangular core area of the Netherlands. Amsterdam, the constitutional capital, remains the focus of the Netherlands, with a bustling commercial center, a busy port, and a variety of light manufactures. Rotterdam, Europe's busiest port, is the shipping gateway to Western Europe, commanding the entries to both the Rhine and Meuse (Maas) rivers. Its modern development mirrors that of Germany's Ruhr industrial region and the adjacent Rhine Valley; thus ongoing manufacturing decline in the Rhine hinterland as well as increased competition from other European ports is eroding Rotterdam's historic situational advantage. The third city in the triangle, The Hague, is the seat of the Dutch government and the home of the United Nations' World Court.

Collectively, these three cities of the triangular core have spawned a **conurbation**, the term geographers use to define the huge multimetropolitan complexes formed by the coalescence of two or more major urban areas. This particular coalescence, known as *Randstad*, has created a ring-shaped urban complex that surrounds a still-rural center (the literal translation of *rand* is edge or margin). A more precise labeling of the conurbation, however, would be Randstad-Holland because Holland (meaning "hollow country") refers specifically to the Dutch heartland that faces the North Sea in these lowest-lying western provinces of the Netherlands.

23

Belgium, the Netherlands, and Luxembourg, with their limited space and modest resource base, have turned to managing international trade as a productive economic activity. Brussels, Belgium's capital, also has acquired a significant political role, serving (with Strasbourg) as one of the co-capitals of the European Union, headquarters of NATO, and administrative center for numerous international economic organizations. Hundreds of multinational corporations have their offices here, enhancing Brussels' role as a financial center and commercial-industrial complex.

Belgium, however, remains a three-region amalgam with an uncertain political future. From time to time devolutionary forces threaten to sever Dutch- (Flemish-) speaking Flanders (with 55 percent of the population) from French-speaking Wallonia (33 percent). In the early 1990s, such a prospect led to a politico-geographical reorganization of the country, establishing three coequal regions (the third: Brussels, in a quasi-federal arrangement). Flemish nationalism, however, remains a potent force.

The Alpine States

Switzerland and Austria share a landlocked situation and the mountainous topography of the Alps—and little else. Austria is a member of the European Union; Switzerland is not. Austria is a monolingual society; Switzerland is multilingual. Austria has a large primate city; multicultural Switzerland does not. Austria has a substantial range of domestic raw materials (including oil); Switzerland does not. Austria is twice as large as Switzerland and has a larger population than its neighbor to the west; Austria also has much more cultivable land than Switzerland.

And yet, as Table I-2 indicates, Switzerland, not Austria, is the leading state in Western Europe's Alpine subregion in terms of per-capita GNP and other indices. The world map of physical landscapes (Fig. I-1) suggests that mountainous countries face certain limits on development; mountainous terrain, and its frequent corollary, **landlocked location**, tend to inhibit the dissemination of ideas and innovations, obstruct circulation, constrain agriculture, and divide cultures. Tibet, Afghanistan, and Ethiopia seem to prove the point. That is why Switzerland is such an important lesson in human geography. The Swiss, through their skills and abilities, have overcome a seemingly restrictive environment and converted it into an asset that keeps them in the forefront of prosperous European countries.

24

At various times in their history, the Swiss employed their Alpine environments to act as middlemen in interregional trade, to use the water cascading from their mountains to build hydroelectric power plants in order to support highly specialized industries, and to attract millions of tourists. Meanwhile, Swiss farmers perfected ways to maximize the productivity of mountain pastures and valley soils. True, the Swiss speak German in the north, French in the west, Italian in the southeast, and even a bit of Rhaeto-Romansch in remote highlands (Fig. 1-8), but economic success has overcome such cultural heterogeneity. Neutrality, security, and stability made Switzerland a haven during war, a world banking giant, a global magnet for money.

The financial center of Zürich (Switzerland's largest city, located in the German-speaking sector), the international city of Geneva (in the far southwest and thus in the French sector), and the capital of Bern (near the French-German linguistic divide) each has its own precision industries, some of historic vintage. Swiss quality, whether in watches, instruments, or specialty foods, is an international standard.

Switzerland is therefore a Western European country, part and parcel of the European core—but even the prosperous Swiss confront problems in this changing realm. Cross traffic between Western and Mediterranean Europe, coupled with a relentlessly growing stream of tourists, is damaging Alpine environments. Switzerland's traditional neutrality and aloofness—it remains the only Western European country that is not yet a member of the EU—is causing concern that in the future the Swiss could lose their competitive edge in various fields.

Austria, by contrast, joined the European Union in 1995, following a referendum in which a sizeable minority opposed the move. German-speaking, overwhelmingly Catholic, and a remnant of the Austro-Hungarian Empire, Austria has a historical geography that is far more reminiscent of unstable Eastern Europe than Switzerland's. Even Austria's physical geography seems to demand that the country look eastward: it is at its widest, lowest, and most productive in the east, where the eastward-flowing Danube links it to Hungary, its old ally in the anti-Muslim campaigns of the past.

Vienna, by far the Alpine subregion's largest city, also lies on the country's eastern perimeter. One of the world's most expressive primate cities, Vienna is an outdoor museum of monumental art and architecture, replete with palaces, castles, mansions, public gardens, statues, magnificent churches, and grand public buildings. Many of these structures commemorate the victories of this outpost of Western culture over the Ottoman Turks at a time when much of Eastern Europe was under Islamic rule. Today, Vienna represents an outpost of another sort: this easternmost city of Western Europe is situated on the doorstep of transforming Eastern Europe, much as Singapore sits at the threshold of Southeast Asia's economies. Corporations doing business in the uncertain commercial climates of Eastern Europe are establishing headquarters in Vienna, using the city and its excellent infrastructure and comparatively uncorrupted institutions as a base for business with its ex-communist neighbors.

As the eastern branch of the Germanic culture that dominates the heart of Western Europe, Austria is part and parcel of this region. Its cultural landscapes are those of the West, its economy is closely tied to the region (more than 40 percent of exports and imports are with Germany), and its political system is representative of the region's democracies. When Austria joined the EU, it met the Maastricht criteria without difficulty. But joining the EU made Austria's eastern boundary something more than an ordinary interregional border: here now lies the great divide between Europe's "ins" and the "outs" that desire to join but are not yet eligible. Becoming part of the EU changed Austria's relative location in ways we cannot yet completely analyze.

THE BRITISH ISLES

Off the coast of mainland Western Europe lie two major islands, surrounded by a constellation of tiny ones, that constitute the British Isles, a discrete region of the European realm (Fig. 1-17). The larger of the two major islands, which also lies nearest to the mainland (a mere 21 miles, or 34 km, at the closest point), is the island called *Britain*; its smaller neighbor to the west is *Ireland*.

The names attached to these islands and the countries they encompass are the source of some confusion. They still are called the British Isles, even though British dominance over most of Ireland ended in 1921. The nation-state that occupies Britain and a corner of northeastern Ireland is officially called the United Kingdom of Great Britain and Northern Ireland—United Kingdom for short and UK by abbreviation. But this country often is referred to simply as Britain, and its people are known as the British. The nation-state of Ireland officially is the Republic of Ireland (*Eire* in Irish Gaelic), but it does not include the whole island of Ireland.

How convenient it would be if physical and political geography coincided! Unfortunately, the two do not. During the long British occupation of Ireland, which is overwhelmingly Catholic, many Protestants from northern Britain settled in northeastern Ireland. In 1921 when British domination ended, the Irish were set free—except in that corner in the north, where London kept control to protect the area's Protestant settlers. That is why the country to this day is officially known as the United Kingdom of Great Britain and Northern Ireland.

Northern Ireland (Fig. 1-17) was home not only to Protestants from Britain, but also to a substantial population of Irish Catholics who found themselves on the wrong side of the border when Ireland was liberated. Ever since, conflict has intermittently engulfed Northern Ireland and spilled over into Britain and even into Western Europe.

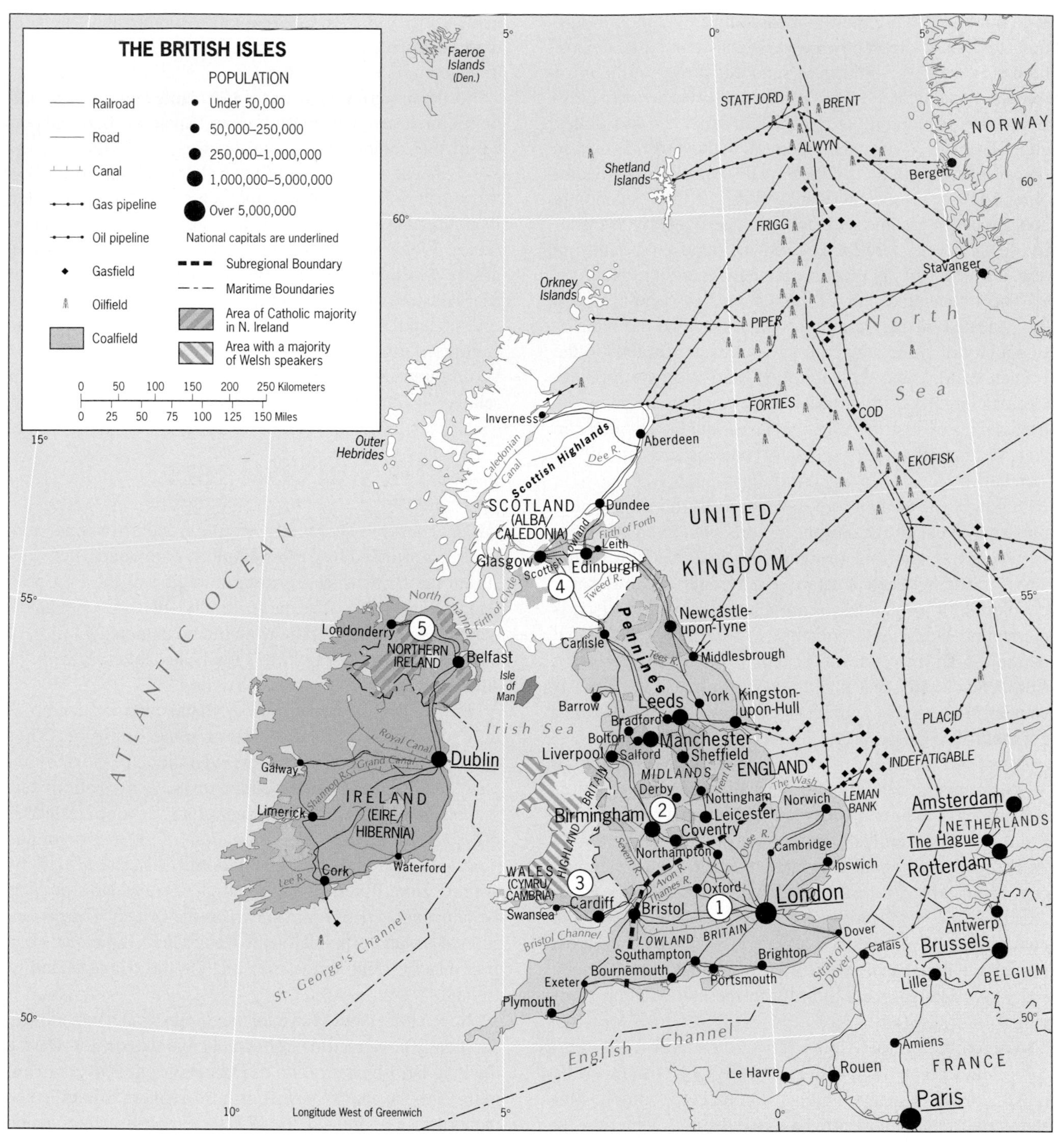

FIGURE 1-17

It has been and remains one of Europe's most costly struggles.

Although all of Britain lies in the United Kingdom, political divisions exist here as well. England is the largest of these units, the center of power from which the rest of the region was originally brought under unified control. The English conquered Wales in the Middle Ages, and Scotland's link to England, cemented when a Scottish king ascended the English throne in 1603, was ratified by the Act of Union of 1707. Thus England, Wales, Scotland, and Northern Ireland became the United Kingdom.

The British Isles form a distinct region of Europe for several reasons. Britain's insularity provided centuries of security from turbulent Europe, protecting the evolving British nation as it achieved a system of parliamentary government that had no peer in the Western world. Having united the Welsh, Scots, and Irish, the British set out to forge what would become the world's largest colonial empire. An era of mercantilism and domestic manufacturing (the latter based on water power from streams flowing off the Pennines, Britain's mountain backbone) foreshadowed the momentous Industrial Revolution, which transformed Britain—and much of the world. British cities became synonyms for specialized products as the smokestacks of factories rose like forests over the urban scene. London on the Thames River anchored an English core area that mushroomed into the headquarters of a global political, financial, and cultural empire. As recently as World War II, the narrow English Channel ensured the United Kingdom's impregnability against German attack, giving the British time to organize their war machine. When the United Kingdom emerged from that conflict as a leading power among the victorious Allies, it seemed that its role in the postwar era was assured.

Two unexpected developments changed that prospect: the collapse of colonialism and the resurgence of mainland Europe. In just a few decades, an empire of centuries was lost and the United Kingdom became, if not merely another European country, then certainly a lesser power in the world. And in just a few years mainland Europe's economies, fueled by the Marshall Plan, rose to challenge Britain's. When the supranationalism movement gathered momentum on the mainland, the United Kingdom, still enmeshed in Commonwealth ties to its former colonies, remained aloof. As we noted earlier, the British tried to counter the old EEC by taking a leading role in EFTA, but eventually (in the early 1970s) the United Kingdom opted to join what has become the European Union.

Almost always, the British have worked to restrain EU moves toward tighter integration and tougher regulations that would come with it. In the late 1990s, when European Monetary Union (EMU) approached, London decided to delay its participation. While the German mark and the French franc are slated to disappear by 2002, the British pound will endure—at least temporarily. The British want the EU to impose only a minimum of rules; for them, monetary union is questionable and federalism is out of the question. In this respect, Britain's historic, insular standoffishness continues.

The United Kingdom

As we noted earlier, the British Isles as a region consists of two political entities: the United Kingdom and Ireland. The UK, with an area about the size of Oregon and a population of 59.9 million (Oregon: 3.5 million), is by European standards a large country. As we will see, the UK has considerable spatial variation. Based on a combination of physiographic, cultural, economic, and political criteria, the United Kingdom can be divided into five subregions: (1) Southern England centered on the London area, (2) Northern England including Manchester, Liverpool, and Birmingham, historic cities at the heart of the Industrial Revolution, (3) Wales to England's west, (4) Scotland in Britain's far north, and (5) Northern Ireland across the Irish Sea (Fig. 1-17).

The United Kingdom is divided into counties (regions in Scotland), many with historic names made famous by the Industrial Revolution as well as by emigrants who transplanted these names to many countries, especially the United States and Australia. In the 1990s, the county map was modified to improve administrative efficiency and to adjust to EU planning standards. But many of the old names—Cornwall, Devon, Kent, Norfolk—continue to form Britain's traditional geographic frame of reference.

Figure 1-17 reveals the importance of still another region of Britain: its maritime zone in the North Sea. Not many of our maps show international boundaries at sea as well as on land, but it is essential here. The rock strata beneath the North Sea contain major reserves of oil and natural gas, and it became necessary to determine which coastal country owned what part of these. As Figure 1-17 shows, the United Kingdom and Norway have the largest share of these crucial energy reserves. Oil platforms and submarine pipelines make the North Sea an integral part of the coastal states, and new discoveries continue to be made.

Southern England is the United Kingdom's most affluent subregion, the center of high-technology and service industries focused on the giant London metropolitan area. Financial, communications, engineering, and energy-related industries cluster here. More than one-third of the UK's population is concentrated in this economically and politically dominant area. London exemplifies the momentum generated by centuries of agglomeration; even when the Industrial Revolution shifted Britain's economic focus to the north, London's primacy continued. Today Southern England's superior connections to the European mainland help keep this core subregion far ahead of the rest.

Northern England, as the map shows, really should be called Central, Northern, and Western England. The center, appropriately called the Midlands, was the scene of the Industrial Revolution and the spectacular rise of Manchester, the source of that revolution; the great port, Liverpool; the peerless industrial center, Birmingham; as well as Sheffield, Leeds, and other manufacturing cities. But here, obsolescence has overtaken what was once ultramodern, and a Rustbelt plagued by social and economic problems now prevails. Heavy immigration from South Asia, the Caribbean region, and Africa, coupled with high unemployment, pose a difficult challenge for the Midland cities. Birmingham, the country's second-largest city, also is the most racially mixed city in Britain, with more than 20 percent foreign-born residents. Immigration peaked just as employment in the automobile and engineering industries fell by half. Now these old industrial cities are trying to become service centers—Liverpool by regaining its transport function, Manchester by attracting financial-service facilities, and all of them by luring tourists and conventions to re-

Among the Realm's Great Cities . . .

London

Sail westward up the meandering Thames River toward the heart of London, and be prepared to be disappointed. London does not overpower or overwhelm with spectacular skylines or beckoning beauty. It is, rather, an amalgam of towns—Chelsea, Chiswick, Dulwich, Hampstead, Islington—each with its own social character and urban landscape. Some of these towns come into view from the same river the Romans sailed 2000 years ago: Silvertown and its waterfront urban renewal; Greenwich with its famed Observatory; Thamesmead, the model modern commuter community. Others somehow retain their identity in the vast metropolis that remains (Fig. 1-9), in many ways Europe's most civilized and cosmopolitan city. And each contributes to the whole in its own way: every part of London, it seems, has its memories of empire, its memorials to heroes, its monuments to wartime courage.

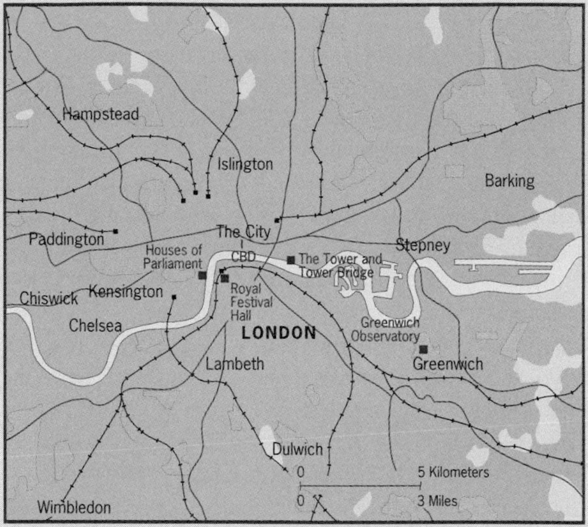

Along the banks of the Thames, London displays the heritage of state and empire: the Tower and the Tower Bridge, the Houses of Parliament (officially known as the Palace of Westminster), the Royal Festival Hall. Step ashore, and you find London to be a memorable mix of the historic and (often architecturally ugly) modern, of the obsolete and the efficient, of the poor and the prosperous. Public transportation, by world standards, is excellent; traffic, however, often is chaotic, gridlocked by narrow streets. Recreational and cultural amenities are second to none. London seems to stand with one foot in the twenty-first century and the other in the nineteenth. Its airports are ultramodern. But when the Eurostar TGV train from Paris emerges from the Channel Tunnel, it has to slow down for the final hour in its three-hour run because the tracks into London's Waterloo Station cannot carry it at its normal high speed.

London remains one of the world's most livable metropolises for its size (7.6 million) largely because of farsighted urban planning that created and maintained, around the central city, a so-called Greenbelt set aside for recreation, farming, and other nonresidential, noncommercial uses (Fig. 1-9). Although London's growth eroded this Greenbelt, in places leaving only "green wedges," the design preserved crucial open space in and around the city, channeling suburbanization toward a zone at least 25 miles (40 km) from the center. The map reveals the design's continuing impact.

furbished portions of downtown. In the shadow of London's dominance, however, recovery will be slow.

Wales, the third of the United Kingdom's five subregions, lies to the west of Northern England. It is a nearly rectangular, rugged territory where ancient Celtic peoples found refuge from invaders. In its western counties, more than half the people still speak Welsh, but in the counties bordering England, it is rarely heard (Fig. 1-17). Because of its high-quality coal reserves, Wales, too, was engulfed by the Industrial Revolution, but when these resources were exhausted and coal production costs mounted, Wales's fortunes declined. Cardiff, the Welsh capital, was once the world's leading coal exporter; today the city shows signs of decline and stagnation. Many Welsh citizens left their historic homeland in search of opportunities elsewhere, but among the 3 million who remained the flame of Welsh nationalism was not extinguished. In 1997 the British government allowed the Welsh population to vote on the creation of a Welsh Assembly to administer public services in Wales, a first devolutionary step. By a slender majority, voters approved.

As the map shows, *Scotland* is much larger than Wales. Its territorial connection with England lies in the narrow northern neck of Britain, in the area where the Roman emperor Hadrian built a wall around 120 AD to keep the barbarians out of Roman England. Scotland is no small outpost: it is nearly twice as large as the Netherlands and has more than 5 million inhabitants, about as many as Denmark. Most of this population is concentrated in the Scottish Lowland anchored by Glasgow in the west and Edinburgh in the east, where coalfields and iron ores enabled the Scots to participate in Britain's Industrial Revolution, creating (among other manufacturing) a world-renowned shipbuilding industry. When decline and obsolescence overtook the area, an alternative emerged: as Figure 1-17 shows, the seafloor beneath the North Sea off Scotland's east coast contains major reserves of oil. Edinburgh's port of Leith, once a fishing port and then a shipbuilding center, now became a major focus for North Sea oil drilling and pipeline equipment. Aberdeen, the old fishing port, became the maritime oil industry's primary service center and saw its economy transformed. Farther north, Inverness specialized in the manufacture and repair of oil platforms.

Meanwhile, some high-tech development, notably in the hinterland of the old industrial city of Glasgow (in a corridor popularly called Silicon Glen), is compensating somewhat for the decline of heavy manufacturing. But none of this growth has been enough to halt a rising sense among the Scots that, in the United Kingdom and in the greater European Union sphere, they are economically and politically disadvantaged. When the British government put the option of a Scottish parliament to the voters in 1997, 74 percent approved, and for many Scots this was just the first stage of a process they hope will lead to total independence. Whether the London government will be able to inhibit the devolutionary forces it has set in motion remains to be seen.

Northern Ireland remains London's most serious domestic political problem. With a population of 1.8 million occupying the northeastern one-sixth of the island of Ireland, here we observe the troubled legacy of British colonial rule. About 55 percent of the people in Northern Ireland trace their ancestry to Scotland or England and are Protestants; some 44 percent are Roman Catholics, who share their Catholicism with virtually the entire population of the Irish Republic on the other side of the border. Although Figure 1-17 suggests that there are majority areas of Protestants and Catholics in Northern Ireland, no clear separation exists; mostly they live in clusters throughout the territory, including walled-off neighborhoods in the major cities of Belfast and Londonderry. Partition is no solution to a conflict that has raged for over three decades at a cost of thousands of lives; Catholics accuse London as well as the local Protestant-dominated administration of discrimination, whereas Protestants accuse Catholics of seeking union with the Republic of Ireland. The cost of Northern Ireland to the UK, in terms of policing measures and economic assistance, has been enormous, and until 1998 no investment was enough to stabilize the situation. In that year, a combination of developments, including negotiations led by a for-

FROM THE FIELD NOTES

"From a ferry on the Mersey River, the skyline of downtown Liverpool seemed to mirror the troubles of this city, the nucleus of a metropolitan area with almost one million inhabitants. The dominant buildings in the CBD remain those built during Liverpool's heyday a century ago, when Liverpool's docks outnumbered even London's, and the city thrived on its trade with America and Africa as Merseyside's manufacturing industries prospered. After World War II, Liverpool declined as its port functions dwindled, its industries weakened, and unemployment soared; this view reveals no modern, glass-sided buildings towering over the older cityscape, symbols of investment and confidence in the future. Once the apex of what was called the *Liverpool Triangle* linking this port to West Africa and eastern America, the Merseyside area now lies remote from the centers of British and European action. Some investment has been made in waterfront redevelopment, but the city's stagnation is evident throughout its historic center (where memories of the famed rock group, the Beatles, have developed into a minor industry). 'There may be ambivalence about the European Union in other parts of Britain,' said my colleague of the Department of Geography at the University of Liverpool as we walked the streets, 'but not here. At least we can see more tangible results of EU investment, for which Liverpool is eligible.' But what will lift Merseyside out of its doldrums?"

mer U.S. Senator, George Mitchell, and two terrorist acts that revulsed all sides in the conflict, produced a complicated settlement that created a Northern Ireland Assembly to which powers would be devolved from London. All parties in Northern Ireland elected representatives to this new Assembly, and for the first time bitter opponents sat down with each other to share power. In addition, the settlement created two councils involving parties beyond those in Northern Ireland alone: a British-Irish Council, which will bring together representatives from Scotland, Wales, as well as Belfast to discuss issues with members of the British and Irish governments, and a North-South Council, which focuses on cross-border matters between Northern Ireland and Ireland itself.

Only time will tell whether this arrangement can overcome the historic animosities of so many of the participants, but the geographic obstacles are immense. The local resource base is poor, unemployment remains high, and economic prospects remain bleak so long as political stability is not assured. As in the Middle East and elsewhere in the world, a few armed extremists can destroy a peacemaking process favored by the vast majority (in Northern Ireland, more than 3600 people have been killed during what locals call "the Troubles"). Keeping Northern Ireland's devolutionary forces on a peaceful track will remain difficult.

The Republic of Ireland

What might have been in Northern Ireland is shown by the Irish Republic itself: a growing, booming service-based economy, the fastest-growing in all of Europe since the turn of the century. Burgeoning cities and towns (and rising real-estate prices), mushrooming industrial parks, bustling traffic and construction everywhere reflect this new era. For the first time ever, workers of Irish descent are returning from foreign places to take jobs at home; non-Irish immigrants also are arriving, posing some new social problems for a closely knit, long-isolated society.

The Republic of Ireland fought itself free from British colonial rule just three generations ago. Its island is shaped like a wide-rimmed saucer tilted toward the east, where the interior lowlands open toward the sea and the capital, Dublin, is located (Fig. 1-17). Ireland's climate is notoriously moist, and its cool environment was well suited to potato farming when this crop was introduced from America in the 1600s, so that the potato quickly took the place of other crops. But even the potato could not withstand the large amount of precipitation that fell year after year during the 1840s, causing blight and rotting the harvest. By then, the Irish had come to depend on the potato as their staple food, and the resulting famines took over a million lives. Another 2 million people emigrated, most in desperation; Ireland's population, which in the 1830s had reached about 8 million, declined by more than one-third. Today, the entire island of Ireland has just 5.6 million residents, and the Republic of Ireland only 3.8 million.

Stagnation continued into the mid-1990s as limited economic opportunity and the conservatism of Irish cultural and religious life generated a steady stream of emigrants. But then the situation changed. European telecommunications service industries saw labor and locational opportunities in Ireland, and soon Ireland was Europe's leading call center for the realm's rapidly expanding toll-free telephone market. Other service industries followed suit, benefiting from Ireland's well-educated but not highly paid labor pool. By the late 1990s Ireland was booming, with its economy producing annual surpluses and its image sparkling as the *Celtic Tiger*.

Meanwhile, a growing tourist trade buoyed Ireland's economy: the country's legendary beauty attracts a large number of visitors every year. The prosperity of the south of England also influences Ireland's regional development. Southeastern Ireland, notably the area centered on Waterford, enjoys a somewhat sunnier climate than the remainder of the island. Lower real-estate prices (compared to those in England and Western Europe), proximity, and service-industry growth have created a Sunbelt phenomenon that is transforming villages into vacation and retirement sites and towns into office parks. All of this is facilitated by Ireland's fully committed EU membership: Ireland, unlike the United Kingdom, is a strong supporter of the EU and has joined EMU.

NORTHERN (NORDIC) EUROPE

North of Europe's Western European core area lies a disconnected group of six countries that exemplify core-periphery contrasts. Northern Europe is a region of difficult environments: generally cold climates, poorly developed soils, limited mineral resources, long distances. Together, the six countries in this region—Sweden, Norway, Denmark, Finland, Estonia, and Iceland—contain just over 25 million inhabitants, which is a lower total population than that of Benelux and only one-seventh that of Western Europe. The overall land area, on the other hand, is almost the size of the entire European core. Here in peripheral *Norden*, as the people call their northerly domain, national core areas lie in the south: note the location of the capitals of Helsinki, Stockholm, and Oslo at approximately the same latitude (Fig. 1-18).

Northern Europe's peripheral situation is more than environmental. As viewed from Europe's core area, Norden is on the way to nowhere. No major shipping lanes lead from Western Europe past Norden to other productive areas of the world, limiting interaction of the sort that ties the British

FROM THE FIELD NOTES
"I had heard of the huge high-technology complex developing near the Dublin (Ireland) airport, and wanted to get a first-hand look. It wasn't easy. When I stopped at this gate guarding the entry to a major computer manufacturer, I asked permission simply to drive to the main building for a perspective of the operation's size. Not only was entry refused, but the bar came down and I was forced to back up in order to turn around. It was not the only experience of its kind: these corporations based in the new 'Celtic Tiger' appeared to guard their privacy with considerable energy. . . In Dublin, where real-estate prices are soaring and the new economy is changing the cityscape, not everyone is favorably disposed toward its inevitable impact."

Isles to the mainland. Moreover, except for relatively small Denmark and Estonia, all of Norden lies separated from the European mainland by water. At all levels of spatial generalization, isolation is pervasive in Norden.

Northern Europe's remoteness, isolation, and environmental severity also have had positive effects for this region. The countries of the Scandinavian Peninsula lay removed from the wars of mainland Europe (although Norway was overrun by Nazi Germany during World War II). The three major languages—Danish, Swedish, and Norwegian—are mutually intelligible, which creates one of the criteria delimiting this region. Another regional criterion is the overwhelming adherence to the same Lutheran church in each of the Scandinavian countries (Norway, Sweden, and Denmark), Iceland, and Finland. Furthermore, democratic and representative governments emerged early, and individual rights and social welfare have long been carefully protected. Women participate more fully in government and politics here than in any other region of the world.

Sweden is the largest Nordic country in terms of both population and territory. Most Swedes live south of the zone where the evergreen forest yields to mixed deciduous and needleleaf forests, in what is climatically the most moderate part of the country (Fig. 1-18). Here lie the capital, core area, and, as Figure 1-7 shows, the main industrial districts; here, too, are the main agricultural areas that benefit from the lower relief, better soils, and milder climate.

Sweden long exported raw or semifinished materials to industrial countries, but today the Swedes are making finished products themselves, including automobiles, electronics, stainless steel, furniture, and glassware. Much of this production is based on local resources, including a major iron ore reserve at Kiruna in the far north (there is a steel mill at Luleå). Swedish manufacturing, in contrast to that of several Western European countries, is based in dozens of small and medium-sized towns specializing in particular products.

Sweden faces the Baltic, not the North Sea, and its ancient crystalline rocks create varied topography but do not contain oil or natural gas. Sweden was a pioneer in developing nuclear power to supplement its hydroelectric sources, and by the late 1970s more than half of the country's electricity came from a dozen nuclear facilities. But in a referendum in 1980, voters decided that the risks of nuclear power were too great and that the industry should therefore be shut down. The first nuclear plant was closed in 1998 despite warnings from business, industry, trade unions, and politicians that this move would hurt Sweden's economy in the years ahead.

Norway does not need a nuclear power industry to supply its energy needs. It has found its economic opportunities on, in, and beneath the sea. Norway's fishing industry, now augmented by highly efficient fish farms, long has been a cornerstone of the economy, and its merchant marine spans the world. But since the 1970s Norway's economic life has been transformed by the bounty of oil and natural gas discovered in its sector of the North Sea. In 2000, Norway ranked as the second-biggest exporter of oil in the world.

FIGURE 1-18

With its limited patches of cultivable soil, high relief, extensive forests, frigid north, and spectacularly fjorded coastline, Norway has nothing to compare to Sweden's agricultural or industrial development. Its cities, from the capital Oslo and the North Sea port of Bergen to the historic national focus of Trondheim as well as Arctic Hammerfest, lie on the coast and have difficult overland connections. Norway has been described as a necklace, its beads linked by the thinnest of strands. But this has not constrained national development. Norway in 2000 had the second-lowest unemployment rate in Europe (after tiny Luxembourg). In terms of income per capita, Norway is the third-richest country in the world.

Norwegians have a strong national consciousness and a spirit of independence. In 1994, when Sweden and Finland voted to join the European Union, the Norwegians again

Greenland and Europe

Greenland, the world's largest island, lies much closer to North America than to Europe. But its historical geography links it to Europe—specifically, Denmark. Peopled originally by Inuits (formerly called Eskimos) whose communities clustered along the southwestern coast, Greenland was reached by Norwegian Vikings before AD 1000 and became a Norwegian dependency in 1261. After Norway united with Denmark in 1380, the frigid island came under Danish rule. When the union ended in 1814, Greenland continued under the Danish crown. In 1953, its status changed from colony to province, and in 1979 Greenland won home rule as a "Self-Governing Overseas Administrative Division" of Denmark with an Inuit name: *Kalaallit Nunaat*.

Greenland's 59,000 inhabitants, 85 percent of whom are Inuits or native-born Greenlanders of European ancestry, depend almost entirely on fishing, a declining industry. Mining operations (lead, zinc) failed, and tourism

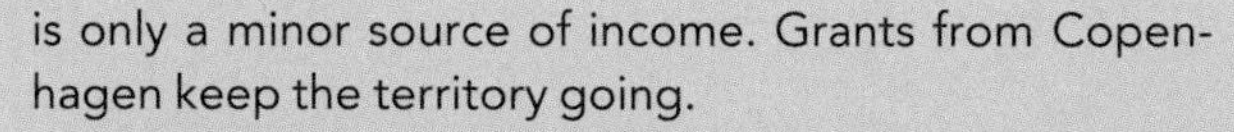

is only a minor source of income. Grants from Copenhagen keep the territory going.

Economic problems and dependence on Denmark have not stifled Greenlanders' political liberties. When Denmark joined the (then) European Economic Community in 1973, Greenland entered the EEC as part of the Danish state. After achieving home rule in 1979, Greenlanders voted to secede, taking their territory out of the budding European Union. They were the first (but possibly not the last) to do so. Still, the Scandinavian imprint on Greenland's Arctic cultural landscapes is indelible, as reflected by the capital, Nuuk (Godthab), with its Danish government buildings and brightly colored wooden houses, located on the comparatively moderate southwestern coast. The realm boundary shown in Figure I-2 confirms that while Greenland lies just a few miles east of Canada's Ellesmere Island, its functional ties are across the Atlantic, not across Baffin Bay.

said no. They did not want to trade their economic independence for the regulations of a larger, even possibly safer, Europe.

Denmark, territorially small by Scandinavian standards, has a population of 5.3 million, which ranks it second in the Nordic region after Sweden. The country consists of the Jutland Peninsula and adjacent eastern islands between Sweden and Western Europe (Denmark also has ties to Greenland—see box titled "Greenland and Europe"). Denmark has a comparatively mild, moist climate; level land and good soils help sustain intensive agriculture over 75 percent of its area. Denmark exports dairy products, meats, poultry, and eggs, mainly to its chief trading partners, Germany, Sweden, and the United Kingdom.

Denmark's capital, Copenhagen, has long been a center where large quantities of goods are collected, stored, and transshipped. This is because it lies at the **break-of-bulk point** [25] where many oceangoing vessels are prevented from entering the shallow Baltic Sea; conversely, ships with smaller tonnages ply the Baltic and bring their cargoes to this pivotal collecting station. Thus Copenhagen is an **entrepôt** [26] whose transfer functions maintain the city's position as the lower Baltic's leading port. Today these spatial advantages are further enhanced by the opening of the first

The lights of Copenhagen line the western horizon in this twilight view from above the Øresund Fixed Link, the new ten-mile-long bridge/tunnel/artificial island complex that connects Denmark's primate city to the southern Swedish city of Malmö. This first direct link between the Scandinavian Peninsula and the main body of Europe greatly strengthens the internal and external economic linkages of the Øresund functional region, focused on the newly united metropolis of 3.2 million people at its core. The most far-reaching changes, however, are likely to occur in Malmö's hinterland, the southernmost Swedish province of Scania. In Europe, where memories are long, many Scanians today still think of themselves as former Danes because their province was forcibly annexed to Sweden in the mid-seventeenth century. Following his study of attitudes on both sides of the waterway, a social scientist at Scania's University of Lund reported that "for a lot of people in Scania, the bridge is their exit from the Swedish nation-state. They feel that they're going home to Denmark." The bottom line here, of course, is yet another example of *devolution*: the Øresund Fixed Link is already contributing to a growing sense of regionalism in southern Sweden and represents an intensifying centrifugal force that will need countering by the Stockholm government. That is why we have added Scania to the map (p. 61) that shows Europe's foci of devolutionary pressures.

direct link to Sweden in 2000, the Øresund bridge-tunnel that crosses the narrow strait east of the Danish capital and connects it by road and rail to Sweden's thriving southernmost province, Scania (see photo, p. 85).

Finland, territorially almost as large as Germany, has only 5.2 million residents, most of them concentrated in the triangle formed by the capital, Helsinki, the textile-producing center, Tampere, and the shipbuilding center, Turku (Fig. 1-18). A land of evergreen forests and glacial lakes, Finland has an economy that has long been sustained by wood and wood product exports. But the Finns, being a skillful and productive people, have developed a diversified economy in which the manufacture of machinery and telecommunications equipment as well as the growing of staple crops are important.

Finland lost crucial lower-latitude territory (by Finnish standards) to the Soviet Union, now incorporated as a Russian "republic" under Moscow's control. Vyborg once was a Finnish town, but Finland avoided the fate of the Baltic states on the other side of the Gulf of Finland, which were incorporated into the Soviet communist empire.

As in Norway and Sweden, environmental challenges and relative location have created Nordic cultural landscapes in Finland, but the Finns are not a Scandinavian people; their linguistic and historic links are instead with the Estonians across the Gulf of Finland. As we will see in Chapter 2, ethnic groups speaking Finno-Ugric languages are widely dispersed across what is today western Russia.

Estonia, northernmost of the three "Baltic states," is part of Nordic Europe by virtue of its ethnic and linguistic ties to Finland. But during the period of Soviet control from 1940 to 1991, Estonia's demographic structure changed drastically: today about 30 percent of its 1.4 million inhabitants are Russians, most of whom came there as colonizers. When Estonia became independent in 1991 after the Soviet collapse, its government instituted new rules for citizenship that many Russians could not meet. This damaged relations with its most important neighbor.

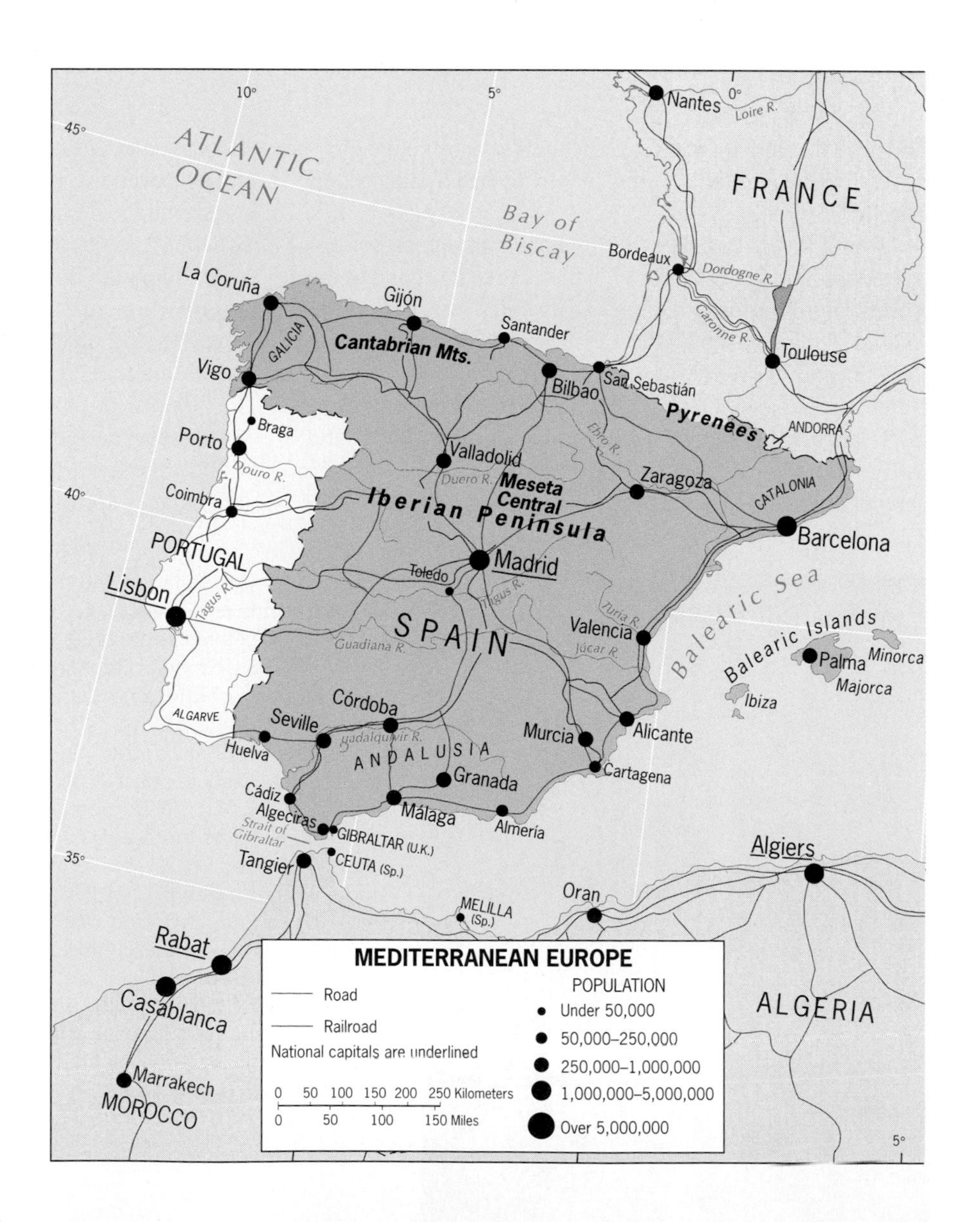

FIGURE 1-19

Today Estonia is following new directions, as reflected by the busy traffic across the Gulf of Finland between Tallinn, the capital, and Helsinki, and by the new free-trade zone at Muuga Harbor to facilitate transit trade with Russia. Relations with the Russian minority improved after the citizenship laws were modified, and Estonia is negotiating with the European Commission for entry into the European Union.

Iceland, the volcanic, glacier-studded island in the frigid waters of the North Atlantic just south of the Arctic Circle, is the sixth country in this region. Inhabited by people with Scandinavian ancestries (population: 285,000), Iceland and its small neighboring archipelago, the Westermann Islands, are of special scientific interest because they lie on the Mid-Atlantic Ridge, where the Eurasian and North American tectonic plates of the Earth's crust are diverging and new land can be seen forming.

Iceland's population is almost totally urban, and the capital, Reykjavik, contains about half the country's inhabitants. The nation's economic geography is almost entirely oriented to the surrounding waters, whose seafood harvests give Iceland one of the world's highest standards of living—but at the risk of overfishing. Disputes over fishing grounds and fish quotas have intensified in recent decades; the Icelanders argue that, unlike the Norwegians or the British, they have little or no alternative economic opportunities.

MEDITERRANEAN EUROPE

On the southern side of Europe's core lie the six countries that constitute the Mediterranean region: Italy, Spain, Portugal, Greece, and the small island countries of Cyprus and Malta (Fig. 1-19). With this shift from near-polar to near-tropical Europe, we should expect strong contrasts, and there are many—but similarities also exist. Once again, this is a region of peninsulas. Like Northern Europe, Mediterranean Europe is a discontinuous region. Moreover, there

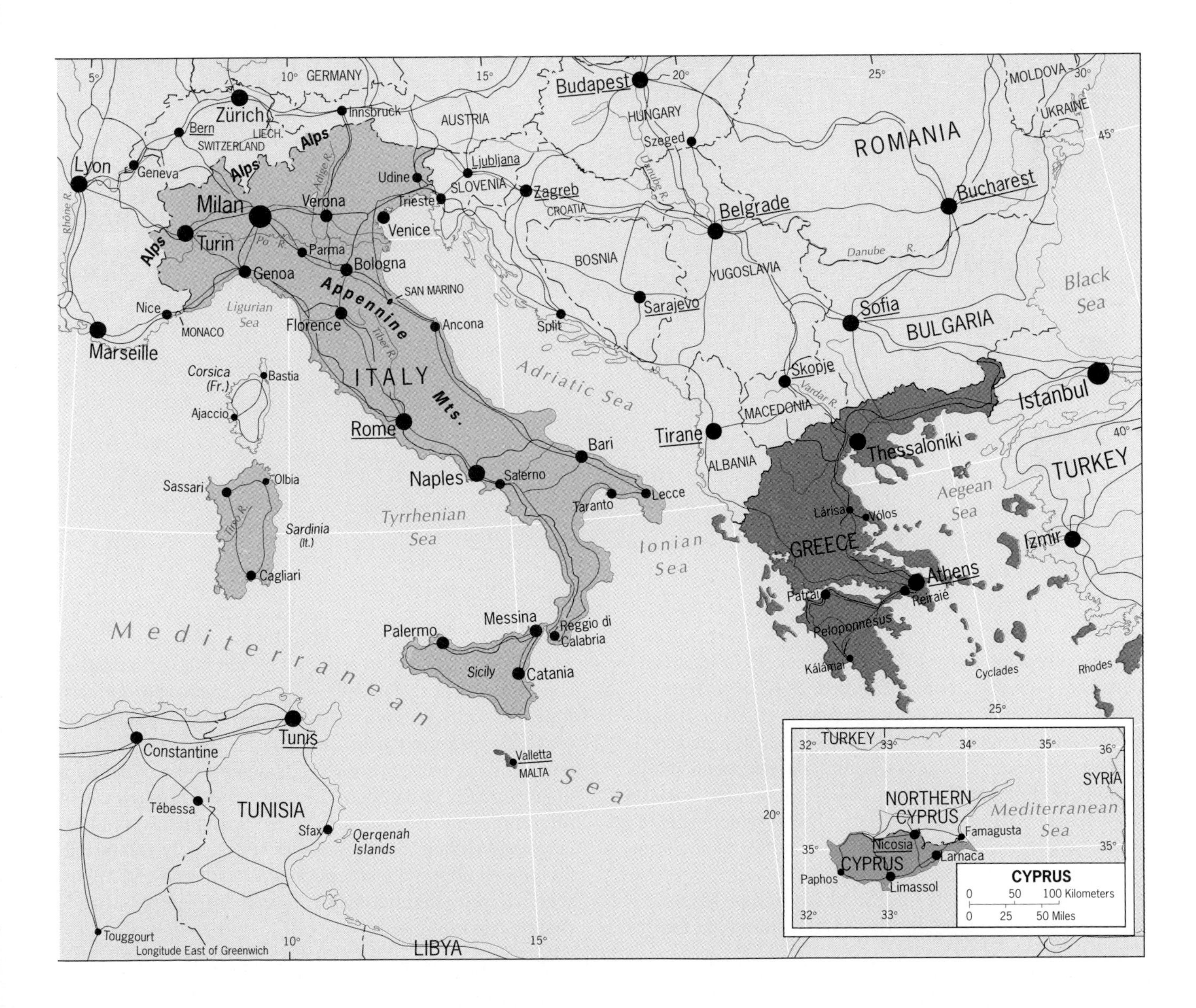

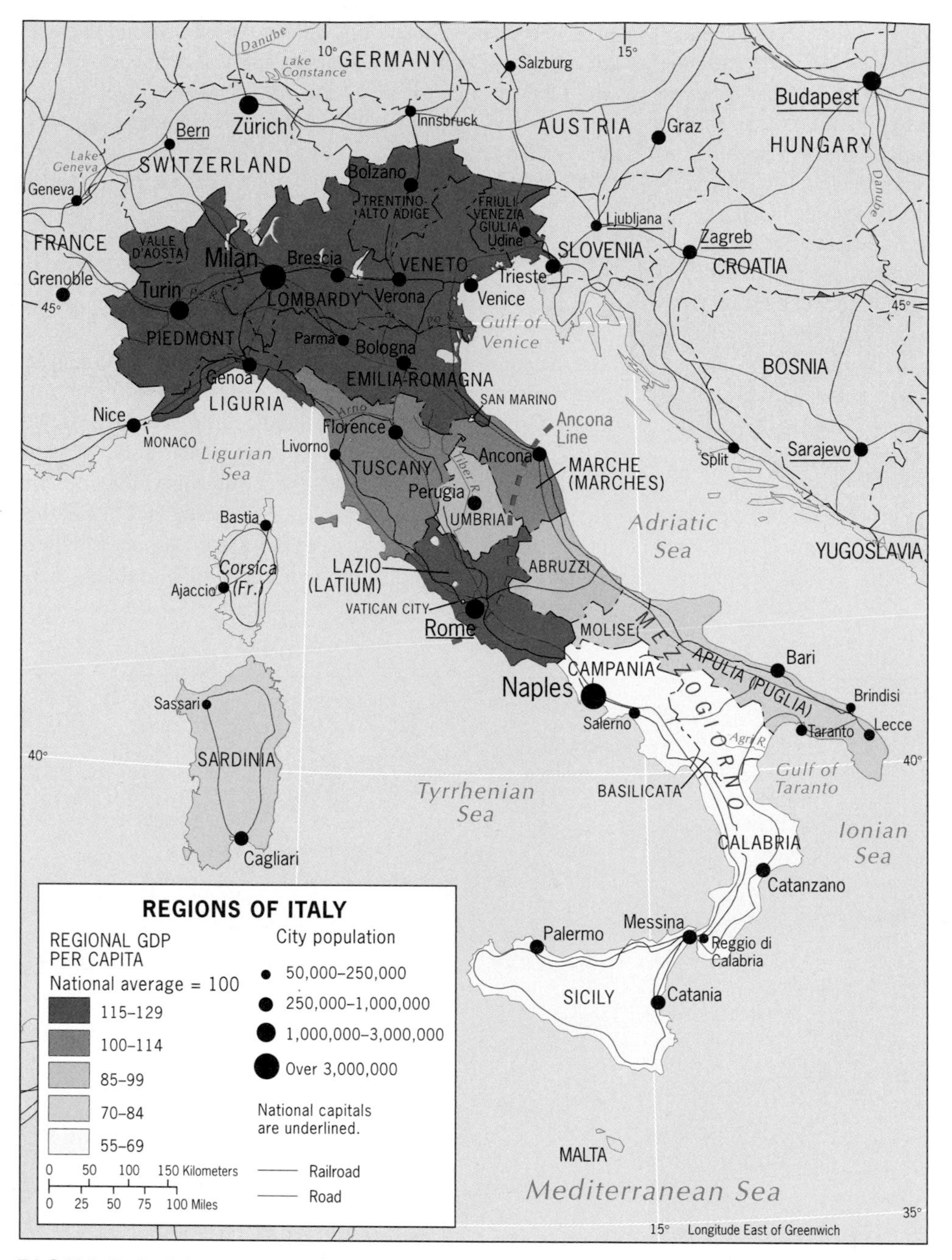

FIGURE 1-20

(again) is separation from the European core, so that core-periphery contrasts are manifest here as well. A degree of cultural continuity, dating from Greco-Roman times, marks the region's religions and languages, lifeways, and landscapes. As Figure 1-2 shows, natural environments in this region are dominated by a climatic regime that bears its very name—Mediterranean. Dry, hot summers are the norm, so that moisture supply often is problematic during the growing season.

In terms of raw materials, Southern Europe (as the region is also called) is not nearly as well endowed as the European core, a situation reflected by the map of industrial complexes (Fig. 1-7). Only northern Spain and northern Italy (the latter through massive imports of coal and iron ore) have become part of the core area's economic geography. Even then, Spain exports much of the minerals it mines directly to Western Europe. Forest resources also offer a discouraging picture: unlike Scandinavia and Finland, the Mediterranean region has been largely deforested. Despite its often rugged topography, a limited and highly seasonal water supply constrains Southern Europe's hydroelectric opportunities.

A key contrast between Northern and Southern Europe lies in their populations. Although territorially smaller than Northern Europe, Mediterranean Europe has nearly five times as many people (119 million in 2002). Population distribution continues to reflect the agricultural bases of the preindustrial era, with large concentrations in coastal lowlands and fertile river basins, although the growth of major industrial centers, notably in northern Italy and northern Spain, has superimposed a new mosaic. Still, urbanization in Southern Europe is far below that in Western or Northern Europe, or in the British Isles (Portugal's 48 percent is one of the lowest in the realm). Other data indicate that living standards in Mediterranean Europe also lag behind (Table I-2). This was, and remains, one of the challenges the European Union must confront as its leaders strive for equality among the 15 members.

Italy

Centrally located in the Mediterranean region, most populous of the Mediterranean states, best connected to the European core, and economically most advanced is Italy (57.7 million), a charter member of Europe's Common Market.

Administratively, Italy is organized into 20 regions, many with historic roots dating back centuries (Fig. 1-20).

Among the Realm's Great Cities . . .

Rome

From a high vantage point, Rome seems to consist of an endless sea of tiled roofs, above which rise numerous white, ochre, and gray domes of various sizes; in the distance, the urban perimeter is marked by highrises fading in the urban haze. This historic city lives amid its past as perhaps no other as busy traffic encircles the Colosseum, the Forum, the Pantheon, and other legacies of Europe's greatest empire.

Founded about 3000 years ago at an island crossing point on the Tiber River about 15 miles (24 km) from the sea, Rome had a high, defensible site. A millennium later, with a population some scholars estimate as high as 1 million, it was the capital of a Roman domain that extended from Britain to the head of the Persian Gulf and from the shores of the Black Sea to North Africa. Rome's emperors endowed the city with magnificent, marble-faced, columned public buildings, baths, stadiums, obelisks, arches, and statuary; when Rome became a Christian city, the domes of churches and chapels added to its luster.

It is almost inconceivable that such a city could collapse, but that is what happened after the center of Roman power shifted eastward to Constantinople (now Istanbul). By the end of the sixth century, Rome probably had fewer than 50,000 inhabitants, and in the thirteenth, a mere 30,000. Papal rule and a Renaissance revival lay ahead, but in 1870, when Rome became the capital of newly united Italy, it still had a population of only 200,000.

Now began a growth cycle that eclipsed all previous records. As Italy's political, religious, and cultural focus (though not an industrial center to match), Rome grew to 1 million by 1930, to 2 million by 1960, and subsequently to 2.7 million where it has leveled off today. The religious enclave of Vatican City, Roman Catholicism's headquarters, makes Rome a twin capital; the Vatican functions as an independent entity and has a global influence that Italy cannot equal.

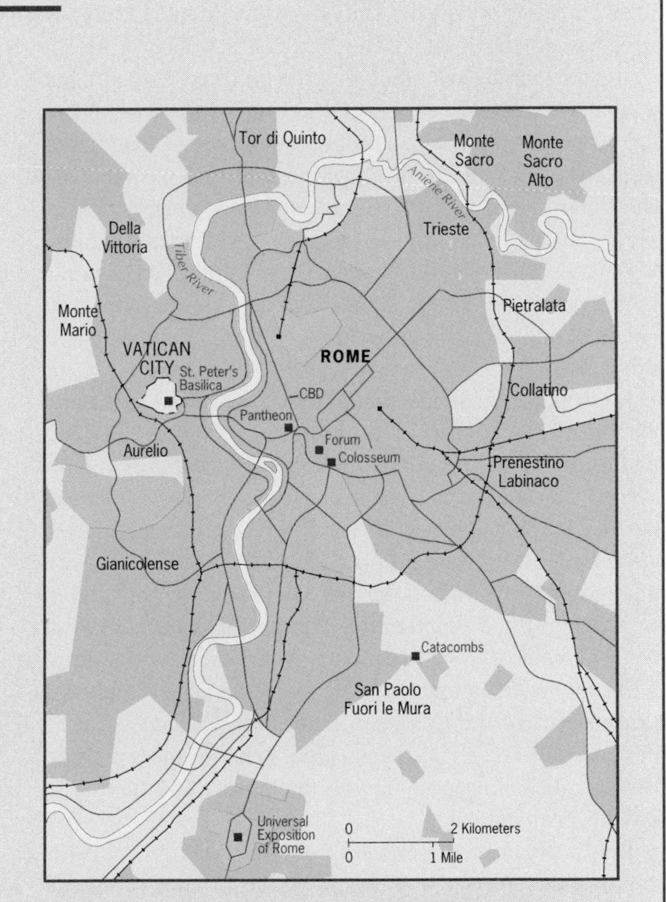

Rome today remains a city whose economy is dominated by service industries: national and local government, finance and banking, insurance, retailing, and tourism employ three-quarters of the labor force. The new city sprawls far beyond the old, walled, traffic-choked center where the Roman past and the Italian future come face to face.

Several of these regions have become powerful economic entities centered on major cities, such as Lombardy (Milan) and Piedmont (Turin); others are historic hearths of Italian culture, including Tuscany (Florence) and Veneto (Venice). These regions in the northern half of Italy stand in strong social, economic, and political contrast to such southern regions as Calabria (the "toe" of the Italian "boot") and Italy's two major Mediterranean islands, Sicily and Sardinia. Not surprisingly, Italy is often described as two countries—a progressive north and a stagnant south, or *Mezzogiorno*.

North and south are bound by the ancient headquarters, Rome, which lies astride the narrow transition zone between Italy's contrasting halves. This zone is referred to as the *Ancona Line*, after the town on the Adriatic coast where the zone reaches the other side of the peninsula (Fig. 1-20). Whereas Rome remains Italy's capital and cultural focus, the functional core area of Italy has shifted northward into Lombardy in the basin of the Po River. Here lies Southern Europe's leading manufacturing complex, in which a large, skilled labor force and ample hydroelectric power from Alpine and Appennine slopes combine with a host of imported raw materials to produce a wide range of machinery and precision equipment. The Milan–Turin–Genoa triangle exports appliances, instruments, automobiles, ships, and many specialized products. Meanwhile, the Po Basin, lying on the margins of the region's dominant Mediterranean climatic regime, enjoys a more even pattern of rainfall distribution throughout the year, making it a productive agricultural zone as well.

Metropolitan Milan embodies the new, modern Italy. Not only is Milan (at 4.3 million) Italy's largest city and leading manufacturing center—making Lombardy one of Europe's Four Motors—but it also is the country's financial and service-industry headquarters. Privately owned factories here tend to be small operations, but they are so well managed and automated that they have a global reputation for efficiency and adaptability. Today, the Milan area, with just 9 percent of Italy's population, accounts for one-third of the country's national income. Indeed, northern Italy has become part of the European core. Italy's success can be read in Table I-2: its per-capita GNP is by far the highest in Mediterranean Europe and rivals that of several countries in the European core.

Italy's overall economic progress is all the more remarkable because the south shows few signs of improvement. (Note, in Fig. 1-20, the much lower GDP of regions lying southeast of the Ancona Line.) The highly advanced industrial north stands in ever sharper contrast to the stagnant, mainly agricultural south, where outdated state-owned enterprises still operate. As Italy's north-south regional disparity grows, social and political problems associated with it increase. Taxpayers in the north object to having their money poured into the "bottomless pit" of the Mezzogiorno. Worse, extremist political reaction in the north has even led to a secessionist movement hoping to make northern Italy an independent country to be called *Padania*. Many northerners see the south as refusing to modernize, unwilling to rid itself of organized crime clans that control village and countryside, and unable to seize the opportunities that have transformed other once-poor parts of Europe.

But Italy's south faces some particular problems stemming from its relative location. Northerners complain of the "Africanization" of the Mezzogiorno as the south forms a conduit from North Africa to the jobs of the European core; but the south does not have the means to stem this tide. During the continuing disintegration of the former Yugoslavia, thousands of refugees sailed across the Adriatic Sea, most of them to southern Italy. More recently the struggle in Kosovo and the economic collapse in Albania generated further refugee movements, straining relief facilities in the very part of Italy least able to handle them.

Is there any hope for the Mezzogiorno? In 1998, Italy announced the discovery of what was described as one of the largest oil reserves found anywhere in the world during the 1990s—in the Agri River Valley of Basilicata (Fig. 1-20). The oil was reported to be of high quality, its volume huge even by North Sea standards. Plans are to drill wells and lay a pipeline to the coast. But to what extent will a privately owned, extractive, highly automated industry help alleviate Basilicata's 30 percent unemployment or reduce the region's poverty? Even an energy bonanza may not be enough to lift the Mezzogiorno out of its dormancy.

Spain and Portugal

At the western end of Southern Europe lies the Iberian Peninsula, separated from France and Western Europe by the rugged Pyrenees and from North Africa by the narrow Strait of Gibraltar (Fig. 1-19). Spain (population: 39.5 million) occupies most of this compact Mediterranean landmass. Portugal lies in its southwestern corner.

Iberia was a cornerstone of the Roman Empire; the Romans founded ports, built roads, and planted crops, including grapevines and olive trees. Later the Muslim Moors invaded and colonized much of the peninsula from North Africa, endowing some cities with magnificent architecture. In the aftermath of the defeat and ouster of the Muslims in 1492, a succession of kingdoms arose on the peninsula. One of these was Portugal, which was founded as a separate state in the twelfth century. Iberia has been divided ever since (except for a brief period between 1580 and 1640), despite the cultural bonds between the Spanish and Portuguese. From Spain and Portugal sailed colonizers of much of the world from the Philippines to Brazil, but in Europe the two countries fell behind after colonialism's golden age. Persistent dictatorial rule and pervasive economic stagnation cast doubt on their future in modern Europe.

Today, however, both countries are democracies, and their economies—especially Spain's—are thriving. As Figure 1-21 shows, Spain followed the leads of Germany and

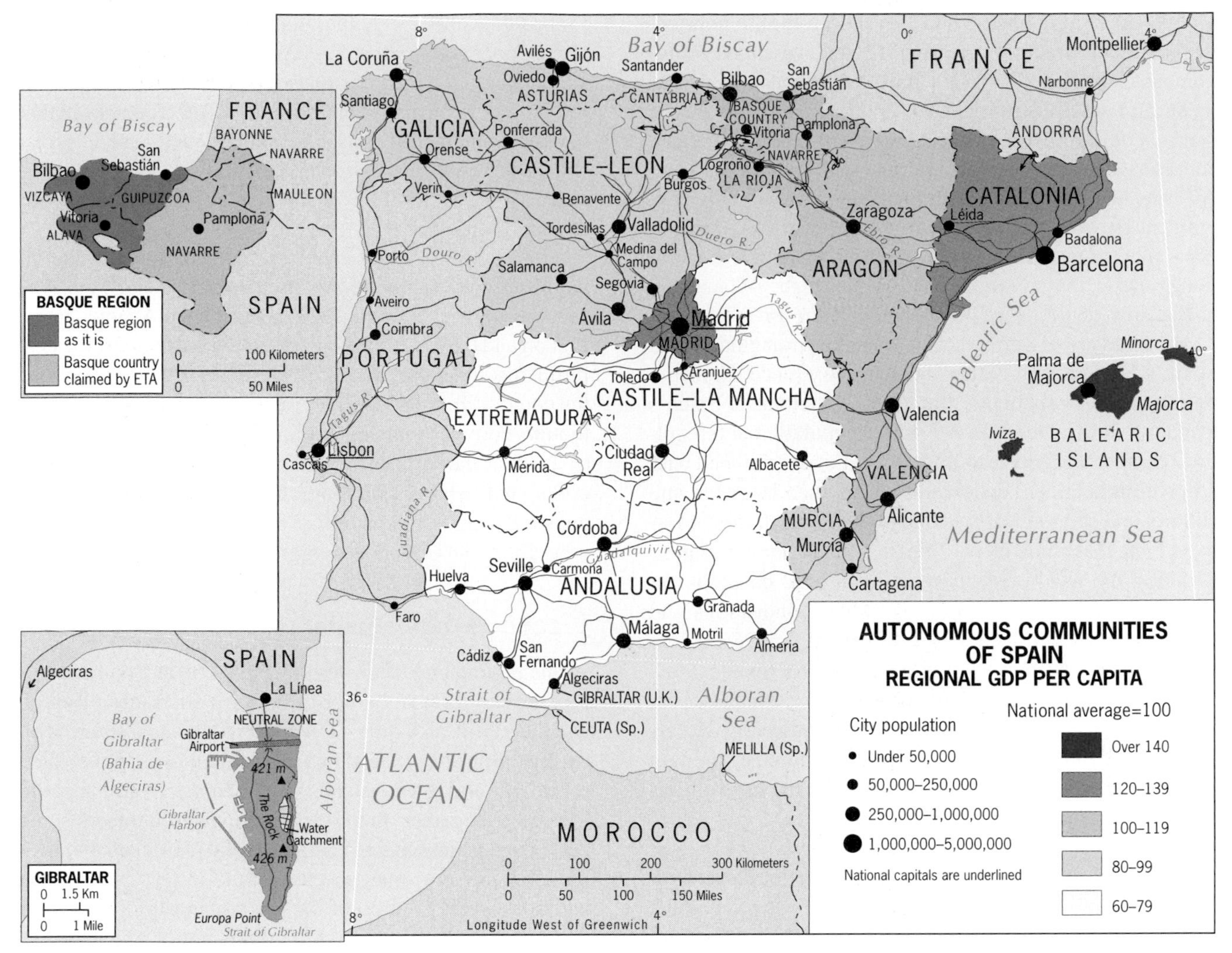

FIGURE 1-21

FROM THE FIELD NOTES

"Every major Spanish city has its *plaza*, a central square flanked by its most impressive public and religious buildings. They tend to look the same, but the details tend to have local or regional significance. This one, in the city of La Coruña in the northwestern corner of Spain, displayed a feature I had seen along many streets: to the classic architecture had been added small, white-framed and windowed balconies called *miradores*. I assumed that this innovation had some environmental cause, and when I asked about it I was told that La Coruña is the windiest city in Spain, and that the specially glazed windows were installed to cope with this. High-velocity winds, of course, are a feature of Mediterranean climates; as the map shows, La Coruña lies especially exposed. 'But please realize,' said the local teacher I had asked about this, 'that there's more to La Coruña than interesting baroque architecture and *miradores*. This was a Roman port, then held by the Moors, then the Portuguese, now Madrid.' Now Madrid? 'Yes, this is Galicia, not just another division of Spain, and we're not only closer to Western Europe than any other part of Iberia, but we're on the direct route between Western Europe and eastern South America. We want more autonomy, more identity.' Here we go again, I thought as I took my notes. Identity. Autonomy. Devolution."

France and reorganized and decentralized its politico-geographical structure, creating 17 regions called Autonomous Communities (ACs). Each AC has its own parliament and administration that control planning, public works, cultural affairs, education, environmental matters, and even, to some extent, international commerce. Each AC can negotiate its own degree of autonomy with the government in Madrid. This new system, however, has not been enough to defuse Spain's most problematic devolutionary issue, that involving the Basques in the AC mapped as Basque Country (Fig. 1-21, especially the top inset map).

Especially noteworthy is Catalonia, the triangular AC in Spain's northeastern corner, adjacent to France. Centered on prosperous, productive Barcelona, Catalonia is Spain's leading industrial area, an AC with a population of 6.5 million (comparable in size to Switzerland's). It is imbued with a fierce nationalism, endowed with its own language and culture, and economically is one of the Four Motors of Europe. Note that while most of Spain's industrial raw materials lie in the northwest, its major industrial development has taken place in the northeast, where innovations and skills propel a high-technology-driven regional economy. In recent years, Catalonia—with 6 percent of Spain's territory and 17 percent of its population—has annually produced 25 percent of all Spanish exports and nearly 40 percent of its industrial exports. Such economic strength translates into political power, and in Spain the issue of Catalonian separatism is never far from the surface.

As Figure 1-21 indicates, Spain's capital and largest city, Madrid, lies near the geographic center of the state. It also lies along an economic-geographic divide. Catalonia and Madrid are Spain's most prosperous ACs; the contiguous group of five ACs between the Basque Country and Valencia rank next. Tourism (notably along the Mediterranean coast) and winegrowing (especially in La Rioja) contribute importantly here. Ranking below this cluster of ACs are the four such units of the northwest: Galicia, Castile-Leon, and industrialized Asturias and Cantabria. Incomes in these regions are below the national average because industrial obsolescence, dwindling raw material sources, and emigration have plagued their economies. Worst off, however, are the three large ACs to the south of Madrid: Extremadura, Castile-La Mancha, and especially Andalusia. Drought, inadequate land reform, scarce resources, and remoteness from Spain's fast-growing northeast are among the factors that inhibit development here. Low local wages have encouraged some companies to set up factories in Seville, Córdoba, and Málaga, but so far the hopes raised by Spain's participation in the European Union have not materialized in these ACs. In the mid-1990s, a spurt of growth was followed by a slowdown that has intensified the regional disparities that weaken the fabric of the state. Even as EU members seek to reduce the differences among themselves, individual countries face the geographic consequences of focused growth that deepens internal divisions.

The state of Portugal (population: 10 million), a comparatively poor country that has benefited enormously from its admission to the EU, occupies the southwestern corner of the Iberian Peninsula. One rule of EU membership is that the richer members assist the poorer ones, and Portugal shows the results in a massive renovation project in the capital, Lisbon, as well as in the modernization of surface transport routes.

Unlike Spain, which has major population clusters on its interior plateau as well as its coastal lowlands, the Portuguese are concentrated along and near the Atlantic coast. Lisbon and the second city, Porto, are coastal cities; the best farmlands lie in the moister western and northern zones of the country. But the farms are small and inefficient, and although Portugal remains dominantly rural, it must import as much as half of its foodstuffs. Exporting textiles, wines, corks, and fish, and running up an annual deficit, the indebted Portuguese economy remains a far cry from those of other European countries of similar dimensions.

Greece and Cyprus

At the eastern end of the region we define as Southern Europe lies Greece, an outlier of both Mediterranean Europe and the European Union. Greece's land boundaries are with Turkey, Bulgaria, Macedonia, and Albania; as Figure 1-19 reveals, it also owns islands just offshore from mainland Turkey. Altogether, the Greek archipelago numbers some 2000 islands ranging in size from Crete (3218 square miles [8335 sq km]) to small specks of land in the Cyclades. In addition, Greeks represent the great majority on the now-divided island of Cyprus, the southern part of which aspires to EU membership.

Ancient Greece was a cradle of Western civilization; later it was absorbed by the expanding Roman Empire. For some 350 years beginning in the mid-fifteenth century, Greece was under the sway of the Ottoman Turks. Greece regained independence in 1827, but not until nearly a century later, through a series of Balkan wars, did it acquire its present boundaries. During World War II, Nazi Germany occupied and ravaged the country, and in the postwar period the Greeks have quarreled with the Turks, the Albanians, and the newly independent Macedonians. This remains a volatile part of the world.

Modern Greece is a nation of 10.6 million centered on historic Athens, one of the realm's great cities. With its port of Piraeus, metropolitan Athens contains about 40 percent of the Greek population, making it one of Europe's most congested and polluted urban areas. Athens is the quintessential primate city; the monumental architecture of ancient Greece still dominates its cultural landscape. The Acropolis and other prominent landmarks attract a steady stream of visitors; tourism is one of Greece's leading sources of foreign revenues, and Athens is only the beginning of what the country has to offer.

Among the Realm's Great Cities . . .

Athens

Take a map of Greece. Draw a line to encompass its land area as well as all the islands. Now find the geographic center of this terrestrial and maritime territory, and you will find a major city nearby. That city is Athens, capital of modern Greece, culture hearth of ancient Greece, source of Western civilization.

Athens today is only part of a greater metropolis that sprawls across, and extends beyond, a mountain-encircled, arid basin that opens southward to the Bay of Phaleron, an inlet of the Aegean Sea. Here lies Piraeus, Greece's largest port, linked by rail and road to the adjacent capital. Other towns, from Keratsinion in the west to Agios Dimitrios in the east, form part of an urban area that not only lies at, but *is*, the heart of modern Greece. With a population of 3.1 million, most of the country's major industries, its densest transport network, and a multitude of service functions ranging from government to tourism, Greater Athens is a well-defined national core area.

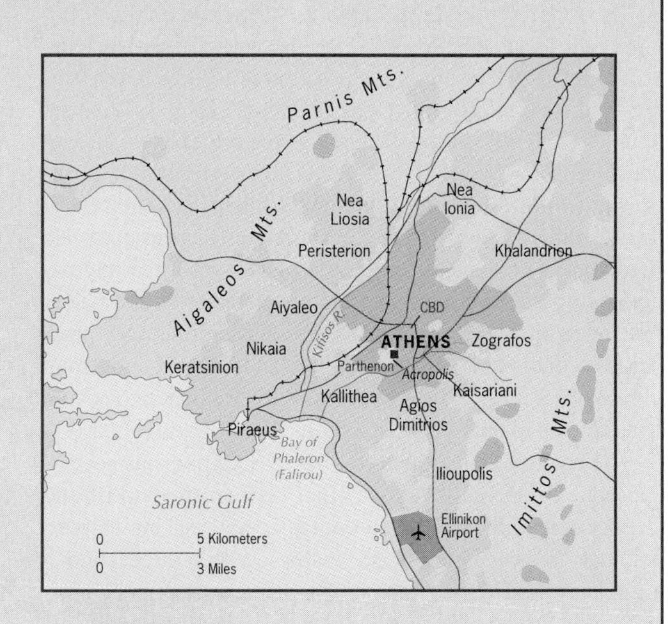

Arrive at Ellinikon Airport south of Athens and drive into the city, and you can navigate on the famed Acropolis, the 500-foot (150-m) high hill that was ancient Athens' sanctuary and citadel, crowned by one of humanity's greatest historic treasures, the Parthenon, temple to the goddess Athena. It is only one among many relics of the grandeur of Athens 25 centuries ago, now engulfed by mainly low-rise urbanization and all too often obscured by smog generated by factories and vehicles and trapped in the basin. Conservationists decry the damage air pollution has inflicted on the city's remaining historic structures, but in truth the greatest destruction was wrought by the Ottoman Turks, who plundered the country during their occupation of it, and the British and Germans, who carted away priceless treasures to museums and markets in Western Europe.

Travel westward from the Middle East or Turkey, and Athens will present itself as the first European city on your route. Travel eastward from the heart of Europe, however, and your impression of its cultural landscape may differ: behind the modern avenues and shopping streets are bazaar-like alleys and markets where the atmosphere is unlike that of any other in this geographic realm. And beyond the margins of Greater Athens lies a Greece that lags far behind this bustling, productive coreland.

Greece's limestone-dominated topography, under Mediterranean climatic conditions, has been sculpted into stark, often barren scenery that forms a dramatic backdrop for the whitewashed towns and villages that dot the countryside. In ancient Greece, extensive forests draped the landscape, but centuries of exploitation have denuded all the Mediterranean countries. None of these conditions makes farming easier because soils have been washed away and arable land amounts to less than 30 percent of the total area. Still strongly agrarian, Greece is self-sufficient in staple foods and imports only livestock products; Mediterranean fruits (grapes, olives, figs, citrus) go to EU markets. Although EU membership has brought many benefits, Greece remains the poorest EU country in terms of GNP and suffers from a huge and persistent trade deficit. Relations with the EU have at times been troubled by the government's failure to act on the economic reforms and loan conditions required by EU membership, but the EU recognizes Greece's importance as a stabilizing force in a fractious part of the realm.

Cyprus lies in the far northeast corner of the Mediterranean Sea, much closer to Turkey than to Greece (Fig. 1-12), but is peopled dominantly by Greeks rather than Turks. In 1571 the Turks conquered Cyprus, then ruled by Venice, and controlled it until 1878 when the British took over. But when the British were ready to give the island independence after World War II, the 80-percent Greek majority preferred union with Greece. The 1950s were years of much ethnic violence; Greek and Turkish communities lay dispersed throughout the island, and friction abounded. Still, in 1960 the British granted Cyprus independence under a complicated constitution that prescribed majority rule but guaranteed minority rights.

This fragile order broke down in 1974, and civil war engulfed the island. Turkey sent in troops and massive dislocation followed, resulting in the partition of Cyprus into northern Turkish and southern Greek sectors (Fig. 1-19, inset map). In 1983 the 40 percent of Cyprus under Turkish control, with about 100,000 inhabitants (and some 30,000 Turkish soldiers), declared itself the independent Turkish Republic of Northern Cyprus. Only Turkey recognizes this ministate (which now contains a population of 180,000); the international community recognizes the government on the Greek side as legitimate. With 875,000 residents, a relatively prosperous economy based on agriculture and tourism, and strong links with Europe, the Greek-side government is seeking membership in the European Union.

The potential for serious conflict over Cyprus has not disappeared. In effect, the "Green Line" between the Turkish and Greek communities constitutes not just a regional border but a boundary between geographic realms. With a large Turkish army in the north, the south has sought security through arms purchases, which provokes Turkish threats of preemptive action. As a result of partition, two ancient adversaries confront each other in dangerous times.

EASTERN EUROPE

As Figure 1-12 shows, Eastern Europe is not only territorially the largest region in the European realm: it also contains more countries (17) than any other European region. Almost all of Eastern Europe lies outside the core, and the problems of the periphery affect many of its countries. From the North European Lowland in Poland to the rugged highlands of the south, this is a region of physiographic, cultural, and political fragmentation. Open plains, major rivers, strategic mountains, isolated valleys, and crucial corridors all have influenced Eastern Europe's tumultuous migrations, epic battles, foreign invasions, and imperial episodes. Illyrians, Slavs, Turks, Hungarians, and other peoples converged on this region from near and far. Ethnic and cultural differences have kept them in chronic conflict.

Geographers call this region a **shatter belt**, a zone of persistent splintering and fracturing. Geographic terminology uses several expressions to describe the breakup of established order, and these tend to have their roots in this part of the world. One of them is **balkanization**. The southern half of Eastern Europe is referred to as the Balkans or Balkan Peninsula, after the name of a mountain range in Bulgaria. Balkanization denotes the recurrent division and fragmentation of this part of Eastern Europe, and it is now applied to any place where such processes take place. A more recent term is *ethnic cleansing*—the forcible ouster of entire populations from their homelands by a stronger power bent on taking their territories. The term may be new, but the process is as old as Eastern Europe itself.

Each episode in the historical geography of Eastern Europe has left its legacy in the cultural landscape. Twenty centuries ago the Roman Empire ruled much of it (Romania is a cartographic reminder of this period); during the past half-century, the Soviet Empire controlled almost all of it. In the intervening two millennia, Christian Orthodox church doctrines diffused from the southeast, and Roman Catholicism advanced from the northwest. Turkish (Ottoman) Muslims invaded and created an empire that reached the environs of Vienna. By the time the Austro-Hungarian Empire ousted the Turks, millions of Eastern Europeans had been converted to Islam. Albania today remains a dominantly Muslim country. In the twentieth century, Eastern Europe became a battleground between superpowers, and the complicated map reflects the results through 1991 (Fig. 1-22).

The subsequent collapse of the Soviet Union freed several of Russia's neighbors, and with only one exception—Belarus—these countries turned their gaze from Moscow to the west, specifically to the European Union and its economic promise. This changed the map of Eastern Europe, repositioning the realm boundary eastward and adding five countries to the region (Fig. 1-23): Latvia, Lithuania, Belarus, Moldova, and Ukraine. Meanwhile, however, two of Eastern Europe's established states fell apart: Czechoslovakia peacefully, Yugoslavia violently.*

The Geographic Framework

As Figure 1-23 shows, Eastern Europe extends from the Baltic Sea in the north to the Black Sea and the Adriatic Sea in the south, and from the German border in the west to the Russian boundary in the east. To help us understand this complex region, we group its 17 countries by geographic subregion:

1. *Countries facing the Baltic Sea.* Poland is the dominant state in this group, which also includes Lithuania and Latvia. Relative location would suggest that Belarus belongs here too, but Belarus has sought reunification with

*Between 1919 and 1991, the name *Yugoslavia* referred to the multicultural state shown in Figure 1-22. When civil war destroyed the old order, parts of Yugoslavia became independent states, including Slovenia, Croatia, Bosnia, and Macedonia. The largest component, dominated by the Serbs, kept the name Yugoslavia. In this chapter we adhere to this usage. In addition to Serbia, the new Yugoslavia also includes Montenegro, Vojvodina, and Kosovo (Fig. 1-23). Following the revolutionary end of the Milosevic regime in the autumn of 2000, the new president of Yugoslavia stated that he was open to suggestions in regard to a name change for his country, but as of Spring 2001 the name Yugoslavia still prevailed.

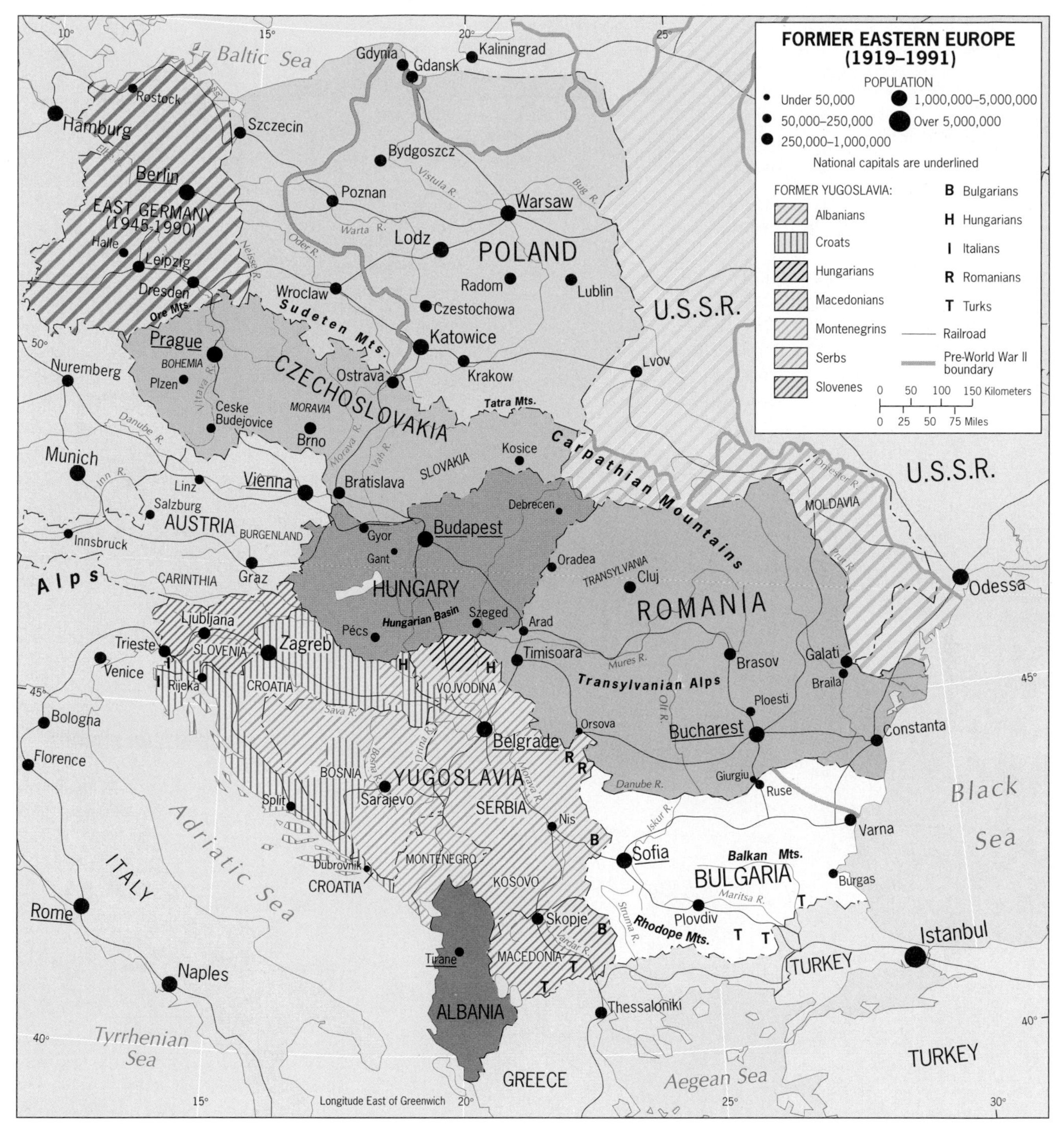

FIGURE 1-22

Russia. Should reunification occur, the current realm boundary shown on Figure 1-23 will move westward.

2. *The landlocked center.* These are the countries at the heart of the old Eastern Europe: the Czech Republic (westernmost and most Western of the region's states), Slovakia, and Hungary.

3. *Countries facing the Black Sea.* This cluster of countries consists of Romania and Bulgaria of the old Eastern Europe, and Moldova and dominant Ukraine (largest territorial state in Europe) of the new.

4. *Countries facing the Adriatic Sea.* Until a decade ago, this grouping would have involved two countries: the former Yugoslavia and Albania. Today there are six, including Albania, Slovenia, Croatia, Bosnia, Macedonia, and the new Yugoslavia; there will be seven if Kosovo's separate status becomes permanent.

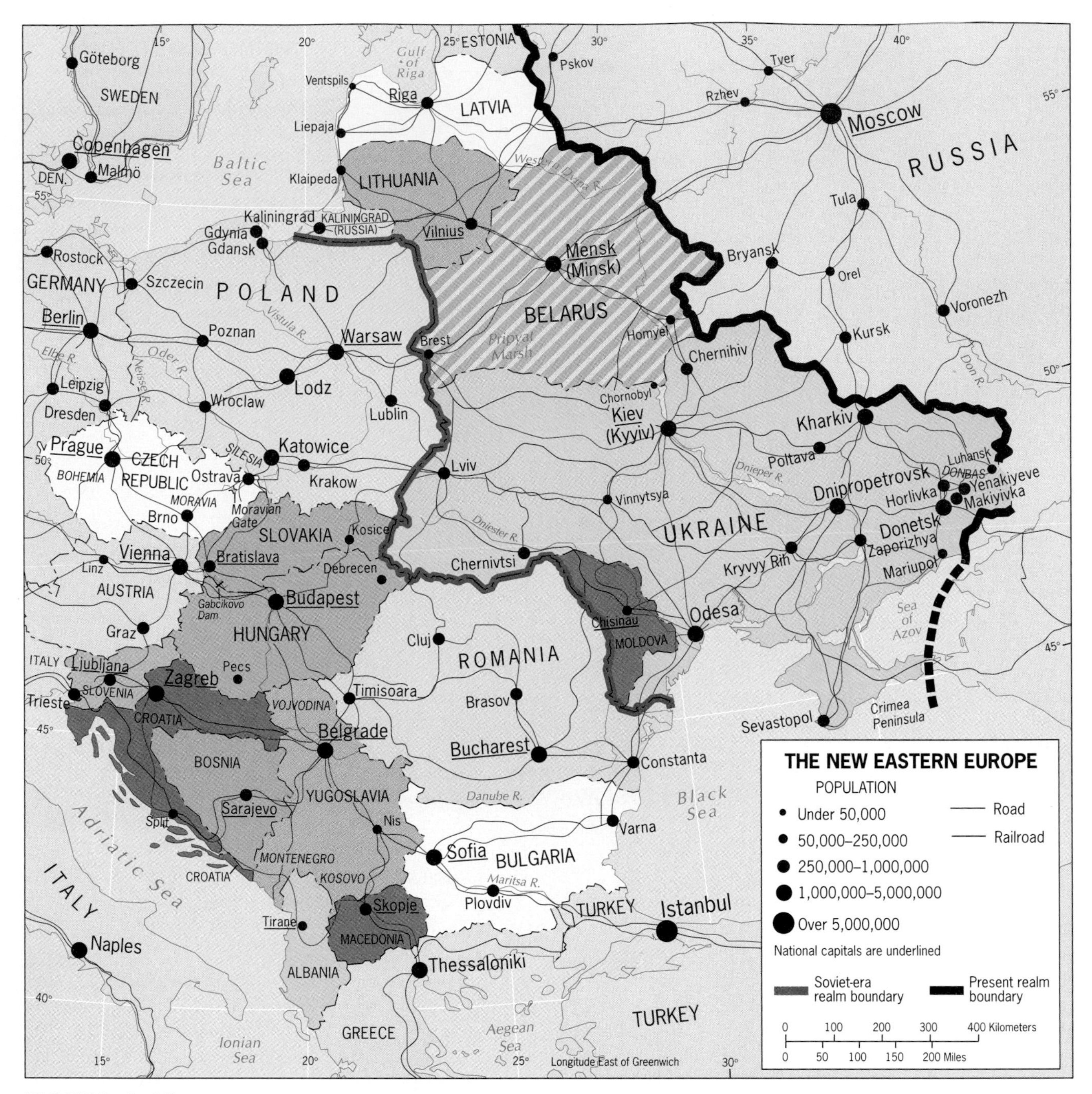

FIGURE 1-23

Countries Facing the Baltic Sea

Poland dominates the northwestern corner of Eastern Europe. Situated on the North European Lowland, Poland has traditionally been an agrarian country, but during the communist period major industrialization occurred in the south. Silesia became Poland's industrial heartland, and Katowice, Wroclaw, and Krakow came to symbolize that new industrial age. But unchecked environmental degradation including air, stream, and groundwater pollution accompanied this transformation of the south. And when the Soviet Union collapsed, Poland was left with outdated, inefficient factories.

Poland's best farmlands also lie in the south, where farming is intensive and wheat is the leading crop. In central Poland, thin and often rocky soils that developed on glacial deposits support rye and potatoes; farther north, the Baltic rim has such poor soils that cultivation gives way to pastureland and moors. Farming in Poland suffered severely during the communist period. The communist planners attempted to collectivize all farms but without making investments comparable to those made in state-owned fac-

FROM THE FIELD NOTES
"Turbulent history and prosperous past are etched on the cultural landscapes of Poland's port, Gdynia, and the old Hanseatic city of Gdansk nearby. Despite major wartime destruction, much of the old architecture survives, and we found restoration underway throughout these twin cities. Gdansk was the stage for the 1980s rebellion of the Solidarity labor union against Poland's communist regime; today city and country struggle through the difficult transition to a new economic and social order. Attracting foreign visitors is one way to raise revenues, and the historic old city on the Baltic is becoming a tourist draw."

tories. This created a severe economic-geographic imbalance, so that Polish farming today has many problems. Farmers still work with tools and equipment that in any Western European country would be housed in an historical museum.

The historic capital and primate city, Warsaw, lies at the head of navigation on the Vistula River in a productive agricultural hinterland. As Figure 1-23 shows, Warsaw lies at the hub of a radiating network of transport routes that reach all parts of the country. Water, rail, and road link the capital to the port of Gdansk on the Baltic Sea coast. The port and commercial center of Szczecin is the Baltic outlet for the northwest of the country.

In their struggle to build a democratic and economically healthy state, the 38.6 million Poles have several advantages. Poland is a nation-state with only minuscule minorities. One religion (Roman Catholicism) prevails overwhelmingly; one language unites Poland. There are major problems, however: the communist legacy, a weak infrastructure, and rural backwardness. Poland's new era is evident in its prospering cities, where modern manufacturing, commercial activity, and service industries thrive. But its urban-rural contrasts and strong regional disparities form long-term challenges.

Lithuania (population: 3.7 million) borders Poland to the north, the Russian exclave of Kaliningrad separating the two along the Baltic coast (Fig. 1-23). The Grand Duchy of Lithuania once dominated Eastern Europe from the Baltic to the Black Sea, a vast empire later reduced by Russian, German, and Polish encroachment. The Soviets annexed Lithuania in 1940, and in 1945, at the Potsdam Conference at the end of World War II, Moscow's hegemony was confirmed—the southwestern corner centered on Königsberg was awarded to Moscow outright (see box titled "Kaliningrad: Russia on the Baltic"). This left Lithuania with a mere 50 miles (80 km) of coastline and no real outlet: the small port of Klaipeda was not even connected by rail to the interior. As a Soviet "republic," Lithuania looked inward, toward Moscow, not to the Baltic Sea and beyond. Hence the capital, Vilnius, is situated near the Belarus border rather than on the coast.

Independence in 1991 produced a difficult transition period and tense relations with Russia, but by the turn of the millennium Russia once again was Lithuania's leading trade partner, providing fuels and buying textiles and farm products. A boundary dispute with Russia was settled by treaty in 1997, and Lithuania still held a trump card: overland access into Russia's Kaliningrad exclave. Ties with Poland also improved in the late 1990s. At the Helsinki summit in 1999, Lithuania was invited to begin preliminary discussions with the European Commission on prospective EU membership.

Latvia, Eastern Europe's northernmost state, has just 2.4 million inhabitants, of whom about one-third are Russians, a result of 50 years of Soviet control (1940 to 1991). Indeed, Latvians make up only about 54 percent of the population, and the Latvian government made it difficult for Russians to become Latvian citizens. This worsened relations between Moscow and Riga until the policy was eased in 1999.

Economic problems plague Latvia as well. Without significant resources other than its good soils and forests, Latvia was industrialized by the Soviets to produce specialized products such as television sets, refrigerators, and railroad cars; now the Latvian government must try to sell these often obsolete factories to private companies. Production has plummeted, but Latvia has few options. One trump card is the oil pipeline from Russia to the port of Ventspils, the major northern route for Russian oil exports. Riga, the capital, also has become a banking center for Russians

Kaliningrad: Russia on the Baltic

29

The Russian **exclave** of Kaliningrad lies wedged between Lithuania and Poland, facing the Baltic Sea through its gigantic naval port (Fig. 1-23). Soon after the Soviets acquired this German base in 1945, virtually all ethnic Germans were expelled. Not much was left of their cultural landscape: relentless British bombing had devastated the place.

Russians replaced the departed Germans, and today Kaliningrad's population approaches 1 million people, 90 percent of whom are Russian. In the Soviet political scheme of things, Kaliningrad had the status of an oblast (see Chapter 2), and it was made part of the Russian Republic—as it remains today.

Uncertainty now prevails in Kaliningrad. Independence is not an issue, but that could change. Despite the almost total destruction of seven centuries of German heritage, the German past still can be felt: the renowned philosopher Immanuel Kant (1724–1804) lived and worked here and lies buried in the ruins of the German cathedral. Even the city's German name, Königsberg, appears now and then in advertising (e.g., for German tourists).

Kaliningrad's post-Soviet years have been difficult. Rumors flew—for example, that the exclave would be used to resettle ethnic Germans from Central Asia, that Russia's president had offered it for sale to Germany. Locals hoped to establish a free-trade area, envisioning Kaliningrad as a future Hong Kong of the Baltic. Rising nationalism and militarism in Russia put an end to all these aspirations. In 2001, Kaliningrad was reverting to its old role as a crucial military base for a country with few warm-water ports—a naval bulwark, potentially, in an Eastern Europe into which Western military alliances, chiefly NATO, are spreading.

seeking a safe haven for their money. All this has rekindled Russian influence and has brought with it organized crime. Although Latvia is in pre-accession discussions with the European Commission, its prospects for membership remain remote.

As Figure 1-23 shows, Belarus borders all three of the other states in this subregion, but Belarus has no coast. Centered on the capital and largest city, Mensk (Minsk), Belarus has 9.9 million citizens, of whom about 80 percent are Belarussians ("White" Russians). Like the Poles and the Slovaks, the Belarussians are a West Slavic people. Only 13 percent are (East Slavic) Russians.

During World War II German armies inflicted terrible damage on Belarus, killing one-quarter of the entire population. In the postwar period, Belarus was one of Moscow's most loyal satellites, although its economy was always weak: its domestic resource base is limited to peat (for fuel), potash (for fertilizers), and some oil. The Soviets made Belarus a major machine-fabricating center, and Mensk a large industrial city producing tractors, trucks, and machine tools.

Belarus's relative location may be its greatest asset. The Druzba oil pipeline and the Northern Lights gas pipeline cross this country on their way to other parts of Eastern Europe. Belarussian territory reaches to within 65 miles (105 km) of Russia's Kaliningrad exclave, raising the possibility of a corridor from Moscow to this window on the relatively warm Baltic. Lately, Belarus's leaders have been pressing the Russians for closer association, even a formal union between Moscow and Mensk. Because of its own economic and political problems, Russia has not responded eagerly—but the map suggests that Belarus may yet become attractive to a less introverted Russian state. Belarus is the sole Eastern European country whose orientation is to the east, not the west.

The Landlocked Center

The Czech Republic and Hungary, two of the three countries in this subregion, are cornerstones of the new Eastern Europe: they are members of NATO and in the process of qualifying for European Union membership. Both countries have overcome the disadvantages of their relative locations to rise to the top of Eastern Europe's hierarchy.

The Czech Republic is the western part of the former state of Czechoslovakia. In 1993 the Czechs and Slovaks decided to divide their country along dominant ethnic lines, resulting in a political "velvet divorce." The Czechs got the better part of the deal; Slovakia is this subregion's poorest country by far.

The Czech Republic centers on Bohemia, the mountain-enclosed core area that contains the historic capital, Prague. This province always has been cosmopolitan and Western in its exposure, outlook, development, and linkages. Prague lies in the basin of the Elbe River, Bohemia's traditional outlet through northern Germany to the North Sea. It is a classic primate city, its cultural landscape faithful to Czech traditions; but it also is an industrial center. The encircling mountains contain many valleys with small towns that specialize, Swiss-style, in fabricating high-quality goods. In Eastern Europe the Czechs always were the leaders in technology and engineering; even during the communist period their products found markets in foreign countries near and far.

As Figure 1-23 shows, the Czech Republic incorporates an eastern province named Moravia, linked to Poland's Silesia by the Moravian Gate, a gap between the Sudeten Mountains and the Carpathians. This gap forms a passage between the Danube Valley to the south and the North European Lowland. During the communist period Moravia became an important industrial zone centered on Ostrava and Brno, but today many of those inefficient state-supported industries are obsolete and uncompetitive. Reviving the economy of Moravia is a major challenge for the Czech government. In the late 1990s economic growth was slow, worsened by labor problems and buffeted by the floods of 1997, the worst of the twentieth century.

In the Czech Republic (population: 10.3 million), the main ethnic issue centers on several hundred thousand gypsies, or Roma, who have historically been victims of discrimination. Slovakia, on the other hand, has a major ethnic minority: about 11 percent of its 5.4 million inhabitants are Hungarians, most of them concentrated in a southern zone along the Danube River (Fig. 1-24). Slovak-Hungarian tensions have risen repeatedly since independence as the Slovak-dominated government has tried to restrict use of the Hungarian language and has disadvantaged the Hungarians in other ways. Prospects of applying for EU membership, however, impelled the government to soften its policies toward the Hungarians.

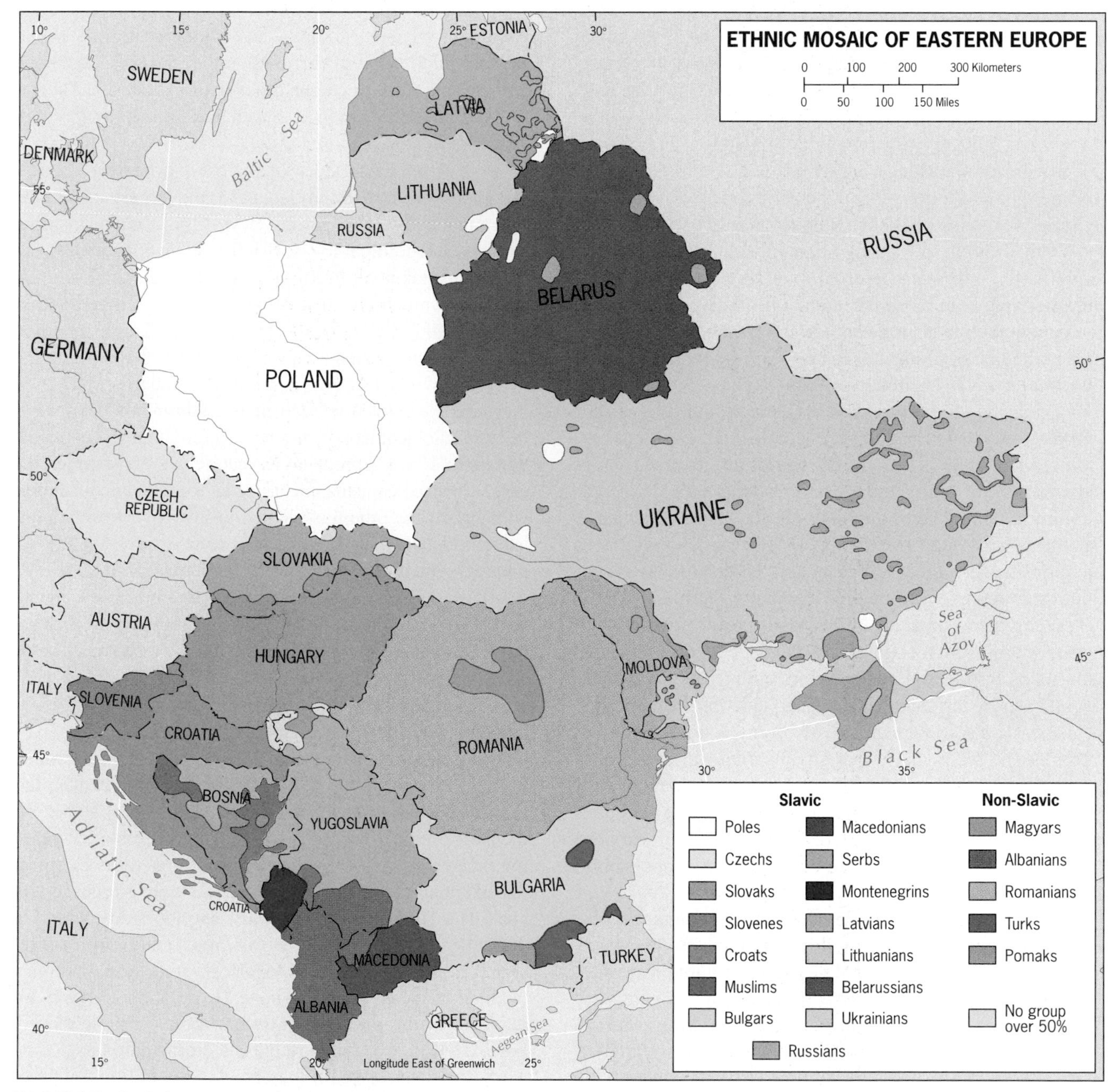

FIGURE 1-24

Slovakia negotiated its secession from the Czechs not only on ethnic and cultural grounds but also on economic bases. The Slovaks did not share the Czechs' determination to pursue rapid economic reforms. In their division of assets, the Slovaks got about one-third of the wealth. Bratislava, the Slovak regional headquarters, now became a national capital.

Slovakia always was the least developed, most rural part of the former Czechoslovakia; it was more cooperative with Soviet authorities and more closely tied to the Soviet economic system than Bohemia. In today's independent Slovakia (population: 5.4 million), vestiges of communist-style authoritarianism, corruption, and mismanagement still infect affairs of state.

To the south of Slovakia lies Hungary, in the middle basin of the Danube River, flanked by Austria to the west and Romania to the east. Hungary is a nation-state of just under 10 million, but Hungarians also live in Slovakia, Romania, Slovenia, Croatia, and Yugoslavia, all remnants of a time when Hungary ruled much of this region (Fig. 1-24). From time to time Hungarian governments express support for these ethnic cohorts in neighboring countries, and they, in turn, look to the mother country for reassurance.

When a nation shows supportive interest in cross-border 30 cohorts, the practice is referred to as **irredentism**, a term that derives from a similar situation involving Italy and Austria in the nineteenth century. Italy wanted to incorporate an Italian-speaking area in neighboring Austria, calling it *Italia Irredenta* (Unredeemed Italy). Today, with numerous minorities located across borders in all parts of the world, irredentism is a global geographic phenomenon.

Eastern Europe's great river, the Danube, touches many countries in the region but serves mostly as a divider, not a unifier. In its middle basin, however, the Danube formed a spatial bond for the Hungarians (Magyars) who migrated here more than a thousand years ago. Of distant Asian origin, the Magyars have neither Slavic nor Germanic roots. They converted their fertile lowland into a thriving state while retaining their cultural and linguistic identity, eventually rising to imperial power in the region. The twin-cities capital astride the Danube, Buda and Pest (better known as Budapest), is a primate city nearly 10 times the size of the next largest town in Hungary—reflecting the continuing rural character of much of the country. After Hungary was freed of Soviet domination in the early 1990s, agricultural reforms provoked strikes and roadblocks. Still, Hungary remains self-sufficient in basic foods and is the region's only major food exporter.

Hungary's transition to democracy and a market economy has been difficult, but political stability and democratic transfers of power (in 1994 the former communists were voted back into office under a socialist banner!) allowed Hungary to address its economic problems. Still a world-class exporter of bauxite (for aluminum production), Hungary also has a diminishing coal reserve used for part of its electricity production, oil and natural gas under the Great Hungarian Plain in the east, and uranium ore near Mecsek to supply its nuclear power plants. Other raw materials range from manganese to copper, but Hungary's future lies in the skills of its well-educated labor force. The old state-founded factories still produce iron and steel (from ores imported by barge up the Danube), but thousands of smaller companies now export precision equipment ranging from microwaves to television sets. In 2001, Hungary's prospects for joining the European Union were bright.

Countries Facing the Black Sea

Four countries form Eastern Europe's Black Sea quadrant: Ukraine, Romania, Moldova, and Bulgaria (Fig. 1-23). All except Moldova have coastlines on the Black Sea, but none of their core areas or capital cities lies on the coast. This reflects an inward orientation that characterizes the subregion as a whole.

Ukraine is Eastern Europe's most populous country (48.9 million); territorially, it is also the largest state in the entire realm. Its capital, Kiev (Kyyiv), is a major historic, cultural, and political focus. Briefly independent before the communist takeover in Russia, Ukraine regained its sovereignty as a much-changed country in 1991. Once a land of farmers tilling its famously fertile soils, Ukraine emerged from the Soviet period with a huge industrial complex in its east—and with a large (22 percent) Russian minority.

Ukraine's boundaries also changed during the Soviet era. In 1954 a Soviet dictator, as a reward for Ukraine's contribution to the USSR's productivity, capriciously transferred the entire Crimea Peninsula, including its Russian inhabitants, to Ukraine. At the time this seemed to involve a mere administrative detail; both Russia and Ukraine were Soviet "republics." But when the Soviet Union collapsed, Crimea's Russians found themselves under the Ukrainian government. After prolonged negotiations, Ukraine retained the Crimea but leased to Russia the rights to use two Crimean naval bases at Sevastopol and South Bay.

The Dnieper River forms a useful geographic reference to understand Ukraine's regional division (Fig. 1-23). To its west lies agrarian, rural, mainly Roman Catholic Ukraine; in its great bend and to the east lies industrial, urban, Russified (and Russian Orthodox) Ukraine. The economic dimension of this division had already evolved by the end of the nineteenth century, when coal mining in the east produced about 90 percent of the czarist empire's high-quality coal. The Donets Basin (*Donbas,* for short) lay less than 200 miles (320 km) from the vast high-quality iron ore reserves at Kryvyy Rih (Krivoy Rog), and the communist planners made eastern Ukraine a Soviet Ruhr of heavy industries. In the process, eastern Ukraine became closely linked to Russia: oil, natural gas, and raw materials flowed from Russia in return for massive exports of finished products. This arrangement created a dependence, especially on energy supplies, that costs Ukraine heavily today.

Moscow's intervention in Ukraine's economic geography generated Russian immigration to Donetsk, Dnipropetrovsk, and other Donbas-area cities; it also brought mismanagement, corruption, and profound environmental destruction. In 1986, Ukraine came to world attention when the Soviet-built nuclear power plant at Chornobyl (Chernobyl) caught fire, wafting deadly radiation as far away as Poland and Norway.

Unlike Poland, the Czech Republic, or Hungary, Ukraine has suffered from political instability and severe economic downturns during its first decade of post-Soviet independence. The national debt is huge (some countries have impounded Ukrainian ships in their ports to secure payments due); the economy has faltered; and social welfare is eroding (life expectancy has actually declined). Corruption and crime have reduced the flow of foreign investment. So discouraged are many Ukrainians that nearly half of them, in recent polls, have expressed a desire for reunion with Russia and a return to the "old days."

And yet Ukraine has assets that other Eastern European countries can only dream about: outlets to the sea and international shipping lanes; ports (including Odesa); massive farm production; experienced, educated, and skilled labor; a large domestic market; and considerable raw materials. But a legacy of dependence, dictatorial rule, and mismanagement has so far prevented Ukraine from fulfilling its potential.

As Figure 1-23 shows. Romania is Ukraine's large neighbor to the southwest, but a small country intervenes—Moldova. Barely (but purposefully) landlocked, Moldova was a Romanian province seized by the Soviets in 1940. In recognition of its Romanian majority, the Soviet regime made Moldova a separate "republic" of the USSR. Undoubtedly, no Soviet planner foresaw that this territory with its (now) 4.3 million people, agrarian economy, and small capital of Chisinau would some day become an independent country. Nor did the Russians (13 percent of the population) and Ukrainians (14 percent) who moved into Moldova, clustering in the strip between the Ukrainian border and the Dniester River, anticipate that they would come under a Romanian-dominated government. A secession movement in "Transdniestra" has simmered ever since independence; meanwhile, Moldova's economy has declined precipitously.

In the early 1990s, when the map of Eastern Europe was being redrawn, it seemed that Moldova might be reattached to Romania, but the Moldovans voted overwhelmingly against such a union. One reason was the prospect of independence and the fruits it might bear; another was Romania's wretched condition at the time. In Romania, communist totalitarianism had reached its destructive extreme in the person of dictator Nicolae Ceausescu, who squandered Romania's wealth, destroyed much of its cultural heritage, and plunged the state into a social abyss from which it will take generations to recover.

The world knew little about the horrific pollution of underground water supplies, streams, soils, and air during the Soviet Union's rapid industrialization. In the former communist colonies as well as Russia itself, this was environmental degradation on a huge scale. Since the end of communist rule, successful efforts have been made to mitigate the problem, but the task is massive—and help must come from countries that have their own record of pollution. This coal mine stands on a bend in the Mius River in Ukraine; the air smells foul, pollutants seep into the soil through piles of waste, and an ugly scum covers the water. Multiply this scene several thousand times, and you have an idea what the countries of the former Soviet Union confront.

Romania has a wide window on the Black Sea and a long border on the Danube River (although the capital, Bucharest, lies on neither and owes its location to its role in the pre-Romanian principality of Walachia). As Figure 1-1 shows, the country has a varied physiography. The resource-rich Carpathian Mountains sweep into Romania from the northwest in a broad arc, and the Transylvanian Alps define the lower Danube Basin. Along their southern flank lie the oil (now nearly exhausted) and gas reserves that long made Romania an energy exporter. The core area lies in the plains of the southeast; Romania's substantial Hungarian minority (2 million, or about 9 percent of Romania's population of 22.4 million) is concentrated in the central north (Fig. 1-24).

Romania's economic geography remains wretched, the result primarily of political infighting and instability, factionalism, corruption, and crime. The Ceausescu regime

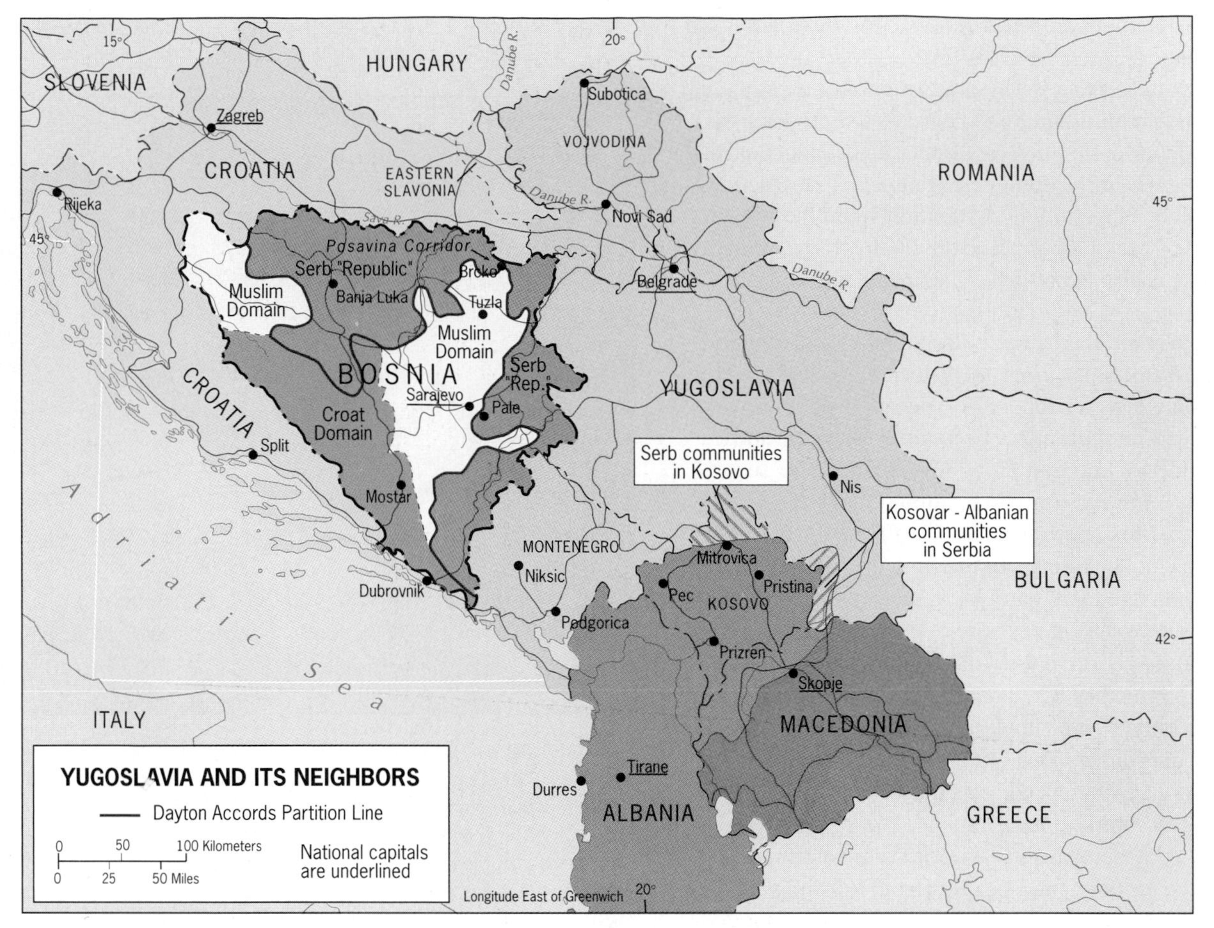

FIGURE 1-25

squandered the country's income on grandiose development and urbanization schemes; Western countries, seeing a hint of independence from Moscow in these programs, provided generous loans. When Ceausescu's schemes failed and the loans came due, Romania exported oil and food to pay the bills while Romanians went without adequate fuel and nourishment. Add to this the obsolescence of the heavy industries built by the state, and Romania's road to reform is longer than that of most other ex-communist countries. Privatization, one key to such reforms, has progressed more slowly here than almost anywhere else in Eastern Europe. Once known as Eastern Europe's most civilized society, its capital the Paris of the region, Romania today is the basket case of the Balkans.

This subregion's southernmost country is Bulgaria, bordered by Turkey, Greece, Macedonia, and Yugoslavia. About the size of Tennessee, Bulgaria is dominated by rugged topography except in the Danube lowland it shares with Romania in the north. The Balkan Mountains form Bulgaria's physiographic backbone and separate the Danube and Maritsa drainage basins. As the map shows, Bulgaria has a strategic location: near the entrance to the Black Sea, flanking the lower Danube River, and on key land routes between Europe and Southwest Asia.

A Bulgarian state did not appear in the Balkans until 1878, when the Russian czar's armies drove the Turks out of this area. The Slavic Bulgars, who form 86 percent of the population of 8.1 million, never forgot who their liberators were, and during the Soviet period they were compliant allies of Moscow. Meanwhile, they did not treat the remaining Turkish minority (9 percent) kindly. In 1974 the government prohibited use of the Turkish language in teaching and closed the country's 1300 mosques. In 1984 the Turks in Bulgaria were forced to give up their Turkish names and take Slavic ones. This discrimination led more than 300,000 Turks to emigrate to Turkey

The Soviets transformed Bulgaria from a peasant to an industrial society in less than a half-century, collectivizing agriculture and making the capital, Sofia, a manufacturing center. The economic picture reflects Moscow's lingering influence here: Russia remains Bulgaria's leading trade partner. But Bulgaria needs economic reform, which is proving especially difficult. The government has been unable to reach a consensus on crucial issues such as land ownership in the rural areas and the privatization of saleable industries. As a result (as Table I-2 indicates), Bulgaria remains one of Eastern Europe's least promising post-Soviet economies.

Countries Facing the Adriatic Sea

As recently as 1990, only two Eastern European countries fronted the Adriatic Sea: Yugoslavia and Albania (Fig. 1-22). Albania survives, but the former Yugoslavia has splintered into five countries—and its disintegration may not be over.

We turn first to the former Yugoslavia ("Land of the South Slavs"), a country extending from Austria to Greece, thrown together in 1918 after World War I, containing 7 major and 17 smaller ethnic and cultural groups (Fig. 1-24). The Slovenes and Croats in the north were Roman Catholics; the Serbs in the south adhered to the Serbian Orthodox church. Several million Muslims also formed part of the cultural mosaic. Two alphabets were in use in separate regions of the country. At first the Royal House of Serbia dominated Yugoslavia; after 1945, a communist dictatorship personified by World War II hero Marshal Tito held Yugoslavia together. But when the communist system collapsed in Eastern Europe and the Soviet Union disintegrated, so did the Yugoslav state.

Communist social planners laid the groundwork for the disaster that befell Yugoslavia. They divided the country into six internal "republics" based on the Soviet model, each dominated (except Bosnia) by one major group. Since all these "republics" inevitably incorporated minorities, the state guaranteed their rights, albeit through autocratic means.

When the communist system collapsed, individual "republics" proclaimed their independence—that is, the majorities in these entities did so. What remained of the machine of state, still dominated by the Serbs, tried to halt this disintegration. That effort soon failed, and new countries named Slovenia, Croatia, Bosnia, Macedonia, and, by subtraction, Serbia appeared on the map. Minorities in these new states, however, objected and, in Croatia and Bosnia, rose against those who were advocating statehood. The result was a catastrophic conflict just when grandiose Euro-unification schemes were under way. Europe, which twice during the twentieth century had plunged the world into war and had vowed that genocide would "never again" cloud its horizons, failed this first test of its declarations. EU members, led by Germany, quickly accorded official recognition to the post-Yugoslavian states, even before the concerns of the frightened minorities could be considered. Then, European powers stood by as more than 250,000 people were killed, perhaps a million more were injured, ethnic cleansing and mass executions depopulated entire areas, refugees streamed into neighboring countries, and historic treasures were demolished. Meanwhile, the conferees in Maastricht and Brussels brought out the champagne to herald the progress of the European Union.

As we will see, the devolutionary process continued into the twenty-first century. Today, the still-evolving map arising from the former Yugoslavia consists of five countries: Slovenia, Croatia, Bosnia, the new (Serbian) Yugoslavia, and Macedonia.

Slovenia was the first "republic" to secede and proclaim independence. Ethnically the most homogeneous but territorially small with only 2 million of the former Yugoslavia's 23 million citizens, Slovenia lay farthest from the Serbian power core and seized its opportunity. This Alpine-mountain state, with the most productive economy among the republics even before independence, today has this subregion's highest per-capita GNP by far. It is among the leading candidates to join the EU in this decade.

Croatia's crescent-shaped territory, with prongs along the Hungarian border and along the Adriatic coast, only partially reflects the distribution of Croats in the former Yugoslavia (Fig. 1-24). About 85 percent of Croatia's 4.6 million people are Croats, but another 800,000 live in a broad zone of southern Bosnia. Croatia's Serb minority, about 12 percent when independence came, has dwindled under Croatian pressure. Under the communists' planned economy, Croatia was the former Yugoslavia's second most prosperous republic (after Slovenia), but civil strife and dislocation did much damage. The former authoritarian, now more democratic leadership in the capital, Zagreb, has been unable as yet to achieve the reforms Croatia needs. Compared to Bosnia, however, Croatia's prospects are bright. During Croatia's brief fight for independence the Serb-led Yugoslav army bombed Zagreb, and when Serb minority areas in Croatia refused to acknowledge Croatian authority, they were forcibly occupied and many Serbs were ousted.

Bosnia, however, suffered a far greater loss of life. Multicultural, landlocked, and positioned between the Croatian state to the west and the powerful Serbian stronghold to the east, Bosnia fell victim to calamitous conflict. Among its 3.8 million people, nearly 50 percent were Muslims, 31 percent Serbs, and 17 percent Croats. The devastated capital, Sarajevo, is a holy place to Muslims and Orthodox Serbs alike and lies on the eastern margin of a wedge-shaped territory where Muslims remain in the majority (Fig. 1-25). The Muslim-dominated government declared independence in 1992, resulting in Serbian rebellion supported by neighboring Serbia.

The civil war that followed claimed an estimated 250,000 lives, dislocated 2 million people, caused a desperate refugee flow, and destroyed much of Bosnia's cultural heritage and infrastructure. In 1995, a U.S. diplomatic initiative resulted in a truce that partitioned Bosnia into two (effectively four) sectors (Fig. 1-25). Signed at Dayton, Ohio by all parties, the Dayton Accords required an international force to police it. To the west and south lies the so-called Muslim-Croat "Federation," and to the north and east lie two quite different parts of

Changes in government in Europe tend to result from well-organized, democratic elections—but not, in October 2000, in Yugoslavia. Elections had been held, but the loser, President Milosevic, refused to concede and schemed, initially with the help of party cadres and elements of the armed forces and police, to schedule a "runoff" vote later. This brought crowds of protesters into the streets and, in scenes not seen since revolutions overthrew dictators generations ago, they stormed the federal parliament building in Belgrade, drove out the occupants, and set fire to offices. Milosevic was forced to step down, and President Kostunica took office soon thereafter. A new day dawned for the Serbian people.

a Serb "Republic" called Srpska. The Serbs, with their original 31 percent of the population, were awarded 48 percent of Bosnia's territory, the Muslims 27 percent, and the Croats 17 percent. As the map shows, the two wings of the Serb "republic" are connected by a narrow link, the Posavina Corridor, just 3 miles wide in a zone that lies in the hinterland of the river port of Brcko. This area was Muslim-dominated before the civil war and witnessed some of the worst ethnic cleansing in the entire country; the Muslims have not forgotten this atrocity. Brcko remained unassigned under the Dayton Accords, but in 1999 it was placed under joint jurisdiction. Both the Serbs, who see this as a vital part of the link between their two Bosnian entities, and the Muslims, to whom the city offers hope for an outlet to the sea, regarded this decision as a defeat.

The geographic name Yugoslavia now refers to what is left of the domain once ruled or dominated by the Serbs: *Serbia*, centered on the old Serb capital of Belgrade on the Danube River; *Vojvodina* in the north, part of which was once Hungarian territory and where a Hungarian minority numbers over 350,000; *Montenegro* on the Adriatic coast, once strongly supportive of the Serbs but recently trying to distance itself from Belgrade; and *Kosovo* in the south, one of the historic Muslim strongholds in the subregion, now under NATO administration following NATO's ouster of the repressive Serb authorities.

With over 6 million inhabitants, an autocratic regime, and a still-potent armed force, Serbia dominates this four-part Yugoslavia (total population: 10.7 million), although it has lost control over Kosovo and faces problems in both Vojvodina and Montenegro. The ethnic Hungarian minority in Vojvodina, comprising one-fifth of the population there, have been better off than the Kosovo Muslims ever were, but some were reluctant to join the Yugoslav army during the campaign in Kosovo. In Hungary, meanwhile, conservative politicians have called for the redrawing of the Hungary-Yugoslav boundary to put the northern third of Vojvodina back in Hungary. Montenegro, with just 650,000 citizens, has sought to distance itself from Serbia, even adopting the German mark as its official currency while forging its own diplomatic and commercial ties with European countries. About one-third of Montenegrins are pro-Serb, and Yugoslavia maintains an armed force in Montenegro. Another crisis may be in the making here.

But the tragedy of Kosovo remained the most serious problem in the region in 2001. This Maryland-sized corner of Yugoslavia, with an Albanian, mainly Muslim majority among its 2 million people, once had considerable autonomy. Here, in 1389, the Serbs lost a decisive battle against advancing Muslim armies of Turkey, and for more than five centuries the Muslims ruled this area. When, in 1912, the Serbs finally recaptured Kosovo, they had come to view the Field of the Blackbirds near Pristina, where that 1389 battle took place, as hallowed ground. But the Serb population of Kosovo never exceeded 10 percent of the total, and Kosovo remained a Muslim society under Serb administration. It was an uneasy, but workable situation—until 1989, when the Serbian dictator Slobodan Milosevic revoked Kosovo's autonomy and began a campaign to suppress Kosovo's Muslim culture.

Kosovo was always a poor, rural territory with one of Europe's lowest standards of living, but Serbia's oppression had predictable effect. Soon armed resistance arose, and Serbs in Kosovo began to take casualties. Thousands of Serbs left Kosovo even as the first ethnic cleansing by Serbian forces began. This, in turn, produced a flood of refugees heading for neighboring Albania, Macedonia, and even that other autonomous Serbian province, Montenegro.

As Kosovo's towns and villages burned and Kosovars were killed in growing numbers, European and U.S. mediators sought to persuade the Serbs and the Kosovars to sign an agreement that would return to Kosovo its autonomy and would guarantee further negotiations three years hence. Although the Kosovars did sign such an agreement, the Serbs refused. This brought on the first military action by NATO forces since that organization's inception. In March 1999 a campaign of massive, punitive bombing commenced, destroying much of Serbia's military capability and severely damaging the country's infrastructure.

Meanwhile, however, Serbian forces continued to destroy people and property in Kosovo, driving hundreds of thousands of Kosovars out of their province and across its borders. By early April, this stream of refugees already exceeded 500,000, creating major humanitarian crises within and outside Kosovo. Thousands of civilians were killed by Serbian action. After the catastrophe in Bosnia this was the worst human tragedy in Europe since World War II, and again Europe had failed to prevent what, after the end of that war, Europeans vowed would never happen again.

As of the winter of 2001, Kosovo's long-term status remains unclear. The Serb population has dwindled to less than 5 percent; United Nations forces maintain control; rebuilding with foreign aid is underway. But Kosovo was never formally detached from Yugoslavia, and it is not an independent entity.

Compared to events in the rest of the former Yugoslavia, the southernmost former Yugoslav "republic" of Macedonia has been fortunate despite its multicultural population (66 percent Macedonian, 24 percent Albanian, 4 percent Turkish) and its extremely sensitive location. When Macedonia declared independence in 1991, Greece warned that the name *Macedonia* was Greek property and that its adoption by an independent state was unacceptable to Athens. The Greeks insisted on *FYROM* (Former Yugoslav Republic of Macedonia), and when Macedonia refused to comply, the Greeks closed the port of Thessaloniki to Macedonian traffic. This action damaged an already weak, mainly subsistence economy in Macedonia, the poorest of all the former components of Yugoslavia. UN troops were stationed in the capital, Skopje, to guard against spillover from conflicts among its neighbors; but no force could have prepared this small country (population: 2 million) for the refugee stream that crossed its borders in March and April of 1999. Macedonia already battled the effects of a stagnant economy coupled with its own ethnic and cultural divisions (Fig. 1-24). The disaster in Kosovo posed a grave risk to this young state.

It is appropriate to save our discussion of Albania for last because even in this turbulent region Albania is an anachronism. It is the only dominantly Muslim state in Europe; some 70 percent of Albanians are nominally Muslims. Today, Albania is Europe's poorest country, ranking lowest on most indices of well-being. It has by far the fastest rate of population growth in the realm. Most Albanians subsist on livestock herding and farming on the one-fifth of this mountainous, rocky, earthquake-prone country that can be used for agriculture. Not surprisingly, thousands of Albanians have tried to reach Italy across the Adriatic Sea, hoping to secure a better life in the EU.

But Albania is not an unimportant country. It is potentially the Muslim gateway into Eastern Europe. Its population of 3.5 million has ethnic affinities with large contiguous Muslim minorities in Kosovo and northwestern Macedonia. Meanwhile, Albania's fragile government in Tirane cannot afford irredentism. Albania has its own cultural divide between the Gegs in the north and the somewhat better-off Tosks in the south (where there also is a small Greek minority), and since 1990 it has suffered from dangerous civil unrest. Will Albania avoid being swept up by the devolutionary forces that have dismembered its coastal neighbor?

With 583 million inhabitants in 39 countries, including some of the world's highest-income economies, a politically stable and economically integrated Europe would be a superpower in the twenty-first century. But Europe's political geography is anything but stable, as devolutionary forces and cultural conflict continue to trouble the realm. Moreover, economic integration so far involves less than half its countries, a process that will become more difficult as the European Union confronts applications from marginally qualifying states in the east. Europe always has been a realm of revolutionary change, and it remains so today.

chapter 2

Russia

FIGURE 2-1

nt © Rand McNally, 2000

CONCEPTS, IDEAS, AND TERMS

1 Continentality
2 Climatology
3 Weather
4 Tundra
5 Taiga
6 Permafrost
7 Forward capital
8 Colonialism
9 Imperialism
10 Federation
11 Russification
12 Collectivization
13 State planning
14 Unitary state system
15 Distance decay
16 Population decline
17 Heartland theory
18 Core area

REGIONS

RUSSIAN CORE AND PERIPHERIES
EASTERN FRONTIER
SIBERIA
RUSSIAN FAR EAST

The name *Russia* evokes cultural-geographic images of a stormy past: terrifying czars, conquering Cossacks, Byzantine bishops, rousing revolutionaries, clashing cultures. Russians repulsed the Tatar (Mongol) hordes, forged a powerful state, colonized a vast contiguous empire, defeated Napoleon, succumbed to communism, and, when the communist system failed, lost most of their imperial domain. Today, Russia, still possessing a huge nuclear arsenal from the Cold War, is in a tenuous—not to say dangerous—transition.

DEFINING THE REALM

Russia is a land of vast distances (11 time zones separate Sakhalin Island in the Pacific and St. Petersburg on the Baltic arm of the Atlantic), bitter cold, remote frontiers, and isolated outposts. Its core area (sometimes called European Russia) lies west of the Ural Mountains, centered on Moscow. To the north, south, and east lie distant, difficult peripheries where economies falter and political systems fail. Under dictatorial communism, those peripheries could be kept in check. Today, they oppose and even rebel against Moscow. The pain of economic change has led to strikes, railroad blockages, and demands for more regional autonomy. Russia's problems form a major obstacle to any New World Order.

Precommunist Russia may be described as a culture of extremes. It was a culture of strong nationalism, resistance to change, and despotic rule. Enormous wealth was concentrated in a small elite. Powerful rulers and bejeweled aristocrats perpetuated their privileges at the expense of millions of peasants and serfs who lived in dreadful poverty. The Industrial Revolution arrived late in Russia, and a middle class was slow to develop. Yet the Russian nation gained the loyalty of many of its citizens, and its writers and artists were among the world's greatest. Authors such as Tolstoy and Dostoyevsky chronicled the plight of the poor; composers celebrated the indomitable Russian people, and, as Tchaikovsky did in his *1812 Overture*, commemorated their victories over foreign foes.

Under the czars, Russia grew from nation into empire. The czars' insatiable demands for wealth, territory, and power sent Russian armies across the plains of Siberia, through the deserts of interior Asia, and into the mountains along Russia's rim. Russian pioneers ventured even farther, entering Alaska, traveling down the Pacific coast of North America, and planting the Russian flag near San Francisco in 1812. But as Russia's empire expanded, its internal weaknesses gnawed at the power of the czars. Peasants rebelled. Unpaid (and poorly fed) armies mutinied. When the czars tried to initiate reforms, the aristocracy objected. The empire at the beginning of the twentieth century was ripe for revolution, which began in 1905 (see box titled "The Soviet Union, 1924–1991").

The last czar, Nicholas II, was overthrown in 1917, and civil war followed. The victorious communists led by V. I. Lenin soon swept away much of the Russia of the past. The Russian flag disappeared. The czar and his family were executed. The old capital of Russia, St. Petersburg, was renamed Leningrad in honor of the revolutionary leader. Moscow, in the interior of the country, was chosen as the

◆ Major Geographic Qualities of Russia

1. Russia is the largest territorial state in the world. Its area is nearly twice as large as that of the next-ranking country (Canada).
2. Russia is the northernmost large and populous country in the world; much of it is cold and/or dry. Extensive rugged mountain zones separate Russia from warmer subtropical air, and the country lies open to Arctic airmasses.
3. Russia was one of the world's major colonial powers. Under the czars, the Russians forged the world's largest contiguous empire; the Soviet rulers who succeeded the czars took over and expanded this empire.
4. For so large an area, Russia's population of under 150 million is comparatively small. The population remains heavily concentrated in the westernmost one-fifth of the country.
5. Development in Russia is concentrated west of the Ural Mountains; here lie the major cities, leading industrial regions, densest transport networks, and most productive farming areas. National integration and economic development east of the Urals extend mainly along a narrow corridor that stretches from the southern Urals region to the southern Far East around Vladivostok.
6. Russia is a multicultural state with a complex domestic political geography. Twenty-one internal Republics, originally based on ethnic clusters, function as politico-geographical entities.
7. Its large territorial size notwithstanding, Russia suffers from land encirclement within Eurasia; it has few good and suitably located ports.
8. Regions long part of the Russian and Soviet empires are realigning themselves in the postcommunist era. Eastern Europe and the heavily Muslim Southwest Asia realm are encroaching on Russia's imperial borders.
9. The failure of the Soviet communist system left Russia in economic disarray. Many of the long-term components described in this chapter (food-producing areas, railroad links, pipeline connections) broke down in the transition to the postcommunist order.
10. Russia long has been a source of raw materials but not a manufacturer of export products, except weaponry. Few Russian (or Soviet) automobiles, televisions, cameras, or other consumer goods reach world markets.

FROM THE FIELD NOTES

"Walking the streets of St. Petersburg in 1994, more than 30 years after our first trip to the city, brought home the magnitude of the changes after 1991. Not only Nevsky Prospekt, the city's main street seen here, but other streets as well, are crowded with cars and pedestrians. Advertisements are everywhere. Small makeshift stalls selling goods ranging from books to family memorabilia line the sidewalks. Luxury goods and services have made their appearance, and the newly wealthy shop at recently opened, swank stores. Also in evidence are the newly destitute and desperate; just down the street from this scene a little girl stood playing the violin all day, her father leaning against the wall nearby, waiting to collect the coins passersby gave her. Crime has escalated as well, and extortion threatens many new small businesses. Like other Russian cities, St. Petersburg (*Peter* as locals call it) is in post-Soviet ferment."

The Soviet Union, 1924-1991

For 67 years Russia was the cornerstone of the *Soviet Union*, the so-called Union of Soviet Socialist Republics (USSR). The Soviet Union was the product of the Revolution of 1917, when more than a decade of rebellion against the rule of Czar Nicholas II led to his abdication. Russian revolutionary groups were called soviets ("councils"), and they had been active since the first workers' uprising in 1905. In that year, thousands of Russian workers marched on the czar's palace in St. Petersburg in protest, and soldiers fired on them. Hundreds were killed and wounded. Russia descended into chaos.

A coalition of military and professional men forced the czar's abdication in 1917. Then Russia was ruled briefly by a provisional government. In November 1917, the country held its first democratic election—and, as it turned out, its last for more than 70 years.

The provisional government allowed the exiled activists in the Bolshevik camp to return to Russia (there were divisions among the revolutionaries): Lenin from Switzerland, Trotsky from New York, and Stalin from internal exile in Siberia. In the ensuing political struggle, Lenin's Bolsheviks gained control over the revolutionary soviets, and this ushered in the era of communism. In 1924, the new communist empire was formally renamed the Union of Soviet Socialist Republics, or Soviet Union in shorthand.

Lenin the organizer died in 1924 and was succeeded by Stalin the tyrant, and many of the peoples under Moscow's control suffered unimaginably. In pursuit of communist reconstruction, Stalin and his adherents starved millions of Ukrainian peasants to death, forcibly relocated entire ethnic groups, and exterminated "uncooperative" or "disloyal" peoples. The full extent of these horrors may never be known. Many of the country's most creative people were executed.

On December 25, 1991, the inevitable occurred: the Soviet Union ceased to exist, its economy a shambles, its political system shattered, the communist experiment a failure. The last Soviet president, Mikhail Gorbachev, resigned, and the Soviet hammer-and-sickle flag flying atop the Kremlin was lowered for the last time and immediately replaced by the white, red, and blue Russian tricolor. A new and turbulent era had begun, but Soviet structures and systems will long cast their shadows over transforming Russia.

new capital for a country with a new name, the *Soviet Union*. Eventually this Union consisted of 15 political entities, each a Soviet Socialist Republic. Russia was just one of these republics, and the name Russia disappeared from the international map.

Russia remained first among equals, however. Not for nothing was the communist revolution known as the Russian Revolution. The Soviet Empire was the legacy of the czars' expansionism, and the new communist rulers were Russians first and foremost. As the new Union was laid out, Russians moved by the millions to the fringes of the empire—the fringes where lay the "republics" formed from the czars' colonies (or were conquered later by the Red Army). The Russification of the Soviet Empire proceeded just as the British and French and Belgians and Portuguese were also moving in large numbers to their colonies. The Soviet Union was a Russian colonial empire.

The world's great colonial empires did not endure, and (as we predicted in earlier editions of this book) neither did the Soviet Union. Non-Russian peoples stirred in opposition to Moscow, their nationalism or ethnic consciousness mobilized by memories of a long-suppressed past. Lithuanians, Ukrainians, Georgians, and other peoples enmeshed in the Soviet Empire moved to throw off the communist yoke. And in Russia itself, nationalism also stirred—in opposition not to other peoples of the empire but to the communist system that for nearly 70 years had bound them together.

Like the last czar long ago, the last communist dictator, Mikhail Gorbachev, tried to control and channel the forces of change that were sweeping through his country. He, too, failed as events overtook him. (Unlike Nicholas II, however, Gorbachev lived to tell the tale in books and lectures.) And so today, Russia's national flag flies once again over the Kremlin, the empire is gone, and the challenge facing Russia involves political stability and economic redirection, not Cold War politics and superpower competition.

SOVIET DEVOLUTION

Under the communists, the colonized pieces of the Soviet Empire outside Russia were given the status of *Soviet Socialist Republics (SSRs)*, and when the Soviet system collapsed, leaders in Russia and in some of the SSRs hoped that a form of voluntary association among these republics would replace it. The Commonwealth of Independent States (CIS) was designed to continue the links forged during communist times. But the CIS never was much more than an acronym. The SSRs soon went their own ways (Table 2-1); only Belarus has sought to rejoin Russia in a formal union.

Table 2-1
WHAT IS HAPPENING TO THE OTHER 14 EX-SOVIET REPUBLICS?

Name	Population (Millions)	Percent Russians	Religion	Official Language	Status
Armenia	3.8	2	Armenian Orthodox (Christian) 48%	Armenian	Armenia is in conflict with neighboring Azerbaijan over the exclave of Nagorno-Karabakh. Public opinion favors closer association with Russia.
Azerbaijan	7.8	6	Shi'ite Muslim 93%	Azerbaijani	Azerbaijan is seeking settlement with Armenia over the Nagorno-Karabakh exclave, and trying to reduce dependence on Russia for oil-export routes.
Belarus	9.9	14	Belarussian Orthodox, Roman Catholic 49%	Belarussian; Russian	Belarus persuaded Russia to sign a "union" agreement in 1997. Political repression, human-rights violations, stagnating economy.
Estonia	1.4	29	Estonian Orthodox, Estonian Lutheran 34%	Estonian	After a period of discrimination against the Russian minority, conditions are improving along with a booming economy. Relations with Nordic neighbors, the European Union, and Russia strengthening.
Georgia	5.5	6	Georgian, Russian Orthodox 47%; Sunni Muslim 11%	Georgian	Georgia has unfriendly relations with Russia and accuses Russia of encouraging separatist elements in the province of Abkhazia. Russian military bases on Georgian soil, and Russian border closures are seen as threats to Georgian sovereignty.
Kazakhstan	15.0	35	Sunni Muslim 47%; Russian Orthodox 10%	Kazakh	Kazakhstan has difficult relations with Russia over treatment of Russians in the northern oblasts, over the use of Cossack troops to patrol the northern border, over Russian oil-company rights in the Caspian area, and over Russian weapons-testing sites inside Kazakhstan.
Kyrgyzstan	5.0	16	Sunni Muslim 71%	Kyrgyz; Russian	An increasingly authoritarian government is impairing the democratization process by imprisoning critics and stifling the press. The Kyrgyz Communist Party has demanded a Belarus-type "union" with Russia.
Latvia	2.4	32	Christian Churches 40%	Latvian	Relations between the government and the large Russian minority have been difficult but improving; naturalization slow. First talks on joining EU have started.
Lithuania	3.7	8	Roman Catholic 79%	Lithuanian	Relations with Russia improved in 1998 with the signing of an agreement settling a boundary dispute. EU turned down Lithuania's plans for membership application.
Moldova	4.3	13	Eastern Orthodox 49%	Romanian	Relations with Russia have suffered from the attempted breakaway by the pro-Russian trans-Dniester region.
Economy					in trouble.
Tajikistan	6.6	7	Sunni Muslim 80%	Tajik	Civil war has pitted the ruling post-communist regime against United Tajik Opposition (UTO) led by the Islamic Renaissance Party.
Turkmenistan	5.4	6	Sunni Muslim 87%	Turkmen	Turkmenistan's government accuses Russia of impeding the flow of natural gas to its customers; an alternative pipeline via Iran is in the works. Problems continue with Azerbaijan over Caspian Sea oil and gas reserves.
Ukraine	48.9	22	Ukrainian Orthodox 61%	Ukrainian	A large and potentially powerful country, Ukraine has settled several disputes with Russia including the rights of its large Russian minority and the disposition of the Soviet Black Sea fleet as well as two Black Sea ports. Pro-Russian sentiment in Ukraine appears to be rising.
Uzbekistan	25.7	6	Sunni Muslim 88%	Uzbek	Repeated setbacks to the post-Soviet democratization process include stifling of opposition and press. Fear of Islamic fundamentalism has produced repressive laws and disputes with Russia as well as the United States and European countries.

Loosening ties with Russia has been easier for some former SSRs than for others. The forces of economic geography cannot be redirected overnight, and as Table 2-1 indicates, sizeable Russian minorities remain in several of the former colonies. Even as they try to stabilize relations with Russia, most of the westernmost SSRs are looking to Europe as they seek a new place in the international community. The five former Soviet Central Asian republics now are part of a distinct geographic region, Turkestan, where Islam is reviving and links to the Muslim world are growing stronger. But in the southwest, between the Black and Caspian seas, the situation is more complicated.

Of the three former SSRs in this area, known as *Transcaucasia*, two have had to remain close to Russia: Christian Armenia, and to a lesser degree, Georgia. The third, Muslim Azerbaijan, is trying to reduce its dependence on Moscow for the transit of its key exports, oil and natural gas. Transcaucasia has one of the world's most complicated cultural mosaics, a transition zone where geographic realms and regions, civilizations and cultures meet (see Fig. 6-20 on p. 328). For reasons we will make clear in the pages ahead, the geographic framework of this area remains in transition.

DIMENSIONS AND DISTRIBUTIONS

In territorial size, Russia is by far the largest country in the world. It extends from the Bering Sea near Alaska to the Gulf of Finland in Europe and from the shores of the Black Sea to the Sea of Japan (Fig. 2-1). From well inside the Arctic Circle, Russia reaches southward to the latitude of Chicago. It is a land of almost unimaginable dimensions, almost twice as large as the United States or China.

Russia also is a land of vast empty spaces. The country's population of 143.2 million is small compared to China's 1.3 billion and India's 1.0 billion. And, as Figure I-9 reveals, Russia's population remains strongly concentrated in "European" Russia—Russia west of the Ural Mountains, which divide the west from Siberia (Fig. 2-1). The name *Siberia*, which means "sleeping land," aptly describes a vast area in which the population still lives in isolated clusters and along discontinuous ribbons. The Russia of both the czars and the Soviet Union wanted to populate the east, to strengthen the Russian presence in that remote frontier. Over time an eastward vanguard of settlements did emerge, the largest ones along two railroads: the Trans-Siberian Railroad built by the czars before the revolution and the other, the Baykal-Amur Mainline (BAM) Railroad, constructed by the communists. The suitably named city of Novosibirsk is such a center, one of many trans-Ural places whose names start with *novo* (meaning "new").

THE PHYSICAL ENVIRONMENT

Russia's eastward march was hampered not only by distance but also by the natural environment. As the northernmost populous country on Earth, Russia has virtually no natural barriers against the onslaught of Arctic air. Moscow lies farther north than Edmonton, Canada, and St. Petersburg lies at latitude 60° North—the latitude of the southern tip of Greenland. Winters are long, dark, and bitterly cold in most of Russia; summers are short and growing seasons limited. Many a Siberian frontier outpost was doomed by cold, snow, and hunger.

To make matters worse, precipitation totals range from modest to minimal because the warm, moist air carried across Europe from the North Atlantic Ocean loses much of its warmth and moisture by the time it reaches Russia. Figures I-7 and I-8 reveal the consequences. Russia's climatic **continentality** (inland climatic environment remote from moderating and moistening maritime influences) is expressed by its prevailing *Dfb* and *Dfc* conditions. Compare the Russian map to that of North America, and you note that, except for a small corner off the Black Sea, Russia's climatic conditions resemble those of the U.S. Upper Midwest and Canada. Along its entire north, Russia has a zone of *E* climates, the most frigid on the planet. In these Arctic latitudes originate the polar airmasses that dominate its environments.

1

By studying Russia's climates we can begin to understand what the map of population distribution (Fig. I-9) shows. **Climatology** is a field of geography that investigates not only the distribution of climatic conditions over the Earth's surface but also the processes that cause this spatial arrangement. The Earth's atmosphere traps heat received in the form of radiation from the Sun, but this *greenhouse effect* varies in space—and over time. As we noted in the Introduction, much of Russia was in the grip of a glaciation until the onset of the warmer Holocene. Today, a natural warming cycle is in progress, perhaps augmented by human action. This enhanced global warming threatens some societies (for example, those with extensive coastal plains that would be inundated by rising sea levels), but Russia may be one of the winners in this process of environmental change. Warmer temperatures and increased moisture may expand the frontiers of settlement shown in Figure I-9.

2

Climate and weather (there is a distinction: *climate* is a long-term average, whereas **weather** refers to existing atmospheric conditions at a given place and time) have always challenged Russia's farmers. Conditions are most favorable in the west, but even there temperature extremes, variable and undependable rainfall, and short growing seasons make farming difficult. During the Soviet period, fer-

3

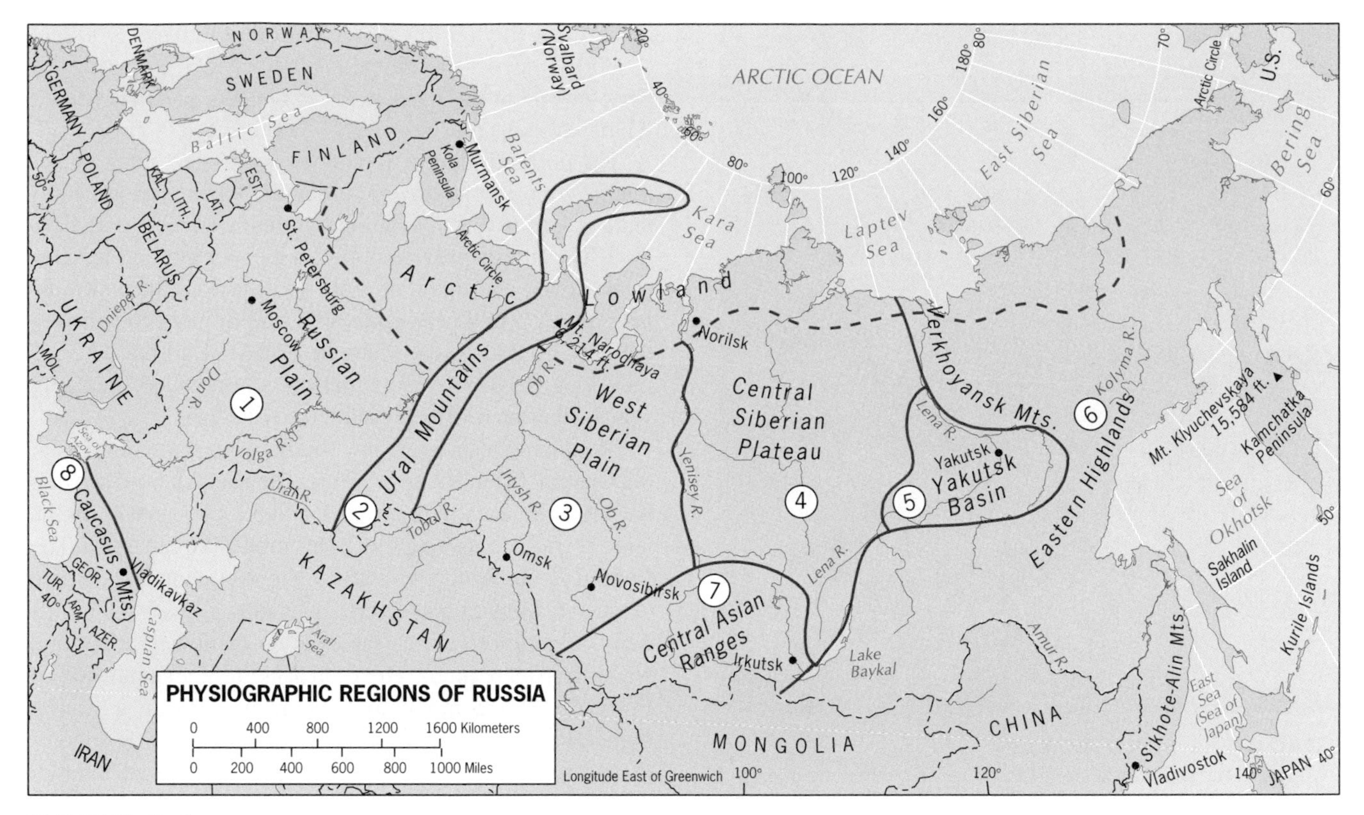

FIGURE 2-2

tile and productive Ukraine supplied much of Russia's food needs, but even then Russia often had to import grain. Soviet rulers wanted to reduce their country's dependence on imported food, and their communist planners built major irrigation projects to increase crop yields in the colonized republics of Central Asia. As we will see in Chapter 6, some of these attempts to overcome nature's limitations spelled disaster for the local people.

Physiographic Regions

The forces shaping Russia's harsh environments are revealed in the map of its physiography (Fig. 2-2). Note how mountains and deserts encircle Russia: the Caucasus in the southwest ⑧; the Central Asian Ranges in the center ⑦; the Eastern Highlands facing the Pacific from the Bering Sea to the East Sea (Sea of Japan) ⑥. The Kamchatka Peninsula has a string of active volcanoes in one of the world's most earthquake-prone zones (Fig. I-5). Warm subtropical air thus has little opportunity to penetrate Russia, while cold Arctic air sweeps southward without impediment. Russia's Arctic north is a gently sloping lowland broken only by the Urals and Eastern Highlands.

Russia's vast and complex physical stage can be divided into eight physiographic regions, each of which, at a larger scale, can be subdivided into smaller units. In the Siberian region, one criterion for such subdivision would be the vegetation. The Russian language has given us two terms to describe this vegetation: **tundra**, the treeless plain along the Arctic shore where mosses, lichens, and some grasses survive, and **taiga**, the mostly coniferous forests that begin to the south where the tundra ends, and extend over vast reaches of the "sleeping land."

4

5

The Russian Plain ① is the eastward continuation of the North European Lowland, and here the Russian state formed its first core area. Travel north from Moscow at its heart, and the countryside soon is covered by needleleaf forests like those of Canada; to the south lie the grain fields of southern Russia and, beyond, those of Ukraine. Note the Kola Peninsula and Barents Sea in the far north: warm water from the North Atlantic comes around northern Norway and keeps the port of Murmansk ice free most of the year. The Russian Plain is bounded on the east by the Ural Mountains ②; though not a high range, it is topographically prominent because it separates two extensive plains. The range of the Urals is more than 2000 miles (3200 km) long and reaches from the shores of the Kara Sea to the border with Kazakhstan. It is not a barrier to east-west transportation, and its southern end is densely populated. Here the Urals yield minerals and fossil fuels.

East of the Urals lies Siberia. The West Siberian Plain ③ has been described as the world's largest unbroken low-

FROM THE FIELD NOTES

"One of the most arduous trips in memory, this ride in an old bus along the Georgian Military Highway from Yerevan in Armenia via Tbilisi in Georgia and on to Pyatigorsk in Russia, all across the Caucasus Mountains. Near the Georgia-Russia border we stopped in sight of Mount Elbrus (18,510 feet; 5642 m), tallest peak in the Caucasus and sometimes mistakenly called the highest mountain in Europe. An Eastern Orthodox Church stood on a high mountain nearby. Along the route were numerous small villages, isolated and hidden away in valleys, some with Islamic crescents painted on doors."

land; this is the vast basin of the Ob and Irtysh rivers. Over the last 1000 miles (1600 km) of its course to the Arctic Ocean, the Ob falls less than 300 feet (90 m). In Figure 2-2, note the dashed line that extends from the Finnish border to the East Siberian Sea and offsets the Arctic Lowland. North of the line, water in the ground is permanently frozen; this 6 **permafrost** creates another obstacle to permanent settlement. Looking again at the West Siberian Plain, we see that the north is permafrost-ridden and the central zone is marshy. The south, however, has such major cities as Omsk and Novosibirsk, within the corridor of the Trans-Siberian Railroad.

East of the West Siberian Plain the country begins to rise, first into the central Siberian Plateau ④, another sparsely settled, remote, permafrost-affected region. Here winters are long and cold, and summers are short; the area remains barely touched by human activity. Beyond the Yakutsk Basin ⑤, the terrain becomes mountainous and the relief high. The Eastern Highlands ⑥ are a jumbled mass of ranges and ridges, precipitous valleys, and volcanic mountains. Lake Baykal lies in a trough that is over 5000 feet (1500 m) deep—the deepest rift lake in the world (see Chapter 7). On the Kamchatka Peninsula, volcanic Mount Klyuchevskaya reaches nearly 15,600 feet (4750 m).

The northern part of region ⑥ is Russia's most inhospitable zone, but southward along the Pacific coast the climate is less severe. Nonetheless, this is a true frontier region. The forests provide opportunities for lumbering, a fur trade exists, and there are gold and diamond deposits. To help develop this promising area, the Soviets constructed the new Baykal-Amur Mainline (BAM) Railroad, a 2200 mile-long (3540 km) route that roughly parallels the aging Trans-Siberian line to the south (see Fig. 2-6).

Mountains also mark the southern margins of Russia: the Central Asian Ranges ⑦ from the Kazakh border in the west to Lake Baykal in the east, and the Caucasus ⑧ in the land corridor between the Black and Caspian seas. The Central Asian Ranges rise above the snow line and contain extensive mountain glaciers. The annual melting brings down alluvium to enrich the soils on the lower slopes and water to irrigate them. The Caucasus form an extension of Europe's Alpine mountain system and exhibit a similar high *relief* (range of elevations), as shown in the photograph on this page. No convenient passes facilitate interaction. Here Russia's southern border is sharply defined by *topography* (surface configuration).

As the physiographic map suggests, the more habitable terrain in Russia becomes latitudinally narrower from west to east; beyond the southern Urals, settlement becomes a discontinuous ribbon. As we will note later in this chapter, isolated towns did develop in interior Siberia (such as Yakutsk on the Lena River and Norilsk near the Yenisey River). Yet, as in Canada, the population is clustered markedly in the southern, most livable zone of the country, a narrow belt that widens only on Russia's Far Eastern, Pacific rim.

EVOLUTION OF THE RUSSIAN STATE

Four centuries ago, when European kingdoms were sending fleets to distant shores to search for riches and capture colonies, there was little indication that the largest of all empires would one day center on a city in the forests halfway between Sweden and the Black Sea. The plains of Eurasia south of the taiga had seen waves of migrants sweep from east to west: Scythians, Sarmatians, Goths, Huns, and others came, fought, settled, and were absorbed or driven off. Eventually, the Slavs, heirs to these Eurasian infusions, emerged as the dominant culture in what is today Ukraine, south of Russia and north of the Black Sea.

From this base on fertile, productive soils, the Slavs expanded their domain, making Kiev (Kyyiv) their southern headquarters and establishing Novgorod on Lake Ilmen in the north. Each was the center of a state known as a *Rus*, and both formed key links on a trade route between the German-speaking Hanseatic ports of the Baltic Sea and the trading centers of the Mediterranean. During the eleventh and twelfth centuries, the Kievan Rus and the Novgorod Rus united to form a large and comparatively prosperous state astride forest and steppe (short-grass prairie).

The Mongol Invasion

Prosperity attracts attention, and far to the east, north of China, the Mongol Empire had been building under Genghis Khan. In the thirteenth century, the Mongol (Tatar) "hordes" rode their horses into the Kievan Rus, and the state fell. By mid-century, Slavs were fleeing into the forests, where the Mongol horsemen had less success than on the open plains of Ukraine. Here the Slavs reorganized, setting up several Russes. Still threatened by the Tatars, the ruling princes paid tribute to the Mongols in exchange for peace.

Moscow, deep in the forest on the Moscow River, was one of these Russes. Its site was remote and defensible. Over time, Moscow established trade and political links with Novgorod and became the focus of a growing area of Slavic settlement. When the Mongols, worried about Moscow's expanding influence, attacked near the end of the fourteenth century and were repulsed, Moscow emerged as the unchallenged center of the Slavic Russes. Soon its ruler, now called a Grand Duke, took control of Novgorod, and the stage was set for further growth.

Grand Duchy of Muscovy

Moscow's geographic advantages again influenced events. Its centrality, defensibility, links to Slavic settlements including Novgorod, and open frontiers to the north and west, where no enemies threatened, gave Moscow the potential to expand at the cost of its old enemies, the Mongols. Many of these invaders had settled in the basin of the Volga River and elsewhere, where the city of Kazan had become a major center of Islam, the Tatars' dominant religion. During the rule of Ivan the Terrible (1547–1584), the Grand Duchy of Muscovy became a major military power and imperial state. Its rulers called themselves czars and claimed to be the heirs of the Byzantine emperors. Now began the expansion of the Russian domain (Fig. 2-3), marked by the defeat

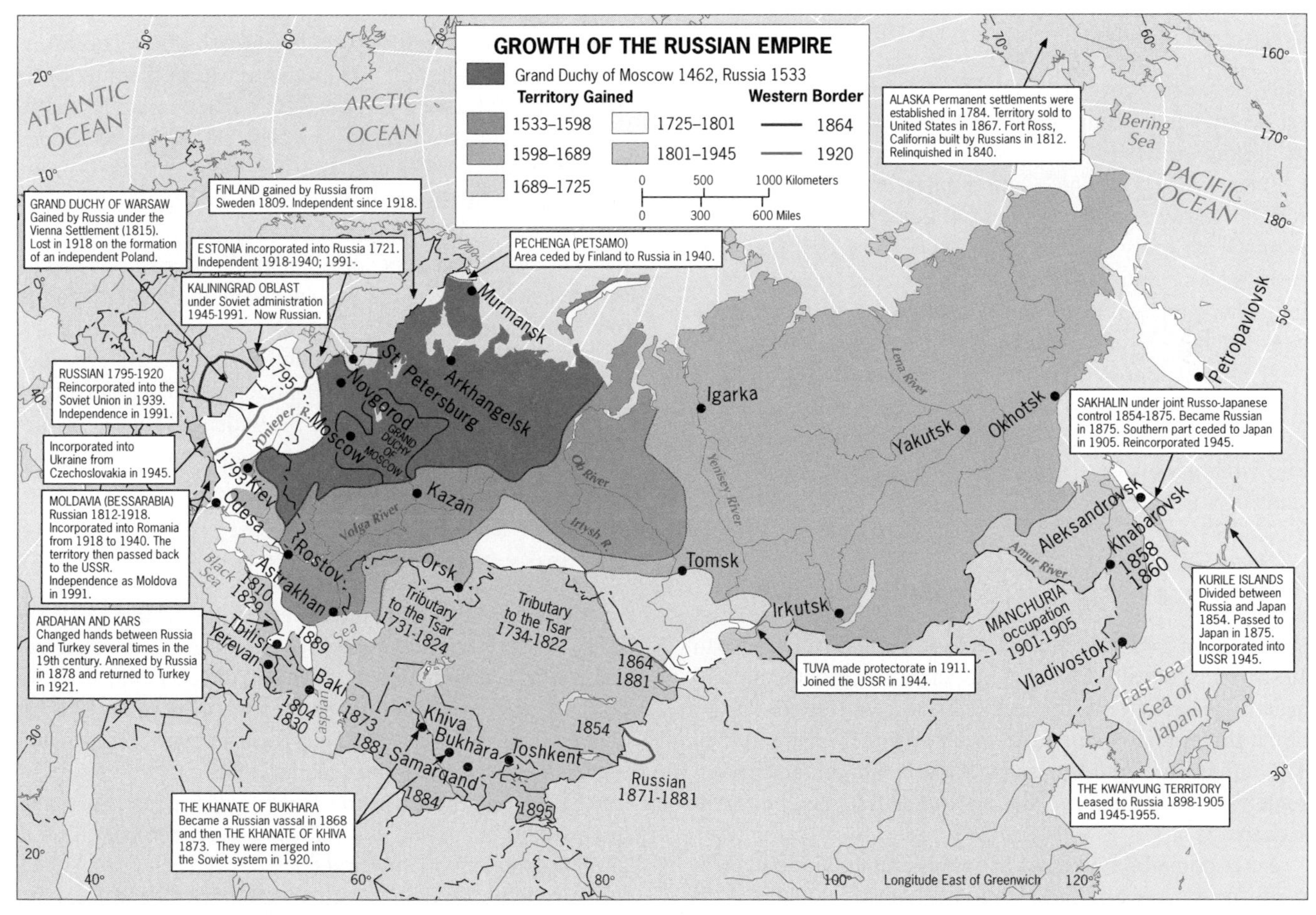

FIGURE 2-3

Russians in North America

The first white settlers in Alaska were Russians, not Western Europeans, and they came across Siberia and the Bering Strait, not across the Atlantic and North America. Russian hunters of the sea otter, valued for its high-priced pelt, established their first Alaskan settlement at Kodiak Island in 1784. Moving south along the North American coast, the Russians founded additional villages and forts to protect their tenuous holdings until they reached as far as the area just north of San Francisco Bay, where they built Fort Ross in 1812.

But the Russian settlements were isolated and vulnerable. European fur traders began to put pressure on their Russian competitors, and St. Petersburg found the distant settlements a burden and a risk. In any case, American, British, and Canadian hunters were decimating the sea otter population, and profits declined. When U.S. Secretary of State William Seward offered to purchase Russia's holdings in 1867, St. Petersburg quickly agreed—for $7.2 million. Thus Alaska, including its lengthy southward coastal extension—the Alaskan "panhandle"—became U.S. territory and, eventually, the forty-ninth State. Although Seward was ridiculed for his decision—Alaska was called "Seward's Folly" and "Seward's Icebox"—he was vindicated when gold was discovered there in the 1890s. The twentieth century proved the wisdom of Seward's action, strategically as well as economically. At Prudhoe Bay off Alaska's northern Arctic slope, large oil reserves were tapped in the 1970s and are still being exploited. And like Siberia, Alaska probably contains yet-unknown riches.

of the Tatars at Kazan and the destruction of hundreds of mosques.

The Cossacks

This eastward expansion of Russia was spearheaded by a relatively small group of seminomadic peoples who came to be known as Cossacks and whose original home was in present-day Ukraine. Opportunists and pioneers, they sought the riches of the eastern frontier, chiefly fur-bearing animals, as early as the sixteenth century. By the mid-seventeenth century they reached the Pacific Ocean, defeating Tatars in their path and consolidating their gains by constructing *ostrogs*, strategic fortified waystations along river courses. Before the eastward expansion halted in 1812 (Fig. 2-3), the Russians had moved across the Bering Strait to Alaska and down the western coast of North America into what is now northern California (see box titled "Russians in North America").

Czar Peter the Great

When Peter the Great became czar (he ruled from 1682 to 1725), Moscow already lay at the center of a great empire—great, at least, in terms of the territories it controlled. The Islamic threat had been ended with the defeat of the Tatars. The influence of the Russian Orthodox Church was represented by its distinctive religious architecture and powerful bishops.

Peter consolidated Russia's gains and hoped to make a modern, European-style state out of his loosely knit country. He built St. Petersburg as a **forward capital** on the doorstep of Swedish-held Finland, fortified it with major military installations, and made it Russia's leading port.

7

Czar Peter the Great, an extraordinary leader, was in many ways the founder of modern Russia. In his desire to remake Russia—to pull it from the forests of the interior to the waters of the west, to open it to outside influences, and to relocate its population—he left no stone unturned. Prominent merchant families were forced to move from other cities to St. Petersburg. Ships and wagons entering the city had to bring building stones as an entry toll. The czar himself, aware that to become a major power Russia had to be strong at sea as well as on land, went to the Netherlands to work as a laborer in the famed Dutch shipyards to learn the most efficient method for building ships. Meanwhile, the czar's forces continued to conquer people and territory: Estonia was incorporated in 1721, widening Russia's window to the west, and major expansion soon occurred south of the city of Tomsk (Fig. 2-3).

Czarina Catherine the Great

Under Czarina Catherine the Great, who ruled from 1760 to 1796, Russia's empire in the Black Sea area grew at the expense of the Ottoman Turks. The Crimean Peninsula, the port city of Odesa (Odessa), and the whole northern coastal zone of the Black Sea fell under Russian control. Also during this period, the Russians made a fateful move: they penetrated the area between the Black and Caspian seas, the mountainous Caucasus with dozens of ethnic and cultural groups, many of which were Islamized. The cities of Tbilisi (now in Georgia), Baki (Baku in Azerbaijan), and Yerevan (Armenia) were captured. Eventually, the Russian push toward an outlet on the Indian Ocean was halted by the British, who held sway in Persia (modern-day Iran), and the Turks. But Catherine the Great had made Russia a colonial power.

The Nineteenth Century

Russian expansionism was not over. While extending the empire southward, the Russians also took on the Poles, old enemies to the west, and succeeded in taking most of what is today the Polish state, including the capital of Warsaw. To the northwest, Russia took over Finland from the Swedes in 1809. During most of the nineteenth century, however, the Russians were preoccupied with Central Asia—the region between the Caspian Sea and western China—where Toshkent (Tashkent) and Samarqand (Samarkand) came under St. Petersburg's control (Fig. 2-2). The Russians here were still bothered by raids of nomadic horsemen, and they sought to establish their authority over the Central Asian steppe country as far as the edges of the high mountains that lay to the south. Thus Russia gained many Muslim subjects, for this was Islamic Asia they were penetrating. Under czarist rule, these people retained some autonomy. Much farther to the east, a combination of Japanese expansionism and a decline of Chinese influence led Russia to annex from China several provinces to the east of the Amur River. Soon thereafter, in 1860, the port of Vladivostok on the Pacific was founded.

Now began the events that were to lead, after five centuries of almost uninterrupted expansion and consolidation, to the first setback in the Russian drive for territory. In 1892, the Russians began building the Trans-Siberian Railroad in an effort to connect the distant frontier more effectively to the western core. As the map shows (p. 115), the most direct route to Vladivostok was across northeastern China (Manchuria). The Russians wanted China to permit the construction of the last link of the railway across their territory, but the Chinese resisted. Taking advantage of the Boxer Rebellion in China in 1900 (see Chapter 9), Russia occupied Manchuria, which threatened Japanese interests. This brought on the Russo-Japanese War of 1904–1905, in which the Russians were disastrously defeated; Japan even took possession of southern Sakhalin Island (which they called Karafuto).

THE COLONIAL LEGACY

Thus Russia, like Britain, France, and other European powers, expanded through **colonialism**. Yet whereas the other European powers expanded overseas, Russian influence traveled overland into Central Asia, Siberia, China, and the Pacific coastlands of the Far East. What emerged was not the greatest empire but the largest territorially contiguous

8

Just west of Khabarovsk, as a tug pushes a bargeload of trucks upstream, a train that has just left the city crosses the Amur River on its long journey inland toward Irkutsk and points much farther west. This bridge not only carries the Trans-Siberian Railroad across the Russian Far East's mightiest river, but also represents the final link in this transcontinental route. It was here, in 1916, that the 25-year-long railbuilding effort came to an end with the completion of a Moscow-Valdivostok line that ran entirely through Russian territory (replacing an earlier shortcut route that cut across Northeast China between Vladivostok and the area east of Lake Baykal).

empire in the world. At the time of the Japanese war, the Russian czar controlled more than 8.5 million square miles (22 million sq km), just a tiny fraction less than the area of the Soviet Union after the 1917 Revolution. Thus the communist empire, to a large extent, was the legacy of St. Petersburg and European Russia, not the product of Moscow and the socialist revolution.

The czars embarked on their imperial conquests in part because of Russia's relative location: Russia always lacked warm-water ports. Had the Revolution not intervened, their southward push might have reached the Persian Gulf or even the Mediterranean Sea. Czar Peter the Great envisaged a Russia open to trading with the entire world; he developed St. Petersburg on the Baltic Sea into Russia's leading port. But in truth, Russia's historical geography is one of remoteness from the mainstreams of change and progress, as well as one of self-imposed isolation. Not even a string of warm-water ports would have been likely to transform Russia into an outward-looking, trading state. The czars' objectives were primarily strategic, not economic.

An Imperial, Multinational State

Centuries of Russian expansionism did not confine itself to empty land or unclaimed frontiers. The Russian state became an imperial power that annexed and incorporated many nationalities and cultures. This was done by employing force of arms, by overthrowing uncooperative rulers, by annexing territory, and by stoking the fires of ethnic conflict. By the time the ruthless Russian regime began to face revolution among its own people, czarist Russia was a hearth of **imperialism**, and its empire contained peoples representing more than 100 nationalities. The winners in the ensuing revolutionary struggle—the communists who forged the Soviet Union—did not liberate these subjugated peoples. Rather, they changed the empire's framework, binding the peoples colonized by the czars into a new system that would in theory give them autonomy and identity. In practice, it doomed those peoples to bondage and, in some cases, extinction.

9

When the Soviet system failed and the Soviet Socialist Republics became independent states, Russia was left without the empire that had taken centuries to build and consolidate—and that contained crucial agricultural and mineral resources. No longer did Moscow control the farms of Ukraine and the oil and natural gas reserves of Central Asia. But look again at Figure 2-3 and you will see that, even without its European and Central Asian colonies, Russia remains an empire. Russia lost the "republics" on its periphery, but Moscow still rules over a domain that extends from the borders of Finland to North Korea. Inside that domain Russians are in the overwhelming majority, but many subjugated nationalities, from Tatars to Yakuts, still inhabit ancestral homelands. Accommodating these many indigenous peoples is one of the challenges facing the Russian Federation today.

On the basis of physiographic, ethnic, historic, and cultural criteria, therefore, Russia constitutes a geographic realm, although transition zones rather than sharp boundaries mark its limits in some areas. Encircled by mountains and deserts, ruled by climatic continentality, unified by a dominant culture, and unmatched on Earth in terms of dimensions, Russia stands apart—from Europe to the west, China to the east, Central Asia to the south. But Russia, as we will see, is a society in transition. The realm's boundaries are unstable, still changing. To the west, Belarus at the opening of the twenty-first century was redirecting its interests from Europe toward Moscow. In the Caucasus, Armenia and Georgia effectively were in the Russian realm's orbit. In Central Asia, millions of Russians live across the border in northern Kazakhstan.

In the 1990s, Russia reorganized in the aftermath of the collapse of the Soviet Union. This reorganization cannot be understood without reference to the seven decades of Soviet communist rule that went before. We turn next to this crucial topic.

THE SOVIET LEGACY

The era of communism may have ended in the Soviet Empire, but its effects on Russia's political and economic geography will long remain. Seventy years of centralized planning and implementation cannot be erased overnight; regional reorganization toward a market economy cannot be accomplished in a day.

While the world of capitalism celebrates the failure of the communist system in the former Soviet realm, it should remember why communism found such fertile ground in the Russia of the 1910s and 1920s. In those days Russia was infamous for the wretched serfdom of its peasants, the cruel exploitation of its workers, the excesses of its nobility, and the ostentatious palaces and riches of the czars. Ripples from the Western European Industrial Revolution introduced a new age of misery for those laboring in factories. There were workers' strikes and ugly retributions, but when the czars finally tried to better the lot of the poor, it was too little too late. There was no democracy, and the people had no way to express or channel their grievances. Europe's democratic revolution passed Russia by, and its economic revolution touched the czars' domain only slightly. Most Russians, and tens of millions of non-Russians under the czars' control, faced exploitation, corruption, starvation, and harsh subjugation. When the people began to rebel in 1905, there was no hint of what lay in store; even after the full-scale Revolution of 1917, Russia's political future hung in the balance.

The Russian Revolution was no unified uprising. There were factions and cliques; the Bolsheviks ("Majority") took their ideological lead from Lenin, while the Mensheviks

("Minority") saw a different, more liberal future for their country. The so-called "Red" army factions fought against the "Whites," while both battled the forces of the czar. The country stopped functioning; the people suffered terrible deprivations in the countryside and in the cities. Most Russians (and other nationalities within the empire as well) were ready for radical change.

That change came when the Revolution succeeded and the Bolsheviks bested the Mensheviks, most of whom were exiled. In 1918, the capital was moved from Petrograd (as St. Petersburg had been renamed in 1914, to remove its German appellation) to Moscow. This was a symbolic move, the opposite of the forward-capital principle: Moscow lay deep in the Russian interior, not even on a major navigable waterway (let alone a coast), amid the same forests that much earlier had protected the Russians from their enemies. The new Soviet Union would look inward, and the communist system would achieve with Soviet resources and labor the goals that had for so long eluded the country. The chief political and economic architect of this effort was the revolutionary leader who prevailed in the power struggle: V. I. Lenin (born Vladimir Ilyich Ulyanov).

The Political Framework

Russia's great expansion had brought many nationalities under czarist control; now the revolutionary government sought to organize this heterogeneous ethnic mosaic into a smoothly functioning state. The czars had conquered, but they had done little to bring Russian culture to the peoples they ruled. The Georgians, Armenians, Tatars, and residents of the Muslim states of Central Asia were among dozens of individual cultural, linguistic, and religious groups that had not been "Russified." In 1917, however, the Russians themselves constituted only about one-half of the population of the entire country. Thus it was impossible to establish a Russian state instantly over this vast political region, and these diverse national groups had to be accommodated.

The question of the nationalities became a major issue in the young Soviet state after 1917. Lenin, who brought the philosophy of Karl Marx to Russia, talked from the beginning about the "right of self-determination for the nationalities." The first response by many of Russia's subject peoples was to proclaim independent republics, as they did in Ukraine, Georgia, Armenia, Azerbaijan, and even in Central Asia. But Lenin had no intention of permitting the Russian state to break up. In 1923, when his blueprint for the new Soviet Union went into effect, the last of these briefly independent units was fully absorbed into the sphere of the Moscow regime. Ukraine, for example, declared itself independent in 1917 and managed to sustain this initiative until 1919. But in that year the Bolsheviks set up a provisional government in Kiev, the Ukrainian capital, thereby ensuring the incorporation of the country into Lenin's Soviet framework.

The Bolsheviks' political framework for the Soviet Union was based on the ethnic identities of its many incorporated peoples. Given the size and cultural complexity of the empire, it was impossible to allocate territory of equal political standing to all the nationalities; the communists controlled the destinies of well over 100 peoples, both large nations and small isolated groups. It was decided to divide the vast realm into Soviet Socialist Republics (SSRs), each of which was delimited to correspond broadly to one of the major nationalities. At the time, Russians constituted about

Russia is a vast country that incorporates a wide range of environmental conditions and numerous ethnic groups or "nationalities." In the far northeast, the reindeer herders of Chukotskiy have historic affinities with the Inuits; in the southern periphery, linkages are with Turks and Persians. Here, in the barren northlands of Yamalo-Nenetskiy Autonomous Region, where the Ob River flows into the Kara Sea and the taiga gives way to tundra, these reindeer sustain a sparse, mobile population far removed from the political and social life of Moscow and the center.

The Soviet Federation and Its Legacy

When the Soviet Union was originally planned and delimited, its name implied what it was to be—a federation. As we note in our discussion of federalism (p. 540), a **federation** involves the sharing of power between a country's central government and its political subdivisions (provinces, States, or, in the Soviet case, "Socialist Republics"). In theory, any Soviet Republic was free to leave the Union. Study the map of the former Soviet Union (Fig. 2-5), and an interesting geographic corollary emerges: every one of the 15 Soviet Republics had a boundary with a non-Soviet neighbor. Not one was spatially locked within the others. This seemed to give geographic substance to the notion that a republic could opt out of the USSR if it so desired. Reality, of course, was different: Moscow's control over the republics made the Soviet Union a federation in theory only.

Nevertheless, the term *federal* appeared even on official Soviet maps. The former Russian Republic's official name was the Russian Soviet Federative Socialist Republic (RSFSR), as any pre-1992 political map should show. The other 14 republics, however, did not have "federal" or "federative" in their names. Why not?

The answer lies not only in Russia's central and paramount position among the republics of the former Soviet

half of the developing Soviet Union's population, and, as Figure 2-4 shows, they also were (and still are) the most widely dispersed ethnic group in the realm. The Russian Republic, therefore, was by far the largest designated SSR, comprising just under 77 percent of total Soviet territory.

Within the SSRs, smaller minorities were assigned political units of lesser rank. These were called Autonomous Soviet Socialist Republics (ASSRs), which in effect were republics within republics; other areas were designated Autonomous Regions or other nationality-based units. It was a complicated, cumbersome, often poorly designed framework, but in 1924 it was launched officially under the banner of the Union of Soviet Socialist Republics (USSR).

Eventually, the Soviet Union came to consist of 15 SSRs (shown in Fig. 2-5), including not only the original republics of 1924 but also such later acquisitions as Moldova (formerly Moldavia in Romania), Estonia, Latvia, and Lithuania. The internal political layout often was changed, some-

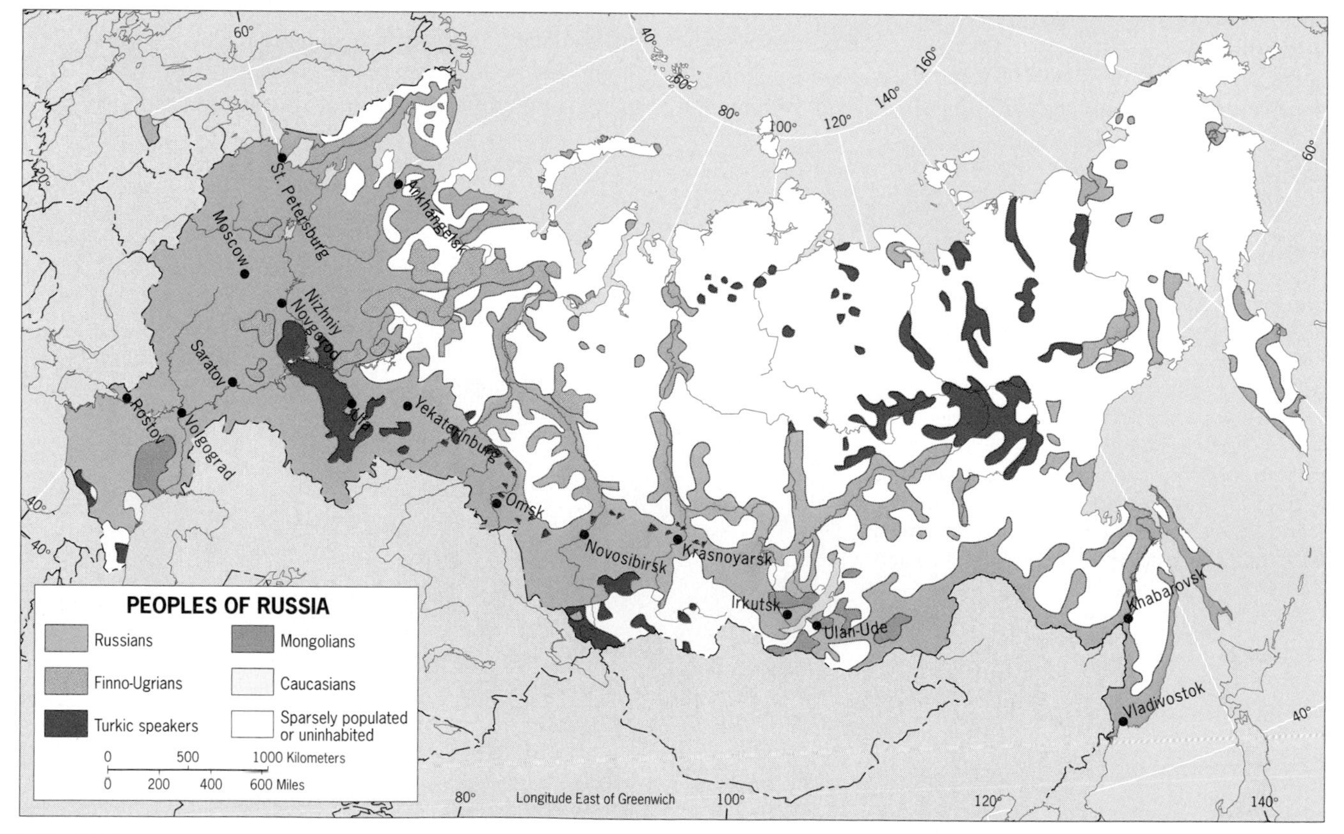

FIGURE 2-4

Union but also in the enormous politicogeographical complexity of the Russian Republic itself. Today, after the devolution of the Soviet Union, the term *Russian Federation* is in common use in Russia, signifying (again theoretically) the relationships between the central authority of Moscow and the governments and administrations of 21 Republics and nearly 70 other jurisdictions that lie within the borders of the Russian state. As we note in the next section of the chapter, this relationship often is strained. Subnational entities want greater autonomy; certain Republics have notions of independence that would match the sovereignty of former Soviet components such as Latvia or Ukraine.

A federation is not simply proclaimed. After half a century of supranational negotiation, there still is no European federation. Federation involves the voluntary sharing of power, the willing association of sometimes disparate peoples who understand that yielding some sovereignty will reward all concerned. The years of Soviet rule did little to spread the ideas and ideals of federalism to the former empire's diverse peoples. Present-day Russia needs those lessons if it is to be a true federal state, but they may be lost on those forging the new system.

times at the whim of the communist empire's dictators. But no communist apartheid-like system of segregation could accommodate the shifting multinational mosaic of the Soviet realm. The republics quarreled among themselves over boundaries and territory. Demographic changes, migrations, war, and economic factors soon made much of the layout of the 1920s obsolete. Moreover, the communist planners made it Soviet policy to relocate entire peoples from their homelands in order to better fit the grand design, and to reward or punish—sometimes capriciously. The overall

FIGURE 2-5

The rolling countryside of Bashkortostan, north of Kazakhstan at the southern end of the Ural Mountains, reveals a republic doing relatively well in the Russian Federation. Modernized, mechanized grain cultivation yields important harvests for a food-deficient country, and oil and natural gas (see the wells at the fringe of the grain field) occur throughout. Add to this a variety of minerals and (see background) extensive pastures, and this republic's diversified economy has given it a strong voice in regional affairs.

effect, however, was to move minority peoples eastward and to replace them with Russians. As time went on, the **Russification** of the Soviet Empire also generated substantial ethnic Russian minorities in all the non-Russian republics (see Table 2-1).

The Russian Republic, though only one of the 15 SSRs, was the Soviet Union's dominant entity—the centerpiece of a tightly controlled federation (see box titled "The Soviet Federation and Its Legacy"). With half the population, the capital city, the realm's core area, and over three-quarters of the Soviet Union's territory, Russia was the empire's nucleus. In other republics, "Soviet" often was simply equated with "Russian"—it was the reality with which the lesser republics lived. Russians came to the other republics to teach (Russian was taught in the colonial schools), to organize (and often dominate) the local Communist Party, and to implement Moscow's economic decisions. This was colonialism, but somehow the communist disguise—how could socialists, as the communists called themselves, be colonialists?—and the contiguous spatial nature of the empire made it appear to the rest of the world as something else. Indeed, on the world stage, the Soviet Union became a champion of oppressed peoples, a force in the decolonization process. It was an astonishing contradiction that would, in time, be fully exposed.

The Soviet Economic Framework

The geopolitical changes that resulted from the founding of the Soviet Union were accompanied by a gigantic economic experiment: the conversion of the empire from a czarist autocracy with a capitalist veneer to communism. From the early 1920s onward, the country's economy would be centrally planned—the communist leadership in Moscow would make all decisions regarding economic planning and development. Soviet planners had two principal objectives: (1) to accelerate industrialization and (2) to **collectivize** agriculture. To accomplish these objectives, the entire country was mobilized, with a national planning commission (Gosplan) at the helm. For the first time ever on such a scale, and for the first time in accordance with Marxist-Leninist principles, an entire country was organized to work toward national goals prescribed by a central government.

The Soviet planners believed that agriculture could be made more productive by organizing it into huge state-run enterprises. The holdings of large landowners were expropriated, private farms were taken away from the farmers, and the land was consolidated into collective farms. Initially, all such land was meant to be part of a *sovkhoz*, literally a grain-and-meat factory in which agricultural efficiency, through maximum mechanization and minimum labor requirements, would be at its peak. But many farmers opposed the Soviets and tried to sabotage the program in various ways, hoping to retain their land.

The farmers and peasants who obstructed the communists' grand design suffered a dreadful fate; their tragedy is one of the factors that nurtured anti-Russian nationalism in the now-independent republics. In the 1930s, for instance, Stalin confiscated Ukraine's agricultural output and then ordered part of the border between the Russian and Ukrainian republics sealed—thereby leading to a famine that killed millions of farmers and their families. In the Soviet Union under communist totalitarianism, the ends justified the means, and untold hardship came to millions who had already suffered under the czars. (In his book *Lenin's Tomb*, David Remnick estimates that between 30 and 60 million people lost their lives from imposed starvation, political purges, Siberian exile, and other causes.)

It was an incalculable human tragedy, but it was hidden from the world by the secretive character of Soviet officialdom. Only after Mikhail Gorbachev initiated his *glasnost* (openness) campaign in the late 1980s did meaningful data on such events emerge. One result has been the rehabilitation of the reputations of many prominent Russians and others who protested these actions and were executed during the dictatorial rule of Josef Stalin, Lenin's successor as head of the Communist Party (1929–1953).

While dissidents were ruthlessly eliminated, Soviet planners finally realized that a smaller-scale collective farm, the *kolkhoz*, would be more acceptable and efficient than the sovkhoz. By 1939, just before the outbreak of World War II, about 90 percent of all farms, ranging from large estates to peasant holdings, had been collectivized.

Productivity, however, did not increase as the Soviet planners had hoped. Farmers like to tend their own land, and they take better care of it than they do the land of the state. Resentment over the harsh imposition of the program undoubtedly played a role also, but ultimately the two persistent problems were poor management and weak incentives for farmers to do their best. Add to these the recurrent weather problems caused by the country's relatively disadvantageous location, and it is not surprising that Soviet farm yields were below expectations—and below what the country needed to feed itself. Throughout the three generations of communist rule, the Soviets had to import food from abroad.

Even as collectivization proceeded, the Soviets brought millions of acres of new land into cultivation through ambitious irrigation schemes. During the 1950s they launched the Virgin and Idle Lands Program in Kazakhstan, turning pasturelands into wheatfields. The program transformed parts of the Kazakh SSR, but it came at the cost of environmental disaster as diverted streams dried up, pesticides poisoned groundwater, and many people became ill. Again, in trying to solve old problems, the grand design created new ones.

The Soviet planners hoped that collectivized and mechanized agriculture would free hundreds of thousands of workers to labor in the industries the Soviets wanted to establish. Enormous amounts of money were allocated to develop manufacturing; the Soviets knew that national power would be based on the country's factories. The transport networks were extended; a second rail route to the Far East (the BAM) was built in the 1970s and 1980s, and such remote places as Almaty (formerly Alma-Ata) and Qaraghandy (formerly Karaganda) were connected to the system (Fig. 2-6). Development of energy was given priority; compared to agriculture, the USSR's industrialization program had good results. Productivity rose rapidly, and when World War II engulfed the empire in 1941, the Soviet manufacturing sector was able to generate the equipment and weapons needed to repel the German invaders.

Yet even in this context, the Soviet grand design held liabilities for the future. Because they could ignore market pressures and certain cost factors, Soviet **state planning** assigned the production of particular manufactured goods to particular places, often disregarding economic geography. For example, the manufacture of railroad cars might be assigned (as indeed it was) to a factory in Latvia. No other factory anywhere else would be permitted to produce this equipment—even if supplies of raw materials would make it cheaper to build them near, say, Volgograd 1200 miles away. Such practices made manufacturing in the USSR extremely expensive, and the absence of competition made managers complacent and workers less productive then they could be.

13

The Soviet system tightly bound the economic geography of the republics to the center—and to each other. Each republic and other administrative entity in the Soviet Union depended for raw materials, energy, or other needs on another part. Assignment by Moscow, rather than market forces, determined the development of territories and places. The city of Bratsk, for example, began as a penal colony in Siberia; once dams and roads had been built, it was made a factory town and became a major regional center of state enterprises. The lack of viability of many such enterprises in the post-Soviet period was a serious impediment to economic reform.

The devolution of the Soviet Empire began even before the reform-minded Gorbachev took office in 1985. A costly military campaign in Afghanistan had failed, and this "Soviet Vietnam" took its toll on Russian society. Soviet economic, political, and social systems were showing signs of failure. Democracy was a beginning to stir in communist-dominated Eastern Europe. Gorbachev knew that the Soviet communist experiment was ending in failure, but he tried to control the forces that were making it collapse. That the dissolution of the old order occurred with minimal loss of life will be a permanent monument to the leader who opened Soviet society to the world—and to itself.

When the USSR dissolved, Russia lost more than territory and population (about 20 million Russians still live in the 14 republics that arose from the former empire); it also lost mineral and fuel resources and faced a spatially dismembered economy. Before the Soviet collapse, the USSR was the world's leading oil producer (most of the oil was consumed at home, so that the Soviet Union never was the leading exporter). But many of the reserves lie outside Russia, and since 1991 Russian leaders have been trying to control the exploitation and export of oil and natural gas from the former Soviet republics. Soviet investments built the industry's infrastructure, they argue, giving Moscow a legitimate claim to a share of the proceeds. Russia also puts pressure on the governments of former Soviet republics not to allow the laying of pipelines to avoid Russian territory. Like other colonial powers, the Russians do not want to let go of their former possessions without maintaining an economic presence.

THE RUSSIAN REALM
POPULATION
Under 50,000
50,000-250,000
250,000-1,000,000
1,000,000-5,000,000
Over 5,000,000
Railroad built before 1917
Railroad built 1917-1991
Road
National capitals are underlined
0 200 400 600 800 1000 Kilometers
0 100 200 300 400 500 600 Miles
ARCTIC
Svalbard (Nor.)
Franz Josef Land
Barents Sea
Novaya Zemlya
Kara Sea
NORWAY
SWEDEN
FINLAND
Murmansk
White Sea
Arkhangelsk
Petrozavodsk
Lake Onega
Ladoga
Helsinki
Tallinn
ESTONIA
LATVIA
Riga
LITHUANIA
Vilnius
Baltic Sea
KALININGRAD (Russia)
POLAND
Warsaw
GERMANY
DEN.
BELARUS
Mensk
St. Petersburg
Novgorod
Lake Peipus
Pechora R.
Vorkuta
N. Dvina R.
Volga R.
Ivanovo
Yaroslavl
Moscow
Tula
Bryansk
Ryazan
Kursk
UKRAINE
Kiev
Dnieper R.
Dniester R.
Chisinau
MOLDOVA
Odesa
Kharkiv
Voronezh
Penza
Saratov
Nizhniy Novgorod
Kazan
Vyatka
Vychegda R.
Kama R.
Izhevsk
Perm
Yekaterinburg
Nizhniy-Tagil
Tolyatti
Samara
Ufa
Magnitogorsk
Orenburg
Ural R.
Chelyabinsk
Volgograd
Rostov
Don R.
Sea of Azov
Crimean Pen.
Sevastopol
Black Sea
Novorossiysk
Krasnodar
Astrakhan
Grozny
GEORGIA
Tbilisi
ARMENIA
Yerevan
AZERBAIJAN
Baki (Baku)
Caspian Sea
TRANS-CAUCASIAN TRANSITION ZONE
TURKEY
IRAQ
IRAN
Tehran
Ob R.
Norilsk
Igarka
Urengoy
Yenisey R.
Sergino
RUSSIAN
Tobol R.
Ishim R.
Irtysh R.
Petropavl
Qostanay (Kustanay)
Kökshetau (Kokchetav)
Omsk
Novosibirsk
Tomsk
Novokuznetsk
Barnaul
Pavlodar
Astana
Semey (Semipalatinsk)
Qaraghandy
Öskemen (Ust-Kamenogorsk)
RUSSIAN TRANSITION ZONE
KAZAKHSTAN
Aral Sea
Amu Darya
Syr Darya
UZBEKISTAN
TURKMENISTAN
Ashgabat
Toshkent
L. Balqash
Almaty
Bishkek
KYRGYZSTAN
TAJIKISTAN
CHINA
Longitude East of Greenwich
50° 60° 70° 80° 90°

FIGURE 2-6

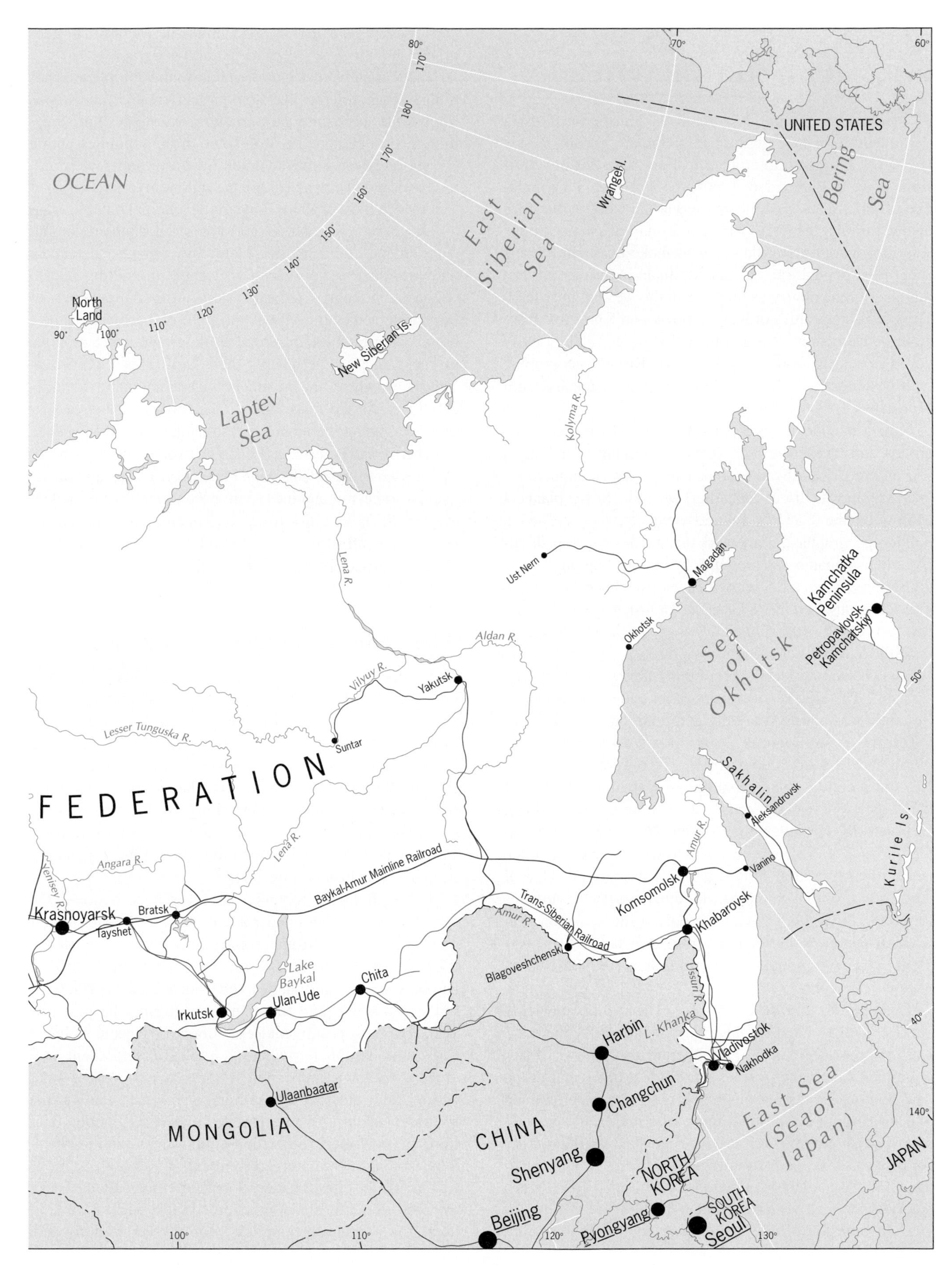

OCEAN
North Land
East Siberian Sea
Laptev Sea
Bering Sea
UNITED STATES
Wrangel I.
New Siberian Is.
Kolyma R.
Lena R.
Aldan R.
Vilyuy R.
Lesser Tunguska R.
Angara R.
Yenisey R.
FEDERATION
Ust Nern
Magadan
Okhotsk
Sea of Okhotsk
Kamchatka Peninsula
Petropavlovsk-Kamchatskiy
Yakutsk
Suntar
Sakhalin
Aleksandrovsk
Kurile Is.
Amur R.
Vanino
Komsomolsk
Khabarovsk
Baykal-Amur Mainline Railroad
Trans-Siberian Railroad
Blagoveshchensk
Krasnoyarsk
Tayshet
Bratsk
Lake Baykal
Irkutsk
Ulan-Ude
Chita
Ussuri R.
L. Khanka
Harbin
Vladivostok
Nakhodka
Changchun
Ulaanbaatar
MONGOLIA
CHINA
Shenyang
East Sea (Sea of Japan)
JAPAN
NORTH KOREA
SOUTH KOREA
Pyongyang
Seoul
Beijing

RUSSIA'S CHANGING POLITICAL GEOGRAPHY

Political maps sometimes show a "European" Russia (bounded on the east by the Urals) and an "Asian" Russia beyond it. No geographic justification exists for this division. True, as Figure 2-6 shows, most of Russia's population of 143 million is concentrated in the western one-fifth of the national territory. But as the map of ethnic groups (Fig. 2-4) reveals, Russians—who make up about 83 percent of the population—are as dominant in the vast eastern regions of the country as they are in the western heartland. The Far Eastern cities of Komsomolsk and Khabarovsk are no less "European" Russian than Rostov or Nizhniy Novgorod (formerly Gorkiy). In short, the Russian geographic realm extends from St. Petersburg to Vladivostok and from Murmansk to Transcaucasia.

Russia is nonetheless a patchwork of nations and ethnic groups, and despite nearly 75 years of communist domination and enforced Russification, many non-Russian cultures remain vibrant. As discussed earlier, when the Soviet planners confronted the USSR's cultural diversity, they created a politico-geographical framework that would, they hoped, satisfy ethnic demands and pressures by establishing smaller political units within the Soviet Socialist Republics. Within the Transcaucasus republic of Georgia, for instance, they laid out two so-called Autonomous Soviet Socialist Republics (ASSRs) and one Autonomous Region.

The most complicated part of this framework lay in Russia itself. The USSR's Russian Soviet Federative Socialist Republic was administratively and politically divided into 6 Territories, 49 Regions, 16 ASSRs, 5 Autonomous Regions, and 10 Autonomous Areas. The 6 Territories and 49 Regions were designed as administrative units, mainly for the purposes of national economic planning, census taking, and so forth. However, the original 16 ASSRs and 5 Autonomous Regions were designed to recognize some of the nearly 40 substantial minorities on Russian soil. One of these regions, for example, was the Jewish Autonomous Region in the remote Soviet Far East, a rugged and nearly isolated corner of the country not much larger than Vermont (and never populated by more than 30,000 Russians of Jewish descent).

Even before the Soviet Union fell apart, this unwieldy politico-geographical framework was failing. While the 5 Autonomous Regions held a higher status than the 10 Autonomous Areas, the peoples and leaders of several of the Autonomous Areas had begun to play a more prominent role in national affairs than those of some of the regions. (The Jewish Region, for instance, was never important.) But these issues were submerged when the devolution of the whole Soviet Union gathered momentum. When Russia emerged as a newly independent republic at the end of 1991, it took the official name Russian Federation in acknowledgment of its inherited, multi-tiered political geography and the diversity of its cultural geography.

The Russian Federation was heir to the complicated administrative structure the Soviets had created, and it was impossible to change that structure overnight. Yet something had to be done because some of the republics within Russia, and even some other administrative units, were agitating for independence or greater autonomy.

In 1992, most of Russia's nearly 90 internal republics, autonomous regions, oblasts, and krays (all of them components of the administrative hierarchy) signed a document known as the Russian Federation Treaty, committing them to cooperate in the new federal system. At first a few units refused to sign, including Tatarstan, scene of Ivan the Terrible's brutal conquest more than four centuries ago, and a republic in the Caucasus periphery, then known as Chechenya-Ingushetia, where Muslim rebels waged a campaign for independence. As the map shows, Chechnya-Ingushetiya split into two separate republics whose names were then spelled Chechenya and Ingushetia (Fig. 2-7). Eventually only Chechnya refused to sign the Russian Federation Treaty, and subsequent Russian military intervention led to a prolonged and violent conflict, with disastrous consequences for Chechnya's people and infrastructure (the capital, Groznyy, was completely destroyed [see photo, p. 127]) and for Russia's political leaders.

The Federal Framework of Russia

The spatial framework of the still-evolving Russian Federation is as complex as that of the Russian Federative Socialist Republic of communist times. As the twenty-first century opened, the Federation consisted of 89 entities: 2 Autonomous Federal Cities, 21 Republics, 11 Autonomous Regions (Okrugs), 49 Provinces (Oblasts), and 6 Territories (Krays). Moscow and St. Petersburg are the two Autonomous Federal Cities. The 21 Republics, recognized to accommodate substantial ethnic minorities in the population, lie in several clusters (Fig. 2-7).

Although the 21 Republics are listed near the top in the hierarchy of Russian federal components, some are far more important (or troublesome) than others. A cluster of them lies in the Russian core, while others are located in Russia's internal peripheries: the Caucasus, far north, Central Asian south, and Siberian east. During the past few years, the most prominent Republics were Chechnya in the Caucasian periphery, where a political settlement still eluded Moscow; Tatarstan in the Russian core, where Tatar nationalism and (Muslim) religious revival were on the rise; and Sakha in the eastern periphery, a Republic one-third the size of Canada with a minuscule population but a huge storehouse of natural resources.

During the 1990s, several entities that had not been awarded Republic status proclaimed their intent to seek it, hoping to increase their influence in the Federation while

Chechnya was the only republic that rejected incorporation into the new Russian Federation, causing a series of armed conflicts that devastated its infrastructure. The capital, Groznyy, was totally destroyed; not a single building or structure survived the wars unscathed. Today the city's remaining population survives in basements and in the remnants of homes and shops. These four women sit on a makeshift bench in front of the ruins of several homes; look up and down the street, and all of it looks like this. The women on the left have piled some wood in the hope of selling it as firewood to other survivors.

reducing Moscow's power over local people and resources. In Tatarstan and in Primorskiy Kray (Maritime Region) on the Pacific coast, there was even talk of secession. In part, these efforts reflected the weakness and corruption of the center, the ineffectiveness of the administration, and the opportunism of local leaders. In 2000, however, President Boris Yeltsin, who had come to symbolize this state of affairs, was succeeded by Vladimir Putin, who appeared determined to weaken the power of the Regions and Republics and their local kingpins. The stage was set for still another revision of the Russian map.

Moscow and the Federal Components

When Russia lost control over its external Soviet Republics and had to go it alone after the communist collapse, its leaders faced a massive problem. A multinational, multicultural state that had been accustomed to authoritarian rule and government control over virtually everything—from factory production to everyday life—now had to be governed in a new way. Democratization of the political system, transition to a market economy, sale of state-owned industries (privatization), and liberation of the press and other media must all take place in an orderly way, or the country would, literally, fall apart.

Russia's leaders knew that their options were limited. They could continue to hold as much power as possible at the center, making decisions in Moscow that would apply to all the Republics, Regions, and other subdivisions of the state. Such a **unitary state system**, with its centralized government and administration, marked authoritarian kingdoms of the past and serves totalitarian dictatorships of the present. Or they could share power with the Republics and Regions, allowing elected regional leaders to come to Moscow to represent the interests of their people. This is the federal system Russia chose as the only way to accommodate the country's economic and cultural diversity.

14

In a *federal system*, the national government usually is responsible for matters such as defense, foreign policy, and foreign trade. The Regions (or provinces, States, or other subdivisions) retain authority over affairs ranging from education to transportation. A federal system does not create unity out of diversity, but it does allow diverse components of the state to coexist, their common interests represented by the national government and their regional interests by their local administrations. Some countries owe their survival as coherent states to their federal frameworks. India and Australia are cases in point.

But to maintain a generally acceptable balance of power between the center and the Regions (or States) is difficult. Disputes over "States' rights" continue to roil the American political scene more than two centuries after the Constitution was adopted. Russia, which also owes its continuity to a grand federal experiment, is barely more than ten years old. Power shifts continue, and the framework of the Russian Federation will undoubtedly change over time. This is one advantage of federation: change can be accommodated without threatening the whole system.

RUSSIA AND ITS INTERNAL DIVISIONS
Russian administrative units
Autonomous Regions within administrative units
Proclaimed internal republics
Peripheries in transition
Each internal republic is colored separately
National capital is underlined
0 200 400 600 800 Kilometers
0 100 200 300 400 500 Miles
ARCTIC
Barents Sea
Kara Sea
NORWAY
SWEDEN
FINLAND
DENMARK
GERMANY
POLAND
LITHUANIA
LATVIA
ESTONIA
BELARUS
UKRAINE
MOLDOVA
KALININGRAD REPUBLIC (Proclaimed)
Baltic Sea
White Sea
Svalbard (Nor.)
Franz Josef Land
Novaya Zemlya
MURMANSK
KARELIYA
ARKHANGELSK
NENETSKIY
KOMI
YAMALO-NENETSKIY
KHANTY-MANSIYSKIY
TYUMEN
VOLOGDA REPUBLIC (Proclaimed)
MOSCOW
NIZHE-GOROD
MORDOVIYA
MARIY-EL
CHUVASHIYA
TATARSTAN
UDMURTIYA
KOMI-PERMYATSKIY
PERM
SVERDLOVSK (Urals Republic) (Proclaimed)
BASHKORTOSTAN
RUSSIA
ADYGEYA
KARACHAYEVO-CHERKESIYA
KABARDINO-BALKARIYA
NORTH OSSETIA
KALMYKIYA
DAGESTAN
CHECHNYA
INGUSHETIYA
GEORGIA
ARMENIA
AZERBAIJAN
TURKEY
IRAQ
IRAN
TRANS CAUCASIAN TRANSITION ZONE
Caspian Sea
Black Sea
Sea of Azov
Crimean Pen.
RUSSIAN TRANSITION ZONE
KAZAKHSTAN
UZBEKISTAN
KYRGYZSTAN
CHINA
KHAKASIYA
ALTAY
ALTAYA
L. Balqash
Lake Peipus
Lake Ladoga
Lake Onega
Volga R.
Don R.
Dnieper R.
Dniester R.
N. Dvina R.
Pechora R.
Vychegda R.
Kama R.
Ural R.
Tobol R.
Ishim R.
Irtysh R.
Ob R.
Yenisey R.
Syr Darya
Longitude East of Greenwich
RUSSIAN ADMINISTRATIVE UNITS (named after their capitals)
1. Astrakhan
2. Belgorod
3. Bryansk
4. Chelyabinsk
5. Ivanovo
6. Kaluga
7. Kemerovo
8. Kostroma
9. Krasnodar
10. Kurgan
11. Kursk
12. Lipetsk
13. Moscow
14. Novgorod
15. Novosibirsk
16. Omsk
17. Orenburg
18. Orël
19. Penza
20. Pskov
21. Rostov
22. Ryazan
23. St. Petersburg
24. Samara
25. Saratov
26. Smolensk
27. Stavropol
28. Tambov
29. Tomsk
30. Tula
31. Tver
32. Ulyanovsk
33. Vladimir
34. Volgograd
35. Voronezh
36. Vyatka
37. Yaroslavl

FIGURE 2-7

OCEAN
North Land
UNITED STATES
Wrangel I.
East Siberian Sea
Bering Sea
CHUKOTSKIY
New Siberian Is.
Laptev Sea
Kolyma R.
TAYMYRSKIY
MAGADAN
KAMCHATKA
Lena R.
SAKHA (YAKUTIYA)
Aldan R.
Vilyuy R.
Sea of Okhotsk
KRASNOYARSK
Lesser Tunguska R.
EVENKIYSKIY
FEDERATION
KHABAROVSK
SAKHALIN
Kurile Is.
Angara R.
Yenisey R.
Lena R.
AMUR OBLAST (Amur Republic [Proclaimed])
Amur R.
YEVREYSKAYA (JEWISH)
IRKUTSK
BURYATIYA
CHITA
UST-ORDYNSKIY
AGINSKIY BURYATSKIY
Ussuri R.
PRIMORSKIY KRAY (Primorye Republic [Proclaimed])
L. Khanka
TYVA
East Sea (Sea of Japan)
MONGOLIA
CHINA
JAPAN
NORTH KOREA
SOUTH KOREA

Russia's federal framework is subject to many pressures. One of these arises from the very origins of the federal map: it had to be built on the Soviet framework with its confusing regional hierarchy (oblasts, krays, okrugs, republics). The Republics had special status before and after 1991, provoking envy among leaders of other Regions. While the leader of Tatarstan spoke publicly of self-determination and autonomy for his Republic, representatives from Chelyabinsk Region were taking orders from Moscow. A federation of unequal parts tends to have problems (witness Canada), and the inherited inequality of Russia's federal components causes difficulties.

Another problem arises from Russia's sheer size. Geographers refer to the principle of **distance decay** to explain how increasing distances between places tend to reduce interactions among them. Because Russia is the world's largest country, distance forms a significant ingredient in the relationships between the capital and the outlying areas. Furthermore, Moscow lies in the far west of the giant country, half a world away from the shores of the Pacific. Not surprisingly, one of the most obstreperous Regions has been Primorskiy, the Region of Vladivostok. In 1997, the Russian government sought to dismiss the elected governor of Primorskiy because, among other things, he refused to implement federal decisions and would not forward the Region's tax revenues to Moscow.

Still another problem is reflected by Figure 2-7: the enormous size variation among the Republics (and to a lesser extent, the Regions). Whereas the (territorially) smallest Republics are concentrated in the Russian core area, the largest lie far to the east, where Sakha is nearly a thousand times as large as Ingushetiya. On the other hand, the populations of the huge eastern Republics are tiny compared to those of the smaller ones in the core. Such diversity spells administrative difficulty.

But perhaps the most serious current problem for Russia is the growing social disharmony between Moscow and the subnational political entities. Almost wherever one goes in Russia today, Moscow is disliked and often berated by angry locals. The capital is seen as the privileged playground for those who have benefited most from the post-Soviet transition—bureaucrats and hangers-on whose economic policies have driven down standards of living, whose greed and corruption have hurt the economy, whose actions in Chechnya have been a disaster, who have allowed foreigners (prominently the United States) to erode Russian power and prestige, who fail to pay the wages of workers toiling in industries still owned by the state, and who do not represent the Russian people. Complaints about the capital are not uncommon in free countries, but in Russia the mistrust between capital and subordinate areas has become so serious that it constitutes a key challenge for the

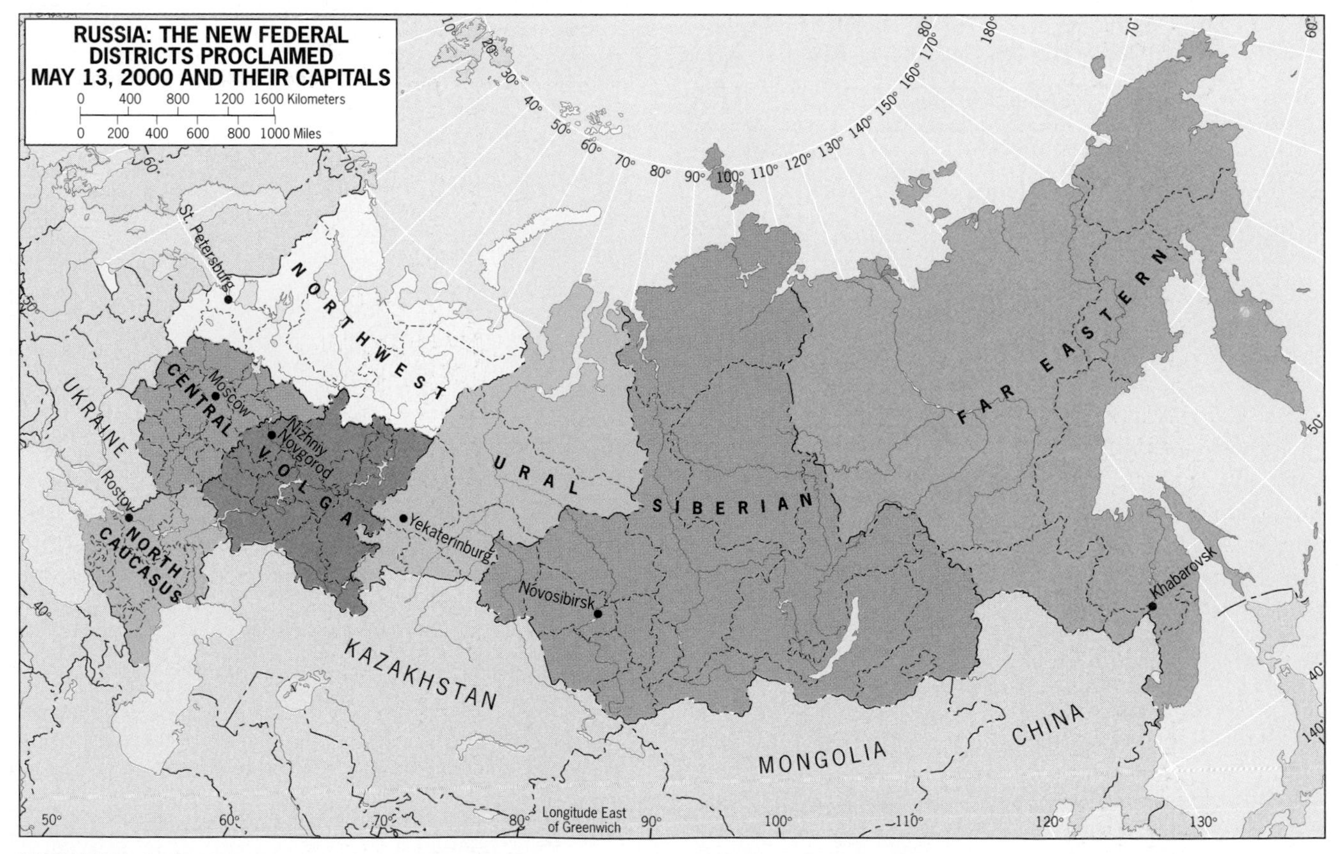

FIGURE 2-8

Putin government—and for any future administrators of the Russian state.

In mid-2000, the Putin administration moved to diminish the influence of the Regions by creating a new spatial framework that combines the 89 Regions, Republics, and other entities into 7 new administrative units—not to enhance their influence in Moscow, but to increase Moscow's authority over them (Fig. 2-8). Each of these new federal administrative districts has its capital city, which will become the conduit for Moscow's "guidance," as the official plan puts it. This framework will counter some powerful vested interests in the Republics and Regions, however, and it will reverse a trend toward ever-greater local self-determination. It is far from certain that the system will work.

Capitalism under the watchful eye of the architect of state communism! Here, on a brilliant spring afternoon near the southwestern edge of Moscow, shoppers browse through an open-air market selling shoes, garments, and sporting goods in front of Lenin Stadium (site of the 1980 Olympics and 1988 World Youth Games). Look closely, and you will spot some familiar brand names. It is a scene unthinkable little more than a decade ago.

CHANGING SOCIAL GEOGRAPHIES

Look at the population data for Russia (Table I-2), and you have firm evidence that the decade of post-Soviet transition has been traumatic for the Russian people. Russia's 2002 population of 143.2 million is not only small for so huge a country: it is also shrinking—and not from emigration. Russian families are having fewer children than are needed to keep the population growing.

In Chapter 1 we noted that the populations of several European countries also are declining as a result of urbanization, economic prosperity, marriage at later ages, and family planning options. But in Russia, studies show that the causes of **population decline** lie in the dislocation, turmoil, fears, and uncertainties arising from the post-Soviet transition. After World War II, during which Russians suffered millions of casualties, the Soviet regime encouraged families to have many children, and women who gave birth to ten children were declared heroes of the Soviet Union and given medals. For a time the Soviet Union had one of the world's faster-growing populations. Now, the statistics reveal shock and anguish. Male life expectancy has dropped from 66 in the mid-1960s to 61 in the early 2000s, and for women it has declined from 74 to 73. The gap between men and women in this respect is the largest in the world, and male alcoholism, suicide, and other manifestations of social disorder are to blame. The uncertainties of life, the unprecedented loss of jobs and security, the competitiveness of a market economy, and related factors have sapped Russia's national confidence, and the demographic data reveal the extent of the damage.

Currently, the Russian population is declining by nearly one million per year. According to medical geographers, this reflects not just the effects of growing alcoholism, drug abuse, heavy smoking, and poor diets (including even malnutrition); it also results from rising rates of disease. Tuberculosis is taking a heavy toll, especially on children; HIV and AIDS cases are multiplying rapidly; sexually transmitted diseases are spreading fast. Even as life expectancies decline, so many people will be sick in the later years of their shortened lifetimes that the health crisis in Russia, if it is not somehow reversed, will overshadow all else, endangering the country's prospects in every sphere.

Coupled with this distress is the decline of national institutions. Even the Communist Party, still strongly present in much of the country, is in disarray, riven by factionalism. The armed forces, until recently one of the two most powerful in the world, are in disrepair, dispirited and humiliated after a calamitous campaign in breakaway Chechnya. Major universities lack money and modern facilities. Leading industries and mining companies still owned by the state force workers to strike because they have not been paid; privatized factories confront economic failure. Legal systems, banking operations, and property regulations are not sufficiently developed to ensure orderly procedures. Tax collection is inadequate.

Such a situation breeds corruption and cronyism, and the areas outside the capital feel that Moscow is thoroughly corrupt. Simultaneously, organized and petty crime have become serious problems throughout Russia. Some observers describe organized crime as Russia's most successful industry, extorting millions of rubles from businesses and individuals, dealing in contraband, and setting up foreign bases through which to channel and launder their gains.

The dangers inherent in Russia's difficult transition are numerous. Frustration is widespread—among underpaid and undersupplied members of the armed forces, among striking workers, among citizens unable to make ends meet even as they cope with petty criminals. Yet Russia remains a formidable nuclear power and retains a vast arsenal of weapons. Concern that nuclear materials and weapons might find their way into illicit arms markets is rising. In addition, Russia's voters may express their frustration by electing leaders who will reverse the nation's experiment with democratic government. It would not be the first time such a reversal has happened.

RUSSIA'S PROSPECTS

When Peter the Great began to reorient his country toward Europe and the outside world, he envisaged a Russia with warm-water ports, a nation no longer encircled by Swedes, Lithuanians, Poles, and Turks, and a force in European affairs. Core-periphery relationships always have been crucial to Russia, and they remain important today. With 20 million Russians living in the former Soviet Republics on Russia's rim, without the effective protection from Moscow that they enjoyed during the communist period, and with conflicts raging among several of these neighbors from Georgia to Tajikistan, Russia worries anew about its periphery. The Russians argue that the collapse of their Soviet Union notwithstanding, they have the right to be involved, and if necessary to intervene, in their former colonial empire.

Heartlands and Rimlands

Russia's relative location has long been the subject of study and conjecture by geographers. Nearly a century ago, the British geographer Sir Halford Mackinder (1861–1947) argued in a still-discussed article entitled "The Geographical Pivot of History" that western Russia and Eastern Europe enjoyed a combination of natural protection and resource wealth that would someday propel its occupants to world power. The protected core area, he reasoned, overshadowed the exposed periphery. Eventually, this pivotal interior region of Eurasia, which he later called the *heartland*, would become a stage for world domination.

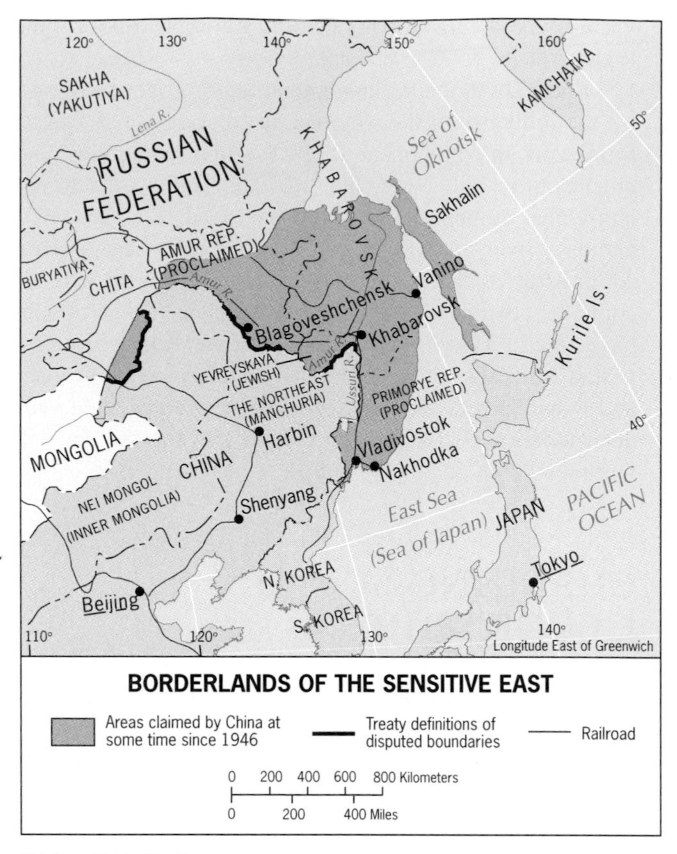

FIGURE 2-9

17

Mackinder's **heartland theory** was published when Russia was a weak, economically backward society ripe for revolution, but Mackinder stuck to his guns. When Russia, now in control of Eastern Europe and with a vast colonial empire, emerged from World War II as a superpower, Mackinder's conjectures seemed prophetic.

But not all geographers agreed. Probably the first scholar to use the term *rimland* (today *Pacific Rim* is part of everyday language) was Nicholas Spykman, who in 1944 countered Mackinder by calculating that Eurasia's periphery, not its core, held the key to global power. Spykman foresaw the rise of rimland states as superpowers and viewed Japan's emergence to wealth and power as just the beginning of that process.

Russia's Role in the World

Even as Russia's leaders seek to address the enormous internal problems that beset this giant country, Moscow also tries to maintain a prominent international role. Russia faces an encroaching NATO in Europe, a strong Muslim presence in Central Asia, a rising China along its eastern flanks. During the 1990s, the Russians settled more than a dozen lingering border disputes with China (some remain unresolved). But for the future, the larger question is likely to involve China's "lost" territories. As Figure 2-9 shows,

Russia's eastern edge just touches North Korea, virtually landlocking China's historic industrial heartland. Russians took these lands from China more than a century ago, and in the future this may again become an issue. Meanwhile, the Russian government in 2001 faced several other, more immediate international challenges and objectives. These included:

1. *Russian interest in the oil and natural gas reserves now owned or shared by the former Soviet Republics.* Moscow seeks to assert its rights to such reserves in the Caspian Basin by negotiating maritime borders in the Caspian Sea with its ex-Soviet neighbors. Russian leaders also want to ensure that oil and gas pipelines from Caspian Sea neighbors cross Russian territory to Black Sea terminals, rather than cross Turkey or Iran.
2. *Russian concern over the fate of the approximately 20 million Russians living in former Soviet Republics.* Moscow has alerted the international community to alleged human-rights violations in several republics, including Estonia and Latvia, and discrimination against Russians in Central Asian countries. As Table 2-1 indicates, Russian remains an official language in only 2 of the 14 former colonies, creating difficulties for Russian expatriates in education and other areas.
3. *Russian determination to remain a champion of Slavic interests in Eastern Europe.* This has led Moscow to support a Serbian regime in Yugoslavia against Western European and U.S. efforts to constrain Belgrade's role as a troublemaker in the region. Moscow has also approved a contractual "union" with neighboring Belarus. Russia again played the Slavic card in its objections to NATO expansion eastward to include Poland, Hungary, and the Czech Republic, which newspapers in Moscow described as a growing threat to "Slavic" interests.
4. *Russian resolve to project what remains of its power onto the international stage by thwarting U.S. and European designs and by negating Western initiatives.* For example, Russia has been reluctant to support UN and U.S. policies toward Iraq, thus creating diplomatic problems for the alliance that opposes Baghdad and signaling support for rogue regimes. In 1999, Russia refused to endorse the NATO campaign against its Eastern European ally, Serbia (the new Yugoslavia), and sought to establish itself as Serbia's protector in the conflict.
5. *Russian problems in the Caucasian periphery range from repeated policy failures in Chechnya (which nominally is part of Russia) to difficult relations with Georgia (which is not).* This area is important to Russia for several reasons: here lie oil and natural gas reserves and the pipelines to transport them; here Russia's multicultural mosaic reaches its greatest complexity; here the Russian realm borders major Islamic neighbors.

Bounded by 14 countries from Norway in the west to North Korea in the east, occupying the heart of the world's greatest landmass and sharing it with the world's two most populous states, China and India, Russia remains a major, if diminished, force in international affairs. The nature of a New World Order now in the making will rely substantially on events in the land the czars forged.

REGIONS OF THE REALM

So vast is Russia, so varied its physiography, and so diverse its cultural landscape that regionalization requires a small-scale perspective and a high level of generalization. Figure 2-10 outlines a four-region framework: the Russian Core and its Peripheries west of the Urals, the Eastern Frontier, Siberia, and the Far East. As we will see, each of these massive regions contains major subregions.

RUSSIAN CORE AND PERIPHERIES

18 The heartland of a state is its **core area**. Here much of the population is concentrated, and here lie its leading cities, major industries, densest transport networks, most intensively cultivated lands, and other key components of the country. Core areas of long standing strongly reflect the imprints of culture and history. The Russian core area, broadly defined, extends from the western border of the Russian realm to the Ural Mountains in the east (Fig. 2-10). This is the Russia of Moscow and St. Petersburg, of the Volga River and its industrial cities, of farms and forests. Here, the Muscovy Russians asserted their power and began the formation of the Russian Empire, forerunner to the now-defunct Soviet Union.

Central Industrial Region

At the heart of the Russian Core lies the Central Industrial Region (Fig. 2-11). The precise definition of this subregion varies, for all regional definitions are subject to debate. Some geographers prefer to call this the Moscow Region, thereby emphasizing that for over 250 miles (400 km) in all

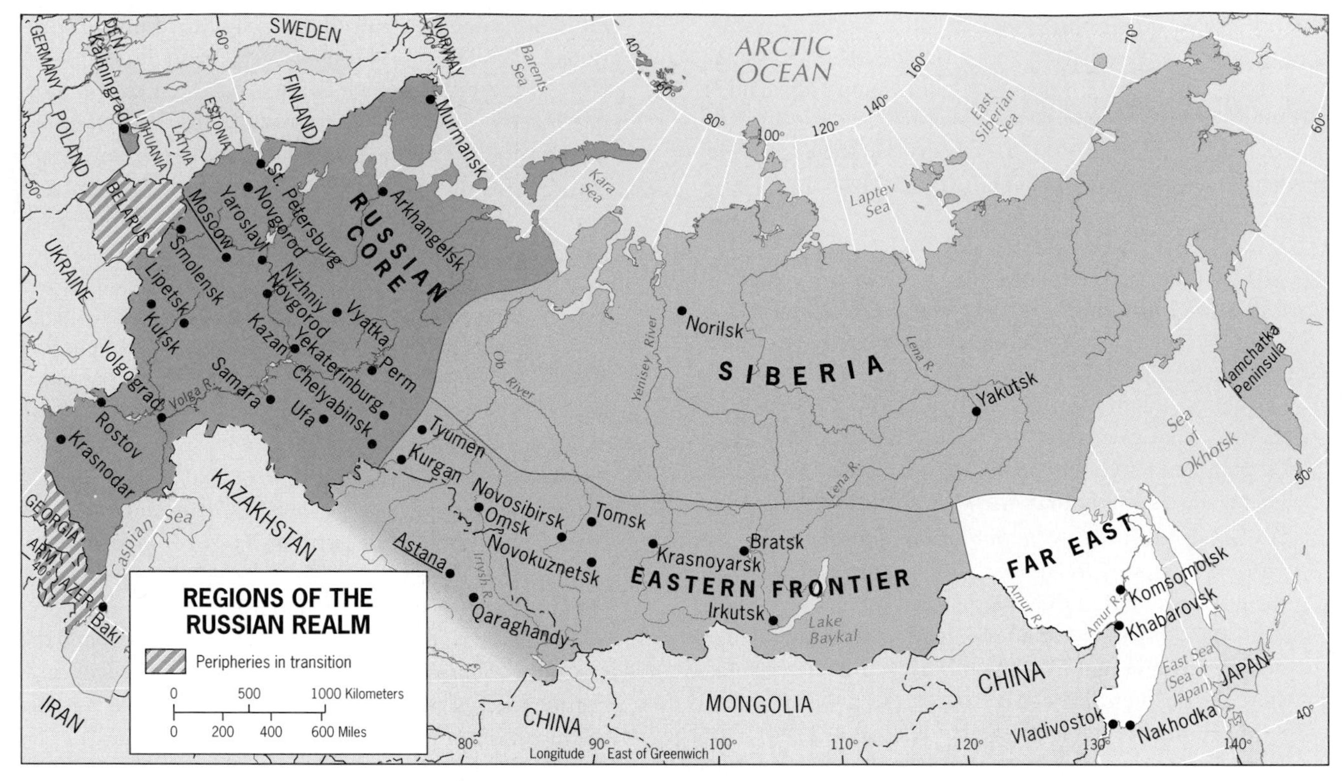

FIGURE 2-10

directions from the capital, everything is orientated toward this historic focus of the state. As both Figures 2-1 and 2-6 show, Moscow has maintained its decisive centrality: roads and railroads converge in all directions from Ukraine in the south; from Mensk (Belarus) and the rest of Eastern Europe in the west; from St. Petersburg and the Baltic coast in the northwest; from Nizhniy Novgorod (formerly Gorkiy) and the Urals in the east; from the cities and waterways of the Volga Basin in the southeast (a canal links Moscow to the Volga, Russia's most important navigable river); and even to the subarctic northern periphery that faces the Barents Sea (see box titled "Facing the Barents Sea").

Moscow itself is a transforming metropolis (2002 population: 9.3 million), with high-rise apartment complexes increasingly dominating the residential landscape. Although this construction has helped alleviate the capital's severe housing shortage, Moscow—like nearly all Russian cities—remains badly overcrowded, with most people forced to accept unreasonably cramped personal living spaces.

Moscow is also the focus of an area that includes some 50 million inhabitants (more than one-third of the country's total population), many of them concentrated in such major cities as Nizhniy Novgorod, the automobile-producing "Soviet Detroit"; Yaroslavl, the tire-producing center; Ivanovo, the heart of the textile industry; and Tula, the mining and metallurgical center where lignite (brown coal) deposits are worked.

St. Petersburg (the former Leningrad) remains Russia's second city, with a population of 5.1 million. Under the

MAJOR CITIES OF THE REALM

City	Population* (in millions)	City	Population* (in millions)
Baki (Baku), Azerbaijan	2.0	St. Petersburg, Russia	5.1
Irkutsk, Russia	0.7	Tbilisi, Georgia	1.3
Kazan, Russia	1.1	Vladivostok, Russia	0.8
Moscow, Russia	9.3	Volgograd, Russia	1.0
Nizhniy Novgorod, Russia	1.5	Yekaterinburg, Russia	1.4
Novosibirsk, Russia	1.5	Yerevan, Armenia	1.3

*Based on 2002 estimates.

Facing the Barents Sea

North of the latitude of St. Petersburg, the region we have named the Russian Core takes on Siberian properties. However, the Russian presence is much stronger in this remote northern periphery than it is in Siberia proper. Two substantial cities, Murmansk and Arkhangelsk, are road- and rail-connected outposts in the shadow of the Arctic Circle.

Murmansk lies on the Kola Peninsula not far from the border with Finland (Fig. 2-6). In its hinterland lie a variety of mineral deposits, but Murmansk is particularly important as a naval base. During World War II, Allied ships brought supplies to Murmansk; the city's remoteness shielded it from German occupation. After the war, it became a base for nuclear submarines. This city is also an important fishing port and a container facility for cargo ships.

Arkhangelsk is located near the mouth of the Northern Dvina River where it reaches an arm of the White Sea (Fig. 2-6). Ivan the Terrible chose its site during Muscovy's early expansion; the czar wanted to make this the key port on a route to maritime Europe. Yet Arkhangelsk, mainly a port for lumber shipments, has a shorter ice-free season than Murmansk, whose port can be kept open with the help of the warm North Atlantic Drift ocean current (and by icebreakers if necessary).

Nothing in Siberia east of the Urals rivals either of these cities—yet. But their existence and growth prove that Siberian barriers to settlement can be overcome.

czars, St. Petersburg was the focus of Russian political and cultural life, and Moscow was a distant second city. Today, however, St. Petersburg has none of Moscow's locational advantages, at least not with respect to the domestic market. It lies well outside the Central Industrial Region near the northwestern corner of the country, 400 miles (650 km) from Moscow. Neither is it better off than Moscow in terms of resources: fuels, metals, and foodstuffs must all be

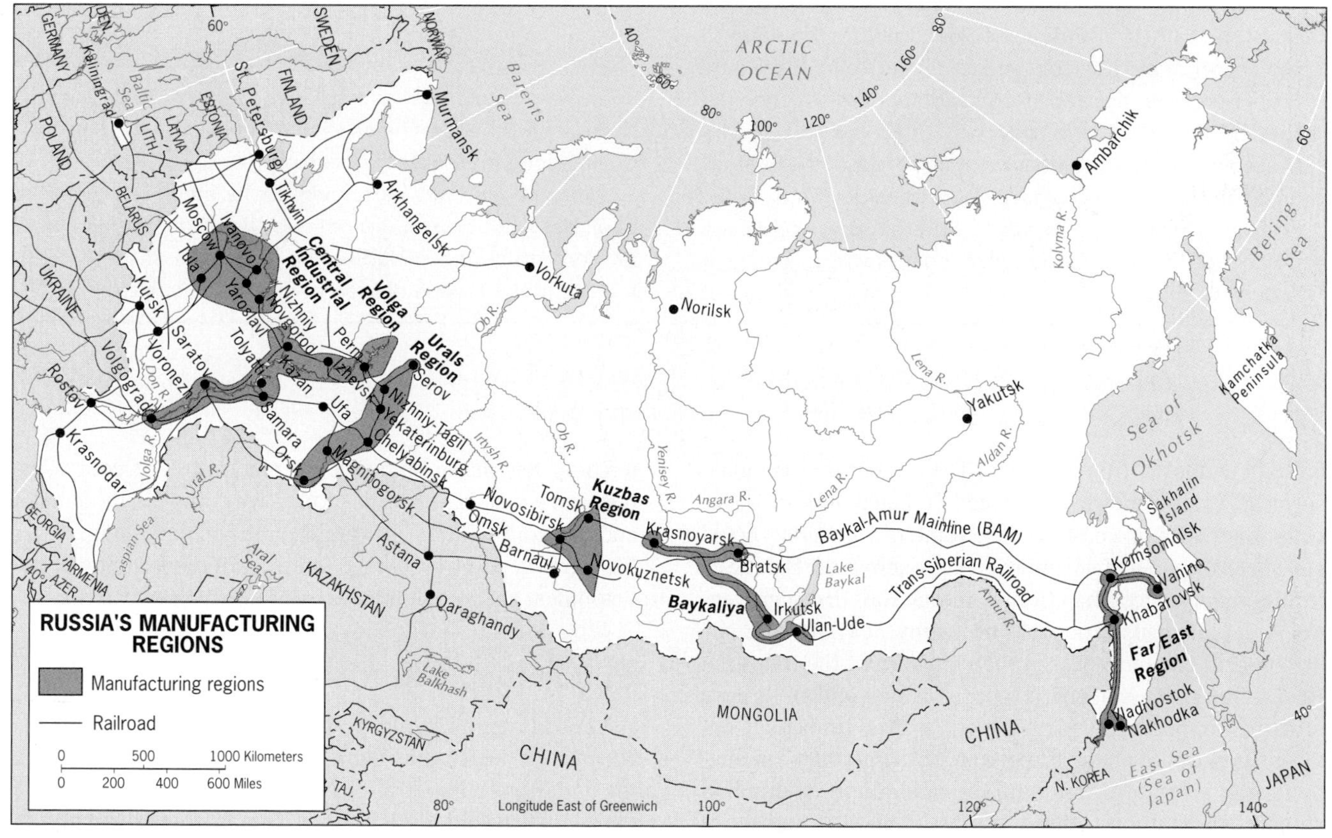

FIGURE 2-11

Among the Realm's Great Cities . . .

Moscow

In the vastness of Russia's territory, Moscow, capital of the Federation, seems to lie far from the center, close to its western margin. But in terms of Russia's population distribution, Moscow's centrality is second to none among the country's cities. Moscow (9.3 million) lies at the heart of Russia's primary core area and at the focus of its circulation systems.

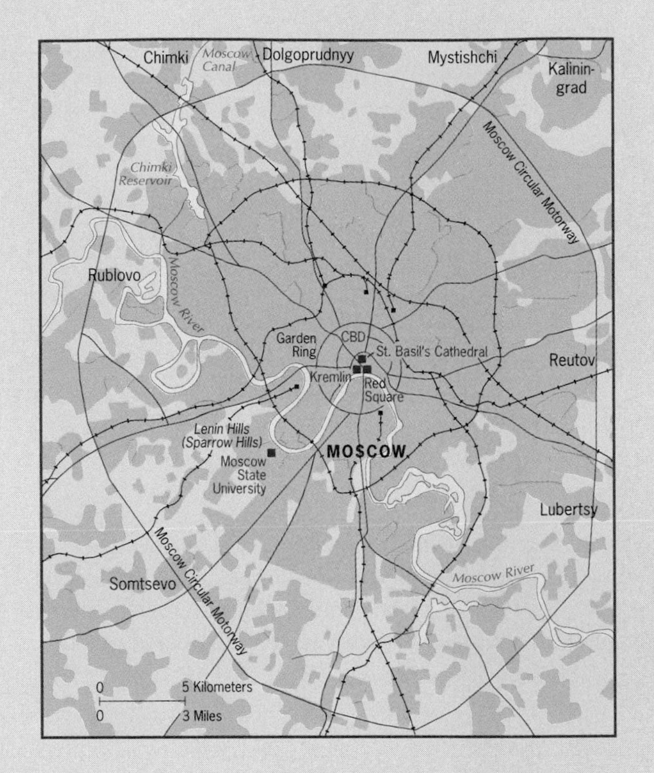

On the banks of the meandering Moscow River, Moscow's skyline of onion-shaped church domes and modern buildings rises from the forested flat Russian Plain like a giant oasis in a verdant setting. Archeological evidence points to ancient settlement of the site, but Moscow enters recorded history only in the middle of the twelfth century. Forest and river provided defenses against Tatar raids, and when a Muscovy force defeated a Tatar army in the late fourteenth century, Moscow's primacy was assured. A huge brick Kremlin (citadel, fortress), with walls more than a mile in length and with 18 towers, was built to ensure the city's security. From this base Ivan the Terrible expanded Muscovy's domain and laid the foundations for Russia's vast empire.

The triangular Kremlin and the enormous square in front of it (Red Square of revolutionary times), flanked by St. Basil's cathedral and overlooking the Moscow River, is the center of old Moscow and is still the heart of the city. From here, avenues and streets radiate in all directions to the Garden Ring and beyond. Until the 1970s, the Moscow Circular Motorway encircled most of the built-up area, but today the city sprawls far outside this beltway. For all its size, Moscow never developed a world-class downtown skyline. Communist policy was to create a "socialist city" in which neighborhoods—consisting of apartment buildings, workplaces, schools, hospitals, shops, and other amenities—would be clustered so as to obviate long commutes into a high-rise city center.

Moscow may not be known for its architectural appeal, but the city contains noteworthy historic as well as modern structures, including the Cathedral of the Archangel and the towers of Moscow State University.

brought in, mostly from far away. The former Soviet emphasis on self-sufficiency even reduced St. Petersburg's asset of being on the Baltic coast because some raw materials could have been imported much more cheaply across the Baltic Sea from foreign sources than from domestic sites in distant Central Asia (only bauxite deposits lie nearby, at Tikhvin).

Yet St. Petersburg was at the vanguard of the Industrial Revolution in Russia, and its specialization and skills have remained important. Today, the city and its immediate environs contribute about 10 percent of the country's manufacturing, much of it through fabricating high-quality machinery. In addition to the usual association of industries (metals, chemicals, textiles, and food processing), St. Petersburg has major shipbuilding plants and, of course, its port and naval station. This productive complex, though not enough to maintain its pre-Revolution advantage over Moscow, kept the city in the forefront of modern Soviet development and will continue to do so in the new Russia.

Povolzhye: The Volga Region

A second region lying within the Russian Core is the *Povolzhye*, the Russian name for an area that extends along the middle and lower valley of the Volga River. It would be appropriate to call this the Volga Region, for that greatest of Russia's rivers is its lifeline and most of the cities that lie in

Among the Realm's Great Cities . . .

St. Petersburg

Czar Peter the Great and his architects transformed the islands of the Neva Delta, at the head of the Gulf of Finland, into the Venice of the North, its palaces, churches, waterfront façades, bridges, and monuments giving St. Petersburg a European look unlike that of any other city in Russia. Having driven out the Swedes, Peter laid the foundations of the Peter and Paul fortress on the bank of the wide Neva River in 1703, and the city was declared the capital of Russia in 1712.

Peter's "window on Europe," St. Petersburg (named after the saint, not the czar) was to become a capital to match Paris, Rome, and London. During the eighteenth century, St. Petersburg acquired a magnificent skyline dominated by tall, thin spires and ornate cupolas, and graced by baroque and classical architecture. The Imperial Winter Palace and the adjoining Hermitage Museum at the heart of the city are among a host of surviving architectural treasures.

Revolution and war lay in St. Petersburg's future. In 1917, the Russian Revolution began in the city (then named Petrograd), and following the communist victory it lost its functions as a capital to Moscow and its name to Lenin. As Leningrad, it suffered through the 872-day Nazi siege during World War II, holding out heroically through endless bombardment and starvation that took nearly 1 million lives and severely damaged many of its buildings.

The communist period witnessed the neglect and destruction of some of Leningrad's most beautiful churches, the intrusion and addition of crude monuments and bleak apartment complexes, and the rapid industrialization and growth of the city (which now totals 5.1 million). Immediately after the collapse of the Soviet Union, it regained its original name and a new era opened. The Russian Orthodox Church revived, building restoration went forward, and tourism boomed. But the transformation from communist to capitalist ways has been accompanied by social problems that include crime and poverty.

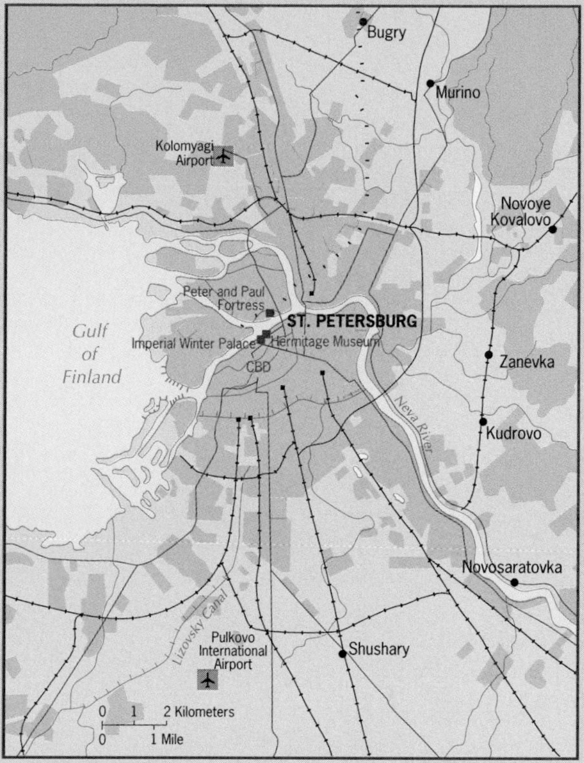

the Povolzhye are on its banks (Fig. 2-11). In the 1950s, a canal was completed to link the lower Volga with the lower Don River (and thereby the Black Sea).

The Volga River was an important historic route in old Russia, but for a long time neighboring regions overshadowed it. The Moscow area and Ukraine were far ahead in industry and agriculture. The Industrial Revolution that came late in the nineteenth century to the Moscow Region did not have much effect in the Povolzhye. Its major function remained the transit of foodstuffs and raw materials to and from other regions.

This transport function is still important, but the Povolzhye has changed. First, World War II brought furious development because the Volga River, located east of Ukraine, was far from the German armies that invaded from the west. Second, in the postwar era the Volga-Urals Region proved to be the greatest source of petroleum and natural gas in the entire Soviet Union. From near Volgograd (formerly Stalingrad) in the southwest to Perm on the Urals' flank in the northeast lies a belt of major oilfields (Fig. 2-12). These deposits were once believed to constitute the former Soviet Union's largest reserve, but as the map shows, later discoveries in western Siberia indicate that even more extensive oil- and gasfields lie beyond the Urals. Still, because of their location and size, the Volga Region's fossil-fuel reserves remain important.

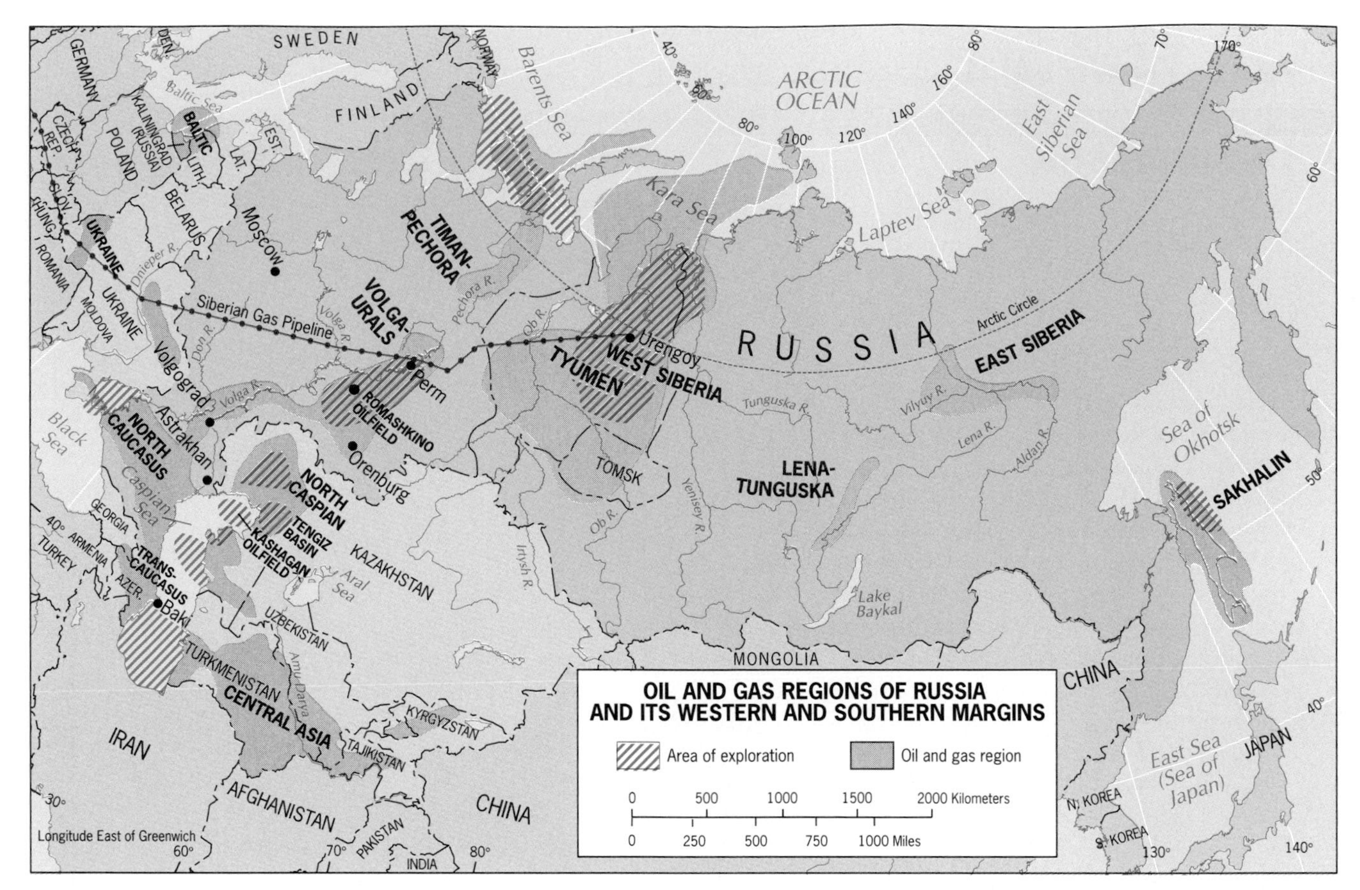

FIGURE 2-12

Third, the transport system has been greatly expanded. The Volga-Don Canal directly connects the Volga waterway to the Black Sea; the Moscow Canal extends the northern navigability of this river system into the heart of the Central Industrial Region; and the Mariinsk canals provide a link to the Baltic Sea. Today, the Volga Region's population exceeds 25 million, and the cities of Samara (formerly Kuybyshev), Volgograd, Kazan, and Saratov all have populations between 1.0 and 1.3 million. Manufacturing has also expanded into the middle Volga Basin, emphasizing more specialized engineering industries. The huge Fiat-built auto assembly plant in Tolyatti, for example, is one of the world's largest of its kind.

The Urals Region

The Ural Mountains form the eastern limit of the Russian Core. They are not particularly high; in the north they consist of a single range, but southward they broaden into a hilly zone. Nowhere are they an obstacle to east-west transportation. An enormous storehouse of metallic mineral resources located in and near the Urals has made this area a natural place for industrial development. Today, the Urals Region, well connected to the Volga and Central Industrial Region, extends from Serov in the north to Orsk in the south (Fig. 2-11).

The Urals Region became prominent during World War II and its aftermath when its remoteness from the German invaders allowed its factories to support the war effort without threat of destruction. Coal had to be imported, but the discovery of oilfields in the zone lying between the Urals and the Volga Region (Fig. 2-12) has relieved the problem of energy supply here. The more recent West Siberian oil and gas exploitation (from fields lying northeast of the Urals) increasingly reinforces development in the Urals.

The Central Industrial, Volga, and Urals regions form the anchors of the Russian core area. For decades they have been spatially expanding toward one another, their interactions ever more intensive. These regions of the Russian Core stand in sharp contrast to the comparatively less developed, forested, Arctic north and the remote upland to the south between the Black and Caspian seas. Thus even within this Russian coreland, frontiers still await growth and development.

The Internal Southern Periphery

Earlier we noted the tier of ethnic-minority republics that occupies the northern flanks of the Caucasus Mountains and extends nearly all the way from the Caspian Sea to the Black Sea (Fig. 2-13). Here lie several of Russia's most dif-

SOUTHERN RUSSIA: INTERIOR AND EXTERIOR PERIPHERIES
POPULATION
Under 50,000
50,000-250,000
250,000-1,000,000
1,000,000-5,000,000
Over 5,000,000
National capitals are underlined
Railroad
Road
Canal
Oil pipeline
Proposed oil pipeline
0 50 100 150 200 250 300 350 Kilometers
0 50 100 150 200 Miles
KAZAKHSTAN
RUSSIA
KALMYKIYA
Volgograd
Rostov
Mariupol
Sea of Azov
Astrakhan
Elista
Tikhoretsk
Kropotkin
Krasnodar
Novorossiysk
Armavir
Stavropol
Maykop
ADYGEYA
Budennovsk
Mineral'nyye Vody
Cherkessk
Pyatigorsk
KARACHAYEVO-CHERKESIYA
KABARDINO-BALKARIYA
Nalchik
INGUSHETIYA
Grozny
CHECHNYA
Makhachkala
Vladikavkaz
Nazran
NORTH OSSETIA
Caspian Sea
Sochi
ABKHAZIA
Sokhumi
Black Sea
CAUCASUS MOUNTAINS
DAGESTAN
SOUTH OSSETIA
Tskhinval
Samtredia
Kutaisi
Supsa
Gori
Batumi
AJARIA
GEORGIA
Tbilisi
Rustavi
Borcka
Lake Mingacevir
AZERBAIJAN
Sumgait
Baki (Baku)
Trabzon
Torul
Gyumri
ARMENIA
Gänjä
L. Sevan
Yerevan
NAGORNO-KARABAKH
Xankändi (Stepanakert)
Askale
Agri
TURKEY
NAXCIVAN AZER.
Armenian controlled
AZERBAIJAN PROVINCE
Lake Van
Elâzig
Tatvan
Van
Marand
Tabriz
IRAN
Ardabil
Lake Urmia
Ceyhan
Don R.
Volga River
Tsimlyansk Res.
Rioni R.
Kura R.
Aras R.
Longitude East of Greenwich

FIGURE 2-13

ficult politico-geographical problems, prominently including rebellious Chechnya. Russia's internal southern periphery is a zone of political instability, social tensions, and economic stagnation. The high-relief topography of much of this zone creates hideouts for insurgents and makes policing difficult.

A combination of geographic factors causes the problems Moscow confronts in this peripheral zone. Here the Russians (and later the Soviets) met and stalled the advance of Islam. Here the cultural mosaic is as intricate as anywhere in the Federation: Dagestan, the southernmost republic fronting the Caspian Sea, has 2 million inhabitants comprising some 30 ethnic groups speaking about 80 languages. And here Russia and its Caspian holdings penetrate the reservoirs of oil and natural gas which this huge, energy-rich basin contains. Groznyy, the capital of Chechnya which adjoins Dagestan to the west, was a crucial oil-industry center, pipeline junction, and service hub during the Soviet era.

But Chechnya also contains a sizeable Muslim population, and fiercely independent Muslim Chechens used Caucasus mountain hideouts to resist Russian colonization during the nineteenth century. Accused of collaboration with the Nazis during World War II, on Stalin's orders the Soviets exiled the entire Chechen population to Central Asia, with much loss of life. Rehabilitated and allowed to return by Khrushchev in the 1950s, the Chechens seized their opportunity in 1991 after the collapse of the Soviet Union and fought the Russian army to a stalemate. But Moscow never granted Chechnya independence (one quarter of the population of 1.2 million was Russian), and an uneasy standoff continued. Meanwhile Chechen hit-and-run attacks on targets in neighboring republics continued, and in 1999 three apartment buildings in Moscow were bombed, resulting in 230 deaths. Russian authorities blamed Chechnyan terrorists for these bombings and ordered a full-scale attack on Chechnya's Muslim holdouts. Groznyy, already severely damaged in the earlier conflict, was totally devastated and the rebels were driven into the mountains, but the Russian armed forces, still taking substantial losses, were unable to establish unchallenged control of all of the Republic.

Russia's costly problems in Chechnya illustrate the fractious nature of its internal southern periphery, from Buddhist Kalmykiya to also-Muslim Ingushetiya. Here the still-evolving federal framework will be put to the test for many years to come.

The External Southern Periphery

Beyond Russia's tier of internal republics lies still another periphery, legally outside the Russian sphere but closely tied to it nevertheless. This is *Transcaucasia*, historically a battleground for Christians and Muslims, Russians and Turks, Armenians and Persians. Today it is a subregion containing three former Soviet Socialist Republics: landlocked Armenia, coastal Georgia on the Black Sea, and Azerbaijan on the land-encircled Caspian Sea. Although these three countries are independent states today, they are so closely bound up with Russia that they are, functionally, part of the Russian realm (Fig. 2-13).

Armenia

As Figure 2-1 shows, landlocked Armenia occupies some of the most rugged and mountainous terrain in the earthquake-prone Transcaucasus. The Armenians are an embattled people who adopted Christianity 17 centuries ago and for more than a millennium sought to secure their ancient homeland here on the margins of the Muslim world. During World War I, the Ottoman Turks massacred much of the Christian Armenian minority and drove the survivors from eastern Anatolia and what is now Iraq into the Transcaucasus. At the end of that war in 1918, an independent Armenia arose, but its autonomy lasted only two years. In 1920, Armenia was taken over by the Soviets; in 1936, it became one of the 15 constituent republics of the Soviet Union. The collapse of the Soviet Empire gave Armenia what it had lost three generations earlier: independence.

Or so it seemed. Soon afterward, the Armenians found themselves at war with neighboring Azerbaijan over the fate of some 150,000 Armenians living in Nagorno-Karabakh, a pocket of territory surrounded by Azerbaijan. Such a separated territory is called an *exclave*, and this one had been created by Soviet sociopolitical planners who, while acknowledging the cultural (Christian) distinctiveness of this cluster of Armenians, nevertheless gave (Muslim) Azerbaijan jurisdiction over it.

That was a recipe for trouble: the arrangement was made to work under authoritarian Soviet rule, but once this rule ended, the Christian Armenians encircled by Muslim Azerbaijan felt insecure and appealed to Armenia for help. In the ensuing conflict, Armenian troops entered Azerbaijan and gained control over the exclave, even ousting Azerbaijanis from the zone between the main body of Armenia and Nagorno-Karabakh (Fig. 2-13). The international community, however, has not recognized Armenia's occupation, and officially the territory remains a part of Azerbaijan. After the turn of the twenty-first century, the matter still remained unresolved.

The impact of the war on Armenia helped make public opinion more favorable toward Moscow. Azerbaijan cut the pipelines that were sending oil and natural gas from the Caspian coast to Armenian consumers; the Armenian economy's subsequent collapse reminded Armenians of their vulnerability in their remote corner between historic (Turkish) and current (Azerbaijani) Muslim enemies.

Another factor that inclined Armenians to look to Moscow was Armenia's Soviet experience. The senior author of this book, on his first field trip to Transcaucasia in 1964, was struck by the relatively relaxed form of communism that seemed to prevail here. Life seemed to have a vitality and color not seen in Russia or even in neighboring Georgia. At the university in Yerevan, Armenian faculty were remarkably candid in their criticism of Moscow—but also acknowledged the Soviets' role in securing their Armenian homeland in this turbulent region. The Armenian Apostolic Church continued to function, and small markets and private farms were common. The repressive atmosphere experienced in Russia, Belarus, and Ukraine seemed absent.

Only the size of Maryland and with a population totalling 3.8 million, Armenia needs Russian support, even guardianship. The revival of the Christian churches in Russia and Armenia has helped renew linkages broken when the Soviet Union collapsed.

Mineral-rich and producing a variety of subtropical fruits in well-watered valleys, Armenia has economic potential, enhanced by plentiful hydroelectric power derived from Soviet-era installations. In 1998, Armenia and Russia signed an agreement whereby Russia will provide natural gas to Armenia and Armenia will allow its territory to be used for a pipeline to export gas from Russia to Turkey. Meanwhile, a political group advocating an official union with Russia of the kind signed by Russia and Belarus managed to gather more than 1 million signatures in Armenia, proof that in this former colony, at least, Moscow's image remains favorable.

Georgia

Of the three former Soviet Republics in Transcaucasia, only Georgia has a Black Sea coast and thus an outlet to the wider world. Smaller than South Carolina, Georgia is a country of high mountains and fertile valleys. Its social and political geographies are complicated. The population of 5.5 million is more than 70 percent Georgian but also includes Armenians (8%), Russians (6%), Ossetians (3%), and Abkhazians (2%). The Georgian Orthodox Church dominates the religious community, but about 10 percent of the people are Muslims.

Unlike Armenia and Azerbaijan, Georgia has no exclaves, but its political geography is problematic nonetheless (Fig. 2-13). Within Georgia's borders lie three minority-based autonomous entities: the Abkhazian and Ajarian Autonomous Republics, and the South Ossetian Autonomous Region.

Sakartvelos, as the Georgians call their country, has a long and turbulent history. Tbilisi, the capital for 15 centuries, lay at the core of an empire around the turn of the thirteenth century, but the Mongol invasion ended this era. Next, the Christian Georgians found themselves in the path of wars between Islamic Turks and Persians. Turning northward for protection, the Georgians were annexed by the Russians in 1800, who were looking for warm-water ports. Like other peoples overpowered by the czars, the Georgians took advantage of the Russian Revolution to reassert their independence; but the Soviets reincorporated Georgia in 1921 and proclaimed a Georgian Soviet Socialist Republic in 1936. Josef Stalin, the communist dictator who succeeded Lenin, was a Georgian.

Georgia is renowned for its scenic beauty, warm and favorable climates, agriculture (especially tea), timber, manganese, and other products. Georgian wines, tobacco, and citrus fruits are much in demand. The diversified economy could support a viable state.

Unfortunately, Georgia's political geography is loaded with centrifugal forces. After Georgia declared its independence in 1991, factional fighting destroyed its first elected government. But worse was to come. In the Autonomous Region of South Ossetia, which borders the Russian internal republic of North Ossetia, conflict broke out over local demands for self-determination; many of Georgia's Ossetians preferred unification with Russia's Ossetians over minority status in independent Georgia. Even as this costly civil strife continued, one of Georgia's two internal republics, Abkhazia (in the country's northwest corner), proclaimed its independence, ousting Georgians and setting up a separatist regime.

In time, this Abkhazian uprising would have dire consequences for Georgia. Initially, Russian army leaders supported the separatist movement, for reasons having less to do with strategy than with comfort and spite: comfort because the Abkhazian Black Sea coast had long been a favorite resort for Russian (Soviet) army brass, and spite because of a lingering animosity toward the Georgian leader, Eduard Shevardnadze, a former Soviet foreign minister who was blamed for contributing to the breakup of the Soviet Empire. With Russian help, the Abkhazians, a minority in their own republic, drove 250,000 Georgians from their homes.

Then Russian attitudes changed—but at a price. In return for permanent Russian military bases on Georgian soil, Moscow ended its support for the Abkhazians and even blockaded the common border between Abkhazia and Russia. When Abkhazia's rebel leaders stood fast, Georgia lost on two counts: unable to regain control over Abkhazia, it lost sovereignty as Russian troops were stationed within its borders.

In the meantime, Georgia's promising economy suffered severely; the tourist industry was moribund, and the country's famed subtropical health resorts (those not destroyed in the conflict) were empty. If there was a positive side to the political geography of Georgia, it was that the southwestern republic within its borders, Ajaria, made no secession attempt. More than 80 percent of Ajaria's population is Georgian, and of the remainder, 10 percent are Russians and 5 percent Armenians.

In the winter of 2001, Georgia was becoming embroiled in the Russian conflict with the Chechens. In the rugged mountains that mark the border area between Chechnya and Georgia, Chechen fighters were seeking refuge in the Geor-

This is the terminal of the oil pipeline that extends all the way from Baki (Baku) in Azerbaijan via Dagestan and Chechnya to Novorossiysk on the Black Sea (Fig. 2-13). This port and petroleum transfer point has become one of Russia's most important economic foci, bringing oil from restive Transcaucasia via several peripheral republics to a facility that gives Russia a competitive maritime outlet to world markets. That is why Russians object to plans allowing Azeri oil to be exported via Georgia to a Turkish Mediterranean port, or from Georgia's own Black Sea outlet at Supsa. The geopolitics of energy create some intense rivalries.

gian territory; Russians violated Georgian space in hot pursuit, and Russia accused Georgia of giving sanctuary to Muslim rebels. The Georgian government warned the Russians against using Georgian territory in their campaign against the Chechens.

Meanwhile, Georgian leaders harbor the hope that, some day, their country may join the European Union. As a Tbilisi newspaper stated in an editorial, if Turkey can be considered for membership, then surely Georgia must be eligible. But given its instability, its corrupt and violent politics, its difficult relations with Russia, and its internal fragmentation, Georgia's European aspirations will not soon be fulfilled.

Azerbaijan

Azerbaijan is the name of an independent state *and* of a province in neighboring Iran. The Azeris (short for Azerbaijanis) on both sides of the border have the same ancestry: they are a Turkish people divided between the (then) Russian and Persian Empires by a treaty signed in 1828. By that time, the Azeris had become Shi'ite Muslims, and when the Soviet communists laid out their grand design for the USSR, they awarded the Azeris their own republic. On the Persian side, the Azeris were assimilated into the Persian Empire, and their domain became a province. Today the former Soviet Socialist Republic is the independent state of Azerbaijan (population: 7.8 million), and the 10 million Azeris to the south live in the Iranian province.

During the brief transition to independence and at the height of their war with the Armenians, the dominantly Muslim Azeris tended to look southward, toward Iran. But geographic realities dictate a more practical orientation. Azerbaijan possesses huge reserves of oil and natural gas; under the Soviets it was one of Moscow's chief regional sources of fuels. The center of the oil industry is Baki (Baku), the capital on the shore of the Caspian Sea—but the Caspian Sea is a lake. To export its oil, Azerbaijan needs pipelines. And those of Soviet vintage link Baki to Russia's Black Sea terminal of Novorossiysk. During the mid-1990s, when Azerbaijan's leaders announced plans to build new pipelines via Georgia or Iran, Russia objected and even meddled in Azeri politics to stymie such plans.

A look at the map (Fig. 2-13) shows Azerbaijan's options. A pipeline could be routed from Baki through Georgia to the Black Sea coast near Batumi, avoiding Russian territory altogether. Another route could run through Georgia and across Turkey to the Mediterranean terminal at Ceyhan. Still another option would run the pipeline across Iran to a terminal on the Persian Gulf. Moscow objects to such alternatives, but Azerbaijan now has powerful international allies. American, French, British, and Japanese oil companies are developing Azerbaijan's offshore Caspian reserves, and the U.S. government prefers the Turkish route.

All these alternate routes are in the future, however; for the present, Azerbaijan's oil must flow across Russian territory, and Azerbaijan must cooperate with Moscow. Given its Muslim roots and its location, the time may come when Azerbaijan turns toward the Islamic world. For the moment, however, this country is inextricably bound up with its neighbors in Transcaucasia and the north.

THE EASTERN FRONTIER

From the eastern flanks of the Ural Mountains to the headwaters of the Amur River, and from the latitude of Tyumen to the northern zone of neighboring Kazakhstan, lies Russia's vast Eastern Frontier Region, product of a gigantic experiment in the eastward extension of the Russian Core (Fig. 2-10). As the maps of cities and surface communications suggest, this eastern frontier is more densely peopled and more fully developed in the west than in the east; at the longitude of Lake Baykal, settlement has become linear, marked by ribbons and clusters along the east-west railroads. Two subregions dominate the geography: the Kuznetsk Basin in the west and the Lake Baykal area in the east.

The Kuznetsk Basin (Kuzbas)

Some 900 miles (1450 km) east of the Urals lies another of Russia's primary regions of heavy manufacturing resulting from the communist period's national planning: the Kuznetsk Basin, or *Kuzbas* (Fig. 2-11). In the 1930s, it was opened up as a supplier of raw materials (especially coal) to the Urals, but that function became less important as local industrialization accelerated. The original plan was to move coal from the Kuzbas west to the Urals and allow the returning trains to carry iron ore east to the coalfields. However, good-quality iron ore deposits were subsequently discovered near the Kuznetsk Basin itself. As the new resource-based Kuzbas industries grew, so did its urban centers. The leading city, located just outside the region, is Novosibirsk, which stands at the intersection of the Trans-Siberian Railroad and the Ob River as the symbol of Russian enterprise in the vast eastern interior. To the northeast lies Tomsk, one of the oldest Russian towns in all of Siberia, founded in the seventeenth century and now caught up in the modern development of the Kuzbas Region. Southeast of Novosibirsk lies Novokuznetsk, a city that produces steel for the region's machine and metal-working plants and aluminum products from Urals bauxite.

Impressive as the concentration of resources may be in the Kuznetsk Basin, the industrial and urban development that has taken place there must largely be attributed again to the ability of the communist state and its planners to promote this kind of expansion, notwithstanding what capitalists would consider excessive investment. With no financial constraints, the state planners pushed the country toward industrialization, hoping that certain areas would reach "takeoff" levels. Then these areas would need fewer direct investments because growth would become self-perpetuating. The Kuzbas, for instance, was expected to become one of the Soviet Union's leading industrial agglomerations, with its own important market and with a location favorable if not to the Urals and points west, then at least to the developing markets of the Far East.

The Lake Baykal Area (Baykaliya)

East of the Kuzbas, development becomes more insular, and distance becomes a stronger adversary. North of the Tyva Republic and eastward around Lake Baykal, larger and smaller settlements cluster along the two railroads to the Pacific coast (Fig. 2-11). West of the lake, these rail corridors lie in the headwater zone of the Yenisey River and its tributaries. A number of dams and hydroelectric projects serve the valley of the Angara River, particularly the city of Bratsk. Mining, lumbering, and some farming sustain life here, but isolation dominates it. The city of Irkutsk, near the southern end of Lake Baykal, is the principal service center for a vast Siberian region to the north and for a lengthy east-west stretch of southeastern Russia.

During the Soviet period the exploitation of coal and other resources in the Kuzbas transformed this area into one of the USSR's industrial heartlands, a megacomplex that grew explosively. In the process, enormous environmental damage was inflicted on an already-fragile countryside. This coal mine not far from Tomsk has been described as an ecological disaster, its environs a jumble of corroding structures, the atmosphere polluted from hundreds of smokestacks.

Beyond Lake Baykal, the Eastern Frontier really lives up to its name: this is southern Russia's most rugged, remote, forbidding country. Settlements are rare, many mere camps. The Buryat Republic (Fig. 2-7) is part of this zone; the territory bordering it to the east was taken from China by the czars and may become an issue in the future. Where the Russian-Chinese boundary turns southward, along the Amur River, the region called the Eastern Frontier ends and Russia's Far East begins.

SIBERIA

Before we assess the potential of Russia's Pacific Rim, we should remember that the ribbons of settlement just discussed hug the southern perimeter of this giant country, avoiding the vast Siberian region to the north (Fig. 2-10). Siberia extends from the Ural Mountains to the Kamchatka Peninsula—a vast, bleak, frigid, forbidding land. Larger than the conterminous United States but inhabited by only an estimated 15 million people, Siberia quintessentially symbolizes the Russian environmental plight: vast distances, cold temperatures worsened by strong Arctic winds, difficult terrain, poor soils, and limited options for survival.

But Siberia also has resources. From the days of the first Russian explorers and Cossack adventurers, Siberia's riches have beckoned. Gold, diamonds, and other precious minerals were found. Later, metallic ores including iron and bauxite were discovered. Still more recently, the Siberian interior proved to contain sizeable quantities of oil and natural gas (Fig. 2-12) and began to contribute significantly to Russia's energy supply.

As the physiographic map (Fig. 2-2) shows, major rivers—the Ob, Yenisey, and Lena—flow gently northward across Siberia and the Arctic Lowland into the Arctic Ocean. Hydroelectric power development in the basins of these rivers has generated electricity used to extract and refine local ores, and in the lumber mills that have been set up to exploit the vast Siberian forests. Soviet planners even considered grandiose schemes to reverse the flow of Siberian streams (to divert Siberian water onto parched farmlands far to the southwest)—ideas their Russian successors are likely to discard.

The human geography of Siberia is fragmented, and much of the region is virtually uninhabited (Fig. 2-4). Ribbons of Russian settlement have developed; the Yenisey River, for instance, can be traced on this map of Soviet peoples (a series of small settlements north of Krasnoyarsk), and the upper Lena Valley is similarly fringed by ethnic Russian settlement. Yet hundreds of miles of empty territory separate these ribbons and other islands of habitation. During the 1930s and 1940s, in the worst excesses of the Stalinist period (and long afterward as well), dissidents and criminals were exiled to Siberian mining and lumbering camps to do hard labor under the harshest conditions; many never returned.

The political geography of eastern Siberia is marked by the growing identity of Sakha (the Yakut Republic). As additional resources are discovered here (including oil and natural gas), this Republic, centered on the capital, Yakutsk, will become more important.

Siberia, Russia's freezer, is stocked with goods that may become mainstays of future national development. Already, precious metals and mineral fuels are bolstering the Russian economy. In time, we may expect Siberian resources to play a growing role in the economic development of the Eastern Frontier and the Russian Far East as well. One step in that process was already taken during Soviet times: the completion of the BAM (Baykal-Amur Mainline) Railroad in the 1980s. This route, lying north of and parallel to the old Trans-Siberian Railroad, extends 2200 miles (3540 km) eastward from Tayshet (near the important center of Krasnoyarsk) directly to the Far East city of Komsomolsk (Fig. 2-6). In the post-Soviet era, the BAM Railroad has been beset by equipment breakdowns and workers' strikes. Nonetheless, it is a key element of the infrastructure that will serve the Eastern Frontier's economic growth in the twenty-first century.

THE RUSSIAN FAR EAST

Imagine this: a country with 5000 miles (8000 km) of Pacific coastline, two major ports, major interior cities nearby, huge reserves of resources ranging from minerals to fuels to timber, directly across from one of the world's largest economies—all this at a time when the Asian Pacific Rim was the world's fastest-growing economic region. Would not that country have burgeoning cities, busy harbors, growing industries, and expanding trade?

In the Russian Far East (Fig. 2-10), the answer is—no. Activity in the port of Vladivostok is a shadow of what it was during the Soviet era, when it was the communists' key naval base. The nearby container terminal at Nakhodka suffers from breakdowns and inefficiencies. The railroad to western Russia carries just a fraction of the trade it did during the 1970s and 1980s. Cross-border trade with China is minimal. Trade with Japan is inconsequential. The region's cities are grimy, drab, moribund. Utilities are shut off for hours at a time because of fuel shortages and system breakdowns. Outdated factories are shut, their workers dismissed. Political relations with Moscow are poor. There is potential here, but little of it has been realized.

As a region, the Russian Far East consists of two parts: the mainland area extending from Vladivostok to the Stanovoy Mountains and the large island of Sakhalin (Figs. 2-1; 2-14). This is cold country: icebreakers have to keep the ports of Vladivostok and Nakhodka open throughout the

FROM THE FIELD NOTES
"Standing in an elevated doorway in the center of Vladivostok, you can see some of the vestiges of the Soviet period: the omnipresent, once-dominant, GUM department store behind the blue seal on the left, and the communist hammer-and-sickle on top of the defunct hotel on the right. But in the much more colorful garb of the people, and the private cars in the street, you see reflections of the new era. Once a city closed to foreigners, Vladivostok now throbs with visitors—and has become a major point of entry for contraband goods."

winter. Winters here are long and bitterly cold; summers are brief and cool. Although the population is small (about 7 million), food must be imported because not much can be grown. Most of the region is rugged, forested, and remote. Vladivostok, Khabarovsk, and Komsomolsk are the only cities of any size. Nakhodka and the newer railroad terminal at Vanino are smaller towns; the population of the whole island of Sakhalin is about 700,000 (on an island the size of Caribbean Hispaniola [15.4 million]). Offshore lie productive fishing grounds, and Russian fleets from Vladivostok and points north catch salmon, herring, cod, and mackerel to be frozen or canned and shipped to local and distant markets.

The Soviet regime rewarded people willing to move to this region with housing and subsidies. The communists, like the czars before them, realized the importance of this frontier (Vladivostok means "We Own the East"), and they used every possible incentive to develop it and link it ever closer to Russia's distant western core. Freight rates on the Trans-Siberian Railroad, for example, were about 10 percent of their real costs; the trains were always loaded in both directions. Vladivostok was a military base and a city closed to foreigners, and Moscow invested heavily in its infrastructure. Komsomolsk in the north and Khabarovsk near the region's center were endowed with state-owned industries using local resources: iron ore from Komsomolsk, oil from Sakhalin, timber from the ubiquitous forests. The steel, chemical, and furniture industries sent their products westward by train, and they received food and other consumer goods from the Russian heartland.

For several reasons the post-Soviet transition has been especially difficult here in the Far East. The new economic order has canceled the region's communist-era advantages: the Trans-Siberian Railroad now must charge the real cost of transporting products from the Eastern Frontier to the Russian Core. State-subsidized industries must compete on market principles, their subsidies having ended. The decline of Russia's armed forces has hit Vladivostok hard. The fleet lies rusting in port; service industries have lost their military markets; the shipbuilding industry has no government contracts. Coal miners in the Bureya River Valley (a tributary of the Amur) go unpaid for months and go on strike; coal-fired power plants do not receive fuel shipments, and cities and towns go dark.

Locals put much of the blame for their region's failure on Moscow, and with reason. As Figure 2-7 shows, the Far East contains only five regions: Primorskiy Kray, Khabarovsk Kray, Amur Oblast, Sakhalin Oblast, and Yevreyskaya, originally the Jewish Autonomous Region. This does not add up to much political clout, and in the capital the Far East has not been given the help it would need to realize its potential. A particularly costly failure involves Japan: the Soviets and the Japanese never signed a peace treaty after World War II, and now Moscow and Tokyo are at odds over the fate of a group of small islands the Soviets occupied during the war and never relinquished. The Japanese want the return of these islands, located in the Kurile chain northeast of Hokkaido (Fig. 2-14); Russian leaders have refused to yield them. In the early 1990s, Soviet President Gorbachev and Russian President Boris Yeltsin tried to resolve the matter but failed. Japan, then at the height of its economic power, offered a $26 billion deal for the islands, a proposal that included exploring the Russian Far East and eastern Siberia for mineral resources, the purchase of raw materials, and assistance in developing the region's infrastructure. But President Yeltsin felt unable to return "one square meter" of Russian territory to the Japanese, and the deal fell through. Thus the Far East lost a major Pacific Rim opportunity.

At the turn of our new century, all eyes are on Sakhalin. Here a new era may be dawning because major oil and natural gas reserves have been discovered, and foreign companies, with Russian partners, are making large investments to exploit them. European, American, and Japanese investments are already having an impact beyond the island itself: in Komsomolsk more than 1000 workers are building the specially reinforced foundations for the huge oil rigs that will be positioned offshore from Sakhalin and that will have to withstand the rough and cold Pacific seas. Japan is

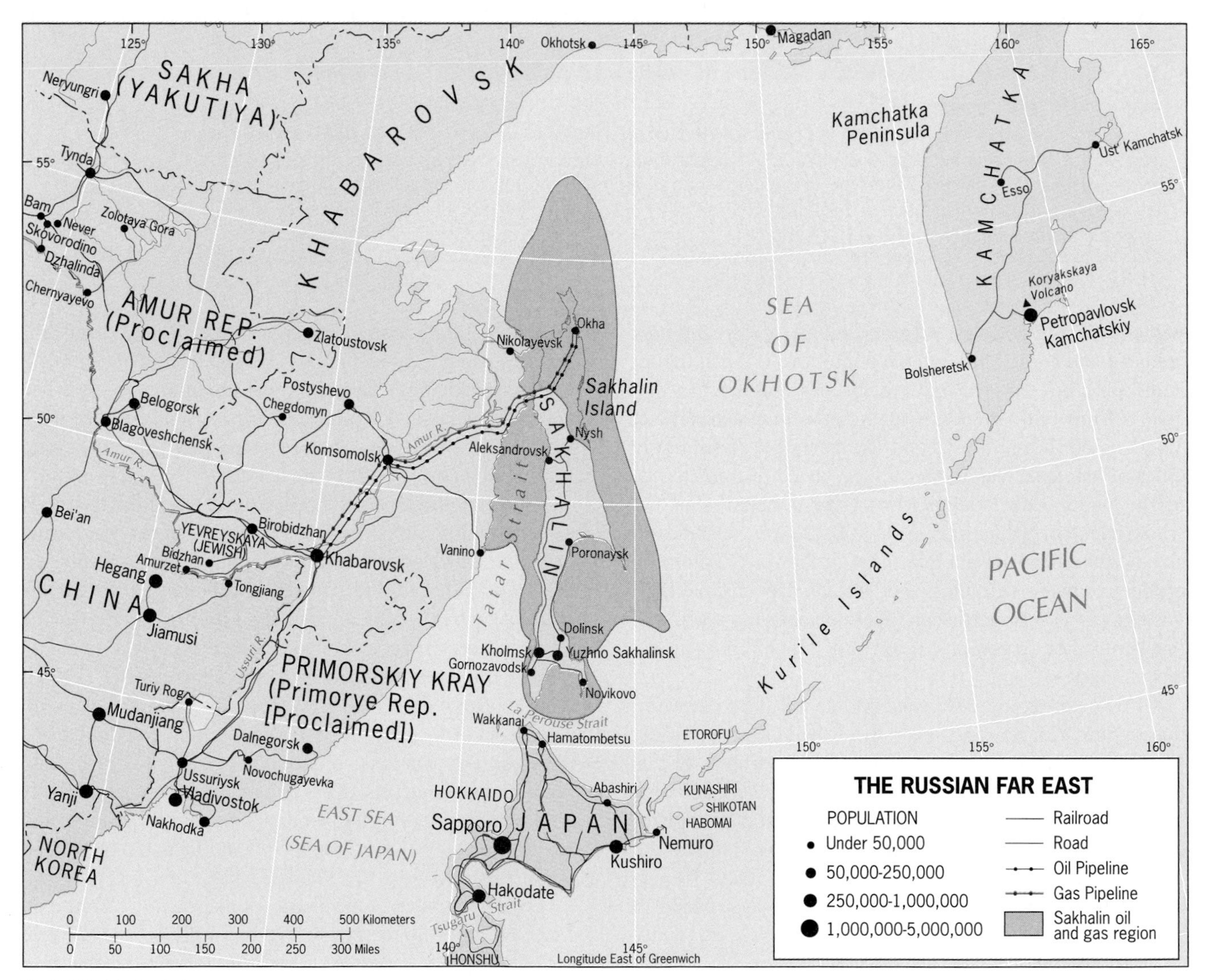

FIGURE 2-14

The Kamchatka Peninsula constitutes one of the Pacific Rim's most geologically active zones, with more than 125 volcanoes of which more than 20 are active. Sudden massive eruptions can endanger airliners on the northern Pacific route between North America and East Asia, which skirts Kamchatka. The region's largest city, Petropavlovsk-Kamchatskiy, lies at the foot of the 11,400-foot (3456-m) high Koryakskaya Volcano near the Pacific coast in southeastern Kamchatka. Soviet-era apartment buildings dominate the cultural landscape of this bleak fishing port founded during Russia's eighteenth-century eastward expansion. Note that, in this *Dfc* climate, the countryside is much greener (with deciduous as well as coniferous vegetation) than it is toward the north and the Siberian interior.

especially interested in this venture because it currently depends on far more distant sources of oil and gas.

If the Sakhalin reserves are to usher in a new age for the Russian Far East, Moscow will have to cooperate. Substantial revenues will have to go directly to the local regions and not to Moscow. And that may be an obstacle. The Far East regions, especially rebellious Primorskiy, do not have high standing in the capital.

Meanwhile, the Russian Far East today is sometimes compared to the American Wild West of another era. Thievery, violent crime, smuggling of stolen cars, corruption, and cronyism make this a difficult place to do business, but some companies are willing to risk it all because they understand the potential. Nevertheless, it will be some time before this region becomes a Pacific Rim success. Mismanagement and incompetence can negate the geographic advantages of relative location and resource wealth, as this lagging region proves.

In any future world order, much will depend on the relationships between Russia and its numerous neighbors. None of these relationships may eventually matter as much as the one between Russia and China. And that one is likely to depend largely on events in Russia's easternmost outpost.

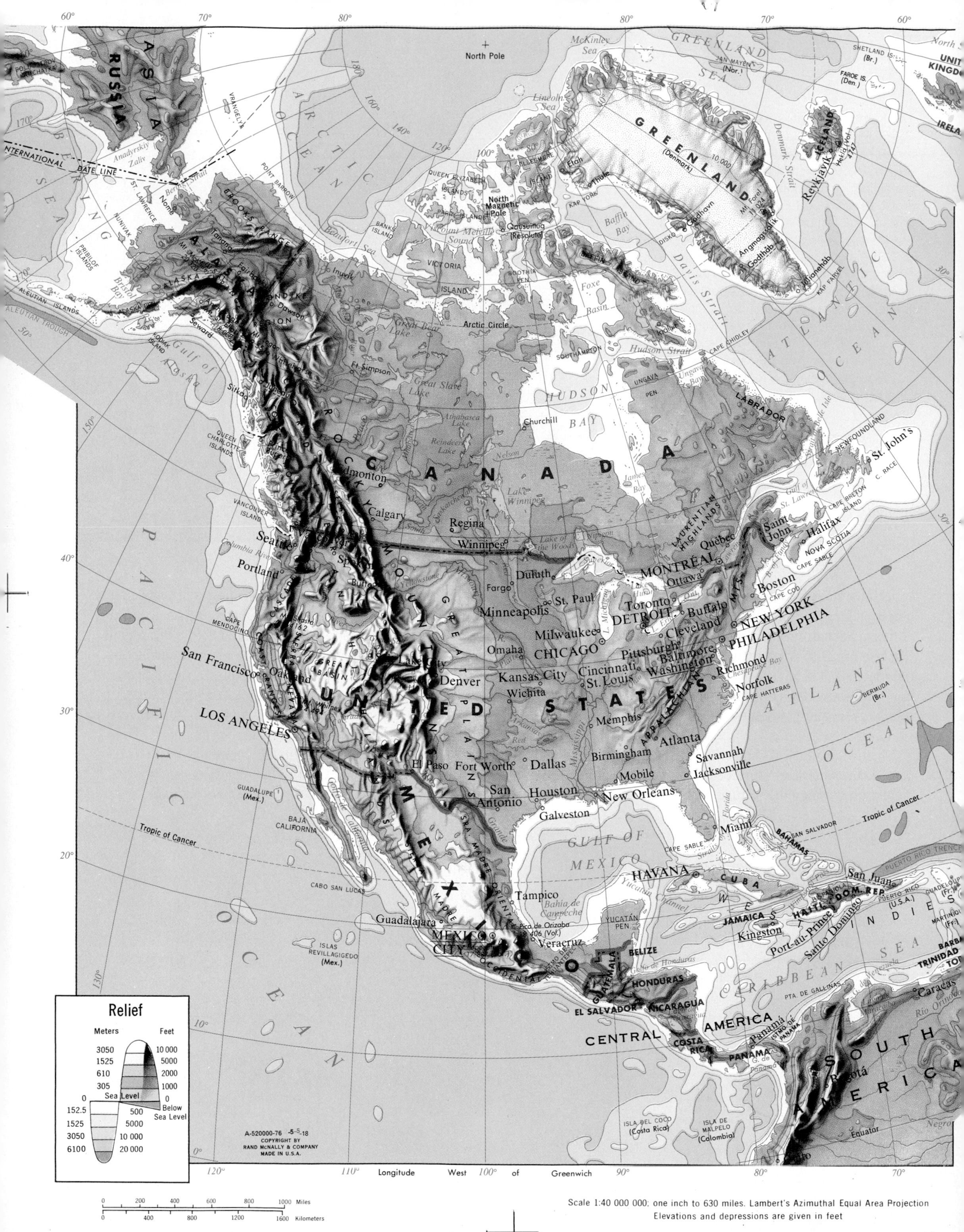

Copyright © Rand McNally, 2000

chapter 3

North America

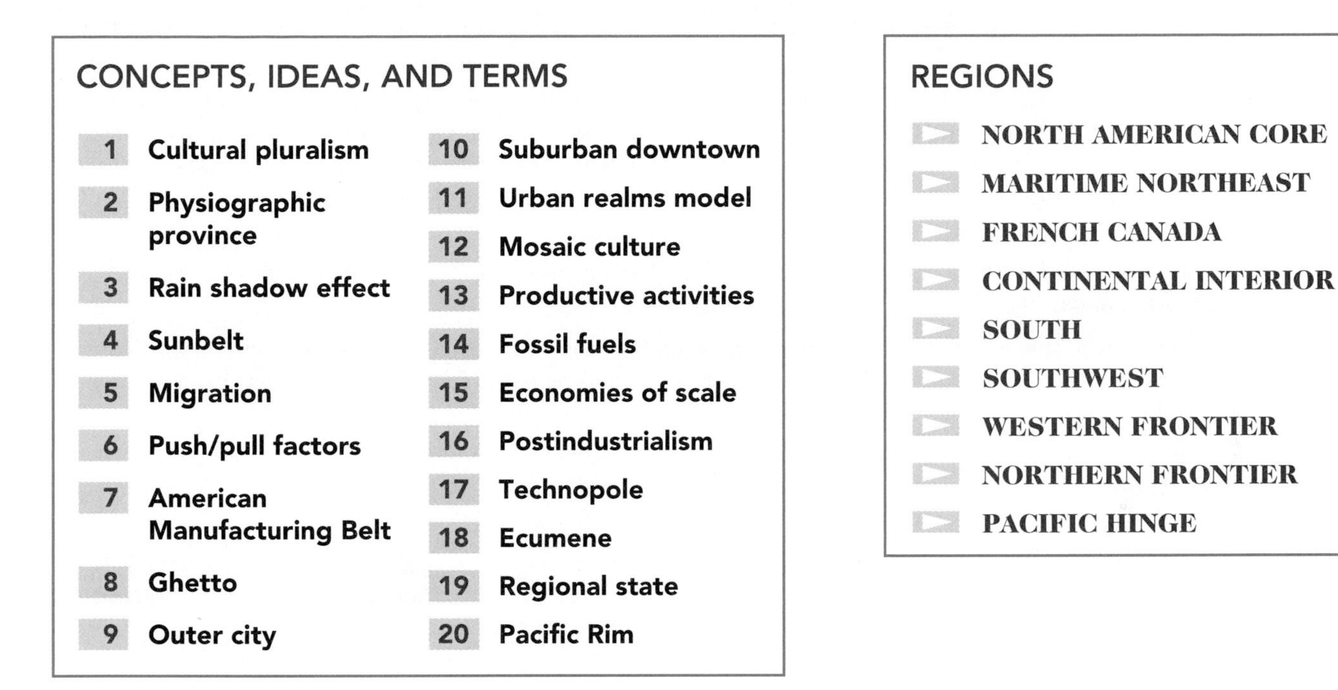

CONCEPTS, IDEAS, AND TERMS

1 Cultural pluralism
2 Physiographic province
3 Rain shadow effect
4 Sunbelt
5 Migration
6 Push/pull factors
7 American Manufacturing Belt
8 Ghetto
9 Outer city
10 Suburban downtown
11 Urban realms model
12 Mosaic culture
13 Productive activities
14 Fossil fuels
15 Economies of scale
16 Postindustrialism
17 Technopole
18 Ecumene
19 Regional state
20 Pacific Rim

REGIONS

- NORTH AMERICAN CORE
- MARITIME NORTHEAST
- FRENCH CANADA
- CONTINENTAL INTERIOR
- SOUTH
- SOUTHWEST
- WESTERN FRONTIER
- NORTHERN FRONTIER
- PACIFIC HINGE

The North American realm consists of two countries, the United States and Canada, which are alike in many ways. Culturally, most (but not all) of the people trace their ancestries to various European countries, and the realm is often called *Anglo-America*, with English serving as the dominant language of the United States and officially sharing equal status with French in Canada. Economically, both rank among the world's most highly advanced by every measure of national development, and they continue to benefit from a still-intensifying trade relationship that functions across the longest open international boundary on Earth. On the global scene, the realm's highly developed societies have clearly achieved a leadership role, which arose from a combination of history and geography. Presented with a rich abundance of natural and human resources over the past 150 years, North Americans have skillfully converted these productive opportunities into continentwide affluence and worldwide influence.

FIGURE 3-1

With more than 75 percent of the population of the United States and Canada residing in towns and cities,* North America's societies have become the world's most highly urbanized. Indeed, nothing symbolizes them as strongly as the skyscrapered panoramas of New York, Toronto, or Chicago—and, increasingly, the booming new suburban business complexes that are transforming their outer metropolitan rings. North Americans also are hypermobile, with networks of superhighways, commercial air routes, and state-of-the-art telecommunications media efficiently interconnecting the

*The population and activity agglomerations we call cities are changing throughout the world as the urban areas they anchor continue to grow. The term *city* applies to the older core or central city; the term *metropolitan* (also *metropolis*) refers to the entire urban complex consisting of (1) the core city and (2) its ring of newer surrounding suburbs that in North America and elsewhere are expanding in size and especially function. As indicated in the Preface, all urban population data provided in this book are metropolitan-area totals unless otherwise specified.

◆ Major Geographic Qualities of North America

1. North America encompasses two of the world's biggest states territorially (Canada is the second largest in size; the United States is third).
2. Both Canada and the United States are federal states, but their systems differ. Canada's is adapted from the British parliamentary system and is divided into ten provinces and three territories. The United States separates its executive and legislative branches of government, and it consists of 50 States, the Commonwealth of Puerto Rico, and a number of island territories under U.S. jurisdiction in the Caribbean Sea and the Pacific Ocean.
3. Both Canada and the United States are plural societies. Although ethnicity is increasingly important, Canada's pluralism is most strongly expressed in regional bilingualism. In the United States, major divisions occur along racial lines.
4. A large number of Quebec's French-speaking citizens supports a strong movement that seeks independence for the province. The movement's high-water mark may have been reached in the 1995 referendum in which (minority) non-French speakers were the difference in the narrow defeat of separation. The prospects for a break-up of the Canadian state have diminished considerably since 2000.
5. North America's population, not large by international standards, is the world's most highly urbanized and mobile. Largely propelled by a continuing wave of immigration, the realm's population total is expected to grow by more than 40 percent over the next half-century.
6. By world standards, this is a rich realm where high incomes and high rates of consumption prevail. North America possesses a highly diversified resource base, but nonrenewable fuel and mineral deposits are consumed prodigiously.
7. North America is home to one of the world's great manufacturing complexes. The realm's industrialization generated its unparalleled urban growth, but a new postindustrial society and economy are rapidly maturing in both countries.
8. The two countries heavily depend on each other for supplies of critical raw materials (e.g., Canada is the leading source of U.S. energy imports) and have long been each other's chief trading partners. Today, the North American Free Trade Agreement (NAFTA), which also includes Mexico, is linking all three economies ever more tightly as barriers to international trade and investment flow are steadily being dismantled.
9. North Americans are the world's most mobile people. Although plagued by recurrent congestion problems, the realm's networks of highways, commercial air routes, and cutting-edge telecommunications are the most efficient on Earth.

realm's far-flung cities and regions. Commuters stream into and out of suburban business centers and central-city downtowns by the millions each working day; most of them drive cars, whose numbers have multiplied more than six times faster than the human population since 1970. Moreover, each year nearly one out of every six individuals changes his or her residence.

As the twenty-first century opens, North America has entered a new age, the third since the arrival of Columbus in the New World more than 500 years ago. Agriculture and rural life dominated the first four centuries; the second age—industrial urbanization—spanned the now-ended twentieth century. In its aftermath, the United States and Canada are today experiencing the maturation of a *postindustrial society and economy*, which is dominated by the production and manipulation of information, skilled services, and high-technology manufactures, and operates within an ever more globalized framework of business interactions. As dynamic new locational forces surface, North America's human geography is undergoing a parallel transformation. At the national level, new regions emerge while the older ones are hard-pressed to reinvent themselves; at the intrametropolitan scale, central cities turn inside-out and provide major new opportunities for their surrounding suburbs (which by themselves contain more than half the U.S. population) as well as the urbanizing countryside beyond. Whatever the outcome, the winners and losers in the current scramble to adapt to the changing spatial infrastructure will shape the geography of these two leading countries far into the future.

Although Canada and the United States share many historical, cultural, and economic qualities, they also differ in significant ways, as can be seen on the map. The United States, somewhat smaller territorially than Canada, occupies the heart of the North American continent and, as a result, encompasses a greater environmental range. The U.S. population is dispersed across most of the country, forming major concentrations along both the (north-south-trending) Atlantic and Pacific coasts. The overwhelming majority of Canadians, however, live in an interrupted east-west corridor that lies across southern Canada, mainly within 200 miles (320 km) of the U.S. border. The United States also encompasses North America's northwestern extension, Alaska. (Offshore Hawai'i, however, belongs in the Pacific Realm.) Thus, unlike Canada, the United States is a *fragmented state*, a discontinuous country whose national territory consists of two or more individual parts separated by foreign territory and/or international waters.

Differences between the two countries also become apparent when we examine the population characteristics that form the basis for important internal social contrasts. The population of the United States in 2002 was 285.4 million; Canada's was 31.0 million, just over one-tenth as large. Although comparatively small, Canada's population is divided by culture and tradition, and this division has a pronounced regional expression. About 84 percent of Canada's citizens speak English, 31 percent speak French, and as many as 10 percent other languages (multilinguality affects the percentage total: for example, 17 percent speak both English and French). *First Nations* (indigenous peoples of Amerindian descent, whose U.S. counterparts are called Native Americans) and *Inuit* (peoples of the Arctic zone, formerly called Eskimos) make up approximately 2 percent of the total.

The strong spatial clustering of *Francophones* (French speakers) in Quebec, the second most populous of the country's ten provinces, accentuates Canada's social division along ethnic and linguistic lines. Nearly 85 percent of this province's population is French Canadian, and Quebec is the historic, traditional, and emotional focus of French culture in Canada. Over the past few decades a strong nationalist movement has emerged in Quebec and today demands outright separation from the rest of Canada (independence was narrowly defeated in the latest referendum held in 1995). With more than 90 percent of Canada's Francophones living in Quebec, French speakers in the other nine provincial populations form minorities that become quite small to the west of Ontario (Quebec's western neighbor and the country's core area and most populous province). The weakness of the French cause in western Canada further contributes to internal regionalism because the federal government's recent effort to keep Quebec in the Canadian federation by constitutionally recognizing it as a "distinct society" deeply offended the large English-speaking ma-

The photo at the left, dominated by North America's tallest skyscraper (Sears Tower, capped by a pair of white spires), shows the heart of the central business district (CBD) of Chicago—the epitome of city-building during the era of industrial urbanization that peaked in the mid-twentieth century. But we now live in the twenty-first century, and the postindustrial transformation of urban America is all but complete. The landscape of the latter is symbolized by the low-rise headquarters of the General Foods Corporation shown in the photo above, the centerpiece of an elegant suburban office campus located in Rye Brook/Port Chester, New York about 20 miles north of Manhattan's Times Square. These scenes represent the most far-reaching change in the history of the American city, and the forces of the past three decades that turned it inside-out and replaced its single-centered with a multi-centered spatial structure.

jorities in the Prairie Provinces (Manitoba, Saskatchewan, and Alberta) and Pacific-bordering British Columbia.

No multilingual divisions affect the unity of the realm's other federation, but **cultural pluralism** of another kind prevails in the United States. More persistent in the U.S. cultural mosaic than language or ethnicity is the division between peoples of European descent (more than 75 percent of the population) and those of African origin (12 percent). Despite the significant progress of the modern civil rights movement, which decidedly weakened *de jure* racial segregation in the public domain, whites so rarely share their living space with blacks that *de facto* residential segregation is all but universal. Deprivation, too, is surprisingly widespread, with notable spatial concentrations of poverty inside the inner-ring slums of central cities and on rural reservations where Native Americans or Inuit reside.

1

DEFINING THE REALM

What geographers call the southern Rocky Mountains is one of the world's most clearly demarcated *physiographic* provinces, wedged between the Great Plains to the east and the Intermontane Basins and Plateaus to the west, and displaying a distinctive natural landscape of high relief and spectacular scenery. Glacial erosion carved this topography and sculpted valleys such as the Aspen Valley shown here; forests drape countrysides and snow covers the peaks. Gold finds once drew settlers here, but today the leading industry is tourism. The slopes and facilities of Aspen are world-renowned.

NORTH AMERICA'S PHYSICAL GEOGRAPHY

Before we examine the human geography of the United States and Canada more closely, we need to consider the physical setting in which they are rooted. The North American continent extends from the Arctic Ocean to Panama (Fig. 3-1), but we will confine ourselves here to the territory north of Mexico—a geographic realm that still stretches from the near-tropical latitudes of southern Florida and Texas to subpolar Alaska and Canada's far-flung northern periphery. The remainder of the North American continent comprises a separate realm, *Middle America*, which is covered in Chapter 4.

Physiography

North America's physiography is characterized by its clear, well-defined division into physically homogeneous regions called **physiographic provinces**. Each region is marked by considerable uniformity in relief, climate, vegetation, soils, and other environmental conditions, resulting in a scenic sameness that comes readily to mind. For example, we identify such regions when we refer to the Rocky Mountains, the Great Plains, and the Appalachian Highlands. However, not all the physiographic provinces of North America are so easily delineated.

2

Figure 3-2 maps the complete layout of the continent's physiography and includes a cross-sectional terrain profile along the 40th parallel. The most obvious aspect of this map of North America's physiographic provinces is the north-south alignment of the continent's great mountain backbone, the Rocky Mountains, whose rugged topography dominates the western segment of the continent from Alaska to New Mexico. The major feature of eastern North America is another, much lower chain of mountain ranges called the Appalachian Highlands, which also trend approximately north-south and extend from Canada's Atlantic Provinces to Alabama. The orientation of the Rockies and Appalachians is important because, unlike Europe's Alps, they do not form a topographic barrier to polar or

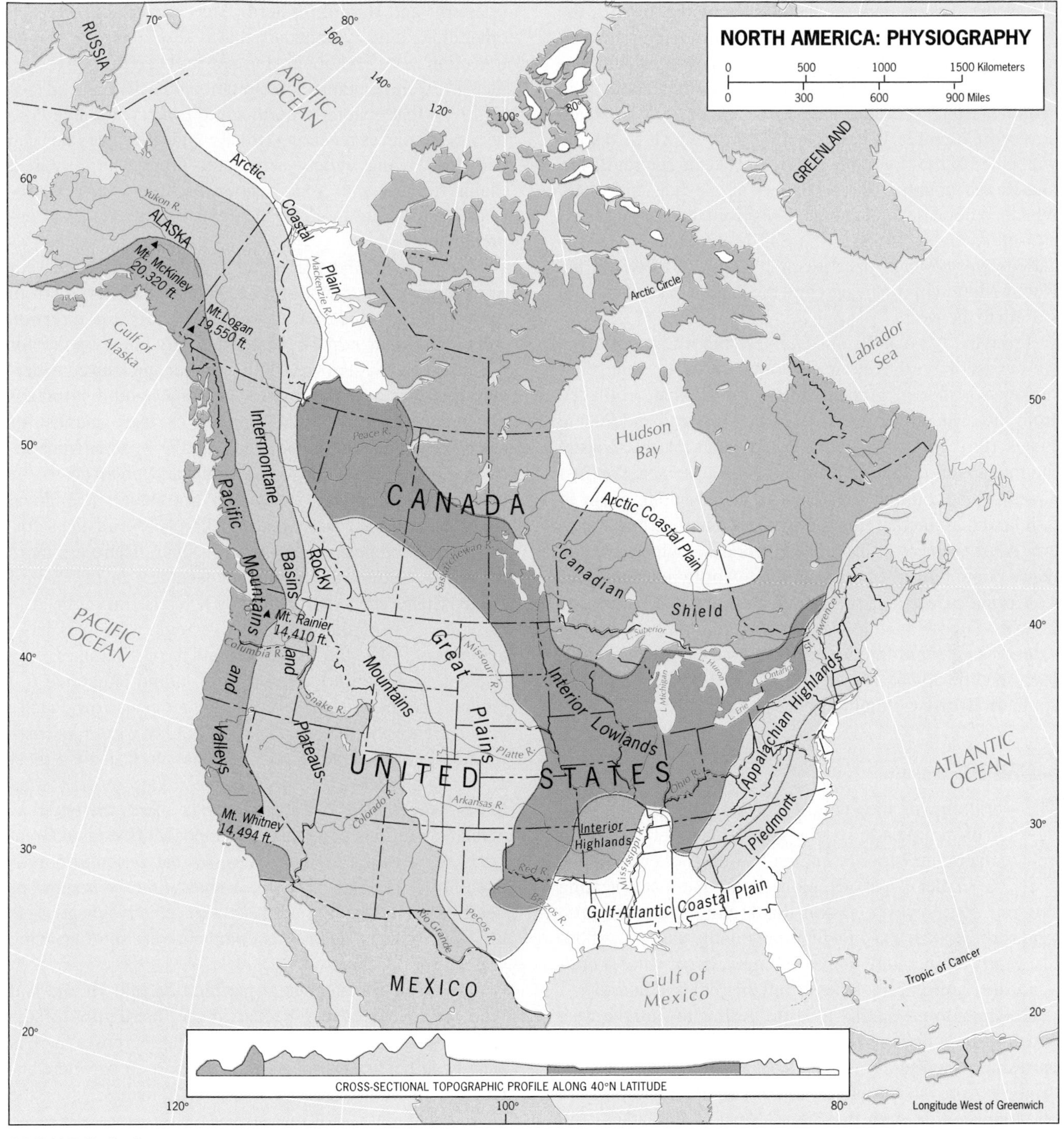

FIGURE 3-2

tropical airmasses flowing southward or northward, respectively, across the continent's interior.

Between the Rocky Mountains and the Appalachians lie North America's vast interior plains, which extend from the Mackenzie Delta on the Arctic Ocean to the Gulf of Mexico. We can subdivide these into several provinces: (1) the great Canadian Shield, which is the geologic core area containing North America's oldest rocks; (2) the Interior Lowlands, covered largely by glacial debris laid down by ice, meltwater, and wind during the Pleistocene glaciation; and (3) the Great Plains, the extensive sedimentary surface that rises slowly westward toward the Rocky Mountains. Along the southern margin, these interior plainlands merge into the Gulf-Atlantic Coastal Plain, which extends from southern Texas along the seaward margin of the Appalachian Highlands and the neighboring Piedmont until it ends at Long Island just to the east of New York City.

On the western side of the Rocky Mountains lies the zone of Intermontane Basins and Plateaus. Within the conterminous United States, this physiographic province includes: (1) the Colorado Plateau in the south, with its thick sediments and spectacular Grand Canyon; (2) the lava-covered Columbia Plateau in the north, which forms the watershed of the Columbia River; and (3) the central Basin-and-Range country (Great Basin) of Nevada and Utah, which contains several extinct lakes from the glacial period as well as the surviving Great Salt Lake. This province is called *intermontane* because of its position between the Rocky Mountains to the east and the Pacific coast mountain system to the west.

From the Alaskan Peninsula to Southern California, the west coast of North America is dominated by an almost unbroken corridor of high mountain ranges that originated from the contact between the North American and Pacific Plates (Fig. I-4). The major components of this coastal mountain belt include California's Sierra Nevada, the Cascades of Oregon and Washington, and the long chain of highland massifs that line the British Columbia and southern Alaska coasts. Three broad valleys—which contain dense populations—are the only noteworthy interruptions: California's Central (San Joaquin-Sacramento) Valley; the Cowlitz-Puget Sound lowland of Washington State, which extends southward into western Oregon's Willamette Valley; and the lower Fraser Valley, which slices through southern British Columbia's coast range.

Climate

The world climate map (Fig. I-8) clearly depicts the various climatic regimes and regions of North America. In general, temperature varies latitudinally—the farther north one goes, the cooler it gets. Regional land-and-water-heating differentials, however, distort this broad pattern. Because land surfaces heat and cool far more rapidly than water bodies, yearly temperature ranges are much larger where *continentality* (interior remoteness from the sea) is greatest.

Precipitation generally tends to decline toward the west (except for the Pacific coastal strip itself) as a result of the 3 **rain shadow effect**. This occurs because Pacific airmasses, driven by prevailing winds, carry their moisture onshore but soon collide with the Sierra Nevada–Cascades wall, forcing them to rise—and cool—in order to crest these mountain ranges. Such cooling is accompanied by major condensation and precipitation, so that by the time these airmasses descend and warm along the eastern slopes to begin their journey across the continent's interior, they have already deposited much of their moisture. Thus the mountains produce a downwind "shadow" of dryness, which is reinforced whenever eastward-moving air must surmount other ranges farther inland, especially the massive Rockies. Indeed, this semiarid (and in places truly arid) environment extends so deeply into the central United States that a broad division can be made between Arid (western) and Humid (eastern) America, which face each other along a fuzzy boundary that is best viewed as a wide transition zone. Although the separating criterion of 20 inches (50 cm) of annual precipitation is easily mapped (see Fig. I-7), that generally north-south *isohyet* (the line connecting all places receiving exactly 20 inches per year) can and does swing widely across the drought-prone Great Plains from year to year because highly variable warm-season rains from the Gulf of Mexico come and go in unpredictable fashion.

On the other hand, precipitation in Humid America is far more regular. The prevailing westerly winds (blowing from west to east—winds are always named for the direction *from* which they come), which normally come up dry for the large zone west of the 100th meridian, pick up considerable moisture over the Interior Lowlands and distribute it throughout eastern North America. A large number of storms develop here on the highly active weather front between tropical Gulf air to the south and polar air to the north. Even if major storms do not materialize, local weather disturbances created by sharply contrasting temperature differences are always a danger. There are more tornadoes (nature's most violent weather) in the central United States each year than anywhere else on Earth. And in winter, the northern half of this region receives large amounts of snow, particularly around the Great Lakes.

Figure I-8 shows the absence of humid temperate (*C*) climates from Canada (except along the narrow Pacific coastal zone) and the prevalence of cold in Canadian environments. East of the Rocky Mountains, Canada's most *moderate* climates correspond to the *coldest* of the United States. Nonetheless, southern Canada shares the environmental conditions that mark the Upper Midwest and Great Lakes areas of the United States, so that agricultural productivity in the Prairie Provinces and in Ontario is substantial. Canada is a leading food exporter (chiefly wheat), as is the United States, despite its comparatively short growing season.

The broad environmental partitioning into Humid and Arid America is also reflected in the distribution of the realm's soils and vegetation. For farming purposes there is usually sufficient soil moisture to support crops where annual precipitation exceeds the critical 20 inches; where the yearly total is less, soils may still be fertile (especially in the Great Plains), but irrigation is often necessary to achieve their full agricultural potential. As for vegetation, the Humid/Arid America dichotomy is again a valid generalization: the natural vegetation of areas receiving more than 20 inches of water annually is *forest*, whereas the drier climates give rise to a *grassland* cover.

Hydrography (Surface Water)

Surface water patterns in North America are dominated by the two major drainage systems that lie between the Rockies and the Appalachians: (1) the five Great Lakes (Superior, Michigan, Huron, Erie, and Ontario) that drain into the St. Lawrence River and (2) the mighty Mississippi-Missouri river network, fed by such major tributaries as the Ohio, Tennessee, and Arkansas rivers. Both are products of the last episode of Pleistocene glaciation, and together they amount to nothing less than the best natural inland waterway system in the world. Human intervention has further enhanced this network of navigability, mainly through the building of canals that link the two systems as well as the St. Lawrence Seaway.

Elsewhere, the northern east coast of the continent is well served by a number of short rivers leading inland from the Atlantic. In fact, many of the major northeastern seaboard cities of the United States—such as Washington, D.C., Baltimore, and Philadelphia—are located at the waterfalls that marked the limit to tidewater navigation (hence their designation as *Fall Line cities*). Rivers in the Southeast and west of the Rockies at first offered little practical value because of their orientation and the difficulty of navigating them. In the far west, however, the Colorado and Columbia rivers have become supremely important as suppliers of drinking and irrigation water as well as hydroelectric power.

THE UNITED STATES

The broad outline of North America we have just sketched will be useful in developing the regionalization scheme that appears in the second half of this chapter. But to fully appreciate the realm's internal regional organization, we first need to examine in some detail the changing human geography of each country. We begin with the United States, and it may be helpful to take a few moments to review the map of its basic contents (Fig. 3-3).

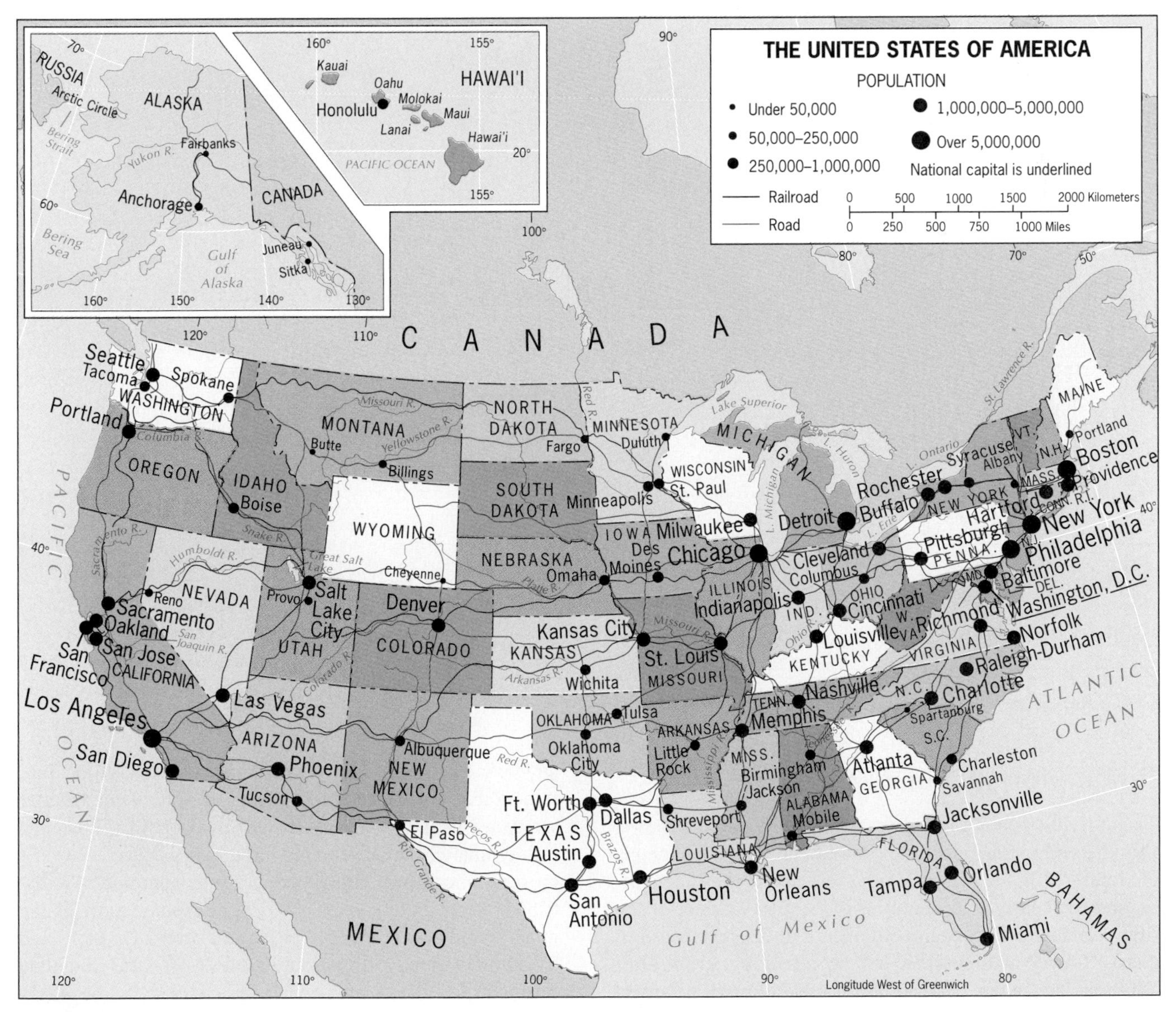

FIGURE 3-3

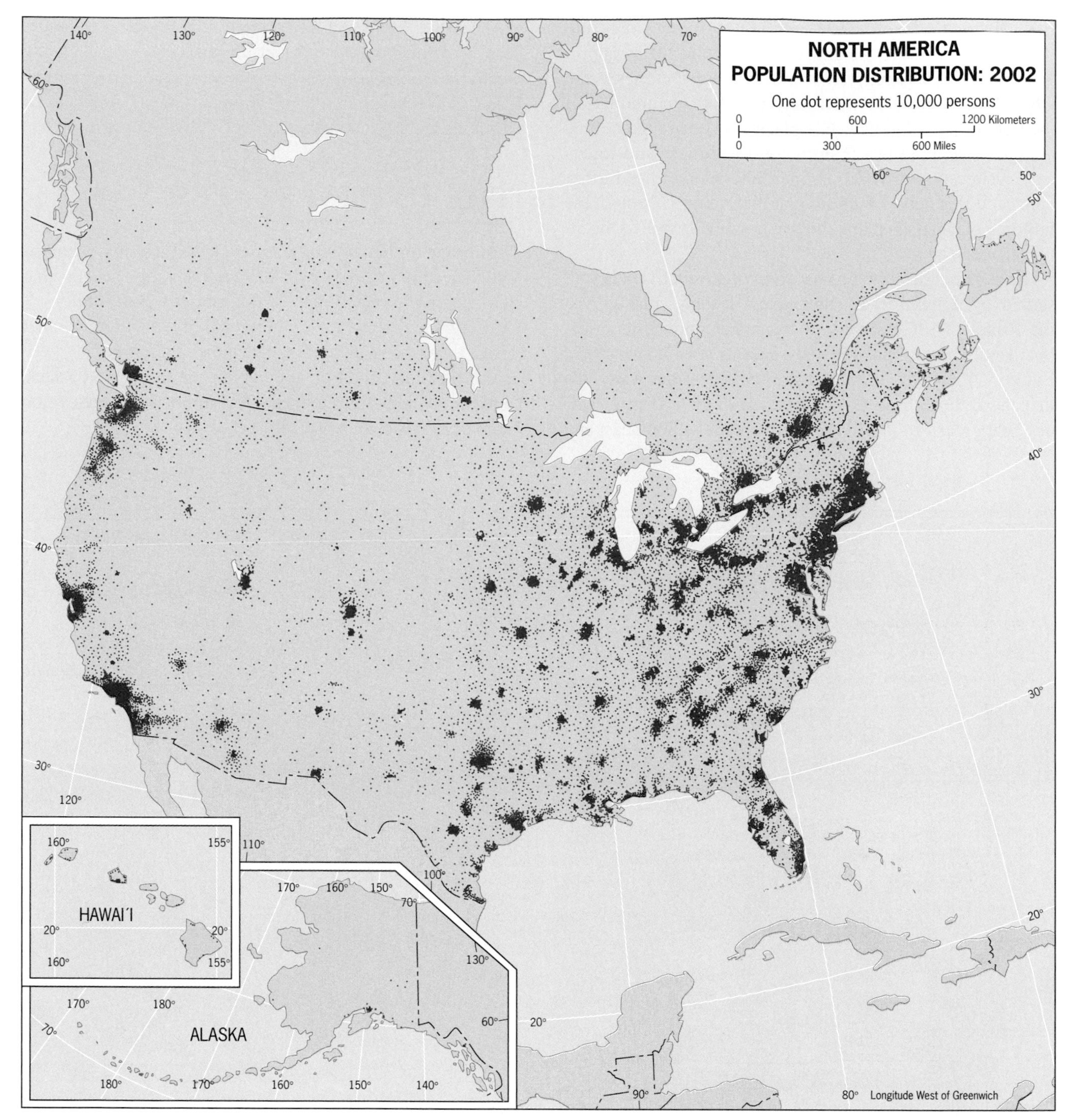

FIGURE 3-4

Population in Time and Space

The current population distribution of the United States is shown in Figure 3-4. It is important to note that this map is the latest "still" in a motion picture, one that has been unreeling for nearly four centuries since the founding of the first permanent European settlements on the northeastern coast. Slowly at first, then with accelerating speed after 1800, as one major transportation breakthrough followed another, Americans (and Canadians) took charge of their remarkable continent and pushed the settlement frontier westward to the Pacific. The swiftness of this expansion was dramatic, but Americans have long been the world's most mobile people. In fact, migrations continue to redistribute population in the United States today, perhaps the most significant being the persistent drift of people and livelihoods toward the South and West (the so-called **Sunbelt**).

4

The Migration Process

The United States as well as Canada is the product of **migration** (a change in residential location intended to be permanent). After tens of thousands of years of native settlement, European explorers reached the shores of North America beginning with Columbus in 1492 (and probably centuries before that). The first permanent colonies were established in the early 1600s, and from them evolved the modern United States of America. The Europeanization of North America doomed the continent's aboriginal societies, but this was only one of many areas around the world where local cultures and foreign invaders came face to face. Between 1835 and 1935, perhaps as many as 75 million Europeans departed for distant shores—most of them bound for the Americas (Fig. 3-5). Some sought religious freedom; others escaped poverty and famine; still others simply hoped for a better life.

Studies of the *migration decision* indicate that migration flows vary in size with (1) the perceived degree of difference between one's home, or source, and the destination; (2) the effectiveness of the information flow, that is, the news about the destination that migrants sent to those who stayed behind waiting to decide; and (3) the distance between the source and the destination (shorter moves attract many more migrants than longer ones). More than a century ago, the British social scientist Ernst Georg Ravenstein studied the migration process, and many of his conclusions remain valid today. For example, every migration stream from source to destination produces a counter-stream of returning migrants who cannot adjust, are unsuccessful, or are otherwise persuaded to go back home. Studies of migration also conclude that several factors are at work in the migration process. **Push factors** motivate people to move away; **pull factors** attract them to new destinations. To those early Europeans, the United States was a new frontier, a place where one might acquire a piece of land and some livestock. The opportunities were reported to be unlimited.

That perception has never changed, and immigration continues to significantly shape the human-geographic complexion of the United States. Today's immigrants account for no less than one-third of the country's annual population growth. Important details about this latest stage of the international migration process are discussed on page 167.

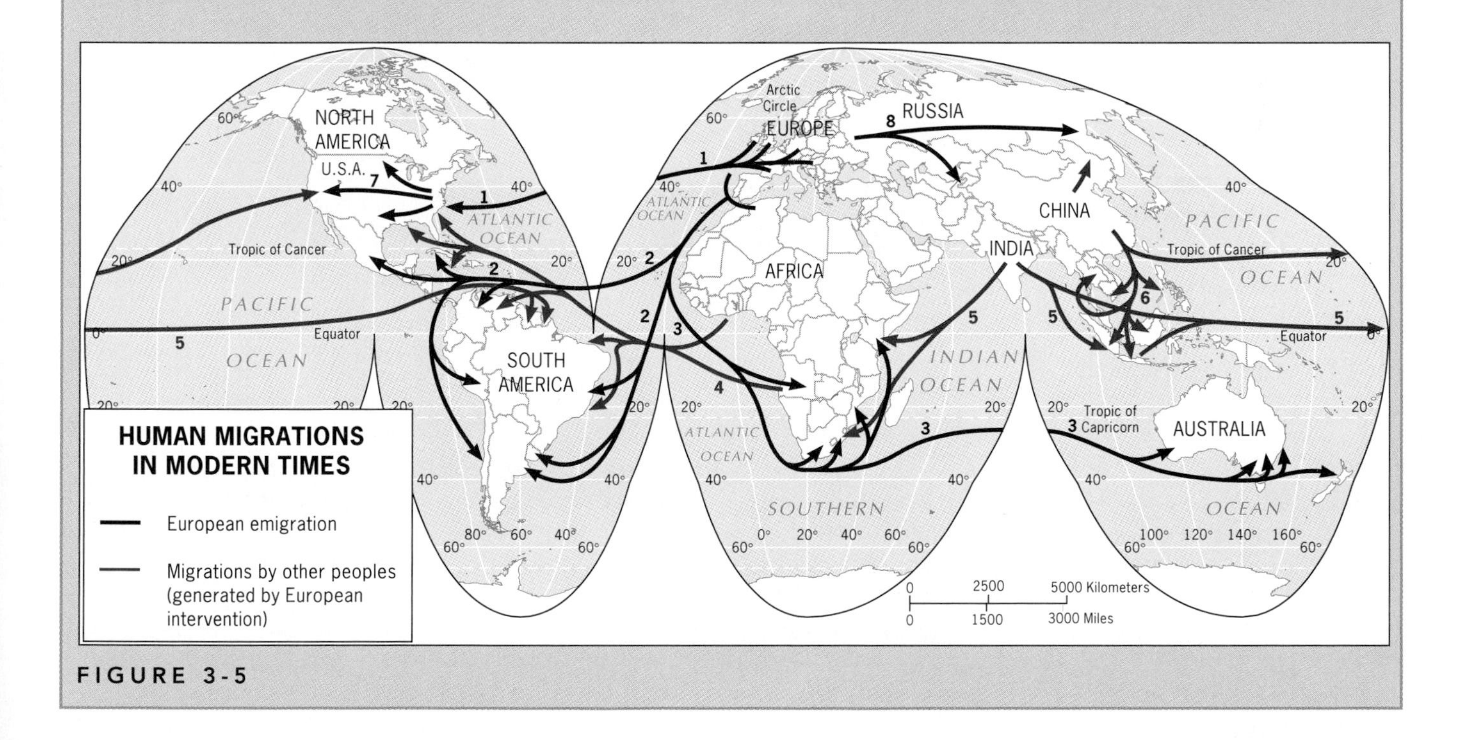

FIGURE 3-5

To understand the contemporary population map, we need to review the major forces that have shaped, and continue to shape, the distribution of Americans and their activities. Since its earliest days, the United States has attracted a steady influx of immigrants who were rapidly assimilated into the societal mainstream (see box titled "The Migration Process"). Within the country, people have sorted themselves out to maximize their proximity to existing economic opportunities, and they have shown little resistance to relocating as the nation's evolving economic geography has successively favored different sets of places over time.

During the past century these transformations spawned a number of major migrations: (1) the still-continuing westward shift but now with that southerly, Sunbelt deflection; (2) the rapid growth of metropolitan areas, first triggered by the late-nineteenth-century Industrial Revolution, which since the 1960s has been largely rechanneled from the central cities to the suburban ring; (3) the movement of African Americans from the rural south to the urban north, which

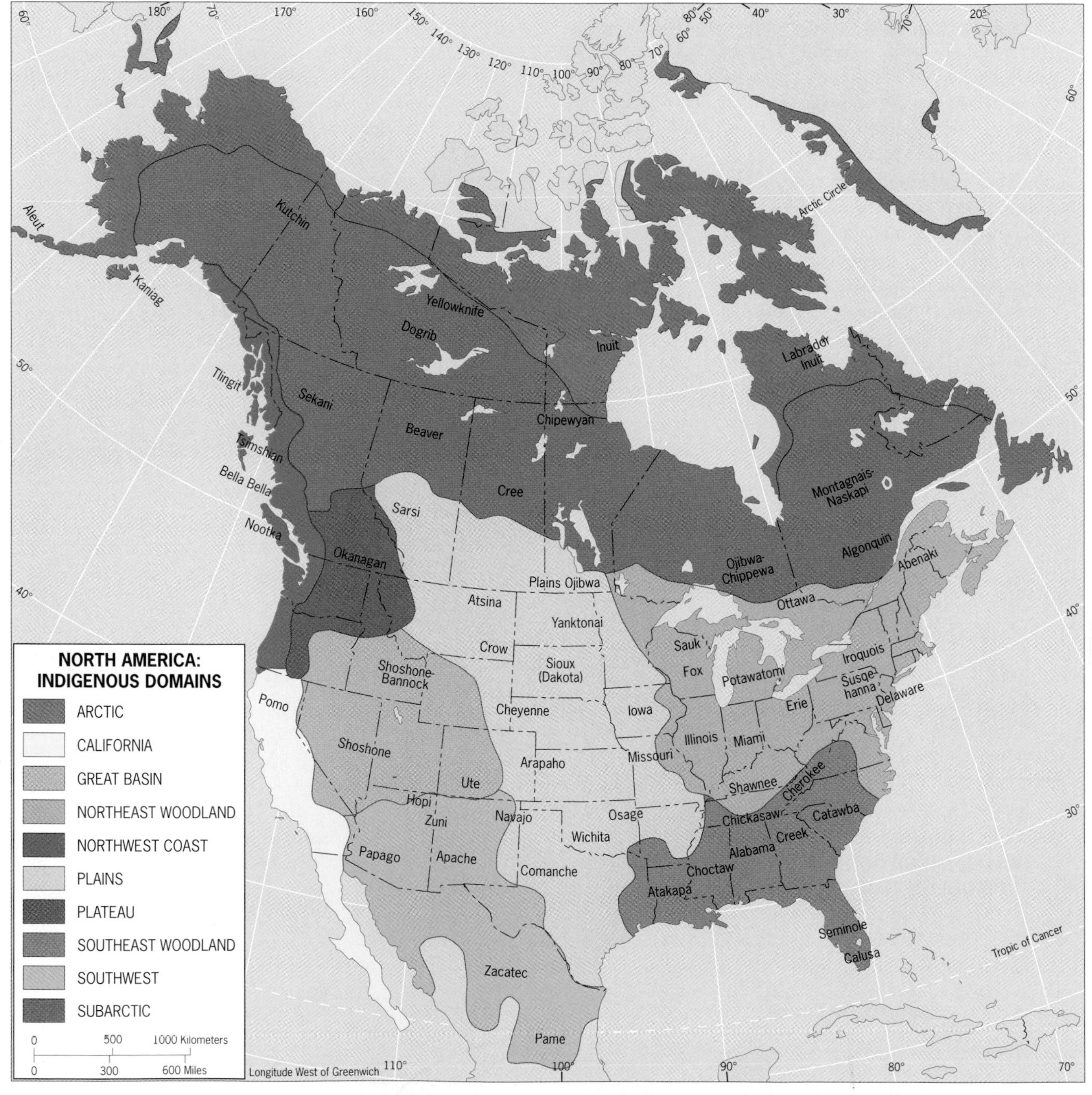

FIGURE 3-6

since the 1970s has become a stronger return flow, particularly among middle-income blacks; and (4) the influx of immigrants from outside the United States, mostly European before 1960 but now dominated by a new wave from Middle America and eastern Asia that is directed, respectively, toward the zone adjacent to the southern border and the urban areas of the Pacific coast.

Let us now look more closely at the historical geography of these changing population patterns, considering first the initial rural influence and then the decisive impacts of industrial urbanization.

Pre-Twentieth-Century Population Patterns

Indigenous North America When the first Europeans set foot on North American soil, the continent was occupied by millions of people whose ancestors had reached the Americas from Asia, via Alaska and probably also across the Pacific, more than 13,000 years (and possibly as long as 30,000 years) ago. In search of Asia, the Europeans misnamed them "Indians," but the historic affinities of these earliest Americans were with the peoples of eastern and northeastern Asia, not India. In North America these Native Americans or First Nations, as they are now called in the United States and Canada respectively, had organized themselves into hundreds of nations with a rich mosaic of languages and a great diversity of cultures (Fig. 3-6). Farmers grew crops the Europeans had never seen; other nations depended chiefly on fishing, herding, hunting, or some combination of these. Elaborate houses, efficient watercraft, effective weaponry, decorative clothing, and wide-ranging art forms distinguished the aboriginal nations. Certain nations had formulated sophisticated health and medical practices; ceremonial life was complex and highly developed; and political institutions were mature and elaborate.

The eastern nations first bore the brunt of the European invasion. By the end of the eighteenth century, ruthless, land-hungry settlers had driven most of the Native American peoples living along the Atlantic and Gulf coasts from their homes and lands, beginning a westward push that was to devastate indigenous society. The U.S. Congress in 1789 proclaimed that "Indian . . . land and property shall never be taken from them without their consent," but in fact this is just what happened. One of the sorriest episodes in American history involved the removal of the eastern Cherokee, Chickasaw, Choctaw, Creek, and Seminole from their homelands in forced marches a thousand miles westward to Oklahoma. One-fourth of the entire Cherokee population died along the way from exposure, starvation, and disease, and the others fared little better. Again, Congress approved treaties that would at least protect the native peoples of the Plains (Fig. 3-6) and those farther to the west, but after the mid-nineteenth century the white settlers ignored those guarantees as well. A half-century of war left what remained of North America's nations with about 4 percent of U.S. territory in the form of mostly impoverished reservations.

In what is today Canada, too, the comparatively small First Nations population was overwhelmed by the numbers and power of European settlers, and decimated by the diseases they introduced. Efforts at restitution and recognition of First Nations rights, however, have gone farther in Canada than in the United States.

Colonial and Nineteenth-Century Development The current spatial distribution of the U.S. population is rooted in the colonial era of the seventeenth and eighteenth centuries that was dominated by England and France. The French sought mainly to organize a lucrative fur-trading network, while the English established settlements along the coast of what is today the northeastern U.S. seaboard. These British colonies quickly became differentiated in their local economies, a diversity that was to endure and later shape American cultural geography. The northern colony of New England (Massachusetts Bay and environs) specialized in commerce; the southern Chesapeake Bay colony (Tidewater Virginia and Maryland) emphasized the plantation farming of tobacco; the Middle Atlantic area lying in between (southeastern New York, New Jersey, eastern Pennsylvania) was home to a number of smaller, independent-farmer colonies.

These neighboring colonies soon thrived and yearned to expand, but the British government responded by closing the inland frontier and tightening economic controls. By 1783 this move had led to colonial unification, British defeat in the Revolutionary War, and independence for the newly formed United States of America. The western frontier of the fledgling nation now swung open, and the zone north of the Ohio River was promptly settled following the discovery that the soils (and climate) of the Interior Lowlands were more favorable for farming than those of the Atlantic Coastal Plain and Piedmont. This triggered the rapid growth of trans-Appalachian agriculture and the widening of seaboard-interior trading ties, and the new interregional complementarities signified that U.S. spatial organization was assuming national-scale proportions.

By the time the westward-moving frontier swept across the Mississippi Valley in the 1820s, the three former seaboard colonies (**A**, **B**, and **C** in Fig. 3-7) had become separate *culture hearths*—primary source areas and innovation centers from which migrants carried cultural traditions into the central United States (as the arrows in Fig. 3-7 show). The northern half of this vast interior space soon became well unified as its infrastructure steadily improved following the introduction of the railroad in the 1830s. The American South, however, did not wish to integrate itself economically with the North, preferring to export tobacco and cotton from its plantations to overseas markets; its insistence on preserving slavery to support this system soon led the South into secession, disastrous Civil War (1861–1865), and a dismal aftermath that took a full century to overcome.

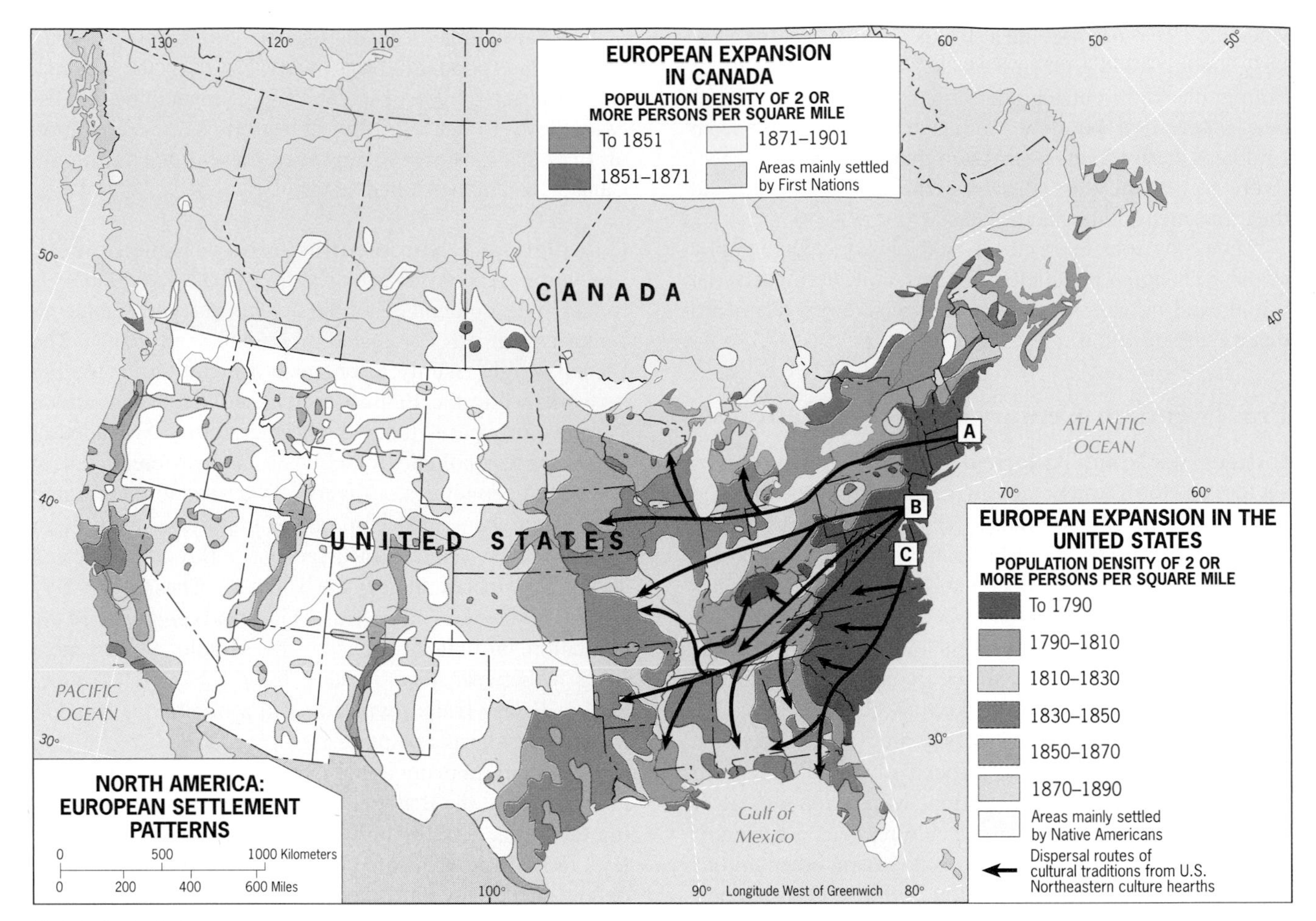

FIGURE 3-7

The second half of the nineteenth century saw the frontier cross the western United States, and by 1869 agriculturally booming California was linked to the rest of the nation by transcontinental railroad (these same steel tracks also opened up the bypassed, semiarid Great Plains). When the American frontier closed in the 1890s, today's rural settlement pattern was firmly in place, anchored to a set of enduring national agricultural regions (discussed later in this chapter). By then, however, the exodus of rural Americans toward the burgeoning cities had begun in response to the Industrial Revolution that had taken hold after 1870.

Post-1900 Industrial Urbanization

The U.S. Industrial Revolution occurred almost a century later than in Europe, but when it finally did cross the Atlantic in the 1870s, it took hold so successfully and advanced so robustly that only 50 years later America was surpassing Europe as the world's mightiest industrial power. The impact of industrial urbanization occurred simultaneously at two levels of spatial generalization. At the national or *macroscale*, a system of new cities swiftly emerged, specializing in the collection, processing, and distribution of raw materials and manufactured goods, linked together by an efficient web of railroad lines. Within that urban network, at the local or *microscale*, individual cities prospered in their new roles as manufacturing centers, generating an internal structure that still forms the geographic framework of most of the central cities of America's large metropolitan areas. We now examine the urban trend at both of these scales.

Evolution of the U.S. Urban System The rise of the national urban system in the late nineteenth century was based on the traditional external role of cities: providing goods and services for their hinterlands in exchange for raw materials. Because people (both as laborers and consumers), commercial activities, investment capital, and transport facilities were already agglomerated in existing (preindustrial) cities, the emerging industrialization movement tended to favor such locations. Their growing incomes, in turn, permitted industrially intensifying cities to invest in a bigger local infrastructure of private and public services as well as housing—and thereby convert each round of industrial expansion into a new stage of urban development.

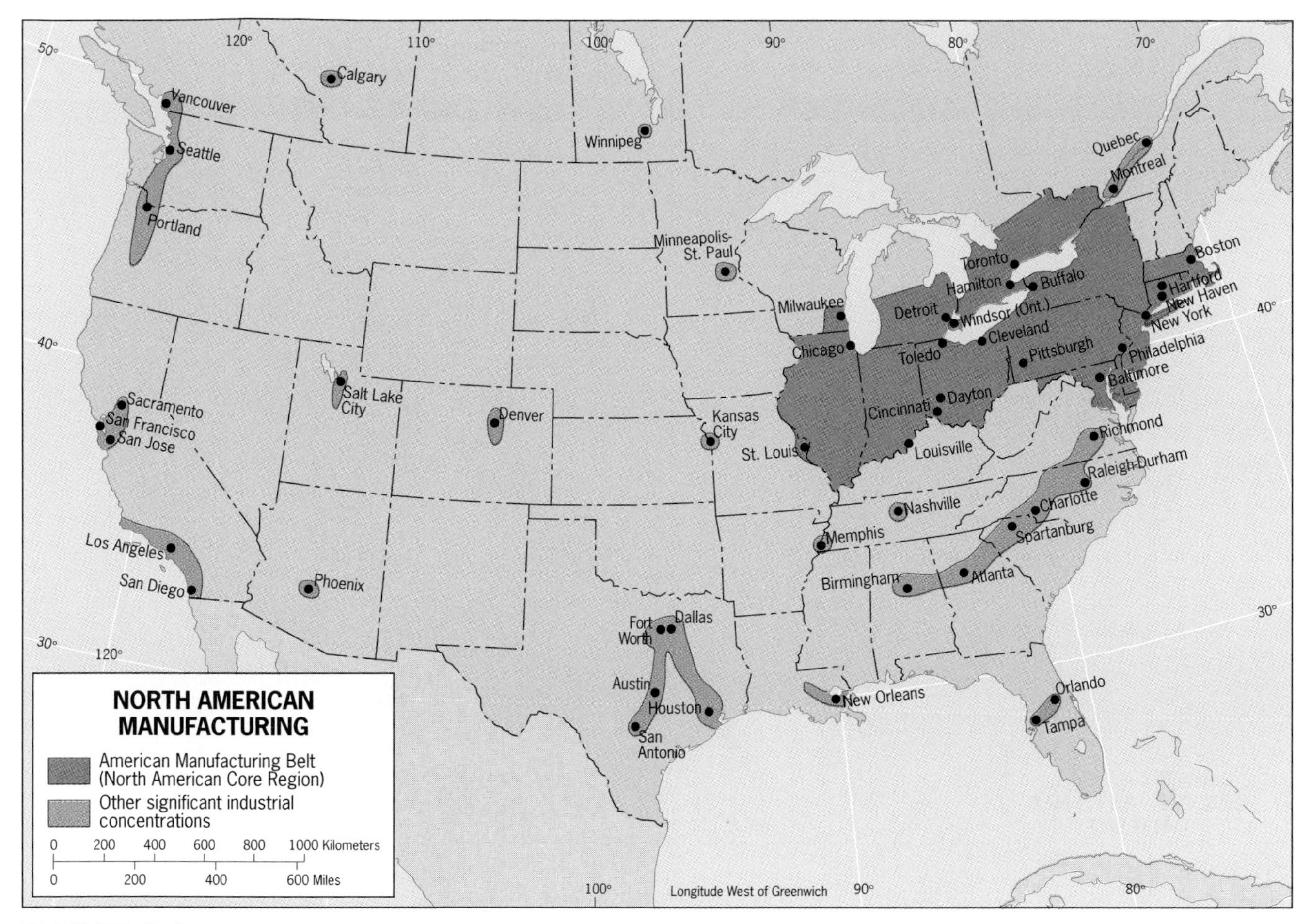

FIGURE 3-8

The national urban system had actually been in the process of formation long before its flowering after 1870. Its evolution over the past two centuries has been studied by John Borchert, who identified **five epochs of metropolitan evolution** based on transportation technology and industrial energy. (1) *The Sail-Wagon Epoch* (1790–1830), marked by primitive overland and waterway circulation; the leading cities were the northeastern ports, which were more heavily oriented to the European overseas trade than to their barely accessible western hinterlands. (2) *The Iron Horse Epoch* (1830–1870), dominated by the arrival and spread of the steam-powered railroad; a nationwide transport system had now been forged, and the national urban system began to take shape, with New York emerging as the primate city by 1850. (3) *The Steel-Rail Epoch* (1870–1920), which spanned the Industrial Revolution and saw the full establishment of the national metropolitan system; new forces shaping growth were the increasing scale of manufacturing, the rise of the steel and auto industries in Midwestern cities, and the introduction of steel rails that enabled trains to travel faster and haul heavier cargoes. (4) *The Auto-Air-Amenity Epoch* (1920–1970), which encompassed the later stage of U.S. industrial urbanization and the maturation of the national urban hierarchy; its key elements were the automobile and the airplane, the expansion of white-collar services jobs, and the growing locational pull of *amenities* (pleasant environments) that increasingly stimulated the urbanization of the suburbs and selected Sunbelt locales. (5) *The Satellite-Electronic-Jet Propulsion Epoch* (1970–), shaped by the newest advancements in information management, computer technologies, global communications, and intercontinental travel; it favors globally oriented metropolises (now often called *world cities*), particularly those along the Pacific and Atlantic coasts that function as international gateways.

Industrialization and the accompanying growth of the urban system reconfigured the realm's economic landscape. The most notable regional transformation was the emergence of the North American Core, or **American Manufacturing Belt**, which contained the lion's share of industrial activity in both the United States and Canada. As Figure 3-8 shows, the geographic form of the Core Region—which includes southern Ontario—was a near-rectangle whose four corners were Boston, Milwaukee, St. Louis, and Baltimore. However, because manufacturing is such a spatially concentrated activity, less than 1 percent of

7

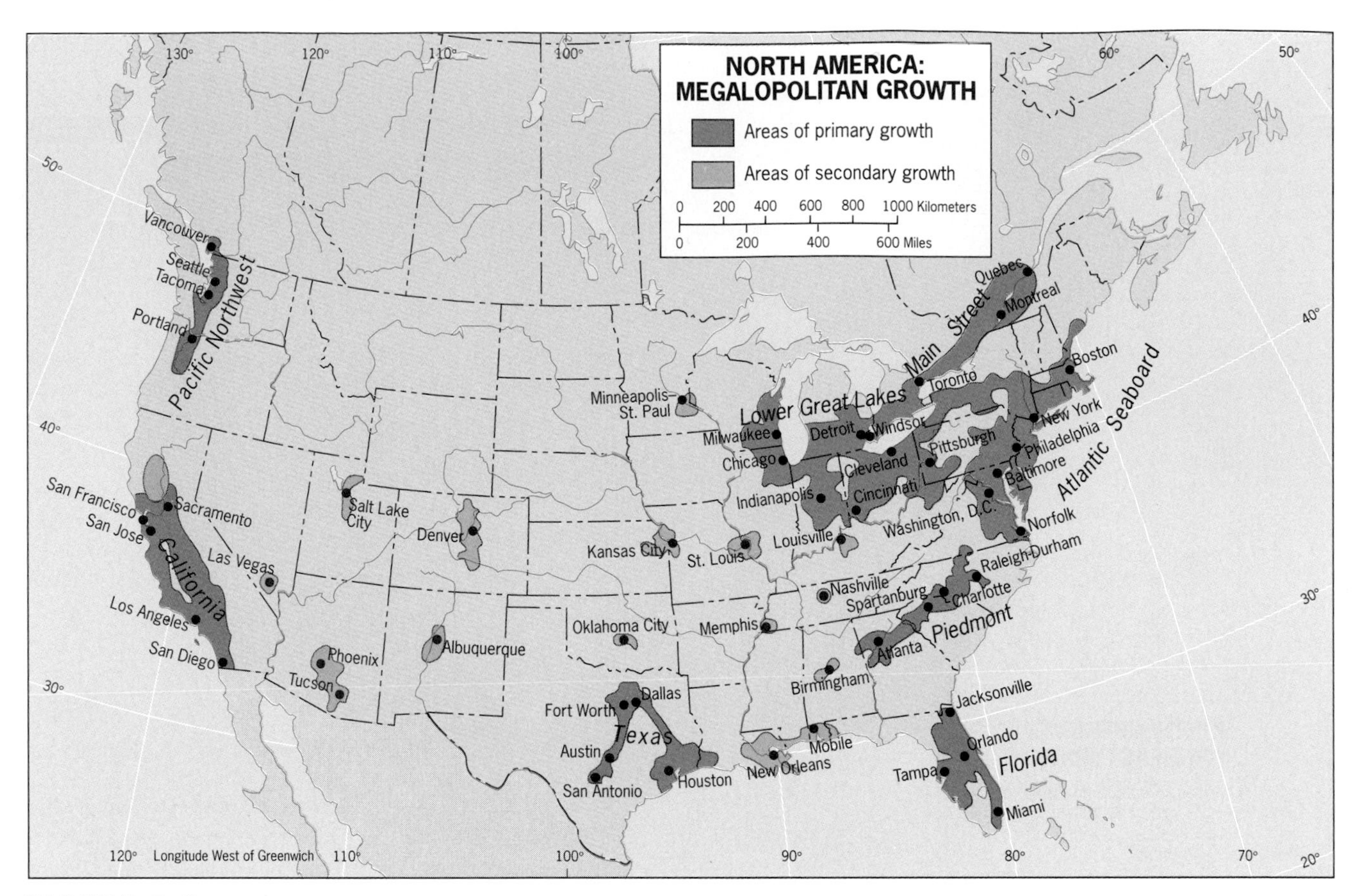

FIGURE 3-9

the territory of the Belt is devoted to industrial land use: most of its factories are tightly clustered into a dozen districts centered on the cities mapped in Figure 3-8.

At the subregional scale, as transportation breakthroughs permitted progressive urban decentralization and *megalopolitan growth*, the expanding peripheries of major cities soon coalesced to form a number of conurbations. The most important of these by far is the *Atlantic Seaboard Megalopolis* (Fig. 3-9), the 600-mile (1000-km) urbanized northeastern coastal strip extending from southern Maine to Virginia that contains metropolitan Boston, New York, Philadelphia, Baltimore, and Washington. This was the economic heartland of the Core; the seat of U.S. government, business, and culture; and the trans-Atlantic trading interface between much of North America and Europe. Six other primary conurbations have also emerged: *Lower Great Lakes* (Chicago-Detroit-Cleveland-Pittsburgh), *Piedmont* (Atlanta- Charlotte-Raleigh/Durham), *Florida* (Jacksonville-Tampa-Orlando-Miami), *Texas* (Houston-Dallas/Ft. Worth-San Antonio), *California* (San Diego-Los Angeles-San Francisco), and the *Pacific Northwest* (Portland-Seattle-Vancouver). Note that the last spills across the border into Canada, which has also spawned its own nationally predominant conurbation—*Main Street* (Windsor-Toronto-Montreal-Quebec City).

The Changing Structure of the U.S. Metropolis

The internal structure of the metropolis reflected the same mixture of forces that shaped the national urban system, especially transportation technology. Rails—in this case, lighter street-rail lines—once again shaped spatial organization as horse-drawn trolleys were succeeded by electric streetcars in the late nineteenth century. The mass introduction of the automobile after World War I changed all that, and America steadily turned from building compact cities to the widely dispersed metropolises of the post–World War II highway era. By 1970, the new intraurban expressway network had equalized location costs throughout the metropolis, setting the stage for suburbia to swiftly transform itself from a residential preserve into a complete outer city with amenities and new prestige that proved highly attractive to the business world. As the newly urbanized suburbs increasingly captured major economic activities, many large cities saw their status diminish to that of coequal. Their once thriving central business districts (CBDs) were now all but reduced to serving the less affluent populations that increasingly dominated the central city's neighborhoods.

This growth process was conceptualized into a four-stage model by John Adams, who identified the four **eras of intraurban structural evolution** that are diagrammed in

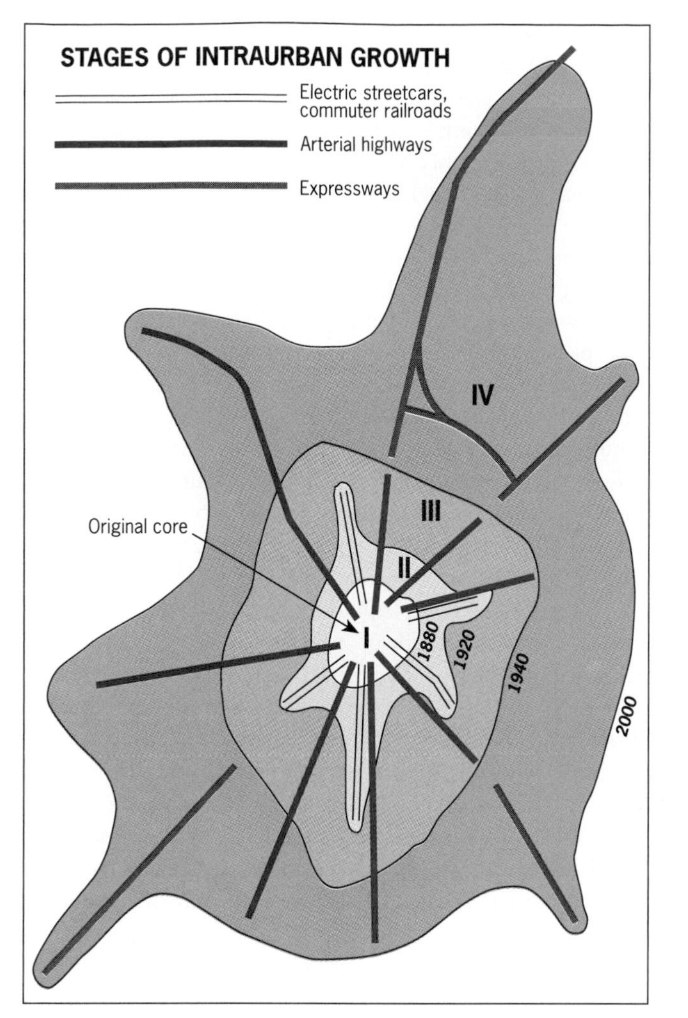

FIGURE 3-10

Figure 3-10. Stage I, prior to 1888, was the *Walking-Horse-car Era*, which produced a compact pedestrian city in which everything had to be within a 30-minute walk, a layout only slightly augmented when horse-drawn trolleys began to operate after 1850. The 1888 invention of the electric traction motor launched Stage II, the *Electric Streetcar Era* (1888–1920); higher speeds enabled the 30-minute travel radius and the urbanized area to expand considerably along new outlying trolley corridors; in the older core city, the CBD, industrial, and residential land uses differentiated into their modern form. Stage III, the *Recreational Automobile Era* (1920–1945), was marked by the initial impact of cars and highways that steadily improved the accessibility of the outer metropolitan ring, thereby launching a wave of mass suburbanization that further extended the urban frontier; during this era, the still-dominant central city experienced its economic peak and the partitioning of its residential space into neighborhoods sharply defined by income, ethnicity, and race. Stage IV, the *Freeway Era* (under way since 1945), saw the full impact of automobiles, with the metropolis turning inside-out as expressways pushed suburban development more than 30 miles (50 km) from the CBD.

The social geography of the evolving industrial metropolis has been marked by the development of a residential mosaic that exhibited the congregating of ever-more-specialized groups. The electric streetcar, which introduced "mass" transit that every urbanite could afford, allowed the heterogeneous, immigrant-dominated city population to sort itself into ethnically uniform neighborhoods. When the United States sharply curtailed foreign immigration in the 1920s, industrial managers discovered the large African American population

FROM THE FIELD NOTES

"Monitoring the urbanization of U.S. suburbs for the past three decades has brought us to Tyson's Corner, Virginia on many a field trip and data-gathering foray. It is now hard to recall from this late-1990s view that only 40 years ago this place was merely a near-rural crossroads. But as nearby Washington, D.C. steadily decentralized, *Tyson's* capitalized on its unparalleled regional accessibility (its Capital Beltway location at the intersection with the radial Dulles Airport Toll Road) to attract a seemingly endless parade of high-level retail facilities, office complexes, and a plethora of supporting commercial services. We would rank it today among the three largest and most influential suburban downtowns (or 'edge cities') in North America."

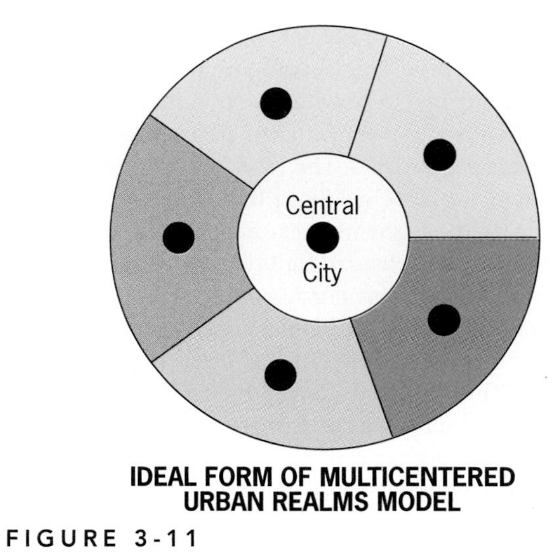

IDEAL FORM OF MULTICENTERED URBAN REALMS MODEL

FIGURE 3-11

of the rural South—increasingly unemployed there as cotton-related agriculture declined—and began to recruit these workers by the thousands for the factories of Manufacturing Belt cities. This influx had an immediate impact on the social geography of the industrial city because whites were unwilling to share their living space with the racially different newcomers. The result was the involuntary segregation of these newest migrants, who were channeled into geographically separate, all-black areas. By the 1950s, these mostly inner-city areas became large expanding **ghettos**, speeding the departure of many white central-city communities and reinforcing the trend toward a racially divided urban society. 8

The huge suburban component of the intraurban residential mosaic is not only home to the more affluent residents of the metropolis: over the past three decades it received so massive an infusion of nonresidential activities that it has been transformed into a full-fledged **outer city**. As its ties to the central city loosened, the outer city's growing independence was accelerated by the rise of major new suburban nuclei (particularly near key freeway interchanges) to serve the new local economies. These multipurpose activity nodes often developed around large regional shopping centers, whose prestigious images attracted scores of industrial parks, office campuses and high-rises, hotels, restaurants, entertainment facilities, and even major league sports stadiums and arenas, which together formed burgeoning new **suburban downtowns** that are an automobile-age version of the CBD (see photo, p. 163). As suburban downtowns flourish, they attract tens of thousands of local residents to organize their lives around them—offering workplaces, shopping, leisure activities, and all the other elements of a complete urban environment. 9 10

These newest spatial elements of the contemporary metropolis are assembled in the model displayed in Figure 3-11. The rise of the outer city has today produced a *multicentered* metropolis consisting of the traditional CBD as well as a set of increasingly coequal suburban downtowns, with each

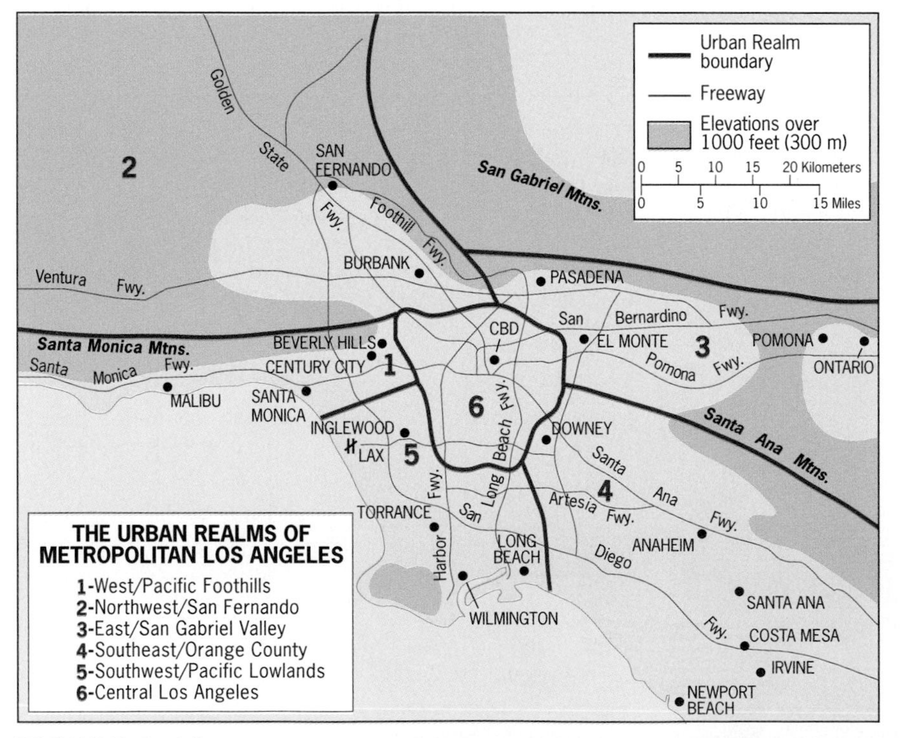

FIGURE 3-12

This photo of a portion of northern Manhattan island was taken in Fort sLee, New Jersey (see map p. 181), looking eastward from atop the Palisades escarpment that lines the west bank of the Hudson River. Located at the base of the Palisades in the foreground is a cluster of upscale Japanese restaurants, groceries, department stores, and electronics and myriad other retailers. This luxurious shopping and entertainment complex is Yaohan Plaza–New York, which serves the tens of thousands of Japanese who work for the New York branch offices of their global corporations. Most live in high-income suburbs north and west of New York City, for which Fort Lee is especially convenient because three expressways and the George Washington Bridge (just off the left of the photo) converge there. The slice of Manhattan seen across the river, extending from about 160th Street on the left to 140th Street on the right, is a totally different social world. This is not the skyscrapered, international-class CBD south of 59th Street, but the highly stressed inner city of the southern end of Washington Heights. The towers that form the skyline backdrop at the upper right are in the heart of the adjacent neighborhood of Harlem. These high-rises are not commercial buildings but the deteriorating—and thoroughly depressing—low-income housing projects that grew out of the sterile urban renewal programs of the post–World War II era.

activity center serving a discrete and self-sufficient surrounding area. James Vance defined these tributary areas as **urban realms**, recognizing in his studies that each such realm maintains a separate, distinct economic, social, and political significance and strength. Figure 3-12 applies the urban realms model to Los Angeles; we could easily draw a similar regionalization scheme for other large U.S. metropolises.

The position of the central city within the new multinodal metropolis of realms is eroding. No longer the dominant metropolitanwide center for goods and services, the CBD increasingly serves the less affluent residents of the innermost realm and those working downtown. As manufacturing employment declined precipitously, many large cities adapted successfully by shifting toward service industries. Accompanying this switch is downtown commercial revitalization, but in many cities for each shining new skyscraper that goes up several old commercial buildings are abandoned. Residential reinvestment has also occurred in many downtown-area neighborhoods but usually requires the displacement of established lower-income residents, an emotional issue that has sparked many conflicts. Beyond the CBD zone, the vast inner city remains the problem-ridden domain of low- and moderate-income people, with most forced to reside in ghettos.

Cultural Geography

In the United States, over the past two centuries, the contributions of a wide spectrum of immigrant groups have shaped—and continue to shape—a rich and varied cultural complex. Great numbers of these newcomers were willing to set aside their original cultural baggage in favor of assimilation into the emerging culture of their adopted homeland, which itself was a hybrid nurtured by constant infusions of new influences. For most upwardly mobile immigrants, this plunge into the much-touted "melting pot" promised a ticket for acceptance into mainstream American society.

American Cultural Bases

As the hybrid culture of the United States matured, it built on a set of powerful values and beliefs: (1) love of newness; (2) desire to be near nature; (3) freedom to move; (4) individualism; (5) societal acceptance; (6) aggressive pursuit of goals; and (7) a firm sense of destiny. Brian Berry has discerned these cultural traits in the behavior of people throughout the evolution of urban America. A "rural ideal" has prevailed throughout U.S. history and is still expressed in a strong bias against residing in cities. When industrialization made urban living unavoidable, those able to afford it soon moved to the emerging suburbs (*newness*) where a form of country life (*close to nature*) was possible in a semiurban setting. The fragmented metropolitan residential mosaic, composed of myriad slightly different neighborhoods, encouraged frequent *unencumbered mobility* as middle-class life revolved around the *individual* nuclear family's *aggressive pursuit* of its aspirations for *acceptance into the next higher stratum of society*. To most Americans these accomplishments confirmed that they could attain their goals through hard work and perseverance, and that they had the ability to realize their *destiny* by achieving the "American Dream" of homeownership, affluence, and total satisfaction.

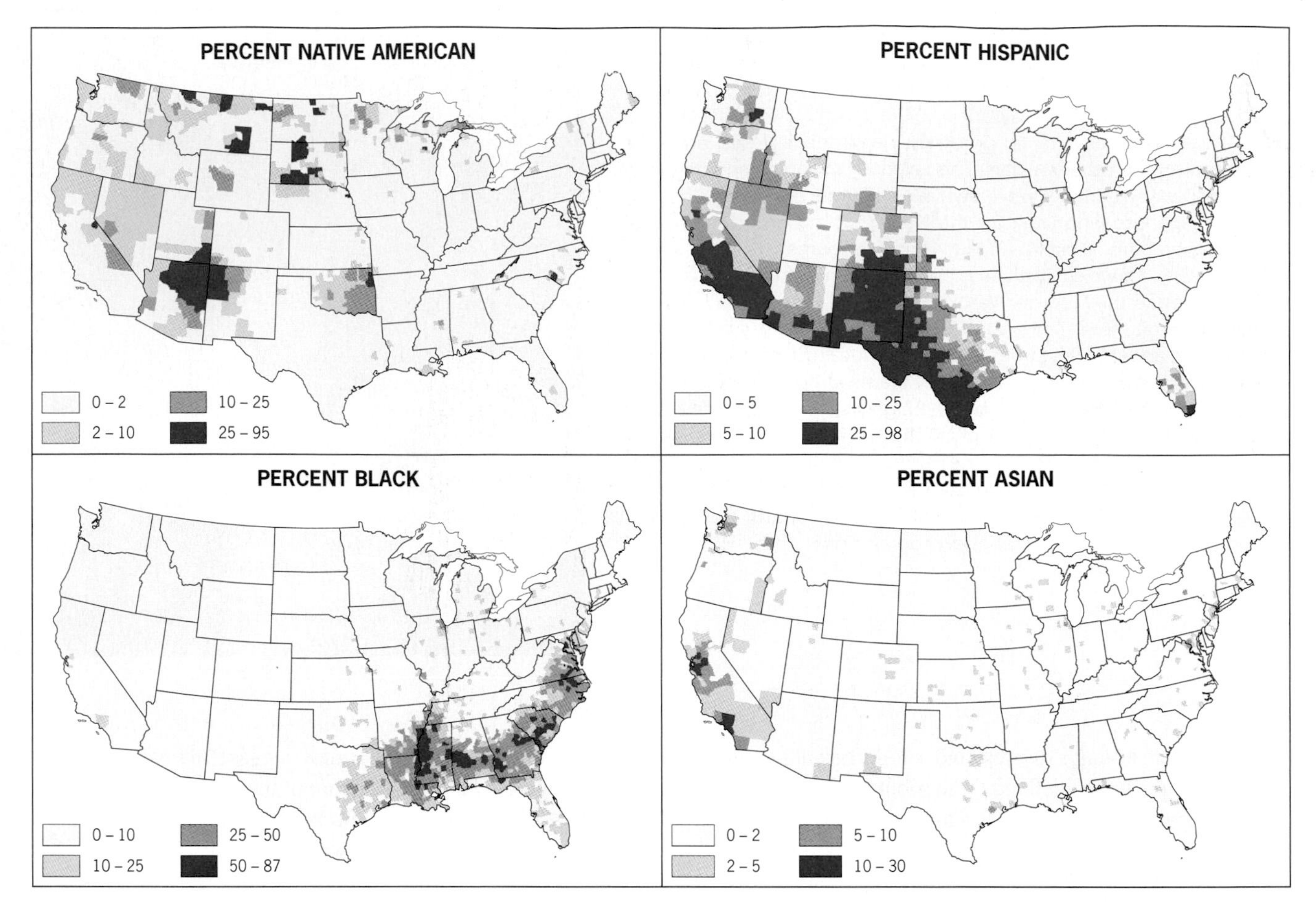

FIGURE 3-13

Language and Religion

Although linguistic variations play a far more important role in Canada, more than one-eighth of the U.S. population speaks a primary language other than English. Differences in English usage are also evident at the subnational level in the United States, where regional variations (*dialects*) can still be noted despite the recent trend toward a truly national society. The South and New England immediately come to mind as areas that still possess distinctive accents.

North America's Christian-dominated kaleidoscope of religious faiths contains important spatial variations. Many major Protestant denominations are clustered in particular regions, with Southern Baptists localized in the southeastern quadrant of the United States, Lutherans in the Upper Midwest and northern Great Plains, and Mormons west of the Rockies but focused on Utah. Roman Catholics are most visibly concentrated in Manufacturing Belt metropolises, New England, and the Mexican borderland zone. Judaism is the nation's most highly agglomerated major religious group, whose largest congregations are clustered in the cities and suburbs of Megalopolis, Southern California, South Florida, and the Midwest.

Ethnic Patterns

Ethnicity (national ancestry) has always played a key role in American cultural geography. Today, whites of European background no longer dominate the increasingly diverse U.S. ethnic tapestry, with ethnics of color and non-European origin comprising nearly 30 percent of the population—a proportion that demographers predict will rise to 50 percent by 2050. In 2000, African Americans constituted the single largest minority group at just over 12 percent of the U.S. total; but Hispanic Americans (also 12+ percent) were growing at a faster rate, and will become the nation's leading minority by 2003.

The spatial distribution of the four largest ethnic minorities is mapped in Figure 3-13. African Americans are regionally concentrated in the South, a legacy of slavery and the plantation economy, but blacks are also a major presence in the central cities of the Manufacturing Belt and West Coast (most urban areas do not register prominently on these small-scale maps). Hispanics are regionally clustered as well, with about 60 percent residing in California, Texas, New Mexico, and South Florida; nonetheless, population geographers expect this rapidly expanding minority to continue to disperse across the country, especially in the West and along the eastern seaboard. Asian Americans are the most agglomerated of the leading ethnic minorities, their distribution dominated by urban-based clusters along the Pacific coast (entry points for the large number who are recent immigrants). The main concentrations of Native Americans are found in the West, where they largely occupy tribal lands on reservations ceded by the federal government, but many live in communities widely scattered across the nation.

The changing ethnic complexion of the United States has long been influenced by immigration, and that still holds true. At the end of the 1990s, nearly 1 million immigrants annually entered the country, perhaps three-quarters of them legally. The source areas, however, have changed dramatically over the past half-century. During the 1950s, just over 50 percent came from Europe, 25 percent from Middle and South America, and 15 percent from Canada; today, approximately 50 percent come from Middle and South America, 30 percent from Asia, and only about 15 percent from Europe and Canada. Although still overwhelmingly directed at urban areas, the geography of immigrant destinations is also changing. The once-dominant Manufacturing Belt States have slipped to less than 25 percent, with most of the newcomers now heading for either metropolitan New York or Chicago. Not surprisingly, given the ongoing growth of Asian and Hispanic Americans (Fig. 3-13), the Sunbelt States have taken up the slack, with California, Texas, and Florida attracting more than 50 percent of all U.S. immigrants.

The Emerging Mosaic Culture

American cultural geography continues to evolve. What is now taking place is a new fragmentation into the emerging 12 nationwide **mosaic culture**, an increasingly heterogeneous complex of separate, uniform "tiles" that cater to more specialized groups than ever before. No longer based solely on such broad divisions as income, race, and ethnicity, today's residential communities of interest are also forming along the dimensions of age, occupational status, and especially lifestyle. Their success and steady proliferation reflect an obvious satisfaction on the part of most Americans. Yet such balkanization—fueled by people choosing to interact only with others exactly like themselves—threatens the survival of important democratic values that have prevailed throughout the evolution of U.S. society.

The Changing Geography of Economic Activity

The economic geography of the United States today is the product of all of the foregoing, as bountiful environmental, human, and technological resources have cumulatively blended together to create one of the world's most advanced economies. Perhaps the greatest triumph was overcoming the tyranny of distance, as people and activities were organized into a continentwide spatial economy that took maximum advantage of agricultural, industrial, and urban development opportunities. Yet, despite these past achievements, American economic geography at the outset of the twenty-first century is once again in the throes of restructuring as the transition is being completed from industrial to postindustrial society.

Major Components of the Spatial Economy

Economic geography is mainly (though not exclusively) concerned with the locational analysis of **productive activities**. Four major sets may be identified: 13

- **Primary activity**: the extractive sector of the economy in which workers and the environment come into direct contact, especially in *mining* and *agriculture*.
- **Secondary activity**: the *manufacturing* sector in which raw materials are transformed into finished industrial products.
- **Tertiary activity**: the *services* sector, including a wide range of activities from retailing to finance to education to routine office-based jobs.
- **Quaternary activity**: today's dominant sector, involving the collection, processing, and manipulation of *information*; a subset, sometimes referred to as **quinary activity**, is the managerial activity associated with decision-making in large organizations.

Historically, each of these activities has successively dominated the American labor force for a time over the past 200 years, with the quaternary sector now dominant. Agriculture dominated until late in the nineteenth century, giving way to manufacturing by 1900. The steady growth of services after 1920 finally surpassed manufacturing in the 1950s but now shares a dwindling portion of the limelight with the still-rising quaternary sector. The approximate breakdown by major sector of employment in the U.S. labor force today is agriculture, 2 percent; manufacturing, 15 percent; services, 18 percent; and quaternary, 65 percent (with about 10 percent in the quinary sector). We now proceed to review these major productive components of the spatial economy in the following coverage of resource use, agriculture, manufacturing, and the postindustrial revolution.

Resource Use

The United States (and Canada) was blessed with abundant deposits of mineral and energy resources. Fortunately, these resources were usually concentrated in sufficient quantities to make long-term extraction an economically feasible proposition, and most of the richest raw material sites are still

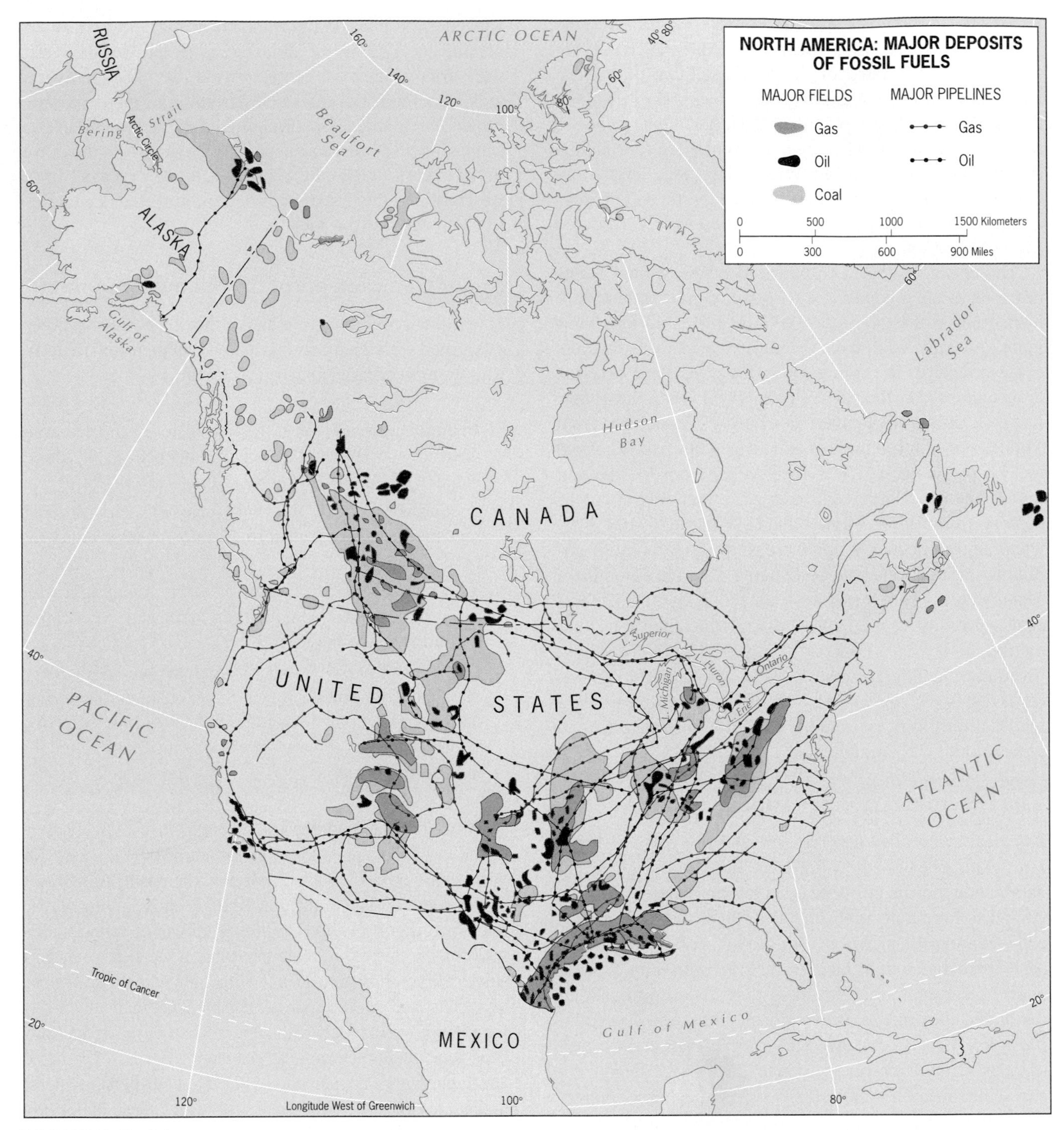

FIGURE 3-14

the scene of major drilling or mining operations. Moreover, the continental and offshore mineral/fuel storehouse may yet contain significant undiscovered resources for future exploitation.

Mineral Resources North America's rich mineral deposits are localized in three zones: the Canadian Shield north of the Great Lakes, the Appalachian Highlands, and scattered areas throughout the mountain ranges of the West. The Shield's most noteworthy minerals are iron ore, nickel, gold, uranium, and copper. Besides vast deposits of soft (bituminous) coal, the Appalachian region also contains hard (anthracite) coal in northeastern Pennsylvania and iron ore in central Alabama. The western mountain zone contains significant deposits of coal, copper, lead, zinc, molybdenum, uranium, silver, and gold.

Fossil Fuel Energy Resources The realm's most strategically important resources are its petroleum (oil), natural gas, and coal supplies—the **fossil fuels**, so named because they were formed by the geologic compression and transformation of plant and tiny animal organisms that lived hundreds of millions of years ago. These energy supplies are mapped in Figure 3-14, which reveals abundant deposits and far-flung distribution networks.

14

The leading *oil*-production areas of the United States are located along and offshore from the Texas-Louisiana Gulf Coast; in the Midcontinent district, extending through western Texas-Oklahoma-eastern Kansas; and along Alaska's central North Slope facing the Arctic Ocean. (Canada's major oilfields lie in a wide crescent curving southeastward from northern Alberta to southern Manitoba and beneath the waters around Newfoundland in the extreme east.) The distribution of *natural gas* deposits generally resembles the geography of oilfields because petroleum and energy gas are usually found in similar geologic formations (the floors of ancient shallow seas). The realm's *coal* reserves rank among the greatest on Earth, and the U.S. portion alone contains at least a 400-year supply; the main producing coalfields are found in Appalachia, the northern U.S. Great Plains/southern Alberta, and southern Illinois/western Kentucky.

Agriculture

Despite the post–1900 emphasis on developing the nonprimary sectors of the spatial economy, agriculture remains an important element in America's human geography. Because it is the most space-consuming economic activity, vast expanses of the U.S. (and Canadian) landscape are clothed with fields of grain. In addition, great herds of livestock are sustained by pastures and fodder crops because this wealthy realm can afford the luxury of feeding animals from its farmlands to meet huge demands for red meat in its diet. The growing application of high-technology mechanization to farming has steadily increased both the volume and value of total agricultural production. It also has been accompanied by a sharp reduction in the number of those actively engaged in agriculture, and today only slightly more than 1.5 percent of the U.S. population still lives on farms.

The regionalization of U.S. agricultural production is shown in Figure 3-15, its spatial organization developed largely within the framework of the *von Thünen model* (see pp. 49–51). As in Europe (Fig. 1-6), the early-nineteenth-century, original-scale model of town and hinterland expanded outward (driven by constantly improving transportation technology) from a locally "isolated state" to

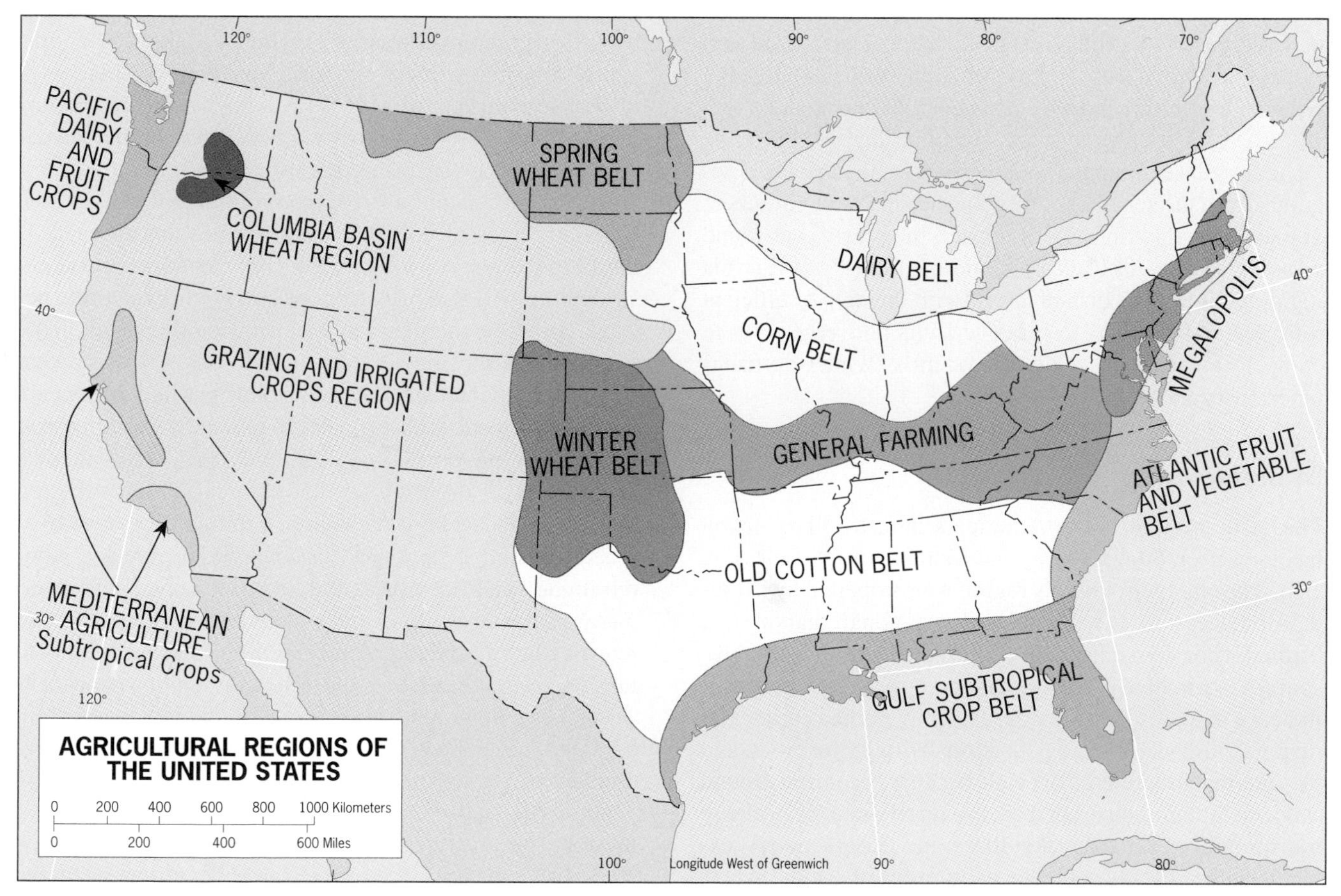

FIGURE 3-15

encompass the entire continent by 1900. As this macrogeographic structure formed, the greatly enlarged original Thünian production zones (Fig. 1-5) were modified: (1) the first ring was now differentiated into an inner fruit/vegetable zone and a surrounding belt of dairying; (2) the forestry ring was displaced to the outermost limits of the regional system because railroads could now transport wood quite cheaply; (3) the field crops ring subdivided into an inner mixed crop-and-livestock ring to produce meat (the Corn Belt, as it came to be called) and an outer zone that specialized in the mass production of wheat grains (the Wheat Belts); and (4) the ranching area remained in its outermost position, a grazing zone that supplied young animals to be fattened in the meat-producing Corn Belt as well as supporting an indigenous sheep-raising industry. The "supercity" anchoring this macro-Thünian regional system was the northeastern Megalopolis, well on its way toward coalescence and already the dominant food market and transport focus of the entire country.

Although the circular rings of the model are not apparent in Figure 3-15, many spatial regularities can be observed (remember that von Thünen, too, applied his model to the real world and thereby distorted the theoretically ideal pattern). Most significant is the sequence of farming regions as distance from the national market increased, especially westward from Megalopolis toward central California, which was the main directional thrust of the historic inland penetration of the United States. The Atlantic Fruit and Vegetable Belt, Dairy Belt, Corn Belt, Wheat Belts, and Grazing Region are indeed consistent with the model's logical structure, each zone successively farther inland astride the main transcontinental routeway. We can attribute deviations from the scheme to irregularities in the environment or to unique conditions. For example, the nearly year-round growing seasons of California and the Gulf Coast-Florida region permit those distant areas (with the help of efficient refrigerated transport) to produce fruits and vegetables in competition with the regional system's (winter-dormant) innermost zone.

Manufacturing

The geography of North America's industrial production has long been dominated by the Manufacturing Belt (Fig. 3-8). The emergence of this region was propelled by (1) superior access to the Megalopolis national market that formed its eastern edge, and (2) proximity to industrial resources, particularly iron ore and coal for the pivotal steel industry that arose in its western half. We noted earlier that manufacturers had a strong locational affinity for cities, and the internal structure of the Belt became organized around a dozen urban-industrial districts interlinked by a dense transportation network. As these industrial centers expanded, they swiftly achieved **economies of scale**, savings accruing from large-scale production in which the cost of manufacturing a single item was further reduced as individual companies mechanized assembly lines, specialized their workforces, and purchased raw materials in massive quantities.

15

This efficient production pattern served the nation well throughout the remainder of the industrial age. Because of the effects of *historical inertia*—the need to continue using hugely expensive manufacturing facilities for their full, multiple-decade lifetimes to cover initial investments—the Manufacturing Belt is not going to disappear anytime soon. But as its aging facilities are retired, a process well under way, the distribution of American industry will change. As transportation costs equalize among U.S. regions, as energy costs now favor the south-central oil- and gas-producing States, as high-technology manufacturing advances reduce the need for lesser skilled labor, and as locational decision-making intensifies its attachment to noneconomic factors, industrial management has increasingly demonstrated its willingness to relocate to regions it perceives as more desirable in the South and West.

Parts of the Manufacturing Belt have been resisting this trend, particularly the industrial Midwest, whose prospects have greatly improved since the "Rustbelt" days of the 1970s and 1980s. Shedding that label in recent years, Midwestern manufacturers are successfully reinventing their operations by rooting out inefficiencies, investing in cutting-edge factories and technologies, and not only beating back foreign competition in the United States but significantly expanding their exports. Nonetheless, there is a downside to this progress because as high-tech manufacturing proliferates, older factory workers discover that the demand for their lower-tech skills shrinks accordingly.

As the U.S. manufacturing map continues to change, the economic lives of countless communities are affected. If the 1990s were a sign of things of come, entrepreneurial restructuring will also play an increasingly important role because corporate mergers and acquisitions often result in the geographic reshuffling of thousands of jobs. Another recent trend likely to intensify is the economic warfare that occurs when employers seek to locate major new facilities and deliberately pit cities, counties, and even States against one another in order to get the best deal. Often exhorted by politicians seeking reelection, communities offer ever greater incentive packages that include tax breaks, labor retraining, and infrastructural improvements. However, many such competitions can fail to produce a decent rate of return: when Alabama recently went to extreme lengths to lure Mercedes-Benz to build a new assembly plant near Tuscaloosa, it wound up costing the State government $200,000 per expected job. (The cost to Tennessee's government to attract the Saturn automaking complex a few years earlier had been one-tenth that figure.) Moreover, these business strategies can be worked in reverse; New York City not long ago offered a $60 million incentive package to keep two Wall Street brokerage firms from moving to the suburbs, and Disneyland got its city and State to spend hundreds of millions on local road improvements after the company threatened to cancel expansion plans and perhaps even move the theme park out of Southern California.

The Postindustrial Revolution

16 The signs of **postindustrialism** are visible throughout the United States (and much of Canada) today, and they are popularly grouped under such labels as "the computer age" or "the new economy." The term *postindustrial* by itself, of course, tells us mainly what the theme of the American economy no longer is. Yet many social scientists also use the term to refer to a set of societal traits that signal an historic break with the past (for example, work experiences now focus mainly on person–to–person interaction rather than person-product or person-environment contact). Many of the urban spatial expressions of the ongoing societal transformation have already been highlighted; here we focus on some broader economic-geographical patterns.

High-technology, white-collar, office-based activities are the leading growth industries of the postindustrial economy. Most are relatively footloose and are therefore responsive to such noneconomic locational forces as geographic prestige, local amenities, and proximity to recreational opportunities. Northern California's *Silicon Valley*—the world's leading center for computer research and development and the headquarters of the U.S. microprocessor industry—epitomizes the blend of locational qualities that attract a critical mass of high-tech companies to a given locality. These include: (1) a world-class research university (Stanford); (2) technological know-how in the form of a large pool of highly educated, highly skilled labor; (3) close proximity to a cosmopolitan urban center (San Francisco); (4) abundant venture capital; (5) a local economic climate and entrepreneurial culture that supports risk-taking and forgives failure by both smaller and big companies; (6) a locally based network of global business linkages; and (7) a high-amenity environment in the form of top-quality housing, pleasant weather, scenic countrysides, and year-round recreational opportunities.

The development of Silicon Valley is so significant to the new postindustrial era that regional-planning scholars Manuel Castells and Peter Hall have conceptualized it as

17 the first **technopole**. Technopoles are planned techno-industrial complexes that innovate, promote, and manufacture the hardware and software products of the new informational economy. On the landscape of the outer suburban city, where almost all of these complexes are located, the signature of a technopole is a low-density cluster of ultramodern, low-rise buildings laid out as a campus—an image as symbolic of today's spatial economy as the smoke-belching factory was of the industrial age a century ago. From Silicon Valley, technopoles have spread in all directions—from San Diego in southwesternmost California to the Route 128 corridor around Boston and from North Carolina's Research Triangle to the lakeside suburbs of Seattle—and many of these technopoles will be noted in the regional section that concludes this chapter. These high-tech complexes are also becoming a global phenomenon as they spring up in other geographic realms. They often arise as joint ventures between government and the private sector; among the local names they go by are science city (Japan; Russia), technopolis (France; Japan), science park (Taiwan)—and even Silicon Glen (Scotland).

CANADA

Like the United States, Canada is a federal state, but it is organized differently. Canada is divided administratively into ten provinces and three territories (Fig. 3-16). The provinces—where 99.7 percent of all Canadians live—range in territorial size from tiny, Delaware-sized Prince Edward Island to sprawling Quebec, more than twice the area of Texas. Beginning in the east, the four Atlantic Provinces are Nova Scotia, New Brunswick, Prince Edward Island, and Newfoundland. To their west lie Quebec and Ontario, Canada's two biggest provinces. Most of western Canada (which is what the Canadians call everything west of Lake Superior) is covered by the three Prairie Provinces—Manitoba, Saskatchewan, and Alberta. In the far west, beyond the Canadian Rockies and facing the Pacific, lies the tenth province, British Columbia.

The three territories—Yukon, the Northwest Territories, and Nunavut—together occupy a massive area half the size of Australia but are inhabited by only about 100,000 people. Nunavut is the newest addition to Canada's political map and deserves special mention. Created in 1999, this new territory is the outcome of a major aboriginal land claim agreement between the Inuit people (formerly called Eskimos) and the federal government, and encompasses all of Canada's eastern Arctic. Nunavut covers fully one-fifth of Canada's land, an area larger than that of any other province or territory. Since 80 percent of its 28,000 residents are Inuit, Nunavut—which means "our land"—is the first territory to have a substantial degree of native self-government (in addition to outright ownership of an area half the size of Texas).

In population size (in descending order), Ontario (11.8 million) and Quebec (7.3 million) are again the leaders; British Columbia ranks third with 4.2 million; next come the three Prairie Provinces with a combined total of 5.2 million; the Atlantic Provinces are the four smallest, together containing 2.4 million. The sparsely populated territories in the far north contain barely 100,000 residents. Canada's total population of 31.0 million is only slightly larger than one-tenth the size of the U.S. population. Spatially, as noted

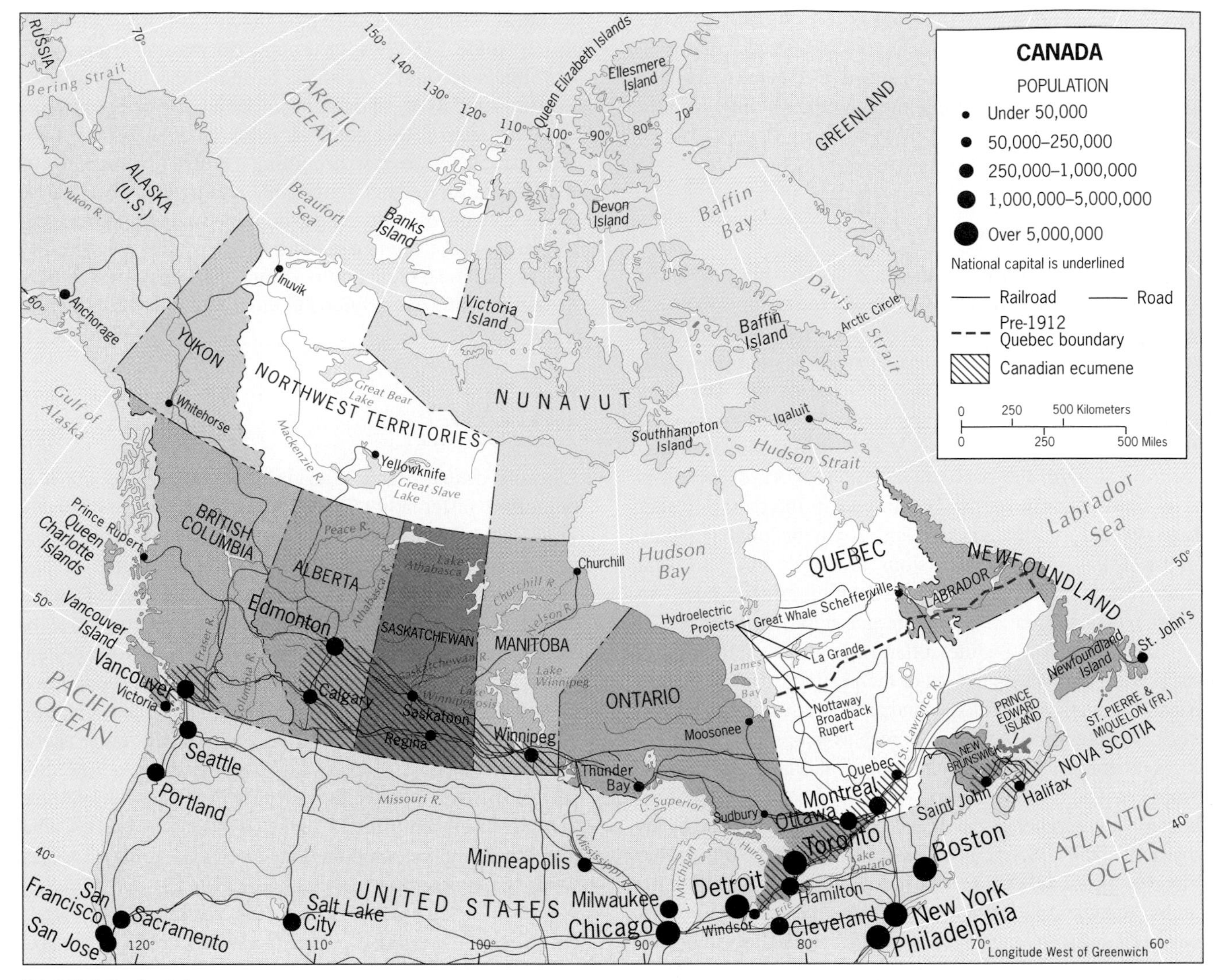

FIGURE 3-16

at the outset of the chapter, Canadians overwhelmingly reside within 200 miles (320 km) of the U.S. border.

Population in Time and Space

The map showing the distribution of Canada's population (Fig. 3-4, p. 156) reveals that only about one-eighth of this enormous country can be classified as its **ecumene**—the inhabitable zone of permanent settlement. As Figure 3-16 indicates, the Canadian ecumene is dominated by a discontinuous strip of population clusters that lines the southern border. We can identify four such clusters on the map, the largest by far being Main Street (home to more than six out of every ten Canadians). As noted earlier (Fig. 3-9), *Main Street* is the conurbation that stretches across southernmost Quebec and Ontario, from Quebec City on the lower St. Lawrence River southwest through Montreal and Toronto to Windsor on the Detroit River. The three lesser clusters are: (1) the Saint John-Halifax crescent in central New Brunswick and Nova Scotia; (2) the prairies of southern Alberta, Saskatchewan, and Manitoba; and (3) the southwestern corner of British Columbia, focused on Canada's third-largest metropolis, Vancouver.

Pre-Twentieth-Century Canada

Canada's population map evolved more slowly than that of the United States. As Figure 3-7 shows, compared to European expansion in the United States, penetration of the Canadian interior lagged several decades behind. In 1850, for example, when the eastern United States and much of the Pacific coast had already been settled, Canada's frontier had only reached Lake Huron. In terms of political geography, Canada did not unify before the last third of the nineteenth century—and then only reluctantly because of fears the United States was about to expand in a northerly direction. Moreover, the Canadian push to the west was delayed by major physical obstacles, including the barren Canadian Shield north and west of the Great Lakes, and the rugged highlands separating the Great Plains from the Pacific (which lacked convenient mountain passes).

The evolution of modern Canada—as well as its contemporary cultural geography—is also deeply rooted in the bicultural division discussed at the outset of this chapter. Its origin lies in the fact that it was the French, not the British, who were the first European colonizers of present-day Canada, beginning in the 1530s. *New France*, during the seventeenth century, grew to encompass the St. Lawrence Basin, the Great Lakes region, and the Mississippi Valley. In the 1680s a series of wars between the English and French began, ending with France's defeat and the cession of New France to Britain in 1763. By the time London took control of its new possession, the French had made considerable progress in their North American domain. French laws, the French land-tenure system, and the Roman Catholic Church prevailed, and substantial settlements (including Montreal on the St. Lawrence) had been established. The British, anxious to avoid a war of suppression and preoccupied with problems in their other American colonies, gave former French Quebec—the territory extending from the Great Lakes to the mouth of the St. Lawrence—the right to retain its legal and land-tenure systems as well as freedom of religion.

After the American War for Independence, London was left with a region it called British North America (the name *Canada*—derived from an aboriginal word meaning settlement—was not yet in use), but whose cultural imprint still was decidedly French. The Revolutionary War drove many thousands of English refugees northward, and soon difficulties arose between them and the French in British North America. In 1791, heeding appeals by these new settlers, the British Parliament divided Quebec into two provinces: Upper Canada, the region upstream from Montreal centered on the north shore of Lake Ontario, and Lower Canada, the valley of the St. Lawrence. Upper and Lower Canada became, respectively, the provinces of Ontario and Quebec (Fig. 3-16). By parliamentary plan, Ontario would become English-speaking and Quebec would remain French-speaking.

This earliest cultural division did not work well, and in 1840 the British Parliament tried again. This time it reunited the two provinces through the Act of Union under which Upper and Lower Canada would have equal representation in the provincial legislature. That, too, failed, and efforts to find a better system finally led to the 1867 British North America Act that established the Canadian federation (consisting initially of Upper and Lower Canada, New Brunswick, and Nova Scotia, later, between 1870 and 1999, to be joined by the other provinces and territories). Under the 1867 Act, Ontario and Quebec were again separated, but this time Quebec was given important guarantees: the French civil code was left unchanged, and the French language was protected in Parliament and in the courts.

Canada Since 1900

By the close of the nineteenth century, the fledgling Canadian federation was making major strides toward regional development and the spatial integration of a continentwide economy. In 1886, the transcontinental Canadian Pacific Railway had finally been completed to Vancouver, a condi-

At first glance, you might mistake this city for Denver, Colorado—a high-rise CBD in the shadow of the Rocky Mountains. But this is Alberta's Calgary, and unlike Denver it lies at the confluence of two rivers, the Bow and the Elbow. Calgary started as a fort on the western edge of Canada's Great Plains; the settlement got a boost when the Grand Pacific Railway reached it in 1883. Later a second railroad, followed by the Trans-Canada Highway, further improved its linkages. But Calgary's biggest shot in the arm was the discovery of oil- and gasfields in its hinterland, which put petroleum refining and petrochemical industries ahead of flour milling and meat packing as industrial mainstays. Today Calgary also is a service center for a large region, with an important high-tech sector evincing its modernization.

tion of British Columbia's entry into the federation 15 years earlier. This new lifeline soon spawned the settlement of not only the far west but also the fertile Prairie Provinces, whose wheat-raising economy expanded steadily as immigrants from the east and abroad poured in during the early twentieth century. Industrialization also began to stir, and by 1920 Canadian manufacturing had surpassed agriculture as the leading source of national income. As noted earlier (Fig. 3-8), the dominant zone of industrial activity is the Toronto-Hamilton-Windsor corridor of southern Ontario (which also functions as one of the 12 districts of the American Manufacturing Belt). That has been the case since the Industrial Revolution took hold in Canada during World War I (1914–1918), when the country heavily supported Britain's war effort on the European mainland.

As industrial intensification proceeded, it was accompanied by the expected parallel maturation of the national urban system. Along the lines of Borchert's five epochs of American metropolitan evolution, Maurice Yeates has constructed a similar multi-stage model of urban-system development that divides the telescoped Canadian experience into three eras. The initial *Frontier-Staples Era* (prior to 1935) encompasses the century-long transition from a frontier-mercantile economy to one oriented to staples (production of raw materials and agricultural goods for export), with increasing manufacturing activity in the budding industrial heartland. By 1930, Montreal and Toronto, reflecting their different cultural constituencies, had emerged as the two leading cities atop the national urban hierarchy (thus Canada has no single primate city).

Next came the *Era of Industrial Capitalism* (1935–1975), during which Canada achieved U.S.-style prosperity. However, most of this development—involving the massive growth of manufacturing, tertiary activities, and urbanization—took place after 1950 because the Great Depression lasted until World War II, which was followed by a lengthy recession. A major stimulus was the investment of U.S. corporations in Canadian branch-plant construction, especially in the automobile industry in Ontario near the automakers' Detroit-area headquarters. In western Canada, the rapid growth of oil and natural gas production fueled Alberta's urban development, but new agricultural technologies reduced farm labor needs and sparked a rural outmigration in neighboring Saskatchewan and Manitoba. The postwar period also saw the ascent of Main Street, which on less than 2 percent of Canada's land quickly came to contain more than 60 percent of its people, contributed two-thirds of its national income, and had nearly 75 percent of its manufacturing jobs.

The third stage, ongoing since 1975, is the *Era of Global Capitalism*, signifying the rise of additional foreign investment from the Asian Pacific Rim and Europe. This, of course, is also the era of transformation into a postindustrial economy and society, and in the process Canada is experiencing many of the same upheavals as the United States. Interestingly, while U.S. urban growth has recently leveled off at 75 percent of the total population, Canada still continues to urbanize (reaching 78 percent in 2002). Most of this development is occurring in the form of new suburbanization. This represents a departure from the recent past because the pre-1990 Canadian metropolis had experienced far less automobile-generated deconcentration than the United States, with the central city retaining much of its middle-income population and economic base. But today Canada's large cities are turning inside-out, and the new intraurban geography is increasingly symbolized by the suburban downtowns that anchor the ultramodern business complexes along the Highway 401 freeway north of Toronto and Alberta's West Edmonton megamall, the world's biggest shopping center.

Cultural/Political Geography

The historic cleavage between Canada's French- and English-speakers, supposedly resolved by the British North America Act, has resurfaced in the past three decades to dominate the country's cultural and political geography. By the time the Canadian federation observed its centennial in 1967, it had become evident that Quebecers regarded themselves as second-class citizens; they believed that bilingualism meant that French-speakers had to learn English but not vice versa; and they perceived that Quebec was not getting its fair share of the country's wealth. Since the 1960s, the intensity of ethnic feelings in Quebec has risen in surges despite the federal government's efforts to satisfy the province's demands. During the 1970s, while a separatist political party came to power in Quebec, a new federal constitution was drawn up in Ottawa. In 1980, Quebec's voters solidly rejected sovereignty when given that choice in a referendum. But the new constitution did *not* satisfy the Quebecers, and throughout the 1980s and early 1990s the Ottawa government struggled unsuccessfully to devise a plan, acceptable to all the provinces, that would keep Quebec in the Canadian federation.

By 1995, with Canada's interest in constitutional reform exhausted, a second referendum on Quebec's sovereignty could no longer be put off. With the reenergized separatist party again leading the way, the Francophone-dominated electorate almost approved independence (which attracted an astonishing 49.4 percent of the vote), an outcome now regarded by Canadians as their country's "near-death experience." Despite calls by many separatists for a follow-up vote that might turn narrow defeat into victory, the years since 1995 have been marked by delay in efforts to organize a third referendum. The main reason is that public opinion polls in Quebec show that support for secession has faded to below 45 percent, and in 2001 the strongest separatist leader (Lucien Bouchard) suddenly quit politics.

Seizing the initiative after 1995, the federal government responded by asking Canada's Supreme Court to review the constitutionality of Quebec's separation, and in 1998 the court ruled that the province had no right to unilateral secession under international law. However, the ruling also declared that Canada is not indivisible; therefore, should a clear majority of Quebec's electorate vote to secede, the Ottawa government and the other provinces would be obliged to negotiate the terms of separation as if the matter were a new amendment to Canada's constitution. Armed by this legal interpretation, the federal government moved to pass legislation that formalizes the court's rulings—and makes the process of secession for a breakaway province far more cumbersome. First, the lower house of the Canadian Parliament will determine the clarity of any secessionist question approved in a provincial referendum (both the 1980 and 1995 votes in Quebec were based on vaguely worded questions that would have received less support had their language been more specific and hard line). Second, that body is also empowered to decide what constitutes a "clear majority" of voters, which is likely to be well beyond a total of 50-percent-plus-one. And third, the legislation lays the groundwork for the court-mandated negotiations over the terms of separation, in which a breakaway province would have to reach settlement with Ottawa and the nine other provinces on such issues as borders, guaranteeing the rights of linguistic and aboriginal minorities, and paying off its share of the Canadian national debt. Should the Quebec situation ever advance to this stage, it is easy to envision that certain provinces (particularly British Columbia and Alberta) would continue their longstanding hard line and insist that Quebec pay a very high price for its independence. Given this likelihood, and all that has gone into the search for a resolution to the Quebec crisis since 1980, negotiations might well end in an impasse—with unknown consequences for the future of northern North America.

Among the issues that an exiting Quebec would have to negotiate, most significant from a geographic standpoint is that the French linguistic region does *not* coincide with the province's territorial boundaries. As can be seen in the map on page 179, "French Canada" is mainly confined to the area centered on the middle and lower St. Lawrence Valley, with some spillover into northern New Brunswick. Moreover, as the distribution of the "no" vote on separation in the 1995 referendum strongly suggests (Fig. 3-17; areas colored pink), there are numerous non-French communities located within the Francophone region—especially along the Ontario and U.S. borders as well as in metropolitan Montreal. Dozens of English-speaking municipalities in this southern periphery of Quebec have spearheaded a partitionist movement by declaring their intention to stay in Canada, arguing that they have the same right to secede from Quebec that Quebec has to secede from Canada. Even though this claim is vehemently opposed by the separatist provincial government (which realizes that a dismembered independent Quebec would be less attractive to voters in any future referendum), the polls show that perhaps as many as half the French speakers agree with it.

On the wider Canadian scene, the culture-based Quebec crisis has stirred the ethnic feelings of the country's 1.2 million native (First Nations and Inuit) peoples. Their assertions have also received a sympathetic hearing in Ottawa, and by the end of the 1990s breakthroughs were being achieved with the creation of Nunavut and a treaty for limited tribal self-government in northern British Columbia. A leading concern among these peoples is that their aboriginal rights be protected by the federal government against the provinces. This is especially true for the First Nations of Quebec's northern frontier, the Cree, whose historic domain covers more than half of the province of Quebec as it appears on current maps. Administration of the Cree was assigned to Quebec's government in 1912, a responsibility that a move toward independence may well invalidate. In any case, the Cree would also likely be empowered to seek independence. This would leave the French-speaking remnant of Quebec with only about 45 percent of the province's present territory (which would be further diminished if the Anglophone partitionist movement succeeded). As Figure 3-16 shows, the territory of the Cree is no unproductive wilderness: it contains vital facilities of the James Bay Hydroelectric Project, a massive scheme of dikes, dams, and artificial lakes that has transformed much of northwestern Quebec and generates electrical power for a huge market within and outside the province.

Finally, the events of the past quarter-century have increasingly impacted Canada's political landscape, and not only in Quebec. Regionalism has also intensified in the west, whose leaders (some even to the point of threatening the region's own secession) oppose federal concessions to Quebec and insist that the equal treatment of all ten provinces is a basic principle that precludes the designation of special status for anyone. The most recent federal elections have clearly revealed the emerging fault lines that surround Quebec and set the western provinces (British Columbia, Alberta, and Saskatchewan) off from the rest of the country. Even the remaining Ontario-led center and the eastern bloc of Atlantic Provinces voted divergently, and today the politico-geographical hypothesis of "Four Canadas" may well be on its way toward becoming reality. Thus, with its national unity under pressure, Canada today confronts the coalescing forces of *devolution* that threaten to transform the new fault lines into permanent fractures.

Economic Geography

As in the United States, the growth of Canada's spatial economy has been supported by a diversified, high-quality *resource base*. We noted earlier that the Canadian Shield is

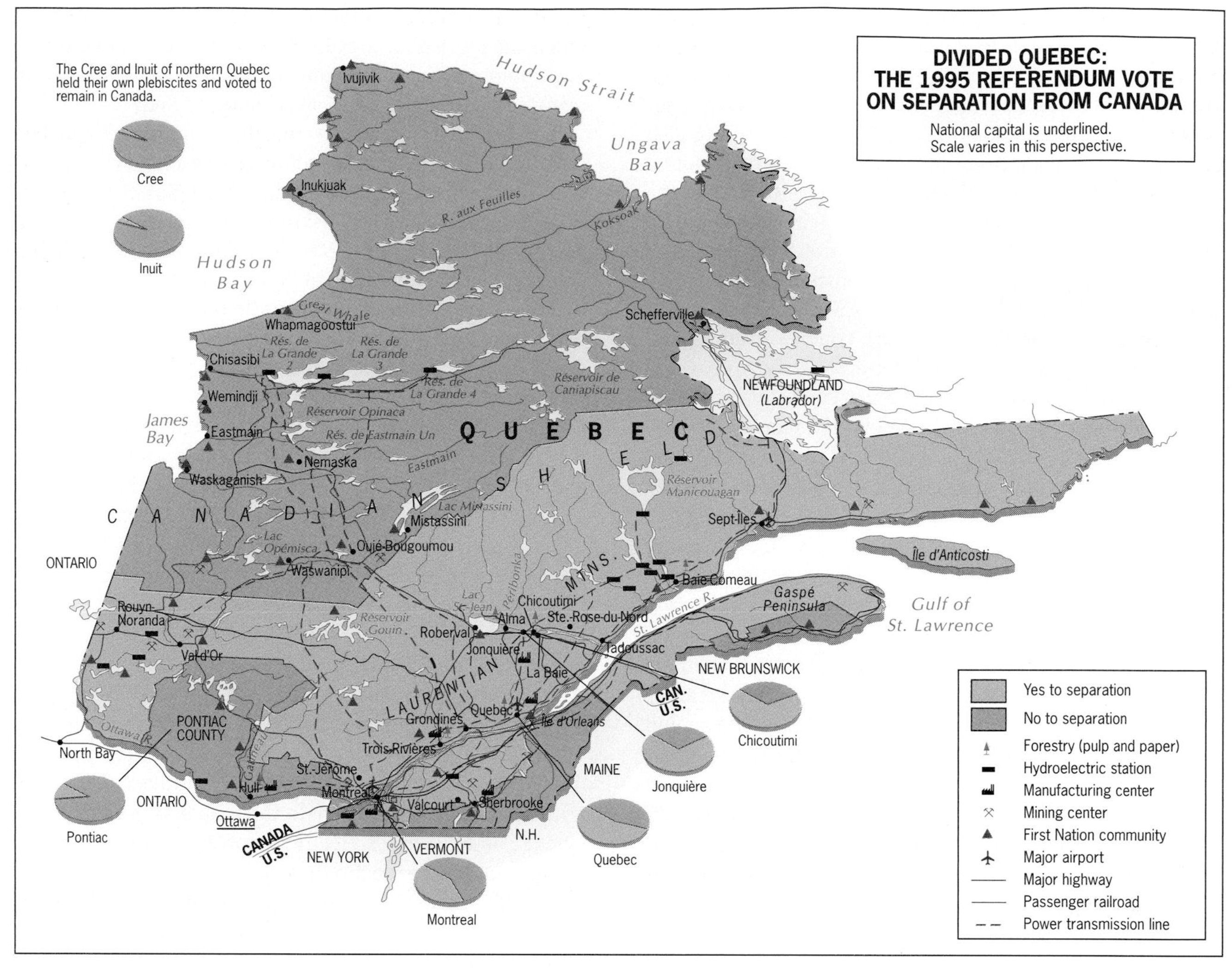

FIGURE 3-17

endowed with rich mineral deposits and that oil and natural gas are extracted in sizeable quantities in Alberta to the southwest of the Shield. Canada has long been a leading *agricultural* producer and exporter, especially of wheat and other grains from its breadbasket in the Prairie Provinces; new technologies keep productivity high, but labor requirements continue to diminish and farm workers now account for only about 2 percent of the national workforce. Postindustrialization has caused substantial employment decline in the *manufacturing sector*, with Southern Ontario's industrial heartland being the most adversely affected area. On the other hand, Canada's robust *tertiary and quaternary sectors* (which today employ more than 70 percent of the total workforce) are creating a host of new economic opportunities. Fortunately, Southern Ontario is benefiting from this postindustrial transformation because the country's leading high-technology, research-and-development complex has grown up around Waterloo and Guelph, a pair of university towns about 60 miles (100 km) west of Toronto.

Canada's economic future is also going to be strongly affected by the continuing development of its trading relationships. The landmark United States–Canada Free Trade Agreement, signed in 1989, initiated a now-completed phasing out of all tariffs and investment restrictions between the two countries, whose annual cross-border flow of goods and services is the largest in the international trade arena. (At the end of the 1990s more than four-fifths of Canada's exports went to the United States, from which it also derived about two-thirds of its imports.) In 1994 these economic linkages were further tightened by the implementation of the North American Free Trade Agreement (NAFTA), which consolidated the gains of the 1989 pact and opened major new opportunities for both countries by adding Mexico to the trading partnership (see box titled "North American Free Trade Agreement [NAFTA]").

Because these free-trade agreements increasingly impact the Canadian spatial economy, they are likely to weaken domestic east-west linkages and strengthen international north-south ties. Since many local cross-border linkages built on geographical and historical commonalities are already well developed, they can be expected to intensify in the future: the Atlantic Provinces with neighboring New England; Quebec (even if independent) with New York State; Ontario with Michigan and adjacent Midwestern States; the Prairie Provinces with the Upper Midwest; and, above all, British Columbia with the (U.S.) Pacific Northwest. Such functional reorientations, of course, constitute yet another set of powerful devolutionary forces confronting the Ottawa government because most of these potential economic fault lines coincide with those that politically demarcate the "Four Canadas."

The rising importance of this framework of transnational regions straddling the U.S.-Canadian border recalls Kenichi Ohmae's **regional state** concept introduced in Chapter 1. A regional state is a "natural economic zone" that defies old borders, and it is shaped by the global economy of which it is a part; its leaders deal directly with foreign partners and negotiate the best terms they can with the national governments under which they operate. Writing about Canada recently, Ohmae identified both a Pacific Northwest (the Seattle-Vancouver axis) and a Great Lakes regional state (the remarkably intertwined Ontario-Michigan industrial complex), and warned that the manner in which Ottawa's leaders dealt with these new economic entities was critical to the survival of the Canadian state. Today his warning is more meaningful than ever because the Pacific Northwest and Great Lakes have been joined by most of the other areas listed above. Indeed, these growing international interactions are increasingly evident in the overall regional configuration of North America, to which we now turn in the concluding part of this chapter.

19

Among the Realm's Great Cities . . .

Toronto

Toronto, the capital of Ontario and Canada's largest metropolis (4.9 million), is the historic heart of English-speaking Canada. The landscape of much of its center is dominated by exquisite Victorian-era architecture and surrounds a healthy downtown that is one of Canada's leading economic centers. Landmarks abound in this CBD, including the Skydome stadium, the famous City Hall with its facing pair of curved high-rises, and mast-like CN Tower, which at 1,815 feet (553 m) is the world's tallest freestanding structure. Toronto also is a major port and industrial complex, with miles of facilities lining the shoreline of Lake Ontario.

Livability is one of the first labels Torontonians apply to their city, which has retained more of its middle class than central cities of its size in the United States. *Diversity* is another leading characteristic because this is North America's richest urban ethnic mosaic. Among the largest of more than a dozen thriving ethnic communities are those dominated by Italians, Portuguese, Chinese, Greeks, and Ukrainians; overall, Toronto now includes residents from 169 countries who speak more than 100 languages; and the immigrant inflow continues strongly, with those born outside Canada constituting more than 40 percent of the city's current pop-ulation.

Toronto has worked well in recent decades, thanks to a metropolitan government structure that fostered central city–suburban cooperation. But that relationship is becoming stressed as the outer city gains a critical mass of population, economic activity, and political clout while the central city was recently forced to accommodate an unwanted amalgamation with five of its nearest suburban neighbors.

North American Free Trade Agreement (NAFTA)

The economies of Canada, the United States, and Mexico formalized their interactions under the *North American Free Trade Agreement (NAFTA)*, which took effect in 1994. This economic alliance has forged the world's largest trading bloc, a (U.S.) $8-trillion-plus market of 413 million consumers (surpassing the European Union's 377 million). Between now and its completion date of 2009, NAFTA will integrate its constituent economies through a number of steps. Thousands of individual tariffs, quotas, and import licenses have been or are scheduled to be rolled back to eliminate most trade barriers for agricultural and manufactured goods as well as tertiary and quaternary services. Moreover, restrictions on the flow of investment capital across international borders are being lifted, and regulations concerning foreign ownership of productive facilities continue to be liberalized.

The architects of NAFTA cite the opportunities the maturing pact will bring. Business investment will surge as unimpeded exports flow throughout the vast new marketplace. At the same time, intensified competition will give consumers access to a wider variety of quality products at reduced prices. Canada—which joined NAFTA mainly to safeguard the concessions it received in its 1989 Free Trade Agreement with the United States—faced the most skeptical citizenry at the outset, but those doubts evaporated as Canadian exports (and new jobs) under NAFTA have risen substantially since 1993. Trade with the United States, already heightened under the bilateral 1989 agreement, has propelled much of this growth, especially the southward export of goods whose tariffs had been lowered.

This progress notwithstanding, NAFTA's evolution continues to be criticized. Organized labor has objected to the redistribution and losses of jobs in a number of industries, dislocations which it perceives stem largely from the implementation of the trade agreement. Environmental activists complain that the pact's architects and component governments have not adequately addressed their concerns. And a number of social scientists and public policy planners argue that NAFTA should be doing much more to reduce regional income disparities and uneven development within its trade area.

As NAFTA realizes its goals, it will require each member country to concentrate on the production of those goods and services for which it has a comparative advantage (in research and development, technological skills, managerial know-how, and the like). As workers move from less productive to more productive industries, the geography of employment will be restructured—a process that would be occurring anyway as the three countries continue to adapt to continental and global economic change.

NAFTA's success has attracted the intense interest of all the other countries in the Western Hemisphere. Although most of them are involved in their own regional efforts aimed at greater economic integration (as we shall see in Chapters 4 and 5), the great majority desire to become part of an enlarged NAFTA—which triggered its own expansion by inviting Chile to join in 1995 (an initiative so far thwarted by the U.S. Congress). To that end, ambitious plans were launched at summit meetings in the mid-1990s to create the *Free Trade Area of the Americas (FTAA)*, a hemisphere-encompassing free-trade zone that would establish a single market of nearly 850 million consumers. If FTAA is implemented by its target date of 2005, it should accelerate the completion of NAFTA—as well as NAFTA's transformation into a much larger supranational organization.

REGIONS OF THE REALM

The ongoing transformation of North America's human geography is fully reflected in its internal regional organization. As new forces uproot and redistribute people and activities, old locational rules no longer apply. However, the varied character of the realm's physical, cultural, and economic landscapes ensures that meaningful regional differences will persist. We will now examine the current areal arrangement of the United States and Canada within a framework of nine regions (Fig. 3-18).

THE NORTH AMERICAN CORE

The Core Region (Fig. 3-18)—synonymous with the American Manufacturing Belt—was introduced earlier in this chapter. Serving as the historic workshop for the linked spatial economies of the United States and Canada, this region was the unquestioned leader and centerpiece during the cen-

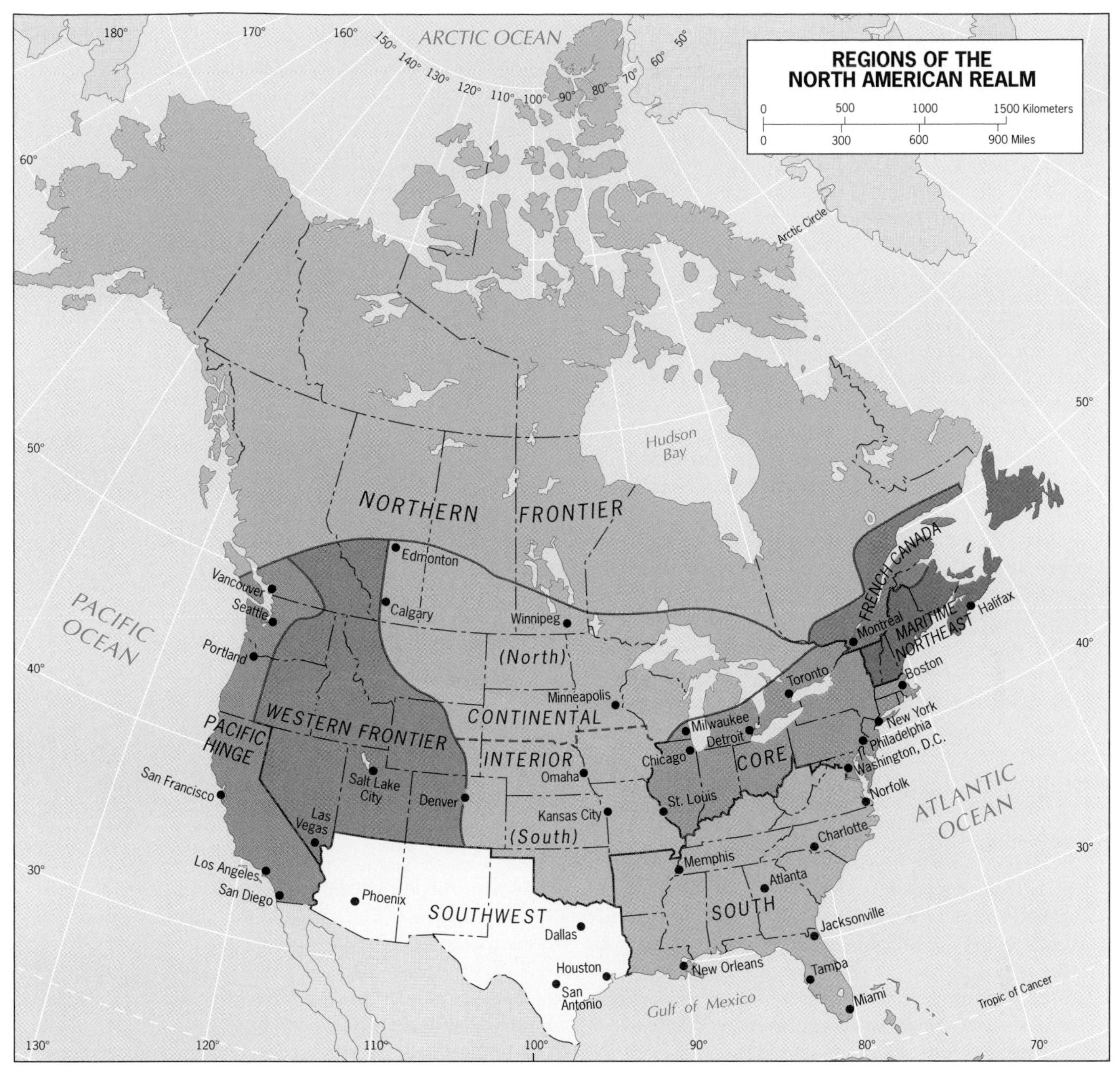

FIGURE 3-18

MAJOR CITIES OF THE REALM

City	Population* (in millions)	City	Population* (in millions)
Atlanta	4.2	New York	20.6
Boston	5.8	Ottawa, Canada	1.2
Chicago	9.0	Philadelphia	6.1
Dallas-Ft. Worth	5.3	San Diego	3.0
Denver	2.6	San Francisco	7.2
Detroit	5.4	Seattle	3.7
Houston	4.8	Toronto, Canada	4.9
Los Angeles	16.8	Vancouver, Canada	2.1
Montreal, Canada	3.4	Washington, D.C.	5.0

*Based on 2002 estimates.

Among the Realm's Great Cities . . .

Chicago

Chicago lies near the southern end of Lake Michigan not far from the geographic center of the conterminous United States. Its centrality and crossroads location were evident to its earliest indigenous settlers, who developed the site as a portage where canoes could be hauled from the lake to the headwaters of a nearby stream that led to the Mississippi River. Centuries later, when the modern U.S. spatial economy emerged, Chi-cago became the leading hub on the continental transport network. As a freight-rail node, it still reigns supreme; and even though most long-distance passengers have switched to jet planes, Chicago has remained the quintessential hub city as O'Hare International became the world's busiest airport more than a generation ago.

Poet Carl Sandburg described Chicago as "the city of big shoulders," underscoring its personality and prowess as the leading U.S. manufacturing center during the century following the Industrial Revolution of the 1870s and 1880s. Among the myriad products that emanated from Chicago's industrial crucible were the first steel-frame skyscraper, elevated railway, refrigerated boxcar, cooking range, electric iron, nuclear reactor, and window envelope. Chicago also became a major commercial center, spawning a skyscraper-dominated central business district second only to New York's in size and influence (see photo, p. 151).

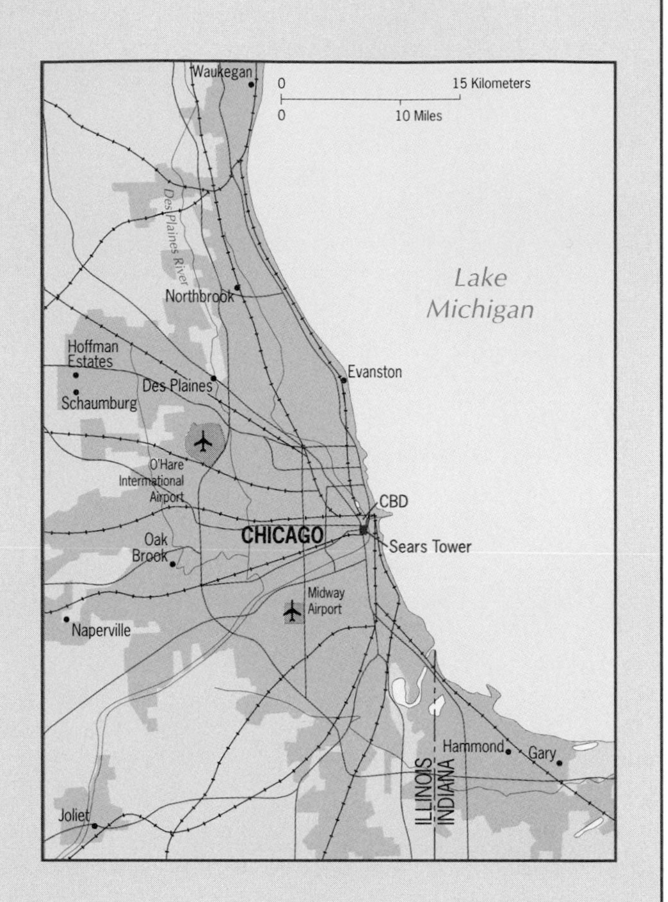

Chicago today (metro population: 9.0 million) is trying hard to make the transition to the postindustrial age. Its diversified economy has proven to be a precious asset in this quest, and the city is determined to reinvent itself as a first-order service center. The new forces it must contend with, however, are symbolized in the fate of the emptying Sears Tower, the city's (and the nation's) tallest building: not long ago Sears & Roebuck abandoned the structure for a new headquarters campus along Interstate-90 in the far northwestern suburb of Hoffman Estates. This shift acknowledges the rise of Chicagoland's thriving outer city, one of the country's largest and most successful. Around O'Hare, a huge and still-expanding suburban office/industrial-park complex has truly become Chicago's "Second City," underscoring that major airports themselves will increasingly function as urban growth magnets in the ever more globalized spatial economy of the twenty-first century.

tury between the Civil War and the close of the industrial age (1865–1970). The rise of postindustrialism over the past three decades has diminished that linchpin regional role, and strengthening challengers to the south and west will continue to siphon key functions away from the Core as this century unfolds. But make no mistake: this is still the geographic heart of North America. Here one finds the largest city and federal capital of both countries. Here, too, are the leading financial markets, corporate headquarters, media centers, cultural facilities, research complexes, and the busiest airports and intercity expressways. Moreover, both the U.S. and Canadian portions of the Core still contain more than one-third of their respective national populations.

Although manufacturing remains highly important within the transformed American economy, productivity and obsolescence problems during the late twentieth century increased production costs in the Core Region, thereby erasing many of its historical competitive advantages over newly emerging industrial areas in the southern and western United States. In parts of the Core (especially its eastern portion), employment dropped sharply in the smokestack industries as factories closed or relocated, speeding the economic decline of numerous mill towns, blue-collar suburbs, and inner central cities. But other areas fought back, none more successfully than the Midwest portion of the Manufacturing Belt, which reinvented itself by aggressively pursuing the high-tech upgrading of its aged industrial base and becoming far more competitive in the international marketplace.

The reconcentration of the pivotal automobile industry since the 1980s is a prime example of the booming Midwest's new manufacturing geography, which foreign as well as domestic carmakers have shaped. A key role was played by Japanese-owned Toyota, which decided to challenge the "Big Three" (GM, Ford, and [then] Chrysler) on its home turf. Toyota's investments were directed at the Core's less-developed southwestern border zone, where a 500-mile (800-km)-long corridor of Interstate-64 between St. Louis and northern West Virginia became the axis of a self-sufficient, state-of-the-art automaking complex centered on the Lexington, Kentucky area. Detroit's Big Three were forced to keep pace and successfully met the challenge by modernizing fabrication technologies and dispersing certain operations away from their traditional base in Michigan to locations elsewhere in the Midwest and as far away as Mexico.

Postindustrial development has also spawned a number of new (or recycled) growth centers in the Core. The major metropolitan complexes of the northeastern Megalopolis—

Among the Realm's Great Cities . . .

New York

New York is much more than the largest city of the North American realm. It is one of the most famous places on Earth; it is the hemisphere's gateway to Europe and the rest of the Old World; it is a tourist mecca; it is the seat of the United Nations; it is one of the globe's most important financial centers—a true "world city" in every sense of that overused term.

New York City consists of five boroughs, centered by the island of Manhattan, which contains the CBD south of 59th Street (the southern border of Central Park). Here is a skyscrapered landscape unequaled anywhere, studded with more fabled landmarks, streets, squares, and commercial facilities than any other cities except London and Paris. This is also the cultural and media capital of the United States, which means the city's influence constantly radiates across the planet thanks to New York-based television networks, newspapers and magazines, book publishers, fashion and design leaders, and artistic trendsetters.

At the metropolitan scale, New York forms the center of a vast urban region, 150 miles (250 km) square (population: 20.6 million), that sprawls across parts of three States in the heart of Megalopolis. That outer city has become a giant in its own right with its population of more than 12 million, massive business complexes, and flourishing suburban downtowns. Thus, despite its global connections, New York is caught in the currents that affect U.S. central cities today. Its ghettos of disadvantaged populations grow steadily; its aging port and industrial base continue to decline; its corporate community is increasingly stressed by the high cost of doing business in ultra-expensive Manhattan.

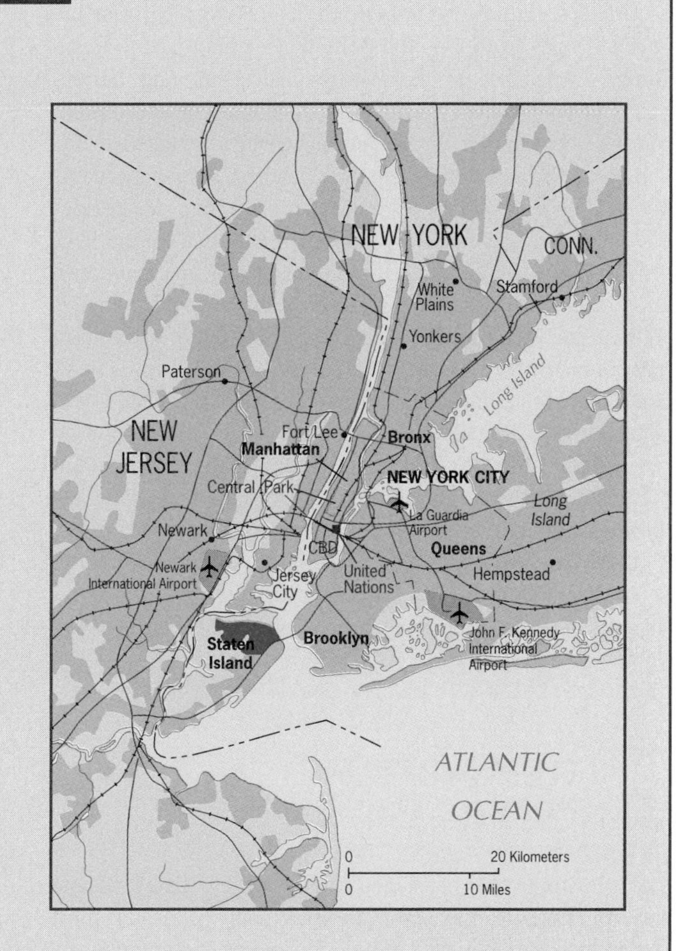

New York has not yet found the solutions but is struggling to overcome these problems by reinventing itself in order to remain a leading economic player in the twenty-first century.

already the scene of much quaternary and quinary economic activity—are adjusting fairly well. The Boston area, well endowed with research facilities, is once again attracting innovative high-tech businesses, mostly toward the Route 128 freeway corridor that girdles the central city (North America's pioneering technopole of the 1950s). New York City remains the national leader in finance and advertising, and is home to significant elements of the broadcast and print media; but even here the city increasingly shares its business leadership with its huge outer suburban city.

The Core Region metropolis that has gained the most from postindustrialization is Washington, D.C. As the information-producing and managerial sectors have blossomed, and as the U.S. federal government extended its ties to the private sector, Washington and its surrounding outer city of affluent Maryland and Virginia suburbs (interconnected by the 66-mile [105-km] Beltway encircling the capital city) has amassed an enormous complex of office, research, high-tech, lobbying, and consulting firms. The most important development has occurred along the outer-suburban expressway corridor between Tysons Corner on the Beltway (see photo, p. 163) and Dulles International, the world's fastest-growing airport. Since the mid-1990s, a large number of major telecommunications and Internet companies have been attracted to the Dulles area, one of the country's leading fiber-optic cable hubs thanks to its local cluster of federal intelligence agencies as well as the corporate headquarters of America Online. This growth is now fanning out from Dulles across much of northern Virginia's sector of the Washington metropolis, which is well on its way to becoming one of the realm's foremost technopoles. Moreover, some are already calling this activity complex the "technological capital" of the United States because more technical workers are employed here than in either Silicon Valley or suburban Boston's Route 128 agglomeration. Clearly, this economic transformation is being felt throughout the Washington region: the area's approximately 350,000 federal government workers are today far outnumbered by the more than 500,000 employed by computer-service, telecom, Internet, aerospace, and biotech companies.

THE MARITIME NORTHEAST

The Maritime Northeast consists of upper New England and the neighboring Atlantic Provinces of easternmost Canada (Fig. 3-18). New England, one of the realm's historic culture hearths, has retained a strong regional identity for almost 400 years. The urbanized southern half of New England has been the northeastern anchor of the Core since the mid-nineteenth century—where it must be regionally classified. However, the six New England States (Maine, New Hampshire, Vermont, Rhode Island, Massachusetts, and Connecticut) still share many common characteristics. Besides this overlap with the Manufacturing Belt (and Megalopolis) in its south, the Maritime Northeast region also extends northeastward across the Canadian border to encompass almost all of the four Atlantic Provinces of New Brunswick, Nova Scotia, Prince Edward Island, and Newfoundland.

A long association based on economic and cultural similarities has tied northern New England to Atlantic Canada. Both have a strong maritime orientation, are rural in character, possess difficult environments with limited land resources, and were historically bypassed in favor of more fertile inland areas. Thus economic growth in upper New England has always lagged behind that of the rest of the realm. Development has centered on primary activities, mainly fishing the (once) rich offshore banks of the nearby North Atlantic, forestry in the uplands, and farming in the few fertile valleys available. Recreation and tourism have boosted the regional economy in recent times, with scenic coasts and mountains attracting millions from the neighboring Core Region. The growth of skiing has also helped, extending the tourist season through the harsh winter months.

Since 1980, New England's roller-coaster economy has experienced spectacular prosperity, hard times, and, most recently, cautious recovery. Weaning itself from its overreliance on the activities that brought on the troubles of a decade ago, New England is developing a more diversified economic base built around telecommunications, financial services, health care, and biotechnology. Nonetheless, the economic revival has not yet taken hold in much of northern New England. Most of the benefits of the recovery are concentrated in the zone closest to metropolitan Boston, not just its immediate suburbs focused on the again-booming Route 128 corridor, but a wide swath of fringe areas that spill over into the southern parts of Maine and particularly New Hampshire. New England's continued growth, however, is by no means assured: apart from the uneven distribution of recovery, the region's costs of doing business are higher than those in the South and Southwest, and the continuing outflow of highly skilled labor raises concerns about the size and quality of the future workforce.

The Atlantic Provinces also experienced hard economic times in the 1990s. Most adversely affected was the groundfish industry as offshore stocks of flounder, haddock, and cod became severely depleted through over-fishing; here, too, disagreements over Canada's maritime fishing boundaries led to the seizure of vessels and heightened tensions with other nations, including the United States. New opportunities, never easy to come by, are most promising in the provinces peripheral to the more heavily populated Nova Scotia and New Brunswick. In tiny Prince Edward Island, tourism was given a boost by the 1997 opening of an 8-mile (13-km) bridge across Northumberland Strait, the island's first direct link to the mainland since it joined the Canadian federation in 1873.

Economic prospects are even brighter in remote Newfoundland, Canada's poorest province, which was the tenth (and last) province to enter the federation just over 50 years ago. Major offshore oil deposits (see Fig. 3-14), centered on the Grand Banks about 200 miles (320 km) east of the capital, St. John's, were discovered in the 1990s just as government-mandated fisheries restrictions were inflicting disaster on the seafood industry. The construction and opening of seabed drilling platforms and coastal support facilities has been swift, additional oil as well as natural gas deposits are being discovered closer to the coast and onshore, and prospects look good that oil can transform the Newfoundland economy along the lines of Norway's over the past quarter-century. (Prospects are also brightening in Nova Scotia, where offshore gas deposits now promise a similar bonanza.)

FRENCH CANADA

Francophone Canada constitutes the effectively settled, southern portion of Quebec, which straddles the central and lower St. Lawrence Valley from where that river crosses the Ontario-Quebec border just upstream from Montreal to its mouth in the Gulf of St. Lawrence. Also included is a sizeable concentration of French speakers, known as the Acadians, who reside just beyond Quebec's provincial boundary in neighboring New Brunswick (Fig. 3-18). The Old World charm of Quebec's cities (see photo below) is matched by an equally unique rural settlement landscape introduced by the French: narrow rectangular farms, known as *long lots*, are laid out in sequence perpendicular to the St. Lawrence, other rivers, and the roads that parallel them, thus allowing each farm access to an adjacent routeway.

The economy of French Canada, however, is no longer rural (although dairying remains a leading agricultural pursuit) and exhibits urbanization rates similar to those of the rest of the country. Industrialization is widespread, supported by cheap hydroelectric energy generated at huge dams in northern Quebec, but relatively little of the region's manufacturing could be classified as high-tech. Tertiary and postindustrial commercial activities are concentrated around Montreal, and tourism and recreation are also important to the regional economy. But the health of these sectors—now improving as the independence movement has stalled—is tied to the resolution of Quebec's political status within Canada.

This uncertainty reminds us that, over the past third of a century, Quebec's heightened nationalism has had a major impact on the province that extends well beyond the economic sphere. During the 1980s, the provincial government enacted new laws that strengthened the French language and culture throughout Quebec. With English domination ended, the French Canadians then channeled their energies into developing Quebec's economy. A byproduct of that ef-

An unmistakably French cultural landscape in North America: the Rue Saint-Louis in the capital of Quebec Province, Quebec. Not an English sign in sight in this center of French/Quebecois culture. The green spires in the background belong to Le Chateau Frontenac, a grand hotel built on the site of the old Fort St. Louis (the Count of Frontenac was a prominent governor of what was then known as New France). French-speaking, Roman Catholic Quebec lies at the core of French Canada.

Among the Realm's Great Cities . . .

Montreal

Montreal lies on a large triangular island in the St. Lawrence River, its historic core laid out along the base of what Montrealers call "the mountain"—flat-topped Mount Royal, from which the city takes its name. Culturally, Montreal is the second largest French-speaking city in the world (after Paris), and therein lie its opportunities and challenges. Without a doubt, this is still one of the realm's most cosmopolitan cities, a strong European flavor pervading its stately CBD, its lively neighborhoods, and its bustling street life. To escape the long, harsh winters here, much of downtown Montreal has become an underground city with miles of handsome passageways lined with shops, restaurants, and movie theaters, all of it served by one of the world's most modern and attractive subway systems.

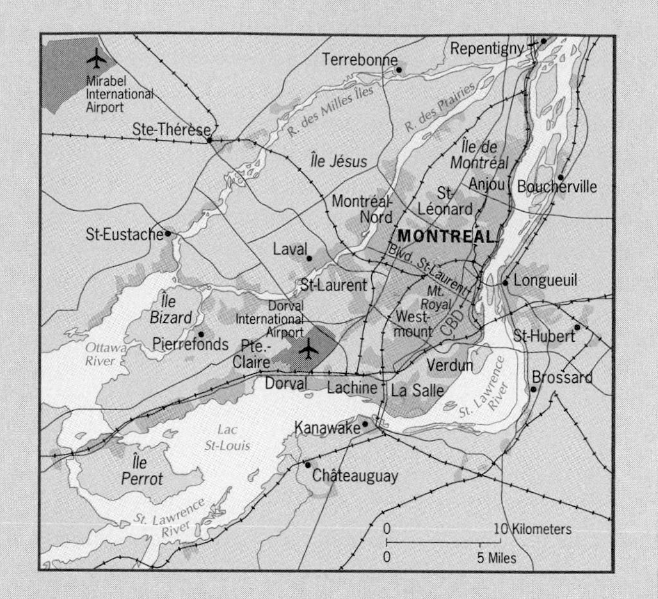

Yet, despite the sparkle and continental atmosphere, Montreal has not been a happy place in recent years. With the escalation of tensions between Francophone and Anglophone Canada over the past three decades, it is understandable that the confrontation should overshadow life in Quebec's largest metropolis (3.4 million), and the city remains divided along ethnolinguistic lines. Spatially, this polarization is especially evident along the Boulevard St. Laurent, which marks the boundary between the French-speakers in Montreal's East End (who constitute about 70 percent of the city's population) and the English speakers who cluster in the West End (the remnant of a larger Anglophone community that was reduced by 40 percent when the French-supremacy language laws were implemented).

Montreal's economy has also suffered. During the 1990s, hundreds of Anglophone firms (and thousands of their workers) fled to Ontario and points west, leaving in their wake a commercial real estate tailspin, rising unemployment, and a noticeable decline in the city's prosperity. Today, with support for secession fading in Quebec, Montreal has begun its comeback. Tourists have returned in droves; a construction boom is reshaping parts of the CBD; new foreign-trading ties, especially with the United States, are reviving the business community; and vibrant suburbs are leading the metropolis to reinvent itself as a hub for information technology, telecommunications, and biopharmaceutical companies.

fort was the rapid urbanization of young families and a loosening of their ties to Roman Catholicism. The outcome was a sharp decline in the provincial birth rate, which the Quebec government tried to offset by encouraging (with cash payments) large families and Francophone immigration (mostly from former colonies of France around the world).

As noted earlier, French Canada also includes Acadia, Canada's largest cluster of Francophones outside Quebec. Here in northernmost New Brunswick, the approximately 250,000 French-speakers constitute one-third of the province's population. The Acadians, however, not only shun the notion of independence for themselves but actively promote all efforts to keep Quebec within the Canadian federation. Unlike the Quebecers, the Acadians have devoted their energies in recent years to accommodation with the Anglophone community. Today, a new relationship (known as "cohabitation") has been worked out between the two groups, based as much on mutual respect for each other's languages as the strict equality of a new, government-mandated system of bilingualism.

THE CONTINENTAL INTERIOR

The Continental Interior extends across the center of both the conterminous United States and the southern tier of Canada (Fig. 3-18). With few exceptions—most notably in the region's northeastern corner where the barren but mineral-rich Canadian Shield meets the Great Lakes—agriculture is the predominant feature of the landscape. Whereas the innermost fruit-and-vegetable and dairying belts in the U.S. macro-Thünian regional system (Fig. 3-15) are in competition with other economic activities in the North American Core to the east, by the time one reaches the Mississippi Valley, meat and grain production prevail for much of the next thousand miles (1,600 km) west to the base of the Rocky Mountains. Because the eastern half of the Continental Interior lies in Humid America and closer to the national food market on the northeastern seaboard, mixed crop-and-livestock farming wins out over less competitive wheat raising, which is relegated to the fertile but semiarid environment of the central and western Great Plains on the dry side of the 100th meridian. The latter area also contains Canada's agricultural heartland north of the 49th-parallel border. Although these fertile Prairie Provinces are less subject to serious drought than the U.S. high plains to the south, they are situated at a more northerly latitude and thus have a shorter growing season.

The distribution of North American corn and wheat farming is mapped in Figure 3-19, and the edges of the most productive zones of these two leading Continental Interior crops coincide with several of the region's boundaries. As we saw earlier in Figure 3-15, the Corn Belt is most heavily focused on Iowa and part of neighboring Illinois, while the Spring and Winter Wheat Belts are respectively centered on North Dakota and western Kansas. This separation of the Wheat Belts is also the basis for subdividing the Continental Interior region, as the dashed line in Figure 3-18 indicates. The North subregion encompasses the Upper Midwest, the northern Great Plains, and the cultivable southern portions of Canada's three Prairie Provinces; the Spring Wheat Belt is its leading agricultural component and is so named because crops are planted in the spring and harvested in late summer or early autumn. The South subregion contains a major part of the Corn Belt in its eastern half and all of the Winter Wheat Belt in its western. The Corn Belt should probably be renamed the *Corn-Soybean Belt* because it also produces huge soybean crops that are grown in rotation with corn (see box titled "Geography of the Mighty Soybean"). The Winter Wheat Belt is centered in the southern Great Plains, and its name is derived from the practice of planting wheat in the fall and harvesting it by late spring; during its infancy winter wheat is able to withstand the milder winter temperatures south of 40°N latitude. The great advantage of this growth cycle is that the crop is harvested before the hot, dry summer takes hold in this semiarid environment.

Throughout the Continental Interior, economic activity is markedly oriented toward farming. Its leading metropolises—Kansas City, Minneapolis–St. Paul, Winnipeg, Omaha, and even Denver—are major processing and marketing points for pork and beef packing, flour milling, and soybean, sunflower, and canola oil production (all of which are increasingly exported). Although the family farm still

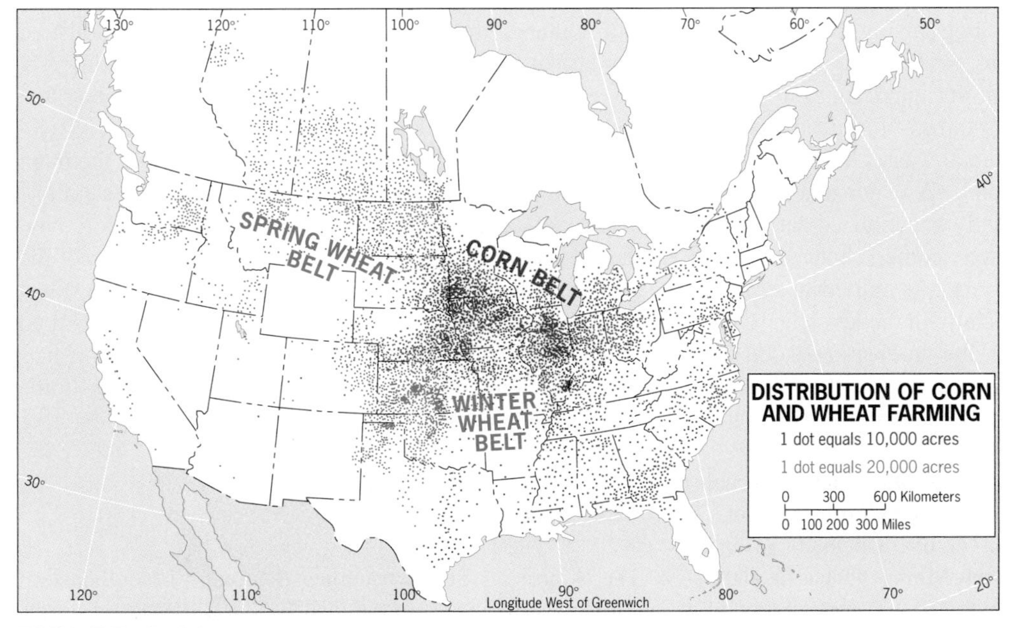

FIGURE 3-19

Geography of the Mighty Soybean

Over the past century, no crop in this realm has expanded faster than the soybean. This hardy legume was one of the earliest plants domesticated, and for nearly 5000 years the Chinese have revered it as a grain vital to life itself. The advantages of soybeans are numerous. They are the world's cheapest and most efficient source of protein, with a level of concentration double that of meat or fish. They are easy to cultivate, adapting to a wide range of environmental conditions, and are less vulnerable to pests and diseases than other temperate-climate food crops. Economically, crop yields are comparatively high, and the demand for soybeans has steadily risen across the world in recent decades.

The dramatic increase of soybean production in the United States has been propelled not only by these advantages but also by this crop's particular suitability to the Corn Belt. It was quickly discovered that soybeans were an ideal crop to rotate with corn because they recondition the soil with nitrogen and other nutrients needed by the latter. In terms of cultivation, soybeans also proved highly profitable to grow on the large, flat fields of the Corn Belt (which continues to account for about 80 percent of all U.S. soybean production) where their farm-machinery requirements were identical to those already in place for raising corn. Moreover, clever marketing constantly widened the versatility of soybeans beyond their use as a nutritious livestock feed, and today they are used in a variety of human foods (ranging from vegetable oil to a plethora of low-calorie products) as well as such industrial goods as fertilizers, paints, and insect sprays.

All of the above also contributed to steadily rising foreign demand, and in many recent years soybeans have ranked as the leading U.S. export commodity. Other producing countries have recognized this opportunity as well, and intensifying global competition after 1990 has resulted in a leveling off of America's share of the world market. Nonetheless, the United States still produces roughly half of the world's annual soybean output. As we note in Chapter 5, most of its competition (another 33 percent of the global supply) comes from Brazil and Argentina in mid-latitude South America, but China is constantly expanding its acreage in a bid to become a major producer. Soybeans will continue to play a key role in North American and international agriculture as the twenty-first century unfolds, buoyed in no small part by this grain's important potential in the struggle to ease the global food shortage.

dominates the rural landscape, the aggressive incursion of large-scale corporate farming now threatens that way of life. This is particularly true for hog raising in the western Corn Belt, where big companies are taking control of the entire production process in their ultramodern, factory-like complexes that gobble up huge swaths of farmland and disgorge appalling quantities of animal wastes that pollute air and groundwater for miles around. (A major new area for such mass-scale hog raising rapidly materialized during the late 1990s in the region's southwest corner, centered on the panhandle of far western Oklahoma.) As individual agricultural opportunities decline, especially in the less intensively farmed Great Plains to the west, the unrelenting exodus of younger people leaves behind an ever more elderly population widely dispersed across a constellation of stressed rural communities struggling to survive the depopulation trend.

Despite the key role that agriculture plays in the regional economy, the nonfarming sectors of the Continental Interior continue to develop and diversify, especially in and around major metropolitan centers. Minnesota has been the most successful in promoting such growth, capitalizing on the reputations of that State's stable economy, highly educated workforce, track record as a product innovator, and "north woods" amenities. In the urban areas of the eastern Great Plains, telemarketing has become a major new pursuit. Omaha, Nebraska is already known as the "toll-free capital" of the United States because so many credit-card, reservations, and catalog-order companies are concentrated there; Gateway 2000, the computer sales giant, has single-handedly put North Sioux City, South Dakota on the map; and Sprint, a global telecommunications leader, has built its new headquarters just southwest of Kansas City. Moreover, the development of energy resources has spurred the growth of the region's northwest; southern Alberta, most notably, is attracting investments from all over the United States to expand production of its massive deposits of natural gas—today's clean-burning fuel of choice for home heating and electrical power generation.

THE SOUTH

The American South occupies the realm's southeastern corner, extending from the lush Bluegrass Basin of northern Kentucky to the swampy bayous of Louisiana's Gulf Coast, and from the knobby hills of West Virginia to the

A port is an interface between the land behind it and the water body in front of it, and some of the world's most productive ports lie at the mouths of major rivers—such as Rotterdam at the North Sea outlet of Western Europe's Rhine. New Orleans, located at the apex of the Mississippi Delta, has benefited from such a favored location since the early nineteenth century. Today, in the era of NAFTA, as the products of the U.S. heartland increasingly enter the international trade arena, southern Louisiana senses a heightened opportunity to tap this goods flow by convincing shippers to export by water via the Mississippi. In order to attract this business, huge investments have recently been made to modernize and expand the Port of New Orleans. Some of the results are visible here along this stretch of the great river just above the Crescent City itself, part of a huge new port complex that will extend more than 20 miles (32 km) upstream from downtown New Orleans (seen in the middle distance).

flat sandy islands off the southernmost extremity of peninsular Florida (Fig. 3-18). Of the realm's nine regions, none has undergone more overall change during the past half-century. Choosing to pursue its own sectional interests practically from the advent of nationhood, the South experienced an even deeper economic and cultural isolation from the rest of the United States during the long and bitter aftermath of the ruinous Civil War. For over a century it languished in economic stagnation, but the 1970s finally witnessed a reassessment of the nation's perception of the region that launched a still-ongoing wave of growth and change unprecedented in Southern history.

Propelled by the forces that created the Sunbelt phenomenon, people and activities streamed into the urban South. Cities such as Atlanta, Charlotte, Miami, and Tampa became booming metropolises practically overnight, and conurbations swiftly formed in such places as southern and central Florida, the Carolina Piedmont, and the western Gulf Coast from Houston (just across the Texas border) through New Orleans to Mobile, Alabama. This "bulldozer revolution" was matched in the most favored rural areas by an agricultural renaissance that stressed such higher-value commodities as soybeans, winter wheat, poultry, and wood products. On the social front, institutionalized racial segregation was dismantled more than three decades ago.

Yet for all the growth that has taken place since 1970, the South remains a region beset by many economic problems because the geography of its development has been decidedly uneven. Although several metropolitan and certain farming areas have benefited, many others containing sizeable populations have not, and the juxtaposition of progress and backwardness is frequently encountered in the Southern landscape. Not surprisingly, the gap between rich and poor is wider here than in any other U.S. region. This economic disparity also carries racial dimensions: compared to whites, blacks have much higher poverty rates and rank well behind in median family income and achievement of middle-class status. Within this checkerboard spatial framework of development, the South's new affluence is largely concentrated in its central-city CBDs and burgeoning suburban rings. At the other end of the income spectrum, its far more widely dispersed struggling areas include less favored agricultural zones (such as the chronically depressed bottomlands that line the lower Mississippi) and myriad rural towns whose only jobs are tied to small-scale plants in locally dying industries such as food processing and garmentmaking. And at the regional scale, yet another geographic pattern of inequality emerges: except for centrally located Atlanta and the Interstate-85 "Boombelt" corridor extending from there northeastward into central North Carolina, a significant proportion of the South's recent economic growth has taken place on its periphery in the Washington suburbs of northern Virginia, on the western Gulf Coast, and along Florida's central (Interstate-4) and southern-coast corridors.

The transformation of the South is also changing the complexion of its demographic mosaic (as was demonstrated so dramatically in Florida in the aftermath of the 2000 presidential election). Racially and economically, the composition of the region's population continues to be reshaped by the inmigration of whites and blacks from the North, with both groups exhibiting higher income levels than their resident Southern counterparts. Ethnically, the rapidly increasing presence of Hispanics and Asians is intensifying the South's cultural heterogeneity—and heightening ethnic tensions as low-skilled newcomers increase the competition for a shrinking number of blue-collar jobs. In age structure, too, the region is diversifying as the influx of retirees steadily increases, not just in Florida but also in Georgia, Virginia, and the Carolinas. Finally, we should remember that the massive construction of new urban landscapes since 1980 has provided ample opportunity for creating fragmented residential environments that typify today's mosaic culture. Thus the filling of the tiles of the Southern population mosaic by an ever more diverse array of social and lifestyle groups is another signal that this once outcast region is completing its convergence with the rest of the country.

This 30-year transition period has ended with the South not only fully absorbed into the *national* economy and society, but also becoming an important player on the *international* stage. During the 1990s, European and Japanese companies were active investors in the region, especially the carmakers who built entire manufacturing complexes. In terms of relative location, the most obvious foreign business opportunities to develop in this era of NAFTA lie to the south in Middle and South America. South Florida has been in the vanguard here for a generation, and Miami appropriately bills itself as "the gateway to Latin America." Lately, other parts of the South have indicated an interest in undercutting this monopoly: Atlanta has opened a number of new nonstop air routes, and New Orleans is seeking to enlarge its share of the lucrative seaborne trade by completing the world's highest technology and longest linear port (24 miles/38 km) (see photo, page 187). And in Memphis plans are proceeding to make this metropolis a global distribution hub, based on the huge air-shipment facilities recently built there by Federal Express and United Parcel Service.

THE SOUTHWEST

Before the late twentieth century, most geographers did not even identify a distinct southwestern region, classifying the U.S.-Mexican borderland zone into separate southward extensions of the Continental Interior and the mountain/plateau country to its west. Today, however, this steadily developing area—constituted by the States of Texas, New Mexico, and Arizona—has earned its place on the regional map of North America (Fig. 3-18). The still-emerging Southwest is also unique in the United States because it is a bicultural regional complex. Atop the crest of the Sunbelt wave, inmigrating Anglo-Americans join a rapidly expanding Mexican-American population anchored to the sizeable long-time resident Hispanic population, which traces its roots to the Spanish colonial territory that once extended from Texas's Gulf Coast to central California. In fact, when we add the large Native American population, it is not inappropriate to recognize the Southwest as a *tricultural* region.

Recent rapid development in Texas, Arizona, and New Mexico is built on a three-pronged foundation: (1) availability of huge amounts of *electricity* to power air conditioners through the long, brutally hot summers; (2) sufficient *water* to supply everything from swimming pools to irrigated crops, so that large numbers may reside in this dry environment; and (3) the *automobile*, so that affluent newcomers may spread themselves out at much-desired low densities. The first and third of these have been rather easily attained because the eastern half of the Southwest is abundantly endowed with oil and natural gas (Fig. 3-14). The future of water supplies, however, is far more problematical. For instance, Phoenix and Tucson could not keep growing at their current swift pace without the massive new Central Arizona Project (CAP) Canal, which transports Colorado River water across hundreds of miles of desert.

The economic growth of the Southwest in recent years has been led by Texas, which during the 1990s weaned itself away from dependence on its oil and gas industries (whose share of the State's GNP is now only about one-quarter of the 1980 level). Following a serious oil-induced recession in the 1980s, the Lone Star State's mix of economic activities was restructured and diversified. Success was not long in coming, and today the eastern Texas triangle formed by Dallas-Ft. Worth, Houston, and San Antonio has become one of the world's most productive postindustrial complexes, specializing in business-information and health-care services and high-technology manufacturing. The metropolis centered on the State capital—Austin, which lies near the heart of the triangle—has become a technopole that is second only to Silicon Valley as a leader in computer research and development as well as a producer of semiconductors and PCs. To the north at the triangle's apex, the Dallas-Ft. Worth (DFW) Metroplex is keeping pace with its own expanding cluster of high-tech industries focused on the *Telecom Corridor* technopole just north of Dallas (the world's largest concentration of telecommunications firms). As Figure 3-20 shows, the geography of employment is increasingly oriented to booming suburban freeway corridors, a spatial pattern intensifying as well in metropolitan Austin, Houston, and San Antonio. The DFW Metroplex is also the leading Texas command post for international trade and NAFTA-spawned development, and it is becoming the northern anchor of a new transnational growth corridor that extends for more than 500 miles (800 km) southward from Dallas through Austin and San Antonio to Monterrey 150 miles (250 km) inside Mexico.

The western flank of the Southwest—whose development is mostly concentrated in southern Arizona—is unique. Equaled only by parts of southern Israel's Negev, this is the most technologically transformed desert environment on Earth. But while the Negev is devoted to agriculture, Arizona's far more intensive development is marked by vast urbanized landscapes focused on two of the largest western U.S. cities, Phoenix and Tucson. During most of the 1990s, metropolitan Phoenix was expanding into the surrounding desert at the rate of an acre per hour, and it still grows at the astonishing yearly rate of roughly 4 percent. Although the forces of the slow-growth movement are gaining momentum, Arizona's conservative political and business leaders have been loath to restrict development. This debate may soon become moot, however, because everyone agrees that retrievable water supplies—the State's lifeblood and key to its future—are reaching their limit.

THE WESTERN FRONTIER

To the north of the western half of the Southwest lies the realm's newest region: the swiftly growing, economically booming Western Frontier. In its east-west extent, this mountain-studded plateau region stretches between (and includes) the Rocky Mountains and the Sierra Nevada-Cascades chain; its longer north-south axis reaches from northern Arizona's Grand Canyon country to the edge of the subarctic snowforest west of the Canadian Rockies (Fig. 3-18). The region's heartland encompasses Utah, Nevada, Idaho, and western Colorado, and it is here that the old intermountain West is most actively reinventing itself. Until recently, such an opportunity would have been unthinkable because of this region's remoteness, dryness, and sparse population. But the Western Frontier's geographic prospects are changing significantly as the postindustrial national economy matures, advances in communications and transportation technologies eliminate interregional dif-

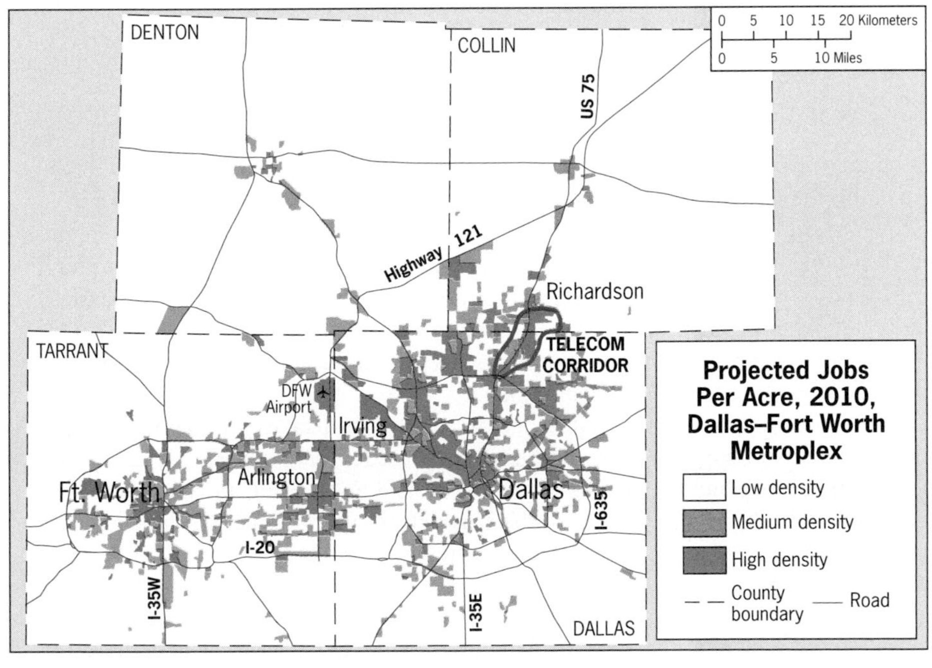

FIGURE 3-20

There is something quite remarkable about the ongoing urbanization of arid valleys in the high desert of the Western Frontier, and nowhere is this phenomenon more starkly on view than in the Las Vegas Valley in Nevada's southernmost extremity. This particular municipality is Henderson, not only Las Vegas's largest suburb but also the fastest-growing urban settlement in the entire United States between 1990 and 2000. Many affluent retirees have also been attracted to this immediate area, and together with their younger counterparts have shaped a classic mosaic-culture urban fabric dominated by a low-density checkerboard of self-contained, socially homogeneous communities. Architectural critics and environmentalists, of course, would have no trouble in declaring the filling of this landscape to be one of the most blatant examples of urban sprawl on view anywhere in North America.

ferences in the costs of doing business, and amenities become the leading locational preference of employers. Undoubtedly, proximity to the increasingly expensive and overcrowded Pacific coast is also a reinforcing factor: well over 1 million disenchanted Californians have moved into the Western Frontier since the late 1980s, lured by its sunny climate, wide-open spaces, spectacular scenery, proliferating jobs, and lower cost of living.

The label *Western Frontier* further emphasizes that the transformation under way is being driven by the influx of people and activities from outside the region. Indeed, the populations of the four States of its heartland all rank among the five fastest-growing in the United States since 1990, with the more than 2 million migrants who arrived during the past decade already constituting more than 20 percent of the Western Frontier's population. In the economic-geographic sphere, the region's identity is being reforged as traditional resource-based industries (mining, timber, livestock grazing) decline while inmigrating and new indigenous companies create thousands of high-tech manufacturing and specialized services jobs. The spatial pattern of this development is strongly focused on the region's mushrooming urban areas and creates an unusual frontier: instead of a single line of advancing settlement, a dispersed set of population clusters is simultaneously expanding around large- and middle-sized cities, high-amenity zones, and interstate freeway corridors.

The hottest growth area in all of the Western Frontier—and one of the fastest-growing metropolises in the United States—is the Las Vegas Valley of southernmost Nevada (see photo above). This boom was triggered by Las Vegas's burgeoning recreation industry (which attracts some 30 million tourists annually), but it is also the product of new economic development dominated by the influx of high-tech and professional services firms that specialize in computers, finance, and health care. Business entrepreneurs have flocked here, and the new employment opportunities sparked a local population explosion, spearheaded by those relocating Californians; relatively cheap land, modest housing prices, low taxes, and the sunny climate further fueled this migration.

Other urban areas are also experiencing accelerated growth, particularly the region's two largest metropolises, centered on Denver and Salt Lake City (which are spawning conurbations extending more than 50 miles north and south of the central city). In both cases the leading amenity is the Rocky Mountains, which soar majestically above the urbanizing flatlands and foothills. Metropolitan Denver has expanded rapidly along (and into) central Colorado's Front Range, building a diverse new, globally oriented economic base dominated by telecommunications, computer software, and financial services. On a somewhat smaller scale, Salt Lake City has followed in Denver's footsteps on the Rockies' western flank and now centers a thriving conurbation that lines northern Utah's Wasatch Front from Ogden in the north to Provo in the south. In fact, the budding technopole of *Software Valley*, the 40-mile (65-km) corridor linking Salt Lake City and Provo, contains one of the world's largest concentrations of information technology companies, built up around two universities, a highly educated workforce, and a postindustrial complex specializing in the production of word-processing and computer-simulation software.

As much as 90 percent of the Western Frontier's development is confined to its widely dispersed urban areas, but the effects of this rapid growth are rippling deeply into the region's vast rural zones. Some of the problems—such as the rising incidence of wildfires as zones bordering forests become ever more populated—are caused by the spillover effects of urbanization on adjacent areas. But many others result from the collision of lifestyles as higher-income, recreation-seeking city dwellers invade a countryside whose long-time, nonaffluent residents are already struggling to preserve a way of life based on declining activities such as mining and extensive agriculture. In the valleys of the Colorado Rockies, ranchers are further squeezed by the upwardly spiralling land values set off by the massive growth of ski resorts—and usually are forced to sell and relocate their operations to less favorable land in even more remote areas. In the small towns of northern Idaho and western Montana, social tensions are inflamed as local right-wing militias increasingly encounter more liberal transplanted Californians. And seemingly everywhere, the plans of developers are blocked by environmentalists—who in the 1990s often received the support of the federal government (the region's biggest landowner). The surprisingly rapid emergence of a new preservation ethic, which increasingly threatens environmentally insensitive land and resource developers, is part of a larger anti-growth movement that is certain to intensify. Throughout this region today, the buzzwords are "managing growth without restricting economic opportunity"—which usually translates into bringing opposing interests together to work out compromises.

Whatever our new century brings, the growth boom of the 1990s that created the realm's newest region has forever altered the economic-geographic and demographic complexion of much of the old intermountain West. Now reborn as the Western Frontier, this no-longer-remote region is putting the often disastrous boom/bust cycles of its mining- and grazing-dominated past behind it, and is making major strides toward establishing itself as a globally connected, technological powerhouse on the inland doorstep of North America's burgeoning sector of the Pacific Rim (see box, p. 193).

THE NORTHERN FRONTIER

The northern half of the realm, lying poleward of roughly 52°N latitude, constitutes the Northern Frontier region (Fig. 3-18). This is by far North America's biggest regional subdivision, covering almost 90 percent of Canada and all of the U.S. State of Alaska. Indeed, if this huge territory were a separate country, it would rank as the world's fifth-largest in size. The latitudinal extent of this region leaves no doubt as to why it is designated as *northern*. In fact, after northern Russia (whose dimensions of continentality are even more pronounced), this is our planet's second greatest expanse of harsh cold environment affecting an inhabited area. As with the Arctic fringes of Eurasia, there are very few physical barrers in northern North America to ameliorate the masses of supercold air that throughout the long winter plunge directly southward from the ice-covered polar ocean. Not surprisingly, most of the Northern Frontier's people and activities are concentrated in the milder, heavily forested southern half of the region.

Unquestionably, this region is also a *frontier*—and far more classically so than the Western Frontier just discussed. The latter's newly activated, dispersed, urbanized development pattern and dramatic population influx are largely attributable to its improved economic-geographic position astride a national axis, within which significant east-west interaction was already taking place. The Northern Frontier, on the other hand, represents a thrust into undeveloped country in an entirely new direction, one marked by the leisurely, long-term advance of a line across parts of Canada's near- and middle-northern periphery that absorbs the areas it crosses into the national spatial economy. Boom and bust cycles since the late nineteenth century have controlled the rate of movement of this development line, which frequently has ebbed backward as much as it has flowed forward. In general, the still sparsely populated Northern Frontier is mostly a region of recent European settlement based on the extraction of newly discovered resources. As we shall see, however, First Nations and other indigenous peoples, who have always regarded this region as a homeland, are today playing a growing role in the affairs of the Northern Frontier and will have much to say about shaping its future.

As noted earlier, the Canadian Shield, which covers the eastern half of the Northern Frontier (Fig. 3-2), is one of the world's richest storehouses of mineral resources. Most of the Shield's developing areas are focused on mining complexes that extract such high-demand metallic ores as nickel, uranium, copper, gold, silver, lead, and zinc. The leading zones of production are located along the Shield's southern margins, but since 1980 the search for additional deposits, the expansion of road and rail networks, and the opening of new mines have all advanced steadily northward. The

Yukon and Northwest Territories have proven bountiful, with gold- and diamond-mining paving the way; and at the opposite, eastern edge of the Shield, near Voisey's Bay on the central Labrador coast of Newfoundland, the largest body of high-grade nickel ores ever discovered is being opened up for full-scale production. Besides metallic minerals, the Northern Frontier is endowed with an abundance of other resources (which include fossil fuels, hydroelectric power opportunities, timber, and fisheries). These are dominated by substantial oil and gas reserves that are concentrated in a wide zone east of the Canadian Rockies between the Yukon and the U.S. border (Fig. 3-14). Most intriguing are the Athabasca Tar Sands of northeastern Alberta, which contain more crude oil than all of Saudi Arabia but deposited as a tar-like substance (bitumen) in near-surface sand formations; recent technological breakthroughs are enabling this energy supply to be extracted ever more profitably, output is steadily rising, and by 2010 the Tar Sands are expected to become Canada's dominant oil-producing zone.

The productive activities of the Northern Frontier are linked together within a far-flung network of mines, oil- and gasfields, pulp mills, and hydropower stations that have spawned hundreds of settlements and thousands of miles of interconnecting transport and communications facilities. Although dominant, this commercial economy constitutes just one component of a dual regional economy. The other component, as Robert Bone points out in his book *The Geography of the Canadian North*, consists of the native economy that "evolved from a subsistence hunting economy to one in which wage employment, transfer payments, and trapping provide cash income while hunting provides country food." This native economy, which adheres to the traditional values of the indigenous peoples (First Nations), has long been tied to the commercial economy as a source of wages. It has also been heavily land-based, and in recent years First Nations throughout the region have become more assertive and filed legal actions to recover lands that Europeans took in the past without negotiating treaties. Both the courts and the federal government have been sympathetic, and the creation of Nunavut in 1999 (see map p. 172) has been the most dramatic breakthrough to date.

The Northern Frontier also contains Alaska, whose geographic opportunities and challenges differ somewhat from those of the Canadian North. Besides being the largest U.S. State, Alaska earns a sizeable income from oil production on its northeastern Arctic slope, and the south-central coast around Anchorage increasingly benefits from its near-midpoint location on the Pacific Rim air route between the "Lower 48" States and East Asia. These advantages have propelled a growth spurt of more than 100 percent since 1970, and today Alaska's total of just over 650,000 accounts for 30 percent of the Northern Frontier's population on less than one-sixth of its land area. Here, too, metallic minerals have long drawn migrants northward. But, unlike Canada to the east, Alaska's recent boom was triggered by the opening of its vast oilfields on the North Slope, one of the hemisphere's leading energy sources since the 800-mile Trans-Alaska Pipeline began operating in 1977 (Fig. 3-14). Dwindling supplies at the main field at Prudhoe Bay are now forcing producers to turn to the huge additional reserves of the Arctic slope, but strong opposition from nature preservationist groups has slowed their plans. To the east lies the Arctic National Wildlife Refuge, where the Bush administration wants to begin drilling. To the west lies the National Petroleum Reserve-Alaska, which the federal government partially opened in 1999 to oil leasing by drillers, who insist that their newly gained sensitivity paired with the latest technologies will minimize the impact of production on the tundra's fragile ecosystems. Another major project that faces a similar battle is the proposal to construct a natural-gas pipeline to parallel the oil pipeline (huge supplies of liquified gas are a by-product of oil drilling). As noted earlier, demand for clean-burning natural gas by electricity producers is rising all across North America.

THE PACIFIC HINGE

The Pacific coastlands of the conterminous United States and southwesternmost Canada, which comprise the realm's ninth region (Fig. 3-18), have been a powerful lure to migrants since the Oregon Trail was pioneered more than 150 years ago. Unlike the remainder of western North America south of 50°N latitude, the strip of land between the Sierra Nevada-Cascade mountain wall and the sea receives adequate moisture. It also possesses a far more hospitable environment, with generally delightful weather south of San Francisco, highly productive farmlands in California's Central Valley, and such scenic glories as the Big Sur coast, Washington State's Olympic Mountains, and the spectacular waters surrounding San Francisco, San Diego, Seattle, and Vancouver. Most of the major development here took place during the post–World War II era, accommodating enormous population and economic growth along this outermost edge of the conterminous United States. But as the twenty-first century opens, it is clear that, in terms of its economic geography, the West Coast no longer represents an end but rather a beginning—a gateway to an abundance of growing opportunities that in recent times have blossomed on many of the distant shores encircling the Pacific Basin. That is why we use the term "hinge," because now this region increasingly forms an interface between North America and the booming, still-emerging *Pacific Rim* (see box titled "The Pacific Rim Connection").

Fifty-plus years of unrelenting growth have taken their toll, and whereas the West Coast States and British Columbia are now savoring the prospects of their Pacific Rim

Los Angeles is infamous for its atmospheric pollution, underscored here in contrasting views of the air-quality extremes over the central business district. Smog shrouds this sprawling city on many days of the year, despite efforts to curb the production of pollutants from motor-vehicle exhaust, factories, and other sources.

location, they must also confront the less pleasant consequences of regional maturity. This is especially true in California, where the massive development of America's most populated and multi-ethnic State (35 million in 2002, 13 percent larger than all of Canada) has been overwhelmingly concentrated in the teeming conurbation extending south from San Francisco through San Jose, the San Joaquin Valley, the Los Angeles Basin, and the southwestern coast into San Diego at the Mexican border. Environmental hazards bedevil this entire corridor, including inland droughts, coastal-zone flooding, mudslides, brushfires, and particularly earthquakes—with the ominous San Andreas Fault practically the axis of megalopolitan coalescence. To all this, humans have added their own abuses of California's fragile habitat, from overuse of water supplies (requiring vast aqueduct systems to import water from hundreds of miles away) to the chronic air pollution of Los Angeles.

These challenges notwithstanding, Californians over the past half-century have built one of the realm's most productive economic machines. Indeed, if the Golden State were an independent country, its economy would today rank as the world's seventh largest. The road to success, however, has often proven to be a bumpy one, as was the case in Southern California during the 1990s. This subregion began the decade struggling as the national recession combined with the unexpected end of the Cold War to deliver a serious blow to the Los Angeles-area economy, which at

The Pacific Rim Connection

Today we continue to witness the rise of a far-flung region born of a string of economic miracles on the shores of the Pacific Ocean. It is a still-discontinuous region, led by its foci on the Pacific's western margins: Japan, coastal China, South Korea, Taiwan, Thailand, Malaysia, Singapore. The regional term **Pacific Rim** has come into use to describe this dramatic development, which over the past quarter-century has redrawn the map of the Pacific periphery—not only in East and Southeast Asia but also in the United States and Canada, and even in such Southern Hemisphere locales as Australia and South America's Chile.

20

The Pacific Rim is also is a superb example of a *functional region*, with economic activity in the form of capital flows, raw-material movements, and trade linkages generating urbanization, industrialization, and labor migration. In the process, human landscapes from Sydney to Santiago are being transformed within the 20,000-mile (32,000-km) corridor that girdles the globe's largest body of water. We will continue to discuss this important regional development and its spatial impacts, in later chapters, wherever the Pacific Rim intersects Pacific-bordering geographic realms.

the time was heavily dependent on its aerospace industry driven by federal government defense expenditures. By the late 1990s, a turnaround was underway as Southern California successfully diversified its economy and developed such new growth sectors as the entertainment industry, foreign trade, and the nation's largest single metropolitan manufacturing complex—whose products range from high-tech medical equipment to the clothing churned out by the low-tech factories of a burgeoning, immigrant-staffed garment industry that continues to outpace New York City's.

The northern portion of the region encompasses Northern California and the Pacific Northwest, the latter extending northward from Oregon's Willamette Valley through the Cowlitz-Puget Sound lowland of western Washington State into the adjoining coastal zone of British Columbia beyond the Canadian border. Northern California, focused on the San Francisco Bay Area, during the 1990s experienced a much smoother path than the southern part of the State. Here an unprecedented period of economic expansion continues, led by Silicon Valley. Within that technopole, no community stands to gain more from this boom than Palo Alto, which is situated at the largest switching point on the Internet that handles more than half of its global traffic. As Internet commerce grows explosively, companies are flocking to Palo Alto to take advantage of its unparalleled access to cyberspace. This is further facilitated by the city's state-of-the-art local fiber-optic network. What is taking shape here is nothing less than an infrastructural hub that will become one of the cornerstones of the so-called New Economy.

Continuing up the coast, the booming Pacific Northwest is also deeply involved in the pioneering and production of

Among the Realm's Great Cities . . .

Los Angeles

It is hard to think of another city that has been in more headlines in the past several years than Los Angeles. Much of this publicity has been negative. But despite its earthquakes, mudslides, brushfires, pollution alerts, showcase trials, and riots, L.A.'s glamorous image has hardly been dented. This is compelling testimony to the city's resilience and vibrancy, and to its unchallenged position as the Western world's entertainment capital.

The plane-window view during the descent into Los Angeles International Airport, almost always across the heart of the metropolis, gives a good feel for the immensity of this urban landscape. It not only fills the huge natural amphitheater known as the Los Angeles Basin, but it also oozes into adjoining coastal strips, mountain-fringed valleys and foothills, and even the margins of the Mojave Desert more than 50 miles (80 km) inland. This quintessential spread city, of course, could only have materialized in the automobile age. In fact, most of it was built rapidly over the past 60 years, propelled by the swift expansion of a high-speed freeway network unsurpassed anywhere in metropolitan America. In the process, Greater Los Angeles became so widely dispersed that today it has reorganized within a geographic framework of six sprawling urban realms (see Fig. 3-12).

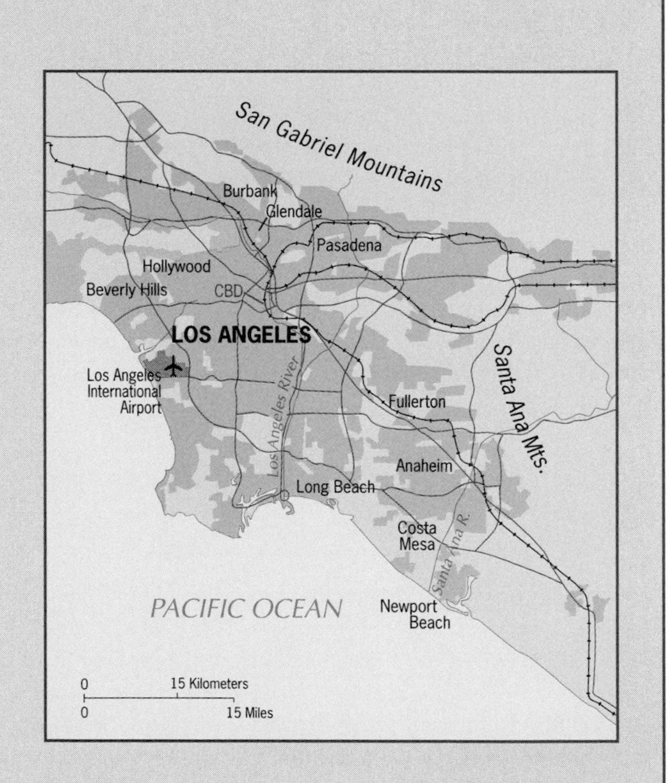

The metropolis as a whole is home to 16.8 million Southern Californians, constitutes North America's second largest urban agglomeration, and forms the southern anchor of the huge California conurbation that lines the Pacific coast from San Francisco southward to the Mexican border (Fig. 3-9). It also is the Pacific Hinge's leading trade, manufacturing, and financial center. In the global arena, Los Angeles is the eastern Pacific Rim's biggest city and the origin of the largest number of transoceanic flights and sailings to Asia.

computer technology, especially in the technopole centered on suburban Seattle's Redmond (Microsoft's headquarters). Originally built on timber and fishing—primary activities that still survive here despite human pressures on the environment—this subregion found its impetus for industrialization in the massive Columbia River dam projects of the 1930s and 1950s, which generated the cheap hydroelectricity that attracted aluminum and aircraft manufacturers. Although Boeing's huge complex outside Seattle is still the area's leading employer, the recent influx of affluent newcomers is fueling the explosive growth of software and Internet companies that is spearheading Seattle's transformation into the prototype metropolis of the New Economy.

North of the Canadian border, the Pacific Hinge's northernmost subregion stands out from the others not only because of its British Columbia location but also because its economic core, Vancouver, has become the most Asianized metropolis in North America. Vancouver lies closer to East Asia on the air and sea routes than the cities to its south, and it got a head start in forging trade and investment linkages to the western Pacific Rim thanks to its large and growing Asian population. Today, ethnic Chinese residents constitute fully 20 percent of the metropolitan population, whose total Asian component now exceeds a remarkable 35 percent. Although a new east-west hybrid culture is trying to establish itself, frictions intensified during the late 1990s as wealthy Asians (many of them expatriates of now-Chinese Hong Kong) took over neighborhoods and replaced traditional homes with much larger houses, surrounded them with walls and fences where none had existed, and cut down most of the trees. More serious is the growing belief by the majority Anglos that the newcomers do not really want to assimilate, preferring instead to avoid the use of English and live, shop, and transact business exclusively within their own communities.

With North America ever more tightly enmeshed in the global economy, nowhere are the realm's international linkages more apparent today than in the Pacific Hinge. From the northern end of this region to the Mexican border, its geographic advantages as a gateway to the burgeoning Pacific Rim are providing opportunities undreamed of barely a decade ago. Asianizing British Columbia now exports more than 40 percent of its goods to the countries of the western Pacific Rim (a proportion far higher than in any other Canadian province or U.S. State). The rest of the Vancouver-Seattle-Portland corridor is following suit, led by its high-technology industries. In California, the foreign trade sector of the State's gigantic economy has tripled since 1990, and today more than 15 percent of its jobs are export-dependent. Both Northern and Southern California are expected to thrive as these trends continue, with metropolitan Los Angeles becoming the financial, manufacturing, and trading capital of the eastern Pacific Rim as well as its transport hub (L.A. overtook New York in the mid-1990s as the leading U.S. seaport in the value of goods passing through it). Thus it is abundantly clear today that the mid-latitude West Coast no longer serves as North America's back door, but forms the hinges of a new front door to the Pacific arena thrown wide open to foster international interactions of every kind.

FIGURE 4-1

Scale 1:16 000 000; one inch to 250 miles. Polyconic Proje
Elevations and depressions are given in feet

chapter 4

Middle America

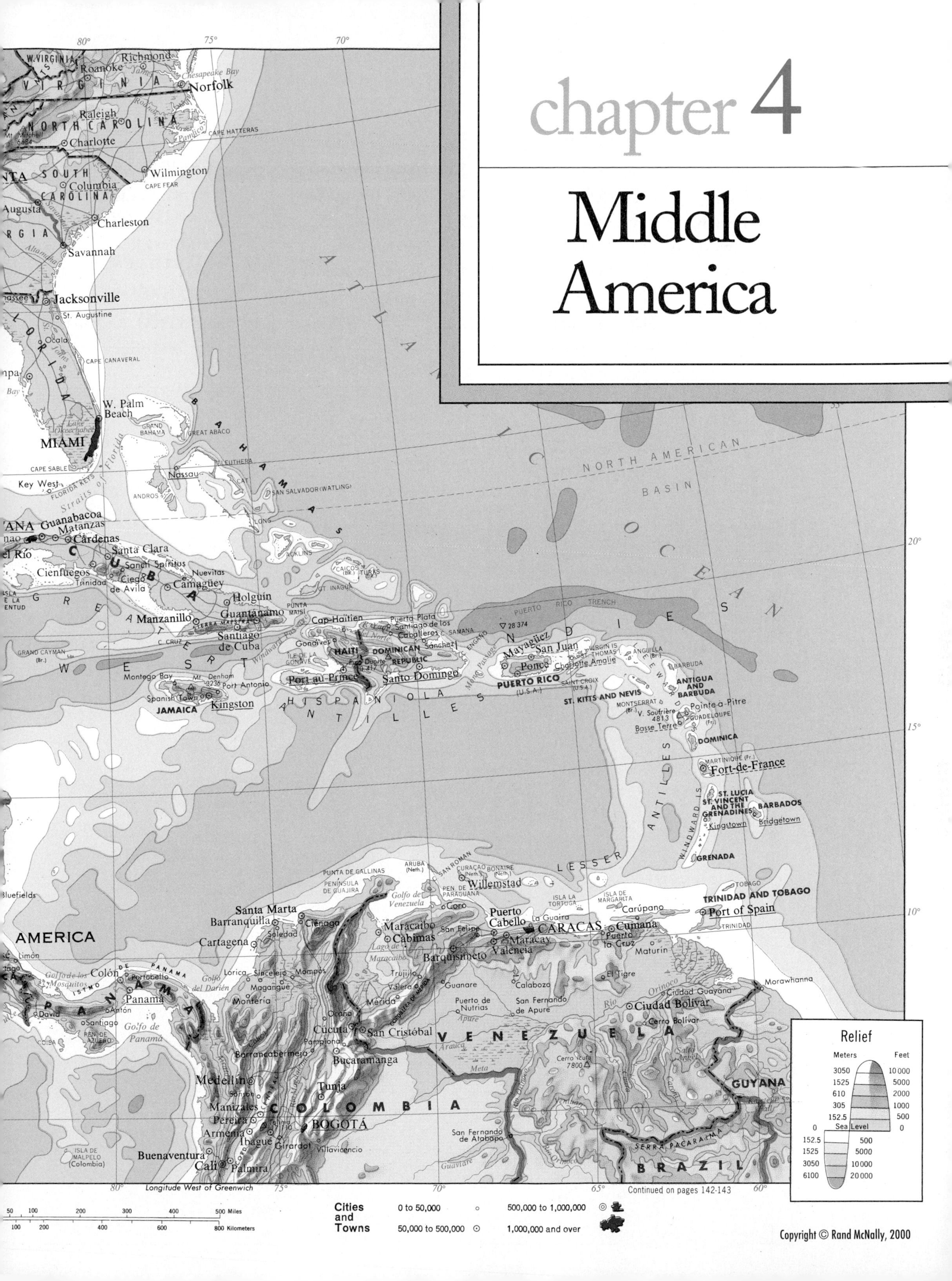

Continued on pages 142-143

Copyright © Rand McNally, 2000

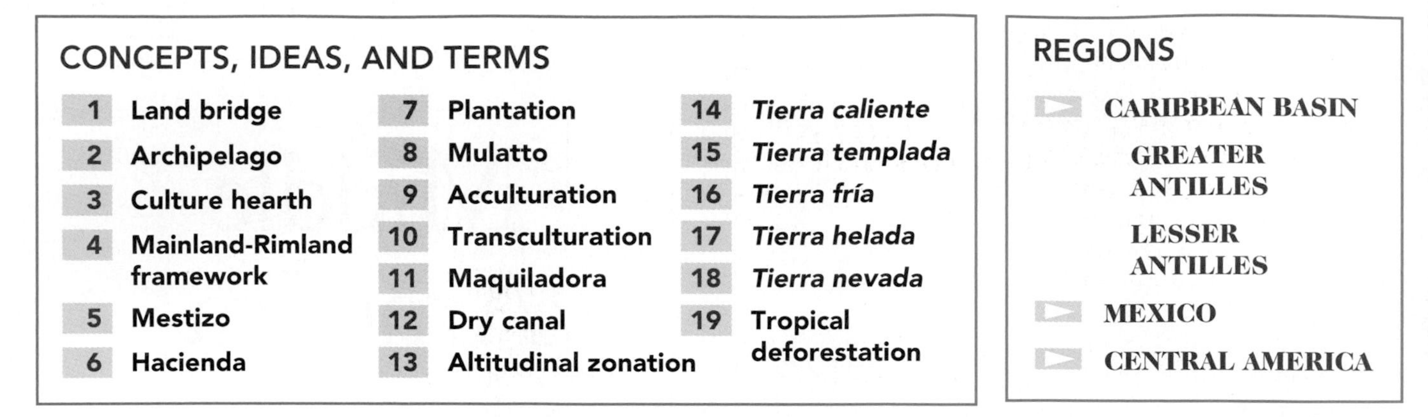

CONCEPTS, IDEAS, AND TERMS

1. Land bridge
2. Archipelago
3. Culture hearth
4. Mainland-Rimland framework
5. Mestizo
6. Hacienda
7. Plantation
8. Mulatto
9. Acculturation
10. Transculturation
11. Maquiladora
12. Dry canal
13. Altitudinal zonation
14. *Tierra caliente*
15. *Tierra templada*
16. *Tierra fría*
17. *Tierra helada*
18. *Tierra nevada*
19. Tropical deforestation

REGIONS

- CARIBBEAN BASIN
  - GREATER ANTILLES
  - LESSER ANTILLES
- MEXICO
- CENTRAL AMERICA

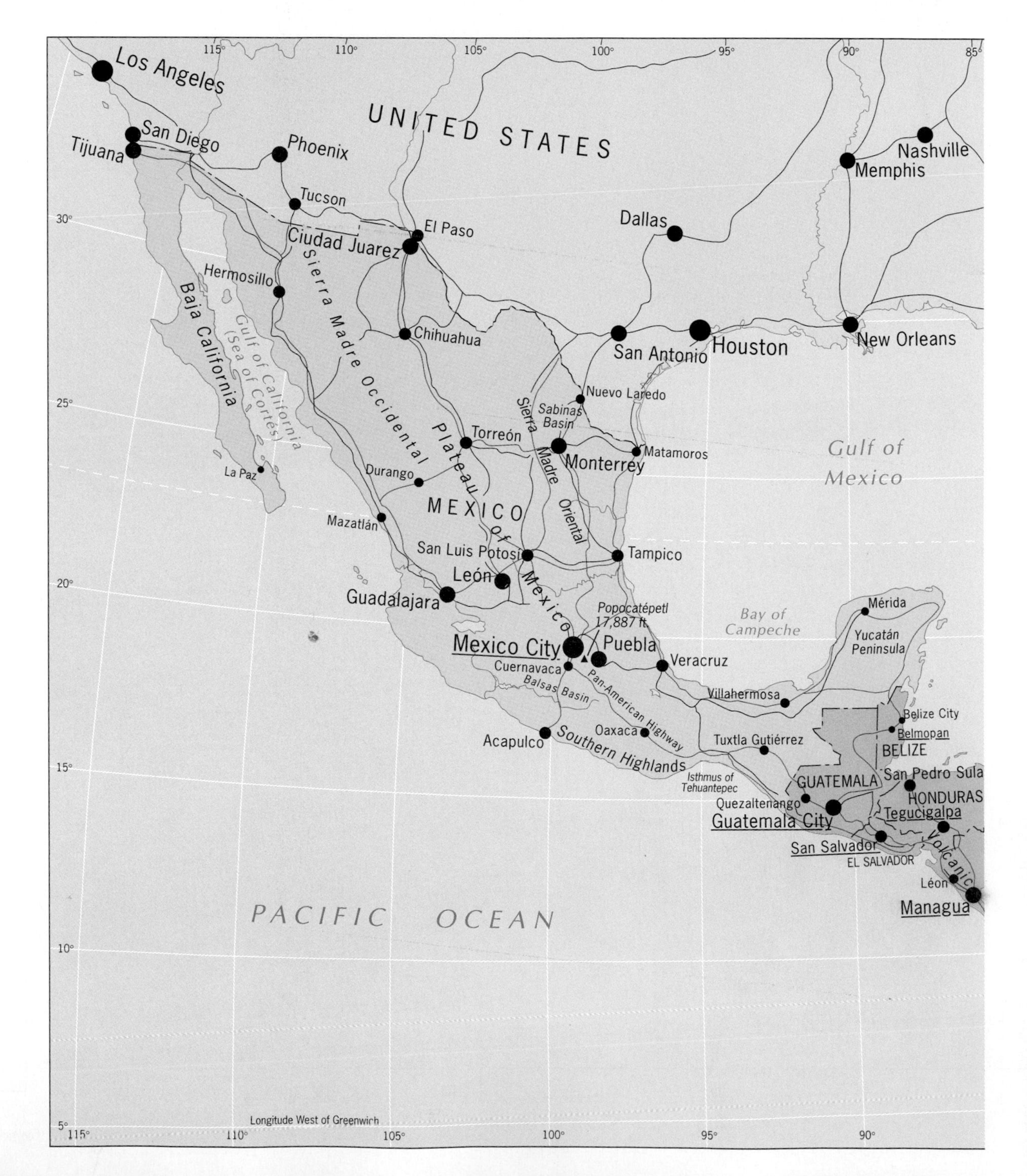

Middle America is a realm of vivid contrasts, turbulent history, continuing political turmoil, and an uncertain future. The realm's diversity, visible in Figure 4-1, encompasses all the lands and islands between the United States to the north and South America to the south. Its regions are mapped in Figure 4-2 and include (1) the substantial landmass of *Mexico*; (2) the narrowing strip of land to its southeast that constitutes *Central America*; and (3) the many large and small islands—respectively known as the *Greater Antilles* and *Lesser Antilles*—of the Caribbean Sea to the east.

Middle America is a realm of soaring volcanoes and forested plains, of mountainous islands and flat coral cays. Moist tropical winds sweep in from the sea, watering windward (wind-facing) coasts while leaving leeward (wind-protected) areas dry. Soils vary from fertile volcanic to desert barren. Spectacular scenery abounds, and tourism is one of the realm's leading industries.

FIGURE 4-2

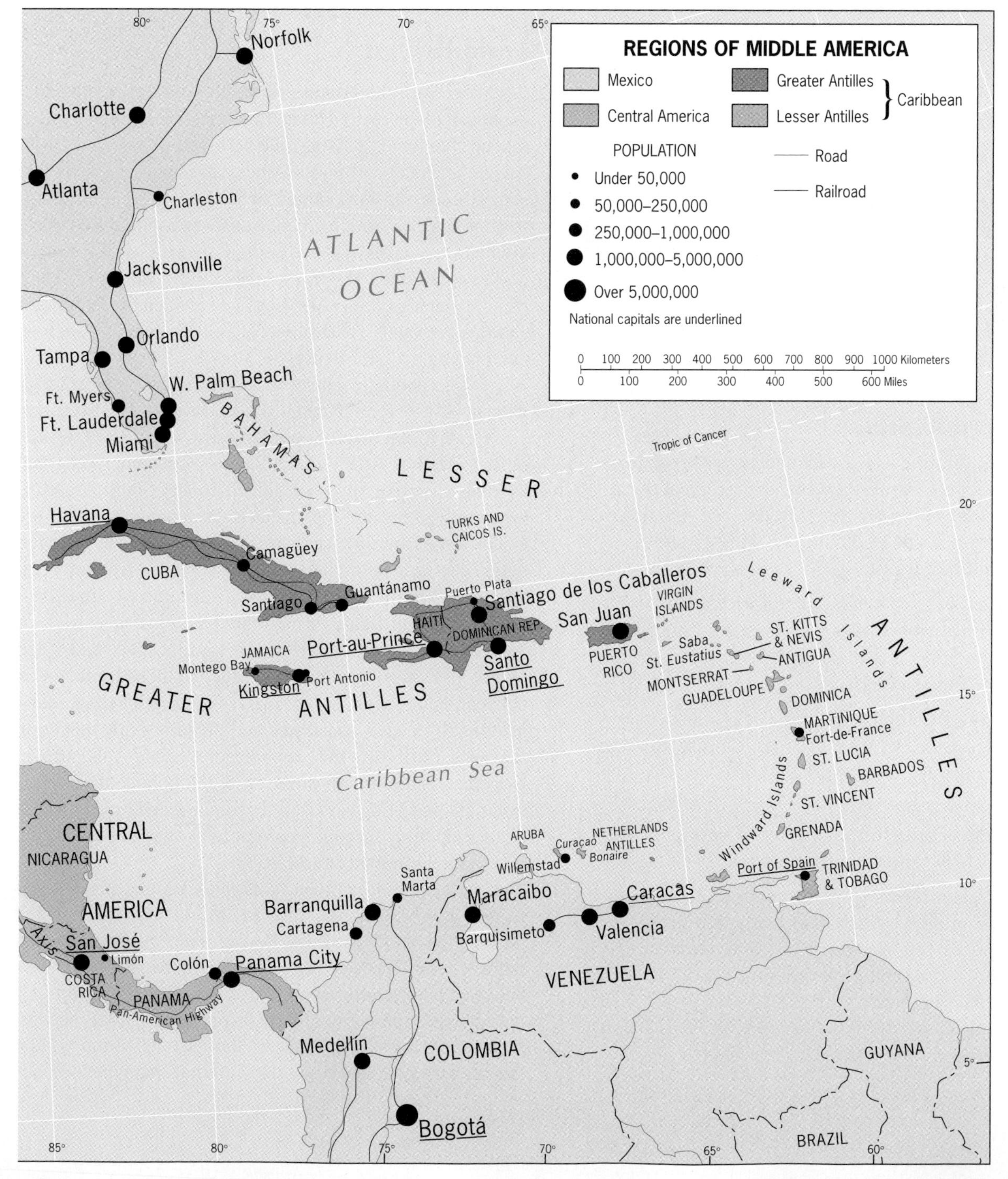

DEFINING THE REALM

Is Middle America a discrete geographic realm? Some geographers combine Middle and South America into a realm they call "Latin" America, citing the dominant Iberian (Spanish and Portuguese) heritage and the prevalence of Roman Catholicism. But these criteria apply more strongly to South America. In Middle America, large populations exhibit African and Asian as well as European ancestries. And nowhere in South America has native, Amerindian culture contributed to modern civilization as strongly as it has in Mexico. The Caribbean Basin is a patchwork of independent states, territories in political transition, and residual colonial dependencies. The Dominican Republic speaks Spanish, adjacent Haiti uses French, Dutch is spoken in Curaçao and on neighboring islands, while English is spoken in Jamaica. Middle America thus gives vivid definition to concepts of cultural-geographical pluralism.

◆ Major Geographic Qualities of Middle America

1. Middle America is a fragmented realm that consists of all the mainland countries from Mexico to Panama and all the islands of the Caribbean Basin to the east.
2. Middle America's mainland constitutes a crucial barrier between Atlantic and Pacific waters. In physiographic terms, this is a land bridge that connects the continental landmasses of North and South America.
3. Middle America is a realm of intense cultural and political fragmentation. The political geography defies unification efforts, but countries and regions are beginning to work together to solve mutual problems.
4. Middle America's cultural geography is complex. African influences dominate the Caribbean, whereas Spanish and Amerindian traditions survive on the mainland.
5. The realm contains the Americas' least-developed territories. New economic opportunities may help alleviate Middle America's endemic poverty.
6. In terms of area, population, and economic potential, Mexico dominates the realm.
7. Mexico is reforming its economy and has experienced major industrial growth. Its hopes for continuing this development are tied to overcoming its remaining economic problems and to expanding trade with the United States and Canada under the North American Free Trade Agreement (NAFTA).

PHYSIOGRAPHY

Land Bridge

Compared to the landmass of South America, the Middle American realm is divided and fragmented. Even its funnel-shaped mainland, a 3800-mile (6000-km) connection between the North and South American realms, narrows to a slim 40-mile (65-km) ribbon of land in Panama. Here this strip of land, or *isthmus*, bends eastward so that Panama's orientation is east-west (with the Panama Canal cutting it northwest-southeast). On the map, continental (mainland) Middle America looks like a bridge between the Americas, and this is exactly what physical geographers call such an isthmian link: a **land bridge**. 1

In their geological history, North and South America were not always connected this way. More than 200 million years ago, Mexico and South America were joined to each other as well as Africa in the Pangaea supercontinent. But as Pangaea broke apart, so did North and South America. Eventually, beneath the ocean that separated them, the Cocos Plate from the southwest collided with and plunged below the Caribbean Plate to the east (Fig. I-4), thereby elevating the Middle American land bridge into its current position. This uplift also formed the wide Plateau of Mexico as volcanoes and fissures in the Earth's crust disgorged enormous flows of molten rock. Much of the volcanism was concentrated in highland corridors that paralleled mainland Middle America's coastlines and steadily built mountain ranges that still form the backbone of the entire land bridge. Nonetheless, these will prove to be temporary surface features: all land bridges exist only for a limited time and then disappear, either through geologic processes or because sea levels rise and submerge them.

If you examine a globe, you can see other present and former land bridges: the Sinai Peninsula between Asia and Africa, the (now-broken) Bering land bridge between northeasternmost Asia and Alaska, and the shallow waters between New Guinea and Australia. Such land bridges have played crucial roles in the dispersal of animals and humans across the planet. But even though continental Middle America forms a land bridge, its internal fragmentation has

Middle America and Central America

Middle America, as we define it, includes all the mainland and island countries and territories that lie between the United States and the continent of South America. Sometimes the term *Central America* is used to identify the same realm, but Central America is actually a region within Middle America (Fig. 4-2). Central America comprises the republics that occupy the strip of mainland between Mexico and Panama: Guatemala, Belize, Honduras, El Salvador, Nicaragua, and Costa Rica. Panama itself is regarded here as belonging to Central America as well. However, it should be noted that many Central Americans still do not consider Panama to be part of their region because that country was, for most of its history, a part of South America's Colombia.

always inhibited movement. Mountain ranges, volcanoes, swampy coastlands, and dense rainforests make contact and interaction difficult, especially where they all combine in most of Central America (see box titled "Middle America and Central America").

Island Chains

As shown in Figure 4-2, the approximately 7000 islands of the Caribbean Sea stretch in a lengthy arc from Cuba and the Bahamas east and south to Trinidad, with numerous outliers outside (such as Barbados) and inside (e.g., the Cayman Islands) the main chain. The four large islands—Cuba, Hispaniola (containing Haiti and the Dominican Republic), Puerto Rico, and Jamaica—are called the *Greater Antilles*. All the remaining smaller islands are called the *Lesser Antilles*.

Once again, the map hints at the physiography. The entire Antillean **archipelago** (island chain) consists of the 2 crests and tops of mountain chains that rise from the floor of

Middle America is the world's most dangerous realm in terms of natural hazards. Hurricanes, volcanic eruptions, earthquakes, and landslides endanger virtually every corner of this realm. Modest homes in the town of San Pedro de Macoris in the Dominican Republic were devastated by the 1998 passage of Hurricane Georges, which killed 70 people in this country and 48 more elsewhere . . . In central Mexico, as seen from the nearby village of Cholula, the Popocatépetl Volcano spews ash in a spectacular eruption during Christmas week of 2000. This 17,945-foot (5470-m) mountain, only 45 miles (72 km) southeast of Mexico City, poses a constant threat to this metropolis, now the world's largest. "El Popo" had been relatively quiescent for centuries before this current episode of activity began in 1994, underscoring that "dormant" is a volcanic category subject to instant change.

FROM THE FIELD NOTES

"We spent Monday and Tuesday upriver at Lamanai, a huge, still mostly overgrown Maya site deep in the forest of Belize. On Wednesday we drove from Belize City to Altun Ha, which represents a very different picture. Settled around 200 BC, Altun Ha flourished as a Classic Period center between AD 300 and 900, when it was a thriving trade and redistribution center for the Caribbean merchant canoe traffic and served as an entrepôt for the interior land trails, some of them leading all the way to Teotihuacán. Altun Ha has an area of about 2.5 square miles (6+ sq km), with the main structures, one of which is shown here, arranged around two plazas at its core. I climbed to the top of this one to get a perspective, and sat down to have my sandwich lunch, imagining what this place must have looked like as a bustling trade and ceremonial center when the Roman Empire still thrived, but a more urgent matter intruded. A five-inch tarantula emerged from a wide crack in the sun-baked platform, and I noticed it only when it was about two feet away, apparently attracted by the crumbs and a small piece of salami. A somewhat hurried departure put an end to my historical-geographical ruminations."

the Caribbean, the result of the same kind of plate action that raised the Middle American land bridge to the west. Some of these crests are relatively stable, but elsewhere they contain active volcanoes. One of the realm's worst disasters took place exactly a century ago on the eastern Caribbean island of Martinique in 1902, when volcanic Mount Pelée erupted with little warning and killed more than 30,000 people living at its base in the port town of St. Pierre. Much more recently, a dormant volcano roared to life in the mid-1990s and forced the temporary abandonment of much of the island-country of Montserrat. And almost everywhere in this realm, earthquakes are an ever-present danger: in the islands as well as on the mainland, the crust is unstable where the Caribbean, Cocos, and North American Plates come together (Fig. I-4). Add the seasonal risk of hurricanes spawned and nurtured by the realm's abundant tropical waters (such as Hurricane Mitch, which in 1998 struck Central America with catastrophic consequences), and the Middle American environment clearly ranks among the world's most hazardous.

LEGACY OF MESOAMERICA

Mainland Middle America was the scene of the emergence of major ancient civilizations. Here lay one of the world's true 3 **culture hearths** (see Fig. 6-3), a source area from which new ideas radiated and whose population could expand and make significant material and intellectual progress. Agricultural specialization, urbanization, and transport networks developed, and writing, science, art, architecture, religion, and other spheres of achievement saw major advances. Anthropologists refer to the Middle American culture hearth as *Mesoamerica*, which extended southeast from the vicinity of present-day Mexico City to central Nicaragua. Its development is especially remarkable because it occurred in very different geographic environments, each presenting obstacles that had to be overcome in order to unify and integrate large areas. First, in the low-lying tropical plains of what is now northern Guatemala, Belize, and Mexico's Yucatán Peninsula, and perhaps simultaneously in Guatemala's highlands to the south, the Maya civilization arose more than 3000 years ago, building cities such as El Mirador between 1000 and 500 BC. Later, far to the northwest on the high plateau in central Mexico, the Aztecs founded a major civilization centered on the largest city ever to exist in pre-Columbian times. Other states with complex cultures arose elsewhere, making Mesoamerica one of the world's crucibles of statecraft and knowledge.

The Lowland Maya

As we note later, the Maya civilization is the only one on the world culture map that arose substantially in the lowland tropics. Its great cities, with stone pyramids, massive temples, and other elaborate buildings, still yield archeological information today. Maya culture over four millennia experienced successive periods of glory and decline, reaching its zenith—the Classic Period—from the third to the tenth centuries AD.

The Maya civilization, forged by a series of city-states, unified an area larger than any of the modern Middle American countries except Mexico. Its population probably totaled between 2 and 3 million; the Maya language—the local *lingua franca*—and some of its related languages are still used in the area to this day. The Maya city-states were marked by hereditary dynastic rule that functioned alongside a powerful, complex religious hierarchy, and the great cities with their imposing structures that today lie in ruins were primarily ceremonial centers (see photo, p. 202). We also know that Maya culture produced skilled artists, writers, mathematicians, and astronomers, but it had a practical side as well. In agriculture and trade, these people achieved a great deal. They grew cotton, created a rudimentary textile industry, and even exported finished cotton cloth by seagoing canoes to other parts of Middle America in return for, among other things, the cacao (raw material of chocolate) they prized so highly that these beans were used as money; obsidian for tools and weapons; and feathers from the quetzal bird for fashion.

The Highland Aztecs

In what is today the intermontane highland zone of Mexico, significant cultural developments were also taking place. Here, just north of present-day Mexico City, lay Teotihuacán, the first true urban center in the Western Hemisphere, founded around the beginning of the Christian era. For nearly seven centuries it thrived, its population reaching perhaps 150,000, its urban landscape dominated by great pyramids, its influence radiating into Maya culture. Later the Toltecs migrated into this area from the north, conquered and absorbed the local Amerindian peoples, and formed a powerful state. The Toltecs' period of hegemony was relatively brief, lasting less than three centuries after their rise to power around AD 900, but they conquered parts of the Maya domain, absorbed many Mayan innovations and customs, and introduced them on the Mexican plateau. When the Toltec state was in turn penetrated by new elements from the north, it was already in decay, but its technology was readily adopted and developed by the conquering Aztecs.

The Aztec state, the pinnacle of organization and power in pre-Columbian Middle America, is thought to have originated in the early fourteenth century with the founding of a settlement on an island in one of the many lakes that lay in the *Valley of Mexico* (the area surrounding what is now Mexico City). This urban complex, a functioning city as well as a ceremonial center, named Tenochtitlán, was soon to become the greatest city in the Americas and the capital of a large and powerful state. Through alliances with neighboring peoples, the Aztecs gained control over the entire Valley of Mexico, the pivotal geographic feature of Middle America that is still the heart of the modern state of Mexico. This 30-by-40-mile (50-by-65-km) region is, in fact, a mountain-encircled basin positioned about 8000 feet (nearly 2500 m) above sea level. Both elevation and interior location affect its climate; for a tropical area, it is quite dry and very cool. The area's lakes formed a valuable means of internal communication, and the Aztecs built canals to connect several of them. This fostered a busy canoe traffic, bringing agricultural produce to the cities, and tribute paid by their many subjects to the headquarters of the ruling nobility.

Throughout the fourteenth century, the Aztec state strengthened its military position, and by the early fifteenth century the conquest of neighboring peoples had begun. The Aztec drive to expand the empire was directed primarily eastward and southward, and they encountered little resistance. The Aztecs' objective was neither to acquire territory nor to spread their culture, but rather to subjugate peoples and towns in order to extract taxes and tribute. As Aztec influence spread throughout Middle America, the goods streaming back to the Valley of Mexico included gold, cacao, and cotton cloth. The state grew ever richer, its population mushroomed, and its cities expanded (Tenochtitlán's population peaked in excess of 100,000). These cities were no longer just ceremonial centers but true cities with a variety of economic and political functions as well as large populations, including specialized, skilled labor forces.

The Aztecs produced a wide range of impressive accomplishments, although they were better borrowers and refiners than they were innovators. They practiced irrigation by diverting water from streams to farmlands, and they built elaborate walls to terrace slopes where soil erosion threatened. Indeed, when it comes to measuring the legacy of Mesoamerica's Amerindians to their successors (and to humankind), the greatest contributions surely came from the agricultural sphere. Corn (maize), the sweet potato, various kinds of beans, the tomato, squash, cacao tree, and tobacco are just a few of the crops that grew in Mesoamerica when the Europeans first made contact.

COLLISION OF CULTURES

We in the Western world all too often believe that history began when the Europeans arrived in some area of the world and that the Europeans brought such superior power to the other continents that whatever existed there previously had little significance. Middle America confirms this misperception: the great, feared Aztec state apparently fell before a relatively small band of Spanish invaders in an incredibly short period of time (1519–1521). But let us not lose sight of a few facts. At first the Aztecs believed the Spaniards were "White Gods" whose arrival had been predicted by Aztec prophecy. Having entered the Aztecs' territory, the earliest Spanish visitors could see that great wealth had been amassed in Aztec cities. Hernán Cortés, for all his 508 soldiers, did not singlehandedly overthrow this powerful empire: he ignited a rebellion by Amerindian peoples who had fallen under Aztec domination and had seen their relatives carried off for human sacrifice to Aztec gods. Led by Cortés with his horses and guns, these indigenous peoples rose against their Aztec oppressors and joined the band of Spaniards headed toward Tenochtitlán. There, thousands of Amerindians died in combat against the Aztec warriors; without their valor Tenochtitlán would not have fallen to the Spaniards so easily.

Spain's defeat of Middle America's dominant indigenous state opened the door to Spanish penetration and supremacy. Nor could any of Middle America's other local states withstand the European invasion. Only some communities protected by geography—deep in the forests or high in the mountains—survived the onslaught for a time. But eventually they too fell victim to the soldiers, the priests, and the Spaniards' diseases.

Effects of the Conquest

In Middle America, the confrontation between Hispanic and native cultures spelled disaster for the Amerindians: a drastic decline in population, rapid deforestation, pressure on vegetation from grazing animals, substitution of Spanish wheat for maize (corn) on cropland, and the concentration of Amerindians into newly built towns. The quick defeat of the Aztec state was followed by a catastrophic decline in population. Middle America had 15 to 25 million native inhabitants when the Spaniards arrived (estimates vary); only a century later, just 2.5 million survived.

The Spaniards were ruthless colonizers, but not more so than other European powers that subjugated other cultures. True, the Spaniards first enslaved the Amerindians and were determined to destroy the strength of indigenous society. But biology accomplished what ruthlessness could not have achieved in so short a time. No native population in the Americas had immunity to the diseases the Spaniards brought: smallpox, typhoid fever, measles, influenza, and mumps. Nor did they have any protection against the tropical diseases that the Europeans introduced through their African slaves, which took enormous tolls on human life in the hot, humid lowlands of Middle America.

Middle America's cultural landscape—its great cities, its terraced fields, its dispersed aboriginal villages—was thus drastically modified. The Amerindian cities withered. Although they destroyed Tenochtitlán, the Spaniards recognized the attributes of its site and situation and rebuilt it as their mainland headquarters. Whereas the Amerindians had used stone almost exclusively as their building material, the Spaniards employed great quantities of wood and used charcoal for heating, cooking, and smelting metal. The onslaught on the forests was immediate, and expanding rings of deforestation quickly formed around the Spanish towns. The Spaniards also introduced livestock, whose numbers multiplied rapidly and made increasing demands not only on the existing grasslands but on the cultivated crops as well. Cattle and sheep quickly became sources of wealth, and the owners of the herds benefited. People and livestock now competed for available food. This competition required the opening up of vast areas of marginal land in higher and

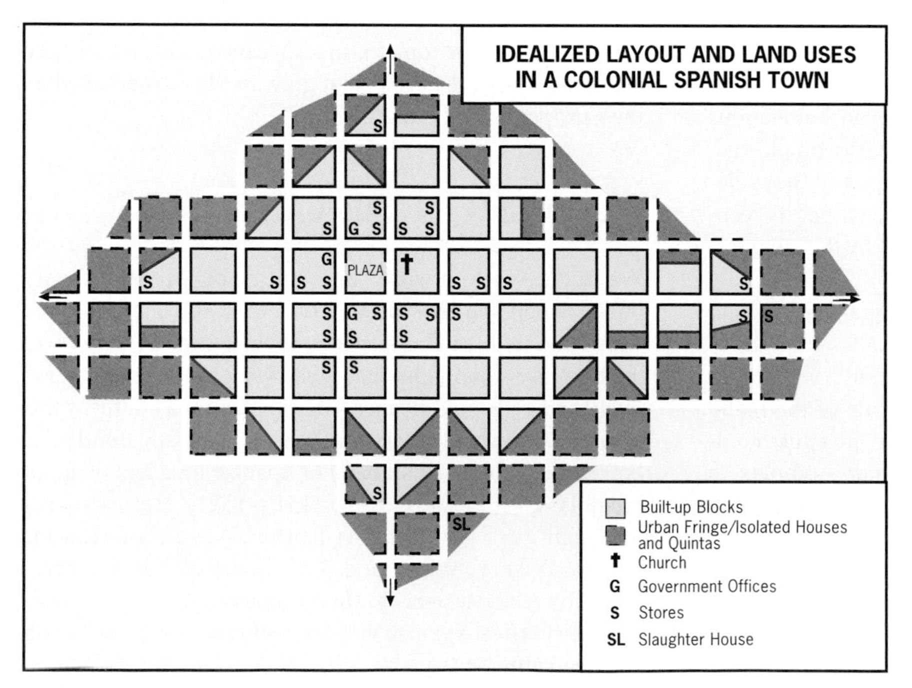

FIGURE 4-3

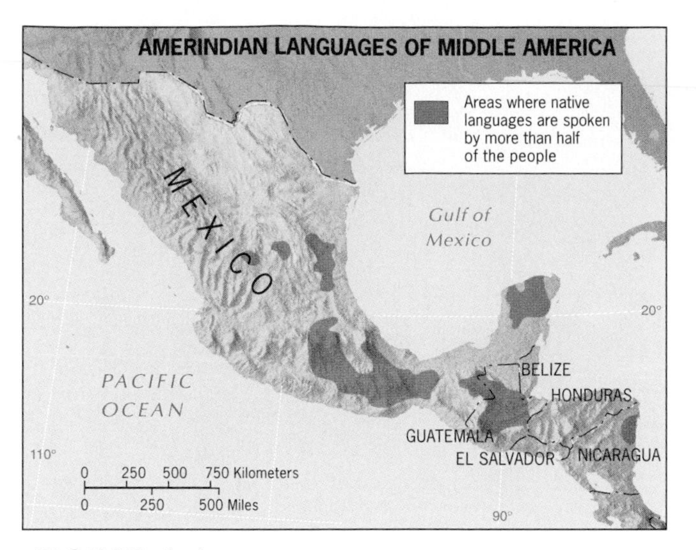

FIGURE 4-4

drier locations, thereby contributing to a major disruption of the region's food-production balance. Moreover, the Spaniards introduced their own crops (notably wheat) and farming equipment, and soon large fields of wheat began to encroach upon the small plots of corn that the natives cultivated. The Spaniards also reconfigured the irrigation systems. To satisfy their need for water for their fields and hydropower for their mills, they took over and modified regional drainage and irrigation networks. As a result, the Amerindians' fields became insufficiently watered, thereby diminishing their food supply even further.

The Spaniards' most far-reaching cultural changes derived from their traditions as town dwellers. To facilitate domination, the Amerindians were moved off their land into nucleated villages and towns the Spaniards established and laid out. In these settlements, the Spaniards could exercise the kind of rule and administration to which they were accustomed (Fig. 4-3). The internal focus of each Spanish town was the central *plaza* or market square, around which both the local church and government buildings were located. The surrounding street pattern was deliberately laid out in *gridiron* form, so that any insurrections by the resettled Amerindians could be contained by having a small military force seal off the affected blocks and then root out the troublemakers. Each town was located near what was thought to be good agricultural land (which was often not so good), so that the Amerindians could go out each day and work in the fields. Packed tightly into these towns and villages, they came face to face with Spanish culture. Here they learned the Europeans' Roman Catholic religion and Spanish language, and paid their taxes and tribute to a new master. Nonetheless, the nucleated indigenous village survived, although its administration was also taken over by the Spaniards and later by their postcolonial successors. Today, the village is still a key feature of remote Amerindian areas in southeastern Mexico and inner Guatemala, where native languages still prevail over Spanish (Fig. 4-4).

Once the indigenous population was conquered and resettled, the Spaniards were able to pursue another primary goal in their New World territory: the exploitation of its wealth for their own benefit. Lucrative trade, commercial agriculture, livestock ranching, and especially mining were avenues to affluence. Gold-mining held the greatest initial promise, and the Spaniards simply took over the existing Amerindian miner workforce. But soon accessible gold deposits diminished, leading Spanish prospectors to begin searching for other valuable minerals. They were quickly successful, finding enormously profitable silver and copper deposits, particularly in a wide frontier zone north of the Valley of Mexico (see Fig. 4-9). Developing these resources produced a host of changes in this part of Middle America because mining towns required labor forces, extractive equipment, food supplies, and the transportation connections to handle the intensifying flows of people, goods, and services. Thus was born a new urban system, one that integrated and organized the Spanish domain in Middle America and also extended effective economic control over some far-flung parts of New Spain. Mining truly became the mainstay of colonial Middle America.

Wherever the Spaniards ruled—in towns, farms, mines, indigenous villages—the Roman Catholic Church was the supreme cultural force transforming Amerindian society. Jesuits and soldiers pushed the Spanish frontier northward and westward, the church acquiescing in the state's often brutal occupation of Amerindian domains. In the wake of conquest, it was the church that controlled, pacified, organized, and acculturated the Amerindian peoples.

MAINLAND AND RIMLAND

In Middle America outside Mexico, only Panama, with its twin attractions of interoceanic transit and gold deposits, became an early focus of Spanish activity (the Spaniards founded Panama City in 1519). Apart from using the approximate route of the modern Panama Canal as an Atlantic-Pacific overland link, the Spaniards' main interest lay on the Pacific side of the isthmus, a base from which Spanish influence steadily extended northwestward into Central America. The leading center of Spanish activity, however, remained in what is today central and southern Mexico.

The major arena of international competition in Middle America lay not on the Pacific side but on the islands and coasts of the Caribbean Sea. Here the British gained a foothold on the mainland, controlling a narrow, low-lying coastal strip that extended southeast from Yucatán to what is now Costa Rica. As the colonial-era map (Fig. 4-5) shows, in the Caribbean the Spaniards faced the British, French, and Dutch, all interested in the lucrative sugar

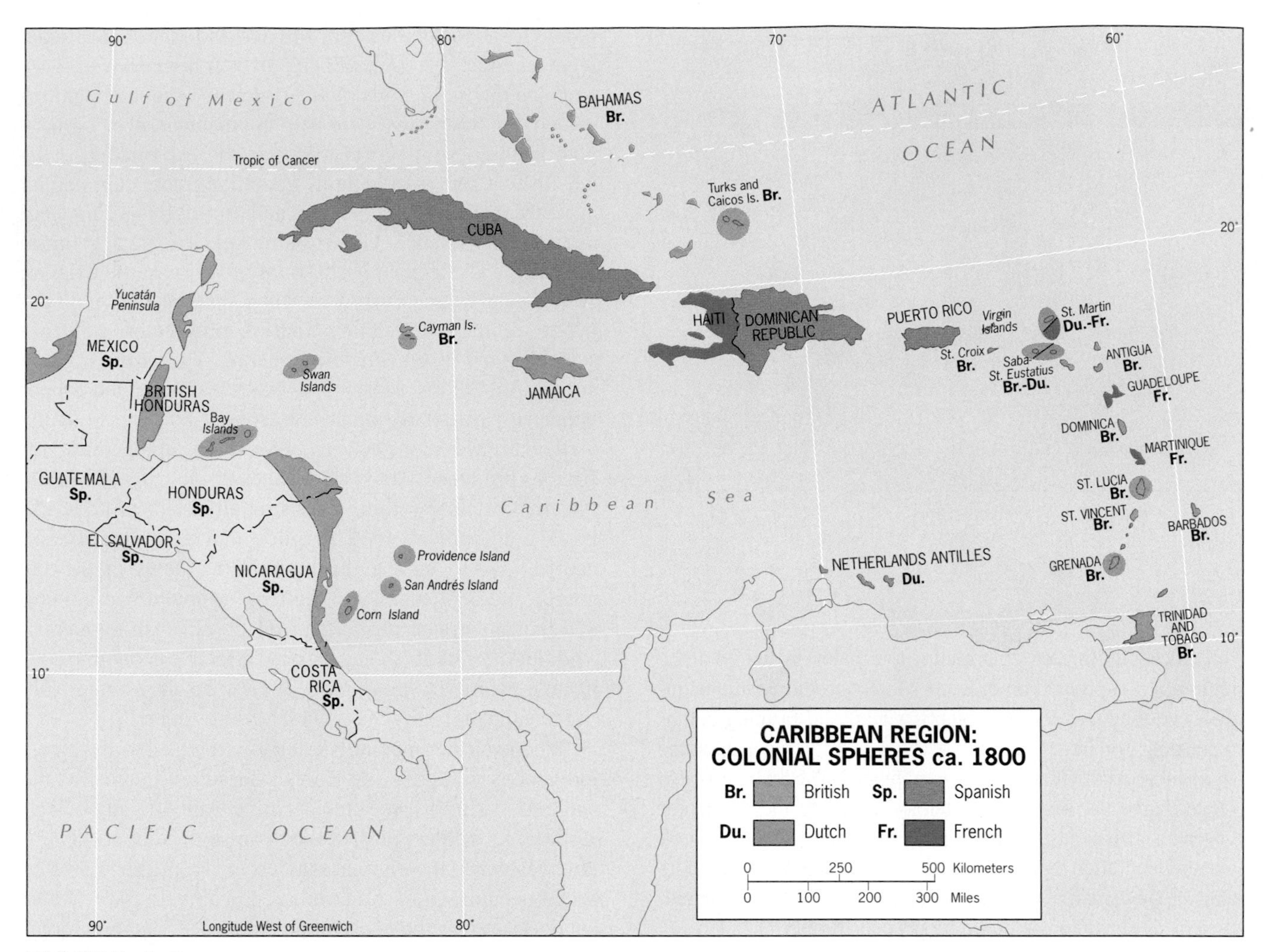

FIGURE 4-5

trade, all searching for instant wealth, and all seeking to expand their empires.

Later, after centuries of European colonial rivalry in the Caribbean Basin, the United States entered the picture and made its influence felt in the coastal areas of the mainland, not through colonial conquest but through the introduction of widespread, large-scale, banana plantation agriculture. The effects of these plantations were as far-reaching as the impact of colonialism on the Caribbean islands. The economic geography of the Caribbean coastal zone was transformed, as hitherto unused alluvial soils in the many riverine lowlands were planted with thousands of acres of banana trees.

Because the diseases the Europeans had brought to the New World had been most rampant in these hot and humid areas, the Amerindian population that survived was too small to provide a sufficient labor force. The comparatively small indigenous populations on the Caribbean islands succumbed to the European's harsh treatment and imported diseases even more rapidly than the mainland Amerindians had. In little more than half a century they were wiped out, creating a labor shortage and generating the trans-Atlantic slave trade from Africa that transformed the Caribbean's demography (see map, p. 349). When labor was needed on continental Middle America's Caribbean coast, tens of thousands of black workers were brought to the mainland from Jamaica and other islands. (Many more came later when the Panama Canal was dug between 1904 and 1914.) Thus the demographic complexion was altered here as well.

These contrasts between the Middle American highlands and the coastal areas and Caribbean islands were conceptualized by John Augelli into the **Mainland-Rimland framework** (Fig. 4-6). Augelli recognized (1) a Euro-Amerindian **Mainland**, which consisted of continental Middle America from Mexico to Panama, with the exception of the Caribbean coastal belt from mid-Yucatán southeastward; and (2) a Euro-African **Rimland**, which included this coastal zone as well as the islands of the

Caribbean. The terms *Euro-Amerindian* and *Euro-African* underscore the cultural heritage of each region. On the Mainland, European (Spanish) and Amerindian influences are paramount, and also include **mestizo** sectors where the two ancestries mixed. In the Rimland, the heritage is European and African.

5

As Figure 4-6 shows, the Mainland is subdivided into several areas based on the strength of the Amerindian legacy. In southern Mexico and Guatemala, Amerindian influences are prominent; in northern Mexico and parts of Costa Rica, those influences are limited; between these areas lie the mestizo sectors, with moderate Amerindian influence. The Rimland is also subdivided. The most obvious division is between the mainland-coastal plantation zone and the islands. But the islands themselves can be classified according to their cultural heritage: a group of islands with Spanish influence (Cuba, Puerto Rico, and the Dominican Republic on old Hispaniola) and another group with other European influences, including the former British West Indies, the various French islands, and the Netherlands Antilles located just off South America on the Caribbean's central southern coast.

Supplementing these contrasts of human habitat are regional differences in outlook and orientation. The Rimland was an area of sugar and banana plantations, of high accessibility, of seaward exposure, and of maximum cultural contact and mixture. The Mainland, being farther removed from these contacts, was an area of greater isolation. The Rimland was the region of the great *plantation*, and its commercial economy was therefore susceptible to fluctuating world markets and tied to overseas investment capital. The Mainland was the region of the *hacienda*, which was more self-sufficient and considerably less dependent on external markets.

The Hacienda

This contrast between plantation and hacienda land tenure in itself constitutes strong evidence for the Rimland-Mainland division. The hacienda was a Spanish institution, but the modern plantation, Augelli argued, was the concept of Europeans of more northerly origin. In the **hacienda**, Spanish landowners possessed a domain whose productivity they might never push to its limits: the very possession of such a

6

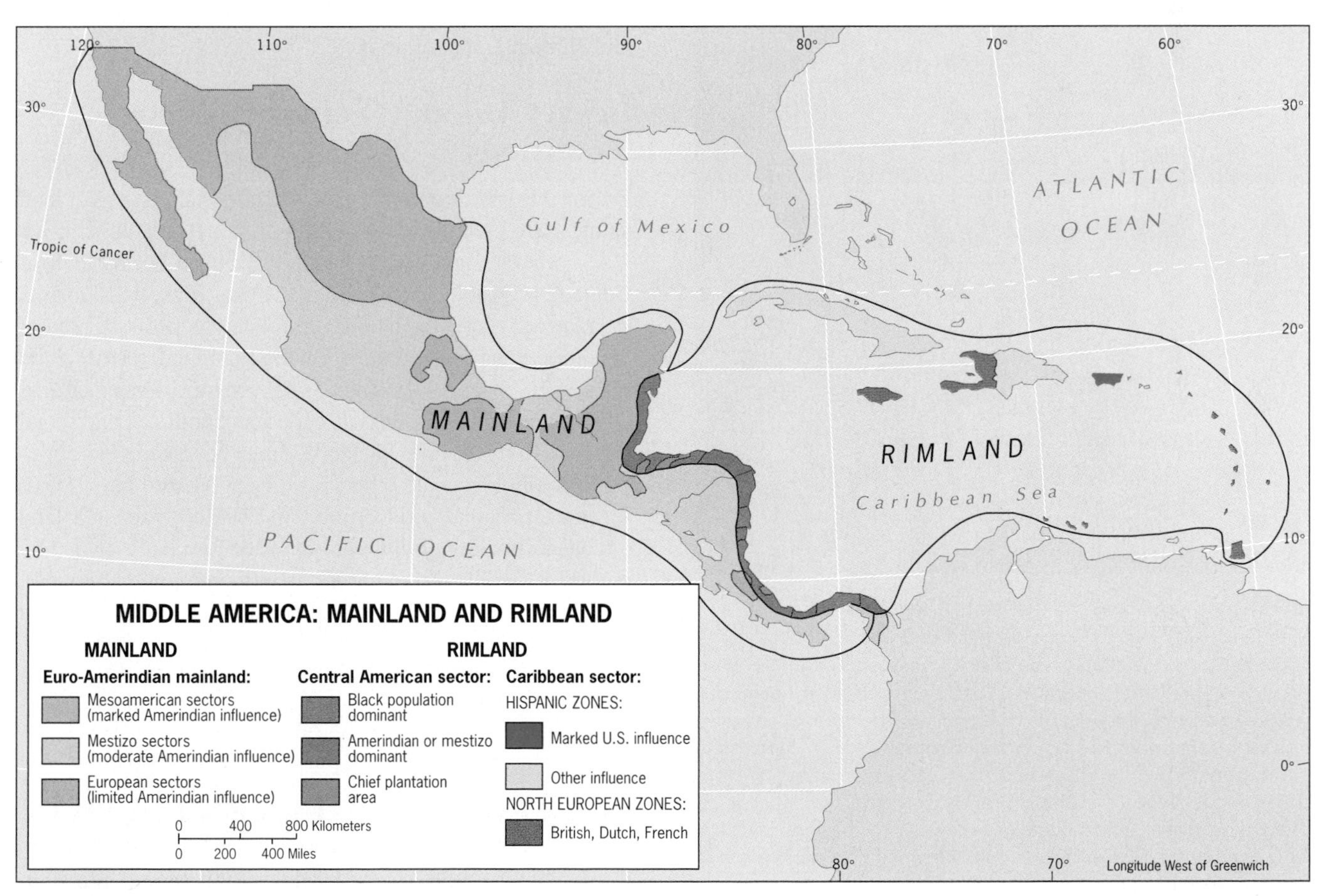

FIGURE 4-6

vast estate brought with it social prestige and a comfortable lifestyle. Native workers lived on the land—which may once have been *their* land—and had plots where they could grow their own subsistence crops. Traditions survived in the Amerindian villages incorporated into the early haciendas, in the methods of farming, and in the means of transporting produce to markets. All this is written as though it is mostly in the past, but the legacy of the hacienda, with its inefficient use of land and labor, is still visible throughout mainland Middle America as well as parts of South America (see photo pair to the left).

Contrasting land uses in the Middle American Rimland and Mainland give rise to some very different rural cultural landscapes. Huge stretches of the realm's best land continue to be controlled by (often absentee) landowners whose haciendas yield export or luxury crops, or foreign corporations that raise fruits for transport and sale on their home markets. The banana plantation shown here lies near the Caribbean coast of Honduras and is owned by United Brands Company. The vast fields of banana plants stand in strong contrast to the lone peasant who ekes out a bare subsistence from small cultivable plots of land, often in high-relief countryside where grazing some goats or other livestock is the only way to use most of the land.

The Plantation

7

The **plantation**, in contrast, was conceived as something entirely different. In their book, *Middle America: Its Lands and Peoples*, Robert West and John Augelli list five characteristics of Middle American plantations that illustrate the differences between hacienda and plantation: (1) plantations are located in the humid tropical coastal lowlands of the realm; (2) plantations produce for export almost exclusively—usually a single crop; (3) capital and skills are often imported so that foreign ownership and an outflow of profits occur; (4) labor is seasonal—needed in large numbers during the harvest period but often idle at other times—and such labor has been imported because of the scarcity of Amerindian workers; and (5) with its "factory-in-the-field" operation, the plantation is more efficient in its use of land and labor than the hacienda. The objective was not self-sufficiency but profit, and wealth rather than social prestige is a dominant motive for the plantation's establishment and operation.

Legacy of Land-Tenure Traditions

During the twentieth century, both traditional systems of land tenure have changed a great deal. In the Rimland, massive U.S. corporate investment in the Caribbean coastal belt of Guatemala, Honduras, Nicaragua, Costa Rica, and Panama transformed that area and brought a new concept of plantation agriculture to the region. On the Mainland, the hacienda system has been under increasing pressure from national governments that view it as an economic, political, and social liability. Indeed, some haciendas have been parceled out to small landholders, and others have been pressed into greater specialization and productivity. In Mexico, hacienda land has been placed in *ejidos*, where it is communally owned by groups of families (though privatization is now underway). With its remnants continuing to break down today, both the hacienda and the plantation have for centuries contributed to the different social and economic directions that gave the Mainland and Rimland their respective and distinct regional personalities.

POLITICAL DIFFERENTIATION

Continental Middle America today is fragmented into eight different countries, all but one of which have Hispanic ori-

FROM THE FIELD NOTES
"Evidently the battle was won."

gins. (In fact, the single exception—Belize, the former British Honduras [see photo above]—is being transformed as thousands of Spanish-speaking immigrants, who arrived from strife-torn neighboring countries over the past two decades, are coming to form the majority of the tiny Belizean population.) Largest of them all is Mexico, the giant of Middle America, whose 737,000 square miles (1,910,000 sq km) constitute more than 70 percent of the realm's entire land area (the Caribbean Basin included), and whose 104 million people outnumber those of all the other countries and islands of Middle America combined.

The cultural variety in Caribbean Middle America is much greater. Here Cuba dominates: its area is almost as large as that of all the other islands put together, and its population of 11.3 million is well ahead of the next-ranking country (the Dominican Republic, with 8.8 million). As we have seen, however, the Caribbean is hardly an area of exclusive Spanish influence. For example, whereas Cuba has an Iberian heritage, its southern neighbor, Jamaica (population 2.7 million, mostly black), has a legacy of British involvement, and in nearby Haiti (6.6 million, overwhelmingly black), the strongest imprints have been African and French. The crowded island of Hispaniola (population 15.4 million) is shared between Haiti and the Dominican Republic, where Hispanic culture prevails. Spanish influence also predominates in the nearby, remaining island of the Greater Antilles, Puerto Rico (4.0 million), a commonwealth of the United States.

The Lesser Antilles, too, exhibits great cultural diversity. This region includes the (once Danish) U.S. Virgin Islands; French Guadeloupe and Martinique; a group of British-influenced islands, including Barbados, St. Lucia, St. Vincent, and Grenada; and the Dutch St. Maarten (shared with the French), Saba, St. Eustatius, and the A-B-C islands of the Netherlands Antilles—Aruba, Bonaire, and Curaçao—off the northwestern Venezuelan coast. Standing apart from the Antillean arc of islands (off northeastern Venezuela) is Trinidad, another former British dependency that, along with its smaller neighbor Tobago, became a sovereign state in 1962.

Independence movements stirred Middle America at an early stage. In the Greater Antilles region, where Spain held Cuba and Puerto Rico and the British controlled Jamaica, Hispaniola's Afro-Caribbean population mounted a rebellion against their French slave masters. They succeeded, as early as 1804, in creating the Republic of Haiti. On the mainland, revolts against Spanish authority (beginning in 1810) achieved independence for Mexico by 1821, and for the Central American republics by the end of the 1820s.

The colonial map of the Caribbean was a patchwork of holdings by British, Dutch, French, Spanish, and Danish interests (Fig. 4-5). All were competitors in the sugar boom made possible by the forced in-migration of African labor. The United States, concerned over European designs in the realm, in 1823 proclaimed the Monroe Doctrine, which was intended to deter any European power from reasserting its authority in newly independent republics or from further expanding its existing domains. The United States as yet had no holdings in Middle America, but the emerging republics on its doorstep presented obvious economic and strategic opportunities.

By the end of the nineteenth century, the United States had become a major force in Middle America. The Spanish-American War of 1898 made Cuba independent and put Puerto Rico under the U.S. flag; soon afterward, the Americans were in Panama constructing the Panama Canal. Meanwhile, U.S. corporations were riding a boom based not on sugar but on huge banana plantations. Central American republics were colonies of the United States in all but name.

Independence came to the Greater and Lesser Antilles in fits and starts. Afro-Caribbean Jamaica as well as Trinidad and Tobago, where the British had brought a large South Asian population, attained full sovereignty from the United Kingdom in 1962; other British islands (Barbados, St. Vincent, Dominica) became independent later. France, however, retains Martinique and Guadeloupe as overseas *départements* of the French Republic, and the Dutch islands are at various stages of autonomy. The U.S. Virgin Islands, purchased from Denmark in 1917, have territorial status with no prospect of independence.

REGIONS OF THE REALM

THE CARIBBEAN BASIN

The 33 island-nations of Caribbean America are crowded with so many people that, as a region (encompassing the Greater and Lesser Antilles), it is the most densely populated part of the Americas. It is also a place of persistent poverty and, in all too many localities, of unrelenting misery. To some degree, at least, communist Cuba and U.S.-affiliated Puerto Rico constitute exceptions to any such generalization made about Caribbean America, and each is profiled in the boxes titled "Awaiting the Post–Castro Era in Cuba" and "Puerto Rico's Uncertain Future." On most of the other islands, however, life for the average person is difficult, often hopeless, and tragically short (e.g., the AIDS crisis of the Caribbean region is proportionately the world's second worst after that of Subsaharan Africa).

All this is in jarring contrast to the early period of riches based on the sugar trade. Yet that initial wealth was gained while an entire ethnic group, the Amerindians, was being wiped off the Caribbean map and while another, the

Awaiting the Post-Castro Era in Frustrated Cuba

Located only 90 miles (145 km) from southernmost Florida, Cuba, the Caribbean's largest island, fell under the dictatorial rule of Fidel Castro in 1959 (following the ouster of the previous dictator, Fulgencio Batista). A convert to Soviet-style communism soon after taking power, Castro spent the next three decades as the Western Hemisphere's pariah, constantly seeking to export his socialist revolution to other parts of the Americas. But all that changed after 1991, following the collapse of the Soviet Union. Cuba's economy was devastated by the loss of most of its former benefactor's market, imported raw materials, and annual subsidy to the tune of about (U.S.) $4 billion. To alleviate this hardship and to shore up his popular support following riots in 1994, Castro was forced to introduce certain free-market reforms—while stubbornly clinging to the "socialist principles" of his revolution of so long ago. Among these changes were massive layoffs of government workers and encouraging the growth of both private businesses and foreign capital investments.

Foreign joint ventures mushroomed during the mid-1990s, and included European-backed tourist resorts and international banks, Japanese automobile dealerships, Israeli textile plants, and a high-tech telephone system developed by Mexican entrepreneurs. With the drive toward economic integration intensifying across the Americas, countries throughout the hemisphere were eager to build trading ties with Cuba because they recognized that the post-Soviet transformation has eliminated Castro as an ideological threat. These new relationships notwithstanding, the United States continued to refuse to deal with Cuba, and the trade embargo it introduced in 1962 (which has long been ignored throughout Middle America) remained in place. Although the embargo is now weakening, it did inflict considerable damage on the Cuban economy over the past 40 years.

Given the island's favorable geographic position and resource base (Fig. 4-7), these developments suggest that Cuba has the potential to become a Caribbean economic tiger. Led by sugar (still the mainstay of the domestic and export economy), the natural conditions for commercial agriculture rank among the most favorable in the Antillean archipelago. Mineral resources include high-quality nickel, manganese, and chromite, and oil deposits may also exist. Skilled manufacturing has been nurtured since 1960, from which a respectable biotechnology industry emerged that successfully pioneered a vaccine for meningitis. Higher-education facilities are the most advanced in the Caribbean Basin, and the tourist industry revived as vacationers (mainly from Europe and Canada) flocked to new beach resorts on the picturesque coast east of the capital city, Havana.

Despite these recent gains, Cuba was not able to sustain its progress beyond the mid-1990s. A major reason was Castro's decision to backtrack on his modest economic reforms in 1997, presumably because they threatened his regime's political stability; more likely, Castro's rigidity and stubborn belief in his failed brand of communism rendered him incapable of changing with the times. With the resources of his police state deployed to prevent civil unrest, the frustrated Cubans can do little but watch as their deepening economic crisis sinks the country into unprecedented levels of poverty. Among the casualties, as Cuba's economic resources diminish, are the regime's achievements in education and health care; the loyalty of the country's youths, who increasingly turn to drug use and crime as their social and economic opportunities evaporate; the output of the pivotal sugar indus-

MAJOR CITIES OF THE REALM

City	Population* (in millions)	City	Population* (in millions)
Ciudad Juárez, Mexico	1.2	Port-au-Prince, Haiti	1.9
Guadalajara, Mexico	4.0	Puebla, Mexico	2.0
Guatemala City, Guatemala	3.5	San José, Costa Rica	1.0
Havana, Cuba	2.3	San Juan, Puerto Rico	1.4
Managua, Nicaragua	1.0	San Salvador, El Salvador	1.5
Mexico City, Mexico	27.0	Santo Domingo, Dominican Rep.	3.8
Monterrey, Mexico	3.5	Tegucigalpa, Honduras	1.0
Panama City, Panama	1.2	Tijuana, Mexico	1.2

*Based on 2002 estimates.

try, now in decline due to shortages of production inputs; tourism, as visitor services deteriorate along with Cuba's reputation as a vacation destination; and the environment for foreign investors (who are now less welcome and encounter more red tape), which limits new ventures and jeopardizes the gains of the 1990s.

With nearly everything in Cuba on hold, the countdown proceeds toward the day that Castro disappears from power. Looking beyond this closing era, it is not difficult to envision that Cuba will join the other recent bastions of communism where capitalist economic forces now hold sway. Interest in Cuba continues on the part of foreign corporations and other investors, particularly in the United States. Thus it is fair to say that, following the changing of the guard, this island-country is likely to take its place as a major player in the economic geography of Middle America.

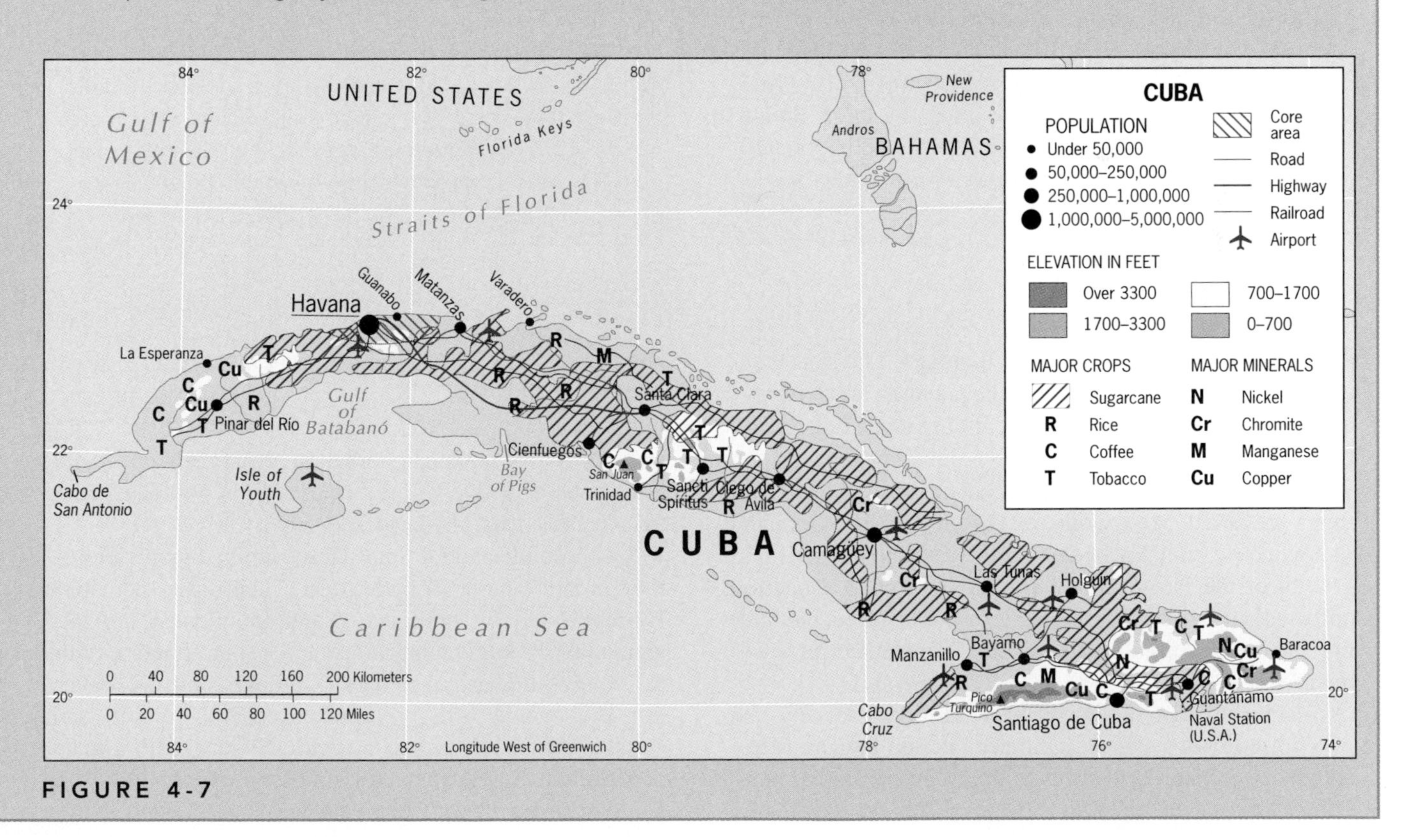

FIGURE 4-7

Puerto Rico's Clouded Future

The largest, most populous U.S. domain in Middle America is Puerto Rico, the easternmost and smallest island of the Greater Antilles chain (Fig. 4-8). This 3400-square-mile (8800-sq km) island, with a population of 4.0 million, is larger than Delaware and more populous than Oregon. It fell to the United States more than a century ago during the Spanish-American War of 1898. Since the Puerto Ricans had been struggling for some time to free themselves from Spanish control, this transfer of power was, in their view, only a change from one colonial power to another. As a result, the first half-century of U.S. administration was difficult, and it was not until 1948 that Puerto Ricans were permitted to elect their own governor.

When the island's voters approved the creation of a Commonwealth in a 1952 referendum, Washington and San Juan, the two seats of government, entered into a complicated arrangement. Puerto Ricans are U.S. citizens but pay no federal taxes on local incomes. The Puerto Rican Federal Relations Act governs the island under the terms of its own constitution and awards it considerable autonomy. Puerto Rico receives large annual subsidies from Washington, totaling about (U.S.) $9 billion per year at the end of the 1990s, which is roughly equivalent to all U.S. foreign aid to the rest of the world combined.

Despite these apparent advantages in the poverty-mired Caribbean, Puerto Rico has not thrived under U.S. administration. Long dependent on a single-crop economy (sugar), the island's industrialization during the 1950s and 1960s was based on its comparatively cheap labor, tax breaks for corporations, political stability, and special access to the U.S. market. As a result, pharmaceuticals, electronic equipment, and apparel top today's list of exports, not sugar or bananas. But this industrialization failed to stem a tide of emigration that carried more than 1 million Puerto Ricans to New York City alone. The same wages that favored corporations kept many Puerto Ricans poor or unemployed. Behind the impressive waterfront skyline of San Juan with its high-rise hotels and tourist attractions lies a landscape of economic malaise. Today, approximately 60 percent of all Puerto Ricans continue to live below the poverty line.

In his recent book, *Puerto Rico: The Trials of the Oldest Colony in the World*, José Trias Monge leaves no doubt as to the consequences of what he considers to be Washington's failed governance of its largest dependency. Puerto Rico has become one of the major points of U.S. entry for the international drug trade; its crime rates are much higher than those in the United States; corruption is rife; and its political debate is bad tempered. The island's newspapers reflect a degree of polarization that should raise warning flags for the future.

During the 1990s, Puerto Ricans were twice asked to vote on their political future in referenda authorized by the U.S. Congress. On each occasion the island's electorate rejected both independence and joining the United States as the 51st State, choosing instead to continue their Commonwealth status. In the first referendum in 1993, the status quo option narrowly defeated Statehood 48.4 to 46.2 percent, and there were subsequent indications that support for joining the U.S. was on the rise. But the 1998 referendum revealed only a minimal shift in opinion as the Commonwealth vote again bested Statehood, this time by 50.2 to 46.5 percent. Those preferring

Africans, was being imported in bondage. The sugar revenues, of course, always went to the planters, not the laborers. Subsequently, the regional economy faced rising competition from other tropical sugar-producing areas; it soon lost its monopoly of the European market, and difficult times prevailed. Meanwhile, just as they did in other parts of the world, the Europeans helped stimulate the rapid growth of the island population by improving sanitation and medical standards. Death rates were lowered, but birth rates remained high and explosive population increases resulted.

When the sugar trade declined, millions were pushed into a life of subsistence, malnutrition, and hunger. Many sought work elsewhere. Tens of thousands of Jamaican laborers went to the plantations of the Rimland coast; large numbers of British West Indians went to England in search of a better life; Puerto Ricans and Dominicans streamed into New York City. Nonetheless, this outflow has failed to stem the tide of regional population growth: almost 40 million people now live on Caribbean islands, a total expected to increase at least another 10 percent by 2010.

Caribbean islanders simply have not had many alternatives in their search for betterment. Their dispersed habitat is fragmented by both water and mountains; the total amount of flat cultivable land is only a small fraction of the far-flung Antillean archipelago. Although some diversification of economic activity has occurred in the region, agriculture remains a significant sector. Sugar is still a major product and continues to head the food export lists of the Dominican Republic, Cuba, and Jamaica; in Haiti, coffee

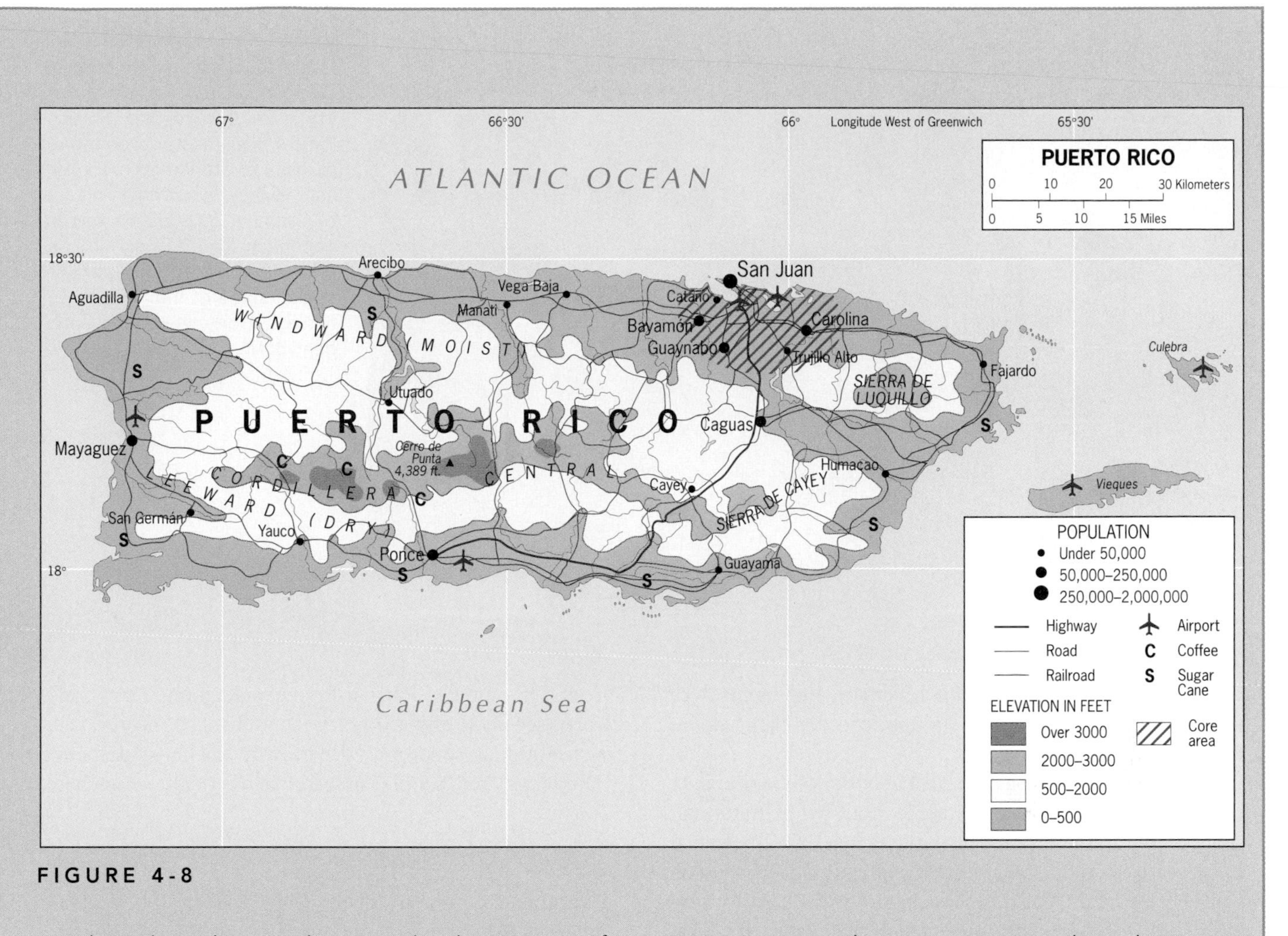

FIGURE 4-8

independence, however, lost ground as the 1993 vote of 4.4 percent shrank to 2.5 percent five years later.

The outcome of the latest referendum will not silence those who, like Trias Monge, argue that any status short of outright sovereignty will serve only to perpetuate Puerto Rico's standing as an American dependency—branding the United States as a colonial power no matter how lofty and democratic its intentions. Thus Puerto Rico will continue to pose a formidable challenge to American statecraft well into the future.

has become the leader among exported primary products. In the Lesser Antilles, however, sugar has retreated markedly, having been supplanted by such crops as bananas, limes, spices, and sea-island cotton (only Barbados and St. Kitts-Nevis retain significant levels of sugar production). And tourism, the region's single growth industry, is a mixed blessing—as we shall discover.

Problems of Widespread Poverty

All the crops grown in the Caribbean—Haiti's coffee, Jamaica's bananas, the Lesser Antilles' fruits, as well as the historically pivotal sugar industry—constantly face severe competition from other parts of the world and have still not become established at a scale that could begin to improve standards of living. Those minerals that do exist in this region—such as Jamaica's bauxite (aluminum ore) and Cuba's nickel—do not support any significant industrialization within the Caribbean Basin itself (only natural-gas-rich Trinidad is making preliminary steps in this direction). As in other lower-income parts of the world, these resources are exported for use elsewhere. And even when postindustrial producers are attracted by an island's labor resources—as in St. Lucia and Barbados, where the population is comparatively well educated—too often it is only because large numbers of workers can be hired at very low wages to perform the most routine data-processing operations (which can quickly be relocated if even cheaper labor supplies become available in another low-income nation). Thus the Caribbean countries, dependent on uncertain for-

The Caribbean is a region of sharp, often searing contrasts. Haiti is the Western Hemisphere's poorest country, its city slums (here in the capital, Port-au-Prince) the most desperate of human environments. Tourism and wealthy waterfront living create another face of the region, generating landscapes of conspicuous consumption—but producing some of the limited economic opportunities in the Caribbean Basin.

eign markets for much of their revenues, find themselves trapped in an international economic order they cannot change.

Given these conditions, the vast majority of the people in this region continue to eke out a precarious living from small plots of ground, are mired in poverty, and are threatened by disease. Food supplies are frequently inadequate on Caribbean islands, with the best land always used to raise cash crops for the export trade rather than staples for local consumption (in fact, many countries—most notably Haiti—are chronically food-deficient). Cultivation methods have undergone little change over the generations. Land inheritance customs have subdivided peasant families' plots until they have become so small that the owner must share-crop some other land or seek work on a plantation in order to supplement the meager harvest. Furthermore, soil erosion constantly threatens: much of the Jamaican countryside is scarred by gulleys and ravines, and Haiti's land has become so ravaged by agricultural overuse and deforestation that almost the entire country has been turned into an ecological wasteland.

With such problems, it would be unlikely for Caribbean America to have many large cities. Nonetheless, the urbanization of the region's population surpasses 60 percent, far exceeding the global average of 45 percent (some agencies estimate that the world's urban population is slightly above 50 percent). And the four leading urban agglomerations—the Dominican Republic's Santo Domingo, Cuba's Havana, Haiti's Port-au-Prince, and Puerto Rico's San Juan—contain nearly 25 percent of the Caribbean's total population. Often, however, towns and cities exhibit even more miserable living conditions than in the poorest rural areas. The slums of Port-au-Prince rank among the world's worst (see photo above), and we should not be surprised that such abysmal conditions drive away the most desperate Haitians as "boat people" in search of a better life elsewhere.

African Heritage

The human geography of the Caribbean islands is also a legacy of Africa, and there are places where the region's cultural landscapes strongly resemble those of West and Equatorial Africa. In the construction of village dwellings, the operation of rural markets, the role of women in rural life, the preparation of certain kinds of food, the methods of cultivation, the nature of the family, artistic expression, and in an abundance of other traditions, the African heritage is visible throughout the Caribbean-American scene.

Nevertheless, in general terms it may still be argued that the average white person is in the best position in this island chain; politically and economically, the **mulatto** (mixed white-black) ranks next; and the black person ranks lowest. In Haiti, for instance, where 95 percent of the population is "pure" black and only 5 percent mulatto, this mulatto minority has held a disproportionate share of power. On neighboring Jamaica, the 8 percent "mixed" sector of the population has played a role of prominence in island politics far out of proportion to its numbers. In the Dominican Republic, the pyramid of power puts the white, Spanish-European sector (16 percent) at the top, the mixed group (73 percent) in the middle, and the black population (11 percent) at the bottom. In Cuba, too, the 11 percent of the population that is black has found itself less favored than the white (37 percent) and mulatto (51 percent) sectors.

8

The composition of the population of the islands is further complicated by the presence of Asians from both China and India. During the nineteenth century, the emancipation of slaves and ensuing local labor shortages brought some far-reaching solutions. Some 100,000 Chinese emigrated to Cuba as indentured laborers (today they still constitute 1 percent of the island's population); and Jamaica, Guadeloupe, Martinique, and especially Trinidad saw nearly 250,000 East Indians arrive for similar purposes. To the African-modified forms of English and French heard in the Caribbean, therefore, can be added several Asian languages. Hindi is particularly strong in Trinidad, whose overall population is now just over 40 percent South Asian. The ethnic and cultural variety of the plural societies of Caribbean America is indeed endless.

Tourism: The Irritant Industry

The resort areas, scenic treasures, and historic locales of Caribbean America attract well over 20 million visitors annually, making this region one of the world's most popular tourist destinations. At least 10 million of these tourists travel on Florida-based cruise ships, and many of these ships' ports of call and nearby coastal strips have enjoyed spectacular growth since 1980. For example, the development of Jamaica's Ocho Rios, Port Antonio, and Montego Bay has boosted the tourist industry so greatly that it now accounts for about one-sixth of the country's gross domestic product and employs more than one-third of the Jamaican labor force.

Certainly, Caribbean tourism is a prospective moneymaker for most of the islands, but it has serious drawbacks. The invasion of overtly poor communities by affluent, sometimes raucous tourists contributes to a rising sense of local anger and resentment. At the same time, tourism can have the effect of debasing local culture, which often is adapted to suit the visitors' tastes (as at hotel-staged "culture" shows). And while tourism does generate income in the Caribbean where alternatives are few, the intervention of island governments and multinational corporations removes opportunities from local entrepreneurs in favor of large operators and major resorts. Today there is evidence, too, that tourism in the region may be approaching its maximum development because the industry grew only sluggishly during the affluent years following the mid-1990s—a combined result of image problems (crime waves; repeated hurricane damage), lagging infrastructures (especially airports and telecommunication systems), rising construction and maintenance costs, and the more effective marketing of competing tourist destinations.

Tourism, then, is a mixed blessing for the developing Caribbean Basin. Given the region's limited options, it provides revenues and jobs where otherwise there would be none. Yet there is a negative cumulative effect that intensifies contrasts and disparities: gleaming hotels tower over substandard housing, luxury liners glide past poverty-stricken villages, opulent meals are served in places where, down the street, children suffer from malnutrition. Clearly, the tourist industry contributes positively to island economies but strains the fabric of the local communities involved.

Regional Cooperation

Many of the world's realms and regions, as we noted in the chapters on Europe and North America, are today exploring new forms of international cooperation. The Caribbean Basin is no exception, the diverse cultural and political backgrounds of its island-nations notwithstanding. Prompted by shared economic interests and a growing sense of regional identity, all 16 independent states in 1994 concluded an agreement to create the Association of Caribbean States (ACS). This initiative was spearheaded by CARICOM (Caribbean Community and Common Market), a three-decade-old customs union of former British colonies, which wanted to expand by building ties to nearby Spanish-speaking countries. These new linkages came to include not only the Dominican Republic and Cuba (as well as Creole-speaking Haiti) in the Caribbean, but all of the realm's mainland countries except Panama, plus Venezuela and Colombia in northern South America. In all, 25 member states constitute the ACS, with the only Caribbean nonparticipants being non-self-governing states affiliated with France, the Netherlands, the United Kingdom, or the United States.

Whereas the ACS marks solid progress toward the broad goal of achieving closer ties among Caribbean-area states, more immediate regional issues also played a role in the organization's formation. Most important was the launching of the North American Free Trade Agreement (NAFTA), which threatened the existing duty-free access of Caribbean exports (especially clothing) to the U.S. market. These privileges had been granted as part of the Caribbean Basin Initiative (CBI) following the U.S. ouster of Grenada's communist regime in 1983, and resulted in substantial job increases in several Caribbean countries as their exports to the United States surged. With its entry into NAFTA in 1994, Mexico became a far more powerful competitor on the U.S. market for these Caribbean goods because the tariffs on Mexican trade were set on a downward course toward elimination; as the new century opened, this competitive advantage was growing, and capital investment projects were continuing to be diverted from the Caribbean Basin to Mexico. By banding together under the new ACS banner, the Caribbean states had initially been able to exert collective pressure on the United States to give them some "NAFTA parity." But the U.S. Congress repeatedly rebuffed these requests during the late 1990s, forcing the ACS countries to confront their intensifying economic problems by themselves as the last of the CBI gains unraveled.

As the widening of the Caribbean's economic spatial organization continues to be pursued, new opportunities are also surfacing at the national level. The most dramatic example occurs at the region's southeastern corner, where Trinidad and Tobago is now riding a natural-gas-driven, industrialization boom that is turning this two-island country into a Pacific Rim-style *economic tiger*. Trinidad has long been an oil producer, but chronically low world prices and dwindling local supplies in recent years forced a reexamination of its natural gas deposits to help counter the economic downturn. That quickly resulted in the discovery of major new supplies, and the country's proven natural gas reserves have tripled over the past ten years. The availability of this cheap and abundant fuel has sparked not only a local gas-production boom (the product can be liquefied and exported by supertanker), but also an influx of energy, chemical, and steel companies from Western Europe, Canada, and even India. Many of the new industrial facilities have agglomerated at the ultramodern Point Lisas Industrial Estate, and they have just catapulted Trinidad to the rank of the world's leading exporter of ammonia and methanol. Natural gas is also an efficient fuel for the manufacturing of metals, and steelmakers as well as aluminum refiners have been attracted to locate here. With Trinidad lying only a few miles from the Venezuelan coast of South America, it is a sea-lane crossroads that is well connected to the vast, near-coastal supplies of iron ore and bauxite that are mined in nearby countries, particularly the Brazilian Amazon.

Elsewhere in the region, as we have noted, development opportunities are far more limited. But at least one other bright spot does exist in the Dominican Republic, whose economy grew the fastest in all of Middle and South America between 1995 and 2000. Attracted by the privatization of formerly government-owned industries, a huge influx of foreign investment spurred the rapid diversification of the country's economic base. The primary sector was a major beneficiary: agriculture has expanded from its traditional export mainstays (sugar, bananas, cacao) into such lucrative new pursuits as fruits, vegetables, and cut flowers, and nickel has become the leading export commodity. Even more important has been the development of tourist-boomled service industries as well as the establishment of a postindustrial sector marked by telecommunications, start-up Internet companies, and clusters of high-tech firms distributed across the country's 43 free-trade zones. Despite these gains, little of the new prosperity has even begun to trickle down to improve the lives of the two-thirds of the population still mired in wretched poverty.

THE CONTINENTAL MOSAIC

Continental Middle America consists of two regions, Mexico and Central America. The former is constituted by a single country, and the latter by seven.

Mexico is Middle America's giant, a region by virtue of its physical size, population total, cultural qualities, resource base, and relative location. Mexico's geographic position adjacent to the United States has been a blessing because it has facilitated economic interaction; it is also a curse because it cost Mexico huge tracts of territory and has led to neighborly friction over massive illegal emigration. Much of that cross-border population outflow has been triggered by economic conditions in Mexico. These conditions may now improve because the North American Free Trade Agreement (NAFTA), despite Mexico's problems, should stimulate further industrial and other forms of regional development.

The seven republics of Central America cannot match Mexico in terms of total population or territory. Nonetheless, they constitute a geographic region—a region plagued by armed conflict, leadership struggles, economic stagnation, rapid population growth, and environmental crisis. We focus first on Mexico and then turn to Central America.

MEXICO

Mexico is the colossus of Middle America. Its 2002 population of 103.6 million exceeds the combined total of all the other countries and islands of the realm by 28 million, and its territory is more than twice as large (Fig. 4-9). Indeed, in all of Spanish-influenced Middle and South America, no other country's population is even half as large as Mexico's. In 2000 Mexico became the eleventh country in the world to have 100 million people. In fact, Mexico has grown so swiftly that its 1970 population of 48.2 million has more than *doubled*; although the growth rate has now moderated, the country is still on track to double in size again before 2040.

The physiography of most of Mexico resembles that of the western United States, although its environments are more tropical. Figure 4-9 shows several prominent features: the elongated Baja (Lower) California Peninsula in the northwest, separated from the mainland by the Gulf of California (which Mexicans call the Sea of Cortés); the far eastern Yucatán Peninsula, jutting out into the Gulf of Mexico; and the Isthmus of Tehuantepec in the southeast, where the Mexican landmass tapers to its narrowest extent. Here in the southeast, Mexico most resembles Central America physiographically; a mountain backbone forms the isthmus, curves southeast into Guatemala, and extends northwest toward Mexico City. Shortly before reaching the capital, this mountain range divides into two chains, the Sierra Madre Occidental in the west and the Sierra Madre Oriental in the east (Figs. 4-2, 4-9). These diverging ranges frame the funnel-shaped Mexican heartland, the center of which consists of the extensive Plateau of Mexico (the Valley of

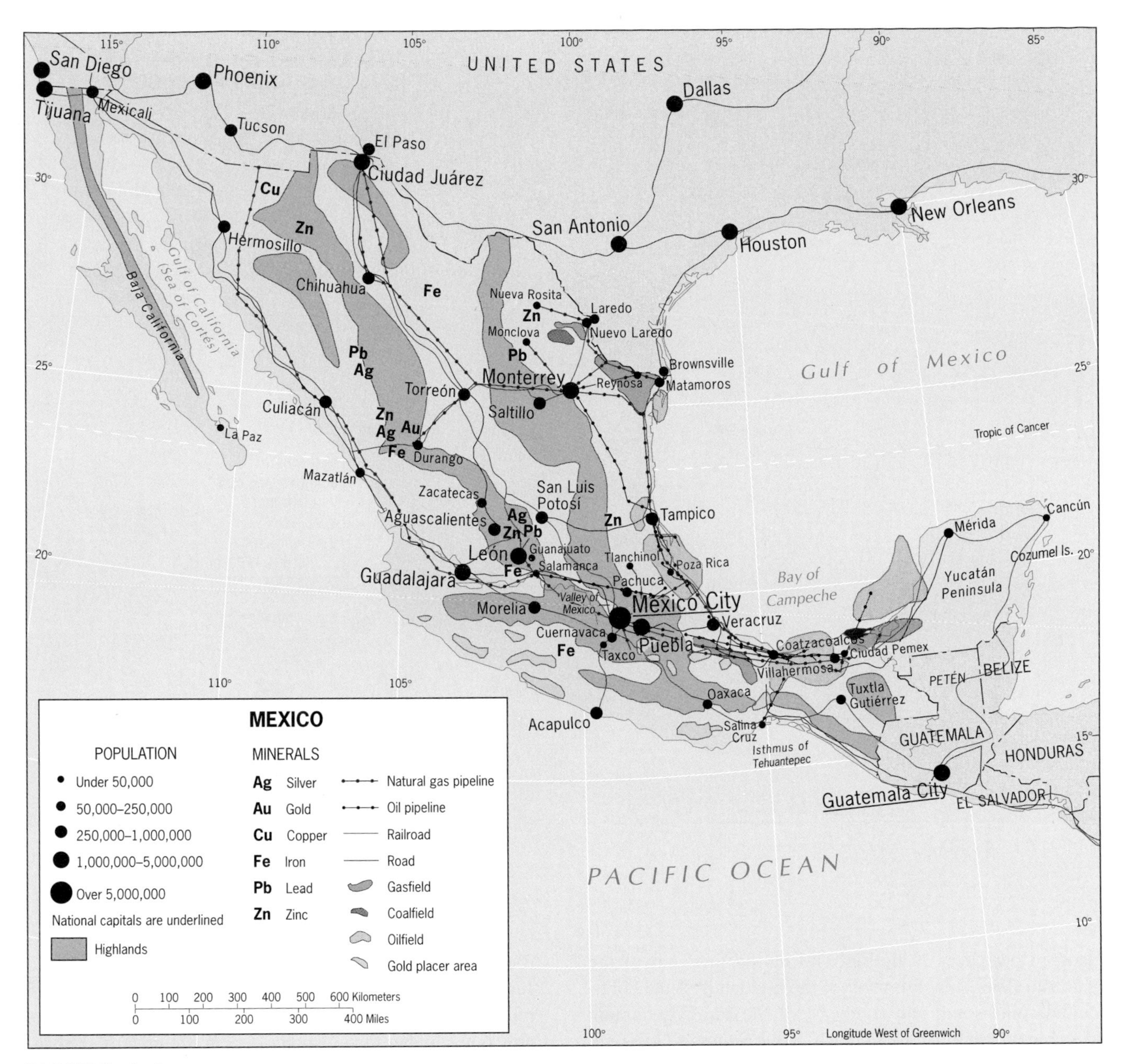

FIGURE 4-9

Mexico lies near its southeastern end). This rugged tableland is some 1500 miles (2400 km) in length and up to 500 miles (800 km) wide. The plateau is highest in the south near Mexico City, where it is about 8000 feet (2450 m) in elevation; from there it gently declines to the north toward the Rio Grande boundary with the United States. As Figure I-8 reveals, Mexico's climates are marked by dryness, particularly in the broad, mountain-flanked North. Most of the better-watered areas lie in the southern half of the country where the major population concentrations have developed.

The distribution of population across Mexico's 31 internal States is shown in Figure 4-10. The largest concentration, containing more than half the Mexican people, extends across the densely populated "waist" of the country from Veracruz State on the eastern Gulf Coast to Jalisco State on the Pacific. The center of this corridor is dominated by the most populous State, Mexico (**3** on the map). At its heart lies the Federal District of Mexico City (**9**), which has just become the world's largest urban agglomeration. In the dry and rugged terrain to the north of this central corridor lie Mexico's least-populated subregions, among them the six States bordering the United States—Tamaulipas, Nuevo León, Coahuila, Chihuahua, Sonora, and Baja California Norte. Southern Mexico also exhibits a sparsely

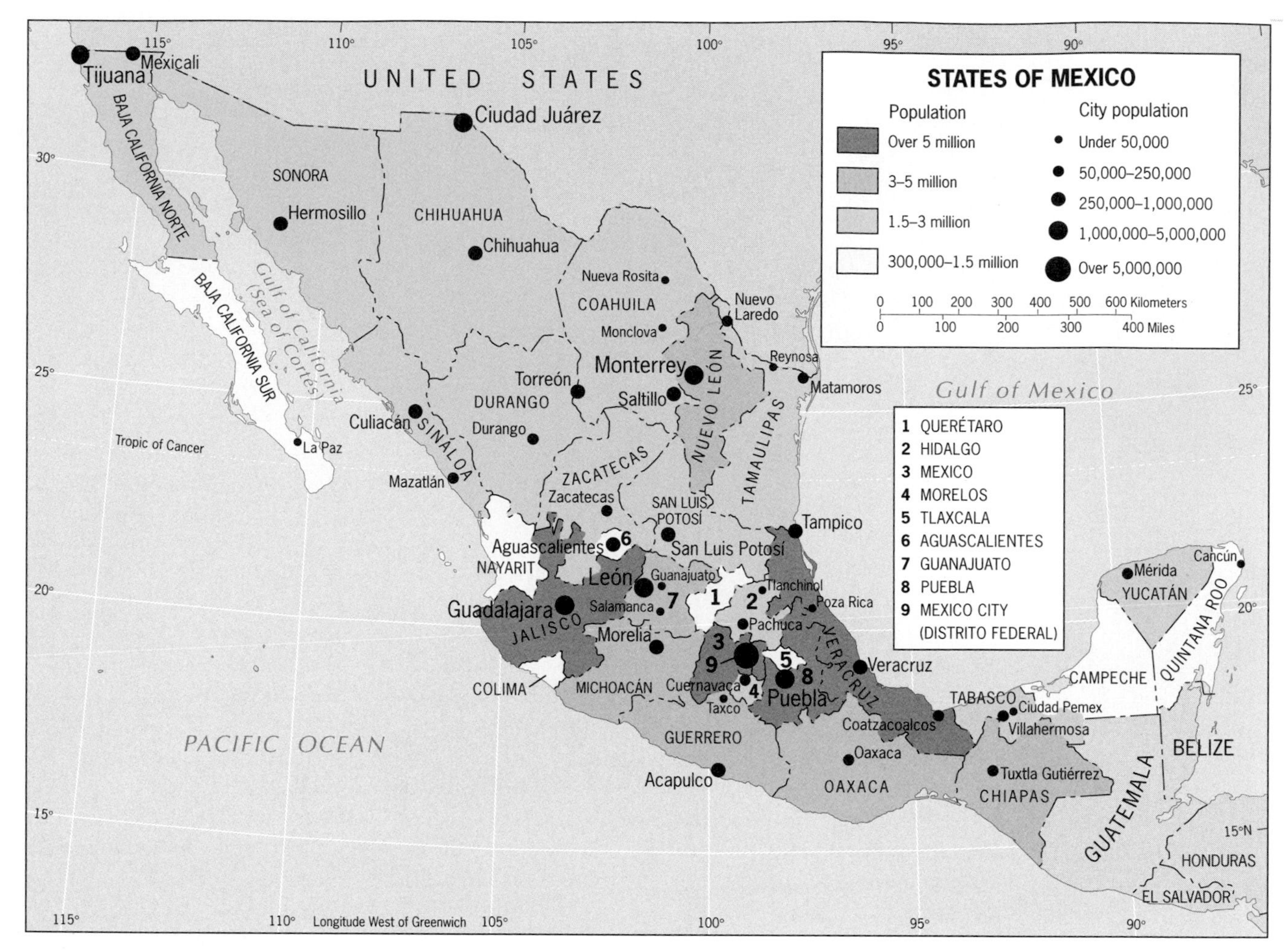

FIGURE 4-10

peopled periphery in the hot and humid lowlands of the Yucatán Peninsula, but most of the highlands of the continental spine south and southeast of Mexico City contain sizeable populations, particularly in the States of Guerrero, Oaxaca, and Chiapas.

Another major feature of Mexico's population map is urbanization, driven by the *pull* of the cities (with their perceived opportunities for upward mobility) in tandem with the *push* of the economically stagnant countryside. Today, no less than 74 percent of the Mexicans reside in towns and cities—up from less than 60 percent three decades ago. This is a surprisingly high proportion for a developing country (the current U.S. figure is 75 percent). Undoubtedly, these numbers are affected by the explosive recent growth of metropolitan Mexico City, which now totals 27 million (making it the largest urban concentration on Earth) and is home to just over 25 percent of the national population. Among the other leading cities in the central population corridor are Guadalajara, Puebla, and León (Fig. 4-10). Outside this heartland, some of the country's fastest-growing cities are found in the U.S. border zone: both Tijuana and Ciudad Juárez recently surpassed 1 million, and Monterrey (at 3.5 million) is booming to the extent that Nuevo León State now ranks higher than its neighbors in Figure 4-10. Urbanization rates at the other end of Mexico, however, are at their lowest in those remote uplands where Amerindian society has been least touched by modernization.

Nationally, the Amerindian imprint on Mexican culture remains quite strong. Today, 60 percent of all Mexicans are mestizos, 20 percent are predominantly Amerindian, and about 10 percent are full-blooded Amerindians; only 9 percent are Europeans. There is a Mexican saying that Mexicans who do not have Amerindian blood in their veins nevertheless have the Amerindian spirit in their minds. Certainly the Mexican Amerindian has been Europeanized, but the Amerindianization of modern Mexican society is so powerful that it would be inappropriate here to speak of one-way, European-dominated **acculturation**. Instead, what took place in Mexico is **transculturation**—the two-way exchange of culture traits between societies in close contact. In the southeastern periphery (Fig. 4-4), several hundred thousand Mexicans still speak only an Amerindian language, and millions more still use these languages in everyday conversation even though they also speak Mexican Spanish. The latter has been strongly influenced by Amerindian languages, but this is only one aspect of Mexican culture that has received an Amerindian impress. Uniquely Mexican modes of dress, foods and cuisine, sculpture and painting, architectural styles, and folkways also vividly reflect the Amerindian contribution. The fusion of these Spanish and Amerindian heritages gives modern Mexico a distinctiveness that it alone possesses in Middle and South America, and is the product of an upheaval that began to reshape the country nearly a century ago.

Revolution and its Aftermath

Modern Mexico was forged in a revolution that began in 1910 and set into motion events that are still unfolding today. At its heart, this revolution was about the redistribution of land, an issue that had not been resolved after Mexico freed itself from Spanish colonial control in the early nineteenth century. As late as 1900, more than 8000 haciendas blanketed virtually all of Mexico's good farmland, and about 95 percent of all rural families owned no land whatsoever and toiled as *peones* (landless, constantly indebted serfs) on the haciendas. Led by the legendary Emiliano Zapata, the triumphant revolution produced a new constitution in 1917 that launched a program of expropriation and parceling out of the haciendas to rural communities.

Since 1917, more than half the cultivated land of Mexico has been redistributed, mostly to peasant communities consisting of 20 families or more. Such lands, as noted earlier, are called *ejidos*; the government holds title to the land, and use rights are parceled out to villages and then individuals for cultivation. Most of the nearly 30,000 *ejidos* carved out of haciendas lie in central and southern Mexico, where Amerindian traditions of land ownership and cultivation survived and where the adjustments were most successfully made. Although this land reform program built an enduring political base for the Institutional Revolutionary Party (PRI) that ruled Mexico from 1929 to 2000, the new landholdings it created were too finely fragmented to provide a real development opportunity. (Belatedly, in the 1990s, the government began to privatize *ejido* lands, but with such small holdings its farmers cannot afford to invest in new technologies to compete against the more efficient, larger-scale agricultural producers of the Mexican North and the United States [whose food products increasingly flow into Mexico as import restrictions vanish under NAFTA].)

Despite an understandable temporary decline in agricultural productivity during the transition, the miracle is that land reform was carried off without major dislocation and that the power of the wealthy land-owning aristocracy could be broken without ruin to the state. Although considerable malnutrition and poverty persisted in the countryside, it was widely recognized that Mexico, alone among the realm's countries with large Amerindian populations, had made significant strides toward solving the land question. Just how much farther Mexico has to go, however, became evident in 1994 when the issue resurfaced at the heart of a rebellion that broke out in southeasternmost Chiapas State, deliberately timed to make global headlines on the very day that NAFTA went into operation.

Chiapas, the poorest of the 31 States, lies wedged against the Guatemalan border and is more reminiscent of Central America than Mexico. The complexion of its rapidly growing population of 4.3 million is heavily Amerindian (largely of Mayan background) and is dominated by families of peasant farmers who eke out a precarious existence by cultivating tiny, hardscrabble plots of land in the rainforested hills. The more productive valley soils have for centuries been incorporated into the estates of the large landholders, a system that endures here virtually unaffected by the land redistribution that reshaped so many other parts of rural Mexico. In response to this lack of development, the Mexican government had pledged to introduce new services and job-creation programs in Chiapas, but its main effort was a coffee-raising scheme in the 1980s that soon came to naught when world coffee prices plummeted. This was followed by the passage of a constitutional amendment that permitted the privatization of Amerindian land (held communally until then), a move the government saw as accelerating the incorporation of indigenous peoples into the national economy, but one resisted by most Amerindians here. These events intensified the longstanding bitterness of the Chiapans, who have experienced almost none of the benefits of Mexico's twentieth-century revolution. Quietly, a radical group of Mayan peasant farmers began to organize and build support to resume the historic struggle of Amerindian *peones* to gain land and fair treatment, as well as escape from their extreme poverty and marginalized position in Mexican society.

On January 1, 1994, this organization, now calling itself the Zapatista National Liberation Army (ZNLA), ignited a guerrilla war with coordinated attacks on several Chiapan towns. By timing this uprising to coincide with the launching of NAFTA and by taking the legendary name of Zapata, the ZNLA (who number only a few thousand but claim a half-million sympathizers) cleverly ensured maximum publicity for its agenda that included land reform, access to greater economic opportunity, heightened cultural identity, and local autonomy. The Mexican military reacted strongly to this offensive and quickly drove the insurgents out of the towns. Although their heavy-handed pursuit ended abruptly when the international press exposed a number of human-rights violations, the army has maintained a major Chiapan presence ever since with up to 50,000 troops. Undeterred, the guerrillas regrouped in the mountains and

Among the Realm's Great Cities . . .

Mexico City

Middle America has only one great metropolis: Mexico City. With 27 million inhabitants, Mexico City is home to just over one-fourth of Mexico's population and grows by about a half-million each year. Even more significantly, in early 2002 Mexico City surpassed Tokyo, Japan (26.4 million) to become the world's largest urban agglomeration.

Lakes and canals marked this site when the Aztecs built their city of Tenochtitlán here seven centuries ago. The conquering Spaniards made it their headquarters, and following independence the Mexicans made it their capital. Centrally positioned and well connected to the rest of the country, Mexico City, hub of the national core area, became the quintessential primate city.

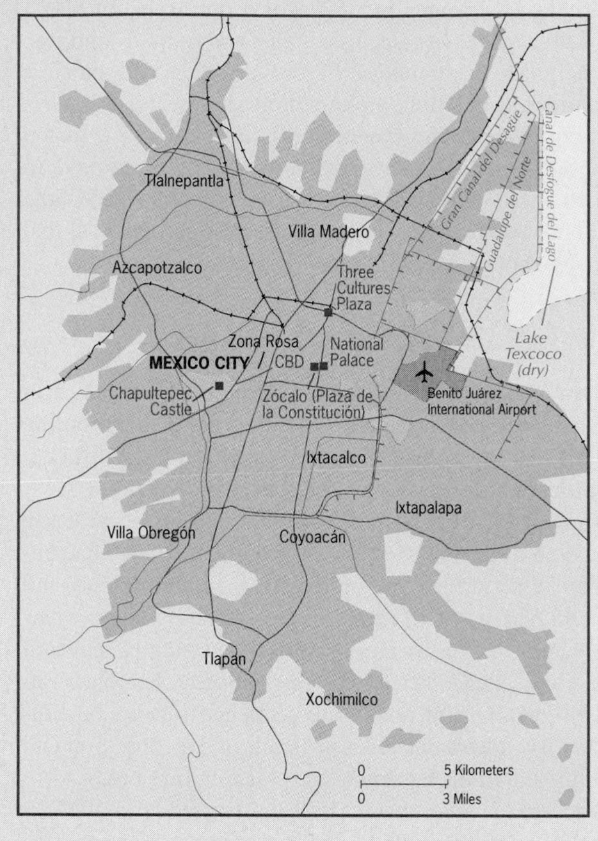

Vivid social contrasts mark the cityscape. Historic plazas, magnificent palaces, churches, villas, superb museums, ultramodern skyscrapers, and luxury shops fill the city center. Beyond lies a zone of comfortable middle-class and struggling, but stable, working-class neighborhoods. Outside this belt, however, lies a ring of more than 500 slums and countless, even poorer *ciudades perdidas*—the "lost cities" where newly arrived peasants live in miserable poverty and squalor (these squatter settlements contain no less than *one-third* of the metropolitan area's population). Mexico City's more affluent residents have also been plagued by problems in recent years as the country's social and political order came close to unraveling. Rampant crime is the most immediate concern, much of it associated with corrupt police as the control of the country's long-time ruling party (the PRI) weakened and collapsed.

Environmental crises parallel the social problems. Local surface waters have long since dried up, and groundwater supplies are approaching depletion; to meet demand, the metropolis must now import a third of its water by pipeline from across the mountains (with about 40 percent lost through leakages in the city's crumbling waterpipe network). Air pollution here is probably the world's worst as more than 4 million motor vehicles and 40,000 factories churn out smog that in Mexico City's thin, high-altitude air sometimes reaches *100 times* the acceptably safe level. And add to all this a set of geologic hazards: severe land subsidence as underground water supplies are overdrawn; the ever-present threat of earthquakes that can wreak havoc on the city's unstable surface (the last big one occurred in 1985); and the risk of volcanic activity as nearby Mount Popocatépetl (see photo, p. 201) may be ending centuries of dormancy.

In spite of it all, the great city continues to beckon, and every year hundreds of thousands of the desperate and the dislocated arrive with hope—and little else.

have remained a political, if not a military, force to be reckoned with as negotiations soon replaced armed conflict. These peace talks collapsed in 1996 when the Zapatistas accused the national government of reneging on its written agreement to give greater recognition to the rights and cultural traditions of Chiapas's indigenous peoples. With the federal authorities making no further concessions but unwilling to face the consequences of wiping out the ZNLA, the stalemate dragged on into the new administration that took over in Mexico City in 2000. Despite the campaign claim of candidate Vicente Fox that he could resolve the Zapatista problem in a few minutes, upon taking office

President Vicente Fox immediately discovered that the challenge is a far more difficult one.

The Chiapas rebellion raises a number of broader issues that have important implications for Mexico's future. First, it emphasizes that certain areas of the country do not participate in the ongoing development thrust. In spatial terms, the Chiapas situation is a classic case of core-periphery confrontation: the population of this poorest outlying State demands its share of the growing national pie at precisely the time (1995–1998) when government statistics show the number of Mexicans living in extreme poverty increased by an astonishing 53 percent. A second issue, one that now affects many of the world's countries, is devolution. The ZNLA demand for "autonomy" is modeled after Spain's Autonomous Communities (see p. 90-92), and involves not secession but the decentralization of powers from the federal to the State government that allows the latter more local control, particularly over cultural affairs.

Another issue raised by the Chiapas conflict is vital to Mexico's social geography. The ZNLA crusade elicited wide sympathy among indigenous populations in five other southern States (Guerrero, Oaxaca, Puebla, Michoacán, and Mexico—each the scene of its own sporadic guerrilla activity since the mid-1990s) and could well spark a nationwide civil rights movement for Amerindians. As in Caribbean societies, darkness of skin color is directly related to a person's social status, a linkage that in Mexico further extends to an individual's degree of "Indianness." Throughout the modern Mexican revolution, the country's ruling elite has been ambivalent in its behavior toward the Amerindians. On one hand, it celebrates the Aztec and Maya legacies as vital components of Mexican identity; on the other, the lagged development of Amerindians is seen as a blot on the modernization movement, and efforts to aid these native peoples are aimed at their greater assimilation into the national culture. Large numbers of predominantly Amerindian Mexicans (a group that totals 20 percent of the Mexican population) have made social gains during the past century. But the remaining 10 percent who are full-blooded Amerindians, and choose to preserve their pre-Hispanic cultural traditions, have been shunned by a racist mainstream society whose discrimination continues to render them second-class citizens.

The Changing Geography of Economic Activity

In recent decades, Mexico has made important progress in several productive sectors. During the early 1990s, its economy was further transformed by the boom preceding the implementation of NAFTA, which over the next few years will bind the economies of Mexico, the United States, and Canada into a single free-trade zone and market of more than 400 million people. Mexico will gain the most because this affiliation is expected to close the economic gap with its wealthy northern neighbor, which still produces some striking contrasts along their 1936-mile (3115-km)-long border (see photo below).

The launching of NAFTA in 1994 was followed by a number of unexpected shocks to the Mexican political establishment and economic system. The most serious of these shocks was a substantial devaluation of the peso in

The border between Mexico and the United States occasionally provides some stunning contrasts. Here, looking westward, we observe the opposing economic geographies that mark the border landscape of the Imperial Valley east of the urban area formed by Mexicali, Mexico and Calexico, California. The larger town of Mexicali (on the left) sprawls eastward along the border, with a cluster of *maquiladora* assembly plants, visible in the left foreground, surrounded by high-density, poor-quality housing. On the U.S. side, lush irrigated croplands blanket the otherwise dry countryside, fed by the All-American Canal that taps the waters of the Colorado River before they cross into Mexico. At this point, the canal swings northward to bypass border-hugging Calexico (right rear) before rejoining the international boundary beyond the town's western edge.

late 1994. As a result, Mexico plunged into an economic recession that lingered through the end of the 1990s and tested the faith of its workforce, business leaders, and foreign investors. The recession also challenged the government to correct the problems and restart the boom, an effort blunted by the political upheaval that accompanied the post–1997 disintegration of the power of the PRI Party, which had ruled Mexico for 71 consecutive years and was finally ousted in the election of 2000. We now review Mexico's changing economic geography against the background of these events.

Agriculture

Although traditional subsistence agriculture and the output of the inefficient *ejidos* have not changed a great deal in the poorer areas of rural Mexico, larger-scale commercial agriculture has diversified during the past three decades and made major gains with respect to both domestic and export markets. The southern half of the Plateau of Mexico is geared mainly to the domestic production of food crops, but in the country's arid northern tier sizeable irrigation projects have been built on streams flowing down from the interior highlands. Along the booming northwest coast of the mainland, which lies within a day's drive of Southern California, mechanized large-scale cotton production now supplies an increasingly profitable export trade. Here, too, wheat and winter vegetables are grown, with fruit and vegetable cultivation attracting foreign investors, particularly to such crops as bananas and sugarcane. Cattle-raising is another leading pursuit and continues to expand onto the Gulf Coast lowlands from its long-time base in the northern interior.

Energy Resources

Although Mexico's metal-mining industries are less important today than they once were, since 1970 the country has enjoyed the advantages and suffered the problems of being a major petroleum producer. Huge oilfields centered on the southern Gulf Coast's Bay of Campeche around Villahermosa in Tabasco State brought Mexico abundant revenues when the world oil price was high in the late 1970s and serious economic difficulties when the price fell after 1980. These discoveries of massive oil and natural gas reserves have made Mexico self-sufficient in energy, adding to already substantial reserves located in oilfields along the Gulf Coast to the northwest and inner Yucatán to the northeast (Fig. 4-9). The high petroleum prices of the 1970s stimulated the beginnings of Mexico's economic-geographic transformation, but the oil crash of the 1980s thwarted its momentum for a decade. This crash occurred because the government found itself without the expected oil revenues it needed to pay interest on the huge foreign loans it had taken to finance domestic development programs. Such a disaster could recur if a future oil boom again tempts Mexican leaders to take such risks.

Industrialization

Manufacturing is the centerpiece of Mexico's latest development episode, but this economic sector actually got its start a century ago. Blessed with a wide range of raw materials, many of which are located in the North (Fig. 4-9), the country began to industrialize in 1903 with the completion of an iron and steel plant (now abandoned) in the northeastern city of Monterrey. A second steel complex was built at nearby Monclova in the 1950s, a period that saw the spreading of factories across many parts of central Mexico, particularly in and around the capital. The industrial sector since that time has seen steady growth and for the past quarter-century has employed at least one-seventh of the Mexican labor force.

The most significant recent development in Mexico's manufacturing geography is the growth of maquiladora plants in the northern border zone. **Maquiladoras** are factories (about half are U.S.-owned) that assemble imported, duty-free components and raw materials into finished industrial products. At least 80 percent of these goods are then re-exported to the United States, whose import tariffs are limited to the value added to products during their Mexican fabrication stage (under NAFTA, these tariffs are scheduled to end by 2003). While the foreign owners largely benefit from the lower costs of doing business (e.g., Mexican wage rates average about one-third of those north of the border), this industrial system offers a wide array of economic advantages for Mexico. Among these advantages are the creation and expansion of job opportunities, increased foreign investment, heightened access to export markets, and the transfer of new technologies into the country. Unfortunately, there is a downside as well: factory workers are usually exploited through long work weeks, receive minimal if any fringe benefits, and too many have no choice but to reside in the squalid slums and shacktowns that often surround the plants (see photo, p. 221).

11

Although this development program was initiated in the 1960s, the number of maquiladoras grew only to a modest 588 (with 122,000 employees) by 1982. Suddenly, however, this innovation took flight, and by the early 1990s about 1800 assembly plants were employing some 500,000 workers; among the goods now being assembled were electronic equipment, electrical appliances, auto parts, clothing, plastics, and furniture. Today, ten years later, nearly 4000 maquiladoras with over 1.4 million employees are operating all along the northern border zone in the urban areas mapped in red in Figure 4-11 and account for nearly 30 percent of Mexico's industrial jobs as well as 52 percent of its manufacturing exports (and 45 percent of its total exports).

The Mexican government is capitalizing on the success of the maquiladora program by promoting industrial growth

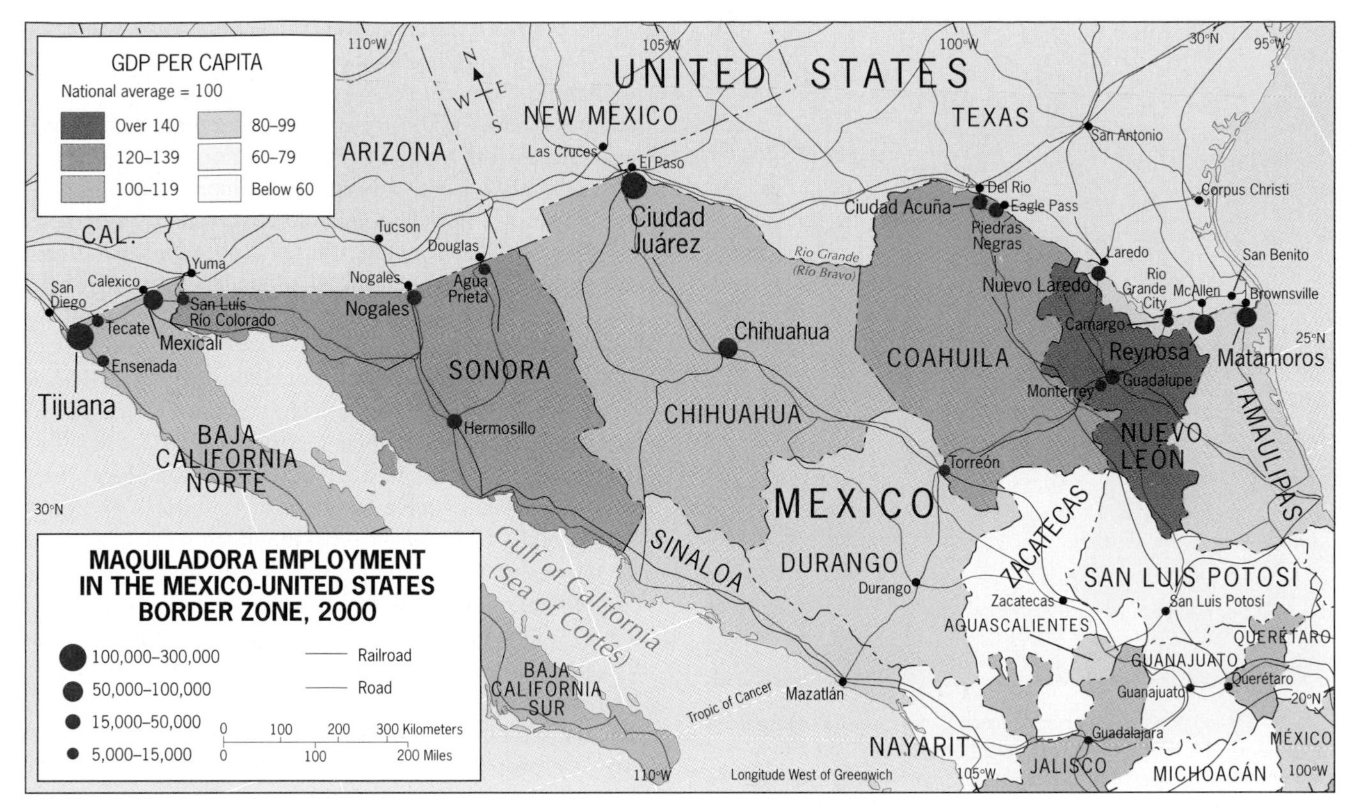

FIGURE 4-11

in other parts of the country. Indeed, the phenomenon itself is gradually expanding southward on its own, and such factories in the interior now account for about one-third of the country's total maquiladora employment. But what development planners desire most is that industrial firms create complete manufacturing complexes within the heart of Mexico rather than limit their investments to assembly plants that hug the U.S. border. In recent years, a number of multinational corporations have undertaken such ventures, most notably the three leading U.S. automakers and four of their Japanese and German competitors. (The automobile industry now ranks among the leaders in Mexico's manufacturing and export sectors, and General Motors' Delphi parts subsidiary is the country's largest private employer.)

To speed the opening of the rest of the country, the government is giving priority to programs aimed at upgrading Mexico's infrastructure to world-class standards, especially its telecommunications, superhighway, and electrical-power networks. Individual megaprojects designed to boost regional development are also being pursued, most ambitiously the **dry canal** across the 150-mile (250-km) Tehuantepec isthmus to move containerized goods between the Pacific port of Salina Cruz and the Gulf port of Coatzacoalcos; this would be not only an overland trade route to rival the aging Panama Canal, but also an ultramodern road/rail corridor lined with factories that specialize in the manufacture and assembly of goods in transit between the affluent countries of the Asian Pacific Rim and the North Atlantic Basin (construction of the railway was being completed in 2001).

Uneven Regional Development

Despite such efforts, an ominous economic divide is deepening between the southern and northern halves of Mexico. South of the capital, technological and social development lags ever farther behind the rest of the country as traditional productive activities (low-output farming, mining, logging) continue to dominate. North of Mexico City—which sits at the apex of an upside-down, "dynamic triangle" whose base is the entire U.S. border—growth corridors are stirring that increasingly exhibit the landscape of the global economy in the form of new manufacturing facilities, technology training academies, and even the beginnings of a "Silicon Valley of Mexico" (built around a complex of electronics plants that have recently clustered near Guadalajara in Jalisco State). Importantly, the rise of the Mexican North is also being expressed in politico-geographical terms because it was here that the swing votes were generated to overthrow the long-entrenched PRI Party in the 2000 election. To capture the forces that are transforming northern Mexico, we focus on two of its leading urban centers, Monterrey and Ciudad Juárez/El Paso.

Likened to a glowing hot bar of freshly cast steel plucked from a furnace, the *Faro del Comercio* (Lighthouse of Commerce) shoots skyward in the historic heart of Monterrey. Saddle Mountain in the background is a reminder that this northeastern Mexican metropolis is situated at a strategic point where the desert mets the great Sierra Madre Oriental range, in the heart of an area so well endowed with raw materials that Mexico's iron and steel industry was born here a century ago. Ever since, Monterrey has been an economic leader, and its new landmark signifies that no city more boldly confronts the Mexican future. Today, building on the opportunities offered by NAFTA, Monterrey is busy forging a role for itself on the international circuit. Capitalizing on its function as the southern anchor of the booming growth corridor that stretches northward to Texas' Dallas–Ft. Worth Metroplex, Monterrey has built a new downtown business complex that includes a convention center booked solidly with trade shows at least two years ahead. And in 1999, the city even hosted the National League's season-opening baseball game (in which the Colorado Rockies outscored the San Diego Padres, 8-2).

Burgeoning Monterrey—150 miles (250 km) inside Mexico, yet close enough to the Texas border to have benefited from all the recent development trends—is frequently singled out as a model of what a future Mexican growth center should be. Among this city's assets are a highly educated and well-paid labor force, a stable international business community, and a thriving high-technology complex of ultramodern industrial facilities that has attracted blue-chip multinational companies (see photo above). In addition, a new expressway link to the Rio Grande is helping to forge an international growth corridor between Monterrey and Dallas-Ft. Worth, which (as noted in Chapter 3) is becoming a primary axis of cross-border trade with Texas. For Mexico as a whole, however, there may also be a negative side to Monterrey's success. This city and its surrounding State of Nuevo León have now become so prosperous (see its standout GDP in Fig. 4-11) that there is widening talk of secession. Similar to the devolutionary movement in northern Italy's Lombardy (p. 90), affluent Nuevo León—which receives back less than 20 centavos in services for every tax peso (100 centavos) it sends to Mexico City—has increasingly complained that the federal government tries to develop the impoverished South by taking more money from the industrial North.

The binational metropolis of Ciudad Juárez/El Paso (1.9 million) sits squarely astride the Rio Grande border between Chihuahua State and extreme West Texas, close to the midpoint of the Mexico-U.S. border at a place where north-south routes have bundled for centuries. As Figure 4-11 reveals, *Paseo del Norte* (the name its residents use for the urban area) contains the single largest concentration of maquiladoras, with upwards of 400 plants employing 275,000 workers. This is a leading element of the *Paseo*'s prosperity and has helped drive an employment boom underway since the mid-1990s. The line separating these twin cities also constitutes the world's busiest border crossing, propelled by the increasing interaction of the two countries and the fact that as much as 25 percent of their annual trade passes through here. But the vast majority of daily movements across the Rio Grande are local as complementary economies continue to emerge: although about 40,000 El Pasoans commute to their jobs in Ciudad Juárez, thousands of residents of the latter cross the river each day to account for nearly half of El Paso's retail sales. Of considerable importance, too, is a rising awareness of the *Paseo* as a single community as evinced in the new cooperative bodies set up to handle highway planning, air and water quality, and other metropolitanwide concerns. And, as in Monterrey, affluence has also triggered thoughts of greater independence, including calls for an open international border and the carving out of a *Paseo* city-state from the State of Chihuahua. These notions are reminiscent of trends along the United States' northern border (see pp. 176-178), where the formation of *regional states* and the intensifying impact of NAFTA are shaping new economic-geographic relationships.

NAFTA and Continuing Challenges

The launching of the North American Free Trade zone in 1994 was supposed to mark an economic turning point for Mexico. Instead, NAFTA's early years were plagued by an unending string of crises that included the Chiapas rebellion, monetary devaluation, economic recession, crime waves associated with uncontrolled drug trafficking, and political scandals and transitions. Nonetheless, trade with the United States increased substantially: more than 85 percent of Mexican exports now go to the U.S., which is also the source for about 75 percent of all imported goods. At the same time, Mexico has become a leading U.S. trade partner, in 1997 dislodging Japan as the United States' second-largest export market after Canada. Thus, when the peso devaluation crisis struck in the mid-1990s, the United States unhesitatingly played a pivotal role in the effort to rescue the Mexican economy.

As Mexico completes its economic recovery and the resumption of its growth boom, we should keep in mind that NAFTA is the first reciprocal trade agreement between a high-income and an upper-middle-income country. As events since 1993 have demonstrated, there are bound to be internal as well as external problems during the transition to the new supranational economic order. Clearly, Mexico's development has been accelerated and enhanced by NAFTA, although the geographic distribution of the benefits to date lopsidedly favors the increasingly affluent northern States.

As Mexico strives to join the ranks of the world's advantaged states, the movements of its inhabitants indicate that the country is still a wobbly giant. About 1 million Mexicans annually migrate to the U.S. border zone, lured by the promise of jobs in agriculture and manufacturing. Most migrants would prefer to keep on going north, and hundreds of thousands do indeed manage to cross the U.S. boundary each year. Large numbers of frustrated Mexicans also illegally attempt to leave the country but are intercepted and deported by U.S. authorities. Undoubtedly, many more get through. In recent years, this illegal immigration has triggered increased public support in the United States for tougher measures to curb such activity; as a result, police patrols have been beefed up, and new barriers have been built along many stretches of the border.

Mexico today is at a crossroads. Its most recent (and truly democratic) national election has installed a new regime that is expected to produce new initiatives aimed at reducing the divisiveness of the U.S. border, devolving power to the States, resolving the Chiapas conflict, and continuing the revitalization of the economy. Expectations are high throughout the country, and results should not be long in coming.

THE CENTRAL AMERICAN REPUBLICS

Crowded onto the narrow segment of the Middle American land bridge between Mexico and the South American continent are seven countries collectively known as the Central American republics (Fig. 4-12). Territorially, they are all quite small; only one, Nicaragua, is substantially larger than the Caribbean island of Cuba. Populations range from Guatemala's 13.5 million down to Panama's 3.0 million in the six Hispanic republics, whereas the sole former British territory, Belize, has only about 280,000 inhabitants.

The narrowing land bridge on which these republics are located consists of a highland belt flanked by coastal lowlands on both the Caribbean and Pacific sides (Fig. 4-1). From the earliest times, the people have been concentrated in the upland *tierra templada* (temperate) zone. Here tropical temperatures are moderated by elevation (see box titled "Altitudinal Zonation" on p. 227), and rainfall is adequate for the cultivation of a variety of crops (Fig. 4-13). As noted earlier, the Middle American highlands are studded with volcanoes, and local areas of fertile volcanic soils are scattered throughout the region. The old Amerindian agglomerations were located in these more fertile parts of the highlands, and this human distribution persisted during the Spanish colonial period.

The distribution of population within Central America, apart from its concentration in the region's uplands, also exhibits greater densities toward the Pacific than toward the Caribbean coastlands (Fig. I-9). El Salvador, Belize, and (to some degree) Panama are exceptions to the rule that people in continental Middle America are concentrated in the *templada* zone. Most of El Salvador is tropical *tierra caliente*, and the majority of its 6.6 million people are crowded (824 to the square mile—a population density approaching India's) onto the intermontane plains lying less than 2,500 feet (750 m) above sea level. In Nicaragua, too, the Pacific-side areas are the most densely populated; the early Amerindian centers lay near Lake Managua, Lake Nicaragua, and in the adjacent highlands. The frequent activity of volcanoes in this Pacific zone is accompanied by the emission of volcanic ash, which settles over the countryside and quickly weathers into fertile soils.

By contrast, the Caribbean coastal lowlands—hot, wet, and awash in leached soils—support comparatively few people. In the most populous republic, Guatemala, the heartland also has long been in the southern highlands. Although the large majority of Costa Rica's population is concentrated in the Central Valley around San José, the Pacific lowlands have been the scene of major in-migration since banana plantations were first established there. Even in Panama there is a strong Pacific orientation. More than half of all Panamanians (and this means about two-thirds of the rural population) live in the southwestern lowlands and on adjoining mountain slopes; another 25 percent live and work in the Panama Canal corridor; and of the remainder, a majority (many of them descendants of black immigrants from the Caribbean) live on the Caribbean, Rimland side of the isthmus.

Middle America's smaller republics face the same problems as the less developed parts of Mexico, only more so; they also share many of the difficulties confronting the poorer Caribbean islands. No present or future challenge, however, is greater than Central America's overpopulation. The region's population explosion began a half-century ago, expanding from a base of 9.3 million people in 1950 to a total of 38.9 million in 2002. Unlike Mexico, which has

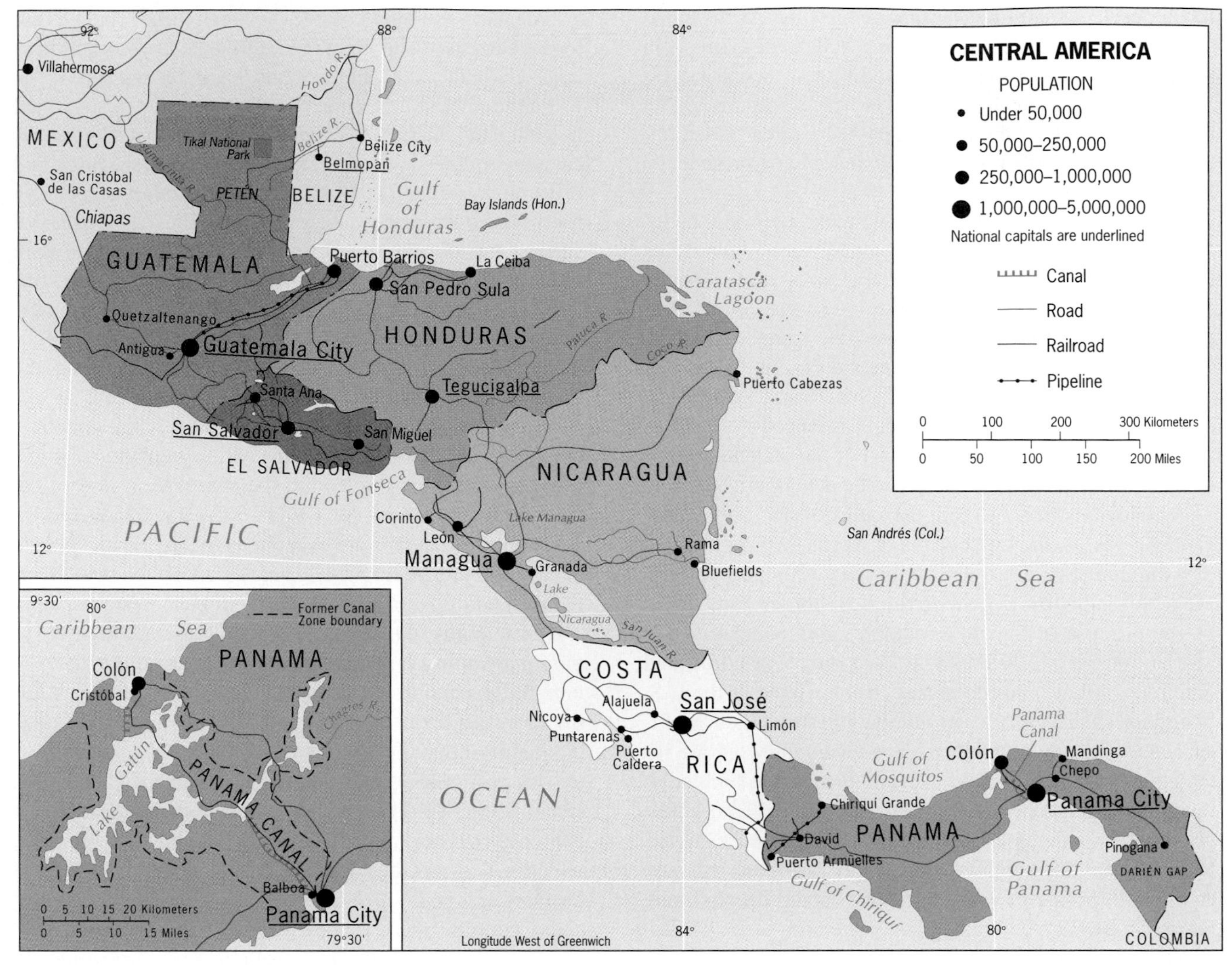

FIGURE 4-12

substantially reduced its rate of natural increase since 1980, Central America (except for Costa Rica and Panama) is on a course that will see a doubling of today's population to 78 million by 2028. This amounts to nothing less than an onrushing demographic catastrophe in a region already unable to cope with most of its social, economic, and natural-resource problems. (To cite just one example, beleaguered Honduras produced 27 percent less food per capita in 1996 than in 1980—and that was *before* the country was ravaged by monstrous Hurricane Mitch in 1998, which set the economy back for years.)

Emergence From a Turbulent Era

Devastating inequities, repressive governments, external interference, and the frequent unleashing of armed forces have destabilized Central America for much of its modern history. The roots of these upheavals are old and deep, and today the region continues its struggle to emerge from a period of turmoil that lasted through the 1980s into the mid-1990s.

Central America is not a large region, but because of its physiography it contains many isolated, comparatively inaccessible locales. Conflicts between Amerindian population clusters and mestizo groups are endemic to the region, and contrasts between the privileged and the poor are especially harsh. Dictatorial rule by local elites followed authoritarian rule by Spanish colonizers, and the latest episode of violent confrontation was simply another manifestation of this persistent polarization. One unprecedented side effect of these recent conflicts has been an intermittent but enormous flow of refugees. These people, by the tens of thousands, were forced to flee from combat and "death-squad" terrorism, often leaving behind broken families and shattered lives.

Despite the challenges that remain, battle-scarred Central America's prospects are finally brightening. The expanding opportunities of individual republics are highlighted in the country profiles that follow. An important breakthrough is also occurring at the supranational level, where a new spirit of cooperation is forging a sense of regional identity that has barely existed in the past. What began in 1993 as an exploratory effort to resuscitate intraregional trade within the framework of the 30-year-old Central American Common Market soon escalated into a series of pacts to create a more meaningful economic union. At the same time, free-trade agreements were negotiated with several nearby countries outside Central America, and all the republics except tiny Belize are working as a group to obtain full membership in NAFTA.

Putting these new concepts into daily practice, however, requires that the Central American republics overcome some major challenges. Foremost among them, if this cooperative development thrust is to succeed, is the liability of a ramshackle regional infrastructure that has long hindered internal as well as cross-border movement and interaction. (The Pan-American Highway, largely constructed during the middle decades of the twentieth century, does link all the republics except Belize, but this aging road has been

Altitudinal Zonation

Continental Middle America and the western margin of South America are areas of high relief and strong local contrasts. People live in clusters in hot tropical lowlands, in temperate intermontane valleys, and even on high plateaus just below the snow line in South America's Andes. In each of these various zones, distinct local climates, soils, vegetation, crops, domestic animals, and modes of life prevail. Such **altitudinal zones**, which are diagrammed in Figure 4-13, are known by specific names as if they were regions with distinguishing properties—as, in reality, they are.

The lowest vertical zone, from sea level to 2500 feet (about 750 m), is known as the ***tierra caliente***, the "hot land" of the coastal plains and low-lying interior basins where tropical agriculture (including banana plantations) predominates. Above this lowest zone lie the tropical highlands containing Middle and South America's largest population clusters, the ***tierra templada*** of temperate land reaching up to about 6000 feet (1800 m). Temperatures here are cooler; prominent among the commercial crops is coffee, while corn (maize) and wheat are the staple grains. Still higher, from about 6000 feet to nearly 12,000 feet (3600 m), is the ***tierra fría***, the cold country of the higher Andes where hardy crops such as potatoes and barley are mainstays. Only small parts of the Middle American highlands reach into the *fría* zone, but in South America this environment is much more extensive in the Andes. Above the tree line, which marks the upper limit of the *tierra fría*, lies the ***tierra helada***; this fourth altitudinal zone, extending from about 12,000 to 15,000 feet (3600 to 4500 m), is so cold and barren that it can support only the grazing of sheep and other hardy livestock. The highest zone of all is the ***tierra nevada***, a zone of permanent snow and ice that reaches to the peaks of the loftiest Andean mountains.

These elevation ranges are for highlands lying in the equatorial latitudes. Of course, as one moves poleward of the tropics, north or south of 15 degrees of latitude, the sequence of five vertical zones is ratcheted downward, with the breaks occurring at progressively lower altitudes.

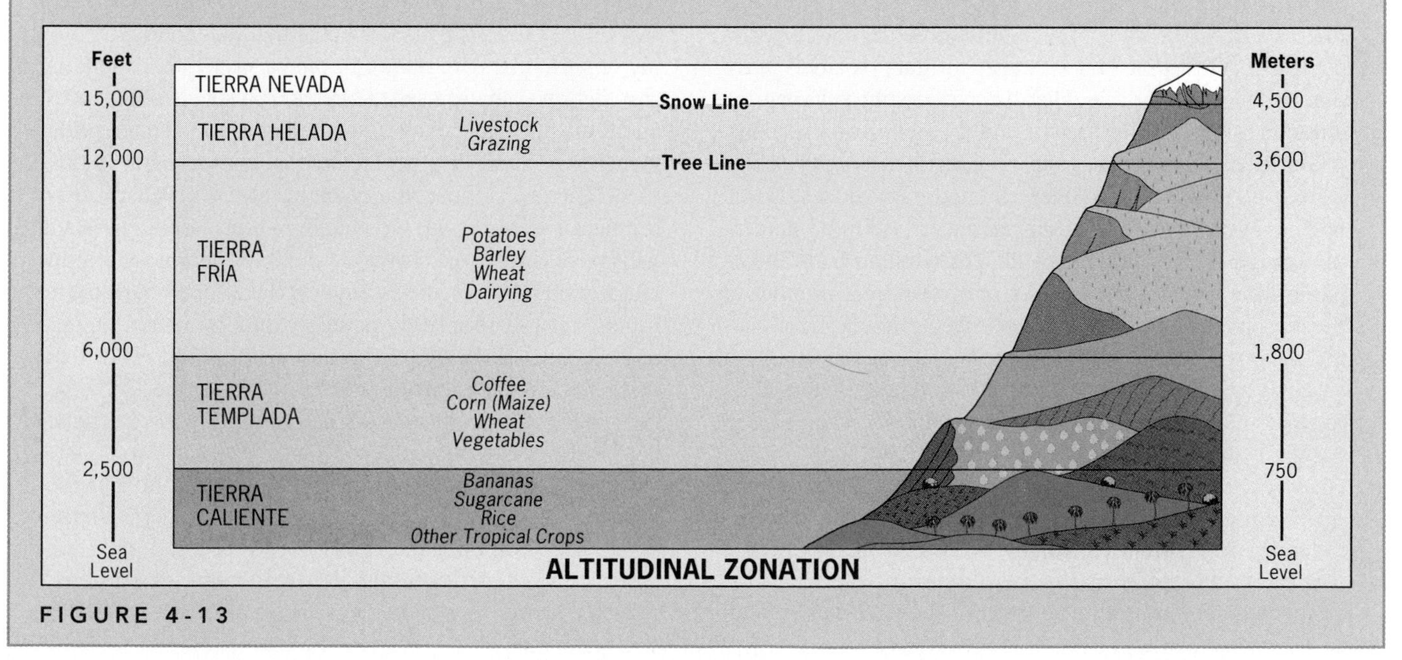

FIGURE 4-13

plagued by constant disruptions and badly needs to be modernized.) As the new century opened, all seven countries had begun the costly process of significantly upgrading their transport, telecommunications, and electrical power networks, but it will be years until the poorer republics can even approach world-class standards. Other obstacles are more easily surmounted; one example is the implementation of procedures smoothing the flow of goods across international boundaries that until recently did not even have adjacent customs houses open at the same time. At any rate, the new spirit of cooperation has already produced results, as both the region's economy and the export trade among Central America's countries have grown steadily since the return of relative peace in the late 1990s.

Guatemala: A Phantom Peace?

Westernmost of Central America's seven republics, Guatemala has more land neighbors than any other. Straight-line boundaries across the tropical forest mark much of the border with Mexico, creating the box-like region of Petén between Chiapas State on the west and Belize on the east (Fig. 4-12). Also to the east lie Honduras and El Salvador.

Guatemala, heart of the ancient Maya Empire and still strongly infused by Amerindian culture and tradition, has just a small window on the Caribbean but a longer Pacific coastline. It was still part of Mexico when the Mexicans threw off the Spanish yoke, and although independent from Spain after 1821, Guatemala did not become a separate republic until 1838. Mestizos, not the Amerindian majority, secured the country's independence.

Most populous of all seven Central American republics with 13.5 million inhabitants (mestizos are now in the [shrinking] majority with 56 percent, Amerindians 44 percent), Guatemala has seen much conflict. Repressive regimes made deals with U.S. and other foreign economic interests that stimulated development, but at a high social cost. Over the past half-century, military regimes have dominated political life. The deepening split between the wretchedly poor Amerindians and the better-off mestizos, who here call themselves *ladinos,* generated a civil war that started in 1960 and has since claimed more than 200,000 lives as well as 50,000 "disappearances" (with 93 percent of these casualties inflicted by the Guatemalan army and its paramilitary death squads). An overwhelming number of the victims have been of Mayan descent; the mestizos control the government, army, and land-tenure system.

Because the insurgents were able to take refuge in Guatemala's remote, rainforested plains on the Mexican border, the war was unwinnable for the ladinos based in the incipient core area of the southern highlands. Rising international outrage over nearly 40,000 human-rights violations perpetrated by the armed forces finally forced the government to the negotiating table, and in 1996 a peace agreement was concluded. The central provisions of the pact were the rebuilding of a civil society that fully included the rebels, and a major reduction in the role and influence of the military. For more than a year the political violence ceased, and initial steps toward stability were taken. But in the spring of 1998 the hopeful new atmosphere was shattered when one of the country's leading human-rights advocates, Bishop Juan José Gerardi, was murdered in typical death-squad fashion two days after he released a report that documented many of the war's atrocities. Today the bishop's assassination still casts a deep shadow over the country as Guatemalans wonder if the peace process will ever be completed. More ominously, the army continues to wield excessive power, and the shaky transition to democracy remains stalled as the indigenous minority awaits constitutional recognition and access to the full rights of citizenship.

The tragedy of Guatemala is that its economic geography has considerable potential but has long been shackled by the unending internal conflicts that have kept the income level of 80 percent of the population below the poverty line. The country's mineral weath is not fully known, but it includes nickel in the highlands and oil in the lower-lying north. Agriculturally, because elevations in the south's mountain backbone reach over 13,000 feet (4000 m), soils are fertile and moisture is ample over areas large enough to produce a wide range of crops including excellent coffee. Ironically, this promising upland zone is now an area of out-migration for a horde of mostly indigenous, land-hungry peasants who are streaming northward into the infertile tropical rainforests of the Petén, thereby doubling the population of the Guatemalan north over the past decade from a quarter- to more than a half-million. The push factors in this migration are the increasing pressures on traditional subsistence farmlands propelled by the realm's highest birth rate, the return of tens of thousands of Amerindians from Mexican exile following the civil war, and ineffective land reform; the pull factors are the (unjustified) beliefs that the land is rich and job opportunities are plentiful, plus a savvy understanding that the government uses the Petén as a social safety valve and looks the other way when it comes to squatter settlements as well as the enforcement of conservation regulations. The last is most significant because the Petén's environmentally sensitive woodlands—especially the Maya Biosphere Preserve (covering the northernmost tier of Guatemala north of the latitude of Tikal National Park in Figure 4-12—are already being trashed by illegal loggers and careless oil drillers. The new settlers follow in their wake and torch the remaining trees, believing they are clearing lands for crop-raising (an activity that will never succeed at low technological levels on these leached soils). Not surprisingly, the widespread fires take a heavy toll and contribute mightily to Central America's tropical deforestation crisis (discussed on p. 232).

Peace and security would also invigorate a major tourist industry based on Guatemala's incomparable Mayan heritage (both past and present), spectacular scenery, and magnificent beaches. And, most of all, stability would enable Guatemala to build the modern infrastructure it so urgently needs (in 2000, the entire country contained barely 3000 miles [5000 km] of paved roads). Even the area around the capital, Guatemala City, still lies poorly connected to the outside world. From there the overland trip to Puerto Barrios on the Caribbean coast is tortuous and slow, and many travelers have been waylaid by criminals. Conditions at least as bad also plague the routes to the south, where only minimal port development has occurred on the Pacific coast. If the challenge of ethnic division can be overcome, the rewards for Guatemala will be substantial.

Belize: Changing Identity

Strictly speaking, Belize is not a Central American republic in the same tradition as the other six. Until 1981, this country, a wedge of land between northern Guatemala, Mexico's Yucatán Peninsula, and the Caribbean (Fig. 4-12), was a dependency of the United Kingdom known as British Honduras. For much of the postindependence period, Guatemala refused to recognize Belize's existence, claiming that an 1859 treaty with Britain had "stolen" the territory from Guatemala. Although tensions have abated in recent years, following Guatemala's 1991 abandonment of its claim to much of Belize, the dispute has not been fully resolved because the Guatemalans still insist they own Belize's southernmost corner. Slightly larger than Massachusetts and with a minuscule population of only 280,000 (many of African descent), Belize has been more reminiscent of a Caribbean island than of a continental Middle American state.

Today, all that is changing as the migrations of the past decade reshape the demographic complexion of Belize. Thousands of residents of African descent (the *Creoles*, predominantly English speakers) have recently emigrated. Most went to the United States, and their departure precipitated a drop in the proportion of Belize's Creole population from 40 percent in 1980 to less than 25 percent today. They have been replaced by tens of thousands of Spanish-speaking immigrants—mostly escapees from strife in nearby Guatemala, El Salvador, and Honduras who saw *their* proportion of the Belizean population rise from 33 to nearly 50 percent between 1980 and 2002. Within the next few years the newcomers will be in the majority, Spanish will become the *lingua franca*, and Belize's cultural geography will exhibit an expansion of the Mainland at the expense of the Rimland.

The Belizean transformation extends to the economic sphere as well. No longer just an exporter of sugar and bananas, Belize is producing new commercial crops and has thriving seafood-processing and clothing industries that have become major revenue earners. Also important is tourism, which annually lures more than 150,000 vacationers to the country's Mayan ruins, resorts, and newly legalized casinos. A growing specialty is ecotourism, based on the natural attractions of the country's near-pristine environment whose coastal zone includes the world's second-largest barrier reef. Belize is also developing a reputation as one of the realm's centers of *offshore banking*—a financial haven for foreign companies and individuals who want to avoid paying taxes in their home countries. This is one of the stimuli spurring the growth of coastal Belize City, which served as the capital until 1970. The current capital, interior Belmopan (home to a mere 7000 inhabitants), became the country's headquarters after Belize City was repeatedly devastated by hurricanes during the 1960s.

Honduras: Deluged by Disaster

We noted at the outset of this chapter that hurricanes rank among the most dangerous of the natural hazards that repeatedly bedevil Middle America. Unfortunately, Honduras took the brunt of one of the worst of these tropical cyclones in 1998, and everything else is on hold while this country spends the opening years of the new century struggling to get back on its feet and rebuild what was already the third

The text reveals the details of the horrifying impact of Hurricane Mitch on Honduras in late 1998. Yet, despite the devastating blow to their already struggling country, the resilience of the Hondurans amazed international disaster-response teams. The gargantuan reconstruction effort began almost immediately, particularly at the grass roots level. With 35 percent of the population living outside urban centers, rural education reform is a high priority, and many of the hardest-hit communities have received aid to rebuild their schools. Nueva Mendez was one of these villages, and here a group of its children proudly pose before their schoolhouse, newly rebuilt on higher ground, only seven months after the storm.

poorest economy in the Americas (after Haiti and Nicaragua). The hurricane, named Mitch, rampaged all across northern Central America and proved to be the costliest disaster in the modern history of the Western Hemisphere. Honduras was hit the hardest, battered for more than two days with winds exceeding 150 mph, four *feet* of rain, and 12-foot tidal surges all along the Caribbean coast. The consequences were catastrophic as massive floods and mudslides were unleashed throughout the country, killing 9200 people, demolishing more than 150,000 homes, destroying 21,000 miles (34,000 km) of roadway and 335 bridges, and rendering 2 million—nearly one-third of the total population—temporarily homeless. Also devastated was the critical agricultural sector that employed two-thirds of Honduras's labor force, accounted for nearly a third of its gross domestic product, and earned more than 70 percent of its foreign revenues. (The key to the rebuilding of the commercial farming sector is reinvestment by the large companies that control banana plantations. Even though the replanting of damaged fields was completed in 2000, this was often accompanied by the installation of more modern irrigation and packing equipment that are expected to reduce labor needs by at least 10 percent.)

With 6.4 million inhabitants, about 90 percent mestizo, struggling Honduras still has years to go in its recovery, especially the permanent repair of its shattered infrastructure—even to the barely functional level that existed before Mitch. Agriculture, livestock, forestry, and some mining of lead and zinc formed the mainstays of the pre-1998 economy, with the familiar Central American products—bananas, coffee, shellfish, apparel—earning most of the external income (coffee is easiest to restore because this crop grows on hillsides above the mud-choked valleys, which upon drying left huge quantities of sand that harm the quality of the soils they blanketed). There has also been some growth of maquiladora-type light industry around San Pedro Sula near the northwestern coast, dominated by a complex of sweatshops that finish clothing designed and mostly produced elsewhere. Foreign investors, however, have been reluctant to risk additional funds to expand manufacturing, first due to regional strife and lately because of the country's reputation as one of the world's most corrupt (U.S.-bound drug smuggling flourishes here, and prostitution is propelling an intensifying AIDS outbreak that has reached epidemic proportions). Nonetheless, Honduras has a democratically elected government, although the military wield considerable power.

Honduras, in direct contrast to Guatemala, has a lengthy Caribbean coastline and a small window on the Pacific (Fig. 4-12). The country also occupies a critical place in the political geography of Central America, flanked as it is by Nicaragua, El Salvador, and Guatemala—all continuing to grapple with the aftermath of years of internal conflict and, most recently, natural disaster. The road back to economic viability is a long and difficult one, but once traversed will still leave four out five Hondurans deeply mired in poverty and the country with little overall improvement in its development prospects. This is yet another tragedy because social and ethnic divisions, which so strongly mark Honduras's neighbors, are not serious here, and the gap between rich and poor, though evident, is not as wide.

El Salvador: Postwar Reconstruction

El Salvador is Central America's smallest country territorially, smaller even than Belize, but with a population about 25 times as large (6.6 million) it is the most densely peopled. Again, like Belize, it is one of only two continental republics that lack coastlines on both the Caribbean and Pacific sides (Fig. 4-12). El Salvador adjoins the Pacific in a narrow coastal plain backed by a chain of volcanic mountains. The country's heartland lies behind those mountains in the interior, where the capital, San Salvador, is located. North of this core area, along the Honduran border, lies another zone of mountains, which has always contained areas beyond effective governmental control.

Unlike neighboring Guatemala, El Salvador has a quite homogeneous population (94 percent mestizo and just 5 percent Amerindian). Yet ethnic homogeneity has not translated into social or economic equality or even opportunity. Whereas other Central American countries were called banana republics, El Salvador was a coffee republic, and the coffee was produced on the huge landholdings of a few landowners and on the backs of a subjugated peasant labor force. The military supported this system and repeatedly suppressed violent and desperate peasant uprisings.

From 1980 to 1992, El Salvador was torn by a devastating civil war that was worsened by outside arms supplies from the United States (supporting the government) and Nicaragua (aiding the Marxist rebel forces). Following the negotiated end to that war, in which some 75,000 people, mostly peasants, were killed, an effort has been underway to prevent a recurrence. But El Salvador, its historical geography one of division and disintegration, is having difficulty overcoming its legacy of searing inequality.

Perversely, the civil war had one positive result: wealthy citizens who left the country and did well in the United States and elsewhere send substantial funds back home (more than U.S. $1 billion annually), which now provide the largest single source of foreign revenues. This boost to the economy has helped stimulate such urban industries as apparel and footwear manufacturing, as well as food processing. The clothing industry in particular has grown rapidly since the mid-1990s. At the same time, however, the country has experienced such a sharp rise in its crime rate (attributed to the availability of weapons following the war) that new foreign investment remains inhibited.

Geographically, with the heartland economy recovering, El Salvador's biggest problems lie in its marginal interior zones, where peasant families on overpopulated lands feel forgotten in the national reconstruction effort. A major stumbling block to revitalization of the agricultural sector has been land reform because efforts to redistribute properties have become embroiled in countless disputes over landownership. El Salvador's future still hangs in the balance.

Nicaragua: Mired in Misfortune

When studying Nicaragua, one does well to look again at the map (Fig. 4-12), which underscores the country's pivotal position in the heart of Central America. The Pacific coast follows a southeasterly direction, but the Caribbean coast is oriented north-south so that Nicaragua forms a triangle of land with its lakeside capital, Managua, located in a great longitudinal valley on the mountainous, earthquake-prone, Pacific side (the city was heavily damaged by quakes in 1885, 1931, and 1972). Indeed, the core area of Nicaragua has always been in this western zone. The Caribbean side, where the mountains and valleys give way to a coastal plain of (disappearing) rainforest, pine savanna, and swampland, has for centuries been home to Amerindian peoples such as Miskito who have been relatively remote from the focus of national life.

Until the end of the 1970s, Nicaragua was the typical Central American republic, ruled by a dictatorial government and exploited by a wealthy land-owning minority, its export agriculture dominated by the huge plantations owned by foreign corporations. It was a situation ripe for insurgency, and in 1979 the leftist Sandinista rebels overthrew the government in Managua. Sandinista rule, however, quickly produced its own excesses, resulting in civil war for most of the 1980s. The conflict ended in 1990, and in the free elections that followed, more democratic, anti-Sandinista regimes were successively voted into office. Nonetheless, Sandinistas continue to wield considerable influence in running the country (especially the military), and Nicaraguans remain so divided that they are unable to resolve most economic and political issues.

Nicaragua's economy has been a leading casualty of this turmoil, and for the past two decades it has ranked as continental Middle America's poorest. Still mired in the difficult aftermath of Hurricane Mitch, the country today challenges Haiti for the dubious distinction of being the realm's—and therefore the hemisphere's—poorest. Even though it was not as widely devastated as neighboring Honduras, Nicaragua's struggle to rebuild much of the national infrastructure is yet another massive challenge piled upon its already formidable burdens. With the exception of coffee, which grows on hillsides largely immune to the effects of flooding, most of northern Nicaragua's crops (which account for 75 percent of the gross national product) were wiped out by the ferocious 1998 storm. The agricultural recovery effort remains a top priority in both Nicaragua and Honduras for two reasons: (1) the products of commercial agriculture are leading foreign-income earners; and (2) if farming opportunities are not promptly restored, emigration from the countryside to the badly overcrowded towns and cities will greatly accelerate, thereby swamping prospects for urban economic development. Moreover, in the months before Mitch, the Nicaraguans had finally resolved major property-ownership issues stemming from Sandinista land redistribution. Since redistribution affected more than 200,000 families and at least one-third of the country's arable farmland, continued national stability necessitates that this important social reform get back on track as soon as possible.

When "normal" conditions return, Nicaragua will still possess few comparative advantages in the competition for foreign investment capital against the region's more prosperous republics. Among the more promising development possibilities is the construction of a *dry canal* across the flatlands of southern Nicaragua, with high-speed trains ferrying freight-filled containers between Caribbean and Pacific ports that can be directly loaded onto ultramodern ships too large to fit through the isthmus's aging Panama Canal. Even though the Nicaraguans were the first to discuss this idea, as we saw earlier Mexico is already implementing plans to complete such a facility across its southern Isthmus of Tehuantepec. Perhaps Central America can support additional interocean transit corridors, but here Nicaragua will confront other competitors who are proposing such projects in Colombia, jointly in El Salvador and Honduras, and even in Panama itself. Economic opportunity, however, must also be weighed against looming demographic disaster: unless Nicaragua can reduce the very high natural increase rate in its population of 5.4 million (the most rapid growth in the Americas, exceeding even Haiti's by an astounding 76 percent), the country's living standards will not have a chance to improve after hurricane recovery is complete.

Costa Rica: Durable Democracy

As if to confirm what was said about Middle America's endless variety and diversity, Costa Rica differs in significant ways from its neighbors and from the norms of Central America as well. Bordered by two volatile countries (Nicaragua to the north and Panama to the east), Costa Rica is a nation with an old democratic tradition and, in this cauldron, no standing army! The country is, in fact, the oldest democracy in Middle and South America, enjoying a freely elected government, except for two brief periods, since 1889. Although the country's Hispanic imprint is similar to that found elsewhere on the Mainland, its early independence, its good fortune to lie remote from regional strife (which fostered an enduring neutrality), and its leisurely pace of settlement allowed Costa Rica the luxury of concentrating on its economic development and on public education—which is universal, free, and produces a literacy rate of 95 percent in a region where the average is only 75

Tropical Deforestation

Before the Europeans arrived, two-thirds of continental Middle America was covered by tropical rainforests. The clearing and destruction of this precious woodland resource to make way for expanding settlement frontiers and the exploitation of new economic opportunities began in the sixteenth century during the Spanish colonial era, and the practice has continued systematically ever since. In recent decades, however, the pace of 19 **tropical deforestation** in Central America has accelerated alarmingly, and since 1950 close to 85 percent of the region's forests have been decimated. Now, about 3 million acres of Central American and Mexican woodland disappear each year, an area equivalent to one-third the size of Belgium. El Salvador has already lost 99 percent of its forests, and most of the six other republics will soon reach that stage.

The causes of tropical deforestation are related to the persistent economic and demographic problems of disadvantaged countries. In Central America, the leading cause has been the need to clear rural lands for cattle pasture as many countries, especially Costa Rica, became meat producers and exporters. The price of this environmental degradation has been enormous, although some gains have been recorded. Because tropical soils are so nutrient-poor, newly deforested areas can function as pastures for only a few years at most. These fields are then abandoned for other freshly cut lands and quickly become the ravaged landscape seen in the photo in this box. Without the protection of trees, local soil erosion and flooding immediately become problems, affecting still-productive nearby areas (a sequence of events that reached catastrophic dimensions all across Honduras and northern Nicaragua when Hurricane Mitch struck in 1998). A second cause of deforestation is the rapid logging of tropical woodlands as the timber industry increasingly turns from the exhausted forests of the mid-latitudes to harvest the rich tree resources of the equatorial zones, responding to accelerating global demands for new housing, paper, and furniture. The third major contributing factor is related to the region's population explosion: as more and more peasants are required to extract a subsistence from inferior lands, they have no choice but to cut down the remaining forest for both firewood and additional crop-raising space, and their intrusion prevents the trees from regenerating (Haiti is the extreme example of this denudation process).

Although deforestation is a depressing event, tropical pastoralists, farmers, and timber producers do not consider it to be life-threatening, and perhaps it even seems to offer some short-term economic advantages. Why, then, should there be such an outcry from the scientific community? And why should the World Resources Institute call this "the world's most pressing land-use problem"? The answer is that unless immediate large-scale action is taken, by the middle of this century the world's tropical rainforests will be reduced to two disappearing patches—the western Amazon Basin of northern South America and the middle Congo Basin of Equatorial Africa.

The tropical forest, therefore, must be a very important part of our natural world, and indeed it is. Biologically, the rainforest is by far the richest, most diversified arena of life on our planet: even though it covers only about (a shrinking) 3 percent of the Earth's land area, it contains about three-quarters of all plant and animal species. Its loss would cause not only the extinction of millions of species, but also what ecologist Norman Myers calls "the death of birth" because the evolutionary process that produces new species would be terminated. Because tropical rainforests already yield countless valuable medicinal, food, and industrial products, many potential disease-combatting drugs or new crop varieties to feed undernourished millions will be irretrievably lost if they disappear.

This scene in Costa Rica shows how badly the land can be scarred in the wake of deforestation. Without roots to bind the soil, tropical rains swiftly erode the unprotected topsoil.

percent. Perhaps most important, internal political stability has prevailed over much of the past 175-odd years. The last brush with conflict more than 50 years ago left the nation resolved to avoid further violence, and the armed forces were abolished in 1948 (along with a military establishment, which is so often the source of trouble throughout Central America).

Like its neighbors, Costa Rica is divided into environmental zones that parallel the coasts. The most densely settled is the central highland zone, lying in the cooler *tierra templada*. Volcanic mountains predominate in much of this zone, but the heartland is the Valle Central (Central Valley), a fertile 40-by-50-mile (65-by-80-km) basin that contains the country's main coffee-growing area and the leading population cluster focused on the capital city. The capital, San José, is atypical of Middle America: clean and virtually slumless, it is the most cosmopolitan urban center between Mexico City and the primate cities of northern South America.

To the east of the highlands are the hot and rainy Caribbean lowlands, a sparsely populated segment of Rimland where many plantations have now been abandoned and replaced by subsistence farmers. Between 1930 and 1960, the U.S.-based United Fruit Company shifted most of the country's banana plantations from the crop-disease-ridden Caribbean littoral to Costa Rica's third zone—the plains and gentle slopes of the Pacific coastlands. This move gave the Pacific zone a major boost in economic growth, although banana production continues to drift to Panama, where wages are lower. The Pacific zone is now an area of diversifying and expanding commercial agriculture (often requiring irrigation) and of successful colonization in previously undeveloped valleys and basins. The new container port recently opened at Puerto Caldera is spurring additional Pacific coast development.

The long-term development of Costa Rica's economy has given it the region's highest standard of living, literacy rate, and life expectancy. Agriculture continues to dominate (with bananas, coffee, seafood, and tropical fruits the leading exports), and tourism is expanding steadily as the country now attracts more than a million foreign visitors a year. But the most impressive economic gains are being generated by a single new industrial complex that the Intel Corporation opened just outside San José in 1998. Here, the world's leading maker of semiconductors has built one of its largest assembly/test facilities in which silicon chips manufactured at fabrication plants are transformed into the processors that constitute the brains of personal computers. Although the Intel complex has sharply boosted the Costa Rican economy (its products account for about 40 percent of the country's exports—the largest such national share by one corporation anywhere on Earth), some of the nation's economists are concerned that this represents too great a dependency on a single company. Downplaying these worries, the country's developers point up Costa Rica's transition from a banana republic to a high-tech republic in which the growing pool of highly skilled workers will launch their own businesses, attract many more foreign firms, and create one of the realm's leading technopoles.

These advances notwithstanding, the country's veneer of development cannot mask serious problems. In terms of social structure, about one-quarter of the population of 3.7 million is trapped in an unending cycle of poverty, and the huge gap between the poor and the affluent is constantly widening. In the environmental sphere, recent progress has come at the expense of the tropical forest: even though the rate of woodland destruction is now slowing, it is too late to avert an ecological disaster because more than 70 percent of Costa Rica's forest has vanished (see box titled "Tropical Deforestation").

Politically, Costa Rica remains quite stable. Despite its proximity to the region's trouble spots, it has resisted involvement because the overwhelming majority of its peace-loving people prefer the country to maintain its neutrality as "the Switzerland of Central America." One of the few clouds on the horizon is a longstanding boundary dispute with neighboring Nicaragua over the waters and banks of the San Juan River, a quarrel that extends westward to include the southern shore of Lake Nicaragua (Fig. 4-12). If this corridor were to change hands, many observers believe that Costa Rica would become yet another entrant in the contest to build a new interocean routeway across Central America.

Panama: Strategic Canal, Reorganizing Corridor

Other republics may *talk* about building canals, but Panama owes its very existence to the idea of an artificial waterway connecting the Atlantic and Pacific oceans to avoid the lengthy circumnavigation of South America. In the 1880s, when Panama was still an extension of neighboring Colombia, a French company tried and failed to build such a canal here; thousands of workers died of yellow fever, malaria, and other tropical diseases, and the company went bankrupt. By the turn of the twentieth century, U.S. interest in a Panama canal (which would shorten the sailing distance between the East and West Coasts by 8000 nautical miles) rose sharply, and the United States in 1903 proposed a treaty that would permit a renewed effort at construction across Colombia's Panamanian isthmus. When the Colombian Senate refused to go along, Panamanians rebelled, and the United States supported this uprising by preventing Colombian forces from intervening. The Panamanians, at the behest of the United States, declared their independence from Colombia, and the new republic immediately granted the United States rights to the Canal Zone, averaging about 10 miles (16 km) in width and just over 50 miles (80 km) in length.

FROM THE FIELD NOTES

"The Panama Canal remains an engineering marvel nearly 90 years after it opened in August 1914. The parallel lock chambers each are 1000 feet long and 110 feet wide, permitting vessels as large as the Queen Elizabeth II to cross the isthmus. Ships are raised by a series of locks to Gatún Lake, 85 feet above sea level. We watched as tugs helped guide the QEII into the Gatún Locks, a series of three locks leading to Gatún Lake, on the Atlantic side. A container ship behind the QEII is sailing up the dredged channel leading from the Limón Bay entrance. The lock gates are 65 feet wide and seven feet thick, and range in height from 47 to 82 feet. The motors that move them are recessed in the walls of the lock chambers. Once inside the locks, the ships are pulled by powerful locomotives called *mules* that ride on rails that ascend and descend the system. It was still early morning, and a major fire, probably a forest fire, was burning near the city of Colón, where land clearing was in progress. This was the beginning of one of the most fascinating days ever."

Soon canal construction commenced, and this time the project succeeded as American engineering and technology—and medical advances—triumphed over a formidable set of obstacles. The Panama Canal (see inset map, Fig. 4-12; photo above) was opened in 1914, a symbol of U.S. power and influence in Middle America. The Canal Zone was held by the United States under a treaty that granted it "all the rights, powers, and authority" in the area "as if it were the sovereign of the territory." Such language might suggest that the United States held rights over the Canal Zone in perpetuity, but the treaty nowhere stated specifically that Panama permanently yielded its own sovereignty in that transit corridor. In the 1970s, as the canal was transferring more than 14,000 ships per year (that number is now only slightly lower, but the cargo tonnage is up significantly) and generating hundreds of millions of dollars in tolls, Panama sought to terminate U.S. control in the Canal Zone. Delicate negotiations began. In 1977, an agreement was reached on a staged withdrawal by the United States from the territory, first from the Canal Zone and then from the Panama Canal itself (a process completed on the final day of 1999).

Panama today reflects some of the usual geographic features of the Central American republics. Its population of 3.0 million is more than two-thirds mestizo and also contains substantial Amerindian, white, and black minorities. Spanish is the official (and majority) language, but English is also widely used. Ribbon-like and oriented east-west, Panama's topography is mountainous and hilly, with some peaks reaching higher than 10,000 feet (3000 m). Eastern Panama, especially Darien Province adjoining Colombia, is densely forested, and here is the only remaining gap in the intercontinental Pan American Highway. (Darien today is a spillover battleground in the conflict between Colombian guerrillas and their government, so completion of the highway remains a dream.) Most of the rural population lives in the uplands west of the canal; there, Panama produces bananas, shrimp and other seafood, sugarcane, coffee, and rice. Much of the urban population (which constitutes 56 percent of the national total) is concentrated in the vicinity of the waterway, anchored by the cities at each end of the canal, Colón and Panama City.

The Panama Canal, despite its age and its inability to accommodate the largest 20 percent of the vessels in today's global merchant fleet, remains the country's focus, its lifeline, and—with a significant proportion of the world's cargoes moving through the waterway each year—its future. To help assure that future astride this crucial international trading artery, the canal's new Panamanian owners are busily pursuing opportunities they hope will transform their trading hub into a full-fledged Central American tiger (Singapore is most often mentioned as the model). Their initial task, now successfully completed, was to demonstrate to the international community that they are fully capable of operating and maintaining the canal without sharply increasing its tolls. They have also begun to attract major new foreign investments that are being channeled into a number of maritime, railroad, manufacturing, and tourism projects located on the prime, canalside real estate vacated by U.S. facilities that were handed over to Panama.

One of the anchors of this development is Colón, the port city at the Caribbean end of the canal. Here, in 1948, a free-trade zone was opened that has since become the world's second-largest such facility after Hong Kong. Most of the zone's activities—thanks to Colón's centrality in the Western Hemisphere—involved the import and distribution of electronic appliances, clothing, jewelry, and other merchandise bound for South America. Today, however, this warehousing operation is being challenged by the rise of competitors in Brazil, Paraguay, and Chile. In order to maintain their supremacy, Colón's resourceful leaders recognized the need to upgrade their goods-handling technology to accommodate containers (whose importance in global commerce was underscored in our earlier discussions of *dry canal* projects in Mexico and Nicaragua). The result was the late-1990s construction of the huge Manzanillo International Terminal, adjacent to the free-trade zone, a state-of-the-art port facility capable of transshipping more than 1000 containers a day.

Today, the Pacific end of the canal is an even more active center of economic development because of its greater number of former U.S. facilities as well as proximity to the capital and largest city, Panama City. One of the most prominent projects is a new tourist port at the site of the abandoned U.S. military base at Fort Amador; here, adjacent to both the canal's entrance and the capital, a complex of shopping centers, casinos, golf courses, and marinas opened in 2001 to lure the thousands of passengers aboard the 300-plus cruise ships that, until now, traversed the waterway each year without stopping. Not surprisingly, many of the industrial parks, ship terminals, and other new cargo-related facilities going up in this immediate area are financed by Chinese, Japanese, Korean, and Taiwanese interests—for whose export-dependent economies the canal has acquired strategic importance as an indispensable link in their global trading networks that span the oceans on both sides of the Isthmus of Panama. As for Panama City itself (the only coastal capital in continental Middle America), an expanding cluster of downtown highrises unlike any other in the realm suggests Miami rather than San Salvador or San José. What generates this world-class skyline that towers over an urban area anchoring a country of only 3 million people? The official answer: international banking. The geographic answer suggests the power of relative location, including a new, eyebrow-raising linkage to drug-plagued Colombia, the country on which the Panamanians a century ago turned their backs.

FIGURE 5-1
Copyright © Rand McNally, 20

chapter 5

South America

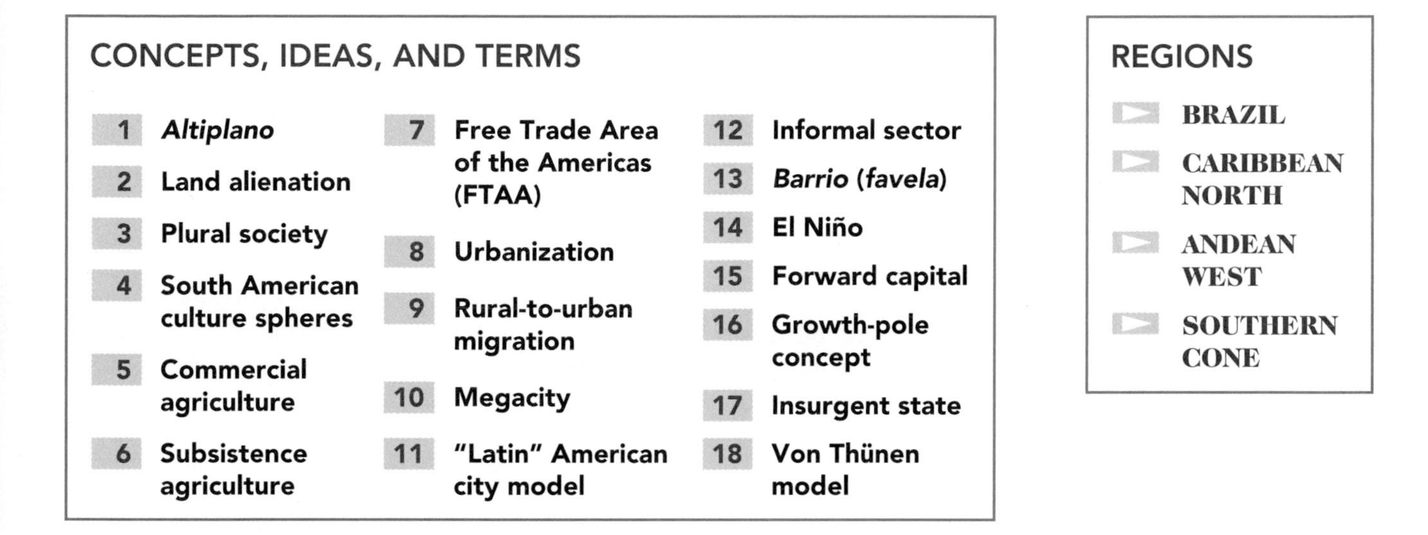

CONCEPTS, IDEAS, AND TERMS

1. *Altiplano*
2. Land alienation
3. Plural society
4. South American culture spheres
5. Commercial agriculture
6. Subsistence agriculture
7. Free Trade Area of the Americas (FTAA)
8. Urbanization
9. Rural-to-urban migration
10. Megacity
11. "Latin" American city model
12. Informal sector
13. *Barrio* (*favela*)
14. El Niño
15. Forward capital
16. Growth-pole concept
17. Insurgent state
18. Von Thünen model

REGIONS

- BRAZIL
- CARIBBEAN NORTH
- ANDEAN WEST
- SOUTHERN CONE

Of all the continents, South America has the most familiar shape—that giant triangle connected by mainland Middle America's tenuous land bridge to its sister continent in the north. What we realize less often about South America is that it lies not only south but also mostly east of its northern counterpart as well. Lima, the capital of Peru—one of the continent's westernmost cities—lies farther east than Miami, Florida. Thus South America juts out much more prominently into the Atlantic Ocean than does North America, and South American coasts are much closer to Africa and even to southern Europe than are the coasts of Middle and North America. But lying so far eastward means that South America's western flank faces a much wider Pacific Ocean than does North America; the distance from its west coast to Australia is nearly twice that from California to Japan.

As if to reaffirm South America's northward and eastward orientation, the western margins of the continent are rimmed by one of the world's longest and highest mountain ranges, the Andes, a gigantic wall that extends unbroken from Tierra del Fuego near the southern tip of the triangle to Venezuela in the far north (Fig. 5-1). Every map of world physical geography clearly reflects the existence of this mountain chain—especially in the alignment of isohyets (lines connecting places of equal precipitation totals) in Figure I-7 and in the elongated zone of highland climate in Figure I-8. Moreover, as Figure I-9 reveals, South America's largest population clusters are located along the eastern and northern coasts, overshadowing those of the Andean west.

The other major physiographic feature of South America dominates its central north: the Amazon Basin. This vast humid-tropical amphitheater is drained by one of Earth's mightiest rivers, the Amazon, which (as Fig. 5-1 shows) is fed by more than a half-dozen major tributaries, each a world-class river all by itself. Much of the remainder of the continent can be classified as plateau, with the most important components being the extensive Brazilian Highlands that cover most of Brazil southeast of the Amazon Basin, the Guiana Highlands located north of the lower Amazon Basin, and the cold Patagonian plateau that blankets the southern third of Argentina. Figure 5-1 also reveals two noteworthy river basins beyond Amazonia: the Paraná-Paraguay Basin of south-central South America, which contains Argentina's fertile Pampa and empties into the Rio de la Plata; and the Orinoco Basin in the far north that drains the interiors of both Colombia and Venezuela.

DEFINING THE REALM

Long characterized by regional disparities, political turmoil, and developmental inertia, South America has entered a new era of opportunity. Its major countries, heretofore accustomed to going their separate ways, are discovering the benefits of forging closer individual and multinational ties, especially through free-trade pacts. Juntas and their remnants have largely been swept off the political landscape, with military forces retreating to their barracks in favor of more democratic forms of government. New transport routes are opening new settlement frontiers in many once-remote parts of the continent. South America now leads the world in new mining ventures: coal and oil in Colombia; iron ore and gold in the wide fringes of the vast Amazon Basin; copper in southern Peru and northern Chile; silver, zinc, and lead in the Andes of Peru. The perception now taking hold is that, finally, things are improving and this realm is at the threshold of a period of unprecedented economic growth.

This optimism, however, must be tempered by the recognition that serious challenges remain to be confronted and overcome. As certain areas of South America make real and lasting progress, too many others continue to be plagued by infrastructure shortcomings, inefficiency and corruption, and endless rounds of no-gain, boom/bust cycles. Most importantly, throughout the realm the gulf between the rich and the poor is steadily widening: since 1980, the proportion of the continent's population living below the poverty level has jumped from 27 to 35 percent. Put another way, according to the World Bank, the richest 20 percent of the population controls *70 percent* of South America's wealth, while the poorest 20 percent controls only (an astonishingly small) 2 percent. These numbers reveal the greatest gap between affluence and poverty to be found in any geographic realm. They also remind us—recent advances notwithstanding—that South America, above all else, is still a continent of stupendous contrasts.

THE HUMAN SEQUENCE

Although modern South America's largest populations are situated in the east and north, during the height of the Inca Empire the Andes Mountains contained the most densely peopled and best organized state on the continent. Although the origins of Inca civilization are still shrouded in mystery, it has become generally accepted that the Incas were descendants of ancient peoples who came to South America via the Middle American land bridge (possibly following earlier migrations from Asia to North America across the Bering land bridge that existed west of Alaska). Thus for thousands of years before the Europeans arrived in the sixteenth century, indigenous Amer-indian communities and societies had been developing in South America.

About a thousand years ago, a number of regional cultures thrived in Andean valleys and basins and at places along the Pacific coast. The llama had been domesticated; religions flourished; sculpture, painting, and other art forms were practiced. From their headquarters in the Cuzco Basin of the Peruvian Andes, the Incas extended their authority over these cultures. Beginning in the late twelfth century, they began to forge the greatest empire in the Americas prior to the coming of the Europeans. Nothing to compare with these cultural achievements of the central Andean zone existed anywhere else in South America.

◆ Major Geographic Qualities of South America

1. South America's physiography is dominated by the Andes Mountains in the west and the Amazon Basin in the central north. Much of the remainder is plateau country.
2. Half of the realm's area and half of its population are concentrated in one country—Brazil.
3. South America's population remains concentrated along the continent's periphery. Most of the interior is sparsely peopled, but sections of it are now undergoing significant development.
4. Interconnections among the states of the realm are improving rapidly. Economic integration has become a major force, particularly in southern South America.
5. Regional economic contrasts and disparities, both in the realm as a whole and within individual countries, are strong.
6. Cultural pluralism exists in almost all of the realm's countries, and is often expressed regionally.
7. Rapid urban growth continues to mark much of the South American realm, and the urbanization level overall is today on a par with the levels in the United States and Europe.

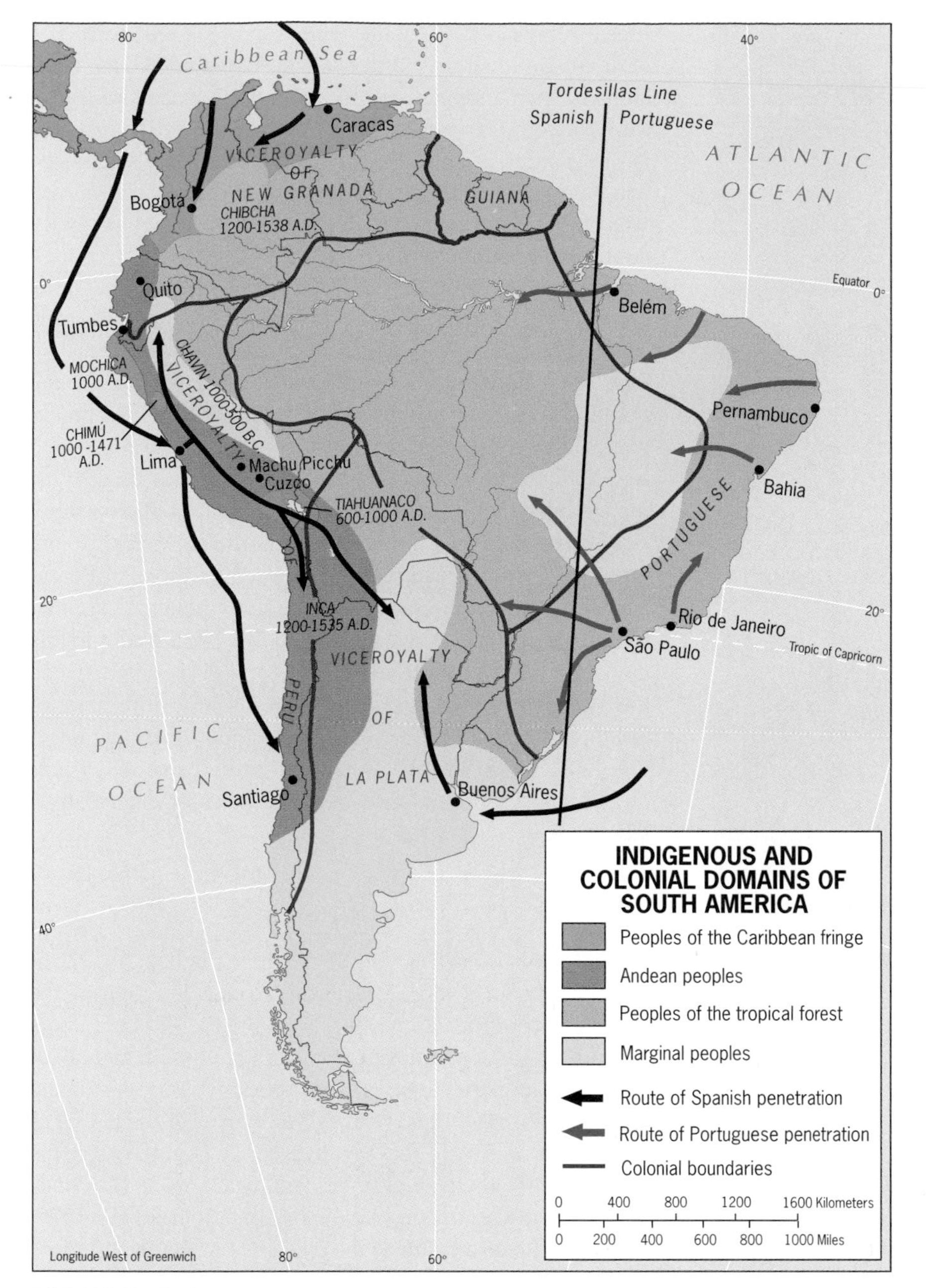

FIGURE 5-2

The Inca Empire

When the Inca civilization is compared to that of ancient Mesopotamia, Egypt, the old Asian civilizations, and the Mexican Aztec Empire, it quickly becomes clear that this civilization was an unusual achievement. Everywhere else, rivers and waterways provided avenues for interaction and the circulation of goods and ideas. Here, however, an empire was forged from a series of elongated basins (called 1 ***altiplanos***) in the high Andes, created when mountain valleys between parallel and converging ranges filled with erosional materials from surrounding uplands. These *altiplanos* are often separated from one another by some of the world's most rugged terrain, with high snowcapped mountains alternating with precipitous canyons. Individual *altiplanos* accommodated regional cultures; the Incas themselves were first established in the intermontane basin of Cuzco (Fig. 5-2). From that hearth, they conquered and extended their authority over the peoples of coastal Peru and other *altiplanos*.

More impressive than the Incas' military victories was their subsequent capacity to integrate the peoples and regions of the Andean domain into a stable and efficiently functioning state. The odds would seem to have been against them

FROM THE FIELD NOTES
"From this high vantage point I got a good perspective of a valley near Pisac in the Peruvian Andes, not far from Cuzco. This was part of the Incan domain when the Spaniards arrived to overthrow the empire, but the terraces you can see actually predate the Inca period. Human occupation in these rugged mountains is very old, and undoubtedly the physiography here changed over time. Today these slopes are arid and barren, and only a few hardy trees survive; the stream in the valley bottom is all the water in sight. But when the terrace builders transformed these slopes, the climate may have been more moist, the countryside greener."

because as they progressed their domain became ever more elongated, making effective control much more difficult. The Incas, however, were expert road and bridge builders, colonizers, and administrators, and in an incredibly short time they consolidated these new territories—which extended from southernmost Colombia southward to central Chile (brown zone, Fig. 5-2).

At its zenith, the Inca Empire may have counted more than 20 million subjects. The Incas themselves were always in a minority in this huge state, and their position became one of a ruling elite in a rigidly class-structured society. The Incas, representative of the emperor in Cuzco, formed a caste of administrative officials who implemented the decisions of their monarch by organizing all aspects of life in the conquered territories. This bureaucracy of Inca administrators strictly controlled the life of the empire's subjects, and there was little personal freedom. So highly centralized was the state and so complete the subservience of its tightly controlled population that a takeover at the top was enough to gain power over the entire empire—as the Spaniards quickly proved in the 1530s.

The Inca Empire, which had risen to greatness so rapidly, disintegrated abruptly under the impact of the Spanish invaders. Perhaps the swiftness of its development contributed to its fatal weakness, but the empire was also ripe for internal revolt in the early sixteenth century as the vaunted administrative framework was increasingly stressed by the Incas' push into the northern Andes. At any rate, apart from spectacular ruins such as those at Peru's Machu Picchu, the empire left behind social values that have remained a part of Amerindian life in the Andes to this day and still contribute to fundamental divisions between the Hispanic and Amerindian population in this part of South America. For example, the Inca state language, Quechua, was so firmly rooted that millions of Amerindians living in the highlands of Peru, Ecuador, and Bolivia still speak it.

The Iberian Invaders

In South America as in Middle America, the location of indigenous peoples largely determined the direction of the thrusts of European invasion. The Incas, like Mexico's Maya and Aztec peoples, had accumulated gold and silver at their headquarters, possessed productive farmlands, and constituted a ready labor force. Not long after the defeat of the Aztecs in 1521, the Spanish conquerors crossed the Panamanian isthmus and sailed southward along the continent's northwestern coast. On his first journey in 1527, Francisco Pizarro heard of the existence of the Inca Empire and soon withdrew to Spain to organize its overthrow. Four years later he returned to the Peruvian coast with 183 men and two dozen horses at a time when the Incas were preoccupied with problems of royal succession and strife in the northern provinces. The events that followed are well known, and in 1533 the party rode victorious into Cuzco.

At first, the Spaniards kept the Incan imperial structure intact by permitting the crowning of an emperor who was in fact under their control. But soon the land- and gold-hungry invaders were fighting among themselves, and the breakdown of the old order began. The new order that eventually emerged in western South America placed the indigenous peoples in serfdom to the Spaniards. Great haciendas were formed by **land alienation** (the takeover of former Amerindian lands), taxes were instituted, and a forced-labor system was introduced to maximize the profits of exploitation.

2

As in Middle America, most of the Spanish invaders had little status in Spain's feudal society, but they brought with them the values that prevailed in Iberia: land meant power and prestige, gold and silver meant wealth.

Lima, the west coast headquarters of the Spanish conquerors, was founded by Pizarro in 1535, about 375 miles (600 km) northwest of the Andean center of Cuzco. Before long Lima was one of the richest cities in the world, its wealth based on the exploitation of vast Andean silver deposits. The city quickly became the capital of the viceroyalty of Peru, as the authorities in Spain integrated the new possession into their colonial empire (Fig. 5-2). Subsequently, when Colombia and Venezuela came under Spanish control and, later, when Spanish settlement began to expand in the coastlands of the Rio de la Plata estuary in what is now Argentina and Uruguay, two additional viceroyalties were added to the map: New Granada in the north and La Plata in the south.

Meanwhile, another vanguard of the Iberian invasion was penetrating the east-central part of the continent, the coastlands of present-day Brazil. This area had become a Portuguese sphere of influence because Spain and Portugal had agreed in the Treaty of Tordesillas (1494) to recognize a north-south line (drawn by Pope Alexander VI) 370 leagues west of the Cape Verde Islands as the boundary between their New World spheres of influence. This border ran approximately along the meridian of 50°W longitude, thereby cutting off a sizeable triangle of eastern South America for Portugal's exploitation (Fig. 5-2).

A brief look at the political map of South America (Fig. 5-3) shows that the 1494 treaty did not succeed in limiting Portuguese colonial territory to the east of the agreed-upon

FIGURE 5-3

50th meridian. True, the boundaries between Brazil and its northern and southern coastal neighbors (French Guiana and Uruguay) both reach the ocean near 50°W; but then the Brazilian boundaries bend far inland to include almost the entire Amazon Basin as well as a good part of the Paraná-Paraguay Basin in the south. Thus Brazil came to be only slightly smaller in territorial size than all the other South American countries combined (and today also accounts for just under half the total population of the realm). The enormous, successful, westward thrust was the work of many Brazilian elements, including missionaries in search of converts and explorers in search of quick wealth. No group, however, did more to achieve this penetration than the so-called *Paulistas*, the settlers of São Paulo who needed Amerindian slave labor to run their highly profitable plantations.

FROM THE FIELD NOTES

"Walking the narrow streets of the old town of Salvador, one of Brazil's oldest cities, long its capital, and the seat of government of Bahia State, revealed the damage done by weathering in this tropical environment. Magnificent churches, public buildings, and mansions were built during the time when this was, by many measures, the most important city in the Southern Hemisphere. The slave trade, the whaling industry, and fortune-making plantation agriculture concentrated enormous wealth here, some of which went into the construction of an opulent city center. But fortunes change, and Salvador lost its political primacy as well as its major sources of income, and the city fell into the decay we see here. During the second half of the twentieth century, though, recovery began, and the United Nations designated the old city as a World Heritage site, making it eligible for rehabilitation. The mansion shown here in the 1980s is now restored, as are hundreds of other buildings in this historic old town."

The Africans

As Figure 5-2 shows, the Spaniards initially got very much the better of the territorial partitioning of South America—not just quantitatively but qualitatively as well. There were no rich Amerindian states to be conquered and looted east of the Andes, and no productive agricultural land was under cultivation. The comparatively few eastern Amerindians constituted no long-term, usable labor force. It has been estimated that the entire area of present-day Brazil was inhabited by no more than 1 million aboriginal people.

When the Portuguese finally began to develop their New World territory, they turned to the same lucrative activity that their Spanish rivals had pursued in the Caribbean—the plantation cultivation of sugar for the European market. And they, too, found their labor force in the same source region, as millions of Africans were brought in slavery to the tropical Brazilian coast north of Rio de Janeiro (see map, p. 349). Not surprisingly, Brazil now has South America's largest black population, which is still heavily concentrated in the country's poverty-stricken northeastern States. Today, with the overall population of Brazil at 175 million, 6 percent (more than 10.5 million) of the people are "pure" black and another 38 percent (67 million) are of mixed African, white, and Amerindian ancestry. Africans, then, definitely constitute the third major immigration of foreign peoples into South America.

Longstanding Isolation

Despite their common cultural heritage (at least insofar as their European-mestizo population is concerned), their adjacent location on the same continent, their common language, and their shared national problems, the countries that arose out of South America's Spanish viceroyalties have existed in a considerable degree of isolation from one another. Distance and physiographic barriers reinforced this separation, and the realm's major population agglomerations still adhere to the coast, mainly the eastern and northern coasts (Fig. I-9). Compared with other geographic realms, South America may be described as underpopulated not just because of its modest total (357 million) for a landmass of its size but also because of the resources available for or awaiting development. This continent never drew as large an immigrant European population as did North America. The Iberian Peninsula could not provide the numbers of people that western and northwestern Europe did, and Spanish colonial policy had a restrictive effect on the European inflow.

The New World viceroyalties existed primarily for the purpose of extracting riches and filling Spanish coffers. In Iberia there was little interest in developing the American lands for their own sake. Only after those who had made Spanish and Portuguese America their permanent home and who had a stake there rebelled against Iberian authority did things begin to change, and then very slowly. South America was saddled with the values, economic outlook, and social attitudes of eighteenth-century Iberia—not the best tradition from which to begin the task of forging modern nation-states.

Independence

Certain isolating factors had their effect even during the wars for independence. Spanish military strength was always concentrated at Lima, and those territories that lay farthest from their center of power—Argentina and Chile—were the first to establish their independence from Spain (in 1816 and 1818, respectively). In the north, Simón Bolívar led the growing independence movement, and in 1824 two decisive military defeats there spelled the end of Spanish power in South America. Thus, in little more than a decade, the Hispanic countries had fought themselves free. But this joint struggle did not produce unity: no fewer than nine countries emerged from the three former viceroyalties.

It is not difficult to understand why this fragmentation took place. With the Andes intervening between Argentina and Chile and the Atacama Desert between Chile and Peru, overland distances seemed even greater than they really were, and these obstacles to contact proved quite effective. Hence the countries of South America began to grow apart—a separation process sometimes heightened by uneasy frontiers. Friction and even wars have been frequent there, and dozens of boundary disputes remain unresolved to this day.

Brazil attained independence from Portugal at about the same time that the Spanish possessions in South America were struggling to end overseas domination, but the sequence of events was quite different. In Brazil, too, there had been revolts against Portuguese control. But the early 1800s, instead of witnessing a decline in Portuguese authority, actually brought the Portuguese government (headed by Prince Regent Dom João) from Lisbon to Rio de Janeiro! Incredibly, Brazil in 1808 was suddenly elevated from colonial status to the seat of empire. It owed its new position to Napoleon's threat to overrun Portugal, which was then allied with the British. Although many expected that a new era of Brazilian progress and development would follow, things soon fell apart. By 1821, with the Napoleonic threat removed, Portugal decided to demote Brazil to its former colonial status. But Dom Pedro, Dom João's son and the new regent, defied his father and led a successful struggle for Brazilian independence.

The postindependence relationships between Brazil and its Spanish-influenced neighbors remained similar to the limited interactions among the individual Hispanic republics themselves. Distance, physical barriers, and cultural contrasts served to inhibit positive contact and interaction. Thus Brazil's orientation toward Europe, like that of the other republics, remained stronger than its involvement with the countries on its own continent. Only within the past ten years have the countries of the South American realm finally begun to recognize the mutual advantages of increasing cooperation, thereby ushering in a new era.

CULTURAL FRAGMENTATION

When we speak of the "orientation" or "interaction" of South American countries, it is important to keep in mind just who does the orienting and interacting, for there is a tendency to generalize the complexities of these countries away. The fragmentation of colonial South America into ten individual republics and the nature of their subsequent relationships were the work of a small minority of the people in each country. At the time of independence, people of African descent in Brazil had little or no voice in the course of events; the Amerindians in Peru, numerically a vast majority, could only watch as their European conquerors struggled with each other for supremacy. It would not even be true to say that the European minorities *in toto* governed and made policy: the wealthy, landholding, upper-class elite determined the posture of the state.

So complex and heterogeneous are the societies and cultures of Middle and South America that practically every generalization has to be qualified. Take the one so frequently used—the term *Latin* America. Apart from the obvious exceptions that can be read from the map, such as Jamaica, Guyana, and Suriname, which are clearly not "Latin" countries, it may be improper even to identify some of the Spanish-influenced republics as "Latin" in their cultural milieu. Certainly the white, wealthy upper classes are of Latin European stock, and they have the most influence at home and are most visible abroad; they are the prominent politicians, businesspeople, writers, and artists. Their cultural environment is made up of the Spanish (and Portuguese) language, the Roman Catholic Church, and the picturesque Mediterranean-style architecture of Middle and South America's cities and towns; these things provide them with a common bond and strong ties to Iberian Europe. But in the mountains and villages of Ecuador, Peru, and Bolivia are millions of people to whom the Spanish language is still alien, to whom the white people's religion is another unpopular element of acculturation, and to whom decorous Spanish styles of architecture are meaningless when a decent roof and a solid floor are still unattainable luxuries.

South America, then, is a continent of **plural societies**, 3 where Amerindians of different cultures, Europeans from Iberia and elsewhere, blacks from western tropical Africa, and Asians from India, Japan, and Indonesia cluster in adjacent areas but do not mix. The result is a cultural kaleidoscope of almost endless variety, whose internal divisions are also reflected in the realm's economic landscape (see box titled "South America's Divided Agricultural Map"). Certainly, calling this human spatial mosaic "Latin" America is not very useful. Is there a more meaningful approach to a regional generalization that would better represent and differentiate the continent's cultural and economic spheres? John Augelli, who also developed the Mainland-Rimland concept for Middle America, made such an attempt. His map (Fig. 5-5) shows that five **South American culture spheres**—internal cultural regions—blanket the realm. This regionalization scheme is quite useful if we remember that South America is undergoing economic change in certain areas today and that these culture spheres are generalized and subject to further modification as well.

Tropical-Plantation Region

The first culture sphere, the *tropical-plantation* region, in many ways resembles the Rimland of the Middle American realm. It consists of several separated areas, of which the

South America's Divided Agricultural Map

The map of South America's dominant livelihood, agriculture (Fig. 5-4), reveals an unusual pattern. Here **commercial** (for-profit) and **subsistence** (minimum-life-sustaining) farming exist side by side to a greater degree than elsewhere in the world, where one normally geographically dominates the other. This, of course, does not represent a planned "balance" between the two but is yet another expression of the continent's deep human-spatial divisions.

The commercial agricultural side of South America is expressed in (1) a huge cattle-ranching zone near the coasts (Map Category 9) that stretches southwestward from northeastern Brazil to Patagonia; (2) Argentina's wheat-raising Pampa (Category 3), which is comparable to the U.S. Great Plains; (3) a Corn Belt-type crop and livestock zone (Category 2) in northeastern Argentina, Uruguay, southern Brazil, and south-central Chile; (4) a number of seaboard tropical plantation strips (Category 6) located in eastern Brazil, the Guianas, Venezuela, Colombia, and Peru; and (5) a Mediterranean-type agricultural zone (Category 5) in central Chile.

In stark contrast to these commercial systems, subsistence farming blankets the rest of South America's arable land. Primitive shifting cultivation (Category 8) occurs in the rainforested Amazon Basin and its hilly perimeter; rudimentary sedentary cultivation (Category 7) dominates the Andean plateau country from Colombia in the north to the Bolivian Altiplano in the south; and a ribbon of mixed subsistence farming (Category 4) courses through most of east-central Brazil between the coastal plantation and interior grazing zones.

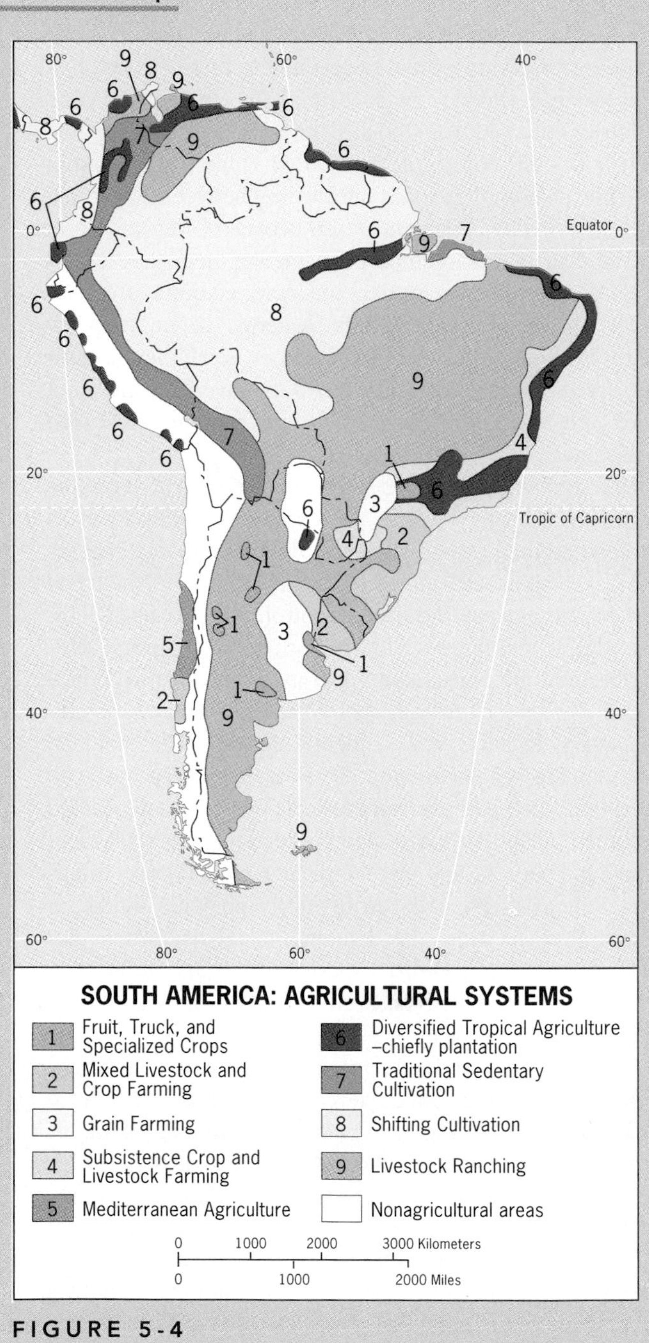

FIGURE 5-4

largest lies along the northeastern Brazilian coast, with four others along the Atlantic and Caribbean coastlands of northern South America. Location, soils, and tropical climates favored plantation crops, especially sugar. The fact that the indigenous population was small led to the introduction of millions of African slave laborers, whose descendants today continue to dominate the racial makeup and strongly influence the cultural expression of these areas. The plantation economy later failed, soils became exhausted, slavery was abolished, and the people were largely reduced to poverty and subsistence—socioeconomic conditions that now dominate most of the region mapped as tropical-plantation.

European-Commercial Region

The second region on Augelli's map, identified as *European-commercial*, is perhaps the most truly "Latin" part of South America. Argentina and Uruguay, each with a population that is at least 85 percent "pure" European and with a strong Hispanic cultural imprint, constitute the bulk of the European-commercial region. Two other areas also lie within it: most of Brazil's core area and the heartland of central Chile. Southern Brazil shares the temperate grasslands of the Argentine Pampa and Uruguay, and this area is important as a zone of livestock raising as well as corn production. Middle Chile is an old Spanish settlement zone, home to the approximately one-sixth of the Chilean population who claim pure Spanish ancestry. Here, in an area of Mediterranean climate (Fig. I-8), cattle and sheep raising as well as mixed farming are practiced. In general, then, the European-commercial region is economically more advanced than the rest of the continent. A commercial economy rather than a subsistence way of life prevails (see box, on p. 244), living standards are better, literacy rates are higher, transportation

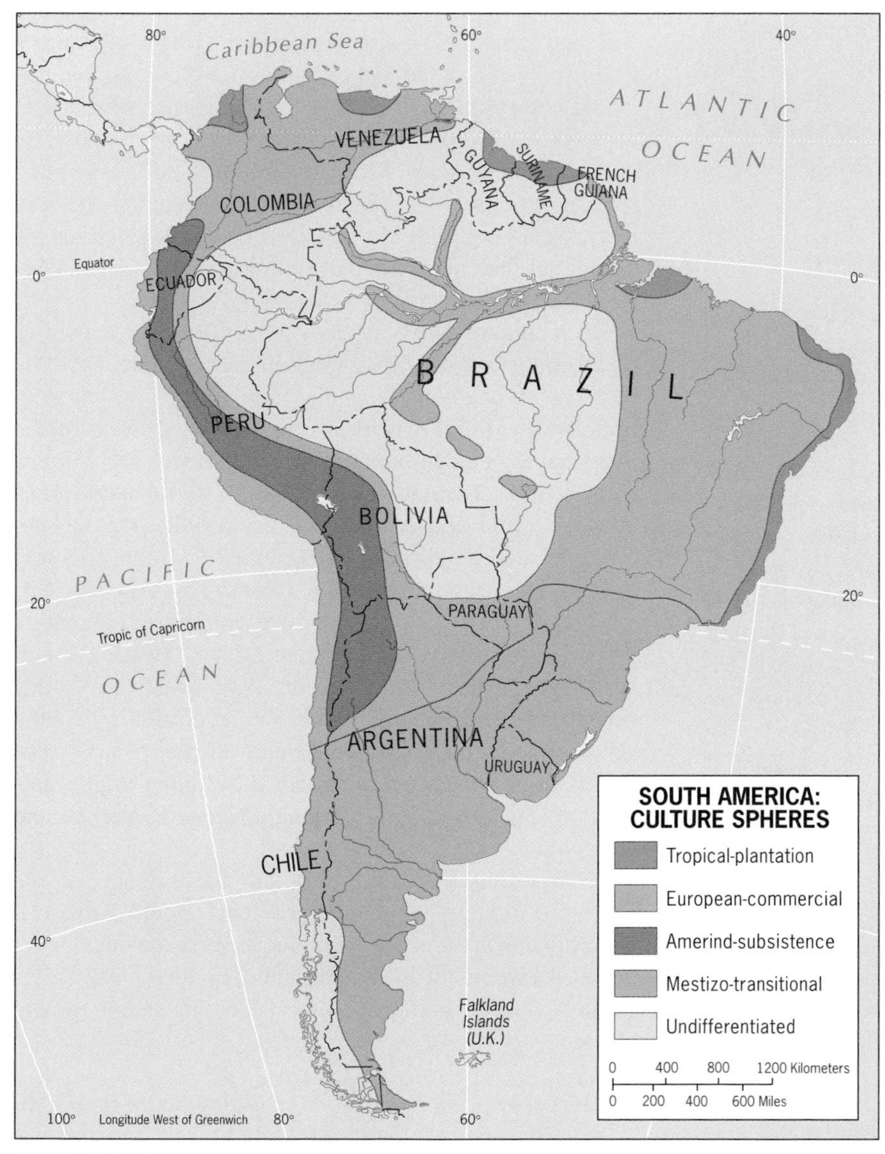

FIGURE 5-5

networks are superior, and (as Augelli has pointed out) the overall development of this region surpasses that of parts of Europe itself.

Amerind-Subsistence Region

The third region is identified as *Amerind-subsistence*, and it forms an elongated zone along the length of the central Andes from southern Colombia to northern Chile/northwestern Argentina, closely approximating the area occupied by the old Inca Empire. The feudal socioeconomic structure that was established here by the Spanish conquerors still survives. The Amerindian population forms a large, landless peonage, living by subsistence or by working on haciendas (or smaller farms called *minifundias*) far removed from the Spanish culture that forms the primary force in the national life of their country. This region includes some of South America's poorest areas, and whatever commercial activity exists tends to be in the hands of whites or mestizos. The Amerindian heirs to the Inca Empire live, often precariously, at high elevations (as much as 12,500 feet/3800 m) in the Andes. Poor soils, uncertain water supplies, high winds, and bitter cold make farming a constantly difficult proposition.

Mestizo-Transitional Region

The fourth region, *mestizo-transitional*, surrounds the Amerind-subsistence region, covering coastal and interior Peru and Ecuador, most of Colombia and Venezuela, much of Paraguay, and large parts of Argentina, Chile, and especially Brazil (including the developing corridors of the Amazon Basin). This is the zone of mixture between European and Amerindian—or African in coastal sections of Venezuela, Colombia, and northeastern Brazil. The map thus reminds us that such countries as Bolivia, Peru, and Ecuador are dominantly Amerindian and mestizo. In Ecuador, for instance, these two groups make up fully 80 percent of the total population, and only about 10 percent can be classified as European. The term *transitional* has an economic connotation, too, because (as Augelli puts it) this region "tends to be less commercial than the European sphere but less subsistent in orientation than dominantly [Amerindian] areas."

Undifferentiated Region

The fifth region on the map is marked as *undifferentiated* because its characteristics are hard to classify. Some of the Amerindian peoples in the interior of the Amazon Basin had remained almost completely isolated from the momentous changes in South America since the days of Columbus. Although remoteness and lack of change are still two notable aspects of this subregion, the ongoing development of Amazonia is reversing that situation. The most remote Amazon backlands as well as the Chilean and Argentinean southwest also are sparsely populated and exhibit only very limited economic development; inaccessibility continues to contribute to the unchanging nature of these areas.

The framework of the five culture spheres just described is necessarily a broad generalization of a rather complex geographic reality. But it does underscore the diversity of South America's peoples, cultures, and economies—a quality that lately has been overshadowed by a still-emerging, continentwide spirit of international cooperation.

ECONOMIC INTEGRATION

All across South America today, the separatism that has so long characterized international relations is giving way as countries discover the benefits of forging new partnerships with one another. The distribution of the realm's resources has always presented opportunities for building interregional complementarities, and the resurgence of civilian, democratic rule is finally making that development possible on a continental scale. More importantly, in the economic sphere, virtually all of South America's republics have in recent years replaced old policies that protected domestic economies with new ones that embrace market-oriented reform and the expansion of trading ties.

With mutually advantageous trade the catalyst, international cooperation is blossoming in every sphere. Periodic flareups of boundary disputes now rarely escalate into prolonged conflicts—witness the 1998 resolution of the longstanding confrontation between Ecuador and Peru or the peaceful settlement of more than 20 territorial disputes between Chile and Argentina since 1990. Cross-border rail, road, and pipeline projects, stalled for years, are now multiplying steadily. In southern South America, five formerly contentious nations are developing the *Hidrovia*, a system of river locks that is opening most of the Paraná-Paraguay Basin to barge transport. And investments today flow freely from one country to another, particularly in the agricultural sector: Brazilian farmers already raise most of Paraguay's soybean crop and are developing huge tracts of fertile land in Brazil-adjoining portions of Bolivia and Uruguay.

Recognizing that free trade can solve many of the realm's economic-geographic problems, South American governments are actively pursuing several avenues of regional economic integration. In 2002, South America's republics were affiliating with the following major trading blocs:

Mercosur Launched in 1995, this Southern Cone Common Market established a free-trade zone and customs union linking Brazil, Argentina, Uruguay, and Paraguay.

Andean Community Formed as the Andean Pact in 1969 but restarted in 1995 as a customs union with common tariffs for imports, its members are Venezuela, Colombia, Peru, Ecuador, and Bolivia.

Group of Three (G-3) Launched in 1995, this free-trade agreement involves Mexico, Venezuela, and Colombia, and aims to eliminate all internal tariffs by 2005.

North American Free Trade Agreement (NAFTA) Launched by the United States, Canada, and Mexico in 1994, NAFTA has sought to expand into South America to include Chile; it aims to phase out all internal tariffs and complete the formation of a free-trade area before 2010.

These organizations, however, represent only an intermediate step toward a much grander goal: the creation of the 7 **Free Trade Area of the Americas (FTAA)**, a single market of more than 800 million consumers that would extend from the Arctic shores of Alaska and Canada to Chile's southernmost Cape Horn. This hemisphere-spanning free-trade area—by far the world's largest—is to be formed by combining 24 existing multinational agreements (including the four listed above) into one. Negotiators will undoubtedly face some formidable obstacles in assembling this gigantic trading bloc. Nonetheless, its architects still remain optimistic that FTAA can meet its target completion date of 2005.

URBANIZATION

As in most other less advantaged countries, residents of South America's republics are leaving the land and moving 8 to the cities. This **urbanization** process intensified sharply after mid-century, and it persists so strongly that South America's urban population percentage now ranks with those of Europe and the United States. In 1925, about one-third of South America's peoples lived in cities and towns, and as recently as 1950 the percentage was just over 40. But by 1975 the continentwide figure had surpassed 60 percent, and today 78 percent of the South American population resides in urban areas. Of course, these percentages mask the actual numbers, which are even more dramatic. Between 1925 and 1950, the realm's towns and cities grew by about 40 million residents as the urbanized percentage rose from 33 to 42. Then between 1950 and 1975, more than 125 million people crowded into the teeming metropolitan areas—more than *three times* the total for the previous quarter-century—and an additional 140 million since 1975 have swelled the continental total to its current 277.5 million.

Nowhere is South America's population of 357 million (28 percent greater than that of the United States) increasing faster than in the towns and cities. We usually assume that the populations of rural areas grow more rapidly than urban areas because farm families traditionally have more children than city dwellers. Yet, overall, the urban population of South America has grown annually by about 5 percent since 1950, while the concomitant increase in rural areas was less than 2 percent. These figures underscore the dimensions of the **rural-to-urban migration** from the 9 countryside to the cities—still another migration process that affects modernizing societies throughout the world.

The generalized spatial pattern of South America's urban transformation is exhibited in Figure 5-6, which shows a

FROM THE FIELD NOTES

"Look one way between the towering skyscrapers of waterfront Rio de Janeiro, and you see one of the world's most famous beaches. Look landward, and luxury gives way to deprivation. Rio's *favelas* are vast and poor. Still, this photograph gives rise to hope. Ten years ago this was a miserable shantytown whose residents lived in shacks. Most of those dwellings have been replaced by more permanent housing; sanitary conditions may not be adequate, but there is electricity, walls are bricked, windows and doors the norm. Some *favela* residents have done well enough to build substantial homes. That is the hope for the future: that stability and rising incomes among *favela* residents will transform shantytown into suburb."

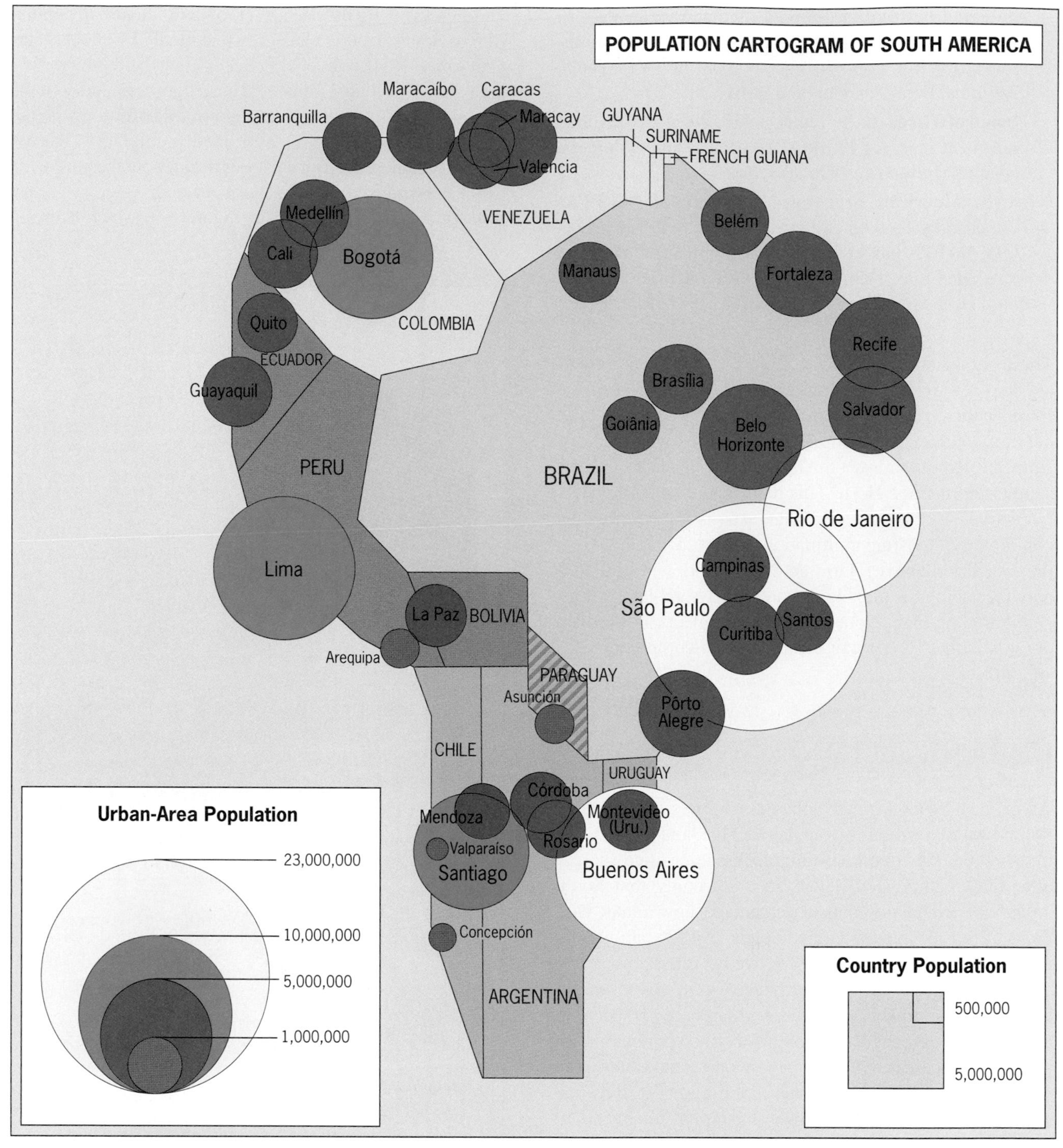

FIGURE 5-6

cartogram of the continent's population.* Here we see not only the realm's 13 countries in population-space relative to each other, but also the proportionate sizes of individual large cities within their total national populations.

*A cartogram is a specially transformed map in which countries and cities are represented in proportion to their populations. Those containing large numbers are "blown up" in population-space, while those containing lesser numbers are "shrunk" in size accordingly. This technique is used to map the countries of the world in Figure I-10.

Regionally, the realm's highest urban percentages occur in southern South America. Today in Argentina, Chile, and Uruguay, at least 85 percent of the people reside in cities and towns (statistically on a par with Europe's most urbanized countries). Ranking next among the most heavily urbanized populations is Brazil at 78 percent (3 percent higher than the United States), now growing so rapidly that it records an increase of 1 percent each year. That may not sound very significant, but once again percentages mask the real numbers: in a population of 175 million, 1 percent indicates that an additional 1.75 million people annually are somehow squeezed into Brazil's badly overcrowded urban centers.

The next highest group of countries, averaging 76 percent urban, border the Caribbean in the north. Here, Venezuela leads with 86 percent, and Colombia follows with 71 percent. Not surprisingly, the Amerind-subsistence-dominated Andean countries constitute the realm's least urbanized zone. Peru, because of its strong Spanish imprint, is that region's leader by far, with 72 percent of its population agglomerated in towns and cities. Ecuador, Bolivia, and transitional Paraguay, on the other hand, lag well behind the rest of the continent, with all three exhibiting urban proportions in the 52 to 63 percent range. Figure 5-6 also tells us a great deal about the relative positions of major metropolises in their countries. Six of them—Brazil's São Paulo and Rio de Janeiro, Argentina's Buenos Aires, Peru's Lima, Colombia's Bogotá, and Chile's Santiago—rank among the world's **megacities** (whose populations exceed 5 million).

10

In South America, as in Middle America, Africa, and Asia, people are attracted to the cities and driven from the poverty of the rural areas. Both *pull* and *push factors* are at work. Rural land reform has been slow in coming, and every year tens of thousands of farmers simply give up and leave, seeing little or no possibility for economic advancement. The cities lure them because they are perceived to provide opportunity—the chance to earn a regular wage. Visions of education for their children, better medical care, upward social mobility, and the excitement of life in a big city draw hordes to places such as São Paulo and Caracas. Road and rail connections continue to improve, so that access is easier and exploratory visits can be made. City-based radio and television stations beckon the listener to the locale where the action is.

But the actual move can be traumatic. Cities in developing countries are surrounded and often invaded by squalid slums, and this is where the uncertain urban immigrant most often finds a first—and frequently permanent—abode in a makeshift shack without even the most basic amenities and sanitary facilities. Many move in with relatives who have already made the transition but whose dwelling can hardly absorb yet another family. And unemployment is persistently high, often exceeding 25 percent of the available labor force. Jobs for unskilled workers are hard to find and pay minimal wages. But still the people come, the overcrowding in the shantytowns worsens, and the threat of epidemic-scale disease (and other disasters) rises. It has been estimated that in-migration accounts for over 50 percent of urban growth in some disadvantaged countries, and in South America the poorest urban populations exhibit high natural-increase rates as well.

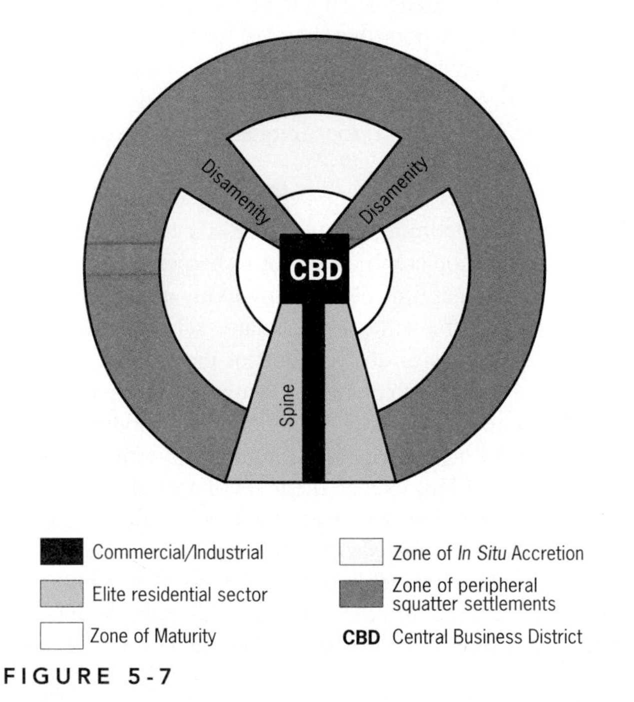

FIGURE 5-7

The "Latin" American City Model

Although the urban experience in the South and Middle American realms has varied because of diverse historical, cultural, and economic influences, there are many common threads that have prompted geographers to search for useful generalizations. One is the model of the intraurban spatial structure of the **"Latin" American city** proposed by Ernst Griffin and Larry Ford (Fig. 5-7), which may have wider application to cities of developing countries in other geographic realms as well.

11

The basic spatial framework of city structure, which blends traditional elements of South and Middle American culture with modernization forces now reshaping the urban scene, is a composite of radial sectors and concentric zones. Anchoring the model is the thriving CBD, which, like its European counterpart, remains the primary business, employment, and entertainment focus of the surrounding metropolitan agglomeration. The landscape of the CBD contains many modern high-rise buildings but also mirrors its colonial beginnings. As shown in Figure 4–3, when the Spanish colonizers laid out their New World cities they created a central square, or *plaza*, dominated by a church and flanked by imposing government buildings. Santiago's *Plaza de Armas* (see photo, p. 250), Bogotá's *Plaza Bolívar*, and Buenos Aires' *Plaza de Mayo* are classic examples. Early in the South American city's development, the plaza formed the hub and focus of the city, surrounded by shopping streets and arcades. Eventually the city outgrew its old center, and new commercial districts formed elsewhere within the CBD, leaving the plaza to serve as a largely ceremonial link with the past.

FROM THE FIELD NOTES

"From a high vantage point overlooking Santiago's *Plaza de Armas* you realize how these great center-city spaces have become preservation centers for colonial and historic architecture. As elsewhere in Hispanic South America, this plaza is flanked by religious and public buildings: in the foreground we can see, from left to right, the municipal offices of the city, the National Historical Museum, and the Main Post Office. Just beyond are the Metropolitan Cathedral and the Library, as well as a glimpse of the Art Museum. In the distance stand modern highrises of the kind that crowd the centers of North American cities, kept at a distance by the rules that protect the plaza's ambience."

Emanating outward from the urban core along the city's most prestigious axis is the commercial *spine*, which is surrounded by the *elite residential sector* (shown in green in Fig. 5-7). This widening corridor is essentially an extension of the CBD, featuring offices, shopping, high-quality housing for the upper and upper-middle classes, restaurants, theaters, and such amenities as parks, zoos, and golf courses that give way to wealthy adjoining suburbs which carry the elite sector beyond the city limits.

The three remaining concentric zones are home to the less fortunate residents of the city (who comprise the great majority of the urban population), with income levels and housing quality decreasing markedly as distance from the city center increases. The *zone of maturity* in the inner city contains the best housing outside the spine sector, attracting middle-class urbanites who invest sufficiently to keep their solidly built but aging dwellings from deteriorating. The adjacent *zone of in situ accretion* is one of much more modest housing interspersed with unkempt areas, representing a transition from inner-ring affluence to outer-ring poverty. The residential density of this zone is usually quite high, reflecting the uneven assimilation of its occupants into the social and economic fabric of the city.

The outermost *zone of peripheral squatter settlements* is home to the impoverished and unskilled hordes that have recently migrated to the city from rural areas. Although housing in this ring consists mainly of teeming, high-density shantytowns, residents here are surprisingly optimistic about finding work and eventually bettering their living conditions. Many achieve their first economic success by becoming part of the **informal sector**. This primitive form of capitalism is now common in many lower-income countries, taking place beyond the control—and especially the taxation—of the government. Participants are unlicensed sellers of homemade goods (such as arts and crafts, clothing, and food specialties) and services (auto repair, odd jobs, and the like). Their willingness to engage in this hard work has transformed many a shantytown into a beehive of activity that can propel resourceful residents toward a middle-class existence elsewhere in the metropolis.

12

A final structural element of many South American cities is the *disamenity sector* that contains relatively unchanging slums, known as **barrios** or **favelas**. The worst of these poverty-stricken areas often include sizeable numbers of people who are so poor that they are forced literally to live in the streets. Thus the realm's cities present sharp contrasts between poverty and affluence, squalor and comfort—a harsh juxtaposition all too frequently observed in the cityscape (see photo, p. 247).

13

As with all geographic models, this particular construct has certain shortcomings. Among the criticisms directed at this generalization are the need for a sharper sectoring pattern and a stronger time element involving a multiple-stage approach. (On the other hand, many scholars view this work as a successful abstraction not only of the "Latin" American city but also of cities in any disadvantaged country.) To address such concerns and keep his model current, Larry Ford recently proposed a number of additions to it. These modifications exceed the level of spatial complexity for our purposes here (which is why they are not included in Fig. 5-7), but Ford stresses that they do not overshadow the usefulness of the original model. Many of the changes incorporate elements of contemporary U.S. suburbanization (which may now be spreading around the world) discussed in Chapter 3: a budding suburban downtown located at the outer end of the spine, focused on a large shopping mall and flanked by new middle-class housing tracts; also added is an industrial park on the opposite edge of the city, which is connected to the suburban downtown by a circumferential highway that runs along the border between the model's yellow- and red-colored rings. Ford also modified some of the inner areas of the city by subdividing the CBD into traditional and modern components, introducing a narrow manufacturing corridor linking the CBD to the industrial park, and adding a small area of gentrification (residential redevelopment) to the zone of maturity.

REGIONS OF THE REALM

We turn now to the countries of South America and the principal characteristics of their regional geography. In general terms, it is possible to group the realm's countries into regional units (Fig. 5-3) because several have qualities in common. To begin with, Brazil by itself constitutes a geographic region because it contains half the continent's land and people. The northernmost Caribbean countries, Venezuela and Colombia, form a second regional unit that includes neighboring Guyana, Suriname, and French Guiana. On the basis of their Amerindian cultural heritage, Andean physiography, and modern populations, the western republics of Ecuador, Peru, and Bolivia constitute a third regional entity. Finally, in South America's southern cone, Argentina, Uruguay, and Chile have a common regional identity as the realm's mid-latitude, most strongly Europeanized states; Paraguay, a transitional country between the Andean west and these southern republics, is also included with the latter.

Before we examine the continent's 13 countries within this regional framework, it is useful to take another look at the spatial distribution of South America's culture spheres (Fig. 5-5). This map clearly indicates that, with the lone exception of tiny Uruguay, cultural fragmentation marks the human geography of *every* country—a pattern that prevails in national economic landscapes as well (Fig. 5-4). Thus central governments throughout the realm are also confronted with the centrifugal forces that accompany plural societies as they struggle to manage and resolve the formidable developmental, political, and environmental problems they face.

MAJOR CITIES OF THE REALM

City	Population* (in millions)
Asunción, Paraguay	1.4
Belo Horizonte, Brazil	4.4
Bogotá, Colombia	6.6
Buenos Aires, Argentina	12.9
Caracas, Venezuela	3.2
Guayaquil, Ecuador	2.5
La Paz, Bolivia	1.6
Lima, Peru	7.9
Manaus, Brazil	1.5
Montevideo, Uruguay	1.2
Quito, Ecuador	1.9
Rio de Janeiro, Brazil	10.8
Santiago, Chile	5.8
São Paulo, Brazil	23.2

*Based on 2002 estimates.

BRAZIL: GIANT OF SOUTH AMERICA

By any measure, Brazil is South America's giant. It is so large that it has common boundaries with all the realm's other countries except Ecuador and Chile (Fig. 5-3). Its tropical and subtropical environments range from the equatorial rainforest of the Amazon Basin to the humid temperate climate of the far south. Territorially, Brazil occupies just under 50 percent of South America and is exceeded on the global stage only by Russia, Canada, the United States, and China. In population size, it also accounts for just below half the continental total and is again the world's fifth-largest country (surpassed only by China, India, the United States, and Indonesia). The Brazilian economy is now the eighth largest on Earth, and its modern industrial base is the seventh largest. Moreover, the country is likely to continue advancing in the international ranks and is poised to become a world force as the twenty-first century unfolds.

Population Patterns

Brazil's population of 175.2 million is as diverse as that of the United States. In a pattern quite familiar in the Americas, the indigenous inhabitants of the country were decimated following the European invasion (less than 200,000 Amerindians now survive deep in the Amazonian interior). Africans came in great numbers, too, and today there are 8.6 million blacks in Brazil. Significantly, however, there was also much racial mixing, and 70 million Brazilians have combined European, African, and minor Amerindian an-

FROM THE FIELD NOTES

"Near the waterfront in Belém lie the now-deteriorating, once-elegant streets of the colonial city built at the mouth of the Amazon. Narrow, cobblestoned, flanked by tiled frontages and arched entrances, this area evinces the time of Dutch and Portuguese hegemony here. Mapping the functions and services here, we recorded the enormous diversity of activities ranging from carpentry shops to storefront restaurants and from bakeries to clothing stores. Dilapidated sidewalks were crowded with shoppers, workers, and people looking for jobs (some newly arrived, attracted by perceived employment opportunities in this growing city of 1.7 million). The diversity of population in this and other tropical South American cities reflects the varied background of the region's peoples and the wide hinterland from which these urban magnets have drawn their inhabitants."

cestries. The remaining 96 million—now barely in the majority at 54 percent—are mainly of European origin, the descendants of immigrants from Portugal, Italy, Germany, and Eastern Europe. The complexion of the population was further diversified by the arrival of Lebanese and Syrians—and, since 1908, the growth of the largest community of ethnic Japanese outside Japan. The Japanese, now numbering about 1 million, are concentrated in São Paulo State where they have risen to the topmost ranks of Brazil's easygoing immigrant society as farmers, urban professionals, business leaders—and traders with Japan.

Brazilian society, to a greater degree than is true elsewhere in the Americas, has made progress in dealing with its racial divisions. To be sure, blacks are still the least advantaged among the country's major population groups, and community leaders continue to complain about discrimination. But ethnic mixing in Brazil is so pervasive that hardly any group is unaffected, and official census statistics about "blacks" and "Europeans" are meaningless. What the Brazilians do have is a true national culture, expressed in an adherence to the Catholic faith (this is the world's largest Roman Catholic country), in the universal use of a modified form of Portuguese as the common language, and in a set of lifestyles in which vivid colors, distinctive music, and a growing national consciousness and pride are fundamental ingredients.

One of the most noteworthy trends in Brazil's population geography has been the recent lowering of the country's high rate of natural increase. Over the past three decades, this rate has been more than halved (to 1.5 percent), and the average number of children born to a Brazilian woman plummeted from 4.4 in 1980 to 2.4 in 2000. Surprisingly, this remarkable slowdown in population increase took place in the absence of an active birth control policy by the Brazilian government—and in direct contradiction to the teachings of the decreasingly influential Catholic Church. Demographers believe that the decline is the result of three factors: a rapid spread in contraceptive usage; an extended period of economic uncertainty that lasted through most of the 1980s and 1990s; and the advent of near-universal access to television, a major force in reshaping the attitudes and aspirations of the Brazilian people.

Development Prospects

Brazil is richly endowed with mineral resources, including enormous iron and aluminum ore reserves, extensive tin and manganese deposits, and sizeable oil- and gasfields (Fig. 5-8). Other significant energy developments involve massive new hydroelectric facilities and the successful substitution of sugarcane-based alcohol (gasohol) for gasoline—allowing well over half of Brazil's cars to use this fuel instead of costly imported petroleum. Besides these natural endowments, Brazilian soils sustain a bountiful agricultural output that makes the country a global leader in the production and export of coffee, soybeans, and orange juice concentrate. Commercial agriculture, in fact, is now the fastest growing economic sector, propelled by mechanization and the opening of a major new farming frontier in the fertile grasslands of southwestern Brazil.

As already noted, Brazil ranks seventh among the world's industrial powers. Much of the momentum for this continuing development was unleashed in the early 1990s after the government opened the country's long-protected industries to international competition and foreign investment. These new policies are proving to be effective because productivity has risen by over a third since 1990 as Brazilian manufactures attain world-class quality. Industrial goods also surpassed (in value) the agricultural sector during the mid-1990s to become the nation's export leader. Trade with

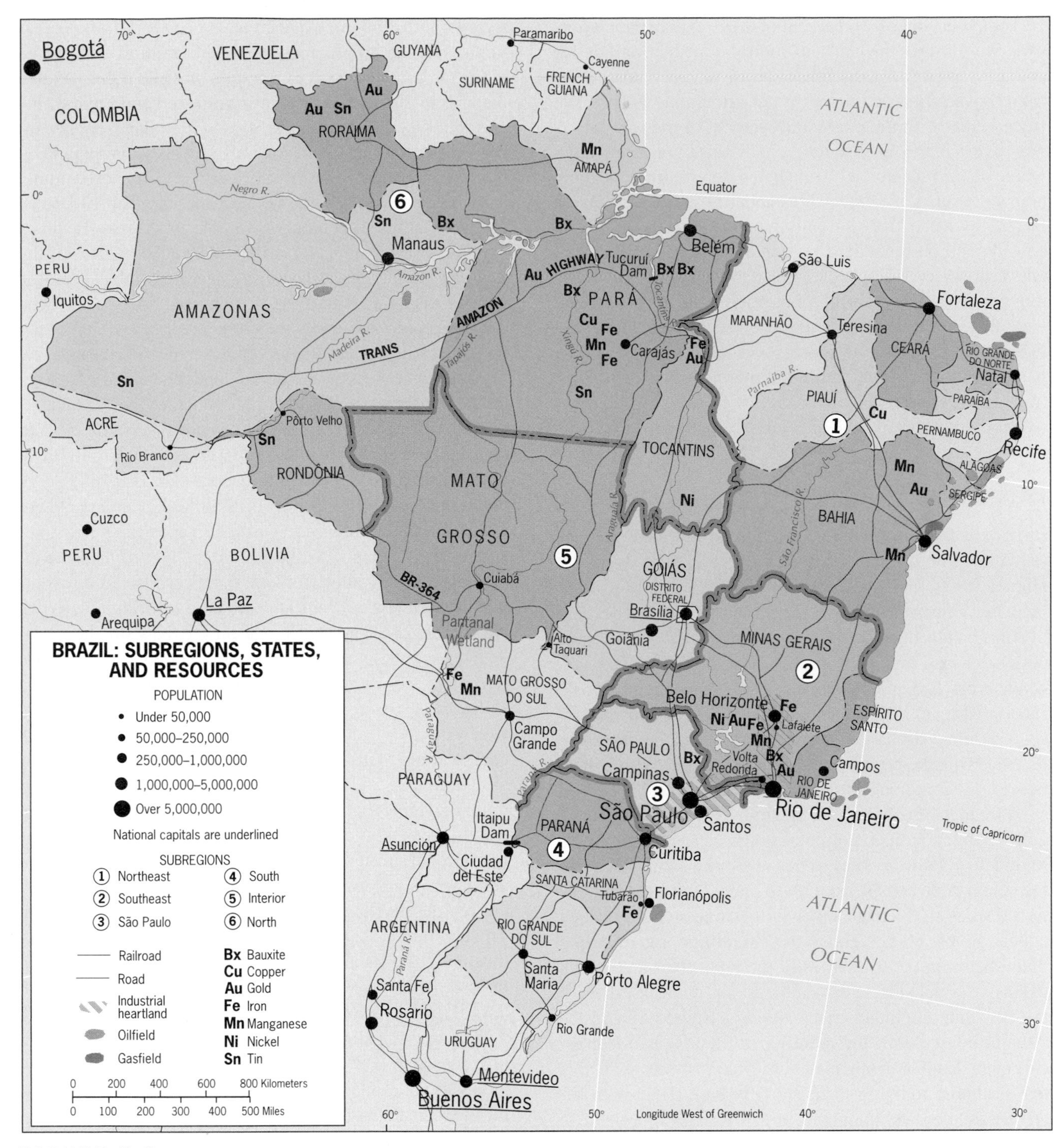

FIGURE 5-8

Argentina—now one of Brazil's primary trading partners—is at the forefront, a relationship expected to intensify as Mercosur (which the Portuguese-speaking Brazilians call *Mercosul*) reaches full operation.

Brazil's current economic performance notwithstanding, the country's overall development during recent decades has been marked by a roller coaster ride of boom and bust. The prospects for keeping economic growth on a more stable footing are heightened by the continent's new free-trade environment, but serious internal problems of social and regional inequality must be mitigated as well. In a realm exhibiting the world's sharpest division between affluence and poverty, Brazil stands out with the widest income gap in all of South America.

Today, the richest 10 percent of the Brazilian population own two-thirds of the land and control more than half of the country's wealth. At the other end of the economic spectrum, poverty has increased by an astounding 50 percent since 1980. With the poorest 20 percent of the population now living in the most squalid conditions to be found anywhere on the planet, too little is being done to improve their housing, education, and health standards (at least half of all Brazilians suffer from chronic malnutrition). Even though considerable regional development is taking place, the benefits are overwhelmingly channeled toward already affluent individuals and powerful corporations, thereby further widening that enormous gulf between modern and traditional Brazil. Unless the country's resources and access to its opportunities are redistributed more equitably, Brazil's forward progress will increasingly be clouded by the negative consequences of such uneven development.

Brazil's Subregions

Brazil is a federal republic consisting of 26 States and the federal district of the capital, Brasília (Fig. 5-8). As in the United States, and for similar reasons, the smallest States lie in the northeast and the larger ones farther west. The State of Amazonas is the biggest, twice the size of Texas with over 600,000 square miles (1.6 million sq km), but its huge area contains barely 3 million people. At the other extreme, Rio de Janeiro State (17,000 square miles/44,000 sq km) on the southeastern coast has a population of almost 15 million. But the State with the largest population by far is São Paulo, now exceeding 37 million and growing phenomenally.

Although Brazil is about as large as the 48 contiguous United States, it does not possess the clear physiographic regionalism familiar to us. It has no Andes Mountains, and the countryside consists mainly of plateau surfaces and low hills (Fig. 5-1). Even the lower-lying Amazon Basin, which covers almost 60 percent of the country, is not entirely a plain: between the tributaries of the great river lie low but extensive tablelands. In the southeast, the plateau surface of the Brazilian Highlands rises slowly eastward toward the Atlantic coast. Along the shoreline itself, there is a steep escarpment leading from plateau surface to sea level that leaves almost no living space along its base. Thus, with relatively little coastal plainland, Brazil is indeed fortunate to possess several very good natural harbors. Given the country's physiographic ambiguity, the six subregions discussed next have no absolute or even generally accepted boundaries. In Figure 5-8, those boundaries have been drawn to coincide with the borders of States, making identifications easier.

The Northeast

The Northeast was Brazil's source area, its culture hearth. The plantation economy took root here at an early date, attracting Portuguese planters who soon imported the country's largest group of African slaves to work in the sugar fields. But the ample and dependable rainfall that occurs along the coast soon gives way to lower and more variable patterns in the interior, which is home to nearly half of the region's 50 million people. This drier inland backcountry—called the *sertão*—is not only seriously overpopulated but also contains some of the worst poverty to be found anywhere in Middle and South America. The Northeast produces barely 15 percent of Brazil's gross domestic product, but its inhabitants constitute almost 30 percent of the national population. Given this staggering imbalance, it is not surprising that the region contains fully half of the country's poor, a literacy rate 20 percent below that of the rest of Brazil, and an infant mortality rate twice the national average that reflects some of the hemisphere's lowest standards of nutrition and health care. Much of the Northeast's misery is rooted in its grossly unequal system of land tenure. Farms must be at least 250 acres to be profitable in the hardscrabble *sertão*, a threshold size that only large landowners can afford (most peasant farmers are unable to assemble landholdings greater than 75 acres).

On top of all this, the Northeast is plagued by a monumental environmental problem: the recurrence of devastating droughts, which are at least partly attributable to **El Niño** (periodic sea-surface-warming events off the continent's northwestern coast). Indeed, extended dry periods occur so regularly that the region has become known as the *Polygon of Drought*. Ironically, there is ample groundwater, but the *sertão*'s huge population of peasant farmers cannot afford to drill the necessary wells. Without such irrigation, human and animal pressures on the land are combining to deplete the natural vegetation and thereby hasten the encroachment of aridity. Understandably, these conditions propel substantial emigration toward the coastal cities and, increasingly, out of the Northeast to the more prosperous regions that line Brazil's Atlantic seaboard to the south. To stem that human tide, the Brazilian government has in recent years pursued a program to purchase underutilized farmland for the settling of landless peasants. Although tens of thousands of families have benefited, overall barely a dent has been made in the Northeast's massive rural poverty crisis.

14

Sugar still remains the Northeast's chief crop along the moister coast, and livestock herding prevails in the *sertão*, with beef cattle on large ranches in the better grazing zones and goats everywhere else. The comparatively small areas of successful commercial agriculture—cotton in Rio Grande do Norte, sisal in Paraíba, and sugarcane in Pernambuco—stand in sharp contrast to the countless patches of rudimentary subsistence agriculture located nearby. Water is always the key, and when agribusiness or government can afford the investment, the land can become highly productive. For example, the lower São Francisco Valley in the States of Bahia, Sergipe, and Alagoas has in recent years been transformed into South America's fastest-growing area of irrigated farming, specializing in tropical fruits.

The Northeast today is Brazil's great contradiction. In the cities of Recife and Salvador (the Portuguese colonial capital from 1549 to 1763), the architecture still bears the imprint of an earlier age of wealth (see photo, p. 242), but thousands of peasants without hope—driven from the land by deteriorating conditions—constantly arrive to expand the usual surrounding shantytowns. As yet, few of the generalizations about emerging Brazil apply here, but there are some bright spots beyond the irrigated-agriculture projects. A huge petrochemical complex was built near Salvador in the late 1980s, creating thousands of jobs, luring further (including foreign) investment, and boosting the region's industrial base in general. Tourism has begun to boom along the entire Northeast coast, whose new beachside resorts are attracting hordes of vacationing Europeans (flying time to Western Europe is only about seven hours). Recife has nurtured a budding software industry and developed a reputation as a major medical center. Most promising of all is the ongoing development of Fortaleza and surrounding Ceará State: thriving new clothing and shoe industries have already put the city on the global economic map, and the social programs of local government have improved education, health care, and employment so substantially that they are cited as a model for all of urban Brazil.

The Southeast

In the State of Bahia, a transition occurs toward the south. The coastal escarpment becomes more prominent, the plateau higher, and the terrain more varied; annual rainfall increases and is seasonally more dependable. The Southeast has been modern Brazil's *core area*, with its major cities and leading population clusters. Gold first drew many thousands of settlers, and other mineral finds also contributed to the influx (with Rio de Janeiro itself serving as the terminus of the "Gold Trail"), but ultimately the region's agricultural possi-

Among the Realm's Great Cities . . .

Rio de Janeiro

Say "South America" and the first image most people conjure up is Sugar Loaf Mountain, Rio de Janeiro's landmark sentinel that guards the entrance to beautiful Guanabara Bay. Nicknamed the "magnificent city" because of its breathtaking natural setting, *Rio* replaced Salvador as Brazil's capital in 1763 and held that position for almost two centuries until the federal government shifted its headquarters to interior Brasília in 1960. Rio de Janeiro's primacy suffered yet another blow in the late 1950s: São Paulo, its hated urban rival 250 miles (400 km) to the southwest, surpassed Rio to become Brazil's largest city—a gap that has been widening ever since. Although these events triggered economic decline, Rio (10.8 million) remains a major entrepôt, air travel and tourist hub, and center of international business. As a focus of cultural life, however, Rio de Janeiro is unquestionably Brazil's leader, and the *cariocas* (as Rio's residents call themselves) continue to set the national pace with their entertainment industries, universities, museums, and libraries.

On the darker side, this city's reputation is increasingly tarnished by the widening abyss between Rio's affluent and poor populations—symbolizing inequities that rank among the world's most extreme (see photo, p. 247). All great cities experience problems, and Rio de Janeiro is currently bedeviled by the drug use and crime waves emanating from its most desperate hillside *favelas* (slums) that continue to grow explosively.

Nonetheless, Rio's planners recently launched a major project (known as "Rio-City") to improve urban life for all residents. This ambitious scheme is designed to reshape nearly two dozen of the aging city's neighborhoods, introduce an ultramodern crosstown expressway to relieve nightmarish traffic congestion, and—most importantly—bring electrical power, paved streets, and a sewage-disposal network to the beleaguered *favelas*.

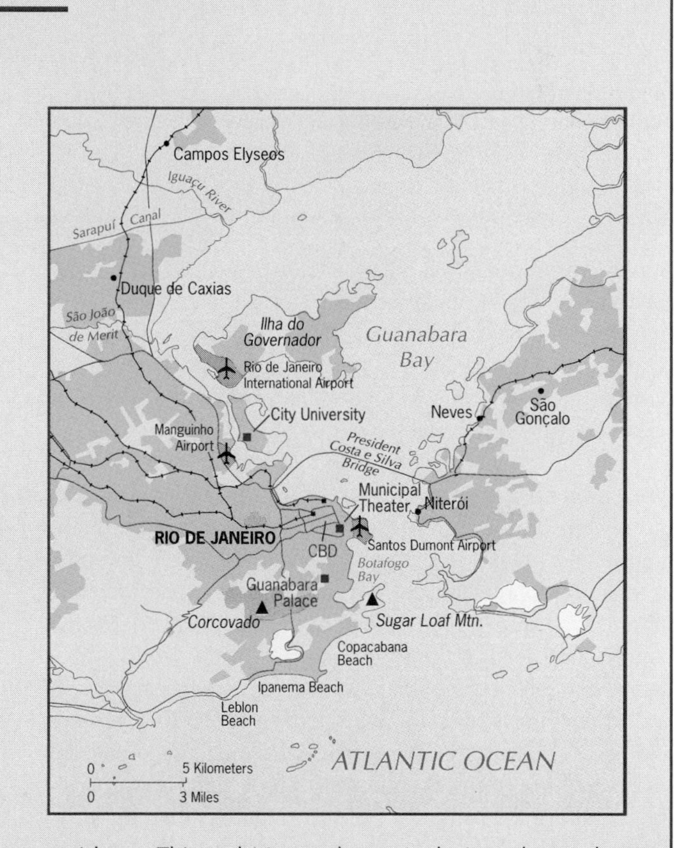

Brazil is the leading citrus producer, these days averaging 25 percent of the world's output and well surpassing the 15-percent share of number two, the United States. Contributing mightily to this Brazilian output is the country's thriving orange-juice-concentrate industry, which is based in São Paulo State. This ultramodern processing plant in the heart of the State's citrus belt is located next to Matão, a railroad town about 250 miles (400 km) northwest of Santos (the city of São Paulo's large outport, gateway to the global sea-lane network).

bilities ensured its stabilization and growth. The mining towns needed food, and prices for foodstuffs were high; farming was stimulated, and many farmers came (with their slave workers) to Minas Gerais State to till the soil. Eventually a pastoral industry came to predominate, with large herds of beef cattle grazing on planted pastures.

The third quarter of the twentieth century brought another mineral age to the region, based not on gold or diamonds but on the iron ores around Lafaiete and the manganese and limestone carried to the steel-making complex at Volta Redonda (Fig. 5-8). Iron-mining soon became one of Brazil's leading economic activities, and since 1990 the country has been in the topmost ranks of the world's producers and exporters. Minas Gerais (the name means "General Mines") produces about 75 percent of Brazil's iron ore and nearly 50 percent of its steel. From this base, industrial diversification in the Southeast has proceeded apace, with the metallurgical center of Belo Horizonte paving the way. That booming city has also become the endpoint of a rapidly developing manufacturing corridor that stretches 300 miles (500 km) southwest to metropolitan São Paulo—which during most of the 1990s experienced growth rates on a par with the highest in the Pacific Rim. At the same time, with an economy equivalent in size to Chile's in the late 1990s, Minas Gerais surpassed Rio de Janeiro to become Brazil's second largest State in industrial output.

São Paulo

The leading industrial producer is the State of São Paulo, the focus of ongoing Brazilian development. This economic-geographic powerhouse accounts for nearly half of the country's gross domestic product and more than a third of its manufacturing activity. With an economy that now matches Argentina's in overall size, burgeoning São Paulo State has been a magnet for migrants (especially from the Northeast). Not surprisingly, it is growing phenomenally and already contains more than 20 percent of Brazil's entire population.

The wealth of São Paulo State was built on its coffee plantations (known as *fazendas*), and up to a generation ago coffee accounted for at least half of Brazil's exports. Although it remains an important crop (Brazil is still the world's leading producer), coffee today claims less than 5 percent of the exports and has been eclipsed by other farm commodities. One of them is orange juice concentrate (here, too, Brazil leads the world); São Paulo State now produces more than double the annual output of Florida, thanks to a climate all but devoid of winter freezes, ultramodern processing plants, and a fleet of specially equipped tankers to ship the concentrate to markets in the United States—and more than 50 other countries ranging from Japan to Russia. Another leading pursuit is soybeans (a crop discussed in the box on p. 186), in which Brazil ranks second among the world's producers. The central and western portions of São Paulo State have benefited the most from this recent agricultural expansion, which also includes the raising of sugarcane for gasohol production, cotton, and rubber trees.

Matching this agricultural prowess is the State's industrial strength. The revenues derived from the coffee plantations provided the necessary investment capital, ores from Minas Gerais supplied the vital raw materials, hydroelectric power from the slopes of the coastal escarpment produced the needed energy, the nearby outport of Santos facilitated access to the ocean, and immigration from Europe, Japan, and other parts of Brazil contributed the increasingly skilled labor force. As the capacity of the domestic market grew, the advantages of central location and agglomeration secured São Paulo's primacy—a position reinforced by the subsequent growth of international trade. This resulted in metropolitan São Paulo becoming the country's—and South America's—leading industrial complex (and megacity). That distinction has continued into the twenty-first century, but, as the box on p. 257 indicates, new challenges and opportunities loom as the forces unleashed by postindustrialism (discussed in Chapter 3) begin to impact manufacturing centers all across the world.

Among the Realm's Great Cities . . .

São Paulo

São Paulo, which lies on a plateau 30 miles (50 km) inland from its Atlantic outport of Santos, possesses no obvious locational advantages. Yet here on this site we find the third largest metropolis on Earth, whose population has multiplied so uncontrollably that São Paulo has more than doubled in size (from 11 to 23 million) over the past quarter-century.

Founded in 1554 as a Jesuit mission, modern São Paulo was built on the nineteenth-century coffee boom. It has since grown steadily as both an agricultural processing center (soybeans, orange juice concentrate, and sugar besides coffee) and a manufacturing complex (accounting for about half of all of Brazil's industrial jobs). It has also become Brazil's primary focus of commercial and financial activity. Today São Paulo's distinctive, high-rise CBD is the very symbol of urban South America and attracts the continent's largest flow of foreign investment as well as the trade-related activities that befit the city's rise as the business capital of Mercosur (Mercosul in Portuguese).

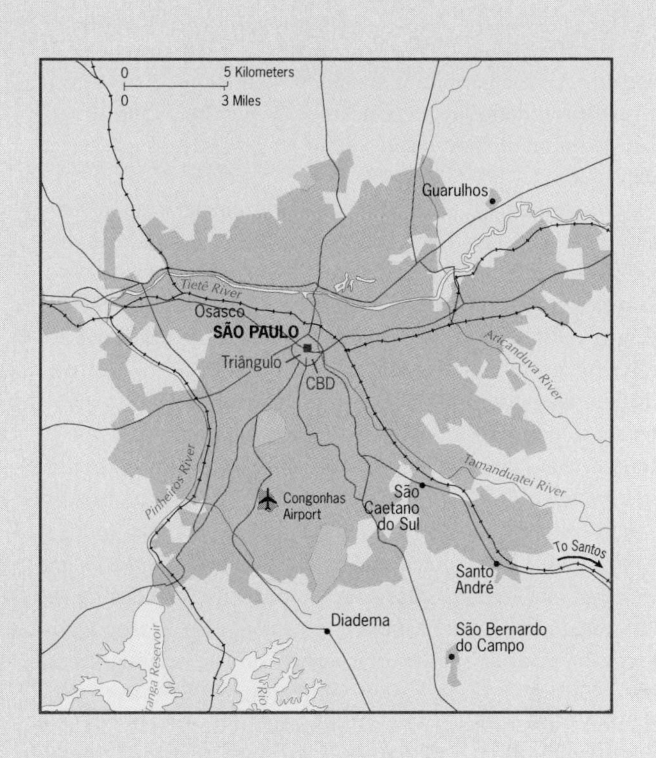

Nonetheless, even for this biggest industrial metropolis of the Southern Hemisphere, the increasingly global tide of postindustrialism is rolling in and São Paulo is being forced to adapt. To avoid becoming a Detroit-style Rustbelt, the aging automobile-dominated manufacturing zone on the central city's southern fringes is today attracting new industries. Internet companies, particularly, have flocked here in such numbers that they already constitute Brazil's largest cluster of e-commerce activity—and prompt many to now call this area *Silicon Village.*

Elsewhere in São Paulo's vast urban constellation—whose suburbs now sprawl outward up to 60 miles (100 km) from the CBD—additional opportunities are being exploited. In the outer northeastern sector, new research facilities as well as computer and telecommunications-equipment factories are springing up. And to the west of central São Paulo, lining the ring road that follows the Pinheiros River, is South America's largest suburban office complex replete with a skyline of ultramodern high-rises.

The huge recent growth of this megacity has also been accompanied by massive problems of overcrowding, pollution, and congestion. Traffic jams here are among the world's worst, with more than twice as many gridlocked motor vehicles on any given day than in Manhattan. Staggering poverty—on a par with Mexico City's—is the most pressing crisis as the ever-expanding belt of shantytowns tightens its grip on much of the metropolis. With its prodigious growth rate of the late twentieth century expected to persist through the foreseeable future, can anything prevent Greater São Paulo from approaching an unimaginable population of *50 million* less than 25 years from today?

The South

Three States, whose combined population has just surpassed 25 million, make up the southernmost Brazilian subregion: Paraná, Santa Catarina, and Rio Grande do Sul. The contribution of recent European immigrants to the agricultural development of southern Brazil, as in neighboring Uruguay and Argentina, has been considerable. Many came not to the coffee country of São Paulo State but to the available lands farther south. Here the newcomers introduced their advanced farming methods to several areas. Portuguese rice farmers clustered in the valleys of Rio Grande do Sul; the Germans, specialists in raising grain and cattle, occupied the somewhat higher areas to the north and in Santa Catarina; and the Italians selected the highest slopes, where they established thriving vineyards. These fertile lands proved highly productive, and with growing markets in the large urban areas to the north, this tristate region became Brazil's most affluent corner. Since 1990, the South's prosperity has been heightened by the expansion of European and other foreign markets, particularly for its tobacco; today, as much as half of Rio Grande do Sul's tobacco acreage is dedicated to raising highly popular, super-nicotine plants (genetically altered varieties created for U.S. cigarette-makers and then shipped secretly to Brazil in violation of export laws), which helps the Brazilians to maintain their position as the world's leading tobacco exporter.

European-style standards of living match the diverse European heritage that is reflected in the towns and countryside, where the German and Italian languages are still spoken alongside Portuguese. In some places, zealous preservation of Old World cultural life has led to hostility against non-European Brazilians. Many communities now actively discourage poor, job-seeking migrants from the north by offering to pay return bus fares or even blocking their household-goods-laden vehicles. Moreover, several extremist groups have arisen to openly espouse the secession of the South from Brazil. The largest, calling itself "The South Is My Country," promotes the creation of a *Republic of the Pampas* and even displays its intended flag. Accusing the separatist movement of subversion and racism, State governments have moved to thwart its progress. They claim the central issue is an economic one, complaining to Brasília that the South receives only about a 60 percent return on the federal taxes it pays. But with polls showing that at least one-third of the population is sympathetic to the idea of secession, officialdom may well have a longer-term struggle on its hands in preventing the Brazilian South from succumbing to the devolutionary forces that are weakening nation-states around the globe.

Economic development in the South is not limited to the agricultural sector. Coal from Santa Catarina and Rio Grande do Sul, shipped north to the steel plants of Minas Gerais, was a crucial element in Brazil's industrialization. Local manufacturing is growing as well, especially in Pôrto Alegre and Tubarão, which contains one of South America's largest steel-making complexes. During the 1990s, an internationally significant center of the computer software industry was established in Florianópolis, the island city and State capital just off Santa Catarina's coast. Known as *Tecnópolis*, this budding technopole continues to grow by capitalizing on its seaside amenities, skilled labor force, superior air-travel and global communications linkages, and government and private-sector incentives to support new companies. Only the sparsely populated inland portion of the South has lagged behind the rest of the region, despite the new presence of massive Itaipu Dam in westernmost Paraná State (the dam supplies São Paulo and surroundings with huge amounts of hydroelectric power but has not yet stimulated the growth of its local area). Further to the north, however, interior Brazil has begun to awaken with a vengeance.

The Interior

The subregion of Interior Brazil—constituted by the three States of Goiás, Mato Grosso, and Mato Grosso do Sul—is also known as the Central-West (or *Centro-Oeste*). This is the region that Brazil's developers have long sought to make a part of the country's productive heartland, and in the 1950s the new capital of Brasília was deliberately situated on its margins (Fig. 5-8). By locating the new capital city in the untapped wilderness 400 miles (650 km) inland from its predecessor, Rio de Janeiro, the nation's leaders dramatically signaled the opening of Brazil's age of development and a new thrust toward the west.

Brasília is also noteworthy in another regard because it represents what political geographers call a **forward capital**. A state will sometimes relocate its capital to a sensitive area, perhaps near a peripheral zone under dispute with an unfriendly neighbor, in part to confirm its determination to sustain its position in that contested zone. An example was Pakistan's decision in the 1960s to move its capital from coastal Karachi to northern Islamabad, a "forward" position near the disputed territory of Kashmir lying in the border zone with India. Brasília does not lie near a contested area, but Brazil's interior was an internal frontier to be conquered by a growing nation. Spearheading that drive, the new capital occupied a decidedly forward position.

15

Despite the subsequent growth of Brasília since 1960—from zero to 2.1 million inhabitants today—it was not until the 1990s that the tristate Interior became economically integrated with the rest of Brazil. The catalyst was the exploitation of the vast *cerrado*—the fertile savannas that blanket the Central-West and make it one of the world's most promising agricultural frontiers (at least two-thirds of its arable land still awaits development). As with the U.S. Great Plains, the flat terrain of the *cerrado* is one of its main advantages because it facilitates the large-scale mechanization of farming with a minimal labor force. Another advantage is rainfall, which is more prevalent than in the Great Plains or Argentine Pampa (Fig. I-7). The leading crop is soybeans, whose output per acre here exceeds even that of the U.S. Corn Belt. Other foodcrops and cotton are also expanding across the farmscape of the Centro-Oeste, but the current pace of regional development is inhibited by a serious accessibility problem that is only now being resolved. Unlike the Great Plains and Pampa, the growth of an efficient transportation network did not accompany the opening of this farming frontier. As a result, the Interior's products must travel along poor roads and intermittent railroads to reach the markets and ports of the Atlantic seaboard. Today several projects are underway to alleviate these bottlenecks. Most important is the *Ferronorte* (popularly called the Soy Railroad), a new privately financed railway that links São Paulo-Santos to the southeastern corner of Mato Grosso State; a northwestward extension to Mato Grosso's capital, Cuiabá, is already being planned, and from there a line may someday run northward to the central Amazon Basin (Fig. 5-8).

The Amazonian North

The territorially largest and most rapidly developing subregion—whose population has more than doubled since 1980 to 13.5 million today—is also the most remote from the core of Brazilian settlement: the seven States of the Amazon Basin (Fig. 5-8). This was the scene of the great rubber boom a century ago, when the wild rubber trees in the *selvas* (tropical rainforests) produced huge profits and the central Amazon city of Manaus enjoyed a brief period of wealth and splendor. The rubber boom ended in 1910, however, when plantations elsewhere (notably in Southeast Asia) began to produce rubber more cheaply, efficiently, and accessibly. For most of the seven decades that followed, Amazonia was a stagnant hinterland, but all that changed dramatically during the 1980s. New development began to stir throughout this awakening region, which currently is the scene of the world's largest migration into virgin territory as over 200,000 new settlers arrive each year. Most of this activity is occurring south of the Amazon River, in the tablelands between the major waterways and along the Basin's wide rim.

Two ongoing development schemes are especially worth noting because they are quintessential expressions of what is occurring here. The first is the *Grande Carajás Project* in southeastern Pará State, a huge multifaceted scheme centered on one of the world's largest known deposits of iron ore in the hills around Carajás (Fig. 5-8). In addition to a vast mining complex, other new construction here includes the Tucuruí Dam on the nearby Tocantins River and a 535-mile (850-km) railroad to the Atlantic port of São Luis. This ambitious development project also emphasizes further mineral exploitation (including bauxite, manganese, and copper), cattle-raising, crop farming, and forestry. What is taking place is an application of the **growth-pole concept**. The term is almost self-explanatory: a growth pole is a location where a set of industries, given a start, will expand and generate widening ripples of development in the surrounding area. Growth plans for Grande Carajás are aimed at just such a stimulation of its immediate hinterland, and if they come to fruition the scheme could one day cover one-sixth of all Amazonia.

16

Understandably, tens of thousands of settlers have descended on this part of the Amazon Basin in recent years. Those seeking business opportunities have been in the vanguard, but they have been followed by masses of lower-income laborers and peasant farmers in search of jobs and land ownership. The initial stage of this colossal enterprise has boosted the fortunes of many towns, most notably the spectacular revival of Manaus northwest of Carajás. An entire industrial complex has now emerged in the free-trade zone that adjoins this city, specializing in the production of TVs, VCRs, and myriad other electronic goods. The advantages for manufacturers here are cheap assembly costs, outstanding air-freight operations at an ultramodern airport, and central location with respect to all of northern South America (further enhanced by the late-1990s completion of an all-weather highway north to Caracas, Venezuela).

Many problems have also arisen as the tide of pioneers rolled across central Amazonia. None has suffered a more tragic plight than the Yanomami, one of the hemisphere's last remaining groups of indigenous people practicing a Stone Age culture. Following the 1987 discovery of gold deposits beneath their lands in Roraima State north of Manaus, the Yanomami homeland was overrun by thousands of claim-stakers who triggered numerous violent confrontations. The government was slow to intervene, and by the time it acted to halt the bloodshed, the Yanomami's fragile way of life had fallen victim to the disruption and pollution of its habitat, the ravages of newly introduced diseases (especially tuberculosis), and the inevitable modernization forced upon it by contact with other Brazilians.

The second leading development scheme, known as the *Polonoroeste Plan*, is located about 500 miles (800 km) to the southwest of Grande Carajás in the 1500-mile-long Highway BR-364 corridor that parallels the Bolivian border and connects the western Brazilian towns of Cuiabá, Pôrto Velho, and Rio Branco (Fig. 5-8). Although the government had planned for the penetration of western Amazonia to proceed via the east-west Trans-Amazon Highway, the migrants of the 1980s and 1990s preferred to follow BR-364 and settle within the Basin's southwestern rim zone, mostly in Rondônia State. Agriculture has been the dominant activity here, attracting affluent growers and ranchers from the South, plantation workers from São Paulo and Paraná States displaced by agrarian mechanization, and subsistence farmers from all over the country. The common denominator has been the quest for land; bitter conflicts continue to break out between peasants and landholders as the Brazilian government cautiously pursues the persistent and volatile issue of land reform.

The usual pattern of settlement in this part of the Brazilian North is something like this. As main and branch highways are cut through the wilderness, settlers (enticed by cheap land) follow and move out laterally to clear spaces for farming (see photo, p. 260). Crops, usually corn or upland rice, are planted, but within three years the heavy equatorial-zone rains leach out soil nutrients and accelerate surface erosion. As soil fertility declines, pasture grasses are then planted, and the plot of land is soon sold to cattle ranchers (most of them associated with agribusiness corporations based in Rio or São Paulo). The peasant farmers then move on to newly opened areas, clear more land for planting, and the cycle repeats itself. As long as open spaces remain, this is a profitable pursuit for all parties, but it ensures the widespread establishment of a low-grade land use that will ultimately concentrate most of the earnings in the hands of

This is what the Amazon's equatorial rainforest looks like from an orbiting satellite after the human onslaught in preparation for settlement. The colors on this Landsat image emphasize the destruction of the trees, with the dark green of the natural forest contrasted against the pale green and pinks of the leveled forest. The linear branching pattern of deforestation here in Rondônia State's Highway BR-364 corridor is explained in the text. But farming here is not likely to succeed for very long, and much of the cleared land is likely to be abandoned. Then the onslaught will resume to clear additional land—as an entire ecosystem comes ever closer to failing forever.

large landowners. It also requires the burning and clearing of enormous stands of tropical woodland: for at least the past decade, more than 22,000 square miles (57,000 sq km) of rainforest has been disappearing *annually* in Amazonia—an area almost the size of Ohio. Moreover, Amazon woodland destruction has much wider implications because it accounts for over half of all tropical deforestation, an environmental crisis of global significance (see box, p. 232).

THE NORTH: CARIBBEAN SOUTH AMERICA

As another look at Figure 5-5 confirms, the countries of South America's northern tier have something in common besides their coastal location: each has a tropical-plantation zone signifying early European plantation development, the in-migration of black laborers, and the absorption of this African element into the population matrix. Not only did black workers arrive; many thousands of South Asians also came to the realm's northern shores as contract laborers. The pattern is familiar: in the absence of large local labor sources, the colonists turned to slavery and indentured workers to serve their lucrative plantations. Between Spanish Venezuela and Colombia on the one hand, and the non-Hispanic "three Guianas" on the other, there is this difference: in Venezuela and Colombia, the population center of gravity soon moved inland and the plantation phase was followed by a totally different economy. In the Guianas, coastal settlement and the plantation economy still predominate in the twenty-first century (see box titled "The Guianas").

Figure 5-9 reveals that Venezuela and Colombia have assets that the Guianas lack: much larger territories and populations; more varied natural environments, including a share of the Andes Mountains (Colombia's being larger); and greater economic opportunities, based on oil reserves in each country that rank among the world's largest deposits. Yet each of these geographic advantages is constrained by negative qualities that continue to prevent the realization of their full potential. Physiographically and demographically, both countries exhibit a pronounced clustering of often-isolated populations in their highland zones, and share a huge, near-empty, and relatively low-lying interior. Economically, both depend on a small number of products (subject to worrisome price fluctuations on the world market) for the bulk of their export revenues—while the vast majority of their inhabitants practice subsistence agriculture and labor under the social and economic inequalities common to most of Iberian America. And politically, both remain under a dark cloud of instability, particularly violence-riddled Colombia, a state bedeviled by ongoing insurgencies and a society that has all but collapsed.

The Guianas

On the north coast of South America, three small countries lie east of Venezuela and adjacent to northernmost Brazil (Fig. 5-9). In "Latin" South America, these territories are anomalies of a sort: their colonial heritage is Western European, and formerly they were known as British, Dutch, and French Guiana—and called *the Guianas*. But two of the three have long been independent: Guyana, Venezuela's neighbor, and Suriname, the Dutch-influenced country in the middle. The easternmost territory, French Guiana, still continues under the colonial rule of France's government, while a local movement to upgrade its political status seeks to gain strength.

None of the three countries has a population over 1 million. Guyana is the largest with 740,000, Suriname has 440,000 inhabitants, and French Guiana a mere 185,000. Culturally and spatially, patterns here are Caribbean: peoples of South Asian and African descent are in the majority, with whites forming a small minority. In Guyana, Asians make up just under half the population, blacks and mixed Africans 44 percent, and the others, including Europeans, are tiny minorities. In Suriname, the ethnic picture is even more complicated, for the colonists brought not only Indians from South Asia (now 37 percent of the population) and blacks (31 percent) but also Indonesians (15 percent) to the territory in servitude. The remaining 10 percent of the Surinamese population consists of a separate category of black communities, peopled by descendants of African slaves (known as Bush blacks) who escaped from the coastal plantations and fled into the forests of the interior. French Guiana is the most European of the three countries. About three-quarters of its small population speak French, and the ethnic majority (66 percent) is of mixed African, Asian, and European ancestry—the *creoles*.

French Guiana, an overseas *département* of France, is the least developed of the three Guianas. The European Space Agency's launch complex at Kourou accounts for fully half of the country's modest productive activity. Most food, however, must be brought in from abroad.

Suriname progressed more rapidly after independence in 1975, but unremitting political instability soon ensued. More than 100,000 residents emigrated to the Netherlands, and during the late 1990s the stagnating economy neared bankruptcy. Nonetheless, Suriname not only remains self-sufficient in rice production but also exports surpluses of that staple crop along with bananas and seafood. In most years, the leading Surinamese income earner is bauxite (aluminum ore), mined in a zone across the middle of the country. That output may soon be surpassed by oil production, however, because significant reserves have been discovered just offshore.

British-influenced Guyana became independent in 1966 amid internal conflict that basically pitted people of African origins against people of Asian descent. Such problems continue to plague this country, in which the great majority of the people live in small villages near the coast. Bauxite, gold from mineral-rich western Guyana, sugar, rice, shrimp, and molasses earn most of the country's annual revenues. Oil is also likely to become a major economic activity in the near future because the newly discovered offshore oilfield that adjoins Suriname extends westward into Guyanese waters. The two countries, however, do not agree on the location of their maritime boundary, and that is certain to increase tensions between these two already-uneasy neighbors.

Today a major environmental controversy engulfs all three countries. With the economies of the two largest Guianas starved for funds, timber companies from Asia's Pacific Rim are offering to pay handsomely to exploit the most precious natural resource—the vast tracts of tropical rainforest that blanket the landscape beyond the populated coastal strips. Concessions (land development rights) have already been granted, but the battle has now been joined by conservationists. The latter, who aim to purchase their own concessions to preserve as much of the rainforest as possible, are only too well aware that the loggers they oppose have ravaged similar woodland environments across many parts of Southeast Asia.

Venezuela

Much of what is important in Venezuela is concentrated in the northern and western parts of the country, where the Venezuelan Highlands form the eastern spur of the north end of the Andes system. Most of Venezuela's 25.2 million people are concentrated in these uplands, which include the capital of Caracas, its early rival Valencia, and the commercial/industrial centers of Barquisimeto and Maracay.

The Venezuelan Highlands are flanked by the Maracaibo Lowlands and Lake Maracaibo to the northwest and by a vast plainland of savanna country, known as the *Llanos*, in the Orinoco Basin to the south and east (Fig. 5-9). The Maracaibo Lowlands, once a disease-infested, sparsely peopled coastland, today constitute one of the world's leading oil-producing areas. Much of the oil is drawn from reserves that lie beneath the shallow waters of the lake itself. Actually, Lake Maracaibo is a misnomer, for

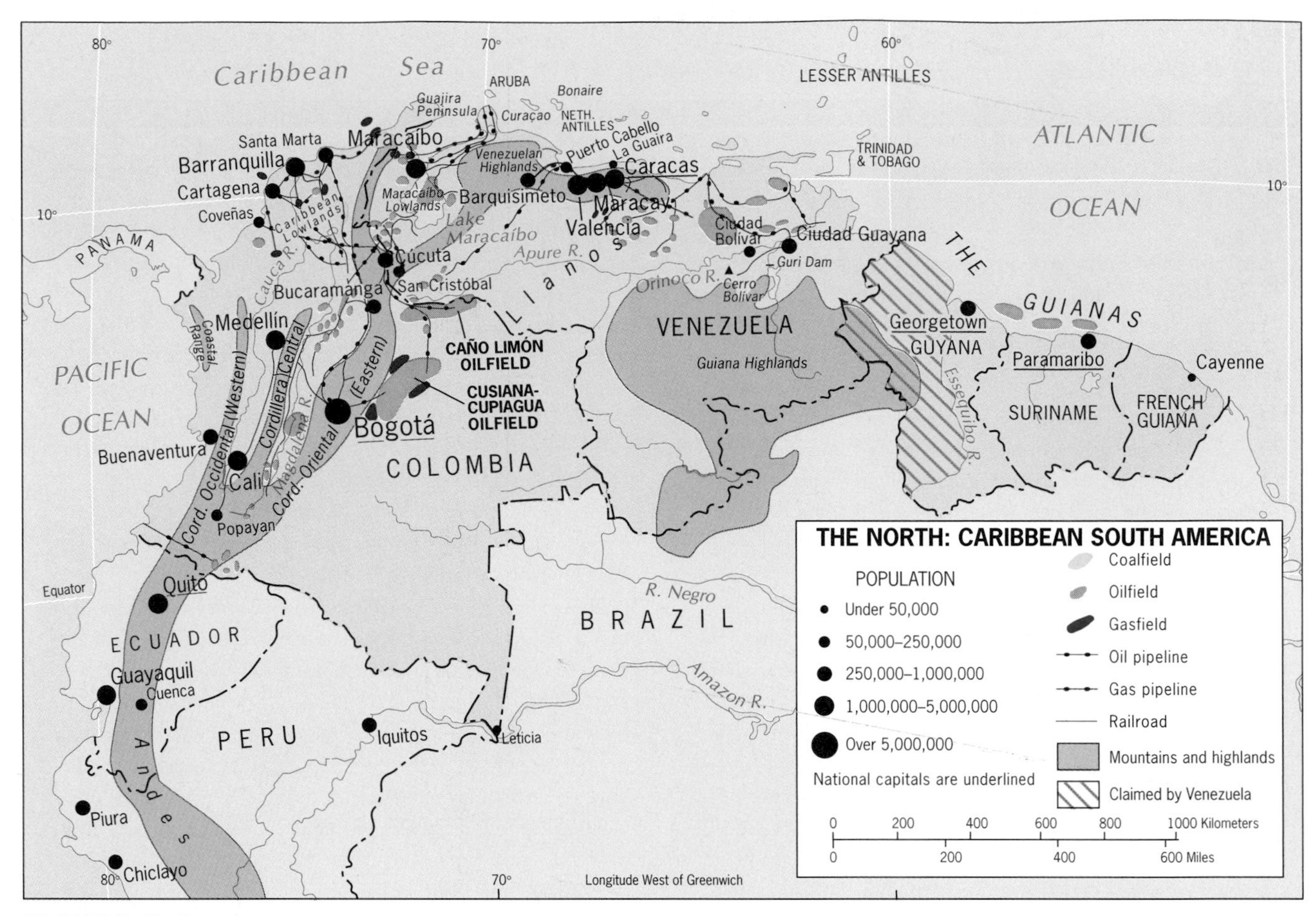

FIGURE 5-9

the "lake" is open to the ocean and is in fact a gulf with a very narrow entry. The country's third-largest city, Maracaibo, is the focus of the petroleum industry that transformed the Venezuelan economy in the 1970s; however, as we shall see, since then oil has been more of a curse than a blessing.

The *Llanos* on the southern side of the Venezuelan Highlands and the Guiana Highlands in the country's southeast, like much of Brazil's Interior, are in an early stage of development. The 200- to 400-mile (300- to 650-km)-long *Llanos* slope gently from the base of the Andean spur to the Orinoco River. Its mixture of savanna grasses and scrub woodland support cattle grazing on higher ground, but widespread wet-season flooding of the more fertile lower-lying areas has thus far inhibited the *Llanos*' commercial agricultural potential (much of the development of the *Llanos* to date has been limited to the exploitation of its substantial oil reserves). Crop-raising conditions are more favorable in the *tierra templada* areas (see box titled "Altitudinal Zones" on p. 227) of the Guiana Highlands. Economic integration of this more remote interior zone with the rest of Venezuela has been spearheaded by the discovery of rich iron ores on the northern flanks of the Guiana Highlands southwest of Ciudad Guayana. Local railroads now connect with the Orinoco, and from there ores are shipped directly to foreign markets.

Despite these opportunities, Venezuela since 1998 has been in upheaval as longstanding economic and social problems finally intensified to the point where the electorate decided to push the country into a new era of radical political change. We noted above that, since the euphoric 1970s, oil had not bettered the lives of most Venezuelans. A major reason was that the government unwisely acquired the habit of living off oil profits, forcing the country to suffer the consequences of the long global oil depression that began in the early 1980s. Now Venezuela found itself heavily burdened by a huge foreign debt it had incurred in borrowing against future oil revenues that were not materializing fast enough. By the mid-1990s, the government was required to sharply devalue the currency, and a political crisis ensued that resulted in a severe recession and widespread social unrest. With more and more Venezuelans enraged at the way their oil-rich country was approaching bankruptcy without making progress toward the more equitable distribution of the national wealth, voters resoundingly turned in an extreme direction in the 1998 presidential election. Expressing their disgust with Venezuela's ruling elite of both political parties, they elected Hugo Chavez, a former colonel who in 1992 had led a failed military coup. By choosing this independent candidate, the voters gave him a mandate to be a strongman type of leader.

This is exactly the course Chavez has taken since entering office in 1999, sweeping aside Congress, the Supreme Court, both major political parties, and supervising the rewriting of the Venezuelan constitution in his own image. Further consolidating his hold on office by engineering a landslide reelection to another term that runs until 2006, Chavez now proclaims himself the leader of a "peaceful leftist revolution" that will transform the country. Although he professes his deepest ties are to "the people" and says social equality ranks highest on his agenda, Chavez has not hesitated to stir up racial divisions by actively promoting mestizos (67 percent of the population) over those of European background (20 percent). Heightening the societal role of the armed forces while criticizing every Venezuelan elite from the corporate community to the Roman Catholic Church, Chavez has clearly signalled that he is a classic military strongman of the type that many dismayed South Americans had thought they would never see again. And in the international arena, Chavez sparks controversy at every turn: angering the government of neighboring Colombia by expressing his "neutrality" in its confrontation with cocaine-producing insurgent forces; unsettling another neighbor, Guyana, by aggressively reviving a century-old territorial claim to the western two-thirds of that country (the striped zone in Fig. 5-9); condemning capitalism and befriending Cuba's Fidel Castro; and harassing the United States—which imports more oil from Venezuela than from any other country—by pressing for higher fuel prices within the OPEC petroleum cartel and warmly courting such rogue leaders as Iraq's Saddam Hussein and Libya's Muammar Gadhafi. All this rhetoric and posturing aside, the new regime has not yet weaned the government from its habit of living off oil profits; and with a simultaneously worsening economy between 1999 and 2001, the country's enthralled electorate is in no mood to accept more years of patience while the Chavez revolution tries to deliver on its promises of a better life for all Venezuelans.

Colombia

Imagine a country more than twice the size of France, not at all burdened by overpopulation, with a physical geography so varied that it can produce crops ranging from the temperate to the tropical, possessing world-class reserves of oil and natural gas, plus 2000 miles of coastline on Atlantic and Pacific waters. Situated in the northwestern corner of South America, closer than any of its neighbors to the markets of the north, with a single language and one dominant religion, would such a country not assuredly thrive amidst the burgeoning economic geography of its hemisphere? The answer is no. Colombia today is a country as troubled as any in the world, riven by violence, its economy imperiled, its politics destabilized, its future a gigantic question mark. Colombia's cultural uniformity has not produced social cohesion. The same physical geography that diversifies its environments also divides its population clusters, which are so tenuously interconnected that the country totals less than 250 miles (400 km) of paved four-lane highway. Its proximity to U.S. markets is a curse as well as a blessing: at the root of Colombia's current problems lies its role in the international drug trade.

The oil reserves beneath the Lake Maracaibo Basin have transformed the economic geography of this shallow inland sea. Drilling platforms stud the lake; pipelines link the wells to shoreside refineries and containers. In one way or another, the lives of virtually the entire population of the town of Cabimas (in the background) are dominated or affected by this industry. Shown here are the storage and export facilities near Cabimas, and a tanker being loaded; Venezuela remains the United States' leading foreign source of petroleum, and oil is the mainstay of the Venezuelan economy.

The Geography of Cocaine

Any geographical discussion of northwestern South America today must take note of one of its most widespread activities: the production of illegal narcotics. Of the enormous flow of illicit drugs that enter the United States each year, the most widely used substance undoubtedly is cocaine—all of which comes from South America, mainly Bolivia, Peru, and Colombia (the sources of more than 75 percent of the total world supply). Within these three countries, cocaine annually brings in billions of (U.S.) dollars and "employs" tens of thousands of workers, constituting an industry that functions as a powerful economic force. Those who operate the industry have accumulated considerable power through bribery of politicians, threats of terrorism, and alliances with guerrilla groups in outlying zones beyond governmental control. The cocaine industry itself is structured within a tightly organized network of territories that encompass the various stages of this drug's production.

The first stage of cocaine production is the extraction of coca paste from the coca plant, a raw-material-oriented activity that is located near the areas where the plant is grown. The coca plant was domesticated in the Andes by the Incas centuries ago; millions of their descendants still chew coca leaves for stimulation and brew them into coca tea, the leading beverage of highland South America. The main zone of coca-plant cultivation is along the eastern slopes of the Andes and in adjacent tropical lowlands in Bolivia, Peru, and Colombia.

Today, five areas dominate in the growing of coca leaves for narcotic production (Fig. 5-11): Bolivia's Chaparé district in the marginal Amazon lowlands northeast of the city of Cochabamba; the Yungas Highlands north of the Bolivian capital, La Paz; north-central Peru's Huallaga Valley; south-central Peru's Apurimac Valley, east of Huancayo; and the green-colored zones around the FARC-controlled territory of southern Colombia (Fig. 5-10). These areas, which produce as many as nine crops per year, are especially conducive to high leaf yields thanks to optimal local climatic and soil environments that also allow plants to develop immunity to many diseases and insect ravages. Despite harassment by (U.S.-supported) government efforts to aerially spray herbicides on them, all of these areas continue to thrive as coca cultivation has become a full-fledged cash crop. Operations in guerrilla-controlled south-central Colombia have now reached the scale of plantations, inducing

Colombia's current disorder is not its first. In the past, civil wars between conservatives and liberals (on Roman Catholic religious issues) developed into conflicts pitting rich against poor, elites against workers. In Colombia today, people still refer to the last of these wars as *La Violencia*, a decade of strife beginning in 1948 during which as many as 200,000 people died. When, in the late 1950s, the two sides—the Conservative and Liberal Parties—called a truce and agreed on constitutional reform to preclude a recurrence, Colombia at last seemed able to realize its economic potential. Exports of coffee, for which the country was famous, soared. Oil exports from reserves in the northern part of the country increased. Minerals, fruits, and cut flowers flowed to foreign markets, half by value to the United States, a quarter to Europe.

But in the 1970s, disaster struck. In remote parts of the country, groups opposed to the power-sharing monopoly between the political parties began a campaign of terrorism, damaging the developing infrastructure and destroying confidence in the future. Simultaneously, the U.S. market for narcotics expanded rapidly, and many Colombians got involved in the drug trade. Powerful and wealthy drug cartels formed in major cities such as Medellín and Cali, with networks that influenced all facets of Colombian life from the peasantry to the politicians. The fabric of Colombian society unraveled.

As Figure 5-9 shows, western Colombia is mountainous while the east lies in the Orinoco and Amazon Basins. Most of the country's 41.6 million people (the largest population in all of Spanish-speaking South America) live in the western and northern parts of Colombia. Physiography has contributed to the concentration of this population into more than a dozen separate clusters. Some of them are located in the Caribbean Lowlands; others lie in the Magdalena, Cauca, and lesser valleys between the parallel ranges of the Andes (including Medellín and Cali); and still others are found in basins within the mountains themselves, the most important being the cluster around the capital city of Bogotá (elevation: 8500 feet [2700 m]). Colombia's huge coffee crop—second only to Brazil's on the continent and in the world—comes from the numerous *templada* zones on the Andean slopes. But the more sparsely populated east and south have become the production centers for the coca crop that sustains the country's illicit narcotics industry (see box titled "The Geography of Cocaine").

Colombia's belt of oil and natural gas reserves flanks the border with Venezuela from Lake Maracaibo south to the Caño Limón oilfield, then continues onward along the Andean front through the country's largest petroleum deposits (the Cusiana-Cupiagua oilfield), and terminates at the newly discovered Guando oilfield in the upper Magdalena Valley southwest of Bogotá (Fig. 5-9). Pipelines connect thousands of nearby subsistence farmers (sometimes at gunpoint) to join the more lucrative ranks of the field workforce. In Peru and Bolivia, the specialized coca-cultivation zones have lured an even larger peasant-farmer population to abandon the less profitable production of food crops, thereby further reducing the capability of these nutrition-poor countries to feed themselves. The coca leaves harvested in these eastern Andean source areas make their way to nearby centers—located at the convergence of rivers and trails—where coca paste is extracted and prepared.

The second stage of production involves refining that coca paste (about 40 percent pure cocaine) into cocaine hydrochloride (more than 90 percent pure), a lethal concentrate that is diluted with substances such as sugar or flour before being sold on the streets to consumers. Cocaine refining requires sophisticated chemicals, carefully controlled processes, and a labor force skilled in their supervision, and here Colombia has predominated. Most of this activity takes place in ultramodern processing centers located in the rebel-held territory of the lowland central-south and east, well beyond the reach of the Bogotá government. Interior Colombia also possesses the geographic advantage of intermediate location, lying between the source areas to the south (as well as locally) and the U.S. market to the north, a short hop for smugglers by air or sea across the Caribbean.

The final stage of production entails the distribution of cocaine to the U.S. marketplace, which depends on an efficient (but clandestine) transportation network. Most common is the use of private planes that operate directly out of remote airstrips near the refineries in areas beyond effective government control. The transport of cocaine to the United States is often a two-step process, making use of such intermediate Middle American transshipment points as the Bahamas, Dominican Republic, Haiti, Puerto Rico, Panama, and particularly Mexico. Although South Florida is reputed to be a leading port of illegal entry, other coastal points in the U.S. Southeast as well as cities along the Mexican border in the Southwest have recorded increases in cocaine trafficking since 1990. Judging by the size of recent drug seizures in the Southwest as well as Southern California, the overland route through mainland Middle America may now have become the smugglers' pathway of choice.

the oilfields to the Caribbean ports of Cartagena, Barranquilla, Santa Marta, and Coveñas. Beneath that coastal lowland, Colombia also has substantial reserves of coal, a product that now ranks high on the list of exports. With its fuel, mineral, and agricultural productivity, Colombia's economy might have been one of South America's most prosperous. But the rise of the narcotics industry, coupled with the legacy of *La Violencia*, has dealt the country a devastating blow.

Today, Colombia is in combat. In the cities, narcoterrorists have committed acts of violence ranging from an invasion of the Supreme Court to the downing of a passenger-filled airliner. In the countryside, drug-financed armies of the political "left" fight equally vicious paramilitary forces of the "right." Rebels sabotage pipelines; oil companies hire and arm personnel to protect them. Civilians by the thousands die in the crossfire of all these belligerent groups, and countless more are dislocated.

Parts of Colombia are beyond the control of its government and its national armed forces, and are essentially ruled by rebels who "protect" the local citizenry against Bogotá's efforts to reestablish authority there. In effect, insurgents in several parts of Colombia have created their own domains, and some of their rulers have even negotiated their status with the legitimate government in the capital. What is happening here is a process long studied by political scientists. It involves three stages: (1) *contention*, the period of initial rebellion of the kind that marked the decade of *La Violencia*; (2) *equilibrium*, when the rebels gain effective control over some of the national territory, as has happened in parts of Colombia; and (3) *counteroffensive*, in which the government and the rebels, usually following the breakdown of negotiations, engage in armed conflict that will decide the future of the country.

The geographer Robert McColl translated this sequence into spatial terms, suggesting that the equilibrium stage represents the emergence of an **insurgent state**. The formation of an insurgent state is the key stage in the process because it creates an informally bounded territory, usually with a distinct core area centered on a captured town, establishes a parallel government, installs a form of administration, and provides some social services such as schools and medical clinics that substitute for those the formal state may have provided. McColl pointed to Cuba and South Vietnam as countries within which insurgent states had developed before the final stage of decisive counteroffensive occurred.

17

Importantly, not all insurgent states succeed. They did in Cuba and Vietnam, but not in Peru or Nigeria. In some cases, the process never advances beyond the initial stage of contention. In Chapter 8 we examine the case of South Asia's Sri Lanka, where an insurgent state has existed for two decades and the counteroffensive has been in progress for years—producing stalemate, not decision.

At the beginning of the twenty-first century, Colombia contained not just one, but several enclaves that were taking on the properties of insurgent states (Fig. 5-10). This is because Colombia confronts not just one but several different guerrilla forces, each operating in particular areas of the country. Thus the M-19 (April 19) Movement focuses on the Urabá area in the north, near the Panamanian border. The Popular Liberation Army is active in the Cauca Valley. The National Liberation Army (ELN) operates in the Arauca area on the Venezuelan border, where it attacks oil facilities, and in the middle Magdalena Valley. And the Revolutionary Armed Forces of Colombia (FARC), by far the most powerful insurgent group, has most recently concentrated on the central-south and east.

In 1999, as an incentive to negotiate, FARC compelled the government to demilitarize a large zone south of Bogotá (now popularly called *Farclandia*), in effect leaving the insurgents there in control of an area the size of Switzerland (red-striped zone in Fig. 5-10). Also important was FARC's simultaneous initiative to take over the far eastern town of Mitú and its environs, an area whose remoteness made it particularly difficult for the Colombian army to regain control. And although the insurgents claimed they were suppressing drug production in the zones under their control, the evidence indicates otherwise. According to the latest official estimates and aerial imagery, coca production has skyrocketed in the FARC-held insurgent state.

Colombia's twin scourges, narcotics and guerrillas, have set the country back immeasurably and pose a colossal threat to its future as a coherent state. In Chapter 1 we discuss the centrifugal forces that affect all countries to some degree, and the centripetal forces that must override them for the state to function. In Colombia, the divisive forces grow ever stronger as the future of the state becomes more and more uncertain. Guerrilla movements that literally and/or virtually control large parts of Colombia no longer base their actions on political philosophies: despite a superficial veneer of socialist ideology, they have evolved from "leftist," "Marxist," or radical religious groups into criminal gangs that specialize in terror, kidnapping, extortion, and racketeering. They obstruct foreign investment and destroy facilities, organize and tax coca farmers, operate airstrips and charge takeoff and landing fees, run sophisticated communications networks, and control hundreds of towns and even entire provinces. In the process, they have driven well over 1 million Colombian farmers off their land and into

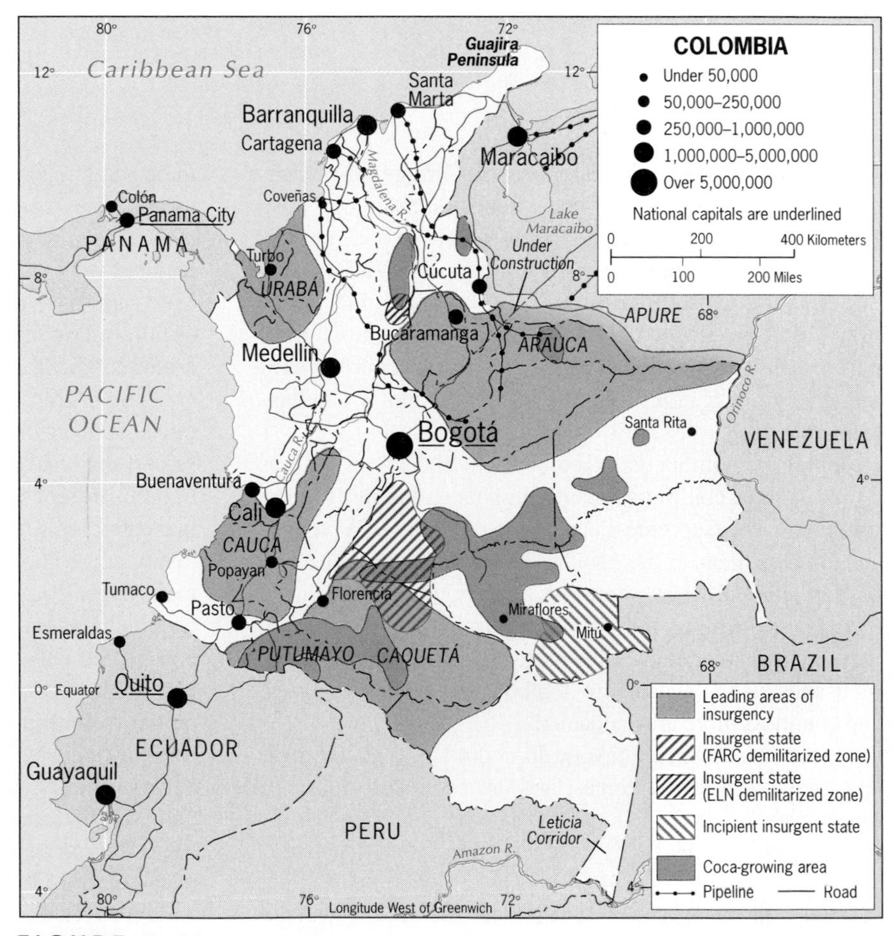

FIGURE 5-10

the cities as refugees, and convinced an additional million (mostly of middle-class status) to flee to neighboring countries and the United States. On the confiscated real estate the rebels establish their own ranches, drug laboratories, and command posts. As the cash from the narcotics trade and from ransoms flows in (estimates for FARC range as high as U.S. $500 million a year), they buy sophisticated weapons that allow them to stalemate the national army.

As the rebels' hold over their domains strengthens, local governments (if they are allowed to remain in place), companies, and individuals are forced to pay "protection" money by direct deposit into guerrilla-controlled bank accounts. Such funds are used, among other things, to buy legitimate businesses such as hotels, restaurants, and stores. Thus the drug trade, directly or indirectly, infects every corner of what remains of Colombian society.

The geographic dimensions of the centrifugal forces these activities represent are particularly worrisome. As Figure 5-10 reveals, the activity spheres of the guerrilla movements place extensive parts of the country beyond the reach of law and order. Colombia's international borders continue to be represented on maps (such as those in this chapter), but the reality on the ground is quite different. When any country confronts disruptive forces that put portions of the national territory beyond government control, the very foundations of the state are threatened.

THE WEST: ANDEAN SOUTH AMERICA

The third regional grouping of South American states—the Andean West (Fig. 5-11)—encompasses Peru, Ecuador, and Bolivia.* The map of culture spheres (Fig. 5-5) shows the Amerind-subsistence region extending along the Andes Mountains, indicating that these countries have large Amerindian components in their populations. In Bolivia about 55 percent of the people are of Amerindian stock, in Peru about 45 percent, and in Ecuador about 25 percent; but keep in mind that these percentages are only approximate because it is often impossible to distinguish between Amerindian and "mixed" people of strong Amerindian background. There are, however, other similarities among the three countries: their incomes are low; they are comparatively unproductive; and, unhappily, they exemplify the grinding poverty of the landless peonage. As the twenty-first century opens, these desperate conditions are no longer being passively tolerated, and uprisings continue to emerge throughout this region—especially among the most disadvantaged indigenous populations.

*Paraguay also exhibits many cultural characteristics of the Andean West but is today turning southward in its economic-geographic orientation. We treat Paraguay as a transitional state and discuss it in the next regional section, the Southern Cone.

Peru

Peru, which straddles the Andean spine for more than 1000 miles (1600 km), is the largest of the region's three republics in terms of territory as well as population (28.3 million). Its half-million square miles (1.3 million sq km) divide both physiographically and culturally into three subregions: (1) the desert coast, the European-mestizo region; (2) the Andean highlands or *Sierra*, the Amerindian region; and (3) the eastern slopes and adjoining *montaña*, the sparsely populated Amerindian-mestizo interior (Fig. 5-11). It is symptomatic of the cultural division still prevailing in Peru that Lima, the capital whose population constitutes more than one-quarter of the national total, is not located centrally in a basin of the Andes but in the peripheral coastal zone.

At Lima in the heart of the desert coastal strip, the Spaniards avoided the greatest of the Amerindian empires, choosing a site a few miles inland from a suitable anchorage that became the modern outport of Callao. From an economic point of view, the Spaniards' choice of a headquarters on the Pacific coast proved to be sound, for the coastal subregion has become commercially the most productive part of the country. A thriving fishing industry based on the cool productive waters of the Peru (Humboldt) Current offshore contributes significantly to the export trade. Irrigated agriculture in some 40 oases distributed all along the arid coast produces cotton, sugar, rice, vegetables, fruits, and wheat, much of which is exported as well.

The Sierra (Andean) subregion occupies about a third of the country and contains the majority of Peru's Amerindian peoples, most of them Quechua-speaking—a legacy of the Inca Empire. Despite the size of its territory and population (nearly half of all Peruvians reside here), the political influence of this subregion is slight, as is its economic contribution (except for the mines). In the high valleys and intermontane basins, the Amerindian population is clustered either in isolated villages, around which people practice a precarious subsistence agriculture, or in the more favorably located and fertile areas where they are tenants, peons on white- or mestizo-owned haciendas. Most of these people never receive an adequate daily caloric intake or balanced diet of any sort. The wheat produced around Huancayo, for example, is mainly exported and would in any case be too expensive for the Amerindians themselves to buy. Potatoes, barley, and corn are among the subsistence crops grown here in the *tierra fría* zone, and in the *tierra helada* of the higher basins the Amerindians graze their llamas, alpacas, cattle, and sheep (these environmental zones are discussed in the box on p. 227). The major mineral products from the Sierra are copper, zinc, lead, and several other metallic minerals, with the largest mining complex centered on Cerro de Pasco.

Of Peru's three subregions, the *Oriente*, or East—the inland slopes of the Andes and the Amazon-drained, rain-

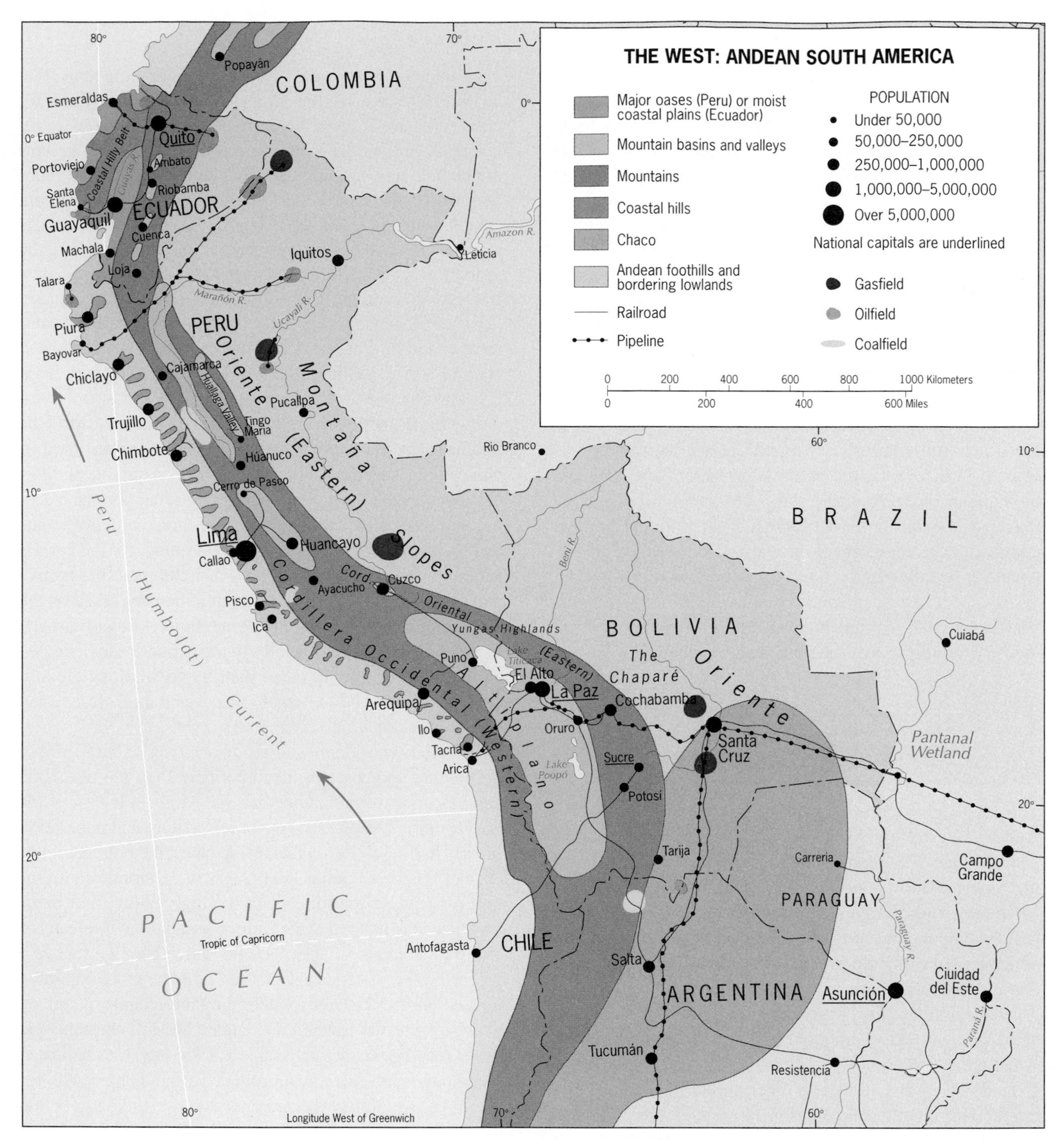

FIGURE 5-11

forest-covered *montaña*—is the most isolated. The focus of the eastern subregion, in fact, is Iquitos, a city that looks east rather than west and can be reached by ocean-going vessels sailing 2300 miles (3700 km) up the Amazon River across northern Brazil. Iquitos grew rapidly during the Amazon wild-rubber boom of a century ago and then declined; now it is finally growing again and reflects Peruvian plans to open up the eastern interior. Petroleum was discovered west of Iquitos in the 1970s, and since 1977 oil has flowed through a pipeline built across the Andes to the Pacific port of Bayovar. As Figure 5-11 shows, the eastern interior also contains major natural gas deposits, and their exploitation is a leading priority in Peru's current development plans.

Peru today is emerging from two decades of instability during which its government was besieged by well-organized guerrilla movements. The most serious threat, which for a time reached the level of an insurgent state, came from the radical-communist *Sendero Luminoso* (Shining Path). Spilling out of its territorial base in the central Andes, this insurgency at its height even threatened Lima. But in the end the government responded strongly, and since the mid-1990s the guerrillas have been in retreat—their leadership in disarray, their territorial control reduced to a few remote localities, and their offensive diminished to sporadic terrorist acts. As the revolutionary threat subsided, investors and tourists returned, Peru's flagging industries revived, and the economic picture brightened. Maintaining such growth beyond the short-term period of recovery, however, is not the greatest socioeconomic challenge now facing Peru's leaders. Looming even larger is the central cause of the guerrilla uprisings: the persistent, desperate poverty of the Amerindians. Shockingly, the latest evidence reveals that the portion of the Peruvian population living in poverty has grown from 46 percent in 1995 to over 50 percent today, a still-widening gap between the affluent and the poor surpassed, among South America's large countries, only by Brazil.

Ecuador

On the map, Ecuador, smallest of the three Andean West republics, appears to be just a corner of Peru. But that would be a misrepresentation because Ecuador possesses a full range of regional contrasts (Fig. 5-11). It has a coastal belt; an Andean zone that may be narrow (under 150 miles or 250 km) but by no means of lower elevation than elsewhere; and an *Oriente*—an eastern subregion that is as sparsely populated and as economically marginalized as that of Peru. As in Peru, nearly half of the people of Ecuador are concentrated in the Andean intermontane basins and valleys, and the most productive region is the coastal strip. Here, however, the similarities end.

Ecuador's Pacific coastal zone consists of a belt of hills interrupted by lowland areas, of which the most important lies in the south between the hills and the Andes, drained by the Guayas River and its tributaries. Guayaquil—the country's largest city, main port, and leading commercial center (but not the capital)—forms the focus of this subregion. Ecuador's coastal lowland, moreover, is not a desert: it consists of fertile tropical plains not afflicted by excessive rainfall (except when periodic El Niño events, which originate along this edge of

Among the Realm's Great Cities . . .

Lima

Even in a realm replete with primate cities, Lima (7.9 million) stands out. Here reside 28 percent of the Peruvian population—who produce over 70 percent of the country's gross domestic product, 90 percent of its collected taxes, and 98 percent of its private investments. Economically, at least, Lima *is* Peru.

Lima began as a modest oasis in a narrow coastal desert squeezed between the cold waters of the Pacific and the soaring heights of the nearby Andes. Near here the Spanish conquistadors discovered one of the best natural harbors on the western shoreline of South America, where they founded the port of Callao; but they built their city on a site 7 miles (11 km) inland where soils and water supplies proved more beneficial. This city was named Lima and quickly became the Spaniards' headquarters for all their South American territories.

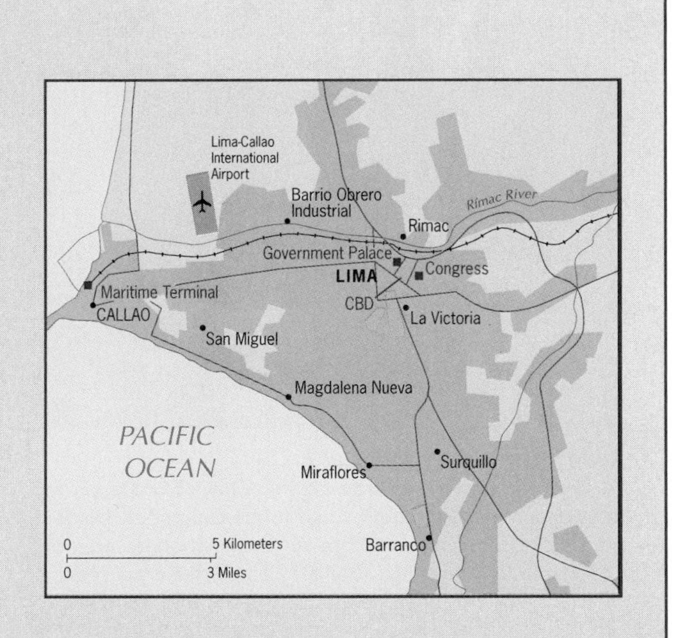

Since independence, Lima has continued to dominate Peruvian national life. The city's population growth was manageable through the end of the 1970s, but the past 20 years have witnessed a disastrous doubling in its size. As usual, the worst problems are localized in the peripheral squatter shantytowns that now house close to one-half of the metropolitan population. Lima's CBD, however, is worlds removed from this squalor, its historic landscape dotted with five-star hotels, new office towers, and prestigious shops as foreign investments have poured in since 1990.

the Pacific Basin, cause a temporary reversal of the moisture regime). The products of this western subregion also differ from those of its Peruvian counterpart. Seafood is a primary industry, but, unlike Peru, whose main product is fishmeal (a fertilizer and cattle food), Ecuador now harvests such vast quantities of shrimp that it has become the world's second-leading producer. Agriculturally, bananas, sugarcane, rice, and cacao are the most important lowland commercial crops; in addition, cattle-raising is a major pursuit here on lands possessing less fertile soils, and lucrative coffee is grown on hillsides as well as in the Andean *templada* zones.

FROM THE FIELD NOTES

"I walked through the street market in Otavalo, Ecuador, on the Andean road north from Quito to the Colombian border. This is Inca country, and at these elevations (Mount Cayambe at nearly 19,000 feet [5800 m] lies nearby) the staple is the potato. Beets and turnips are also for sale, and many other crops—common and unusual—are cultivated in this area. The people here, both sellers and buyers, seem quite prosperous. They were well dressed against the cool mountain weather and business was brisk. Comparing this scene to what I had seen in the Ecuadorian port city of Guayaquil helped me understand this small country's intense regionalism better. Mestizo coast and Amerindian interior are worlds apart."

Ecuador's coastal subregion is also far less Europeanized than Peru's Pacific-facing lowland because the white component of the national population of 13.1 million is only about 10 percent. A sizeable proportion of whites are engaged in administration and hacienda ownership in the central Andean zone, where most of the 25 percent of the Ecuadorians who are Amerindian also reside—and, not surprisingly, where land-tenure reform is an explosive issue. (Of the country's remaining population, 10 percent is black and mulatto, and 55 percent is mestizo—many with a stronger Amerindian ancestry.) The differing interests of the Guayaquil-dominated coastal lowland and the Andean-highland subregion focused on the capital (Quito) have long fostered a deep regional cleavage between the two. This schism has intensified in the past few years as Guayas Province, which contains Guayaquil, led a rising chorus of lowlander complaints that the tax revenues sent to Quito from this wealthiest part of Ecuador were being squandered by national government bureaucrats. The Guayas legislative council even went as far as organizing a referendum on autonomy (which the voters resoundingly supported)—a potentially significant step down the slippery slope of devolution in an era in which such centrifugal forces are weakening nation-states across the world.

In the jungles of the *Oriente* subregion, petroleum production has reached substantial levels over the past quarter-century, and oil tops the country's export list. A trans-Andean pipeline was opened in 1972 to connect the rich interior oilfields to the port of Esmeraldas, but poor management contributed to a continuing pollution nightmare of repeated oil spills and toxic-waste dumping along parts of the 300-mile (480-km) route; despite recent cleanup efforts, this is likely to remain one of the continent's worst environmental disasters for years to come. Stimulating the wider economic development of the interior remains a leading priority of the Ecuadorian government. That policy was given a major boost in 1998 by the settlement of a long-running dispute with neighboring Peru over Ecuador's southeastern border, a sporadically violent confrontation that on at least three occasions since 1980 had escalated to the brink of war. Although Ecuador did not gain sovereignty over the territory it claimed, the Peruvians made several concessions to foster increased trade and the integration of the border zone. The most important of these was to grant the Ecuadorians navigation rights on the Amazon and its northern Peruvian tributaries as well as special access to build two trading-center ports on this riverine network.

These and other growth opportunities notwithstanding, most economic development is now on hold in Ecuador because the country is in the vanguard of a regionwide social movement, led by indigenous peoples, that is threatening to destabilize the Andean West. This movement, grounded in the longstanding exclusion of Amerindians within each of the region's three national societies, has since the late 1990s spawned a number of uprisings as frustrations can no longer be contained in the face of repeatedly failed promises by governments that have raised expectations for social and economic improvement. Ongoing geographic change further fuels this unrest because the poorest people increasingly urbanize and come to witness close-up how their second-class standing and lack of education bar them from access to the knowledge-based labor market driven by the forces of economic globalization. In Ecuador, the widely supported Confederation of Indigenous Nationalities of Ecuador (CONAIE) has spearheaded the protest movement, demanding a multiethnic state in which the Amerindians will be granted full rights of citizenship, better access to landownership, greater political involvement, and an end to government neglect which ensures that indigenous communities end up with inadequate educational systems and infrastructural linkages.

Bolivia

From Ecuador southward through Peru, the Andes broaden until in Bolivia they reach a width of some 450 miles (720 km). In both the eastern and western Andean cordilleras (ranges), peak elevations in excess of 20,000 feet (6000 m) are recorded. Between these two great ranges lies the *Altiplano* proper (Fig. 5-11; as noted earlier, an altiplano is an elongated, high-altitude basin). On the boundary between Peru and Bolivia, freshwater Lake Titicaca—the highest large lake on Earth—lies at 12,507 feet (3700 m) above sea level. Here, in its west, is the heart of modern Bolivia; here, too, lay one of the centers of Inca civilization, and indeed of pre-Inca cultures. Bolivia's capital, La Paz, is also located on the *Altiplano* at 11,700 feet (3570 m), and as such is one of the highest cities in the world.

Lake Titicaca helps make the *Altiplano* livable, for this large body of water ameliorates the coldness in its vicinity, where the snow line lies just above the plateau surface. On the surrounding cultivable land, grains have been raised for centuries in the Titicaca Basin to the extraordinary elevation of 12,800 feet (3850 m). To this day, this area of Peru and Bolivia supports a major cluster of Amerindian subsistence farmers.

Modern Bolivia is the product of the European impact, an influence that bypassed many of the Amerindian population clusters. These indigenous Bolivians no more escaped the loss of their land than did their Peruvian or Ecuadorian counterparts, especially east of the *Altiplano*. What made the richest Europeans in Bolivia wealthy, however, was not land but minerals. The town of Potosí in the eastern cordillera became a legend for the immense deposits of silver in its vicinity; tin, zinc, copper, and several ferroalloys were also discovered there. Over most of the past century, Bolivia's tin deposits, which ranked among the world's richest, yielded much of the country's annual export income; but since 1980, declining tin reserves and falling world prices have forced much of the industry to shut down. Today zinc has replaced tin as the leading metallic mineral export, and natural gas and oil are accounting for a growing share of foreign revenues. Recently affiliating with the Mercosur trading bloc as an associate member, Bolivia now exports large quantities of gas via a new pipeline to Brazil and Argentina. In return, Brazil is assisting in the development of the southeastern lowlands, where commercial agriculture—notably the production of soybeans, now Bolivia's leading export—is steadily expanding in the fertile savanna around Santa Cruz, which has become the country's second-largest city.

Bolivia has had a turbulent history. Apart from endless internal struggles for power, the country first lost its corridor to the Pacific coast during the 1880s in a disastrous conflict with Chile; then in 1903 it lost its northern territory of Acre to Brazil in a dispute involving the rubber boom in the Amazon Basin; and finally, it lost 55,000 square miles (140,000 sq km) of southeastern Gran Chaco territory to Paraguay in the 1930s. The most critical loss by far was that of its outlet to the sea—which may yet be regained from Chile if longstanding negotiations can be concluded successfully.

Although Bolivia has rail connections to the Chilean ports of Arica and Antofagasta, it remains severely disadvantaged by its landlocked situation. Because the western Andean cordillera and the *Altiplano* form the country's inhospitable western margins, one might suppose that Bolivia would look eastward and that its *Oriente* would be somewhat better developed than that of Peru or Ecuador. As we noted, this interior subregion is now awakening, but its limited infrastructure still confines economic activity and population growth to a few pockets. The densest settlement clusters within Bolivia's dispersed population of 8.6 million are in the valleys and basins of the eastern cordillera, where agricultural opportunities are more favorable than on the barren *Altiplano* to the west.

Regional development here in South America's second-poorest country (only tiny Guyana has a lower per capita GNP) is also inhibited because an appalling 70 percent of all Bolivians still live in poverty. The vast majority of the poor are the long-neglected Amerindian peoples, who finally began to be recognized in Bolivian society during the 1990s through such limited reforms as bilingual education programs (less than half of the country's citizens speak Spanish as a first language). The insufficiency of this government response—and the now-dashed hopes of continued improvement raised by these minimal reforms—are evident in Bolivia's strong participation in the social movements that today besiege the Andean region. After a number of local confrontations in the late 1990s, the bitter feelings of the indigenous peoples burst into the open all across the *Altiplano* with a week-long protest in early 2000 that blockaded roads and brought the country to a standstill. This explosion was triggered by a riot in defiance of increased water rates in the city of Cochabamba to cover the privatization of the municipal waterworks, a sell-off perceived to be financed by the lower-income masses for the economic benefit of the elite. Choosing not to react in a heavy-handed way to violent protests that had quickly spread to other parts of the country, the La Paz government capitulated after a few days. But almost nothing was done to address the social inequalities that cause the discontent experienced by so many Bolivians, which can only be mitigated by integrating the impoverished masses into the mainstream of national life.

THE SOUTHERN CONE: MID-LATITUDE SOUTH AMERICA

South America's four southern countries—Argentina, Uruguay, Paraguay, and Chile—constitute the realm's final regional grouping (Fig. 5-12). Because the continent resembles an overstuffed ice cream cone on the map, this region also goes by its popular name — the *Southern Cone*. We noted earlier that the Southern Cone Common Market (Mercosur) began operating here in 1995 as the hemisphere's second-largest trading bloc after NAFTA. Mercosur encompasses Argentina, Uruguay, Paraguay, and Brazil; Chile and Bolivia participate as associate members.

Regional ties in the Southern Cone are built on its modern history as the heart of the European-commercial culture sphere, which also spills across the Brazilian border to include much of that country's core area (Fig. 5-5). Today Mercosur not only perpetuates those linkages, but has also drawn peripheral Paraguay into its orbit. Paraguay may be a less-developed state and transitional with respect to the adjoining Andean West region, but its neighbors to the south and east unhesitatingly welcomed Paraguay into their new free-trade zone and customs union. Two of the largest dams on Earth (Yacyretá and Itaipu) already operate on the Paraná River along Paraguay's eastern border, and one of Mercosur's biggest ventures calls for the completion of the 2150-mile (3450-km)-long *Hidrovia* waterway that would open most of the Paraná-Paraguay river basin to barge transport. Intraregional ties are also being strengthened by infrastructure improvements; many are already underway or in the planning stage. These include the construction of new routes over and under the Andes, the world's longest bridge across the Rio de la Plata between Buenos Aires and Uruguay, and hundreds of miles of new superhighways and railroads.

Argentina

The largest Southern Cone country by far is Argentina, whose territorial size ranks second only to Brazil in this geographic realm; its population of 37.8 million ranks third after Brazil and Colombia. Argentina exhibits a great deal of physical-environmental variety within its boundaries, and the vast majority of the Argentines are concentrated in the physiographic subregion known as the *Pampa* (a word meaning "plain"). Figure I-9 underscores the degree of clustering of Argentina's inhabitants on the land and in the cities of the Pampa. It also shows the relative emptiness of the other six subregions (mapped in Fig. 5-12): the scrub-forest *Chaco* in the northwest; the mountainous *Andes* in the west, along whose crestline lies the boundary with Chile; the arid plateaus of *Patagonia* south of the Rio Colorado; and the undulating transitional terrain of intermediate *Cuyo, Entre Rios* (also known as "Mesopotamia" because it lies between the Paraná and Uruguay rivers), and the *North*.

The Argentine Pampa is the product of the past 150 years. During the second half of the nineteenth century, when the great grasslands of the world were being opened up (including those of the interior United States, Russia, and Australia), the economy of the long-dormant Pampa began to emerge. The food needs of industrializing Europe grew by leaps and bounds, and the advances of the Industrial Revolution—railroads, more efficient ocean transport, refrigerated ships, and agricultural machinery—helped make large-scale commercial meat and grain production in the Pampa not only feasible but also highly profitable. Large haciendas were laid out and farmed by tenant workers; railroads radiated ever farther outward from the booming capital of Buenos Aires and brought the entire Pampa into production.

Over the decades, within the framework of the **von Thünen model** (see pp. 49-51), several specialized agricultural areas appeared on the Pampa. As we would expect, a zone of fruit and vegetable production became established near the huge Buenos Aires conurbation located beside the estuary of the Rio de la Plata. To the southeast is the predominantly pastoral district, where beef cattle and sheep are raised. In the drier west, northwest, and southwest, wheat becomes the important commercial grain crop, but half the land remains devoted to grazing. Among the exports, cereals have usually led by value, followed by meat products, animal feed, and vegetable oils. Yet even with all this output the Pampa has hardly begun to fulfill its productive potential, which could easily double with more intensive agricultural practices. One problem is that the land inheritance system too often fragments farms into inefficiently small units; another has been the underutilization of fertilizers.

18

Today all that is changing as awareness grows of new global market opportunities driven by rising food demands. Argentina's government paved the way by reducing agricultural export taxes in 1990, launching a boom that would see farm exports double by the end of the decade. More recently, foreign investors have been attracted by the Pampa's relatively modest land prices and costs of doing business. Cattle-raising in particular is being reinvigorated, and production has risen with the spread of U.S.-style feedlots that turn out larger quantities of high-quality, grain-fed animals. New crops such as soybeans (Argentina is now the world's third-largest producer) are appearing as well, and more farmers than ever before are seeking to maximize out-

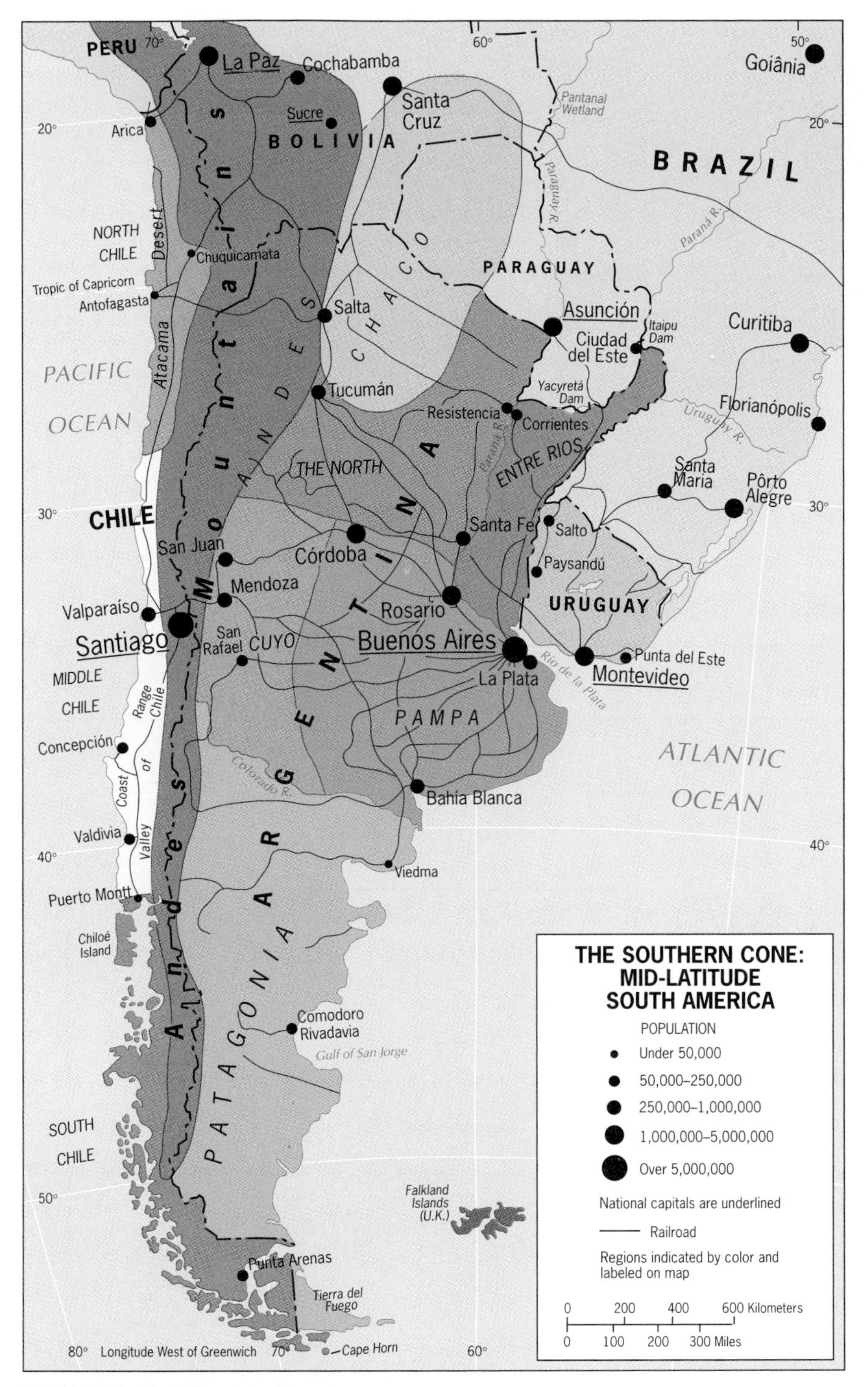

FIGURE 5-12

puts by increasing their usage of pesticides, fertilizers, and anti-erosion methods. Not surprisingly, these developments are enlarging the scale of production, and as the sizes of farms and ranches expand, hundreds of thousands of smaller operators find themselves under intensifying pressure to keep up.

Argentina's affluence is reflected in its bustling cities, which epitomize the European-commercial cultural character of the realm's southernmost countries. In fact, no less than 90 percent of the Argentine population may be classified as urbanized, a proportion well above the average of Western Europe's countries. Just under 13 million (34 percent) reside in metropolitan Buenos Aires, which also contains most of the industries, many of them managed by Italians, Spaniards, and other immigrants. Córdoba, Argentina's second city with a much smaller population of 1.5 million, is another focus of industrial growth (here, in the late 1990s, U.S., European, and Japanese automakers built state-of-the-

FROM THE FIELD NOTES

"Whether you arrive in Buenos Aires by air or by sea, this city does not present a memorable site. It lies on the western side of the Plata estuary, not far south from the delta of the Paraná River. Sail upstream, and the low, flat banks of the Rio de la Plata narrow, presenting a string of coastal villages and some industry; then the highrises of Buenos Aires come into view, seemingly standing in the muddy water (silting is a problem in the channel). As the photo shows, Buenos Aires also lacks a dramatic CBD skyline; for so large a city, its vertical development is limited. Its areal sprawl, however, is considerable; there are few obstacles to expansion. Unremarkable architecturally, Buenos Aires has an Ibero-Italian atmosphere with some colonial remnants—in places. Historic preservation has not been a preoccupation here."

Among the Realm's Great Cities . . .

Buenos Aires

Its name means "fair winds," which first attracted European mariners to the site of Buenos Aires alongside the broad estuary of the muddy Rio de la Plata. The shipping function has remained paramount (see photo, above), and to this day the city's residents are known throughout Argentina as the *porteños* (the "port dwellers"). Modern Buenos Aires was built on the back of the nearby Pampa's grain and beef industry. It is often likened to Chicago and the Corn Belt in the United States because both cities have thrived as interfaces between their immensely productive agricultural hinterlands and the rest of the world.

Buenos Aires (12.9 million) is yet another classic South American primate metropolis, housing more than one-third of all Argentines, serving as the capital since 1880, and functioning as the country's economic core. Moreover, Buenos Aires is a cultural center of global standing, a monument-studded city that contains the world's widest boulevard (*Avenida 9 de Julio*).

During the half-century between 1890 and 1940, the city was known as the "Paris of the South" for its architecture, fashion leadership, book publishing, and performing arts activities (it still has the world's biggest opera house, the *Teatro Colón*). With the recent restoration of democracy, Buenos Aires is now trying to recapture its golden years. Besides reviving these cultural functions, the city has added a new one: the leading base of the hemisphere's motion picture and television industry for Spanish-speaking audiences.

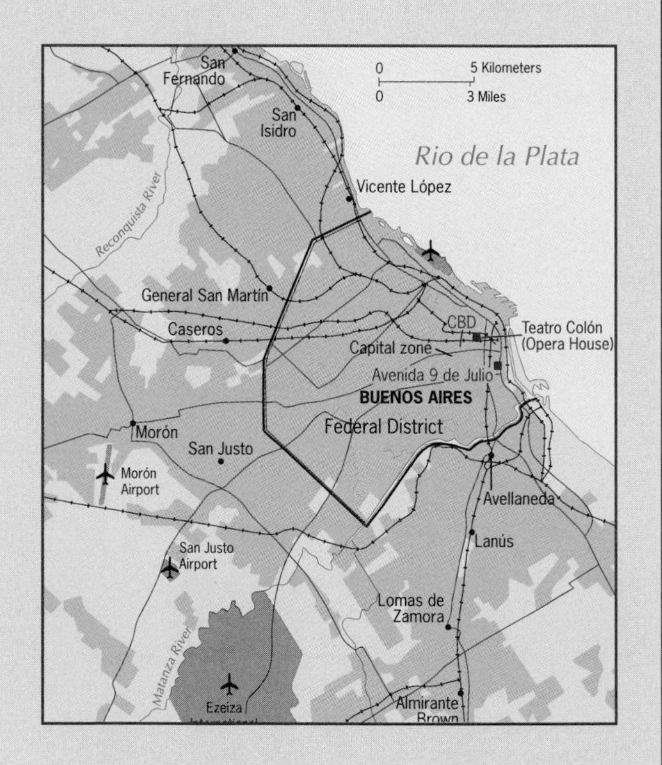

art component and assembly plants to tap into the growing Mercosur market for cars, especially neighboring Brazil's). Much of the manufacturing in the larger cities is associated with the processing of Pampa products and the production of consumer goods for the domestic market. Nearly one out of every three wage earners in the country is engaged in manufacturing—another indication of Argentina's somewhat advanced economic standing. Nonetheless, a severe recession plagued the country at the opening of the twenty-first century, undermining confidence in the future and leading many younger Argentines to emigrate to southern Europe.

Besides its high degree of clustering, Argentina's population also exhibits a decidedly peripheral distribution. The Pampa subregion covers only a little more than 20 percent of Argentina's territory, but with nearly half of the people concentrated here the rest of the country cannot be densely populated. Outside the Pampa, pastoralism is an almost universal pursuit, but the quality of the cattle is much lower than in the Pampa; in near-arid Patagonia sheep are raised, but ecological problems caused by overgrazing have resulted in a sharp decline since 1990. Some of the more distant areas contain potentially significant resources: in Patagonia an oilfield is in production near coastal Comodoro Rivadavia, and extensive reserves may also exist in the northern Chaco border zone with Paraguay. Yerba maté, a local tea, is produced in the North and Entre Rios subregions; and the quebracho-tree extract (for tanning leather) that both Paraguay and Argentina export comes from the central Paraguay-Paraná Basin. Northeastern Argentina is also an area targeted for industrial development, sparked by the massive Yacyretá Dam on the Paraná River. In both the North and Cuyo subregions, streams flowing off the Andes provide opportunities for irrigated agriculture; Tucumán is a major sugar-producing center, and Mendoza is the heart of a productive district of vineyards and fruit orchards. But despite these sizeable near-Andean outposts and their good rail connections, effective Argentina remains the area within a radius of 350 miles (560 km) of Buenos Aires.

Uruguay

Uruguay, unlike Argentina or Chile, is compact, small, and rather densely populated. This buffer state of old became a fairly prosperous agricultural country, in effect a smaller-scale Pampa (though possessing less favorable soils and topography). Figures 5-1 and I-8 show the similarity of physical conditions on the two sides of the Plata estuary. Montevideo, the coastal capital, contains more than 35 percent of the country's population of 3.3 million; from here, railroads and roads radiate outward into the productive agricultural interior. In the immediate vicinity of Montevideo lies Uruguay's major farming area, which produces vegetables and fruits for the metropolis as well as wheat and fodder crops. Most of the rest of the country is used for grazing cattle and sheep, with beef products, wool and textile manufactures, and hides dominating the export trade. Tourism is another major economic activity as Argentines, Brazilians, and other visitors increasingly flock to the Atlantic beaches at Punta del Este and other thriving resort towns.

FROM THE FIELD NOTES

"Arrived in Mendoza (Argentina) yesterday to begin a week's study of the wine industry here. Stayed at an old hotel still showing cracks in the walls from the 1861 earthquake. Was invited to the local opera performing *Carmen*, starting time: 11:30 PM. Mendoza is so hot in the daytime that people here use the night to work as well as play, and I was told that the city is 'dead' from about 1 to 6 PM. This morning's guided excursion produced some remarkable viticultural vistas, but nothing topped the drama of the mountains and the famous 'Christ of the Andes' statue at the foot of what is actually the Andean range called the *Sierra de Los Paramillos*. Beyond the mountains in this westward view lies Chile, and during the movement for independence from Spain, General José de San Martin planned and executed an expedition to liberate Chile in 1817 from his headquarters in Mendoza."

Uruguay's area of 67,500 square miles (176,000 sq km)—less territory even than Guyana—does not leave much room for population clustering. Nevertheless, a special quality of the land area of Uruguay is that it is rather evenly peopled right up to the boundaries with Brazil and Argentina. And of all the countries in South America, Uruguay is the most truly European, notably lacking the racial minorities found even in Chile and Argentina, but with a sizeable non-Spanish European component in its population. As for future development prospects as Mercosur matures (Montevideo is the bloc's administrative capital), Uruguay is in an excellent position to capitalize on its location between the Pampa to the south and Brazil's most dynamic areas to the north.

Paraguay

As Figure 5-3 indicates, Paraguay is one of those countries (Kazakhstan is another) that lies across a regional transition zone. In terms of economic development, Paraguay's per capita GNP is more typical of the Andean West than the Southern Cone. Ethnically, too, the pattern is more characteristic of the West: about 95 percent of Paraguay's 5.8 million people are mestizo, but with so pervasive an Amerindian influence that any white ancestry is almost totally submerged. As for languages, Amerindian Guaraní is so widely spoken alongside Spanish that the country is surely one of the world's most completely bilingual. The physiography, however, is decidedly non-Andean because all of Paraguay lies to the east of the mountains in the center of the Paraná-Paraguay Basin; from here, the outlet to the sea has always been to the south. As the Basin's waterways are improved—and as Mercosur drives the economic integration of the Southern Cone countries—that regional orientation is steadily strengthening.

Paraguay's landlocked position has had much to do with its modest economic development to date. Opportunities for pastoral and agricultural industries have long existed but went largely unrealized because exports must be shipped through Buenos Aires—a long haul from the capital of Asunción via the (as-yet-unimproved) Paraguay-Paraná waterway. Nonetheless, soybean products, cotton, timber, vegetable oils, and hides reach foreign markets. Grazing in the dry Chaco to the west of the Paraguay River is an important commercial activity, but here cattle generally do not compare favorably to those of Argentina. A prosperous Mennonite agricultural colony in the remote heart of this zone is more successful, producing a large peanut crop and about half of the country's dairy products. In the northeast border zone, many Brazilians have crossed the Paraná River to purchase fertile farmlands and raise lucrative soybeans (making Paraguay the sixth-largest producer, whose annual crop accounts for 2 percent of the world's supply). Further downstream, two ultramodern dams present possibilities for the growth of manufacturing (Fig. 5-12). Here, too, located on the river where Paraguay, Argentina, and Brazil meet, is Ciudad del Este. This notorious frontier city, so conveniently situated next to spectacular Iguaçu Falls, draws its hordes of visitors because of its reputation as a bargain retailer of consumer goods—cheap knockoff goods, that is, because the largely clandestine local economy thrives as South America's gateway for counterfeit products and contraband ranging from imitation athletic shoes to pirated CDs to illegally trafficked weapons.

Chile

For 2500 miles (4000 km) between the crestline of the Andes and the coastline of the Pacific lies the narrow strip of land that is the Republic of Chile. On average just 90 miles (150 km) wide (and only rarely over 150 miles or 250 km in width), Chile is the world's quintessential example of what *elongation* means to the functioning of a state. Accentuated by its north-south orientation, this severe territorial attentuation not only results in Chile extending across numerous environmental zones; it has also contributed to the country's external political, internal administrative, and general economic problems. Nonetheless, throughout most of their modern history, the Chileans have made the best of this potentially disastrous centrifugal force: from the beginning, the sea has constituted an avenue of longitudinal communication; the Andes Mountains continue to form a barrier to encroachment from the east; and when confrontations loomed at the far ends of the country, Chile proved to be quite capable of coping with its northern rivals, Bolivia and Peru, as well as Argentina in the extreme south.

As Figures I-8 and 5-12 indicate, Chile is a three-subregion country. About 90 percent of its 15.6 million people are concentrated in what is called Middle Chile, where Santiago, the capital and largest city, and Valparaíso, the chief port, are located. North of Middle Chile lies the Atacama Desert, which is wider, drier, and colder than the coastal desert of Peru. South of Middle Chile, the coast is broken by a plethora of fjords and islands, the topography is mountainous, and the climate—wet and cool near the Pacific—soon turns drier and colder against the Andean interior. South of the latitude of Chiloé Island there are no permanent overland routes, and there is hardly any settlement. These three subregions are also clearly apparent on the map of culture spheres (Fig. 5-5), which displays a mestizo north, a European-commercial zone in Middle Chile, and an undifferentiated south. In addition, a small Amerind-subsistence zone in northern Chile's Andes is shared with Argentina and Bolivia.

Some intraregional differences exist between northern and southern Middle Chile, the country's core area. Northern Middle Chile, the land of the hacienda and of Mediterranean climate with its dry summer season, is an area of (usually irrigated) crops that include wheat, corn, vegetables, grapes, and other Mediterranean products. Livestock raising and fodder crops also take up much of the productive land but are continuing to give way to the more efficient and profitable cultivation of fruits for export. Southern Middle Chile, into which immigrants from both the north and Europe (especially Germany) have pushed, is a better-watered area where raising cattle has predominated. But here, too, more lucrative fruit, vegetable, grain, and other expanding food crops are changing the area's agricultural specializations.

Prior to the 1990s, the arid Atacama region in the north accounted for more than half of Chile's foreign revenues. The Atacama contains the world's largest exploitable deposits of nitrates. At first these provided the country's economic mainstay, but this mining industry declined after the discovery of methods of synthetic nitrate production early in the twentieth century. Subsequently, copper became the chief export (Chile possesses the world's largest reserves). It is found in several places, but the main concentration lies on the eastern margin of the Atacama Desert near the town of Chuquicamata, not far from the port of Antofagasta.

Chile today is emerging from a development boom that transformed its economic geography during the 1990s, a growth spurt that established its reputation as South America's greatest success story. Following the withdrawal of its repressive military dictatorship in 1990, Chile embarked on a program of free-market economic reform that brought stable growth (of up to 8 percent a year), dramatically lowered inflation and unemployment, reduced the poverty rate by more than 20 percent, and attracted massive foreign investment. The last is of particular significance because these new international connections enabled the export-led Chilean economy to diversify and develop in some badly needed new directions. Copper—whose prices have fallen steadily over the past several years—remains the single leading export, but many other mining ventures (including gold) have been launched. In the agricultural sphere, fruit and vegetable production for export has soared because Chile's harvests coincide with the winter farming lull in the high-income countries of the Northern Hemisphere. Other primary economic activities that are increasing their foreign earnings are seafood and wood products, especially paper. Industrial expansion is occurring as well, though at a more leisurely pace, and new manufactures include a modest array of goods that range from basic chemicals to computer software.

Chile's newly internationalized economy has propelled the country to forge a prominent role for itself on the global trading scene. Even the Japanese expressed interest in building ties, and for a time during the 1990s Japan was its leading trading partner. This foothold in the Pacific Rim arena notwithstanding, Chile's trade linkages are now more heavily oriented toward its own hemisphere. The Chileans quickly realized that regional economic integration offered the best opportunities for sustained progress and eagerly accepted the 1994 invitation of the United States, Canada, and Mexico to join the North American Free Trade Agreement. Although the U.S. Congress continues to delay ratification, Chile remains first in line to join NAFTA. In the meantime, responding to warm overtures received from the countries beyond its Andean border, Chile has affiliated with Mercosur as an associate member; however, despite exhortations from its Mercosur partners, Chile in 2000 decided against seeking full membership, preferring instead to negotiate its own free-trade agreement directly with the United States.

Chile ended the twentieth century on a less encouraging note as its still-dominant copper sector dragged the national economy down to a growth standstill. With this commodity earning more than a third of all foreign revenues, there was little the Chileans could do in 1998 when the Asian economic downturn stifled demand and caused a worldwide collapse in the already low price of copper. This was followed by a period of painful readjustment throughout Chile but little loss of confidence in the economy because the financial system retained its strength. Chile's planners also learned some valuable lessons from this experience. One was that the country's economic diversification away from its reliance on copper had neither been as comprehensive nor as rapid as first believed. Among the other lessons were that Chile remains too heavily dependent on its natural resources, urgently needs to expand the skills of its workforce, and must work harder to develop its manufacturing, services, and quaternary sectors.

Chile's recent transformation is widely touted as the "economic model" for all of Middle and South America to emulate. That is almost certainly an overstatement because few countries in these geographic realms can match the Chilean combination of natural and human resources and development circumstances. Yet all of those countries strongly aspire to follow in Chile's footsteps and for the first time in their modern histories are cooperating toward making that a reality. Their goal is the economic integration of all the Americas within the next few years—an accomplishment that could lead to a better life for every inhabitant of the Western Hemisphere.

chapter 6

North Africa/Southwest Asia

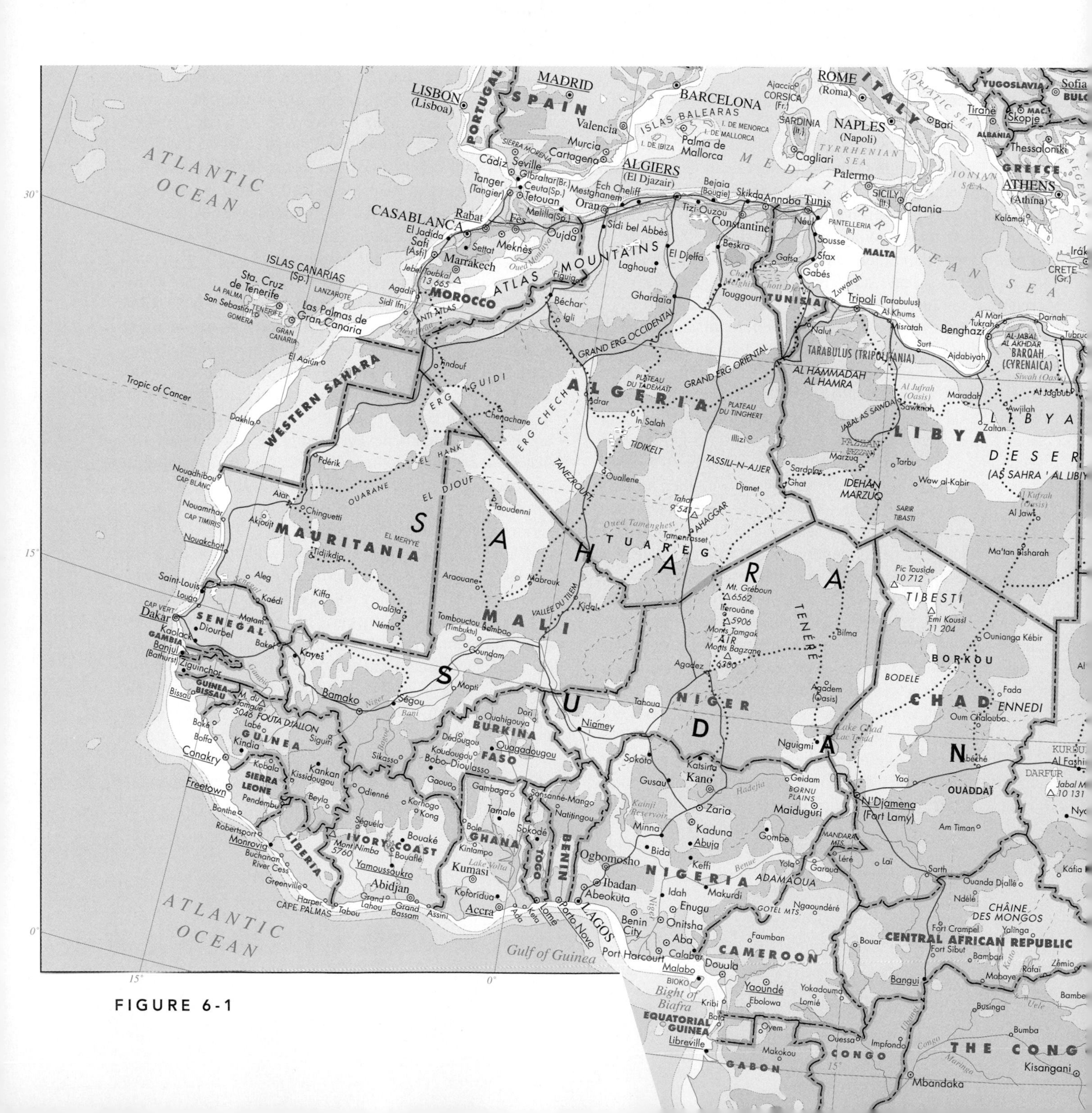

FIGURE 6-1

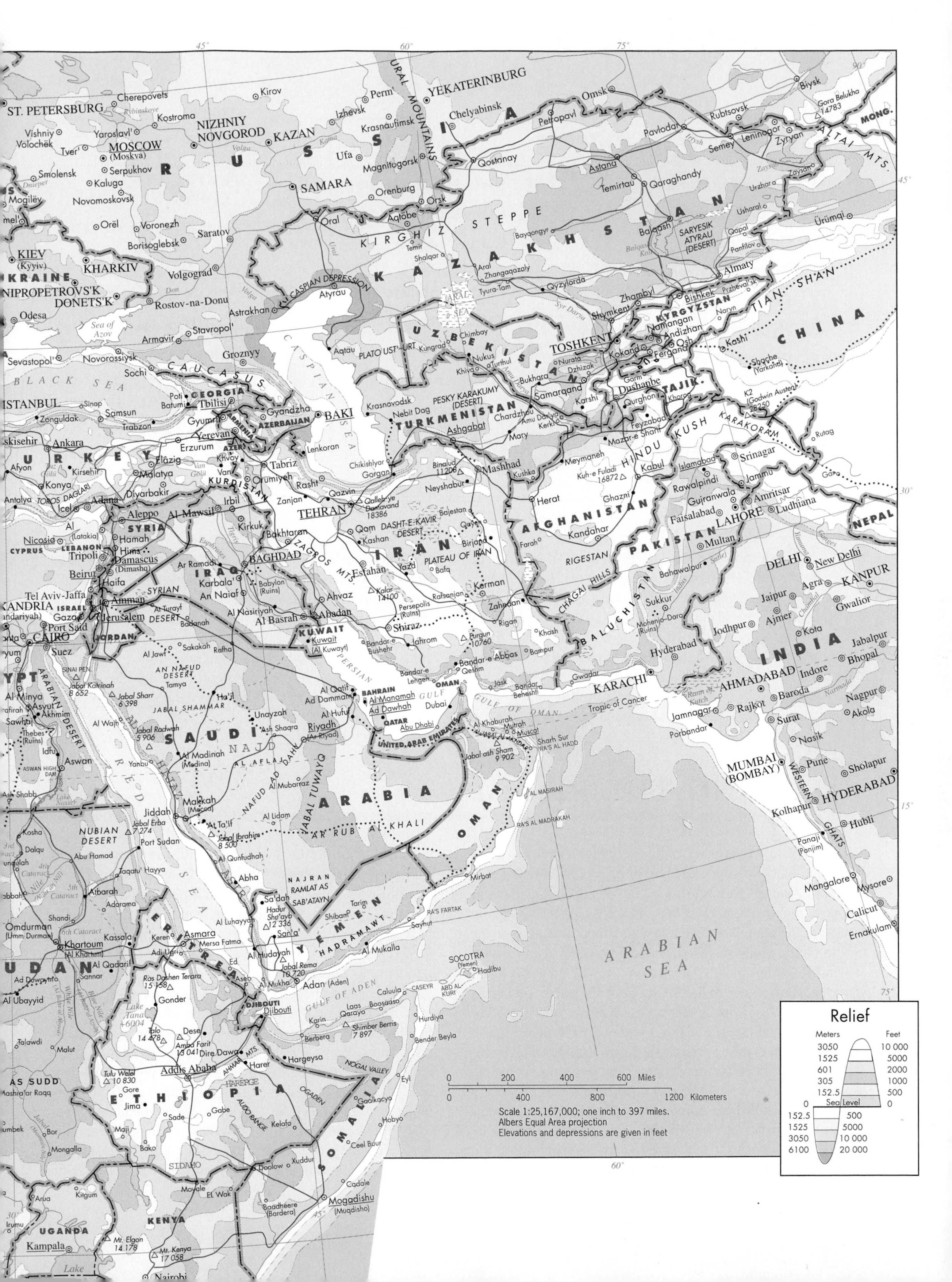

45°
60°
75°
90°
45°
30°
15°
75°
60°
45°
ST. PETERSBURG
Cherepovets
Rybinskoye
Kirov
Perm'
YEKATERINBURG
URAL MOUNTAINS
Omsk
Biysk
Gora Belukha 14783
Vishniy Volochëk
Yaroslavl'
Kostroma
NIZHNIY NOVGOROD
KAZAN
Izhevsk
Krasnoufimsk
Chelyabinsk
Petropavl
Rubtsovsk
MONG.
Tver'
MOSCOW
(Moskva)
R
U
S
S
I
A
Kama
Ufa
Magnitogorsk
Qostanay
Pavlodar
Irtysh
Semey
Leninogor
Zyryan
ALTAI MTS.
Smolensk
Serpukhov
Kaluga
Dnieper
SAMARA
Orenburg
Orsk
Astana
Temirtau
Qaraghandy
Zaysan
Mogilëv
Novomoskovsk
Orël
Voronezh
Saratov
Oral
Aqtöbe
STEPPE
KIRGHIZ
Urzhar
Usharal
Ürümqi
Borisoglebsk
Ural
Temir
Bayqongyr
Balqash
SARYESIK ATYRAU (DESERT)
Qapal
Panfilov
KIEV
(Kyyiv)
KHARKIV
Volgograd
Shalqar
Aral
Zhangaqazaly
Balqash Köli
KRAINE
NIPROPETROVS'K
DONETS'K
Don
Volga
CASPIAN DEPRESSION
Atyrau
K
A
Z
A
K
H
S
T
A
N
Tyura-Tam
Qyzylorda
Almaty
Rostov-na-Donu
ARAL SEA
Syr Darya
Zhambyl
Bishkek
Przheval'sk
Odesa
Sea of Azov
Astrakhan
Shymkent
KYRGYZSTAN
TIAN SHAN
Armavir
Stavropol'
Aqtau
PLATO UST-URT
UZBEKISTAN
Chimbay
Kungrad
Nukus
TOSHKENT
Namangan
Andizhan
Naryn
CHINA
Sevastopol'
Novorossiysk
Groznyy
Khiva
Turtkul
Nurata
Dzhizak
Kokand
Fergana
Osh
Kashi
Shache (Yarkand)
BLACK SEA
Sochi
CAUCASUS
CASPIAN SEA
Amu Darya
Bukhara
Samarqand
Dushanbe
Garm
TAJIK.
ISTANBUL
Sinop
Samsun
Poti
GEORGIA
Batumi
Tbilisi
ARMENIA
Krasnovodsk
Nebit Dag
PESKY KARAKUMY (DESERT)
TURKMENISTAN
Karshi
Qurghon
Khorog
K2 (Godwin Austen) 28250
Zonguldak
Trabzon
Gyandzha
BAKI
Ashgabat
Chardzhou
Amu Darya
Kerki
Feyzabad
Gyumri
AZERBAIJAN
Mary
Mazar-e Sharif
HINDU KUSH
KARAKORAM
Rutag
Eskisehir
Ankara
Erzurum
Yerevan
AZER.
Lenkoran
Meymaneh
TURKEY
Elâzig
Khvoy
Tabriz
Chikishlyar
Binalud 11208
Mashhad
Kuh-e Fuladi 16872
Kabul
Islamabad
Srinagar
Afyon
Van Gölü
Kirsehir
Malatya
Orumiyeh
Gorgan
Kushka
Rawalpindi
Jammu
Konya
TOROS DAGLARI
Diyarbakir
KURDISTAN
Rasht
Neyshabur
Gujranwala
Amritsar
Gar
Antalya
Adana
Icel
Irbil
Zanjan
Qazvin
Qolleh-ye Damavand 18386
Herat
Ghazni
Faisalabad
Ludhiana
Aleppo
Al Mawsil
TEHRAN
Bajestan
AFGHANISTAN
LAHORE
NEPAL
Al (Latakia)
SYRIA
Tigris
Kirkuk
Qom
DASHT-E-KAVIR DESERT
Qayen
Kandahar
Farah
PAKISTAN
Multan
Nicosia
CYPRUS
LEBANON
Hamah
Bakhtaran
Kashan
Birjand
Euphrates
Hims
Tripoli
Damascus
(Dimashq)
BAGHDAD
ZAGROS MTS.
IRAN
PLATEAU OF IRAN
RIGESTAN
Sutlej
DELHI
New Delhi
Yamuna
Ganges
Beirut
Ar Ramadi
IRAQ
Esfahan
Yazd
Bafq
CHAGAI HILLS
BALUCHISTAN
Bahawalpur
Haifa
Karbala'
Babylon (Ruins)
Kerman
KANPUR
Tel Aviv-Jaffa
Amman
SYRIAN
An Najaf
Kalar 14100
Rafsanjan
Zahedan
Jaipur
Agra
KANDRIA
ISRAEL
Al Turayf
DESERT
Badanah
Al Nasiriyah
Ahvaz
Persepolis (Ruins)
Sukkur
Chambal
Ajmer
Gwalior
Gaza
Jerusalem
Abadan
Al Basrah
Shiraz
Rigan
Mohenjo-Daro (Ruins)
Indus
Jodhpur
Port Said
JORDAN
KUWAIT
Kuwait
(Al Kuwayt)
Khash
Kota
CAIRO
Suez
Sakakah
Rafha
Bandar-e Bushehr
Jahrom
Purgun 10760
Hyderabad
INDIA
Jabalpur
Al Jawf
PERSIAN
Bampur
Bhopal
SINAI PEN.
Jabal Katrinah 8 652
AN NAFUD DESERT
Tamya
Bandar-e Abbas
Qeshm
Gwadar
KARACHI
AHMADABAD
Indore
Ha'il
Bandar-e Lengeh
OMAN
Jask
Bandar Beheshti
Al Minya
Asyut
ARABIAN DESERT
Jabal Sharr 6 398
JABAL SHAMMAR
Al Qatif
BAHRAIN
Ad Dammam
Al Manamah
Ad Dawhah
GULF
Dubai
GULF OF OMAN
Tropic of Cancer
Rann of Kutch
Baroda
Narmada
Nagpur
Akhmim
Sawhaj
Al Wajh
Jabal Radwah 5 906
Unayzah
Ash Shaqra
Al Hufuf
QATAR
Abu Dhabi
Riyadh
(Ar Riyad)
Jamnagar
Rajkot
Surat
Akola
Thebes (Ruins)
Idfu
SAUDI
NAJD
Al Khaburah
Mutrah
Muscat
Porbandar
Nasik
Aswan
ASWAN HIGH DAM
Yanbu
Al Madinah (Medina)
AL AFLAJ
UNITED ARAB EMIRATES
Jabal ash Sham 9 902
Sharh Sur
RA'S AL HADD
Pune
Sholapur
MUMBAI (BOMBAY)
Abu Shabb
Lake Nasser
RED
AL HIJAZ
NAFUD AD DAHY
Al Mubarraz
JABAL TUWAYQ
ARABIA
WESTERN
Makkah (Mecca)
AL MASIRAH
Kolhapur
HYDERABAD
Jiddah
At Ta'if
Al Lidam
OMAN
RA'S AL MADRAKAH
Hubli
Kosha
NUBIAN DESERT
Jabal Erba 7 274
Jabal Ibrahim 8 500
AR RUB AL KHALI
GHATS
Panaji (Panjim)
Dalqu
Abu Hamad
Port Sudan
Al Qunfudhah
4th Cataract
Taqatu' Hayya
Abha
NAJRAN
RAMLAT AS SAB'ATAYN
Mirbat
Mangalore
Mysore
Nile (Nahr an Nil)
5th Cataract
Atbarah
Adarama
Sa'dah
Hadur Shu'ayb 12 336
Tarim
YEMEN
RA'S FARTAK
Shandi
Al Luhayyah
Shibam
HADRAMAWT
Sayhut
Calicut
Omdurman
(Umm Durman)
6th Cataract
Kassala
ERITREA
Keren
Asmara
Mersa Fatma
San'a'
Al Mukalla
Ernakulam
Khartoum
(Al Khartum)
Adi Ugri
Al Hudaydah
Jabal Rema 10 720
SOCOTRA (Yemen)
Hadibu
ARABIAN SEA
SUDAN
Al Qadarif
Sannar
Ed
Aseb
Al Mukha
Adan (Aden)
Ad Duwaym
White Nile
Blue Nile
Ras Dashen Terara 15 158
Gonder
DJIBOUTI
Djibouti
GULF OF ADEN
Caluula
CASEYR
RAS AL KURI
Al Ubayyid
Lake Tana 6004
Laas Qoray
Boosaaso
Karin
Shimber Berris 7 897
Hurdiya
Talo 14 478
Dese
Amba Farit 13 041
Dire Dawa
Berbera
Bender Beyla
Talawdi
Malut
AHMAR MTS.
Harer
Hargeysa
NOGAL VALLEY
AS SUDD
Tulu Welel 10 830
Addis Ababa
Eyl
Mashra' ar Raqq
HARERGE
Gore
Jima
ETHIOPIA
OGADEN
Gaalkacyo
AUDO RANGE
Gabe
Sade
Kelafo
Hobyo
Maji
Bor
Mongalla
Bako
SOMALIA
Ceel Buur
SIDAMO
Doolow
Xuddur
Moyale
El Wak
Arua
Kitgum
Cadale
Baadheere (Bardera)
Mogadishu
(Muqdisho)
KENYA
UGANDA
Mt. Elgon 14 178
Kampala
Mt. Kenya 17 058
Lake
Nairobi
0 200 400 600 Miles
0 400 800 1200 Kilometers
Scale 1:25,167,000; one inch to 397 miles.
Albers Equal Area projection
Elevations and depressions are given in feet
Relief
Meters
Feet
3050
10 000
1525
5000
601
2000
305
1000
152.5
500
0
Sea Level
0
152.5
500
1525
5000
3050
10 000
6100
20 000

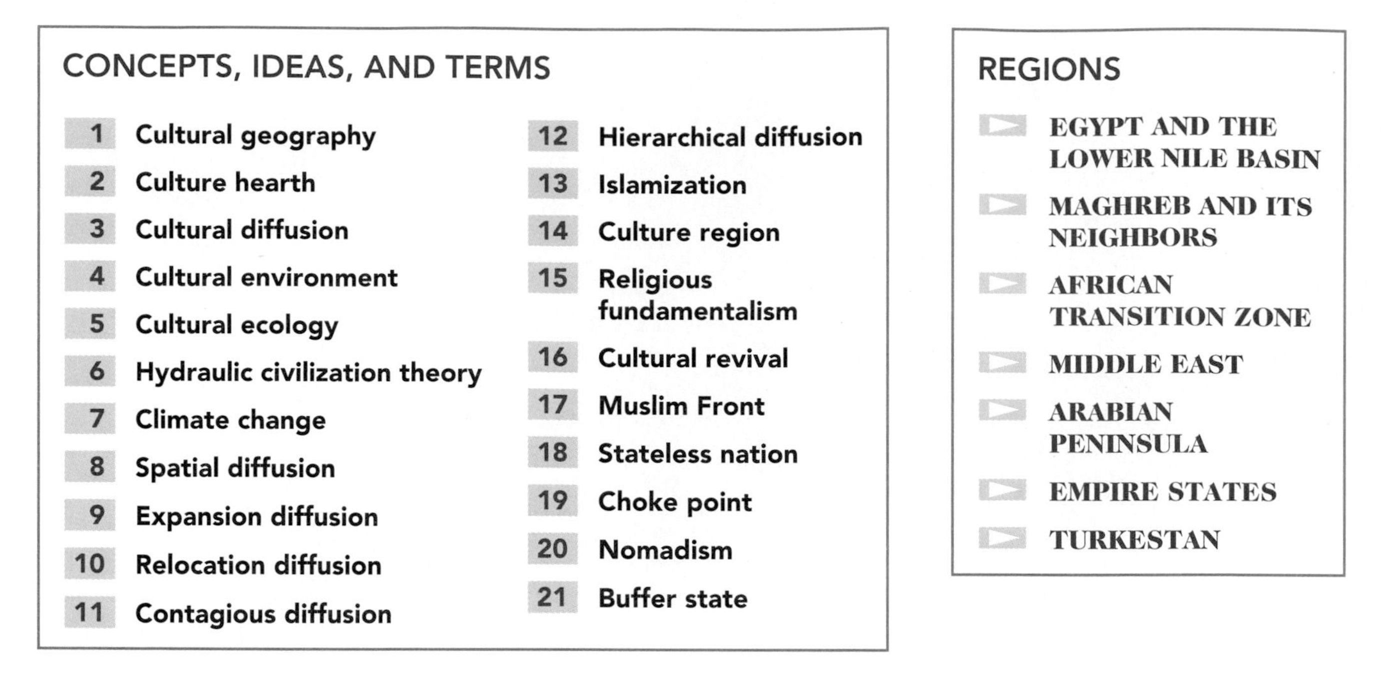

CONCEPTS, IDEAS, AND TERMS

1. Cultural geography
2. Culture hearth
3. Cultural diffusion
4. Cultural environment
5. Cultural ecology
6. Hydraulic civilization theory
7. Climate change
8. Spatial diffusion
9. Expansion diffusion
10. Relocation diffusion
11. Contagious diffusion
12. Hierarchical diffusion
13. Islamization
14. Culture region
15. Religious fundamentalism
16. Cultural revival
17. Muslim Front
18. Stateless nation
19. Choke point
20. Nomadism
21. Buffer state

REGIONS

- EGYPT AND THE LOWER NILE BASIN
- MAGHREB AND ITS NEIGHBORS
- AFRICAN TRANSITION ZONE
- MIDDLE EAST
- ARABIAN PENINSULA
- EMPIRE STATES
- TURKESTAN

From Morocco on the shores of the Atlantic to the mountains of Afghanistan, and from the Horn of Africa to the steppes of inner Asia lies a vast geographic realm of enormous cultural complexity. It stands at the crossroads where Europe, Asia, and Africa meet, and it is part of all three (Fig. 6-1). Throughout history, its influences have radiated to these continents and to practically every other part of the world as well. This is one of humankind's primary source areas. On the Mesopotamian Plain between the Tigris and Euphrates rivers (in modern-day Iraq) and on the banks of the Egyptian Nile arose several of the world's earliest civilizations. In its soils, plants were domesticated that are now grown from the Americas to Australia. Along its paths walked prophets whose religious teachings are still followed by hundreds of millions. And at the opening of the twenty-first century, the heart of this realm is beset by some of the most bitter and dangerous conflicts on Earth.

DEFINING THE REALM

It is tempting to characterize this geographic realm in a few words and to stress one or more of its dominant features. It is, for instance, often called the "Dry World," containing as it does the vast Sahara as well as the Arabian Desert. But most of the realm's people live where there is water—in the Nile Delta, along the Mediterranean coastal strip (or *tell*) of northwesternmost Africa, along the Asian eastern and northeastern shores of the Mediterranean Sea, in the Tigris-Euphrates Basin, in far-flung desert oases, and along the lower mountain slopes of Iran south of the Caspian Sea and of Turkestan to the northeast. We know this world region as one where water is almost always at a premium, where peasants often struggle to make soil and moisture yield a small harvest, where nomadic peoples and their animals circulate across dust-blown flatlands, where oases are islands of sedentary farming and trade in a sea of aridity. But it also is the land of the Nile, the lifeline of Egypt, the crop-covered tell of northwestern Africa, the verdant coasts of Turkey, the meltwater-fed valleys of Central Asia.

Before we investigate this realm further, we need to look again at Figure I-8, the map of the world climates. Note the high degree of coincidence between the extent of the *B* (desert and steppe) climates and the limits of both the North African and Southwest Asian segments. In North Africa, the border between this realm and Subsaharan Africa comes close to matching the southern limits of the *BSh* zone. In Southwest Asia, deserts, steppes, and mountain climates dominate. Except for the Mediterranean coasts, this is a realm of low and highly variable annual precipitation, of searing daytime heat and chilling nighttime cold, of strong winds and dust-laden air. Soils are thin; mountain slopes carry little vegetation. To all this, water brings exception—not only along coasts and rivers but also in oases and *qanats*, tunnels dug into water-bearing rock strata at an

◆ Major Geographic Qualities of North Africa/Southwest Asia

1. North Africa and Southwest Asia were the scene of several of the world's great ancient civilizations, based in its river valleys and basins.
2. From this realm's culture hearths diffused ideas, innovations, and technologies that changed the world.
3. The North Africa/Southwest Asia realm is the source of three world religions: Judaism, Christianity, and Islam.
4. Islam, the last of the major religions to arise in this realm, transformed, unified, and energized a vast domain extending from Europe to Southeast Asia and from Russia to East Africa.
5. Drought and unreliable precipitation dominate natural environments in this realm. Population clusters exist where water supply is adequate to marginal.
6. Certain countries of this realm have enormous reserves of oil and natural gas, creating great wealth for some but doing little to raise the living standards of the majority.
7. The boundaries of the North Africa/Southwest Asia realm consist of volatile transition zones in several places in Africa and Asia.
8. Conflict over water sources and supplies is a constant threat in this realm, where population growth rates are high by world standards.
9. The Middle East, as a region, lies at the heart of this realm; and Israel lies at the center of the Middle East conflict.
10. Religious, ethnic, and cultural discord frequently cause instability and strife in this realm.

angle so that the water drains to the surface. The population map (Fig. I-9) reminds us how scattered and isolated the population clusters are—as a matter of necessity.

An "Arab World"?

North Africa/Southwest Asia is also often referred to as the Arab World. This term implies a uniformity that does not actually exist. First, the name Arab is applied loosely to the peoples of this area who speak Arabic and related languages, but ethnologists normally restrict it to certain occupants of the Arabian Peninsula—the Arab "source." In any case, the Turks are not Arabs, and neither are most Iranians or Israelis. Moreover, although the Arabic language prevails from Mauritania in the west across all of North Africa to the Arabian Peninsula, Syria, and Iraq in the east, it is not spoken in other parts of this realm. In Turkey, for example, Turkish is the major language, and it has Ural-Altaic rather than Arabic's Semitic or Hamitic roots. The Iranian language belongs to the Indo-European linguistic family. Other "Arab World" languages that have separate ethnological identities are spoken by the Jews of Israel, the Tuareg people of the Sahara, the Berbers of northwestern Africa, and the peoples of the transition zone between North Africa and Subsaharan Africa to the south.

An "Islamic World"?

Yet another name given to this realm is the World of Islam. The prophet Muhammad (Mohammed) was born in Arabia in AD 571, and in the centuries after his death in 632, Islam spread into Africa, Asia, and Europe. This was the age of Arab conquest and expansion. Their armies penetrated southern Europe, their caravans crossed the deserts, and their ships plied the coasts of Asia and Africa. Along these routes they carried the Muslim (Islamic) faith, converting the ruling classes of the states of the West African savanna, threatening the Christian stronghold in the highlands of Ethiopia, penetrating the deserts of inner Asia, and pushing into India and even the island extremities of Southeast Asia. Islam was the religion of the marketplace, the bazaar, the caravan. Where necessary, it was imposed by the sword, and its protagonists aimed directly at the political leadership of the communities they entered. Today, the Islamic faith with its more than 1 billion followers extends well beyond the limits of the realm discussed here (Fig. 6-2). It is the major religion in northern Nigeria, in Pakistan, and in Indonesia; it is influential in East Africa; and it still prevails in its old strongholds of Albania, Kosovo, and Bosnia in Eastern Europe. Moreover, in South Asia more than 120 million Muslims live in Hindu-dominated India, the world's largest religious minority.

On the other hand, the World of Islam is not entirely Muslim either. In Israel, Judaism is the prevailing faith; Christianity remains strong in Lebanon; and ancient Coptic Christian churches still exist in Egypt. Thus the connotation of World of Islam when applied to North Africa/Southwest Asia is far from satisfactory: the religion prevails far beyond these areas, and within the realm there is more religious diversity than it implies.

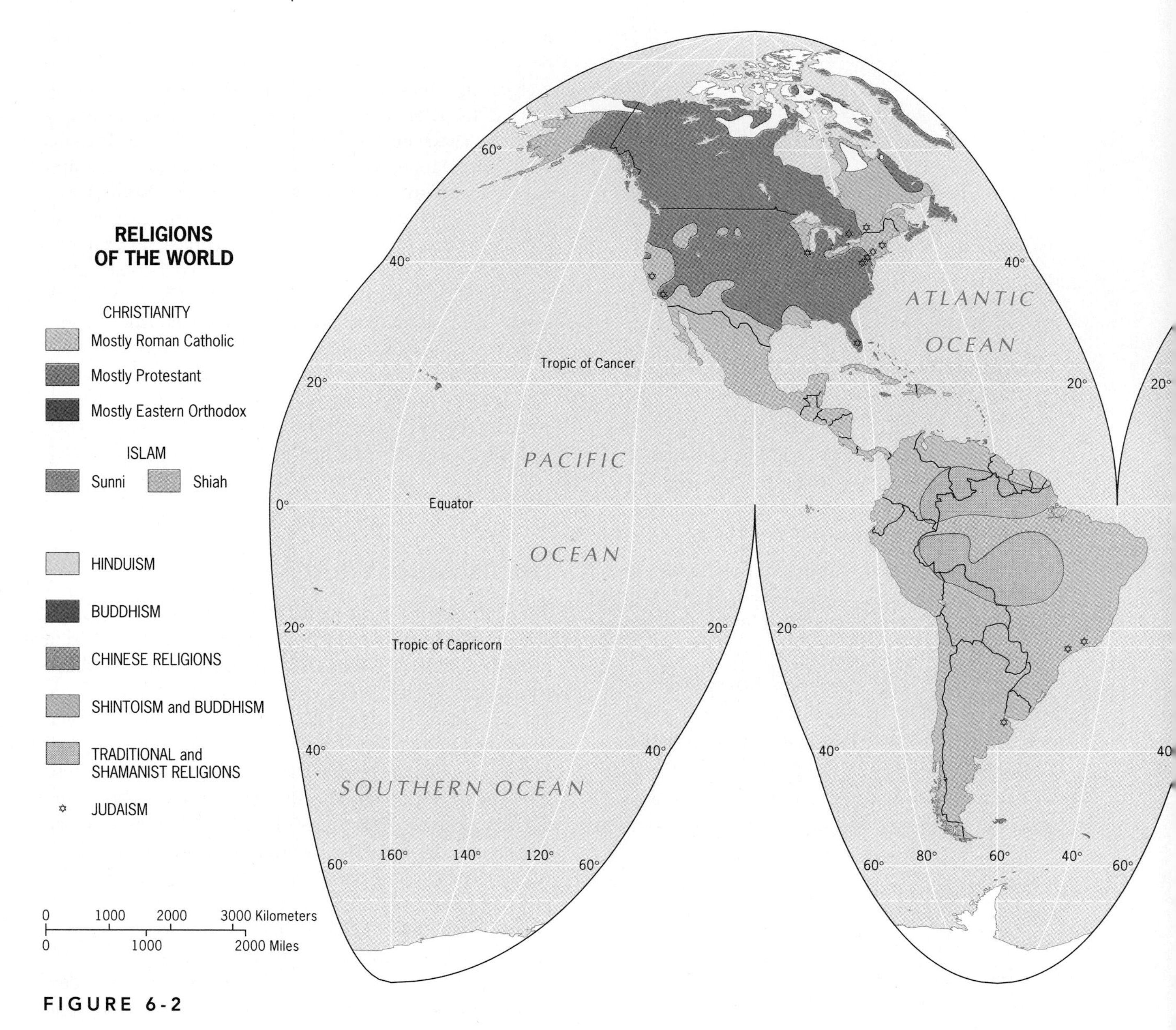

FIGURE 6-2

"Middle East"?

Finally, this realm is frequently called the Middle East. That must sound odd to someone in, say, India, who might think of a Middle West rather than a Middle East! The name, of course, reflects the biases of its source: the "Western" world, which saw a "Near" East in Turkey, a " Middle" East in Egypt, Arabia, and Iraq, and a "Far" East in China and Japan. Still, the term has taken hold, and it can be seen and heard in everyday usage by scholars, journalists, and members of the United Nations. In view of the complexity of this realm, its transitional margins, and its far-flung areal components, the name *Middle East* need be faulted only for being imprecise—it does not make a single-factor region of North Africa/Southwest Asia, as do the terms *Dry World*, *Arab World*, and *World of Islam*. In this chapter we do use this name—but only for one of its regions, not for the realm as a whole.

HEARTHS OF CULTURE

This geographic realm occupies a pivotal part of the world: here Eurasia, crucible of human cultures, meets Africa, source of humanity itself. A million years ago, the ancestors of our species walked from East Africa into North Africa and Arabia and spread from one end of Asia to the other. One hundred thousand years ago, *Homo sapiens* crossed these lands on the way to Europe, Australia, and, eventually, the Americas. Ten thousand years ago, human com-

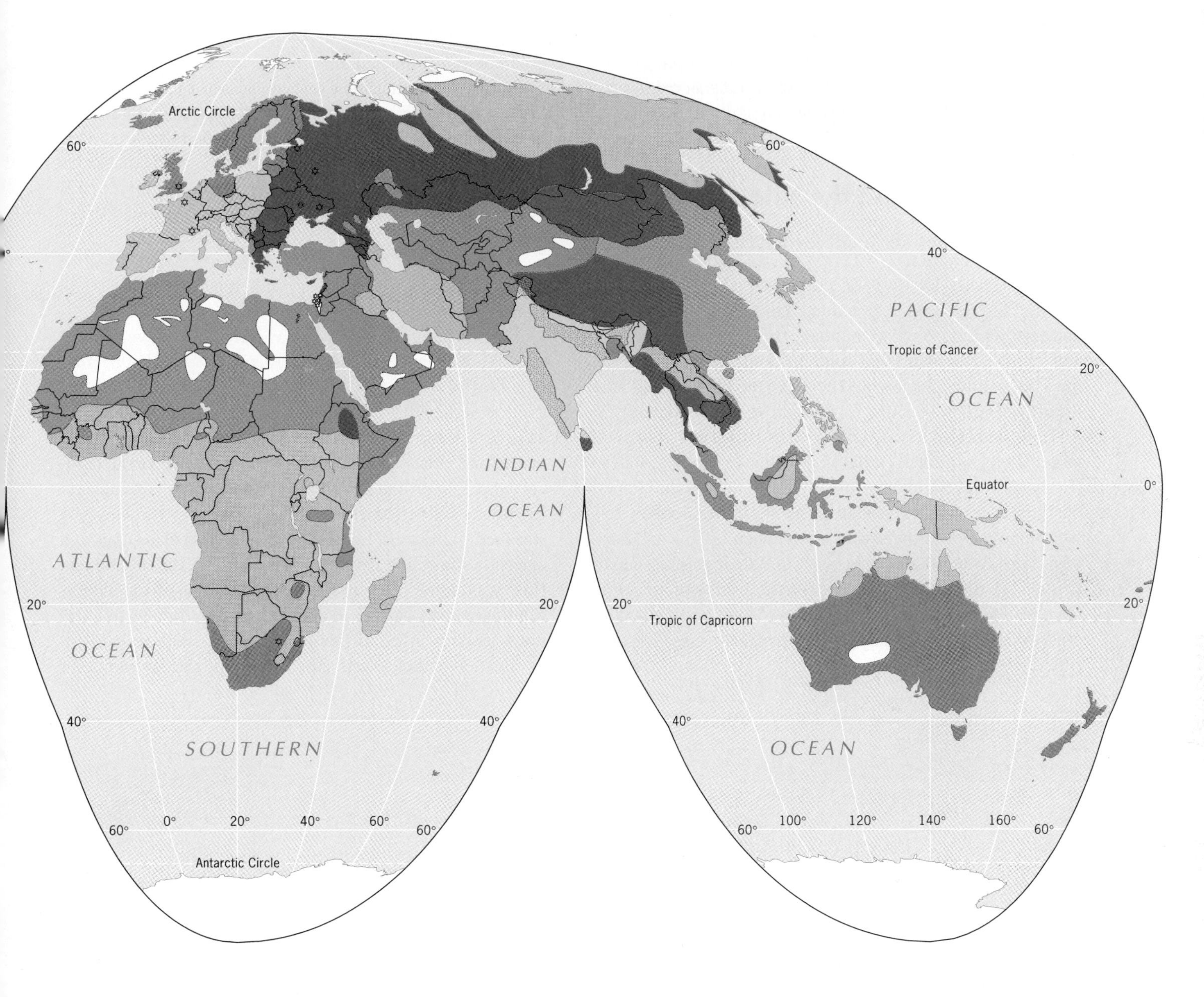

munities in what we now call the Middle East began to domesticate plants and animals, learned to irrigate their fields, enlarged their settlements into towns, and formed the earliest states. One thousand years ago, the heart of the realm was stirred and mobilized by the teachings of Muhammad and the Quran (Koran), and Islam was on the march from North Africa to India. Today this realm is a cauldron of religious and political activity, weakened by conflict but empowered by oil, plagued by poverty but fired by a wave of religious fundamentalism.

In the Introduction (pp. 16-17) we discussed the concept of culture and its regional expression in the cultural landscape. **Cultural geography**, we noted, is a wide-ranging and comprehensive field that studies spatial aspects of human cultures, focusing not only on cultural landscapes but also on **culture hearths**, the crucibles of civilization, sources of ideas, innovations, and ideologies that changed regions and realms. Those ideas and innovations spread far and wide through a set of processes that we study under the rubric of **cultural diffusion**. Because we understand these processes better today, we can reconstruct ancient routes by which the knowledge and achievements of culture hearths spread (that is, diffused) to other areas. Another topic of cultural geography, also relevant in the context of the North African/Southwest Asian realm, is the **cultural environment** that a dominant culture creates. Human cultures exist in long-term accommodation with (and adaptation to) their natural environments, exploiting opportunities that these environments present and coping with the extremes they can impose. The study of the relationship between human societies and natural environments has become a separate branch of cultural geography called **cultural ecology**. As we will see, the North African/Southwest Asian realm presents many opportunities to investigate cultural geography in regional settings.

5

Mesopotamia and the Nile

In the basins of the major rivers of this realm (the Tigris and Euphrates of modern-day Turkey, Syria, and Iraq, and the Nile of Egypt) lay two of the world's earliest culture hearths (Fig. 6-3). Mesopotamia, "land amidst the rivers," had fertile alluvial soils, abundant sunshine, ample water, and animals and plants that could be domesticated. Here, in the Tigris-Euphrates lowland between the Persian Gulf and the uplands of present-day Turkey, arose one of humanity's first culture hearths, a cluster of communities that grew into larger societies and, eventually, into the world's first states. (Early state development probably was going on simultaneously in East Asia's river basins as well.) Mesopotamians were innovative farmers who knew when to sow and harvest crops, water their fields, and store their surplus. Their knowledge diffused to villages near and far, and a *Fertile Crescent* evolved extending from Mesopotamia across southern Turkey into Syria and the Mediterranean coast beyond (Fig. 6-3).

Irrigation was the key to prosperity and power in Mesopotamia, and urbanization was its reward. Among many settlements in the Fertile Crescent, some thrived, grew, enlarged their hinterlands, and diversified socially and occupationally; others failed. What determined success? One theory, the **hydraulic civilization theory**, holds that cities that could control irrigated farming over large hinterlands held power over others, used food as a weapon, and thrived. One such city, Babylon on the Euphrates River, endured for nearly 4000 years (from 4100 BC). A busy port, its walled and fortified center endowed with temples, towers, and palaces, Babylon for a time was the world's largest city.

6

Egypt's cultural evolution may have started even earlier than Mesopotamia's, and its focus lay upstream from (south of) the Nile Delta and downstream from (north of) the first of the Nile's series of rapids, or cataracts (Fig. 6-3). This part of the Nile Valley is surrounded by inhospitable desert, and unlike Mesopotamia (which lay open to all comers), the Nile provided a natural fortress here. The ancient Egyptians converted their security into progress. The Nile was their highway of trade and interaction; it also supported agriculture through irrigation. The Nile's cyclical ebb and flow was much more predictable than that of the Tigris-Euphrates river system. By the time Egypt finally fell victim to outside invaders (about 1700 BC), a full-scale urban civilization had emerged. Ancient Egypt's artist-engineers

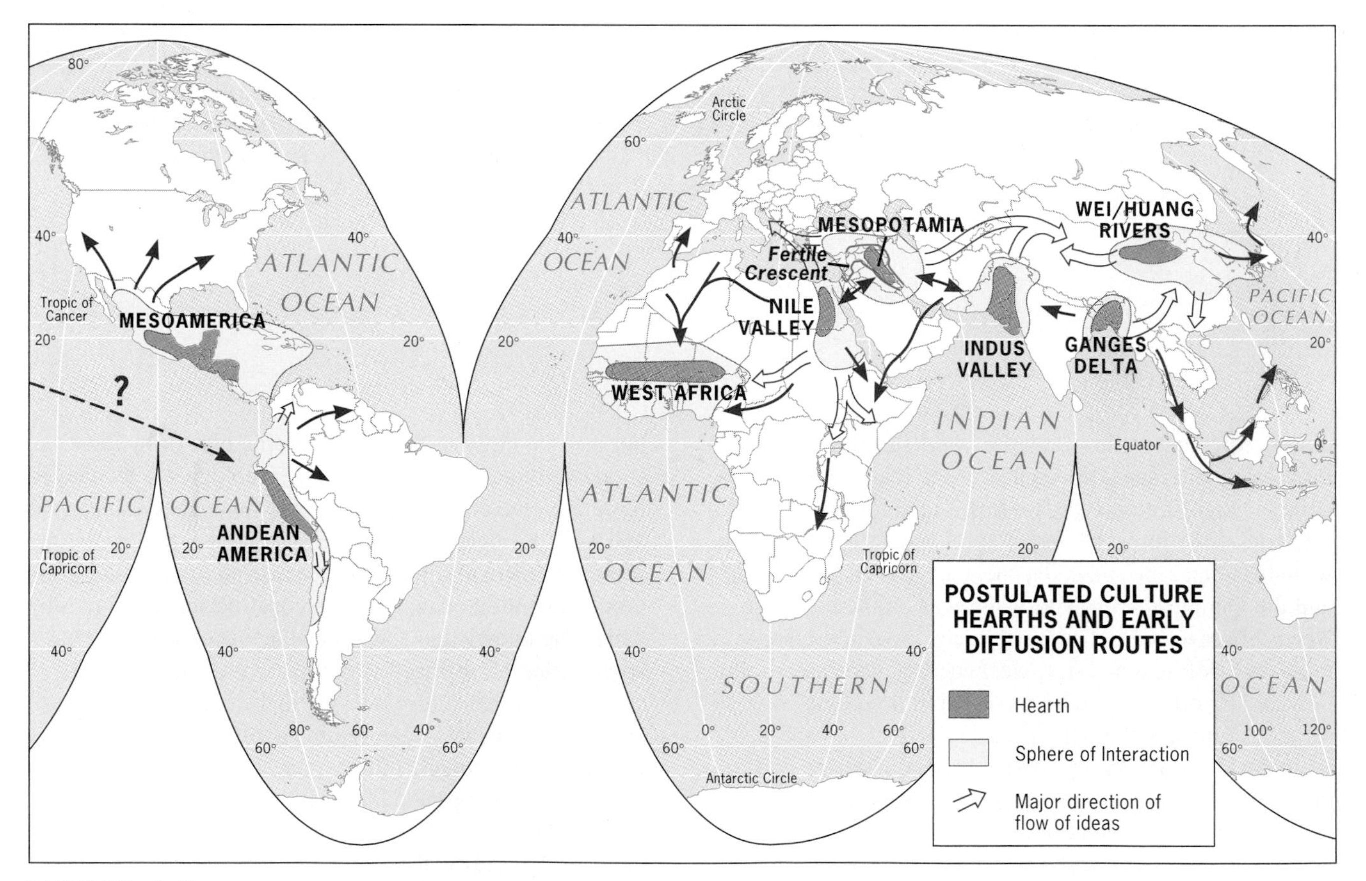

FIGURE 6-3

The ancient Romans transformed the North African countryside by building cities, laying out irrigation systems, and farming fields that had been pastures for the herds of nomads. Their structures, from colonnaded avenues to tiered amphitheaters and from cobbled roadways to covered aqueducts, were built so well that parts of them endure to this day. These are the surviving remnants of a covered aqueduct that carried water for many miles across the undulating landscape of northeastern Tunisia. A cross-section of the "tube" atop the columned structure can be seen at the upper left; the extent of the system can be gauged at the right. Roman engineers sloped the aqueduct in such a way that water could be drawn off to various fields along its course. Not only was the system functional; the structure itself was designed to convey beauty as well as strength and effectiveness. Arched openings between the columns allowed strong Mediterranean winds to pass through, reducing the danger of weakening and collapse. It worked for centuries.

left a magnificent legacy of massive stone monuments, some of them containing treasure-filled crypts of god-kings called Pharaohs. These tombs have enabled archeologists to reconstruct the ancient history of this culture hearth.

To the east, separated from Mesopotamia by more than 1200 miles (1900 km) of mountain and desert, lies the Indus Valley (Fig. 6-3). By modern criteria, this eastern hearth lies outside the realm under discussion here; but in ancient times it had cultural and commercial ties with the Tigris-Euphrates region. Mesopotamian innovations reached the Indus region early, and eventually the cities of the Indus became power centers of a civilization that extended far into present-day northern India.

Today, the world continues to benefit from the accomplishments of the ancient Mesopotamians and Egyptians. They domesticated cereals (wheat, rye, barley), vegetables (peas, beans), fruits (grapes, apples, peaches), and many animals (horses, pigs, sheep). They also advanced the study of the calendar, mathematics, astronomy, government, engineering, metallurgy, and a host of other skills and technologies. In time, many of their innovations were adopted and then modified by other cultures in the Old World and eventually in the New World as well. Europe was the greatest beneficiary of these legacies of Mesopotamia and ancient Egypt, whose achievements constituted the foundations of "Western" civilization.

Decline and Decay

Many of the early cities of this realm's culture hearths are archeological curiosities today. In some places, new cities have been built on the sites of the old, but the great cultural traditions of this realm eventually went into deep decline after many centuries of continuity. Several of these ancient urban centers are located in what is now desert. If we presumed that they were not built in the middle of these drylands, we could postulate a hypothesis that also could form an alternative to the hydraulic civilization theory: that of climate change.

7

Climate change, not a monopoly over irrigation techniques, may have given certain cities in the ancient Fertile Crescent an advantage over others; and climate change, associated with shifting environmental zones after the last Pleistocene glacial retreat, may have destroyed the last of the old civilizations. Perhaps overpopulation and human destruction of the natural vegetation contributed to the process. Indeed, some cultural geographers suggest that the momentous innovations in agricultural planning and irrigation technology were not "taught" by the seasonal flooding of the rivers but were forced on the inhabitants as they tried to survive changing environmental conditions.

The scenario is not difficult to imagine. As outlying areas began to fall dry and farmlands were destroyed, people congregated in the already crowded river valleys—and made every effort to increase the productivity of the land that could still be watered. Eventually overpopulation, destruction of the watershed, and perhaps reduced rainfall in the rivers' headwater areas dealt the final blow. Towns were abandoned to the encroaching desert; irrigation canals filled with drifting sand; croplands dried up. Those who could migrated to areas that were still reputed to be productive. Others stayed, their numbers dwindling, increasingly reduced to subsistence.

As old societies disintegrated, power emerged elsewhere. First the Persians, then the Greeks, and later the Romans imposed their imperial designs on the tenuous lands and disconnected peoples of North Africa/Southwest Asia. Roman technicians converted North Africa's farmlands into irrigated plantations whose products went by the boatload to Roman Mediterranean shores (see photo, p. 285). Thousands of people were carried as slaves to the cities of the new conquerors. Egypt was quickly colonized, as was the area we now call the Middle East. One region that lay distant, and therefore remote from these invasions, was the Arabian Peninsula, where no major culture hearth or large cities had emerged and where the turmoil had not affected Arab settlements and nomadic routes.

STAGE FOR ISLAM

Islam today dominates the cultural geography of this realm, but long before Islam arose, other religions had emerged: Zoroastrianism in what is today Iran, and Judaism (and later Christianity) in the area of modern Israel. However, the teachings of Zoroaster remained confined to Persia; Judaism was devastated by the Babylonians and later by the Romans; and Christianity became an alien faith as its center moved to Rome. Until the early seventh century AD, none of the religions that had their origins in this realm became dominant within it.

Muhammad the Prophet

In a remote place on the Arabian Peninsula, where the foreign invasions of the Middle East had had little effect on the Arab communities, an event occurred early in the seventh century that was to change history and affect the destinies of people in many parts of the world. In a town called Mecca (Makkah), about 45 miles (70 km) from the Red Sea coast in the Jabal Mountains, a man named Muhammad in the year AD 611 began to receive revelations from Allah (God). Muhammad (571–632) was then in his early forties and had barely 20 years to live. Convinced after some initial self-doubt that he was indeed chosen to be a prophet, Muhammad committed his life to fulfilling the divine commands he believed he had received. Arab society was in social and cultural disarray, but Muhammad forcefully taught Allah's lessons and began to transform his culture. His per-

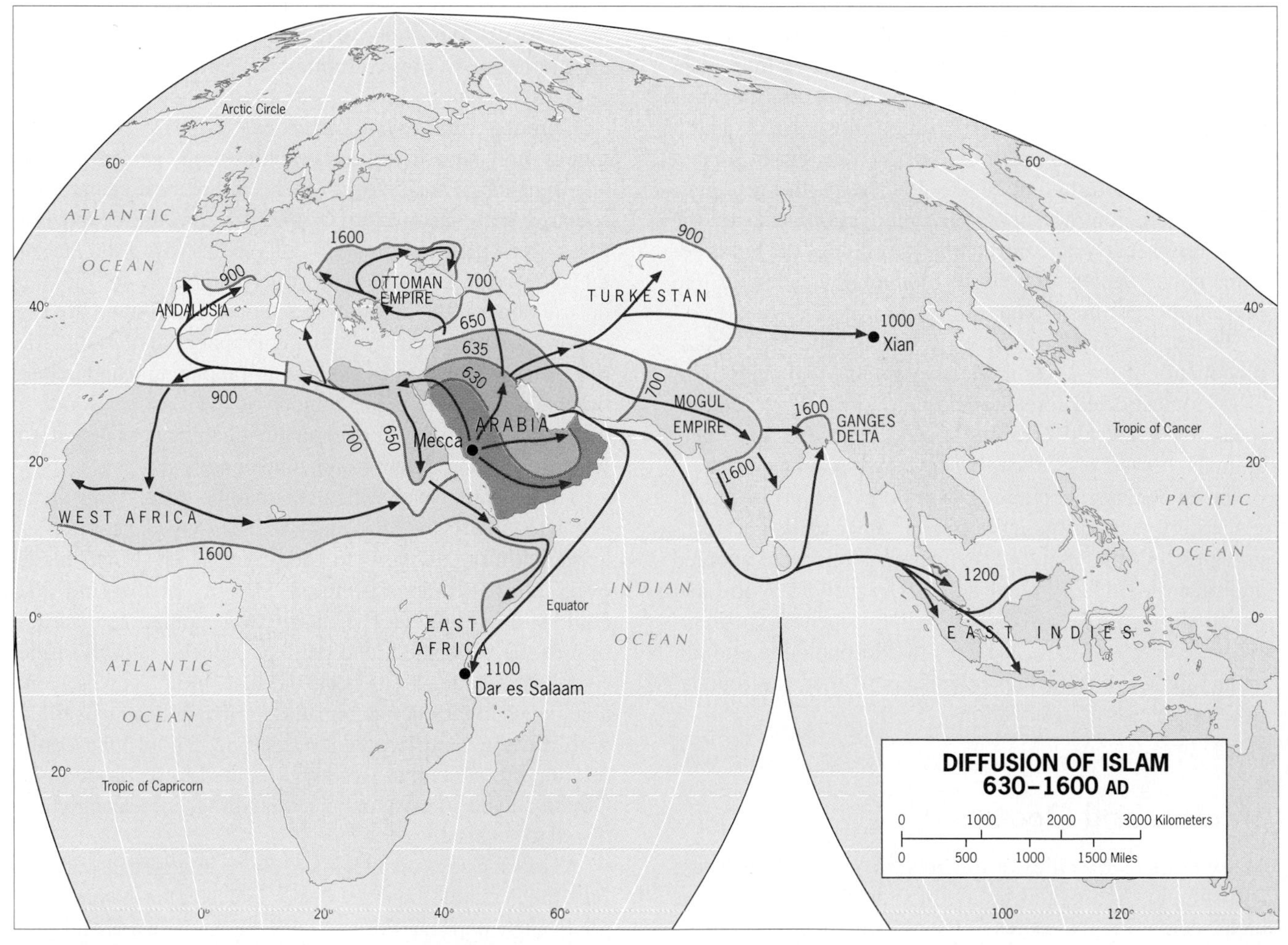

FIGURE 6-4

Diffusion Processes

One of the most interesting fields in geography studies how ideas, inventions, practices, and other phenomena spread through a population across space and time. A new musical style, a word or phrase, a new way of doing things, new fashions in clothing disseminate by a process called **spatial diffusion**. So do technological innovations, diseases, religious practices, and political ideas.

8

Understanding diffusion processes allows us to reconstruct the dispersal of cultural and technological ideas in the past. It also helps us predict such things as the impact of epidemics or the probability that an invention will find a market to justify the cost of producing it. Spatial diffusion has many practical applications.

In 1952, the Swedish geographer Torsten Hägerstrand published a fundamental study of spatial diffusion entitled *The Propagation of Diffusion Waves.* Diffusion takes place in two types of processes: **expansion diffusion**, when propagation waves originate in a strong and durable source area and spread outward, affecting an ever larger area and population; and **relocation diffusion**, in which an idea, or innovation (or, as with AIDS, a virus) is carried—usually by migrants—to a distant location and diffuses from there.

9

10

Islam, as Figure 6-4 shows, initially spread by rapid expansion diffusion from its western Arabian source. It affected virtually every village, town, and nomadic community through a form of expansion diffusion shaped by local proximity; geographers call this process **contagious diffusion**. But Islam was also disseminated through the conversion of kings, chiefs, and other high officials, who in turn propagated it through their bureaucracies. This is another form of expansion diffusion, called **hierarchical diffusion**, which has many variations; its main spatial expression is the downward filtering of whatever is being diffused from larger to smaller places within a national- or continental-scale urban hierarchy.

11

12

The map also shows that Islam spread far and wide by relocation diffusion, notably to the Ganges Delta and to the East Indies, where seafarers established remote new source areas. In East Africa Islam arrived through both expansion and relocation diffusion. Today the diffusion of Islam continues in many areas, including North America, where it arrived by relocation diffusion and now spreads by expansion diffusion.

sonal power soon attracted enemies, and in 622 he fled from Mecca to the safer haven of Medina (Al Madinah), where he continued his work. This moment, the *hejira* ("migration"), marks the starting date of the Muslim era, Year 1 on Islam's calendar. Mecca, of course, later became Islam's holiest place.

The precepts of Islam in many ways constituted a revision and embellishment of Judaic and Christian beliefs and traditions. All of these faiths have but one god, who occasionally communicates with humankind through prophets; Islam acknowledges that Moses and Jesus were such prophets but considers Muhammad to be the final and greatest prophet. What is Earthly and worldly is profane; only Allah is pure. Allah's will is absolute; Allah is omnipotent and omniscient. All humans live in a world that Allah created for their use but only to await a final judgment day.

Islam brought to the Arab World not only the unifying religious faith it had lacked but also a new set of values, a new way of life, a new individual and collective dignity. Islam dictated observance of the Five Pillars: (1) repeated expressions of the basic creed, (2) the daily prayer, (3) a month each year of daytime fasting (Ramadan), (4) the giving of alms, and (5) at least one pilgrimage in each Muslim's lifetime to Mecca. And Islam prescribed and proscribed in other spheres of life as well. It forbade alcohol, smoking, and gambling. It tolerated polygamy, although it acknowledged the virtues of monogamy. Mosques appeared in Arab settlements, not only for the (Friday) sabbath prayer, but also as social gathering places to knit communities closer together. Mecca became the spiritual center for a divided, widely dispersed people for whom a collective focus was something new.

The Arab-Islamic Empire

Muhammad provided such a powerful stimulus that Arab society was mobilized almost overnight. The prophet died in 632, but his faith and fame spread like wildfire. Arab armies carrying the banner of Islam formed, invaded, conquered, and converted wherever they went. As Figure 6-4 shows, by AD 700 Islam had reached far into North Africa, into Transcaucasia, and into most of Southwest Asia. In the centuries that followed, it penetrated Southern and Eastern Europe, Central Asia's Turkestan, West Africa, East Africa, and South and Southeast Asia, even reaching China by AD 1000.

At the heart of this religious realm lay an Arab Empire. Its original capital was at Medina in Arabia, but as it expanded, its headquarters were moved first to Damascus in Syria and later to Baghdad on the Tigris River. Meanwhile,

The Flowering of Islamic Culture

The conversion of a vast realm to Islam was not only a religious conquest: it was also accompanied by a glorious explosion of Arab culture. In science, the arts, architecture, and other fields, Arab society far outshone European society. While the Western European remnants of the Roman Empire languished, Arab energies soared.

When the wave of Islamic diffusion reached the Maghreb (western North Africa), the Arabs saw, on the other side of the narrow Strait of Gibraltar, an Iberia ripe for conquest and ready for renewal. An Arab-Berber alliance, called the Moors, invaded Spain in 711 and controlled all but northern Castile and Catalonia before the end of the eighth century.

It took seven centuries for Catholic armies to recapture all of the Arabs' Iberian holdings, but by then the Muslims had made an indelible imprint on the Spanish and Portuguese cultural landscape. The Arabs brought unity and imposed the rule of Baghdad, and their works soon overshadowed what the Romans had wrought. Al-Andalus, as this westernmost outpost of Islam was called, was endowed with thousands of magnificent mosques, castles, schools, gardens, and public buildings. The ultimately victorious Christians destroyed most of the less durable art (pottery, textiles, furniture, sculpture) and burned the contents of Islamic libraries, but the great Islamic structures survived, including the Alhambra in Granada, the Giralda in Seville, and the Great Mosque of Córdoba, three of the world's greatest architectural achievements. While Spanish and Portuguese culture became Hispanic-Islamic culture, the Muslims were transforming their cities from Turkestan to the Maghreb in the image of Baghdad. The lost greatness of a past era still graces those townscapes today.

FROM THE FIELD NOTES

"Seville was a Muslim capital of Iberia, and walking the streets of the old city center is an adventure in cultural and historical geography. Seville fell to the Muslim invaders in 711, and the Muslims built a large mosque in the heart of town. Later the Christians, who ousted the Muslims in 1248, destroyed most of the mosque and built a huge cathedral on the site, but they saved the ornate minaret, made it the bell tower, and called it the *Giralda* (left). This minaret had been built by the Muslims to match the 220-foot (67-m) minaret of the Koutoubia Mosque in Marrakech (see page 305). But the finest Islamic legacy in Seville surely is the Alcazar Palace (above), begun in the late twelfth century and embellished and finished by the Christians after their victory. Its arched façades and intricately carved walls remain a monument to the Muslim architects and artists who designed and created them."

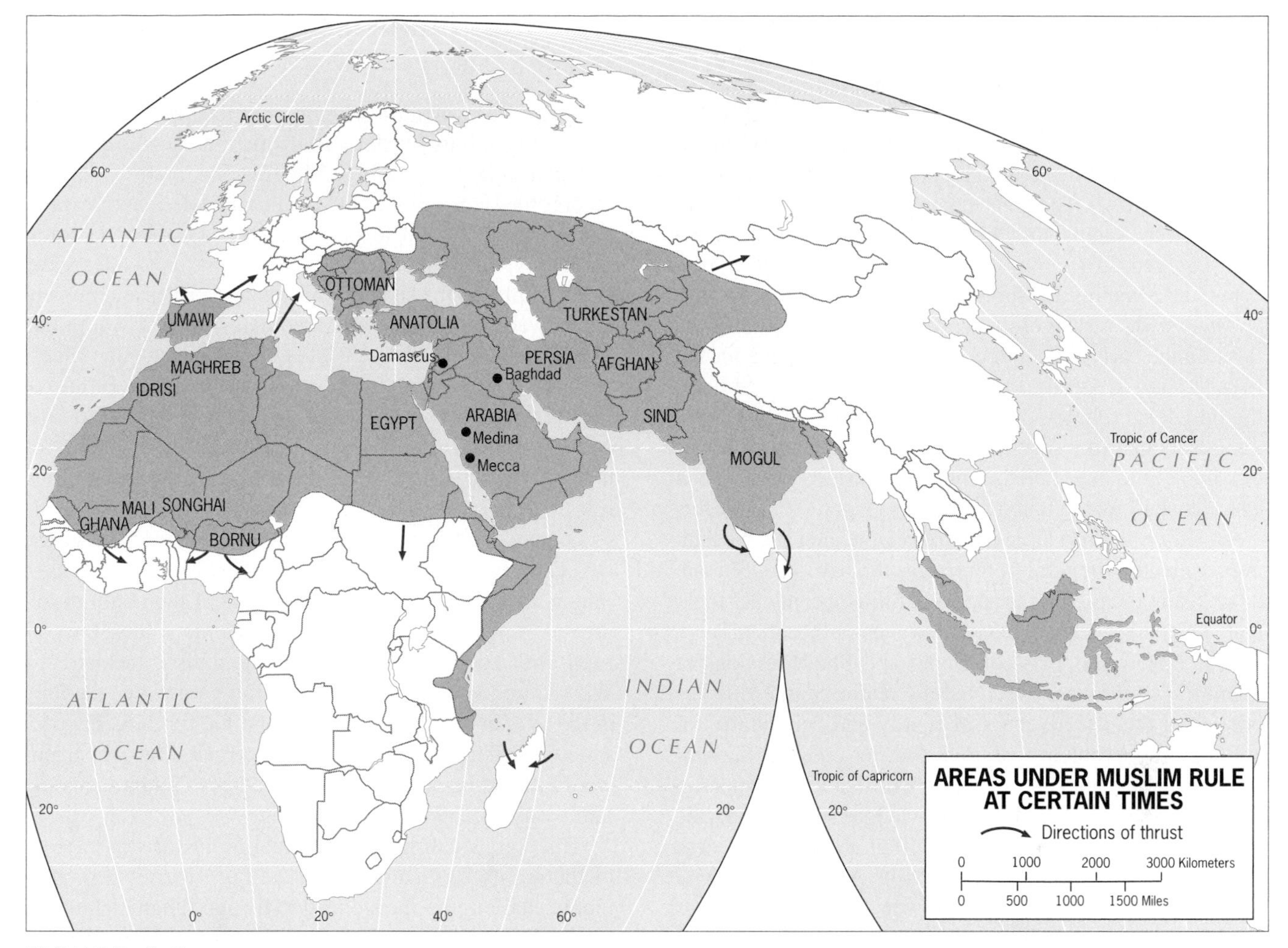

FIGURE 6-5

Islam was carried ever farther outward by camel caravan and pilgrim, by sailor, scholar, and sultan. Here was a manifestation of the process of *spatial diffusion* (see box titled "Diffusion Processes"). And while Islam expanded, the Arab Empire that lay at its source matured and prospered. In architecture, mathematics, and science, the Arabs overshadowed their European contemporaries. The Arabs established institutions of higher learning in many cities including Baghdad, Cairo, and Toledo (Spain), and their distinctive cultural landscapes united their vast domain (see box titled "The Flowering of Islamic Culture"). Non-Arab societies in the path of the Muslim drive were not only Islamized, but also Arabized, adopting other Arab traditions as well. Islam had spawned a culture; it still lies at the heart of that culture today.

As we noted, Islam's expansion eventually was checked in Europe, Russia, and elsewhere. But a map showing the total area under Muslim sway in Eurasia and Africa reveals the enormous dimensions of the domain affected by 13 **Islamization** at one time or another (Fig. 6-5). Islam continues to expand, now mainly by relocation diffusion. There are Islamic communities in cities as widely scattered as Vienna, Singapore, and Cape Town, South Africa; Islam is also growing rapidly in the United States. With more than 1.1 billion adherents today, Islam is a vigorous and burgeoning cultural force around the world.

ISLAM DIVIDED

For all its vigor and success, Islam still fragmented into sects. The earliest and most consequential division arose after Muhammad's death. Who should be his legitimate successor? Some believed that only a blood relative should follow the prophet as leader of Islam. Others, a majority, felt that any devout follower of Muhammad was qualified. The first chosen successor was the father of Muhammad's wife (and thus not a blood relative). But this did not satisfy those who wanted to see a man named Ali, a cousin of Muhammad, made *caliph* (successor). When Ali's turn came, his followers, the Shi'ites, proclaimed that Muhammad finally had a legitimate successor. This offended the Sunnis, those who did not see a blood relationship as necessary for the succession. From the beginning of this disagreement, the numbers of Muslims who took the Sunni side far exceeded those who regarded themselves as Shiah (followers) of Ali. The great expansion of Islam was largely propelled by Sunnis; the Shi'ites survived as small minorities scattered throughout the realm. Today, about 85 percent of all Muslims are Sunnis.

But the Shi'ites vigorously promoted their version of the faith. In the early sixteenth century their work paid off: the royal house of Persia (modern-day Iran) made Shi'ism the only legal religion throughout its vast empire. That domain extended from Persia into lower Mesopotamia (modern Iraq), into Azerbaijan, and into western Afghanistan and Pakistan. As the map of religions (Fig. 6-2) shows, this created for Shi'ism a large **culture region** and gave the faith unprecedented strength. Iran remains the bastion of Shi'ism in the realm today, and the appeal of Shi'ism continues to radiate into neighboring countries and even farther afield.

14

The differences between Sunni and Shiah versions of Islam have profoundly affected the realm. Sunni Muslims believe in the effectiveness of family and community for solving life's problems, whereas Shi'ites consider their infallible *imams* (mosque officials who lead worshipers in prayer) the sole source of true knowledge. Sunnis tend to be comparatively reserved; Shi'ites often are passionate and emotional. For example, the death of Ali's son, Husayn, who would have been caliph and was killed by Sunnis, is commemorated annually with intense processions during which the marchers beat themselves with chains and cut themselves with sharp metal instruments. To many Sunnis, this behavior is unseemly and excessive.

During the late twentieth century, Shi'ism gained unprecedented influence in the realm. In its heartland, Iran, a *shah* (king) tried to secularize the country and to limit the power of the imams; he provoked a revolution that cost him the throne and made Iran an Islamic republic—in fact, a Shi'ite republic. Before long, Iran was at war with neighboring, Sunni-dominated Iraq, and Shi'ite parties and communities elsewhere were invigorated by the newfound power of Shi'ism. From Arabia to Africa's northwestern corner, Sunni-ruled countries warily watched their Shi'ite minorities, newly imbued with religious fervor. Mecca, the holy place for both Sunnis and Shi'ites, became a battleground during the week of the annual pilgrimage, and for a time the (Sunni) Saudi Arabian government denied entry to Shi'ite pilgrims. Recently that schism has healed somewhat, but intra-Islamic sectarian differences run deep.

Smaller Islamic sects further diversify this realm's religious landscape. Some of these sects play a disproportionately large role in the societies and countries of which they are a part, as we will see in our regional discussion.

Religious Fundamentalism in the Realm

Another cause of intra-Islamic conflict lies in the resurgence of **religious fundamentalism**, or as Muslims refer to it, *religious revivalism*. In Shi'ite Iran, the imams wanted to reverse the shah's moves toward liberalization and secularization: they wanted to (and did) recast society in traditional, revivalist Islamic molds. An *ayatollah* (leader under Allah) replaced the shah in 1979; Islamic rules and punishments were instituted. Urban women, many of whom had been considerably liberated and educated during the shah's regime, resumed more traditional Islamic roles. Vestiges of Westernization, encouraged by the shah, disappeared. The war against Iraq (1980–1990) began as a conflict over territory but became a holy war that cost more than a million lives.

15

Islamic fundamentalism did not rise in Iran alone, nor was it confined to Shi'ite communities. Many Muslims—Sunnis as well as Shi'ites—in all parts of the realm disapproved of the erosion of traditional Islamic values, the corruption of society by European colonialists and later by Western modernizers, and the declining power of the faith in the secular state. As long as economic times were good, such dissatisfaction remained submerged. But when jobs were lost and incomes declined, a return to fundamental Islamic ways became more appealing.

This set Muslim against Muslim in all the regions of the realm. Revivalists fired the faith with a new militancy, challenging the status quo from Afghanistan to Algeria. The militants forced their governments to ban "blasphemous" books, to resegregate the sexes in schools, to enforce traditional dress codes, to legitimize religious-political parties, and to heed the wishes of the *mullahs* (teachers of Islamic ways). Militant Muslims proclaimed that democracy inherited from colonialists and adopted by Arab nationalists was incompatible with the rules of the Quran.

In Algeria, the growing power of the fundamentalists led to a crisis. In 1991, during democratic elections, the so-called Islamic Salvation Front (ISF) showed great strength. It was clear that the ISF would defeat the government in the second round of voting scheduled some weeks later; leaders of the ISF stated that they would transform Algeria into an Islamic republic. The army then intervened, compelled the government to resign, and the military forces took control. Algeria was plunged into a long crisis of confrontation and violence, ruining its social fabric, devastating its economy, killing tens of thousands, and spilling over into France, Algeria's former colonial ruler and home to a large Algerian minority.

Already, Sudan, Afghanistan, and of course Iran are officially Islamic republics. To many orthodox Sunnis who prefer to keep mosque and state separate, what has happened in Sudan is frightening. In 1989, a democratically elected government there was overthrown in a military coup. The army allowed the leaders of the National Islamic Front to institute Islamic laws; "nonbelievers" were purged, and the *Sharia* criminal code was introduced. (The Sharia laws prescribe corporal punishment, amputations, stoning, and lashing for both major and minor offenses.) Algeria's fundamentalists often referred to Sudan as their model as the Algerian elections approached because Sudan proved that an Islamic state could indeed be established in a Sunni-dominated country. The specter of a Sudan-style regime in Algiers helped precipitate the crisis that began in 1991.

The rift between more liberal Muslims and fundamentalists poses a major challenge for the future. Militant Muslims confront the governments of Egypt, Jordan, Tunisia, and even Turkey. These governments have reacted in various ways (in Egypt, a combination of appeasement and containment; in Tunisia and Algeria, repression; in Jordan, co-option). Other countries, such as Morocco and Saudi Arabia, clearly are vulnerable. Hanging in the balance is the future of the realm.

Fundamentalism and religious militancy are not exclusively Islamic phenomena. They also infect Christian, Judaic, Hindu, and Buddhist societies. Nowhere, however, does the fundamentalist drive exhibit the intensity and vigor it displays in the Islamic realm.

Islam and Other Religions

Two other major faiths had their sources in the *Levant* (the area extending from Greece eastward around the Mediterranean coast to northern Egypt), and both were older than Islam. Islam's rise submerged many smaller Jewish communities, but the Christians, not the Jews, waged centuries of holy war against the Muslims, seeking, through the Crusades, not only to drive Islam back but to reestablish Christian communities where they had dominated before the Islamic expansion. The aftermath of that campaign still marks the realm's cultural landscape today. A substantial Christian minority (about one-fourth of the population) remains in Lebanon, and Christian minorities also survive in Israel, Syria, Egypt, and Jordan. Strained relations between

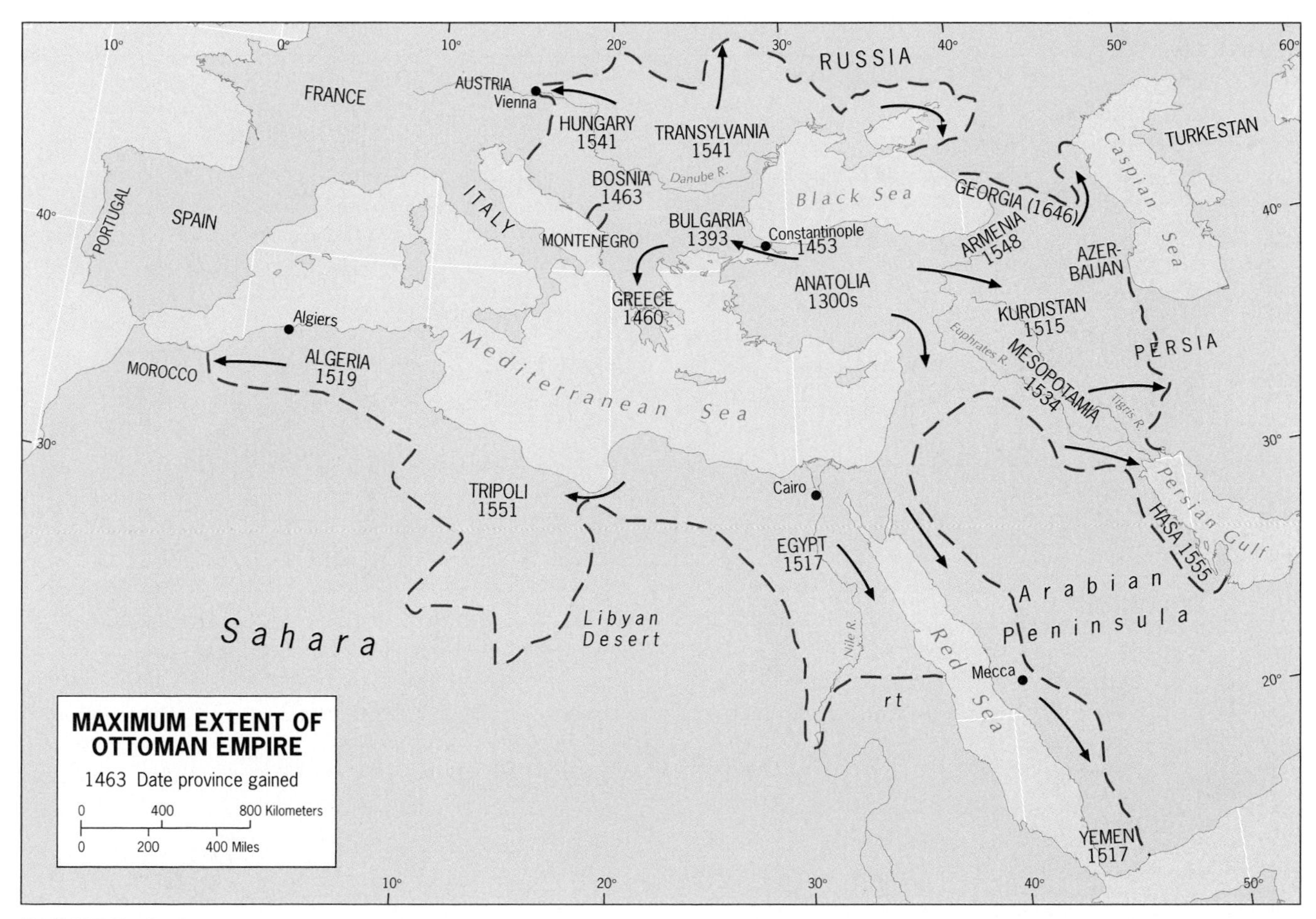

FIGURE 6-6

the long-dominant Christian minority in Lebanon and the Muslim majority (itself divided into five sects) contributed to the disastrous armed struggle that engulfed this country in the 1970s and 1980s.

But the most intense conflict in modern times has pitted the Jewish state, Israel, against its Islamic neighbors near and far. Israel's United Nations–sponsored creation in 1948 precipitated more than a half-century of intermittent strife and attempts at mediation; it also caused friction among Islamic states in the region. Jerusalem—holy city for Judaism, Christianity, and Islam—lies in the crucible of this confrontation.

The Ottoman Aftermath

It is a geographic twist of fate that Islam's last great advance into Europe resulted in the European occupation of Islam's very heartland. The Ottomans (named after their leader, Osman I), based in what is today Turkey, conquered Constantinople (now Istanbul) in 1453 and pushed into Eastern Europe. Soon Ottoman forces were on the doorstep of Vienna; they also invaded Persia, Mesopotamia, and North Africa (Fig. 6-6). The Ottoman Empire under Suleyman the Magnificent, who ruled from 1522 to 1560, was the most powerful state in western Eurasia.

The Ottoman Empire survived for more than four centuries, but it lost territory as time went on, first to the Hungarians, then to the Russians, and later to the Greeks and Serbs until, after World War I, the European powers took over its provinces and made them colonies—colonies we now know by the names of Syria, Iraq, Lebanon, and Yemen (Fig. 6-7). As the map shows, the French and the British took large possessions; even the Italians annexed part of the Ottoman domain.

The boundary framework that the colonial powers created to delimit their holdings was not satisfactory. As Figure I-9 reminds us, this realm's population of more than 500 million is clustered, fragmented, and strung out in river valleys, coastal zones, and crowded oases. The colonial powers laid out long stretches of boundary as ruler-straight lines across uninhabited territory; they saw no need to adjust these boundaries to cultural or physical features in the landscape. Other boundaries, even some in desert zones, were poorly defined and never marked on the ground. Later, when the colonies had become independent states, such boundaries led to quarrels, even armed conflicts, among neighboring Muslim states.

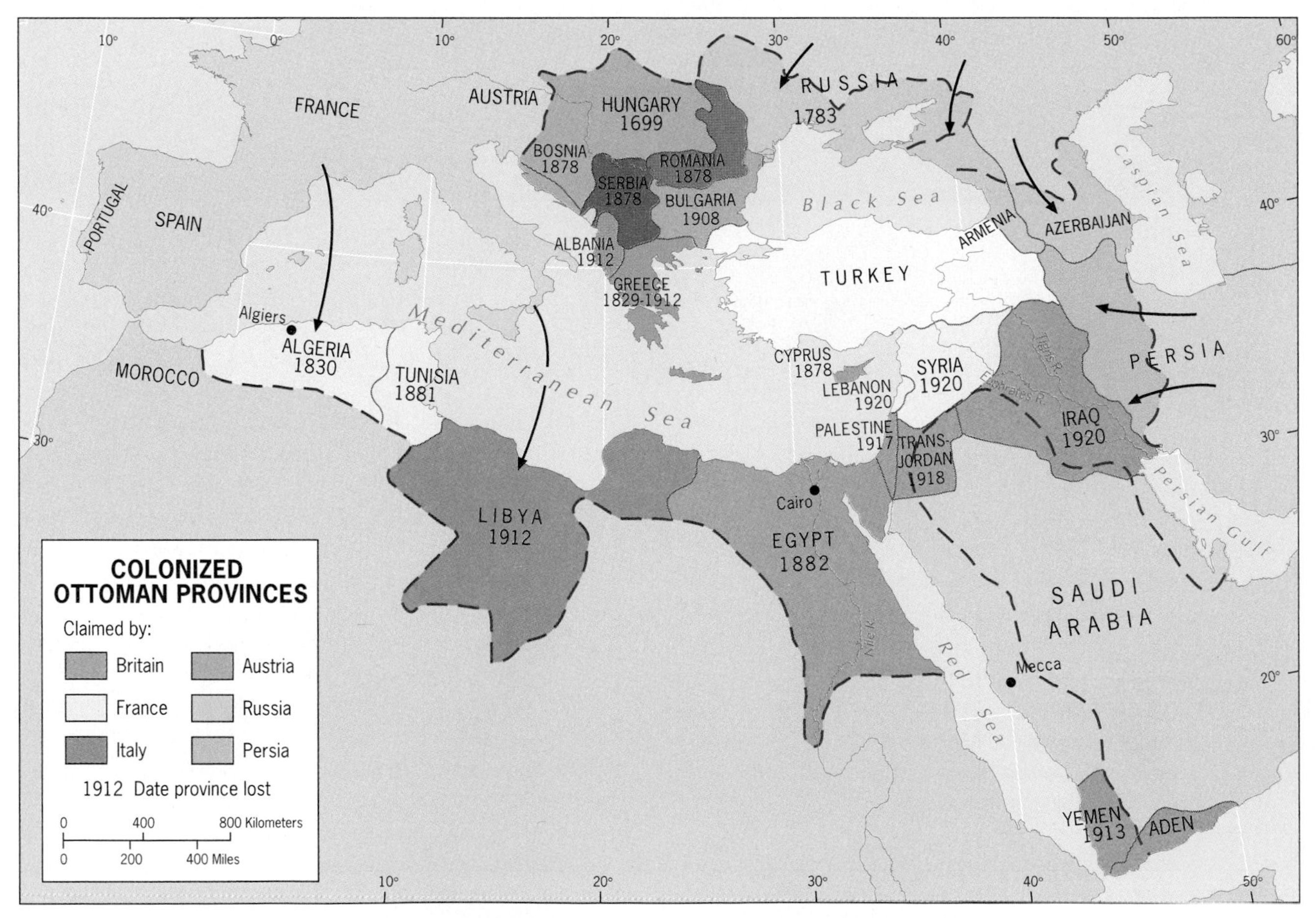

FIGURE 6-7

THE POWER AND PERIL OF OIL

Travel through the cities and towns of North Africa and Southwest Asia, talk to students, shopkeepers, taxi drivers, and migrants, and you will hear the same refrain: "Leave us in peace, let us do things our traditional way. Our problems—with each other, with the world—seem always to result from outside interference. The stronger countries of the world exploit our weaknesses and magnify our quarrels. We want to be left alone."

That wish might be closer to fulfillment were it not for two relatively recent events: the creation of the state of Israel and the discovery of some of the world's largest oil reserves. We will discuss the founding of Israel in the regional section of this chapter. Here we focus on the realm's most valuable export product: oil.

It is sometimes said that oil is the realm's most valuable resource. That is not accurate, however. Most of the people here remain farmers, so that the realm's most important resources are actually water and tillable soil. Oil is indeed in demand throughout the world. But there was a Muslim world long before the first barrel of oil was taken from the ground, and there are tens of millions of people whose lives, to this day, are only indirectly affected by oil revenues. For example, Syria and Jordan must import oil, Morocco has produced almost none, and Turkey also depends heavily on oil imports.

Location of Known Reserves

In general terms, oil (and associated natural gas) exists in this realm in three discontinuous zones (Fig. 6-8). The most productive of these zones extends from the southern and southeastern part of the Arabian Peninsula northwestward around the rim of the Persian Gulf, reaching into Iran and continuing northward into Iraq, Syria, and southeastern Turkey, where it peters out. The second zone lies across North Africa and extends from north-central Algeria eastward across northern Libya to Egypt's Sinai Peninsula, where it ends. The third zone begins on the margins of the realm in eastern Azerbaijan, continues eastward under the Caspian Sea into Turkmenistan and Kazakhstan, and also reaches into Uzbekistan, Tajikistan, Kyrgyzstan, and Afghanistan.

The search for oil goes on, in this realm and in many other areas of the world. As a result, estimates of the realm's reserves are subject to change as new discoveries are announced. Current assessments suggest that more than 65 percent of the world's known oil reserves lie in the North Africa/Southwest Asia realm.

In terms of production, Saudi Arabia long has been among the world leaders, and it is the undisputed top exporter. Two countries that rival Saudi Arabia as producers, the United States and Russia, consume most of their own output. The United States, in fact, is the world's leading importer. Thus the oil production in Southwest Asia and North Africa is crucial to the rest of the world. As Figure I-11 indicates, oil wealth has elevated several of this realm's countries into the upper-middle and high-income categories. Petroleum wealth also has enmeshed these Islamic societies in world affairs, so that a threat to stability in an oil-rich country risks foreign intervention.

When the colonial powers laid down the boundaries that partitioned this realm among themselves, no one knew about the riches that lay beneath the ground. A few wells had been drilled, and production in Iran had begun as early as 1908 and in Egypt's Sinai Peninsula in 1913. But the major discoveries came later, in some cases after the colonial powers had already withdrawn. Some of the newly independent countries, such as Libya, Iraq, and Kuwait, found themselves with wealth undreamed of when the Turkish Ottoman Empire collapsed. As Figure 6-8 shows, however, others were less fortunate. A few countries had (and still have) potential. The smaller, weaker emirates and sheikdoms on the Arabian Peninsula always feared that powerful neighbors would try to annex them (Kuwait faced this prospect in 1990 when Iraq invaded it). The unevenly distributed oil wealth, therefore, created another source of division and distrust among Islamic neighbors.

A Foreign Invasion

The oil-rich countries of the realm had a coveted energy source, but they lacked the equipment or skills to exploit it. These had to come from the outside world and entailed what many tradition-bound Muslims feared most: penetration by the vulgarities of Western ways. In his book *The Middle East*, geographer William Fisher reports that Saudi Arabia had two major cultural forces: Islam and Aramco, the joint Arab-American oil company. Yet while some Western ideas and practices were diffused, Islamic societies proved resistant to foreign cultural influences. Most schools, for example, remain segregated by sex; the role of women in society varies by country but remains traditional in many countries. While women in Turkey enjoy considerable freedom and opportunity, Afghan women are compelled to adhere to the strictest Islamic orthodoxy.

In geographic terms, the impact of oil, its production and sale, in the exporting countries of this realm can be summarized as follows:

1. *High Incomes*. When oil prices on international markets were high, several countries in this realm ranked as the highest-income societies in the world. Even when oil prices declined, virtually all the petroleum-exporting states remained in the upper-middle-income category (Fig. I-11).

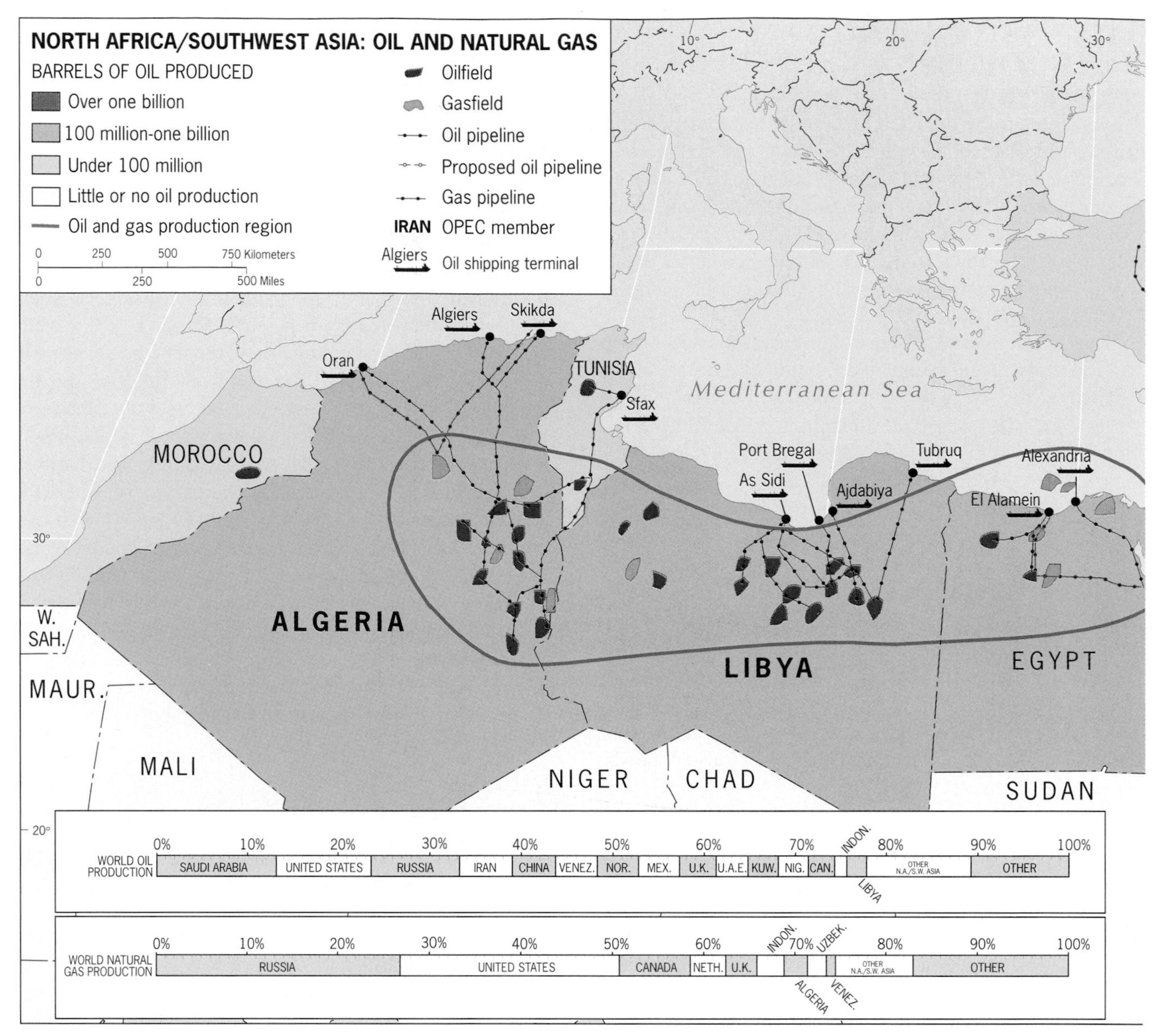

FIGURE 6-8

FROM THE FIELD NOTES

"The impact of oil wealth is nowhere better illustrated than it is along the waterfront of the famous Creek of Dubai (adjoining the city of the same name, largest in the United Arab Emirates). We stopped on the bridge that links the two sectors of Dubai to observe the old *dhows* (most of these wooden boats now with motors as well as triangular sail) still at their centuries-old moorings, overlooked not by traditional Arab buildings but by modern skyscrapers. Oil may have transformed the local economy, but don't count the role of the dhows out just yet. They still carry trade and contraband. We went to the dock and asked what was being transported. 'Jeans and cassettes to Iran,' we were told, 'and caviar and carpets from Iran. It's a good business.' So the traditional role of the dhows goes on, even in the age of oil."

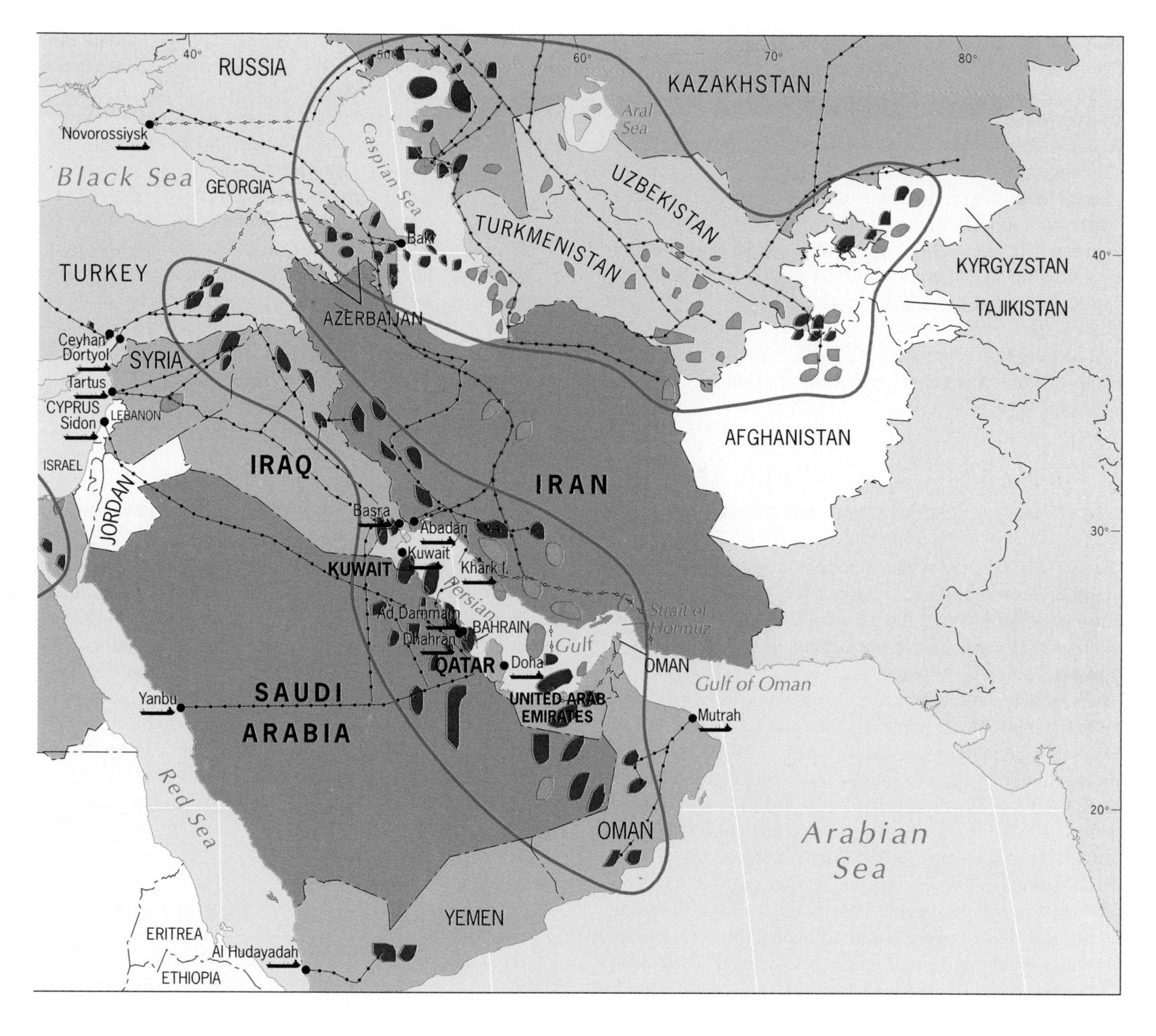

2. *Modernization.* Huge oil revenues transformed cultural landscapes throughout the realm, producing a façade of modernization in the cityscape of ports and capitals. Gleaming glass-encased office buildings towered over mosques; superhighways crossed ancient camel paths; state-of-the-art port facilities handled oil and oil-funded trade.
3. *Industrialization.* Farsighted governments among those with oil wealth, realizing that petroleum reserves will not last forever, have invested some of their income in industrial plants that will outlast the era of oil. Petrochemical industries, plastics fabrication, and desalinization plants are among these facilities.
4. *Intra-Realm Migration.* The oil wealth has attracted millions of workers from less favored parts of the realm to work in the oilfields, in the ports, and in many other, mainly menial capacities. This has brought many Shi'ites to the countries of eastern Arabia; hundreds of thousands of Palestinians also work there as laborers. Saudi Arabia, with a population of 23 million, has more than 5 million foreign workers.
5. *Inter-Realm Migration.* The willingness of workers from such countries as Pakistan, India, and Sri Lanka to work for wages even lower than those the oil industry pays has attracted a substantial flow of temporary immigrants from outside the realm. These workers serve mostly as domestics, gardeners, refuse collectors, and the like.
6. *Regional Disparities.* Oil wealth and its manifestations in the cultural landscape create strong contrasts with areas not directly affected. The ultramodern east coast of Saudi Arabia is a world apart from large areas of its inte-

rior, where it becomes a land of desert, oasis, and camel, of vast distances, slow change, and isolated settlements. To some degree, this phenomenon affects all oil-rich countries.

7. *Foreign Investment.* Governments and Arab businesspeople have invested oil-generated wealth in foreign countries. These investments have created a network of international involvement that links many of this realm's countries not only to the economies of foreign states, but also to growing Arab (and thus Islamic) communities in those states.

The map (Fig. 6-8) contains a warning that came home to the oil-rich states (not only in this realm but elsewhere as well) in the 1980s when, after a period of high oil prices and huge revenues, oil prices fell and incomes plummeted. Even the power of an 11-member cartel, OPEC (Organization of Petroleum Exporting Countries), could not recover the lost advantage. Figure 6-8 shows a system of oil and gas pipelines that strongly resembles the exploitative interior-to-coast railroad lines in a mineral-rich colony of the past. Such a pattern spells disadvantage for the exporter, whether colony or independent country. Markets, not raw-material exporters, dominate international trade. Oil brought this realm into contact with the outside world in ways unforeseen just a century ago. Oil has strengthened and empowered some of its peoples; it has dislocated and imperiled others. It has truly been a double-edged sword.

REGIONS OF THE REALM

Identifying and delimiting regions in this vast geographic realm is a considerable challenge. Not only are population clusters widely scattered, but also cultural transitions—internal and external—make it difficult to discern a regional framework.

As we have noted in earlier discussions, the world's regional geographic framework is subject to change. When Columbus sailed for the New World in 1492, the entire Balkan region of Eastern Europe was under the sway of the Ottoman Empire, and Islam cast its shadow over Vienna and Venice. In those times, this was not just a North African/Southwest Asian realm but an Eastern European one as well. Little more than two centuries ago, after the Austrian Empire had wrested the upper Danube Basin from the Turks, the Muslims still ruled over much of Romania, Bulgaria, Serbia, Bosnia, Albania, and Greece. Not until the second decade of the twentieth century did the Ottomans lose the last of their European holdings, which finally erased the European Muslim region from the map. Ever since, Christian Greece and Islamic Turkey have been antagonistic neighbors.

In Asia, too, expanding Islam suffered setbacks. The Muslims spread northward from Persia into the Transcaucasian corridor between the Black and Caspian seas, reaching the northern slopes of the Caucasus Mountains and converting such peoples as the Ingush and the Chechens to Islam. But Russian armies stopped the advance and conquered the entire Muslim frontier. East of the Caspian Sea lay another arena of Islamic expansion, and much of Central Asia became a mosaic of Muslim societies, a vast but sparsely peopled region where Islam traveled by caravan and took root in the oases. Again, the Russian czars had other plans for this region, and the scattered khanates and emirates were no match for St. Petersburg's armies. Later, when the Soviet communists inherited the Russian Empire, they divided Central Asia into five colonies and proceeded to try to extinguish Islam in favor of communism's official atheism. In their final act of attempted conversion, the Soviets in 1979 invaded Afghanistan to support a Marxist reform government that had seized control there with Soviet aid. It was a fateful enterprise, leading to a guerrilla war in which the United States poured weapons and money into the anti-Soviet camps. Thousands of refugees fled into neighboring Pakistan and Iran. These developments had long-term consequences that still endure and affect not just Afghanistan but the world at large.

MAJOR CITIES OF THE REALM

City	Population* (in millions)
Algiers, Algeria	2.0
Almaty, Kazakhstan	1.3
Baghdad, Iraq	5.1
Beirut, Lebanon	2.1
Cairo, Egypt	11.1
Casablanca, Morocco	3.8
Damascus, Syria	2.5
Istanbul, Turkey	10.2
Jerusalem, Israel	0.7
Khartoum, Sudan	3.0
Riyadh, Saudi Arabia	3.7
Tehran, Iran	7.4
Tel Aviv, Israel	2.3
Toshkent, Uzbekistan	2.2
Tunis, Tunisia	2.0

*Based on 2002 estimates.

Islam's setbacks had different results in different regions. In Eastern Europe, Ottoman imprints were soon expunged by the revival of old Christian traditions except in Albania, Kosovo, and Bosnia, where remnants of it survived, and in Bulgaria, where small Turkish minorities remained. But in the Caucasus area and in the five former Soviet republics of Central Asia (the region called Turkestan), Islam proved more durable. After the Soviet collapse in 1991, Islam quickly reasserted itself, and a wave of Muslim fervor engulfed these old Islamic frontiers. Here was a classic case of **cultural revival**, the regeneration of a long-dormant culture through internal renewal and external infusion.

16

Of all the geographic realms we discuss in this book, the North African/Southwest Asian realm has seen the most territorial change. It has expanded, contracted, and expanded again in Europe and Asia, as well as Africa; its margins remain in flux. Islam, we noted earlier, is only one dimension of it, but today it is Islam that energizes its borders—in West and northeastern Africa, in the Caucasus area, in northern Turkestan, in northwestern Pakistan. Unlike the Russian, North American, South Asian, or East Asian realms, this geographic realm has no dominant, anchoring state. It is not bounded by oceans or rivers or mountain ranges. Islam dominates and invigorates it, but several Islamic countries are not part of it. We have studied *countries* with multiple cores; here is a *realm* with many core areas, linked by the tenets of an ancient civilization that was infused by the prophet Muhammad fourteen centuries ago.

The following are the regional components of this far-flung realm today (Fig. 6-9):

1. **Egypt and the Lower Nile Basin**. This region in many ways constitutes the heart of the realm as a whole. Egypt (together with Iran and Turkey) is one of the realm's three most populous countries. It is the historic focus of this part of the world and a major political and cultural force. It shares with its southern neighbor, Sudan, the waters of the lower Nile River.
2. **The Maghreb and Its Neighbors**. Western North Africa (the Maghreb) and the areas that border it also form a region, consisting of Algeria, Tunisia, and Morocco at the center and Libya, Chad, Niger, Mali, and Mauritania along the broad periphery. The last four of these countries also lie astride or adjacent to the broad transition zone where the Arab-Islamic realm of northern Africa merges into Subsaharan Africa.
3. **The African Transition Zone**. From southern Mauritania in the west to Somalia in the east, across the entire African landmass at its widest extent, the realm dominated by Islamic culture interdigitates with that of Subsaharan Africa. No sharp dividing line can be drawn here: people of African ethnic stock have adopted the Muslim faith and Arabic language and traditions. As a result, this is less a region than a broad zone of transition.
4. **The Middle East**. This region includes Israel, Jordan, Lebanon, Syria, and Iraq. In effect, it is the crescent-like zone of countries that extends from the eastern Mediterranean coast to the head of the Persian Gulf.
5. **The Arabian Peninsula**. Dominated by the large territory of Saudi Arabia, the Arabian Peninsula also includes the United Arab Emirates, Kuwait, Bahrain, Qatar, Oman, and Yemen. Here lies the source and focus of Islam, the holy city of Mecca; here, too, lie many of the world's greatest oil deposits.
6. **The Empire States**. Two of the realm's giants, states with imperial histories and majestic cultures, dominate this region: Turkey and Iran. Turkey today is the realm's most secular state (most Turks are Sunni Muslims); Shi'ite Iran is one of its Islamic republics. To the north, Azerbaijan, former Soviet republic and once a part of Persia's empire, lies in the turbulent, Muslim-infused Transcaucasian Transition Zone. To the south, the northern part of the island of Cyprus is under Turkey's thrall.
7. **Turkestan**. Turkish influence ranged far and wide in Southwest and Central Asia, and following the Soviet collapse that influence proved to be durable and strong. In the five former Soviet republics the strength and potency of Islam vary, and the new governments deal warily (sometimes forcefully) with Islamic revivalists. Boundaries of this region, as in the African Transition Zone, do not always coincide with national borders. Kazakhstan is the most prominent case, but regional cultural influences also radiate into China and Pakistan. Afghanistan also forms part of this large, landlocked, turbulent region.

EGYPT AND THE LOWER NILE BASIN

Egypt occupies a pivotal location in the heart of a realm that extends over 6000 miles (9600 km) longitudinally and some 4000 miles (6400 km) latitudinally. At the northern end of the Nile and of the Red Sea, at the eastern end of the Mediterranean Sea, in the northeastern corner of Africa across from Turkey to the north and Saudi Arabia to the east, adjacent to Israel, to Islamic Sudan, and to militant Libya, Egypt lies in the crucible of this realm. Because it owns the Sinai Peninsula (recently lost to and regained from Israel), Egypt, alone among states on the African continent, has a foothold in Asia, a foothold that gives it a coast overlooking the strategic Gulf of Aqaba (the northeasternmost arm of the Red Sea). Egypt also controls the Suez Canal, vital link between the Indian and Atlantic oceans and lifeline of Europe. We hardly need to further justify Egypt's designation (together with northern Sudan) as a discrete region.

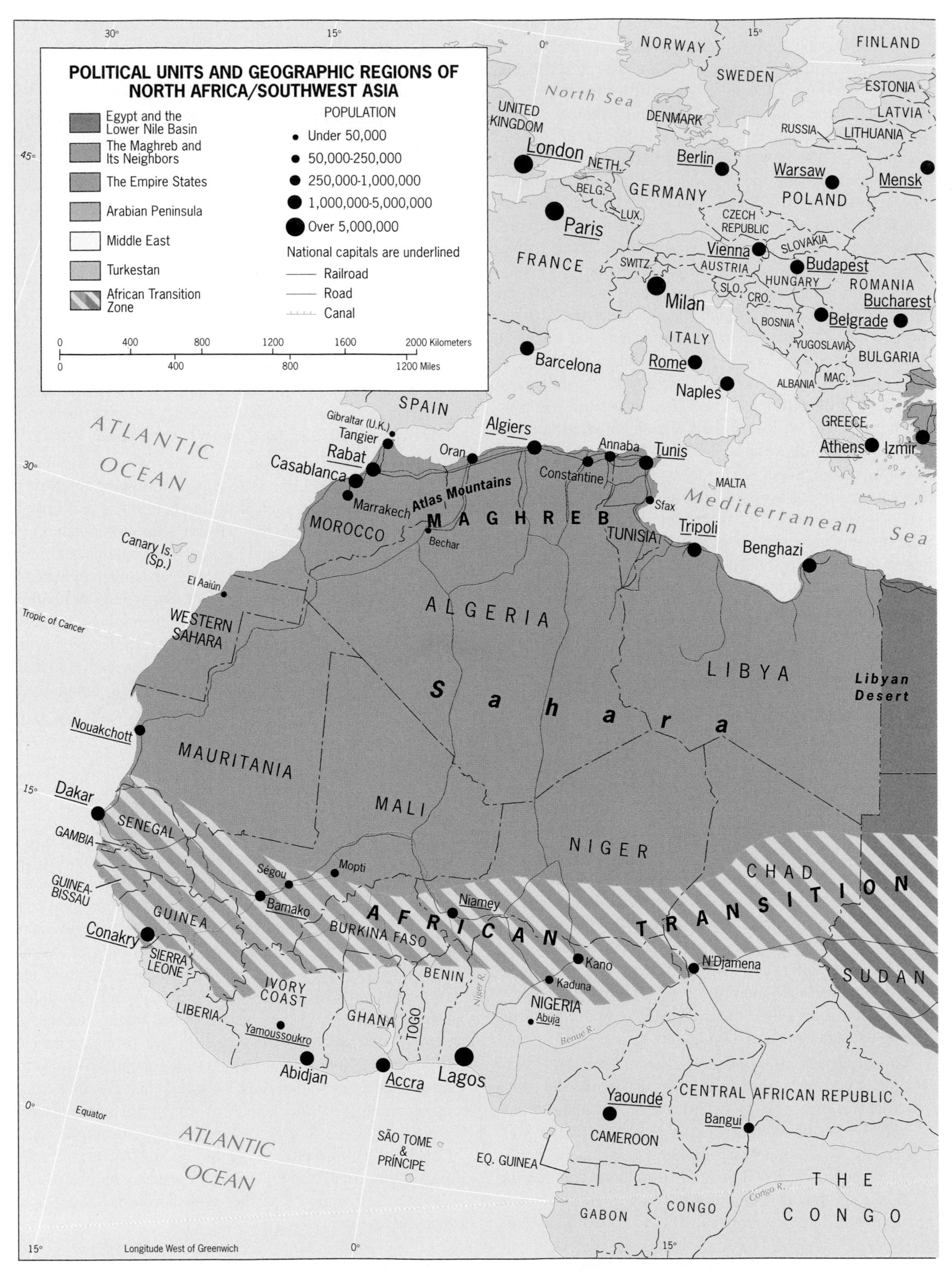
POLITICAL UNITS AND GEOGRAPHIC REGIONS OF NORTH AFRICA/SOUTHWEST ASIA
Egypt and the Lower Nile Basin
The Maghreb and Its Neighbors
The Empire States
Arabian Peninsula
Middle East
Turkestan
African Transition Zone
POPULATION
Under 50,000
50,000-250,000
250,000-1,000,000
1,000,000-5,000,000
Over 5,000,000
National capitals are underlined
Railroad
Road
Canal
0 400 800 1200 1600 2000 Kilometers
0 400 800 1200 Miles
ATLANTIC OCEAN
North Sea
Mediterranean Sea
NORWAY
SWEDEN
FINLAND
ESTONIA
LATVIA
LITHUANIA
RUSSIA
DENMARK
UNITED KINGDOM
London
NETH.
BELG.
LUX.
GERMANY
Berlin
Warsaw
POLAND
Mensk
CZECH REPUBLIC
SLOVAKIA
Paris
FRANCE
SWITZ.
AUSTRIA
Vienna
Budapest
HUNGARY
ROMANIA
Bucharest
SLO.
CRO.
BOSNIA
Belgrade
YUGOSLAVIA
BULGARIA
Milan
ITALY
Rome
Naples
Barcelona
ALBANIA
MAC.
GREECE
Athens
Izmir
SPAIN
Gibraltar (U.K.)
Tangier
Rabat
Casablanca
Marrakech
Oran
Algiers
Annaba
Tunis
Constantine
Sfax
MALTA
Atlas Mountains
MAGHREB
MOROCCO
Bechar
TUNISIA
Tripoli
Benghazi
Canary Is. (Sp.)
El Aaiún
Tropic of Cancer
WESTERN SAHARA
ALGERIA
LIBYA
Libyan Desert
Sahara
Nouakchott
MAURITANIA
MALI
NIGER
CHAD
Dakar
SENEGAL
GAMBIA
GUINEA-BISSAU
Mopti
Ségou
Bamako
Niamey
GUINEA
Conakry
BURKINA FASO
AFRICAN TRANSITION
SIERRA LEONE
LIBERIA
IVORY COAST
Yamoussoukro
Abidjan
GHANA
TOGO
BENIN
Niger R.
Accra
Lagos
Kano
Kaduna
NIGERIA
Abuja
Benue R.
N'Djamena
SUDAN
Equator
ATLANTIC OCEAN
SÃO TOME & PRÍNCIPE
EQ. GUINEA
Yaoundé
CAMEROON
CENTRAL AFRICAN REPUBLIC
Bangui
GABON
CONGO
Congo R.
THE CONGO
30° 15° 0° 15°
45° 30° 15° 0°
Longitude West of Greenwich

FIGURE 6-9

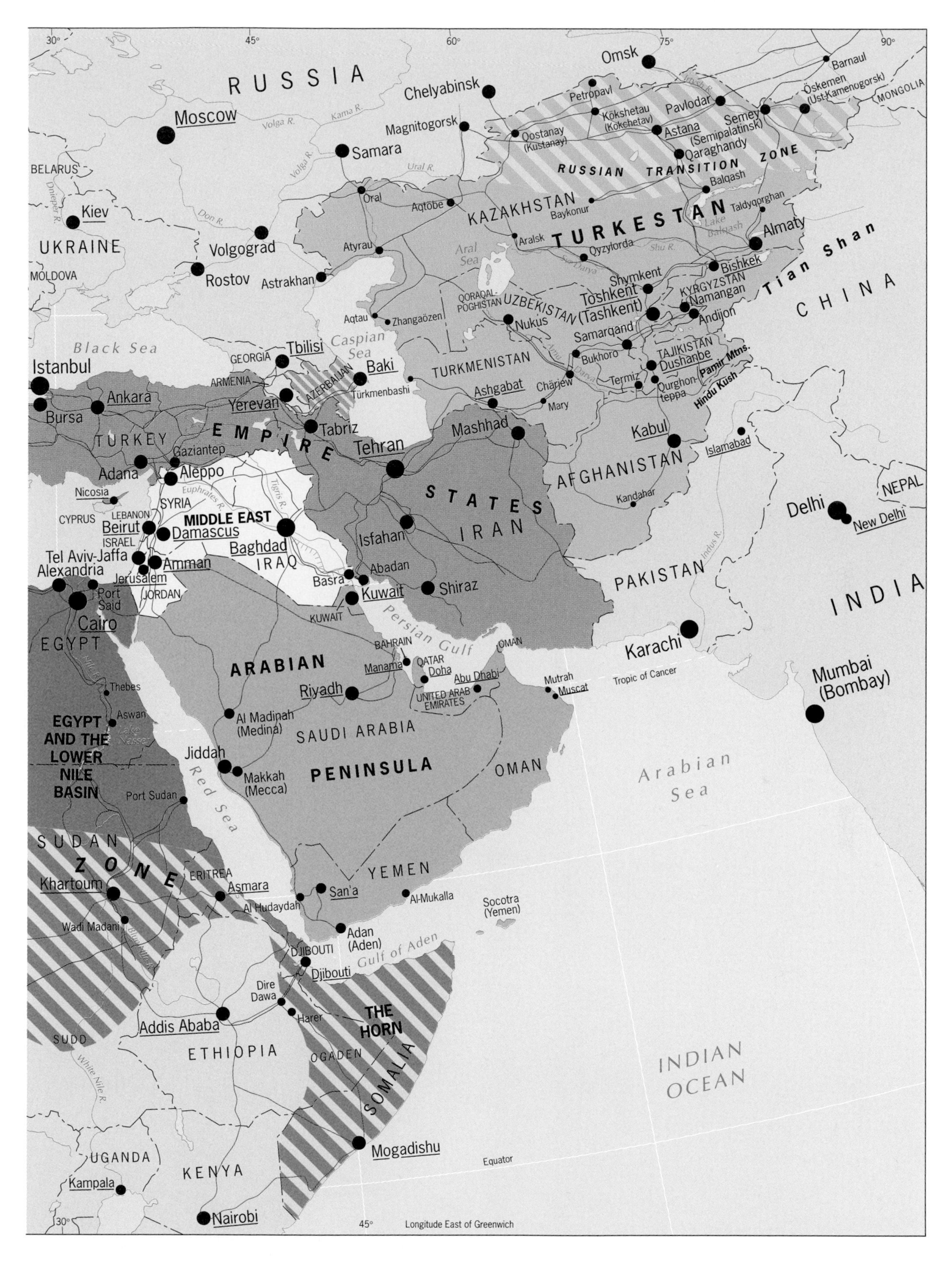
RUSSIA
Moscow
Omsk
Chelyabinsk
Magnitogorsk
Samara
Volgograd
Rostov
Astrakhan
Kiev
UKRAINE
BELARUS
MOLDOVA
KAZAKHSTAN
TURKESTAN
RUSSIAN TRANSITION ZONE
Astana
Pavlodar
Almaty
Bishkek
KYRGYZSTAN
UZBEKISTAN
Toshkent (Tashkent)
Samarqand
TAJIKISTAN
Dushanbe
TURKMENISTAN
Ashgabat
Tian Shan
CHINA
Black Sea
Caspian Sea
Istanbul
Ankara
TURKEY
EMPIRE STATES
Tbilisi
Yerevan
Baki
Tabriz
Tehran
Mashhad
IRAN
Isfahan
Shiraz
AFGHANISTAN
Kabul
Islamabad
PAKISTAN
Karachi
INDIA
Delhi
New Delhi
NEPAL
Mumbai (Bombay)
MIDDLE EAST
SYRIA
Damascus
Beirut
Baghdad
IRAQ
Amman
Jerusalem
JORDAN
Kuwait
Persian Gulf
ARABIAN PENINSULA
SAUDI ARABIA
Riyadh
Jiddah
Makkah (Mecca)
OMAN
Muscat
YEMEN
San'a
Adan (Aden)
Arabian Sea
Alexandria
Cairo
EGYPT
EGYPT AND THE LOWER NILE BASIN
Red Sea
SUDAN ZONE
Khartoum
ERITREA
Asmara
Djibouti
Addis Ababa
ETHIOPIA
THE HORN
SOMALIA
Mogadishu
KENYA
Nairobi
UGANDA
Kampala
INDIAN OCEAN
Tropic of Cancer
Equator
Longitude East of Greenwich

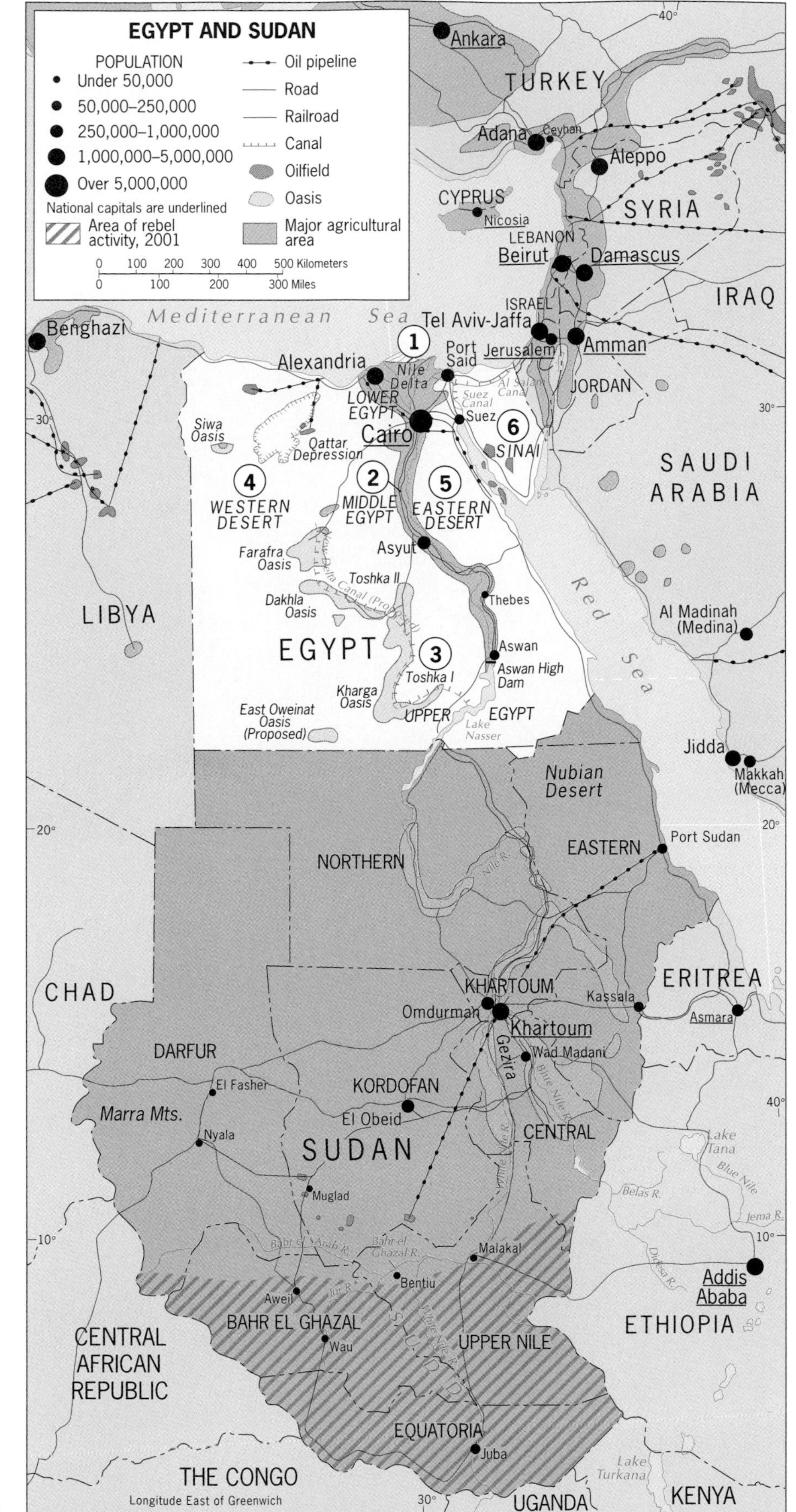
EGYPT AND SUDAN
POPULATION
Under 50,000
50,000–250,000
250,000–1,000,000
1,000,000–5,000,000
Over 5,000,000
National capitals are underlined
Area of rebel activity, 2001
Oil pipeline
Road
Railroad
Canal
Oilfield
Oasis
Major agricultural area
0 100 200 300 400 500 Kilometers
0 100 200 300 Miles
Ankara
TURKEY
Adana
Ceyhan
Aleppo
SYRIA
CYPRUS
Nicosia
LEBANON
Beirut
Damascus
IRAQ
ISRAEL
Tel Aviv-Jaffa
Jerusalem
Amman
JORDAN
Mediterranean Sea
Benghazi
Alexandria
Nile Delta
Port Said
Al Salam Canal
Suez Canal
Suez
LOWER EGYPT
Cairo
SINAI
Siwa Oasis
Qattar Depression
WESTERN DESERT
MIDDLE EGYPT
EASTERN DESERT
SAUDI ARABIA
Asyut
Farafra Oasis
Toshka II
Thebes
Dakhla Oasis
LIBYA
EGYPT
Aswan
Aswan High Dam
Toshka I
Kharga Oasis
East Oweinat Oasis (Proposed)
UPPER EGYPT
Lake Nasser
Red Sea
Al Madinah (Medina)
Jidda
Makkah (Mecca)
Nubian Desert
Port Sudan
EASTERN
NORTHERN
Nile R.
CHAD
KHARTOUM
Kassala
ERITREA
Asmara
Omdurman
Khartoum
Gezira
Wad Madani
DARFUR
El Fasher
KORDOFAN
Marra Mts.
El Obeid
Nyala
SUDAN
CENTRAL
Lake Tana
Blue Nile
Belas R.
Muglad
Jema R.
Malakal
Addis Ababa
Bentiu
Aweil
BAHR EL GHAZAL
Wau
UPPER NILE
CENTRAL AFRICAN REPUBLIC
ETHIOPIA
EQUATORIA
Juba
THE CONGO
Lake Turkana
UGANDA
KENYA
Longitude East of Greenwich
1
2
3
4
5
6
40°
30°
20°
10°

FIGURE 6-10

The Nile River

In the fifth century BC, the Greek scholar Herodotus described Egypt as the gift of the Nile, but Egypt also was a product of natural protection. In the protective isolation of their culture hearth, the ancient Egyptians converted security into stability and progress. The great stone sculptures and pyramids in modern Egypt's cultural landscape bear witness to the flowering of a continuous civilization that is now more than 5000 years old.

Egypt's Nile is the aggregate of two great branches upstream: the White Nile, which originates in the streams that feed Lake Victoria in East Africa, and the Blue Nile, whose source lies in Lake Tana in Ethiopia's highlands. The two Niles converge at Khartoum in modern-day Sudan. Fly northward from Khartoum along the Nile toward Egypt, and you will realize how small and vulnerable this ribbon of water looks in the vast wasteland of the Sahara. And on this ribbon depend the lives of tens of millions of people.

About 95 percent of Egypt's 71.1 million people live within a dozen miles (20 km) of the great river's banks or in its delta (Fig. 6-10). It has always been this way: the Nile rises and falls seasonally, watering and replenishing soils and crops on its banks. In April and May, the river usually is at its lowest level, a trickle through the desert. Then during the summer months it rises, to reach flood stage at Cairo, the capital, in October. Then the Nile may be as much as 20 feet (6 m) above its lowest stage.

The ancient Egyptians used *basin irrigation*, building fields with earthen ridges and trapping the floodwaters with their fertile silt, to grow their crops. That practice continued for thousands of years until, during the nineteenth century, the construction of permanent dams made it possible to irrigate Egypt's farmlands year round. These dams, with locks for navigation, controlled floods, expanded the country's cultivable area and allowed the farmers to harvest more than one crop per year on the same field. In a single century, all of Egypt's farmland was brought under *perennial irrigation*.

The greatest of all Nile projects, the Aswan High Dam (which began operating in 1968), creates Lake Nasser, one of the world's largest artificial lakes (Fig. 6-10). As the map shows, the lake extends into Sudan, where 50,000 people had to be resettled to make way for it. The Aswan High Dam increased Egypt's irrigable land by nearly 50 percent and today provides the country with about 40 percent of its electricity. But, as is so often the case with megaprojects of this kind, the dam also produced serious problems. Snail-carried schistosomiasis and mosquito-transmitted malaria thrived in the dam's standing water, afflicting hundreds of thousands of people living nearby. By blocking most of the natural fertilizers in the Nile's annual floodwaters, the dam necessitated the widespread use of artificial fertilizers, very costly to small farmers and damaging to the natural environment. And the now fertilizer- and pesticide-laden Nile no longer supports the fish fauna offshore, reducing the catch and depriving coastal populations of badly needed proteins.

Egypt's elongated oasis along the Nile, just 3 to 15 miles (5 to 25 km) wide, broadens north of Cairo across a delta anchored in the west by the great city of Alexandria and in the east by Port Said, gateway to the Suez Canal. The delta contains extensive farmlands, but it is a troubled area today. The ever more intensive use of the Nile's water and silt upstream is depriving the delta of much needed replenishment. And the low-lying delta is geologically subsiding, raising fears of salt-water invasion from the Mediterranean Sea that would damage soils here.

Egypt's millions of subsistence farmers, the *fellaheen*, still struggle to make their living off the land, as did the peasants of the Egypt of five millennia past. Rural landscapes seem barely to have changed; ancient tools are still used, and dwellings remain rudimentary. Poverty, disease, high infant mortality rates, and low incomes prevail. In the

FROM THE FIELD NOTES

"On the eastern edge of central Cairo we saw what looked like a combination of miniature mosques and elaborate memorials. Here lie buried the rich and the prominent of times past in what locals call the 'City of the Dead.' But we found it to be anything but a dead part of the city. Many of the tombs here are so large and spacious that squatters have occupied them. Thus the City of the Dead is now an inhabited graveyard, home to at least one million people. The exact numbers are impossible to determine; indeed, whereas metropolitan Cairo in 2002 has an official population of just over 11 million, many knowledgeable observers believe that 16 million is closer to the mark."

Among the Realm's Great Cities . . .

Cairo

Stand on the roof of one of the high-rise hotels in the heart of Cairo, and in the distance you can see the great pyramids, monumental proof of the longevity of human settlement in this area. But the present city was not founded until Muslim Arabs chose the site as the center of their new empire in AD 969. Cairo became and remains Egypt's primate city, situated where the Nile River opens into its delta, home today to almost one-sixth of the entire country's population.

Cairo at 11.1 million ranks among the world's 20 largest urban agglomerations, and it shares with other cities of the poorer world the staggering problems of crowding, inadequate sanitation, crumbling infrastructure, and substandard housing. But even among such cities, Cairo is noteworthy for its stunning social contrasts. Along the Nile waterfront, elegant skyscrapers rise above carefully manicured, Parisian-looking surroundings. But look eastward, and the urban landscape extends gray, dusty, almost featureless as far as the eye can see. Not far away, more than a million squatters live in the sprawling cemetery known as the City of the Dead (p. 301). On the outskirts, millions more survive in overcrowded shantytowns of mud huts and hovels.

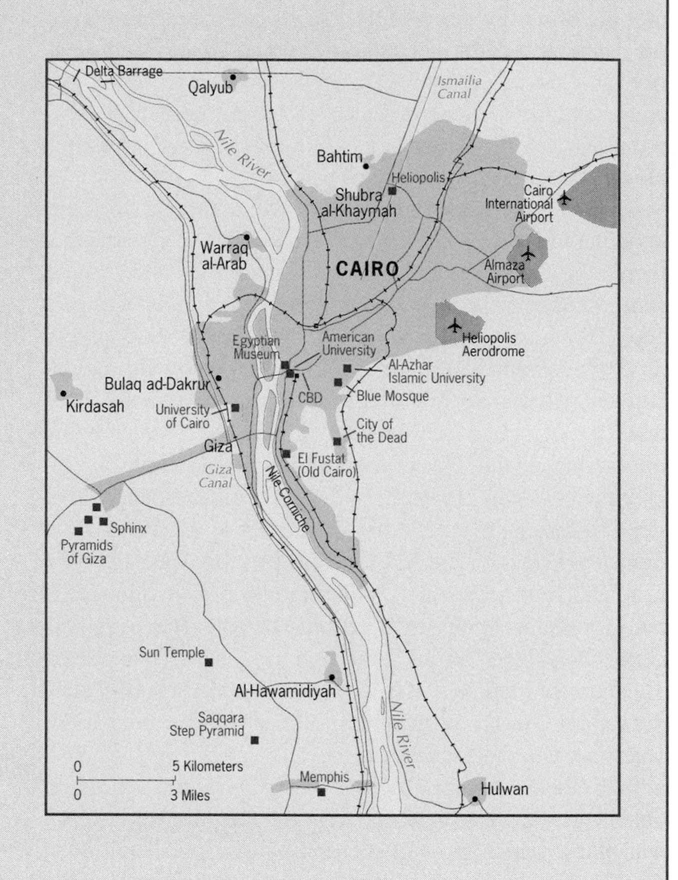

And yet Cairo is the dominant city not only of Egypt but for a wider sphere; it is the cultural capital of the Arab World, with centers of higher learning, splendid museums, world-class theater and music, and magnificent mosques and Islamic learning centers. Although Cairo always has been primarily a center of government, administration, and religion, it also is a river port and an industrial complex, a commercial center, and, as it sometimes seems, one giant bazaar. Cairo is the heart of the Arab World, a creation of its geography, and a repository of its history.

1990s, the Egyptian government intensified its drive to expand irrigated farmlands. The newly dug Al-Salam (Peace) Canal sends water from the eastern part of the Nile Delta into the Sinai Peninsula. A much larger project, the two-phased Toshka, to be built over the next decade, will take 10 percent of the Nile's flow from Lake Nasser and divert it into the Western Desert along a string of oases (Fig. 6-10). This more than 300-mile (500-km) canal will bring some 1.5 million acres (over 600,000 hectares) of new farmland into production. And the proposed East Oweinat Project in western Upper Egypt will be watered from local aquifers. But these efforts should be viewed in the context of Egypt's relatively high rate of population growth (2.0 percent at the turn of the century). Without modernization and greater efficiency in the existing farmlands, no amount of expansion can keep pace with the needs of this burgeoning population.

Egypt's Regional Geography

Egypt has six subregions, mapped in Figure 6-10. Most Egyptians live and work in Lower (i.e., northern) and Middle Egypt (regions ① and ②), the country's core area anchored by Cairo and flanked by the leading port and second industrial center, Alexandria. The economy has benefited from further oil discoveries in the Sinai (region ⑥) and in the Western Desert (region ④), so that Egypt now is self-sufficient and even exports some petroleum. Cotton and textiles are the other major source of external income, but the important tourist industry has been hurt by Islamic extremists. As the population mushrooms, the gap between food supply and demand widens, and Egypt must import grain. Since the late 1970s, Egypt has been a major recipient of U.S. foreign aid.

Egypt at the opening of this century is at a crossroads in more ways than one. Its planners know that reducing the high birth rate would improve the demographic situation, but revivalist Muslims object to any programs that promote family planning. Its accommodation with Israel helps ensure foreign aid but divides the people. Its government faces a fundamentalist challenge. Egypt's future, in this crucial corner of the realm, is uncertain.

Northern, Arabized Sudan

As Figure 6-10 shows, Egypt is flanked by two countries that have posed challenges to its leadership: Sudan to the south and Libya to the west. Sudan, upstream along the Nile, is the more formidable and problematic neighbor. Apart from its obvious location of control over Nile River waters before they reach Lake Nasser, Sudan also is a beacon for Islamic militants in the region. Sudan is officially an Islamic Republic; its government has been accused by the United States of supporting international terrorism. (In 1998 the United States attacked one of Sudan's chemical factories with rockets, alleging that weapons were being manufactured there.) During the 1991 Gulf War, when Egypt and other Arab countries sided with the United States, Sudan backed Iraq.

More than twice as large as Egypt and with a population of just under 31 million, Sudan lies centered on the confluence of the White Nile (from Uganda) and the Blue Nile (from Ethiopia). Here the twin capital, Khartoum-Omdurman, anchors a large agricultural area where cotton was planted during colonial times. The British administration combined northern Sudan, which was Arabized and Islamized, with a large area to the south, which was African and where many villagers had been Christianized. After the British left, the regime in Khartoum wanted to impose its Islamic rule on the south, and a bitter civil war ensued. The cost in human lives and dislocation is incalculable; in 2001 an estimated 4 million people remained refugees in their own land.

Port Sudan on the Red Sea is Sudan's maritime outlet, but the country's economy is symbolic of the periphery. Its per-capita income is one of the world's lowest. Trade with Saudi Arabia, Sudan's main partner, exchanges sheep and cotton for oil. But change may be on the way. The pipeline that used to transfer oil from Port Sudan to Khartoum is about to be supplemented by one whose oil will move the other way: a major reserve has been discovered east of Muglad, about 400 miles (640 km) southwest of Khartoum but still within the northern, Arabized sector of Sudan (Fig. 6-10). An underground pipeline nearly 1000 miles (1600 km) long to Port Sudan via Khartoum will bring self-sufficiency and make the country an oil exporter.

Meanwhile, the contrasts between Egypt and Sudan could hardly be stronger: a moderate secular Arab country adjoins a militant Arabized African state at war with its own people. Water, oil, ideology, and international involvement make for a risky political geography here.

THE MAGHREB AND ITS NEIGHBORS

The countries of northwestern Africa are collectively called the *Maghreb*, but the Arab name for them is more elaborate than that: *Djezira-al-Maghreb*, or "Isle of the West," in recognition of the great Atlas Mountain range rising like a huge island from the Mediterranean Sea to the north and the sandy flatlands of the immense Sahara to the south.

The countries of the Maghreb (sometimes spelled *Maghrib*) are Morocco, last of the North African kingdoms; Algeria, a secular republic beset by the religious-political problems we noted earlier; and Tunisia, smallest and most Westernized of the three (Fig. 6-11). Libya, facing the Mediterranean between the Maghreb and Egypt, is unlike any other North African country: an oil-rich desert state whose population is almost entirely clustered in settlements along the coast.

Whereas Egypt is the gift of the Nile, the Atlas Mountains form the nucleus of the settled Maghreb. These high ranges wrest from the rising air enough orographic rainfall to sustain life in the intervening valleys, where good soils support productive farming. From the vicinity of Algiers eastward along the coast into Tunisia, annual rainfall averages more than 30 inches (75 cm), a total more than three times as high as that recorded for Alexandria in Egypt's delta. Even 150 miles (240 km) inland, the slopes of the Atlas still receive over 10 inches (25 cm) of rainfall. The effect of the topography can be read on the world map of precipitation (Fig. I-7): where the highlands of the Atlas terminate, desert conditions immediately begin.

The Atlas Mountains are structurally an extension of the Alpine system that forms the orogenic backbone of Europe, of which the Alps and Italy's Appennines are also parts. In northwestern Africa, these mountains trend southwest-northeast and begin in Morocco as the High Atlas, with elevations close to 13,000 feet (4000 m). Eastward, two major ranges dominate the landscapes of Algeria proper: the Tell Atlas to the north, facing the Mediterranean, and the Saharan Atlas to the south, overlooking the great desert. Between these two mountain chains, each consisting of several parallel ranges and foothills, lies a series of intermontane basins (analogous to South America's Andean *altiplanos* but at lower elevations), markedly drier than the northward-facing slopes of the Tell Atlas. In these valleys, the rain-shadow effect of the Tell Atlas is reflected not only in the steppe-like natural vegetation but also in land-use

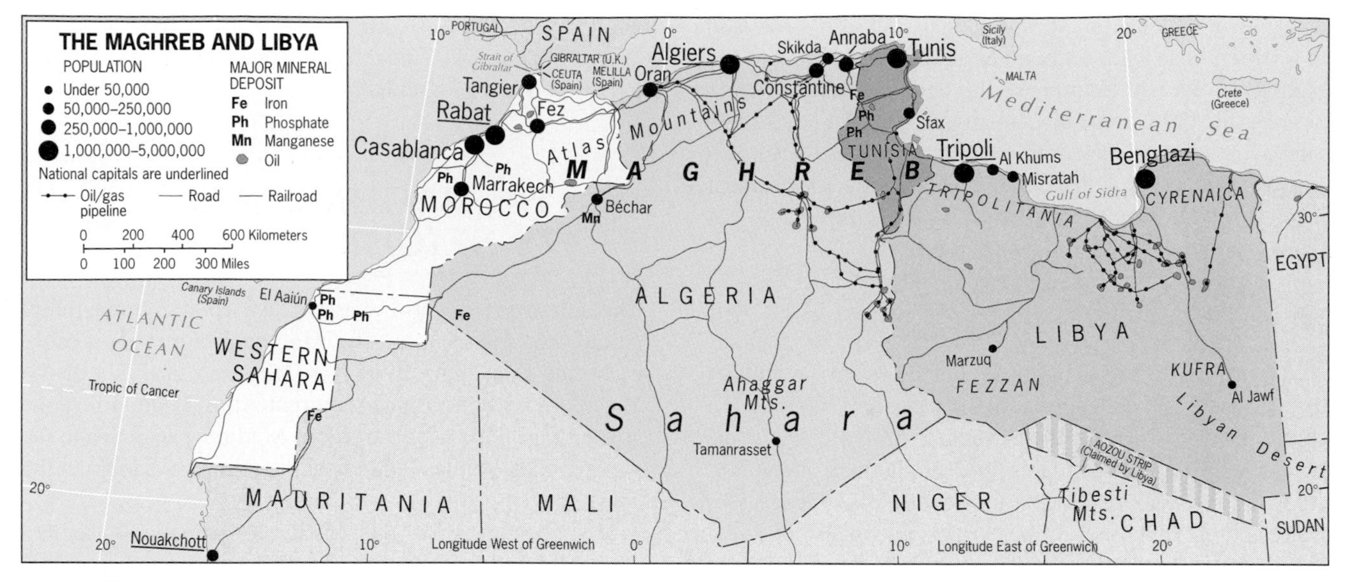

FIGURE 6-11

patterns: pastoralism replaces cultivation, and stands of short grass and bushes blanket the countryside.

During the colonial era, which began in Algeria in 1830 and lasted until the early 1960s, well over a million Europeans came to settle in North Africa—most of them French, and a large majority bound for Algeria—and these immigrants soon dominated commercial life. They stimulated the renewed growth of the region's towns; Casablanca, Algiers, Oran, and Tunis became the urban foci of the colonized territories. Although the Europeans dominated trade and commerce and integrated the North African countries with France and the European Mediterranean world, they did not confine themselves to the cities and towns. They recognized the agricultural possibilities of the favored parts of the tell (the lower Tell Atlas slopes and narrow coastal plains that face the Mediterranean) and established thriving farms. Not surprisingly, agriculture here is Mediterranean. Algeria soon became known for its vineyards and wines, citrus groves, and dates; Tunisia has long been one of the world's leading exporters of olive oil; and Moroccan oranges went to many European markets.

Oil and Emigrants

The fortunes of the Maghreb states since independence (four decades ago) have varied. Morocco became embroiled in a territorial conflict involving Western Sahara to its south, a former Spanish possession. A small but resilient group of Western Saharan residents resisted Morocco's absorption of this desert area, and Algeria and Mauritania gave them support. This strained relations between Morocco and its neighbors, but in Morocco itself the campaign had a unifying effect at a time when the king faced growing antimonarchist sentiment. But even without Western Sahara (the issue remained unresolved in 2001), Morocco is a large and important state with a population of almost 30 million. A conservative kingdom in a revolutionary region, Morocco acquired a two-chamber legislature as recently as 1997; in 1999 power passed from King Hassan II to his son, King Muhammad VI. The country's economic problems continue: it is not self-sufficient in food, had no significant energy resources until the recent discovery of the Talsint oilfield, and must survive on specialty farm exports, sales of unfinished minerals and ores (iron, manganese, phosphates), and tourism. The core area lies in the north, where the four major cities are located; Marrakech is the focus of the center of the country, a traditional center in sight of the Atlas and now a significant tourist attraction.

Close inspection of Figure 6-11 reveals that Spain possesses two exclaves on Moroccan shores: Ceuta and Melilla, holdovers from the time when France controlled this part of the country, and Spain the southwest. When Morocco achieved independence in 1956, Spain held on to Ceuta and Melilla, and today these small coastal cities are Spain's problem. They form Europe's Moroccan "back door," through which tens of thousands of Moroccan illegal emigrants have made their way to Spain and into the European Union. Would-be migrants swim or boat from Morocco to Ceuta and Melilla, and once there, are on Spanish soil. Spanish and Moroccan authorities have tried to limit this traffic, as well as the direct cross-water movement of rowboats and fishing vessels.

It is only 8 miles (13 km) from Morocco to Spain across the Strait of Gibraltar, and not only Moroccans but also migrants from Subsaharan Africa make the crossing, sometimes losing their lives. But unemployment is high in Morocco, opportunities are few, and controls are strict, so the countries of the European Union beckon and Spain is the threshold.

Algeria's political and economic geographies have been overshadowed by the events of the early 1990s, when, in the first free elections since independence in 1962, the Islamic Salvation Front—which advocated making Algeria an Islamic republic—won overwhelmingly in provincial and local votes. The army intervened, national elections were postponed, the president was assassinated, and a civil war began that claimed an estimated 100,000 lives.

FROM THE FIELD NOTES

"Having seen the *Giralda* in Seville, Spain (photo, p. 288), I was interested to view the tower of the Koutoubia Mosque in Marrakech, Morocco, also built in the twelfth century and closely resembling its Andalusian counterpart. Some 220 feet (67 m) tall, the Koutoubia minaret was built by Spanish slaves and is a monument to the heyday of Islamic culture of the time. I soon discovered that it is not as easy as it looks to get a clear view of the tower, and a boy offered his help, guiding me around the block on which the mosque is situated. He was well informed on the building's history and on the varying fortunes of his home town. It was European mispronunciation of the name 'Marrakech' that led to Morocco being called Morocco, he reported. And Marrakech was Morocco's capital for a very long time, and still should be. 'But back to the mosque,' I said. 'What does the sign say?' He smiled. 'You'll just have to learn some Arabic, just as we know French,' he said. Obviously a teacher in the making, I thought."

Algiers is Algeria's primate city, centrally situated along the Mediterranean coast, and home to 2 million of the country's 33 million inhabitants. Not even the urban centers were safe from the strife that engulfed the country, however. Indeed, the conflict spilled over into France, home to nearly 3 million North Africans, a majority of them Algerians. Meanwhile, Algeria's oil and natural gas industries, under protection from the army, expanded in the east (Fig. 6-11) and transformed the economy. A country that once earned its outside income from Mediterranean crops and some minerals now gets 95 percent of its foreign exchange from oil and gas.

As Table I-2 shows, the highest GNP per capita in the Maghreb is earned in Tunisia, smallest of the North African countries territorially but with a sizeable population of 10 million. Successive governments have severely repressed Muslim radicals, and comparative stability has allowed the relatively diversified economy, ranging from textiles and farm products to mineral production and tourism, to expand and prosper. Tunis, the historic primate city in the northeastern corner, dominates this most urbanized of the Maghreb states.

The Desert Coast

Between the eastern terminus of the Atlas range (in Tunisia) and Egypt's Nile Delta, the Sahara reaches the shores of the Mediterranean Sea. Here lies Libya, small in population (5.4 million), large in area (nearly three times bigger than Texas), and rich in oil.

Almost rectangular in shape, Libya is a country whose four corners matter most (Fig. 6-11). What limited agricultural possibilities exist lie in the northwest in Tripolitania, centered on the capital, Tripoli, and in the northeast in Cyrenaica, where Benghazi is the urban focus. Between these two coastal clusters, which are home to 90 percent of the Libyans, lies the Gulf of Sidra, a deep Mediterranean bay. Libya has claimed this gulf as its sea, but other countries have not accepted this claim. Libya's two southern corners are the desert Fezzan, a mountainous area near the southwestern border with Algeria and Niger, and the sparsely populated Kufra oasis in the southeast. Despite its huge size and tiny population, Libya has claimed a sector (the Aozou Strip) of its southern neighbor, Chad (Fig. 6-11).

Libya's long-time ruler, Muammar Gadhafi, used the huge revenues from major oil reserves in the country's northeast to build megaprojects, including the 600-mile (950-km) pipeline that carries water from aquifers beneath the Sahara to the parched provinces of Cyrenaica and Tripolitania; to promote revolutionary Arab-Islamic causes abroad; and, more recently, to boost pan-Africanism. Ordinary Libyans benefited from the oil wealth, but not comparably to the citizens of the oil-rich countries of the Arabian Peninsula. Still, hundreds of thousands of Africans from Nigeria, Ghana, and other countries came to Libya to work

FROM THE FIELD NOTES
"As you walk toward the entryways of Marrakech's *medina* (which means 'ancient Moorish town') you become aware that, whereas the *medinas* of most other Islamic cities tend to be white, this one is reddish. In fact, I learned, locals call it the 'red city,' because this is the color of the area's clay. 'They whitewash the buildings elsewhere,' said my Arab companion, 'but we rather like this distinctive hue.' The *medina* is a maze of narrow streets and alleys with hundreds of shops and stalls selling goods ranging from candles to carpets and sandals to spices. Most of the shoppers are locals, but Marrakech (Morocco's third-largest city) has become a tourist attraction, and business spills over into the square outside."

for the Libyans at wages higher than they could earn at home. Then, in the autumn of 2000, an economic downturn provoked massive anti-African attacks, killing hundreds and driving tens of thousands out of the country. Arab-African, Muslim-Christian conflict formed the roots of this tragedy, little noticed around the world—but spelling the end of Gadhafi's pan-African campaign.

THE AFRICAN TRANSITION ZONE

Islam diffused by land and by water. It crossed the Mediterranean Sea and the Indian Ocean; it gained footholds on many coasts from East Africa to Indonesia. In North Africa, Islam also crossed the desert. It came up the Nile by boat and across the Sahara by caravan, and it reached the peoples of the interior steppes and savannas on the southern side of the desert. There, along a wide stretch of Africa known today as the *Sahel* (an Arabic word for "border" or "margin"), Islam's proselytizers converted millions to the Muslim faith. As we will discover in the next chapter, the zone between the desert and the forest was one of Africa's culture hearths, and here lay large and durable states. Their rulers and subjects adopted Islam, and soon huge throngs of West Africans marched along the savanna corridor eastward on their annual pilgrimages to Mecca.

Then came Europe's colonial powers, and the modern political map of Africa took shape. Muslims and non-Muslims were thrown together in countries not of their making. Whereas the countries of North Africa are more than 90 percent Muslim, those of coastal West Africa, from Liberia to Benin, are well below 50 percent Muslim (Fig. 6-12). The north-to-south transition is especially clear in the heart of the continent: Libya is 97 percent Muslim, Chad 54, Central African Republic 15, and The Congo 1.

In some countries of the African Transition Zone Muslims and non-Muslims live side by side, as in Eritrea in the east and Guinea-Bissau in the west, but more often the transition has a regional dimension. As Figure 6-12 shows, northern Nigeria, northern Chad, and northern Sudan are dominantly Muslim, while in each of these countries the south is mainly Christian and animist (that is, traditional forms of belief). This does not mean that no Christians live in the north or Muslims in the south, but it does create a regional contrast expressed in cultural landscapes, social norms, legal practices, and a host of other ways. This regional divergence has led to severe political tension in Nigeria and to open conflict in Chad and Sudan. In Ethiopia, which is mapped as 31 percent Muslim, the mainly Somali Muslims are concentrated in the east and the non-Muslims to the west of the line shown extending from Djibouti to eastern Kenya. Eastern Ethiopia is ethnically Somali and culturally Islamic, and in its Ogaden area cross-border affinities with fellow Somalis (living in Somalia to the east) are far stronger than internal relations with the rest of Ethiopia. Harer, once the capital of a Muslim state and Ethiopia's only Muslim-style, walled city, is worlds removed from the capital, Addis Ababa, founded by the Christian Amhara whose rulers conquered the Muslim east.

At the easternmost end of the African Transition Zone lies Somalia, where some 7.7 million people (virtually all Muslim) live pastoral and farming lives at the mercy of a desert-dominated climate. Somalia came to world attention during the 1990s, when droughts killed thousands, the clan-based social order collapsed, and the United Nations tried to help—with disastrous results. Clan warfare demolished what remained of the social system and devastated the capital, Mogadishu. In 1991 the northern part of the country proclaimed itself independent under the name of Somaliland, but no other states recognized it. In 2000, Somalia still had no central government, but a meeting of several thousand representatives of the country's hundreds of clans, convened in Djibouti, appointed 225 members of a transitional National Assembly. Again there is hope that a Somali state system will emerge.

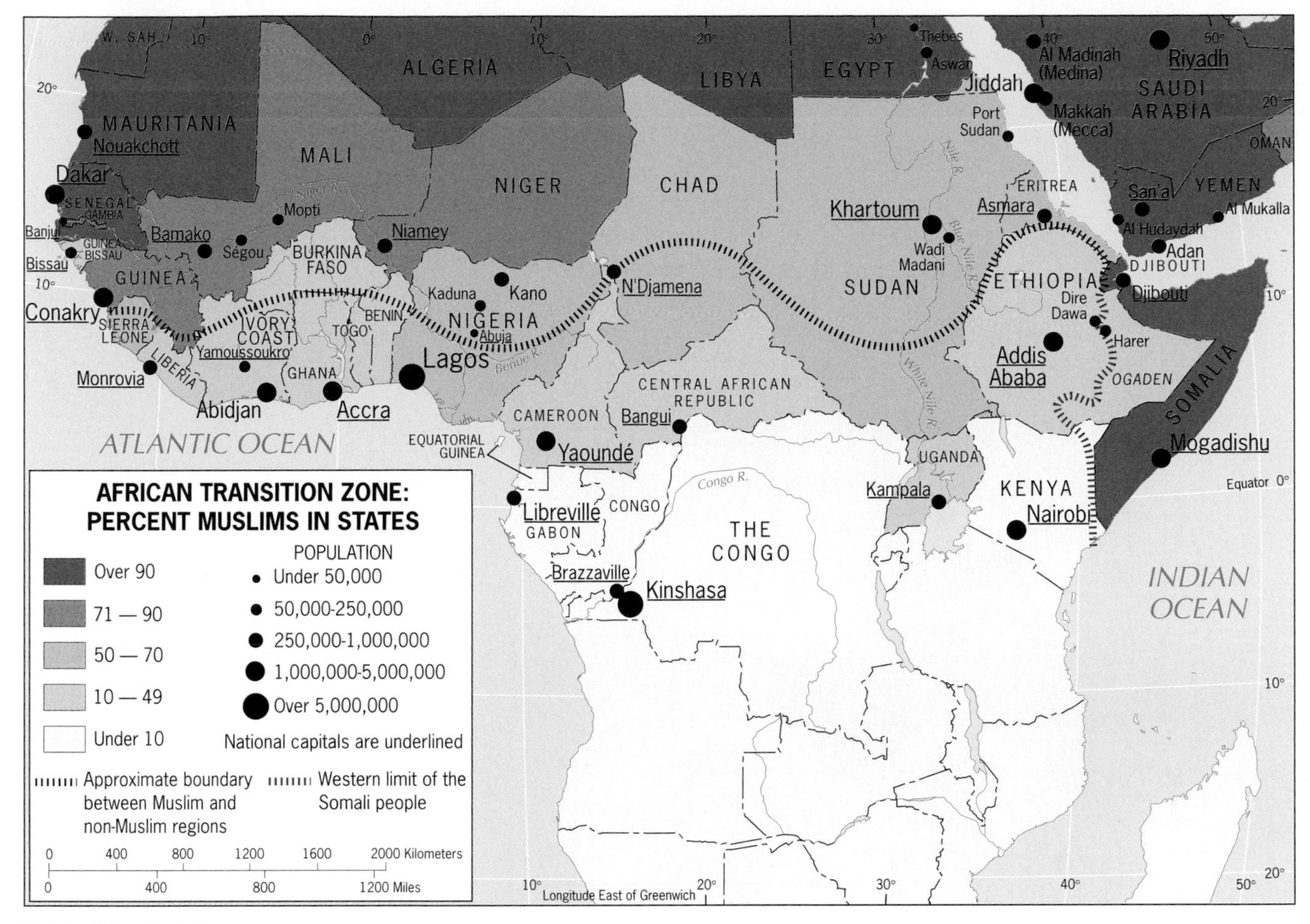

FIGURE 6-12

We should note that the figures on the map are estimates only. The number of Muslim and non-Muslim adherents in the African Transition Zone cannot be known with any certainty, but researching this issue suggests a trend: as the years pass, the estimated number of Muslim believers increases. Some observers refer to the African Transition Zone as Africa's **Muslim Front**, a zone where Islam continues to expand, a religious frontier affecting two dozen African countries from Guinea in the west to Ethiopia in the east.

17

THE MIDDLE EAST

The regional term *Middle East*, we noted earlier, is not satisfactory, but it is so common and generally used that avoiding it creates more problems than it solves. It originated when Europe was the world's dominant realm and when places were "near," "middle," and "far" from Europe: hence a Near East (Turkey), a Far East (China, Japan, Korea, and other countries of East Asia), and a Middle East (Egypt, Arabia, Iraq). If you check definitions used in the past, you will see that the terms were applied inconsistently: Syria, Lebanon, Palestine, even Jordan sometimes were included in the "Near" East, and Persia and Afghanistan in the "Middle" East.

Today, the geographic designation *Middle East* has a more specific meaning. And at least half of it has merit: this region, more than any other, lies at the middle of the vast Islamic realm (Fig. 6-9). To the north and east of it, respectively, lie Turkey and Iran, with Muslim Turkestan beyond the latter. To the south lies the Arabian Peninsula. And to the west lie the Mediterranean Sea and Egypt, and the rest of North Africa. This, then, is the pivotal region of the realm, the very heart of it.

Five countries form the Middle East (Fig. 6-13): Iraq, largest in population and territorial size, facing the Persian Gulf; Syria, next in both categories and fronting the Mediterranean; Jordan, linked by the narrow Gulf of Aqaba to the Red Sea; Lebanon, whose survival as a unified state has come into question; and Israel, Jewish nation in the crucible of the Muslim world.

Iraq

California-sized Iraq (24.4 million) comprises nearly 60 percent of the total area of the Middle East and has 42 per-

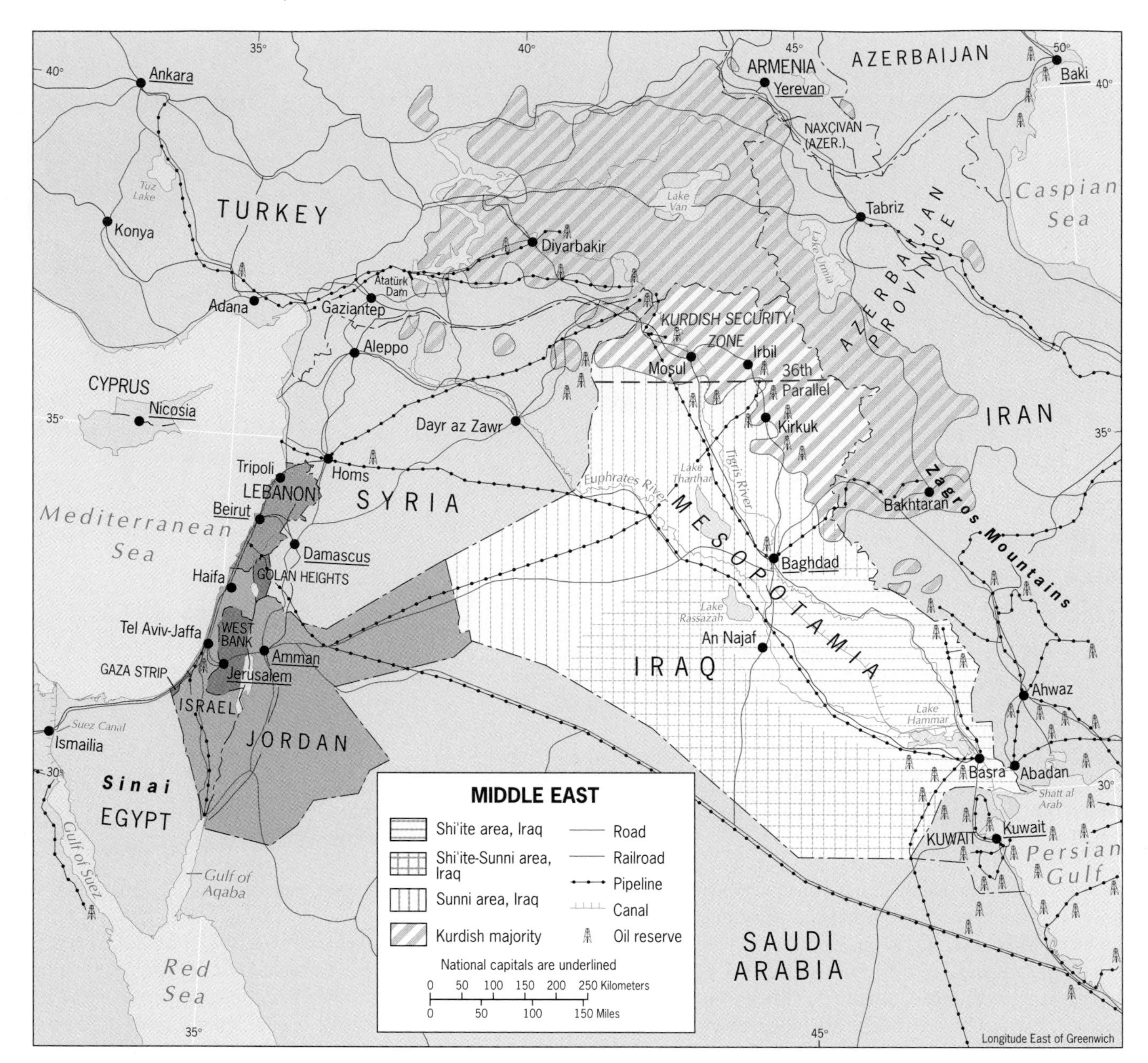

FIGURE 6-13

cent of the region's population. With major oil reserves and large areas of irrigated farmland, Iraq also is best endowed with natural resources. It is heir to the early Mesopotamian states and empires that emerged in the basin of the Tigris and Euphrates rivers, and the country is studded with archeological sites.

Today, Iraq is bounded by six neighbors and has recently had adversarial relationships with most of them. To the north lies Turkey, source of both of Iraq's vital rivers. To the east lies Iran, target of a decade of destructive war during the 1980s. At the head of the Persian Gulf is Kuwait, invaded by Iraq's armies in 1990. To the south lies Saudi Arabia, ally of Iraq's adversaries. And to the west, Iraq is adjoined by Jordan and Syria. Jordan afforded Iraq an outlet to the sea via the port of Aqaba during the 1991 Gulf War in defiance of UN sanctions, but relations have since soured. Syria lies functionally most remote from Iraq despite its long desert border with its embattled neighbor. Note, however, that the Euphrates River traverses Syria before entering Iraq. Water is a potent source of tension in this arid zone.

The regional geography of Iraq reveals the divisions within the country (Fig. 6-13). The heart of Iraq is the area around the capital, Baghdad, on the Tigris River amid the productive farmlands of the Tigris-Euphrates Plain. Here most of the people are Sunni Muslims, who dominate the core area and the country's political machine.

As many as 10 million of Iraq's 24 million citizens, however, are not Sunnis but Shi'ites, concentrated in the populous south. Here lie some of Shiah's holiest places, and here the Ayatollah Khomeini, escaping persecution by the Shah of Iran, was first given refuge before moving to Paris and eventually returning to Tehran. But Iraq's Shi'ites do not have proportional representation in the country's government, and the 1991 Gulf War spelled disaster for them and their provinces. Fearing a Shi'ite rebellion after the Iraqi army was ousted from Kuwait, Baghdad suppressed a Shi'ite uprising in which towns, villages, and religious shrines were heavily damaged and untold casualties inflicted.

A close look at Figure 6-13 reveals the importance of this southern zone of Iraq. The Tigris and Euphrates join to become the Shatt al Arab, Iraq's water outlet to the Persian Gulf. Over its last 50 miles (80 km) or so, the Shatt al Arab waterway also becomes the boundary between Iraq and Iran. Iraq's territorial claim to land on the Iranian side of the waterway precipitated the long war between the two countries that began in 1980.

Also note the situation of Iraq's most important southern city, Basra, and the position of the boundary with Kuwait. In this area lie several major oilfields; one, the Rumailah oilfield, extends from beneath Iraqi soil into Kuwait. In 1990, Iraq claimed that Kuwait was draining oil from this field by drilling slanted wells under Iraqi territory; that the boundary between the two countries was not agreed upon; and that Kuwait was failing to adhere to OPEC rules on oil production and pricing. The Shatt al Arab waterway still was filled with the wreckage of the just-ended Iran-Iraq War, so that the annexation of Kuwait would also give Iraq a new outlet to the Persian Gulf. This combination of justifications and potential returns persuaded Iraq to embark on its disastrous invasion of Kuwait in mid-1990, which triggered the Gulf War early the following year.

The core area, centered on Baghdad, and the Shi'ite south are two of Iraq's major subregions; a third subregion lies in the north. Here most of the people are Sunni Muslims, but they are not Arabs: this is the land of the Kurds. Fewer than 4 million Kurds live in the mainly mountainous areas of northern Iraq, and they constitute perhaps 15 percent of the total Kurdish population in the realm. The Kurds are minorities in all the countries they inhabit, and certain governments like to undercount their minorities. Geographers estimate that there may be as many as 25 million Kurds, with the largest number in Turkey, the next largest minority in Iran, then in Iraq, Syria, and small clusters even in Armenia and Azerbaijan (Fig. 6-13).

As the map shows, Iraq's Kurds occupy a sensitive area of the country: the huge oil reserves on which the city of Kirkuk is situated lie on the margins of the Kurdish domain. During the Iran-Iraq War and again during the Gulf War, the Kurds rose against their Iraqi rulers and briefly succeeded in controlling parts of their homeland, but Iraq's response always ended such efforts, often with great loss of life. In the late 1980s, Baghdad even used cyanide and mustard gas against Kurdish villagers. At the end of the Gulf War, the United Nations established a security zone between the 36th parallel and Iraq's northern border to encourage refugees who had crossed into Turkey and Iran to return. However, as Figure 6-13 shows, much of Iraq's Kurdish domain lies outside this safety zone. Once again, the Kurds are what they have always been: victims of more powerful neighbors (see box titled "A Future Kurdistan?").

A Future Kurdistan?

Political maps of the region do not show it, but where Turkey, Iraq, and Iran meet, the cultural landscape is not Turkish, Iraqi, or Iranian. Here live the Kurds, a fractious and fragmented nation of at least 25 million (their numbers are uncertain). More Kurds live in Turkey than in any other country (perhaps as many as 14 million); possibly as many as 8 million in Iran; about half that number in Iraq; and smaller numbers in Syria, Armenia, and even Azerbaijan (see Fig. 6-13).

The Kurds have occupied this isolated, mountainous, frontier zone for over 3000 years. They are a nation, but they have no state; nor do they enjoy the international attention that peoples of other **stateless nations** (such as the Palestinians) receive. Turkish and Iraqi repression of the Kurds, and Iranian betrayal of their aspirations, briefly make the news but are soon forgotten. Relative location has much to do with this: its remoteness and the obstacles created by the ruling regimes inhibit access to their landlocked domain.

18

Many Kurds dream of a day when their fractured homeland will be a nation-state. Most would agree that the city of Diyarbakir, now in southeastern Turkey, would become the capital. It is the closest any Kurdish town comes to a primate city, although the largest urban concentration of Kurds today is in the shantytowns of Istanbul, where more than 3 million have migrated. In their heartland, meanwhile, the Kurds, their shared goals notwithstanding, are a divided people whose intense disunity has thwarted their objectives; and as in the case of Europe's Basques, a small minority of extremists has tended to use violence in pursuit of their aims.

The Kurds—without oil reserves, without a seacoast, without a powerful patron, without a global public-relations machine—are victims of a conspiracy between history and geography. One of the world's largest stateless nations hopes for a deliverance that is unlikely to come.

Iraq's infrastructure and economy were shattered during the Gulf War, but Iraq had already wasted much of its potential on the earlier conflict with Iran—and on mismanagement, corruption, and inefficiency. With its good agricultural land and its enormous oil income, Iraq should be one of the economic success stories of the entire realm. Instead, its failed leadership has made it one of the world's tragedies.

Syria

If one issue can galvanize virtually the entire realm (other than Islam and oil), it is Israel. During the Gulf War, Iraq sought to draw Israel into the conflict by aiming missiles at it. Getting Israel into the fray would have been easier if Iraq had what Syria does: a common border with the Jewish state. Iraq, which once tried to build nuclear weapons to threaten Israel, is the only Middle Eastern country without territorial contact with the Israelis.

Syria has historic proof of its proximity: it lost a piece of territory, the Golan Heights, to Israel in 1967. More than 30 years later, negotiators were still trying to find a way to arrange the return of all or part of this area to Syria, but there are many problems. Jewish settlers have occupied parts of the Golan Heights; opposition to returning the Golan is strong in Israel, a democratic society where voters can oust representatives whose policies they do not approve. On the Syrian side is the problem of continuity. Syria is not a democracy; since 1963 it has been a republic under a military regime. Moreover, while Syria's population of 17.4 million is about 75 percent Sunni Muslim, the ruling elite comes from a smaller Islamic sect based there, the Alawites. Leaders of this powerful minority have retained control over the country for decades, at times by ruthless suppression of dissent. In 2000, president-for-life Hafez al-Assad died and was succeeded by his son Bashar, signaling a continuation of the political status quo.

Like Lebanon and Israel, Syria has a Mediterranean coastline where crops can be raised without irrigation. Behind this densely populated coastal zone, Syria has a much larger interior than its neighbors, but its areas of productive capacity are widely dispersed. Damascus, in the southwest corner of the country, was built on an oasis and is considered to be the world's oldest continuously inhabited city. It is now the capital of Syria, with a population of 2.5 million.

The far northwest is anchored by Aleppo, the focus of cotton- and wheat-growing areas in the shadow of the Turkish border. Here the Orontes River is the chief source of irrigation water, but in the eastern part of the country the Euphrates Valley is the crucial lifeline. It is in Syria's interest to develop its eastern provinces, and recent discoveries of oil there will speed that process. Syria is self-sufficient in staple grains and earns substantial revenues from its cotton exports, but most of its modest income comes from oil. Syria is also a country where agriculture still has room to expand. This expansion would improve the cohesion of the state and bring its separate subregions into a tighter spatial framework. But what Syria needs most is modernization; in a globalizing world, it remains one of the most stagnant and isolated societies.

Jordan

None of this can be said for Jordan (5.4 million), the desert kingdom that lies east of Israel and south of Syria. It, too, was a product of the Ottoman collapse, but more than any other Arab state, it suffered when Israel was created. First, Jordan's trade used to flow through Haifa, now an Israeli port. Today Jordan has to depend on destabilized Lebanon's harbors or the tedious route via the Gulf of Aqaba in the far south. Second, when Jordan became independent in 1946 its population was only about 400,000, including nomads, peasants, villagers, and a modest percentage of urban dwellers. Then, with the partition of Palestine and the creation of Israel in 1948, Jordan received more than half a million Arab refugees. Soon it also became responsible for another half a million Palestinians who, though living on the western side of the Jordan River (now known as the *West Bank*), were incorporated into the state. Thus refugees outnumbered residents by more than two to one, and internal political problems were added to external ones—not to mention the economic difficulties of beginning national life as a poor country.

Jordan has survived with U.S., British, and other aid, but its problems have hardly diminished. Many Jordanian residents are still only minimally committed to the country, do not consider themselves its citizens, and give little support to the hard-pressed monarchy. Dissatisfied groups constantly threaten to drag the country into another conflict with Israel. The 1967 war was disastrous for Jordan, which lost both the West Bank (its claim was formally surrendered to the Palestine Liberation Organization in 1988) and its sector of Jerusalem (the kingdom's second largest city). Where hope for progress might lie—for example, in the development of the Jordan River Valley—political conflicts intrude. The capital city, Amman, reflects the limitations and poverty of the country. Without oil, without much farmland, without unity or strength, and overwhelmed with refugees, Jordan presents one of the bleaker economic pictures in the Middle East. Its survival, and the disproportionately large role Jordan has played in the political life of the region, may be attributed substantially to the 46-year reign of its King Hussein. That era of leadership—and the strong centripetal force it represented—came to an end with the king's death in 1999, and the country now faces greater challenges than ever before under the leadership of his son, King Abdullah II.

Lebanon

Lebanon, Israel's northern coastal neighbor on the Mediterranean Sea, is one exception to the rule that the Middle East is the world of Islam: one-fourth of the population of 4.3 million are Christian. Less than one-eighth the territorial size of Jordan (and only half the size of Israel), Lebanon has a long history of trade and commerce, beginning with the ancient Phoenicians who were based here more than a thousand years before the birth of Christ. Lebanon must import much of its staple food, wheat. The coastal belt below the mountains, though intensively cultivated, normally cannot produce enough grain to feed Lebanon's population. In the interior, the mountain-flanked, fertile Bekaa Valley, watered by the Litani River, produces a wide range of fruits and vegetables.

Lebanon fell apart in 1975 when a civil war broke out between Muslims and Christians. This was a conflict with many causes. Lebanon for several decades had a political system created by the French colonialists in the 1920s that divided power between these two leading communities, but the basis for that system had become outdated. In the 1930s, Muslims and Christians in Lebanon were at approximate parity; but since then, the Muslims have increased at a much faster rate than the more urbanized and generally wealthier Christians (many of whom have emigrated from Lebanon in recent decades). The Muslims' displeasure with this outdated political arrangement was expressed in several rebellions before the full-scale civil war. By then, Lebanon had also become a base for over 300,000 Palestinian refugees. These people, many of them living in squalid camps, were never satisfied with Lebanon's moderate posture toward Israel. When the first fighting between Muslims and Christians broke out in the northern coastal city of Tripoli, the Palestinians joined the conflict on the Muslim side. In the process, Lebanon was wrecked.

Beirut, the capital, once a city of great architectural beauty and often described as the Paris of the Middle East, was heavily damaged as Christian, Sunni Muslim, Shi'ite Muslim, Druze (an Islamic sect), and Palestinian militias—as well as the Lebanese and Syrian armies—fought for control. By the end of the 1980s, Beirut had been almost destroyed, and only the poorest 150,000 war-ravaged residents remained of the 1.5 million who had lived there in the late 1970s. Since 1990, the conflict has abated. Beirut is being rebuilt, but it remains a divided city. East Beirut is a Christian enclave, whereas West Beirut is almost exclusively Muslim. Even so, the city is reviving and now contains 2.1 million people.

As the Muslims' strength intensified in Lebanon, the Christians concentrated in an area along the coast between Beirut and Tripoli (Fig. 6-13). In the meantime, Israel had occupied a security zone in the south, and various competing Muslim factions controlled other parts of the country. In 1976, Syria sent its army to try to pacify Lebanon, but the Syrians, too, eventually fell victim to Lebanon's fractious character. By backing various client factions, Syria helped reduce the violence outside Beirut but then found itself inextricably enmeshed in Lebanon's troubles. Meanwhile, other countries, notably Iran, supported anti-Israeli factions in Lebanon, complicating matters still more.

In 2000, the Israeli government decided to abandon the security zone it had maintained in southern Lebanon, raising hopes that life in the border area (now under UN supervision) could return to normal.

Israel

Israel lies at the heart of the Arab World (Fig. 6-9). Its neighbors are Lebanon and Syria to the north and northeast, Jordan to the east, and Egypt to the southwest—all of whom still resent the creation of the Jewish state in their midst. Since 1948, when Israel was created as a homeland for the Jewish people on the recommendation of a United Nations commission, the Arab-Israeli conflict has overshadowed all else in the Middle East.

Indirectly, Israel was the product of the collapse of the Ottoman Empire. Britain gained control over Palestine, and British policy supported the aspirations of European Jews for a homeland in the Middle East, embodied in the concept of Zionism—a return to Zion, ancient Israel. In 1946, the British granted independence to the territory lying east of the Jordan River, and "Transjordan" (now the state of Jordan) came into being. Shortly afterward, the territory west of the Jordan River was partitioned by the United Nations, and the Jewish people got slightly more than half of it—including, of course, land that had long been occupied by Arabs. Jews actually owned only about 8 percent of Palestine's land, but they made up more than one-third of its population.

As soon as the Jewish people declared the independent state of Israel on May 14, 1948, the new country was attacked by its Arab neighbors, who rejected Israel's right to exist. In the ensuing battle, Israel not only held its own but also gained crucial territory in central and northern Palestine and in the Negev Desert to the south (Fig. 6-14). At the end of this first Arab-Israeli War, in 1949, the Jewish population controlled 80 percent of what had been Palestine west of the Jordan River.

As Figure 6-14 shows, Israel now owned not only the territory allocated to it by the United Nations but additional areas facing Egypt, Lebanon, and the West Bank (the area west of the Jordan River shown on the map). During the conflict, troops from Transjordan (the independent Arab kingdom east of the Jordan River) invaded this West Bank area, and in 1950 the king formally annexed it to his country, which he renamed Jordan.

This early conflict was only the first in a series of wars between Israel and its Arab neighbors. In 1967, a week-long military conflict resulted in a major Israeli victory: Israel took the Golan Heights from Syria, the West Bank from Jordan, and the Sinai Peninsula (up to the Suez Canal itself) from Egypt. In 1973, another brief war led to Israel's withdrawal from the Suez Canal to truce lines in the Sinai Peninsula. In

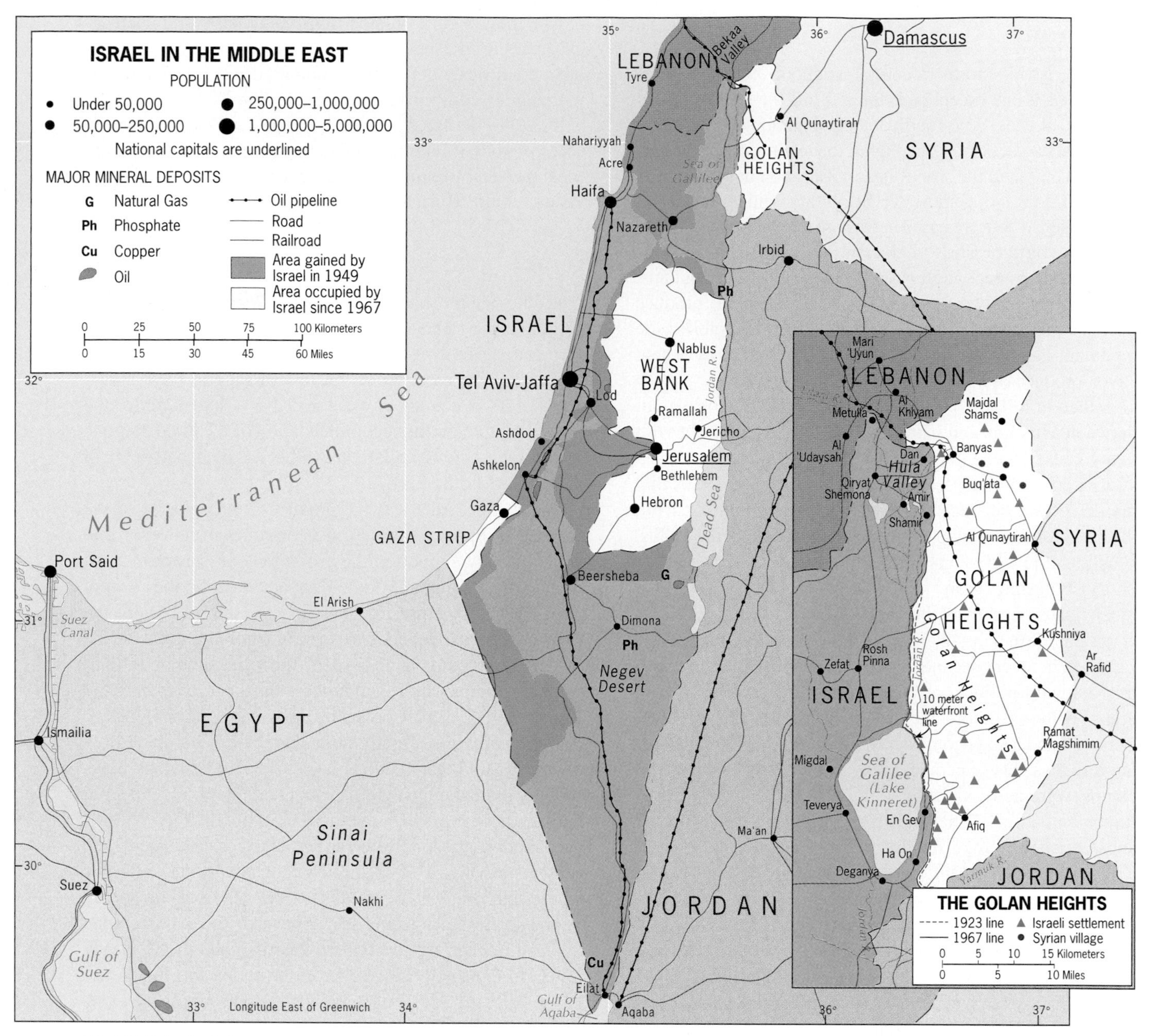

FIGURE 6-14

1978, Israel made peace under U.S. auspices, and Israel later returned all of the conquered Sinai Peninsula to Egypt.

Since then, a fragile peace has been sustained, and in the 1990s some progress was made on key issues involving the relationships between Israel and its Arab neighbors and between Israel and the leadership of the Palestinians, who have been under Israeli control in the Gaza Strip and in the West Bank. Agreements normalizing relationships with Egypt and Jordan were major steps on this difficult road, and the establishment of the Palestinian Authority (PA) to govern Gaza as well as several West Bank cities was another achievement. Many obstacles still stand in the way, however, including the following:

1. *The Golan Heights.* The return of the Golan Heights to Syria may be a precondition for normalizing relations with Syria, but the political climate in Israel itself may make ceding the Golan impossible.
2. *The West Bank.* Even after its capture by Israel in 1967, the West Bank might have become a Palestinian homeland (and possibly a state), but Jewish immigration to the area made such a future difficult. In 1977 only 5000 Jews lived on the West Bank; by 2000 there were almost 200,000, making up about 10 percent of the population and creating a seemingly inextricable jigsaw of Jewish and Arab settlements (Fig. 6-15).

FIGURE 6-15

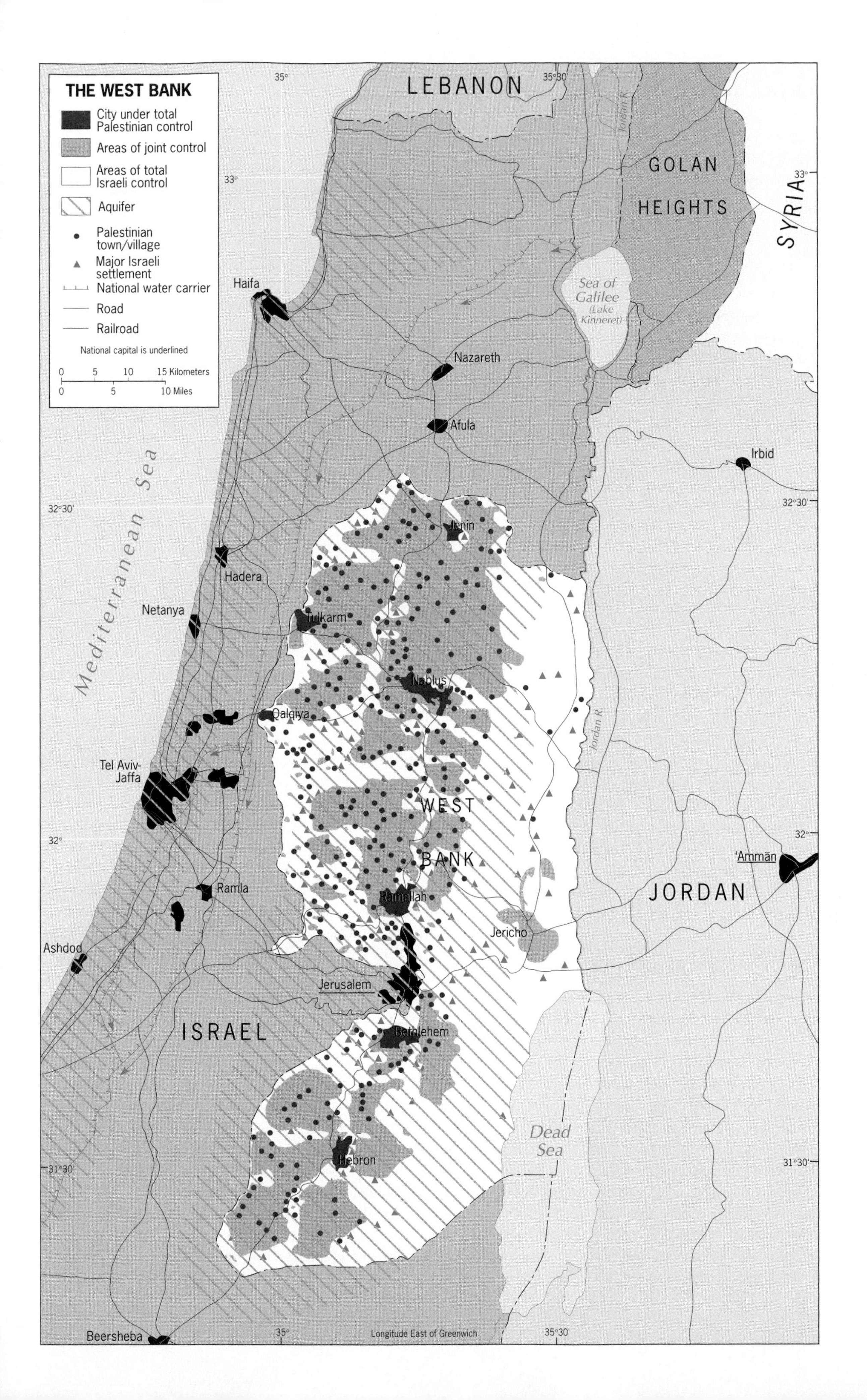

THE WEST BANK
City under total Palestinian control
Areas of joint control
Areas of total Israeli control
Aquifer
Palestinian town/village
Major Israeli settlement
National water carrier
Road
Railroad
National capital is underlined
0 5 10 15 Kilometers
0 5 10 Miles
LEBANON
GOLAN HEIGHTS
SYRIA
Sea of Galilee (Lake Kinneret)
Jordan R.
Haifa
Nazareth
Afula
Irbid
Mediterranean Sea
Jenin
Hadera
Netanya
Tulkarm
Nablus
Qalqiya
Tel Aviv-Jaffa
WEST BANK
'Ammān
JORDAN
Ramla
Ramallah
Jericho
Ashdod
Jerusalem
ISRAEL
Bethlehem
Dead Sea
Hebron
Beersheba
35°
35°30'
33°
32°30'
32°
31°30'
Longitude East of Greenwich

The Palestinian Dilemma

Ever since the creation of Israel in 1948 in what had been the British Mandate of Palestine, Arabs who had called Palestine their homeland for centuries have lived as refugees in neighboring countries. Many have been assimilated into the societies of Israel's neighbors, but others still live in refugee camps. Israel is now more than a half-century old, and most Palestinians were born after the partition of Palestine in 1948.

The Palestinians call themselves a nation without a state (much as the Jews did before Israel was founded), although they and their descendants make up the majority of Jordan's population today. They demand that their grievances be heard and that a Palestinian state be created. The first steps toward such a state have been taken as part of the Arab-Israeli peace process, and negotiations continue.

Current estimates of Palestinian populations in the realm are as follows:

Israel and Occupied Territories		4,715,000
Israel	1,240,000	
West Bank	2,270,000	
Gaza Strip	1,205,000	
Jordan		2,540,000
Lebanon		500,000
Syria		443,000
Saudi Arabia		334,000
Iraq		87,000
Egypt		72,000
Kuwait		35,000
Libya		31,000
Other Arab States		570,000
TOTAL		9,327,000

3. *The Palestinians.* Although Palestinian issues tend to focus on Gaza and the West Bank, and on Arabs who call themselves Palestinians in countries that neighbor Israel (principally Jordan), more than 1 million Palestinian Arabs live in Israel proper (see box titled "The Palestinian Dilemma").
4. *Jerusalem.* Holy city for Jews, Christians, and Muslims, Jerusalem was to become an international city under the UN blueprint for Palestine. It was captured along with the West Bank by Israel in the 1967 war, and Israel now considers Jerusalem its capital despite the Arabs' objections. This city is so important for the future of the region that we focus on it in the following section.

A Challenge for the Twenty-First Century

The Jewish faith was founded by Abraham and was adopted in the Sinai area during the exodus from Egypt to the "promised land" between the Jordan River and the Mediterranean Sea, then inhabited by the Philistines—the name from which *Palestine* is probably derived. Here the Jews founded a state, and Jerusalem was its capital. Inhabited for nearly 40 centuries, Jerusalem (Yerushalayim) was the site of the First Temple, where the faith took form. After King Solomon's time, the kingdom split up and was conquered in turn by the Assyrians, Babylonians, Persians, and Greeks. The First Temple was destroyed by the Babylonians (586 BC), but it was rebuilt as the Second Temple during Persian rule. A Jewish state was reborn during a revolt against Greek rule, but the Romans overpowered it. Under Roman rule, Christianity, a messianic religion based on the teachings of Jesus Christ, arose in the Jewish heartland. When the Jews rebelled against the Romans, the Roman armies destroyed the Second Temple in AD 70, leaving only a part of its western retaining wall standing. Today, this is the most revered site in Jewish life, drawing an unending stream of the faithful to bemoan the loss of the Temple.

Under Roman domination, Jerusalem and Palestine became a focus for Christian pilgrimage and conversion, and Christian churches arose in the city. In the seventh century AD, the expansion of Islam from Mecca and Arabia endowed Jerusalem, which they called Al-Quds, the holy, with still another spiritual dimension. Despite repeated destructions, Jerusalem today remains a city of holy places: the Western Wall; the Church of the Holy Sepulchre, where Jesus was buried; and the El Aqsa Mosque and the Mosque of Omar (the Dome of the Rock) on the Temple Mount, from which the Prophet Muhammad is believed to have ascended to heaven.

When the United Nations delimited the modern state of Israel in 1947, Jerusalem posed a formidable challenge. Look again at Figure 6-14: the UN delimitation of Israel (colored tan) did not include Jerusalem. By Palestine Partition Resolution (No. 181) that created Israel, Jerusalem was designated an international city, open to Jews, Muslims, and Christians alike. As Figure 6-14 shows, the ensuing war enlarged the Jewish state far beyond its UN dimensions, pushing back the borders of the Arab-controlled West Bank and creating a corridor linking Jerusalem to Israeli territory.

The United Nations intended Tel Aviv to be Israel's capital, and so it was, but Israel's territorial gains during the 1948–1949 war allowed it to transfer its government to Jerusalem. In 1950, despite vehement Arab opposition, Israel declared Jerusalem to be its capital, making this, in effect, a *forward capital* at the apex of a West Bank wedge.

In truth, however, Israel did not conquer all of Jerusalem in 1948–1949. As Figure 6-16 shows, the cease-fire line that eventually became the recognized West Bank boundary crossed through the heart of the city. The Old City, with its holy sites including the Western Wall, remained in Jordanian hands, as did East Jerusalem. The New City to the west fell to the Israelis, who set about modernizing that sector and populating it with Jews. However, Jews could not pass through the Old City to the Western Wall, so that they were separated from their most important pilgrimage site.

The 1967 war and its aftermath changed the situation once more. Israel now conquered the West Bank, including all of Jerusalem, and the Old City was again connected to the New. Soon afterward, Israel extended the municipal boundaries of Jerusalem far to the north, south, and east of what had been East Jerusalem (Fig. 6-16). This involved annexing a substantial chunk of West Bank territory, complicating any future solution of the West Bank question.

Until then, West Jerusalem was predominantly Jewish, and East Jerusalem mainly Arab. The United Nations and the international community continued to hope that the existing division of Jerusalem might yet, in time, form the basis for a permanent settlement in which East Jerusalem would be part of a Palestinian state, West (New) Jerusalem would remain part of Israel, and the Old City would be internationalized. But Israel had other plans. In 1980, it reaffirmed that Jerusalem was its capital and called on all nations that had diplomatic relations with Israel to move their embassies from Tel Aviv (where most still were—and are) to the holy city. Meanwhile, Israel began to construct Jewish housing in the zone of municipal expansion. This housing program aimed to create a ring of Jewish neighborhoods around the old East Jerusalem (Fig. 6-16) that would end the distinction between an Arab east and a Jewish west.

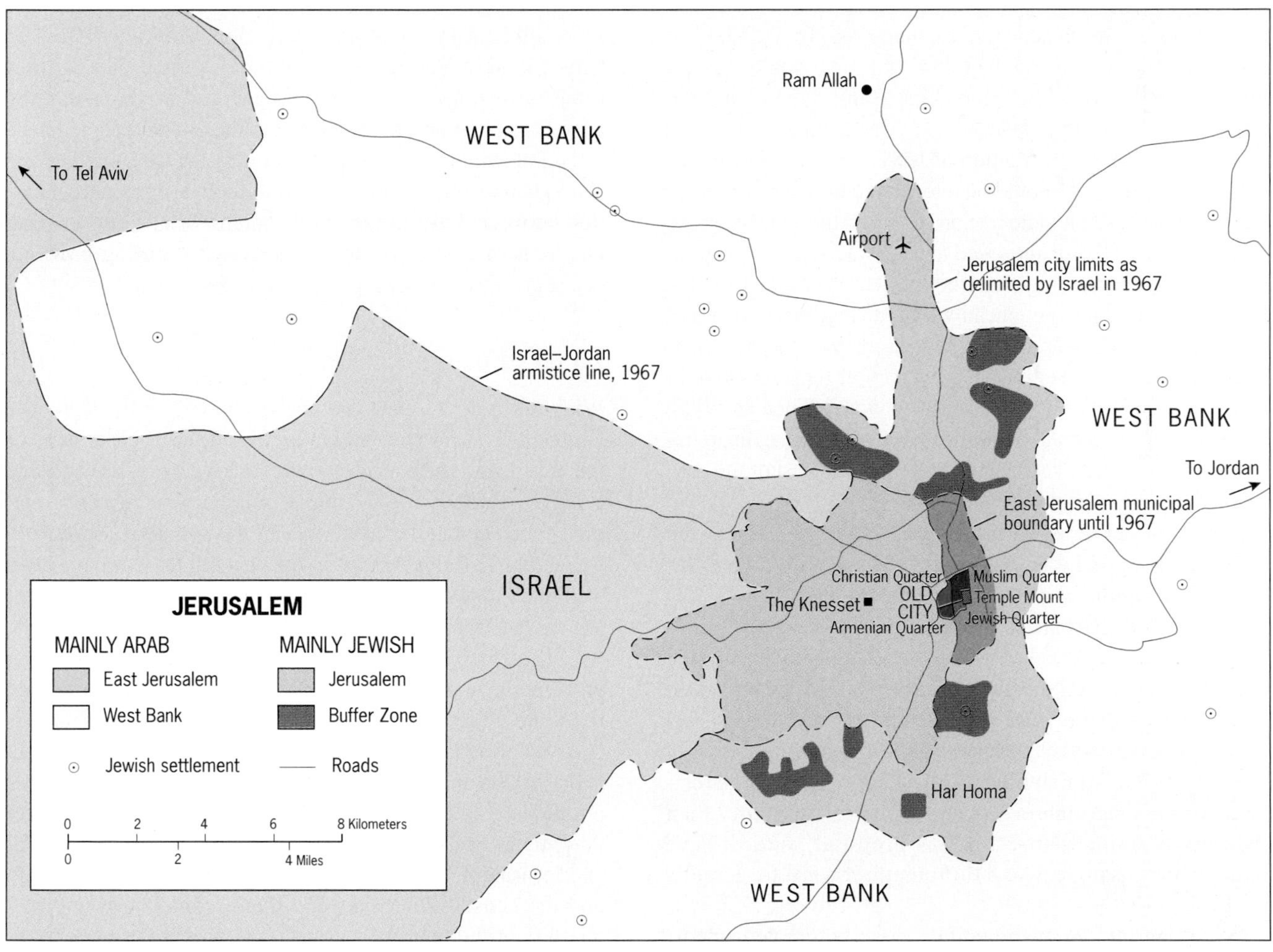

FIGURE 6-16

FROM THE FIELD NOTES

"Cultural contrast pervades Jerusalem's urban landscape; social tension is palpable. From almost any high vantage point you can see the places of worship and the holy ground that means so much to Jew, Christian, and Muslim: synagogues, churches, and mosques, walled cemeteries, sacred shrines, historic sites, all juxtaposed as in this panorama. In the distance, in another direction, you can see the Jordan Valley; Jerusalem is where the West Bank meets Israel. When the Israeli government decided to make Jerusalem its political headquarters, it was a move designed to proclaim the nation's dominance here: Jerusalem would not become an international city. This makes Jerusalem another example of a forward capital, the vanguard of the state in potentially contested territory."

Furthermore, the presence of as many as 200,000 Jewish residents in areas beyond the Old City limits would create a buffer zone between the Arab West Bank beyond and the Arab parts of the city inside.

Until 1993, the construction of Jewish housing in the area shown in Figure 6-16 could be interpreted as a legitimate design to consolidate a victory won in war. But the Oslo Peace Accords of 1993 were supposed to end Israeli land expropriation and further housing construction. However, in 1997 the Israeli government went ahead with plans to begin construction on the Har Homa project in Jerusalem's southeast corner. This enraged the Palestinians, whose chief representative in Jerusalem called it a "declaration of war." Then in 1998, Jerusalem's municipal boundaries were enlarged again, to incorporate several Jewish communities to the west of the city, further altering its ethnic balance.

Any comprehensive settlement between Israel, the Palestinians, and Israel's Arab neighbors will have to include Jerusalem. In mid-2000, then-President Bill Clinton convened the parties at Camp David near Washington, D.C. to seek a solution to the dilemma, but the negotiations failed. Shortly thereafter, Clinton announced his support for the relocation of the United States embassy from Tel Aviv to Jerusalem, a move long deferred because of Palestinian objections. According to the president, Israeli negotiators had shown more flexibility than their Arab counterparts, and this proposal, sure to have major repercussions, was a diplomatic reward for Israel's posture.

What are the options remaining for a future Jerusalem? One is the internationalization of the entire city under UN auspices, a prospect likely to be unacceptable to Arabs and Jews alike. Another is the internationalization of the Old City, a kind of Vaticanization of the historic center. A third is the expansion of Jerusalem along its eastern flank, thereby annexing a part of the West Bank to the city and designating it the Palestinian capital. But events in the aftermath of the Camp David negotiations overtook such designs. Street battles between Palestinians and Israelis broke out, costing hundreds of casualties and immeasurably setting back the peace process—and the prospects for Jerusalem.

Israel's Future

With the help of massive foreign aid (mainly from the United States), large remittances from Jews in other parts of the world, and the energies of its settlers, Israel has become a high-income society amidst comparative poverty. (Israel's per-capita GNP is almost 15 times that of Jordan and more than 12 times that of Egypt.) Irrigated areas have been expanded, dryland cultivation has succeeded with technological advances that are now disseminated around the world, oil is bought from Egypt under the normalization agreement, an iron and steel plant at Tel Aviv is maintained for strategic reasons, and the country has become a world-class technopole built on a major pool of scientific talent and a base of more than 2000 high-tech companies. Israel's population of 6.4 million is highly urbanized, with just over 90 percent living in such cities as Tel Aviv, Jaffa, Haifa, and Jerusalem. The core area, including Tel Aviv and Haifa and the coastal zone between them, contains over three-quarters of the country's population, by far the most modernized cluster in any non-oil-producing country anywhere in the realm.

As the peace process continued, new obstacles constantly arose. One involved water. Over 30 percent of Israel's water supply comes from aquifers that lie under the West Bank. Of this, about one-fifth is allocated to the 2.3 million Palestinians living on the West Bank, and a smaller quantity goes to the nearly 200,000 Jewish settlers. The rest is consumed in Israel proper, mainly Tel Aviv and Jerusalem. When the next stage in the transfer of land from Israeli to PA control occurs, some of the most important pumping stations will come under Palestinian jurisdiction. The Israelis do not want to change existing patterns and systems of water use, and they want the Palestinians to promise that no new wells will be drilled that would divert water from critical aquifers. Modern societies like Israel consume large amounts of water; Israelis on average use four times as much water as Palestinians. The Palestinians argue that they should not have to support this huge disparity. Israelis insist that without a separate agreement on water use, they will not favor further autonomy for Palestinian areas on the West Bank.

But Israel's future depends on reaching a satisfactory settlement with its Palestinian neighbors, as well as on normalizing relations with moderate and secular Arab states. In the heart of the realm, Israel is at the focus of an animosity that has a 13-century history. Now, in the age of nuclear, chemical, and biological weapons and long-range missiles, Israel's search for an accommodation with its neighbors is a race against time.

THE ARABIAN PENINSULA

The regional identity of the Arabian Peninsula is clear: south of Jordan and Iraq, the entire peninsula is encircled by water. This is a region of old-style emirates and sheikdoms made wealthy by oil; it also contains the site where Islam originated.

As a region, the Arabian Peninsula is environmentally dominated by a desert habitat and politically dominated by the Kingdom of Saudi Arabia (Fig. 6-17). With 830,000 square miles (2,150,000 sq km), Saudi Arabia is the realm's fourth biggest state; only Kazakhstan, Algeria, and Sudan are larger. On the peninsula, Saudi Arabia's neighbors (moving clockwise from the head of the Persian Gulf) are Kuwait, Bahrain, Qatar, the United Arab Emirates, the Sultanate of Oman, and the Republic of Yemen (created in 1990 through the unification of former North Yemen and South Yemen). Together, these countries on the eastern fringes of the peninsula contain about 27 million inhabitants; the largest by far is Yemen, with 18 million. The interior boundaries of several of these states, however, are still inadequately defined here in one of the world's last remaining frontier-dominated zones.

Saudi Arabia

Saudi Arabia itself has only 22.9 million inhabitants in its vast territory, but we can see the kingdom's importance in Figure 6-8: the Arabian Peninsula contains the Earth's largest concentration of known petroleum reserves. Saudi Arabia occupies most of this area and by some estimates may possess as much as one-quarter of the world's oil deposits. As Figure 6-17 shows, these reserves lie in the eastern part of the country, particularly along the Persian Gulf coast and in the Rub al Khali (Empty Quarter) to the south.

The national state that is Saudi Arabia was only consolidated in the 1920s by King Ibn Saud. At the time, it was a mere shadow of its former greatness as the source of Islam and the heart of the Arab World. Apart from some permanent settlements along the coasts and in scattered oases, there was little to stabilize the country; most of it is desert, with annual rainfall under 4 inches (10 cm). The land surface rises generally from east to west, so that the Red Sea is fringed by mountains that reach nearly 10,000 feet (3000 m). Here the rainfall is slightly higher, and there are some farms (coffee is a cash crop). These mountains also contain gold, silver, and other metals, and the Saudis hope to diversify their exports by adding minerals from the west to the oil from the east.

Figure 6-17 reveals that most economic activities in Saudi Arabia are concentrated in a wide belt across the "waist" of the peninsula, from the boomtown of Dhahran on the Persian Gulf through the national capital of Riyadh in the interior to the Mecca–Medina area near the Red Sea. A modern transportation and communications network has recently been completed. But in the more remote zones of the interior, Bedouin nomads still ply their ancient caravan routes across the vast deserts. For decades, Saudi Arabia's royal families were virtually the sole beneficiaries of their country's incredible wealth, and there was hardly any impact on the lives of villagers and nomads. When the oil boom arrived in the 1950s, foreign laborers were brought in (today there are between 4 and 5 million, many of them Shi'ite) to work in the oilfields, ports, factories, and as servants. The east boomed, but the rest of the country lagged behind.

That is now changing, however. Agriculture in particular is receiving major government investments because the Saudis want to increase the country's ability to feed itself in order to prevent the food weapon from being used against them (as they themselves have occasionally wielded the oil weapon). Widespread well drilling has enhanced water supplies to support crops. These achievements have been enormously expensive, for they involve pumping water from aquifers far below the desert surface and constructing extensive center-pivot irrigation systems. Despite these efforts, the Saudis are beginning to realize that their goal of attaining self-sufficiency in food production is elusive. To begin with, the underground water supplies are a one-time-only, nonrenewable resource. Desalinization plants serve

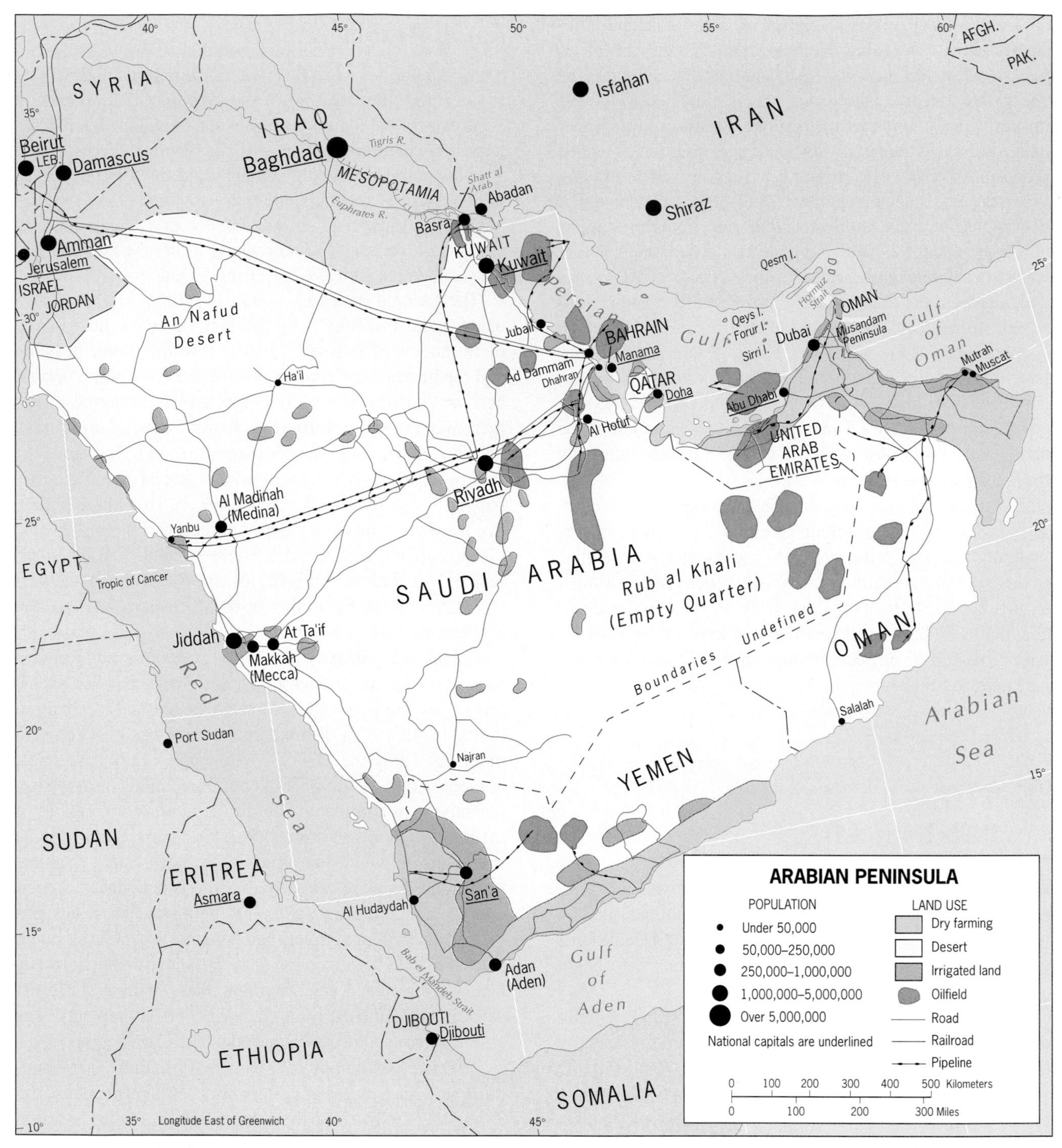

FIGURE 6-17

some areas, but even an oil-rich kingdom cannot afford to raise its crops with desalinized sea water. Even more importantly, the Saudi population is growing at a much faster pace (3.0 percent, thereby doubling in just 23 years) than the rate at which the domestic food supply can be increased.

The country's rulers also have improved housing, health care, and education, and since 1970 they have spent hundreds of billions of dollars on national development programs. The collapse of oil prices in the 1980s and again in the 1990s slowed the progress of these programs, but overall living standards have improved significantly. Industrialization has also been stimulated. The new planned city of Jubail, north of Dhahran on the Persian Gulf, has become an industrial center with state-of-the-art petrochemical and metal fabrication plants. Similarly, Yanbu, north of Jidda on the Red Sea coast, lies at the end of a trans-Arabian oil pipeline, an incipient industrial outpost for the west.

Saudi Arabia's conservative monarchism, official friendship with the West, and social contrasts resulting from its economic growth have raised political opposition, for which no adequate channels exist. Long-term stability may be threatened in this, the region of Islam's birth.

On the Periphery

Five of Saudi Arabia's six neighbors on the Arabian Peninsula face the Persian and Oman gulfs (Fig. 6-17) and are monarchies in the Islamic tradition. All five also derive substantial revenues from oil. Their populations range from 0.6 to 2.9 million in addition to hundreds of thousands of foreign workers; these are not strong or (OPEC aside) influential states. They do, however, display considerable geographic diversity. Kuwait, at the head of the Persian Gulf, almost cuts Iraq off from the open sea, an issue the Gulf War did not settle. Bahrain is an island state, a tiny territory, with dwindling oil reserves. Its approximately 680,000 people are 50 percent Shi'ite and only 35 percent Sunni; nearly two-thirds of its labor force is foreign. Neighboring Qatar consists of a peninsula jutting out into the Persian Gulf, a featureless, sandy wasteland made habitable by oil (now declining in importance) and natural gas (rising simultaneously). The United Arab Emirates (UAE), a federation of seven emirates, faces the Persian Gulf between Qatar and Oman. The reigning sheik is an absolute monarch in each of the emirates, and the seven sheiks together form the Supreme Council of Rulers. In terms of oil revenues, however, there is no equality: two emirates—Abu Dhabi and Dubai—have most of the reserves. But the UAE's economy is unusual in the region because its growth is less dependent on oil as a result of heavy investments in other projects. Two of these projects are Dubai's Port Rashid shipyards and Jebel Ali Free Zone, where hundreds of foreign companies are based.

The eastern corner of the Arabian Peninsula is occupied by the Sultanate of Oman, another absolute monarchy, centered on the capital, Muscat. Figure 6-17 shows that Oman consists of two parts: the large eastern corner of the peninsula and a small but critical cape to the north, the Musandam Peninsula, that protrudes into the Persian Gulf to form a narrow **choke point**—the Hormuz Strait (Iran lies on the opposite shore). Tankers that leave the other Gulf states must negotiate this narrow channel at slow speed, and during politically tense times warships have had to protect them. Iran's claim of several small islands near the Strait that are owned by the UAE is a potential source of dispute. 19

This leaves what is, in many ways, Saudi Arabia's most substantial and potentially most difficult neighbor: Yemen. Our chart (see Table I-2) states that Yemen has an area of 203,900 square miles, but that is an estimate: the boundary between Yemen and Saudi Arabia has never been properly defined. In the early 1990s, following the unification of the two Yemens, the Yemeni government authorized foreign

FROM THE FIELD NOTES

"The port of Mutrah, Oman, like the capital of Muscat nearby, lies wedged between water and rock, the former encroaching by erosion, the latter crumbling as a result of tectonic plate movement. From across the bay one can see how limited Mutrah's living space is, and one of the dangers here is the frequent falling and downhill sliding of large pieces of rock. It took about five hours to walk from Mutrah to Muscat; it was extremely hot under the desert sun but the cultural landscape was fascinating. Oil also drives Oman's economy, but here you do not find the total transformation seen in Kuwait or Dubai. Townscapes (as in Mutrah) retain their Arab-Islamic qualities; modern highways, hotels, and residential areas have been built, but not at the cost of the older and the traditional. Oman's authoritarian government is slowly opening the country to the outside world after long-term isolation."

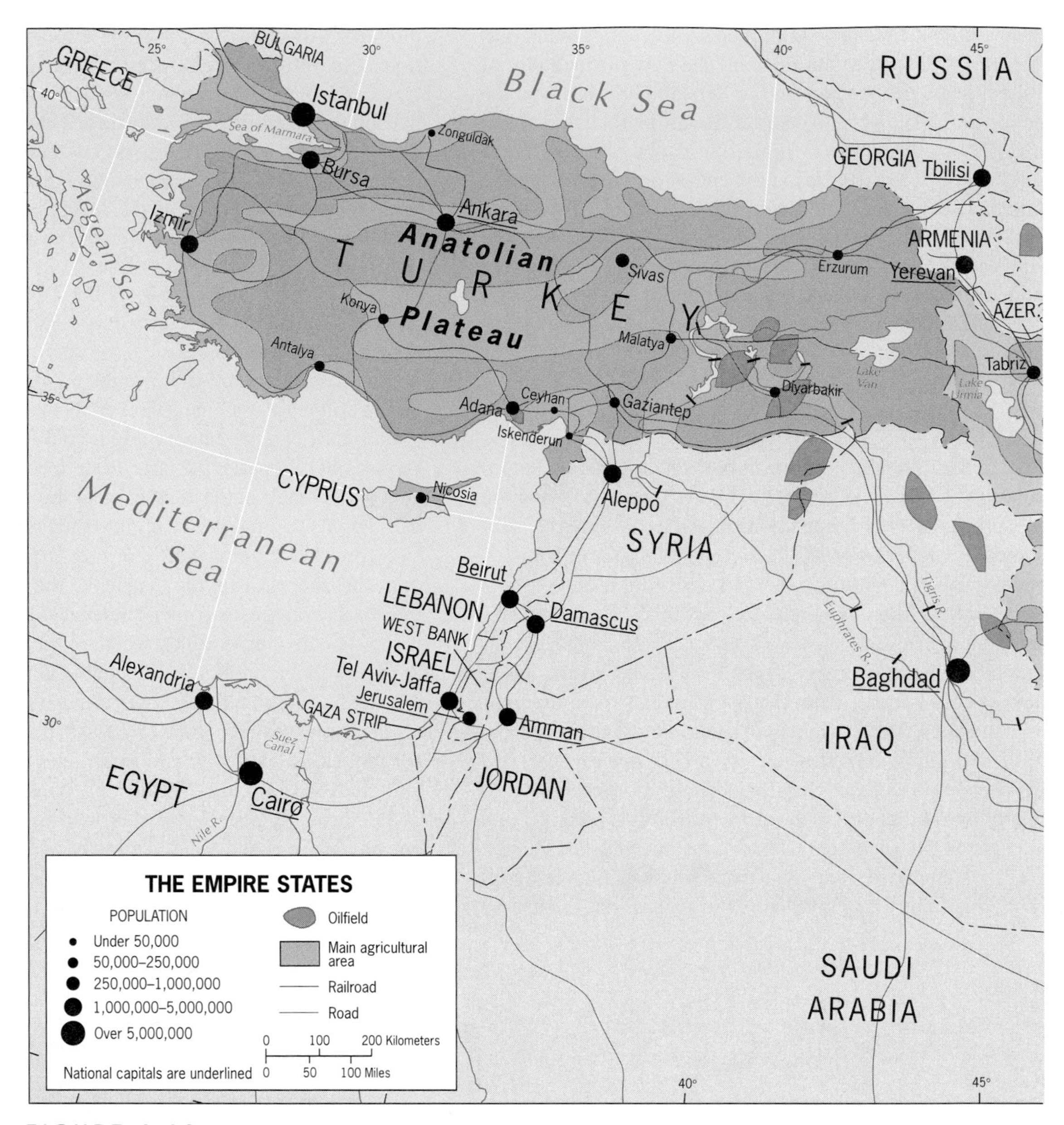

FIGURE 6-18

companies to explore for oil in the promising borderland. Saudi Arabia immediately demanded that it be stopped until the boundary had been defined and delimited. (The Saudis' relationship with the Yemenis, never friendly, deteriorated after Yemen chose Iraq's side during the 1991 Gulf War; Saudi Arabia expelled some 1 million Yemenis working in the kingdom.)

The size of Yemen's population also is difficult to pin down: the best current estimate is 18 million. Significantly, nearly half the Yemenis are Shi'ites. San'a, formerly the capital of North Yemen, became the headquarters of the unified country, and Adan (Aden) is the only port of consequence other than Al Hudaydah on the Red Sea coast. (This is one reason why American warships stopped here to refuel until terrorists attacked one in October, 2000.) The new Yemen was born as a multiparty, secular, democratic state, the only one in the region. But its economy is by far the weakest, with limited oil production.

Yemen occupies a strategic part of the peninsula. Between Yemen and Djibouti in Africa's Horn, the mouth of the Red Sea narrows to the confined Bab el Mandeb Strait, another choke point on Arabia's periphery. The immediate prospects for instability, however, are on land, not at sea. Monarchical, Sunni, oil-rich, modernizing Saudi Arabia stands in stark contrast to democratic, strongly Shi'ite, agricultural, poor, underdeveloped Yemen. Such a contrast, against a backdrop of territorial disagreement, spells trouble.

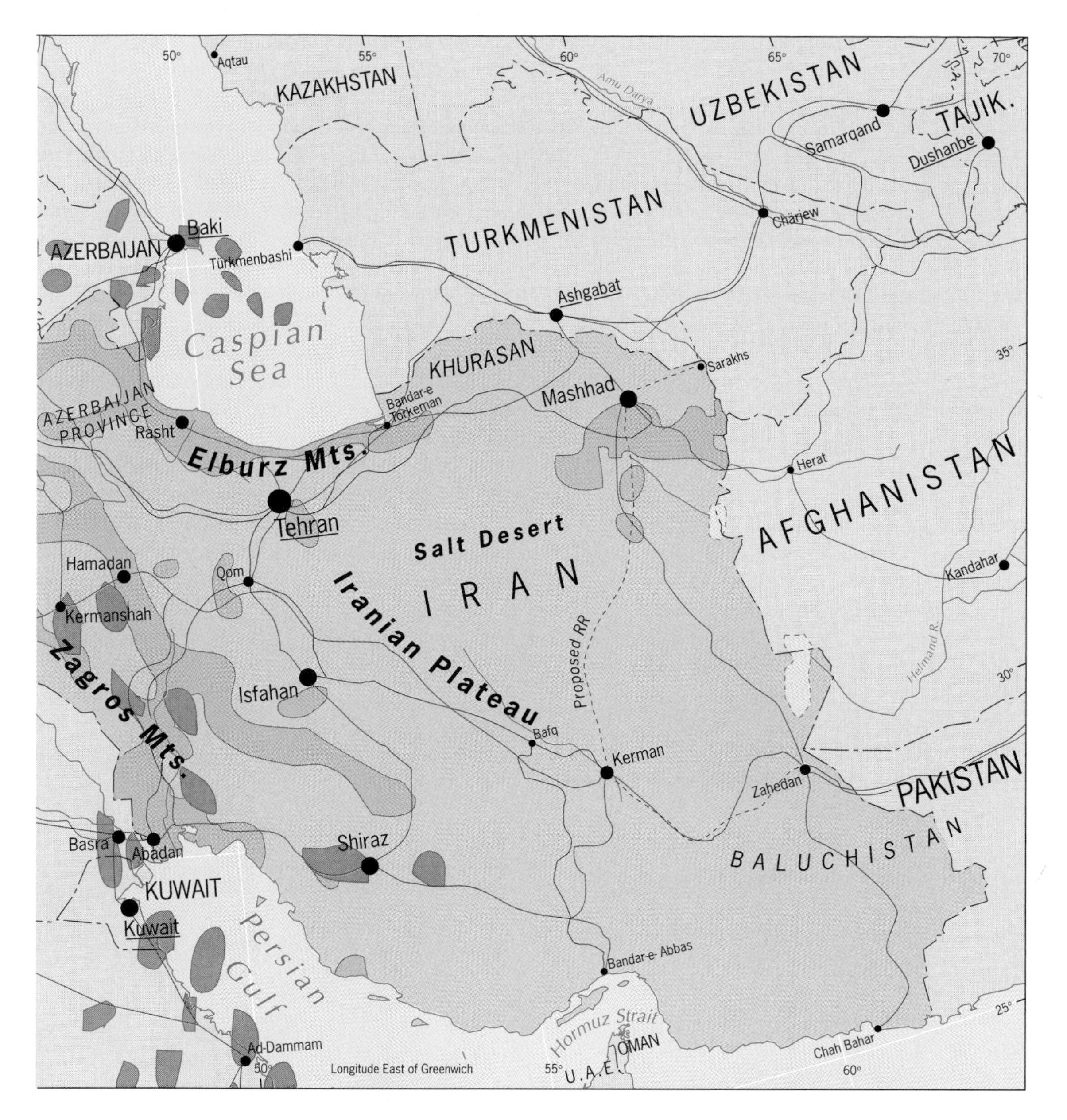

THE EMPIRE STATES

North of the eastern Mediterranean Sea, the Middle East, and the Persian Gulf lie a tier of states that connect this realm's turbulent present to an imperial past. Two states dominate this region, where Arab ethnicity gives way but Islamic culture endures: Turkey and Iran. A third country, Azerbaijan, has historic ties with it because it was a province of the Persian Empire before Russia absorbed it in the nineteenth century (Fig. 6-18). Today, still-dominantly Shi'ite Azerbaijan has closer links with Russia than with Iran, but the Transcaucasian Transition Zone in which it lies is a changeable region. Orientations, linkages, and regional boundaries may well change here, bringing Azerbaijan into this (rather than the Russian) sphere.

The boundaries of this region, in fact, are already changing. At the turn of the twenty-first century, the northern part of the Mediterranean island of Cyprus was functionally a part of it because Cyprus remained partitioned into a Turkish, Islamic north and an independent Greek, Christian south. This division may not be permanent, but at present it creates a Turkish outpost in the Mediterranean, and a realm boundary fractures Cyprus. Elsewhere, the boundary between Iran and Iraq was contested throughout the 1980s, and in the east Iran's borders are under stress due to Afghanistan's chronic instability.

So many things change along the border between the Middle East and these Empire States that we seem to be entering a different realm here. Gone is the Arabic language we heard from Morocco to Oman and from Saudi Arabia to the Horn of Africa. Gone, too, is the Arab nationalism that so strongly influences political geography from the Maghreb to the Middle East. But here to stay is the overarching imprint of Islam, in cities, villages, and countryside, not only in the cultural landscapes but also in ways of life and views of community and the world. And when we get beyond the plateaus of Turkey and the mountains of Iran, we find that this part of the realm, too, has its deserts, oases, caravans, and clustered populations. We have crossed a regional boundary, but we have not left the realm.

Turkey

Who are the Turks, the people who have transformed their pivotal peninsula into a secular, modernizing state? Historical geographers report that the Chinese, as early as the third century, gave the name *Tukiu* (or *Tuchueh*) to the scattered, nomadic peoples then living on the steppes of their Siberian frontier. By the sixth century AD, these nomadic "Turkic" peoples had created a loosely knit domain extending from Mongolia to the Black Sea. Expanding westward and southward, they were energized by their contact with Islam. One of the Turkic peoples, the Oguz, entered Afghanistan and Iran, there confronting the eastern flank of the Byzantine Empire, which was centered in Anatolia (modern-day Turkey) and Greece. In a decisive battle late in the eleventh century they defeated the Byzantine armies, opening the door to a massive migration into Anatolia. There they became the founders of the Turkish nation, and one of their rulers, the sultan Osman I, founded the dynasty that would eventually conquer much of Eastern Europe under the banner of the Ottoman Empire.

After the Turkic expansion came the Mongols, and in Asia numerous Turkic peoples were conquered or driven into remote areas: the Uyghurs, Kazakhs, Kyrgyz, Uzbeks, Turkmen, and others. Their Turkish ancestry was diluted, but Turkic culture was not extinguished. Nor did their allegiance to Islam end.

Among the Realm's Great Cities . . .

Istanbul

From both shores of the Sea of Marmara, northward along the narrow Bosporus toward the Black Sea, sprawls the fabulous city of Istanbul, known for centuries as Constantinople, headquarters of the Byzantine Empire, capital of the Ottoman Empire, and, until Atatürk moved the seat of government to Ankara in 1923, capital of the modern Republic of Turkey as well.

Istanbul's site and situation are incomparable. Situated where Europe meets Asia and where the Black Sea joins the Mediterranean, the city was built on the requisite seven hills (as Rome's successor), rising over a deep harbor that enters the Bosporus from the west, the famous Golden Horn. A sequence of empires and religions endowed the city with a host of architectural marvels that give it, when approached from the water, an almost surreal appearance—and what is seen today is a mere remnant of history's accretion, survivals of earthquakes, fires, and combat.

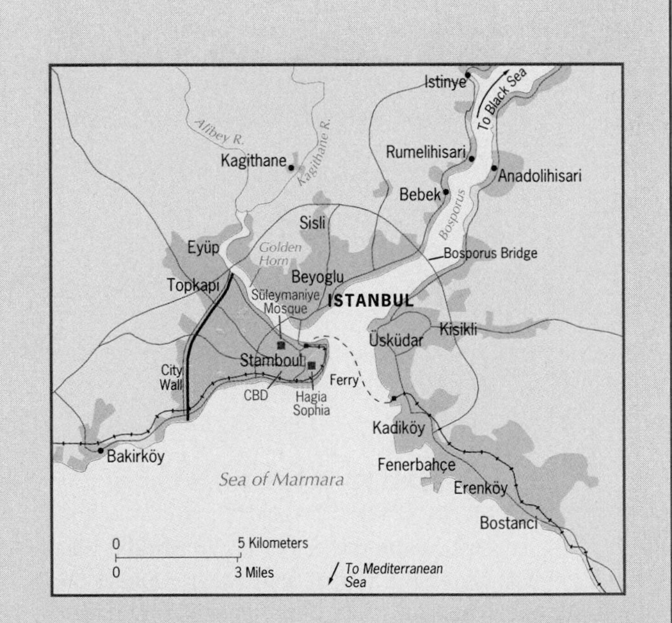

Turkey's political capital may have moved to Ankara, but Istanbul remains its cultural and commercial headquarters. It also is the country's leading urban magnet, luring millions from the poverty-stricken countryside. In the heart of the city known as Stamboul and in the "foreign" area north of the Golden Horn called Beyoglu, modern buildings vie for space and harbor views, blocking the vistas that long made Istanbul's cultural landscape unique. On the outskirts, shantytowns emerge so rapidly that Istanbul's population is doubling every 13 years at current rates, having topped 10 million in 2002. Istanbul's infrastructure is crumbling under this unprecedented influx, its growth is virtually without planning controls, and its legacy of two millennia of cultural landscapes is threatened.

Earlier in this chapter we chronicled the historical geography of the Ottoman Empire, its expansion, cultural domination, and collapse. By the beginning of the twentieth century, the country we now know as Turkey lay at the center of this decaying and corrupt state, ripe for revolution and renewal. This occurred in the 1920s and thrust into prominence a leader who became known as the father of modern Turkey: Mustafa Kemal, known after 1933 as Atatürk, meaning "Father of the Turks."

The ancient capital of Turkey was Constantinople (now Istanbul), located on the Bosporus, part of the strategic straits connecting the Black and Mediterranean seas. But the struggle for Turkey's survival had been waged from the heart of the country, the Anatolian Plateau, and it was here that Atatürk decided to place his seat of government. Ankara, the new capital, possessed certain advantages: it would remind the Turks that they were (as Atatürk always said) Anatolians; it lay nearer the center of the country than Istanbul; and it could therefore act as a stronger unifier. Istanbul lies on the threshold of Europe, with the minarets and mosques of this largest and most varied Turkish city rising above a townscape that resembles Eastern Europe.

Although Atatürk moved the capital eastward and inward, his orientation was westward and outward. To implement his plans for Turkey's modernization, he initiated reforms in almost every sphere of life within the country. Islam, formerly the state religion, lost its official status. The state took over most of the religious schools that had controlled education. The Roman alphabet replaced the Arabic. A modified Western code supplemented Islamic law. Symbols of old—growing beards, wearing the fez—were prohibited. Monogamy was made law, and the emancipation of women was begun. The new government emphasized Turkey's separateness from the Arab World, and it has remained aloof from the affairs that engage other Islamic states.

Atatürk's government naturally faced opposition from religious leaders, notably in the rural areas. Turkey was, and remains, largely an agricultural country, with tens of millions of villagers living in remote farming areas, where resistance to change, especially rapid change, was to be expected. But there were other issues to confront as well. One of these continues to agitate Turkey today: its history of mistreating minorities. Soon after the outbreak of World War I, the (pre-Atatürk) regime decided to expel all the Armenians living in its national territory, most of them in the northeast. Nearly 2 million Turkish Armenians were uprooted and brutally forced out; as many as 600,000 were killed in a campaign that still arouses anti-Turkish emotions among Armenians today. Many fled to Russia, where the Soviet communist regime later established an Armenian republic—now Turkey's Christian neighbor.

In modern times Turkey has faced another minority problem, evident from Figure 6-13: its large and regionally concentrated Kurdish population. About one-fifth of Turkey's population of 67.3 million is Kurdish, and successive Turkish governments have mishandled relationships with this nation, even prohibiting the use of Kurdish speech and music in public places at one time (see box, p. 309). Seeking to suppress a small but violent extremist group among the Kurds, the Turks violated the human rights of the entire minority, an issue that not only roils national politics but also affects Turkey's place in the community of nations. Turkish aspirations to join the European Union are imperiled by human-rights questions arising from the Kurdish situation.

Turkey is a mountainous country of generally moderate relief and, as Figure I-8 indicates, considerable environmental diversity ranging from steppe to highland. On the dry Anatolian Plateau villages are small, and subsistence farmers grow cereals and raise livestock. Coastal plains are not large, but they are productive and densely populated. Textiles (from home-grown cotton) and farm products dominate the export economy, but Turkey also has substantial mineral reserves, some oil in the southeast, massive dam-building projects on the Tigris and Euphrates rivers, and a small steel industry based on domestic raw materials. Normally self-sufficient in staples, Turkey would seem to have a bright future.

But at the turn of the present century, Turkey faced several serious challenges. Potentially the most consequential was the rise of Islamic fundamentalism, which was partly the result of the Kurdish problem. In the shantytowns around Istanbul and Ankara, several million displaced Kurds form a ready market for revivalist leaders who recruit them to their anti-secular, anti-Western causes. By providing bread and basic accommodations to homeless and unemployed urban Kurds, a political party that called itself the Welfare Party (now the Virtue Party) won local elections in Istanbul and Ankara, vowing to reverse the emancipation of women, to reinstitute gender segregation in education, and to separate the sexes in public transport. The military, always a force in Turkish politics, seeing itself as the guardian of Atatürk's policies, then demanded the Welfare Party's suppression, and the party was declared illegal in 1998. The wisdom of this action was debated outside as well as inside Turkey; it came under heavy criticism in the Islamic world and also in Europe.

Another cultural issue in modern Turkey involves one of Islam's many sects, in this case the Alevis. Most of Turkey's Muslims are Sunnis, and some are orthodox and strict. The Alevis, who number about 15 million (nearly 20 percent of the population), are Shi'ites—but they practice a rather nonchalant version of the faith. While orthodox Sunnis worship in mosques, live by Islam's daily rules, and fast during the daytime during Ramadan, Alevis do not gather in mosques and do not follow the rules or the Ramadan fast. This has led to serious cultural conflict, including murder and arson. The historic center of Turkey's Alevi sect lies in

Divided Cyprus

In the northeastern corner of the Mediterranean Sea, just 50 miles (80 km) from Turkey but 275 miles (440 km) from the nearest Greek territory, lies the island of Cyprus. Greek since ancient times but conquered by the Ottomans in 1571, the population of just under 900,000 remains about 80 percent Greek today. During the British colonial period between 1878 and 1960, Greeks and Turks lived intermixed throughout much of the island, but after independence intercultural relations deteriorated and in 1974 a civil war resulted in partition. The northern 40 percent of Cyprus became an almost exclusively Turkish domain, and the south remained Greek (see inset map, p. 87). In 1983, the Turkish community declared itself the independent Turkish Republic of Northern Cyprus (TRNC).

The international community continues to recognize the government of the Greek side of Cyprus as the only legitimate one, and as we noted in Chapter 1 (pp. 93-94), the south has prospered economically. The poorer north depends heavily on Turkish subsidies, and contrasts between the two sectors are deepening. A field trip across Cyprus would leave you in no doubt that a realm boundary exists here: thriving, interconnected, globalizing, tourist-dependent southern Cyprus stands in stark contrast to the isolated, stagnant TRNC. In the now-Muslim north, Greek Orthodox churches and public art have been destroyed; in the now-Christian south, mosques are being converted to other uses or stand abandoned in decay. In the north, infrastructures are deteriorating and incomes are low; in the south, modernization continues and incomes are high.

International efforts at mediation have failed to move Cypriot Greeks and Turks toward reintegration, although polls indicate that many people on both sides of the dividing "Green Line" desire it. But time is hardening, not softening, the divergence of north and south.

Sivas, in the east-central part of the country northwest of the Kurdish zone. Like the Kurds, many Alevis have left their cultural base, where they are under pressure from orthodox Sunnis, for the more cosmopolitan cities.

Coupled with Europe's reluctance to take the first steps that would lead to Turkey's admission to the European Union, these developments cast an ominous shadow over Turkey's future and, by extension, over the future of the area at the heart of which it lies. Turkey has signaled its interest in joining the European Union, but as already noted, the human-rights issue and Greece's objections are diminishing its prospects. Ultimately, however, the potential advantages of Turkey's membership in the European economic framework may override these obstacles; for one, Turkey's admission would prove that Europe is not an exclusive "Christian club." Islamic fundamentalists in Turkey hope that the drive for admission will be stalled long enough for them to take power in Ankara. In any case, the delays, in their eyes, prove that Europe is anti-Islamic. If Europe closes the door against Turkey, Islamic extremists will benefit enormously. Should Turkey experience what happened in neighboring Iran during the 1970s and more recently in Algeria, the consequences would be disastrous.

Compared to such prospects, Turkey's other problems seem minor: the question of still-divided Cyprus (see box titled "Divided Cyprus"), relations with neighboring Iraq and Syria over dam building and oil pipelines, a boundary quarrel with Syria, and a dispute with Greece over maritime boundaries off the west coast. Turkey's relative location gives it an importance far greater than its dimensions, population, economy, and social fabric would otherwise command. Here turbulent Eastern Europe meets the world of Islam; here lies the choke point between the Black Sea and the Mediterranean; here lie the sources of rivers that form the lifelines of neighbors; here is the gateway to Transcaucasia and Turkestan. In this world of growing ethnic and cultural consciousness, Turkey lies at the most crucial of crossroads.

Iran

Iran, Turkey's neighbor to the east, also has a history of imperial conquest. In 1971, the then-reigning shah and his family celebrated the 2500th anniversary of Persia's first monarchy with unmatched royal splendor. But by 1979, revolution had engulfed Iran, and Shi'ite fundamentalists drove the shah from power. The monarchy was replaced by an Islamic republic, and a frightful wave of retribution followed.

As Figure 6-18 shows, Iran also occupies a critical area in this turbulent realm. It controls the entire corridor between the Caspian Sea and the Persian Gulf. To the west it adjoins Turkey and Iraq, both historic enemies. To the north (west of the Caspian Sea) Iran borders Azerbaijan and Armenia, where once again Muslim confronts Christian. To the east Iran meets Pakistan and Afghanistan, and east of the Caspian Sea lies volatile Turkmenistan.

In many places, Iran seems to spill over into its neighbors, a legacy of its expansionist days. As the map shows, Azerbaijan is the name of an Iranian province as well as an independent republic and would be part of Iran today were it not for Russian-Soviet intervention (see p. 142). Similarly, there is an Iranian Baluchistan and a Pakistani Baluchistan; the border marks an Iranian retreat enforced by the British when they ruled India and what is now Pakistan. The war of the 1980s with Iraq started as a territorial conflict over Iraq's access to the Persian Gulf.

Iran, as Figure 6-18 demonstrates, is a country of mountains and deserts. The heart of the country is an upland, the Iranian Plateau, that lies surrounded by even higher mountains, including the Zagros in the west, the Elburz in the north along the Caspian Sea coast, and the mountains of Khurasan to the northeast. The Iranian Plateau therefore is actually a huge highland basin marked by salt flats and wide expanses of sand and rock. The highlands wrest some moisture from the air, but elsewhere only oases break the arid monotony—oases that for countless centuries have been stops on the area's caravan routes.

In eastern Iran, neighboring Pakistan, and Afghanistan, people still move with their camels, goats, and other livestock along routes that are almost as old as the human history of this realm. Usually they follow a seasonal and annual cycle, visiting the same pastures year after year, pitching their tents near the same stream. It is a lifestyle especially associated with this realm: **20** **nomadism**. In Iran as elsewhere, nomads are not aimless wanderers across boundless plains. They know their terrain intimately, and they carefully judge how long to linger and when to depart based on many years of experience along the route. Nor are nomads simply scavengers. A nomadic community has its division of labor in which some are skilled at crafts and make leather and metal objects for sale or trade; others are herders; and still others are farmers who may grow a crop when the group stops long enough during the summer at a suitable site. Caravans of nomads still travel the trails of Southwest Asia and North Africa, but their numbers are dwindling. Their lands are being penetrated by roads, the borders they used to cross freely are being sealed, and their environments are changing as desertification claims the steppe.

In ancient times, Persepolis in southern Iran (located near the modern city of Shiraz) was the focus of a powerful Persian kingdom, a city dependent on *qanats*, underground tunnels carrying water from moist mountain slopes to dry flatland sites many miles away. Today, Iran's population of 69.4 million is 63 percent urban, and the capital, Tehran, lies far to the north, on the southern slopes of the Elburz Mountains. This mushrooming metropolis of 7.4 million, lying at the heart of modern Iran's core area, still depends in part on the same kinds of qanats that sustained Persepolis more than 2000 years ago. As such, Tehran symbolizes the internal contradictions of Iran: a country in which modernization has taken hold in the cities, but little has changed in the vast countryside, where the *mullahs* led their peasant followers in a revolution that overthrew a monarchy and installed a theocracy.

The last shah of Iran, Muhammad Reza Pahlavi, hoped to be the Atatürk of his country, a modernizer and reformer who used Iran's enormous income from oil to promote industrialization, update agriculture, improve health-care services, emancipate women—and to build a huge security apparatus to contain any opposition and the armed forces to establish his regional power base. In the end, Pahlavi was no Atatürk. His policies ran counter to the Islamic traditions of most of his people. Among the Muslim leaders who opposed him was an exiled *ayatollah* (Allah-ordained) named Khomeini. In time, this fanatical religious figure became the symbol of the Islamic revolution that exploded after he returned to Iran from exile in 1979 and continued to dominate Iran beyond his death 10 years later.

As Figure 6-8 shows, Iran has a large share of the oil riches in this part of the world. Petroleum and petroleum products provide about 90 percent of the country's income. The reserves lie in a zone along the southwestern periphery of Iran's territory, and Abadan became its "oil capital" near the head of the Persian Gulf. But Iran is a large and populous country, and the wealth oil generated could not transform it the way the last shah intended, a transformation that had it occurred might have staved off the revolution. Modernization remained but a veneer: in the villages away from Tehran's polluted air, the holy men continued to dominate the lives of ordinary Iranians. As elsewhere in the Muslim world, urbanites, villagers, and nomads remained enmeshed in a web of production and profiteering, serfdom, and indebtedness that has always characterized traditional society here. The revolution swept this system away, but it did not improve the lot of Iran's millions. A devastating war with Iraq (1980–1990), into which Iran ruthlessly poured hundreds of thousands of its young men, sapped both the coffers and energies of the state. When it was over, Iran was left poorer, weaker, and aimless, its revolution spent on unproductive pursuits.

In the 1990s, new challenges confronted Iran. Internally, the Islamic conservative factions that controlled national affairs since the revolution suffered several electoral setbacks. Iranians demanded more liberal policies and elected a progressive, Muhammad Khatami, as president in 1997. In Tehran and other cities, the mood for social change was evident in opposition to the strict Islamic dress code and to government constraints on economic activity. But by the end of 2000, it was clear that the conservative Islamists were thwarting the reformers: journalists were imprisoned, newspapers shut down, students arrested, books impounded. Once again, Iranian society appeared headed for civil disorder.

Externally, Iran faces problems in several directions. Iraq's mistreatment of its Shi'ite communities produces tensions and revives memories of the costly 1980s war. In the Caspian Sea area, Iran has a stake not only in oil reserves but in the competition with Turkey, Georgia, Azerbaijan, and Russia for the location of pipelines for export. And to the east lies turbulent Afghanistan where, in the late 1990s, a fundamentalist Sunni Islamic (Taliban) movement took control that made Tehran's regime look liberal by comparison. Events in Afghanistan sent thousands of refugees into Iran, and the Iranians supported the forces opposed to the Afghani fundamentalists. Nowhere is the Shiah-Sunni schism spatially sharper than at this realm's eastern margin, along the boundary between Iran and Afghanistan.

Revolutionary Iran disavowed its Persian imperial memories and ambitions, but that does not mean that its national interests now stop at its borders. Tehran still has regional objectives and global ambitions ranging from the repudiation of Israel to the promotion of Islam. It is acquiring and building missiles and is suspected of developing weapons of mass destruction. Iran remains a potent force in the realm and beyond; its revolution may have ended a monarchy, but it did not extinguish all ties to an imperial past.

TURKESTAN

For centuries Turkish (Turkic) peoples held sway over a vast Central Asian domain that extended from Mongolia and Siberia to the Black Sea. Propelled by population growth and energized by Islam, they penetrated Iran, defeated the Byzantine Empire, and colonized much of Eastern Europe. Eventually, their power declined as Mongols, Chinese, and Russians invaded their strongholds. But these conquerors could not expunge them, as the names on the modern map prove (Fig. 6-19). The latest conquerors, the Russian czars and their communist successors, created Soviet Socialist Republics named after the majority peoples within their borders. Thus the Kazakhs, Turkmen, Kyrgyz, and other Turkic peoples retained some geographic identity in what was Soviet Central Asia.

Turkestan is a still-changing region. In some areas, the cultural landscapes of neighboring realms overlap into it, for example, in northern (Russian) Kazakhstan. In other areas, Turkestan extends into adjacent realms, as in Uyghur-influenced western China (Xinjiang). Some areas once penetrated by Turkic peoples are no longer dominated by them, for instance, Afghanistan. And certain peoples now living in Turkestan are not of Turkic ancestry, notably the Tajiks. This is a fractious region in sometimes turbulent transition.

The underlying cultural-geographic reason for this contentiousness—not only in Turkestan but in neighboring regions in Southwest and South Asia as well—is illustrated in Figure 6-20. This detailed map of ethnolinguistic groups actually is a generalization of an even more complex mosaic of peoples and cultures. Every cultural domain on the map also includes minorities that cannot be mapped at this scale, so that people of different faiths, languages, and ways of life rub shoulders everywhere. Often this results in friction, and at times such friction escalates into ethnic conflict.

As we define it, Turkestan includes most or all of six states: (1) Kazakhstan, territorially larger than the other five combined but situated astride an ethnic transition zone; (2) Turkmenistan, with important frontage on the Caspian Sea and bordering Iran and Afghanistan; (3) Uzbekistan, the most populous state and situated at the heart of the region; (4) Kyrgyzstan, wedged between powerful neighbors and chronically unstable; (5) Tajikistan, regionally and culturally divided as well as strife torn; and (6) Afghanistan, engulfed in war almost continuously since it was invaded by Soviet forces in 1979.

During their hegemony over Central Asia the Soviets tried to suppress Islam and install secular regimes (this was their objective when they invaded Afghanistan as well), but today Islam's revival is one of the defining qualities of this region. From Almaty to Samarqand, mosques are being repaired and revived, and Islamic dress again is part of the cultural landscape. In other ways, too, Turkestan reflects the norms of this realm: in its dry-world environments and the clustering of its population, its mountain-fed streams irrigating farms and fields, its sectarian conflicts, its oil-based economies. It also is a region where democratic government remains an elusive goal.

Kazakhstan

The territorial giant of the region, Kazakhstan, has five neighbors of which two are giants themselves: China and Russia (Fig. 6-19). To the east, Kazakhs live across the border in western China's Xinjiang region. In the north, nearly 5 million Russians inhabit a wide zone within the republic. The cultural contrast between the Russified north and the Turkestan south is a key challenge for Kazakhstan's future.

During the communist period, Almaty was the capital of Kazakhstan. Its location in the far southeast of the country therefore put Soviet administrators in the heart of the Kazakh sector. But after independence, the agenda changed. Now, Kazakh leaders worried that the Russified north might someday separate from Kazakhstan and join Russia. One way to forestall potential separatism, they reasoned, was to move the capital to this area. In 1998, Astana (formerly called Aqmola) became Kazakhstan's capital following the hasty construction of essential facilities. Unlike historic and scenic Almaty, Astana lies on a windswept plain where summers are hot and winters are bitter, and where the bleak communist-era architecture does little to raise the spirits. But Astana (population: 390,000) is 70 percent Russian and 30 percent Kazakh, and it puts the Kazakh government squarely in the middle of the country's Russified zone.

It may not be enough. In November 1999, Kazakh authorities arrested 22 ethnic Russians and accused them of

FIGURE 6-19

plotting to seize government buildings in Öskemen (formerly Ust-Kamenogorsk) in order to establish an independent Russian republic there. Eastern Kazakhstan Province, of which Öskemen is the capital, remains about 80 percent Russian. Cossacks from Siberia first settled here more than two centuries ago, and until the mid-1930s it was considered part of the Siberian Eastern Frontier. Then the communist regime made it part of the Soviet Republic of Kazakhstan. Now it is a still-largely-Russian society under Kazakh rule.

Kazakhstan's north still has the best surface communications, the largest cities, and most industries, all resulting

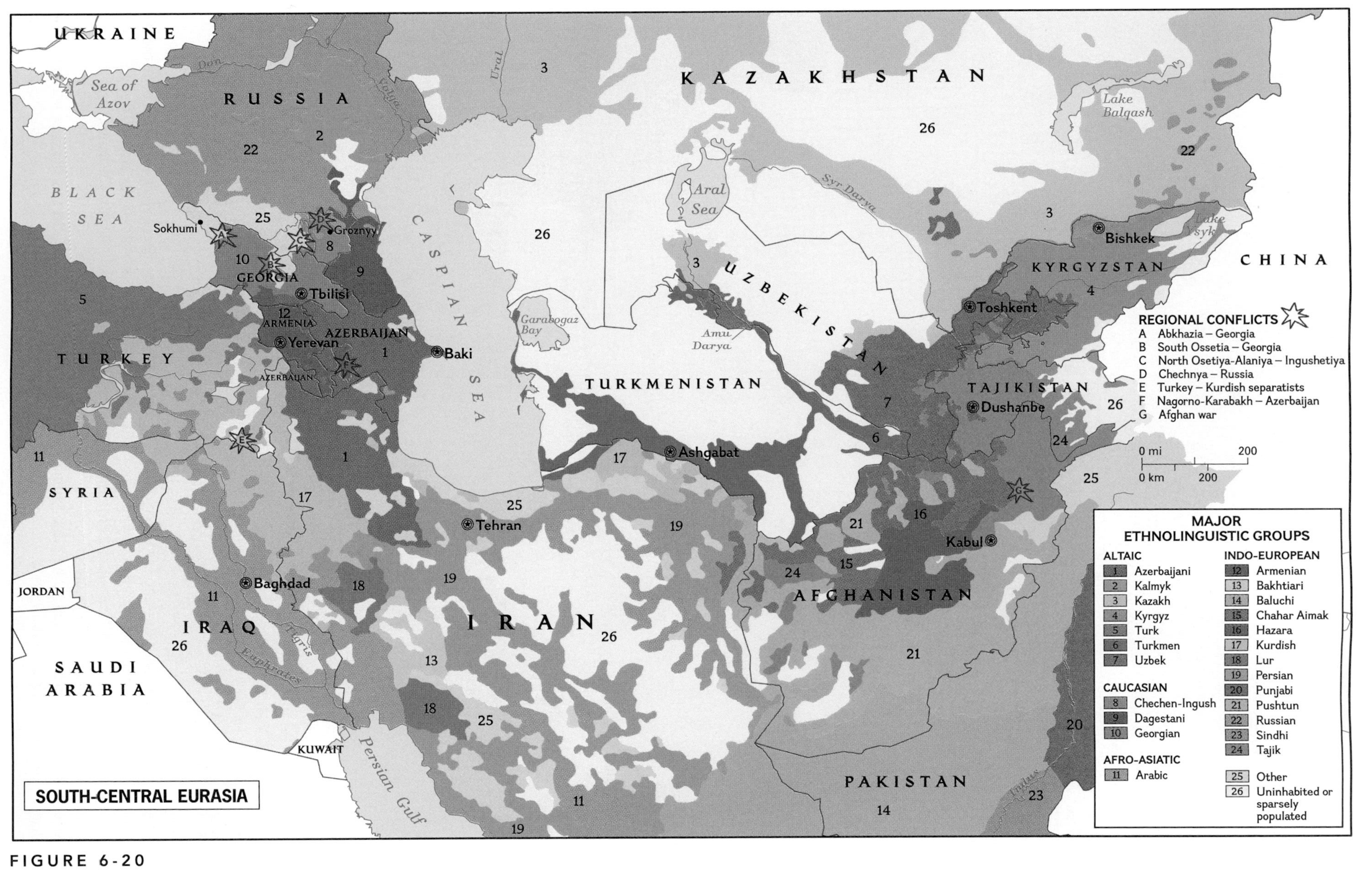
SOUTH-CENTRAL EURASIA
UKRAINE
RUSSIA
KAZAKHSTAN
UZBEKISTAN
TURKMENISTAN
KYRGYZSTAN
TAJIKISTAN
CHINA
AFGHANISTAN
PAKISTAN
IRAN
IRAQ
TURKEY
SYRIA
JORDAN
SAUDI ARABIA
KUWAIT
GEORGIA
ARMENIA
AZERBAIJAN
BLACK SEA
CASPIAN SEA
Sea of Azov
Aral Sea
Garabogaz Bay
Lake Balqash
Lake Issyk
Persian Gulf
Don
Volga
Ural
Syr Darya
Amu Darya
Tigris
Euphrates
Indus
Sokhumi
Groznyy
Tbilisi
Yerevan
Baki
Tehran
Baghdad
Ashgabat
Toshkent
Bishkek
Dushanbe
Kabul
REGIONAL CONFLICTS
A Abkhazia – Georgia
B South Ossetia – Georgia
C North Osetiya-Alaniya – Ingushetiya
D Chechnya – Russia
E Turkey – Kurdish separatists
F Nagorno-Karabakh – Azerbaijan
G Afghan war
0 mi 200
0 km 200
MAJOR ETHNOLINGUISTIC GROUPS
ALTAIC
1 Azerbaijani
2 Kalmyk
3 Kazakh
4 Kyrgyz
5 Turk
6 Turkmen
7 Uzbek
CAUCASIAN
8 Chechen-Ingush
9 Dagestani
10 Georgian
AFRO-ASIATIC
11 Arabic
INDO-EUROPEAN
12 Armenian
13 Bakhtiari
14 Baluchi
15 Chahar Aimak
16 Hazara
17 Kurdish
18 Lur
19 Persian
20 Punjabi
21 Pushtun
22 Russian
23 Sindhi
24 Tajik
25 Other
26 Uninhabited or sparsely populated

FIGURE 6-20

The Caspian Sea area is becoming an arena for competition—a contest between governments and corporations over the oil and natural gas resources this part of the former Soviet Union contains. Among the countries with Caspian Sea coasts (and therefore with offshore as well as land-based potential), Kazakhstan is in a strong position. One of Kazakhstan's most promising oilfields is the Tengiz Basin (see map, p. 138), and the Kazakh government has issued leases and permits for foreign companies to exploit it. This huge installation, recently built by Chevron, brings American technology to this remote but profitable part of Turkestan.

from the Soviet expansion that made this zone, essentially, part of Russia's Eastern Frontier (see Figs. 2-6, 2-10). Even the Soviet space program and missile development plants were based here; the space program has continued after Kazakhstan's independence by agreement with the Russians. But Kazakhstan's west, where it borders the Caspian Sea, will determine the country's economic future. Some estimates indicate that the oil reserves in the Tengiz Basin at the northeast end of the Caspian Sea (now being developed under contract with a U.S. oil company, as shown in the photo above) may rank among the largest in the world. And because of its lengthy Caspian Sea coastline, Kazakhstan can claim any additional petroleum reserves found offshore on its side of the sea (including the especially promising Kashagan oilfield [Fig. 6-19]). The Soviets built large refineries and, near Atyrau, a nuclear power plant. This remote part of Kazakhstan will grow in importance during the twenty-first century.

Kazakhstan's relative location is also important. Kazakhstan, as the map shows, forms a corridor between the Caspian oil reserves and China, which is likely to become one of the fastest-growing consumers of oil in the decades ahead. Pipelines across Kazakhstan could eliminate, or greatly reduce, China's dependence on oil brought in by tanker along distant sea lanes. Already, Kazakhstan has awarded China the sole rights to explore and exploit the Uzen reserve south of the Tengiz. This puts Kazakhstan in a crucial situation in Central Asia.

Kazakhstan's political and social future is another matter. The Kazakhs constitute just 45 percent of the population of 15 million; the Russian minority is about 31 percent and shrinking. Where the Kazakhs are in the majority, in the south, they share their land with many minorities large and small. The last census taken by the Soviets enumerated nearly 100 minorities, including Ukrainians, Germans, Uzbeks, Tatars, and Uyghurs. Less than 50 percent of the people adhere to Sunni Islam, and the Russian Orthodox Church is strong in the north.

Such divisions tend to cause problems, especially during periods of economic difficulty, and Kazakhstan remains a country where potential outweighs reality. Mismanaged during Soviet times, when irrigation schemes disrupted natural drainage lines and pesticides polluted groundwater (all to produce cotton), Kazakhstan faces ecological disaster in the Aral Sea area (Fig. 6-19; photos p. 330). The Soviets also plundered Kazakhstan's mineral resources, notably at Qaraghandy, a city built to mine a huge coalfield. In its decade of independence, the country has suffered a series of setbacks, and its workers are restive. But if it can avoid partition, Kazakhstan's future may yet be bright.

Turkmenistan

In the southwest of Turkestan lies the desert republic of Turkmenistan, with just 5.4 million inhabitants. Extending eastward from the shore of the Caspian Sea to the boundary with Afghanistan, and bordering Iran for over 700 miles (1100 km), this area was part of the old Muslim Turkestan before it became a Soviet Republic in 1925. As large as Nevada and Utah combined, Turkmenistan was the frontier domain of many nomadic peoples when Soviet efforts to modernize the area began. Although 76 percent of the republic's inhabitants

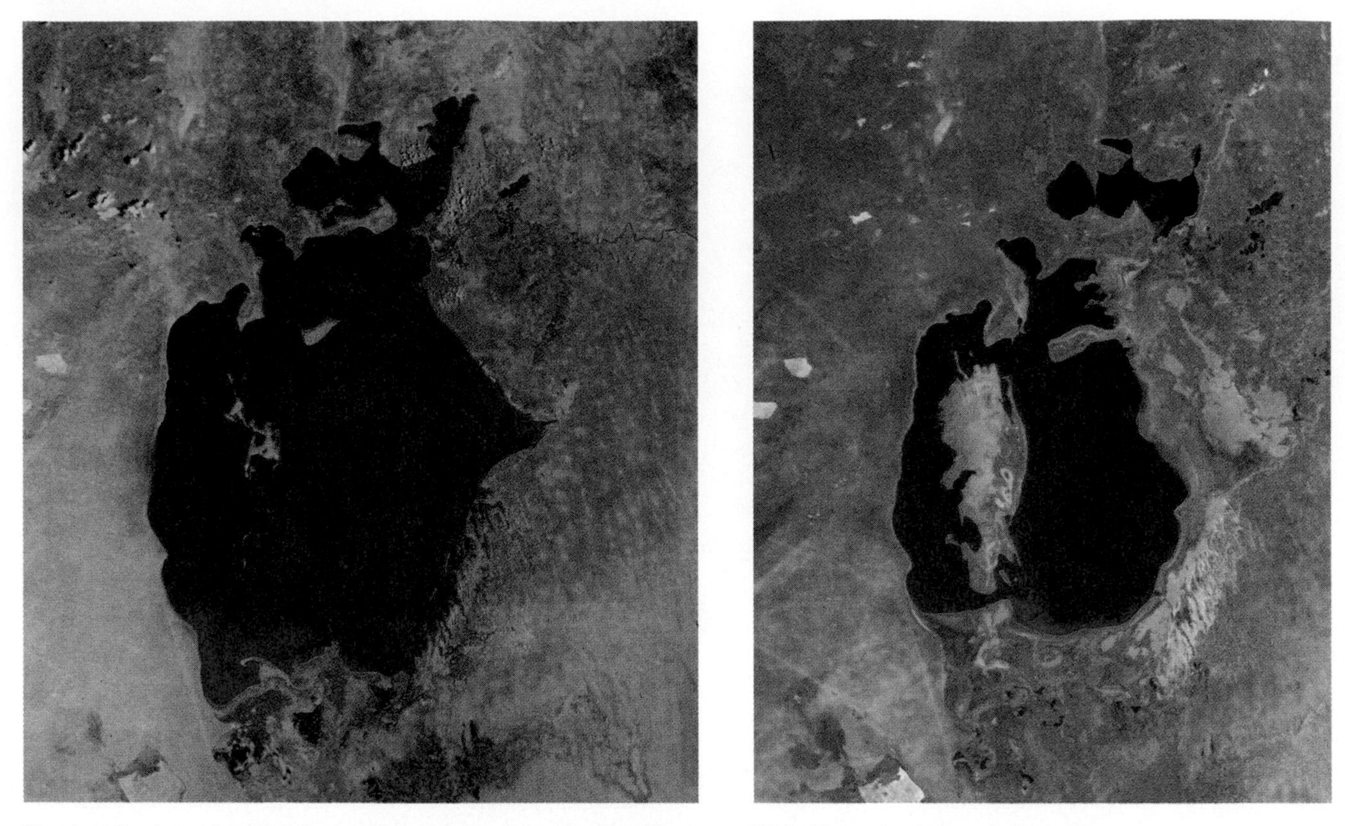

The Aral Sea is a saltwater lake on the boundary between Kazakhstan and Uzbekistan, both former Soviet Socialist Republics. During the communist period, the lake's two inflowing rivers were diverted for irrigation, resulting in the dramatic shrinkage illustrated by these two Landsat images. The image on the left was taken in the 1960s, when the Aral Sea covered an area of about 26,300 square miles (68,000 sq km). Thirty years later, the lake had been cut in two and the total water area combined was less than 13,000 square miles (33,700 sq km). The surface had dropped about 50 feet (15 m) and a large island in what is now known as the "Greater Sea" (as opposed to the "Lesser Sea" to the north) threatened to become another land bridge. The fate of the Aral Sea is but one chapter in Soviet-era mismanagement of the environment in the headlong rush to increase agricultural productivity. The Kazakh and Uzbek governments, with assistance from international agencies, are now in the process of restoring the flow of the two feeder rivers to arrest the Aral Sea's decline, but it is unlikely that a significant rise in the water level can be achieved.

are Turkmen, many ethnic divisions still fragment the predominantly Muslim population. There also are Russians (7 percent, mostly in the towns), Uzbeks (another 10 percent), Kazakhs, Ukrainians, Tatars, and Armenians.

Soviet efforts to stabilize the population and to make it more sedentary centered on a massive project: the Garagum (Kara Kum) Canal, begun in the 1950s to bring water from mountains to the east into the heart of the desert. The canal eventually will reach the Caspian Sea; today, it is more than 700 miles (1100 km) long, and about 3 million acres have been brought under cultivation. But by diverting water from the Amu Darya River (Fig. 6-19), the engineers who planned the Garagum Canal contributed greatly to the drying up of the Aral Sea.

Farmers in the irrigated lands grow cotton, corn, vegetables, and fruits, but many Turkmen still herd sheep and sell Astrakhan fur (from the pelts of young lambs) for export. Turkmenistan's real hope, however, lies in its sector of the Caspian Basin, where oil and natural gas abound. They already rank at the top of the country's exports by value, but foreign companies are involved in major new developments. A key problem is egress: how to send the products to markets overseas. Iran, always interested in building influence in Turkestan, has offered to support construction of a pipeline from western Turkmenistan to its Persian Gulf port of Abadan. Other options would involve shipping the oil from the Turkmenistan port of Türkmenbashi (Balkhan) to Baki (Baku) in Azerbaijan, and thence by pipeline across the Caucasus region and Turkey to a Mediterranean port (Ceyhan), or via Russia to a Black Sea port (Novorossiysk). Turkmenistan does not have a favorable location.

Uzbekistan

Populous Uzbekistan (25.7 million) occupies the heart of Turkestan, bordering every state in the region. California-sized and sharing the Aral Sea Basin (as well as the ecological damage the Soviets did there), Uzbekistan suffered even more soil and groundwater pollution and associated high infant mortality and cancer rates than Kazakhstan.

Straight-line boundaries across the desert mark western Uzbekistan, but the eastern border is another story. Here lie the capital, Toshkent, and several other cities, creating a marginal core area nearly encircled by Kyrgyzstan and Tajikistan. The overcrowded Farghona Valley forms a key part of this core area, but the valley extends into Tajikistan to the south and into Kyrgyzstan to the north. Not only does it produce fruits and vegetables as well as cotton (another Soviet legacy), but its mines and wells yield coal, mercury, antimony, and oil.

The boundaries on the map do not conform to cultural divisions here. Just over 75 percent of the population is Uzbek, with Russian, Tajik, Kyrgyz, and other minorities, but the real issue is the Uzbek population beyond Uzbekistan's borders. Nearly 25 percent of Tajikistan's population, 14 percent of Kyrgyzstan's, and 10 percent of Turkmenistan's are Uzbek. Moreover, about 6 percent of Afghanistan's population is also Uzbek. The government-controlled press in Uzbekistan often refers to a "greater Uzbekistan" in the future, and when Uzbeks are threatened in neighboring countries, as in Afghanistan, irredentist feelings surface. Conflict in Tajikistan and Afghanistan, in particular, has aroused Uzbek nationalism.

Another problem in Uzbek (and other Central Asian) eyes is *Wahhabism*, a particularly virulent form of Sunni fundamentalism that has taken hold in the former Soviet Republics of Central Asia. The Wahhabis derive from an eighteenth-century puritanical movement inspired by an Arabian Muslim theologian, al-Wahhab. It demands literal belief in the Quran (Koran), and it insists that political entities be Muslim states based on Islamic law. The Wahhabis (who in Arabic call themselves "unitarians") might have faded from view had the royal family of Saudi Arabia not adopted their beliefs. When the Saudi Arabian kingdom was formed in the 1930s, Wahhabism became its religious foundation.

When Islam revived in post-Soviet Central Asia, several Islamic sects vied for influence. The new governments, with former communist members in influential positions, formed a ready target for Wahhabi opponents when they tried to restrict Islamic militants of all kinds. Memories of the atheist past, economic failures, widespread corruption, and Soviet-style restrictions on free speech made Turkestan a fertile recruiting ground for Wahhabism. This challenge will roil the cultures of Turkestan for years to come.

Kyrgyzstan

In southeastern Turkestan, topography and political geography create a situation reminiscent of the Caucasus. Uzbekistan's eastward wedge of territory lies intertwined with Kyrgyzstan and Tajikistan. Tajikistani exclaves lie within Kyrgyzstan. Cultural geographies are complex and fragmented. Boundaries separate people of similar ethnic background and unite groups that have little in common. Mountain chains isolate communities living in remote valleys. Under such circumstances, nation-building is difficult.

The great mountain ranges of the Tian Shan dominate the center of Kyrgyzstan. Soviet planners created this entity as a Socialist Republic, but today the indigenous inhabitants, the Turkic people known as the Kyrgyz, constitute barely 60 percent of the population of 5 million. Russians, their numbers declining, make up about 15 percent, Uzbeks 14 percent. Numerous smaller minorities create a complex cultural geography. About 70 percent of the people profess allegiance to Islam, and Wahhabism has gained a strong foothold here. The town of Osh is often referred to as the headquarters of the movement in Turkestan.

Kyrgyzstan has been plagued by economic problems, political failures (it has one of the region's most authoritarian regimes), and social disorder. The capital, Bishkek, lies on the Tian Shan's north slope, near the border with Kazakhstan. Surface communications between northern Kyrgyzstan and Kazakhstan and Uzbekistan are better than they are within Kyrgyzstan itself.

Most of the people are pastoralists; in addition to sheep and cattle, the Kyrgyz raise yaks for meat and milk on high-altitude pastures unsuitable for other livestock. Irrigated valleys and lowlands yield wheat, fruits, and vegetables. Industries are mainly processing and packaging plants for locally produced textiles and foods. The economy is typical for a low-income, peripheral country. Here is how statistics can deceive: in the late 1990s, Kyrgyzstan reported the highest percentage increase in industrial production among the states in the former Soviet orbit. The reason was the output of a gold refinery, a joint Canadian-Kyrgyz venture. But its impact on the overall living conditions of the people was minimal.

At the opening of the century Kyrgyzstan faced a growing problem because of its relative location: illegal drug operatives were using it as a conduit between sources in Afghanistan and Tajikistan and consumers in Europe.

Tajikistan

Iowa-sized Tajikistan borders Afghanistan and China. The gigantic Pamirs dominate its eastern reaches where high, glacier-sustaining mountains form the source of irrigation water for the neighboring desert republics of Uzbekistan and Turkmenistan. The two tallest peaks in this range, both over 23,000 feet (7000 m), tower above some of the world's most spectacular highland scenery.

The Tajiks are a separate people in the region; they are of Persian (Iranian), not Turkic, origin and speak a Persian (and thus Indo-European) language. They constitute 62 percent of the population of 6.6 million, but many Tajiks live in neighboring Afghanistan and China. The Tajiks share their land with Uzbeks in the west and northwest (25 percent), Russians (a shrinking 7 percent), Armenians, and others. Most Tajiks, despite their Persian affinities, are Sunni Muslims, not Shi'ites.

The capital of Tajikistan, Dushanbe, lies in the far west. When the Tadzhik Soviet Socialist Republic (as the communists spelled it) was established in 1929, it was an area of semi-nomadic herding and handicraft industries. Soviet rule and economic planning brought large industrial enterprises, mining, and irrigated agriculture, but most Tajiks remained farmers and herders.

After independence, former communist political managers remained in key government positions but did not have Moscow to back them up. Soon Tajikistan plunged into civil war as Islamic groups attacked government installations, army bases, and even members of the administration. The conflict had a regional as well as cultural dimension: the government in Dushanbe was at odds with the country's northern region, hotbed not only of Islamic fundamentalism but also of anti-Tajik, Uzbek activism. In the late 1990s, when the government finally negotiated a peace accord with the United Tajik Opposition representing the main Islamic groups, many smaller groups and bands of armed rebels were left out of the process. They rejected the accord and continued their campaign of violence against any and all authorities. Meanwhile, Tajikistan's economy and social order collapsed. After the turn of the century, the country's future remained clouded.

Afghanistan

The southernmost country in this region would have become a part of Soviet Central Asia if Moscow's invasion of 1979 had succeeded. Instead, the Soviets started a cycle of conflicts that took an unimaginable toll on the peoples of this landlocked, divided state.

Afghanistan exists because the British and Russians, competing for hegemony in this area during the nineteenth century, agreed to tolerate it as a cushion, or **buffer state**, between them. This is how Afghanistan acquired the narrow extension leading from the main territory eastward to the Chinese border—the Wakhan Corridor (Fig. 6-19). As the colonialists delimited it, Afghanistan adjoined the domains of the Turkmen, Uzbeks, and Tajiks to the north, Persia (now Iran) to the west, and the western flank of British India (now Pakistan) to the east.

21

Geography and history seem to have conspired to divide Afghanistan. As Figure 6-1 shows, the towering Hindu Kush range dominates the center of the country, creating three broad environmental zones: the relatively well-watered and fertile northern plains and basins; the rugged, earthquake-prone central highlands; and the desert-dominated southern plateaus. Kabul, the capital, lies on the southeastern slope of the Hindu Kush, linked by narrow passes to the northern plains and by the Khyber Pass to Pakistan.

Across this variegated landscape moved countless peoples: Greeks, Turks, Arabs, Mongols, and others. Some settled here, their descendants today speaking Persian, Turkic, and other languages. Others left archeological remains or no trace at all. The present population of Afghanistan (28.1 million, plus 4 million refugees living outside the country) has no ethnic majority. This is a country of minorities in which the Pushtuns (or Pathans) of the east are the most numerous but make up less than 40 percent of the total (substantially less if those across the borders are subtracted). The second-largest minority are the Tajiks, a world away across the Hindu Kush, concentrated in the zone near Afghanistan's border with Tajikistan. The Hazara of the central highlands and the south, the Uzbeks and Turkmen in the northern border areas, the Baluchi of the southern deserts, and other, smaller groups scattered across this riven country create one of the world's most complex cultural mosaics (Fig. 6-20).

Episodes of conflict have marked the history of Afghanistan, but none was as costly as its involvement in the Cold War. Following the Soviet intervention of 1979, the United States supported the Muslim opposition, the *Mujahideen* ("strugglers"), with modern weapons and money, and the Soviets were forced to withdraw. Soon the factions that had been united during the anti-Soviet campaign were in conflict, delaying the return of some 4 million refugees who had fled to Pakistan and Iran. The situation resembled the pre-Soviet past: a feudal country with a weak and ineffectual government in Kabul.

In 1994, what at first seemed to be just another warring faction appeared on the scene: the so-called Taliban ("students of religion") from religious schools in Pakistan. Their avowed aim was to end Afghanistan's chronic factionalism by instituting strict Islamic law. Popular support in the war-weary country, especially among the Pushtuns, led to a series of successes, and by 1996 the Taliban had taken Kabul. Afghanistan's physical geography and northern opposition then slowed the Taliban's drive, but in 1998 they penetrated the Uzbek and Tajik strongholds and captured Mazar-e-Sharif, a key northern city.

The Taliban's imposition of Islamic law was so strict and severe that Islamic as well as non-Islamic countries objected. Restrictions on the activities of women ended their professional education, employment, and freedom of movement, and had a devastating impact on children as well. Public amputations and stonings enforced the Taliban's code.

While Pakistan, source of the movement, remained supportive of the new Sunni regime in Kabul, Iran expressed support for the Shi'ite Hazara minority and for the Persian-speaking Tajiks in the north.

Data-reporting agencies often list Afghanistan as the most blighted country in the world, with minimal incomes, grave social dislocation, high infant and child mortalities, and short life expectancies, harsh realities in this once-colonial, now-ideological Islamic frontier.

Region and Realm

When independence came to Turkestan's republics, it was welcomed by several sectors of the population: by former Soviets who anticipated that they would seize control in their respective countries, by Islamic groups hoping to regain the right to worship and spread the faith, by long-suppressed minorities that envisaged a better future, by intellectuals who hoped for democracy and greater freedom. These expectations were not always compatible. The Soviets had suppressed intraregional rivalries and ethnic disputes; their departure lifted those controls and stirred conflict. Furthermore, the end of Soviet rule led to migrations and associated instability. Peoples who had been exiled to Soviet Central Asia during the brutal communist regime saw their opportunity to return home. Others, fearful or uncertain about the future in the new Turkestan, left for Ukraine, Germany, or other destinations.

The end of Soviet rule nevertheless seemed to open new opportunities: for Western oil companies in search of joint ventures, for mining firms hoping to exploit known and yet-unknown reserves, for industrial corporations eager to employ low-cost labor. Then there were those who hoped to expand their nation's or their faith's influence: secular representatives from Turkey, Islamic proselytizers from Iran, merchants from Pakistan, managers from India.

A decade later, the level of expectation—and competition—is much lower. Government authoritarianism on the Soviet pattern continues; corruption, inefficiency, and the suppression of dissent arouse fears of instability that deter foreign involvement. Islamic fundamentalism in the form of the Wahhabi and Taliban movements further cloud the investment climate. Thus Turkestan, in the opening years of the twenty-first century, is still another component of this far-flung geographic realm to suffer from ethnic and cultural strife, its borders under pressure and its peoples divided. It confirms a central principle of regional geography: frameworks are temporary, change is permanent.

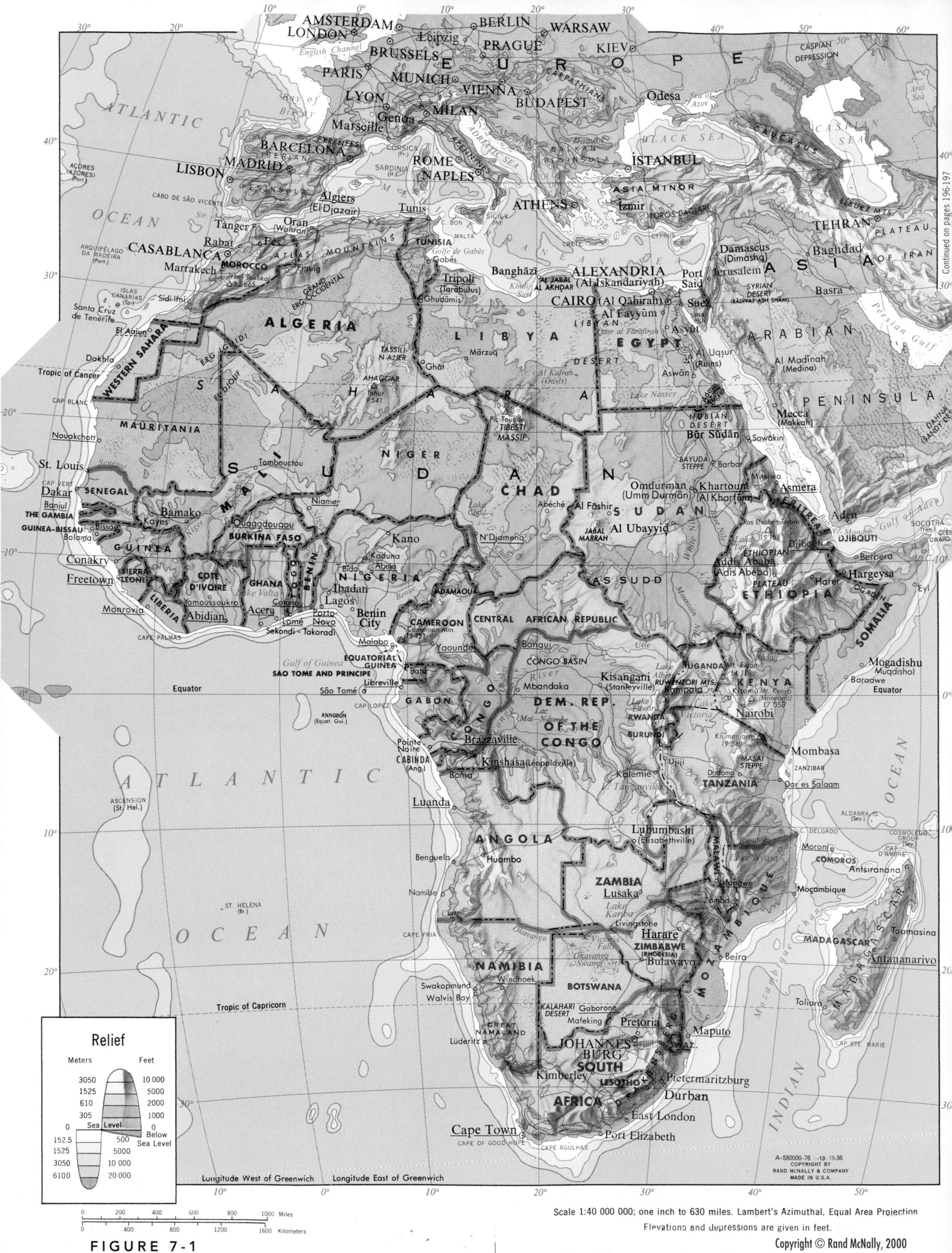

FIGURE 7-1

Copyright © Rand McNally, 2000

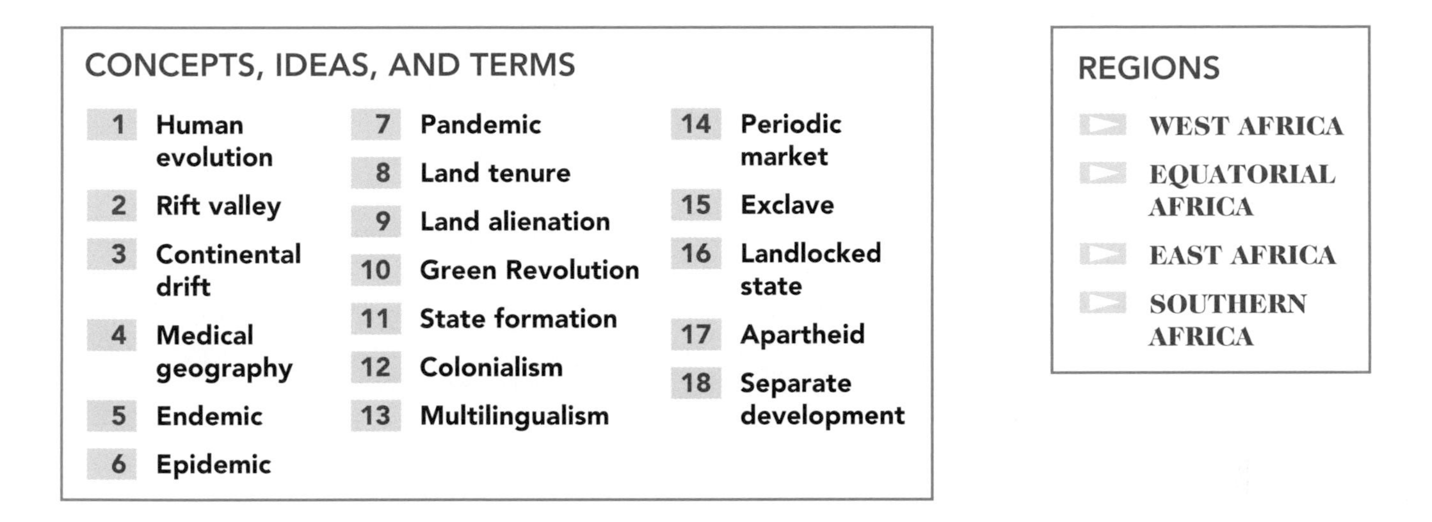

The African continent occupies a special place in the world. Beneath its surface lie some of the oldest parts of the planet's crust. Today Africa lies at the heart of the Earth's landmasses, a relative location that may yet become one of its greatest assets. And current archeological research indicates that our species, *Homo sapiens*, emerged in Africa. We may have spread around the globe, but at the source we are all Africans.

For millions of years, Africa was the stage for the great drama of **human evolution**. The first tools ever made were used by our Hominid ancestors in Africa. Language probably originated in Africa. The first sedentary human communities may well have formed in Africa. Our first artistic expression likely took place somewhere in Africa.

More recently, North Africa's Nile Valley was the scene of one of the earliest durable and creative civilizations, a culture hearth whose innovations radiated in all directions, Subsaharan Africa included. Ancient Egypt was to Africa what, thousands of years later, ancient Greece was to Europe: a source of knowledge and ideas. When West Africa's earliest states formed, their rulers modeled the political system on the Egyptian example.

In those times, more than 2000 years ago, the African continent was a single geographic realm. The partition of Africa, through the Islamization and Arabization of the north, the European colonization, and the desiccation of the Sahara, came later. Today, Africa is a continent of two geographic realms: the north, a part of the greater Islamic realm that connects it to Southwest Asia and beyond; and the rest—Subsaharan Africa—an Africa defined by languages, modes of life, and cultural landscapes.

DEFINING THE REALM

AFRICA'S PHYSIOGRAPHY

Before we investigate Subsaharan Africa's human geography, we should take note of the continent's unique physical geography (Fig. 7-1). To begin with, Africa has an unusual location. No other landmass is so squarely positioned astride the equator, reaching almost as far to the north as to the south. This location has much to do with Africa's vegetation, soils, agricultural potential, and population distribution. The continent lies entirely in tropical and subtropical latitudes.

◆ Major Geographic Qualities of Subsaharan Africa

1. Physiographically, Africa is a plateau continent without a linear mountain backbone, with a set of Great Lakes, variable rainfall, generally low-fertility soils, and mainly savanna and steppe vegetation.

2. Dozens of nations, hundreds of ethnic groups, and many smaller entities make up Subsaharan Africa's culturally rich and varied population.

3. Most of Subsaharan Africa's peoples depend on farming for their livelihood.

4. Health and nutritional conditions in Subsaharan Africa need improvement as the incidence of disease remains high and diets are often unbalanced. The AIDS pandemic began in Africa and has become a major health crisis in this realm.

5. Africa's boundary framework is a colonial legacy; many boundaries were drawn without adequate knowledge of or regard for the human and physical geography they divided.

6. The realm is rich in raw materials vital to industrialized countries, but much of Subsaharan Africa's population has little access to the goods and services of the world economy.

7. Patterns of raw-material exploitation and export routes set up in the colonial period still prevail in most of Subsaharan Africa. Interregional and international connections are poor.

8. During the Cold War, great-power competition magnified conflicts in several Subsaharan African countries, with results that will be felt for generations.

9. Severe dislocation affects many Subsaharan African countries, from Liberia to Rwanda. This realm has the largest refugee population in the world today.

10. Government mismanagement and poor leadership afflict the economies of many Subsaharan African countries.

Africa also accounts for about one-fifth of the Earth's entire land surface. The north coast of Tunisia lies 4800 miles (7700 km) from the south coast of South Africa. Coastal Senegal, on the extreme western *Bulge* of Africa, lies 4500 miles (7200 km) from the tip of the *Horn* in easternmost Somalia. These distances have environmental implications. Much of Africa is far from maritime sources of moisture. In addition, as Figure I-8 shows, large parts of the landmass lie in latitudes where global atmospheric circulation systems produce arid conditions. The Sahara in the north and the Kalahari in the south form part of this globe-girdling desert zone. Water supply is one of Africa's great problems.

Mountains

When we examine Africa's topographic map, however, it becomes clear that the adjective "unique" describes the continent's physiography. Take, for example, the distribution of the world's linear mountain ranges. Every major landmass has at least one mountainous backbone: South America's Andes, North America's Rocky Mountains, Europe's Alps, Asia's Himalayas. Yet Africa, covering one-fifth of the land surface of the Earth, has nothing comparable. The Atlas Mountains of the far north occupy a mere corner of the landmass, and the Cape Ranges of the far south are not of continental dimensions. And where Africa does have high mountains, as in Ethiopia and South Africa, these are really deeply eroded plateaus—or, as in East Africa, high snowcapped volcanoes. Missing in Africa are those elongated, parallel ranges of the Andes or Alps.

Great Lakes and Rift Valleys

This discovery stimulates us to look for other unusual features of Africa's physiographic map and to summarize them on an outline map that highlights its distinctive properties (Fig. 7-2). What else is noteworthy about Africa's physical geography? As Figure I-1 shows, Africa is one of the only two continents that contain a cluster of Great Lakes. Those of North America owe their origins to the work of glaciers; those of Africa, congregated in the east, result from powerful tectonic forces in the Earth's crust.

Except for Lake Victoria, East Africa's lakes are remarkably elongated, from Lake Malawi in the south to Lake Turkana in the north. What causes this elongation and the persistent north-south alignment that we can observe in these lakes? The lakes occupy parts of deep trenches that cut through the East African Plateau, trenches that extend well beyond the lakes themselves. Northeast of Lake Turkana, such a trench cuts the Ethiopian Highlands into two sections, and the entire Red Sea looks much like a north-

FIGURE 7-2

ward continuation of it. On both sides of Lake Victoria, smaller lakes lie in similar trenches, of which the western one runs into Lake Tanganyika and the eastern one extends completely across Kenya, Tanzania, and Malawi (Fig. 7-2).

2 The technical term for these trenches is **rift valleys**. As the name implies, they are formed when huge parallel cracks or faults appear in the Earth's crust and the strips of crust between them sink or are pushed down to form great linear valleys. Altogether, these rift valleys stretch more than 6000 miles (9600 km) from the north end of the Red Sea to Swaziland in Southern Africa. In general, the rifts from Lake Turkana southward are between 20 and 60 miles (30 and 90 km) wide, and the walls, sometimes sheer and sometimes step-like, are well defined.

River Courses

Next, we must note Africa's unusual river systems (Fig. 7-2). Africa has several great rivers, with the Nile and Congo ranking among the most noteworthy in the world. The Niger rises in the far west of Africa, on the slopes of the Futa Jallon Highlands, but first flows inland toward the Sahara. Then, after forming an interior delta, it suddenly elbows southeastward, leaves the desert, and plunges over falls as it cuts through the plateau area of Nigeria, creating another large delta at its mouth. The Congo River begins as the Lualaba River on the boundary between The Congo* and Zambia; for some distance, it flows northeast before turning north, then west and southwest, finally cutting through the Crystal Mountains to reach the ocean. Note that the upper courses of these first two rivers appear to be unrelated to the continent's coasts where they eventually exit. For the Zambezi River, whose headwaters lie in Angola and northwestern Zambia, the situation is the same; the river first flows south, toward the inland delta known as the Okavango Swamp, and then turns northeast and southeast, eventually to reach its delta south of Lake Malawi. Finally, there is the famed erratic course of the Nile River, which braids into numerous channels in the Sudd area of southern Sudan and, in its middle course, actually reverses direction and flows southward before resuming its flow toward the Mediterranean delta in Egypt. With so many peculiarities among Africa's river courses, could all of them have been affected by the same event at some time in the continent's history? Perhaps—but first let us look further at the map.

Plateaus and Escarpments

All continents have low-lying areas—witness the Gulf-Atlantic Coastal Plain of North America or the riverine lowlands of Eurasia and Australia. But as Figure 7-1 shows, coastal lowlands are few and of limited extent in Africa. In fact, we can call Africa a *plateau continent*; except for coastal Moçambique and Somalia and along the northern and western coasts, almost the entire continent lies above 1000 feet (300 m) in elevation, and fully half of it is over 2500 feet (800 m) high. Even the Congo Basin, Equatorial Africa's huge tropical lowland, lies well over 1000 feet above sea level in contrast to the much lower-lying Amazon Basin across the Atlantic.

Although Africa is mostly plateau, its surface is not completely flat and unbroken. In the first place, rivers have been eroding that surface for millions of years and have made some fairly good cuts in it. For example, Victoria Falls on the Zambezi is 1 mile (1.6 km) wide and over 300 feet (90 m) high. Volcanoes and other types of mountains, some of them erosional leftovers, stand well above the landscape in many areas. The Sahara, where both the Ahaggar and Tibesti mountains reach about 10,000 feet (3000 m) in elevation, is no exception. In several places the plateau has sagged under the weight of accumulating sediments. In the Congo Basin, rivers transported sand and sediment downstream for tens of millions of years and dropped their erosional loads into what was once a gigantic lake the size of an interior sea. Today the lake is gone, but the thick sediments that press this part of the African surface into a giant basin prove that it was there. And, as Figure 7-2 shows, this was not the only inland sea. To the south, the Kalahari Basin was filling with sediments that now constitute that desert's sand; far to the north, in the Sahara, three similar basins lie centered on Sudan, Chad, and what today is Mali (the Djouf Basin).

The margins of Africa's plateau are significant, too. Much of the continent, because it is a plateau, is surrounded by an escarpment. In Southern Africa, where this feature is especially pronounced, the Great Escarpment (as it is called there) marks the plateau's edge for hundreds of miles; here the land drops precipitously from more than 5000 feet (1500 m) in elevation to a narrow, hilly coastal belt. From Congo to Swaziland and intermittently on or near most of the African coastline, a scarp bounds the interior upland. Such escarpments also are found in other parts of the world: in Brazil at the eastern margins of the Brazilian Highlands, and in India at the western edge of its Deccan Plateau. But Africa, even for its size, has a disproportionately large share of this topographic phenomenon.

Continental Drift and Africa

Africa's remarkable and unusual physiography was one piece of evidence that the geographer Alfred Wegener, early in the twentieth century, used to construct his hypothesis of **continental drift**. According to this idea (first introduced on pp. 8-9), all the landmasses on Earth were assembled into one giant continent named *Pangaea*; the southern continents constituted *Gondwana*, the southern part of this supercontinent (Fig. 7-3). After a long (geologic) period of unity, this huge landmass began to break up more than 200 million years ago. Africa, which lay at the heart of Gondwana, attained the approximate shape we see today when North and South America, Antarctica, Australia, and India drifted radially away. Later, geophysicists gave the name *plate tectonics* to this process, and we now understand that the breakup of Pangaea was only the most recent in a series of continental collisions and separations that spans much of the Earth's history (see Fig. I-4).

Africa's situation at the heart of the supercontinent explains much of what we see in its landscapes today. The Great Escarpment is a relic of the gigantic faults (fractures) that formed when the neighboring landmasses split off. The rift valleys are only the most recent evidence of the pulling

*Two countries in Africa have the same short-form name: Congo. In this book, we use *The Congo* for the larger Democratic Republic of the Congo and *Congo* for the smaller Republic of Congo.

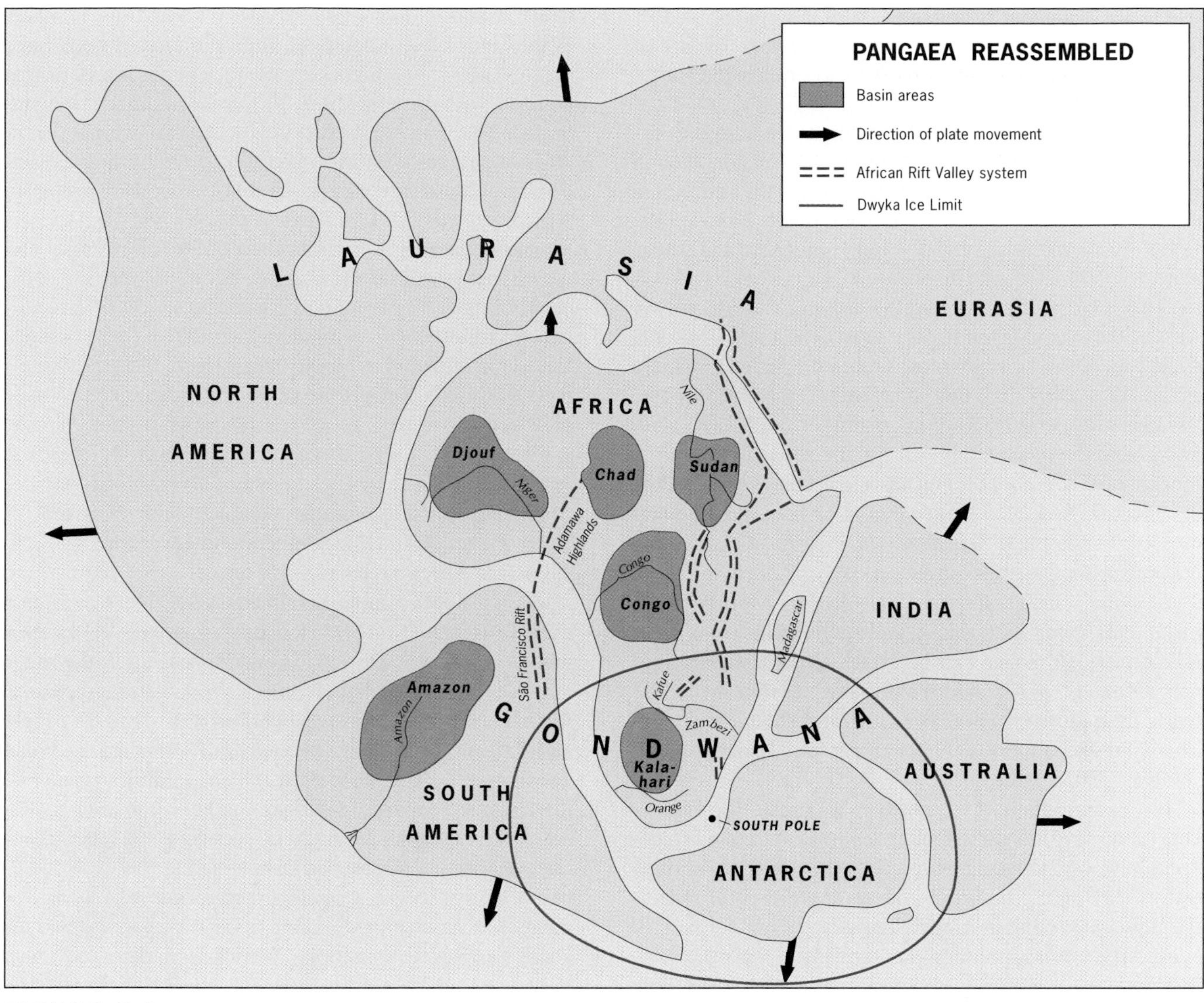

FIGURE 7-3

forces that affect the African Plate; the Red Sea is an advanced stage of such rifts, and we may expect East Africa to separate from the rest of Africa much as Arabia did earlier (and Madagascar before that). Africa's rivers once filled lakes within the continent, but they did not reach the distant ocean. Today, the lakes are drained by rivers that eroded inland when the oceans began to wash Africa's new shorelines.

And why does Africa lack those mountain chains that led us to look at the map more closely? The answer may lie in the direction and distance of plate motion. South America moved far westward, its plate colliding with the plate under the waters of the Pacific Ocean; the Andes crumpled up, accordion-like, in the collision. India moved northeastward, wedging into the Asian landmass; similarly, the Himalayas rose upward. But Africa moved comparatively little, being more affected by pulling, tensional forces that create rifts than by the converging pressures of plate collision that generate the uplifting of mountains. The physiographic map decidedly reveals more than just the location of rivers, plains, and lakes.

NATURAL ENVIRONMENTS

Africa lies astride the equator. All of Africa south of the Sahara, except its southernmost tip, lies in the tropics. There, elevations are low and temperatures are relentlessly high. At higher elevations, temperatures moderate, but much of Africa's interior plateau lies far from the ocean or is under the influence of the Earth's mid-latitude high-pressure belts. That combination of factors subjects much of this continent to variable weather and frequent droughts. Figures I-7 and I-8 provide the evidence: note how much of this realm lies under *B*-climate conditions. And remember that high equatorial temperatures reduce the efficacy of higher precipitation there. Forty inches of rain in the hot savanna is not much.

As Figure I-8 shows, Africa's climatic regions are distributed almost symmetrically about the equator, though more so in the center of the landmass than in the east, where elevation changes the picture. The hot, rainy climate of the Congo Basin merges gradually, both northward and southward, into climates with distinct winter-dry seasons. "Winter," however, is marked more by drought than by cold. In parts of the area mapped *Aw* (savanna), the annual seasonal cycle produces two rainy seasons, often referred to locally as the "long rains" and the "short rains," separated by two "winter" dry periods. As you go farther north and south, away from the moist Congo Basin, the dry season(s) grow longer and the rainfall diminishes and becomes less and less dependable. Annual averages shown on maps are virtually meaningless: a place that reports a 15-inch yearly rainfall may reach this "average" in successive years showing 11, 5, 29, 23, and 7 inches during a half-decade. Most Africans make their living by farming, and such variability of rainfall in marginal zones can be catastrophic. Compare Figures I-8 and I-7 to get an idea of the steep decline in precipitation from more than 80 inches (200 mm) at the equator in The Congo to a mere 4 inches in parts of Namibia in the south and Chad in the north.

In our Introduction we noted that climate and natural vegetation are linked; climatic regions, therefore, sometimes have vegetative names (such as the equatorial rainforest or *Af* climate). In Africa, dense rainforest still reflects the prevailing climate in the Congo Basin and in parts of West Africa, but population pressure and associated deforestation are shrinking it. Flying over the canopy you now see great swaths of cleared land, where the trees have been felled, the stumps burned, and crops planted. Lumbering also takes its toll as Africa's magnificent hardwoods are exported to the economically richer world.

Beyond what remains of the rainforest, where precipitation declines and dry seasons appear, the forest yields to the tree-studded grasslands of the *savanna*, a sometimes parklike vegetation with open grasslands and flat-topped shade trees. Here the farmers mix agriculture with pastoralism, but diseases endemic to the savanna's herds of wildlife have kept the livestock population down. Travel farther north (or south), and you will see the grass become shorter and thinner, patches of bare earth appear, trees yield to bush. Now the savanna gives way to the *steppe*, the sparse grassland that is vulnerable to overgrazing. Here, on the margins of the desert, ecologies are fragile, but human needs continue to grow. Where there is not enough pasture for cattle, goats roam the countryside, surviving the dry seasons but causing desertification when the rains fail.

Africa's rainforests and savannas form the world's last refuges for wildlife ranging from primates to wildebeests. Gorillas and chimpanzees survive in dwindling numbers in threatened forest habitats, while millions of herbivores range in great herds across the savanna plains where people compete with them for space. European colonizers, who introduced hunting as a "sport" (a practice that was not part of African cultural traditions) and who brought their capacities for mass destruction to animals as well as people in Africa, helped clear vast areas of wildlife and push species to near-extinction. Later they laid out game reserves and other types of conservation areas, but these were not sufficiently large or well enough connected to allow herd animals to follow their seasonal and annual migration routes. The same climatic variability that affects farmers also affects wildlife, and when the rangelands wither, the animals seek better pastures. When the fences of a game reserve wall them off, they cannot survive. When there are no fences, the wildlife invades neighboring farmlands and destroys crops, and the farmers retaliate. After thousands of years of equilibrium, the competition between humans and animals in Africa has taken a new turn. It is the end of an era.

As we note in more detail later, the fate of Africa's wildlife is bound up with the rate of growth of Africa's human population. During the middle decades of the twentieth century, Subsaharan Africa's population increased faster than that of any other world realm. This set people and wildlife on an unprecedented collision course. African governments did much to protect local wildlife by buttressing and even expanding national parks, combating poaching, protecting wildlife-migration corridors, and other means (Kenya, for example, banned all hunting for a time). Wildlife-based tourism became an important source of income for many African countries, adding a sense of purpose and urgency to conservation efforts. Means were devised to give local people, those actually living in contact with the wildlife, a stake in preservation.

And yet Subsaharan Africa, given its enormous size, is not a densely populated realm by world standards. All its countries *combined* have a population only half that of China alone. Why is space such a problem? One answer lies—again—in the natural environment. Travel across Africa, and you will be impressed by its vast open spaces. But stop at a road cut and examine the reddish-yellow soil, and you know one of the reasons: the infertility of the soils many Africans must depend on. The lack of long-term fertility of rainforest-area soils is well known: cut down the forest, and crops will grow for only a year or two before the farmer must move on and repeat the process. Tropical forest soils are excessively leached and support the forest only because the forest supports itself, using nutrients from its own decaying biomass. Nor are savanna soils, also loaded with iron and aluminum but few nutrients, easily farmed. The sun bakes them hard and dry; the farmer must water and fertilize and turn them to ensure a crop. But water is scarce and fertilizer expensive, so yields tend to be small. Food production in Subsaharan Africa has declined over the past several decades, and soil exhaustion is one of the causes.

Africa does have areas of good soils, ample water, and high productivity: the volcanic soils of Mount Kilimanjaro and those of the highlands around the Western Rift Valley, the soils in the Ethiopian Highlands, those in the moister areas of higher-latitude South Africa and parts of West Africa yield good crops when social conditions do not disrupt the farming communities. But such areas are small in the vastness of Africa. This realm has nothing to compare to the vast alluvial basins of India and China, not even a Nile Valley and Delta, where comparatively small areas of fertile, irrigated soils can support tens of millions of people. Except for small (in some cases experimental) patches of rice and wheat, Africa is the land of corn (maize), millet, and root crops, far less able to provide high per-acre yields. Africa's natural environment poses a formidable challenge to the millions who depend directly on it.

ENVIRONMENT AND HEALTH

Africa's natural environment creates other difficulties as well. High temperatures and humidity, where these prevail, create breeding grounds for organisms that carry disease. Mosquitoes, flies, fleas, worms, even snails can transmit such diseases. From birth, Africans (especially the people living in *A* climates) are exposed to a wide range of dangerous diseases. Over the centuries they have suffered from repeated outbreaks of killer plagues, some of them related to the wildlife that shares their habitat.

4 **Medical geography** studies people's health in spatial context. Medical geographers employ methods of spatial analysis to track disease outbreaks, predict their diffusion, identify their sources, detect their carriers, and prevent their repetition. Today GIS (geographic information systems) and other modern methods are employed to achieve these goals, and the alliance between doctors and geographers has already yielded significant results. Doctors know how a disease ravages the body; geographers know how climatic conditions such as wind direction or variations in river flow can affect the distribution and effectiveness of disease carriers. This collaboration helps protect vulnerable populations.

Tropical Africa, the source of many serious illnesses, is the focus of much of medical geography's work. Not only the carriers (*vectors*) of infectious diseases but also cultural traditions that facilitate transmission, such as sexual practices, food selection and preparation, and personal hygiene, play their role—and all can be mapped. Comparing medical, environmental, and cultural maps can lead to crucial evidence that helps combat the scourge.

In Africa today, hundreds of millions of people carry one or more maladies, often without knowing exactly what ails them. A disease that infects many people (the *hosts*) in a kind of equilibrium, without causing rapid and widespread deaths, is said to be **endemic** to the population. People affected may not die suddenly or dramatically, but their health deteriorates, energy levels fall, and the quality of life declines. In tropical Africa hepatitis, venereal diseases, and hookworm are among the public health threats in this category. 5

When a disease outbreak has local or regional dimensions, it is called **epidemic**. It may claim thousands, even tens of thousands of lives, but it remains confined to a certain area, perhaps one defined by the range of its vector. In tropical Africa, trypanosomiasis, the disease known as sleeping sickness and vectored by the tsetse fly, has regional proportions. The great herds of savanna wildlife form the *reservoir* of this disease, and the tsetse fly transmits it to livestock and people. It is endemic to the wildlife, but it kills cattle, so Africa's herders try to keep their animals in tsetse-free zones. African sleeping sickness appears to have originated in a West African source area during the fifteenth century, and it diffused throughout much of tropical Africa (Fig. 7-4). Its epidemic range was limited by that of the tsetse fly: where there are no tsetse flies, there is no sleeping sickness. More than anything else, the tsetse fly has kept Subsaharan Africa's savannalands largely free of livestock and open to wildlife. Should a remedy be found, livestock would replace the great herds on the grasslands. 6

When a disease spreads worldwide, it is described as **pandemic**. Africa's and the world's most deadly vectored disease is malaria, transmitted by a mosquito and killer of as many as 1 million children each year. Whether malaria has an African origin is not known, but it is an ancient affliction. 7

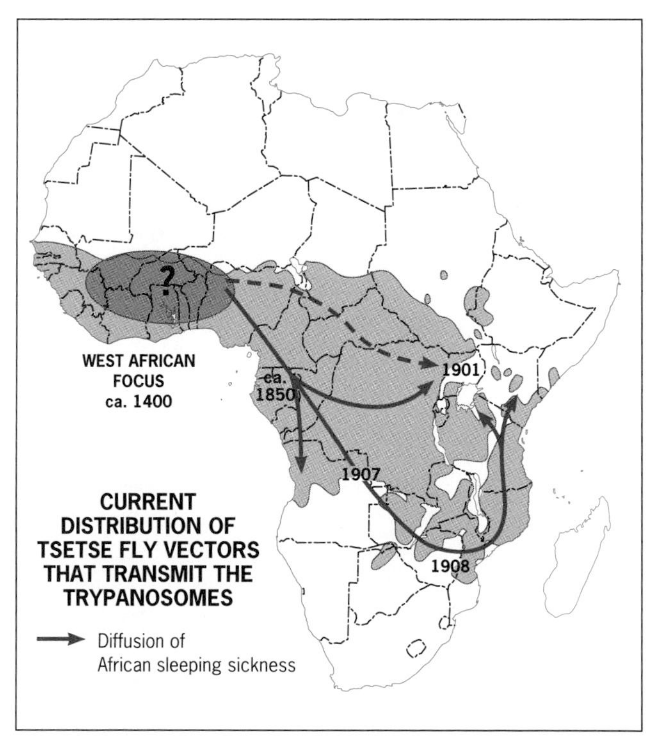

FIGURE 7-4

AIDS in Subsaharan Africa

AIDS—Acquired Immune Deficiency Syndrome—spread in Africa during the 1980s and became a pandemic during the 1990s. Its impact on Africa is devastating.

Persons infected by HIV (human immunodeficiency virus) do not immediately or even soon display symptoms of AIDS. In the early stages, only a blood test will reveal infection, and then only by indicating that the body is mobilizing antibodies to fight HIV. People can carry the virus for years without being aware of it; during that period they can unwittingly transmit it to others. For that reason, official statistics of AIDS cases lag far behind the reservoir of those infected, especially in countries where populations cannot be thoroughly tested.

At the opening of the twenty-first century, more than 32 million people were known to be infected by HIV worldwide, and of these about 27 million lived in 34 tropical African countries according to United Nations sources. In the early 1990s, the worst-hit countries were in Equatorial Africa in what was called the AIDS Belt from (then) Zaïre to Kenya. But by the end of the decade the most severely afflicted countries lay in Southern Africa. In Zimbabwe and Botswana, more than 25 percent of all persons 15–49 were infected with HIV; in Zambia, nearly 20 percent; and in South Africa, the region's most populous country, about 13 percent. These are the official numbers; medical geographers estimate that 20 to 25 percent of the *entire* population of several tropical African countries is infected.

As the number of deaths from AIDS rises, African countries' vital statistics are showing the demographic impact. In Botswana, life expectancy in 1994 was 60 years; in 1999 it had declined to 39, and, UN specialists predict, it will drop further during the first decade of the twenty-first century. In Zimbabwe, population growth was 3.3 percent per year in the early 1980s; now it is 1.0 percent. Another UN report said that more than 10 million people had died of AIDS in Africa between 1990 and 1998.

Geographer Peter Gould, in his 1993 book *The Slow Plague*, called Africa a "continent in catastrophe," pointing out that the official data on which these depressing reports are based understate the magnitude of the calamity. African governments, Gould wrote, conceal the real dimensions of the tragedy for various reasons: a sense of shame, concern that the real figures will drive investors out and keep tourists away, and the arrogance of military rulers used to denying reality. Many doctors in Africa, he added, say that AIDS cases are 80 to 90 percent underreported.

No part of tropical Africa is spared. West African countries, too, are reporting growing numbers of AIDS cases. Ivory Coast, reporting that more than 10 percent of adults are infected there, may provide the most accurate data. In Nigeria, AIDS incidence among adults is more than 4 percent (which, in that populous country, means that more than 5 million people are ill with AIDS).

The misery the AIDS pandemic has created is beyond measure, but its economic impact is beyond doubt. Workers are sick on the job; but jobs are scarce and medical help is unavailable or unaffordable for many, so the workers stay on the job until they collapse. In a realm where unemployment is high, new workers immediately fill the vacancies, but they, too, are unwell. Corporations report skyrocketing AIDS-related costs. In Botswana, whose economy has been prospering due to increased diamond production, highly trained diamond sorters are being lost to AIDS, and training replacement workers is depressing profits. Medical expenses, multiplying death benefits, funeral payments, and recruiting and training costs are reducing corporate incomes from Kenya's sugar plantations to South Africa's gold mines. Thus the impact of AIDS is recorded in the continent's economic statistics. The pattern shown in Figure I-11, where Africa is concerned, is due in substantial part to the AIDS pandemic.

In countries of the global core, progress has been made in containing AIDS and prolonging the lives of those afflicted. This has taken AIDS off the medical center stage, but such a turnaround is not in prospect for Africa. The cost of the medications that combat AIDS is too high, and private drug companies will not dispense their products at prices affordable to African patients. When former President Clinton, during his 2000 visit to Nigeria, proposed a $1 billion loan to help the fight against AIDS, Africans argued that the last thing Africa needed was more debt. Thus the best short-term strategy is for governments to attempt to alter public attitudes and behavior, notably by making condoms available as widely and freely as possible. Uganda, once one of tropical Africa's worst-afflicted countries, managed to slow the rate of AIDS dispersal by these methods—and through a vigorous advertising campaign on billboards, in newspapers, and on radio and television.

Even so, Uganda's adult infection rate in 2000 still was 9.5 percent officially. Medical geographers and others are predicting that Africa's worst-affected countries may see population declines of as much as 10 to 20 percent, a disaster comparable only to Europe's bubonic plague of the fourteenth century and the collapse of Native American populations after the introduction of smallpox by the European invaders in the sixteenth. Once again environmental, historical, and cultural factors have combined to torment the peoples of Subsaharan Africa.

Hippocrates, the Greek physician of the fifth century BC, mentions it in his writings. Apes, monkeys, and several other species also suffer from it. Fever attacks, anemia, and enlargement of the spleen are its symptoms. Malaria has diffused around the world and prevails not only in tropical but also in temperate areas. Eradication campaigns against the mosquito vector have had some success, but always the carrier has come back with renewed vigor. At present as many as 300 million people annually are recorded as having malaria, but the actual number probably is much greater because in much of tropical Africa malaria is, simply, a way of life affecting entire populations. The short life expectancies for tropical Africa reported in Table I-2 reflect infant and child mortality from malarial infection.

Another pandemic disease with African origins is yellow fever, also vectored by a mosquito. African monkeys form the reservoir for this scourge, which once diffused worldwide but is now largely confined to Africa and South America. From its source area in West Africa, yellow fever not only reached South America but diffused into Central America and even the United States; its defeat in Panama nearly a century ago made possible the construction of the Panama Canal. In Africa it remains a threat: Senegal had an outbreak during the 1960s that claimed more than 20,000 lives.

During the 1970s, Africa was afflicted by still another malady that appears to have started in the equatorial forest, became an epidemic, and grew into a pandemic: AIDS (see box titled "AIDS in Subsaharan Africa"). Monkeys may have activated this disease in humans, but whatever the source, Africa has been the most severely affected by it. In 2000, there were an estimated 32 million AIDS cases in the world; more than 27 million were in Africa. Again, these official calculations probably conceal a much worse crisis, and the population growth rates for many African countries are declining as a result. During the early part of the twenty-first century, some countries will experience actual population decreases. Life expectancies are dropping to levels not seen for centuries.

The great incubator of vectored and nonvectored diseases in humid equatorial Africa continues to threaten local populations. Sudden outbreaks of Ebola fever in Sudan during the 1970s, in The Congo in the 1990s, and in Uganda in 2000-2001 projected the risks that many Africans live with onto the world's television screens. But Africa's woes are soon forgotten. Only the fear that an African epidemic might evolve into a global pandemic mobilizes world attention. Africans cope with the planet's most difficult environments, but they are least capable of combating its hazards because their resources are so limited.

Even when international efforts are made to assist Africa, the unintended results may be catastrophic. Among Africa's endemic diseases is schistosomiasis, also called bilharzia. The vector is a snail, and the transmitted parasites enter via body openings when people swim or wash in slow-moving water where the snails thrive (fast-moving streams are relatively safe). When development projects designed to help African farmers dammed rivers and streams and sent water into irrigation ditches, farm production rose—but so did the incidence of schistosomiasis, which causes internal bleeding and fatigue, although it is rarely fatal. Medical geographers' maps showed the spread of schistosomiasis around the dam projects, and soon the cause was clear: by slowing down the water flow, the engineers had created an ideal environment for the snail vector.

Poverty is a powerful barrier to improving health, and major interventions in the natural environment carry costs as well as benefits. Africa needs improved medical services, more effective child-immunization programs, mobile clinics for remote areas, and similar remedies. While 71 percent of its people continue to live in rural areas, the natural environment will weigh heavily on a vulnerable population.

LAND AND FARMING

In their penetrating book, *Geography of Sub-Saharan Africa*, editor Samuel Aryeetey-Attoh and his colleagues focus on the issue of land tenure, crucial in Africa because most Africans are farmers. **Land tenure** refers to the way people own, occupy, and use land. African traditions of land tenure are different from those of Europe or the Americas. In most of Subsaharan Africa, communities, not individuals, customarily hold land. Occupants of the land have temporary, custodial rights to it and cannot sell it. Land may be held by large (extended) families, by a village community, or even by a traditional chief who holds the land in trust for the people. His subjects may house themselves on it and farm it, but in return they must follow his rules.

8

When the European colonizers took control of much of Subsaharan Africa, their land ownership practices clashed head-on with those of Africa. Africans believed that their land belonged to their ancestors, the living, and the yet-unborn; Europeans saw unclaimed space and felt justified in claiming it. What Africans called **land alienation**—the expropriation of (often the best) land by Europeans—changed the pattern of land tenure in Africa. By the time the Europeans withdrew, private land ownership was widely dispersed and could not be reversed. Postcolonial African states tried to deal with this legacy by nationalizing all the land and doing away with private ownership, reverting in theory to the role of the traditional chief. But this policy has not worked well. In the rural areas, the government in the capital is a remote authority often seen as unsympathetic to the plight of farmers. Governments keep the price of farm products low on urban markets, pleasing their supporters but frustrating farmers. Large landholdings once owned by Europeans seem now to be occupied by officials or those the government favors. The colonial period's approach to land tenure left Africa with a huge social problem.

9

Shifting agriculture in Africa's tropical forest zones still forms the mainstay for millions. Here in the Ituri Forest in the northeastern corner of The Congo, trees have been felled, the land cleared, and between stumps and rotting trunks the people farm cassava, yams, and bananas. The soil will sustain these crops for some years, but then yields will diminish and the land will be abandoned, so that another plot must be cleared. Women do the traditional farming in Africa; here three women from the village of Ngodi are weeding the field while two youngsters look on.

Rapid population growth makes the problem worse. Traditional systems of land tenure, which involve subsistence farming in various forms ranging from shifting cultivation to pastoralism, work best when population is fairly stable. Land must be left fallow to recover from cultivation and pastures kept free of livestock so the grasses can revive. A population explosion of the kind Africa experienced during the mid-twentieth century destroys this equilibrium. Soils cannot rest, and pastures are overgrazed. As land becomes degraded, yields decline.

This is the situation in much of Subsaharan Africa, where more than 70 percent of the people depend on farming for their livelihood. That percentage is slowly declining, but it remains the highest for any realm in the global periphery. Subsaharan Africa is sometimes thought of as a storehouse of minerals and fuels and great underground riches, but farming is the key to its economy. And not just subsistence farming: agricultural exports also produce foreign exchange revenues needed to pay for essential imports. The European colonizers identified suitable environments and laid out plantations to grow cocoa, coffee, tea, and other luxury crops for the markets of the core. Irrigation schemes enabled the export of cotton, groundnuts (peanuts), and sesame seeds. Today, national governments and some private owners continue to operate these projects.

But most African farmers are subsistence farmers who grow grain crops (corn, millet, sorghum) in the drier areas and root crops (yams, cassava, sweet potatoes) where moisture is ample. Shifting cultivation still occurs in remote parts of The Congo and neighboring locales in Equatorial Africa, but this practice is dwindling as the forest shrinks. Others are pastoralists, driving their cattle and goats along migratory routes to find water and pasture, sometimes clashing with sedentary farmers whose fields they invade. Whatever their principal mode of farming, however, all African farmers today feel the effects of population pressure and must cope with dwindling space and damaged ecologies. By *intercropping*, planting several types of crops in one cleared field, some of which resemble those in the forest, shifting cultivators extend the life of their plots. In *compound farming*, usually near a market center, farmers combine the use of compound (village) and household waste as fertilizer with intensive care of the crops to produce vegetables, fruits, and root crops as well as eggs and chickens for urban consumers. Other subsistence farmers use *diversification* to cope with such problems; they combine their cultivation of subsistence crops with nonfarm work, including full-time jobs in nearby towns.

Notwithstanding all these adaptations, African farm production is declining for many reasons, including government mismanagement and corruption. When Nigeria became independent in 1960, its government embarked on costly industrialization, and agriculture was given a lower priority. In that year, Nigeria was the world's largest producer and leading exporter of palm oil; today, Nigeria must import palm oil to meet its domestic demand. Inadequate rural infrastructures create another obstacle. Roads from the rural areas to the market centers are in bad and worsening shape, often impassable during the rainy season and rutted when dry. Ineffective links between producers and consumers create economic losses. Storage facilities in the villages tend to be inadequate, so that untold quantities of grains, roots, and vegetables are lost to rot, pests, and theft. Electricity supply in rural areas also requires upgrading: without power, pumps cannot be run for water supply, and the processing of produce cannot be mechanized. Rural farmers have inadequate access to credit: banks would rather lend to urban homeowners than to distant villagers. This is especially true for women who, according to current estimates, produce 75 percent of the food in Subsaharan Africa.

Along the banks of Africa's rivers and on the shores of the Great Lakes fishing augments local food supplies, but Africa's fishing industries are still underdeveloped and contribute only modestly to overall nutrition. During the postcolonial period, Subsaharan Africa's population has more than doubled; food production has declined. Agricultural exports have also decreased, but food imports have mushroomed. The situation is especially clear in terms of per-capita output: in 1999 it was 10 percent below that in 1989. Even advances in biotechnology that closed the global gap between food production and demand largely passed Africa by (see box titled "A Green Revolution for Africa?")

A Green Revolution for Africa?

The **Green Revolution**—the development of more productive, higher-yielding types of grains—has narrowed the gap between world population and food production. Where people depend mainly on rice and wheat for their staples, the Green Revolution pushed back the specter of hunger. The Green Revolution has had less impact in Subsaharan Africa, however. In part, this relates to the realm's high rate of population growth (which is substantially higher than that of India or China). Other reasons have to do with Africa's staples: rice and wheat support only a small part of the realm's population. Corn (maize) supports many more, along with sorghum, millet, and other grains. In moister areas, root crops, such as the yam and cassava, and the plantain (similar to the banana) supply most calories. These crops were not priorities in Green Revolution research.

Lately there have been a few signs of hope. Even as terrible famines struck several parts of the continent and people went hungry in places as widely scattered as Moçambique and Mali, scientists worked toward two goals: first, to develop strains of corn and other crops that would be more resistant to Africa's virulent crop diseases, and second, to increase the productivity of those strains. But these efforts faced serious problems. An average African acre planted with corn, for example, yields only about half a ton of corn, whereas the world average is 1.3 tons. A virus-resistant corn variety has now been developed that also yields more than the old variety, and it has been introduced throughout Subsaharan Africa. In Nigeria, where it was first distributed, yields rose significantly, contributing to a near doubling of farm production in that country since 1980 according to UN data. Hardier types of root crops also raised yields significantly.

But Africa needs even more than this to reverse the cycle of food deficiency. (Nor is the Green Revolution an unqualified remedy: the poorest farmers, who need the help most, can least afford the more expensive higher-yielding seeds and also cannot pay for the pesticides that may be required.) It has been estimated that, for the realm as a whole, food production has fallen about 1 percent annually even as population grew by as much as 3 percent. A lack of capital, inefficient farming methods, inadequate equipment, soil exhaustion, male dominance, apathy, and devastating droughts contributed to this decline. The seemingly endless series of civil conflicts (in Uganda, The Congo, Liberia, Angola, and Rwanda) also reduced farm output. The Green Revolution may narrow the gap between production and need, but the battle for food sufficiency in Africa is far from won.

AFRICA'S HISTORICAL GEOGRAPHY

Africa is the cradle of humanity. Archeological research has chronicled 7 million years of transition from Australopithecenes to Hominids to *Homo sapiens*. It is therefore ironic that we know comparatively little about Subsaharan Africa from 5000 to 500 years ago—that is, before the onset of European colonialism. This is partly due to the colonial period itself, during which African history was neglected, many African traditions and artifacts were destroyed, and many misconceptions about African cultures and institutions became entrenched. It is also a result of the absence of a written history over most of Africa south of the Sahara until the sixteenth century—and over a large part of it until much later than that. The best records are those of the savanna belt immediately south of the Sahara, where contact with North African peoples was greatest and where Islam achieved a major penetration.

The absence of a written record does not mean, as some scholars have suggested, that Africa does not have a history as such prior to the coming of Islam and Christianity. Nor does it mean that there were no rules of social behavior, no codes of law, no organized economies. Modern historians, encouraged by the intense interest shown by Africans generally, are now trying to reconstruct the African past, not only from the meager written record but also from folklore, poetry, art objects, buildings, and other such sources. Much has been lost forever, though. Almost nothing is known of the farming peoples who built well-laid terraces on the hillsides of northeastern Nigeria and East Africa or of the communities that laid irrigation canals and constructed stone-lined wells in Kenya; and very little is known about the people who, perhaps a thousand years ago, built the great walls of Zimbabwe. Porcelain and coins from China, beads from India, and other goods from distant sources have been found in Zimbabwe and other points in East and Southern Africa, but the trade routes within Africa itself—let alone the products that circulated on them and the people who handled them—still remain the subject of guesswork.

FROM THE FIELD NOTES

"Got up before dawn this date in the hope of being the first visitor to the Great Zimbabwe ruins, a place I have wanted to see ever since I learned about it in a historical geography class. Climbed the hill and watched the sun rise over the great elliptical 'temple' below, then explored the maze of structures of the so-called fortification on the hilltop above. Next, for an incredible 90 minutes I was alone in the interior of the great oval walls of the temple. What history this site has seen—it was settled for perhaps six centuries before the first stones were hewn and laid here around AD 750, and from the 11th to the 15th centuries this may have been a ceremonial or religious center of a vast empire whose citizens smelted gold and copper and traded them via Sofala and other ports on the Indian Ocean coast for goods from South Asia and even China. No cement was used: these walls are built with stones cut to fit closely together, and in the valley they rise over 30 feet (nearly 10 m) high. And what secrets Great Zimbabwe still conceals—where are the plans, the tools, the quarries? The remnants of an elaborate water-supply system suggest a practical function for this center, but a decorated conical tower and adjacent platform imply a religious role. Whatever the answers, history hangs heavily, almost tangibly over this African site."

African Genesis

Africa on the eve of the colonial period was a continent in transition. For several centuries, the habitat in and near one of the continent's most culturally and economically productive areas—West Africa—had been changing. For 2000 years, probably more, Africa had been innovating as well as adopting ideas from outside. In West Africa, cities were developing on an impressive scale; in central and Southern Africa, peoples were moving, readjusting, sometimes struggling with each other for territorial supremacy. The Romans had penetrated to southern Sudan, North African peoples were trading with West Africans, and Arab *dhows* were sailing the waters along the eastern coasts, bringing Asian goods in exchange for gold, copper, and a comparatively small number of slaves.

Consider the environmental situation in West Africa as it relates to the past. As Figures I-7 and I-8 indicate, the environmental regions in this part of the continent exhibit a decidedly east-west orientation. The isohyets (lines of equal rainfall totals) run parallel to the southern coast (Fig. I-7); the climatic regions, now positioned somewhat differently from where they were two millennia ago, still trend strongly east-west (Fig. I-8); the vegetation pattern also reflects this situation, with a coastal forest belt yielding to savanna (tall grass with scattered trees in the south, shorter grass in the north) that gives way, in turn, to steppe and desert.

We know that African cultures had been established in all these environmental settings for thousands of years. One of these, the Nok culture, endured for over eight centuries on the Benue Plateau (north of the Niger-Benue confluence in modern Nigeria) from about 500 BC to the third century AD. The Nok people made stone as well as iron tools, and they left behind a treasure of art in the form of clay figurines representing humans and animals. But we have no evidence that they traded with distant peoples. The opportunities created by environments and technologies still lay ahead.

Early Trade

West Africa, over a north-south span of a few hundred miles, displayed an enormous contrast in environments, economic opportunities, modes of life, and products. The peoples of the tropical forest produced and needed goods that were different from the products and requirements of the peoples of the dry, distant north. For example, salt is a prized commodity in the forest, where the humidity precludes its formation, but it is plentiful in the desert and steppe. This enabled the desert peoples to sell salt to the forest peoples in exchange for ivory, spices, and dried foods. Thus there evolved a degree of *regional complementarity* between the peoples of the forest and those of the drylands. And the savanna peoples—those located in between—found themselves in a position to channel and handle the trade (which is always economically profitable).

The markets in which these goods were exchanged prospered and grew, and cities arose in the savanna belt of West Africa. One of these old cities, now an epitome of isolation, was once a thriving center of commerce and learning and one of the leading urban places in the world—Timbuktu. Others, predecessors as well as successors of Timbuktu, have declined, some of them into oblivion. Still other savanna cities, such as Kano in the northern part of Nigeria, are still important.

Early States

Strong and durable states arose in the West African culture hearth (see Fig. 6-3). The oldest state we know anything about is Ghana. Ancient Ghana was located to the northwest of the modern country of Ghana. It covered parts of present-day Mali, Mauritania, and adjacent territory. Ghana lay astride the upper Niger River and included gold-rich streams flowing off the Futa Jallon Highlands, where the Niger has its origins. For a thousand years, perhaps longer, old Ghana managed to weld various groups of people into a stable state. The country had a large capital city complete with markets, suburbs for foreign merchants, religious shrines, and, some miles from the city center, a fortified royal retreat. Taxes were collected from the citizens, and tribute was extracted from subjugated peoples on Ghana's periphery; tolls were levied on goods entering Ghana, and an army maintained control. Muslims from the northern drylands invaded Ghana in about AD 1062, when it may already have been in decline. Even so, the capital was protected for 14 years. However, the invaders had ruined the farmlands and destroyed the trade links with the north. Ghana could not survive. It finally broke into smaller units.

In the centuries that followed, the focus of politico-territorial organization in the West African culture hearth shifted almost continuously eastward—first to ancient Ghana's successor state of Mali, which was centered on Timbuktu and the middle Niger River Valley, and then to the state of Songhai, whose focus was Gao, a city on the Niger that still exists. This eastward movement may have been the result of the growing influence and power of Islam. Traditional religions prevailed in ancient Ghana, but Mali and its successor states sent huge, gold-laden pilgrimages to Mecca along the savanna corridor south of the Sahara, passing through present-day Khartoum and Cairo. Of the tens of thousands who participated in these pilgrimages, some remained behind. Today, many Sudanese trace their ancestry to the West Africa savanna kingdoms.

West Africa's savanna region undoubtedly witnessed momentous cultural, technological, and economic developments, but other parts of Africa also progressed. Early states emerged in present-day Sudan, Eritrea, and Ethiopia. Influenced by innovations from the Egyptian culture

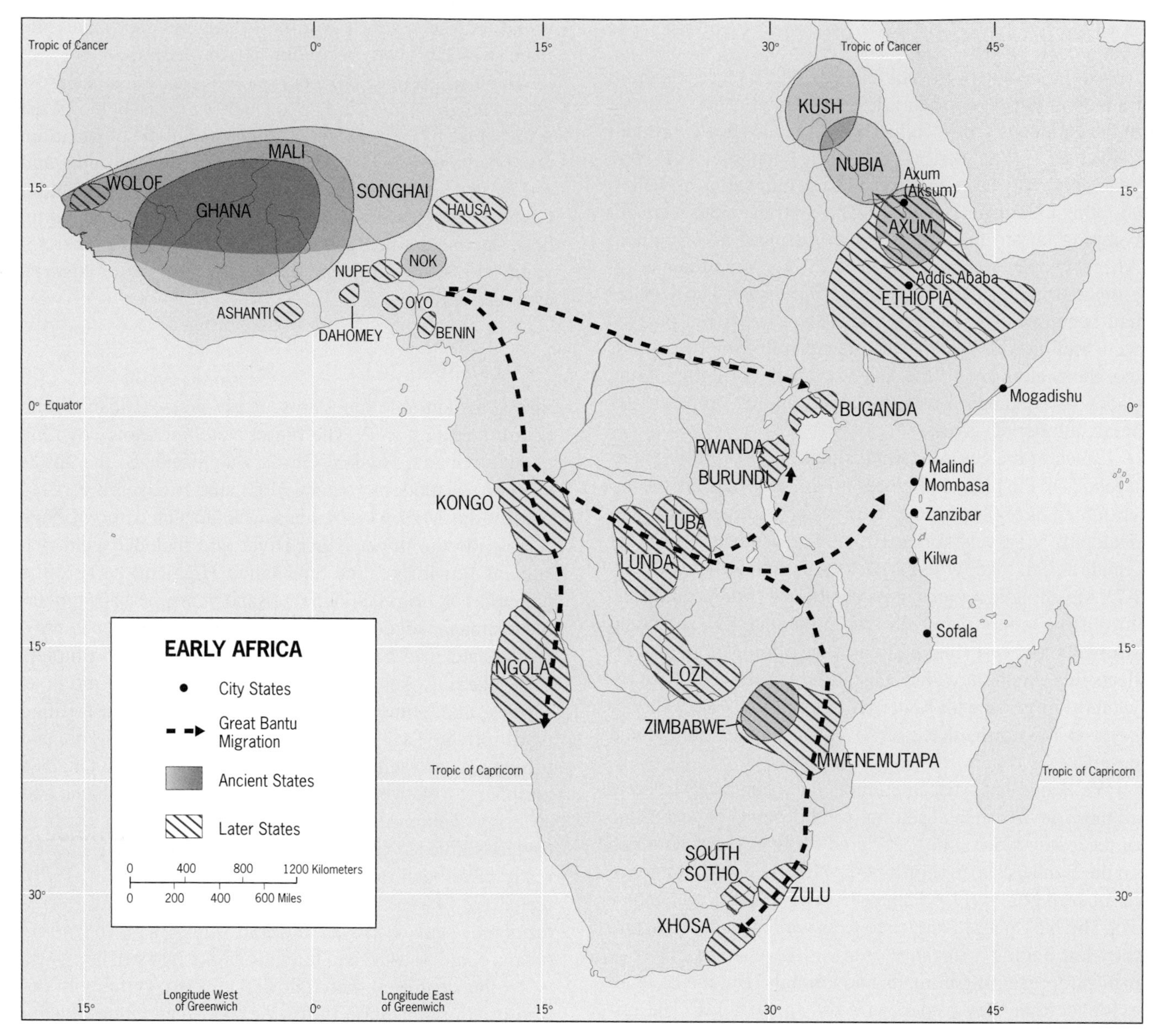

FIGURE 7-5

hearth, these kingdoms were stable and durable: the oldest, Kush, lasted 23 centuries (Fig. 7-5). The Kushites built elaborate irrigation systems, forged iron tools, and built impressive structures as the ruins of their long-term capital and industrial center, Meroe, reveal. Nubia, to the southeast of Kush, was Christianized until the Muslim wave overtook it in the eighth century. And Axum was the richest market in northeastern Africa, a powerful kingdom that controlled Red Sea trade and endured for six centuries. Axum, too, was a Christian state that confronted Islam, but Axum's rulers deflected the Muslim advance and gave rise to the Christian dynasty that eventually shaped modern Ethiopia.

11 The process of **state formation** diffused throughout Africa and was still in progress when the first European contacts occurred in the late fifteenth century. Large and effectively organized states developed on the equatorial west coast (notably Kongo) and on the southern plateau from the southern part of The Congo to Zimbabwe. East Africa had several city-states, including Mogadishu, Kilwa, Mombasa, and Sofala.

A crucial event affected virtually all of Equatorial, West, and Southern Africa: the great Bantu migration from present-day Nigeria-Cameroon southward and eastward across the continent. This migration appears to have occurred in waves starting as long as 5000 years ago, populating the Great Lakes area and penetrating South Africa, where it resulted in the formation of the powerful Zulu Empire in the nineteenth century (Fig. 7-5).

All this reminds us that, before European colonization, Africa was a realm of rich and varied cultures, diverse lifestyles, technological progress, and external trade. But Europe's intervention would forever change its evolving political map.

The Colonial Transformation

European involvement in Subsaharan Africa began in the fifteenth century. It would interrupt the path of indigenous African development and irreversibly alter the entire cultural, economic, political, and social makeup of the continent. It started quietly in the late fifteenth century, with Portuguese ships groping their way along the west coast and rounding the Cape of Good Hope. Their goal was to find a sea route to the spices and riches of the Orient. Soon other European countries were sending their vessels to African waters, and a string of coastal stations and forts sprang up. In West Africa, the nearest part of the continent to European spheres in Middle and South America, the initial impact was strongest. At their coastal control points, the Europeans traded with African middlemen for the slaves who were needed to work New World plantations, for the gold that had been flowing northward across the desert, and for ivory and spices.

Suddenly, the centers of activity lay not with the cities of the savanna but in the foreign stations on the Atlantic coast. As the interior declined, the coastal peoples thrived. Small forest states gained unprecedented wealth, transferring and selling slaves captured in the interior to the European traders on the coast. Dahomey (now called Benin) and Benin (now part of neighboring Nigeria) were states built on the slave trade. When slavery eventually came under attack in Europe, those who had inherited the power and riches it had brought opposed abolition vigorously in both continents.

Although slavery was not new to West Africa, the *kind* of slave raiding and trading the Europeans introduced certainly was. In the savanna, kings, chiefs, and prominent families traditionally took a few slaves, but those slaves' status was unlike anything that lay in store for those who were shipped across the Atlantic. In fact, large-scale slave trading had been introduced in East Africa long before the Europeans brought it to West Africa. African middlemen from the coast raided the interior for able-bodied men and women and marched them in chains to the Arab markets on the coast (Zanzibar was a notorious market). There, packed in specially built *dhows*, they were carried off to Arabia, Persia, and India. When the European slave trade took hold in West Africa, however, its volume was far greater. Europeans, Arabs, and collaborating Africans ravaged the continent, forcing perhaps as many as 30 million persons away from their homelands in bondage (Fig. 7-6). Families were destroyed, as were whole villages and cultures; those who survived their exile suffered unfathomable misery.

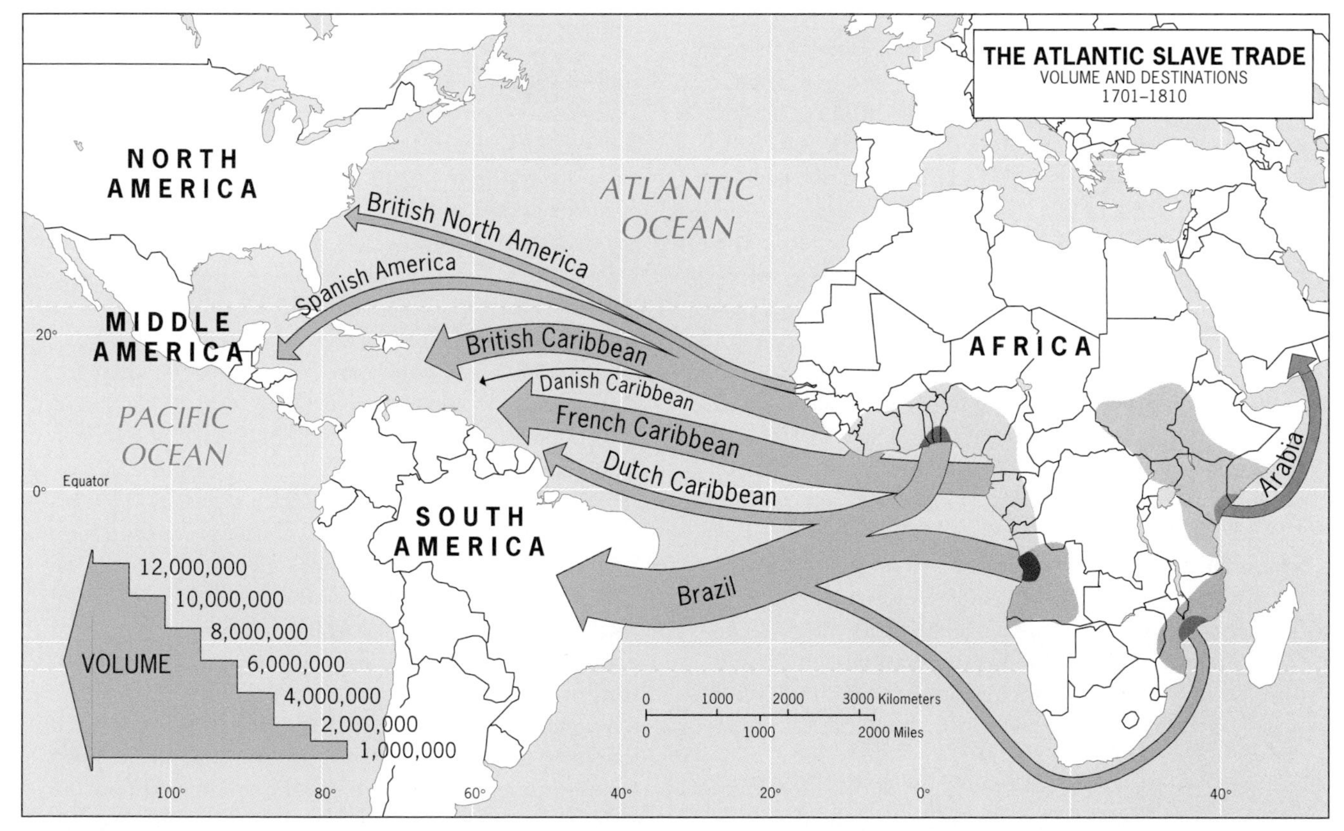

FIGURE 7-6

The European presence on the West African coast completely reoriented its trade routes, for it initiated the decline of the interior savanna states and strengthened the coastal forest states. Moreover, the Europeans' insatiable demand for slaves ravaged the population of the interior. But it did not lead to any major European thrust toward the interior or produce colonies overnight. The African middlemen were well organized and strong, and they held off their European competitors, not just for decades but for centuries. Although the Europeans first appeared in the fifteenth century, they did not carve West Africa up until nearly 400 years later, and in many areas not until after 1900.

In all of Subsaharan Africa, the only area where European penetration was both early and substantial was at its southernmost end, at the Cape of Good Hope. There the Dutch founded Cape Town, a waystation to their developing empire in the East Indies. They settled in the town's hinterland and migrated deeper into the interior. They brought thousands of slaves from Southeast Asia to the Cape, and intermarriage was common. Today, people of mixed ancestry form the largest part of Cape Town's population. Later the British took control not only of the Cape but also of the expanding frontier and brought in tens of thousands of South Asians to work on their plantations. Multiracial South Africa was in the making.

Elsewhere in the realm, the European presence remained confined almost entirely to the coastal trading stations, whose economic influence was strong. No real frontiers of penetration developed. Individual travelers, missionaries, explorers, and traders went into the interior, but nowhere else in Africa south of the Sahara was there an invasion of white settlers comparable to Southern Africa's.

Colonization

In the second half of the nineteenth century, after more than four centuries of contact, the European powers finally laid claim to virtually all of Africa. Parts of the continent had been "explored," but now representatives of European governments and rulers arrived to create or expand African spheres of influence for their patrons. Competition was intense. Spheres of influence began to crowd each other. It was time for negotiation, and in late 1884 a conference was convened in Berlin to sort things out. This conference laid the groundwork for the now-familiar politico-geographical map of Africa (see box titled "The Berlin Conference").

The Berlin Conference

In November 1884, the imperial chancellor and architect of the German Empire, Otto von Bismarck, convened a conference of 14 states (including the United States) to settle the political partitioning of Africa. Bismarck wanted not only to expand German spheres of influence in Africa but also to play off Germany's colonial rivals against one another to the Germans' advantage. The major colonial contestants in Africa were: (1) the British, who held beachheads along the West, South, and East African coasts; (2) the French, whose main sphere of activity was in the area of the Senegal River and north of the Congo Basin; (3) the Portuguese, who now desired to extend their coastal stations in Angola and Moçambique deep into the interior; (4) King Leopold II of Belgium, who was amassing a personal domain in the Congo Basin; and (5) Germany itself, active in areas where the designs of other colonial powers might be obstructed, as in Togo (between British holdings), Cameroon (a wedge into French spheres), South West Africa (taken from under British noses in a swift strategic move), and East Africa (where German Tanganyika broke the British design for a solid block of territory from the Cape north to Cairo).

When the conference convened in Berlin, more than 80 percent of Africa was still under traditional African rule. Nonetheless, the colonial powers' representatives drew their boundary lines across the entire map. These lines were drawn through known as well as unknown regions, pieces of territory were haggled over, boundaries were erased and redrawn, and African real estate was exchanged among European governments. In the process, African peoples were divided, unified regions were ripped apart, hostile societies were thrown together, hinterlands were disrupted, and migration routes were closed off. Not all of this was felt immediately, of course, but these were some of the effects when the colonial powers began to consolidate their holdings and the boundaries on paper became barriers on the African landscape (Fig. 7-7).

The Berlin Conference was Africa's undoing in more ways than one. The colonial powers superimposed their domains on the African continent. By the time Africa regained its independence after the late 1950s, the realm had acquired a legacy of political fragmentation that could neither be eliminated nor made to operate satisfactorily. The African politico-geographical map is thus a permanent liability that resulted from three months of ignorant, greedy acquisitiveness during a period when Europe's search for minerals and markets had become insatiable.

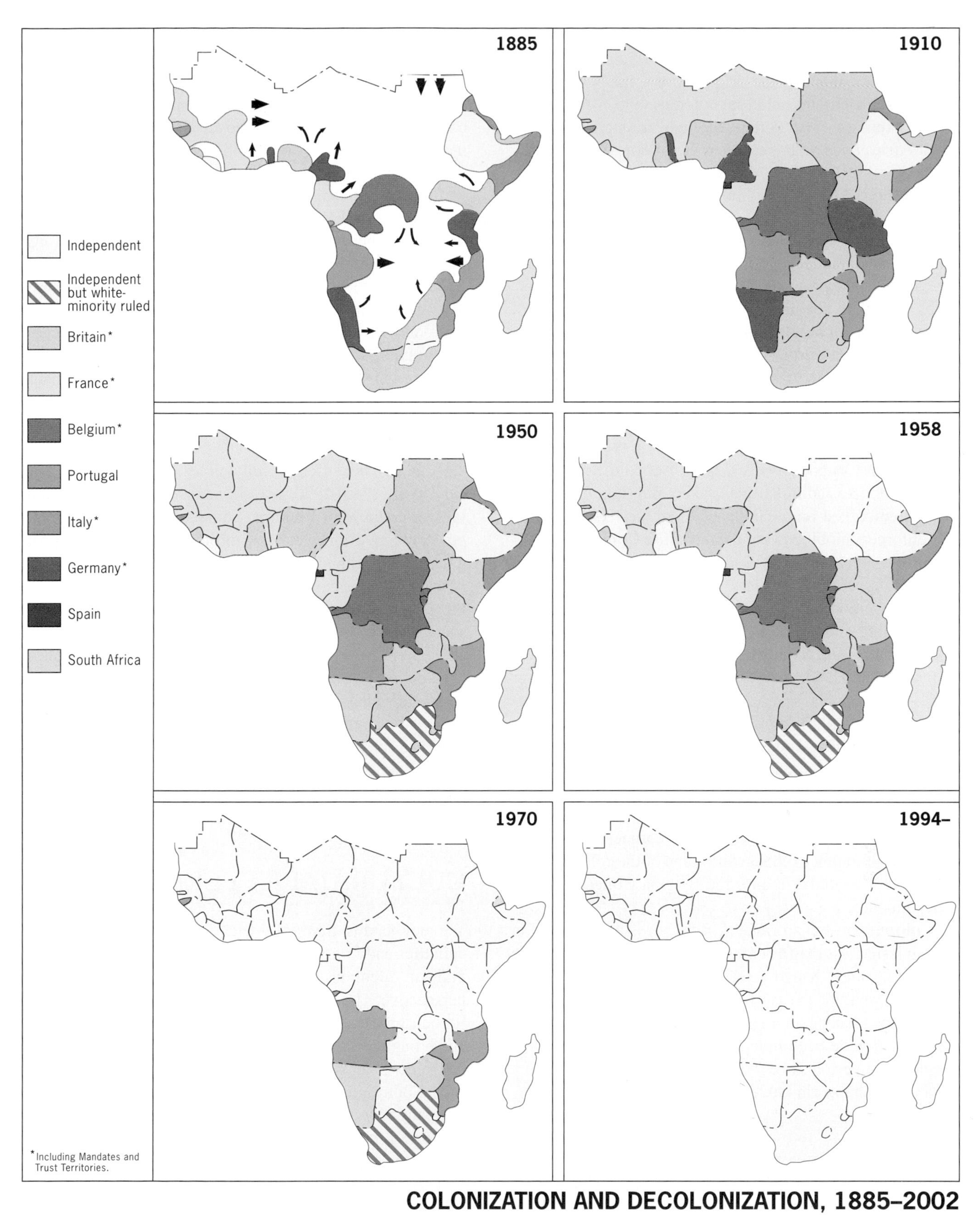

COLONIZATION AND DECOLONIZATION, 1885–2002

FIGURE 7-7

Figure 7-7 shows the result. The French dominated most of West Africa, and the British East and Southern Africa. The Belgians acquired the vast territory that became The Congo. The Germans held four colonies, one in each of the realm's regions. The Portuguese held a small colony in West Africa and two large ones in Southern Africa (see the map dated 1910).

The European colonial powers shared one objective in their African colonies: exploitation. But in the way they governed their dependencies, they reflected their differences. Some colonial powers were themselves democracies (the United Kingdom and France); others were dictatorships (Portugal, Spain). The British established a system of indirect rule over much of their domain, leaving indigenous power structures in place and making local rulers representatives of the British Crown. This was unthinkable in the Portuguese colonies, where harsh, direct control was the rule. The French sought to create culturally assimilated elites that would represent French ideals in the colonies. In the Belgian Congo, however, King Leopold II, who had financed the expeditions that staked Belgium's claim in Berlin, embarked on a campaign of ruthless exploitation. His enforcers mobilized almost the entire Congolese population to gather rubber, kill elephants for their ivory, and build public works to improve export routes. For failing to meet production quotas, entire communities were massacred. Killing and maiming became routine in a colony in which horror was the only common denominator. After the impact of the slave trade, King Leopold's reign of terror was Africa's most severe demographic disaster. By the time it ended, after a growing outcry around the world, as many as 10 million Congolese had been murdered. In 1908 the Belgian government took over, and slowly its Congo began to mirror Belgium's own internal divisions: corporations, government administrators, and the Roman Catholic Church each pursued their sometimes competing interests. But no one thought to change the name of the colonial capital: it was Leopoldville until the Belgian Congo achieved independence.

12 **Colonialism** transformed Africa, but in its post-Berlin form it lasted less than a century. In Ghana, for example, the Ashanti (Asante) Kingdom still was fighting the British in the early years of the twentieth century; by 1957, Ghana was independent again. In a few years, much of Subsaharan Africa will have been independent for half a century, and the colonial period is becoming an interlude rather than a paramount chapter in modern African history.

Legacies

Nevertheless, some colonial legacies will remain on the map for centuries to come. The boundary framework has been modified in a few places, but it is a politico-geographical axiom that boundaries, once established, tend to become entrenched and immovable. The transport systems were laid out to facilitate the movement of raw materials from interior sources to coastal outlets. Internal circulation was only a secondary objective, and most African countries still are not well connected to each other. Moreover, the colonizers founded many of Subsaharan Africa's cities or built them on the sites of small towns or villages.

In many countries, the elites that gained advantage and prominence during the colonial period also have retained their preeminence. This has led to authoritarianism in some postcolonial African states (for details, see the regional discussion below), and to violence and civil conflict in others. Military takeovers of national governments have been a byproduct of decolonization, and hopes for democracy in the struggle against colonialism have too often been dashed. Yet a growing number of African countries, including Ghana, Tanzania, Botswana, South Africa, and, most recently, Nigeria, have overcome the odds and achieved representative government under dauntingly difficult circumstances.

Although not strictly a colonial legacy, the impact of the Cold War (1945–1990) on African states should be noted. In three countries—Ethiopia, Somalia, and Angola—great-power competition, weapons, military advisers, and, in Angola, foreign armed forces magnified civil wars that would perhaps have been inevitable in any case. In other countries, ideological adherence to foreign dogmas led to political and social experiments (such as one-party Marxist regimes and costly farm collectivization) that cost Africa dearly. Today, debt-ridden Africa is again being told what to do, this time by foreign financial institutions. The cycle of poverty that followed colonialism exacts a high price from African societies. As for future development prospects, the disadvantages of peripheral location vis-à-vis the world's core areas continue to handicap Subsaharan Africa.

CULTURAL PATTERNS

We may tend to think of Africa in terms of its prominent countries and famous cities, its development problems and political dilemmas, but Africans themselves have another perspective. The colonial period created states and capitals, introduced foreign languages to serve as the *lingua franca*, and brought railroads and roads. The colonizers stimulated labor movements to the mines they opened, and they disrupted other migrations that had been part of African life for many centuries. But they did not change the ways of life of most of the people. More than 70 percent of the realm's population still live in, and work near, Africa's hundreds of thousands of villages. They speak one of more than a thousand languages in use in the realm. The villagers' concerns are local; they focus on subsistence, health, and safety. They worry that the conflicts over regional power or political ideology will engulf them, as has happened to millions in Liberia, Sierra Leone, Ethiopia, Rwanda, The Congo, Moçambique, and Angola since the 1970s. Africa's largest peoples are major nations, such as the Yoruba of Nigeria and the Zulu of South Africa. Africa's smallest peoples number just a few thousand. As a geographic realm, Subsaharan Africa has the most complex cultural mosaic on Earth.

African Languages

Africa's linguistic geography is a key component of that cultural intricacy. Most of Subsaharan Africa's more than 1000 languages do not have a written tradition, making classification and mapping difficult. Scholars have attempted to delimit an African language map, and Figure 7-8 is a composite of their efforts. One feature is common to all language maps of Africa: the geographic realm begins approximately where the Afro-Asiatic language family (mapped in yellow in Fig. 7-8) ends, although the correlation is sharper in West Africa than to the east.

In Subsaharan Africa, the dominant language family is the Niger-Kordofanian family. It consists of two subfamilies, the tiny Kordofanian concentration in northeastern Sudan and the pervasive Niger-Congo subfamily that extends across most of the realm from West Africa to East and Southern Africa. The Bantu language forms the largest branch in this subfamily, but Niger-Congo languages in West Africa, such as Yoruba and Akan, also have millions of speakers. Another important language family is the Nilo-Saharan family, extending from Maasai in Kenya northwest to Teda in Chad. No other language families are of similar extent or importance: the Khoisan family, of ancient origins, now survives among the dwindling Khoi and San peoples of the Kalahari; the

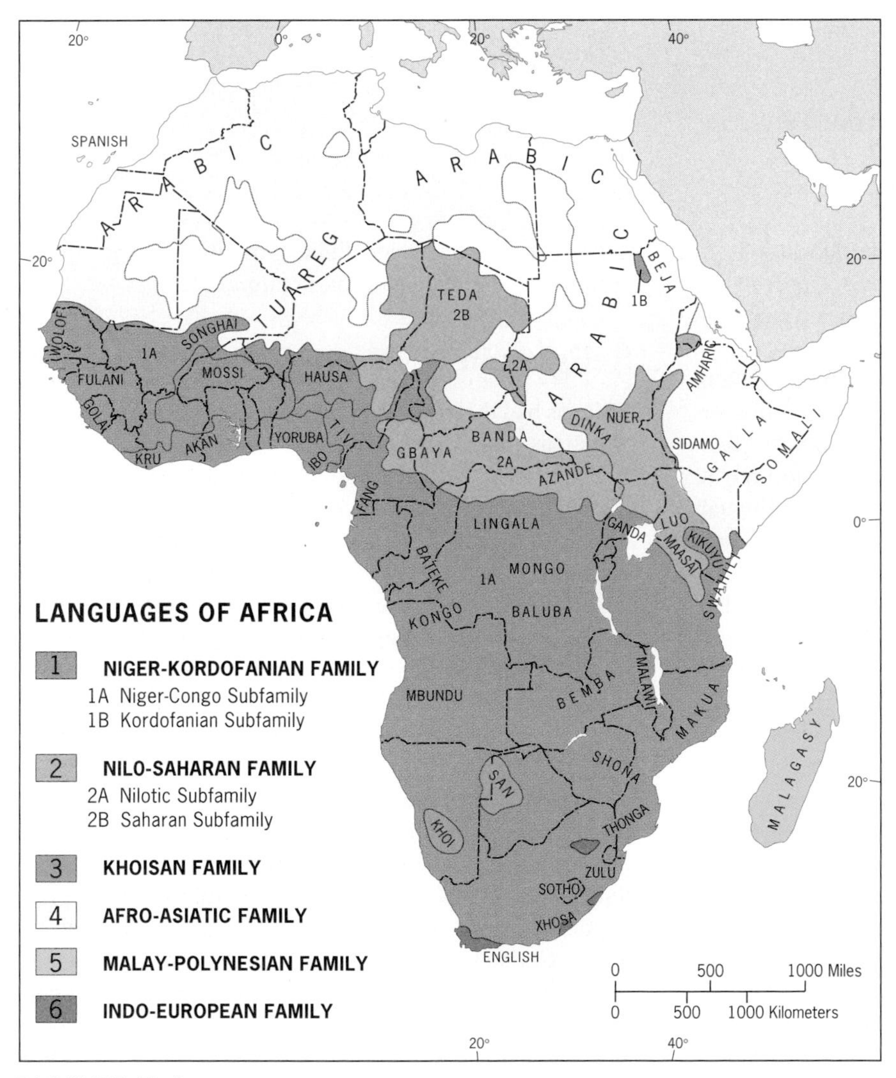

FIGURE 7-8

The faithful kneel during Friday prayers at a mosque in Kano, northern Nigeria. The survival of Nigeria as a unified state is an African success story; the Nigerians have overcome strong centrifugal forces in a multi-ethnic country that is dominantly Muslim in the north, Christian in the south. In the 1990s, some Muslim clerics began calling for an Islamic Republic in Nigeria, and after the death of the dictator Abacha and the election of a non-Muslim president, the Islamic drive intensified. A number of Nigeria's northern States adopted Sharia (strict Islamic) law, which led to destructive riots between the majority Muslims and minority Christians who felt threatened by this turn of events. The northern city of Kaduna saw a near-civil war in which hundreds were killed, churches and mosques burned, and whole neighborhoods devastated. Late in 2000 the Kaduna State government reversed itself and suspended the imposition of Sharia law, but other States persevered despite social unrest. Can Nigeria avoid the fate of Sudan?

small white minority in South Africa speak Indo-European languages; and Malay-Polynesian languages prevail in Madagascar, which was peopled from Southeast Asia before Africans reached it.

About 40 African languages are spoken by 1 million people or more, and a half-dozen by about 10 million or more: Hausa (50 million), Yoruba (23 million), Ibo, Swahili, Lingala, and Zulu. Although English and French have become important *linguae francae* in multilingual countries such as Nigeria and Côte d'Ivoire (where officials even insist on spelling the name of their country—Ivory Coast—in the Francophone way), African languages also serve this purpose. Hausa is a common language across the West African savanna; Swahili is widely used in East Africa. And pidgin languages, mixtures of African and European tongues, are spreading along West Africa's coast. Millions of Pidgin English (called *Wes Kos*) speakers use this medium in Nigeria and Ghana.

13

Multilingualism can be a powerful centrifugal force in society, and African governments have tried with varying success to establish "national" alongside local languages. Nigeria, for example, made English its official language because none of its 250 languages, not even Hausa, had sufficient internal interregional use. But using a European, colonial language as an official medium invites criticism, and Nigeria remains divided on the issue. On the other hand, making a dominant local language official would invite negative reactions from ethnic minorities. Language remains a potent force in Africa's cultural life.

Religion in Africa

Africans had their own religious belief systems long before Christians and Muslims arrived to convert them. And for all of Subsaharan Africa's cultural diversity, Africans had a consistent view of their place in nature. Spiritual forces, according to African tradition, are manifest everywhere in the natural environment, not in a supreme deity that exists in some remote place. Thus gods and spirits affect people's daily lives, witnessing every move, rewarding the virtuous, and punishing (through injury or crop failure, for example) those who misbehave. Ancestral spirits can inflict misfortune on the living. They are everywhere: in the forest, rivers, mountains.

As with land tenure, the religious views of Africans clashed fundamentally with those of outsiders. Monotheistic Christianity first touched Africa in the northeast when Nubia and Axum were converted, and Ethiopia has been a Coptic Christian stronghold since the fourth century AD. But the Christian churches' real invasion did not begin until the onset of colonialism after the turn of the sixteenth century. Christianity's various denominations made inroads in different areas: Roman Catholicism in much of Equatorial Africa mainly at the behest of the Belgians, the Anglican

Church in British colonies, and Presbyterians and others elsewhere. But almost everywhere, Christianity's penetration led to a blending of traditional and Christian beliefs, so that much of Subsaharan Africa is nominally, though not exclusively, Christian. Go to a church in Gabon or Uganda or Zambia, and you may hear drums instead of church bells, sing African music rather than hymns, and see African carvings alongside the usual statuary.

Islam had a different arrival and impact. Long before the colonial invasion, Islam advanced out of Arabia, across the desert, and down the east coast. Muslim clerics converted the rulers of African states and commanded them to convert their subjects. They Islamized the savanna states and penetrated into present-day northern Nigeria, Ghana, and Ivory Coast. They encircled and isolated Ethiopia's Coptic Christians and Islamized the Somali people in Africa's Horn. They established beachheads on the Kenya coast and took over Zanzibar. On the map, the African Transition Zone defines the Muslim Front (Fig. 6-12). In the field, Arabizing Islam and European Christianity competed for African minds, and Islam proved to be a far more pervasive force. From Senegal to Somalia, the population is virtually 100 percent Muslim, and Islam's rules dominate everyday life. The Sunni *mullahs* would never allow the kind of marriage between traditional and Christian beliefs seen in much of formerly colonial Africa. This fundamental contradiction between Islamic dogma and Christian accommodation creates a potential for conflict in countries where both religions have adherents.

MODERN MAP AND TRADITIONAL SOCIETY

Africa's political geography is the product of colonial competition, with relatively minor adjustments during the postcolonial period. International political boundaries, once established, tend to become rapidly entrenched. The colonial interlude may have been brief, but it left an indelible legacy.

The boundaries European powers superimposed on Subsaharan Africa showed little concern for Africa's own evolving political framework. It is often said that the Europeans paid no attention to Africa's cultural mosaic, but that is not true. While their imperial borders were essentially the result of international competition, the internal divisions the Europeans laid out in their colonies did sometimes accommodate existing African patterns. Nevertheless, the European boundary framework often threw together African peoples who had histories of strife; elsewhere it separated peoples with common cultural bonds.

The result was a framework that led to the independence of some 45 African states ranging from Nigeria (131 million today) to São Tomé and Príncipe (175,000) and from The Congo (875,000 sq mi/2,265,000 sq km) to the Seychelles (177 sq mi/455 sq km). The economic prospects of many of these countries are poor, not least because Africa has more landlocked countries than the rest of the world combined. Social conditions in most African countries also are wretched, as shown in Table I-2. Across much of Subsaharan Africa, infant and child mortality and the incidence of disease are high, while life expectancies are low. In a world where the gap between food needs and availability has been closed, Africa still suffers from inadequate nourishment and unbalanced diets in its poorest countries.

All these situations are the result of a combination of factors, some common to the global periphery, others unique to Subsaharan Africa. Certainly the ethnic tensions resulting from the political framework played a role, but even if the map had been laid out more sensibly, conflict between the traditional African political systems and the colonial (and Islamic) systems that outsiders tried to install would have arisen. Africa's newly independent states needed the loyalty of their citizens; instead, they faced split loyalties and parochial attitudes. Postcolonial circumstances gave long-suppressed, historically powerful African societies the opportunity to reassert themselves: the Kikuyu in Kenya, the Ashanti in Ghana, the Ibo in Nigeria, the Shona in Zimbabwe, the Zulu in South Africa (after the end of white rule). Often the interests of these societies ran counter to those of the governments of the newly independent states, and the reaction was to turn inward. The Ashanti of Ghana, for example, consider themselves subjects of their king first, Ghanaians second. The Ibo went further: they mounted a civil war to secure independence from the Nigerian state into which they had been incorporated. Virtually all postcolonial African states were plural societies that confronted such centrifugal forces to some degree, and many found that the political systems the Europeans had helped them establish could not solve the problem. The Europeans had visions of democracy and federalism; the Islamic countries along the northern and eastern margin tried Sharia law and subjugation by force. Neither worked; both provoked strife.

The political map of Subsaharan Africa, therefore, has 45 states but no nation-states (apart from some microstates and ministates in the islands and in the south). Centrifugal forces are powerful, and outside interventions during the Cold War, when communist and anticommunist foreigners took sides in local civil wars, worsened conflict within African states. Colonialism's economic legacy was not much better. In tropical African capitals, core areas, port cities, and transport systems were laid out to maximize profit and facilitate exploitation of minerals and soils; the colonial mosaic inhibited interregional communications except where cooperation enhanced efficiency. Colonial Zambia and Zimbabwe, for example (then called Northern and Southern Rhodesia), were landlocked and needed outlets, so railroads were built to Portuguese-owned ports. But such routes did little to create intra-African linkages. The modern map reveals the results: in West Africa you can travel from the coast into the interior of all the coastal states along railways or adequate roads. But no high-standard roadway was ever built to link these coastal neighbors to each other.

To overcome such disadvantages, African states must cooperate regionally. An important step was taken in the 1990s when 12 countries joined in the Southern African Development Community (SADC). Although the SADC's resources are modest, its goals are crucial. It works to lower tariff barriers among Southern Africa's countries, encouraging regional commerce. It coordinates and supports infrastructural projects, including the construction of highways to link member countries and hydroelectric projects to improve power supplies in rural areas. It assists member governments facing rebellions. In the late 1990s, the organization succeeded in some spheres, mainly economic; it failed in others, especially in its efforts to combat political unrest. When The Congo's then ruler, Laurent Kabila, appealed to the SADC for help against a rebellion in the east of his vast country, the organization could not agree on a joint position. South Africa, its dominant force, wanted negotiation; Zimbabwe, seeing an opportunity to counter South African influence, sent troops. So deep was the disagreement among SADC members over this issue that the organization itself seemed to be at risk. When Lesotho's government asked the SADC for help in quelling post-election violence, South Africa intervened but underestimated the task, with disastrous results. Still, the SADC represents an important step in Subsaharan Africa's politico-geographical evolution.

Population and Urbanization

Subsaharan Africa remains the least urbanized geographic realm, but it also ranks among the world's fastest urbanizing realms today. In 2002, 29 percent of all Africans resided in urban areas, up from 24 percent a decade earlier. From Conakry to Cape Town, rural-to-urban migration keeps growing.

Earlier, we noted that Subsaharan Africa already had urban traditions before Arabs and Europeans introduced theirs. In West Africa and to a lesser extent in coastal East Africa, parts of these indigenous cities have survived, for example, in Ife and Oyo (Nigeria). But the European colonizers founded or modernized most of what are now the realm's leading cities, and modern Africa's urban system is mainly another colonial legacy. During colonial times, cities and towns tended to be segregated and in-migration was to varying degrees controlled. But after independence the new governments could not stem the tide (some tried by evicting homeless newcomers and banishing sidewalk traders). Today many African cities are severely overcrowded, encircled by some of the worst slum developments in the world.

If 29 percent of Subsaharan Africa's population now live in cities and towns, this amounts to about 188 million people. Tens of millions of them have left their rural abodes and moved directly to the nearest city, pushed off the land by crop failures and debt or attracted by imagined opportunities in town. The newcomers will look for ethnic cohorts when they arrive, creating urban outskirts that resemble the cultural mosaic of the countryside. Mombasa, Kenya, for example, has squatter and slum neighborhoods where Luo, Kamba, Kikuyu, and other ethnic groups congregate. In the African spirit of communal assistance, the new arrivals are accommodated while they try to find work.

Africa's cities had often become centers of modest core areas during the colonial period, and of course they were also government headquarters. This *formal sector* of the city used to be the dominant one, with government control and regulations affecting civil service, business, industry, and workers. Today, however, African cities look different. From a distance, the skyline still resembles that of a modern center. But in the streets, on the sidewalks right below the shopwindows, there are hawkers, basket weavers, jewelry sellers, garment makers, wood carvers—a second economy, most of it beyond government control. This *informal sector* now dominates many African cities. It is peopled by the rural immigrants, who also work as servants, apprentices, construction workers, and in countless other menial jobs.

Millions of urban immigrants, however, cannot find work, at least not for months or even years at a time. They live in squalid circumstances, in desperate poverty, and governments cannot assist them. As a result, the squatter rings around (and also within) many of Africa's cities are unsafe—uncomfortable, unhealthy slums without adequate shelter, water supply, or basic sanitation. Garbage-strewn (no solid-waste removal here), muddy and insect-infested during the rainy season, and stifling and smelly during the dry period, they are incubators of disease. Yet few of its residents return to their villages. Every new day brings hope.

In our regional discussion we focus on some of Subsaharan Africa's cities, all of which, to varying degrees, are stressed by the rate of population influx. Despite the plight of the urban poor and the poverty of Africa's rural areas, some of Africa's capitals remain the strongholds of privileged elites who, dominant in governments, fail to address the needs of other ethnic groups. Discriminatory policies and artificially low food prices disadvantage farmers and create even greater urban-rural disparities than the colonial period saw. But today the prospect of democracy brings hope that Africa's rural majorities will be heard and heeded in the capitals.

REGIONS OF THE REALM

On the face of it, Africa seems to be so massive, compact, and unbroken that any attempt to justify a contemporary regional breakdown is doomed to fail. No deeply penetrating bays or seas create peninsular fragments as in Europe. No major islands (other than Madagascar) provide the broad regional contrasts we see in Middle America. Nor does Africa really taper southward to the peninsular proportions of South America. And Africa is not cut by an Andean or a Himalayan mountain barrier. Given Africa's colonial fragmentation and cultural mosaic, is regionalization possible? Indeed it is.

Maps of environmental distributions, ethnic patterns, cultural landscapes, historic culture hearths, and colonial frameworks yield the four-region structure shown in Figure 7-9:

1. *West Africa* includes the countries of the western coast and Sahara margin from Senegal and Mauritania in the west to population-giant Nigeria and Niger (and part of Chad) in the east.
2. *Equatorial Africa* centers on the vast state of The Congo, and also includes Congo, Gabon, Cameroon, and the Central African Republic, a part of Chad, and southern Sudan.
3. *East Africa* also lies astride the equator, but its environments are moderated by elevation. Kenya and Tanzania are

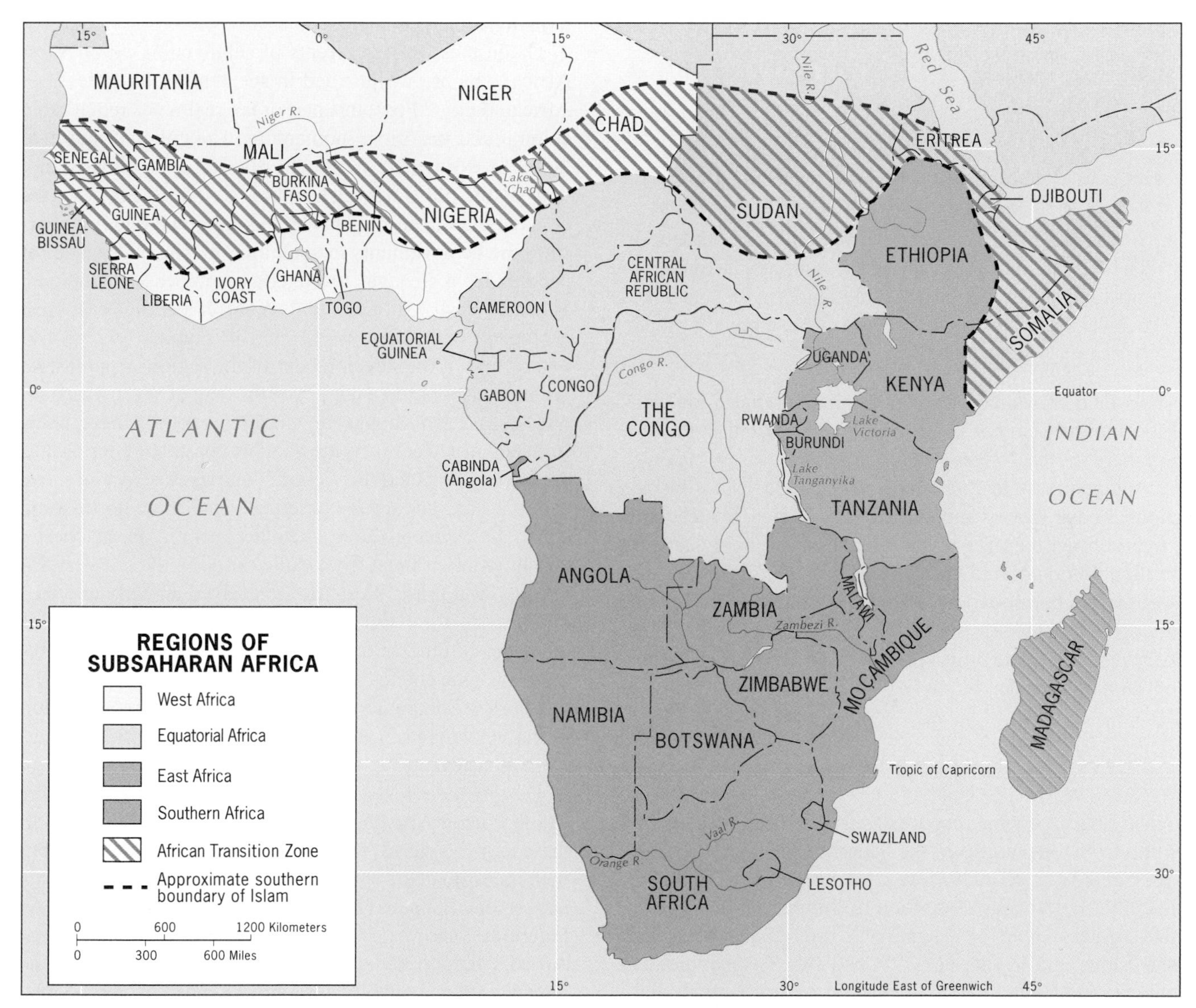

FIGURE 7-9

MAJOR CITIES OF THE REALM

City	Population* (in millions)
Abidjan, Ivory Coast	3.6
Accra, Ghana	2.2
Addis Ababa, Ethiopia	3.0
Cape Town, South Africa	3.1
Dakar, Senegal	2.3
Dar es Salaam, Tanzania	2.6
Durban, South Africa	1.4
Harare, Zimbabwe	1.9
Ibadan, Nigeria	1.9
Johannesburg, South Africa	2.4
Kinshasa, The Congo	5.7
Lagos, Nigeria	15.2
Lusaka, Zambia	1.8
Mombasa, Kenya	0.9
Nairobi, Kenya	2.6

*Based on 2002 estimates.

the coastal states; Uganda, Rwanda, and Burundi are landlocked. Highland Ethiopia also forms part of this region.

4. *Southern Africa* extends from the southern borders of The Congo and Tanzania to the continent's southernmost cape. Ten countries, including Angola and Zimbabwe, form part of this region, whose giant is South Africa.

The island of Madagascar, in the Indian Ocean opposite Moçambique, cannot be incorporated into either East or Southern Africa, for geographic reasons we discuss later.

A gigantic state dominates three of Africa's four regions, in one way or another. In West Africa, the leader is Nigeria by virtue of its huge population of more than 130 million. In Equatorial Africa, The Congo dominates by reason of its enormous territory, larger than the rest of the region combined. In Southern Africa, the giant is South Africa, not because of its size, but because of its economic power and influence.

WEST AFRICA

West Africa occupies most of Africa's Bulge, extending south from the margins of the Sahara to the Gulf of Guinea coast and from Lake Chad west to Senegal (Fig. 7-10). Politically, the broadest definition of this region includes all those states that lie to the south of Western Sahara, Algeria, and Libya and to the west of Chad (itself sometimes included) and Cameroon. Within West Africa, a rough division is sometimes made between the large, mostly steppe and desert states that extend across the southern Sahara (Chad could also be included here) and the smaller, better-watered coastal states.

Apart from once-Portuguese Guinea-Bissau and long-independent Liberia, West Africa comprises four former British and nine former French dependencies. The British-influenced countries (Nigeria, Ghana, Sierra Leone, and Gambia) lie separated from one another, whereas Francophone West Africa is contiguous. As Figure 7-10 shows, political boundaries extend from the coast into the interior, so that from Mauritania to Nigeria, the West African habitat is parceled out among parallel, coast-oriented states. Across these boundaries, especially across those between former British and former French territories, there is only limited interaction. For example, in terms of value, Nigeria's trade with Britain is about 100 times as great as its trade with nearby Ghana. The countries of West Africa are not interdependent economically, and their incomes are largely derived from the sale of their products on the non-African international market.

Given these cross-currents of subdivision within West Africa, why are we justified in speaking of a single West African region? First, this part of the realm has remarkable cultural and historical momentum. The colonial interlude failed to extinguish West African vitality, expressed not only by the old states and empires of the savanna and the cities of the forest, but also by the vigor and entrepreneurship, the achievements in sculpture, music, and dance, of peoples from Senegal to Nigeria's southeastern Iboland. Second, West Africa contains a set of parallel east-west ecological belts, clearly reflected in Figures I-7 and I-8, whose role in the development of the region is pervasive. As the transport-route pattern on the map of West Africa indicates, overland connections within each of these belts, from country to country, are poor; no coastal or interior railroad ever connected this tier of countries. Yet spatial interaction is stronger across these belts, and some north-south economic exchange does take place, notably in the coastal consumption of meat from cattle raised in the northern savannas. And third, West Africa received an early and crucial imprint from European colonialism, which—with its maritime commerce and slave trade—transformed the region from one end to the other. This impact reached into the heart of the Sahara, and it set the stage for the reorientation of the whole area, from which emerged the present patchwork of states.

Despite the effects of the slave trade, West Africa today is Subsaharan Africa's most populous region (Fig. I-9). In these terms, Nigeria (whose census results are in doubt, but with an estimated population of 130.6 million) is Africa's largest state; Ghana (20.4 million) ranks high as well. As Figure I-9 shows, West Africa also constitutes one of Africa's major population clusters. The southern half of the region, understandably, is home to most of the people. Mauritania, Mali, and Niger include too much of the unproductive Sahel's steppe and the arid Sahara to sustain popu-

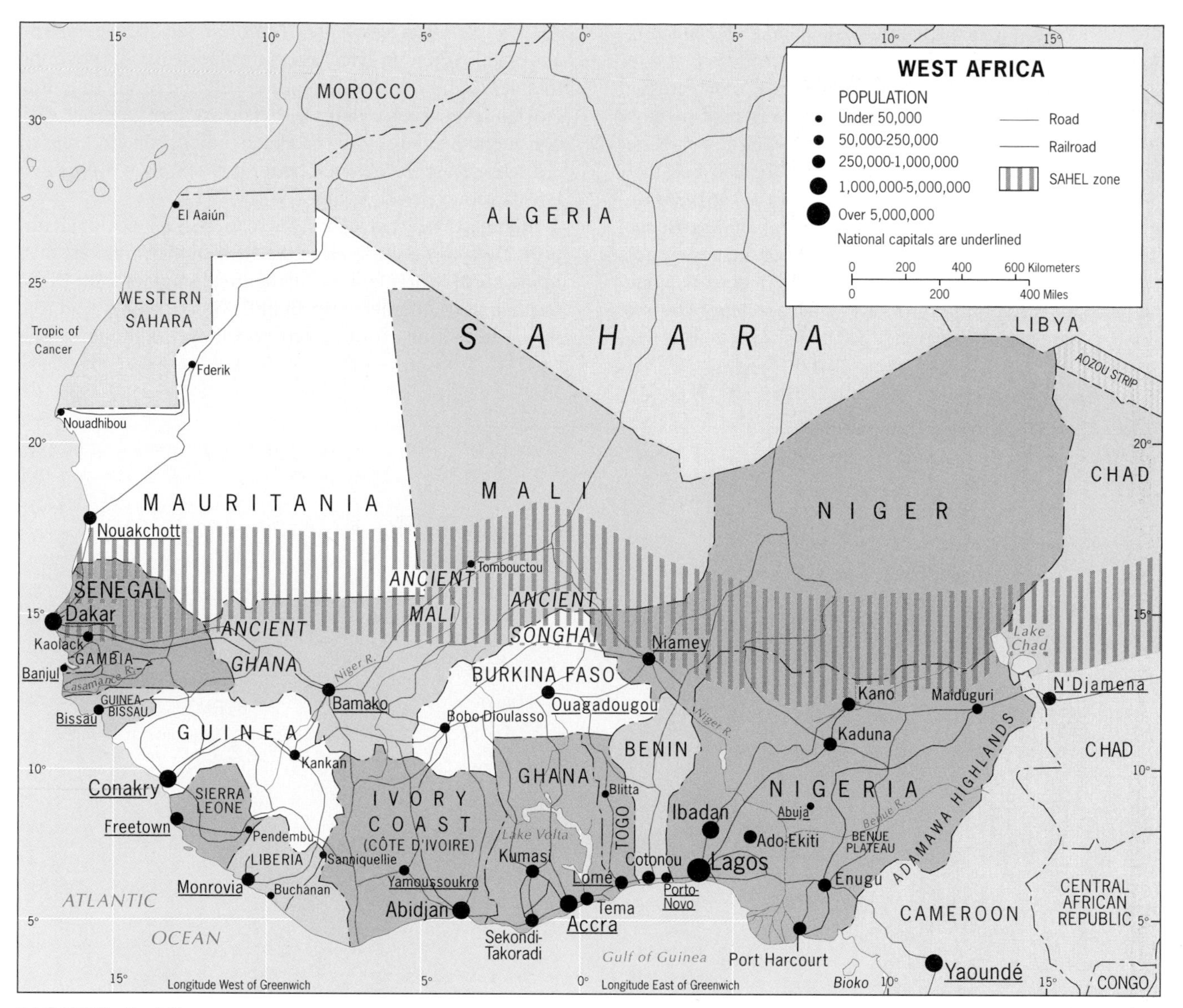

FIGURE 7-10

lations comparable to those of Nigeria, Ghana, or Ivory Coast.

The peoples along the coast reflect the modern era that the colonial powers introduced: they prospered in their newfound roles as middlemen in the coastward trade. Later, they experienced the changes of the colonial period; in education, religion, urbanization, agriculture, politics, health, and many other endeavors, they adopted new ways. In contrast, the peoples of the interior retained their ties with a different era in African history. Distant and aloof from the main theater of European colonial activity and often drawn into the Islamic orbit, they experienced a significantly different kind of change. But the map reminds us that Africa's boundaries were not drawn to accommodate such contrasts. Both Nigeria and Ghana possess population clusters representing the interior as well as the coastal peoples, and in both countries the wide cultural gap between north and south has produced political problems.

Nigeria: West Africa's Cornerstone

When Nigeria achieved full independence from Britain in 1960, its government faced the task of administering a European political creation containing three major nations and nearly 250 other peoples ranging from several million to a few thousand in number. The country had been endowed with a federal political system consisting of three regions, each focused on one of the three leading nations. In the southwest lay the Western Region, the traditional home of the Yoruba, a people with long urban traditions and a complex and highly developed culture. In the southeast lay the Eastern Region, historic home of the Ibo people, less affected than the Yoruba by the colonial impact, less urbanized, but widely dispersed to other parts of Nigeria and even beyond. And in the north, the Northern Region was the largest and most populous, the domain of the Hausa-Fulani cluster, a region of ancient cities, Islam, and Sahelian environments.

The map of Nigeria shows a compact territory crossed by the Niger River and its tributary, the Benue, together forming a Y-shaped system (Fig. 7-11). The lower Niger and its delta separated the Yoruba and the Ibo and their Western and Eastern Regions; the core of the Northern Region was situated between the upper Niger and the Benue. The departing colonial power, recognizing the new state's regional diversity, believed that a federal framework would help Nigeria's government to cope with the resulting stresses. Some progress had been made in the construction of road and railroad links across the "Middle Zone" between north and south, English had been introduced as the common tongue for the educated elite (it is still one of the four official languages today along with Yoruba, Ibo, and Hausa), a considerable civil service had been built up, and in terms of education Nigeria was in a better position than most other African countries or colonies.

For reasons obvious from the map, Britain's colonial imprint always was stronger in the two southern regions than in the north. Christianity became the dominant faith in the south, and southerners, especially Yoruba, took a lead role in the transition from colony to independent state. The

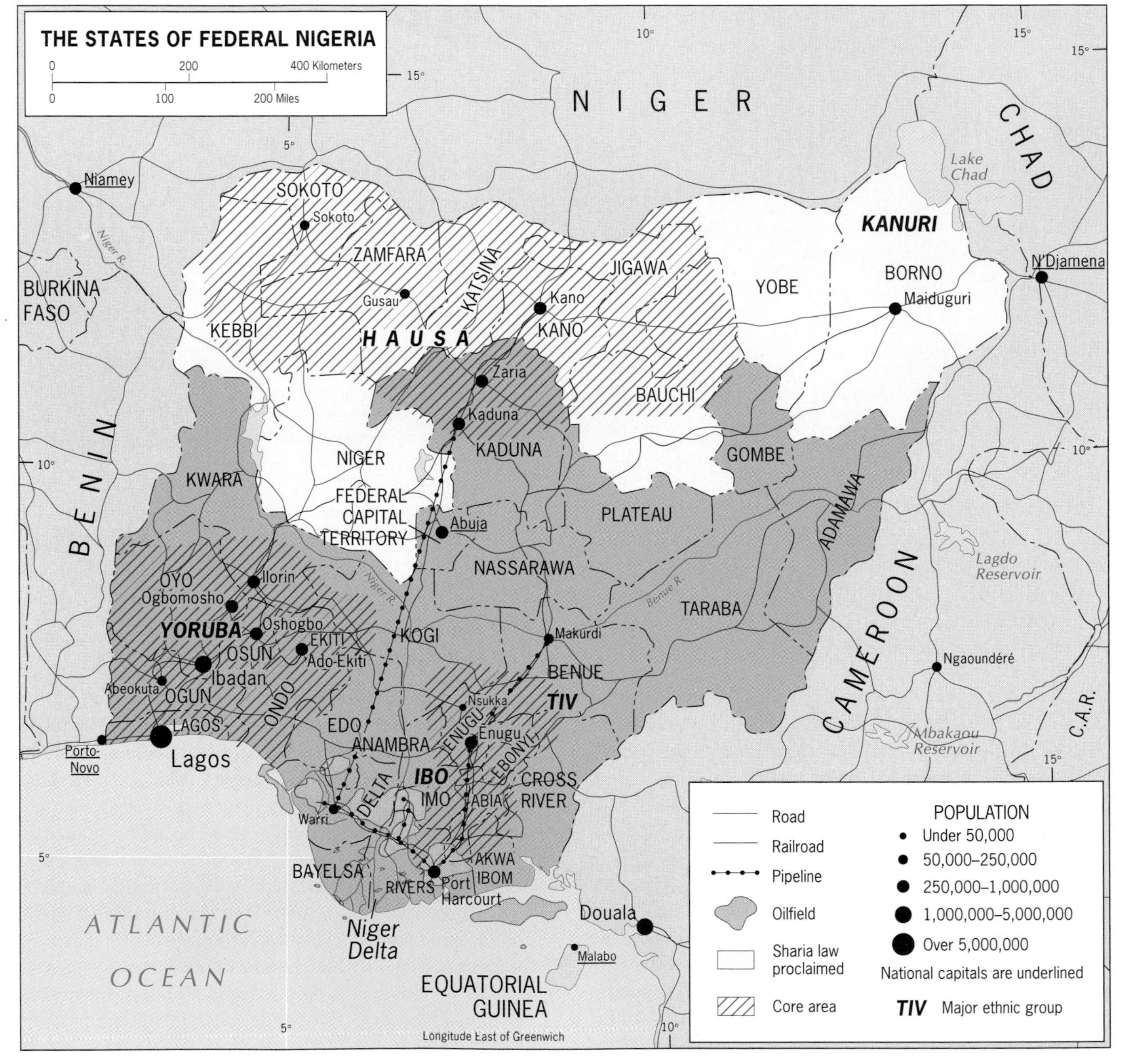

FIGURE 7-11

Among the Realm's Great Cities . . .

Lagos

In a realm that is only 29 percent urbanized, Lagos, former capital of federal Nigeria, is the exception: a teeming metropolis of 15.2 million sometimes called the Calcutta of Africa.

Lagos evolved over the past three centuries from a Yoruba fishing village, Portuguese slaving center, and British colonial headquarters into Nigeria's largest city, major port, leading industrial center, and first capital. Situated on the country's southwestern coast, it consists of a group of low-lying islands and sand spits between the swampy shoreline and Lagos Lagoon. The center of the city still lies on Lagos Island, where the high-rises adjoining the Marina overlook Lagos Harbor and, across the water, Apapa Wharf and the Apapa industrial area. The city expanded southward onto Ikoyi Island and Victoria Island, but after the 1970s most urban sprawl took place to the north, on the western side of Lagos Lagoon.

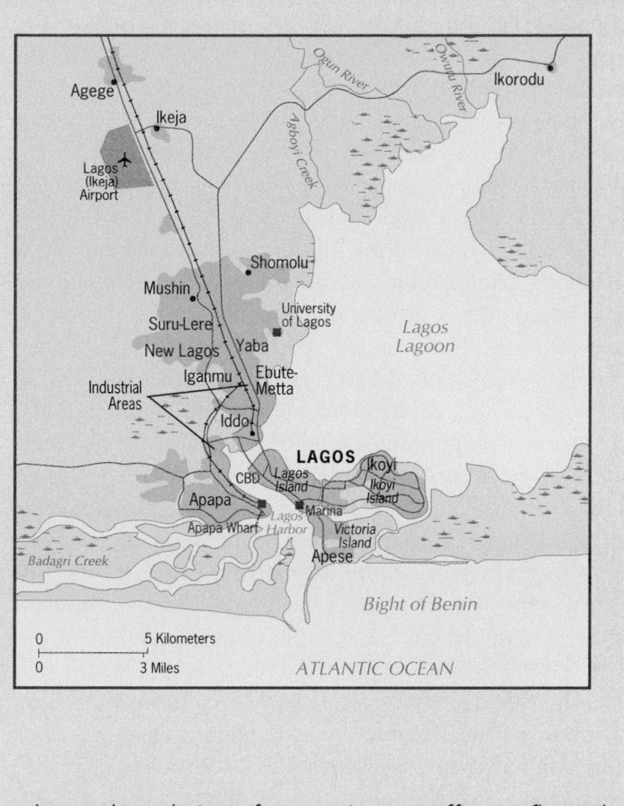

Lagos's cityscape is a mixture of modern high-rises, dilapidated residential areas, and squalid slums. From the top of a high-rise one sees a seemingly endless vista of rusting corrugated roofs, the houses built of cement or mud in irregular blocks separated by narrow alleys. On the outskirts lie the shantytowns of the less fortunate, where shelters are made of plywood and cardboard and lack even the most basic facilities.

By world standards, Lagos ranks among the most severely polluted, congested, and disorderly cities. Mismanagement and official corruption are endemic. Laws, rules, and regulations, from zoning to traffic, are flouted. The international airport is notorious for its inadequate security and for extortion by immigration and customs officers. In many ways, Lagos is a city out of control.

choice of Lagos, the port of the Western Region, as the federal capital (and not one of the cities in the more populous north) reflected British hopes for the country's future.

Nigeria is a large country territorially (about the size of Texas plus Oklahoma), and as Figure 7-10 shows, it extends farther northward than any of the other coastal West African states as far west as Guinea. This has the effect of incorporating much larger populations within its northern borders than its neighbors have. It was an objective of Nigeria's original three-region federal system that the two smaller southern regions would balance the single, larger one in the Islamic north.

But Nigeria's three-region federation did not last long. Initially, it was not the Northern Region but the Eastern Region that mounted a campaign to secede from the federation, although the north did play a role. Many Ibo had moved to northern cities to take jobs northerners were unwilling to do; some became successful and prominent entrepreneurs. These energetic Christians aroused animosities among the majority Muslims, and in September and October 1966 a series of massacres occurred. Surviving Ibos fled from the north in huge numbers, and in May 1967 the government of the Eastern Region declared its independence as the Republic of Biafra. This led to a civil war that lasted three years and cost an estimated 1 million lives. To ensure that such a secession attempt would not happen again, successive Nigerian governments and regimes repeatedly modified the federal map. Today Nigeria has 36 States as well as a new Federal Capital Territory containing the more centrally located capital, Abuja (Fig. 7-11). But during most of its existence as an independent state, Nigeria has not functioned as a federation. Civilian government was repeatedly abrogated by military dictatorships that centralized power, enriched the rulers, and impoverished the country.

Large oilfields were discovered under the Niger Delta during the 1950s, when Nigeria's agricultural sector produced most of its exports (peanuts, palm oil, cocoa, cotton) and farming still had priority in national and State development plans. Soon, revenues from oil production dwarfed all other sources, bringing the country a brief period of prosperity and promise. But before long Nigeria's oil wealth brought more bust than boom. Misguided development plans now focused on grand, ill-founded industrial schemes and costly luxuries such as a national airline; the continuing mainstay of the vast majority of Nigerians, agriculture, fell into neglect. Worse, poor management, corruption, outright theft of oil revenues during military misrule, and excessive borrowing against future oil income led to economic disaster. The country's infrastructure collapsed. In the cities, basic services broke down. In the rural areas, clinics, schools, water supplies, and roads to markets crumbled. In the Niger Delta area, local people beneath whose land the oil was being exploited demanded a share of the revenues and reparations for ecological damage; the military regime under General Abacha responded by arresting and executing nine of their leaders. On global indices of national well-being, Nigeria sank to the lowest rungs even as its production ranked it as high as the world's tenth-largest oil producer, with the United States its chief customer.

In 1999, Nigeria's hopes were raised when, for the first time since 1983, a democratically elected president was sworn into office. But even as President Obasanjo confronted the country's tough realities—idle, dated factories; abandoned farms and plantations; countless unemployed; ingrained habits of corruption and nonpayment of taxes; a fast-growing population that now exceeds 130 million; the deepening AIDS crisis—a cultural imbroglio with catastrophic potential loomed. The return to democracy brought with it a loss of power among the northern elite that had long prevailed in the military, and now northern States, predominantly Muslim, began flexing their muscles in other ways. The State of Zamfara, whose politicians had fallen out of favor in the government of President Obasanjo, proclaimed that Sharia (Muslim) law would henceforth apply not only to matters such as marriage and inheritance, but to all aspects of life. Sharia punishments, as noted in Chapter 6, are severe, including amputations for theft. But it also affects daily life: men and women cannot ride in taxis together, schools are segregated by sex, many jobs are reserved for males. Muslim leaders in Zamfara insisted that Sharia law would affect only Muslims, not Christians and others in the State, but this was not so. Christian women, for example, had to find public taxis driven by Christians, but most taxis are driven by Muslims who no longer pick up either Muslim or Christian women.

Several other northern States followed Zamfara in proclaiming Sharia law, despite the federal government's appeal to slow their initiatives. When Kaduna State imposed Sharia, riots between Muslims and Christians devastated the old capital city of Kaduna. There and elsewhere, the imposition of Sharia led to the departure of thousands of Christians, intensifying the cultural fault line that threatens the Nigerian federation. By the beginning of 2001, ten northern States had proclaimed Sharia, although Kaduna's government repealed its decision, at least temporarily (Fig. 7-11).

All this raises the prospect that Nigeria, West Africa's cornerstone and one of Africa's most important states, may succumb to devolutionary forces arising from its location on the African Transition Zone. This would be a calamitous development. A stable, well-governed, and economically growing Nigeria would be a beacon to region and realm; its collapse could infect Africa's political geography far and wide.

The Coast and Interior

Nigeria is one of 17 states (counting Chad and offshore Cape Verde [not shown in Fig. 7-10]) that constitute the region of West Africa. Four of these countries, comprising a huge territory on the Sahara's margins but containing small populations, are landlocked: Mali, Burkina Faso, Niger, and Chad. Figure I-8 shows clearly how steppe and desert conditions dominate the natural environments of these four interior states. Figure I-9 reveals the concentration of population in the steppe zone and along the ribbon of water the Niger River provides. Scattered oases form the remaining settlements and anchor regional trade.

But even the coastal states do not escape the dominance of the desert over West Africa. Mauritania's environment is almost entirely desert. Senegal, as Figure 7-10 shows, is a Sahel country; and not only northern Nigeria but also northern Benin, Togo, and Ghana have interior steppe zones. The loss of pastures to desertification is a constant worry for the livestock herders there.

West Africa's states share the effects of the environmental zonation depicted in Figures I-7 and I-8, but they also have distinct regional geographies. Benin, Nigeria's neighbor, has a growing cultural and economic link with the Brazilian State of Bahia, where many of its people were taken in bondage and where elements of West African culture have survived. Ghana, once known as the Gold Coast, was the first West African state to achieve independence (1957), with a sound economy based on cocoa exports. Two grandiose postindependence schemes can be seen on its map: the port of Tema, which was to serve a vast West African hinterland, and Lake Volta, which resulted from the region's largest dam project. When neither fulfilled expectations, Ghana's economy collapsed. In the 1990s, stable and democratic government (following a military regime) produced some recovery.

Ivory Coast (which, as we have stated, officially still goes by its French name, *Côte d'Ivoire*) translated three decades of autocratic but stable rule into economic progress that gave it lower-middle-income status based mainly on

Ivory Coast's (*Côte d'Ivoire's*) first and longtime leader, Félix Houphouët-Boigny, always told his people, especially those who had moved to the capital or overseas, to remember their home villages and support them. He set an example by funding public facilities and modernizing infrastructure in his own village, Yamoussoukro. He invested heavily in the modern highway linking Yamoussoukro to Abidjan. But his most extravagant spending created this elaborate replica of St. Peter's in Rome. Even larger than the original and thus the largest church in Christendom, this basilica (built to the finest detail by Italian craftsmen using imported goods ranging from marble to faucets) was consecrated by Pope John Paul II in 1989. Critics argued that *Côte d'Ivoire's* francs should have been spent on other ventures, and Muslim northerners complained about this "cultural excess." Meanwhile the basilica towers over Yamoussoukro today, a nonfunctional memorial to the man who put his village on the map.

cocoa and coffee sales. Continued French involvement in the country's affairs contributed to this prosperity; the capital, Abidjan, reflected its comparative well-being. But in a familiar pattern, Ivory Coast's president-for-life first engineered the transfer of the capital to his home village, Yamoussoukro, and then spent tens of millions of dollars building a Roman Catholic basilica there to rival that of St. Peter's in Rome (see photo above). It was dedicated just as the country's economy was slowing and its social conditions worsened. By the late 1990s, the national economy once again was in the low-income category (Fig. I-11). Political issues with regional overtones roiled Ivoirian society in 2000, when anger over a presidential candidate's ethnic origins grew into a countrywide campaign against residents who had immigrated from Burkina Faso, strengthening the Muslim presence in the north. Thousands were forced to flee, raising the specter of a north-south, Muslim-Christian schism in this long-stable country of 16.7 million people.

All of West Africa has been affected by what has happened in Liberia and Sierra Leone. Liberia, a country founded in 1822 by freed slaves who returned to Africa with the help of American colonization societies, was ruled by their descendants for more than six generations. Rubber plantations and iron mines made life comfortable for the "Americo-Liberians," but among the local peoples, resentment simmered. A military coup in 1980, in which the president was killed, was followed in 1989 by full-scale civil war that pitted ethnic groups against each other and drove hundreds of thousands of refugees into neighboring countries, including Ivory Coast, Guinea, and Sierra Leone. Monrovia, the capital (named after U.S. president James Monroe) was devastated; an estimated 230,000 people, almost 10 percent of the population, perished. In 1997, one of the rebel leaders, Charles Taylor, became president, but his presidency was tainted by allegations of brutality during the war.

One of the countries affected by these events was Sierra Leone, Liberia's coastal neighbor, also founded as a haven for freed slaves, in this case by the British in 1787. Independent since 1961, Sierra Leone went the all-too-familiar route from self-governing Commonwealth member to republic to one-party state to military dictatorship. But in the 1990s a civil war brought untold horror to this small country even as refugees from Liberia were arriving. An organization called the Revolutionary United Front (RUF), another called the "sobels" (soldiers turned rebels), and those supporting the legitimate government engaged in a struggle that devastated town and countryside alike. The RUF mutilated civilians, including children, they did not kill; its leaders traded the interior's diamonds for weapons. Nigerian forces, under the aegis of a regional organization called ECOWAS (Economic Community of West African States), intervened in 1998, but the RUF continued the war. United Nations efforts to stabilize Sierra Leone followed, coupled with British special forces to police Freetown, the capital. Rebel leaders, including those accused of the most heinous acts, were offered roles in a future government in return for an armistice. Meanwhile, UN estimates indicated that about 500,000 refugees had crossed the border into Guinea. In 2000, Sierra Leone ranked dead last on the world's list of nations' well-being.

Why has this particular part of West Africa fallen victim to such violence? The causes are many: ethnic and historic animosities, the concentration of wealth and power in minority hands, coast–interior rivalries, the roles of the "national" armies and local militias, all able to secure weapons. But it also was a matter of scale and location. Small countries such as Liberia and Sierra Leone become realistic takeover targets for ambitious rebels in ways that Ivory Coast and Ghana do not; and when small countries have isolated and disconnected minorities that are not effectively linked to the national scene, these minorities become pawns in the struggle among armed factions. Note that Ivory Coast to the east and Senegal to the northwest were affected by the events just discussed, but they were not destabilized.

Senegal had the advantage of lying separated from Sierra Leone by Guinea, another Francophone country along the West African coast. Its Sahelian environmental problems notwithstanding, Senegal managed to convert its colonial advantage (its capital, Dakar, was the headquarters for France's West African empire) into lasting progress. By no means a rich country, Senegal depends for foreign income on farm exports (chiefly peanuts), fishing (the leading source), phosphate sales, and iron ore production. But its most valuable asset has been its democratic tradition, which has now lasted more than four decades. This has enabled Senegal to weather some storms, including a failed union with its English-speaking enclave, Gambia, and a secessionist movement in the southwestern Casamance District (Fig. 7-10).

Senegal is over 90 percent Muslim, but its population of 10 million is dominated by the Wolof, the largest ethnic group constituting about 37 percent of the total, the Fulani, and the Serer (17 percent each). The Wolof are concentrated in and around the capital and wield most of the power. Senegal's leaders have maintained close relationships with France, which is still Senegal's leading trading partner and financial supporter in times of need. Without oil, diamonds, or other lucrative income sources and with an overwhelmingly subsistence-farming population, Senegal nevertheless managed to achieve GNP levels that ranked among the region's highest (see Table I-2). Here is proof that reasonably representative government and stability are greater assets than gold, liquid or otherwise.

Populous, multicultural, divided West Africa remains a region of farmers and herders (the latter range over the savannas and steppes of the north) along a tier of fast-changing environments between ocean and desert. Local village markets drive the traditional economy, and **periodic markets**—not open every day but operating every three, four, or more days—ensure that all villages participate in the exchange network. Traditions of this kind endure here, even as the cities beckon the farmers and burst at the seams. The region's great challenges are economic survival and nation-building, constrained by a boundary framework that is as burdensome as any in Africa.

14

EQUATORIAL AFRICA

The term *equatorial* is not just locational but also environmental. The equator bisects Africa, but only the western part of central Africa features the conditions associated with the low-elevation tropics: intense heat, high rainfall and extreme humidity, little seasonal variation, rainforest and monsoon-forest vegetation, enormous biodiversity. To the east, beyond the Western Rift Valley, elevations rise, and cooler, more seasonal climatic regimes prevail. As a result, we recognize two regions in these lowest latitudes: Equatorial Africa to the west and East Africa to the east.

Equatorial Africa is physiographically dominated by the giant Congo Basin. The Adamawa Highlands separate this region from West Africa; rising elevations and climatic change mark its southern limits (see the *Cwa* boundary in Fig. I-8). Its political geography consists of eight states, of which The Congo (formerly Zaïre) is by far the largest in both territory and population (Fig. 7-12).

Five of the other seven states—Gabon, Cameroon, São Tomé and Príncipe, Congo, and Equatorial Guinea—all have coastlines on the Atlantic Ocean. The Central African Republic and Chad, the south of which is part of this region, are landlocked. In certain respects, the physical and human characteristics of Equatorial Africa extend even into southern Sudan. This vast and complex region is in many ways the most troubled region in the entire Subsaharan African realm.

During the colonial period, the name *Congo* became attached to two countries on opposite sides of the region's major river: the Belgian Congo and the French Congo. After independence, both countries initially retained this identity, but in 1971 the ruler of the larger (former Belgian) Congo changed its name to Zaïre. In 1997, following a prolonged revolution that started in eastern Zaïre, a new regime took power and renamed the country The Democratic Republic of the Congo. As we noted at the beginning of this chapter, we refer to this country as *The* Congo, and to its smaller neighbor simply as Congo.

The Congo

As the map shows, The Congo has but a tiny window (23 miles; 37 km) on the Atlantic Ocean, just enough to accommodate the mouth of the Congo River. Oceangoing ships can reach the port of Matadi, inland from which falls and rapids make it necessary to move goods by road or rail to the capital, Kinshasa. This is not the only place where the Congo River fails as a transport route. Follow it upstream on Figure 7-12, and you note that other transshipments are necessary between Kisangani and Ubundu, and at Kindu. Follow the railroad south from Kindu, and you reach another narrow corridor of The Congo's territory at the city of Lubumbashi. That vital part of The Congo contains most of its major mineral resources, including copper and cobalt.

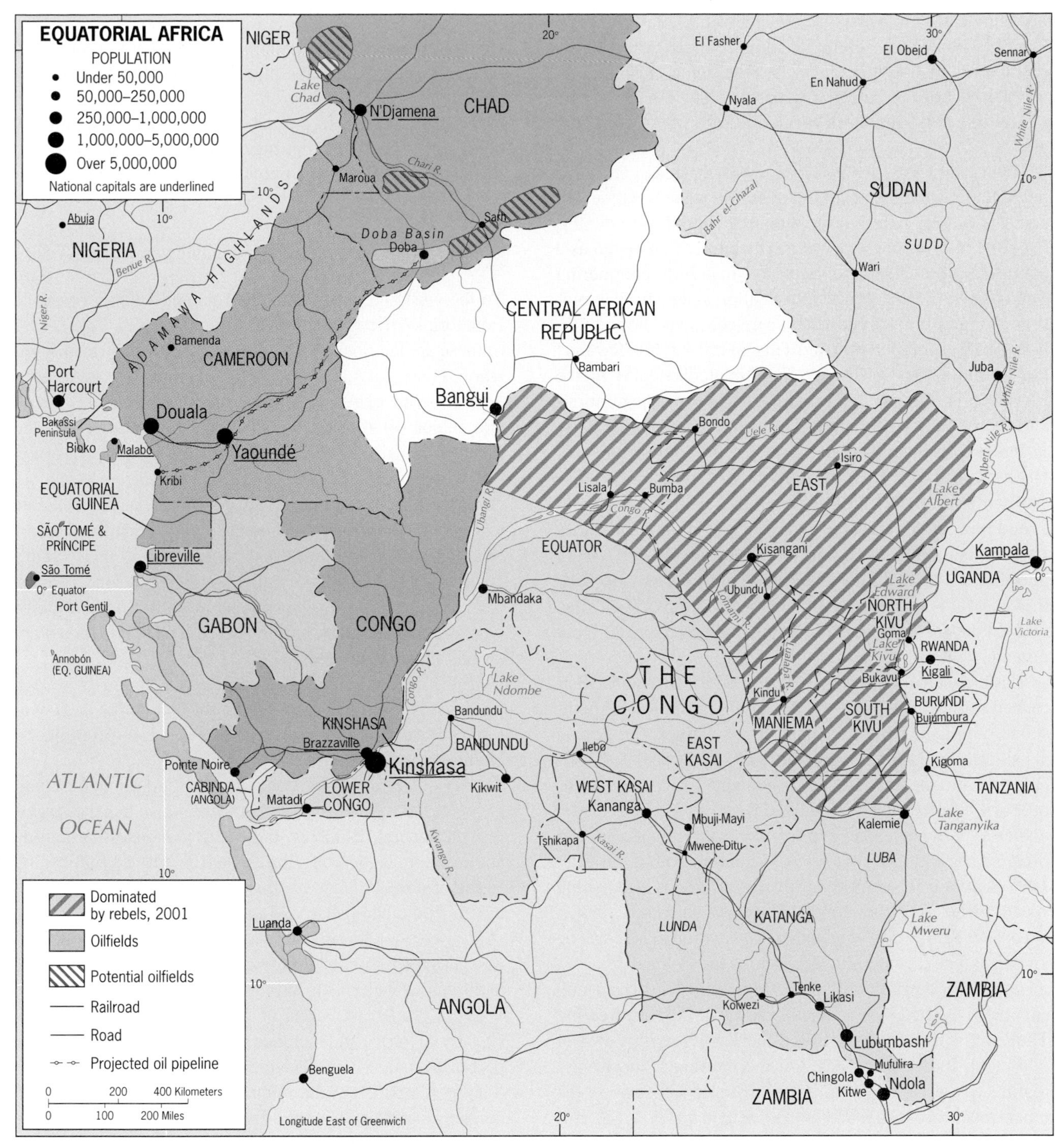

FIGURE 7-12

As the map shows, the most efficient export route from this area, called Katanga, lies by rail across Angola to the port of Benguela. Civil war in Angola, however, disrupted this route.

With a territory not much smaller than the United States east of the Mississippi, a population of 55.4 million, a rich and varied mineral base, and much good agricultural land, The Congo would seem to have all the ingredients needed to lead this region and, indeed, Africa. But strong centrifugal forces, arising from its physiography and cultural geography, pull The Congo apart. The immense forested heart of the basin-shaped country creates communication barriers between east and west, north and south. Many of The Congo's productive areas lie along its periphery, separated by enormous distances. These areas tend to look across the border, to one or more of The Congo's nine neighbors, for outlets, markets, and often ethnic kinship as well.

The Congo's civil wars of the 1990s started in one such neighbor, Rwanda, and spilled over into what was then still Zaïre. Rwanda, Africa's most densely populated country, has for centuries been the scene of conflict that began as a contest between Hutu farmers and invading, conquering Tutsi pastoralists. Colonial intervention and cultural evolution changed the ground rules. The colonists' boundaries hemmed in a core of Hutu and Tutsi in hilly, fertile Rwanda and Burundi, but left many Tutsi-related peoples in the neighboring Belgian Congo. The Tutsi, always the minority and always better off than the Hutu, dominated political life in the postcolonial period. A seemingly endless cycle of Hutu revenge and Tutsi retribution cost hundreds of thousands of lives. In the mid-1990s another such outbreak generated one of the largest refugee streams in human history from Rwanda into Zaïre (and into Tanzania as well). Hutu guerrillas, who had killed hundreds of thousands of Tutsi in Rwanda, now terrorized the refugee camps. Their plan was to take control of eastern Zaïre, creating a greater Hutu domain.

Eastern Zaïre, however, was peopled by Tutsi and Tutsi-related groups, and soon the Hutu aggressors found themselves outgunned. But they had started a larger war. Soon a Tutsi-dominated force, aided not only by Rwanda's Tutsi-led government but also by a sympathetic Uganda, conquered eastern Zaïre and moved westward. Their leader was an experienced rebel commander from Katanga (named Shaba during the Zaïrian period), Laurent Kabila. In May 1997 Kabila's forces, having overrun Zaïre at astonishing speed, swept into the capital, Kinshasa, and took power. Days later Kabila announced that Zaïre would henceforth be known as The Democratic Republic of the Congo.

Ousting a corrupt, collapsing regime in Kinshasa (long supported by the United States) was one thing; resurrecting The Congo was another. Soon, the Kabila regime was at odds with its former allies. On the one hand, the Kabila regime was accused of failing to institute reforms; on the other, rebel forces that had helped him now saw opportunities to achieve their own territorial aims. As The Congo fell into renewed disarray, its breakdown affected other countries within and beyond its region. In the east, Uganda, Rwanda, and Burundi supported an anti-Kabila campaign even as other groups, also opposed to the new regime, fought among themselves. In the west, Kabila had the support of Zimbabwe, Angola, and Namibia, all of whom sent armed forces. But such support came at a price; Zimbabwean generals secured costly business privileges in The Congo's mineral and agricultural areas.

By early 2001 Kabila had been assassinated, and The Congo remained divided roughly as shown on Figure 7-12. Peacemaking efforts have made little headway against the complex and intractable problems of the devolving Congo. More than 200 ethnic groups are contained by a colonial boundary framework that needs reconstruction—and that reconstruction cannot be painless for either The Congo or for Africa.

Across the River

To the west and north of the Congo and Ubangi rivers lie Equatorial Africa's other seven countries (Fig. 7-12). Two of these are landlocked. Chad, straddling the African Transition Zone as well as the regional boundary with West Africa, is one of Africa's most remote countries. But several major oil discoveries in the south and a projected pipeline to Kribi in Cameroon will link Chad more effectively with this region. The Central African Republic, chronically unstable and poverty-stricken, never was able to convert its agricultural potential and mineral resources (diamonds, uranium) into real progress. And one country consists of two small, densely forested volcanic islands: São Tomé and Príncipe, a ministate with a population of only 175,000 and a few exports derived from its cocoa plantations and coconut trees.

The four coastal states present a different picture. All four possess oil reserves and share the Congo Basin's equatorial forests; oil and timber, therefore, rank prominently among their exports. In Gabon, this combination has produced Equatorial Africa's only upper-middle-income economy. Of the four coastal states, Gabon also has the largest proven mineral resources, including manganese, uranium, and iron ore. Its capital, Libreville (the only coastal capital in the region), reflects all this in its high-rise downtown, bustling port, and fast-growing squatter settlements.

Cameroon, less well endowed with oil or other raw materials, has the region's strongest agricultural sector by virtue of its higher-latitude location and high-relief topography. Western Cameroon is one of the more developed parts of Equatorial Africa and includes the capital, Yaoundé, and the port of Douala. In 1997, Cameroon granted China the right to exploit a bauxite (aluminum) reserve, but relations with neighboring Nigeria were difficult. Cameroon and Nigeria quarreled over their joint border on the oil-rich Bakassi Peninsula—which lies in the far west, where the Cameroonians' *lingua franca* is English, not French as it is elsewhere in the country.

With five neighbors, Congo could be a major transit hub for this region, especially for The Congo if it recovers from civil war. Its capital, Brazzaville, lies across the Congo River from Kinshasa and is linked to the port of Pointe Noire by road and rail. But devastating power struggles have negated Congo's geographic advantages.

As Figure 7-12 shows, Equatorial Guinea consists of a rectangle of mainland territory and the island of Bioko, where the capital of Malabo is located. A former Spanish colony that remained one of Africa's least-developed territories, Equatorial Guinea, too, has been affected by the oil business in this area. Petroleum products now dominate its exports, but, as in so many other oil-rich countries, this bounty has not significantly raised incomes for most of the people.

One other territory would seem to be a part of Equatorial Africa: Cabinda, wedged between the two Congos just to the north of the Congo River's mouth. But Cabinda is one of those colonial legacies on the African map—it belonged to the Portuguese and was administered as part of Angola. 15 Today it is an **exclave** of independent Angola, and a valuable one: it contains major oil reserves.

EAST AFRICA

East of the row of Great Lakes that marks the eastern border of The Congo (Lakes Albert, Edward, Kivu, and Tanganyika), central Africa takes on a different character. The land rises from the Congo Basin to the East African Plateau. Hills and valleys, fertile soils, and copious rains mark the transition in Rwanda and Burundi. Eastward the rainforest disappears and open savanna cloaks the countryside. Great volcanoes rise above a rift-valley-dissected highland. At the heart of the region lies Lake Victoria. In the north the surface rises above 10,000 feet (3300 m), and so deep are the trenches cut by faults and rivers there that the land was called, appropriately, Abyssinia (now Ethiopia).

Five countries, as well as the highland part of Ethiopia, form this East African region: Tanzania, Kenya, Uganda, Rwanda, and Burundi. Here the Bantu peoples that make up most of the population met Nilotic peoples from the north, including the Maasai. In the hills of Rwanda and Burundi, as we noted, a stratified society developed in which the minority Tutsi, in their cattle-owning kingdoms, dominated the Hutu peasantry. The coast was the scene of many historic events: the arrival of Islam, the visit of Ming Dynasty Chinese fleets in the 1400s, the quest for power by the Turks, the Arab slave trade, the European colonial competition. Here developed the East African *lingua franca*, Swahili.

Kenya

Kenya is neither the largest nor the most populous country in East Africa, but over the past half-century it has been the dominant state in the region. Its skyscrapered capital at the heart of its core area, Nairobi, is the region's largest city; its port, Mombasa, is the region's busiest. During the 1950s, the Kikuyu nation led a vicious rebellion that hastened the departure of the British from the region.

After independence, Kenya chose a capitalist path of development, aligning itself with Western interests. Without major known mineral deposits, Kenya depended on coffee and tea exports and on a tourist industry based on its magnificent national parks (Fig. 7-13). Tourism became its largest single earner of foreign exchange, and Kenya prospered, apparently proving the wisdom of its capitalist course.

But serious problems arose. Kenya during the 1980s had the highest rate of population growth in the world, and population pressure on farmlands and on the fringes of the

FROM THE FIELD NOTES

"Visiting the village in the Kenya Highlands where a graduate student was doing fieldwork on land reform, I took the long way and drove along the top of the eastern wall of the Eastern Rift Valley. Often the valley wall is not sheer but terraced, and soils on those terraces are quite fertile; also, the west-facing slopes tend to be well-watered. Here African farmers built villages and laid out communal plots, farming these lands in a well-organized way long before the European intrusion."

Among the Realm's Great Cities . . .

Nairobi

Nairobi is the quintessential colonial legacy: there was no African settlement on this site when, in 1899, the railroad the British were building from the port of Mombasa to the shores of Lake Victoria reached it. However, it had something even more important: water. The fresh stream that crossed the railway line was known to the Maasai cattle herders as Enkare Nairobi (Cold Water).

The railroad was extended into the interior, but Nairobi grew. Indian traders set up shop. The British established their administrative headquarters here. When Kenya became independent in 1963, Nairobi naturally was the national capital.

Nairobi owes its primacy to its governmental functions, which ensured its priority through the colonial and independence periods, and to its favorable situation. To the north and northwest lie the Kenya Highlands, the country's leading agricultural area and the historic base of the dominant nation in Kenya, the Kikuyu. Beyond the rift valley to the west lie the productive lands of the Luo in the Lake Victoria Basin. To the east, elevations drop rapidly from Nairobi's 5000 feet (1660 m), so that highland environs make a swift transition to tropical savanna that, in turn, yields to semiarid steppe.

A moderate climate, a modern city center, several major visitor attractions (including Nairobi National Park, on the city's doorstep), and a state-of-the-art airport

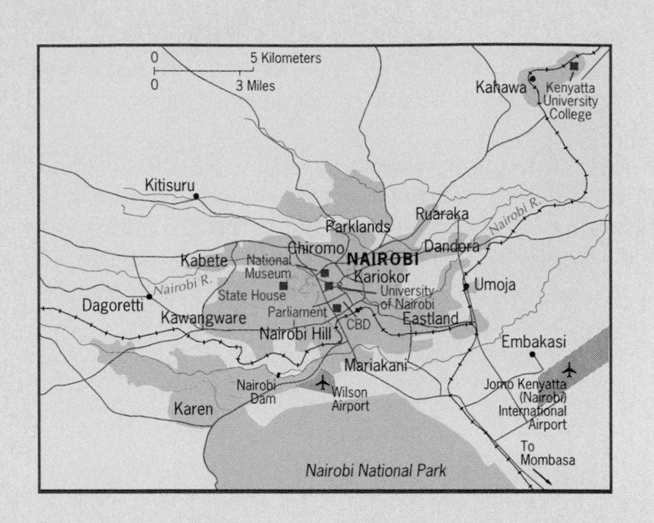

have boosted Nairobi's fortunes as a major tourist destination, though wildlife destruction, security concerns, and political conditions have damaged the industry in recent years.

Nairobi is Kenya's principal commercial, industrial, and educational center. But its growth (to 2.6 million today) has come at a price: its modern central business district stands in stark contrast to the squalor in the shantytowns that house the countless migrants its apparent opportunities attract.

wildlife reserves mounted. Poaching became worrisome, and tourism declined. During the late 1990s, violent weather buffeted Kenya, causing landslides and washing away large segments of the crucial Nairobi-Mombasa Highway. This was followed by a severe drought lasting several years, bringing famine to the interior. Meanwhile, government corruption siphoned off funds that should have been invested. Democratic principles were violated, and relationships with Western allies were strained. The AIDS epidemic brought another setback to a country that, in the early 1970s, had appeared headed for an economic takeoff.

Today, Kenya's prospects are uncertain. Geography, history, and politics have placed the Kikuyu (22 percent of the population of 31.6 million) in a position of power. But there are other major peoples (see Fig. 7-13) and several smaller ones. The Luhya, Luo, Kalenjin, and Kamba together constitute about 50 percent of the population, and on the territorial margins of the country there are peoples such as the Maasai, Turkana, Boran, and Galla. Creating and sustaining a political system that ensures democracy and represents the interests of these disparate peoples is Kenya's unmet challenge.

Tanzania

Tanzania (a name derived from Tanganyika plus Zanzibar) is the biggest and most populous East African country (37.4 million). Its total area exceeds that of the other four countries combined. Tanzania has been described as a country without a core because its clusters of population and zones of productive capacity lie dispersed—mostly on its margins on the east coast (where the capital, Dar es Salaam, is located), near the shores of Lake Victoria in the northwest, near Lake Tanganyika in the far west, and near Lake Malawi in the interior south. This is in sharp contrast to Kenya, which has a well-defined core area in the Kenya Highlands (centered on Nairobi in the heart of the country). Moreover, Tanzania is a country of many peoples, none numerous enough to dominate the state. About 100 ethnic

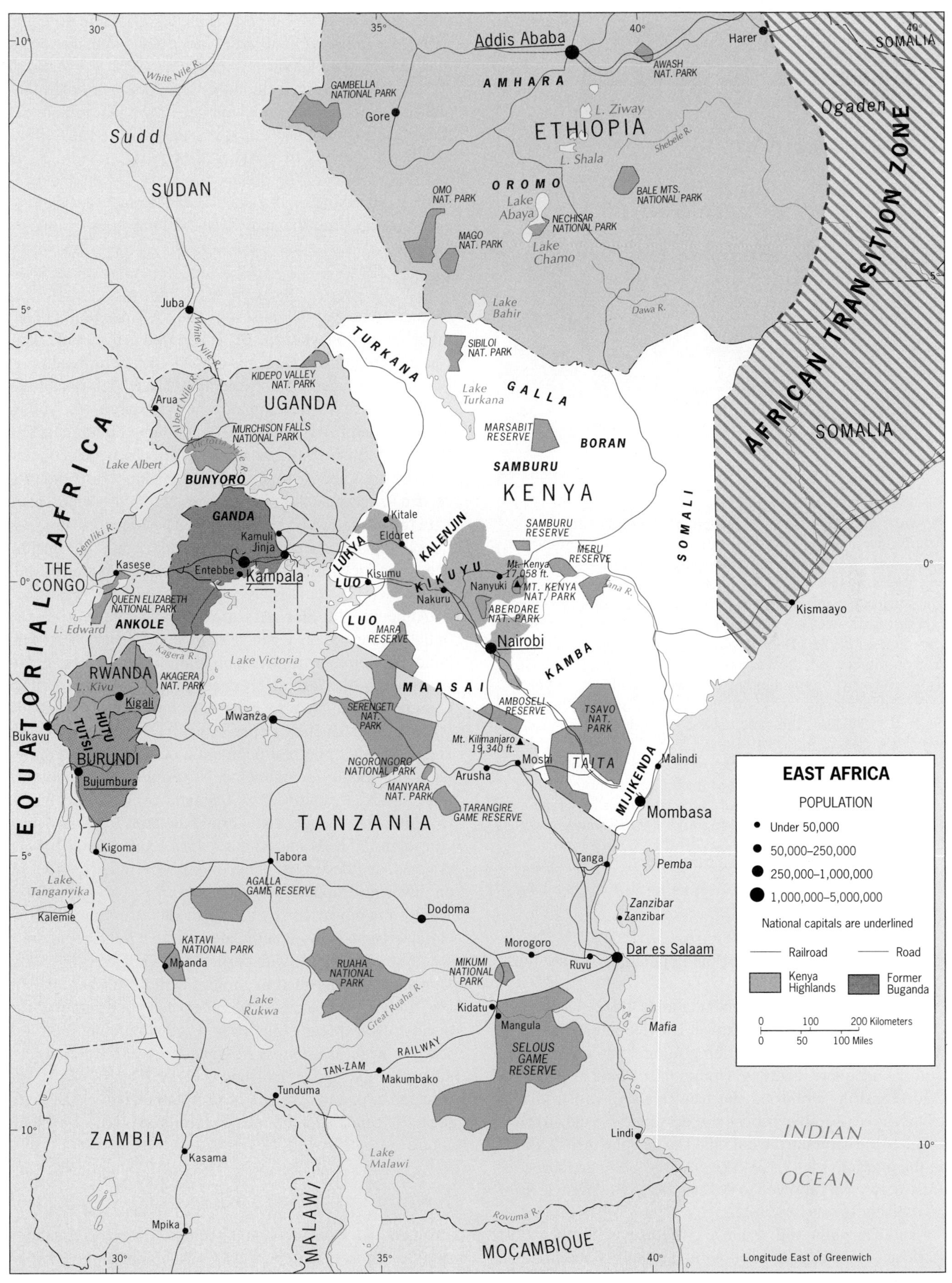

EAST AFRICA
POPULATION
Under 50,000
50,000–250,000
250,000–1,000,000
1,000,000–5,000,000
National capitals are underlined
Railroad
Road
Kenya Highlands
Former Buganda
0 100 200 Kilometers
0 50 100 Miles
Longitude East of Greenwich
Addis Ababa
ETHIOPIA
AMHARA
OROMO
Harer
SOMALIA
Ogaden
AFRICAN TRANSITION ZONE
AWASH NAT. PARK
GAMBELLA NATIONAL PARK
Gore
L. Ziway
L. Shala
Shebele R.
BALE MTS. NATIONAL PARK
OMO NAT. PARK
MAGO NAT. PARK
Lake Abaya
NECHISAR NATIONAL PARK
Lake Chamo
Lake Bahir
Dawa R.
Sudd
SUDAN
White Nile R.
Juba
TURKANA
SIBILOI NAT. PARK
Lake Turkana
GALLA
BORAN
MARSABIT RESERVE
SAMBURU
KENYA
SOMALI
KIDEPO VALLEY NAT. PARK
UGANDA
Arua
Albert Nile R.
MURCHISON FALLS NATIONAL PARK
Victoria Nile R.
Lake Albert
BUNYORO
GANDA
EQUATORIAL AFRICA
Semliki R.
Kamuli
Jinja
Kasese
Entebbe
Kampala
THE CONGO
QUEEN ELIZABETH NATIONAL PARK
ANKOLE
L. Edward
Kitale
Eldoret
KALENJIN
LUHYA
Kisumu
LUO
KIKUYU
Nakuru
Nanyuki
Mt. Kenya 17,058 ft.
MT. KENYA NAT. PARK
SAMBURU RESERVE
MERU RESERVE
ABERDARE NAT. PARK
Kismaayo
MARA RESERVE
Nairobi
KAMBA
Lake Victoria
Kagera R.
RWANDA
L. Kivu
Kigali
AKAGERA NAT. PARK
HUTU
TUTSI
BURUNDI
Bujumbura
Bukavu
Mwanza
MAASAI
SERENGETI NAT. PARK
AMBOSELI RESERVE
TSAVO NAT. PARK
Mt. Kilimanjaro 19,340 ft.
Moshi
Arusha
NGORONGORO NATIONAL PARK
MANYARA NAT. PARK
TARANGIRE GAME RESERVE
TAITA
MIJIKENDA
Malindi
Mombasa
TANZANIA
Kigoma
Tabora
Lake Tanganyika
Kalemie
AGALLA GAME RESERVE
Tanga
Pemba
Zanzibar
Dodoma
KATAVI NATIONAL PARK
Mpanda
Morogoro
Ruvu
Dar es Salaam
RUAHA NATIONAL PARK
MIKUMI NATIONAL PARK
Lake Rukwa
Great Ruaha R.
Kidatu
Mangula
Mafia
SELOUS GAME RESERVE
TAN-ZAM RAILWAY
Makumbako
Tunduma
ZAMBIA
Kasama
Lake Malawi
Lindi
INDIAN OCEAN
MALAWI
Mpika
Rovuma R.
MOÇAMBIQUE

FIGURE 7-13

FROM THE FIELD NOTES

"It was 108 in the shade, but the narrow alleys of Zanzibar's Stone City felt even hotter than that. I spent some time here when I was working on my monograph on Dar es Salaam in the 1960s and had not been back. In those days, African socialism and *uhuru* were the watchwords; since then Tanzania has not done well economically. But here were signs of a new era: a People's Bank in a former government building, and on the old fort's tower a poster saying 'Think Digital Go Tritel.' Nearby, on the sandy beach where I relaxed 35 years ago, was evidence that Zanzibar has not escaped the ravages of AIDS. Now the sand served as a refuge for the sick, who were resting there. 'It's better, bwana, than the corridor of the clinic,' said a young man who could walk only a few steps at a time with the help of a cane, and who breathed with difficulty as he spoke. Here as everywhere in Subsaharan Africa, AIDS has severely strained already-limited medical facilities."

groups co-exist; one-third of the population, mainly those on the coast, are Muslims.

In contrast to Kenya, Tanzania after independence embarked on a socialist course toward development, including a massive farm collectivization program that was imposed without adequate planning. Communist China helped Tanzania construct a railroad, the Tan-Zam Railway, from Dar es Salaam to Zambia; but the project failed. So did an effort to move the capital from colonial Dar es Salaam to Dodoma in the interior (Fig. 7-13). The country's limited tourist infrastructure was allowed to degenerate, giving Kenya virtually the entire tourist market.

But Tanzania did achieve what several other East African countries could not: political stability and a degree of democracy. Thus Tanzania changed economic direction in the late 1980s without social turmoil and embarked on a market-oriented "recovery" program. While Tanzanian coffee, cotton, tobacco, and other farm products sold in increasing amounts on world markets, most Tanzanians remained subsistence farmers. At the turn of the century, Tanzania still had one of the world's poorest economies. In addition, demands for greater autonomy, even independence, were rising on the island of Zanzibar, a devolutionary challenge Tanzania may have difficulty overcoming.

Uganda

Uganda contained the most important African political entity in this region when the British arrived in the 1890s. This was the Kingdom of Buganda (shown in dark brown in Fig. 7-13), which faced the north shore of Lake Victoria, had an impressive capital at Kampala, and was stable—as well as ideally suited for indirect rule over a large hinterland. The British established their headquarters at nearby Entebbe on the lake (thus adding to the status of the kingdom) and proceeded to organize their Uganda protectorate in accordance with the principles of indirect rule. The Baganda (the people of Buganda) became the dominant people in Uganda, and when the British departed, they bequeathed a complicated federal system to perpetuate Baganda supremacy.

16

Although a **landlocked state** dependent on Kenya for an outlet to the ocean, Uganda at independence (1962) had better economic prospects than many other African countries. It was the largest producer of coffee in the British Commonwealth. It also exported cotton, tea, sugar, and other farm products. Copper was mined in the southwest, and an Asian immigrant population of about 75,000 dominated the country's commerce. Nevertheless, political disaster struck. Resentment at Baganda overlordship fueled revolutionary change, and a brutal dictator, Idi Amin, took control in 1971. He ousted the Asians, exterminated his opponents, and destroyed the economy. Eventually, in 1979, an invasion supported by neighboring Tanzania drove Amin from power, but by then Uganda lay in ruins. Recovery has been slow, complicated by the AIDS epidemic, which struck Uganda with particular severity. In the 1990s, Uganda was embroiled in conflict with its northern neighbor, Sudan. Each accused the other of supporting rebellion against itself. Meanwhile, even as economic and political reforms were in progress in Uganda, its government also supported the Tutsi minorities in Rwanda and The Congo. Situated at one of Africa's most volatile crossroads, landlocked Uganda (population: 24.6 million) faces a difficult future.

Rwanda and Burundi

Rwanda and Burundi would seem to occupy Tanzania's northwest corner, and indeed they were part of the German colonial domain conquered before World War I. But during that war Belgian forces attacked the Germans from their Congo bases and were awarded these territories when the conflict was over in 1918. The Belgians used them as labor sources for their Katanga mines.

Rwanda (7.5 million) and Burundi (6.4 million) are physiographically part of East Africa, but their cultural geography is linked to the north and west. Here, as we noted earlier, Tutsi pastoralists from the north subjugated Hutu farmers (who had themselves made serfs of the local Twa [pygmy] population), setting up a conflict that was originally ethnic but became cultural. Certain Hutu were able to advance in the Tutsi-dominated society, becoming to some extent converted to Tutsi ways, leaving subsistence farming behind, and rising in the social hierarchy. These so-called "moderate" Hutu were—and are—often targeted by other Hutus, who resent their position in society, so that today conflict between radical Hutus, on the one hand, and Tutsi and moderate Hutu, on the other, has become a cultural rather than strictly an ethnic one.

This conflict has now had repercussions not only in Rwanda and Burundi but also in The Congo, Uganda, Tanzania, and even Angola and Congo. At the opening of this century, the minority Tutsi and Hutu moderates were in control of the government in both Rwanda and Burundi, where the Tutsi-dominated military had staged a coup. The Hutu majority's leaders, of course, call for democracy, which would put them in power. The situation is one of Africa's most intractable.

Highland Ethiopia

As Figure 7-13 shows, the East African region also encompasses the highland zone of Ethiopia, including the capital, Addis Ababa, the source of the Blue Nile, Lake Tana, and the Amharic core area that was the base of the empire that lost its independence only from 1935 to 1941. Ethiopia, mountain fortress of the Coptic Christians who held their own here, eventually became a colonizer itself. Its forces came down the slopes of the highlands and conquered much of the Islamic part of Africa's Horn, including present-day Eritrea and the Ogaden area, a Somali territory. (Geographically, these are parts of what we have mapped as the African Transition Zone in Fig. 6-9.)

Physiographically and culturally, highland Ethiopia is part of the East African region. But because Ethiopia was not colonized and because its natural outlets are to the Red Sea, not southward to Mombasa, effective interconnections between former British East Africa and highland Ethiopia never developed. But the Amhara and Oromo peoples of Ethiopia are Africans, not Arabs, nor have they been Arabized or Islamized as in northern Sudan and Somalia. The independence and secession of Eritrea in 1993 effectively landlocked Ethiopia, but for a few years there was cooperation and Ethiopia used Eritrean Red Sea ports. In 1998, however, a boundary dispute led to a bitter and costly war, and Ethiopia was forced to turn to Djibouti for a maritime outlet. With adversaries on three sides, Ethiopia is likely to turn increasingly toward East Africa, however tenuous the surface links are today.

SOUTHERN AFRICA

Southern Africa, as a geographic region, consists of all the countries and territories lying south of Equatorial Africa's The Congo and East Africa's Tanzania (Fig. 7-14). Thus defined, the region extends from Angola and Moçambique (on the Atlantic and Indian Ocean coasts, respectively) to South Africa and includes a half-dozen landlocked states. Also marking the northern limit of the region are Zambia and Malawi. Zambia is nearly cut in half by a long land extension from The Congo, and Malawi penetrates deeply into Moçambique. The colonial boundary framework, here as elsewhere, produced many liabilities.

Southern Africa constitutes a geographic region in both physiographic and human terms. Its northern zone marks the southern limit of the Congo Basin in a broad upland that stretches across Angola and into Zambia (the tan corridor extending eastward from the Bihe Plateau in Fig. 7-2). Lake Malawi is the southernmost of the East African rift-valley lakes; Southern Africa has none of East Africa's volcanic and earthquake activity. Most of the region is plateau country, and the Great Escarpment is much in evidence here. There are two pivotal river systems: the Zambezi (which forms the border between Zambia and Zimbabwe) and the Orange-Vaal (South African rivers that combine to demarcate southern Namibia from South Africa).

The social, economic, and political geographies of Southern Africa also confirm the regional definition. Landlocked Zambia, whose economic core area is the Copperbelt northwest of Ndola, always looked southward for its outlets to the sea, electrical power, and fuels. Malawi's core area and outlets lie in its south. Moçambique, with its lengthy Indian Ocean coastline, has served as an exit not only for Malawi but also for Zimbabwe (through the port of Beira) and South Africa (through Maputo). Offshore Madagascar, however, remains a separate entity (see box titled "Distinctive Madagascar").

Southern Africa is the continent's richest region materially. A great zone of mineral deposits extends through the heart of the region from Zambia's Copperbelt through Zimbabwe's Great Dyke and South Africa's Bushveld Basin and Witwatersrand to the goldfields and diamond mines of the Orange Free State and northern Cape Province in the heart of South Africa. Ever since these minerals began to be exploited in colonial times, many migrant laborers have come to work in the mines. The range and volume of miner-

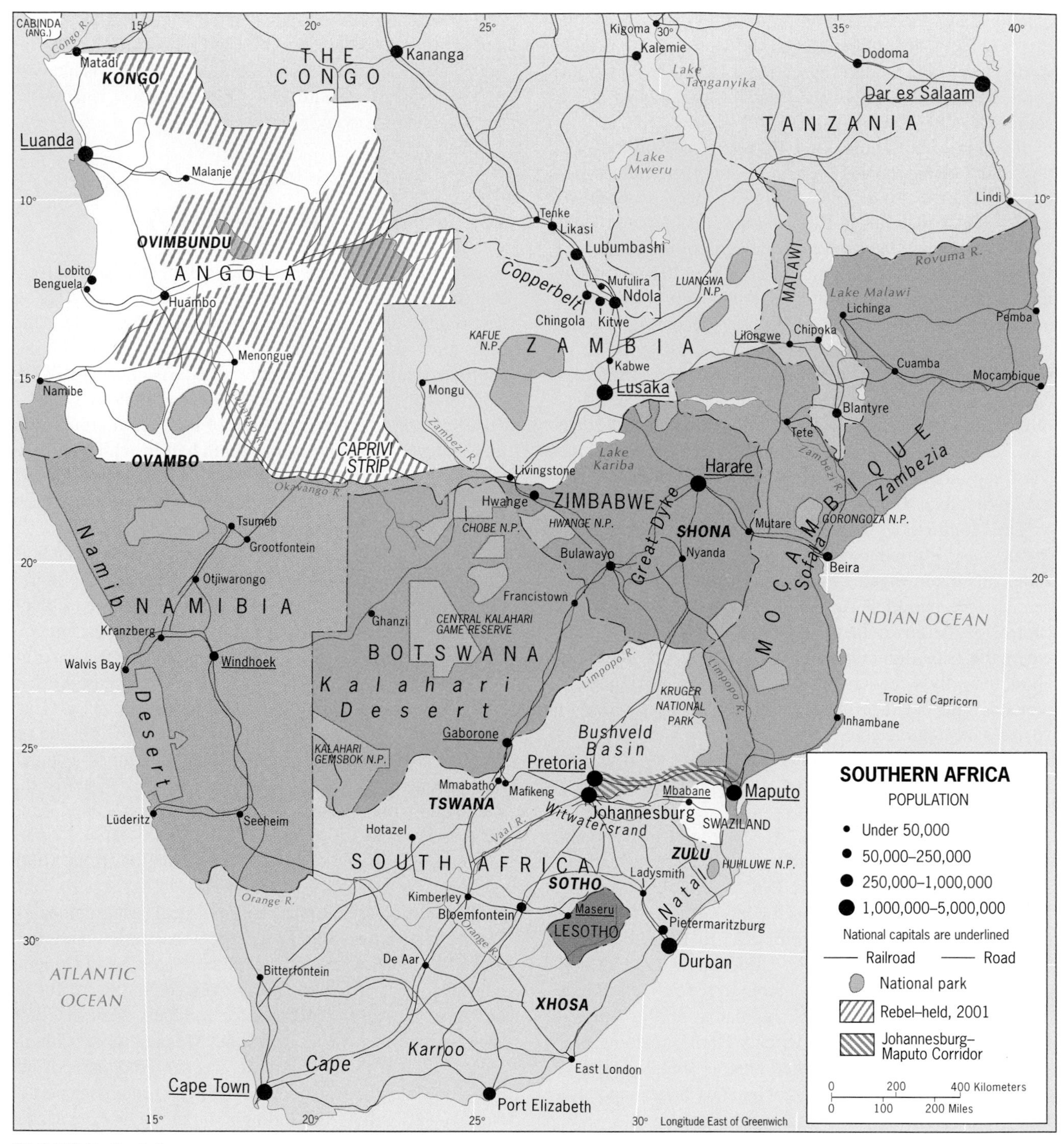

FIGURE 7-14

als mined in this belt are enormous, from the copper of Zambia and the chrome and asbestos of Zimbabwe to the gold, chromium, diamonds, platinum, coal, and iron ore of South Africa. However, not all of Southern Africa's mineral deposits lie in this central backbone. There is coal in western Zimbabwe at Hwange and in central Moçambique near Tete. In Angola, petroleum from oilfields along the north coast heads the export list; diamonds are mined in the northeast, manganese and iron on the central plateau. Namibia produces copper, lead, and zinc from a major mining complex anchored to the town of Tsumeb in the north, and diamond deposits lie in the beaches along the Atlantic Ocean in the south.

Southern Africa's agricultural diversity matches its mineral wealth. Vineyards drape the slopes of South Africa's Cape Ranges; tea plantations hug the eastern escarpment

Distinctive Madagascar

Maps showing the geographic regions of the Subsaharan Africa realm often omit Madagascar. And for good reason: Madagascar differs strongly from Southern Africa, the region to which it is nearest—just 250 miles (400 km) away. It also differs from East Africa, from which it has received some of its cultural infusions. Madagascar is the world's fourth largest island, a huge block of Africa that separated from the main landmass 160 million years ago. About 2000 years ago, the first settlers arrived—not from Africa (although perhaps via Africa) but from Southeast Asia. Malay communities flourished in the interior highlands of the island, which resembles Africa in having a prominent eastern escarpment and a central plateau (Fig. 7-15). Here was formed a powerful kingdom, the empire of the Merina. Its language, Malagasy, of Malay-Polynesian origin, became the indigenous tongue of the entire island (Fig. 7-8).

The Malay and Indonesian immigrants brought Africans to the island as wives and slaves, and from this forced immigration evolved the African component in Madagascar's population of 15.8 million. In all, nearly 20 discrete ethnic groups coexist in Madagascar, among which the Merina (4 million) and Betsimisaraka (2 million) are the most numerous. Like mainland Africa, Madagascar experienced colonial invasion and competition. Portuguese, British, and French colonists appeared after 1500, but the Merina were well organized and resisted the colonial conquest. Eventually Madagascar became part of France's empire, and French became the *lingua franca* of the educated elite.

Because of its Southeast Asian imprint, Madagascar's staple food is rice, not corn. It has some minerals, including chromite, iron ore, and bauxite, but the economy is weak, damaged by long-term political turmoil and burdened by rapid population growth. The infrastructure crumbled; in 1990 the "main road" from the capital to the nearest port (Fig. 7-15) was a potholed 150 miles that took 10 hours for a truck to navigate.

Meanwhile, Madagascar's unique flora and fauna retreated before the human onslaught. Madagascar's long-term isolation kept evolution here so distinct that the island is a discrete zoogeographic realm. Primates living on the island are found nowhere else; 33 varieties of lemurs are unique to Madagascar. Many species of birds, amphibians, and reptiles are also exclusive to this island. Their home, the rainforest, covered 65,000 square miles (168,000 sq km) in 1950, but today only about one-third of it is left. Logging, introduced by the colonists, damaged it; slash-and-burn agriculture is destroying it; and severe droughts in the 1980s and 1990s intensified the impact. Obviously, Madagascar should be a global conservation priority, but funds are limited and the needs are enormous. Malnutrition and poverty are powerful forces when survival is at stake for villages and families.

Madagascar's cultural landscape retains its Southeast Asian imprints, in the towns as well as the paddies. The capital, Antananarivo, is the country's primate city, its architecture and atmosphere combining traces of Asia and Africa. Poverty dominates the townscape here, too, and there is little to attract inmigrants (Madagascar is only 22 percent urbanized). But perhaps the most ominous statistic is Madagascar's population doubling time of just 24 years. In this respect Madagascar resembles Africa, not Southeast Asia.

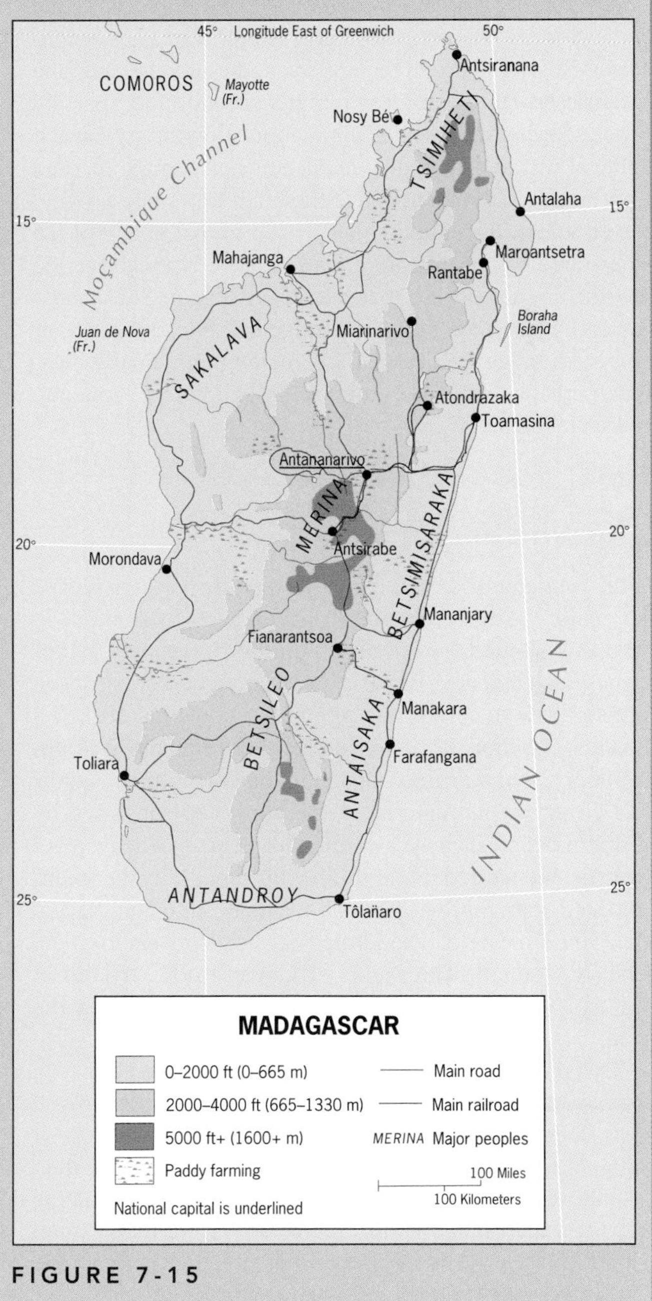

FIGURE 7-15

slopes of Zimbabwe. Before civil war destroyed its economy, Angola was one of the world's leading coffee producers. South Africa's relatively high latitudes and its range of altitudes create environments for apple orchards, citrus groves, banana plantations, pineapple farms, and many other crops. Farther north, tobacco has long been one of Zimbabwe's leading products. And while corn (maize) is the staple grain for most of Southern Africa's farmers, they also grow wheat and other cereals. Even the pastoral industry is varied, with large herds of beef cattle on the grassy highveld (upland), tens of millions of sheep on the southeast's pastures, and dairy farms around the big cities.

Despite this considerable wealth and potential, the countries of Southern Africa have not prospered. As Figure I-11 shows, most remain mired in the low-income category (Moçambique, with a per-capita GNP of only $210, is one of the world's poorest states); only South Africa and neighboring Botswana are in the upper-middle-income rank, the latter desert state being the realm's second most sparsely populated country. Rapid population growth, civil wars, political instability, poor management, corruption, and environmental problems have inhibited economic growth. Nevertheless, the situation is superior to that in any of the other three regions; more countries have risen above the low-income level here than anywhere else in Subsaharan Africa. As the new millennium opened, the resurgence of South Africa gave hope that this region might finally blossom, perhaps to lead the entire realm to a higher economic and social plane.

The Northern Tier

In the four countries that extend across the northern part of the region—Angola, Zambia, Malawi, and Moçambique—problems abound. Angola (13.7 million), formerly a Portuguese dependency, is one of Africa's richest countries in terms of raw materials and agricultural potential. It had a thriving economy at independence in 1975 but was engulfed by civil war that outside involvement made worse. The government, centered in the north and supported by northern peoples, chose a communist course. When a rebel movement emerged in the south with support by the (then) South African regime, Angola's leaders called on the Soviet Union for help. The United States also supported the rebel movement. Thousands of Cuban troops arrived in the capital, Luanda, escalating the war but securing the state.

This unrest devastated the Angolan economy. Farms lay abandoned, fields were land-mined, railroads destroyed, ports damaged. The railroad from Katanga to the port of Benguela, once a major income source, lay idle. Civilian casualties were high. But the government could pursue the war for a reason shown in Figure 7-14. Note that a small part of Angola lies separated from the main body of the country across the Congo River in the extreme northwest. This is the exclave of Cabinda, noted above, which contains oil reserves whose revenues supported the Luanda government during its war.

In the late 1990s, the United Nations brokered an agreement between the government and the rebel movement in the interior that awarded the insurgents a role in the administration. But this arrangement broke down, and as much as half of the country remained under rebel control or in contention. Meanwhile, smugglers carried many millions of dollars' worth of diamonds from Angola's uncontrolled frontier into The Congo and beyond. Political stability could make Angola one of Africa's most prosperous countries, but this is the one commodity this country lacks.

On the opposite coast, Moçambique (population: 19.9 million) was even less fortunate. Also a former Portuguese domain that chose a Marxist course for development, Moçambique had far fewer resources than Angola. Before independence, its chief sources of income came from its cashew and coconut plantations and from its relative location. As the map shows, Moçambique's port of Beira is ideally situated to handle the external trade of Zimbabwe and southern Malawi, and its capital, Maputo, is the closest port to the great mining and industrial complex centered on the Witwatersrand in South Africa. In better times, goods streamed through these ports, and when a hydroelectric plant was built on the Zambezi River at Cabora Bassa, prospects seemed fair. Then, however, the economy collapsed under bad management, a rebel movement aided for a time by South African interests destroyed the social order, famines broke out, and Moçambique descended into chaos. More than a million refugees streamed into Malawi, the hydroelectric project was damaged, the port facilities at Beira ceased functioning, and Maputo's transit role ended.

In the late 1990s, some semblance of stability had returned to Moçambique. The government abandoned its Marxist principles, the rebel movement laid down its arms, refugees returned, and the new South Africa was offering agricultural and technical help rather than arms to insurgents. In 1996, the two countries launched the Maputo Development Corridor, a scheme to revitalize the links (rail, road, and air) between Johannesburg and its nearest port, to stimulate trade and investment, and to boost mining, farming, and tourism along the route (Fig. 7-14). But it will take generations for the country to climb from the depths of impoverishment to which it had fallen.

Between Angola and Moçambique lie three landlocked states that formed part of the British colonial empire: Zambia, Malawi, and Zimbabwe. Zambia (10.0 million) and Malawi (10.8 million) are less developed than Zimbabwe, although Zambia contains the mineral-rich Copperbelt. The decline of world prices for its minerals, as well as the problems and costs associated with their long-distance transportation, have hurt the economy of Zambia. Malawi's economic geography is almost completely agricultural, with a variety of crops including tea, cotton, tobacco, and peanuts. This emphasis on farming has helped cushion the economy against market swings and has meant that more of its labor force has been able to find productive work than in many other African countries. Still, more workers toil in the mines of other countries than are gainfully employed within Malawi itself.

Southern States

Six countries constitute Africa's southernmost tier of states and form a distinct subregion within Southern Africa: Zimbabwe, Namibia, Botswana, Swaziland, Lesotho, and the Republic of South Africa. As Figure 7-14 shows, four of the six are landlocked. Diamond-exporting Botswana occupies the heart of the Kalahari Desert and surrounding steppe; only 1.6 million people, most of them subsistence farmers, inhabit this Texas-sized country. In 2001, no country in Africa was more severely afflicted by the AIDS epidemic than Botswana. South Africa encircles Lesotho (2.2 million) and surrounds most of Swaziland (1.0 million), ancestral home of the Swazi nation. Botswana, Lesotho, and Swaziland depend heavily on the income that workers who labor in South African mines, factories, and fields send home.

Apart from South Africa, the region's giant, the most important southern state undoubtedly is Zimbabwe (11.5 million), landlocked but well endowed with mineral and agricultural resources. Zimbabwe (the country is named after stone ruins in its interior) is mostly an elevated plateau between the Zambezi and Limpopo rivers, with the desert to the west and the Great Escarpment to the east. Its core area is defined by the mineral-rich Great Dyke and its envi-

FROM THE FIELD NOTES

"My second visit to the Kariba Dam on the Zambezi River between Northern and Southern Rhodesia (now renamed Zambia and Zimbabwe) showed the huge dam wall complete and the lake filling up. The four-lane highway across the top of the dam was not yet open for traffic, and construction was still in progress on the turbine housing and other buildings. This was the time when mega-dams were thought to hold the solution to development problems: the Kariba Dam would provide electricity to places as far away as Salisbury (now Harare) and the Copperbelt would support a fishing industry, would enable farming along its shores, would stimulate the tourist industry. It was therefore worth dislocating tens of thousands of people living in the valley and eliminating wildlife refuges (a massive effort was made to rescue wildlife caught on islands as the water rose). But, like other megaprojects of this kind in Africa and elsewhere in the less advantaged world, the Kariba Dam failed to live up to expectations and generated unanticipated problems, medical and otherwise. A hint of the latter can be seen in the extreme left of this photograph: the floating vegetation at the turbine-tunnel entrance. A Brazilian water hyacinth, reportedly introduced by two Brazilian monks who built a small church where the new shoreline would be, clogged the upper reaches of Lake Kariba and spread into Equatorial Africa's rivers."

rons, extending southwest from the vicinity of the capital, Harare, to the country's second city, Bulawayo. Copper, asbestos, and chromium (of which Zimbabwe is one of the world's leading sources) are among its major mineral exports, but Zimbabwe is not just an ore-exporting country. Farms produce tobacco, tea, sugar, cotton, and other crops (corn is the staple). A bitter civil war preceded independence, but many whites, encouraged by the government, stayed on their farms after the black African majority took power. Despite disputes over land policy, Zimbabwe's political transition did not derail the economy as it did in so many other countries.

Zimbabwe's problems mounted during the 1980s, however, when droughts caused African subsistence farmers to abandon their land. As thousands of livestock died, the Mugabe government in 1998 announced that it would take land from white farmers and distribute it to black African families. Social and political tensions rose, especially when farmers charged the government with assigning newly acquired land to cronies rather than those in need. In 2000, the situation worsened when thousands of so-called "war veterans" targeted and invaded white farms, drove out and in several cases killed their owners, destroyed the farm workers' dwellings, and occupied the land. The Mugabe regime tolerated these actions with an eye on upcoming elections, but in the process the Zimbabwean agricultural economy was devastated. The impact of these events will afflict Zimbabwe for years to come.

Two nations form most of Zimbabwe's population: the Shona (71 percent) and the Ndebele (16 percent), the latter concentrated in the southwest of the country. Shona domination produced a government that, by the turn of the century, had been in power for 20 years. The long-term president, Robert Mugabe, commanded a regime marked by deepening corruption, including legislative obstruction of attempts by private Zimbabweans to compete with businesses in which government members had personal financial interest. As attempts to modernize the infrastructure were hampered and the once-promising mixed economy faltered, the government responded to pressure by curbing human rights.

Once in the vanguard of the "frontline" states that battled *apartheid* in South Africa, Zimbabwe's leadership was diminished after South Africa ended white rule and emerged as a regional power. Seeking new opportunities to escape from South Africa's shadow and influence on African affairs, Mugabe intervened in the crisis in The Congo in the late 1990s. He sent troops to help the Kabila regime, although South Africa preferred a diplomatic solution. In the process, Zimbabwe dealt a blow not only to South Africa's primacy but also to the still-formative SADC.

The western tip of Zimbabwe just barely touches the northeastern "Caprivi Finger" of Namibia, the region's youngest independent state. Once a German colony named South West Africa, the country was endowed with that narrow strip of land to connect it to the Zambezi River by the mapmakers in Berlin. When South Africa took control of the colony after World War I, the Caprivi Strip became strategically important; in the 1980s, there were rumors of missile (even nuclear) tests in the area. But Namibia achieved independence in 1990, and South Africa abandoned its nuclear program (nonetheless, in 1999, the Lozi people in the Strip launched an ill-fated secession campaign). In 1994, the last territorial vestige of South Africa was erased when the port of Walvis Bay, a South African exclave, was transferred to Namibia.

Namibia is appropriately named after a desert (the coast-paralleling Namib). Only 1.9 million people, most concentrated in the moister northern border areas, inhabit this huge country, which is about as large as Texas and Oklahoma combined. The capital, Windhoek, is centrally situated and therefore distant from this northern cluster. That is because the major economic activities—mining in the Tsumeb area and ranching across the steppe country of the south—happen elsewhere, not in the subsistence-dominated north.

But that is where the voters are, and Namibia will undoubtedly undergo structural readjustment. Much of its productive capacity remains in foreign hands, and Namibia must import food to supplement what little it produces. For Namibia, the long haul toward true sovereignty has just begun.

South Africa

The Republic of South Africa is the giant of Southern Africa, an African country at the center of world attention, a bright ray of hope not only for Africa but for all humankind.

Long in the grip of one of the world's most notorious racial policies (**apartheid**, or "apartness," and its derivative, **separate development**), South Africa today is shedding its past and building a new future. That virtually all parties to the earlier debacle are now working cooperatively to restructure the country under a new flag, a new national anthem, and a new leadership was one of the great events of the twentieth century. Now, with a new century opened, South Africa is poised to take its long-awaited role as the economic engine for the region—and perhaps beyond.

17

18

South Africa stretches from the warm subtropics in the north to Antarctic-chilled waters in the south. With a land area in excess of 470,00 square miles (1.2 million sq km) and a heterogeneous population of 44.5 million, South Africa is the dominant state in Southern Africa. It contains the bulk of the region's minerals, most of its good farmlands, its largest cities, best ports, most productive factories, and most developed transport networks. Mineral exports from Zambia and Zimbabwe move through South African ports. Workers from as far away as Malawi and as nearby as Lesotho work in South Africa's mines, factories, and fields.

Historical Geography

South Africa's location has much to do with its human geography. On the African continent, peoples migrated southward—first the Khoisan-speakers and then the Bantu peoples—into the South African cul-de-sac. On the oceans, the Europeans arrived to claim the southernmost Cape as one of the most strategic places on Earth, the gateway from the Atlantic to the Indian Ocean, a waystation on the route to Asia's riches. The Dutch East India Company founded Cape Town as early as 1652, and soon the Hollanders began to bring Southeast Asians to the Cape to serve as domestics and laborers. When the British took over about 150 years later, Cape Town had a substantial population of mixed ancestry, the source of today's so-called Coloured sector of the country's citizenry.

The British also altered the demographic mosaic by bringing tens of thousands of indentured laborers from their South Asian domain to work on the sugar plantations of east-coast Natal. Most of these laborers stayed after their period of indenture was over, and today South Africa counts about 1 million Indians among its people. Most are still concentrated in Natal and prominently so in metropolitan Durban.

As we noted earlier, South Africa had been occupied by Europeans long before the colonial "scramble for Africa" gained momentum. The Dutch, after the British took control of the Cape, trekked into the South African interior and, on the high plateau they called the *highveld*, founded their own republics. When diamonds and gold were discovered there, the British challenged the Boers (descendants of the Dutch settlers) for these prizes. In 1899–1902, the British and the Boers fought the Boer War. The British won, and British capitalists took control of South Africa's economic and political life. But the Boers negotiated a power-sharing arrangement and eventually won hegemony. Having long since shed their European links, they now called themselves *Afrikaners*, their word for Africans, and proceeded to erect the system known as apartheid.

These foreign immigrations and struggles took place on lands that Africans had already entered and fought over. When the Europeans reached the Cape, Bantu nations were driving the weaker Khoi and San peoples into less hospitable territory or forcing them to work in bondage. One great contest was taking place in the east and southeast, below the Great Escarpment. The Xhosa nation was moving toward the Cape along this natural corridor. Behind them, in Natal (Fig. 7-14), the Zulu Empire became the region's most powerful entity in the nineteenth century. On the highveld, the North and South Sotho, the Tswana, and other peoples could not stem the tide of European aggrandizement, but their numbers ensured survival (unlike many of Australia's Aboriginal groups, which disappeared).

In the process, South Africa became Africa's most pluralistic and heterogeneous society. People had converged on the country from Western Europe, Southeast Asia, South Asia, and other parts of Africa itself. At the end of the twentieth century, Africans outnumbered non-Africans by about 4 to 1 (Table 7-1).

Table 7-1
DEMOGRAPHIC DATA FOR SOUTH AFRICA

Population Groups	Estimates, 2002 (in millions)
African nations	35.9
Zulu	8.9
Xhosa	7.3
Sotho (N and S)	6.6
Tswana	3.4
Others (6)	9.7
Europeans	3.9
Afrikaners	2.6
English-speakers	1.2
Others	0.1
Mixed (Coloureds)	3.5
African/Europeans	3.3
Malayan	0.2
South Asian	1.2
Hindus	0.8
Muslims	0.4
TOTAL	44.5

Social Geography

Heterogeneity also marks the spatial demography of South Africa. Despite centuries of migration and (at the Cape) intermarriage, labor movement (to the mines, farms, and factories), and massive urbanization, regionalism pervades the human mosaic. The Zulu nation still is largely concentrated in the province the Europeans called Natal. The Xhosa still cluster in the Eastern Cape, from the city of East London to the Natal border and below the Great Escarpment. The Tswana still occupy ancestral lands along the Botswana border. Cape Town still is the core area of the Coloured population; Durban still has the strongest Indian imprint. Travel through South Africa, and you will recognize the diversity of rural cultural landscapes as they change from Swazi to Ndebele to Venda.

This persistent regionalism induced the Afrikaner-led white government of the apartheid period, after 1949, to extend its policies and lock the country in a grand design called separate development. By creating "homelands" that would later become "independent" republics, the white minority could assign all Africans to their national source area, making them foreigners in white South Africa while they worked there. This was a form of what is now called *ethnic cleansing*, and between 1960 and 1980 as many as 3.5 million black Africans were forced from their homes and relocated in their assigned "homelands." The social cost of this program was enormous and will be felt well into the twenty-first century. Under the provisions of separate development, nearly 80 percent of South Africa was designated as white-owned, and barely more than 15 percent was to accommodate the African homelands. There, overcrowding led to excessive pressure on the land, accelerating soil erosion and general environmental deterioration. Crowding also produced conflict among homeland residents, but more serious strife pitted those who were willing to cooperate with the white regime (for example, by serving in government-approved homeland administrations) against those who used all possible means to oppose apartheid and separate development.

For all the misery and dislocation it caused, separate development could not stem the tide of urbanization in South Africa. Millions of workers, jobseekers, and illegal migrants created vast shantytown rings around the cities. The existing "townships" or "locations," as the older African settlements were called, that once were the poorest parts of the cities, now seemed well-off by comparison. Black townships such as Soweto (for South Western Townships) in metropolitan Johannesburg were approved under apartheid laws, even acquiring some services. In the shantytowns, no legalities protected the residents, and the government never stopped trying to eradicate them. Bulldozers would sweep away whole settlements overnight, their residents loaded on trucks and sent to the homelands.

Still the number of both township residents and squatters grew, and eventually the townships became the hubs of opposition to apartheid. Violent uprisings and crippling strikes proved the growing power of the majority and the inability of the state to control it. By the mid-1980s, international attention was focused on South Africa, economic sanctions had been imposed, and the end of the apartheid era was in sight. The South African president, P. W. Botha, began to negotiate with the country's most famous political prisoner, Nelson Mandela. The long-banned African National Congress (ANC) was permitted to engage in political activity again, and exiled leaders began to return home. When, in February 1990, Mandela walked out of prison after 28 years behind bars, he and the last Afrikaner president, F. W. de Klerk, began a process of negotiation and accommodation that led to the transfer of power from the white regime to a government elected by all the voters. In April 1994, a new era began in South Africa. Nelson Mandela, of distinguished Xhosa ancestry, had become president of an ANC-dominated government at Cape Town.

Political Reorganization

The route from prison gate to presidency was not an easy one either for Mandela or for the South Africa he hoped to lead; and again geography played a big role. De Klerk led the white minority, and Mandela the black majority, but there was a third force: the largest single nation in South Africa, the Zulu, whose leader, Chief Mangosuthu Buthelezi, saw a conspiracy between Mandela and de Klerk that would damage Zulu interests. Buthelezi's political movement, Inkatha, demanded equity in the new South Africa and promised to scuttle the conferences that were negotiating the compromise South Africa needed. Inkatha carried out its threats. In the townships of the major cities and in the hostels of the mining companies, Inkatha "warriors" killed thousands, violence that threatened to become a civil war. Buthelezi and his followers talked of secession and independence for Natal if the new and democratic South Africa was not to their liking.

In the end, geography came to South Africa's rescue. What the country needed was an election that would give the ANC a majority, but not one large enough to overwhelm minority interests. To gain Inkatha's cooperation, Inkatha needed to win in Natal. To give substance to the promise of democracy in South Africa, the ANC would also have to lose in one other province.

To accomplish this (not to say to rig the election), the conference that prepared for the power transfer redrew the map, and thus the electoral map, of South Africa. Since 1910, South Africa had had four provinces: the Cape, Natal, the Orange Free State (north of the great Orange River), and the Transvaal ("across the Vaal," the major Orange tributary). These four provinces essentially represented the European occupation of the country: the Cape and Natal were British strongholds, and the Orange Free State and Transvaal were Boer republics until their defeat in the Boer War (Fig. 7-16). The new map (Fig. 7-17) created nine provinces, leaving intact Natal (but calling it Kwazulu-Natal) and the Orange Free State. Three new provinces and part of a fourth were carved out of the Cape: the Western Cape, essentially Cape Town and its hinterland, where the electorate is mainly Coloured and white; the Eastern Cape, dominated by the Xhosa nation; the Northern Cape, by far the least populous, mainly rural province; and the Northwest, part Cape and part Transvaal, where the Tswana nation is concentrated. The remainder of the Transvaal also became three provinces: Gauteng, the heart of the country's core area centered on Johannesburg; Northern Province, home to many Afrikaners who opposed the new South Africa; and Mpumalanga, a mix of wealthy white farmers and densely-peopled Swazi and other African areas.

In the momentous 1994 elections, the ANC won seven of the nine provinces. It lost in Kwazulu-Natal, where an alliance of Zulu, Asian, and white voters prevailed over the ANC, and in the Western Cape, where white and Coloured voters combined to defeat ANC candidates. Nor did the ANC gain the 70 percent majority it would have needed to govern without opposition support. What all parties in South Africa accomplished after more than four decades of oppression and mismanagement was an example to a world quick to use force to reach political goals.

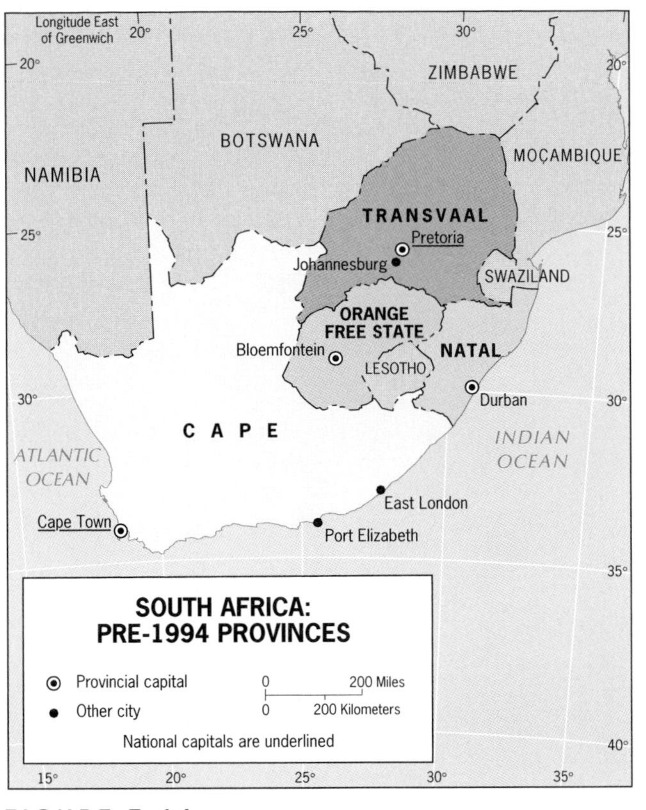

FIGURE 7-16

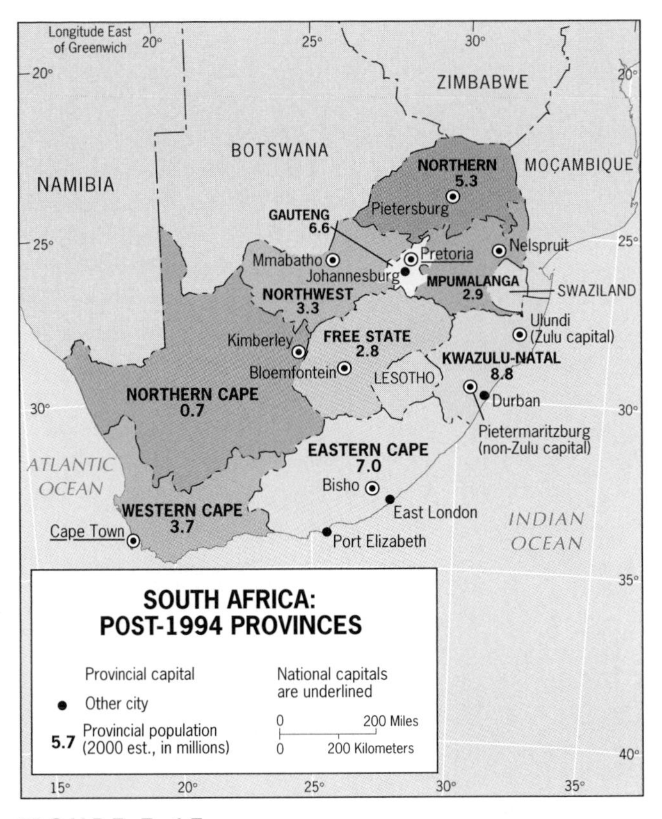

FIGURE 7-17

South Africa's difficult transition to post-apartheid normality and stability, however, is not over. President Mandela provided the leadership the country needed at a crucial time. But now South Africa has entered a post-Mandela era. On June 2, 1999, ANC leader Thabo Mbeki, who had served as President Mandela's deputy, became the country's second popularly elected president. In other African states, the succession from heroic founder-of-the-nation to political inheritor-of-the-presidency has not gone well, but South Africans prepared themselves by negotiating a constitution that contains a structure for presidential succession. That constitution took effect in stages from 1996 to 1999 and may safeguard the country against the problems that have befallen so many other states in the realm.

Soon after assuming the presidency, Mbeki was buffeted by two crises. In 2000, a world conference on AIDS in Durban focused attention on South Africa's own health emergency: estimates of those infected with HIV in the country in the 15-to-49 age group ranged from 13 to 20 percent. The president, in controversial statements before and during the conference, left observers wondering whether he accepted the scientific evidence involved. And in the same year, the farm invasions in Zimbabwe (see page 376) raised fears among white farmers, especially in South Africa's Northern Province adjacent to Zimbabwe. White representatives in the South African Parliament called on the president to rebuke Zimbabwe's leaders, but some black representatives objected to this demand and even expressed support for the Mugabe regime. President Mbeki, in a difficult political bind, did not speak out.

Economic Geography

Undoubtedly the most serious immediate problems for the new South Africa are economic, and this is one of the president's strengths. Ever since diamonds were discovered at Kimberley in the 1860s, South Africa has been synonymous with minerals. The Kimberley finds, made in a remote corner of what was then the Orange Free State (the British soon annexed it to the Cape), set into motion a new economic geography. Rail lines were laid from the coast to the "diamond capital" even as fortune seekers, capitalists, and tens of thousands of African workers, many from as far afield as Lesotho, streamed to the site. One of the capitalists was Cecil Rhodes, of Rhodes Scholarship fame, who used his fortune to help Britain dominate Southern Africa.

Just 25 years after the diamond discoveries, prospectors found what was long to be the world's greatest goldfield on a ridge called the Witwatersrand (Fig. 7-14). This time the site lay in the so-called South African Republic (the Transvaal), and again the Boers were unable to hold the prize. Johannes-

Among the Realm's Great Cities . . .

Johannesburg

Subsaharan Africa displays only one incipient conurbation, and Johannesburg, South Africa lies at the heart of it. Little more than a century ago, Johannesburg was a small (though rapidly growing) mining town based on the newly discovered gold reserves of the Witwatersrand.

Today Johannesburg forms the focus of a megalopolis of about 6 million, extending from Pretoria to the north to Vereeniging to the south, and from Springs in the east to Krugersdorp in the west. In 2002 metropolitan Johannesburg itself had a population of 2.4 million, second only to Cape Town (3.1 million).

Johannesburg's skyline is the most impressive in all of Africa, a forest of skyscrapers reflecting the wealth generated here over the past hundred years. Look southward from a high vantage point, and you see the huge mounds of yellowish-white slag from the mines of the "Rand," the so-called mine dumps, partly overgrown today, interspersed with suburbs and townships. In a general way, Johannesburg developed as a white city in the north and a black city in the south. Soweto, the black township, lies to the south; Houghton and other spacious, upper-class suburbs, once exclusively white residential areas, lie to the north.

Johannesburg has neither the scenery of Cape Town nor the climate and beaches of Durban. The city is a mile above sea level, and its thin air often is polluted from smog created by automobiles, factories, mine-dump dust, and countless cooking fires in the townships and shantytowns that ring the metropolis.

In the past century, the Johannesburg area produced nearly one-half of all the world's gold by value. But today, Johannesburg lies at the heart of an industrial, commercial, and financial complex whose name on the new map of South Africa is *Gauteng*.

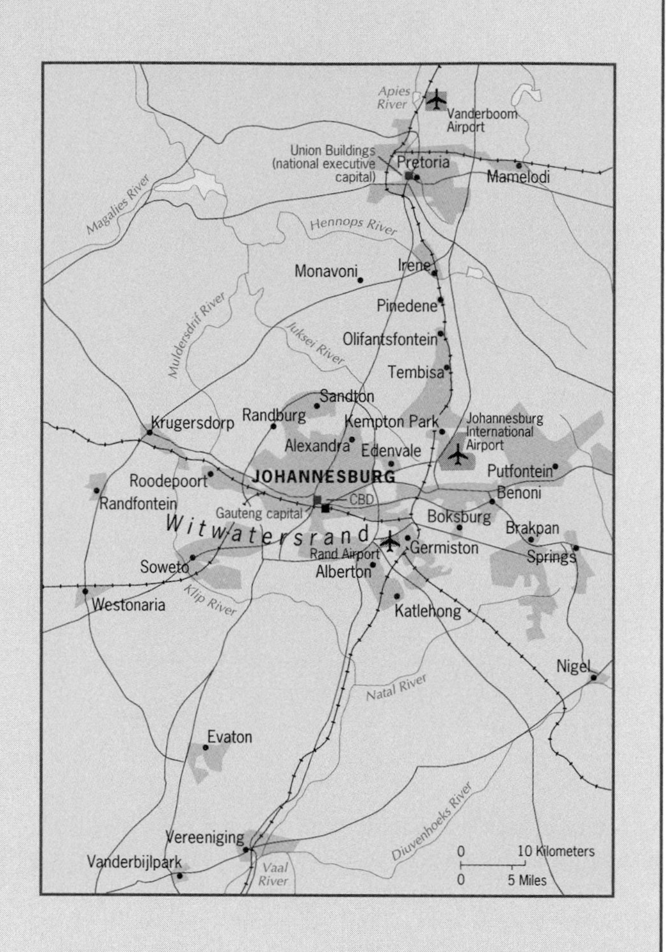

burg became the gold capital of the world, and a new and even larger stream of foreigners arrived, along with a huge influx of African workers. Cheap labor enlarged the profits. Johannesburg grew explosively, satellite towns developed, and black townships mushroomed. The Boer War was only an interlude here on the mineral-rich Witwatersrand.

During the twentieth century, South Africa proved to be even richer than had been foreseen. Additional goldfields were discovered in the Orange Free State. Coal and iron ore were found in abundance, which gave rise to a major iron and steel industry. Other metallic minerals, including chromium and platinum, yielded large revenues on world markets. Asbestos, manganese, copper, nickel, antimony, and tin were mined and sold; a thriving metallurgical industry developed in South Africa itself. Capital flowed into the country, white immigration grew, farms and ranches were laid out, and markets multiplied.

South Africa's cities grew apace. Johannesburg was no longer just a mining town; it became an industrial complex and a financial center as well. The old Boer capital, Pretoria, just 30 miles (50 km) north of the Witwatersrand, became the country's administrative center during apartheid's days. In the Orange Free State, major industrial growth (including oil-from-coal technology) matched the expansion of mining. While the core area developed megalopolitan characteristics, coastal cities expanded as well. Durban's port served not only the Witwatersrand but a wider regional hinterland as well. Cape Town was becoming South Africa's

FROM THE FIELD NOTES

"Looking down on this enormous railroad complex, we were reminded of the fact that almost an entire continent was turned into a wellspring of raw materials carried from interior to coast and shipped to Europe and other parts of the world. This complex lies near Witbank in the eastern Rand, a huge inventory of freight trains ready to transport ores from the plateau to Durban and Maputo. But at least South Africa acquired a true transport network in the process, ensuring regional interconnections; in most African countries, railroads serve almost entirely to link resources to coastal outlets."

largest city; its port, industries, and productive agricultural hinterland gave it primacy over a wide area.

Apartheid ruined these prospects and exposed the economy's weaknesses. The Afrikaner government made huge investments in its separate development policy, which contradicted fundamental principles of economic geography. International sanctions against apartheid hurt the economy, labor unrest weakened it further, and lower prices for those products that did get sold hurt even more. In the black townships, the school systems fell apart under the cry of "liberation before education," creating a huge undereducated (later unemployable) mass of young people.

For South Africa later is now, and the problems these policies created have arrived. As the twenty-first century dawned, South Africa's economy still depended heavily on the export of metals and minerals, but mineworkers demanding higher wages cut into the profits. Manufactured goods figured hardly at all in the export picture. Unemployment stood at high levels even as employed workers sought to raise their wages. In the cities, the whites fled to the suburbs, transforming the downtowns into commuter workplaces and the streets into bazaars. Violent robberies replaced the political crime of the pre-transfer period. Most ominous is the revolution of rising expectations, a phenomenon many African countries experienced after decolonization. Land pressure and housing needs cannot be accommodated overnight, but the newly empowered majority will expect its government to act expeditiously. Joblessness, housing shortages, and land pressure can form a potent mix to destabilize a society, and the South African government faces a daunting challenge.

On maps of development indices, South Africa is portrayed as an upper-middle-income economy, but averages mean little in this country of strong internal core-periphery contrasts. In its great cities, industrial complexes, mechanized farms, and huge ranches, South Africa resembles a high-income economy, much like Australia or Canada. But outside the primary core area (centered on Johannesburg) and beyond the secondary cores and their linking corridors lies a different South Africa, where conditions are more like those of rural Zambia or Zimbabwe. In terms of such indices as life expectancy, infant and child mortality, overall health, nutrition, education, and many others, a wide range marks the population sectors listed in Table 7-1. South Africa has been described as a microcosm of the world, exhibiting in a single state not only a diversity of cultures but also a wide range of human conditions. If South Africa can keep on course, the one-time pariah of apartheid will become a guidepost to a better world.

TOSHKENT
UZBEKISTAN
Namangan
Jalal-Abad
Osh
Andijon
Farghona
Quqon
Bekabad
Khujand
Bukhoro
Navoi
Samarqand
Sary-Tash
Pik Kommunizma 24 590
Charjew
Karshi
TURKMENISTAN
Dushanbe
TAJIKISTAN
Mary
Qurghonteppa
Khorugh
Murghab
PAMIRS
Mazār-e Sharīf
Gushgy
Meymaneh
Kūh-e Fūlādī 16 847
Herāt
PAROPAMISUS
Kābul
AFGHANISTAN
KHYBER PASS
Peshawar
Islāmābād
Rawalpindi
Ghaznī
Farāh
Daryācheh-ye Sīstān
Kandahār
Dera Ismāīl Khān
RĪGESTĀN
Chaman
Loralai
Quetta
Dera Ghāzi Khān
Multan
Bahāwalpur
CHAGAI HILLS
Nushki
Kalat
Khāsh
BALUCHISTAN DESERT
Hamun-i Mashkel
BRAHUI
SULAIMAN RANGE
Shikārpur
Mohenjo Dero (Ruins)
Sukkur
IRAN
KIRTHAR RANGE
SIND
Gwādar
Hyderābād
KARACHI
Tropic of Cancer
RANN OF KUTCH
Bhuj
GUJARAT
Māndvi
Jāmnagar
Rājkot
KATHIAWAR PENINSULA
Porbandar
Junagādh
Verāval
Diu
Damān
Silvassa
PAKISTAN
Gilgit
Chitral
HINDU KUSH
JAMMU AND KASHMIR
LINE OF CONTROL
Srīnagar
Jammu
Jhelum
Sialkot
Gujrānwāla
Faisalābād
LAHORE
Amritsar
Jullundur
Ludhiāna
PUNJAB
Firozpur
Patiāla
Bhatinda
Ambāla
Chandīgarh
Simla
Dehra Dūn
Hardwār
Sahāranpur
HARYANA
Meerut
Morādābād
DELHI
New Delhi
THAR DESERT
Bīkaner
Alwar
Mathura
Rāmpur
Bareilly
Alīgarh
Shāhjahānpur
Āgra
Bharatpur
Jaipur
RĀJASTHAN
Ajmer
Jodhpur
Tonk
Farrukhābād
KANPUR
Lucknow
Gwalior
Sheopur
Kota
Shivpuri
Jhānsi
ARAVALLI RANGE
Udaipur
Jhālawar
Abu Road
Pālanpur
CENTRAL INDIAN PLATEAU
AHMADĀBĀD
Ujjain
Bhopāl
Indore
Baroda
VINDHYA RANGE
MADHYA PRADESH
Sāgar
Jabalpur
Surat
Burhānpur
Amrāvati
Dhule
Akola
Wardha
Nāgpur
Nāsik
Aurangābād
MUMBAI (BOMBAY)
Ahmadnagar
MAHĀRĀSHTRA
Pune
KONKAN COAST
WESTERN GHATS
Sholāpur
Sāngli
Kolhāpur
Belgaum
GOA
Panaji
Hubli
Bellary
Nizāmābād
HYDERABAD
Gulbarga
Raichūr
DECCAN
Warangal
Chandrapur
Vijayawāda
Guntūr
Tenāli
Kurnool
Cuddapah
Nellore
KARNATAKA
Mangalore
Kolār
BANGALORE
Mysore
Mahe
Calicut
Coimbatore
MALABAR COAST
KERALA
Ernākulam
Alleppey
Quilon
Trivandrum
PLATEAU
Salem
TAMIL NADU
Vellore
CHENNAI (MADRAS)
Kānchipuram
Pondicherry
Cuddalore
Kumbakonam
Tiruchchirāppalli
Nāgappattinam
Madurai
Tuticorin
Tirunelveli
Gulf of Mannar
EASTERN GHATS
ANDHRA PRADESH
COROMANDEL COAST
Eluru
Machilīpatnam
Yanam
Rājahmundry
Kākināda
Vishākhapatnam
Vizianagaram
GOLCONDA COAST
Berhampur
Puri
Bhubaneshwar
Cuttack
Jājpur
Balasore
ORISSA
CHHATTISGARH
Raipur
Bhilai
Sambalpur
Bilāspur
Raigarh
Raurkela
Jamshedpur
Rānchi
JHARKHAND
CHOTA NAGPUR PLATEAU
Murwāra
Rewa
Prayagraj (Allāhābād)
Mirzāpur
Vārānasi
UTTAR PRADESH
Gorakhpur
Faizābād
NORTH INDIAN PLAIN
Patna
Gaya
BIHAR
Monghyr
Bhāgalpur
Darbhanga
Giridih
Asansol
Burdwān
Berhampore
WEST BENGAL
Howrah
KOLKATA (CALCUTTA)
Kharagpur
Bhātpāra
Mouths of the Ganges
Hooghly
INDIA
UTTARANCHAL
HIMACHAL PRADESH
Nanda Devi 25 645
Almora
Dhaulagiri 26 810
NEPAL
Kathmandu
Mt. Everest 29 035
Kānchenjunga 28 208
SIKKIM
Lalitpur
Darjeeling
Gangtok
Yatung
Punakha
Thimphu
BHUTAN
ARUNACHAL PRADESH
Sadiya
Sibsāgar
Jorhāt
ASSAM
NAGALAND
Kohīma
Cooch Behār
MEGHALAYA
Shillong
KHASI HILLS
Imphāl
MANIPUR
Silchar
MIZORAM
TRIPURA
Rangpur
Rājshāhi
Nasirābād
BANGLADESH
Dhaka
Comilla
Khulna
Noākhāli
Chittagong
Cox's Bazar
Mt. Victoria 10 018
Sittwe
Kyaukpyu
Yenangyaung
Sandoway
HIMALAYA
Wushi
Aksu
Kashi
Bachu
XINJIANG UYGUR (SINKIANG)
TARIM BASIN
TAKLA MAKAN (DESERT)
Yengisar
Muztagata 24 757
Shache (Yarkand)
Karghalik (Yecheng)
Hotan
Yutian (Keriya)
Qiemo (Qarqan)
ALTUN SHAN
TSAIDAM BASIN
QINGHAI
KUNLUN SHAN
K2 (Godwin Austen) 28 250
KARAKORAM PASS
KARAKORAM
LANAK LA
CHINA
PLATEAU OF TIBET
XIZANG (TIBET)
TANGGULA SHAN
Rutog
Gar
GANGDISE SHAN
NYAINQENTANGLHA SHAN
Nam Co
Lhasa
Xigazê
Gyangzê
Brahmaputra
Yamzho Yumco
Kula Kangri 24 784
ARABIAN SEA
LAKSHADWEEP (LACCADIVE IS.) (India)
LAKSHADWEEP
LACCADIVE SEA
MALDIVES
INDIAN OCEAN
BAY OF BENGAL
ANDAMAN ISLANDS (India)
Palk Strait
Jaffna
Mannar
Trincomalee
Anurādhapura
Puttalam
SRI LANKA (CEYLON)
Kandy
Pidurutalagala 8281
Colombo
Kotte
Galle
Matara
A Area occupied by Pakistan and claimed by India
B Area claimed and occupied by India; status disputed by Pakistan.
C Area occupied by China and claimed by India.
D Area occupied by India and claimed by China.
0 50 100 200 300 400 500 Miles
0 100 200 400 600 800 Kilometers
Scale 1:16,896,000; one inch to 266 miles
Polyconic projection
Elevations and depressions are given in feet
Longitude East of Greenwich
Relief
Meters
3050
1525
610
305
152.5
0
Sea Level
152.5 500
1525 5000
3050 10 000
6100 20 000

chapter 8 South Asia

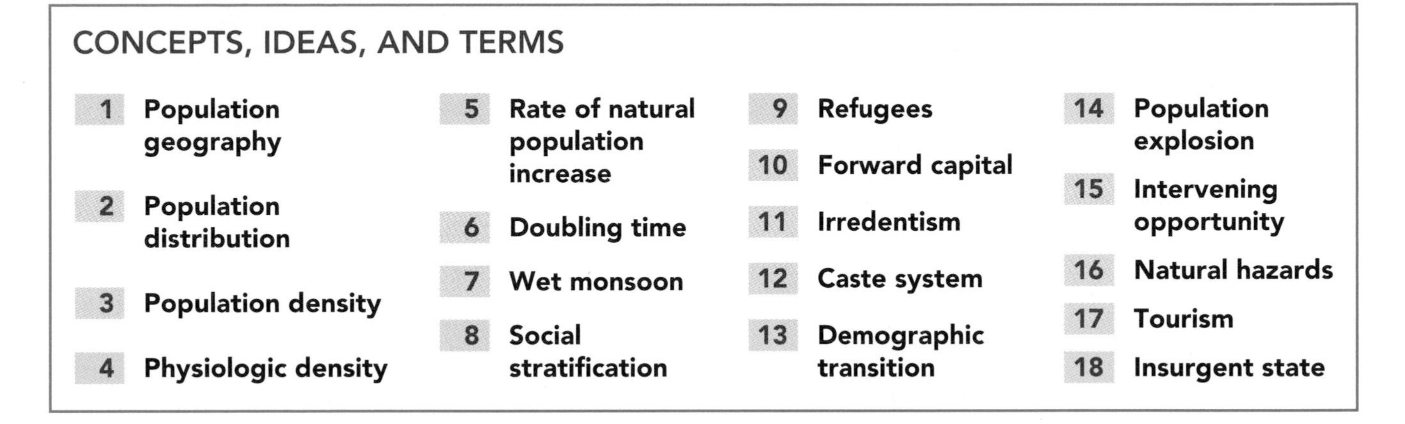

From Iberia to Arabia and from Malaysia to Korea, Eurasia is a landmass fringed by peninsulas. The largest of all is the great triangle of India that divides the northern Indian Ocean into two seas: the Arabian Sea to the west and the Bay of Bengal to the east (Fig. 8-1). The peninsula of India forms the heart of South Asia, a vast, varied, volatile geographic realm.

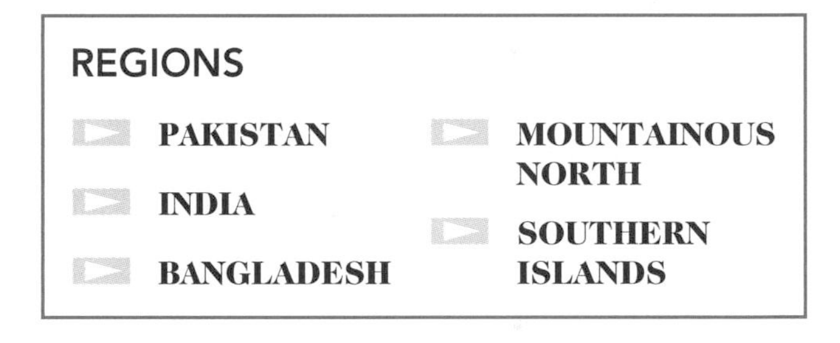

DEFINING THE REALM

Mountains, deserts, and coastlines combine to make South Asia one of the world's most vividly defined physiographic realms. To the north, the Himalaya Mountains create a natural wall between South Asia and China. To the east, mountain ranges and dense forests mark the boundary between South and Southeast Asia. To the west, rugged highlands and expansive deserts separate South Asia from its neighbors. Within these confines lies a geographic realm that is more densely populated than any other. If current population trends continue, it will soon be the most populous realm on Earth as well.

South Asia consists of five regions (Fig. 8-2). Its keystone is India, whose population passed the 1 billion mark in 1999. In the west lies Pakistan. South Asia's eastern flank is centered on Bangladesh. The northern region consists of the mountainous lands of Kashmir, Nepal, and Bhutan. And the southern region includes the islands of Sri Lanka and the Maldives. As the map shows, India divides into several subregions.

South Asia's physiographic boundaries are formidable barriers, but they have not prevented conquerors or proselytizers from penetrating it. As a result, today this realm is a patchwork of religions, languages, traditions, and cultural landscapes. So complex is this mosaic that, remarkably, its political geography numbers just seven states (and only five on the mainland).

Among the cultural infusions was Islam. Today, Pakistan is an Islamic republic, and Islam provides the cement for the state. Pakistan's eastern border with India is a cultural divide in more ways than one: dominantly Hindu India

FIGURE 8-1

Major Geographic Qualities of South Asia

1. South Asia is clearly defined physiographically and is bounded by mountains, deserts, and ocean; the Indian peninsula is Eurasia's largest.
2. South Asia is the world's most poverty-afflicted realm, with low average incomes, low levels of education, poorly balanced diets, and poor overall health.
3. With only 3 percent of the world's land area but 22 percent of its population, more than half of it engaged in subsistence farming, South Asia's economic prospects are bleak.
4. Population growth rates in South Asian countries are among the highest in the world; India's population surpassed the 1 billion mark in 1999.
5. The North Indian Plain, the lower basin of the Ganges River, contains the heart of the world's second largest population cluster.
6. Despite encircling mountain barriers, invaders from ancient Greeks to later Muslims penetrated South Asia and complicated its cultural mosaic.
7. British colonialism unified South Asia under a single flag, but the empire fragmented into several countries along cultural lines after Britain's withdrawal.
8. Pakistan, South Asia's western region, lies on the flanks of two realms: largely Muslim North Africa/Southwest Asia and dominantly Hindu South Asia.
9. India is the world's largest federation and most populous democracy, but its political achievements have not been matched by enlightened economic policies.
10. Religion remains a powerful force in South Asia. Hinduism in India, Islam in Pakistan, and Buddhism in Sri Lanka all show tendencies toward fundamentalism and nationalism.
11. Active and potential boundary problems involve internal areas (notably between India and Pakistan in Kashmir) as well as external locales (between India and China in the northern mountains).

is a secular, not a theocratic, state. Why, then, do we include Pakistan in the South Asian rather than the Southwest Asian/North African realm? One criterion is ethnic continuity, which links Pakistan to India rather than to Afghanistan or Iran. Another is historical geography. Pakistan was part of Britain's South Asian Empire, and it originated from the partition of that domain between Muslim and Hindu majorities. Although Urdu is the official national language of Pakistan, English is the *lingua franca*, as it is in India. Furthermore, the border between India and Pakistan does not signify the eastern frontier of Islam in Asia. About 127 million of India's more than 1 billion citizens are Muslims, and in South Asia's eastern region, Bangladesh (population: 133 million) is more than 85 percent Muslim. Finally, Pakistan and India are locked in a struggle to control a vital mountainous area in the far north, where the British withdrawal left the boundary between them unresolved. Even as Indian and Pakistani cricket teams play each other on sun-baked pitches, their armies face off in a deadly conflict—a conflict that has the potential to unleash a nuclear war.

Events in the wake of the Soviet occupation of Afghanistan (1979–1989) may yet change the political and cultural geography of South and Southwest Asia to such an extent that the regional framework could be modified. Pakistan's orientation toward India and the rest of South Asia was changed by the conflict in Afghanistan, which drove several million Pushtun (Pathan) refugees across its common border. This refugee influx, coupled with a massive infusion of weapons into Afghanistan under U.S. auspices, changed the ideological landscape in Pakistan. Islamic fundamentalism mushroomed, and Pakistan, long preoccupied with issues involving India and South Asia, began to look westward. Islamic schools teaching strict Islamic precepts, approved by Pakistan and supported financially by Saudi Arabia, gave rise to the Taliban movement that swept into power in Afghanistan during the mid-1990s. Today, South Asia's western flank is unstable and in transition.

A REALM OF POVERTY

At the opening of the twenty-first century, South Asia accounts for more than one-fifth of the world's population and two-thirds of its poorest inhabitants. Its literacy rates are among the lowest in the world. Nearly half of the people in this realm earn less than the equivalent of one U.S. dollar per day. It is estimated that half the children in South Asia are malnourished and underweight, most of them girls. South Asia is often called the most deprived realm in the world.

A combination of geographic factors underlies this tragic picture. With 22 percent of the world's population but just 3 percent of its land area, South Asia lacks the natural resources to raise living standards for its hundreds of millions of subsistence farmers. Governmental policies contribute to the problem: while East and Southeast Asia forged ahead by looking outward, encouraging exports and foreign investment, and spending heavily on literacy and technical education, health care, and land reform, South Asian governments tended to adopt bureaucratic controls and state planning. Cultural traditions also play their role. Resistance to change and reluctance by the privileged to open doors of opportunity to the less advantaged inhibit economic advancement for all.

A key factor lies in the rate of population growth. This is an aspect of **population geography**, the spatial view of demography. Figure I-9 reveals the concentration of population in South Asia's major river valleys, underscoring the

1

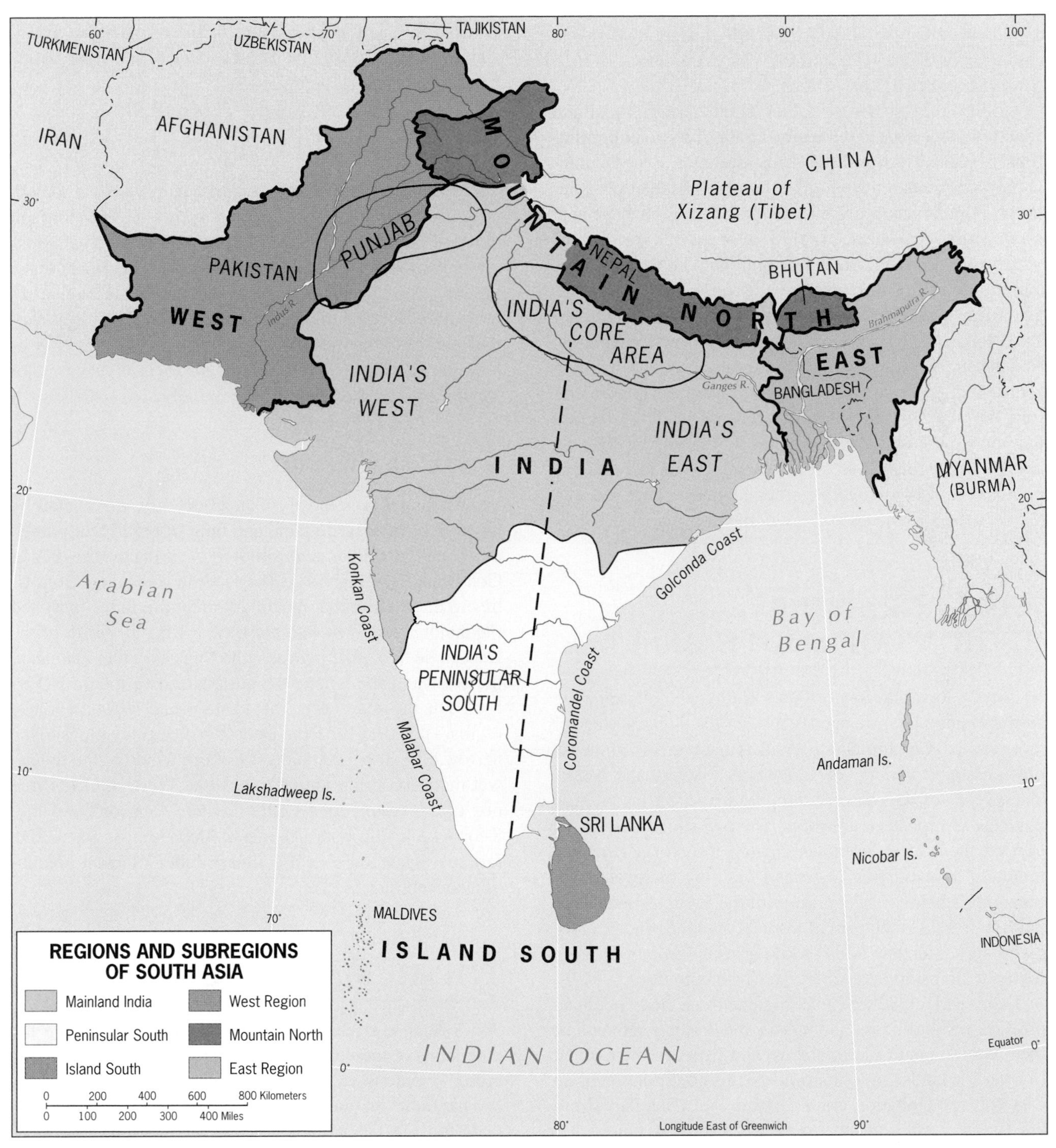

FIGURE 8-2

2 dependence of the majority on these ribbons of life. **Population distribution**, as this map reveals, should be compared to **population density**, a more specific measure. Column 4 of Table I-2 lists population density by country, but Figure I-9 proves that this average needs elaboration. Parts of India, populous as it is, are sparsely peopled; other areas are teeming. The national averages represent the *arithmetic densities* for those countries (905 in India), but they conceal regional variations. A better measure represents the number of people in a country per unit area of agriculturally productive land, and this is called its **physiologic density** (listed in column 5 of Table I-2). In India today, this index is 1615 per square mile (625 per sq km).

3

4

Critical to any assessment of South Asia's social and economic problems is the realm's **rate of natural population increase**. By world standards, these rates, for the individual countries, remain high: around 2.0 percent per year. When populations grow this fast, economic advancement becomes elusive. The needs of the new arrivals absorb most of the gains. From this rate of natural increase we can calculate the **doubling time** of a population. Muslim countries tend to have high growth rates and thus low doubling times. Take Pakistan, whose current population is 159 million. The latest data show its rate of natural increase as 2.8 percent and its doubling time as 25 years; at current rates, therefore, Pakistan will have 318 million people by 2027, barring major changes. India will have 2 billion people by 2039, an almost unimaginable situation. Such rates of increase can only retard, if not altogether stop, economic development. We return to this theme in the regional discussion.

5

6

PHYSIOGRAPHIC REGIONS OF SOUTH ASIA

Before we look into South Asia's complex and fascinating cultural geography, we need to discuss the physical stage of this populous realm. South Asia is a realm of immense physiographic variety, of snowcapped peaks and forest-clad slopes, of vast deserts and broad river basins, of high plateaus and spectacular shores. The collision of two of the Earth's great tectonic plates created the world's highest mountain ranges, their icy crests yielding meltwaters for great rivers below. The workings of the Earth's atmosphere put South Asia in the path of tropical cyclones and produce reversing seasonal windflows known as *monsoons*. This is a realm of almost infinite variety, a world unto itself.

In general terms, we can recognize three clearly defined physiographic zones in South Asia: the northern mountains, the southern peninsular plateaus, and between them a belt of river lowlands. Superimposed on this configuration is an east-west precipitation gradient from wet (Bangladesh) to dry (western Pakistan) that is clearly visible in Figures I-7 and I-8, broken only by the strip of high moisture along India's southwestern Malabar Coast (Fig. 8-1).

Northern Mountains

The northern mountains extend from the Hindu Kush and Karakoram Ranges in the northwest through the Himalayas in the center (Mount Everest, the world's tallest peak, lies in Nepal) to the ranges of Bhutan and the Indian State of Arunachal Pradesh in the east. Dry and barren in the west on the Afghanistan border, the ranges become green and tree-studded in Kashmir, forested in the lower sections of Nepal, and even more densely vegetated in Arunachal Pradesh. Transitional foothills, with many deeply eroded valleys cut by rushing meltwaters, lead to the river basins below.

River Lowlands

The belt of river lowlands extends eastward from Pakistan's lower Indus Valley (the area known as Sind) through the wide plain of the Ganges Valley of India and on across the great double delta of the Ganges and Brahmaputra in Bangladesh (Fig. 8-1). In the east, this physiographic region often is called the North Indian Plain. To the west lies the lowland of the Indus River, which rises in Tibet, crosses Kashmir, and then bends southward to receive its major tributaries from the Punjab ("Land of Five Rivers").

Southern Plateaus

Peninsular India is mostly plateau country, dominated by the massive Deccan, a tableland built of basalt that poured out when India separated from Africa during the breakup of Gondwana (see Fig. 7-3). The Deccan (meaning "South") tilts to the east, so that its highest areas are in the west and the major rivers flow into the Bay of Bengal. North of the Deccan lie two other plateaus, the Central Indian Plateau to the west and the Chota-Nagpur Plateau to the east (Fig. 8-1). On the map, note the Eastern and Western Ghats ("hills") that descend from Deccan plateau elevations to the narrow coastal plains below. Onshore winds of the annual wet monsoon bring ample rain to the Western Ghats (see box titled "South Asia's Life-Giving Monsoon" and Fig. 8-3). As a result, here lies one of India's most productive farming areas and one of southern India's largest population concentrations.

THE HUMAN SEQUENCE

South Asia is a realm of great river basins. Between the mountains of the north and the uplands of the south lie the broad, populous valleys of the Brahmaputra, the Ganges, and the Indus. In one of these valleys, the Indus (see box titled "India"), lies evidence of the realm's oldest civilization, contemporary to and interacting with ancient Mesopotamia. Unfortunately, much of the earliest record of this

South Asia's Life-Giving Monsoon

The dynamics of daily summer seabreezes are well known. The hot sun heats up the land, the air over this surface rises, and cool air from over the water blows in across the beach, replacing it. Teeming beaches and high-cost seafront apartments attest to the efficacy of this natural air conditioning.

Under certain circumstances this kind of circulation can affect whole regions. When an entire landmass heats up, a huge low-pressure system forms over it. This system can draw vast volumes of air from over the oceans onto the land. It is not just a local, daily phenomenon. It takes months to develop, but once it is in place, it also persists for months. When the inflow of moist oceanic air starts over South Asia, the **wet monsoon** has arrived. It may rain for 60 days or more. The countryside turns green, the paddies fill, and another dry season's dust and dirt are washed away. The region is reborn (see photo pair, p. 415).

Not all continents or coasts experience monsoons. A particular combination of topographic and atmospheric circumstances is needed. Figure 8-3 shows how it works in South Asia. It is June. For months, a low-pressure system has been building over northern India. Now, this cell has become so powerful that it draws air from over a large area of the warm, tropical Indian Ocean toward the interior. Some of that moisture-laden air is forced upward against the Western Ghats ①, cooling as it rises and condensing large amounts of rainfall. Other streams of air flow across the Bay of Bengal and get caught up in the convection over northeastern India and Bangladesh ②. Seemingly endless rain now inundates a much larger area, including the whole North Indian Plain. The mountain wall of the Himalayas stops the air from spreading and the rain from dissipating ③. The air moves westward, drying out as it flows toward Pakistan ④.

After persisting for weeks, the system finally breaks down and the wet monsoon gives way to periodic rains and, eventually, another dry season. Then the anxious wait begins for the next year's monsoon, for without it India would face disaster. In recent years, the monsoon has shown signs of irregularity and, as in 1987, partial failure. In India, life hangs by a meteorological thread.

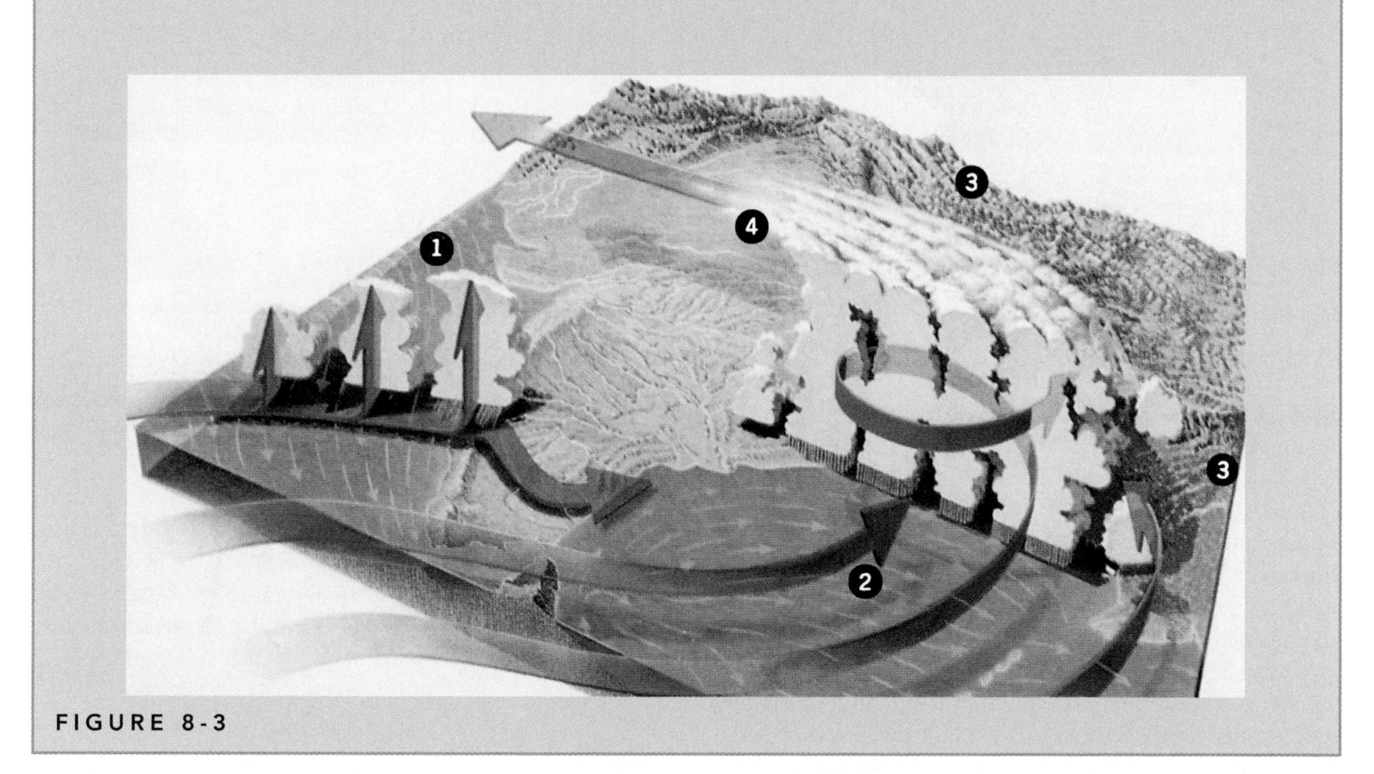

FIGURE 8-3

civilization lies buried beneath the present water table in the Indus Valley, but those archeological sites that have yielded evidence indicate that here was a sophisticated culture with states and large, well-organized cities. As in Mesopotamia and the Nile Valley, this civilization made considerable advances in the technology of irrigation and was based on the productivity of the irrigated soils in the Indus lowlands (Fig. 8-4).

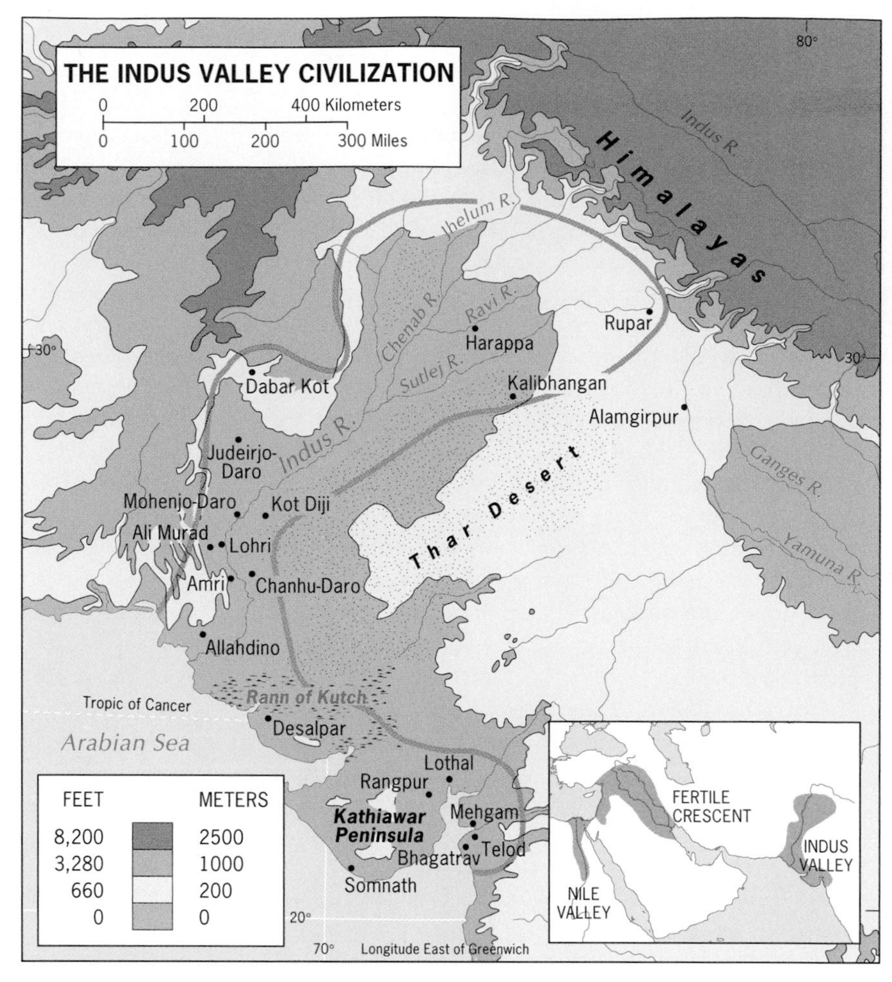

FIGURE 8-4

Development of the River Lowlands

The Indus Valley civilization was to India what the lower Nile culture hearth was to Africa: ideas and innovations diffused from there eastward and southward through the many different communities and societies that coexisted in the peninsula (see Fig. 6-3). But far more than the ancient Egyptians did in Africa, the Indus culture brought coherence to India, and over an even longer period of diffusion. By the time the Indus civilization, anchored by the cities of Harappa and Mohenjo-Daro, began to decline about 4000 years ago, many Indian cultures shared the legacy of the Harappans.

Now a new force entered the scene: the Indo-Europeans (or *Aryans*), who invaded the Indus Valley from Iran, adopted many innovations of the Indus civilization, and pushed their frontier of settlement eastward beyond the valley into the Ganges lowland, where they founded an urban culture of their own. Pushing southward into the peninsula, they conquered and absorbed the tribes they found there. Their language, Sanskrit, began to differentiate into the linguistic complex of modern-day India (Fig. 8-5).

> ### India
>
> Use of the name *India* for the heart of the South Asian realm derives from the Sanskrit word *sindhu,* used to identify the ancient civilizations in the Indus Valley. This word became *sinthos* in Greek descriptions of the area and then *sindus* in Latin. Corrupted to *indus,* which means "river," it was first applied to the region that now forms the heart of Pakistan. Subsequently, it was again modified to *India* to refer generally to the land of river basins and clustered peoples from the Indus in the west to the lower Brahmaputra in the east.

In the centuries following this invasion, Indian culture underwent a period of growth and development. From a formless collection of isolated tribes and their villages, regional organization emerged. Towns developed, arts and crafts blossomed, and trade with Southwest Asia increased. Most important, Hinduism emerged from the beliefs and

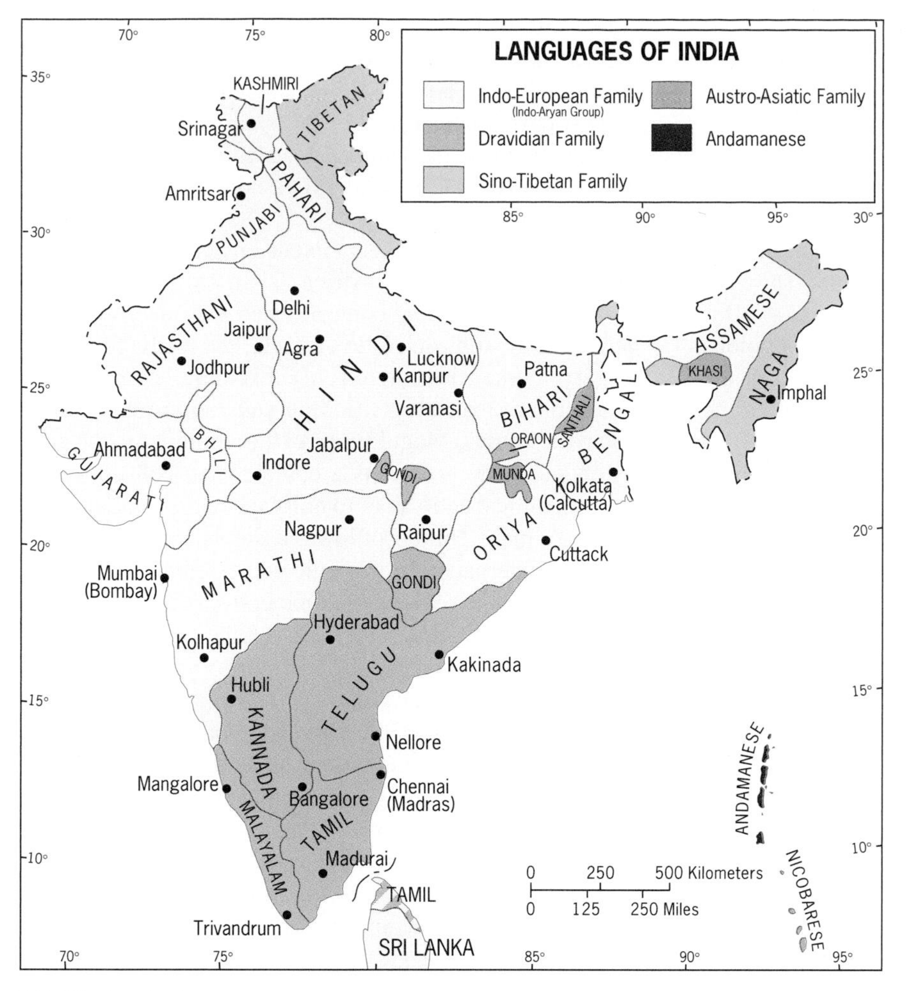

FIGURE 8-5

practices that the Indo-Europeans brought to India, and soon a new way of life, based on this faith, evolved. A multilayered **social stratification** developed, which powerful priests controlled and administered. These Brahmans stood at the head of a complex, bureaucratic hierarchy—a caste system—in which soldiers, artists, merchants, peasants, and all others had their place. But the India of, say, 3000 years ago was not peaceful. Aggressive and expansionist kingdoms arose, competing and struggling with each other for greater power and influence.

It was in one of these kingdoms, located in northeastern India, that Prince Siddhartha, better known as Buddha, was born. His birth in the sixth century BC (more than 2500 years ago) was unremarkable, but his actions were unique: he gave up his princely position to seek salvation and enlightenment through religious meditation. His teachings demanded the rejection of Earthly desires and prescribed a reverence for all forms of life. He walked the length and breadth of India and attracted a substantial following, but his teachings did not have a major impact on Hindu-dominated society during his lifetime. That impact would come later, when the ruler of a powerful Indian state decided to make Buddhism the state religion.

Before this occurred, however, South Asia was buffeted by yet another series of outside penetrations from the west and northwest. First the Persians pushed into the Indus Basin, and next the Greeks, under Alexander the Great, invaded not only the Indus Valley but also the very heartland of India, the Ganges Plain, late in the fourth century BC.

The Southern Peninsula

While all this was happening in northern India, the peninsular south lay comparatively isolated, protected by distance from the arena of cultural innovation, infusion, and conflict. Southern India had been settled by ancient peoples long before the Indus and Ganges civilizations arose—peoples whose historic linkages are not clear. Their physical appearance suggests connections to Africans and indigenous Australians; their languages, too, are distinctive and not related to those of the Indo-European north. As Figure 8-5 indicates, the south is a distinct subregion of India: both the peoples and their languages are known as *Dravidian*. There has, of course, been considerable intermixture with the north, but southern India is still a discrete region. The four major Dravidian languages—Telugu, Tamil, Kanarese (Kannada), and Malayalam—have long literary histories. Today, nearly one-fifth of India's 1 billion citizens speak Telugu or Tamil.

Aśoka's Mauryan Empire

When the Greeks withdrew from the Ganges Basin and the Hindu heartland was once again free, a powerful empire arose there—the first true empire in the realm. This, the Mauryan Empire, extended its influence over India as far west as the Indus Valley (thus incorporating the populous Punjab) and as far east as Bengal (the double delta of the Ganges and Brahmaputra); it reached as far south as the modern city of Bangalore.

This Mauryan Empire was led by a series of capable rulers who achieved stability over a vast domain. Undoubtedly the greatest of these leaders was Aśoka, who reigned for nearly 40 years during the middle of the third century BC. Aśoka was a believer in Buddhism, and it was he who elevated this religion from obscurity to regional and ultimately global importance.

In accordance with Buddha's teachings, Aśoka reordered his government's priorities from conquest and expansion to a Buddhist-inspired search for stability and peace. He sent missionaries to the outside world to carry Buddha's teachings to distant peoples, thereby also contributing to the diffusion of Indian culture. As a result, Buddhism became permanently established as the dominant religion in Sri Lanka (formerly Ceylon), and it established footholds as far afield as Southeast Asia and Mediterranean Europe. Ironically, Buddhism thrived in these remote places even as it declined in India itself. With Aśoka's death, the faith lost its strongest supporter.

The Mauryan Empire represented India's greatest political and cultural achievements in its day, and when the empire collapsed, late in the second century AD, India fragmented into a patchwork of states. Once again, India lay open to infusions from the west and northwest, and across present-day Pakistan they came: Persians, Afghans, Turks, and others driven from their homelands or attracted by the lands of the Ganges.

The Power of Islam

In the late tenth century, Islam came rolling like a giant tide across the subcontinent, spreading from Persia in the west and Afghanistan in the northwest. Of course, the Indus Valley lay directly in the path of this Islamic advance, and virtually everyone was converted. Next the Muslims penetrated the Punjab, the subregion that lies astride the present Pakistan-India border, and there perhaps as many as two-thirds of the inhabitants became converts. Then Islam crossed the bottleneck where Delhi is situated and diffused east- and southeastward into the Gangetic Plain and the subregion known as Hindustan—India's evolving core area. Here Islam's proselytizers had less success, persuading perhaps one in eight Indians to become Muslims. In the meantime, Islam arrived at the Ganges Delta by boat, and present-day Bangladesh became overwhelmingly Islamic. (To the south of the Ganges heartland, however, Islam's diffusion wave lost its energy: Dravidian India never came under Muslim influence.)

Islam's vigorous, often violent onslaught changed Indian society. As in West Africa, Islam often was superimposed through political control: when the rulers were converted, their subjects followed. By the early fourteenth century, a sultanate centered at Delhi controlled more of the subcontinent than even the Mauryan Empire had earlier. Later, the Islamic Mogul Empire (the similarity to the word "Mongol" is by no means a coincidence) constituted the largest political entity ever to unify the realm in precolonial times. To many Hindus of lower caste, Islam represented a welcome alternative to the rigid socio-religious hierarchy in which they were trapped at the bottom. Thus Islam was the faith of the ruling elites and of the disadvantaged, a powerful cultural force in the heartland of Hinduism.

Just as Islam weakened in southern Europe, so its force ultimately became spent in vast and populous India. For all the Muslims' power, they never managed to convert a majority of South Asians. They dominated the northwest corner of the realm (present-day Pakistan), where Lahore became one of Islam's greatest cities. But in all of what is today India, less than 15 percent of the population became and remained Muslim. And throughout the period of Islamic intervention, the struggle for cultural supremacy continued. Placid Hinduism and aggressive Islam did not easily coexist.

The European Intrusion

Into this turbulent complexity of religious, political, and linguistic disunity yet another element began to intrude after 1500: European powers in search of raw materials, markets, and political influence. Because the Europeans profited from the Hindu-Muslim contest, they exploited local rivalries, jealousies, and animosities. British merchants gained control over the trade with Europe in spices, cotton, and silk goods, ousting the French, Dutch, and Portuguese. The British East India Company's ships also took over the intra-Asian sea trade between India and Southeast Asia, which had long been in the hands of Arab, Indonesian, Chinese, and Indian merchants. In effect, the East India Company (EIC) became India's colonial administration.

In time, however, the East India Company faced problems it could not solve. Its commercial activities remained profitable, but it became entangled in a widening effort to maintain political control over an expanding Indian domain. The Company was an ineffective governing agent at a time when the increasing Westernization of India brought to the fore new and intense frictions. Christian missionaries were challenging Hindu beliefs, and many Hindus believed that the British were out to destroy the caste system. Changes also came in public education, and the role and status of women began to improve. Aristocracies saw their positions threatened as Indian landowners had their estates expropriated. Finally, in 1857 the three-month Sepoy Mutiny (named after the Sepoys, Indian troops in the service of the EIC) broke out and changed the entire situation. It took a major military effort to put it down, and from that time the East India Company ceased to function as the government of India. Administration was turned over to the British government, the Company was abolished, and India formally became a British colony—a status it held for the next 90 years, until this *raj* (rule) ended in 1947.

Colonial Transformation

Four centuries of European intervention in South Asia greatly changed the realm's cultural, economic, and political directions. Certainly the British made positive contributions to Indian life, but colonialism also brought serious negative consequences. In this respect there are important differences between the South Asian case and that of Subsaharan Africa. When the Europeans came to India, they found a considerable amount of industry, especially in metal goods and textiles, and an active trade with both Southwest and Southeast Asia in which Indian merchants played a leading role. The British intercepted this trade, changing the whole pattern of Indian commerce.

India now ceased to be South Asia's manufacturer, and soon the country was exporting raw materials and importing manufactured goods—from Europe, of course. India's handicraft industries declined; after the first stimulus, the export trade in agricultural raw materials also suffered as other parts of the world were colonized and linked in trade to Europe. Thus the majority of India's people (who were farmers then as now) suffered an economic setback as a result of the manipulations of colonialism. Although the colonial period brought considerable increases in total volume of trade, the composition of the trade India now supported by no means brought a better life for its people.

Neither did the British manage to accomplish what the Mauryans and the Moguls had tried to do: unify the subcontinent and minimize its internal cultural and political divisions. When the Crown took over from the East India Company in 1857, about 750,000 square miles (nearly 2 million sq km) of Indian territory were still outside the British sphere of influence. Slowly, the British extended their control over this huge unconsolidated area, including several pockets of territory already surrounded but never integrated into the previous corporate administration. Moreover, the British government found itself obligated to support a long list of treaties that the Company's administrators had made with numerous Indian princes, regional governors, and feudal rulers.

These treaties guaranteed various degrees of autonomy for literally hundreds of political entities in India, ranging in size from a few acres to Hyderabad's more than 80,000 square miles (200,000 sq km). The British Crown saw no alternative but to honor these guarantees, and India was carved up into an administrative framework under which there were more than 600 "sovereign" territories in the subcontinent. These "Native States" had British advisors; the large British provinces such as Punjab, Bengal, and Assam had British governors or commissioners who reported to the viceroy of India, who in turn reported to parliament and the monarch in London. In all, this near-chaotic amalgam of modern colonial control and traditional feudalism reflected and in some ways deepened the regional and local disunities of the Indian subcontinent. Although certain parts of India quickly adopted and promoted the positive contribu-

FROM THE FIELD NOTES
"More than a half-century after the end of British rule, the centers of India's great cities continue to be dominated by the Victorian-Gothic buildings the colonists constructed here. Here is evidence of a previous era of globalization, when European imprints transformed urban landscapes. Walking the streets of Mumbai (the British called it Bombay) you can turn a corner and be forgiven for mistaking the scene for London, double-deckered buses and all. One of the British planners' major achievements was the construction of a nationwide railroad system, and railway stations were given great prominence in the urban architecture. I had walked up Naoroji Road, having learned to dodge the wild traffic around the circles in the Fort area, and watched the throngs passing through Victoria Station. Inside, the facility is badly worn, but the trains continue to run, bulging with passengers hanging out of doors and windows."

tions of the colonial era, other areas rejected and repelled them, thereby adding yet another element of division to an increasingly complicated human spatial mosaic.

Colonialism did produce assets for India. The country was bequeathed one of the best transport networks of the colonial domain, especially the railroad system (although the network focused on interior-seaport linkages rather than fully interconnecting the various parts of the country). British engineers laid out irrigation canals through which millions of acres of land were brought into cultivation. Settlements that had been founded by Britain developed into major cities and bustling ports, led by Bombay (now Mumbai), Calcutta (now Kolkata), and Madras (now Chennai). These three cities are still three of India's largest urban centers, and their cityscapes bear the unmistakable imprint of colonialism. Modern industrialization, too, was brought to India by the British on a limited scale. In education, an effort was made to combine English and Indian traditions; the Westernization of India's elite was supported through the education of numerous Indians in Britain. Modern practices of medicine were also introduced. Moreover, the British administration tried to eliminate features of Indian culture that were deemed undesirable by any standards—such as the burning alive of widows on their husbands' funeral pyres, female infanticide, child marriage, and the caste system. Obviously, the task was far too great to be achieved in barely three generations of colonial rule, but independent India itself has continued these efforts where necessary.

Partition

Even before the British government decided to yield to Indian demands for independence, it was clear that British India would not survive the coming of self-rule as a single political entity. As early as the 1930s, the idea of a separate Pakistan was being promoted by Muslim activists, who circulated pamphlets arguing that British India's Muslims were a nation distinct from the Hindus and that a separate state consisting of Sind, Punjab, Baluchistan, Kashmir, and a portion of Afghanistan should be created from the British South Asian Empire in this area. The first formal demand for such partitioning was made in 1940, and, as later elections proved, the idea had almost universal support among the realm's Muslims.

As the colony moved toward independence, a political crisis developed: India's majority Congress Party would not even consider partition, and the minority Muslims refused to participate in any future unitary government. But partition would not be a simple matter. True, Muslims were in the majority in the western and eastern sectors of British India, but Islamic clusters were scattered throughout the realm (Fig. 8-6). Any new boundaries between Hindus and Muslims to create an Islamic Pakistan and a Hindu India would have to be drawn right through areas where both sides coexisted. People by the millions would be displaced.

Nor were Hindus and Muslims the only people affected by partition. The Punjab area, for example, was home to millions of Sikhs, whose leaders were fiercely anti-Muslim. But a Hindu-Muslim border based only on those two groups would leave the Sikhs in Pakistan. Even before independence day, August 15, 1947, Sikh leaders talked of revolt, and there were some riots. But no one could have foreseen the dreadful killings and mass migrations that followed the creation of the boundary and the formation of independent Pakistan and India. Just how many people felt compelled to participate in the ensuing migrations will never be known; 15 million is the most common estimate. It was human suffering on an incomprehensible scale.

Even so large a flow of cross-border **refugees**, however, hardly began to "purify" India of Muslims. After the initial mass exchanges, there still were tens of millions of Muslims in India (Fig. 8-6). Today, the Muslim minority in Hindu-dominated India is almost as large as the whole Islamic population of Pakistan, having more than tripled since the late 1940s. It is the world's largest minority, far more than a mere remnant of the days when Islam ruled the realm. This force will play a growing role in the India of the future.

9

Flight was one response to the 1947 partition of what had been British India, resulting in one of the greatest mass population transfers in human history. Here, two trainloads of eastbound Hindu refugees fleeing (then) West Pakistan arrive at the station in Amritsar, the first city inside India.

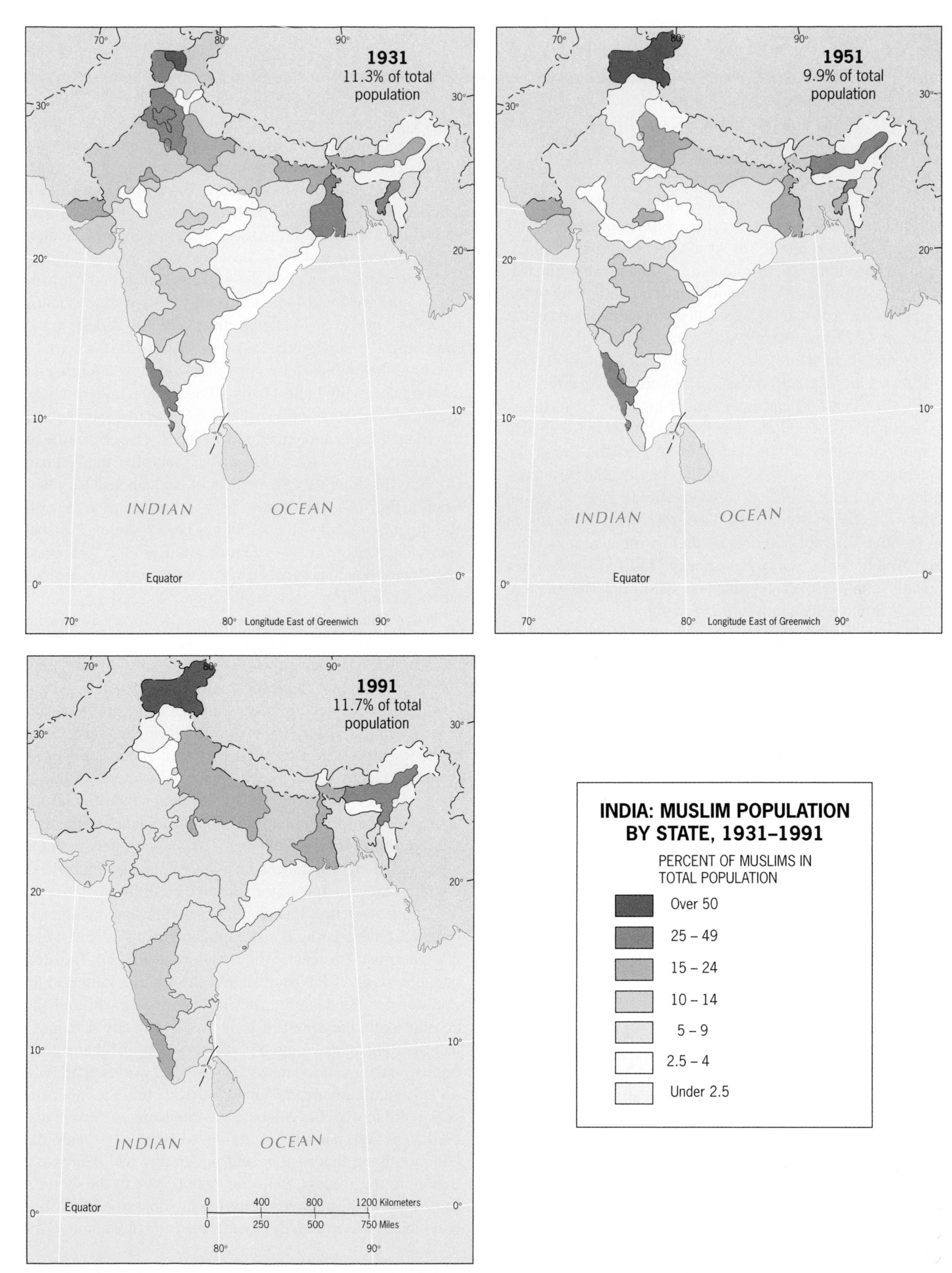
1931
11.3% of total population
1951
9.9% of total population
1991
11.7% of total population
INDIAN OCEAN
Equator
Longitude East of Greenwich
0 400 800 1200 Kilometers
0 250 500 750 Miles
INDIA: MUSLIM POPULATION BY STATE, 1931–1991
PERCENT OF MUSLIMS IN TOTAL POPULATION
Over 50
25 – 49
15 – 24
10 – 14
5 – 9
2.5 – 4
Under 2.5

FIGURE 8-6

REGIONS OF THE REALM

PAKISTAN: ON SOUTH ASIA'S WESTERN FLANK

If India is the dominant entity in South Asia, why focus first on Pakistan? There are several reasons, both historic and geographic. Here lay South Asia's earliest urban civilizations, whose innovations radiated into the great peninsula. Here lies South Asia's Muslim frontier, contiguous to the great Islamic realm to the west and irrevocably linked to the enormous Muslim minority to its east. Pakistan's cultural landscapes bear witness to its transitional location. Teeming, disorderly Karachi is the typical South Asian city; as in India, the largest urban center lies on the coast. Historic, architecturally Islamic Lahore is reminiscent of the scholarly centers of Muslim Southwest Asia. In Pakistan's east, the partition boundary divides a Punjab that is otherwise continuous, a land of villages, wheat fields, and irrigation ditches. In the northwest, Pakistan resembles Afghanistan in its huge migrant populations and its mountainous frontier. And in the far north, Pakistan and India are locked in a deathly conflict over Jammu and Kashmir. The western flank is South Asia's most critical region.

If, as is so often said, Egypt is the gift of the Nile, then Pakistan is the gift of the Indus. The Indus River and its principal tributary, the Sutlej, nourish the ribbons of life that form the heart of this populous country (Fig. 8-7). Territorially, Pakistan is not large by Asian standards; its area is about the same as that of Texas plus Louisiana. But Pakistan's population of 159.2 million makes it one of the world's ten most populous states. Among Muslim countries (officially, it is known as the Islamic Republic of Pakistan) only Southeast Asia's Indonesia is larger, but Indonesia's Islam is much less pervasive than Pakistan's. Pakistan is an active participant in the worldwide resurgence of Islamic fervor; Indonesia is not—at least not yet.

Pakistan lies like a giant wedge between Iran and Afghanistan to the west and India to the east. This wedge extends from the Arabian Sea in the south, where it is widest, to the snowcapped mountains of the north. Here in the far north of Pakistan, the boundary framework is complex and jurisdictions are uncertain. India, China, Pakistan, and, until recently, the former Soviet Union have all vied for control over parts of this mountainous northern frontier. The Soviets, by virtue of their temporary rule over Afghanistan, controlled the Wakhan Corridor (the long narrow segment of land labeled "Afghanistan" in Fig. 8-8). The Chinese have claimed areas to the east. But the major territorial conflict has been between Pakistan and India over the areas of Jammu and Kashmir. This issue has plagued relations between these two countries for decades (see box titled "The Problem of Kashmir").

When Pakistan became an independent state following the partition of British India in 1947, its capital was Karachi on the south coast, near the western end of the Indus Delta. As the map shows, however, the present capital is Islamabad, near the larger city of Rawalpindi in the north, not far from Kashmir. By moving the capital from the "safe" coast to the embattled interior and by placing it on the doorstep of contested territory, Pakistan announced its intent to stake a claim to its northern frontiers. And by naming the city Islamabad, Pakistan proclaimed its Muslim foundation, here in the face of the Hindu challenge. This politico-geographical use of a national capital can be assertive, and Islamabad exemplifies the principle of the **forward capital**. 10

At independence, Pakistan had a bounded national territory, a capital, a cultural core, and a population—but it had few centripetal forces to bind state and nation. The disparate regions of Pakistan shared the Islamic faith and an aversion for Hindu India, but little else. Karachi and the coastal south, the desert of Baluchistan, the city of Lahore and the Punjab, the rugged northwest along Afghanistan's border, and the mountainous far north are worlds apart, and a Pakistani nationalism to match that of India at independence did not exist. Successive Pakistani governments, civilian as well as military, turned to Islam to provide the common bond that history and geography had denied the nation. In the process, Pakistan became one of the world's most theocratic states; its common law, based on the English model, was gradually transformed into a Quranic (Ko-

MAJOR CITIES OF THE REALM

City	Population* (in millions)
Ahmadabad, India	4.4
Bangalore, India	5.9
Chennai (Madras), India	7.0
Colombo, Sri Lanka	1.0
Delhi-New Delhi, India	13.6
Dhaka, Bangladesh	13.8
Hyderabad, India	7.4
Karachi, Pakistan	12.9
Kathmandu, Nepal	0.9
Kolkata (Calcutta), India	13.5
Lahore, Pakistan	6.6
Mumbai (Bombay), India	19.4
Varanasi, India	1.4

*Based on 2002 estimates.

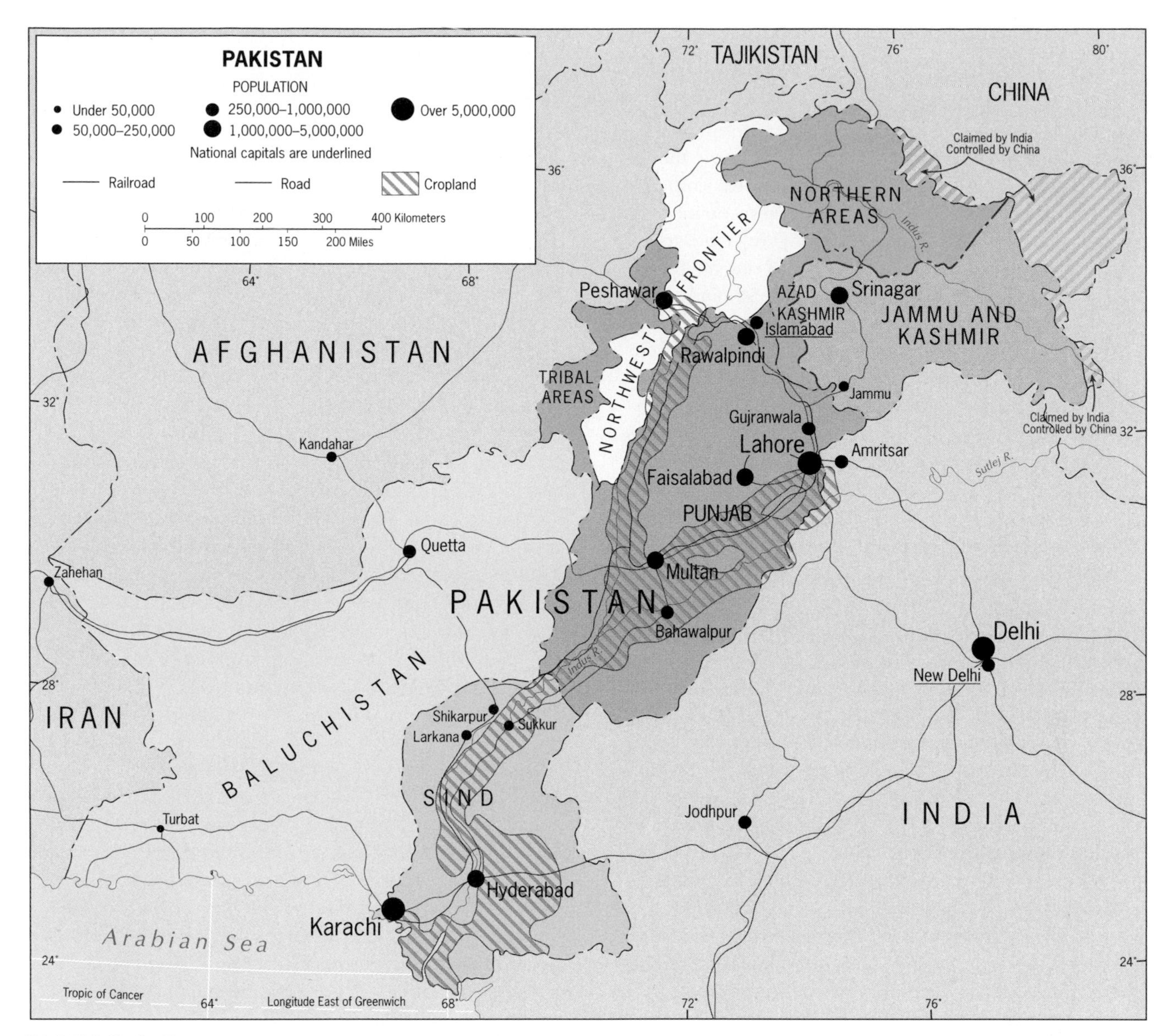

FIGURE 8-7

ranic) system with Islamic Sharia courts and associated punishments.

But even Islam itself is not unified in restive Pakistan. About 80 percent of the people are Sunni Muslims, and the Shia minority numbers about 16 percent. Sunni fanatics intermittently attack Shi'ites, leading to retaliation and creating grounds for subsequent revenge.

Despite the Islamization of Pakistan's plural society, it remains a strongly regionalized country in which Urdu is the official language and English is still the *lingua franca* of the elite. Yet several other major languages prevail in diverse parts, and lifeways vary from nomadism in Baluchistan to irrigation farming in the Punjab to pastoralism in the northern highlands.

Subregions

Punjab

Pakistan's core area is the Punjab, the Muslim heartland across which the postcolonial boundary between Pakistan and India was superimposed. (As a result, India also has a region called Punjab, sometimes spelled Panjab there.) Pakistan's Punjab is home to nearly 60 percent of the country's population. In the triangle formed by the Indus River and its tributary, the Sutlej, live some 90 million people. Punjabi is the language here, and intensive farming of wheat is the mainstay.

The Problem of Kashmir

Kashmir is a territory of high mountains surrounded by Pakistan, India, China, and, along several miles in the far north, Afghanistan (Fig. 8-8). Although known simply as Kashmir, the area actually consists of several political divisions, including the State properly referred to as Jammu and Kashmir (one of the 562 Indian States at the time of independence) and the administrative areas of Gilgit in the northwest and Ladakh (including Baltistan) in the east. The main conflict between India and Pakistan over the final disposition of this territory has focused on the southwest, where Jammu and Kashmir are located.

When partition took place in 1947, the existing States of British India were asked to decide whether they would go with India or Pakistan. In most of the States, the local ruler made this decision, but Kashmir was an unusual case. It had about 5 million inhabitants at that time, nearly half of them concentrated in the basin known as the Vale of Kashmir (where the capital, Srinagar, is located). Another 45 percent of the people were concentrated in Jammu, which leads down the foothill slopes of the Himalayas to the edge of the Punjab. The rest of the population was scattered through the mountains, including Pushtuns in Gilgit and other parts of the northwest. Of these population groups, the people of the mountain-encircled Vale of Kashmir are almost all Muslims, whereas most of Jammu's population is Hindu.

But in the State of Jammu and Kashmir the ruler was Hindu, not Muslim, although the overall population was more than 75 percent Muslim. Thus the ruler had to make a difficult decision in 1947—to go with Pakistan and thereby exclude the State from Hindu India, or to go with India and thereby incur the wrath of most of the people. Hence the maharajah of Kashmir sought to remain outside both Pakistan and India and to retain autonomous status. This decision was followed, after the partitioning of India and Pakistan, by a Muslim uprising against Hindu rule in Kashmir. The maharajah asked for Indian help, and Pakistan's forces came to the aid of the Muslims. After more than a year's fighting and through the intervention of the United Nations, a cease-fire line left Srinagar, the Vale, and most of Jammu and Kashmir (including nearly four-fifths of the territory's population) in Indian hands (Fig. 8-8). Eventually, this line began to appear on maps as the final boundary settlement, and Indian governments have proposed that it be so recognized.

Why should two countries, whose interests would be served by peaceful cooperation, allow a distant mountainland to trouble their relationship to the point of war? There is no single answer to this question, but there are several areas of concern for both sides. In the first place, Pakistan is wary of any situation whereby India would control vital irrigation waters needed in Pakistan. As the map shows, the Indus River, the country's lifeline, crosses Kashmir. Moreover, other tributary streams of the Indus originate in Kashmir, and it was in the Punjab that Pakistan learned the lessons of dealing with India for water supplies. Second, the situation in Kashmir is analogous to the one that led to the partition of the whole subcontinent: Muslims are under Hindu domination. The majority of Kashmir's people are Muslims, so Pakistan argues that free choice would deliver Kashmir to the Islamic Republic. A free plebiscite is what Pakistanis have sought—and the Indians have thwarted. Furthermore, Kashmir's connections with Pakistan before partition were much stronger than those between Kashmir and India, although India has invested heavily in improving its links to Jammu and Kashmir since the military stalemate. In recent years, it did seem more likely that the cease-fire line (now called the Line of Control) would

Three cities anchor this core area: Lahore, the outstanding center of Islamic culture in the realm, Faisalabad, and Multan. Lahore, now home to 6.6 million people, lies close to the India-Pakistan border. Founded about 2000 years ago, Lahore was situated favorably to become a great Muslim center during the Mogul period, when the Punjab was a corridor into India. As a center of royalty, Lahore was adorned with numerous magnificent buildings, including a great fort, multiple palaces, and several mosques displaying superb stonework and marble embellishments. The site of an old university and many majestic gardens, Lahore is the cultural focus of Islam. The city received hundreds of thousands of refugees and grew rapidly after partition. Lahore did lose its eastern hinterland, but its new role in independent Pakistan sustained its growth.

Punjab's relationship with Pakistan's other three provinces is one of the country's weak points. Both the governments and peoples of the other three provinces feel uneasy about the dominance of Punjab, the populous, powerful core of the country.

Sind

South of Punjab lies Sind, centered on the chaotic port city of Karachi on the coast and on Hyderabad upriver. As Figure 8-7 suggests, Sind has three dominant landscapes: arid

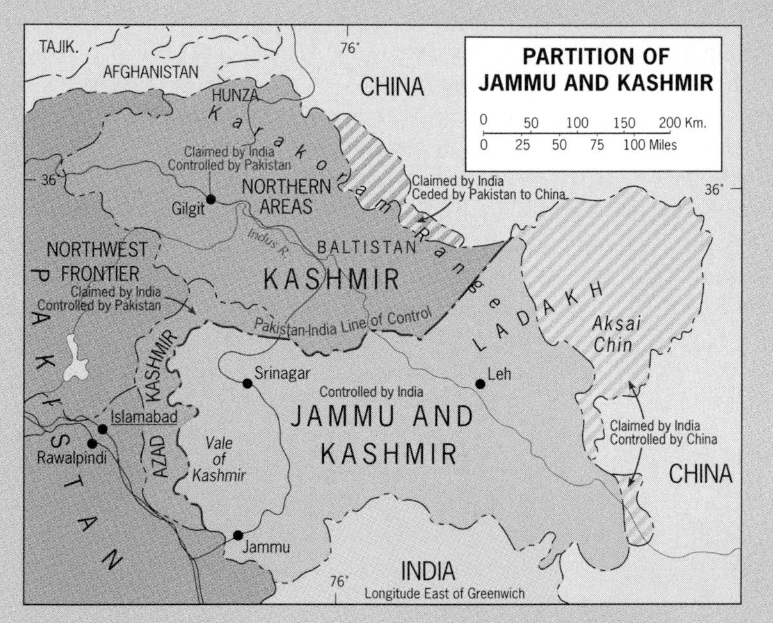

FIGURE 8-8

indeed become a stable boundary between India and Pakistan. (The incorporation of the State of Jammu and Kashmir into the Indian federal union was accomplished as far back as 1975, when India was able to reach an agreement with the State's chief minister and political leader.)

The drift toward stability was reversed in the late 1980s when a new crisis engulfed Kashmir. Extremist Muslim groups demanding independence escalated a long-running insurgency in 1988. That brought a swift crackdown by the Indian military, whose harsh tactics prompted the citizenry to support the separatists far more strongly. In the early 1990s, Pakistan charged the Indians with widespread human-rights violations, and the Indians accused the Pakistanis of inciting secession and supplying arms. Between 1990 and 1995, nearly 10,000 people were killed in the sporadic fighting that continues to afflict this scenic area. In 1995, Muslim extremists began a campaign that involved capturing and killing foreign visitors, a move that was condemned by representatives of both sides. Nevertheless, the memory of three wars between India and Pakistan defied all attempts at compromise. But all this was overshadowed by the implications of the 1998 test explosion of nuclear bombs by India and Pakistan. This development transformed Kashmir from a problem frontier to a potentially calamitous flashpoint.

rocky hills to the west, desert to the east, and between these the fertile, irrigated alluvial valley of the lower Indus River. The British built large irrigation systems here, and improved seed strains have made Sind a Pakistani breadbasket for wheat and rice. Commercially, cotton is king here, supporting major textile industries in the cities and towns.

A flood of immigrants from India moved into Sind during and after partition, and relations between immigrants and locals have not always been good. Karachi has become the focus of conflict between immigrant Muhajirs and their descendants and other groups in this poverty-stricken, crime-ridden megacity of 12.9 million. Karachi grew explosively during and after the refugee influx, and its infrastructure crumbled even as its population increased by millions. The social order broke down, law enforcement deteriorated, and in the mushrooming shantytowns gang battles over turf spilled over into the built-up areas. Large multicultural cities can put severe strain on low-income economies when conditions spin out of control.

Meanwhile, Punjab's domination of the national scene rankles political and business leaders in Sind, and calls for greater autonomy, and even secession, are heard. Certainly Sind's contributions to Pakistan's economy outweigh its influence in distant Islamabad.

Baluchistan

A trip westward from Sind and Karachi into the rocky and sandy deserts makes one wonder whether this is the same country. Baluchistan's arid landscapes extend far into neighboring Iran, and the sparse population of this remote and forbidding territory still moves across the borders into and out of Iran along ancient caravan routes. Pakistan is a Sunni Islamic country, but about 16 percent of the people adhere to Shi'ite principles. Shi'ism arrived in Pakistan along these desert routes from its Iranian stronghold.

Baluchistan may figure more importantly in Pakistan's economic future than its present. The geology indicates considerable mineral potential, and oil and natural gas have been found here.

FROM THE FIELD NOTES

"Driving through the Khyber Pass that links Afghanistan and Pakistan was a riveting experience. From the dry and dusty foothills on the Afghanistan side, with numerous ruins marking historic battles, I traversed the rugged Hindu Kush via multiple hairpin turns and tunnels into Pakistan. This is one of the most strategic passes in the world, but for invading armies it was no easy passage: defensive forces had the advantage in these treacherous valleys. But now the roads and tunnels facilitate the movement of refugees, drugs, and arms, and they are used by militant separatist forces from both sides of the border. Pakistan's Northwest Frontier Province, and especially the city of Peshawar, are hotbeds of these activities."

Northwest Frontier

The fourth subregion of Pakistan is appropriately called the Northwest Frontier. This is the area that faces turbulent and disintegrating Afghanistan, and it has been the scene of massive refugee influxes, local separatist movements, and social dislocation. The Northwest Frontier is dominated by mountain ranges, mountain-encircled basins, and strategic passes. The most famous pass of all is the Khyber Pass to Afghanistan (see photo this page). The Turks invaded the upper Indus Valley by this route; later the Mongols (Moguls) streamed through it on their way to India; and in recent decades millions of war-weary emigrants have walked this road to safety in Pakistan.

The largest city in the Northwest Frontier Province is Peshawar, but after Baluchistan this is Pakistan's least urbanized area. Peshawar lies in an alluvium-filled, fertile valley where fields of wheat and corn cover the countryside. But with its large refugee population (which, during the 1980s, exceeded 4 million), the Northwest Frontier has faced difficult times, which United Nations assistance only partly eased.

The ethnic and cultural backgrounds of the population in the Northwest Frontier vary. Even before the refugee influx from Afghanistan, a large part of the population consisted of Pushtuns (also called Pathans, Pakhtuns, or Pashtuns), who were closely related to the Pushtuns of central Afghanistan. At times, the Pushtuns of Afghanistan urged their kinspeople in Pakistan to demand more autonomy, if not outright independence—a practice we know as **irredentism**. When the Soviets invaded Afghanistan in 1979 and war broke out, millions of Pushtuns moved from Afghanistan to the Northwest Frontier. Pakistan not only accommodated this refugee population but countered the secession movement by improving the integration of the Northwest Frontier Province with the rest of the country through road building, economic aid programs, and enhanced educational opportunities.

At the beginning of the twenty-first century, Pakistan's Northwest Frontier remained tense and strained. It is only 170 miles (270 km) from Peshawar to Afghanistan's capital of Kabul through the Khyber Pass, and events in Afghanistan continue to affect the province. A large refugee population remains, and many are unlikely ever to return home, although they retain links to Afghanistan's factions. Western countries, alleging that Afghanistan is a base for terrorist organizations, have attempted to coerce Pakistan to help combat this terrorism, but Islamic militants warn Islamabad against such collaboration; again the Northwest Frontier is in the middle of this issue. As Figure 8-7 shows, the Northwest Frontier bounds not only Afghanistan but also Jammu and Kashmir, where Pakistani and Indian forces frequently clash. Among Pakistan's disparate subregions, the Northwest Frontier lies in the most difficult location

Livelihoods

For all its size and growing regional influence, Pakistan remains a low-income economy. As the twenty-first century opened, urbanization was just surpassing 30 percent. Population growth was high at 2.8 percent, implying a doubling time of only 25 years. Subsistence farming of food crops still occupied most of the people. Life expectancy for a child born today is just 58 years. Illiteracy still afflicts about 60 percent of the population over age 15.

Nevertheless, Pakistan has made significant economic progress during its five decades-plus of independence. Irrigation has expanded enormously; land reform has progressed; and the cotton-based textile industry has generated important revenues. Despite a limited mineral resource base, manufacturing has grown, including a steel mill at Port Qasim near Karachi.

To the rest of the world, Pakistan sells textiles, carpets, tapestries, leather goods—and rice. Despite its large and growing population, Pakistan during the 1990s was able to export rice (although it did import some wheat), which is a measure of the effect of the Green Revolution as well as its program of farm expansion. In recent years, Pakistan has also become one of the largest producers of heroin in southern Asia, and its opium and hashish enter the international drug trade despite government efforts to interdict it. Afghanistan, too, produces heroin and has failed to eradicate it in the fields.

Pakistan's legitimate economy has shown substantial growth in recent years, but the country faces an enormous public debt and needs external assistance to stay afloat. As Table I-2 shows, Pakistan has the highest per-capita GNP among the major states of mainland South Asia, and a reduction in the high rate of population growth would further improve its prospects. As we will see, these prospects are now endangered by a new threat: a nuclear arms race with its giant neighbor.

Emerging Regional Power

A country's underdevelopment is not necessarily a barrier against its emergence as a power to be reckoned with in international affairs. When Pakistan became independent in 1947, the country was weak, disorganized, and divided. Just five and a half decades later, Pakistan is a major military force and a nuclear power.

To say that Pakistan in 1947 was a divided country is no exaggeration. In fact, upon independence, present-day Pakistan was united with present-day Bangladesh, and the two countries were called West Pakistan and East Pakistan, respectively. The basis for this scheme was Islam: in Bangladesh, too, Islam is the state religion. Between the two Islamic wings of Pakistan lay Hindu India. But there was little else to unify the easterners and westerners, and their union lasted less than 25 years. In 1971, a costly war of secession led to independence for East Pakistan, which took the name Bangladesh; at the same time, West Pakistan became Pakistan.

The loss of Bangladesh was no disaster for Pakistan: in virtually every respect, Bangladesh was (and remains) even more severely impoverished than Pakistan (see Table I-2). Pakistan's challenges lay closer to home—in Kashmir, in the Northwest Frontier, and, most importantly, to the east in India. India had supported independence for Bangladesh and did not resume diplomatic relations with Pakistan until five years later. But the political relationship between India and Pakistan has remained tense. Apart from the conflict in Kashmir, their other differences are numerous. India remained a democracy while Pakistan became a military dictatorship; India's treatment of its Muslim minorities frequently riled Pakistan; India developed a close relationship with the same Soviet Union seen as a threat by Islamabad. During the Cold War, India tilted toward Moscow, while Pakistan, despite frequent disputes, was favored by Washington in order to curb Soviet influence in Afghanistan.

Pakistan was confronted by new challenges across its eastern border in the 1990s when India exploded several nuclear bombs at its test site. Worldwide consternation at this provocation put intense pressure on Pakistan not to follow suit, but Pakistan had little choice. Days later, it set off nuclear explosions at its own atomic test site. The "Islamic Bomb" was now reality, and the prospect of nuclear war, which had seemed to be receding, once again reasserted itself.

Pakistan thus finds itself encircled by adversity and risk. United by Islam but little else, facing domestic and international strife from Karachi to Kashmir, Pakistan is vulnerable to threats from all sides. Its involvement in Afghanistan risks a reverse "Afghanization," Taliban-style, of Pakistani society. Its struggle over Jammu and Kashmir risks nuclear war with a far more powerful neighbor. Even its relations with its near-neighbors in Turkestan are at risk because Turkestan fears Pakistan's power projected via Afghanistan.

No longer merely a newly decolonized, economically disadvantaged country trying to survive, Pakistan has taken a crucial place in the political geography of two neighboring realms in turbulent transition.

INDIA: FIFTY-PLUS YEARS OF FEDERATION

Nearly three-quarters of the great land triangle of South Asia is occupied by a single country—India, the world's most populous democracy and, in terms of human numbers, the world's largest federation. Consider this: India has nearly as many inhabitants as live in all the countries of Subsaharan Africa *plus* North Africa/Southwest Asia combined—75 of them. At present rates of population growth, India not only

FROM THE FIELD NOTES
"We were docked in the port of Mumbai (Bombay), literally surrounded by a fleet of warships including an aircraft carrier (foreground) and a half-dozen submarines. India is often in the news because of its ongoing conflict with Pakistan, but here was evidence of India's other agenda: its projection of strength as a regional power in the Indian Ocean." Note: after this photograph was taken, India as well as Pakistan successfully tested nuclear bombs, and the potential for devastating conflict grows.

will outnumber these two realms before the end of our new century but much sooner will overtake even China. Thus, if India holds together as a single state, it will become the world's most populous country, bar none.

That India has endured as a unified country is a politico-geographical miracle. India is a cultural mosaic of immense ethnic, religious, linguistic, and economic diversity and contrast; it is a state of many nations. The period of British colonialism gave India the underpinnings of unity: a single capital, an interregional transport network, a *lingua franca*, a civil service. Upon independence in 1947, India adopted a federal system of government, giving regions and peoples some autonomy and identity, and allowing others to aspire to such status. Unlike Africa, where federal systems failed and where military dictatorships replaced them, India remained essentially democratic and retained a federal framework in which States have considerable local authority.

This political, democratic success has been achieved despite the presence of powerful centrifugal forces in this vast, culturally diverse country. Relations between the Hindu majority and the enormous Muslim minority, better in some States than in others, have at times threatened to destabilize the entire federation. Local rebellions, demands by some minorities for their own States, frontier wars, even involvement in a foreign but nearby civil war (in Sri Lanka) have buffeted the system—which has bent but not collapsed. India has succeeded where others in the postcolonial world have failed.

This success has not been matched in the field of economics, however. After more than 50 years of independence India remains a very poor country, and not all of this can be blamed on the colonial period or on population growth, although overpopulation remains a strong impediment to improvement of living standards. Much of it results from poor and inconsistent economic planning, too much state ownership of inefficient industries, excessive government control over economic activities, bureaucratic suppression of initiative, corruption, and restraints against foreign investment. As we will see later, a few bright spots in some of India's States contrast sharply to the overwhelming poverty of hundreds of millions.

States and Peoples

The map of India's political geography shows a federation of 28 States, 6 Union Territories (UTs), and 1 National Capital Territory (NCT) (Fig. 8-9). The federal government retains direct authority over the UTs, all of which are small in both territory and population. The NCT, however, includes Delhi, and the capital, New Delhi, and has over 13 million inhabitants.

The political spatial organization shown in Figure 8-9 is mainly the product of India's restructuring following independence from Britain. Its State boundaries reflect the broad outlines of the country's cultural mosaic: as far as possible the system recognizes languages, religions, and cultural traditions. Indians speak 14 major and numerous minor languages, and while Hindi is the official language (and English is the *lingua franca*), it is by no means universal. The map is the product of endless compromise—endless because demands for modifications of it continue to this day; as recently as late 2000, the federal government authorized the creation of three new States. In the northeast lie very small States established to protect the local traditions of small populations; minority groups in the larger States ask why they should not receive similar recognition.

With only 28 States for a national population of 1.039 billion, several of India's States contain more people than many countries of the world (Table 8-1). As Figure 8-9

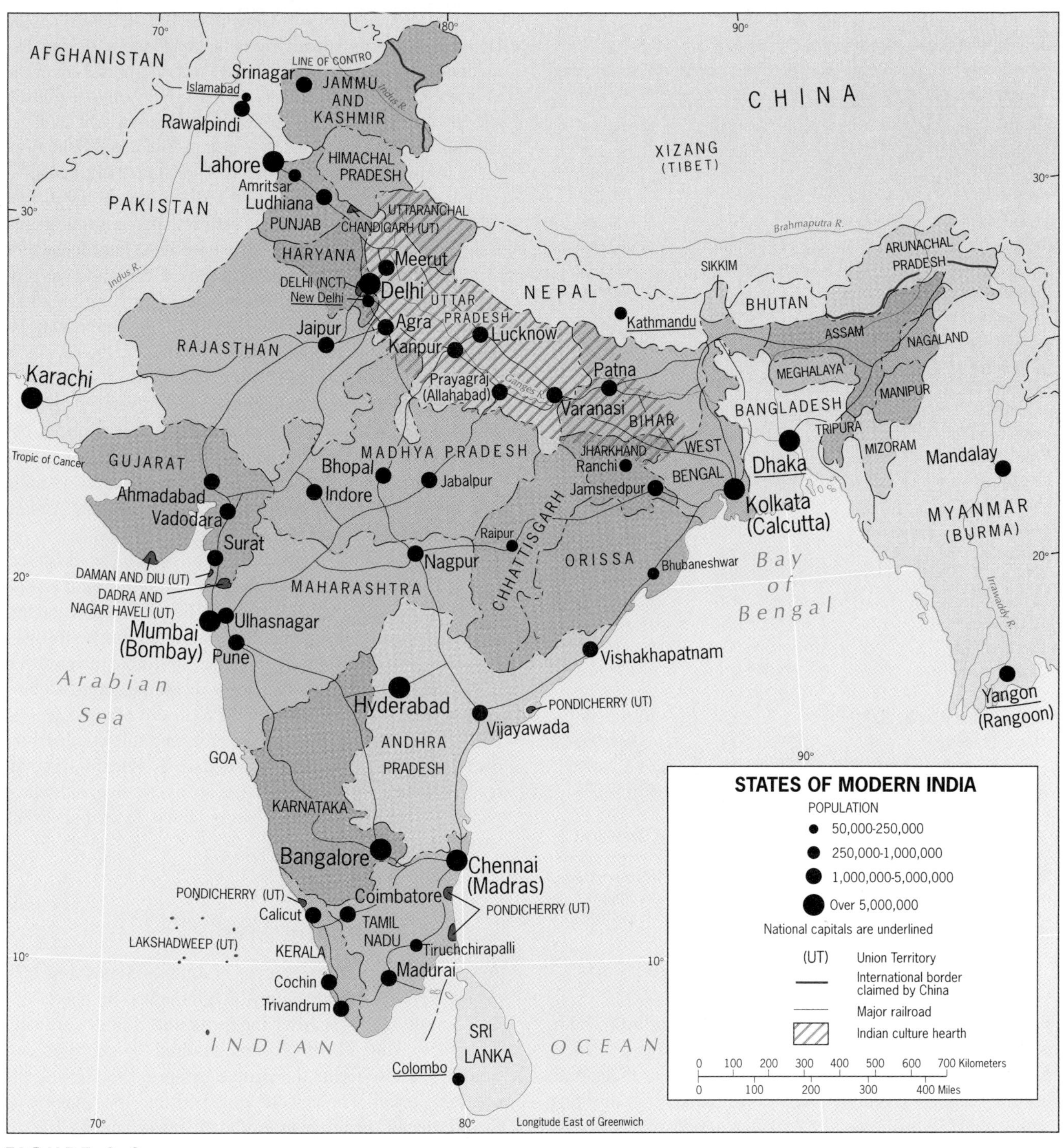

FIGURE 8-9

shows, the (territorially) largest States lie in the heart of the country and on the great southward-pointing peninsula. Uttar Pradesh (just over 160 million) and Bihar (about 77 million) constitute much of the Ganges River Basin and are the core area of modern India (see box titled "Solace and Sickness from the Holy Ganges"). Maharashtra (almost 98 million), anchored by the great coastal city of Bombay (renamed Mumbai in 1996), also has a population larger than that of most countries. West Bengal, the State that adjoins Bangladesh, has more than 84 million residents, 13.5 million of whom live in its urban focus, Calcutta (renamed Kolkata in 2000).

These are staggering numbers, and they do not decline much toward the south. Southern India consists of four States linked by a discrete history and by their distinct Dravidian languages. Facing the Bay of Bengal are Andhra

Table 8-1
INDIA: POPULATION BY STATE

State	1991 Census	2002 Estimate[a]
Andhra Pradesh	66,508,000	81,891,000
Arunachal Pradesh	865,000	1,129,000
Assam	22,414,000	27,866,000
Bihar	86,374,000	76,499,000
Chhattisgarh (Estab. 2000)	n/a	17,600,000
Goa	1,170,000	1,343,000
Gujarat	41,310,000	49,878,000
Haryana	16,464,000	20,800,000
Himachal Pradesh	5,171,000	6,086,000
Jammu and Kashmir	7,719,000	9,855,000
Jharkhand (Estab. 2000)	n/a	29,900,000
Karnataka	44,977,000	53,833,000
Kerala	29,099,000	32,911,000
Madhya Pradesh	66,181,000	64,916,000
Maharashtra	78,937,000	97,876,000
Manipur	1,837,000	2,298,000
Meghalaya	1,775,000	2,365,000
Mizoram	690,000	917,000
Nagaland	1,210,000	1,855,000
Orissa	31,660,000	37,815,000
Punjab	20,282,000	24,034,000
Rajasthan	44,006,000	55,875,000
Sikkim	406,000	505,000
Tamil Nadu	55,859,000	63,943,000
Tripura	2,757,000	3,557,000
Uttaranchal (Estab. 2000)	n/a	10,400,000
Uttar Pradesh	139,112,000	161,916,000
West Bengal	68,078,000	84,355,000
National Capital Territory (Estab. 1993)	n/a	13,585,000
Union Territories	1,998,000	2,697,000

[a]2002 estimates based on projections from 1991 census. Increases calculated on the basis of growth-rate data by individual State. Populations rounded to nearest 000. Populations of the three new States (established in 2000) based on press reports.

Pradesh (82 million) and Tamil Nadu (64 million), both part of the hinterland of the megacity of Madras (renamed Chennai in 1997) and located on the coast near their joint border. Facing the Arabian Sea are Karnataka (54 million) and Kerala (33 million). Kerala, often at odds with the federal government in New Delhi, has long had the highest literacy rate in India and one of the lowest rates of population growth owing to strong local government and strictly enforced policies. "It's a matter of geography," explained a teacher in the Kerala city of Cochin. "We are here about as far away as you can get from the capital, and we make our own rules."

As Figure 8-9 shows, India's smaller States lie mainly in the northeast, on the far side of Bangladesh, and in the northwest, toward Jammu and Kashmir. North of Delhi, India is flanked by China and Pakistan, and physical as well as cultural landscapes change from the flatlands of the Ganges to the hills and mountains of spurs of the Himalayas. In the State of Himachal Pradesh, forests cover the hillslopes and relief reduces living space; only 6 million people live here, many in small, comparatively isolated clusters. Before independence and political consolidation, the colonial government called this area the "Hill States."

But the map becomes even more complex in the distant northeast, beyond the narrow corridor between Bhutan and Bangladesh. The dominant State here is Assam, famed for its tea plantations and important because its oil and gas production amounts to more than 40 percent of India's total. Assam attained full Statehood in 1972, just after Bangladesh became independent from Pakistan. Most of its 28 million residents live in the lowland of the Brahmaputra River, upstream from Bangladesh; illegal migration from Bangladesh into Assam has at times strained relations between India and its eastern Muslim neighbor.

In the Brahmaputra Valley, Assam resembles the India of the Ganges. But in almost all directions from Assam, things change. To the north, in sparsely populated Arunachal Pradesh (1.1 million), we are in the Himalayan offshoots again. To the east, in Nagaland (1.9 million), Manipur (2.3 million), and Mizoram (0.9 million), lie the forested and terraced hillslopes that separate India from Myanmar (Burma). This is an area of numerous ethnic groups (more than a dozen in Nagaland alone) and of frequent rebellion against Delhi's government. And to the south, the States of Meghalaya (2.4 million) and Tripura (3.6 million), hilly and still wooded, border the teeming floodplains of Bangladesh. Here in the country's northeast, where peoples are always restive and where population growth is still soaring, India faces one of its strongest regional challenges.

India's Changing Map

As we noted, the present map of India's States and UTs (Fig. 8-9) is not the one with which India was born as a sovereign state in 1947. After independence, the government first had to contend with several hundred "princely states," fiefdoms whose rights the British had protected during the colonial period. These were absorbed into the States, and the privileged "princely orders" were phased out by 1972.

Next, the Indian government reorganized the country on the basis of its major regional languages (see Fig. 8-5). Hindi, spoken by more than one-third of the population, was designated the country's official language, but 13 other major languages also were given national status by the Indian constitution, including the four Dravidian languages of the south. English, it was anticipated, would become India's common language, its *lingua franca* at government, administrative, and business levels. Indeed, English not only remained the language of national administration but also became the chief medium of commerce in growing

Solace and Sickness from the Holy Ganges

Stand on the banks of the Ganges River in Varanasi, Hinduism's holiest city, and you will see people bathing in the holy water, drinking it, and praying as they stand in it—while the city's sewage flows into it nearby, and the partially cremated corpses of people and animals float past. It is one of the world's most compelling—and disturbing—sights.

The Ganges (*Ganga*, as the Indians call it) is Hinduism's sacred river. Its ceaseless flow and spiritual healing power are earthly manifestations of the Almighty. Therefore, tradition has it, the river's water is immaculate, and no amount of human (or other) waste can pollute it. On the contrary: just touching the water can wash away a believer's sins.

At Varanasi, Prayagraj (Allahabad) and other cities and towns along the Ganges, the river banks are lined with Hindu temples, decaying ornate palaces, and dozens of wide stone staircases called *ghats*. These stepped platforms lead down to the water, enabling thousands of bathers to enter the river. They come from the city and from afar, many of them pilgrims in need of the healing and spiritual powers of the water. It is estimated that more than a million people enter the river somewhere along its 1600-mile (2600-km) course every day. During religious festivals, the number may be ten times as large.

By any standards, the Ganges is one of the world's most severely polluted streams, and thousands among those who enter it become ill with diarrhea or other diseases; many die. In 1986, then-Prime Minister Rajiv Gandhi launched a major scheme to reduce the level of pollution in the river, a decade-long construction program of sewage treatment plants and other facilities. In the mid-1980s, Gandhi was told, nearly 400 million gallons of sewage and other wastes were being disgorged into the Ganges every day. The plan called for the construction of nearly 40 sewage treatment plants in riverfront cities and towns.

Many Hindus, however, opposed this costly program to clean up the Ganges. To them, the holy river's spiritual purity is all that matters. Getting physically ill is merely incidental to the spiritual healing power contained in a drop of the Ganga's water.

The stone steps leading into the Ganges' waters in Varanasi, India. Varanasi is India's holiest city, and millions descend these *ghats* every year.

urban India. English was the key to better jobs, financial success, and personal advancement, and the language constituted a common ground in higher education.

The newly devised framework based on the major regional languages, however, proved to be unsatisfactory to many communities in India. In the first place, many more languages are in use than the 14 that had been officially recognized. Demands for additional States soon arose. As early as 1960, the State of Bombay was divided into two language-based States, Gujarat and Maharashtra. And the pressures were not just linguistic. As recently as 2000, three major changes again transformed the map of India.

After a long struggle that involved railroad blockades, strikes, and bombings as well as political manipulation, 18 southern districts of the State of Bihar achieved Statehood in 2000 as Jharkhand. Here, the prime motives were economic: the tribal peoples in poverty-stricken southern Bihar felt disadvantaged and exploited. Without an effective voice in the Bihar State government, they campaigned for recognition at the federal level—and succeeded.

Another new State was established right next door, in what was eastern Madhya Pradesh. Here the quest for recognition began during colonial times, in the 1930s, and for a time the tribal peoples in northeastern Madhya Pradesh even dreamed of joining Jharkhand, should its campaign succeed. But in 2000 they were rewarded in the form of the State of Chhattisgarh (Fig. 8-9) with a population of nearly 18 million and a considerable, if largely undeveloped, nonagricultural resource base. In terms of social indicators, Chhattisgarh is one of India's poorest States. More than two-thirds of its households have no electricity, more than half the population is illiterate, infant mortality is far higher than the national average, and women are far worse off than average. With direct representation at the federal level, however, there is hope for improvement.

Still another State created in 2000 is Uttaranchal, which split from India's most populous, Ganges-basin, core-area State, Uttar Pradesh. This new State would have qualified for what the colonial power called the "Hill States" category: it lies in the highlands above the populous Ganges lowland, between Nepal to the southeast and Himachal Pradesh to the northwest (Fig. 8-9). This State was authorized in recognition of its environment and ways of life, far different from those prevailing in the teeming Ganges Valley.

For many years India has faced quite a different set of cultural-geographic problems in its northeast, where numerous ethnic groups occupy their own niches in a varied, forest-clad topography. The Naga, a cluster of peoples whose domain had been incorporated into Assam State, rebelled soon after India's independence. A protracted war brought federal troops into the area; after a truce and lengthy negotiations, Nagaland was proclaimed a State in 1961. This led the way for other politico-geographical changes in India's problematic northeastern wing.

A further dilemma involves India's Sikh population. The Sikhs (the word means "disciples") adhere to a religion that was created about five centuries ago to unite warring Hindus and Muslims into a single faith. This faith's principles rejected negative aspects of Hinduism and Islam, and it gained millions of followers in the Punjab and adjacent areas. During the colonial period, many Sikhs supported British administration of India, and by doing so they won the respect and trust of the British, who employed tens of thousands of Sikhs as soldiers and policemen. By 1947, there was a large Sikh middle class in the Punjab. When independence came, many left their rural homes and moved to the cities to enter urban professions. Today, they still exert a strong influence over Indian affairs, far in excess of the less than 2 percent of the population (about 19 million) they constitute.

After independence, the Sikhs demanded that the original Indian State of Panjab (Punjab) be divided into a Sikh-dominated northwest and a Hindu-majority southeast. The government agreed, so that Punjab as now constituted (Fig. 8-9) is India's Sikh stronghold, whereas neighboring Haryana State is mainly Hindu. Unfortunately, this redelimitation did not defuse all the pressures in the area. A militant Sikh minority demanded an even more autonomous Sikh State in the Punjab, to be called Khalistan. Gradually this minority gained strength, and when the radical Sikh leaders held the State in the grip of violence and intimidation, the federal government sent troops to contain them. This led to a disastrous confrontation in 1984 at the Golden Temple in the city of Amritsar, Sikhism's holiest shrine. In its aftermath, Prime Minister Indira Gandhi was assassinated by two of her Sikh bodyguards, and the crisis deepened. To this day, relations between Sikhs and non-Sikhs are fractious; cycles of violence still affect life in the comparatively prosperous Punjab.

These ethnic, cultural, and regional problems are but a sample of the stresses on India's federal framework. There is no Muslim State in India, but India has about 127 million Muslims within its borders—the largest cultural minority in the world. As Figure 8-6 shows, the percentage of Muslims is highest in remote Jammu and Kashmir, but it also is substantial in such widely dispersed States as Kerala, Assam, and Uttar Pradesh. Moreover, the Muslim population today (approximately 12 percent) constitutes a larger percentage than it did after partition (9.9 percent). This Islamic minority also ranks among the most rapidly growing sectors of India's population, and is strongly urbanized as well—nearly one-third of the population of India's largest city, Mumbai (Bombay), is Muslim.

Centrifugal Forces: From India to Hindustan?

In Chapter 1 we introduced the concept of centrifugal and centripetal forces (see box, p. 54), respectively the dividing and unifying forces that continuously affect all states. No country in the world exhibits greater cultural diversity than India, and variety in India comes on a scale unmatched anywhere else on Earth. Such diversity spells strong centrifugal forces, although, as we will see, India also has powerful consolidating bonds.

Among the centrifugal forces, Hinduism's stratification of society into castes remains pervasive. Under Hindu dogma, *castes* are fixed layers in society whose ranks are based on ancestries, family ties, and occupations. The **caste system** may have its origins in the early social divisions into priests and warriors, merchants and farmers, craftspeople and servants; it may also have a racial basis, for the Sanskrit term for caste is color. Over the centuries, its complexity grew until India had thousands of castes, some with a few hundred members, others containing millions. Thus, in

12

city as well as in village, communities were segregated according to caste, ranging from the highest (priests, princes) to the lowest (the untouchables). The term *untouchable* has such negative connotations that some scholars object to its use. Alternatives include *dalits* (oppressed), the common term in Maharashtra State but coming into general use; *harijans* (children of God), which was Gandhi's designation, still widely used in the State of Bihar; and *Scheduled Castes*, the official government label.

A person was born into a caste based on his or her actions in a previous existence. Hence, it would not be appropriate to counter such ordained caste assignments by permitting movement (or even contact) from a lower caste to a higher one. Persons of a particular caste could perform only certain jobs, wear only certain clothes, worship only in prescribed ways at particular places. They or their children could not eat, play, or even walk with people of a higher social status. The untouchables occupying the lowest tier were the most debased, wretched members of this rigidly structured social system. Although the British ended the worst excesses of the caste system, and postcolonial Indian leaders—including Mohandas (Mahatma) Gandhi (the great spiritual leader who sparked the independence movement) and Jawaharlal Nehru (the first prime minister)—worked to modify it, a few decades cannot erase centuries of class consciousness. In traditional India, caste provided stability and continuity; in modernizing India, it constitutes an often painful and difficult legacy.

Today we can discern a geography of caste—a degree of spatial variation in its severity. Cultural geographers estimate that about 15 percent of all Indians are of lower caste, about 40 percent of backward caste (one important rank above the lower caste), and some 18 percent of upper caste, at the top of which are the Brahmans, men in the priesthood. (The caste system does not extend to the Muslims, Sikhs, and other non-Hindus in India, which is why these percentages do not total 100.) The colonial government and successive Indian governments have tried to help the lowest castes. This effort has had more effect in the urban than in the rural areas of India. In the isolated villages of the countryside, the untouchables often are made to sit on the floor of their classroom (if they go to school at all); they are not allowed to draw water from the village well because they might pollute it; and they must take off their shoes, if they wear any, when they pass higher-caste houses. But in the cities, untouchables have reserved for them places in the schools, a fixed percentage of State and federal government jobs, and a quota of seats in national and State legislatures. Mohandas Gandhi, who took a special interest in the fate of the untouchables (harijans) in Indian society, accomplished much of this reform.

The caste system remains a powerful centrifugal force, not only because it fragments society but also because efforts to weaken it often result in further division. Gandhi himself was killed, only a few months after independence, by a Hindu fanatic who opposed his work for the least fortunate in Indian society. Today, India is being swept by a wave of Hindu fundamentalism that is caused, at least in part, by continuing efforts to help the poorest. Higher castes see themselves as disadvantaged, and they take refuge in a "return" to fundamental Hindu values.

The radicalization of Hinduism and the infusion of Hindu nationalism into India's politics loom today as twin threats to the country's unity. Hindu nationalist political parties are polarizing the electorate at the State as well as federal level; in Maharashtra, radical Hindu political leaders managed to rename Bombay, the capital, giving this great international city an old Hindu toponym, Mumbai. If Indian politics fragment along religious lines, the miracle of Indian unity may come to an end.

Centripetal Forces

In the face of all these divisive forces, what bonds have kept India unified for so long? Without question, the dominant binding force in India is the cultural strength of Hinduism, its sacred writings, holy rivers, and influence over Indian life. For most Indians, Hinduism is a way of life as much as it is a faith, and its diffusion over virtually the entire country (Muslim, Sikh, and Christian minorities notwithstanding) brings with it a national coherence that constitutes a powerful antidote to regional divisiveness. Over the long term, however, the key ingredients of this Hinduism have been its gentility and introspection, radical outbursts notwithstanding. Now the specter of Hindu fanaticism threatens this vital bond.

Another centripetal force lies in India's democratic institutions. In a country as culturally diverse and as populous as India, reliance on democratic institutions has been a birthright ever since independence, and democracy's survival—raucous, sometimes corrupt, always free—has been a crucial unifier.

Furthermore, communications are better in much of India than in many other countries in the global periphery, and the continuous circulation of people, ideas, and goods helps bind the disparate state together. Before independence, opposition to British rule was a shared philosophy, a strong centripetal force. After independence, the preservation of the union was a common objective, and national planning made this possible.

India's capacity for accommodating major changes and its flexibility in the face of regional and local demands are also a centripetal force. Boundaries have been shifted; internal political entities have been created, relocated, or otherwise modified; and secessionist demands have been handled with a mixture of federal power and cooperative negotiation. Indians in South Asia have accomplished what Europeans in Yugoslavia could not, and India's history of success is itself a centripetal force.

Finally, no discussion of India's binding forces would be complete without mentioning the country's strong leader-

FROM THE FIELD NOTES

"The streets of India's cities often seem to be one continuous market, with people doing business in open storefronts, against building walls, or simply on the sidewalk. Here the formal and informal sectors of India's economy intermingle. I walked this way in Delhi every morning, and the store selling mattresses and pillows was always open. But the women in the foreground, selling handkerchiefs and other small items from a portable iron rack, were sometimes here, and sometimes not; one time I ran into them about a half-mile down the road. As I learned one tumultuous day, every time the government tries to exercise some control over the street hawkers and sidewalk sellers, massive opposition results and chaos can ensue. A few days later, everything is as it was. Change comes very slowly here."

ship. Gandhi, Nehru, and their successors did much to unify India by the strength of their compelling personalities. For many years, leadership was a family affair: Nehru's daughter, Indira Gandhi, twice took decisive control (in 1966 and 1980) after weak governments, and her son, Rajiv Gandhi (who later, in 1991, was also assassinated), was prime minister in the late 1980s. Since his death, the crucial question of India's leadership has hung in the balance.

The Population Dilemma

The population of the realm's seven countries today totals nearly 1.4 billion—more than one-fifth of all humankind. India alone has 1.039 billion inhabitants, second only to China among the countries of the world, and is on course to overtake it. Such rapid population growth poses a threat to national development. India's federal government and the governments of its States have enacted legislation and implemented programs to reduce the rate of population growth. Although these initiatives have had some effect, India's population at the turn of the twenty-first century was still growing at an annual rate of 1.8 percent (the realm as a whole was expanding by 1.9 percent).

Earlier we noted the implications for India of population growth so rapid that the country's doubling time is only 39 years. Economic gains are being overtaken by growing numbers, and increasing population is creating nutritional problems and risks. The gap between food requirements and farm production narrowed during the Green Revolution, but hundreds of millions of children do not get balanced meals or adequate calories. In Table I-2, note that except for tiny Maldives, the country with the slowest population growth also is the country with the highest GNP: Sri Lanka.

India's Demographic Challenge

On pp. 385-386 we introduced the field of population geography. One dimension of this field involves the global distribution of population growth and its implications. Comparing the map of world economies (Fig. I-11) with the list of population growth rates in Table I-2 reveals a clear pattern: the bulk of rapid population growth is occurring in the lower-income economies. In many of the high-income economies, population growth is small, has leveled off, or is even negative. These higher-income economies have gone through the so-called **demographic transition**, a four-stage sequence that took them from high birth rates and high death rates in preindustrial times to very low birth rates and very low death rates today (Fig. 8-10). Stages 2 and 3 in this model constitute the **population explosion**, a hallmark of the twentieth century: death rates in the industrializing and urbanizing countries dropped, but birth rates

13

14

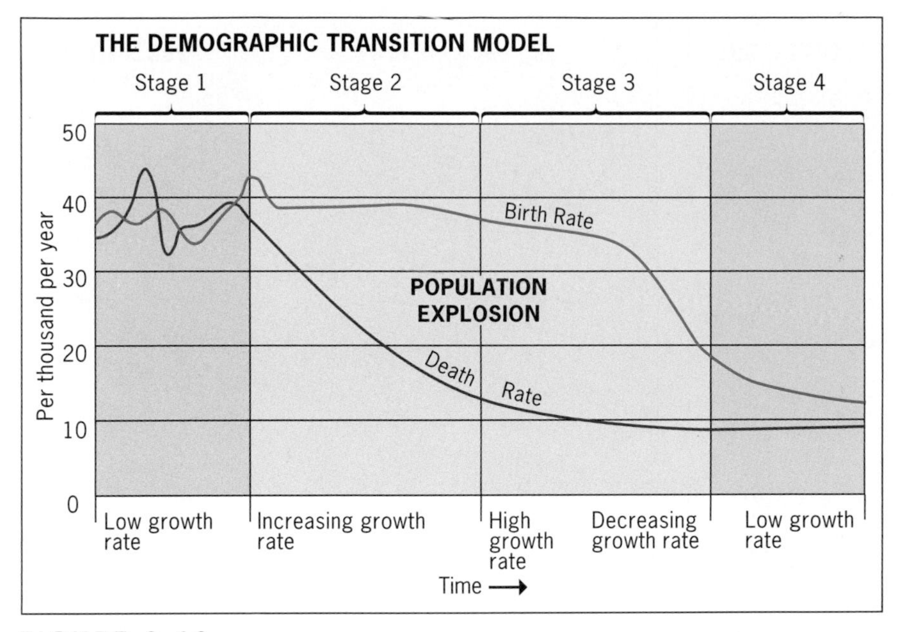

FIGURE 8-10

took longer to decline. In 1900, the world's population was about 1.5 billion; by 2000 it had surpassed 6 billion.

When the British ruled India during the nineteenth century, the country still was in the first stage, with high birth rates and high death rates; the high death rates were caused not only by a high incidence of infant and child mortality but also by famines and epidemics. As Figure 8-10 indicates, the population during Stage 1 does not grow or decline much, but it is not stable. Famines and disease outbreaks kept erasing the gains made during better times. But then India entered the second stage. Birth rates remained high, but death rates declined because medical services improved (soap came into widespread use), food distribution networks became more effective, farm production expanded, and urbanization developed. In the 1920s, India's population still was growing at a rate of only 1.04 percent, but by the 1970s, that rate had shot up to 2.22 percent per year (Fig. 8-11). Note that India gained 28 million people during the 1920s but a staggering 135 million during the 1970s.

Has India entered the third stage, when the death rate begins to level off and birth rates decline substantially, narrowing the gap and slowing the annual increase? The rate of increase suggests it: from 2.22 percent during the 1970s, it dropped to 2.11 percent in the 1980s and to a projected 1.88 percent during the 1990s (Fig. 8-11). But India has another problem. During its population explosion, its numbers grew

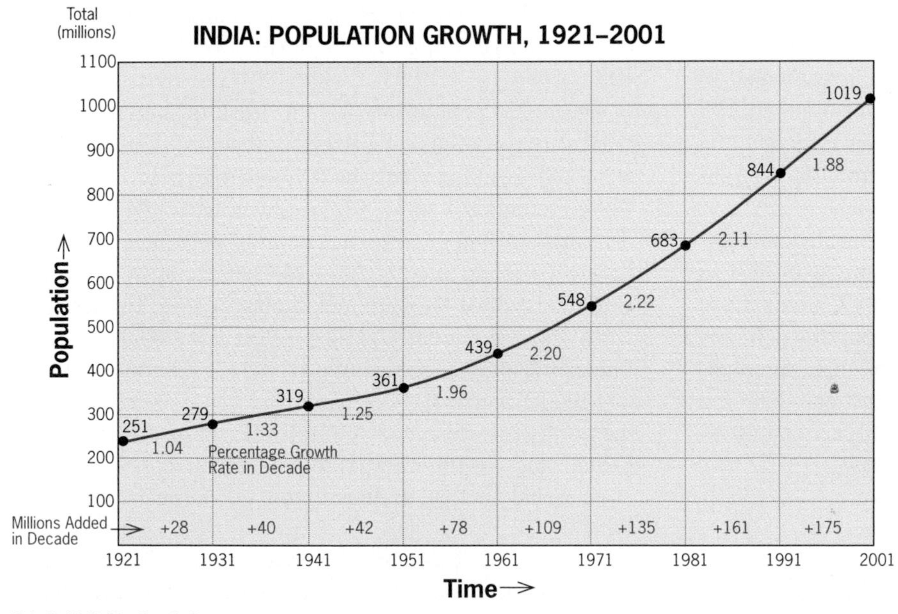

FIGURE 8-11

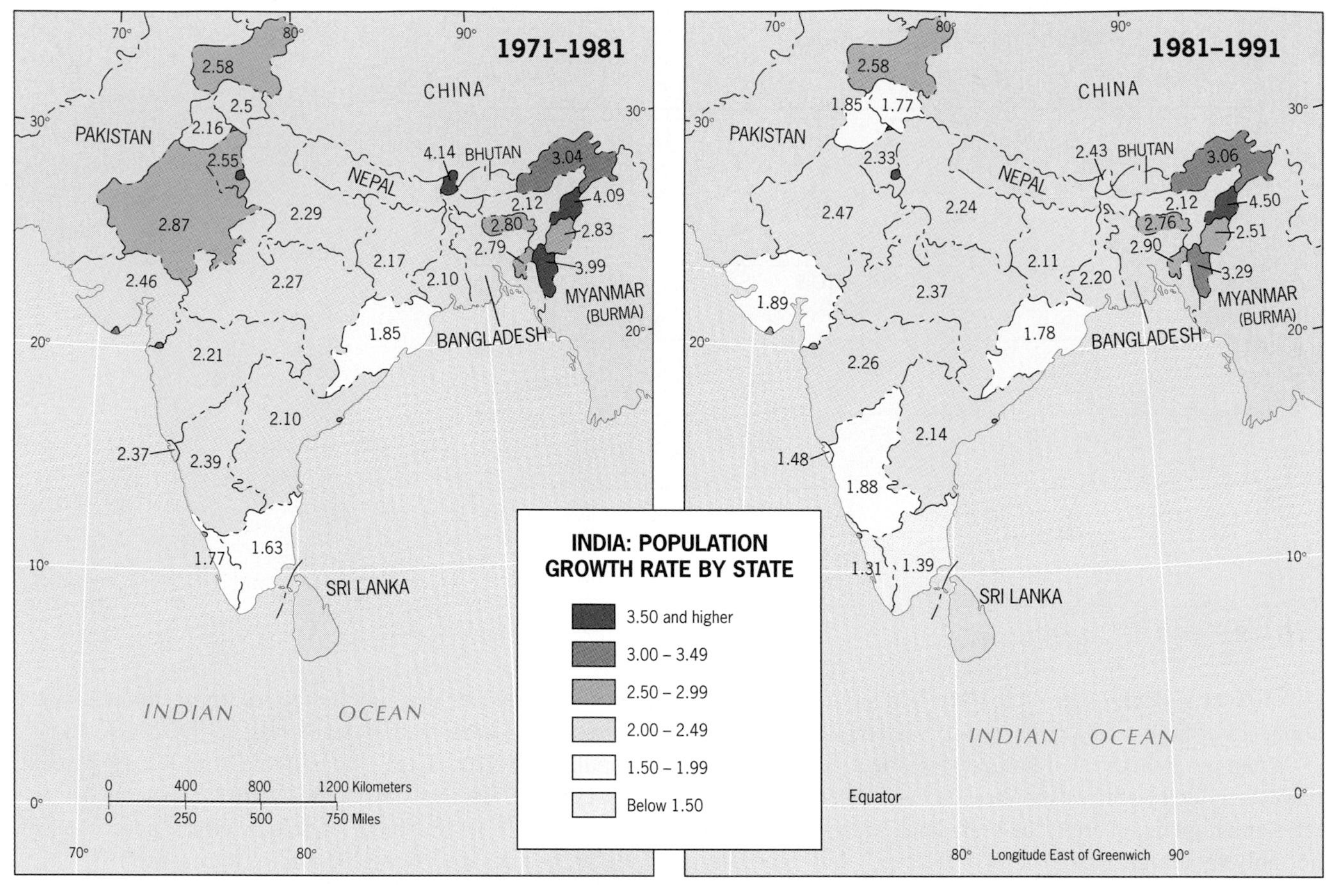

FIGURE 8-12

so large that even a declining rate of natural increase continues to add ever greater numbers to its total. In Figure 8-11, we see that while the decadal rate of increase dropped from 2.22 to 2.11 between 1970 and 1990, the millions added grew from 135 in the 1970s to 161 during the 1980s. The rate of natural increase for the 1990s is estimated to have been 1.88 percent, adding a still larger number: 175 million, taking the total past 1 billion during 1999. If India has indeed entered the third stage of the demographic transition, it will not feel its effects for some time.

Some population geographers theorize that all countries' populations will eventually stabilize at some level, just as Europe's did. Certain governments, notably China's, have instituted regulations to limit family size, but this policy is more easily implemented by dictatorships than democracies. And even if such stabilization were just one doubling time away in India, the country still would have an astronomical total of more than 2 billion residents.

Geography of Demography

Statistics for a country as large as India tend to lose their usefulness unless they are put in geographic context. In its demographic as in so many of its other aspects, there is not just one India but several regionally different and distinct Indias. Figure 8-12 takes population growth down to the State level and provides a comparison between the census periods of 1971–1981 and 1981–1991.

During the period from 1971 to 1981, the highest growth rates were recorded in the States of the northeast and northwest, and only three States had growth rates below 2 percent. But between 1981 and 1991, no fewer than eight States, including Goa, had growth rates below 2 percent. Comparing these two maps tells us that little has changed in India's heartland, where the populous States of Uttar Pradesh and Bihar show slight decreases but West Bengal and Madhya Pradesh display equally slight increases. The most important reductions in rates of natural increase are recorded in the northwest, west, and south. In the tip of the peninsula, Kerala and neighboring Tamil Nadu have growth rates comparable to that of the region's leading country, Sri Lanka.

Figure 8-12 also suggests a broad contrast between eastern and western India. We will soon see that this is not just a matter of demography.

Urbanization

When people move to the cities, they tend to have smaller families. Incentives that cause rural couples to have many children do not prevail in the cities, where living space is highly confined. Urbanization, therefore, contributes to reduced rates of natural increase in populations.

India is not yet a highly urbanized society, but we should not lose sight of the dimensions. Only 28 percent of India's population was urbanized in 2002, but that 28 percent represents more than 290 million people—more than the entire population of the United States.

And India's rate of urbanization is on the upswing. People by the hundreds of thousands are arriving in the already-teeming cities, swelling urban India by about 5 percent annually, almost three times as fast as the overall population growth. Not only do the cities attract as they do everywhere; many villagers are driven off the land by the desperate conditions in the countryside. As villagers manage to establish themselves in Mumbai or Kolkata or Chennai, they help their relatives and friends to join them in squatter settlements that often are populated by newcomers from the same area, bringing their language and customs with them and cushioning the stress of the move.

As a result, India's cities display staggering social contrasts. Squatter shacks without any amenities at all crowd against the walls of modern high-rise apartments and condominiums. Hundreds of thousands of homeless roam the streets and sleep in parks, under bridges, on sidewalks. As crowding intensifies, social stresses multiply. Disorder never seems far from the surface; sporadic rioting, often attributable to rootless urban youths unable to find employment, has become commonplace in India's cities.

India's modern urbanization has its roots in the colonial period, when the British selected Calcutta (Kolkata), Bombay (Mumbai), and Madras (Chennai) as regional trading centers and fortified ports. Madras was fortified as early as 1640; Bombay (1664) had the situational advantage of being the closest of all Indian ports to Britain; and Calcutta (1690) lay on the margin of India's largest

FROM THE FIELD NOTES
"Searing social contrasts abound in India's overcrowded cities. Even in Mumbai (Bombay), India's most prosperous large city, hundreds of thousands of people live like this, in the shadow of modern apartment buildings. Within seconds we were surrounded by a crowd of people asking for help of any kind, their ages ranging from the very young to the very old. Somehow this scene was more troubling here in well-off Mumbai than in Kolkata (Calcutta) or Chennai (Madras), but it typified India's urban problems everywhere."

Among the Realm's Great Cities . . .

Mumbai (Bombay)

Another historic name is disappearing from the map: Bombay. In precolonial times, fishing folk living on the seven small islands at the entrance to this harbor named the place after their local Hindu goddess, Mumbai. The Portuguese, first to colonize it, called it Bom Bahia, "Beautiful Bay." The British, who came next, corrupted both to Bombay, and so it remained for more than three centuries. Now local politics has taken its turn. Leaders of a Hindu nationalist party that control the government of the State of which Bombay is the capital, Maharashtra, passed legislation to change its name back to Mumbai. In 1996, India's federal government approved this change.

Mumbai's 19.4 million people make this India's largest city. Maharashtra is India's economic powerhouse, the State that leads the country in virtually every respect. Locals dream of an Indian Ocean Rim of which Mumbai will be the anchor.

As such, Mumbai is a microcosm of India, a burgeoning, crowded, chaotic, fast-moving agglomeration of humanity (see photo, p. 409). The gothic-Victorian architecture of the city center is a legacy of the British colonial period. Shrines, mosques, temples, and churches evince the pervasive power of religion in this multicultural society. Street signs come in a bewildering variety of scripts and alphabets. Creaking double-decker buses compete with oxwagons and handcarts on the congested roadways. The throngs on the sidewalks—businesspeople, holy men, sari-clad women, beggars, clerks, homeless wanderers—spill over into the streets.

Mumbai is an urban agglomeration of city-sized neighborhoods, each with its own cultural landscape. The seven islands have been connected by bridges and causeways, and the resulting Fort District still is the center of the city, with many of its monuments and architectural landmarks. Marine Drive leads to the wealthy Malabar Hill area, across Back Bay. Northward lie the large Muslim districts, the Sikh neighborhoods, and other ethnic and cultural enclaves. Beyond are some of the world's largest and poorest squatter settlements.

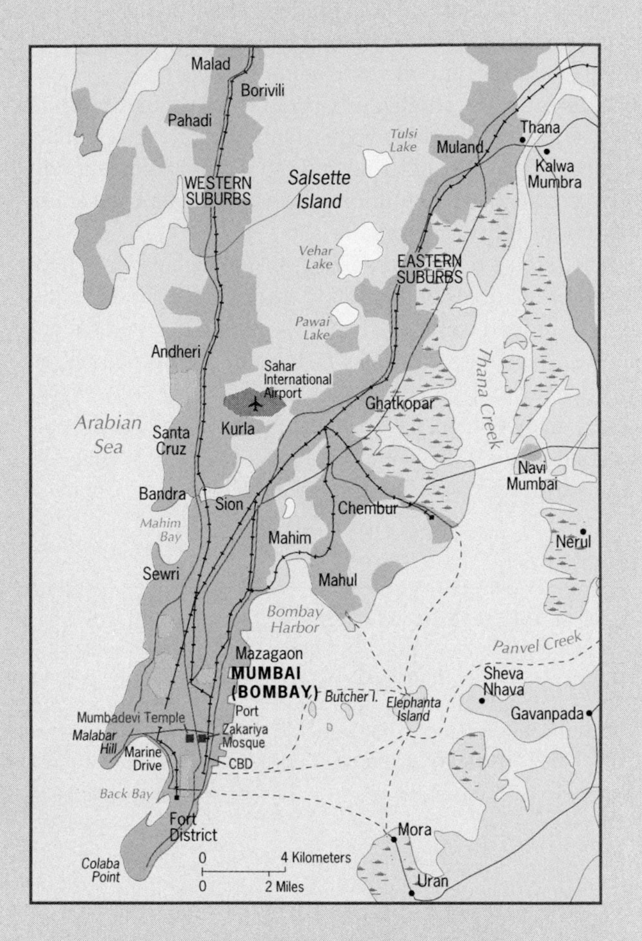

It is situation, not site, that gave Bombay primacy in South Asia. The opening of the Suez Canal in 1869 made Bombay the nearest Indian port to Europe. Today, Mumbai is the focus of India's fastest-growing economic zone—not yet a tiger but on the move.

population cluster and had the most productive hinterland, to which the Ganges Delta's countless channels connected it. This natural transport network made Calcutta an ideal colonial headquarters, but the population of Bengal was often rebellious. In 1912 the British moved their colonial government from Calcutta to the safer interior city of New Delhi, built adjacent to the old Mogul capital of Delhi.

Figure 8-9 displays the distribution of major urban centers in India. Except for Delhi-New Delhi, the largest cities have coastal locations: Kolkata (Calcutta) dominates the east, Mumbai (Bombay) the west, and Chennai (Madras) the south. But urbanization also has expanded in the interior, notably in the core area. The surface interconnections among India's cities remain inadequate (notably the road network), but an Indian urban system is emerging.

Economic Geography

If India has faced problems in its great effort to achieve political stability and national cohesion, these problems are more than matched by the difficulties that lie in the way of economic growth and development. The large-scale factories and power-driven machinery of the colonial powers wiped out a good part of India's indigenous industrial base. Indian trade routes were taken over. European innovations in health and medicine sent the rate of population growth soaring, without introducing solutions for the many problems this spawned. Surface communications improved and food distribution systems became more efficient, but local and regional food shortages occurred (and still do) as droughts frequently caused crop failures. Today, nearly half of India's one-billion-plus people live in abject poverty, and the prospects of reducing that high level of human misery anytime soon are not encouraging. (Yet even with its modest annual per-capita GNP [U.S. $440], the sheer size of India's population has created a big overall economy—the world's sixth largest according to the latest rankings.)

Among the Realm's Great Cities . . .

Kolkata (Calcutta)

Calcutta is synonymous with all that can go wrong in large cities: poverty, dislocation, disease, pollution, crime, corruption. To call a city the Calcutta of its region is to summarize urban catastrophe. West Bengal's leaders had this in mind when they proposed that the city's local name, Kolkata, be restored. The city got its dreadful reputation (as well as its Anglicized name) during colonial times, when plague, malaria, and other diseases claimed countless thousands in legendary epidemics. The British chose the site, 80 miles (130 km) up the Hooghly River from the Bay of Bengal and less than 30 feet (9 m) above sea level, not far from some unhealthy marshes but well placed for commerce and defense. When the British East India Company was granted freedom of trade in the populous hinterland, Calcutta's heyday began; when (in 1772) the British made it their colonial capital, the city prospered. The British sector of the city was drained and raised, and so much wealth accumulated here that Calcutta became known as the "city of palaces." Outside the British town, rich Indian merchants built magnificent mansions. Beyond lay neighborhoods that were often based on occupational caste, whose names are still on the map today (such as Kumartuli, the potters' district). Almost everywhere, on both banks of the Hooghly, lay the huts and hovels of the poorest of the poor. Searing social contrasts characterized Calcutta.

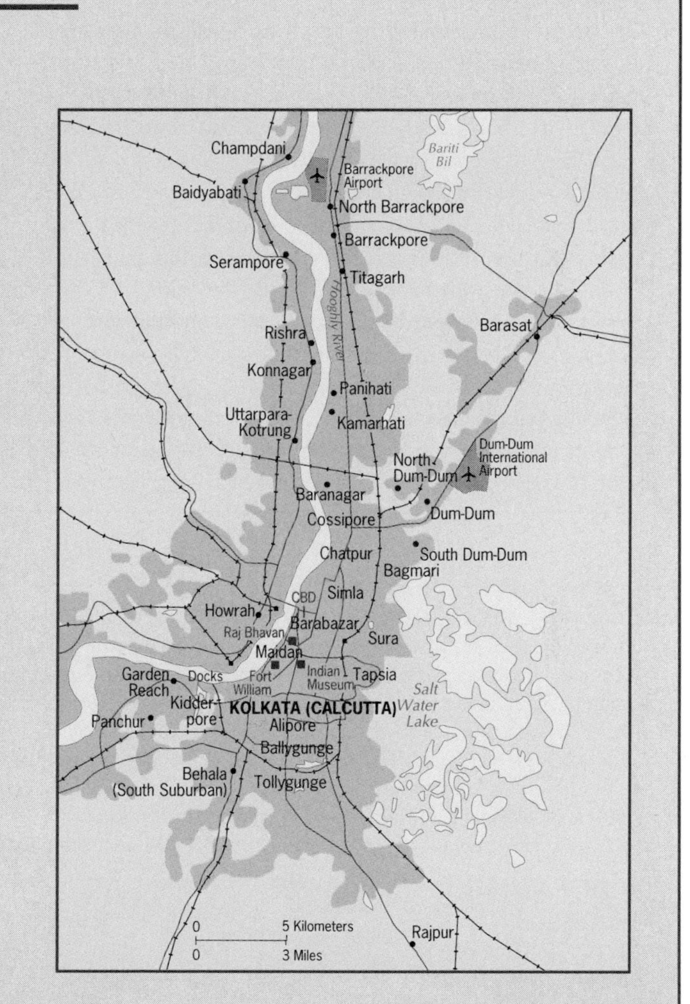

The twentieth century was not kind to Calcutta. In 1912 the British moved their colonial capital to New Delhi. The 1947 partition that created Pakistan also created then-East Pakistan (now Bangladesh), cutting off a large part of Calcutta's hinterland and burdening the city with a flood of refugees. The Indian part of the city had arisen virtually without any urban planning, and the influx created almost unimaginable conditions. Today, Kolkata counts 13.5 million residents, including as many as 500,000 homeless. Beyond the façade of the downtown lies what may be the sickest city of all.

Among the Realm's Great Cities . . .

Delhi New and Old

Fly directly over the Delhi–New Delhi conurbation into the regional airport, and you may not see the place at all. A combination of smog and dust creates an atmospheric soup that can limit visibility to a few hundred feet for weeks on end. Relief comes when the rains arrive, but Delhi's climate is mostly dry. The tail-end of the wet monsoon reaches here during late June or July, but altogether the city gets only about 25 inches (60 cm) of rain a year.

When the British colonial government decided to leave Calcutta and build a new capital city adjacent to Delhi, conditions were different. South of the old city lay a hill about 50 feet (15 m) above the surrounding countryside, on the right bank of the southward-flowing Yamuna tributary of the Ganges. Compared to Calcutta's hot, swampy environment, Delhi's was agreeable. In 1912, it was not yet a megacity. Skies were mostly clear. Raisina Hill became the site of a New Delhi.

This was not the first time rulers chose Delhi as the seat of empire. Ruins of numerous palaces mark the passing of powerful kingdoms. But none brought to the Delhi area the transformation the British did. In 1947 the Indian government decided to keep its headquarters here. In 1970 the metropolitan population exceeded 4 million. By 2002 it was 13.6 million.

Delhi is popular as a seat of government for the same reason as its present expansion: the city has a fortuitous relative location. The regional topography creates a narrow corridor through which all land routes from northwestern India to the North Indian Plain must pass, and Delhi lies in this gateway. Thus the twin cities not only contain the government functions; they also anchor the core area of this populous country.

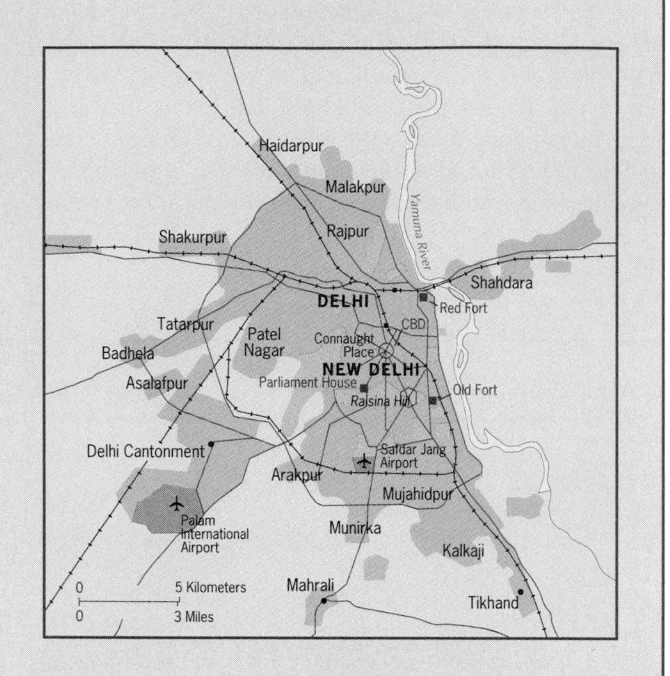

Old Delhi once was a small, traditional, homogeneous town. Today Old and New Delhi form a multicultural, multifunctional urban giant.

FROM THE FIELD NOTES

"Negotiating the traffic in India's chaotic cities is always a challenge as creaking buses, vintage cars, taxis large and small, scooters, and bicycles mingle in a mass of movement that sometimes makes streets look like rivers of humanity. I watched the scene from a vantage point on Nungambakkam Road in Chennai (formerly Madras), impressed that so much high-speed congestion produced no accidents. 'You have to have a sense of humor to be part of all that,' said a man who stopped to chat. 'Just down the street, make a left and then a right, and you'll see that even the authorities do.' He was referring to the sign appealing to drivers to obey the seemingly nonexistent rules. India is in dire need of highway and road improvement; in cities like Chennai, the road system is pretty much the way the British left it in the 1940s when India became independent."

Agriculture

India's underdevelopment is nowhere more apparent than in its agriculture. Traditional farming methods persist, and yields per acre and per worker remain low for virtually every crop grown under this low-technology system. Moreover, the transportation inefficiencies of the traditional farming system hamper the movement of agricultural commodities. In 2001, less than 50 percent of India's 600,000 villages were accessible by motorable road, and today animal-drawn carts still outnumber motor vehicles nationwide.

As the total population grows, the amount of cultivated land per person declines. Today, this physiologic density is 1615 per square mile (625 per sq km). However, this is nowhere near as high as the physiologic density in neighboring Bangladesh, where the figure is more than twice as great (3623 and 1399, respectively). But India's farming is so inefficient that this comparison is deceptive. Fully two-thirds of India's huge working population depends directly on the land for its livelihood, but the great majority of Indian farmers are poor and cannot improve their soils, equipment, or yields. Those areas in which India has substantially modernized its agriculture (as in Punjab's wheat zone) remain islands in a sea of agrarian stagnation.

This stagnation has persisted in large measure because India, after independence, failed to implement a much-needed nationwide land reform program. Roughly one-quarter of India's entire cultivated area is still owned by less than 5 percent of the country's farming families, and little land redistribution was taking place. Perhaps half of all rural families own either as little as an acre or no land at all. Independent India inherited inequities from the British colonial period, but the individual States of the federation would have had to cooperate in any national land reform program. As always, the large landowners retained considerable political influence, so the program never got off the ground.

To make matters worse, much of India's farmland is badly fragmented as a result of local rules of inheritance, thereby inhibiting cooperative farming, mechanization, shared irrigation, and other opportunities for progress. Not surprisingly, land consolidation efforts have had only limited success except in the States of Punjab, Haryana, and parts of Uttar Pradesh, where modernization has gone farthest. Official agricultural development policy, at the federal and State levels, has also contributed to India's agricultural malaise and the uneven distribution of progress. Unclear priorities, poor coordination, inadequate information dissemination, and other failures have been reflected in the country's disappointing output.

It is instructive to compare Figure 8-13, showing the distribution of crop regions and water supply systems in India, with Figure I-7, which shows mean annual precipitation in India and the world. In the comparatively dry northwest, notably in the Punjab and neighboring areas of the upper Ganges, wheat is the leading cereal crop. Here, India has made major gains in annual production through the introduction of high-yielding grain varieties developed under the banner of the Green Revolution, the international research program that played a key role in overcoming the food crises of the 1960s. These "miracle crops" led to the expansion of cultivated areas, the construction of new irrigation systems, and the more intensive use of fertilizer (a mixed blessing, for fertilizers tend to be expensive and the "miracle" crops are more heavily dependent on them).

FROM THE FIELD NOTES

"Travel into rural India, and you soon grasp the realities of Indian agriculture. Human and animal labor predominate; farming methods and equipment are antiquated. At this village I watched the women feed scrawny sticks of sugarcane into a rotating press turned slowly by a pair of bullocks prodded by a boy. In this family enterprise, the grandfather is responsible for boiling the liquid sugar in a huge pan over the fire pit (upper right). The dehydrated sugar is then molded into large round cakes to be sold at the local market. As I observed this slow and inefficient process, I understood better why India is among the world's leaders in terms of sugarcane *acreage*—and among the world's last in terms of *yields*."

Toward the moister east, and especially in the wet-monsoon-drenched areas (Fig. 8-13), rice becomes the dominant staple. About one-fourth of India's total farmland lies under rice cultivation, most of it in the States of Assam, West Bengal, Bihar, Orissa, and eastern Uttar Pradesh and along the Malabar-Konkan coastal strip facing the Arabian Sea. These areas receive over 40 inches (100 cm) of rainfall annually, and irrigation supplements precipitation where necessary.

India devotes more land to rice cultivation than any other country, but yields per acre remain among the world's lowest—despite the introduction of "miracle rice." Nevertheless, the gap between demand and supply has narrowed, and in the late 1980s India actually exported some grain to Africa as part of a worldwide effort to help refugees there.

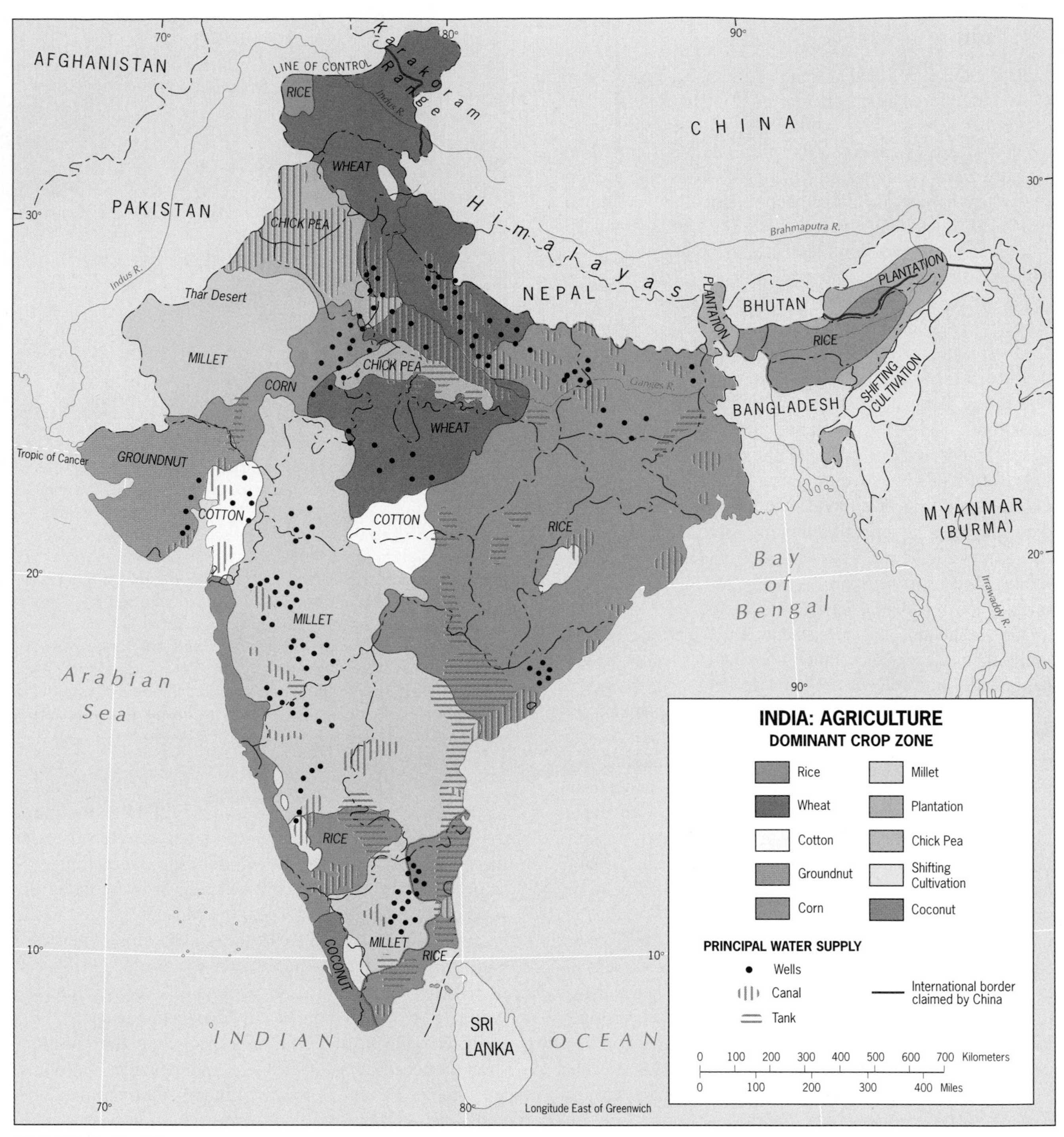

FIGURE 8-13

The situation remains precarious, however. As the population map (Fig. I-9) shows, there is a considerable degree of geographic covariation between India's rice-producing zones and its most densely populated areas. India is just one poor-harvest year away from another food crisis.

Figure 8-13 simplifies India's complex agricultural mosaic. It shows the dominant crop in the country's agricultural regions without additional detail. Obviously, for example, wheat is not grown throughout mountainous Kashmir, but where crops are grown, wheat is the most widespread. Again, rice dominates in the green-colored areas but does not exclude all other crops.

Subsistence remains the fate of tens of millions of Indian villagers who cannot afford fertilizers, cannot cultivate the new and more productive strains of rice or wheat, and cannot escape the cycle of poverty. Perhaps as many as 175

The arrival of the annual wet-monsoon rains transforms the Indian countryside. By the end of May, the paddies lie parched and brown, dust chokes the air and it seems that nothing will revive the land. Then the rains begin, and blankets of dust turn into layers of mud. Soon the first patches of green appear on the soil, and by the time the monsoon ends all is green. The photograph on the left, taken just before the onset of the wet monsoon in the State of Goa, shows the paddies before the rains begin; three months later the countryside looks as on the right.

million of these people do not even own a plot of land and must live as tenants, always uncertain of their fate. This is the enduring reality against which optimistic predictions of improved nutrition in India must be weighed. True, rice and wheat yields have increased at slightly more than the rate of population growth since the Green Revolution. But food security remains elusive, and India continues to face the risks inherent in the ever-growing needs of its burgeoning population.

Industrialization

Despite the problems its farmers face, agriculture must be the foundation for development in India. Agriculture employs approximately two-thirds of the workers, generates most of the government's tax revenues, contributes many of its chief exports by value (cotton textiles, tea, fruits and vegetables, jute products, and leather goods all rank high), and produces most of the money the country can spend in other sectors of the economy. Add to this the compelling need to grow more and more food crops, and we can understand India's heavy investment in agriculture.

In 1947 India inherited the mere rudiments of an industrial framework. After more than a century of British control over the economy, only 2 percent of India's workers were engaged in industry, and manufacturing and mining combined produced only about 6 percent of the national income. Textile and food processing were the dominant industries. Although India's first iron-making plant opened in 1911 and the first steel mill began operating in 1921, the initial major stimulus for heavy industrialization came after the outbreak of World War II. Manufacturing was concentrated in the largest cities: Kolkata (Calcutta) led, Mumbai (Bombay) was next, and Chennai (Madras) ranked third.

The geography of manufacturing still reflects those beginnings, and industrialization in India has proceeded slowly, even after independence (Fig. 8-14). Kolkata now anchors India's eastern industrial region—the Bihar-Bengal District—where jute manufactures dominate, but cotton, engineering, and chemical industries also operate. On the nearby Chota-Nagpur Plateau to the west, coal-mining and iron and steel manufacturing have developed.

On the opposite side of the subcontinent, two industrial areas dominate the western manufacturing region: one is centered on Mumbai and the other on Ahmadabad. This dual region, lying in Maharashtra and Gujarat States, specializes in cotton and chemicals, with some engineering and food processing. Cotton textiles have long been an industrial mainstay in India, and this was one of the few industries to benefit from the nineteenth-century economic order the British imposed. With the local cotton harvest, the availability of cheap yarn, abundant and inexpensive labor, and the power supply from the Western Ghats' hydroelectric stations, the industry thrived, so that today India outranks Britain itself in the volume of its exports.

As we note in the next section, Mumbai and Maharashtra State have emerged as India's first modern, globally connected, export-oriented, service-industry-supported economic region. Its early start, capital accumulation, better educated labor force, and relative location all contributed to this development. Mumbai and environs cannot yet match the economic power of comparable-sized complexes on the Pacific Rim, but it is the vanguard of India's modernization.

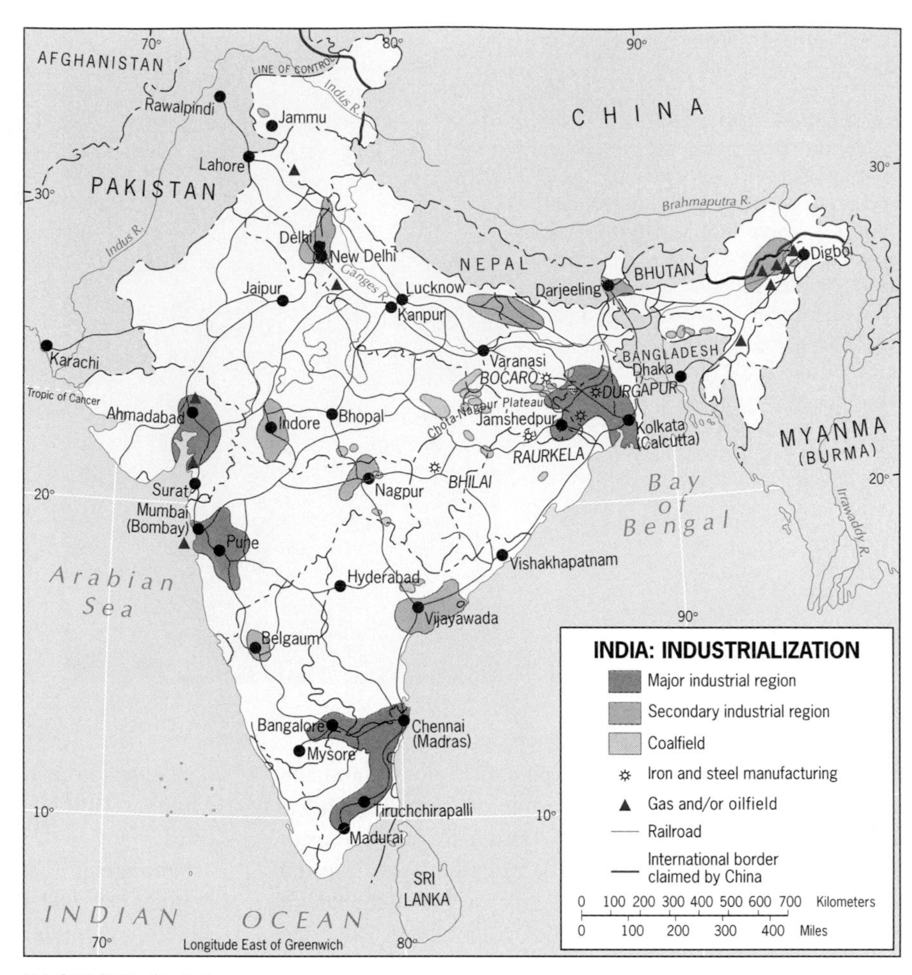

FIGURE 8-14

The southern industrial region consists chiefly of a set of linear, city-linking corridors focused on Chennai, specializing in textile production and light engineering activities. Today, all of India's manufacturing regions are increasing their output of ready-to-wear garments—another legacy of the early development of cotton textiles. Clothing has become India's second-leading export by value; the production of gems and jewelry, another growing specialization, ranks first.

An important development in the south is occurring in and around Bangalore in the State of Karnataka, India's

Bangalore, near the meeting point of the States of Karnataka, Andhra Pradesh, and Tamil Nadu, is the center of what has come to be called India's "Silicon Plateau." This is the heart of the country's high-tech development, and from here many skilled workers have made their way to Europe and the United States, where their abilities earn them more than at home. Even so, Bangalore is one of India's highest-income locations, as can be seen from the appearance of this workforce leaving its shift at a branch of International Instruments. Why Bangalore? The city has great centrality with good connections in all directions; it lies more than 3000 feet (900 m) above sea level, making for tolerable summers; it has long been an educational center and has highly skilled residents; it is a cultural meeting (and competing) place of Kannada-, Tamil-, and Telugu speakers, making English a needed common language; and its State government has encouraged the high-tech sector through critical infrastructure improvements.

"Silicon Plateau." Several hundred software companies are based here, one-third of them foreign with names such as IBM, Texas Instruments, and Motorola (see photo, p. 416). What attracts them, and makes them profitable, is the low cost of India's software engineers, who earn about one-fifth of what their foreign colleagues earn. The emigration of technicians is becoming a problem, but replacements are still plentiful. Bangalore is proof of India's potential in the modern world.

Despite some imbalances and inefficiencies, India's industrial resource base is well endowed. Limited high-quality coal deposits are exploited in the Chota-Nagpur area. In combination with large lower-grade coalfields elsewhere, the country's total output is high enough to rank it among the world's ten leading coal producers. With no known major petroleum reserves (some oil comes from Assam, Gujarat, Punjab, and offshore from Mumbai), India must spend heavily on fuel imports every year. Major investments have been made in hydroelectric plants, especially multipurpose dams that provide electricity, enhance irrigation, and facilitate flood control. India's iron ore deposits in Bihar (northwest of Calcutta) and Karnataka (in the heart of the Deccan) may rank among the largest in the world. Jamshedpur, located west of Calcutta in the eastern industrial region, has become India's steel-making and metals-fabrication center. Yet India still exports iron ore as a raw material to the higher-income industrialized countries, mainly Japan. For low-income, revenue-needy countries, entrenched practices are difficult to break.

India East and West

The most commonly cited, and most clearly evident, regional division of India is between north and south. The north is India's heartland, the south its Dravidian appendage; the north speaks Hindi as its *lingua franca*, the south prefers English over Hindi; the north is bustling and testy, the south seems slower and less agitated.

But there is another, as yet less obvious, but potentially more significant divide across India. In Figure 8-14, draw a line from Lucknow, on the Ganges River, south to Madurai, near the southern tip of the peninsula. To the west of this line, India is showing signs of economic progress, the kind of economic activity that has brought Pacific Rim countries such as Thailand and Indonesia a new life. To the east, India has more in common with less promising countries also facing the Bay of Bengal: Bangladesh and Myanmar (Burma).

As with other regional divides, there are exceptions to our east-west delineation. Indeed, our map seems to suggest that much of India's industrial strength lies in the east. But what the map cannot reveal is the profitability of those industries. True, the east is rich in iron and coal, but the heavy industries built by the state in the 1950s are now outdated, uncompetitive, and in decline. The hinterland of Kolkata now contains India's Rustbelt. The government keeps many industries going but at a high cost. Old industries, such as carpetmaking and cottonweaving, continue to use child labor to remain viable. The State of Bihar represents the stagnation that afflicts much of India east of our line: by several measures it ranks among the poorest of the 28 States.

Compare this to western India. The State of Maharashtra, the hinterland of Mumbai, leads India in many categories, and Mumbai leads Maharashtra. Many smaller, private industries have emerged here, manufacturing goods ranging from umbrellas to satellite dishes and from toys to textiles. Across the Arabian Sea lie the oil-rich economies of the Arabian Peninsula. Hundreds of thousands of workers from western India have found jobs there, sending money back to families from Punjab to Kerala. More importantly, many have used their foreign incomes to establish service industries back home. Outward-looking western India, in contrast to the inward-looking east, has begun to establish other ties to the outside world. Satellite links have enabled Bangalore to become the center of a growing software-producing complex reaching world markets. The beaches of Goa, the small State immediately to the south of Maharashtra, appeal to the tourist markets of Europe. This is, in fact, a classic case of **intervening opportunity** because resorts have sprung up along Goa's coast, and European tourists who once went to the more distant Maldives and Seychelles are coming to Goa. Maharashtra's economic success also has spilled over into Gujarat to the north, and even landlocked Rajasthan (the next State to the north) is experiencing the beginnings of what, by Indian standards, is a boom.

15

The boom has created political problems, however. Not only is Maharashtra State a rising economic power; it also is the base of a strong Hindu nationalist political movement whose leaders object to foreign intrusions and have blocked major development projects and other enterprises. They halted a huge industrial scheme about halfway through and closed a fast-food operation that they deemed incompatible with local culture. Such clashes between foreign interests and domestic traditions are not unique to India, but the rising tide of Hindu fundamentalism has uncertain prospects and unsettles investors to whom India remains a high-risk calculation.

Nevertheless, India's east-west divide shows a growing contrast that puts the west far ahead. The hope is that Maharashtra's success will spread northward and southward along the Arabian Sea coast and will ultimately diffuse eastward as well. But for this to happen, India will have to bring its population spiral under control.

India's Contested North

So pressing are India's social problems, its economic disparities, and its disputes with Pakistan that other politico-geographical issues are sometimes forgotten. Prominent among these other issues are (1) India's support for those Tibetans who have resisted China's absorption of their homeland, and (2) China's claims to Indian territory in India's northwest and northeast. At the turn of the twenty-first century, India accommodated some 300,000 Tibetan refugees and continued to give public backing to the Tibetan Buddhists' ousted leader, the Dalai Lama. This posture damaged India's relations with China but underscored India's position on religious freedom. We can see some (but at this scale, not all) of China's territorial claims to Indian land in Figures 8-8 and 8-9. The Chinese publish maps that show most of the Indian State of Arunachal Pradesh and a corner of Jammu and Kashmir as Chinese territory. India's north once was a colonial frontier; today it is a frontier in a new guise.

BANGLADESH: PERSISTENT POVERTY

On the map of South Asia, Bangladesh looks like another State of India: the country occupies the area of the double delta of India's great Ganges and Brahmaputra rivers, and India almost completely surrounds it on its landward side (Fig. 8-15). But Bangladesh is an independent country, born in 1971 after its brief war for independence against Pakistan, with a territory about the size of Wisconsin. Today it is one of the poorest and least developed countries on Earth, with a population of 133 million that is growing at an annual rate of 1.9 percent.

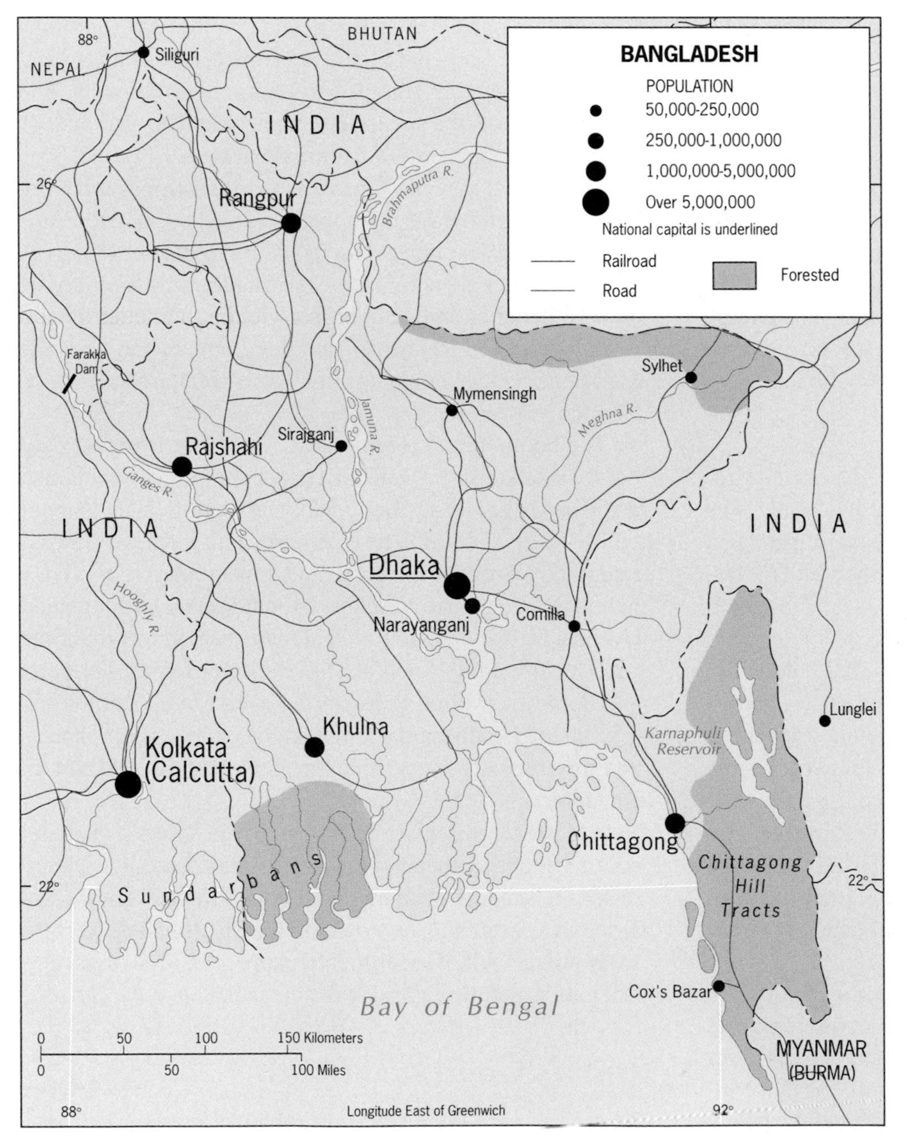

FIGURE 8-15

Natural Hazards

In the spring of 1991, Bangladesh was struck by yet another in an endless series of natural disasters: a devastating hurricane (or cyclone, as these tropical storms are called in this part of the world) that killed as many as 150,000 people (we will never know the exact toll). The storm, on a curving northward path across the Bay of Bengal, pushed a surging wall of water nearly 20 feet high across the islands and flatlands of the delta and swept most of the southeastern port city of Chittagong off the map. The storm surge forced its way well inland along the winding channels of the Ganges-Brahmaputra Delta, causing death and destruction far from the exposed coastlands along the bay. When the waters receded, the bodies of countless people and animals were carried out to sea, later to wash up on the beaches. It was a catastrophe of unimaginable proportions—but it was not, by far, the worst calamity that Bangladesh has suffered. During the twentieth century, eight of the ten deadliest natural disasters in the entire world struck this single country. What makes Bangladesh so vulnerable to these **natural hazards**? 16

Let us look at Figure 8-15 again. The land of Bangladesh lies just barely above sea level; the deltaic plain of the Ganges-Brahmaputra is a labyrinth of stream channels. Only in the extreme east and southeast do these flatlands yield to hills and mountains. The delta's alluvial soils are extremely fertile, and every available patch of it is under crops: rice and wheat for subsistence, jute and tea for cash. The rivers' annual floods bring silt to renew the farmlands' fertility; at the seaward margins of the delta, the silt piles up to form new islands. Even as this new land builds up, people move in to farm it. The crush of ever more mouths to feed compels this migration.

Now consider the inverted-funnel shape of the Bay of Bengal (Fig. 8-1). Tropical cyclones form often in this warm-water, humid-air environment (in contrast to the Arabian Sea, with its drier air and desert coasts, on the western side of India). The cyclones frequently move northward along a rightward-curving track. As they do, the water that piles up ahead of the storm has no place to go: the bay becomes ever narrower and shallower. And so, time and again, storm surges smash into the delta, sweeping people, livestock, and crops from the land. After the storm abates, the returning outrush of water causes added devastation as the delta's normally placid channels become raging torrents.

Unlike the comparatively wealthy Dutch, the Bangladeshis cannot combat their environmental enemy by building huge seawalls and floodgates. Flood and storm warning systems are insufficient; escape plans and routes are inadequate. A program was recently begun to construct concrete, storm-proof shelters on pillars, to which trapped villagers might flee. The available shelters saved few when the 1991 cyclone struck. Even this is too costly for Bangladesh to bear.

Sharing the Ganges

Bangladesh suffers from too much water—and too little. When the mighty Ganges rises during the wet monsoon, the danger is flood. But when the dry season comes, the risk is drought as its distributaries (see Fig. 8-15) run dry. That is why Bangladesh objected to India's construction of the Farakka Dam across the Ganges just before the river leaves India, to divert water toward burgeoning Kolkata. The result of this barrier was a reduction of irrigation and drinking water in the Khulna area, salt-water intrusion from the Bay of Bengal, loss of farmlands and forests—and a major illegal migration of uprooted farmers from Khulna westward into the Indian State of West Bengal. In 1997, Bangladesh and India signed a treaty that stipulates the division of Ganges waters, easing a long-term problem in the region.

Problems in South Asia's Eastern Region

Bangladesh's economic condition is reflected by its GNP per capita (a mere U.S. $350, ranking it among the world's lowest-income economies) and by its very low levels of urbanization (only 20 percent) and industrialization. This is a land of subsistence farmers with one of the highest physiologic densities in the world (3623 people per square mile/1399 per sq km). But higher-yielding varieties of rice and the introduction of wheat in the crop rotation where climate allows it have improved diets and food security. Nevertheless, nutrition is barely adequate and diets remain unbalanced.

Dhaka, the centrally situated capital, and the southeastern port of Chittagong are the only urban centers of consequence in this dominantly rural society. Infrastructure is weak: there are only single road and railway bridges across the Ganges River, both in the west near the Indian border. A railroad bridge across the Brahmaputra (Jamuna) River, which will link Rangpur to Mymensingh and the east, was still under construction in 2001. When you travel from Dhaka to any town some distance away, you should be prepared for crawling road traffic and time-consuming ferry transfers. Indeed, much of the country can be reached only by boat, thousands of which ply the many waterways.

Although population growth is Bangladesh's prime challenge, we must not lose sight of the global context. For example, a child born in Bangladesh will consume, during an equal life span, only about 3 percent of what a U.S.-born child will consume in food, energy, minerals, and other natural resources. Put another way, a single American child will consume what 33 Bangladeshi children do. Yes, Bangladesh faces a population crush, but populous Bangladesh strains the world's resources far less than a high-income country of the same dimensions.

Another challenge for Bangladesh lies in its relations with India, and not just over issues involving water sources. India supported Bangladesh during its 1971 war for independence with Pakistan, but Bangladesh remains a dominantly Muslim country with a sizeable (about 10 percent) Hindu minority. Figure 8-9 shows that Bangladesh adjoins the Indian State of West Bengal along its entire western boundary, and West Bengal shares with Bangladesh the delta of the Ganges River (in fact, the *Bangla* in Bangladesh means Bengal). Bangladesh and India have been at odds over water supplies and cross-border migration, but other problems involve the interior. As the map shows, the territory of Bangladesh leaves but a narrow corridor between the States of northeastern India, including Assam, and the rest of the country. The easiest routes between India and its northeastern extremity lie across northern Bangladesh, but Dakha has refused to allow India such transit. This, in turn, has made India's rail links skirting northern Bangladesh vulnerable to terrorism, which remains a problem in this turbulent eastern region. Bangladesh has repeatedly used this issue as a bargaining chip in its water disputes with India. When India agreed to the 1997 treaty relating to the waters of the Ganges, the government in Dhaka wanted to reward India by permitting Indian road vehicles transit through the north, but opposition parties blocked this initiative.

Even Bangladesh itself has suffered from the social and political instability that is endemic in the hilly and mountainous areas of this eastern region of South Asia. For many years, an independence movement has been active in the Chittagong Hill Tracts, the forested hinterland of Chittagong.

The prospects for Bangladesh are bleak. All the disadvantages of the global periphery afflict this populous, powerless country where survival is the leading industry and all else is luxury.

THE MOUNTAINOUS NORTH

South Asia, as we noted earlier, is one of the world's most clearly defined geographic realms both physically and culturally. Walls of mountains stand between India and China—mountains that defy penetration even in this age of modern highways. And those mountains are more than barriers: fed by melting snow, South Asia's great life-giving rivers rise here, sustaining hundreds of millions in the valleys and plains far below. Control over those source areas has caused centuries of conflict, as reflected on the political map. In addition to the border disputes between India and China relating to Kashmir and Arunachal Pradesh, several smaller highland areas are in contention along the Indo-Chinese boundary between Kashmir and Nepal (see the areas marked Ⓒ on Fig. 8-1).

As Figure 8-1 shows, a tier of landlocked countries and territories lies across this mountainous northern zone. From Afghanistan in the west through Jammu and Kashmir and Nepal to Bhutan in the east, these isolated, remote, vulnerable entities are products of a long and complicated frontier history. The recent misfortunes of one of them, Afghanistan, and the disappearance (as a separate country) of another, Sikkim, underscores their vulnerability. India absorbed Sikkim, wedged between Nepal and Bhutan, in 1975 and made it one of its States. The kingdoms of Nepal and Bhutan, however, retain their independence.

Nepal and Bhutan

Nepal, which lies directly northeast of India's Hindu coreland, is the size of Illinois and contains a population of 25.1 million. It has three geographic zones (Fig. 8-16): a southern, subtropical, fertile lowland called the Terai; a central belt of Himalayan foothills with swiftly flowing streams and deep valleys; and the spectacular high Himalayas themselves (topped by Mount Everest) in the north. The capital, Kathmandu, lies in the east-central part of the country in an open valley of the central hill zone.

Nepal is materially poor but culturally rich. The Nepalese are a people of many sources, including India, Tibet, and interior Asia; about 90 percent are Hindu, and Hinduism is the country's official religion, but Nepal's Hinduism is a unique blend of Hindu and Buddhist ideals. Thousands of temples and pagodas ranging from the simple to the ornate grace the cultural landscape, especially in the valley of Kathmandu, the country's core area. Although over a dozen languages are spoken, 90 percent of the people also speak Nepali, a language related to Indian Hindi.

Nepal's problems are underdevelopment and centrifugal political forces. Limited living space and a large and increasing population have caused environmental degradation. Deforestation is particularly severe—over one-third of Nepal's alpine woodlands have been cut over into wastelands since the 1960s. The growing population of subsistence farmers has been forced to expand into higher-altitude wilderness zones for sufficient crop-raising space (on steep terraces) and to obtain the firewood that supplies most of Nepal's energy needs. But soil quality is poor throughout the uplands, new farms are soon abandoned after a few seasons of declining productivity, and land denudation accelerates. Moreover, the steep slopes and the awesome power of the wet-monsoon rains cause severe soil erosion in treeless areas, and so much silt is now transported out of the Himalayas that, according to some researchers, the process is heightening the risk of flooding in the crowded lower Ganges and Brahmaputra basins. With about half its farmland already abandoned to erosion and with 95 percent of its population engaged in subsistence agriculture (rice, corn, wheat, and millet), Nepal today faces a serious ecological crisis.

With the Himalayan peaks its main attraction, Nepal has a substantial tourist industry. But tourists' spending in

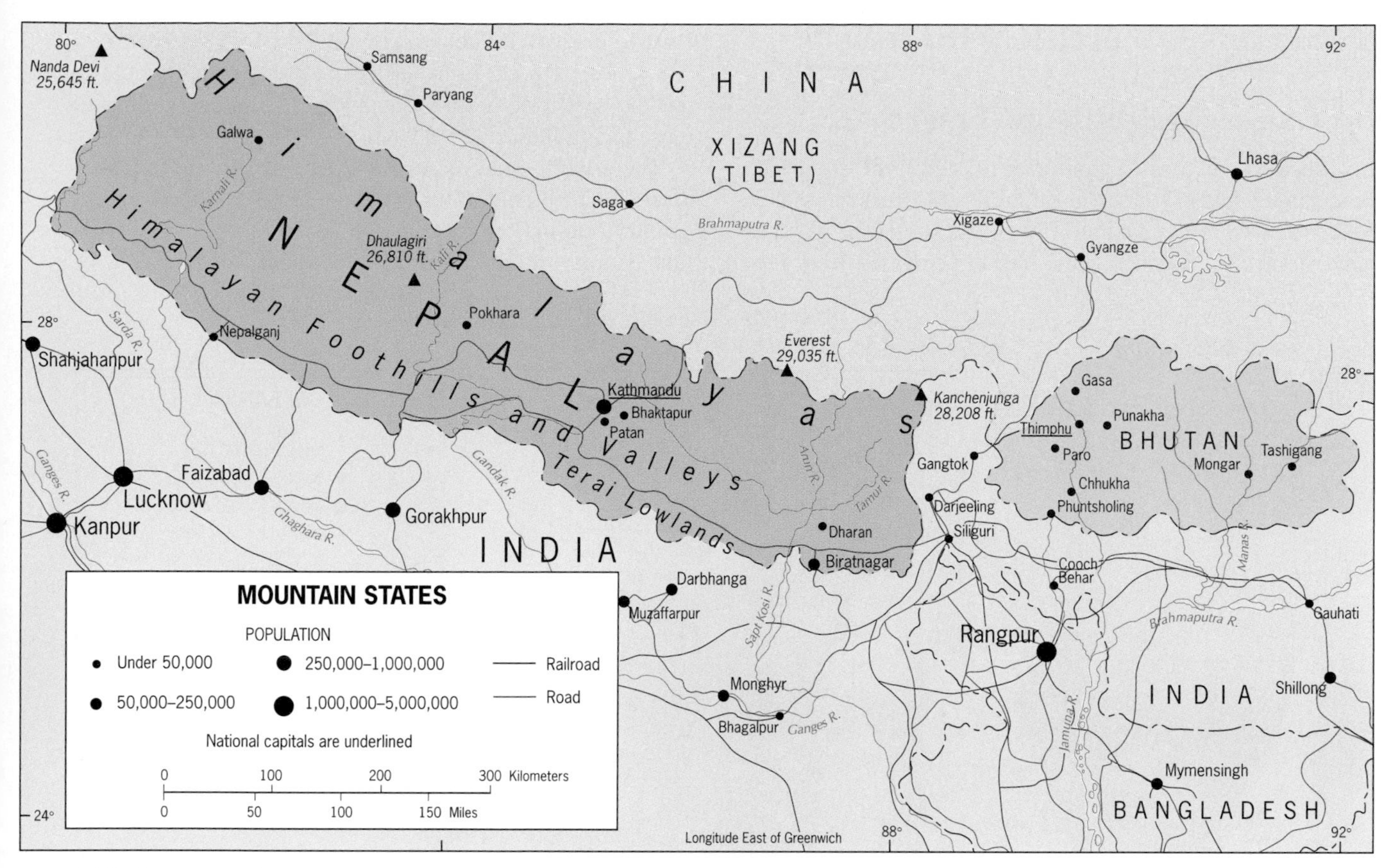

FIGURE 8-16

Nepal is relatively modest; more money is made outside the country by tour operators. Moreover, **tourism** has damaged the ecology: special expeditions have had to remove tons of rubbish left behind by trekkers and mountain climbers.

17

As the data in Table I-2 underscore, Nepal is a severely underdeveloped country; its per-capita GNP (U.S. $210) is the lowest in South Asia, well below even than that of Bangladesh. The country's infrastructure is weak, and regionalism is strong. In terms of political geography, support for the old monarchy in the core area was not enough to forestall a nationwide demand for more democracy that, during the late 1980s, created costly disruption. In 1991, democratic elections ushered in a new era. Although Nepal is now a constitutional monarchy, the end of absolute monarchy did not resolve Nepal's economic woes. Nepal needs integration and improved communications; the southern Terai zone, with its near-tropical lowlands resembling neighboring India, is a world apart from the hills of the central zone. And the peoples of the west have origins and traditions different from those in the east. Nepal is fearful of domination by the giant to the south, but even relations with nearby Bhutan have been problematic, especially because Nepal now has representative government whereas Bhutan continues to be an absolute monarchy. Landlocked, regionally fragmented, economically deteriorating, and culturally splintered, Nepal faces the future with many liabilities and few assets. Survival as a coherent state is its greatest challenge.

As noted earlier, until 1975 Nepal's eastern neighbor, Sikkim, was an independent country. But in that year an overwhelming majority of its people voted to join India. That leaves mountainous Bhutan, wedged between India and China's Tibet, as the only other buffer between Asia's giants.

In landlocked, fortress-like Bhutan, time seems to have stood still. Bhutan is officially a constitutional monarchy, but its king rules the country with virtually absolute power; economic subsistence and political allegiance are the norms of life for most of the population of just under one million. Thimphu, the capital, has about 50,000 inhabitants. The symbols of Buddhism, the state religion, dominate its cultural landscape. Social tensions arise from the large but diminishing Nepalese minority, most of whom are Hindus and some of whom have been persecuted by the dominant Bhutia. In the 1990s, a flow of Nepalese refugees from Bhutan caused tensions between Kathmandu and Thimphu.

Forestry, hydroelectric power, and tourism all have potential here, and Bhutan has considerable mineral resources. But isolation and inaccessibility preserve traditional ways of life in this mountainous buffer state.

THE SOUTHERN ISLANDS

Sri Lanka: South Asian Tragedy

Sri Lanka (known as Ceylon prior to 1972), the compact, pear-shaped island located just 22 miles (35 km) across the Palk Strait from the southern tip of the Indian peninsula, is the fourth independent state to have emerged from the British sphere of influence in South Asia (Fig. 8-17). Sovereign since 1948, Sri Lanka has had to cope with political as well as economic problems, some of them quite similar to those facing India and Pakistan.

There were good reasons to create a separate independence for Sri Lanka. This is neither a Hindu nor a Muslim country; the majority—some 70 percent—of its 19.7 million people are Buddhists. Furthermore, unlike India or

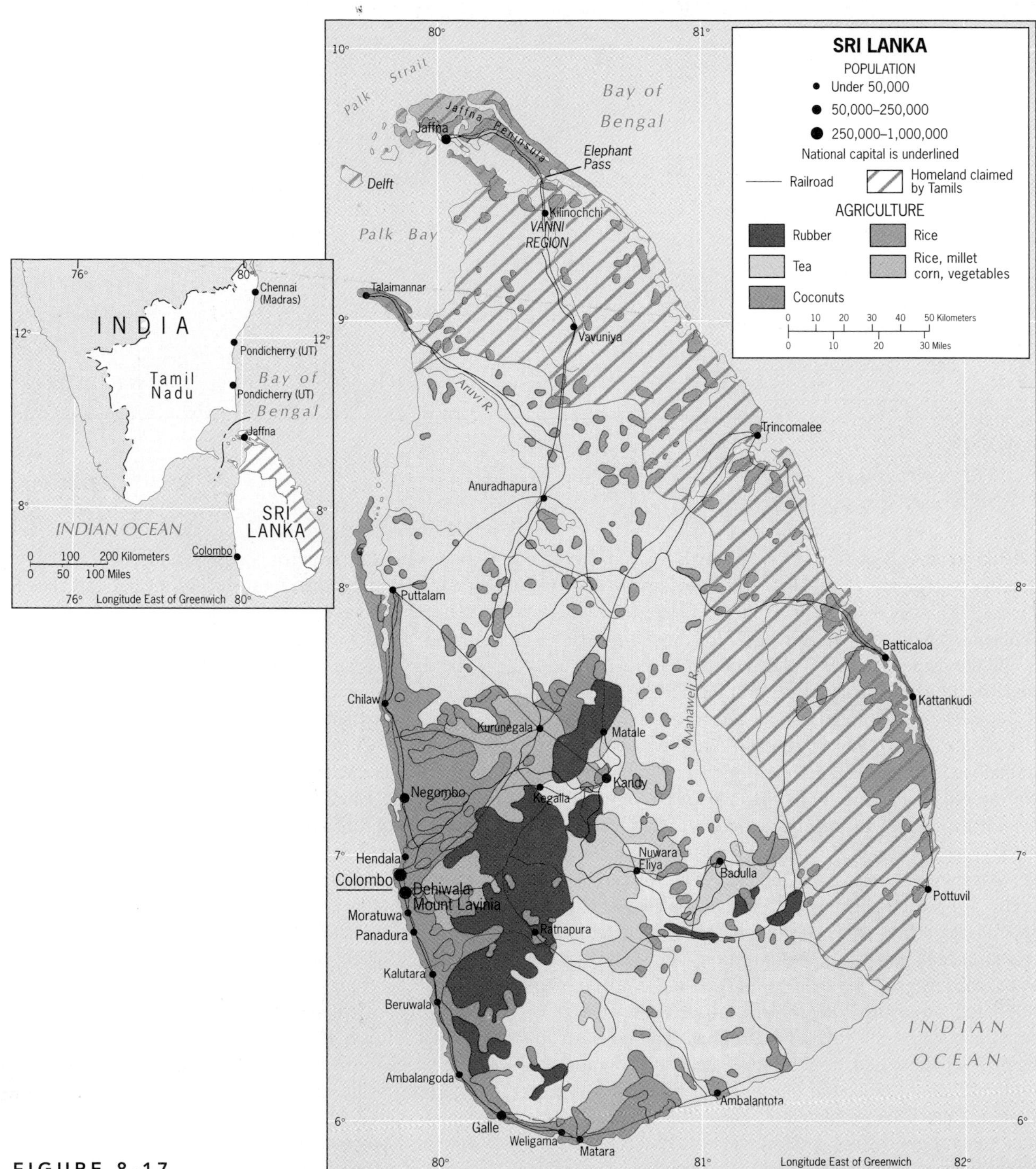

FIGURE 8-17

Pakistan, Sri Lanka is a plantation country (a legacy of the European period), with export agriculture still the mainstay of the external economy.

Most of Sri Lanka's people are not Dravidian but are of Aryan origin with a historical link to ancient northern India. After the fifth century BC, their ancestors began over several centuries to migrate to Ceylon, a relocation that brought to this southern island the advanced culture of the northwestern subcontinent. Part of that culture was the Buddhist religion; another component was a knowledge of irrigation techniques. Today, the descendants of these early invaders, the Sinhalese, speak a language (Sinhala) belonging to the Indo-European linguistic family of northern India.

The Dravidians from southern India never came in sufficient numbers to challenge the Sinhalese. They introduced the Hindu way of life, brought the Tamil language to northern Sri Lanka, and eventually constituted a substantial minority (now 18 percent) of the country's population. Their numbers were markedly strengthened during the second half of the nineteenth century when the British brought hundreds of thousands of Tamils from the adjacent mainland to Ceylon to work on the tea plantations. Sri Lanka later sought to repatriate this ethnic element in its population and even signed an agreement to that effect with India. After independence, successive Sri Lankan governments marginalized the Tamil minority (although Tamil was made a "national language" in 1978). The result was a costly armed rebellion in the island's far north that evolved into a full-scale separatist movement.

Sri Lanka is not a large island (it is about the size of West Virginia), but it is mountainous. The highest uplands, in the south, reach over 8000 feet (2500 m), from where steep, thickly forested slopes lead down to an encircling lowland. The north, including the Jaffna Peninsula, is entirely low-lying. The rivers from the uplands feed the paddies. Rice, not wheat, is the staple crop.

The moist southwest has long been the leading agricultural zone. The plantations introduced by the European colonizers still produce coconuts in the lowlands, rubber at intermediate elevations, and tea (for which Sri Lanka is famous) in the highlands. Tea accounts for one-fourth of the country's exports by value. Rice is another story. In the 1960s, a combination of family-planning policies and malaria eradication measures, coupled with the repopulation of lowlands and the expansion of rice growing, produced self-sufficiency for Sri Lanka. But, as in India, Sri Lankan paddy farming is not known for its efficiency, population growth has again increased, and today Sri Lanka must import rice to meet its domestic demand.

Colombo is the focus for what little industry has developed in Sri Lanka, most of it in food processing and otherwise largely dependent on the small local market. Colombo's cityscape mirrors the fate of the country since independence; an early period of optimism and modernization was interrupted by the civil war involving the Tamil northerners. In the mid-1990s, the prospect of a settlement contributed to some renewed investment; two tall office towers were built by Singaporean investors, the first major change in the city's skyline for many years. But hopes for a permanent truce were dashed as the war heated up again. A sizeable Tamil minority clusters in Colombo's Pettah District, and violence has intermittently struck the city. The tourist industry, once a major revenue earner, has been devastated.

FROM THE FIELD NOTES

"The long civil war involving Tamil demands for a separate state has severely damaged Sri Lanka's economy. Its impact can be seen in the capital, Colombo, where the tourist industry collapsed and new investment dwindled. In the mid-1990s, with hopes for a resolution of the crisis rising, some new building finally began in Colombo: the twin towers seen rising here above the cityscape were being built by Singaporean investors anticipating a resumption of economic progress in this embattled country. But even the capture of Jaffna by government forces did not lower the tension in the capital. We saw armed military checkpoints everywhere." Note: The towers were completed, but badly damaged later by a massive bomb set off by the Tamil Tigers in the business district. A suicide bomber made an attempt on the life of the Prime Minister; the Temple of the Tooth, Sri Lankan Buddhism's holiest shrine, was damaged by a car bomb that killed many. Sri Lanka's future is threatened by a conflict from which there seems to be no escaping.

Fragmentation?

A look at the map suggests that Sri Lanka, off the coast of India, might share some situational advantages with Taiwan, off the coast of China. But Sri Lanka is no economic

FROM THE FIELD NOTES

"Sri Lanka is a dominantly Buddhist country, the religion of the majority Sinhalese. Most areas of the capital, Colombo, have numerous reminders of this in the form of architecture and statuary: shrines to the Buddha, large and small, abound. But walk into the Tamil parts of town, and the cultural landscape changes drastically. This might as well be a street in Chennai or Madurai: elaborate Hindu shrines vie for space with storefronts and Buddhist symbols are absent. The people here seemed to be less than enthused about the Tamil Tigers' campaign for an independent state. 'This would never be a part of it anyway,' said the fellow walking toward me as I took this photograph. 'We're here for better or worse, and for us the situation up north makes it worse.' But, he added, Sri Lankan governments of the past had helped create the situation by discriminating against Tamils."

tiger on an Indian Ocean Rim. Certainly Sri Lanka has opportunities (and despite the war, its per-capita GNP still is the highest in the region except for tiny Maldives). The tragedy of Sri Lanka is that more enlightened political leadership could have prevented what happened. The Tamils of the north and east had long given warning that they could not achieve equal rights in education, employment, land ownership, and linguistic and political representation. When the Sri Lankan government failed to accommodate Tamil demands, an insurrection broke out, and Tamil leaders called for a Cyprus-like partition of the island. Tamil forces (the Tamil Tigers) managed to secure the Jaffna Peninsula and some adjacent areas, and demanded a sovereign state, Eelam (the striped zone in Fig. 8-17).

The sequence of events is depressingly familiar in this devolutionary world. Just as the Turks of northern Cyprus received support from nearby Turkey, so Sri Lanka's Tamils got help from India's more than 50 million Tamils. Sri Lanka asked the government of India to help curb the Tamil uprising, which embroiled India in a foreign conflict with dangerous domestic implications. In 1991, Rajiv Gandhi, the leading Indian politician campaigning for the office of prime minister, was assassinated near the Tamil city of Chennai; his party's role in the Sri Lankan civil war was the apparent cause.

In Chapter 5 we discussed the three-stage model of the 18 evolution of the **insurgent state** (see pp. 265-266). Initial *contention* is followed by temporary *equilibrium*, leading eventually to *counteroffensive*. In the equilibrium stage, the rebels manage to secure a territorial domain that takes on the characteristics of a state-within-a-state. Such an insurgent state has a core area anchoring an informally bounded territory within which an alternative government takes control. This regime may be sufficiently secure and stable to provide social services such as schools and medical clinics. Its headquarters take on the properties of a capital city. Negotiators may even be invited to this capital to bargain for a settlement. Failure of such bargaining may produce the counteroffensive stage.

In Sri Lanka, the equilibrium stage was reached in the early 1990s, when the Jaffna Peninsula (see Fig. 8-17) became the core of an insurgent state. Tamil Eelam headquarters were based in the town of Jaffna, and the equilibrium stage prevailed for several years as Tamil rebels extended their domain to include much of the Vanni territory in the northern part of the island.

The counteroffensive stage began in 1995, when a massive government force attacked the Jaffna Peninsula and drove the Tamil Tigers out of Jaffna town. But the fall of Jaffna alone did not destroy the insurgent state, and in the late 1990s the insurgents established a new headquarters in Kilinochchi. Some 200,000 refugees from the conflict swelled the population of the Vanni region, and the outcome of the struggle remained uncertain. Meanwhile, terrorist attacks on civilian targets in Colombo and other cities caused thousands of casualties, diminishing prospects for a settlement.

FROM THE FIELD NOTES
"Some countries have to take warnings of global warming more seriously than others. As we approached the Maldives, the islands lay like lilypads on the surface of a pond. No part of this country's natural surface lies more than 6 feet (less than 2 meters) above sea level. The upper floors of the buildings in the capital, Maale, form the Maldives's highest points. Almost any rise in sea level would threaten this Indian Ocean outpost of South Asia."

Despite this costly and protracted conflict, the country still had the second-highest GNP per capita in the entire South Asian realm throughout the 1990s. At the turn of the twenty-first century, Sri Lanka remained a regional symbol of opportunities lost.

The Maldives

The Maldives reflect what Sri Lanka has lost. Imagine a country of more than a thousand tiny islands whose combined area is just 115 square miles (less than 300 sq km), whose highest elevation is barely over 6 feet (2 m) above sea level in a region of tropical storms more than 400 miles (650 km) from the nearest continent. Add a population of 295,000, one quarter of it on the capital island named Maale, of Sinhalese and Dravidian extraction and 100 percent Muslim, and we have the essentials of South Asia's tiniest state.

Except for one geographic qualification: The Maldives has translated its tropical, palm-studded, beach-fringed island environment into a tourist mecca that attracts tens of thousands of mainly European visitors annually. Environment and relative location now bring the Maldives per-capita revenues unimagined in mainland South Asia. In Sri Lanka hotels stand empty and beaches see few foreigners. Sri Lanka's loss is the Maldives' gain. If there is a dark cloud on the Maldives' horizon, it is beyond their control: global warming. If the current cycle of global warming continues and sea level rises, the country could disappear.

Figure I-11 reminds us of a troubling reality: South Asia is the only geographic realm on Earth that consists entirely of countries in the lowest-income category—except for the Maldives and troubled Sri Lanka (which together account for only 1.5 percent of both the realm's population and territory). South Asia also contributes by far the largest share in the ongoing global population spiral. It is a realm where Islam meets infidel and where contentious neighbors have nuclear weapons. Nowhere are the problems of the periphery more severe than in this disadvantaged realm, and the countries of the global core cannot ignore the plight of this walled-off, peninsular corner of the planet.

Continued on pages 184-185

Scale 1:16 000 000; one inch to 250 miles. Polyconic Projection
Elevations and depressions are given in feet

Continued on pages 198-199

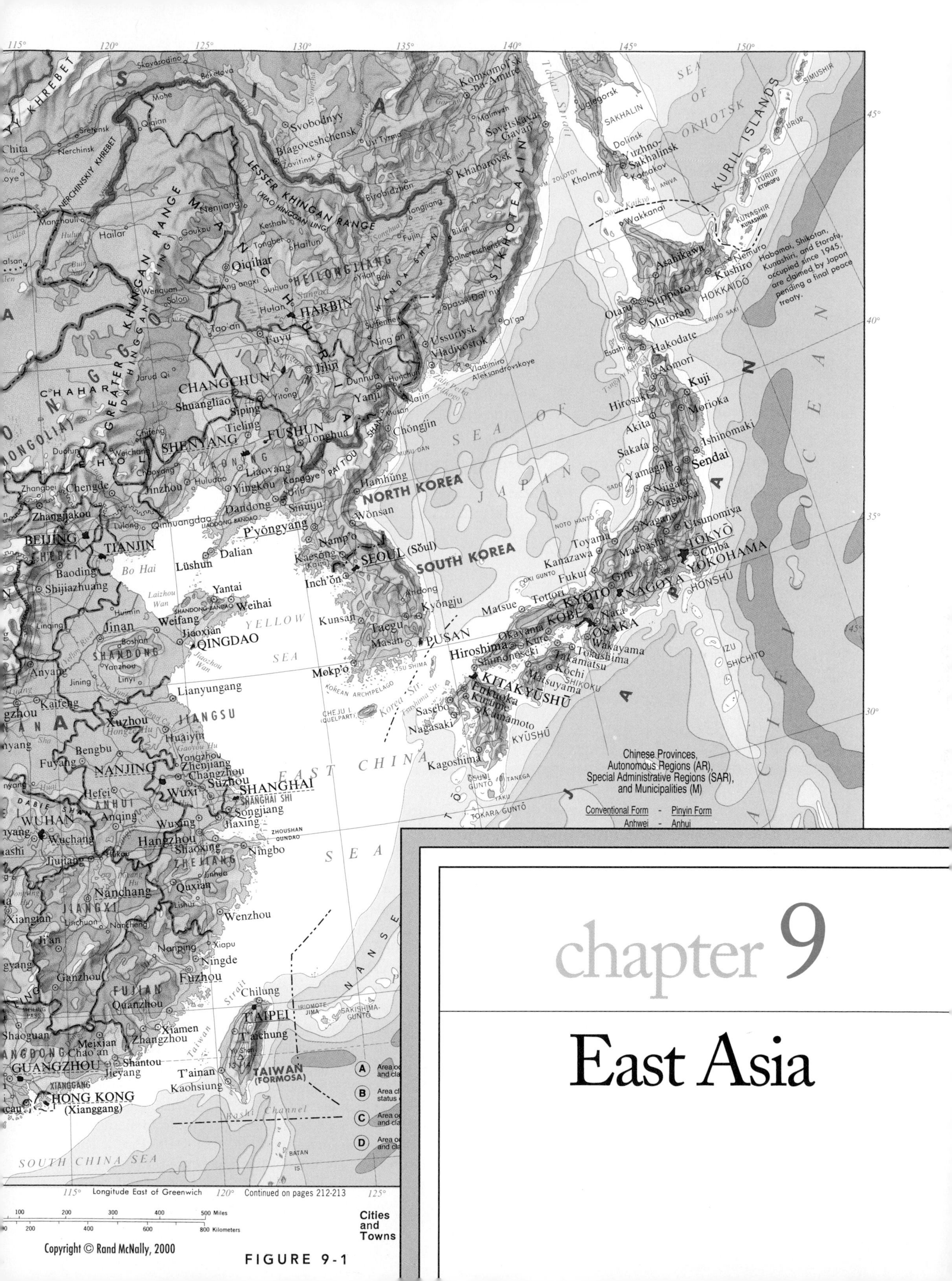

FIGURE 9-1

chapter 9

East Asia

CONCEPTS, IDEAS, AND TERMS

1. Pacific Rim
2. Confucianism
3. Sinicization
4. Extraterritoriality
5. Special Administrative Region (SAR)
6. Economic restructuring
7. Core area
8. Geography of development
9. Overseas Chinese
10. Special Economic Zone (SEZ)
11. Economic tiger
12. Regional state
13. Buffer state
14. Jakota Triangle
15. Modernization
16. Relative location
17. Areal functional organization
18. Regional complementarity
19. State capitalism

REGIONS

- CHINA PROPER
- XIZANG (TIBET)
- XINJIANG
- MONGOLIA
- JAKOTA TRIANGLE (JAPAN-KOREA-TAIWAN)

East Asia is a geographic realm like no other. At its heart lies the world's most populous country. On its periphery lies one of the globe's most powerful national economies. Along its coastline, on its peninsulas, and on its islands an economic boom has transformed cities and countrysides. Its interior contains the world's highest mountains and vast deserts. It is a storehouse of raw materials. The basins of its great rivers produce food that can sustain more than a billion people.

DEFINING THE REALM

The East Asian geographic realm consists of six political entities: China, Mongolia, North Korea, South Korea, Japan, and Taiwan. Note that we refer here to "political entities" rather than "states." In changing East Asia, the distinction is significant. Taiwan, which its government officially calls the Republic of China, functions as a state but is regarded by mainland China (the People's Republic of China) as a temporarily wayward province. North Korea is not a full member of the United Nations, and the division of the Korean Peninsula may be temporary.

As defined here, East Asia lies between the vast expanses of Russia to the north and the populous countries of South and Southeast Asia to the south. This geographic realm extends from the deserts of Central Asia to the Pacific islands of Japan and Taiwan. Environmental diversity is one of its hallmarks.

East Asia also is the hub of the evolving regional phenomenon called the **Pacific Rim**. From Japan to Taiwan and from South Korea to Hong Kong (Xianggang), the Pacific frontage of East Asia is being transformed (see box titled "Names and Places"). Skyscrapers tower over Haikou, capital of the once-dormant southern Chinese island of Hainan. Luxury automobiles from Europe and America ply the streets of Dalian, long the drab port for China's Northeast. A forest of construction cranes marks the emergence of Pudong, a huge industrial zone where Shanghai stakes its entry into the Pacific Rim's booming economy. Millions of people are on the move, abandoning their farms and villages and seeking work in such projects.

Japan was the leader in East Asia's Pacific Rim development. Long before this regional term even came into general use, Japan had built a giant economy with global connections and was the only highly developed country in Asia not plagued by stagnation. Although China lay isolated and South Korea struggled in the aftermath of its terrible war against communism (1950–1953), Japan built a society unlike that of any other in eastern Eurasia and merited recognition as a discrete geographic realm. But Pacific Rim developments are diminishing the contrasts between Japan and its East Asian neighbors. South Korea today challenges

Major Geographic Qualities of East Asia

1. East Asia is encircled by snowcapped mountains, vast deserts, cold climates, and Pacific waters.
2. East Asia was one of the world's earliest culture hearths, and China is one of the world's oldest continuous civilizations.
3. East Asia is the world's most populous geographic realm, but its population remains strongly concentrated in its eastern regions.
4. China, the world's largest nation-state demographically, is the current rendition of an empire that has expanded and contracted, fragmented and unified many times during its long existence.
5. China today remains a mainly rural society, and its vast eastern river basins feed hundreds of millions in a historic pattern that continues today.
6. China's sparsely peopled western regions are strategically important to the state, but they lie exposed to minority pressures and Islamic influences.
7. Along China's Pacific frontage an economic transformation is taking place, affecting all the coastal provinces and creating an emerging Pacific Rim region.
8. Increasing regional disparities and fast-changing cultural landscapes are straining East Asian societies.
9. Japan, the economic giant of the East Asian realm, has a history of colonial expansion and wartime conduct that still affects international relations here.
10. East Asia may witness the rise of the world's next superpower as China's economic and military strength and influence grow—and if China avoids the devolutionary forces that fractured the Soviet Union.
11. The political geography of East Asia contains a number of flashpoints that can generate conflict, including Taiwan, North Korea, and several island groups in the realm's seas.

Names and Places

In 1958, the government of the People's Republic of China adopted the so-called *pinyin* system of standard Chinese, which replaced the Wade-Giles system used since colonial times. Pinyin was adopted not to teach foreigners how to spell and pronounce Chinese names and words but to establish a standard form of the Chinese language throughout China. The pinyin system is based on the pronunciation of Chinese characters in Northern Mandarin, the Chinese spoken in the region of the capital and the north in general.

The new linguistic standard caught on slowly outside China, but today it is in general use. The old name of the capital, Peking, has become Beijing. Canton is now Guangzhou. The Yangtze Kiang (River) is now the Chang Jiang/Yangzi. Tientsin, Beijing's port, is now Tianjin. A few of the old names persist, however. The Chinese call their colony Xizang, but many maps still carry the name Tibet.

Pinyin usage also affected personal names. China's long-time ruler, Mao Tse-tung, is now called Mao Zedong. His eventual successor, Teng Hsiao-ping, was more simply Deng Xiaoping. And remember: when the Chinese write their names, they use the last name first. The current president, Jiang Zemin, is Zemin to friends and Mr. Jiang to others. To the Chinese, it is Bush George and Cheney Dick, not the other way around.

Japan on international markets, selling goods ranging from electronics to automobiles. In southeastern China, Guangdong Province is an economic juggernaut that includes Shenzhen, recently the world's fastest growing city. Japan's own investments in Pacific Rim economies have helped not only to lessen the contrasts, but also to enhance the linkages between Tokyo and the mainland. In short, Japan has again become part of a functional region within the East Asian realm.

No matter how powerful Japan's economy became during the second half of the twentieth century, China remains the colossus of East Asia. One of the world's oldest continuous civilizations, China was a major culture hearth when Japan was an isolated frontier inhabited by the Ainu. Migrations that originated in China advanced through the Korean Peninsula and reached the islands. Philosophies, religions, and cultural traditions (including, for example, urban design and architecture) diffused from China to Korea and Japan, and to other locales on the margins of the Chinese hearth. Over the past millennium, China repeatedly expanded to acquire empires, only to collapse in chaos when its center failed to maintain control. European colonial powers took

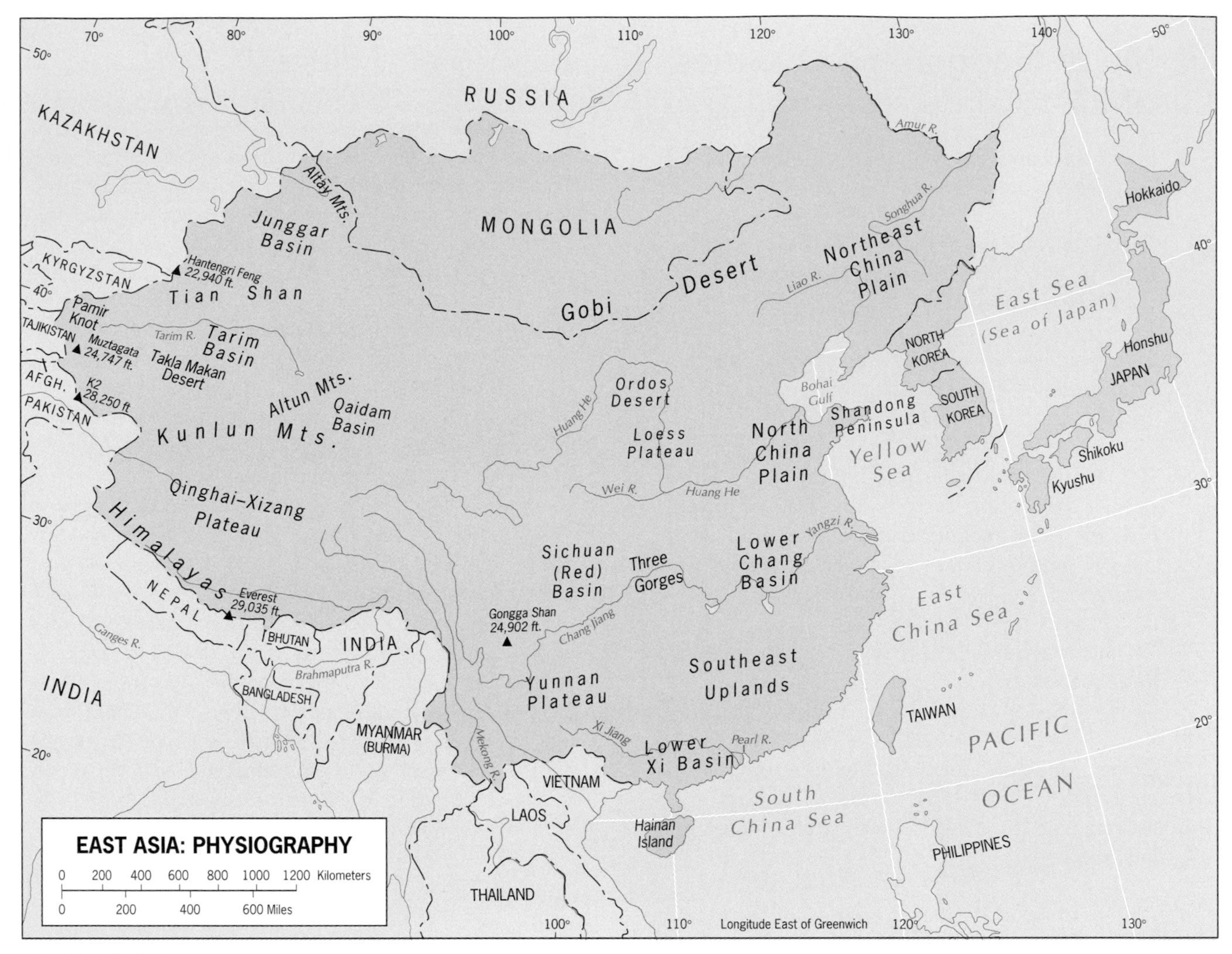

FIGURE 9-2

such opportunities to partition China among themselves. The Russians and the Japanese pushed their empires deep into China. But always China recovered, and today China may again be poised to expand its sphere of influence and, indeed, to take its place as a world power in the twenty-first century.

NATURAL ENVIRONMENTS

Figure 9-1 dramatically illustrates the complex physical geography of the East Asian realm. In the southwest lie ice-covered mountains and plateaus, the Earth's crust in this region crumpled up like the folds of an accordion. A gigantic collision of tectonic plates is creating this landscape as the Indian Plate pushes northward into the underbelly of the Eurasian Plate (Fig. I-4). The result is some of the world's most spectacular scenery, but snow, ice, and cold are not the only dangers to human life here. Earthquakes and tremors occur almost continuously, causing landslides and avalanches. As the map shows, the high mountains and plateaus widen from a relatively narrow belt in the Karakoram to form Xizang's (Tibet's) vast plateau, flanked by the Himalayas to the south. Then, east of Tibet, the mountain ranges converge again and bend southward into Southeast Asia, where they lose their high relief.

As Figure I-9 shows, this Asian interior is one of the world's most sparsely populated areas, but it is nevertheless critical to the lives of hundreds of millions of people. In these high mountains, fed by the melting ice and snow, rise the great rivers that flow eastward across China and southward across Southeast and South Asia. Throughout the Holocene, these rivers have been eroding the uplands and depositing their sediments in the lowlands, in effect creating the alluvium-filled basins that now sustain huge populations. Fertile alluvial soils and adequate growing seasons, combined with ample water and millions of hands to sow the wheat and plant the rice, have allowed the emergence of one of the great population concentrations on Earth.

Figure I-8 records this high-elevation zone of East Asia as category *H*, highland climate, which is by far the largest area of its kind in the world. Here one criterion overpowers the rules of climatic classification: altitude. The mountains wrest virtually all moisture from air masses moving northward and block their path into interior Asia. As Figure I-8 shows, desert conditions prevail in East Asia's northern interior. Figure 9-1 names two of the more famous ones: the Takla Makan of Xinjiang and the Gobi of Mongolia.

Physiography, therefore, has much to do with East Asia's population distribution, but even the more habitable and agriculturally productive east has its limitations. The northeast suffers from severe continentality, with long and bitterly cold winters. High relief encircles the river basins north of the Yellow Sea and dominates much of the northern part of the Korean Peninsula, creating strong local environmental contrasts. South Korea, as Figure I-8 shows, experiences relatively moderate conditions comparable to those of the U.S. Southeast; North Korea has a harsh continental climate like that of North Dakota.

In general, coastal, peninsular, and insular East Asia possess more moderate climates than the interior. Like South Korea, southern Japan and southeastern China have humid temperate climates, and southernmost Taiwan and Hainan Island even have areas of tropical (*A*) conditions. Proximity to the ocean tends to moderate climatic environments, and coastal East Asia proves the point.

Before we focus on East Asia's human geography, it is useful to look at a map of this realm's complex physical stage (Fig. 9-2). From the high-relief interior come three major river systems that have played crucial roles in the human drama. In the north, the Huang He (Yellow River) arises deep in the high mountains, crosses the Ordos Desert and the Loess Plateau, and deposits its fertile sediments in the vast North China Plain, where East Asia's earliest states emerged. In the center, the Chang Jiang (Long River), called the Yangzi downstream, crosses the Sichuan Basin and the Three Gorges, where a huge dam project is under way, and waters extensive ricefields in the Lower Chang Basin. And in the south, the Xi Jiang (West River) originates on the Yunnan Plateau and becomes the Pearl River in its lowest course. Its estuary, flanked by several of China's largest urban-industrial complexes, has become one of the hubs of the evolving Pacific Rim.

Further scrutiny of Figure 9-2 indicates that a fourth river system plays a role in China: the Liao River in the northeast and its basin, the Northeast China Plain. As the map suggests, however, the Liao is not comparable to the great rivers to its south, its course being shorter and its basin, in this higher-latitude area, much smaller. As we will see, China's Northeast was for some time the country's industrial, not agricultural, heartland.

Looking again toward the interior, note the Loess Plateau south of the Ordos Desert, where the Huang He (Yellow River) makes its giant loop. Loess is a fertile, windblown deposit composed of rock pulverized by glaciers. Add water (in this case the middle Huang and its tributaries) and an adequate growing season, and a sizeable population will arise. To the south, deep in the interior, lies the Sichuan (Red) Basin, crossed by the Chang Jiang. This basin has supported human communities for a long time, and, as we noted in the Introduction, you can actually see its current population cluster on the world population map (Fig. I-9). The Sichuan Basin, encircled as it is by mountains, is one of the world's most clearly defined physiographic regions, and the concentration of its approximately 120 million inhabitants reflects that definition.

Still farther to the south lies the Yunnan Plateau, source of the tributaries that feed the Xi River. Much of southeastern China has comparatively high relief; it is hilly and in

FROM THE FIELD NOTES

"From the train, travelling from Beijing to Xian, we had a memorable view of the Loess Plateau. Loess is a fine-grained dust formed from rocks pulverized by glacial action, blown away by persistent winds and deposited in sometimes well-defined locales. Loess covers much of the North China Plain, which is what makes it so fertile, but there it is not as thick as in the Loess Plateau in the middle basin of the Huang He (Yellow River). Here the loess averages 250 feet (75 m) in thickness and in places reaches as much as 600 feet (180 m). Because loess has some very distinctive physical properties, it tends to create unusual landscapes. Through a complicated physical process following deposition, loess develops the capacity to stand upright in walls and columns, and resists collapse when it is excavated. As a result the landscape looks terraced: streams cut deep and steep-sided valleys. The Loess Plateau is a physiographic region, but it is a cultural region too. Hundreds of thousands of people have literally dug their homes into vertical faces of the kind shown here, creating cave-like, multi-room dwellings with wooden exterior doors."

places mountainous. This high relief has helped limit contacts between China and Southeast Asia.

East Asia's Pacific margin is a jumble of peninsulas and islands. The Korean Peninsula looks like a near-bridge from Asia to Japan, and indeed it has served as such in the past. The Liaodong and Shandong peninsulas protrude into the Yellow Sea, which continues to silt up from the sediments of the Huang and Liao rivers. Off the mainland lie the islands that have played such a crucial role in the modern human geography of Asia and, indeed, the world: Japan, Taiwan, and Hainan. Japan's environmental range is expressed by cold northern Hokkaido and warm southern Kyushu, but Japan's core area lies on its main island, Honshu. As Figure 9-1 shows, myriad smaller islands flank the mainland and dot the East and South China Seas. As we will discover, some of these smaller islands have major significance in the human geography of this realm.

HISTORICAL GEOGRAPHY

Consider this: there is no evidence that *Homo sapiens*, modern humans, reached the Americas any earlier than 40,000 years ago, and many paleoanthropologists argue that the only available evidence indicates that humans crossed from Eurasia to Alaska a mere 14,000 to 13,000 years ago. On the opposite side of the Pacific, however, the story of hominid and human settlement spans hundreds of thousands of years—perhaps 1 million.

What took so long? If hominids like *Homo erectus* left their ancestral African homelands and migrated as far away as eastern Asia more than half a million years ago, and if modern humans followed them and reached Australia some 50,000 years ago, what kept the hominids from the Americas altogether, and what delayed the humans?

No satisfactory answers have yet been found to questions like these. The width of the Pacific Ocean may have doomed maritime migrants, although trans-Pacific migration may eventually have occurred. The frigid temperatures along the northern route from Siberia to Alaska may have stopped hominids and humans alike. But the Pleistocene was punctuated by warm interglaciations; why did it take until the early Holocene for humans to make the Bering Strait crossing?

All this implies that for hundreds of thousands of years East Asia was a cul-de-sac, a dead-end for migrants out of Africa. Many archeological sites in the realm have yielded evidence of *Homo erectus*, including what is perhaps the most famous of all: Peking Man. In a cave near the Chinese capital, Beijing, an archeologist in 1927 found a single tooth he recognized as representing a hominid. Later excavations proved him right as parts of more than three dozen skeletons were unearthed. They proved that Peking Man and his associates made stone and bone tools, had a communal culture, controlled the use of fire, hunted wildlife, and cooked their meat. Later arrivals undoubtedly challenged older communities, and the frontier of settlement must have expanded across the river basins of the east. But when *Homo sapiens*, modern humans, arrived in East Asia, the hominids could not compete. Current anthropological theory holds that the better-equipped humans eliminated the hominids, perhaps between 60,000 and 40,000 years ago, after which the human migration into the Americas could begin.

In the context of East Asia we should mention a minority view. Some anthropologists state that the evidence supports the notion that modern humanity developed *not* from one stock in Africa, but from four stocks in widely separated parts of the world: Africa, western Eurasia (the Caucasoids), Australia (the Australoids), and eastern Eurasia (the Mongoloids). According to this idea, today's East Asians trace their ancestry to Peking Man and beyond.

Early Cultural Geography

Whatever the outcome of the debate over human origins in East Asia, it is clear that humans have inhabited the plains and river basins, foothills, and islands of this realm for a very long time. Hunting sustained both the hominids and the early human communities; fishing drew them to the coasts and onto the islands. The first crossing into Japan may have occurred as long as 10,000 to 12,000 years ago, possibly much longer, when the Jomon people, a Caucasoid population of uncertain geographic origins, entered the islands; their modern descendants, the Ainu, spread throughout the archipelago. Today, only about 20,000 persons living in northernmost Hokkaido trace their ancestry to Ainu sources.

About 2300 years ago the Yayoi people, rice farmers who had settled in Korea, appear to have crossed by boat to Kyushu, Japan's southernmost island, from where they advanced northward. The Ainu, who subsisted by fishing, trapping, and hunting, were driven back, but gene-pool studies show that much mixing of the groups took place; they also show that the Yayoi invasion was followed by other incursions from the Asian mainland. By then, powerful dynastic states had already arisen in what is today China, and early Chinese culture traits thus found their way into Japan through the process we know as relocation diffusion.

On the Asian mainland, plant and animal domestication had begun as early as anywhere on Earth. We in the Western world take it for granted that these momentous processes began in what we now call the Middle East and diffused from the Fertile Crescent to other parts of Eurasia and the rest of the world. But the taming of animals and the selective farming of plants may have begun as early, or earlier, here in East Asia. As in Southwest Asia, the fertile alluvial soils of the great river basins and the ebb and flow of stream water created an environment of opportunity, and millet and rice were being harvested between 7000 and 8000 years ago.

Even during this Neolithic period of increasingly sophisticated stone tools, East Asia was a mosaic of regional cultures. Their differences are revealed by the tools they made and the decorations on their bowls, pots, and other utensils. An especially important discovery of two 8000-year-old pots in the form of a silkworm cocoon, from China's Hebei Province, suggests a very ancient origin for one of the region's leading historic industries.

As noted earlier, plant and animal domestication produced surpluses and food storage, enabling population growth and requiring wider regional organization. Here as elsewhere during the Neolithic, settlements expanded, human communities grew more complex, and power became concentrated in a small group, an *elite*.

This process of state formation is known to have occurred in only a half dozen regions of the world, and China was one of these. But evidence about China's earliest states has long been scarce. Today, however, archeologists are focusing on the lower Yi-Luo River Valley in the western part of Henan Province, where the first documented Chinese dynasty, the Xia Dynasty (2200–1770 BC) existed. The capital of this ancient state, Erlitou, has been found, and archeologists now refer to the Xia Dynasty as the Erlitou culture. Secondary centers are being discovered, and Erlitou tools and implements in a wider area prove that the Xia Dynasty represents a substantial state.

All early states were ruled by elites, but China's political history is chronicled in *dynasties* because here the succession of rulers came from the same line of descent, sometimes enduring for centuries. In the transfer of power, family ties counted for more than anything else. Dynasties were overthrown, but the victors did not change this system. Dynastic rule lasted into the twentieth century.

The Xia Dynasty may have been the earliest Chinese state, but it lay in the area where, later, more powerful dynastic states arose: the North China Plain. Here the tenets of what was to become Chinese society were implanted early and proved to be extremely durable. From this culture hearth ideas, innovations, and practices diffused far and wide. From agriculture to architecture, poetry to porcelain making, influences radiated southward into Southeast Asia, westward into interior Asia, and eastward into Korea and Japan. In the North China Plain lay the origins of what was to become the Middle Kingdom, which its citizens considered the center of the world.

Table 9-1 summarizes the sequence of dynasties that may have begun with Xia more than four millennia ago and ended in 1911 when the last emperor of the Qing (Manchu) Dynasty, a boy 6 years old, was forced to abdicate the Chinese throne. As the table shows, every dynasty contributed importantly to the development of Chinese society, sometimes progressively through enlightened policies and efficient administration, at other times regressively when capricious cruelty, authoritarian excesses, and xenophobia prevailed. During those 4000 years, China evolved into the world's most populous nation, always rebounding after disastrous famines and floods, its territory covering more than 80 percent of the East Asian realm. (Fig. 9-3).

Dominant as China is in its regional sphere, there is more to East Asia than the Chinese giant. To the north, the landlocked state of Mongolia is what remains of the time when Mongol armies conquered much of Eurasia (as Table 9-1 shows, even China fell under the Mongol sway). To the east, the Korean Peninsula escaped incorporation into China. Offshore, Japan first surpassed China as a military power, colonizing Manchuria (as it was then called) and penetrating deep into China's heartland. Later, Japan outclassed China as an economic power, achieving world status while China languished under Maoist communism. And as the twenty-first century dawned, Taiwan remained a separate political entity, its detachment from China resulting from the struggle that brought the communists to power in Beijing.

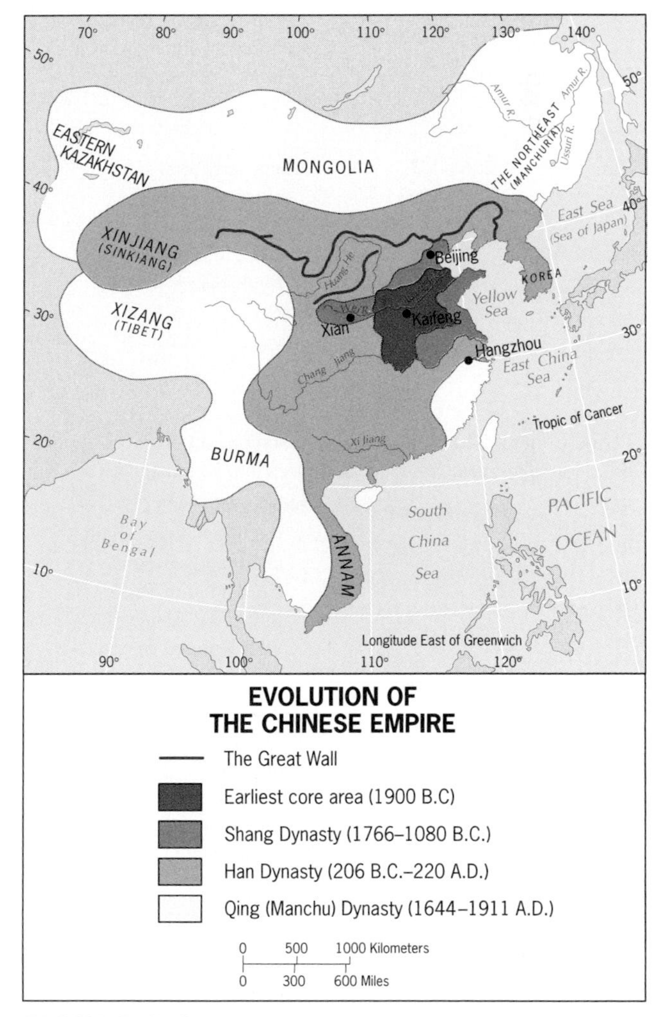

FIGURE 9-3

Table 9-1
THE CHINESE DYNASTIES, 2200 BC to AD 1911

Dynasty Name(s)	Date(s)	Areas Governed	Major Features	Geographic Impact
Xia	c2200–c1770 BC	Small part of Huang River Basin, centered in Yi-Luo tributary.	First dynastic state (?); Neolithic technology, sophisticated stone tools.	Beginnings of stream diversion, irrigation.
Shang (latter part Yin)	c1766–c1080 BC	Huang-Wei River confluence across Shandong and Henan Provinces. Anyang capital.	Neolithic to Bronze Age transition. Timber houses with thatched roofs, beginnings of Chinese writing, moon-cycle-based calendar. Superb bronze vases and other implements.	Unifies large region in lower Huang River Basin.
Zhou	c1027–221 BC Two periods: Spring and Autumn c1027–481 BC; Warring States 475–221 BC	A people centered in Shaanxi Province to the west of the Shang area overthrow the Shang rulers and expand their domain eastward. Warring States is a period of feudalism.	Formative period for China. Irrigation systems, iron smelting, horses, expanded farm production. Major towns develop. Writing system is established. Chopsticks come into use. Taoism and Confucianism arise; Taoism centers on mysticism, individualism, the personal "way", and Confucianism on duties, community standards, respect for government.	Taoism diffuses widely throughout East Asia, reaches South Asia, and influences Buddhism. Confucianism becomes China's guiding philosophy for more than 2000 years. Building of *Great Wall* begins.
Qin (Ch'in)	221–206 BC	Most of North China Plain, south into Lower Chang Basin; first consolidation of large Chinese state.	Time of bureaucratic dictatorship after ferment of Zhou: Confucian Classics burned, Taoism combated. Much of Great Wall is built at huge cost in lives. Cruel despotism, excesses. Emperor's tomb discovered at Xian in 1976 with 6000 terracotta life-sized men and horses.	Ruler Ch'in (Qin)'s name immortalized as *China*.
Han	206 BC–AD 220	Major territorial expansion adds Xinjiang in interior Asia, Vietnam in Southeast Asia.	Second formative dynasty for China; restoration of Confucian principles and a flowering of culture. First use of eunuchs by authoritarian government, planting a long-term weakness in administration. Xian becomes one of the greatest cities of the ancient world. Chinese refer to themselves as the *People of Han*.	About the same time, same size as the Roman Empire. *Silk Road* carries goods from China across inner Asia to Syria and on to Rome.
Period from AD 220–580 witnessed disunion and division. Important developments included the arrival of Buddhism in East Asia and the diffusion of Chinese influence and ideas into Korea and across Korea into Japan.				
Sui	AD 581–618	North and South China reunified, Xian rebuilt and greatly expanded. Conflict with Turks in Central Asia and with Koreans in east.	Brief but important dynasty with revival of Confucian rituals and practices in education. Brutal but modernizing regime conducts a census, establishes a penal code, and engages in massive public works.	Construction of key segments of Grand Canal links Huang and Yangzi Rivers and thus connects northern and southern economies.
Tang	AD 618–907	Defeat of Turks in Central Asia. Campaign to put down rebellions in the south succeeds, though northern frontier remains unstable.	A golden age for China and a third formative dynasty. Enormous and efficient bureaucracy dictates virtually all aspects of life. Xian now the cultural capital and largest city in the world. *Buddhism* thrives, pagodas arise everywhere. The arts, Chinese and foreign, flourish.	Arab and Persian seafarers visit Chinese ports, the Silk Route is loaded with trade, and China seems set for an international era. It is not to be: *Islam* arrives in Central Asia and Tang armies are defeated by Arabs in 751. The Silk Route breaks down. In the east, however, Chinese influences permeate Korea and penetrate Japan.

Table 9-1 (continued)

Dynasty Name(s)	Date(s)	Areas Governed	Major Features	Geographic Impact
The 907–960 period is known as the Five Dynasties because five leaders tried to establish dynasties in northern China in quick succession, each failing in turn. In southern China it is known as the time of the Ten Kingdoms, for ten regimes ruled various parts of the region. Cultural continuity prevails despite political instability.				
Song (Northern (Southern	AD 960–1279 AD 960–1127) AD 1127–1279)	Rulers consolidate the Five (northern) Dynasties, then begin capturing the Ten (southern) Kingdoms. They are unable to control the Liao state north of the Great Wall. Mongols invade, drive Song rulers southward, and end dynasty in 1279.	Preoccupied with organization and administration, Song rulers create a competent, efficient *civil service*. A defensive rather than an expansionist period. An era of rich cultural and intellectual achievement in mathematics, astronomy, mapmaking. Paper and movable type are invented, as is *gunpowder*. Large ships are built. The arts, from porcelain making to poetry, continue to thrive. Practice of footbinding begins. Improved rice varieties feed estimated 100 million inhabitants.	Capital is moved from northern Kaifeng on Huang River to Hangzhou in the south as Song rulers lose control over the north. Mongol invasion is followed by Kublai Khan's choice of Beijing as his capital (1272). Mongols use Chinese authoritarian and bureaucratic traditions to forge their own dynastic rule. Several cities have populations exceeding 1 million.
Yuan	AD 1264–1368	Controls eastern Asia from Siberia in the north to the Vietnam border in the south, including present-day Russian Far East and most of Korea.	Dynasty begins in north before Song ends in south. Repressive regime, social chasms between Mongol rulers and Chinese ruled. Deep and enduring hostility between Chinese and Mongols. Cultural *isolationism* results; Mongols are acculturated to Chinese norms. Marco Polo visits East Asia.	Most of present-day Mongolia is part of the Chinese sphere for the first time. Mongol effort to enter Japan fails. Mongol power over much of Eurasia stimulates trade and flow of information. Beijing grows into major city.
Ming	AD 1368–1644	Rules over all of eastern China from Amur River in north to Red River (Vietnam) in south, as far west as Inner Mongolia and Yunnan. At various times incorporates North Korea, most of Mongolia, even Myanmar (Burma).	Chinese rule again after Mongol Yuan dynasty. Stable but autocratic and inward-looking regime. Major advances in science and technology. In first half of fifteenth century, Chinese oceangoing vessels, larger than any in Europe, explore Pacific and Indian Ocean waters and reach East Africa. Farming expands, silk and cotton industries improve, printing plants multiply. Defensive walls are built in the north and around cities, including the Forbidden City in Beijing, the Ming capital. But mismanagement, infighting, eunuch influence in the palaces, and anti-commercialism weaken Ming rule.	Early Ming *ships* and fleets reach Southeast Asian and Indian Ocean shores long before smaller European vessels do, but the Chinese advantage is lost when Ming rulers turn isolationist and order maritime ventures halted. Population growth causes serious problems of food supply and political control.
Qing (Manchu)	AD 1644–1911	*Largest* China-centered empire ever incorporates Mongolia, much of Turkestan, Xizang (Tibet), Myanmar, Indochina, Korea, Taiwan.	Foreign rule again: Manchu, a people with Tatar links living in present-day Northeast China, seize an opportunity and take control of Beijing. Improbably, a group numbering about 1 million rule a nation of several hundred million. It is done by keeping Ming systems of administration and retaining (and rewarding) Ming officials. Expansion creates an empire. Population pressure and the concentration of land ownership in fewer hands create problems worsened by floods and famines. European powers and Japan force concessions on weakening Qing rulers; defeats in war lead to revolution and collapse.	Manchu (Qing) rulers create the map that today forms the justification for China's claims to Xizang (Tibet), Taiwan, and the central Eurasian interior; at the height of its power, the Qing empire forces the Russians to recognize Manchu authority over China's Northeast as far north as the Argun River. Latent claims to Russia's Far East are based on Manchu predominance there.

REGIONS OF THE REALM

East Asia presents us with an opportunity to illustrate the changeable nature of regional geo°graphy. Our regional delimitation is based on current circumstances, and it predicts ways the framework may change. It is anything but static (Fig. 9-4). At the beginning of the twenty-first century, we can identify five geographic regions in the East Asian realm. These are:

1. *China Proper*. Almost any map of China's human geography—population distribution, urban centers, surface communications, agriculture, industry—emphasizes the strong concentration of Chinese activity in the country's eastern sector. This is the "real" China, where its great cities, populous farmlands, and historic sources are located. Long ago, scholars called this *China Proper*, and it is a good regional designation (Fig. 9-4). But China is a large and complex country, and a number of subregions are nested within China Proper. Some of these, such as the North China Plain and the Sichuan Basin, are old and well-established geographic units. One in particular is new: China's Pacific Rim, still growing, yet poorly defined, and shown as "formative" in Figure 9-4.
2. *Xizang (Tibet)*. The high mountains and plateaus of Xizang, ruled by China but still widely known by its older name of Tibet, form a stark contrast to teeming China Proper. Here, next to one of the world's largest and most populous regions, lies one of the emptiest and, in terms of inhabited space, smallest regions.

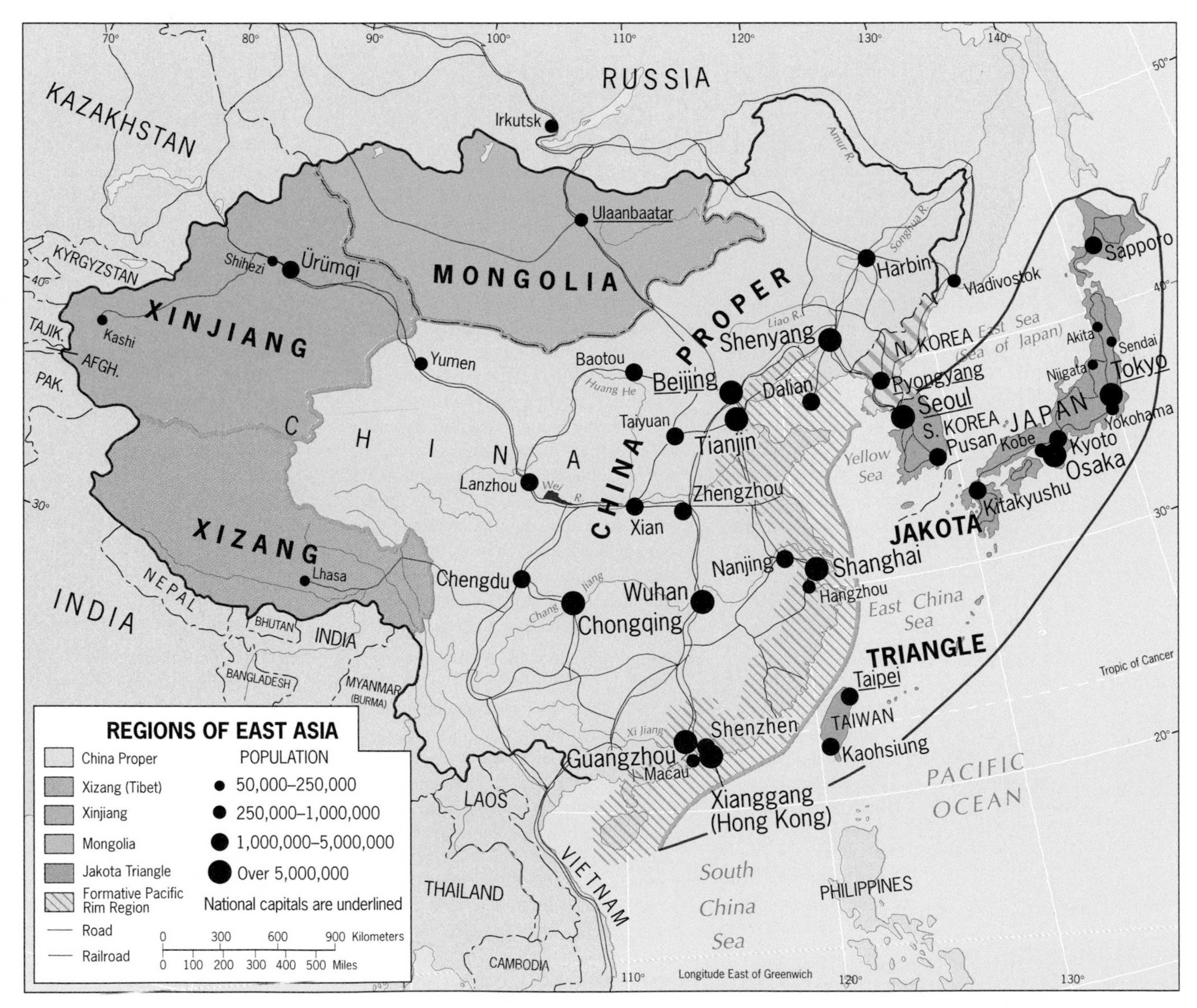

FIGURE 9-4

MAJOR CITIES OF THE REALM

City	Population* (in millions)
Beijing (Shi), China	10.9
Chongqing (Shi), China	30.0†
Guangzhou, China	3.9
Hong Kong (Xianggang) SAR, China	7.1
Macau SAR, China	0.5
Nanjing, China	2.8
Osaka, Japan	11.0
Pusan, South Korea	3.8
Seoul, South Korea	9.9
Shanghai (Shi), China	13.0
Shenyang, China	4.9
Shenzhen, China	3.9
Taipei, Taiwan	2.6
Tianjin (Shi), China	9.3
Tokyo, Japan	26.4
Wuhan, China	5.5
Xian, China	3.2

*Based on 2002 estimates
†See discussion of this estimate on p. 447.

3. *Xinjiang.* The vast desert basins and encircling mountains of Xinjiang form a third East Asian region. Again, physical as well as human geographic criteria come into play: here China meets Islamic Central Asia.
4. *Mongolia.* The desert state of Mongolia forms East Asia's fourth region. Like Tibet, landlocked Mongolia, vast but sparsely peopled, stands in stark contrast to populous China Proper.
5. *Jakota Triangle.* East Asia's fifth region is defined by its economic geography. *Ja*pan, South *Ko*rea, and *Ta*iwan (the name "Jakota" derives from the first two letters of each) were transformed by Pacific Rim economic developments during the second half of the twentieth century. Its recent emergence foreshadows further changes in the decades ahead as Korean unification becomes a possibility and contrasts between the Jakota Triangle and Pacific Rim China diminish.

CHINA PROPER

It is often maintained that the disintegration of the Soviet Union marked the collapse of the world's last remaining great empire. Russian imperialism forged a vast colonial empire that Moscow's communist despots inherited, but now the pieces of that empire are independent states. The age of empires, say some scholars, is over.

From a geographic viewpoint, that verdict is premature. By many measures, an empire far more populous than Russia's ever was still dominates the eastern quadrant of Asia. It is a realm without democracy or multiparty elections. It is a power core that controls territories that are colonies in all but name. It is a land of many disadvantaged minorities. It is a domain that still lays claim to territories beyond its borders. It is a regime that threatens neighbors. It is China, the last of the twentieth century's devolving empires.

China has not managed to shut out the winds of political and economic change that are sweeping the world. China's east is affected by the momentous changes marking the western Pacific Rim; in the west, China lies exposed to the postcolonial transformation of Turkestan (see Chapter 6). And in 1989 a student-led, labor-supported, pro-democracy movement that crested tragically at central Beijing's Tiananmen Square in a bloody suppression struck at the very heart of China. Ever since, human-rights issues have roiled China's international relations. None of this, however, has impelled China's communist rulers to do what Soviet leaders did. The Communist Party's monopoly over the politics of China remains as strong as ever. The world may have turned its back on communist dogma, but China has not. And China, we should remember, contains more than one-fifth of all humankind. Those who now write about a post-communist world are—again—too hasty.

CHINESE PERSPECTIVES

When we in the Western world chronicle the rise of civilization, we tend to focus on the historical geography of Southwest Asia, the Mediterranean, and Western Europe. Ancient Greece and Rome were the crucibles of culture; Mediterranean and Atlantic waters were the avenues of its diffusion. China lay remote, so we believe, barely connected to this Western realm of achievement and progress. When an Italian adventurer named Marco Polo visited China during the thirteenth century and described the marvels he saw there, his work did little to change European minds. Europe was and would always be the center of civilization.

The Chinese, naturally, take a different view. Events on the western edge of the great Eurasian landmass were deemed irrelevant to theirs, the most advanced and refined culture on Earth. Roman emperors were rumored to be powerful, and Rome was a great city, but nothing could match the omnipotence of China's rulers. Certainly the Chinese city of Xian far eclipsed Rome as a center of sophistication. Chinese civilization existed long before ancient Greece and Rome emerged, and it was still there long after they collapsed. China, the Chinese teach themselves, is eternal. It was, and always will be, the center of the civilized world.

We should remember this notion when we study China's regional geography because 4000 years of Chinese culture and perception will not change overnight—not even in a generation. Time and again, China overcame the invasions and depredations of foreign intruders, and afterward the Chinese would close off their vast country against the outside world. Just 30 years ago, in the early 1970s, there were just a few *dozen* foreigners in the entire country with its (then) nearly 1 billion inhabitants. The institutionalization of communism required this insularity, and even the Soviet advisors had been thrown out. But by the early 1970s, China's rulers decided that an opening to the Western world would be advantageous, and so U.S. President Richard Nixon was invited to visit Beijing. That historic occasion, in 1972, ended this latest period of isolation—as always, on China's terms. Since then, China has been open to tourists and business people, teachers and investors. Tens of thousands of Chinese students have been sent to study at American and other Western institutions. Long-suppressed ideas flowed into China, and the pro-democracy movement arose and climaxed in 1989. China's rulers knew that their violent repression of this movement would anger the world, but that did not matter because they deemed foreign condemnation irrelevant. Foreigners in China had done much worse. Moreover, Westerners had no business interfering in China's domestic affairs.

Relative Location and Isolation

As Table 9-1 reminds us, throughout their nation's history, the Chinese have sought to close their country to foreign influences; the most recent episode of exclusion occurred just a few decades ago. Exclusion is one of China's recurrent traditions, made possible by China's relative location and Asia's physiography. In other words, China's "splendid isolation" was made possible by geography.

Earlier we noted the role of relief and desert in encircling the culture hearth of East Asia, but equally telling is the factor of distance. Until recently, China lay far from the modern source areas of innovation and change. True, China—as the Chinese emphasize—was itself such a hearth, but China's contributions to the outside world remained limited, essentially, to finely made arts and crafts. China did interact with Korea, Japan, Taiwan, and parts of Southeast Asia, and eventually millions of Chinese emigrated to neighboring countries. But compare these regional links to those of the Arabs, who ranged worldwide and who brought their knowledge, religion, and political influence to areas from Mediterranean Europe to Bangladesh and from West Africa to Indonesia. Later, when Europe became the center of intellectual and material innovation, China found itself farther removed, by land or sea, than almost any other part of the world.

Today, modern communications notwithstanding, China still is distant from almost anywhere else on Earth. Going by rail from Beijing to Moscow, the capital of China's Eurasian neighbor, involves a tedious journey that takes the better part of a week. Direct surface connections with India are practically nonexistent. Overland linkages with Southeast Asian countries, though improving, remain tenuous.

But for the first time in its history, China now lies near a world-class hearth of technological innovation and financial power: Japan. This proximity to an industrial and financial giant is critical for the momentous economic developments taking place in China's coastal provinces. Japanese investments and business partnerships have transformed the economic landscape of Pacific-coast China. American and European trade links also are important, but Japan's role was crucial. Japan's economic success set the Pacific Rim engine in motion, and Japan's best financial years happened to coincide with China's reopening to foreigners in the 1970s. Geographic and economic circumstances combined to transform the map and made "Pacific Rim" a household word around the world.

Will China's openness to foreign influences be permanent this time? At the turn of the twenty-first century, it seemed inconceivable that China's open door could swing shut again. China now is a member of the WTO (World Trade Organization). Today it is part of a global network of finance and trade that connects it to many countries from South Korea to South Africa. The country also is unified, with little prospect of the kind of regional fragmentation seen so often in the past. Minorities do not threaten national unity. But always in China authoritarian rule has spawned rebellion, and China remains an authoritarian state. Communist dogma and capitalist energy coexist in this favorable economic climate. How long it can last no one knows.

EXTENT AND ENVIRONMENT

China's total area is slightly smaller than that of the United States including Alaska: each country has about 3.7 million square miles (9.6 million sq km). As Figure 9-5 reveals, the longitudinal extent of China and the 48 contiguous U.S. States also is similar. Latitudinally, however, China is considerably wider. Miami, near the southern limit of the United States, lies halfway between Shanghai and Guangzhou. Thus China's lower-latitude southern region takes on characteristics of tropical Asia. In the Northeast, too, China incorporates much of what in North America would be Que-

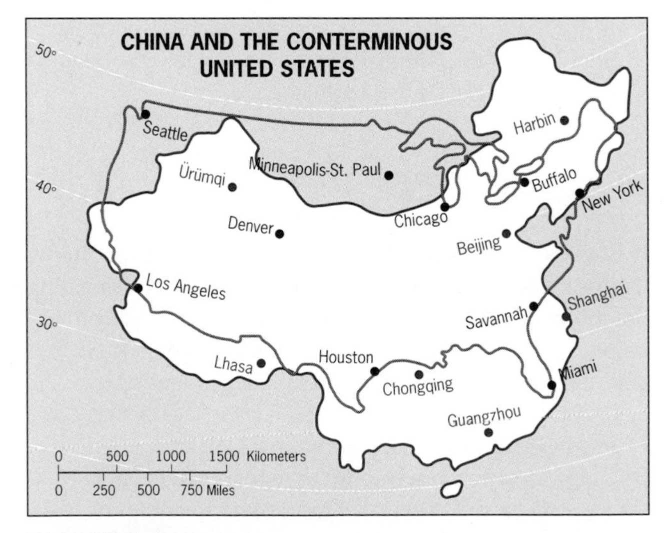

FIGURE 9-5

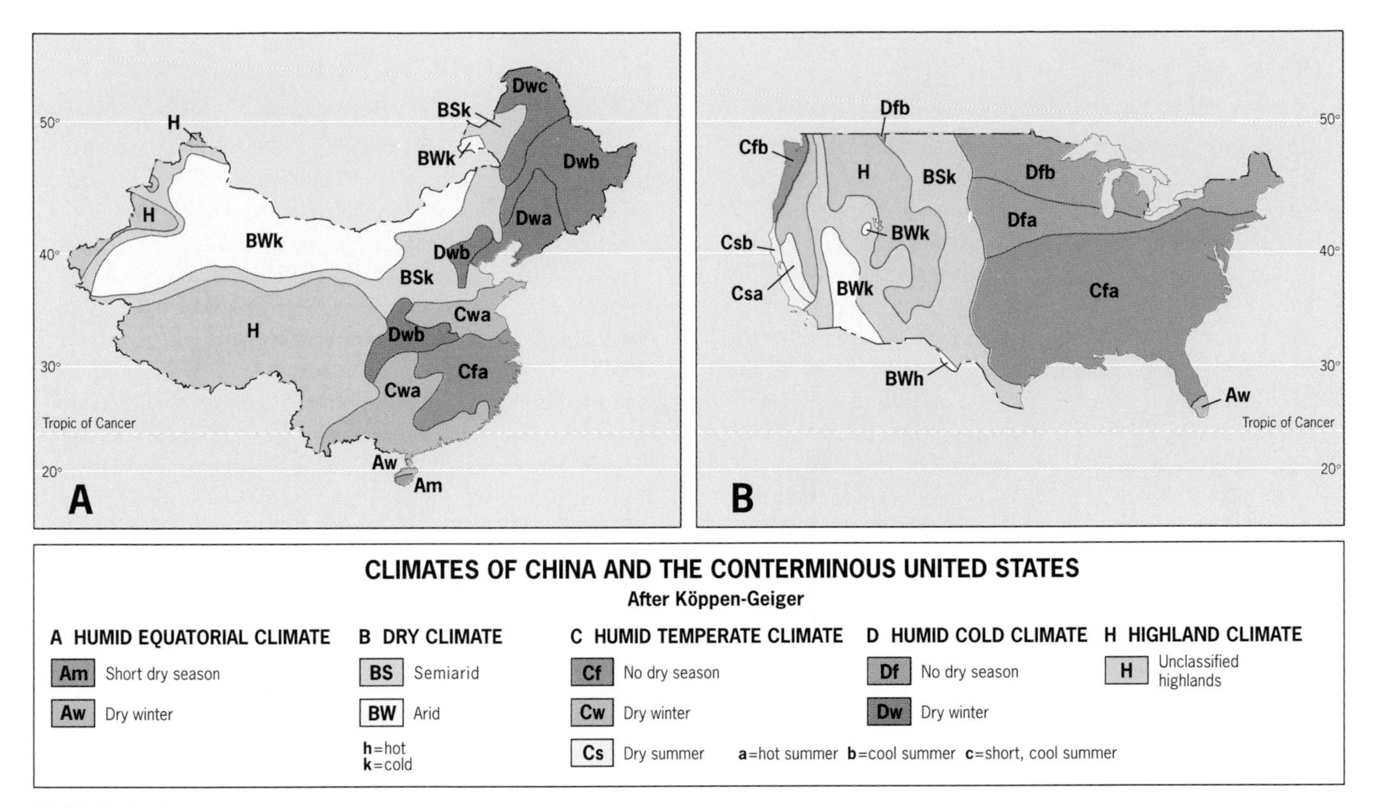

FIGURE 9-6

bec and Ontario. Westward, China's land area becomes narrower and physiographic similarities increase. But, of course, China has no west coast.

Now compare the climate maps of China and the United States in Figure 9-6 (which are enlargements of the appropriate portions of the world climate map in Figure I-8). Note that both have a large southeastern climatic region marked *Cfa* (that is, humid, temperate, warm-summer), flanked in China by a zone of *Cwa* (where winters become drier). Westward in both countries, the *C* climates yield to colder, drier climes. In the United States, moderate *C* climates develop again along the Pacific coast. China, however, stays dry and cold as well as high in elevation at equivalent longitudes.

Note especially the comparative location of the U.S. and Chinese *Cfa* areas in Figure 9-6. China's lies much farther to the south. In the United States, the *Cfa* climate extends beyond 40° North latitude, but in China, cold and generally winter-dry *D* climates take over at the latitude of Virginia. Beijing has a warm summer but a bitterly cold and long winter. Northeast China, in the general latitudinal range of Canada's lower Quebec and Newfoundland, is much more severe than its North American equivalent. Harsh environments prevail over vast regions of China, but, as we will see, nature compensates in spectacular fashion. From the climatic zone marked *H* (for highlands) in the west come the great life-giving rivers whose wide basins contain enormous expanses of fertile soils. Without these waters, China would not have a population more than four times that of the United States.

If you were to travel in China, environments and distances would at times seem familiar. The distance from Shanghai to the capital, Beijing, is not much greater than that from Washington, D.C. to Chicago. Flying cross-country would take about the same amount of time, given similar aircraft. But be prepared: China is not yet a country of modern transport facilities. Generally efficient airports, cross-country superhighways, or bullet trains have yet to make their impact. Still, return visitors are struck by the continuous improvement in China's transport infrastructure. Airports are being built and modernized. Competing airlines have bought and leased a growing fleet of planes. A new superhighway links the port of Tianjin with Beijing, and another (a toll road!) connects Shenzhen to Guangzhou. Developing, modernizing China is on the move.

EVOLVING CHINA

Even if China is not the world's longest continuous civilization (Egypt may claim this distinction), no other state on Earth can trace its cultural heritage as far back as China can. China's fortunes rose and fell, but over more than 40 centuries its people created a society with strong traditions, values, and philosophies. Kongfuzi (Confucius), whose teachings and writing still influence China, lived during the Zhou Dynasty, 2500 years ago (see box titled "Kongfuzi [Confucius]"). The Chinese still refer to themselves as the "People of Han," the dynasty that marks the breakdown of the old feudal order, the rise of military power, the unifica-

Kongfuzi (Confucius)

2

Confucius (*Kongfuzi* or *Kongzi* in pinyin) was China's most influential philosopher and teacher. His ideas dominated Chinese life and thought for over 20 centuries.

Kongfuzi was born in 551 BC and died in 479 BC. Appalled at the suffering of ordinary people during the Zhou Dynasty, he urged the poor to assert themselves and demand explanations for their harsh treatment by the feudal lords. He tutored the indigent as well as the privileged, giving the poor an education that had hitherto been denied them and ending the aristocracy's exclusive access to the knowledge that constituted power.

Kongfuzi's revolutionary ideas extended to the rulers as well as the ruled. He abhorred supernatural mysticism and cast doubt on the divine ancestries of China's aristocratic rulers. Human virtues, not godly connections, should determine a person's place in society, he taught. Accordingly, he proposed that the dynastic rulers turn over the reins of state to ministers chosen for their competence and merit. This was another Kongzi heresy, but in time this idea came to be accepted and practiced.

His Earthly philosophies notwithstanding, Kongfuzi took on the mantle of a spiritual leader after his death. His thoughts, distilled from the mass of philosophical writing (including Daoism) that poured forth during his lifetime, became the guiding principles of the formative Han Dynasty. The state, he said, should not exist just for the power and pleasure of the elite; it should be a cooperative system for the well-being and happiness of the people.

With time, a mass of writings evolved, much of which Kongfuzi never wrote. At the heart of this body of literature lay the Confucian Classics, 13 texts that became the basis for education in China for 2000 years. From government to morality and from law to religion, the Classics were Chinese civilization's guide. The entire national system of education (including the state examinations through which everyone, poor or privileged, could enter the civil service and achieve political power) was based on the Classics. Kongfuzi championed the family as the foundation of Chinese culture, and the Classics prescribe a respect for the aged that was a hallmark of Chinese society.

But Kongfuzi's philosophies also were conservative and rigid, and when the colonial powers penetrated China, Kongzi's Classics came face to face with practical Western education. For the first time, some Chinese leaders began to call for reform and modernization, especially of teaching. Kongzi principles, they said, could guide an isolated China, but not China in the new age of competition. But the Manchu rulers resisted this call, and the Nationalists tried to combine Kongzi and Western knowledge into a neo-Kongzi philosophy, which was ultimately an unworkable plan.

The communists who took power in 1949 attacked Kongzi thought on all fronts. The Classics were abandoned, indoctrination pervaded education, and, for a time, even the family was viewed as an institution of the past. Here the communists miscalculated. It proved impossible to eradicate two millennia of cultural conditioning in a few decades. The fading spirit of Kongfuzi will haunt physical and mental landscapes in China for generations to come.

tion of a large empire, the institution of property rights, the flourishing of architecture, the arts, and the sciences, and the development of trade along the Silk Route (Table 9-1). The Han Dynasty reigned 2000 years ago. The walled city of Xian, then called Ch'angan, was the early Han capital and one of the greatest cities of the ancient world.

As we try to gain a better understanding of China's present political, economic, and social geography, we should not lose sight of the China that endured for thousands of years under a sequence of imperial dynasties as a single entity. Larger and more populous than Europe, China had its divisive feudal periods, but always it came together again, ruled—mostly dictatorially—from a strong center as a unitary state. China was, and remains, a predominantly rural society, a powerless and subservient population controlled by an often ruthless bureaucracy. Retributions, famines, epidemics, and occasional uprisings took heavy tolls in the countryside. But in the cities, Chinese culture flourished, subsidized by the taxes and tribute extracted from the hinterlands. As China grew, it incorporated minorities ranging from Koreans and Mongols to Uyghurs and Tibetans; in the far south, it now includes several peoples with Southeast Asian affinities. Like the Chinese citizenry itself, these minorities have experienced both benevolent government and brutal subjugation. But it has always been China's wish to **Sinicize** them: to endow them with the elements of Chinese culture.

3

A Century of Convulsion

When the European colonialists appeared in East Asia, China long withstood them with a self-assured superiority based on the strength of its culture and the reassuring continuity of the state. There was no market for the British East India Company's rough textiles in a country long used to finely fabricated silks and cottons. There was little interest in the toys and trinkets the Europeans produced in the hope of barter for Chinese tea and porcelain. Even key European inventions, such as the mechanical clock, though considered amusing and entertaining, were ignored and even deprecated as irrelevant to Chinese culture.

Among the Realm's Great Cities . . .

Xian

The city known today as Xian is the site of one of the world's oldest urban centers. It may have been a settlement during the Shang-Yin Dynasty more than 3000 years ago; it was a town during the Zhou Dynasty, and the Qin emperor was buried here along with 6000 life-sized terracotta soldiers and horses, reflecting the city's importance. During the Han Dynasty the city, then called Ch'angan, was one of the greatest centers of the ancient world, the Rome of ancient China. Ch'angan formed the eastern terminus of the Silk Route, a storehouse of enormous wealth. Its architecture was unrivalled, from its ornamental defensive wall with elaborately sculpted gates to the magnificent public buildings and gardens at its center.

Situated on the fertile loess plain of the upper Wei River, Ch'angan was the focus of ancient China during crucial formative periods. After two centuries of Han rule, political strife led to a period of decline, but the Sui emperors rebuilt and expanded Ch'angan when they made it their capital. During the Tang Dynasty, Ch'angan again became a magnificent city with three districts: the ornate Palace City; the impressive Imperial City, which housed the national administration; and the busy Outer City containing the homes and markets of artisans and merchants.

After its Tang heyday the city again declined, although it remained a bustling trade center. During the Ming Dynasty it was endowed with some of its architectural landmarks, including the Great Mosque marking the arrival of Islam; the older Big Wild Goose Pagoda dates from the influx of Buddhism. After the Ming period, Ch'angan's name was changed to Xian (meaning "Western Peace"), then to Siking, and in 1943 back to Xian again.

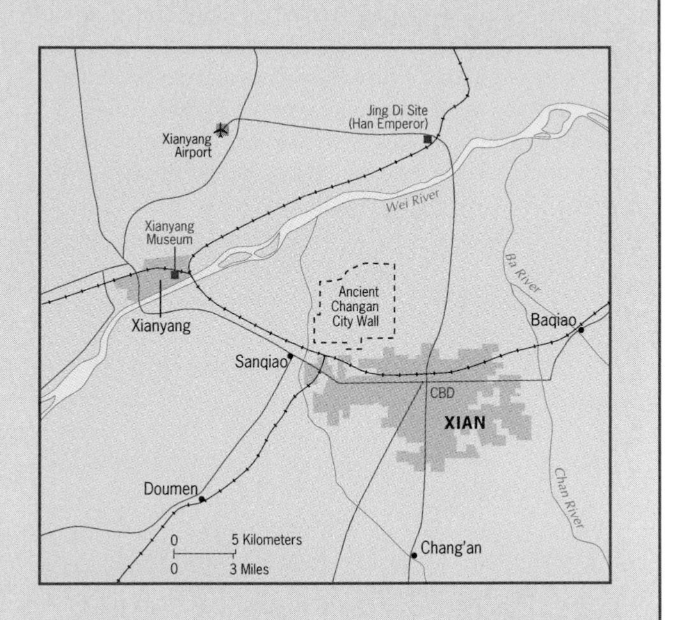

Having been a gateway for Buddhism and Islam, Xian in the 1920s became a center of Soviet communist ideology. The Nationalists, during the struggle against the Japanese, moved industries from the vulnerable east to Xian, and when the communists took power, they enlarged Xian's industrial base still further. The present sprawling city (population: 3.2 million) lies southwest of the famed tombs, its cultural landscape now dominated by a large industrial complex that includes a steel mill, textile factories, chemical plants, and machine-making facilities. Little remains (other than some prominent historic landmarks) of the splendor of times past, but Xian's location on the railroad to the vast western frontier of China sustains its long-term role as one of the country's key gateways.

FROM THE FIELD NOTES

"To pass through the gate from Tiananmen Square and enter the Manchu Emperors' Forbidden City was a riveting experience. Look back now toward the gate, and you see throngs of Chinese visiting what less than a century ago was the exclusive domain of the Qing Dynasty's absolute rulers—so absolute that unauthorized entry was punished by instant execution. The vast, walled complex at the heart of Beijing not only served as the imperial palace; it was a repository of an immense collection of Chinese technological and artistic achievements. Some of this is still here, but most of all you are awed by the lingering atmosphere, the sense of what happened here in these exquisitely constructed buildings that epitomized what high Chinese culture could accomplish. Inevitably you think of the misery of millions whose taxes and tribute paid for what stands here."

The self-confident Ming emperors even beat the Europeans at their own game. In the fourteenth century, great oceangoing vessels sailed the South China Sea; Chinese fleets carrying as many as 20,000 men reached Southeast and South Asia and even East Africa. Several times as large and far more sophisticated than anything built in Europe, China's ships, together with their sailors and the goods they carried, demonstrated the potential of the Middle Kingdom. But suddenly China faced a crisis at home: the impact of the onset of the Little Ice Age that also affected Europe. Cold and drought decimated the wheat harvest on the North China Plain, and boats were needed by the thousands to carry rice from the Chang Basin to the hungry north. That spelled the end of China's long-range maritime explorations; it may have changed the historical geography of the world.

Even when Europe's sailing ships made way for steam-driven vessels and newer and better European products (including weapons) were offered in trade for China's tea and silk, China continued to reject European imports and resisted commerce in general. (When the Manchus took con-

FIGURE 9-7

trol of Beijing in 1644, their bows and arrows proved superior to Chinese-manufactured muskets which were so heavy and hard to load that they were almost useless.) The Chinese kept the Europeans confined to small peninsular outposts, such as Macau, and minimized interaction with them. Long after India had succumbed to mercantilism and economic imperialism, China maintained its established order. This was no surprise to the Chinese. After all, they had held a position of undisputed superiority in their Celestial Kingdom as long as could be remembered, and they had dealt with foreign invaders before.

A (Lost) War on Drugs

All this confidence was shattered during the Manchu (Qing) Dynasty, China's last imperial regime. The Manchus, who had taken control in Beijing as a small minority and who had grafted their culture onto China's, had the misfortune of reigning when the standoff with Europeans ended and the balance of power shifted in favor of the colonialists.

On two fronts in particular, the economic and the political, the European powers destroyed China's invincibility. Economically, they succeeded in lowering the cost and improving the quality of manufactured goods, especially textiles, and the handicraft industries of China began to collapse in the face of unbeatable competition. Politically, the demands of the British merchants and the growing English presence in China led to conflicts. In the early part of the nineteenth century, the central issue was the importation into China from British India of opium, a dangerous and addictive intoxicant. Opium was destroying the very fabric of Chinese culture, weakening the society, and rendering China easy prey for colonial profiteers. As the Manchu government moved to stamp out the opium trade in 1839, armed hostilities broke out, and soon the Chinese found themselves losing a war on their own territory. The First Opium War (1839–1842) ended in disaster: China's rulers were forced to yield to British demands, and the breakdown of Chinese sovereignty was under way.

British forces penetrated up the Chang Jiang and controlled several areas south of it (Fig. 9-7); Beijing hurriedly sought a peace treaty by which it granted leases and concessions to foreign merchants. In addition, China ceded Hong Kong Island to the British and opened five ports, including Guangzhou (Canton) and Shanghai, to foreign commerce. No longer did the British have to accept a status that was inferior to the Chinese in order to do business; henceforth, negotiations would be pursued on equal terms. Opium now flooded into China, and its impact on Chinese society became even more devastating. Fifteen years after the First Opium War, the Chinese again tried to stem the disastrous narcotic tide, and the foreigners who had attached themselves to their country again defeated them. Now the government legalized cultivation of the opium poppy in China itself. Chinese society was disintegrating; the scourge of this drug abuse was not defeated until after the revival of Chinese power in the twentieth century.

But before China could reassert itself, much of what remained of its independence steadily eroded (see box titled "Extraterritoriality"). In 1898 the Germans obtained a lease on

Extraterritoriality

During the nineteenth century, as China weakened and European colonial invaders entered China's coastal cities and sailed up its rivers, the Europeans forced China to accept a European doctrine of international law—**extraterritoriality**. Under this doctrine, foreign states and their representatives are immune from the jurisdiction of the country in which they are based. Today, this applies to embassies and diplomatic personnel. But in Qing (Manchu) China, it went far beyond that.

The European, Russian, and Japanese invaders established as many as 90 *treaty ports*—extraterritorial enclaves in China's cities under unequal treaties enforced by gunboat diplomacy. In their "concessions," diplomats and traders were exempt from Chinese law. Not only port areas but also the best residential suburbs of large cities were declared to be "extraterritorial" and made inaccessible to Chinese citizens. In the city of Guangzhou (Canton in colonial times), Sha Mian Island in the Pearl River was a favorite extraterritorial enclave. A sign at the only bridge to the island stated, in English and Cantonese, "No Dogs or Chinese."

Christian missionaries fanned out into China, their residences and churches fortified with extraterritorial security. In many places, Chinese found themselves unable to enter parks and buildings without permission from foreigners. This involved a loss of face that contributed to bitter opposition to the presence of foreigners—a resentment that exploded in the Boxer Rebellion of 1900.

After the collapse of the Qing Dynasty in 1911, the Chinese Nationalists negotiated an end to all commercial extraterritoriality in China Proper; the Russians, however, would not yield in then-Manchuria. Only Hong Kong and Macau retained their status as colonies.

When China's government in 1980 embarked on a new economic policy that gave major privileges and exemptions to foreign firms in certain coastal areas and cities, opponents argued that this policy revived the practice of extraterritoriality in a new guise. This issue remains a sensitive one in a China that has not forgotten the indignities of the colonial era.

Qingdao on the Shandong Peninsula, and the French acquired a sphere of influence in the far south at Zhanjiang (Fig. 9-7). The Portuguese took Macau; the Russians obtained a lease on Liaodong in the Northeast as well as railway concessions there; even Japan got into the act by annexing the Ryukyu Islands and, more importantly, Formosa (Taiwan) in 1895.

After four millennia of recurrent cultural cohesion, economic security, and political continuity, the Chinese world lay open to the aggressions of foreigners whose innovative capacities China had negated to the end. Now the ships flying European flags lay in the ports of China's coasts and rivers, but China had not learned to manufacture the cannons to blast them out of the water. The smokestacks of foreign factories rose over the townscapes of its great cities, and no Chinese force could dislodge them. The Japanese, seeing China's weakness, invaded Korea. The Russians entered Manchuria, the home base of the Manchus. The foreign invaders even took to fighting among themselves, as Japan and Russia did in Manchuria in 1904. (Today this part of China is called the *Northeast*, the name Manchuria having been—understandably—rejected.)

A New China Rises

In the meantime, organized opposition to the foreign presence in China was gathering strength, and the twentieth century opened with a large-scale revolt against all outside elements. Bands of revolutionaries roamed both cities and countryside, attacking not only the hated foreigners but also Chinese who had adopted Western cultural traits. Known as the Boxer Rebellion (after a loose translation of the Chinese name for these revolutionary groups), this 1900 uprising was put down with much bloodshed by an international force consisting of British, Russian, French, Italian, German, Japanese, and American soldiers. Simultaneously, another revolutionary movement was gaining support, aimed against the Manchu leadership itself. In 1911, the emperor's garrisons were attacked all over China, and in a few months the 267-year-old Qing Dynasty was overthrown. Indirectly, it too was a casualty of the foreign intrusion, and it left China divided and disorganized.

The fall of the Manchus and the proclamation of a republican government in China did little to improve the country's overall position. The Japanese captured Germany's holdings on the Shandong Peninsula, including the city of Qingdao, during World War I. When the victorious European powers met at Versailles in 1919 to divide the territorial spoils, they affirmed Japan's rights in the area. This led to yet another Chinese effort to counter the foreign scourge. Nationwide protests and boycotts of Japanese goods were organized in what became known as the May Fourth Movement. One participant in these demonstrations was a charismatic young man named Mao Zedong.

Nonetheless, China remained badly divided after World War I. By the early 1920s, there were two governments—one in Beijing and another in the southern city of Guangzhou (Canton), where the famous Chinese revolutionary, Sun Yat-sen, was the central figure. Neither government could pretend to control much of China. The Northeast was in chaos, petty states were emerging all over the central part of the country, and the Guangzhou "Parliament" controlled only a part of Guangdong Province in the Southeast. Nevertheless, it was just at this time that the power groups that would ultimately vie for supremacy in China were formed. While Sun Yat-sen was trying to establish a viable Nationalist government in Guangzhou, intellectuals in Shanghai founded the Chinese Communist Party. Several of these intellectuals had been leaders in the May Fourth Movement, and in the early 1920s they received help from the Communist Party of the Soviet Union. Mao Zedong was already a prominent figure in these events.

Initially, the new Communist Party and the Nationalists led by Sun Yat-sen cooperated with each other. The Nationalists were stronger and better organized, and they hoped to use the communists in their anti-foreign (especially anti-British) campaigns. By 1927, the foreigners were on the run; the Nationalist forces entered cities and looted and robbed at will, while aliens were evacuated or sometimes killed. But as the Nationalists continued their drive northward and just as success was within reach, internal dissension arose. Soon, the Nationalists were as busy purging the communists as they were pursuing foreigners. The central figure to emerge among the Nationalists during this period was Chiang Kai-shek. Sun Yat-sen died in 1925, and when the Nationalists established their capital at Nanjing (Nanking) in 1928, Chiang was the country's leader.

Three-Way Struggle

The post-Manchu period of strife and division in China resembled other times in its history when, after a long period of comparative stability under dynastic rule, the country fragmented into rival factions. In the first years of the Nanjing government's hegemony, the campaign against the communists intensified and thousands were killed. Chiang's armies drove the communists ever deeper into the interior (Mao himself escaped the purges only because he was in a remote rural area at the time). For a while, it seemed that Nanjing's armies would break the back of the communist movement in China.

The Long March

A core area of communist peasant forces survived in the zone where the provinces of Jiangxi and Hunan adjoin in southeastern China, and these forces defied Chiang's attempts to destroy them. Their situation grew worse, however, and in 1933 the Nationalist armies were on the verge of encircling this last eastern communist stronghold. The communists decided to avoid inevitable strangulation by leaving. Nearly 100,000 people—armed soldiers, peasants, local leaders—gathered near Ruijin and started to walk westward in 1934. This was a momentous event in modern

China, and among the leaders of the column were Mao Zedong and Zhou Enlai. The Nationalists rained attack after attack on the marchers but never succeeded in wiping them out completely; as the communists marched, they were joined by new sympathizers.

The Long March (see the route in Fig. 9-7), as this drama has come to be called, first took the communists to Yunnan Province, where they turned north to enter western Sichuan. They then traversed Gansu Province and eventually reached their goal, the mountainous interior near Yanan in Shaanxi Province. The Long March covered nearly 6000 miles (10,000 km) of China's more difficult terrain, and the Nationalists' continuous attacks killed an estimated 75,000 of the original participants. Only about 20,000 survived the epic migration, but among them were Mao and Zhou, who were convinced that a new China would arise from the peasantry of the rural interior to overcome the urban easterners whose armies could not eliminate them.

The Japanese

While the Nanjing government was pursuing the communists, foreign interests exploited the situation to further their own objectives in China. The Soviet Union held a sphere of influence in Mongolia and was on the verge of annexing a piece of Xinjiang. Japan was dominant in the Northeast, where it controlled ports and railroads. The Nanjing government tried to resist the expansion of Japan's sphere of influence. When the effort failed, the Japanese set up a puppet state in the region; they appointed a Manchu ruler and called their new possession Manchukuo.

The inevitable full-scale war between the Chinese and the Japanese broke out in 1937. For a while, the Chinese communists and the Nationalists stopped fighting each other in order to concentrate on fighting Japan, but soon their factional war erupted again. Now Chinese communists were fighting Chinese Nationalists while both fought the Japanese. In the process, China broke up into three regions: the Japanese sphere in the north and east (Fig. 9-7), the Nationalists' domain centered on their capital of Chongqing in the Sichuan Basin, and the communist zone in the interior west. The Japanese, by pursuing and engaging Chiang's Nationalist forces, enabled the communists to grow stronger and enhance their prestige in China's western areas.

The Japanese committed unspeakable atrocities in their campaign in China. Millions of Chinese citizens were shot, burned, drowned, subjected to gruesome chemical and biological experiments, and otherwise wantonly victimized. Years later, when China's economic reforms of the 1980s and 1990s led to a renewed Japanese presence in China, the Chinese public and its leaders called for Japan to acknowledge and apologize for these wartime abuses. In Japan, this pitted apologists against strident nationalists, causing a political crisis. In 1992, Emperor Akihito visited China and referred to the war but stopped short of a formal apology. The book is not yet closed on this most sensitive issue.

Communist China Arises

After the U.S.-led Western powers defeated Japan in 1945, the civil war in China quickly resumed. The United States, hoping for a stable and friendly government in China, sought to mediate the conflict but at the same time recognized the Nationalists as the legitimate government. The United States also aided the Nationalists militarily, destroying any chance of genuine and impartial mediation. By 1948, it was clear that Mao Zedong's well-organized militias would defeat Chiang Kai-shek. Chiang kept moving his capital—back to Guangzhou, seat of Sun Yat-sen's first Nationalist government, then back to Chongqing. Late in 1949, after a series of disastrous defeats in which hundreds of thousands of Nationalist forces were killed, the remnants of Chiang's faction gathered Chinese treasures and valuables and fled to the island of Taiwan. There, they took control of the government and proclaimed their own Republic of China.

Meanwhile, on October 1, 1949, standing in front of the assembled masses at the Gate of Heavenly Peace on Beijing's Tiananmen Square, Mao Zedong proclaimed the birth of the People's Republic of China.

CHINA'S HUMAN GEOGRAPHY

After more than a half-century of communist rule, China is a society transformed. It has been said that the year 1949 actually marked the beginning of a new dynasty not so different from the old, an autocratic system that dictated from the top. In that view, Mao Zedong simply bore the mantle of his dynastic predecessors. Only the family lineage had fallen away; now communist "comrades" would succeed each other.

And certainly some of China's old traditions continued during the communist era, but in many other ways Chinese society was totally overhauled. Benevolent or otherwise, the dynastic rulers of old China headed a country in which—for all its splendor, strength, and cultural richness—the fate of landless people and of serfs often was undescribably miserable; in which floods, famines, and diseases could decimate the populations of entire regions without any help from the state; in which local lords could (and often did) repress the people with impunity; in which children were sold and brides were bought. The European intrusion made things even worse, bringing slums, starvation, and deprivation to millions who had moved to the cities.

The communist regime, dictatorial though it was, attacked China's weaknesses on many fronts, mobilizing virtually every able-bodied citizen in the process. Land was taken from the wealthy; farms were collectivized; dams and levees were built with the hands of thousands; the threat of hunger for millions receded; health conditions improved; child labor was reduced. Mao's long tenure (1949–1976) and apparent omnipotence may remind us of the dynastic rulers' frequent longevity and absolutism, but the new China he left behind was vastly different from the old.

In several areas, however, Mao's strict adherence to Marxist dogma had problematic consequences for China. One of these consequences is familiar to anyone who has studied communist planned economies anywhere: government's control over all productive capacity, not only agriculture but also industry. China's industrialization burdened the country with thousands of inefficient, uncompetitive state-owned manufacturing plants. Another of Mao's dictums had to do with population. Like the Soviets (and influenced by a horde of Soviet advisors and planners), Mao refused to impose or even recommend any population policy, arguing that such a policy would represent a capitalist plot to constrain China's human resources. As a result, China's population grew explosively during this rule.

Yet another costly episode of Mao's rule was the so-called *Great Proletarian Cultural Revolution*, launched by Mao Zedong during his last decade in power (1966–1976). Fearful that Maoist communism was being contaminated by Soviet "deviationism" and worried about his own stature as its revolutionary architect, Mao unleashed a campaign against what he viewed as emerging elitism in society. He mobilized young people living in cities and towns into cadres known as Red Guards and ordered them to attack "bourgeois" elements throughout China, criticize Communist Party officials, and root out "opponents" of the system. He shut down all of China's schools, persecuted untrustworthy intellectuals, and encouraged the Red Guards to engage in what he called a renewed "revolutionary experience." The results were disastrous: Red Guard factions took to fighting among themselves, and anarchy, terror, and economic paralysis followed. Thousands of China's leading intellectuals died, moderate leaders were purged, and teachers, elderly citizens, and older revolutionaries were tortured to make them confess to "crimes" they did not commit. As the economy suffered, food and industrial production declined. Violence and famine killed as many as 30 million people as the Cultural Revolution spun out of control. One of those who survived was a Communist Party leader who had himself been purged and then been reinstated—Deng Xiaoping. Deng was destined to lead the country in the post-Mao period of economic transformation.

FIGURE 9-8

Political and Administrative Divisions

Before we investigate the emerging human geography of contemporary China, we should acquaint ourselves with the country's political and administrative framework (Fig. 9-8). For administrative purposes, China is divided into the following units:

4 Central-Government-Controlled Municipalities (*Shi's*)
5 Autonomous Regions
22 Provinces
2 Special Administrative Regions

The four central-government-controlled municipalities are the capital, Beijing; its nearby port city, Tianjin; China's largest metropolis, Shanghai; and the Chang river port of Chongqing, in the interior. These *Shi's* form the cores of China's most populous and important subregions, and direct control over them from the capital entrenches the central government's power.

We should note that the administrative map of China continues to change—and to pose problems for geographers. The city of Chongqing was made a *shi* in 1996, and its "municipal" area was enlarged to incorporate not only the central urban area but a huge hinterland covering all of eastern Sichuan Province. As a result, the "urban" population of Chongqing is officially 30 million, making this the world's largest metropolis—but in truth, the central urban area has no more than about 6 million inhabitants. And because Chongqing's population is officially *not* part of the province that borders it to the west (Fig. 9-8), the official population of Sichuan declined by 30 million when the *Chongqing Shi* was created. Despite what Table 9-2 suggests, therefore, the population within Sichuan Province's outer borders remains the largest of any of China's administrative divisions.

The five Autonomous Regions were established to recognize the non-Han minorities living there. Some laws that apply to Han Chinese do not apply to certain minorities. As we saw in the case of the former Soviet Union, however, demographic changes and population movements affect such regions, and the policies of the 1940s may not work in the twenty-first century. Han Chinese immigrants now outnumber several minorities in their own regions. The five Autonomous Regions (A.R.'s) are: (1) Nei Mongol A.R. (Inner Mongolia); (2) Ningxia Hui A.R. (adjacent to Inner Mongolia); (3) Xinjiang Uyghur A.R. (China's northwest corner); (4) Guangxi Zhuang A.R. (far south, bordering Vietnam); and (5) Xizang A.R. (Tibet).

China's 22 Provinces, like U.S. States, tend to be smallest in the east and largest toward the west. The territorially smallest are the three easternmost provinces on China's coastal bulge: Zhejiang, Jiangsu, and Fujian. The two largest are Qinghai, flanked by Tibet, and Sichuan, China's Midwest.

As with all large countries, some provinces are more important than others. The Province of Hebei nearly surrounds Beijing and occupies much of the core of the country. The Province of Shaanxi is centered on the great ancient city of Xian. In the southeast, momentous economic developments are occurring in the Province of Guangdong, whose urban focus is Guangzhou. When, in the pages that follow, we refer to a particular province or region, Figure 9-8 is a useful locational guide.

In 1997, the British dependency of Hong Kong (Xianggang) was taken over by China and became the country's first **Special Administrative Region (SAR)**. In 1999, Portugal similarly transferred Macau, opposite Hong Kong on the Pearl River Estuary, to Chinese control, creating the second SAR under Beijing's administration.

5

Population Issues

Table 9-2 underscores the enormity of China's population: many Chinese provinces have more inhabitants than do most of the world's countries. With a 2002 population of just over 1.3 billion, China remains the largest nation on

Table 9-2
CHINA: POPULATION BY MAJOR ADMINISTRATIVE DIVISIONS*
(In Millions)

Provinces	(1990 Census)	(2002 Estimate)
Anhui	56.2	64.3
Fujian	30.0	35.3
Gansu	22.4	25.1
Guangdong	62.8	77.9
Guizhou	32.4	37.6
Hainan	6.6	7.5
Hebei	61.1	71.1
Heilongjiang	35.2	39.2
Henan	85.5	96.1
Hubei	54.0	61.3
Hunan	60.7	68.9
Jiangsu	67.1	76.4
Jiangxi	37.7	43.2
Jilin	24.7	27.1
Liaoning	39.5	44.1
Qinghai	4.5	5.3
Shaanxi	32.9	36.3
Shandong	84.4	93.0
Shanxi	28.8	32.7
Sichuan	107.2	90.2
Yunnan	37.0	41.8
Zhejiang	41.5	46.8
Autonomous Regions		
Guangxi Zhuang	42.2	47.6
Nei Mongol	21.5	28.5
Ningxia Hui	4.7	5.6
Xinjiang Uyghur	15.2	18.4
Xizang	2.2	2.3

*Population data for the four centrally-administered municipalities and the Hong Kong (Xianggang) and Macau Special Administrative Regions are shown in the table on p. 437.

Earth. We should view these huge numbers in the context of China's population growth rate. High rates of population growth dim national development prospects in many countries in the periphery, but China has a special problem. Even a modest growth rate—such as 1 percent annually—still adds some 13 million to the total every year.

Mao Zedong, as we have noted, refused to institute policies that would reduce the rate of population increase because he believed (encouraged by Soviet advisors) that this would play into the hands of his capitalist enemies. Fast-growing populations, he said, constituted the only asset many poor, communist, or socialist countries had. If the capitalist world wanted to reduce growth rates, it must be an anti-communist plot.

But other Chinese leaders, realizing that rapid population growth would stymie their country's progress, tried to stem the tide through local propaganda and education. Some of these leaders lost their positions as "revisionists," but their point was not lost on China's future leaders. After Mao's death, China embarked on a vigorous population-control program. In the early 1970s, the annual rate of natural increase was about 3 percent; by the mid-1980s, it was down to 1.2 percent. Families were ordered to have one child only, and those who violated the policy were penalized by losing tax advantages, educational opportunities, and even housing privileges. Today, China's census bureau officially reports a growth rate of 0.9 percent.

Although the one-child policy gained general acceptance, since it was vigorously policed throughout China by Communist Party members, it also had serious negative consequences. Abortions soared, sometimes in the third trimester and coerced by Party enforcers. Female infanticide also rose because families wanted their only child to be a boy. Families tried to hide their second and third children by sending them to family members in other villages; upon

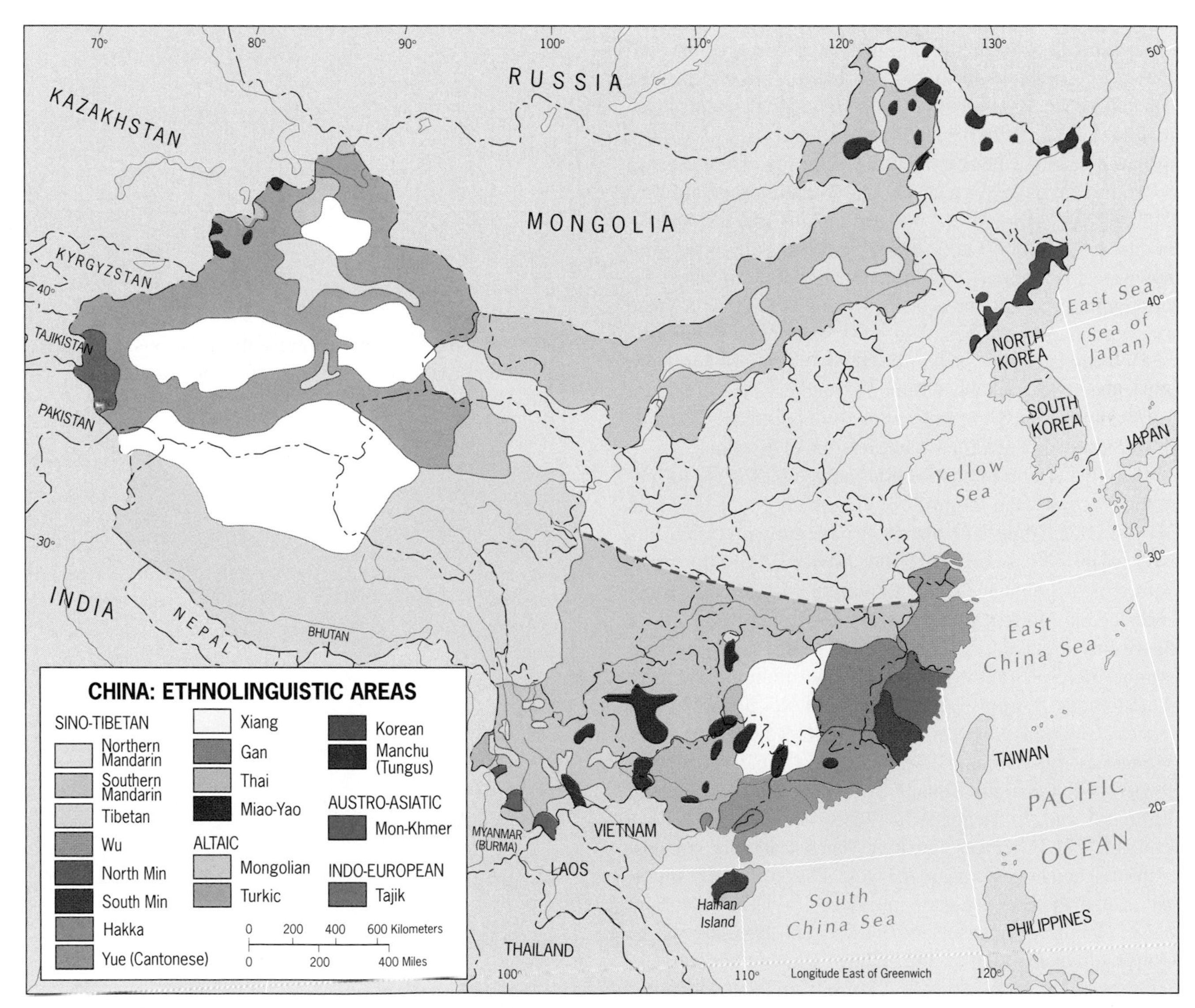

FIGURE 9-9

discovery by Party members, some parents had their houses burned down. When reports of such dreadful byproducts of the population-control campaign leaked to the outside world, the policy was relaxed during the late 1980s. Certain farming and fishing families were allowed to have a second child to fulfill future labor needs, and rural families could request permission to have a second child if their first baby was a girl. Ethnic minority groups were mostly exempted.

Partly because of disobedience and partly because of some relaxation of the one-child policy, China's growth rate may be inching upward again from the officially posted 0.9 percent. The objective of China's demographic planners was to reach 1.2 billion no earlier than the year 2000, but that goal was missed, perhaps by as many as 70 million. This means that an additional 70 million people must be housed, educated, treated medically, and, eventually, employed—a population almost the size of the largest country in Europe, Germany. Against this background, China's wish for rapid economic progress will not easily be fulfilled.

The Minorities

We asserted earlier that China remains an empire; its government controls territories and peoples that are in effect colonized. The ethnolinguistic map (Fig. 9-9) apparently confirms this proposition. This map should be seen in context, however. It does not show local-area majority populations but instead reveals where minorities are concentrated. For example, the Mongolian population in China is shown to be clustered along the southeastern border of Mongolia in the Autonomous Region called Nei Mongol (see Fig. 9-8). But even in that A.R., the Han Chinese, not the Mongols, are now in the majority.

Nevertheless, the map gives definition to the term *China Proper* as the home of the People of Han, the ethnic (Mandarin-speaking) Chinese depicted in tan and light orange in Figure 9-9. From the upper Northeast to the border with Vietnam and from the Pacific coast to the margins of Xinjiang, this Chinese majority dominates. When you compare this map to that of population distribution (to be discussed shortly), it will be clear that the minorities constitute only a small percentage of the country's total. The Han Chinese form the largest and densest clusters.

In any case, China controls non-Chinese areas that are vast, if not populous. The Tibetan group numbers under 3 million, but it extends over all of settled Xizang. Turkic peoples inhabit large areas of Xinjiang. Thai, Vietnamese, and Korean minorities also occupy areas on the margins of Han China. As we will see later, the Southeast Asian minorities in China have participated strongly in the Pacific Rim developments on their doorsteps. Hundreds of thousands have migrated from their Autonomous Regions to the economic opportunities along the coast.

Numerically, Chinese dominate in China to a far greater degree than Russians dominated their Soviet Empire. But territorially, China's minorities extend over a proportionately larger area. The Ming and Manchu rulers bequeathed the People of Han an empire.

PEOPLE AND PLACES OF CHINA PROPER

A map of China's population distribution (Fig. 9-10) reveals the continuing relationship between the physical stage and its human occupants. In technologically advanced countries, we have noted, people shake off their dependence on what the land can provide; they cluster in cities and in other areas of economic opportunity. This depopulates rural areas that may once have been densely inhabited. In China, that stage has not yet been reached. While China has large cities, as in India the great majority of the people (69 percent) still live on—and from—the land. Thus the map of population distribution reflects the livability and productivity of China's basins, lowlands, and plains. Compare Figures 9-6A and 9-10, and China's continuing dependence on soil, water, and warmth will be evident.

The population map also suggests that in certain areas environmental limitations are being overcome. Industrialization in the Northeast, irrigation in the Inner Mongolia Autonomous Region, and oil-well drilling in Xinjiang enabled millions of Chinese to migrate from China Proper into these frontier zones, where they now outnumber the indigenous minorities.

Nevertheless, physiography and demography remain closely related in China. To grasp this relationship, it is useful to compare Figures 9-2 (physiography) and 9-10 (population) as our discussion proceeds. On the population map, the darker the color, the denser the population: in places we can follow the courses of major rivers through this scheme. Look, for example, at China's Northeast. A ribbon of population follows the Liao and the Songhua rivers. Also note the huge, nearly circular population concentration on the western edge of the red zone; this is the Sichuan Basin, in the upper course of the Chang Jiang. Four major river basins contain more than three-quarters of China's nearly 1.3 billion people:

1. The Liao-Songhua Basin or the Northeast China Plain
2. The Lower Huang He (Yellow River) Basin, known as the North China Plain
3. The Upper and Lower Basins of the Chang Jiang (Yangzi River)
4. The Basins of the Xi (West) and Pearl River

Not all the people living in these river plains are farmers, of course. China's great cities also have developed in these populous areas, from the industrial centers of the Northeast to the Pacific Rim upstarts of the South. Both Beijing and

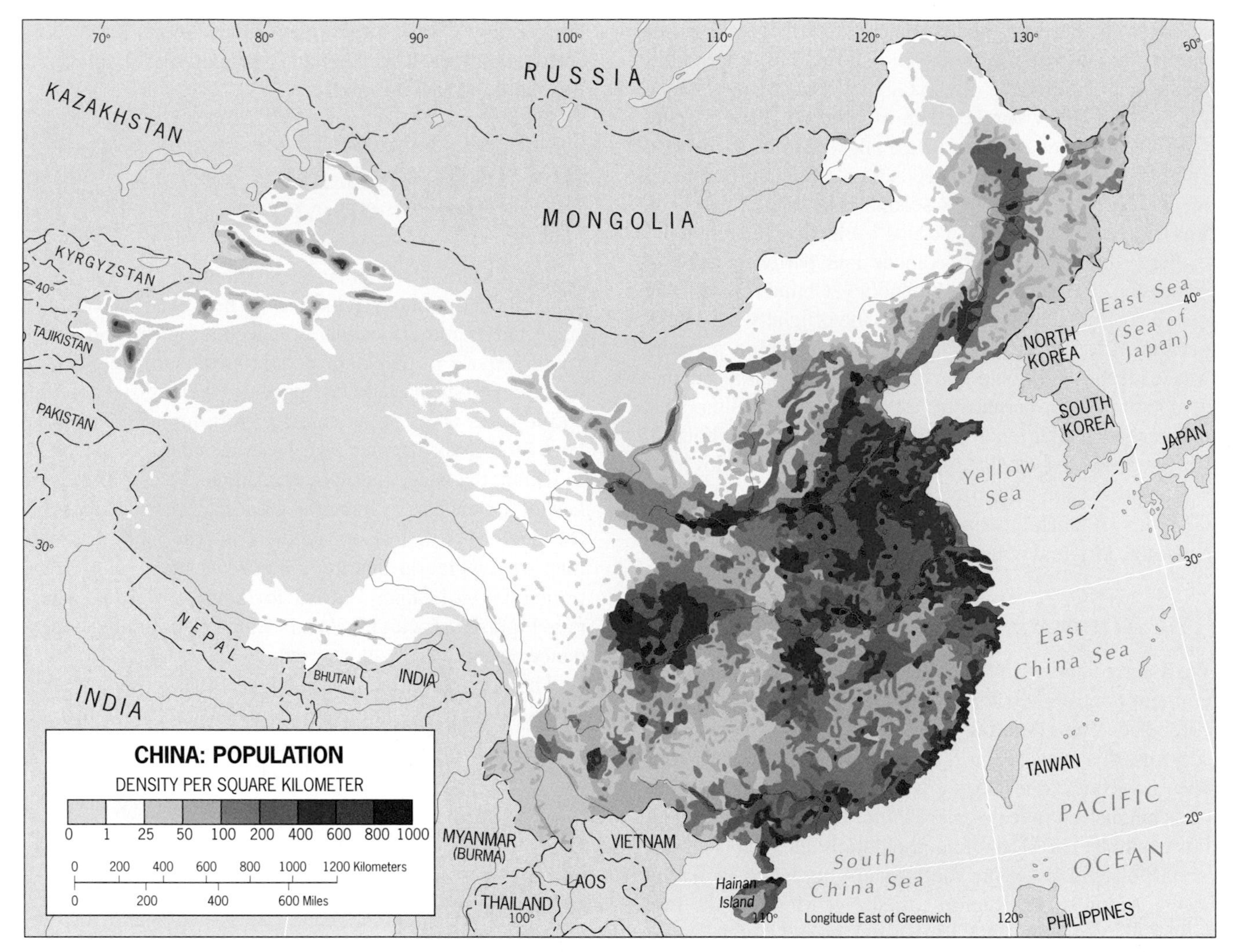

FIGURE 9-10

Shanghai lie in major river basins. And, as the map shows, the hilly areas between the river basins are not exactly sparsely populated either. Note that the hill country south of the Chang Basin (opposite Taiwan) still has a density of 130 to 260 people per square mile (50 to 100 per sq km).

Northeast China Plain

The Northeast China Plain is the heartland of China's Northeast, ancestral home of the Manchus who founded China's last dynasty, battleground between Japanese and Russian invaders, Japanese colony, heartland of communist industrial development, and now losing ground to the industries of the Pacific Rim. This used to be called Manchuria. As the administrative map (Fig. 9-8) shows, there are three provinces here: Liaoning in the south, facing the Yellow Sea; Jilin in the center; and Heilongjiang, by far the largest, in the north.

When you are familiar with the vast farmlands of the North China Plain, the southernmost part of the Northeast China Plain looks so similar that it seems to be a continuation. Near the coast, farms look as they do near the mouth of the great Huang He; to the east, the Liaodong Peninsula looks like the Shandong Peninsula across the water. That impression soon disappears, however. Travel northward, and the Northeast China Plain reveals the effects of cold climate and thin soils. Farmlands are patchy; smokestacks seem to rise everywhere. This is industrial, not primarily farming, country. But the landscape looks like a Rustbelt: equipment often is old and outdated, transport systems are inadequate, buildings are in disrepair. Stream, land, and air pollution is dreadful. The Northeast in some ways resembles industrial zones in former communist Eastern Europe.

The Northeast has experienced many ups and downs. Although Japanese colonialism was ruthless and exploitive, Japan did build railroads, roads, bridges, factories, and other components of the regional infrastructure. (At one time, half of the entire railroad mileage in China was in the Northeast.) After the Japanese were ousted, the Soviets looted the area of machinery, equipment, and other goods. During the late

1940s, the Northeast was a ravaged frontier. But then the communists took power and made the industrial development of the Northeast a priority. From the 1950s until the 1970s, the Northeast led the nation in manufacturing growth. Its population, just a few million in the 1940s, mushroomed to more than 100 million. Towns and cities grew exponentially.

All this growth was based on the Northeast's considerable mineral wealth (Fig. 9-11). Iron ore deposits and coalfields lie concentrated in the Liao Basin; aluminum ore, ferro-alloys, lead, zinc, and other metals are plentiful, too. As the map shows, the region also is well endowed with oil reserves; the largest lie in the Daqing Reserve between Harbin and Qiqihar. The communist planners made Changchun capital of Jilin Province, the Northeast's automobile manufacturing center. Shenyang, capital of Liaoning Province, became the leading steelmaking complex. Fushun and Anshan were assigned other functions, and the communists encouraged Han Chinese to migrate to the Northeast to find employment

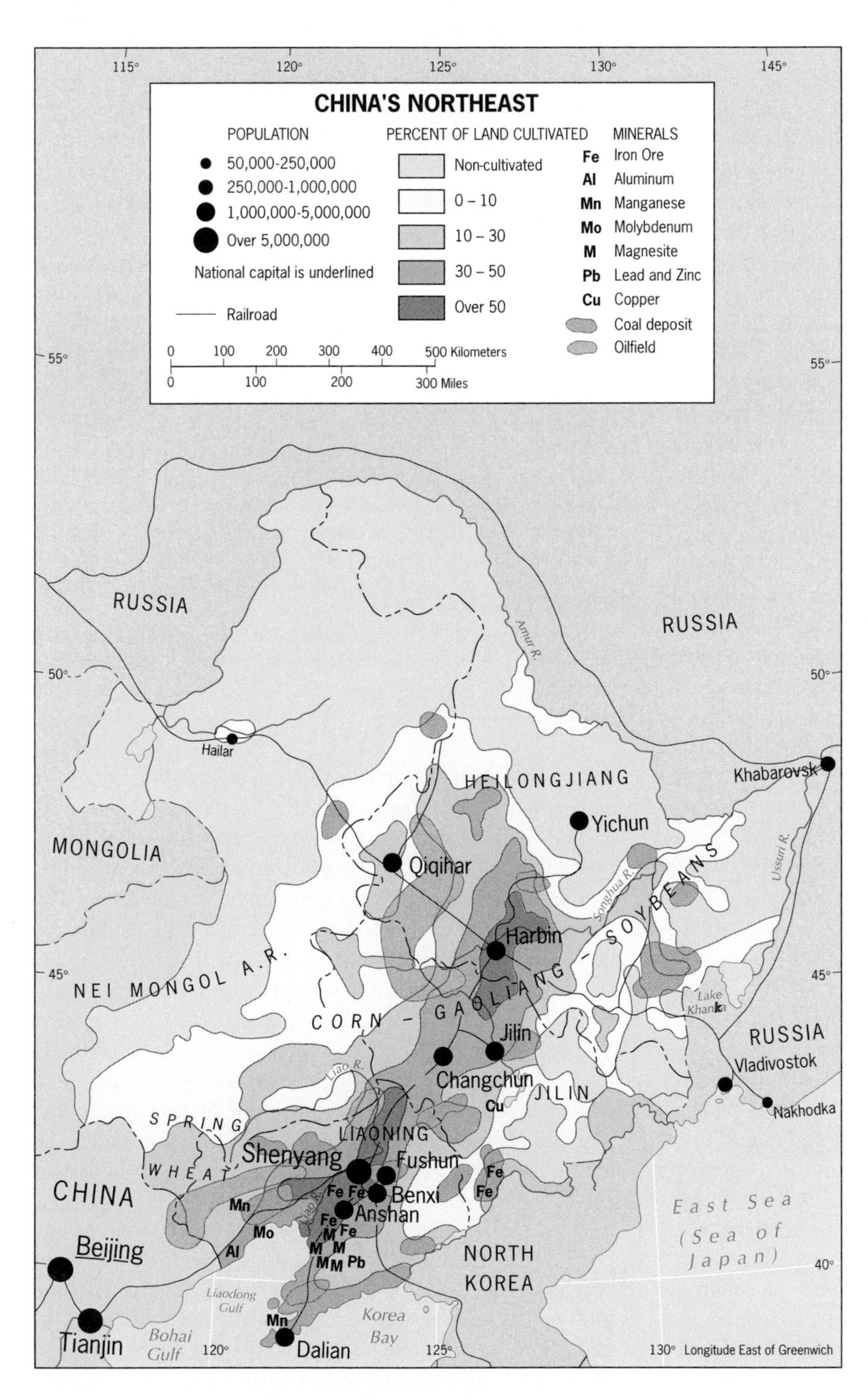

FIGURE 9-11

there. To attract them, the government built apartments, schools, hospitals, recreational facilities, even old-age homes, all to speed the region's industrial development.

For a time, it worked. During the 1970s, the Northeast was responsible for fully one-quarter of the entire country's industrial output. Stimulated by the needs of the growing cities and towns, farms expanded and crops diversified. The Songhua River Basin in the hinterland of Harbin, capital of Heilongjiang Province, became a productive agricultural zone despite the environmental limitations here. The Northeast was a shining example of what communist planning could accomplish.

6 Then came the **economic restructuring** of the post-Mao era. Under the new, market-driven economic order, the state companies that had led China's industrial resurgence were unable to adapt. Suddenly they were aging, inefficient, obsolete. Official policy encouraged restructuring and privatization, but the huge state factories could not (and much of their labor force would not) comply. Today, the Northeast's contribution to China's industrial output is only about 10 percent as the production of factories in the Southeast, from Shanghai to Guangdong, rises annually.

Economic geographers, however, will not write off the prospects of this northeastern corner of China; this area, as we noted, has had its downturns before. The Northeast still contains a storehouse of resources, including extensive stands of oak and other hardwood forests on the mountain slopes that encircle the Northeast China Plain. The locational advantages of the major centers are undiminished. Harbin, for example, lies at the convergence of five railroads and is situated at the head of navigation on the Songhua River, providing a link to the northeast corner of the region and, beyond the Amur River, to Russia's Khabarovsk. Moreover, when the Koreas reunite, producing what may become the western Pacific Rim's second Japan, and when the Russian Far East takes off (Vladivostok and Nakhodka are the natural ports for the upper Northeast), this northern frontier of China may yet reverse its decline and enter still another era of economic success.

North China Plain

Take the new superhighway between the port of Tianjin and the capital, Beijing, or the train from Beijing southward to Anyang, and you see a physical landscape almost entirely devoid of relief and a cultural landscape that repeats itself endlessly (see photo this page). Villages, many showing evidence of the collectivization program of the early communist era, pass at such regular intervals that some kind of geometry seems to control the cultural landscape. Carefully diked canals stretch in all directions. Rows of trees mark fields and farms. Depending on the season, the air is clear, as it is during the moist summer when the wheat crop covers the countryside like a carpet, or choked with dust, which can happen almost any other time of the year. The soil here is a fertile mixture of alluvium and loess (deposits of fine, windborne silt or dust of glacial origin), and the slightest breeze stirs it up. In the early spring the dust hangs like a fog over the landscape, covering everything with a fine powder. Add this to urban pollution and the cities become as unhealthy as any in the world.

FROM THE FIELD NOTES
"We flew over the North China Plain for an hour, and one could not miss the remarkably regular spacing of the villages on this seemingly table-flat surface. Many of these villages became communes during the communist reorganization of China's social order. The elongated buildings (many added since the 1950s) each were designed to contain one or more extended families. After the demise of Mao and his clique the collectivization effort was reversed and the emphasis returned to the small nuclear family. In this village and others we overflew, the long sides of the buildings almost always lay east-west, providing maximum sun exposure to windows and doors, and enhancing summer ventilation from longitudinal breezes. The more distant view looks hazy, because the North China Plain during the spring is wafted by loess-carrying breezes from the northwest. On the ground, a yellow-gray hue shrouds the atmosphere until a passing weather front temporarily clears the air."

For all we see, there are geographic explanations. The geometry of the settlement pattern results from the flatness of the uniformly fertile surface and the distance villagers could walk to neighboring villages and back during daylight. The canals form part of a vast system of irrigation channels designed to control the waters of the Huang He (Yellow River). The rows of trees serve as windbreaks. And while wheat dominates the crops of the North China Plain, you will also see millet, sorghum, soybeans, cotton for the textile industry, tobacco (the Chinese are heavy smokers), and fruits and vegetables for the urban markets.

The North China Plain is one of the world's most heavily populated agricultural areas. Figure 9-10 shows that most of it has a density of more than 1000 people per square mile (400 per sq km), and in some parts the density is twice as high. Here, the ultimate hope of the Beijing government lay less in land redistribution than in raising yields through improved fertilization, expanded irrigation facilities, and the more intensive use of labor. A series of dams on the Huang River, including the Xiaolangdi Dam upstream in Henan Province, now reduce the flood danger, but outside the irrigated areas the ever-present problem of rainfall variability and drought persists. The North China Plain has not produced any substantial food surplus even under normal circumstances; thus when the weather turns unfavorable the situation soon becomes precarious. The specter of famine may have receded, but the food situation is still uncertain in this very critical part of China Proper.

The North China Plain is an excellent example of the concept of a national **core area**. Not only is it a densely populated, highly productive agricultural zone, but it also is the site of the capital and other major cities, a substantial industrial complex, and several ports, among which Tianjin ranks as one of China's largest (Fig. 9-12). Tianjin, on the Bohai Gulf, is linked by rail and highway (less than a two-hour

FIGURE 9-12

Among the Realm's Great Cities . . .

Beijing

Beijing (10.9 million), capital of China, lies at the northern apex of the roughly triangular North China Plain, just over 100 miles (160 km) from its port, Tianjin, on the Bohai Gulf. Urban sprawl has reached the hills and mountains that bound the Plain to the north, a defensive barrier fortified by the builders of the Great Wall. You can reach the Great Wall from central Beijing by road in about an hour.

Although settlement on the site of Beijing began thousands of years ago, the city's rise to national prominence began during the Mongol (Yuan) Dynasty, more than seven centuries ago. The Mongols, preferring a capital close to their historic heartland, endowed the city with walls and palaces. Following the Mongols, later rulers at times moved their capital southward, but the government always returned to Beijing. From the third Ming emperor onward, Beijing was China's focus; it was ideally situated for the Manchus when they took over in 1644. During the twentieth century, China's Nationalists again chose a southern capital, but when the communists prevailed in 1949 they reestablished Beijing as their (and China's) headquarters.

Ruthless destruction of historic monuments carried out from the time of the Mongols to the communists (and now during China's modernization) has diminished, but not destroyed, Beijing's heritage. From the Forbidden City of the Manchu emperors to the fifteenth-century Temple of Heaven, the capital remains an open-air museum and the cultural focus of China. Successive emperors and aristocrats, moreover, bequeathed the city numerous parks and additional recreational spaces that other cities lack. Today Beijing is being transformed as a result of China's new economic policies. A forest of high-rises towers above the retreating traditional cityscape, avenues have been widened, a beltway has been built. A new era dawns in this historic capital.

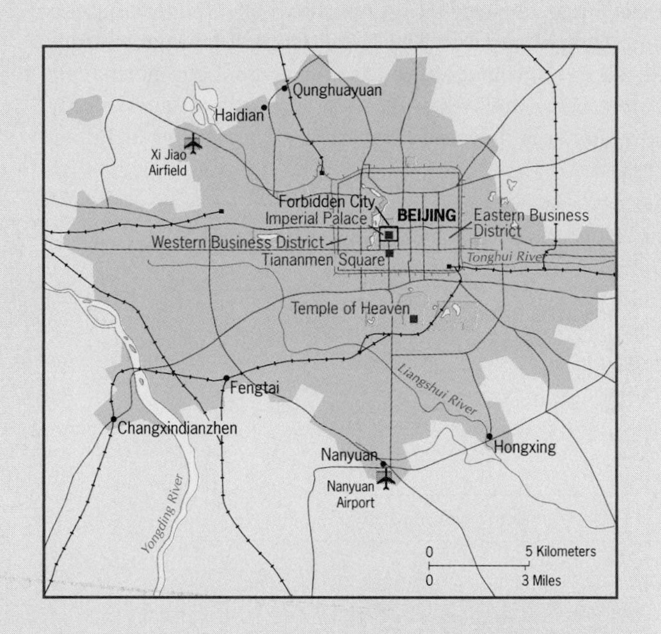

drive) to Beijing. Like many of China's harbors, that of Tianjin's river port is not particularly good; but Tianjin is well situated to serve not only the northern sector of the North China Plain and the capital, but also the Upper Huang Basin and Inner Mongolia beyond (see Fig. 9-1). Tianjin, again like several other Chinese ports, had its modern start as a treaty port, but the city's major growth awaited communist rule. For decades it was a center for light industry and a flood-prone harbor, but after 1949 the communists constructed a new artificial port and flood canals. They also chose Tianjin as a site for major industrial development and made large investments in the chemical industry (in which Tianjin still leads China), iron and steel production, heavy machine manufacturing, and textiles. Today, with a population of 9.3 million, Tianjin is China's fourth largest metropolis and the center of one of its leading industrial complexes.

Northeast of Tianjin lies the smaller city of Tangshan. In 1976, a devastating earthquake, with its epicenter between Tangshan and Tianjin, killed an estimated 750,000 people, the worst natural disaster of the twentieth century. So inept was the government's relief effort in nearby Beijing that the widespread resentment it provoked may have helped end the Maoist period of communist rule in China.

Beijing, unlike Tianjin, is China's political, cultural, and educational center. Its industrial development has not matched Tianjin's. The communist administration did, however, greatly expand the municipal area of Beijing, which (as we noted), is not controlled by the Province of Hebei but is directly under the central government's authority. In one direction, Beijing was enlarged all the way to the Great Wall—30 miles (50 km) to the north—so that the "urban" area includes hundreds of thousands of farmers. Not surprisingly, this has circumscribed an enormous total population, enough to rank Beijing, with its 10.9 million inhabitants, among the world's larger megacities.

Inner Mongolia

Northwest of the core area as defined by the North China Plain, along the border with the state of Mongolia, lies

Inner Mongolia, administratively defined as the Nei Mongol Autonomous Region (see Fig. 9-8, p. 446). Originally established to protect the rights of the approximately 5 million Mongols who live outside the Mongolian state, Inner Mongolia has been the scene of massive immigration by Han Chinese. Today Chinese outnumber Mongols here by nearly four to one. Near the Mongolian border, Mongols still traverse the steppes with their tents and herds, but elsewhere irrigation and industry have created an essentially Chinese landscape. The A.R.'s capital, Hohhot, has been eclipsed by Baotou on the Huang He, which supports a corridor of farm settlements as it crosses the dry land here on the margins of the Gobi and Ordos deserts. With about 29 million inhabitants, Inner Mongolia still cannot compare to the huge numbers that crowd the North China Plain, but recent mineral discoveries have boosted industry in Baotou and livestock herding is expanding. Nei Mongol may retain its special administrative status, but it functions as part of Han China in all but name.

Basins of the Chang/Yangzi

In contrast to the contiguous, flat agricultural-urban-industrial North China Plain defined by the sediment-laden Huang River, the basins and valleys of the Chang Jiang (Long River) display variation in elevation and relief. The Chang River, whose lower-course name becomes the Yangzi, is an artery like no other in China. Near its mouth lies the country's largest city, Shanghai. Part of its middle course is now being transformed by a gigantic engineering project. Farther upstream, the Chang crosses the populous, productive Sichuan Basin. And unlike the Huang, the Chang Jiang is navigable to oceangoing ships for over 600 miles (1000 km) from the coast all the way to Wuhan (Fig. 9-12). Smaller ships can reach Chongqing, even after dam construction. Several of the Chang River's tributaries also are navigable, so that 18,500 miles (30,000 km) of water transport routes serve its drainage basin. Thus the Chang Jiang constitutes one of China's leading transit corridors. With its tributaries it handles the trade of a vast area, including nearly all of middle China and sizeable parts of the north and south. The North China Plain may be the core area of China, but in many ways the Lower Chang Basin is its heart.

Early in Chinese history, when the Chang's basin was being opened up and rice and wheat cultivation began, a canal was built to link this granary to the northern core of old China. Over 1000 miles (1600 km) long, this was the longest artificial waterway in the world, but during the nineteenth century it fell into disrepair. Known as the Grand Canal, it was dredged and rebuilt when the Nationalists controlled eastern China. After 1949, the communist regime continued this restoration effort, and much of the canal is now again open to barge traffic, supplementing the huge fleet of vessels that hauls domestic interregional trade along the east coast (Fig. 9-12).

As Figure 9-13 shows, the Lower Chang Basin is an area of both rice and wheat farming, offering further proof of its pivotal situation between south and north in the heart of China Proper. Shanghai lies at the coastal gateway to this productive region on a small tributary of the Chang (now Yangzi) River, the Huangpu. The city has an immediate hinterland of some 20,000 square miles (50,000 sq km)—smaller than West Virginia—containing more than 50 million people. About two-thirds of this population are farmers who produce food, silk filaments, and cotton for the city's industries.

Travel upriver along the Yangzi/Chang Jiang, and you meet an unending stream of vessels large and small, including numerous "barge trains"—as many as six or more barges pulled by a single tug in an effort to save fuel (they are slow, but time is not the primary concern). The traffic between Shanghai and Wuhan—Wuhan is short for Wuchang, Hanyang, and Hankou, a coalescing conurbation of three cities—makes the Yangzi one of the world's busiest waterways. Smoke-belching boat engines create a more or less permanent plume of pollution here, worsening the regional smog created by the factories of Nanjing and others perched on the waterfront. Here China's Industrial Revolution is in full gear, with all its environmental consequences.

Above Wuhan the river traffic dwindles because the depth of the Chang reduces the size of vessels that can reach Yichang. River boats carry coal, rice, building materials, barrels of fuel, and many other items of trade. But the middle course of the Chang River is becoming more than a trade route. In the northward bend of the great river be-

A decade ago, this was farmland across the Huangpu River from Shanghai's waterfront Bund. Today the name *Pudong* is synonymous with the fastest-paced urban development in the world. A forest of skyscrapers ranging from the ultramodern to the futuristic now overlooks the Bund's Victorian-era architecture. Shanghai, Chinese economic geographers predict, will overtake Hong Kong and become the country's leading business, financial, and service center as well as a powerful industrial giant. Everything in this photograph is new: the buildings, the lake, the pavilion. It is Saturday, the most popular day for weddings, and a newlywed couple heads for their party.

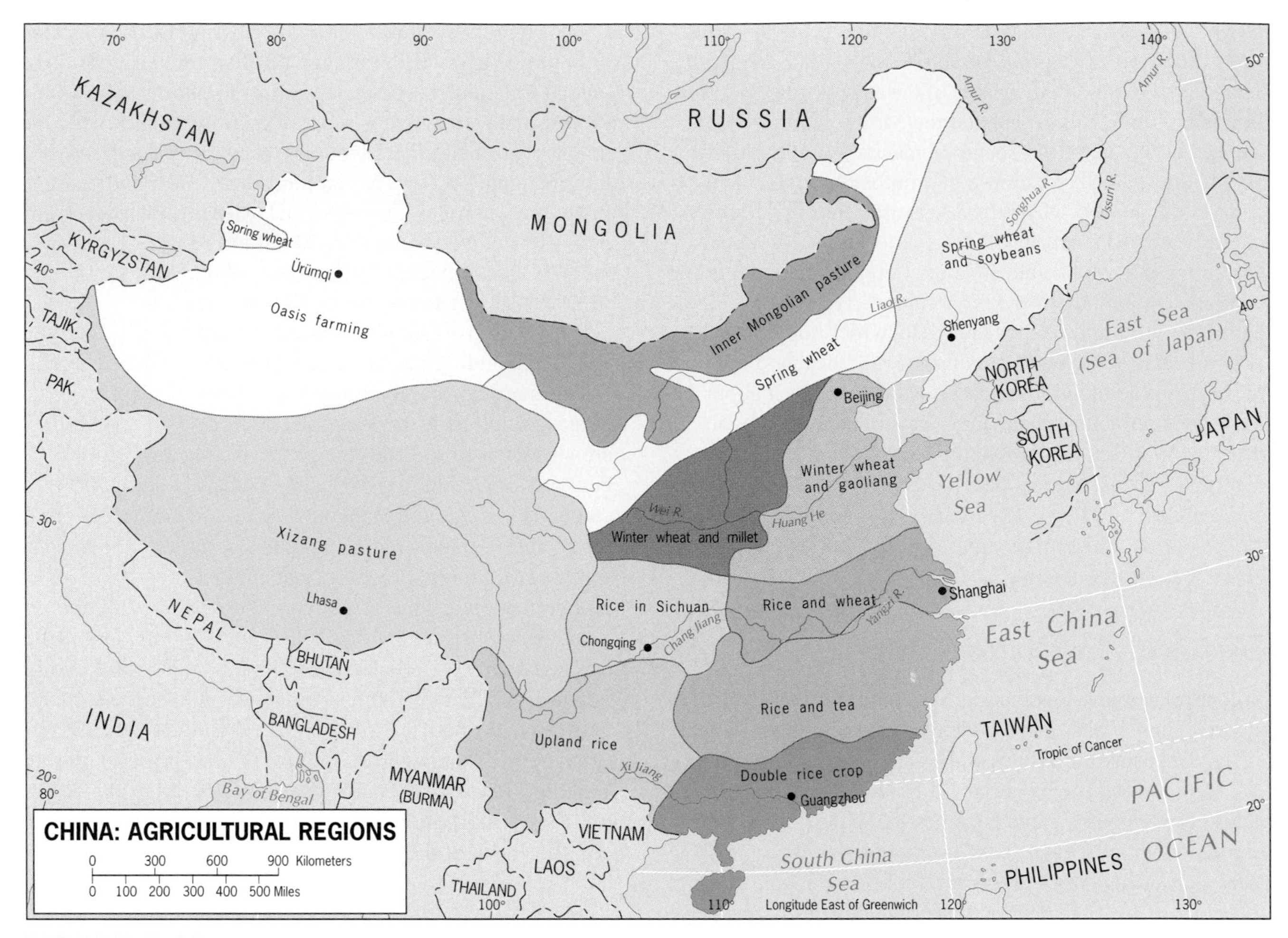

FIGURE 9-13

tween Yichang and Chongqing, where the Chang has cut deep troughs on its way to the coast, one of the world's largest engineering projects is in progress. It has several names: the Sanxia Project, the Chang Jiang Water Transfer Project, the Three Gorges Dam, and the New China Dam. It is most commonly called the Three Gorges Dam because it is here that the great river flows through a 150-mile (240-km) series of steep-walled valleys less than 360 feet (110 m) wide. Near the lower end of this natural trough, the dam will rise to a height of over 600 feet (180 m) above the valley floor to a width of 1.3 miles (2.1 km), creating a reservoir that will inundate the Three Gorges and extend more than 380 miles (over 600 km) upstream. China's project engineers say that the dam will end the river's rampaging flood cycle, enhance navigation, stimulate development along the new lake perimeter, and provide at least one-tenth, and perhaps as much as one-eighth, of China's electrical power supply. As such it will transform the heart of China, which is why many Chinese like to compare this gigantic project to the Great Wall and the Grand Canal, and call it the New China Dam.

The Chinese government approved the massive project in 1992. Work began in 1993, and the Chang Jiang's course was diverted late in 1997, a year ahead of schedule. In 1998, the engineers forecast completion of the scheme on or before the target date, which is mid-2009. All this is happening despite the paucity of foreign investment. Citing its growing doubts about megaprojects of this kind and its concern over the environmental damage, the World Bank declined to provide funds to support the Three Gorges Dam. Other lenders also withheld their money, but China, barring an economic collapse, has the financial resources to keep the project on track. And as construction proceeds, the Chinese will be able to offer foreign companies attractive contracts for the many related projects (power generation and transmission, roads, bridges, high-tech facilities) that will arise from it. Many will participate, whatever their governments' official view of the New China Dam.

It is a measure of the post-Mao changes in China's still-communist society that the Three Gorges project can be criticized—not only in foreign circles, but in China itself. Most Chinese citizens seem to support the initiative, but some scholars as well as nonprofessionals have spoken out against it. In a country where such dissent still tends to be regarded as "anti-communist propaganda," this takes courage: the risks are considerable, and some opponents have been subjected to harsh treatment. Those within China who oppose the project include poets who bemoan the disappearance of one of China's most fabled and inspiring slivers of scenery, archeologists who decry the loss of prehistoric sites, biologists who abhor the ecological damage the dam will cause, and social workers who fear that the nearly 1.5 million people who will be displaced by the rising water will not be adequately protected. Some engineers fear that the silt-laden Chang Jiang will soon clog up both the reservoir and the river channel above it, possibly blocking the very port of Chongqing which the project is supposed to benefit. And there are those who remember the fate of countless thousands who had the misfortune to live below dams constructed by Mao-era builders. They worry about insufficient quality control on the Three Gorges project and fear that an unprecedented catastrophe may lie ahead.

One city that will be (and has already been) strongly affected by the Three Gorges Dam project is Chongqing, upstream from the reservoir. Boats of up to 1000 tons could reach this river port before dam building began, and the city's planners always hoped that this link to the coast would yield economic benefits arising from China's Pacific Rim

Among the Realm's Great Cities . . .

Shanghai

Sail into the mouth of the great Chang/Yangzi River, and you see little to prepare you for your encounter with China's largest city. For that, you turn left into the narrow Huangpu River, and for the next several hours you will be spellbound. To starboard lies a fleet of Chinese warships. On the port side you pass oil refineries, factories, and neighborhoods. Soon, rusty tankers and freighters line both sides of the stream. High-rise tenements tower behind the cluttered, chaotic waterfront where cranes, sheds, boatyards, piles of rusting scrap iron, mounds of coal, and stacks of cargo vie for space. In the river, your boat competes with ferries, barge trains, cargo ships. Large vessels, anchored in midstream, are being offloaded by dozens of lighters tied up to them in rows. The air is acrid with pollution. The noise—bells, horns, whistles—is deafening.

What strikes you is the vastness and sameness of Shanghai's cityscape, until you pass beneath the first of two gigantic suspension bridges. Suddenly, everything changes. To the left, or east, the space-age Oriental Pearl Television Tower looking like a rocket on its launchpad rises above a row of modern, glass-and-chrome skyscrapers that make the Huangpu look like Hong Kong's Victoria Harbor. To the right stands a row of Victorian buildings, monuments to the British colonialists who made Shanghai a treaty port and started the city on its way to greatness. Everywhere, construction cranes rise above the cityscape. Shanghai looks like one vast construction zone—complete with a 1510-foot office tower (called the World Financial Center) that will become the world's tallest building when it opens.

Shanghai is an Open Coastal City, where China invites investment by offering privileges to foreign companies. The government is spending heavily to make Shanghai a tiger on the Pacific Rim, and the scene on the east side of the Huangpu may be unmatched in East Asia. There, an immense business and industrial complex, the Special

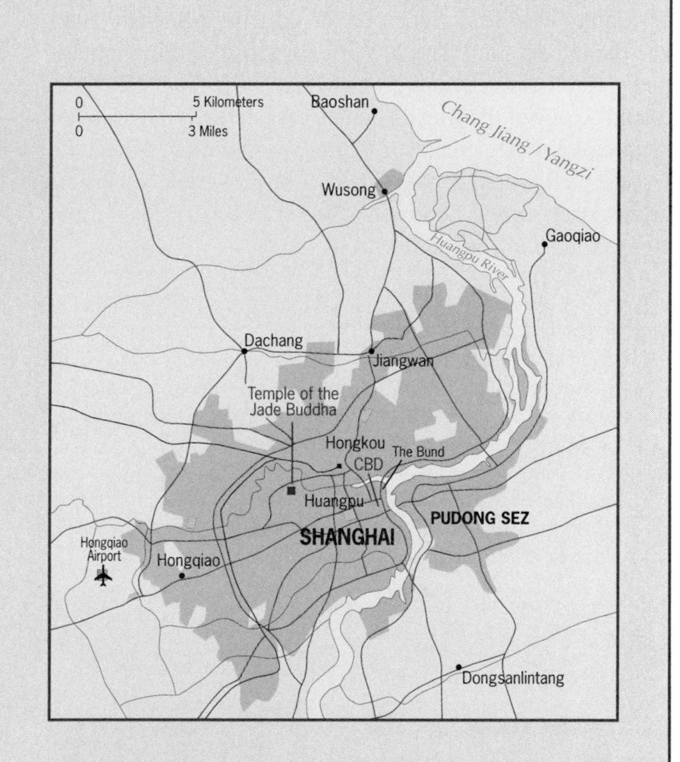

Economic Zone (SEZ) named *Pudong*, has arisen since 1990. Its growth rivals that of Shenzhen. Its future, say locals, will eclipse that of Xianggang (Hong Kong).

Meanwhile, Shanghai's past is being obliterated. Whole neighborhoods in the central city are falling to the wreckers' bulldozers, their occupants sent far way to those tenements we saw along the river. Even as citizens are ousted, jobseekers from the hinterland jam the streets. Nearly 3 million converged on Shanghai when word of the Pudong project spread. About 1 million found work. With 13 million residents today, Shanghai remains China's largest city, and by century's end it ranked among the world's six largest. But Shanghai may have traded its soul for a stock market.

development. Dam construction has disrupted and delayed river traffic, but the diversion channel around the Three Gorges Dam will restore Chongqing's port functions. Still, its relative location would suggest that its greater economic opportunities lie westward, in the fast-developing Sichuan Basin, rather than eastward through its tenuous link with the coast. Having recently been made a *Shi* by the regime in Beijing, its municipal territory enlarged to incorporate some 30 million people, Chongqing now finds itself near the upper end of China's largest reservoir. The last Nationalist capital may yet become the number one growth pole in China's interior.

Sichuan, its basin crossed by the Chang after the river emerges from the mountains of the west, frequently was a problem province for both China's dynastic rulers and the communist regime. Its thriving capital, Chengdu, was one of China's most active centers of dissent during the 1989 pro-democracy movement. With about 120 million inhabitants (including Chongqing) today, Sichuan Province would be one of the world's ten largest states, but it does not even contain 10 percent of China's population. Sichuan's fertile, well-watered soils yield a huge variety of crops: grains (rice, wheat, corn), soybeans, tea, sugarcane, and many fruits and vegetables.

FROM THE FIELD NOTES

"It may not look like much, but this home-made, powered thresher is doing the work of several people in this paddy near Guilin in the Guangxi Zhuang Autonomous Region of southern China. Three people do the bunching, feeding, and bagging! This rural scene led me to wonder: what will happen when China's agriculture becomes more mechanized? How will millions of redundant farmers be employed?"

In many ways Sichuan is a country unto itself, isolated by topography and distance from the government in Beijing. The people here are Han Chinese, but cultural landscapes reveal the area's proximity to Southeast Asia. Along the great riverine axis of the Chang/Yangzi, China changes character time and again.

Basins of the Xi (West) and Pearl River

As Figure 9-12 shows, the Xi River and its basins are no match for the Chang or Huang, not even for the Liao. This southernmost river even seems to have the wrong name: Xi means West! It reaches the coast in a complex area where it forms a delta immediately adjacent to the estuary of the Pearl River. In this subtropical part of China, local relief is higher than in the lowlands and basins of the center and north, so that farmlands are more confined. Especially in the interior areas, water supply is a recurrent problem. On the other hand, its warm climate permits the double-cropping of rice. Regional food production, however, has never approached that of the North China Plain or the Lower Chang Basin. And, as Figure 9-10 shows, the population in this southern part of China is smaller than that in the more northerly heartlands.

Again unlike the Huang and Chang rivers, which rise in snow-covered interior mountains, the Xi is a shorter stream whose source is on the Yunnan Plateau (see Fig. 9-2). Just up the coast from its delta, however, lies one of the most important areas of modern China. As we will note later, the mouth of the Pearl River is flanked by some of China's fastest-growing economic complexes, including Guangzhou, capital of Guangdong Province, Hong Kong, the former British dependency, and adjoining Shenzhen, one of the world's fastest-growing cities.

This is all the more remarkable because, as the maps emphasize, South China is not especially well endowed with geographic advantages, locational or otherwise. The Pearl River is navigable for larger ships only to Guangzhou, and the Xi for small river boats no farther than Wuzhou. Guangdong Province, the key province of southern China, lies far from China's heartlands. Nor is South China blessed with the resources that propelled the industrial development of the Northeast and other parts of China. Indeed, one of the principal objectives of the Three Gorges Dam project is to provide South China with ample electricity. As Figure 9-14 shows, northern and western China have most of the coal, oil, and natural gas resources, not the South.

China Proper, with its populous and powerful subregions, is in many ways the dominant region in the East Asian geographic realm. But China Proper is only one part of China. To the west lie two vast regions with relatively sparse populations: Xizang (Tibet) and Xinjiang. And here, in this remote western periphery adjoining Turkestan, China meets Muslim Central Asia (Fig. 9-15).

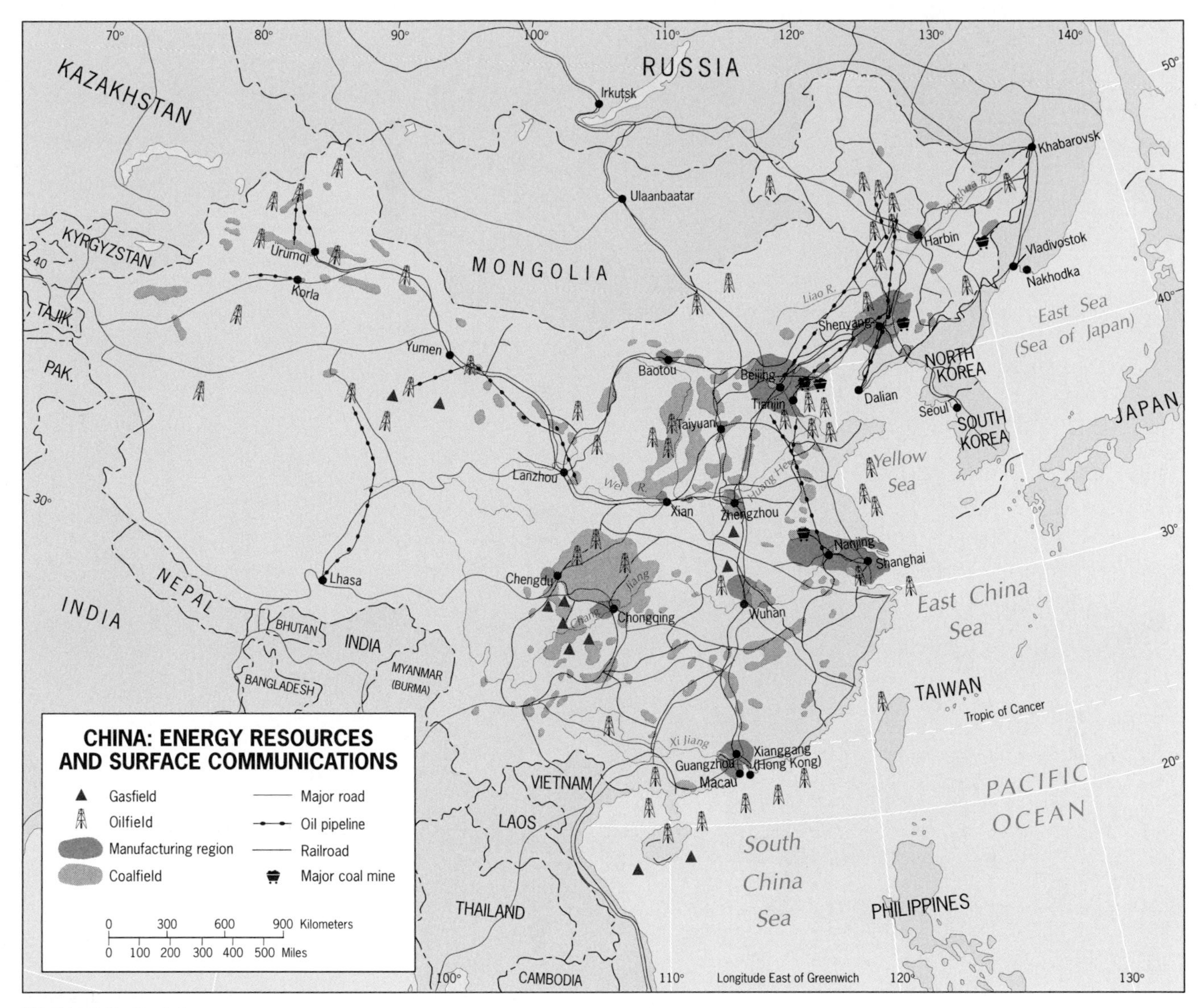

FIGURE 9-14

XIZANG (TIBET)

As the climate and physiographic maps demonstrate, harsh physical environments dominate this colony of China. It is a vast, sparsely peopled region designated as an Autonomous Region, but in fact Xizang (Tibet) is an occupied society. In terms of its human geography, we should take note of two subregions.

The first is the core area of Tibetan culture, between the Himalayas to the south and the Trans-Himalayas lying not far to the north. In this area, some valleys lie below 7000 feet (2130 m); the climate is comparatively mild, and some cultivation is possible. Here lies Tibet's main population cluster, including the crossroads capital of Lhasa. The Chinese government has made investments to develop these valleys, which contain excellent sites for hydroelectric power projects (some have been used for a few light industries) and promising mineral deposits. The second area of interest is the Qaidam Basin in the north (Fig. 9-15). This basin lies thousands of feet below the surrounding Kunlun and Altun Mountains and has always contained a concentration of nomadic pastoralists. Recently, however, exploration has revealed the presence of oilfields and coal reserves below the surface of the Qaidam Basin, and these resources are now being developed.

As Figure 9-3 reminds us, Tibet came under Chinese domination during the Manchu (Qing) Dynasty in 1720, but it regained its separate status in the late nineteenth century. China's communist regime took control after the invasion of 1950; in 1959, China crushed an uprising after Tibetan villagers tried to resist the Chinese presence. Tibetan society had been organized around the fortress-like monasteries of Buddhist monks who paid allegiance to their supreme leader, the Dalai Lama. The Chinese wanted to modernize

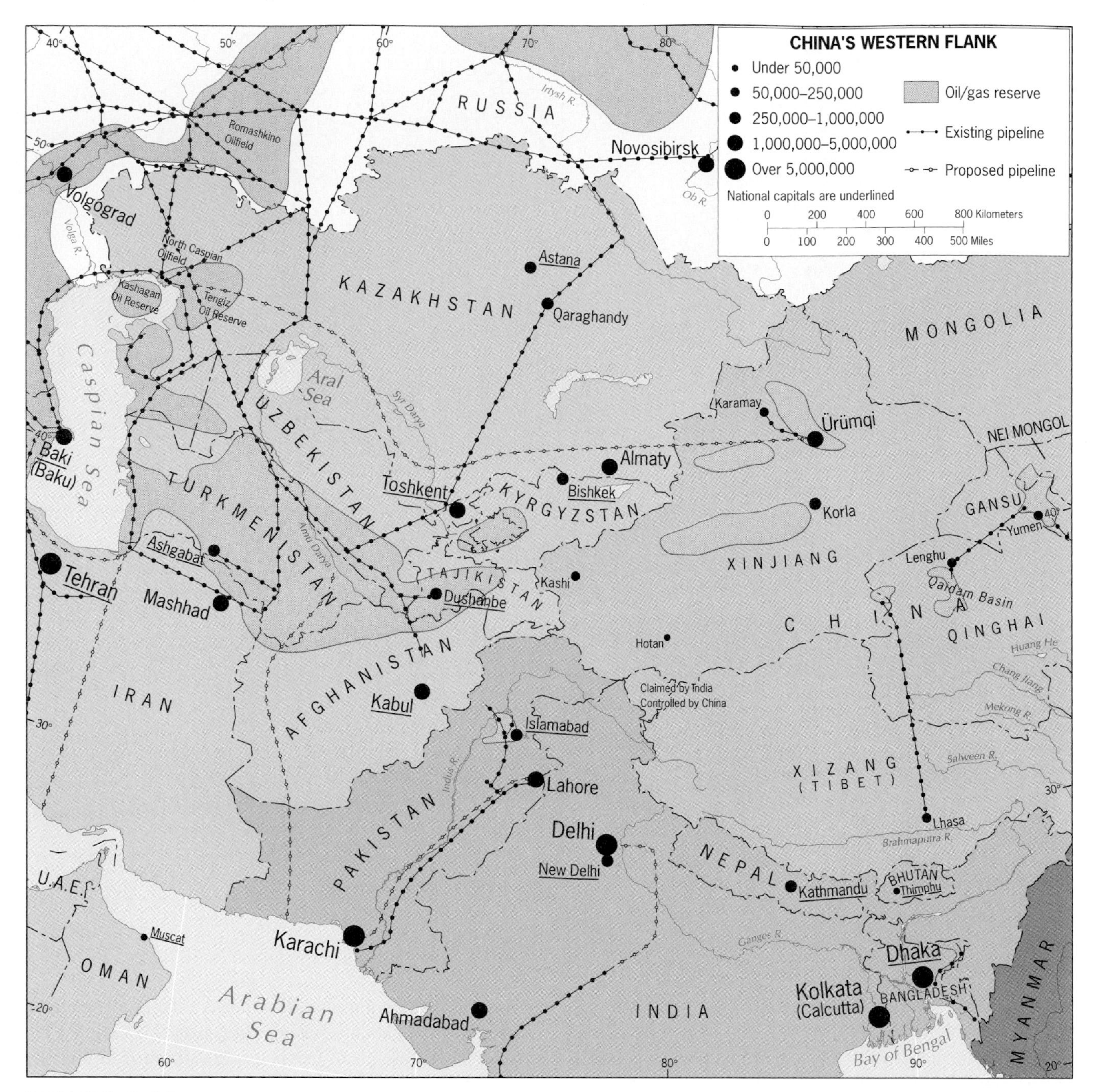

FIGURE 9-15

this feudal system, but the Tibetans clung to their traditions. In 1959, they proved no match for the Chinese armed forces; the Dalai Lama was ousted, and the monasteries were emptied. The Chinese destroyed much of Tibet's cultural heritage, looting its religious treasures and works of art. Their harsh rule devastated Tibetan society, but after Mao's death in 1976 the Chinese relaxed their tight control. Although amends were made (religious treasures were returned to Xizang, monastery reconstruction was permitted, and Buddhist religious life resumed), pro-independence rioting has been frequent since 1987. As a result, the Chinese have again tightened their grip. Since its formal annexation in 1965, Xizang has been administered as an Autonomous Region. Although it is large in area, its population is only slightly more than 2 million.

Visiting Chinese-occupied Tibet is a depressing experience. Tibetan domestic architecture has beauty and expressiveness, but the Chinese have built ugly, gray structures to house the occupiers—often directly in front of Tibetan monasteries and shrines. This juxtaposition of Chinese banality and Tibetan civility seems calculated to perpetuate hostility, an anachronism in this supposedly postcolonial

FROM THE FIELD NOTES

"From the top floor of an office building I got a good view over Ürümqi (Urumchi), the capital of western China's Xinjiang Uyghur Autonomous Region, and part of the Junggar Basin in which it lies. In modern times, Ürümqi had its first growth spurt during the period of Soviet communist involvement, when some of the office and apartment buildings seen here were constructed. Now a new phase of expansion is under way as China seeks to strengthen the Han presence in this still-remote frontier; the Holiday Inn, a U.S.-Chinese joint venture complete with rotating rooftop restaurant, reflects this."

world. During the rule of Mao Zedong, the Chinese officially regarded Tibet as a territorial cushion against India (then allied with the Soviet Union). The breakup of the Soviet Empire has not softened Beijing's stand.

XINJIANG

China's northwestern corner consists of the giant Xinjiang-Uyghur Autonomous Region, constituting over one-sixth of the country's land area and situated in a strategic part of Central Asia (Fig. 9-15). Linked to China Proper via the vital Hexi Corridor in Gansu Province to the east, Xinjiang is China's largest single administrative area.

But even today, after more than a half-century of vigorous Sinicization, Chinese remain a minority here. About 40 percent of the population of 18 million is Chinese (up from 5 percent in 1949), and the Chinese control virtually all aspects of life. The majority are Muslim Uyghurs, who make up approximately half the total, and Kazakhs, Kyrgyz, and others with cultural affinities across the border to all five republics of former Soviet Central Asia.

The physical geography of Xinjiang is dominated by high mountain ranges and vast basins (Fig. 9-2). The southern margin is defined by the Kunlun Shan (Mountains) beyond which lies Xizang (Tibet). Across the north-center of the region lies the mountain wall of the Tian Shan. Between the Kunlun and Tian Shan extends the vast, dry Tarim Basin within which lies the Takla Makan Desert, and north of the Tian Shan lies another depression, the Junggar Basin. Rivers rising on the mountain slopes disappear beneath the sediments of the basins, sustaining a ring of oases where their waters come near the surface or where wells and *qanats* make irrigation possible (Fig. 9-16).

This would not seem to be the most favorable geography, but Xinjiang has other assets that China exploits. The Qing (Manchu) armies that fought the nomadic Muslim warlords here in the 1870s saw the agricultural possibilities and planted the first crops. The communist rulers after 1949 exiled political opponents as well as criminals to notorious prison camps in this remote frontier. Then Xinjiang turned out to have extensive reserves of oil and natural gas (Figs. 9-15; 9-16). Here, too, China could experiment with space technology, rocketry, and atomic weaponry far from densely populated areas (and shielded from foreign spies). And now, as Figure 9-15 reminds us, China's westernmost borders lie relatively close to the great energy reservoirs of the Caspian Sea Basin—and only one neighbor lies in the way. Pipeline construction across Kazakhstan to Xinjiang is in progress.

Xinjiang's evolving human geography is revealed by Figure 9-16. Most of the Han Chinese are concentrated in the north-central cluster of cities and towns between the capital, Ürümqi, and the oil center of Karamay. Of special interest here is the city of Shihezi, founded by the Chinese army and now China's model for the region. Over 90 percent of its half-million population are ethnic Chinese; Shihezi's comparative modernity and orderly life stand in sharp contrast to the poorer, often squalid towns and settlements outside the Chinese heartland. Also centered on the Ürümqi-Karamay area is Xinjiang's largest agricultural zone, where cotton and fruits yield export revenues. As the

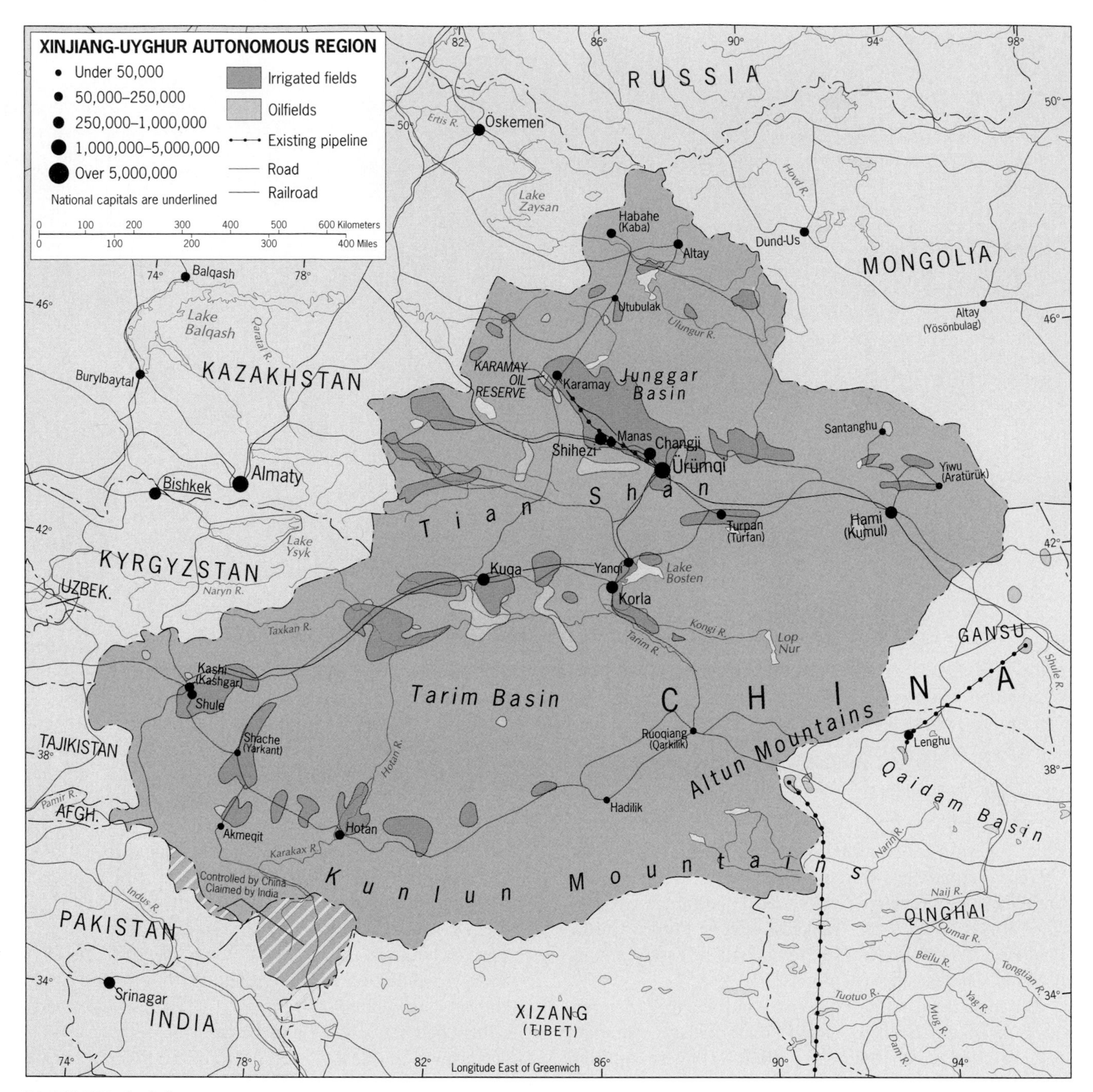

FIGURE 9-16

map shows, a ring of oases and clusters of irrigated agriculture encircles both the Tarim and Junggar Basins.

Beijing has made extensive efforts to integrate Xinjiang into the Chinese state. A railroad and highway link Ürümqi and Lanzhou via the Hexi Corridor. Westward from Ürümqi, a railroad crosses the border with Kazakhstan and connects to the Öskemen-Almaty line; another railroad, opened in 2000, connects the capital to Kashi (Kashgar) in the distant southwest (Fig. 9-16). But vast spaces in Xinjiang remain distant and inaccessible. As the map shows, just one road, recently upgraded, loops around the vast southern and eastern flank of the Tarim Basin from Kashi via Hotan to Korla.

China's western frontier region, therefore, is a place of challenges as well as opportunities. Here a regime that from time to time displays intolerance of "foreign" belief systems meets a majority Muslim population that resists incorporation and whose leaders sometimes call for secession. Here the gap between comparatively well-off Chinese and poorer minorities is large—and spatially vivid. Here, too, China has unresolved boundary disputes with neighbors, including Tajikistan and India. On the other hand, Xinjiang's energy resources (also including some coal), agricultural and industrial potential, and relative location make this a promising frontier, a symbol for the rest of Central Asia of what Chinese administration can accomplish.

CHINA'S PACIFIC RIM: EMERGING REGION?

If we had been studying the geography of China just 30 years ago, we would not have had to pay special attention to a zone fronting the Pacific Ocean. All our generalizations about China—its low level of urbanization and stagnant cities, the government's total control of industry and the inefficiency of its state enterprises, the isolation of its rural areas, the changeless poverty of remote villages, the strict rule of Beijing enforced by the Communist Party—would have applied to all of China Proper. But today, one generation after Mao's demise, China is in a momentous economic and social transition. This process, which Mao's successor Deng Xiaoping set in motion, is changing China as a whole but is nowhere more evident than along the shores of the Pacific Ocean.

From Dalian on the doorstep of the Northeast to Hainan Island off the country's southernmost point, China is booming. In the cities, old neighborhoods are being bulldozed to make way for modern skyscrapers. In the countryside of the coastal provinces, thousands of factories employing millions of workers who used to farm the land are changing life in the towns and villages. New roads, airports, power plants, and dams are being built. It is an uneven, sometimes chaotic process that is affecting certain parts of China's coast more than others. It is penetrating parts of China's interior, but meanwhile social gaps are widening: between advantaged and disadvantaged, well-off and poor, urbanite and villager. It is creating regional economic disparities that China has never seen before. And it is producing what may become a separate geographic region along China's Pacific littoral, a region that has more in common with Japan, South Korea, and Taiwan than with Yunnan or Gansu.

In the Introduction, we discussed issues relating to the geography of development, emphasizing that, in this complex, globalizing world, the concept of developed versus underdeveloped countries (DCs and UDCs) is no longer tenable. By almost any measure, China remains what used to be called a UDC. As Figure I-11 shows, its per-capita GNP still classifies it as a low-income economy, the lowest of categories. But China now has its own core-periphery duality: in its Pacific coast provinces, per-capita incomes today are far above the national average of U.S. $750, while hundreds of millions of villagers in China's remote interior earn much less. In its burgeoning Pacific Rim, China resembles a modernizing economy with "developed" characteristics. In much of its interior, it does not. This contrast is one reason a discrete Pacific Rim region appears to be forming in China Proper.

Geography of Development

Many yardsticks are used to gauge a country's development, including, as we noted, GNP per capita; percentage of workers in farms, factories, or other kinds of employment; amount of energy consumed; efficiency of transport and communications; use of manufactured metals like aluminum and steel in the economy; productivity of the labor force; and social measures such as literacy, nutrition, medical services, and savings rates. All these data are calculated per person, yielding an overall measure that ranks the country among all those providing statistics. These data do not, however, tell us *why* some countries, or parts of countries, exhibit the level of development they do.

The **geography of development** leads us into the study of raw-material distributions, environmental conditions, cultural traditions, historic factors (such as the lingering effects of colonialism), and factors of location. In the emergence of China's Pacific Rim, relative location played a major role, as we will see. 8

Geographers tend to view the development process spatially, whereas other scholars focus on structural aspects. In the 1960s, economist Walt Rostow formulated a global model of the development process that is still much discussed today. This model suggested that all developing countries follow an essentially similar path through five interrelated growth stages. In the earliest of these stages, a *traditional society* engages mainly in subsistence farming, is locked in a rigid social structure, and resists technological change. When it reaches the second stage, *preconditions for takeoff*, progressive leaders move the country toward greater flexibility, openness, and diversification. Old ways are abandoned, workers move from farming into manufacturing, and transport improves. This leads to the third stage, *takeoff*, when the country experiences a type of industrial revolution. Sustained growth takes hold, industrial urbanization proceeds, and technological and mass-production breakthroughs occur. If the economy continues to expand, the fourth stage, *drive to maturity*, brings sophisticated industrial specialization and increasing international trade. Some countries reach a still more advanced fifth stage of *high mass consumption* marked by high incomes, widespread production of consumer goods and services, and most workers employed in the tertiary and quaternary economic sectors.

This model takes little account of core-periphery contrasts within individual countries. In China, takeoff conditions (stage 3) exist in much of the Pacific Rim, but other major areas of this vast country remain in stage 1. China has passed through its transition from tradition-bound to progressive leadership, but the effect of that leadership's reformist policies has not been the same in all parts of the country. The most fertile ground lies along the Pacific, where China has a history of contact with foreign enterprises, where the country is most open to the world, and from where many Chinese departed for Southeast Asian (and other) countries. These **Overseas Chinese** were positioned to play a crucial role in China's Pacific Rim development: many had left kin and community there, and those links facilitated the investment of hundreds of millions of dollars when the opportunity arose. Many of the Pacific Rim's factories were built by means of this repatriated money.

China's Development Under Communism

In 1999, China marked half a century of communist governance after more than 40 centuries of dynastic rule. In 1949, when the communist era began in China, nothing foreshadowed the emergence of an economic juggernaut along the Pacific coast—or anywhere else in the country. China had been wracked by war, leaving its economy in ruins. In addition, famine threatened. The Nationalists had fled but had taken control of the island of Taiwan.

In the 1950s, the communist regime launched massive reform and reconstruction programs, initially with the guidance of Soviet advisors. To speed development, the Soviets recommended, capital derived from the sale of agricultural products should be invested in heavy machinery. This would launch China rapidly into an industrial era, as had happened in the USSR in the 1930s. But the Soviet model did not work in China: there were no agricultural surpluses because the huge population farmed at a subsistence level. Soon the Soviets were told to leave, and the still-new regime looked for other options.

After a fierce debate, China changed course. Now the idea was to achieve rapid development through a complete reorganization of society. All land was expropriated, farming was collectivized, and labor was organized into "brigades" based in communes rather than traditional villages. Men and women were assigned to segregated "production teams," which destroyed millions of families and in effect orphaned their children. Rather than huge factories, China's communist planners now wanted small backyard steel furnaces in all communes, making investment in large industrial plants less necessary.

This draconian program was called the *Great Leap Forward* (1957–1960), and its impact was devastating. Farm labor was diverted into the local industries; farm implements were melted to meet steel quotas in communes; peasants resisted the collectivization program and slaughtered their animals. Agricultural productivity spiraled downward, and famines resulted. It is estimated that between 20 and 30 million people perished as a result.

The failure of the Great Leap Forward led to another intense dispute among China's communist rulers. Some blamed the way the system was implemented, arguing that it had been done with excessive speed and force. Others argued that the system itself was at fault and that farmers should be given back their plots of land and be provided with incentives to grow crops rather than coercion to grow crops. The latter view prevailed, and the commune system began to be dismantled. But this "revisionist" ideological victory was to have fateful consequences. By the middle of the 1960s, Mao Zedong was planning to retaliate by launching his disastrous Great Proletarian Cultural Revolution, whose death toll exceeded even that of the Great Leap Forward.

Mao's death in 1976 left communist China on the verge of civil war, with a power struggle in the capital between a clique of orthodox Marxists and a group who called themselves "pragmatic moderates." Some historical geographers likened the contest to similar ones waged between repressive tyrants and progressive reformers of dynastic times, but none of the latter had an impact on China's cultural geography comparable to Deng Xiaoping, leader of the victorious moderates.

Deng argued that communist political rule could be wedded to capitalist economic practice. "Communism with a Chinese face" would keep the Communist Party in power, the People's Liberation Army in charge, and the Standing Committee of the Politburo (the supreme policy-making body of the Party) in control. He also engineered the opening of China to foreign science and technology, and allowed tens of thousands of Chinese students to attend foreign universities in order to bring back the skills China needed. All this set into motion the events that were to transform Pacific Rim China within a few years. But when the social consequences of his actions produced political demands for democracy and an end to Communist Party hegemony, Deng authorized the violent repression of that movement in 1989. It was a test of strength, and Deng's regime prevailed. In the forward rush of Pacific Rim economic development, it caused hardly a ripple.

Transforming the Economic Map

China's communist era has witnessed the metamorphosis of the world's most populous nation. Despite the dislocations of collectivization, communization, the Cultural Revolution, power struggles, policy reversals, and civil conflicts, China as a communist state in the aftermath of Mao managed to stave off famine and hunger, control its population growth rate, strengthen its military capacity, improve its infrastructure, and enhance its world position. It took over Hong Kong from the British and Macau from the Portuguese, closing the chapter on colonialism. When economic turbulence struck its Pacific Rim and Southeast Asian neighbors, China withstood the onslaught. Its trade with neighbors near and far is growing. Foreign investment in China has mushroomed. And still its government espouses communist dogma.

How could the government manipulate the coexistence of communist politics and market economics? That was the key question confronting Deng Xiaoping and his comrades when they took power in 1979. In terms of ideology, this objective

FIGURE 9-17

would seem unattainable; an economic "open-door" policy would surely lead to rising pressures for political democracy.

But Deng thought otherwise. If China's economic experiments could be spatially separated from the bulk of the country, the political impact would be kept at bay. At the outset, the new economic policies would apply mainly to China's bridgehead on the Pacific Rim, leaving most of the vast country comparatively unaffected. Accordingly, the government introduced a complicated system of **Special Economic Zones (SEZs)**, so-called *Open Cities*, and *Open Coastal Areas*, which would attract technologies and investments from abroad and transform the economic geography of eastern China (Fig. 9-17).

10

In these economic zones, investors are offered many incentives. Taxes are low. Import and export regulations are eased. Land leases are simplified. The hiring of labor under contract is allowed. Products made in the economic zones may be sold on foreign markets and, under some restrictions, in China as well. Even Taiwanese enterprises may operate here. And profits made may be sent back to the investors' home countries.

When Deng's government made the decisions that would reorient China's economic geography, location was a prime consideration. Beijing wanted China to participate in the global market economy, but it also wanted to cause as little impact on interior China as possible—at least in the first stages. The obvious answer was to position the Special Economic Zones along the coast. Initially, the government established four SEZs, all with particular locational properties (Fig. 9-17):

1. *Shenzhen*, adjacent to then-booming British Hong Kong on the Pearl River Estuary in Guangdong Province
2. *Zhuhai*, across from then still-Portuguese Macau, also on the Pearl River Estuary in Guangdong Province
3. *Shantou*, opposite southern Taiwan, a colonial treaty port, also in Guangdong Province, source of many Chinese now living in Thailand

The story of Shenzhen has no parallel in the modern world. Across Hong Kong's once-sealed, now-porous border with China a fishing village was overrun by urban and industrial development so that its population went from under 20,000 to over 3 million in just 15 years. Investment from Hong Kong, Japan, and China, as well as indirectly from Taiwan, and cheap labor from Guangdong and Fujian Provinces as well as the Guangxi-Zhuang Autonomous Region, in a few years produced an industrial and service center that became the prototype of China's Special Economic Zones, where business and industry were given incentives and advantages not available elsewhere. When you visit Shenzhen today, it looks so established and functional that it is easy to forget that this city has, in effect, no history as yet other than its phenomenal growth. . . Macau, on the other hand, which reverted to China at the end of 1999 and became the second Special Administrative Region (SAR), has a centuries-long history of Portuguese colonialism. Long a gambling mecca and rife with organized crime, Macau has a population just one-eighth of Shenzhen and its recent acquisition presents China's leaders with quite a different prospect. A favorable relative location near the Zhuhai SEZ and major infrastructure improvements, including an international airport and huge island-linking bridges, enhance Macau's potential as a participant in the economic growth of the Pearl River Estuary.

4. *Xiamen*, on the Taiwan Strait, also a colonial treaty port (then known as Amoy in the local dialect), in Fujian Province, source of many Chinese now based in Singapore, Indonesia, and Malaysia

In 1988 and 1990, respectively, two additional SEZs were proclaimed:

5. *Hainan Island*, declared an SEZ in its entirety, its potential success linked to its location near Southeast Asia
6. *Pudong*, across the river from Shanghai, China's largest city, different from other SEZs because it was a giant state-financed project designed to attract large multinational companies

China also opened 14 other coastal cities for preferential treatment of foreign investors (Fig. 9-17). Again, most of these cities had been treaty ports in the colonial extraterritoriality period. They were chosen for their size, overseas trading history, links to emigrated Chinese, level of industrialization, and pool of local talent and labor. These 14 Open Cities are, from north to south:

Dalian (Liaoning Province)
Qinhuangdao (Hebei)
Tianjin Shi
Yantai (Shandong)
Qingdao (Shandong)
Lianyungang (Jiangsu)
Nantong (Jiangsu)
Shanghai Shi
Ningbo (Zhejiang)
Wenzhou (Zhejiang)
Fuzhou (Fujian)
Guangzhou (Guangdong)
Zhanjiang (Guangdong)
Beihai (Guangxi-Zhuang A.R.)

When all these Pacific Rim initiatives were implemented, China's market-conscious communist leaders had reason to expect an economic boom in the coastal provinces. (On the map, note that the new economic policy affected every coastal province, as well as the single coastal A.R.) Japan already was a highly developed economy (at the final stage of the Rostow model). South Korea and Taiwan were well beyond takeoff. Hong Kong proved what could be done on this side of the Pacific. Singapore, Southeast Asia's most successful state, is 76 percent Chinese. In truth, China's pragmatic planners were unsure of just what forces they might unleash, which is why foreign investment rules were stricter in the Open Cities and Open Coastal Areas than in the freewheeling SEZs.

Deng's economic initiatives faced opposition within the Communist Party, a thread of dissent that still survives. Some senior members of the Politburo argued that the SEZs and Open Coastal Cities constituted a return to the days of the treaty ports, when foreigners could mistreat Chinese without compunction. When several of the Open Coastal Cities lagged behind, abandoned by foreign investors leaving unemployed workers behind, and when several factory fires killed dozens of workers because of lax safety rules, these opponents made their voices heard. But Deng's personal intervention, visiting SEZs and exhorting their administrators and workers, kept the process on track.

CHINA'S PACIFIC RIM TODAY

When China's planners laid out the SEZs, they held the highest hopes for Shenzhen, immediately adjacent to the thriving British dependency of Hong Kong. They knew that the kinds of attractions the SEZs offered to foreign investors would entice many factory owners paying Hong Kong wages to move across the border to Shenzhen. Successful Chinese living in Southeast Asia could use Hong Kong as a base for investment in Shenzhen; Taiwanese, even during the political standoff between Beijing and Taipei, could channel investments into this favored SEZ.

Those hopes were not disappointed. In the late 1970s, Shenzhen was just a sleepy fishing and duck-farming village of about 20,000. By 2002, its population surpassed 3.9 million—the fastest growth of any urban center on Earth in human history. Thousands of factories, large and small, moved into the SEZ from Hong Kong. Workers by the hundreds of thousands came from Guangdong Province, Guangxi-Zhuang A.R., and beyond to look for jobs. High-rise buildings housing banks, corporate offices, hotels, and other service facilities made Shenzhen look like a new Hong Kong (see photo, p. 466).

Shenzhen's economic impact was far-reaching. Persuading Hong Kong industries to cross the border precipitated a sharp modification of Hong Kong's own economy; attracting investments from Overseas Chinese, mainly in Southeast Asia, opened financial coffers long closed to China; and enabling Taiwanese investors to participate in China's Pacific Rim boom created links with the "wayward province" that have economic as well as political benefit. By its very success (in its 1990s heyday it produced 15 percent of all of China's exports by value), Shenzhen signaled the way for the other SEZs and Open Cities.

Hong Kong (Xianggang) and Its Role

Shenzhen's extraordinary growth was energized by a particular set of geographic circumstances (Fig. 9-18). As a British dependency, Hong Kong (Xianggang) grew into what economic geographers called an **economic tiger** on the Pacific Rim. Here, on the left bank of the Pearl River Estuary, more than 7 million people (97 percent of Chinese ancestry, many of them escapees from communism) crowded onto 400 square miles (1000 sq km) of fragmented, hilly territory under the tropical sun and created an economy larger than that of a hundred countries.

11

Hong Kong (the Cantonese name means "fragrant harbor") had three components during British colonial rule (Fig. 9-18). One, the island of Hong Kong, was the headquarters, containing the capital of Victoria and overlooking the colony's magnificent port. The second part consisted of the other islands (among which Lantau, by far the largest, was chosen as the site of a new international airport) plus the peninsula called Kowloon, directly across the harbor from Victoria. The third was known as the New Territories, the bulk of the mainland. China ceded the first two parts (the islands and the Kowloon Peninsula) permanently to Britain in 1841 and 1860, respectively. Britain leased the New Territories for 99 years in 1898.

When that lease expired in 1997, the British, following difficult negotiations, agreed to yield authority over all three of the dependency's components. China promised to keep Hong Kong's capitalist economy intact, so that there would be "one country, two systems." At midnight on June 30, 1997, Hong Kong became the Xianggang Special Administrative Region (SAR). True to the letter (if not always the spirit) of this commitment, the name "Hong Kong" has continued in general use.

Shenzhen had benefited from Hong Kong's economic success long before the Chinese took over in Victoria, however. Until the 1950s Hong Kong had been just another trading colony—a busy port but little more. Then the Korean War and the United Nations embargo on trade with China cut the colony's connections with its hinterland, necessitating an economic reorientation. Hong Kong possessed no mineral resources, but it did have virtually unlimited human resources—people often desperate to earn a wage, no matter how small. All that was needed was a supply of raw materials, investment in equipment, production know-how, and efficient social organization. Among the Chinese population were many Chinese entrepreneurs from the Shanghai area who had fled communism in 1949, and they played a crucial role now. Within a few years, a huge textile industry and many light manufacturing industries developed. Products were made at low cost and found ready markets throughout the world. Accumulated capital was then used to establish factories making electrical equipment, appliances, and countless other consumer goods. As the flood of immigrants continued, many of them refugees from communist China, labor costs remained low and products stayed competitive. Hong Kong exported products to markets the world over. Meanwhile, the colony served as a back door to closed China, which rewarded it by providing fresh water and staple foods.

As Hong Kong's economy diversified, the city became one of the world's leading financial centers with the Hong Kong dollar's value pegged to the U.S. dollar. During the 1980s, Chinese banks and companies began to enter the Hong Kong economy even as Hong Kong factories moved to Shenzhen to take advantage of the lower wages there. The links between Hong Kong and Shenzhen grew closer as goods were prepared in Hong Kong, assembled in Shenzhen, and exported from the colony. In the 1990s, about 25 percent of China's foreign trade passed through Hong Kong.

FIGURE 9-18

As Shenzhen's industrial economy grew, that of Hong Kong declined; but Hong Kong's financial and service sectors expanded. Tourism earned significant revenue, for example.

By the time the Chinese took over Hong Kong, problems began to plague both Xianggang and the Shenzhen SEZ. Wages, though still low in Shenzhen, were rising, and some companies that might have chosen Shenzhen for their factories now "leapfrogged" the SEZ in pursuit of still-lower labor costs. Meanwhile, an economic crisis occurred in Southeast Asia, where the values of local currencies plummeted. This development had severe repercussions for Hong Kong banks and their Chinese partners. It also affected Overseas Chinese investment in China and battered the service industries, including tourism. Hong Kong's overinflated real estate market collapsed, and the Hong Kong dollar and stock market plummeted. The first several years of Chinese control over Hong Kong were difficult indeed, but not all the problems were of China's making.

On December 20, 1999 China acquired the other European colony of the Pearl River Estuary: Portuguese Macao (see photo, p. 466). With only 8 square miles (21 sq km) of territory and less than a half million inhabitants (97 percent Chinese, about 15 percent Roman Catholic), Macau, as the Chinese prefer to spell it, has none of the economic clout of Hong Kong. Casino-driven tourism and organized crime were its hallmarks before the transfer and will be the new administration's preoccupations in the years ahead. Like Hong Kong, Macau has been designated a Special Administrative Region (SAR), guaranteeing the continuation of its existing social and economic systems for 50 years. As Figure 9-19 shows, Macau adjoins the Zhuhai SEZ; it boasts a major international airport linked to one of its two major islands, Taipa, which is in turn connected to the mainland by a four-lane causeway.

The Other SEZs

Every economy experiences ups and downs, and China's Pacific Rim is no exception. But China's unprecedented investment in regional infrastructure will bear permanent witness to its Deng-inspired economic drive. Even as Hong Kong was turned over to China, for example, Chinese money put the finishing touches on its new international airport (known earlier as Chek Lap Kok), connected to the

Pearl River Hub

When economic geographers measure the contributions China's provinces make to the national economy, one province stands out, year after year: Guangdong, capital Guangzhou (3.9 million), population 78 million. And most of Guangdong's production comes from one vast megalopolis that flanks the Pearl River Estuary (Fig. 9-19).

The triangle formed by Hong Kong, Guangzhou, and Macau, and including Shenzhen and Zhuhai, had all the requirements when China's new economic policy was proclaimed: a huge reservoir of cheap labor, available capital, experienced management, and modern expandable infrastructures in the international centers of the then-British and Portuguese cities. In 1980, most of the estuary's frontage consisted of marshes, tidal flats, ricefields, and duck ponds. Today, its population exceeds 23 million, and the cultural landscape consists of miles of gray box-like buildings, high-rises, multi-lane highways, bridges, parking lots, and loading docks. It is a concrete-and-asphalt desert under a permanent pall of smog. And development continues: the area now has four major international airports. Container-port and bridge-building rushes ahead. If the current growth continues, the Pearl River Hub will soon have three cities with more than 5 million inhabitants, five with more than 1 million, and six with more than a half million. Per-capita incomes here are estimated to be 10 times the national average (in Hong Kong/Shenzhen, 25 times).

The Pearl River Hub exactly fits the model of the **regional state** as defined by the Japanese scholar Kenichi Ohmae. These "natural economic zones" defy old political borders, and they are shaped by the global economy of which they are parts; their leaders deal directly with foreign partners and negotiate the best terms they can with the national governments under which they operate. The Pearl River Megalopolis has moved so far ahead of China as a whole, so far ahead even of its Pacific Rim cohorts, that it is a political as well as an economic force in China.

Guangdong Province always has been a place apart in China. Far from meddlesome Beijing and with strong regional cultural traditions, it has long been entrepreneurial and outward-looking. Today its bustling Pearl River Megalopolis is in the vanguard of the new China.

12

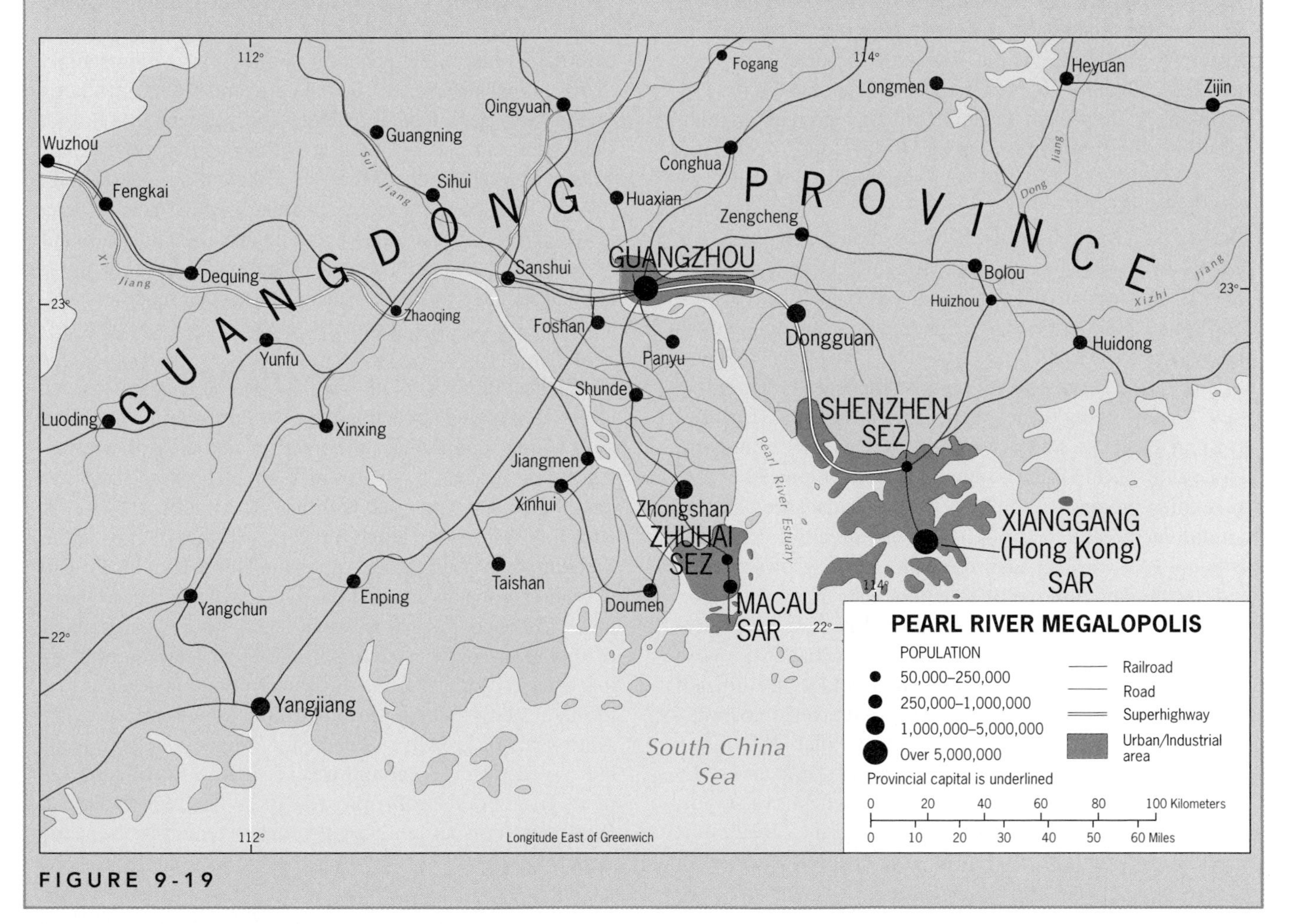

FIGURE 9-19

mainland by the world's longest road-and-rail suspension bridge. Work is now proceeding on another world-class bridge and two tunnels beneath Victoria Harbor in what is being called the largest public works project on Earth.

If this is true, it will have strong competition from Shanghai's gigantic construction scheme at Pudong. From an upper-floor hotel room on the other (Bund) side of the Huangpu River, we counted 117 cranes engaged in the construction of skyscrapers that already include the tallest structure in East Asia and will soon boast the tallest building in the world. The skyline of Pudong has risen from village homes to Hong Kong proportions in one decade, as a wall of glass-and-metal high-rises towers over the Huangpu (see photo, p. 455). This, too, is largely a public works project: it evinces Beijing's determination to shift China's Pacific Rim focus from south to north (see box titled "Pearl River Hub"). In 2001, much of the floor space in Pudong stood unoccupied, reflecting the Asian economic downturn. But the Beijing government is undeterred. Occasional references by public officials to the flight of Chinese entrepreneurs to Hong Kong during the 1949 communist takeover provide a glimpse of the Communist Party's motives in making this enormous financial commitment to this newest SEZ.

None of the other SEZs has come close to what has happened in Shenzhen or Pudong, but development in Xiamen, Shantou, Zhuhai, and Hainan has not been insignificant. Xiamen was established especially to attract Taiwanese investment, and hundreds of Taiwanese factories have been built here. Shantou has benefited from Overseas Chinese investment, mainly from Thailand but also from other countries. However, Shantou is the least developed of the current SEZs because of weak infrastructure and because the city lacks the talent and trading traditions of Xiamen. Zhuhai, which is to Macau what Shenzhen is to Hong Kong, suffers because Macau is no Hong Kong; nevertheless, Zhuhai has expanded on its own merits and, in the mid-1990s, had a higher per-capita income than any other city in the province of Guangdong except Shenzhen. This reflects the rise of the service sector here. In the late 1990s companies "leapfrogging" Shenzhen often established factories in Zhuhai, where wages were below Shenzhen standards. Thousands of factories producing goods ranging from toys to textiles create a pollution-choked landscape; working conditions in the factories often are dreadful. Zhuhai's problem is export facilities: container traffic grows rapidly, and transfer and port facilities cannot keep up.

In many ways the most interesting SEZ is Hainan, where SEZ regulations produced what the Chinese called the Wild South. Hainan's governor established policies he himself described as low and free: low wages and taxes, free movement of people and money. The result was what Willem van Kemenade in his book *China, Hong Kong, Taiwan Inc.* describes as "the lawless Sicily of China, where crime, corruption, piracy, and prostitution were rampant." Real estate speculation went out of control, converting the capital of Haikou into a jungle of high-rises in various stages of completion, their prices sometimes quadrupling in one year. Then the governor got involved in a scheme to import Mercedes automobiles for "export" at high prices to mainland China, and Beijing was forced to intervene. When the effort to establish an SEZ-within-an-SEZ at the port of Yangpu failed, Hainan's tumultuous tenure as a development center under SEZ initiatives came to a halt.

Beijing has not withdrawn Hainan's SEZ designation, probably because the island's relative location is important to China and gives hope for future success. With a population of 7.5 million (including two of China's poorest minorities, the Southeast Asian Li and Miao), Hainan represents China's deepest penetration into the South China Sea. The island is historically part of the Southeast Asian economic sphere, and much contraband, including illicit drugs, enters China via Hainan. In addition, Hainan has become a base for the pirates that now stalk the seas here. Recently, Hainan's government has been trying to exploit its tropical situation and still-unfettered social life by encouraging tourism. But the unfinished high-rises in Haikou are evidence that this SEZ's bubble has burst.

Toward the Interior

Economic growth in the Pacific Rim has created growing regional disparities in the new China. When Deng proclaimed China's new economic policy, he acknowledged that regional differences in incomes and living standards would increase, but he argued that those places able to "get rich" should do so, creating growth poles whose fortunes would eventually benefit the rest of the country. More than two decades later, his successor Jiang Zemin leads a China in which the highest incomes in the urban Pacific Rim are 80 times larger than those in the rural west. Concern is rising that such contrasts may lead to provincial rebellions and a general breakdown of order.

At the turn of the century, plans were being made to channel economic growth from the Pacific Rim into the interior. If SEZs succeeded in the east, they might also thrive near China's borders in the interior, and SEZ-like privileges were granted to an area in Yunnan, bordered by Vietnam, Laos, and Myanmar (Burma). The notion of a Pacific Rim-linked corridor along the Chang/Yangzi River to the Sichuan Basin, benefiting Anhui and Hubei Provinces, also was under consideration. But two obstacles hampered such projects: the central government's reluctance to divest itself of thousands of outdated and inefficient state enterprises that drain the treasury, and the Asian economic downturn of the late 1990s, which slowed China's economic growth. After years of growth at 9 percent or thereabouts, the well-buttressed Chinese economy nevertheless felt the effects of its neighbors' problems. The risks inherent in Pacific Rim-related disparities may therefore worsen in the years ahead.

CHINA: A GLOBAL SUPERPOWER?

Two centuries ago, Napoleon remarked that China was a giant asleep; whoever awakened the Chinese giant, he said, would be sorry. Today China is awake, aware of its power and its potential. As the twenty-first century opens, China is poised to take the world's center stage.

Who awakened China? This momentous development is another consequence of the Cold War between the superpowers, the United States and the Soviet Union. The Cold War worsened many regional and even local conflicts as each superpower supported, and armed, the side it favored. Most of those conflicts are over. But the Soviet communist advisors and technicians to Mao's regime sowed the seeds of China's awakening. Russia and China had long been adversaries in eastern Eurasia, and soon Chinese ideologues and Soviet "revisionists" were at loggerheads. Now China found itself at odds with an immediate neighbor who possessed a vast nuclear arsenal. Mao and his comrades decided that an opening to the capitalist West was in China's long-term interest, and in 1972 U.S. President Nixon and his entourage arrived in Beijing at Mao's invitation. Before the 1970s were over, Mao had died, and those who favored the open-door policy won the ensuing power struggle. The rest is geography. China's cultural and economic landscapes were transformed in just two decades. China not only was awake, it was astir.

Today, China is on the march. Beijing has taken the United Nations seat long occupied by Taiwan's mainlanders. Its leading trading partners are not neighboring Russia or India, but Japan and the United States. (Trade with the United States is boosted by China's Most-Favored-Nation [MFN] status.) Japan is best situated to sell China's more than 400 million city dwellers such products as air conditioners, electronic goods, and automobiles; the United States imports from China a wide range of goods including textiles and toys. China also is a major exporter of weapons. Its global trade network is expanding in ways Mao Zedong and his isolationist colleagues could not have foreseen.

The Chinese rulers who took control in Beijing in the late 1970s, led by Deng Xiaoping, called themselves "pragmatists." That appellation referred especially to their economic policies, but it is applicable to other practices as well. Self-interest, regardless of the views or actions of others in the world community, guides China's use of military power and strategic decision-making. In mid-1989, Chinese armed forces brutally destroyed a student-led democracy movement in the heart of Beijing and in other cities, with a huge loss of life. In the early 1990s, China established diplomatic relations with South Korea, facilitating economic ties but infuriating communist North Korea. But shortly thereafter, when a UN coalition tried to stop North Korea from acquiring the ability to build nuclear weapons, China took North Korea's side. When the United States in 1994 indicated that it might end China's MFN status because of human-rights violations, the Chinese were prepared to pay that price—and the United States "decoupled" the two issues. When the Taiwanese president was allowed to enter the United States for his university class reunion in 1995, China delayed accepting a new U.S. ambassador to Beijing and sold Silkworm missiles to Iran over U.S. objections.

These might be mere incidents in the evolving relationship between China and the rest of the world, but there are more serious portents. Beijing's apparent unwillingness to resolve the Taiwan issue may result in confrontation. China's actions in the South China Sea, which it claims in contravention of United Nations rules, may lead to conflict over the Spratly Islands where China has laid claim to, and built structures on, a reef that lies within the Philippines' Exclusive Economic Zone (a concept discussed in Chapter 12). Furthermore, China has both actual and latent conflicts over boundaries and territories with India, Kazakhstan, and Russia.

What will China's position be in the world of the twenty-first century? China possesses a wide range of natural resources, from fuels and metals to forests and fertile soils. Its human resources are enormous, and its peoples' skills and capacities have been demonstrated both within China and outside its borders. Barring a Soviet-type organizational collapse, China will become the giant of East Asia, and not just demographically.

Thus China is likely to become more than an economic force of world proportions: China also appears on course to achieve global superpower stature. As long as it retains its autocratic form of government (which allowed it to impose draconian population policies and comprehensive economic experiments without having to consult the electorate), China will be able to practice the kind of state capitalism that—in another guise—made South Korea an economic power. And unlike Japan or South Korea, China has no constraints on its military power. Its People's Liberation Army served as security forces during the pro-democracy turbulence of 1989; China's military is the largest standing army in the world, with nearly 3 million soldiers and some 1.2 million reserves. Its equipment, however, is so antiquated that, at the turn of the century, Beijing could not contemplate a military invasion of Taiwan. China's fleet of about 100 naval vessels is aging and includes no aircraft carriers, battleships, or cruisers. Its air force has nearly 5000 planes, but most are over two decades old. However, China has home-made intercontinental and medium-range ballistic missiles, and several hundred nuclear bombs.

During the twentieth century, the United States and the Soviet Union were locked in a 45-year Cold War that repeatedly risked nuclear war. That fatal exchange never happened, in part because it was a struggle between superpowers who understood each other comparatively well. While the politicians and military strategists were plotting, the cultural doors never closed: American audiences listened to Prokofiev and Shostakovich, watched Russian ballet, and read Tolstoy and Pasternak even as the Soviets cheered Van Cliburn, read Hemingway, and lionized American dissidents. In short, it was an intracultural Cold War, which reduced the threat of mutual destruction.

The twenty-first century may witness a far more dangerous geopolitical struggle in which the adversaries may well be the United States and China. U.S. power and influence still prevail in the western Pacific, but it is easy to discern areas where Chinese and American interests will diverge (Taiwan is only one example). American bases in Japan, thousands of American troops in South Korea, and American warships in the East and South China Seas are potential grounds for dispute. All this might generate the world's first *inter*cultural Cold War, in which the risk of fatal misunderstanding is incalculably greater than it was during the last.

How can such a Cold War be averted? Trade, scientific and educational links, and cultural exchanges are obvious remedies: the stronger our interconnections, the less likely is a deepening conflict. We Americans should learn as much about China as we can, to understand it better, to appreciate its cultural characteristics, to recognize the historico-geographical factors underlying China's views of the West. For some, this geographic survey may be a first step toward a lifelong interest in a country whose history is ten times longer than our own.

For more than 40 centuries, China has known authoritarianism of both the brutal and the benevolent kind, has been fractured by regionalism only to unify again, and has depended on communalism to survive environmental and despotic depredations. For 25 centuries Confucianism has guided it. Time and again, China has been opened to, and ruled by, foreigners, and time and again it has retreated into isolationism when things went wrong. Such a reversal may no longer be possible, given what we have seen in this chapter. But now a renewed force is rising in modernizing China: nationalism. A spate of recent books published in China, including one by Song Qiang et al. titled *China Can Say No*, reflects the frustrations of many Chinese over what they view as American arrogance and insensitivity to China's traditions and interests. To these authors (whose books are bestsellers in China), the Americans are only the latest of a series of foreign invaders, none of whom has brought China long-term good. Washington's interference in a quarrel among Chinese (the Taiwan issue) infuriates them, and they warn of dire consequences if the policy persists.

The sequence of events that Mao and Nixon set in motion during the 1970s already has had momentous consequences. Today when we look ahead to a modernizing, mobilizing China as the twenty-first century unfolds, we can hear the echo of Napoleon's words.

MONGOLIA

Between China's Inner Mongolia and Xinjiang to the south and Russia's Eastern Frontier region to the north lies a vast, landlocked, isolated country called Mongolia that constitutes a discrete region of East Asia (Fig. 9-4). With only 2.6 million inhabitants in an area larger than Alaska, Mongolia is a steppe- and desert-dominated vacuum between two of the world's most powerful countries. Its historical geography is tied to Russia and the former Soviet Union as well as to China. Mongolia was part of the Mongol Empire, source of those great waves of horsemen who rode into the Slavic heartland seven centuries ago. The Mongols also invaded China, establishing the short-lived Yuan Dynasty (1264–1368). Eventually, however, the Mongol domain fell under Chinese control. From the late 1600s until 1911, Mongolia was part of the Chinese Empire.

The Mongolians seized independence in 1911 while China was in revolutionary chaos. In those days, present-day Mongolia was known as Outer Mongolia; the ill-defined region to its southeast and east was called Inner Mongolia. As the Chinese tried to reassert their hegemony over Outer Mongolia, the Russian Revolution broke out. This allowed the (Outer) Mongolians to hold off the Chinese. In the early 1920s, the country became a People's Republic on the Soviet model, under Soviet protection. Inner Mongolia remained a Chinese-dominated frontier and became the Nei Mongol Autonomous Region under Beijing.

Today, with its vast deserts, grassy plains, forest-clad mountains, and fish-filled lakes, Mongolia lies uneasily between larger, unstable neighbors, neither of which can permit its incorporation into the other. Despite its historic associations, ethnic affinities, and cultural involvement with China, the Soviets guided Mongolia's development for some 70 years. The capital, Ulaanbaatar, the country's largest city, lies close to the Lake Baykal subregion of Russia's Eastern Frontier. China seems far away, across the forbidding Gobi Desert.

But Mongolia's cultural and economic landscapes are East Asian, despite the Soviet/Russian veneer, and the winds of political change have reached here too. The Mongolian alphabet, abolished under Soviet educational regulations, has been reinstated. Soon after the Soviet collapse, elections brought the Democrats, an old alliance of anti-Soviet activists, to power; in the 2000 elections, the Mongolian People's Revolutionary Party, the party of ex-communists, made a landslide comeback, proving that democracy had firmly taken hold. But the reason for the ex-communist victory was worrisome: in the countryside, Mongolia's 800,000 nomadic herders suffered severe hardship while politicians wrangled in the capital. The price of cashmere, Mongolia's best-known export, had fallen in part because Chinese traders, coming across the border with Inner Mongolia, bought up the best of it, leaving what remained poorer in quality and cheaper on local markets. And Mongolia's herders suffered through the worst winter in memory during 1999–2000, losing livestock and getting inadequate help from the government.

Mongolia's economic opportunities include a major copper mine and other metal production which, in combination, are more valuable than cashmere. But, as Figure I-11 reveals, this still is a low-income economy (per-capita GNP: $380), with very limited foreign investment and inefficient state enterprises slowing development.

More troubling to Mongolians, however, is their country's regional geography. Mongolia remains a landlocked, isolated, vulnerable **buffer state**, its vast, Wyoming-like expanses wedged, Tibet-like, between populous, more powerful neighbors. Mongolia's northern boundary leaves only a narrow Russian corridor south of Lake Baykal. Across Mongolia's southern border lies a Chinese Autonomous Region containing more people of Mongol descent than live in Mongolia itself. Will Mongolia escape the fate of those other inner-Asian buffers, Xizang and Afghanistan?

13

THE JAKOTA TRIANGLE REGION

We turn now to a region that exemplifies the future of East Asia. Along the Pacific Rim, from Japan (through South Korea, Taiwan, China's Guangdong) to Singapore, rapid economic development has transformed community and society. In China, as we saw, a Pacific Rim region is in the making, still discontinuous today but likely to extend all along the coast in the future. In Japan, South Korea, and Taiwan we can observe that future—today. That is why we recognize an East Asian region we call the **Jakota Triangle**, consisting of Japan, Korea (South Korea at present but probably a reunited Korea later), and Taiwan (Fig. 9-20). This is a region of great cities, huge consumption of raw materials from all over the world, voluminous exports, and global financial linkages. It also is a region of social problems, political uncertainties, and economic vulnerabilities.

14

In the discussion that follows, we deal in turn with Japan, South Korea, and Taiwan, the components of the

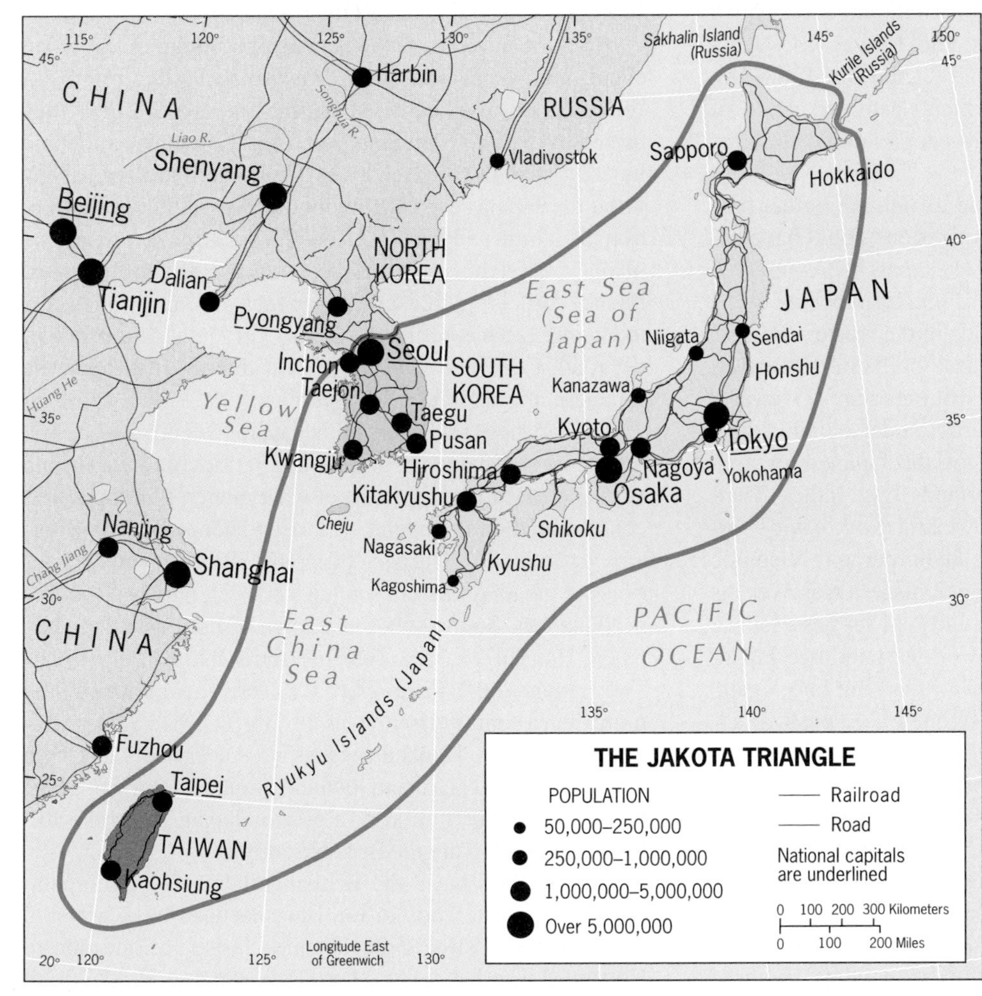

FIGURE 9-20

Jakota Triangle region. This delimitation produces a problem, for North Korea is excluded—not only from this region but, for obvious reasons, from China Proper as well. It is an unusual situation that reflects precisely North Korea's incompatible position in this realm, and indeed in the world today. We discuss North Korea after we have focused on the Jakota region's dominant component, Japan.

JAPAN

When we assess China's prospects of becoming a superpower, we should remember what happened in Japan in the nineteenth century. In 1868, a group of reform-minded modernizers seized power from an old guard, and by the end of the century Japan was a military and economic force. From the factories in and around Tokyo and from urban-industrial complexes elsewhere poured forth a stream of weapons and equipment the Japanese used to embark on colonial expansion. By the mid-1930s, Japan lay at the center of an empire that included all of the Korean Peninsula, the whole of China's Northeast (which the Japanese called Manchukuo), the Ryukyu Islands, and Taiwan, as well as the southern half of Sakhalin Island (called Karafuto). Not even a disastrous earthquake, which destroyed much of Tokyo in 1923 and killed 143,000 people, could slow the Japanese drive.

World War II saw Japan expand its domain farther than the architects of the 1868 "modernization" could have anticipated. By early December 1941, Japan had conquered large parts of China Proper, all of French Indochina to the south, and most of the small islands in the western Pacific. Then, on December 7, 1941, Japanese-built aircraft carriers moved Tokyo's warplanes within striking range of Hawai'i, and the surprise attack on Pearl Harbor underscored Japan's confidence in its war machine. Soon the Japanese overran the Philippines, the (then) Netherlands East Indies, Thailand, and British Burma and Malaya, and drove a wide corridor through the heart of China to the border with Vietnam.

A few years later, Japan's expansionist era was over. Its armies had been driven from virtually all its possessions, and when American nuclear bombs devastated two Japanese cities in 1945, the country lay in ruins. But once again, Japan, aided this time by an enlightened U.S. postwar administration, surmounted disaster.

Japan's economic recovery and its rise to the status of world economic superpower was the success story of the second half of the twentieth century. Japan lost the war and its empire, but it scored many economic victories in a new global arena. Japan became an industrial giant, a technological pacesetter, a fully urbanized society, a political power, and an affluent nation. No city in the world today is without Japanese cars in its streets; few photography stores lack Japanese cameras and film; laboratories the world over use Japanese optical equipment. From microwave ovens to VCRs, from oceangoing ships to camcorders, Japanese goods flood the world's markets.

Japan's brief colonial adventure helped lay the groundwork for other economic successes along the western Pacific Rim. The Japanese ruthlessly exploited Korean and Formosan (Taiwanese) natural and human resources, but they also installed a new economic order there. After World War II, this infrastructure facilitated an economic transition—and soon made both Taiwan and South Korea competitors on world markets.

Spatial Limitations

In discussing the economic geography of Europe, we noted how the comparatively small island of Britain became the crucible of the Industrial Revolution, which gave it a huge head start and advantage over the continent, across which the Industrial Revolution spread decades later. Britain was not large, but its insular location, its local raw materials, the skills of its engineers, its large labor force, and its social organization combined to endow it not only with industrial strength but also with a vast overseas empire.

After the modernizers took control of Japan in 1868—an event known as the *Meiji Restoration* (the return of "enlightened rule" centered on the Emperor Meiji)—they turned to Britain for guidance in reforming their nation and its economy. In the decades that followed, the British advised the Japanese on the layout of cities and the construction of a railroad network, on the location of industrial plants, and on the organization of education. The British influence still is visible in the Japanese cultural landscape today: the Japanese, like the British, drive on the left side of the road. Consider how this affects the effort to open the Japanese market to U.S. automobiles!

The Japanese reformers of the late nineteenth century undoubtedly saw many geographic similarities between Britain and Japan. At that time, most of what mattered in Japan was concentrated on the country's largest island, Honshu (literally, "mainland"). The ancient capital, Kyoto, lay in the interior, but the modernizers wanted a coastal, outward-looking headquarters. So they chose the town of Edo, on a large bay where Honshu's eastern coastline turns sharply (Fig. 9-21). They renamed the place *Tokyo* ("eastern capital"), and little more than a century later it was the largest urban agglomeration in the world. Honshu's coasts were near mainland Asia, where raw materials and potential markets for Japanese products lay. The notion of a greater Japanese empire followed naturally from the British example.

But in other ways, the British and Japanese archipelagoes, at opposite ends of the Eurasian landmass, differed considerably. In total area, Japan is larger. In addition to Honshu, Japan has three other large islands—Hokkaido to the north and Shikoku and Kyushu to the south—as well as numerous small islands and islets, for a total land area of

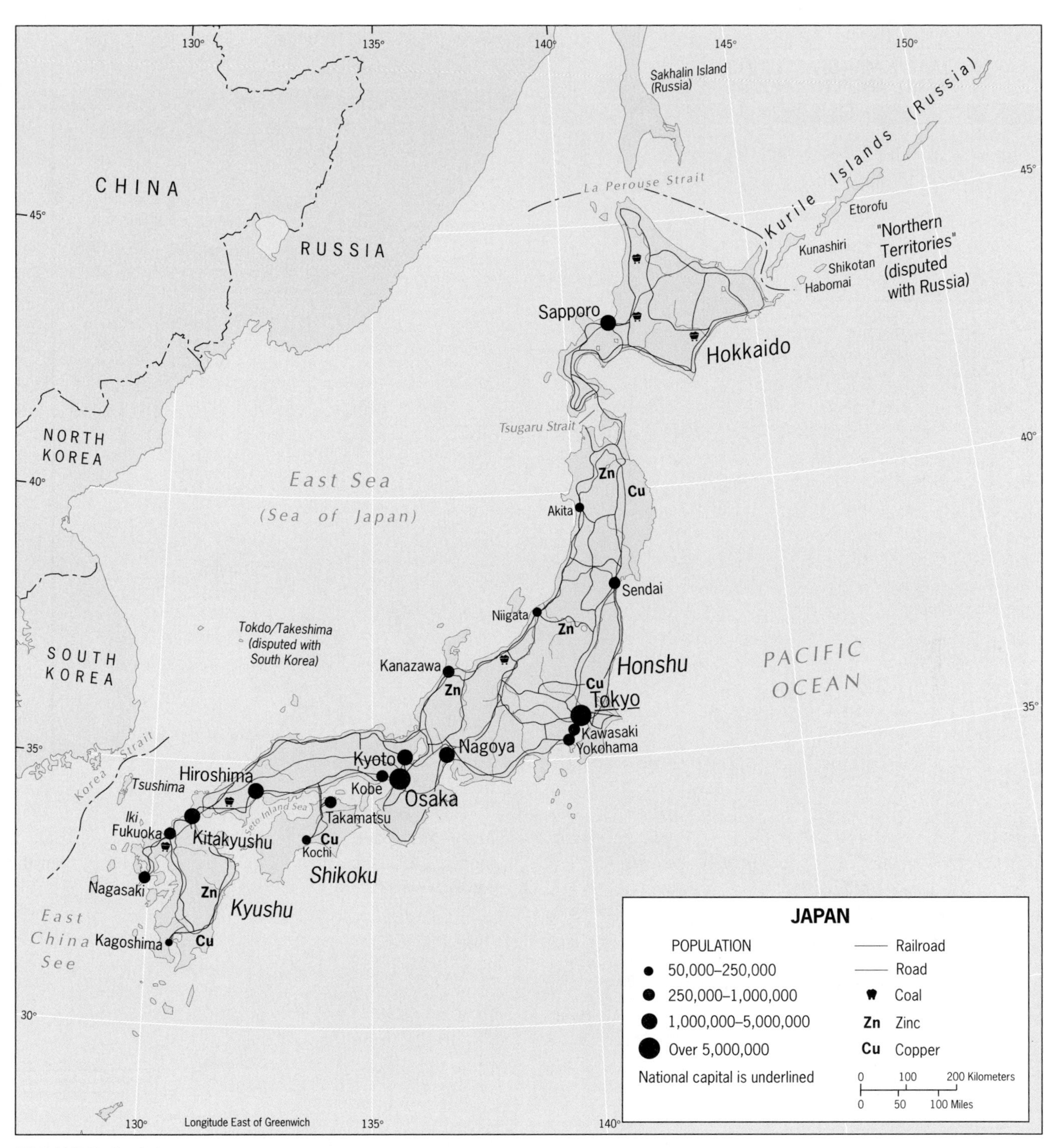

FIGURE 9-21

about 146,000 square miles (377,000 sq km). Much of this area is mountainous and steep-sloped, geologically young, earthquake-prone, and studded with volcanoes. Britain has lower relief, is older geologically, does not suffer from severe earthquakes, and has no active volcanoes. And in terms of the raw materials for industry, Britain was much better endowed than Japan. Self-sufficiency in iron ore and high-quality coal gave Britain the head start that lasted a century.

Japan's high-relief topography has been an ever-present challenge. All of Japan's major cities, except the ancient capital of Kyoto, are perched along the coast, and virtually all lie partly on artificial land claimed from the sea. Sail into Kobe harbor, and you will pass artificial islands designed for high-volume shipping and connected to the mainland by automatic space-age trains. Enter Tokyo Bay, and the refineries and factories to your east and west stand on huge

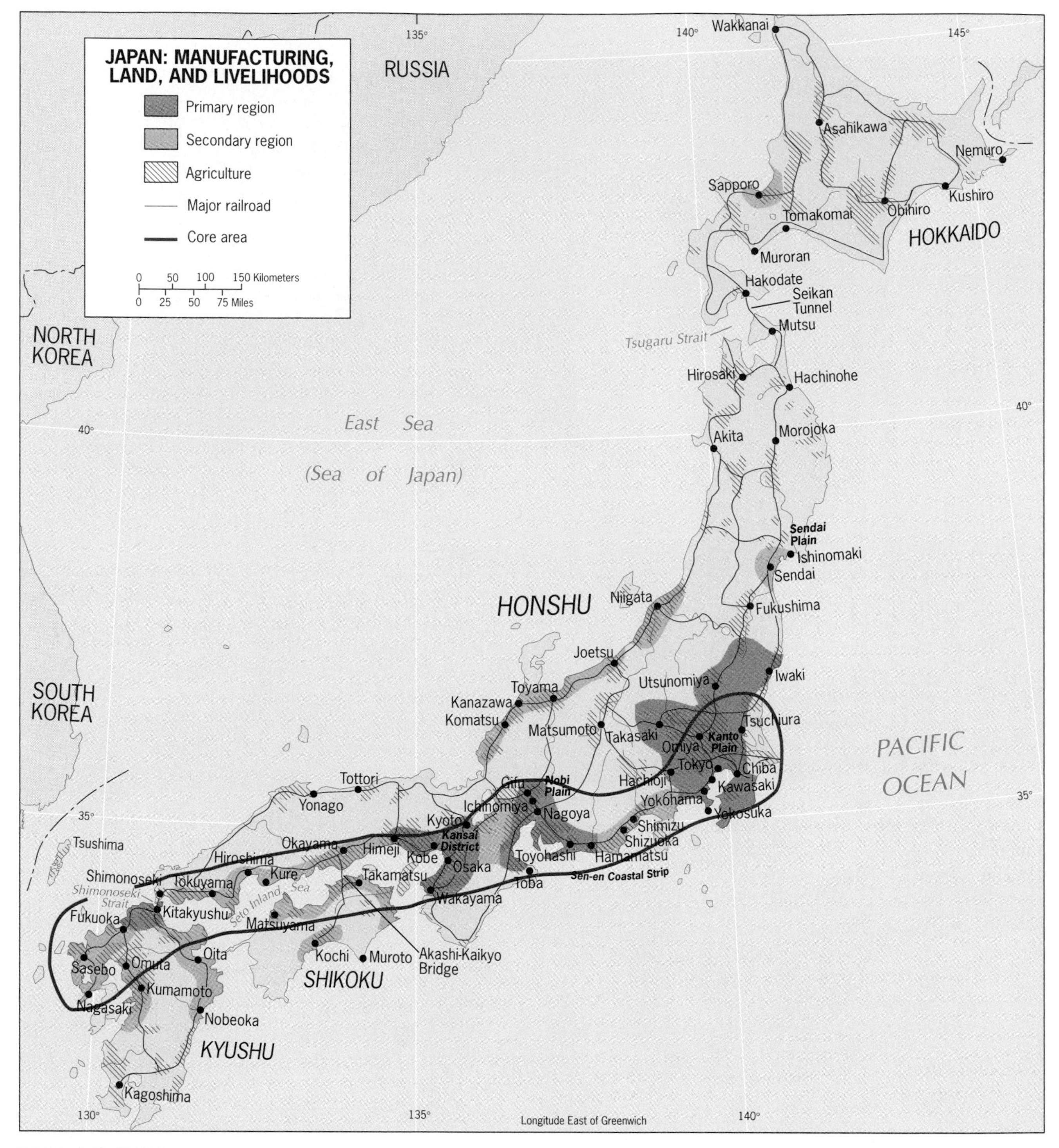

FIGURE 9-22

expanses of landfill that have pushed the bay's shoreline outward. With 127.2 million people, the vast majority (78 percent) living in towns and cities, Japan uses its habitable living space intensively—and expands it wherever possible.

As Figure 9-22 shows, farmland in Japan is both limited and regionally fragmented. Urban sprawl has invaded much cultivable land. In the hinterland of Tokyo lies the Kanto Plain; around Nagoya, the Nobi Plain; and surrounding Osaka, the Kansai District—each a major farming zone under relentless urban pressure. All three of these plains lie within Japan's fragmented but well-defined core area (delimited by the red line on the map), the heart of Japan's prodigious manufacturing complex.

From Realm to Region

For nearly a half-century, Japan's seemingly insurmountable economic lead, its almost unique (for so large a population) ethnic homogeneity, its insular situation, and its cultural uniformity justified its designation as a discrete geographic realm, not a region of a greater East Asia. But rapid change on the rim of the western Pacific—both within Japan and beyond it—has eroded the bases for this distinction. In the mid-1990s, Japan's economic engine faltered while those of its Pacific Rim neighbors accelerated; its social fabric was torn by unprecedented acts of violence, and its foreign relations were soured by disputes over wartime atrocities in China and Korea. Japan stood alone in the vanguard of Pacific Rim developments, but it is no longer discrete.

Early Directions

On the map, the Korean Peninsula seems to extend like an unfinished bridge from the Asian mainland toward southern Japan. Two Japanese-owned islands, Tsushima and Iki, in the Korea Strait almost complete the connection (Fig. 9-21). In fact, Korea was linked to Japan in prehistoric times, when sea levels were lower. Peoples and cultural infusions reached Japan from Asia time and again, even after rising waters inundated the land bridge between them.

The Japanese islands were inhabited even before the ancestors of the modern Japanese arrived from Asia. As we noted earlier, the Ainu, a people of Caucasian ancestry, had established themselves on all four of the major islands thousands of years before. But the Ainu could not hold their ground against the better organized and armed invaders, and they were steadily driven northward until they retained a mere foothold on Hokkaido. Today, the last vestiges of Ainu ancestry and culture are disappearing.

By the sixteenth century, Japan had evolved a highly individualistic culture, characterized by social practices and personal motivations based on the Shinto belief system. People worshiped many gods, venerated ancestors, and glorified their emperor as a divine figure, infallible and omnipotent. The emperor's military ruler, or *shogun*, led the campaign against the Ainu and governed the nation. The capital, Kyoto, became an assemblage of hundreds of magnificent temples, shrines, and gardens. In their local architecture, modes of dress, festivals, theater, music, and other cultural expressions, the Japanese forged a unique society. Handicraft industries, based on local raw materials, abounded.

Nonetheless, in the mid-nineteenth century, Japan hardly seemed destined to become Asia's leading power. For 300 years the country had been closed to outside influences, and Japanese society was stagnant and tradition-bound. When the European colonizers first appeared on Asia's shores, Japan tolerated, even welcomed their merchants and missionaries. But as the European presence in East Asia grew stronger, the Japanese began to shut their doors. Toward the end of the sixteenth century, the emperor decreed that all foreign traders and missionaries should be expelled. Christianity (especially Roman Catholicism) had gained a considerable foothold in Japan, but now this was feared as a prelude to colonial conquest. Early in the seventeenth century, the shogun launched a massive and bloody campaign that practically stamped out Christianity.

Determined not to suffer the same fate as the Philippines, which by then had fallen to Spain, the Japanese henceforth allowed only minimal contact with Europeans. A few Dutch traders, confined to a little island near the city of Nagasaki, were for many decades the sole representatives of Europe in Japan. Thus Japan entered a long period of isolation, lasting past the middle of the nineteenth century.

Japan could maintain its aloofness, while other areas were being enveloped by the colonial tide, because of its strong central government and well-organized military, the natural protection its islands provided, and its remoteness along East Asia's difficult northern coast. Also, Japan's isolation was far less splendid than that of China, whose exquisite silks, prized teas, and skillfully made wares attracted traders and usurpers alike.

When Japan finally came face to face with the new weaponry of its old adversaries, it had no answers. In the 1850s, the steel-hulled "black ships" of the American fleet sailed into Japanese harbors, and the Americans extracted one-sided trade agreements. Soon the British, French, and Dutch were also on the scene, seeking similar treaties. When there was local resistance to their show of strength, the Americans quickly demonstrated their superiority by shelling parts of the Japanese coast. By the late 1860s, even as Japan's modernizers were about to overturn the old order, no doubt remained that Japan's protracted isolation had come to an end.

A Japanese Colonial Empire

When the architects of the Meiji Restoration confronted the challenge to build a Japan capable of competing against powerful adversaries in a changing world, they took stock of the country's assets and liabilities. Material assets, they found, were limited. To achieve industrialization, coal and iron ore were needed. In terms of coal, there was enough to support initial industrialization. Coalfields in Hokkaido and Kyushu were located near the coast; since the new industries also were on the coast, cheap water transportation was possible. As a result, the shores of the (Seto) Inland Sea (Fig. 9-21) became the sites of many factories. But there was little iron ore, certainly nothing like what was available domestically in Britain, and not enough to sustain the massive industrialization the reformers had in mind. This commodity would have to be purchased overseas and imported.

On the positive side, manufacturing—light manufacturing of the handicraft type—already was widespread in Japan. In cottage industries and in community workshops,

FROM THE FIELD NOTES
"The city of Kyoto, chronologically Japan's second capital (after Nara; before Tokyo), is the country's principal center of culture and religion, education and the arts. Tree-lined streets lead past hundreds of Buddhist temples; tranquil gardens provide solace from the bustle of the city. I rode the bullet train from Tokyo and spent my first day following a walking route recommended by a colleague, but got only part of the way because I felt compelled to enter so many of the temple grounds and gardens. And not only Buddhism, but also Shinto makes its mark on the cultural landscape. I passed under a *torii*—a gateway usually formed by two wooden posts topped by two horizontal beams turned up at their ends—which signals that you have left the secular and entered the sacred, and found this beautiful Shinto shrine with its orange trim and olive-green glazed tiles."

the Japanese produced textiles, porcelain, wood products, and metal goods. The small ore deposits in the country were enough to supply these local industries. Power came from human arms and legs and from wheels driven by water; the chief source of fuel was charcoal. Importantly, Japan did have an industrial tradition and an experienced labor force that possessed appropriate manufacturing skills. All this was not enough to lead directly to industrial modernization, but it did hold promise for capital formation. The community and home workshops were integrated into larger units, hydroelectric plants were built, and some thermal (coal-fired) power stations were constructed in critical areas.

For the first time, Japanese goods (albeit of the light manufactured variety) began to compete with Western products on international markets. The Japanese planners resisted any infusion of Western capital, however. While Western support would have accelerated industrialization, it would have cost Japan its economic autonomy. Instead, farmers were more heavily taxed, and the money thus earned was poured into the industrialization effort.

Another advantage for Japan's planners lay in the country's military tradition. Although the shogun's forces had been unable to repel the invasions of the 1850s, this was the result of outdated equipment, not lack of manpower or discipline. So while its economic transformation gathered momentum, the military forces, too, were modernized. Barely more than a decade after the Meiji Restoration, Japan laid claim to its first Pacific prize: the Ryukyu Islands (1879). A Japanese colonial empire was in the making.

When Japan gained control over its first major East Asian colonies, Taiwan and Korea, its domestic raw-material problems were essentially solved, and a huge labor pool fell under its sway as well. High-grade coal, high-quality iron ore, and other resources were shipped to the factories of Japan; and in Taiwan as well as in Korea, the Japanese built additional manufacturing plants to augment production. This, in turn, provided the equipment to sustain the subsequent drive into China and Southeast Asia.

Modernization

The reformers who set Japan on a new course in 1868 probably did not anticipate that three generations later their country would lie at the heart of a major empire sustained by massive military might. They set into motion a process of **modernization**, but they managed to build on, not replace, Japanese cultural traditions. We in the Western world tend to equate modernization with Westernization: urbanization, the spread of transport and communications facilities, the establishment of a market (money) economy, the breakdown of local traditional communities, the proliferation of formal schooling, the acceptance and adoption of foreign innovations. In the non-Western world, the process often is viewed differently. There, "modernization" is seen as an outgrowth of colonialism, the perpetuation of a system of wealth accumulation introduced by foreigners driven by greed. In this view, the local elites who replaced the colonizers in the newly independent states only continue the disruption of traditional societies, not their true modernization. Traditional societies, they argue, can be modernized without being Westernized.

15

In this context, Japan's modernization is unique. Having long resisted foreign intrusion, the Japanese did not achieve the transformation of their society by importing a Trojan horse; it was done by Japanese planners, building on the existing Japanese infrastructure, to fulfill Japanese objectives. Certainly Japan imported foreign technologies and adopted innovations from the British and others, but the Japan that was built, a unique combination of modern and traditional elements, was basically an indigenous achievement.

Relative Location

Japan's changing fortunes over the past century reveal the influence of **relative location** [16] in the country's development. When the Meiji Restoration took place, Britain, on the other side of the Eurasian landmass, lay at the center of a global empire. The colonization and Europeanization of the world were in full swing. The United States was still a developing country, and the Pacific Ocean was an avenue for European imperial competition. Japan, even while it was conquering and consolidating its first East Asian colonies (the Ryukyus, Taiwan, Korea), lay remote from the mainstream of global change.

Then Japan became embroiled in World War II and dealt severe blows to the European colonial armies in Asia. The Europeans never recovered: the French lost Indochina, and the Dutch were forced to abandon their East Indies (now Indonesia). When the war ended, Japan was defeated and devastated, but at the same time the Japanese had done much to diminish the European presence in the Pacific Basin. Moreover, the global situation had changed dramatically. The United States, Japan's trans-Pacific neighbor, had become the world's most powerful and wealthiest country, whereas Britain and its global empire were fading. Suddenly Japan was no longer remote from the mainstream of global action: now the Pacific was becoming the avenue to the world's richest markets. Japan's relative location—its situation relative to the economic and political foci of the world—had changed. Therein lay much of the opportunity the Japanese seized after the postwar rebuilding of their country.

Japan's Spatial Organization

Imagine this: 127.2 million people crowded into a territory the size of Montana (population: 925,000), most of it mountainous, subject to frequent earthquakes and volcanism, with no domestic oilfields, little coal, few raw materials for industry, and not much level land for farming. If Japan today were an underdeveloped country in need of food relief and foreign aid, explanations would abound: overpopulation, inefficient farming, energy shortages.

True, only an estimated 18 percent of Japan's national territory is designated as habitable. And Japan's large population is crowded into some very big cities. Moreover, Japan's agriculture is not especially efficient. But Japan defeated the odds by calling on old Japanese virtues: organizational efficacy, massive productivity, dedication to quality, and adherence to common goals. Even before the Meiji Restoration, Japan was a tightly organized country of some 30 million citizens. Historians suggest that Edo, even before it became the capital and was renamed Tokyo, may have been the world's largest urban center. The Japanese were no strangers to urban life, they knew manufacturing, and they prized social and economic order.

Areal Functional Organization

All this proved invaluable to the modernizers when they set Japan on its new course. The new industrial growth of the country could be based on the urban and manufacturing development that was already taking place. As we noted, Japan does not possess major domestic raw-material sources, so no substantial internal reorganization was necessary. However, some cities were better sited and enjoyed better situations relative to those limited local resources and, more importantly, to external sources of raw materials than others. As Japan's regional organization took shape, a hierarchy of cities developed; Tokyo took and kept the lead, but other cities grew rapidly into industrial centers.

This process was governed by a geographic principle that Allen Philbrick called **areal functional organization**, [17] a set of five interrelated tenets that help explain the evolution of regional organization, not only in Japan but throughout the world. Human activity, Philbrick reasoned, has spatial focus. It is concentrated in some locale, whether a farm or factory or store. Every one of these establishments occupies a particular location; no two of them can occupy exactly the same spot on the Earth's surface (even in high-rises, there is a vertical form of absolute location). Nor can any human activity proceed in total isolation, so that interconnections develop among these various establishments. This system of interconnections grows more complex as human capacities and demands expand. Each system (for example, farmers sending crops to market and buying equipment at service centers) forms a unit of areal functional organization. In the Introduction we referred to functional regions as systems of spatial organization; we can map Philbrick's units of areal functional organization as regions. These regions evolve because of what he called "creative imagination" as people apply their total cultural experience and their technological know-how when they organize and rearrange their living space. Finally, Philbrick suggests, we can recognize levels of development in areal functional organization, a ranking of places and regions based on the type, extent, and intensity of exchange.

In the broadest sense, we can divide regions of human organization into three categories: subsistence, transitional, and exchange. Japan's areal organization reflects the exchange category. Within each of these categories, we can also rank individual places on the basis of the number and kinds of activities they generate. Even in regions where subsistence activities dominate, some villages have more interconnections than others. The map of Japan (Fig. 9-22) showing its resources, urban settlements, and surface communications can tell us much about Japan's economy. It looks just like maps of other parts of the world where an exchange type of areal organization has developed—a hierarchy of urban centers ranging from the largest cities to the tiniest hamlets, a dense network of railways and roads connecting these places, and productive agricultural areas near and between the urban centers.

For Japan, however, the map shows us something else: Japan's external orientation, its dependence on foreign trade. All primary and secondary regions lie on the coast. Of all the cities in the million-size class, only Kyoto lies in the interior (Fig. 9-22). If we deduced that Kyoto does not match Tokyo–Yokohama, Osaka–Kobe, or Nagoya in terms of industrial development, that would be correct—the old capital remains a center of small-scale light manufacturing. Actually, Kyoto's ancient character has been deliberately preserved, and large-scale industries have been discouraged. With its old temples and shrines, its magnificent gardens, and its many workshop and cottage industries, Kyoto remains a link with Japan's premodern past.

Leading Economic Regions

As Figure 9-22 shows, Japan's dominant region of urbanization and industry (along with productive agriculture) is the *Kanto Plain*, which contains about one-third of the Japanese population and is focused on the Tokyo–Yokohama–Kawasaki metropolitan area. This gigantic cluster of cities and suburbs (the world's second largest urban agglomeration), interspersed with intensively cultivated farmlands, forms the eastern anchor of the country's elongated and fragmented core area. Besides its flatness, the Kanto Plain possesses other advantages: its fine natural harbor at Yokohama, its relatively mild and moist climate, and its central location with respect to the country as a whole. (The region's only disadvantage is its vulnerability to earthquakes—see box titled "When the Big One Strikes."). It has also benefited enormously from Tokyo's designation as the modern capital, which coincided with Japan's embarkation on its planned course of economic development. Many industries and businesses chose Tokyo as their headquarters in view of the advantages of proximity to the government's decision makers.

The Tokyo–Yokohama–Kawasaki conurbation has become Japan's leading manufacturing complex, producing more than 20 percent of its annual output. The raw materials for all this industry, however, come from far away. For example, the Tokyo area is among the chief steel producers in Japan, using iron ores from the Philippines, Malaysia, Australia, India, and even Africa; most of the coal is imported from Australia and North America, and the petroleum comes from Southwest Asia and Indonesia. The Kanto Plain cannot produce nearly enough food for its massive resident population. Imports must come from Canada, the United States, and Australia as well as other areas in Japan. Thus Tokyo depends completely on its external trading ties for food, raw materials, and markets for its products, which run the gamut from children's toys to high-precision optical equipment to the world's largest ocean-going ships.

The second-ranking economic region in Japan's core area is the Osaka–Kobe–Kyoto triangle—also known as the *Kansai District*—located at the eastern end of the Inland Sea. The Osaka–Kobe conurbation had an advantageous position with respect to Manchuria (Manchukuo) during the height of Japan's colonial empire. Situated at the head of the Inland Sea, Osaka was the major Japanese base for

Tokyo, at the center of one of the largest metropolises in the world, continues to change. Landfilling and bridge-building in the bay continue; skyscrapers sprout amid low-rise neighborhoods in this earthquake-prone area; traffic congestion worsens. The red-painted Tokyo Tower, a beacon in this part of the city, was modeled on the Eiffel Tower in Paris but, as a billboard at its base announces, is an improvement over the original: lighter steel, greater strength, less weight. Tokyo Bay, part of which can be seen from this vantage point, was the scene of one of history's most costly environmental disasters during the earthquake of 1923. A giant *tsunami* (seismic sea wave) swept up the bay from the epicenter even as landfills liquefied and buildings sank into the mud. The death toll was 140,000; a much larger population than lived in Tokyo in 1923 is now at risk of a repeat.

When the Big One Strikes

The Tokyo–Yokohama–Kawasaki urban area is the second biggest metropolis on Earth (only Mexico City is larger). But Tokyo is more than a large city: it constitutes the most densely concentrated financial and industrial complex in the world. Two-thirds of Japan's businesses worth more than U.S. $50 million are clustered here, many with vast overseas holdings. More than half of Japan's huge industrial profits (averaging about U.S. $100 billion annually between the late 1980s and the downturn of the late 1990s) are generated in the factories of this gigantic metropolitan agglomeration.

But Tokyo has a worrisome environmental history because three active tectonic plates are converging here (Fig. 9-23). All Japanese know about the "70-year" rule: over the past three-and-a-half centuries, the Tokyo area has been struck by major earthquakes roughly every 70 years—in 1633, 1703, 1782, 1853, and 1923. The Great Kanto Earthquake of 1923 set off a firestorm that swept over the city and killed an estimated 143,000 people. Tokyo Bay virtually emptied of water; then a *tsunami* (seismic sea wave) roared back in, sweeping everything before it. That Japan could overcome this disaster was evidence of the strength of its economy.

Today, Tokyo is more than a national capital. It is a global financial and manufacturing center in which so much of the world's wealth and productive capacity are concentrated that an earthquake comparable to the one of 1923 would have a calamitous effect worldwide. Ominously, Tokyo at the outset of the new century is a much more vulnerable place than the Tokyo of the 1920s. True, building regulations are stricter, and civilian preparedness is better. But whole expanses of industries have been built on landfill that will liquefy; the city is honeycombed by underground gas lines that will rupture and stoke countless fires; congestion in the area's maze of narrow streets will hamper rescue operations; and many older high-rise buildings do not have the structural integrity that has lately emboldened builders to erect skyscrapers of 50 stories and more. Add to this the burgeoning population of the Kanto Plain—approaching 30 million on what may well be the most dangerous 4 percent of Japan's territory—and we realize that the next big earthquake in the Tokyo area will not be a remote, local news story.

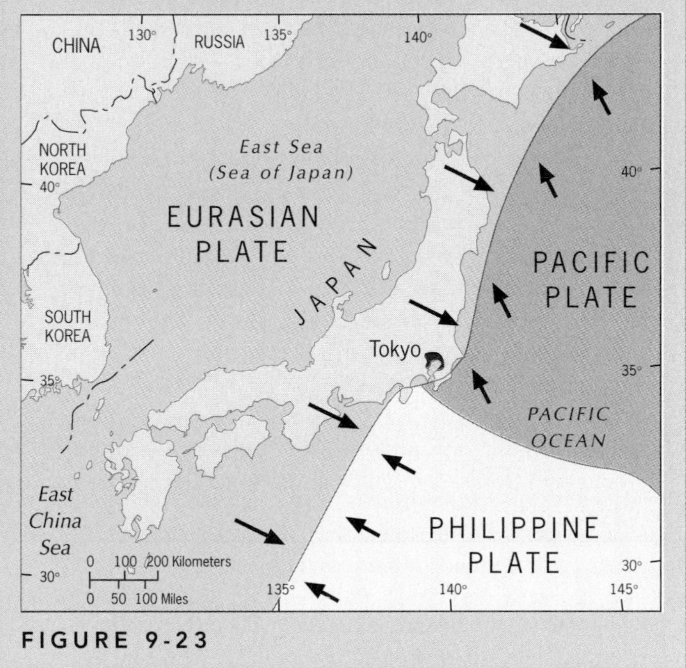

FIGURE 9-23

the China trade and for the exploitation of Manchuria, but it suffered when the empire was destroyed and it lost its trade connections with China after World War II. Kobe (like Yokohama, which is Japan's chief shipbuilding center) has remained one of the country's busiest ports, handling both Inland Sea traffic and extensive overseas linkages. Kyoto, as we noted, remains much as it was before Japan's great leap forward, a city of small workshop industries. The Kansai District is also an important farming area. Rice, of course, is the most intensively grown crop in the warm, moist lowlands, but this agricultural zone is smaller than the one on the Kanto Plain. Here is another huge concentration of people that must import food.

The Kanto Plain and Kansai District, as Figure 9-22 indicates, are the two leading primary regions within Japan's core. Between them lies the *Nobi Plain* (also called the Chubu District), focused on the industrial metropolis of Nagoya, Japan's leading textile producer. The map indicates some of the Nagoya area's advantages and liabilities. The Nobi Plain is larger than the lowlands of the Kansai District; thus its agricultural productivity is greater, though not as great as that of the Kanto Plain. But Nagoya has neither Tokyo's centrality nor Osaka's position on the Inland Sea: its connections to Tokyo are via the tenuous Sen-en Coastal Strip. Its westward connections are better, and the Nagoya area may be coalescing with the Osaka–Kobe conurbation. Still, the quality of Nagoya's port is not nearly as good as Tokyo's Yokohama or Osaka's Kobe, and it has had silting problems.

Westward from the three regions just discussed—which together constitute what is often called the Tokaido mega-

Among the Realm's Great Cities . . .

Tokyo

Many urban agglomerations are named after the city that lies at their heart, and so it is with the second largest of all: Tokyo (26.4 million). Even its longer name—Tokyo–Yokohama–Kawasaki—does not begin to describe the congregation of cities and towns that form this crowded metropolis that encircles the head of Tokyo Bay and continues to grow, outward and upward.

Near the waterfront, some of Tokyo's neighborhoods are laid out in a grid pattern. But the urban area has sprawled over hills and valleys, and much of it is a maze of narrow, winding streets and alleys. Circulation is slow, and traffic jams are legendary. The train and subway systems, however, are models of efficiency—although during rush hours you must get used to the *shirioshi* pushing you into the cars to get the doors closed.

At the heart of Tokyo lies the Imperial Palace with its moats and private parks. Across the street, buildings retain a respectful low profile, but farther away Tokyo's skyscrapers seem to ignore the peril of earthquakes. Nearby lies one of the world's most famous avenues, the Ginza, lined by department stores and luxury shops. In the distance you can see an edifice that looks like the Eiffel Tower, only taller: this is the Tokyo Tower, a multi-purpose structure designed to test lighter Japanese steel, transmit television and radio signals, detect Earth tremors, monitor air pollution, and attract tourists.

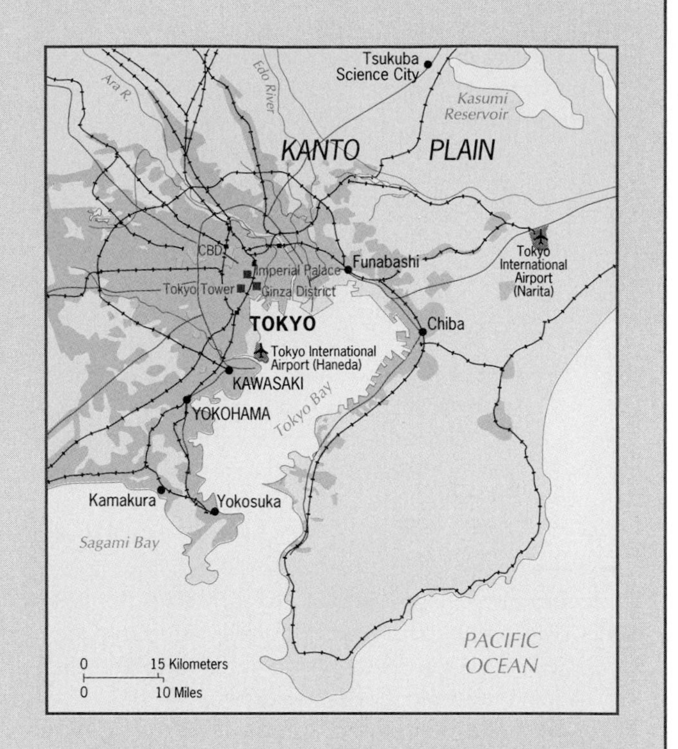

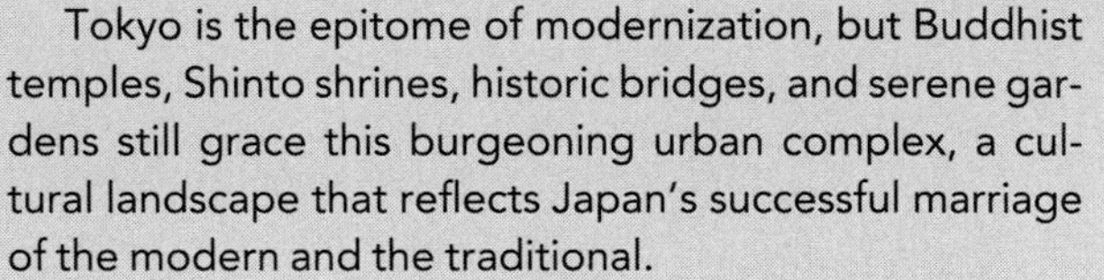
Tokyo is the epitome of modernization, but Buddhist temples, Shinto shrines, historic bridges, and serene gardens still grace this burgeoning urban complex, a cultural landscape that reflects Japan's successful marriage of the modern and the traditional.

lopolis—extends the Inland Sea, along whose shores the remainder of Japan's core area is continuing to develop. The most impressive growth has occurred around the western entry to the Inland Sea (the Strait of Shimonoseki), where *Kitakyushu*—a conurbation of five cities on northern Kyushu—constitutes the fourth Japanese manufacturing complex and primary economic region. Road and railway tunnels connect Honshu and Kyushu, but the northern Kyushu area does not have an urban-industrial equivalent on the Honshu side of the strait. The Kitakyushu conurbation includes Yawata, site of northwest Kyushu's (rapidly declining) coal mines. The first steel plant in Japan was built there on the basis of this coal; for many years it was the country's largest. The advantages of transportation here at the western end of the Inland Sea are obvious: no place in Japan is better situated to do business with Korea and China. As relations with mainland Asia expand, this area will reap many benefits. Elsewhere on the Inland Sea coast, the Hiroshima–Kure urban area has a manufacturing base that includes heavy industry. And on the coast of the Korea Strait, Fukuoka and Nagasaki are the principal centers—the former an industrial city, the latter a center of large shipyards.

Only one major Japanese manufacturing complex lies outside the belt extending from Tokyo in the east to Kitakyushu in the west—the secondary region centered on Toyama on the Sea of Japan (East Sea). The advantage here is cheap power from nearby hydroelectric stations, and the cluster of industries reflects it: paper manufacturing, chemical factories, and textile plants. Figure 9-22 also gives an inadequate picture of the variety and range of industries that exist throughout Japan, many of them oriented to local (and not insignificant) markets. Thousands of manufacturing plants operate in cities and towns other than those shown on this map, even on the cold northern island of Hokkaido, which is connected to Honshu by the Seikan rail tunnel, the world's longest, beneath the treacherous Tsugaru Strait.

The map of Japan's areal organization shows that four primary regions dominate its core area. Each of them is primary because each duplicates to some degree the contents of the others. Each contains iron and steel plants, is served by one major port, and lies in or near a large, productive farming area. What the map does not show is that each also has its own external connections for the overseas acquisition of raw materials and sale of finished products. These linkages may even be stronger than those among the four internal regions of the core area. Only for Kyushu and its coal have domestic raw materials affected the nature and location of heavy manufacturing, and these resources are almost depleted. In the structuring of the country's areal functional organization, therefore, more than just the contents of Japan itself is involved. In this respect Japan is not unique: all countries that have exchange-type organization must adjust their spatial forms and functions to the external interconnections required for progress. But for few countries is this truer than Japan.

East and West, North and South

The previous discussion shows that Japan has its own Pacific Rim: almost all of Honshu's economic expansion is occurring on the east coast. Japanese geographers often call this eastern zone *Omote Nippon* ("Front Japan"), the Pacific-facing, megalopolitan, high-tech, dynamic sector that embodies the modern economic superpower—as opposed to *Ura Nippon* ("Back Japan"), the comparatively underdeveloped periphery of unkempt farms and poor villages bordering the Sea of Japan (East Sea) to the west. In a country that needs food as much as Japan does, the derelict state of farmlands in *Ura Nippon* is of much concern to the country's planners.

The congestion on Honshu is another concern. Efforts to link Hokkaido and southern Shikoku to the "mainland" have produced world-record tunnels and bridges such as the Akashi Kaikyo Bridge, designed to diffuse Honshu's economic primacy to more remote areas. After the Kobe earthquake of 1995, planning began to relocate Tokyo's capital functions to a site between 40 and 200 miles (65 and 320 km) west of Tokyo, not only for safety reasons but primarily to remove governmental offices from the overcrowded and high-cost Tokyo urban area. In 1999, with the Japanese economy slowed by recession, this project was delayed, but Japanese geographers report that the relocation of the national capital remains on the agenda.

Food and Population

Japan's economic modernization so occupies center stage that we can easily forget the country's considerable achievements in agriculture. Japan's planners, who are as interested in closing the food gap as in expanding industries, have created extensive networks of experiment stations to promote mechanization, optimal seed selection and fertilizer use, and information services to distribute knowledge about enhancing crop yields to farmers as rapidly as possible. Although this program has succeeded, Japan faces the unalterable reality of its stubborn topography: it simply lacks sufficient land to farm. Populous Japan has one of the highest physiologic population densities—7950 per square mile (3070 per sq km)—on Earth. (We discuss this concept on pp. 386 and 413.)

The Japanese go to extraordinary lengths to maximize their limited farming opportunities, making huge investments in research on higher-yielding rice varieties, mechanization, irrigation, terracing, fertilization, and other practices. More than 90 percent of farmland is assigned to food crops. Hilly terrain is given over to cash crops, such as tea and grapes (Japan has a thriving wine industry). Vegetable gardens ring all the cities.

But all this costs money, and Japan's food prices are high. Millions of Japanese have moved from the farms to the cities, depopulating the countryside which needs farmers. Japan's political system gives underpopulated rural areas disproportionate influence, and produce prices are kept artificially high, all to induce farmers to stay on the land. It has not worked. Between 1920 and 2000, full-time farmers declined from 50 to under 4 percent of the labor force. Yet the policies persist. In the early 1990s, American farmers proved that they could provide rice to Japanese consumers at one-sixth the price the Japanese were paying for home-grown rice. But the government resisted pressure to open its markets to imports.

With their national diet so rich in starchy foods like rice, wheat, barley, and potatoes, the Japanese need protein to balance it. Fortunately, they can secure enough of it, not by buying it abroad but by harvesting it from rich fishing grounds near the Japanese islands. With their customary thoroughness, the Japanese have developed a fishing industry that is larger than that of the United States or any of the long-time fishing nations of northwestern Europe, and it now supplies the domestic market with a second staple after rice. Although mention of the Japanese fishing industry brings to mind a fleet of ships scouring the oceans and seas far from Japan, most of this huge catch (about one-seventh of the world's annual total) comes from waters within a few dozen miles of Japan itself. Where the warm Kuroshio and Tsushima currents meet colder water off Japan's coasts, a rich fishing ground yields sardines, herring, tuna, and mackerel in the warmer waters and cod, halibut, and salmon in the seas to the north. Japan's coasts have about 4000 fishing villages, and tens of thousands of small boats ply the waters offshore to bring home catches that are distributed to local and city markets. The Japanese also practice *aquaculture*—the "farming" of freshwater fish in artificial ponds and flooded paddy (rice) fields, of seaweeds in aquariums, and of oysters, prawns, and shrimp in shallow bays. They are even experimenting with cultivating algae for their food potential.

When you walk the streets of Japan's cities, you will notice something that Japanese vital statistics are reporting:

Pedestrians crossing the Ginza (Tokyo's Fifth Avenue) on a warm summer day. A generation ago you would have seen some traditional dress among the crowd, but urban Japan's modernization is pretty well complete, and in public the Japanese tend to dress quite conservatively. Suits and school uniforms are the norm.

the Japanese are getting taller and heavier. Coupled with this change is evidence that heart disease and cancer are rising as causes of mortality. The reason seems to lie in the changing diets of many Japanese, especially the younger people. When rice and fish were the staples for virtually all, and the calories available were limited but adequate, the Japanese as a nation were among the world's healthiest and longest-lived. But as fast-food establishments diffused throughout Japan, and Western (especially American) tastes for red meat and fried food spread among children and young adults, their combined impact on the population was soon evident. Coupled with the heavy cigarette smoking that prevails in all the Asian countries of the Pacific Rim, this is changing the region's medical geography.

Japan's Pacific Rim Prospects

As the twenty-first century opens, Japan remains the economic giant on the western Pacific Rim—but a giant with an uncertain future. Japanese products still dominate world markets and Japanese investments span the globe, but in the late 1990s Japan's economy faltered. Its growth rate declined. Many companies scaled back their operations, and some went bankrupt. The banking system proved to be vulnerable. The first Pacific Rim economic tiger was ailing.

Japan's problems stemmed from a combination of circumstances at home and abroad. Overseas, the collapse of Southeast Asian currencies and economies hurt Japanese investments there. From automobile plants to toy factories, long-profitable operations turned into liabilities. In Japan itself, financial mismanagement, long tolerable because of the booming economy, now took its toll. In addition, competition from still-healthy economies, notably from Taiwan but also from struggling South Korea, undercut Japanese goods on world markets.

In the long run, all this may prove to have been a temporary setback, but Japan is unlikely to regain its former economic dominance on the Pacific Rim. Japan's dependence on foreign oil is a potential weakness that could further depress its economy should international supply lines be disrupted. Western competition for Japan's markets is growing. The countless investments Japan made around the world during its time of plenty, from skyscrapers and movie studios to golf courses and hotels, have lost much of their value and must be sold at huge losses. While Japan reorganizes, American and European businesses buy its assets and take advantage of its problems.

Were you to make a field trip to Japan to study its economic geography, none of this might be immediately evident. The cities still bustle with activity, raw materials continue to arrive at the ports, container traffic still stirs the docks. Even in a downturn, this mighty economy outproduces all others on the western Pacific Rim. And, indeed, that observation is relevant to the future. Japan's infrastructure has the capacity to propel an economic rebound. But it needs new opportunities.

One of these opportunities lies in the Russian Far East and beyond, in Siberia. As we noted in Chapter 2, the Russian Far East lies directly across from northern Honshu and Hokkaido and is a storehouse of raw materials awaiting exploitation. For this, the Japanese are ideally located—but political relations with the former Soviet Union, and now with Russia, have stymied all initiatives (see box titled "The Wages of War"). Japan should be one of the major beneficiaries of any substantial oil reserves near Sakhalin (see map, p. 146), but again the necessary connections may not be possible. Japan also has a major potential opportunity in Korea, especially if a unification process takes hold there. But relations with Korea continue to be eroded by Korean memories of Japanese behavior during World War II and by Japan's refusal to acknowledge its misdeeds.

All this must be seen against the backdrop of Japan's changing society. The population of 127 million is aging rapidly, and population geographers expect it to rise only slightly until 2007, when it will begin to decline. This decline will increase quite rapidly, to about 100 million in 2050 and only 67 million in 2100. Already, for some time, Japan has faced a labor shortage (although the recent economic downturn produced rising unemployment, a novelty in the postwar period). Over the long term, Japan will need millions of immigrants to keep its economy going, but for culturally homogeneous, ethnically conscious Nippon, which has historically resisted immigration, this will be a

The Wages of War

Japan and the former Soviet Union never signed a peace treaty to end their World War II conflict. Why? There are four reasons, and all of them are on the map just to the northeast of Japan's northernmost large island, Hokkaido (Fig. 9-21). Their names are Habomai, Shikotan, Kunashiri, and Etorofu. The Japanese call these rocky specks of the Kurile Island chain their "Northern Territories." The Soviets occupied them late in the war and never gave them back to Japan. Now they are part of Russia, and the Russians have not given them back either.

The islands themselves are no great prize. During World War II, the Japanese brought 40,000 forced laborers, most of them Koreans, to mine the minerals there. When the Red Army overran them in 1945, the Japanese were ordered out, and most of the Koreans fled. Today the population of about 50,000 is mostly Russian, many of them members of the military based on the islands and their families. At their closest point the islands are only 3 miles (5 km) from Japanese soil, a constant and visible reminder of Japan's defeat and loss of land. Moreover, territorial waters bring Russia even closer, so that the islands' geostrategic importance far exceeds their economic potential.

Attempts to settle the issue have failed. In 1956, Moscow offered to return the tiniest two, Shikotan and Habomai, but the Japanese declined, demanding all four islands back. In 1989, then-Soviet President Mikhail Gorbachev visited Tokyo in the hope of securing an agreement. The Japanese, it was widely reported, offered an aid-and-development package worth U.S. $26 billion to develop Russia's eastern zone—its Pacific Rim and the vast resources of the eastern Siberian interior. This would have begun the transformation of Russia's Far East, stimulated the ports of Nakhodka and Vladivostok, and made Russia a participant in the spectacular growth of the western Pacific Rim.

But it was not to be. Subsequently, Russian Presidents Yeltsin and Putin also were unable to come to terms with Japan on this issue, facing opposition from the islands' inhabitants and from their own governments in Moscow. And so World War II, 57 years after its conclusion, continues to cast a shadow over this northernmost segment of the western Pacific Rim.

wrenching reality. Still, Japanese leaders will have no choice. The country's "graying" will strain welfare systems and tax burdens.

Modernization, meanwhile, is further stressing Japan's society. The Japanese have a strong tradition of family cohesion and veneration of their oldest relatives, but this custom is breaking down. Younger people are less willing than their parents and grandparents to accept overcrowded housing, limited comforts, and little privacy. In Japan, for all its material wealth, family homes tend to be small, cramped, often flimsily built, and sometimes lacking the basic amenities other high-income countries consider indispensable. With their disposable income, the Japanese during the boom economy came to rank among the world's most-traveled tourists and, having seen how Americans and Europeans of similar income levels live, they returned home dissatisfied.

As we just noted, Japan remains one of the most uniform, culturally homogeneous societies in today's interactive world. The Chinese closed their doors to foreigners intermittently; the Japanese have done so far more effectively. Other than the dwindling Ainu, there are no indigenous minorities. Koreans were brought to Japan during World War II to serve as forced laborers, but otherwise Japanese residency and citizenship imply an almost total ethnic sameness. During periods of labor shortage, the notion to invite foreign workers would not be entertained. When the United States, after the Vietnam War, sought new homes for hundreds of thousands of Vietnamese refugees, Japan accepted a few hundred after much negotiation. Such jealously guarded ethnic-cultural uniformity is unlikely to be viable in the globalizing world of the twenty-first century.

Japan's future in the geopolitical framework of the western Pacific also is uncertain. More than a half-century after the end of World War II, American forces are still based on Japanese territory, guaranteeing Japanese security. China and Korea accept this arrangement, which constrains Japan's rearming, in view of what happened during World War II. But in Japan, where nationalism is rising, domestic military preparedness is a growing political issue. The situation is fraught with potential problems.

Japan is therefore a pivotal country in the western Pacific Rim that stands at economic, social, and geopolitical crossroads. Its future will profoundly influence not only East Asia and the Pacific, but the world as a whole.

KOREA

On the Asian mainland, directly across the Sea of Japan (East Sea), lies the peninsula of Korea (Fig. 9-24), a territory about the size of the State of Idaho, much of it mountainous and rugged, and containing a population of 70 million. Unlike Japan, however, Korea has long been a divided country, and the Koreans a divided nation. For uncounted centuries Korea has been a pawn in the struggles of more powerful neighbors. It has been a dependency of China and a colony of Japan. When it was freed from Japan's oppressive rule at the end of World War II (1945), the victorious Allied powers divided Korea for administrative purposes. That division gave North Korea (north of the 38th parallel) to the forces of the Soviet Union and South Korea to those of the United States. In effect, Korea traded one master for two new ones. The country was not reunited for the rest of the century because North Korea immediately fell under the communist ideological sphere and became a dictatorship in

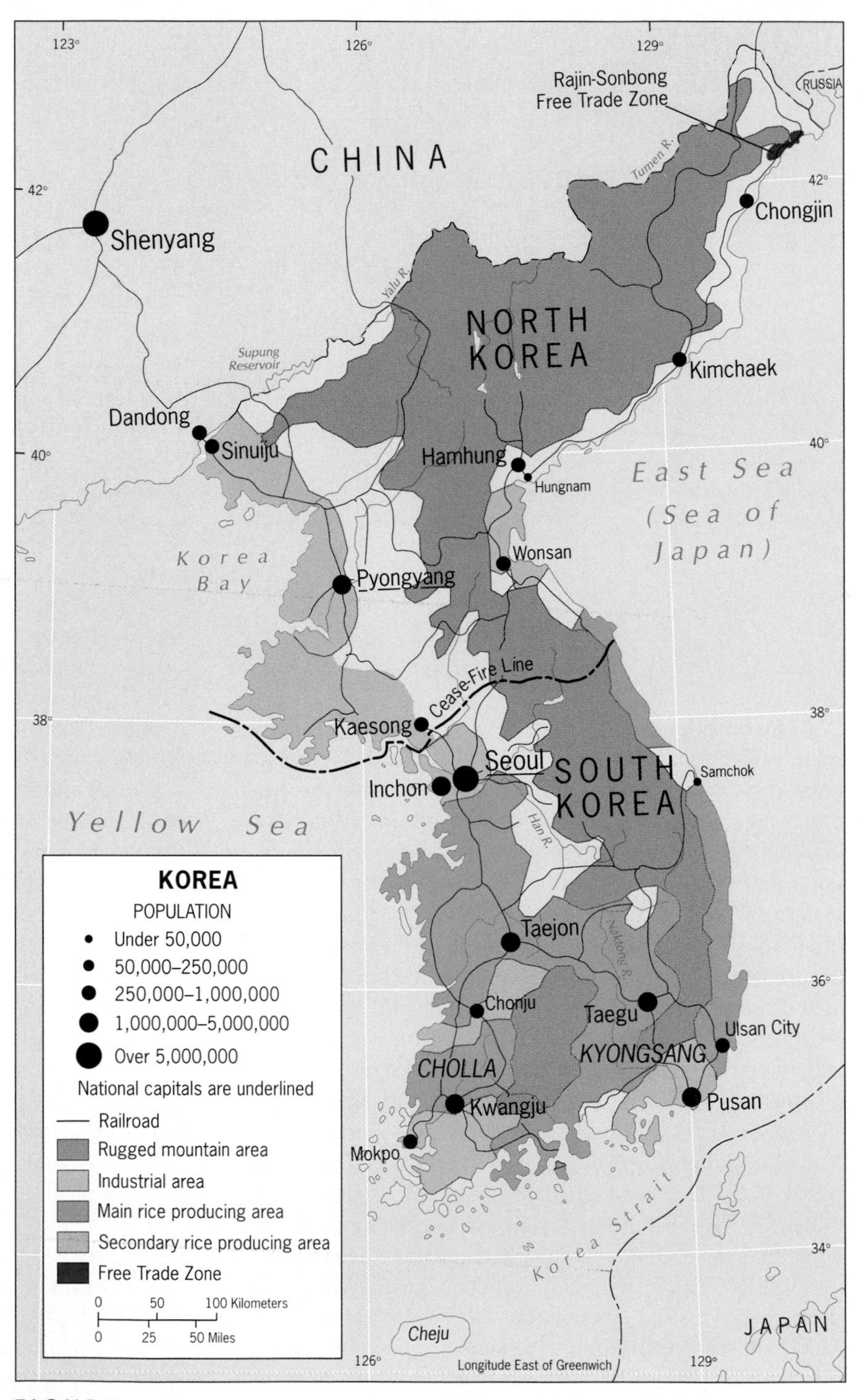

FIGURE 9-24

the familiar (but in this case extreme) pattern. South Korea, with massive American aid, became part of East Asia's capitalist perimeter. Once again, it was the will of external powers that prevailed over the desires of the Korean people.

In 1950, North Korea sought to reunite the country by force and invaded South Korea across the 38th parallel. This attack drew a United Nations military response led by the United States. Thus began the devastating Korean War (1950–1953) in which North Korea's forces pushed far to the south only to be driven back across its own half of Korea almost to the Chinese border. Then China's Red Army entered the war and drove the UN troops southward again. A cease-fire was arranged in 1953, but not before the people and the land had been ravaged in a way that was unprecedented even in Korea's violent past. The cease-fire line shown in Figure 9-24 became a heavily fortified *de facto* boundary. For more than four decades virtually no contact of any kind has occurred across this border, which divides economies and families alike.

As a result, the Jakota Triangle, as a region of East Asia, must include South Korea but exclude North Korea (see box titled "Obsolescent North Korea"). Times may change, as they did in formerly divided Germany, but at the turn of the century South Korea, with 45 percent of Korea's land area but with two-thirds of the Korean population, has emerged as one of the economic tigers on the Pacific Rim. As an undivided country, Korea might have achieved even more because North and South Korea are in what geographers call **regional complementarity**. This condition arises when two adjacent regions complement each other in economic-geographic terms. In this case, North Korea has raw materials that the industries of South Korea need; South Korea produces food the North needs; North Korea produces chemical fertilizers farms in the South need.

18

In the past half-century, however, the two Koreas were cut off from each other, and they developed in opposite directions. North Korea, whose large coal and iron ore deposits attracted the Japanese and whose hydroelectric

Obsolescent North Korea

The Korean Peninsula is physiographically well-defined, and its people—the Korean nation—have occupied it for thousands of years. But internal political division, foreign cultural intrusion, and colonial subjugation have created a turbulent history. Now Koreans are divided again, this time by ideology. Even as the capitalist South became one of the Pacific Rim's economic tigers, North Korea remained an isolated, stagnant state on a Stalinist-communist model abandoned long ago even in the former Soviet Union itself.

For over a half-century, North Korea has been ruled by two men: Kim Il Sung, the "Great Leader," and his son Kim Jong Il, the "Dear Leader." Total allegiance to the dictators is ensured by an omnipotent police and military. Self-imposed isolation is North Korea's hallmark. The state controls all aspects of life in North Korea.

Economic policies under the communist regime have been disastrous. Per-capita incomes in the North are less than one-tenth of those in the South; poverty is rampant. Infant mortality is nearly four times as high. Industrial equipment is 25 years out of date. Exports are negligible. When a combination of failed agricultural policies and environmental extremes created famine in the late 1990s, the regime resisted relief efforts, even from South Korea. As during the Maoist period in China, the world will never know the number of casualties.

North Korea did progress on one front: nuclear capability and associated weaponry. In 1999 it fired a test missile across northern Japan, heightening international concern. The North Koreans used their nuclear threat to negotiate a multi-billion-dollar technical aid program to build reactors for peaceful purposes, with Japan and South Korea among the major contributors.

Simultaneously, signs appeared that North Korea began to seek ways to improve relations with South Korea and to open, however slightly, its doors to the outside world. A free-trade zone was declared at Rajin-Sonbong, in the country's far north (it is not likely to become a growth pole since a double barbed-wire fence encloses it), and cruise ships were allowed to dock there. More significantly, Kim Jong Il sanctioned a series of meetings with South Korean officials during 2000, which resulted in the first reunions for members of families separated by the cease-fire line of 1953. Although these visits lasted just a few hours, their very occurrence was a momentous event.

The reunification of the Koreas, if it is to take place, will entail far greater difficulties than did that of the Germanies. A United States armed force of 37,000 still patrols the most heavily fortified border in the world; reconciliation will involve and affect foreigners as well as Koreans. It will alter the balance of power in the region in unforeseeable ways. Its cost will dwarf that of Germany. Still, the boundary superimposed on the Korean nation is a leftover of the Cold War, and its elimination, and the reintegration of North Korea into the international community, is a laudable goal.

Among the Realm's Great Cities . . .

Seoul

Seoul (9.9 million), located on the Han River, is ideally situated to be the capital of all of Korea, North and South (its name means capital in the Korean language). Indeed, it served as such from the late fourteenth century until the early twentieth, but events in that century changed its role. Today, the city lies in the northwest corner of South Korea, for which it serves as capital; not far to the north lies the tense demilitarized zone (DMZ) that contains the cease-fire line with North Korea. That line cuts across the mouth of the Han, depriving Seoul of its river traffic. Its ocean port, Inchon, has emerged as a result.

Seoul's undisciplined growth, attended by a series of recent accidents including the failure of a major bridge over the Han and the collapse of a six-story department store, reflects the unbridled expansion of the South Korean economy as a whole, as well as the political struggles that carried the country from autocracy to democracy. Central Seoul lies in a basin surrounded by hills to an elevation of about 1000 feet (330 m), and the city has sprawled outward in all directions, even toward the DMZ. An urban plan designed in the early 1960s was overwhelmed by the immigrant flow.

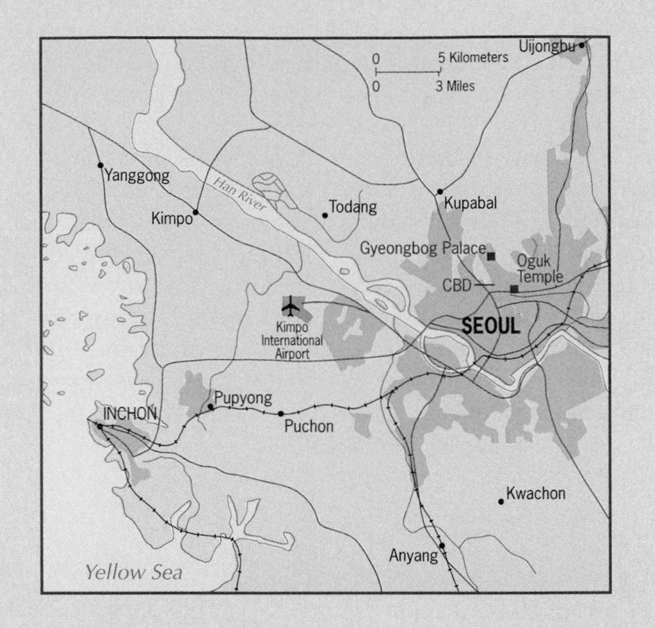

During the period of Japanese colonial control, Seoul's surface links to other parts of the Korean Peninsula were improved, and this infrastructure played a role in the city's later success. Seoul is not only the capital but also the leading industrial center of South Korea, exporting sizeable quantities of textiles, clothing, footwear, and (increasingly) electronic goods. South Korea is still an economic tiger on the Pacific Rim, and Seoul is its heart.

plants produce electricity that the South could use, carried on what limited trade it generated with the Chinese and the Soviets. South Korea's external trade links were with the United States, Japan, and Western Europe.

In the early postwar years, there were few indications that South Korea would emerge as a major economic force on the Pacific Rim and, indeed, on the world stage. Over 70 percent of all workers were farmers; agriculture was inefficient; the country was stagnant. But huge infusions of aid, first from the United States and then Japan, coupled with the reorganization of farming and stimulation of industries (even if they lost money), produced a dramatic turnaround. Large feudal estates were parceled out to farm families in 3-hectare (7.5-acre) plots, and a program of massive fertilizer importation was begun. Production rose to meet domestic needs, and some years yielded surpluses.

The industrialization program was based on modest local raw materials, plentiful and capable labor, ready overseas (especially American) markets, and continuing foreign assistance. Despite corrupt dictatorial rule, political instability, and social unrest, South Korean regimes managed to sustain a rapid rate of economic growth that placed the country, by the late 1980s, among the world's top ten trading powers. To get the job done, the government borrowed heavily from overseas, and it controlled banks and large industries. South Korea's growth resulted from state capitalism rather than free-enterprise capitalism, but it had impressive results. South Korea became the world's largest shipbuilding nation; its automobile industry grew rapidly; and its iron and steel and chemical industries began to thrive. It has placed less emphasis, however, on smaller, high-technology industries, which are the strengths of other Pacific Rim economic tigers. So the future remains uncertain.

Today, South Korea is prospering, and we can see its economic prowess on the map (Fig. 9-24). The capital, Seoul, with 10 million inhabitants, ranks among the world's larger megacities and is the anchor of a huge industrial complex facing the Yellow Sea at the waist of the Korean Peninsula. Hundreds of thousands of farm families migrated to the Seoul area after the end of the Korean War (today only about 20 percent of South Koreans remain on the land). They also moved to Pusan, the nucleus of the

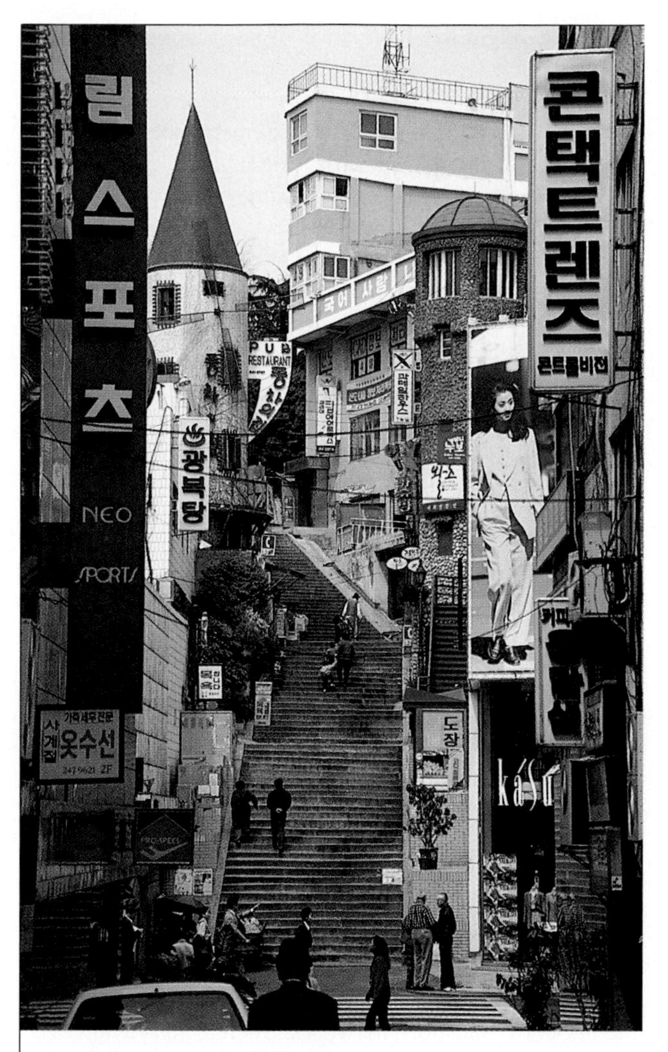

FROM THE FIELD NOTES

"Pusan, located at the southeastern corner of the Korean Peninsula, is a major port, industrial center, commercial hub, and focus for a large agricultural region. Sail or fly into Pusan, however, and you get a geographic lesson on the problems a city's site can present during urban expansion. Pusan lies at the southern end of a rugged mountain chain (see Fig. 9-24) and the city itself sprawls up hillsides, into narrow valleys, and, in the port area, onto reclaimed land. Walking the central city, we found streets sometimes ending in stairways."

country's second largest manufacturing zone, located on the Korea Strait opposite the western tip of Honshu. And the government-supported, urban-industrial drive here continues. Just 30 years ago, Ulsan City, 40 miles (60 km) north of Pusan along the coast, was a fishing center with perhaps 50,000 inhabitants; today its population exceeds 1.4 million, nearly half of them the families of workers in the Hyundai automobile factories and the local shipyards and docks.

The third industrial area shown in Figure 9-24, anchored by the city of Kwangju, has advantages of relative location that will spur its development, but it faces domestic problems. Kwangju lies at the center of the Cholla region, the southwest corner of South Korea; the southeast is called Kyongsang. For more than three decades beginning in the 1960s, the army generals who ruled South Korea all came from Kyongsang, and through state control of banks and other monopolies they bestowed favors on Kyongsang companies while denying access to potential competitors in Cholla. The regime in Seoul also thwarted infrastructure development in Cholla, even to the point of impeding plans for improved railroad links between Kwangju and the capital. This regional friction led to protests in Cholla and violent repression by the military rulers, the aftermath of which is fading only slowly in this part of the now-democratic country. People in Cholla often say that they feel discriminated against in their own country and that it is prejudice, not ability, that has held their region back. They have reasons: the Samsung group of companies, for example, long had a stated policy not to employ any graduates from Cholla's universities.

Thus South Korea's economic growth has not gone smoothly; dictatorial rule and police-state tactics to keep control were part of the transition. Today, South Korea is a more democratic society, but the Asian economic downturn of the late 1990s exposed the weaknesses of its financial and business structure, inherited from the earlier period. Giant corporate conglomerates called *chaebol* conspired to suppress competition; they controlled the banking system and dictated government economic policy. It was perhaps the ultimate version of what economic geographers call **state capitalism**, and it proved to have weaknesses similar to communist planned economies. When South Korea felt the impact of events in other parts of Asia, the system collapsed and a large infusion of funds from international banks was necessary. As the new century began, South Korea's economy had stabilized and was recovering, but circumstances within as well as outside of the country have combined to slow its climb up the ladder already topped by Japan.

TAIWAN

The third component of the region we have called the Jakota Triangle is another, and more durable, success story on the Pacific Rim: Taiwan. Along with the Penghu (Makung) Islands in the Taiwan Strait and the islands of Jinmen Dao (Quemoy) near Xiamen and Matsu Tao (Matsu) near Fuzhou at the coast of mainland China, Taiwan is regarded by Beijing's communist regime as a Chinese province temporarily "wayward," that is, disobedient. After the reunification of Portuguese Macau with China in 1999, Taiwan remained the only major unresolved territorial issue in China's postcolonial, postwar sphere.

Beijing's claim to Taiwan has strengths as well as weaknesses. Han people settled on Taiwan, but the island had been settled thousands of years earlier by Malay-Polynesian groups who had split into mountain (east) and plains (west) inhabitants. When the Chinese emigrated from famine-stricken Fujian Province to Taiwan in the seventeenth century, they soon displaced and assimilated the plains people, but the mountain aborigines held out until the Japanese colonial invasion. Formal Chinese control began in 1683 when the expansionist Qing (Manchu) Dynasty made Taiwan part of Fujian Province. Rice and sugar became important exports to the mainland, and by the 1840s Taiwan's population was an estimated 2.5 million. In 1886 the island was declared a province of China, but by then the British had established two treaty ports there, the French had blockaded it, and the Japanese had sent expeditions to it. Nine years later, in 1895, China ceded Taiwan to Japan in the treaty settlement of the Sino-Japanese War, and it became a Japanese colony.

When China's aging Manchu Dynasty was overthrown following the rebellion of 1911, a Nationalist government in 1912 proclaimed the Republic of China (ROC), led by Sun Yat-sen. Naturally, this government wanted to oust all colonialists, not only Europeans but also Japanese, and not just from the mainland, but from Taiwan as well. On Taiwan, however, the Japanese held on, using the island as a source of food and raw materials as well as a market for Japanese products. To ensure all this, the Japanese launched a prodigious development program involving road and railroad construction, irrigation projects, hydroelectric schemes, mines (mainly for coal), and factories. Farmlands were expanded and farming methods improved. While mainland China was engulfed in conflict, Taiwan remained under Japanese control. Not until 1945, when Japan was defeated in World War II, did Taiwan become an administrative part of China again. But only briefly. As we noted earlier, the Nationalists, under the leadership of Chiang Kai-shek, now faced the rising tide of communism on the mainland, and the ROC government fought a losing war. In 1949, the flag of the Republic of China was taken to a last stronghold: offshore Taiwan. There, Chiang and his followers, with their military might, weapons, and wealth taken from the mainland, established what they (and the world, led by the United States) proclaimed as the legitimate government of all China. On the island, the Nationalists, helped by the United States, began a massive reconstruction of the war-damaged infrastructure. In the international arena, the ROC represented the country; in the still-young United Nations, the Nationalists of Taiwan occupied China's seat.

On the huge mainland, however, it was the communists in Beijing who ruled and called their country the People's Republic of China (PRC). Thus was born the two-China dilemma which still confronts the international community. When the ROC was ousted from the United Nations in 1971 to make way for the PRC, legitimacy was also transferred.

All this might have remained an internal dispute but for two major developments: Taiwan's rise as an economic tiger on the Pacific Rim, coupled with remarkable strides toward democracy made in the ROC; and the emergence on the world political and economic stage of communist-ruled China following decades of isolation. Today, the power of the PRC is growing rapidly, and Taiwan's heirs to the ROC face an uncertain future; but it is the ROC, not the PRC, that has managed to combine economic success with democratization.

Taiwan, as Figure 9-25 shows, is not a large island. It is smaller than Switzerland but has a population much larger (22.6 million), most of it concentrated in an arc lining the western and northern coasts. The Chungyang Mountains, an area of high elevations (some over 10,000 feet [3000 m]), steep slopes, and dense forests, dominate the eastern half of the island. Westward, these mountains yield to a zone of hilly topography and, facing the Taiwan Strait, a substantial coastal plain. Streams from the mountains irrigate the paddyfields, and farm production has more than doubled since 1950 even as hundreds of thousands of farmers left the fields for work in Taiwan's expanding industries.

Today, the lowland urban-industrial corridor of western Taiwan is anchored by the capital, Taipei (Taibei), at the island's northern end and rapidly growing Kaohsiung (Gaoxiong) in the far south.* The Japanese developed Chilung (Jilong), Taipei's outport, to export nearby coal, but now the raw materials flow the other way. Taiwan imports raw cotton for its textile industry, bauxite (for aluminum) from Indonesia, oil from Brunei, and iron ore from Africa. Taiwan has a developing iron and steel industry, nuclear power plants, shipyards, a large chemical industry, and modern transport networks. Increasingly, however, Taiwan is exporting products of its budding high-technology industries: personal computers, telecommunications equipment, and precision electronic instruments. Taiwan has enormous brainpower, and many foreign firms join in the research and development carried on in such places as Hsinchu (Xinzhu) in the north, where the government has helped establish a technopole centered on the microelectronics and personal computer industries. In the south, the "science city" of Tainan specializes in microsystems and information technology.

With less than one-fiftieth of the population of mainland China, Taiwan in the late 1990s still ranked ahead of its giant neighbor as a trading nation. A substantial portion of this trade is with the People's Republic of China; despite more than half a century of animosity, ties between the Taiwanese and the mainlanders remain strong. The British-colonial status of Hong Kong (Xianggang) facilitated this trade; ROC-PRC trade channeled through Hong Kong could be construed as indirect trade.

*The Taiwanese have retained the old Wade-Giles spelling of place names; China now uses the *pinyin* system. Names in parentheses are written according to the pinyin system.

The establishment of China's Special Economic Zones has gone a long way toward solving the economic problems arising from the political discord between China and Taiwan. In the almost-anything-goes environment of the SEZs, Taiwanese business owners are permitted to build or buy factories just as "real" foreigners are, and as we noted earlier, the Xiamen SEZ was situated directly across from Taiwan for just this purpose. As a result, thousands of Taiwanese companies now operate in China even as political tensions continue.

In some ways, Taiwan's emergence as an economic tiger on the Pacific Rim is even more spectacular than Japan's. True, Taiwan received much Western assistance, but it got out of debt faster than anyone expected. Today, per-capita income is well above U.S. $10,000 per year, which is higher than that in many European countries. Taiwan's trade surplus has yielded tens of billions of dollars in reserves, which is being used to further its own development and to invest overseas. Taiwan, for example, is the largest foreign investor in the rebuilding of Vietnam. And unlike Japan and South Korea, Taiwan weathered the economic storm of the late 1990s without crisis.

Taiwan's greatest problem, therefore, is political, not economic. The Nationalists who made the island their base in 1949 established a Republic of China that was authoritarian, but the ROC has evolved into a democratic state. This was not an easy process, but when Taiwan's own democracy movement culminated in 1991, it produced not a massacre, but free elections for a new parliament. Taiwan's leaders now proclaim that the direct and popular election of their president in 1996 was the first such vote not just in Taiwan or the ROC, but in the four-thousand-year history of the Chinese state. Taiwan thus presents itself as an alternative model to the communist-ruled PRC, proving to the world that democracy is compatible with Chinese culture and tradition. Nevertheless, the international community does not recognize Taiwan as an independent state, and a substantial segment of the Taiwanese electorate itself opposes any move toward outright sovereignty. Beijing dictates that Taiwan is off limits to U.S. cabinet members. In 1995, when the United States permitted the ROC's president to visit the American university of which he is a graduate, so that he might attend a class reunion, relations between Washington and Beijing plummeted. In 2000, when Taiwan's voters elected a president who, during the campaign, had proclaimed Taiwan's "equality" with mainland China, Beijing threatened force and massed weaponry across the Taiwan Strait.

The political geography of Taiwan, therefore, is at odds with its economic geography. The ROC stood firm against communism in the 1960s and 1970s; but its long-term ally, the United States, was required to publicly commit itself to a one-China policy, the one being the PRC, the communist giant. Even as goods and capital flow between Taiwan and the mainland, Beijing engages in threatening acts (such as nearby missile tests) when democracy in the ROC takes another step forward.

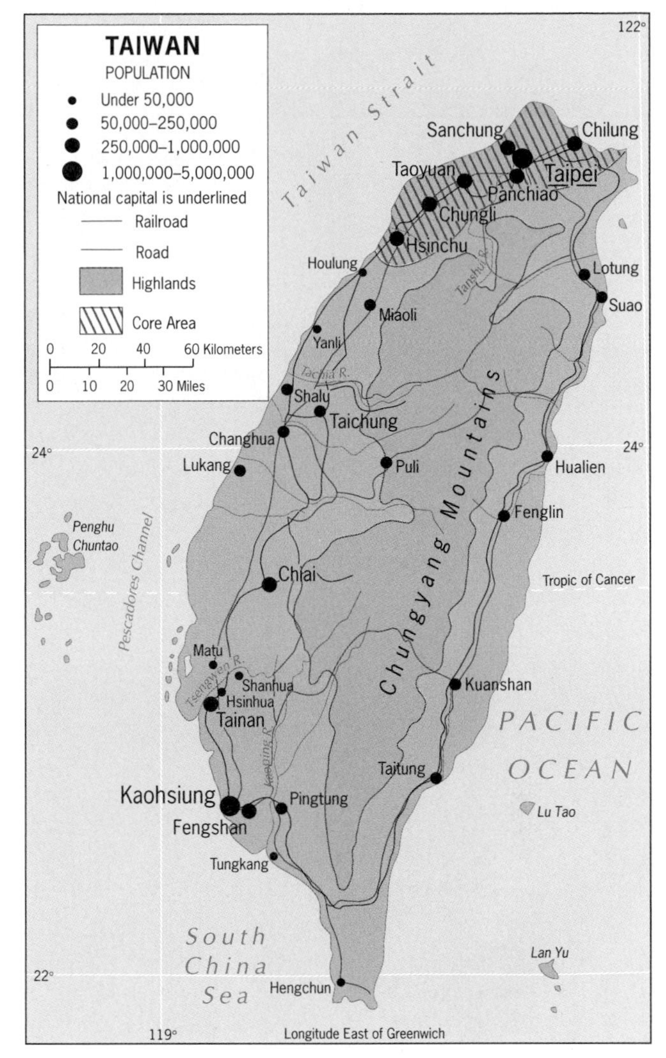

FIGURE 9-25

Today, Taiwan's political options are as limited as its economic horizons are endless. The communist Chinese have made it clear that a move toward independence, based on the likely outcome of any referendum, would lead to intervention. Though militarily well prepared, Taiwan could not long hold off the PRC's enormous armed forces. In any case, independence for Taiwan has philosophical disadvantages for the leaders of the ROC, as it would diminish their claim to legitimacy as the alternative model for a reunified China. In practical terms, conflict on the Pacific Rim would be disastrous for both sides, whatever course it were to take.

And so Taiwan's future hangs in the balance, a wayward province with national aspirations, a quasi-state with global connections, a test of Chinese capacity for compromise and accommodation. In this era of devolutionary forces, the fate of Taiwan will be a harbinger of the world order of the twenty-first century.

Copyright © Rand McNally, 2000

Scale 1:16 000 000; one inch to 250 miles. Polyconic Projection
Elevations and depressions are given in feet

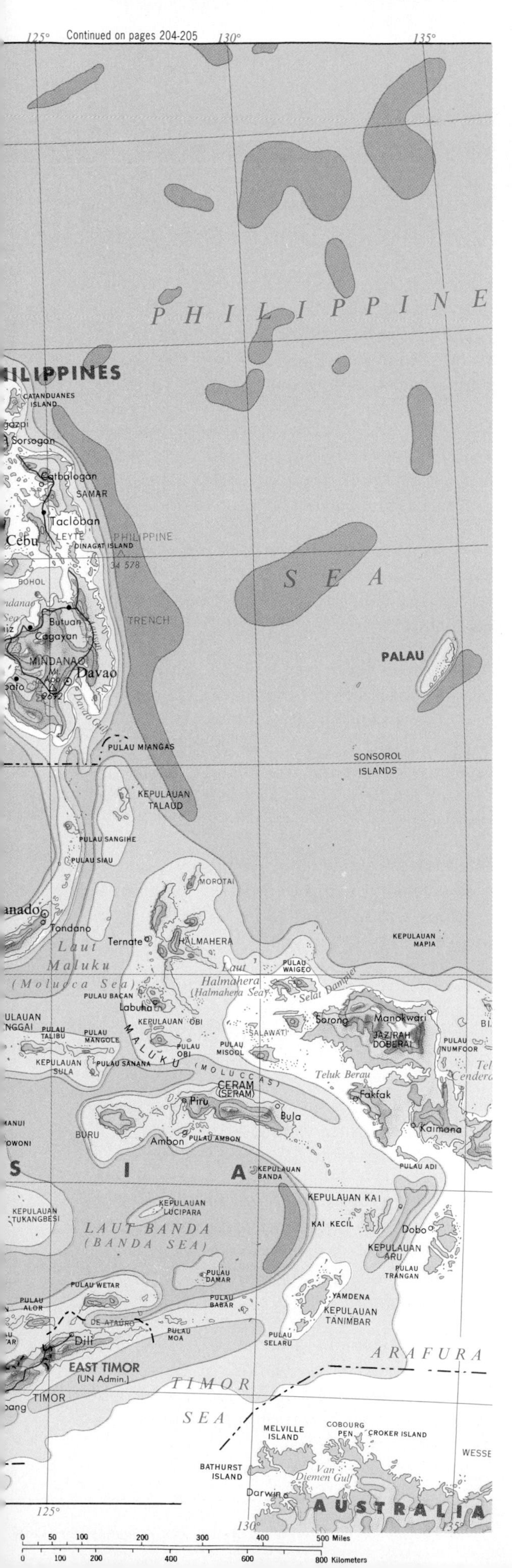

chapter 10

Southeast Asia

CONCEPTS, IDEAS, AND TERMS

1. **Buffer zone**
2. **Shatter belt**
3. **Overseas Chinese**
4. **Organic theory**
5. **State boundaries**
6. **Antecedent boundary**
7. **Subsequent boundary**
8. **Superimposed boundary**
9. **Relict boundary**
10. **State territorial morphology**
11. **Compact state**
12. **Protruded state**
13. **Elongated state**
14. **Fragmented state**
15. **Perforated state**
16. **Domino theory**
17. ***Entrepôt***
18. **Archipelago**
19. **Transmigration**

REGIONS

- MAINLAND SOUTHEAST ASIA
- INSULAR SOUTHEAST ASIA

Southeast Asia . . . . the very name roils American emotions. Here the United States owned its only major colony. Here American forces triumphed over Japanese enemies. Here the United States fought the only war it ever lost. Here Washington's worst Cold War fears failed to materialize. Here American companies invested heavily when the Pacific Rim's economic growth transformed dormant economies into potential Pacific tigers. Southeast Asia, once remote and stagnant, has taken center stage in our globalizing world.

FIGURE 10-1

DEFINING THE REALM

Southeast Asia is a realm of peninsulas and islands, a corner of Asia bounded by India on the northwest and China on the northeast (Fig. 10-1). Its western coasts are washed by the Indian Ocean, and to the east stretches the vast Pacific. From all these directions, Southeast Asia has been penetrated by outside forces. From India came traders; from China, settlers; from across the Indian Ocean, Arabs to engage in commerce and Europeans to build empires; and from across the Pacific, the Americans. Southeast Asia has been the scene of countless contests for power and primacy—the competitors have come from near and far.

Southeast Asia's geography in some ways resembles that of Eastern Europe. It is a mosaic of smaller countries on the periphery of one of the world's largest states. It has been 1 a **buffer zone** between powerful adversaries. It is a **shatter** 2 **belt** in which stresses and pressures from without and within have fractured the political geography. Like Eastern Europe, Southeast Asia exhibits great cultural diversity. It is a realm of hundreds of cultures, numerous languages and dialects, and several major religions.

Figure 10-2 shows the dimensions of the Southeast Asian geographic realm, but note the disconformity between the eastern boundary of the realm and the eastern limits of its most populous state, Indonesia. The easternmost part of Indonesia is the western half of the island of New Guinea, where indigenous cultures are not Southeast Asian but Pacific. Today Indonesia rules what is in effect a Pacific island colony, although Irian Jaya (West Irian) is officially one of its provinces. While we refer in this chapter to Irian Jaya because of its association with Indonesia, including the prospect of a name change for this Papuan territory, we discuss all of New Guinea under the Pacific Realm in Chapter 12.

Because the politico-geographical map (Fig. 10-2) is so complicated, it should be studied attentively. One good way to strengthen your mental map of this realm is to follow the mainland coastline from west to east. The westernmost state in the realm is Myanmar (called Burma before 1989 and still referred to by that name), the only country in Southeast Asia that borders both India and China. Myanmar shares the "neck" of the Malay Peninsula with Thailand, heart of the *mainland* region. The south of the peninsula is part of Malaysia—except for Singapore, at the very tip of it. Facing the Gulf of Thailand is Cambodia. Still moving generally eastward, we reach Vietnam, a strip of land that extends all the way to the Chinese border. And surrounded by its neighbors is landlocked Laos, remote and isolated. This leaves the islands that constitute *insular* Southeast Asia: the Philippines in the north and Indonesia in the south, and between them the offshore portion of Malaysia, situated on the largely Indonesian island of Borneo. Also on Borneo lies the ministate of Brunei, small but,

◆ Major Geographic Qualities of Southeast Asia

1. Southeast Asia extends from the peninsular mainland to the archipelagos offshore. Because Indonesia controls part of New Guinea, its functional region reaches into the neighboring Pacific geographic realm.
2. Southeast Asia, like Eastern Europe, has been a shatter belt between powerful adversaries and has a fractured cultural and political geography shaped by foreign intervention.
3. Southeast Asia's physiography is dominated by high relief, crustal instability marked by volcanic activity and earthquakes, and tropical climates.
4. A majority of Southeast Asia's more than half-billion people live on the islands of just two countries: Indonesia, with the world's fourth-largest population, and the Philippines. The rate of population increase in the insular region of Southeast Asia exceeds that of the mainland.
5. Although the overwhelming majority of Southeast Asians have the same ancestry, cultural divisions and local traditions abound, which the realm's divisive physiography sustains.
6. The legacies of powerful foreign influences, Asian as well as non-Asian, continue to affect the cultural landscapes of Southeast Asia.
7. Southeast Asia's political geography exhibits a variety of boundary types and several categories of state territorial morphology.
8. The Mekong River, Southeast Asia's Danube, has its source in China and borders or crosses five Southeast Asian countries, sustaining tens of millions of farmers, fishing people, and boat owners.
9. The realm's giant in terms of territory as well as population, Indonesia, has not asserted itself as the dominant state because of mismanagement and corruption; but Indonesia has enormous potential.

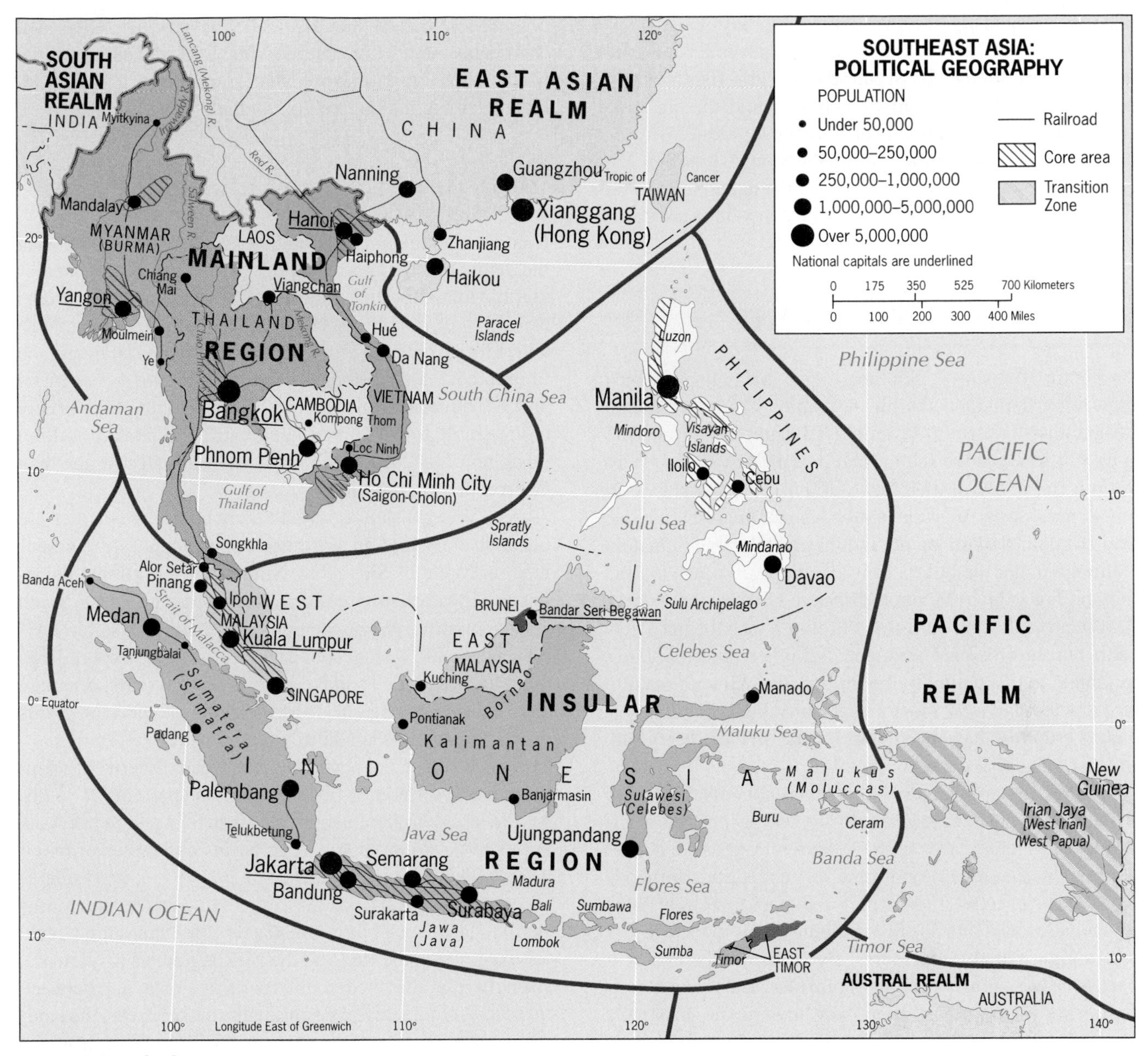

FIGURE 10-2

as we will see, important in the regional picture. And finally, a new state is just appearing on the map in the realm's southeastern corner: East Timor, a former Portuguese colony annexed by Indonesia in 1976 and released to United Nations supervision in 1999.

These are countries of a geographic realm that has no dominant state—no China, no India, no Brazil—although one country, Indonesia, contains 40 percent of its total population and has the potential to emerge as a commanding force. Neither did any single, dominant core of indigenous culture develop here as it did in East Asia. In the river basins and on the plains of the mainland, as well as on the islands offshore, a flowering of cultures produced a diversity of societies whose languages, religions, arts, music, foods, and other achievements formed an almost infinitely varied mosaic—but none of those cultures rose to imperial power. The European colonizers forged empires here, often by playing one state off against another; the Europeans divided and ruled. Out of this foreign intervention came the modern map of Southeast Asia, as only Thailand (formerly Siam) survived the colonial era as an independent entity. Thailand was useful to two competing powers, the French to the east and the British to the west. It was a convenient buffer, and while the colonists carved pieces off Thailand's domain, the kingdom endured.

Indeed, the Europeans accomplished what local powers could not: the formation of comparatively large, multicultural states that encompassed diverse peoples and societies and welded them together. Were it not for the colonial intervention, it is unlikely that the 17,000 islands of far-flung Indonesia would today constitute the world's fourth largest country in terms of population. Nor would the nine sultanates of Malaysia have been united, let alone with the peoples of northern Borneo across the South China Sea. For good or ill, the colonial intrusion consolidated a realm of few culture cores and numerous ministates into less than a dozen countries.

PHYSICAL GEOGRAPHY

As Figure 10-1 shows, Southeast Asia is a realm in which high relief dominates the physiography. From the Arakan Mountains in western Myanmar (Burma) to the glaciers (yes, glaciers!) of the Indonesian part of New Guinea, elevations rise above 10,000 feet (3300 m) in many places. In Myanmar, Mount Victoria is such a peak; in northern Laos, Phu Bia comes close; on the Indonesian island of Sumatera (Sumatra under the old spelling) the tallest volcano is Kerinci; on Jawa (Java) Mounts Slamet and Semeru tower over the countryside; and on Borneo Mount Kinabalu, not a volcano but an erosional remnant, is higher than any other mountain in this realm. In the Philippines, Mount Pinatubo at 4874 feet (1477 m) is not especially high, but this volcano's eruptions have had major impact on the country and, in 1991, on global climate.

The relief map reminds us that this is not only the Pacific Rim but also the Pacific Ring of Fire, where the crust is unstable, earthquakes are common, and volcanoes are active. Among the islands, Borneo is the sole exception. Borneo is a slab of ancient crust, pushed high above sea level by tectonic forces and eroded into its present mountainous topography.

As Figure 10-1 underscores, rivers rise in the highland backbones of the islands and peninsulas, and deposit their sediments as they wind their way toward the coast; the physiography of Sumatera demonstrates this unmistakably. The volcanic hills, plateaus, and better-drained lowlands are fertile and, in the warmth of tropical climates, can yield multiple crops of rice.

On the peninsular mainland we see a pattern that is already familiar: rivers rising in the Asian interior that create alluvial plains and deltas. The Mekong River is the Chang/Yangzi of Southeast Asia: you can trace it all the way from China via Laos, Thailand, and Cambodia into southern Vietnam, where it forms a massive and populous delta. In the west, Myanmar's key river is the Irrawaddy; Thailand's is the Chao Phraya. In the north, the Red River Basin is the breadbasket of northern Vietnam.

No survey of the physical geography of Southeast Asia would be complete without reference to the realm's seas, gulfs, straits, and bays. Irregular and indented coastlines such as these, with thousands of islands near and far, create difficult problems when it comes to drawing maritime boundaries (the islands also form havens for rebels and criminals). As we note later, Southeast Asia has one of the most complex maritime boundary frameworks in the world, and it is rife with potential for conflict.

POPULATION GEOGRAPHY

Compared to the huge population numbers and densities in the habitable regions of South Asia and China, demographic totals for the countries of Southeast Asia, with the exception of Indonesia, seem modest. Again, comparisons with Europe come to mind. Three countries—Thailand, the Philippines, and Vietnam—have populations between 60 and 85 million. Laos, quite a large country territorially (comparable to the United Kingdom), had just 5.5 million inhabitants in 2002. Cambodia, substantially larger than Greece, had 12.7 million.

As Table I-2 shows, arithmetic population densities are not especially high in Southeast Asia except in the urbanized ministate of Singapore. Similarly, physiologic densities are lower than in eastern China or neighboring South Asian countries. When the political situation is stable, several Southeast Asian countries are able to export large quantities of rice. Thailand and, in recent years, Vietnam have ranked among the world's leading exporters in this crucial staple of Asian diets.

Viewed in spatial perspective, one can discern both similarities and differences in the population patterns of Southeast Asia and its giant neighbors. As in East and South Asia, population clusters in the large basins and deltas of major rivers, where densities are high and the countryside is parceled into paddies as far as the eye can see. Numerous small, space-conserving villages dot the landscape which, for example, in the Red and the Mekong deltas in Vietnam, strongly resemble their counterparts in China. But between these low-lying alluvial basins, in the higher areas, the landscape takes on a savanna character, with reddish, leached, less fertile soils and much sparser population. And in the islands of Indonesia and the Philippines, Southeast Asia has environments that are not present at all in India or China—well-watered volcanic soils supporting luxuriant natural vegetation and capable of sustaining intensive agriculture. Whole countrysides have been meticulously terraced to create paddyfields and irrigate them (see cover photo).

People and Land

It is noteworthy that of Southeast Asia's 546 million inhabitants, well over half (303 million, 55 percent) live on the islands of Indonesia and the Philippines, leaving the realm's mainland countries with just 45 percent of the population. With such high population pressure in adjacent realms, why has Southeast Asia not been flooded by waves of immigrants?

In fact, Southeast Asia has received its share of immigrants, but overland invasions have been comparatively limited. Several factors have hindered travel along overland routes into Southeast Asia. First, physical obstacles inhibit such movement. In Chapter 8, we noted the barrier effect of the densely forested hills and mountains along the border between northeastern India and northwestern Myanmar (Burma). North of Myanmar lies forbidding Xizang (Tibet), and northeast of Myanmar and north of Laos is the high, rugged Yunnan Plateau. Transit is easier in the east, between southeastern China and northern Vietnam, and this has indeed been an avenue for contact and migration. As the ethnolinguistic map of China (p. 448) shows, migration from Southeast Asia into China also has occurred in this area.

Second—as reflected in the population distribution map (Fig. I-9)—the highly clustered agricultural opportunities in Southeast Asia lie separated by large stretches of less productive land, and these in turn are crossed by high-relief zones that hinder movement. The map of Southeast Asia reveals several concentrated areas of productive capacity but few corridors that would facilitate migration.

As we noted earlier, immigrants to Southeast Asia over the past millennium have come mainly by sea, not by land. That is true not only for Arabs and Europeans but also for neighboring Chinese. Mostly the Chinese immigrants came not as farmers seeking land but as traders and workers looking for opportunities in the cities and towns. "Chinatowns" form a vital part of most Southeast Asian cities today, especially the coastal ones.

The Ethnic Mosaic

Southeast Asia's peoples come from a common stock just as (Caucasian) Europeans do, but this has not prevented the emergence of regionally or locally discrete ethnic or cultural groups. Figure 10-3 displays the broad distribution of ethnolinguistic groups in the realm, but be aware that this is a generalization. At the scale of this map, numerous small groups cannot be depicted.

The map shows the rough spatial coincidence, on the mainland, between major ethnic group and modern political state. The Burman dominate in the country once known as Burma (Myanmar); the Thai occupy the state once known as Siam (Thailand); the Khmer form the nation of Cambodia and extend northward into Laos; and the Vietnamese inhabit the long strip of territory facing the South China Sea.

Territorially, by far the largest population is classified in Figure 10-3 as Indonesian, the inhabitants of the great archipelago that extends from Sumatera* west of the Malay Peninsula to the Malukus (Moluccas) in the east and from the lesser Sunda Islands in the south to the Philippines in the north. Collectively, all these peoples—the Filipinos, Malays, and Indonesians—shown in Figure 10-3 are known as Indonesians, but they have been divided by history and politics. Note, on the map, that the Indonesians in Indonesia itself include Javanese, Madurese, Sundanese, Balinese, and other large groups; hundreds of smaller ones are not shown. In the Philippines, too, island isolation and

FROM THE FIELD NOTES

"It was a relatively cool day in Hanoi, which was just as well because the drive from Haiphong had taken nearly four hours, much of the time waiting behind oxcarts, and the countryside was hot and humid. On my way to an office along Hang Bai Street I ran into this group of youngsters and their teacher on their way to the Ho Chi Minh memorial. He right away realized the level of my broken French, and explained patiently that he taught history, and took his class on a field trip several times a year. 'They get all dressed up and it is a big moment for them,' he said, 'and there is a lot of history to be visited in Hanoi.' The kids had seen me take this picture and now they crowded around. I told them that I noticed that *rouge* seemed to be a favorite color in their outfits. 'But that should not surprise you, *monsieur*,' said the teacher, smiling. 'This is Vietnam, communist Vietnam, and red is the color that rallies us!' In the city at the opposite end of the country named after Ho Chi Minh (founder of his country's Communist Party, president from 1945 to 1969, victor over the French colonizers, and leader in the war against the United States), my experiences with teachers and students had been quite different, to the point that most called their city Saigon. But here in Hanoi the ideological flame still burned brightly."

*As in Africa, names and spellings have changed with independence. In this chapter, we will use the contemporary spellings, except when we refer to the colonial period. Thus Indonesia's four major islands are Jawa, Sumatera, Kalimantan (the Indonesian part of Borneo), and Sulawesi. The Dutch called them Java, Sumatra, Dutch Borneo, and Celebes, respectively.

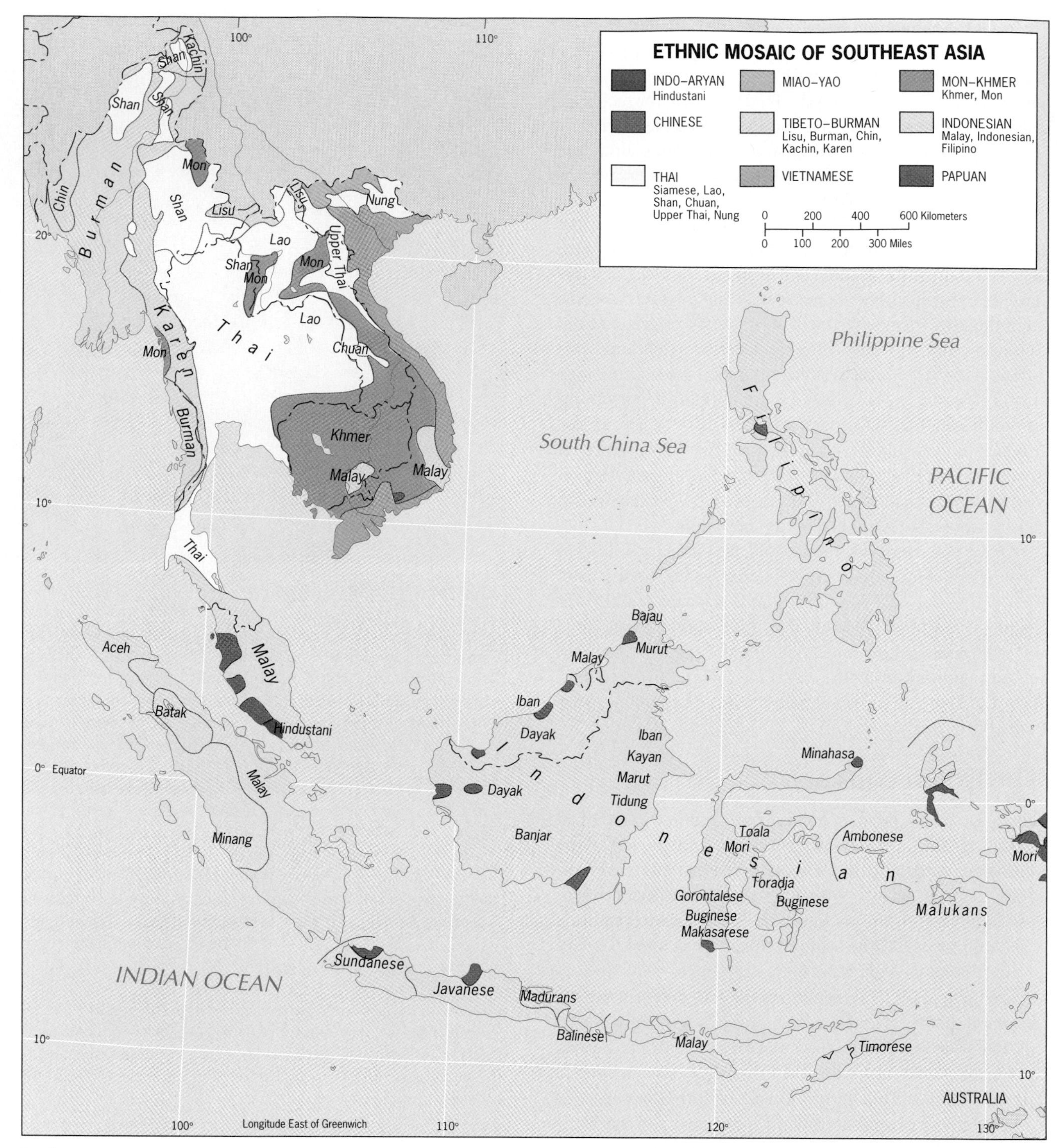

FIGURE 10-3

contrasting ways of life are reflected in the cultural mosaic. Also part of this Indonesian ethnic-cultural complex are the Malays, whose heartland lies on the Malay Peninsula but who form minorities in other areas as well. Like most Indonesians, the Malays are Muslims, but Islam is a more powerful force in Malay society than, in general, in Indonesian culture.

Figure 10-3 also reminds us that (again like Eastern Europe) Southeast Asia has many ethnic minorities. On the Malay Peninsula, note the South Asian (Hindustani) cluster; Hindu communities with Indian ancestries exist in many parts of the peninsula, but here in the southwest they form the majority in a small area. Nearby lies the largest single exception to the rule that Chinese minorities in

Southeast Asia are urban-based: along the west coast of the Malay Peninsula, Chinese are farmers and villagers as well as city dwellers. In the north of the realm, minorities share the lands in which the Burman, Thai, and Vietnamese dominate. Those minorities, as a comparison between Figures 10-2 and 10-3 proves, tend to occupy zones peripheral to the regional core areas, where forests often are dense and difficult to penetrate, where isolation from the power cores prevails, where remoteness promotes detachment from the national state, and where, time and again, ethnic conflicts have raged as the core-based majority tried to control the frontier. During the war in Vietnam, peoples in the highlands (the Montagnards) tended to be sympathetic to the American campaign because of their longstanding hostility to the dominant (lowland) Vietnamese. Today, the Shan and the Karen in Myanmar are in a state of conflict with the regime based in the core area.

HOW THE POLITICAL MAP EVOLVED

The leading colonial competitors in Southeast Asia were the Dutch, French, British, and Spanish (with the last replaced by the Americans in their stronghold, the Philippines). The Japanese had colonial objectives here as well, but these came and went during the course of World War II.

The Dutch acquired the greatest prize: control over the vast archipelago now called Indonesia (formerly the Netherlands East Indies). France established itself on the eastern flank of the mainland, controlling all territory east of Thailand and south of China. The British conquered the Malay Peninsula, gained power over the northern part of the island of Borneo, and established themselves in Burma as well. Other colonial powers also gained footholds but not for long. The exception was Portugal, which held on to its eastern half of the island of Timor (Indonesia) until after the Dutch had been ousted from their East Indies.

Figure 10-4 shows the colonial framework in the late nineteenth century, before the United States assumed control over the Philippines. Note that while Thailand survived as an independent state, it lost territory to the British in Malaya and Burma and to the French in Cambodia and Laos.

The Colonial Imprint

The colonial powers divided their possessions into administrative units as they did in Africa and elsewhere. Some of these political entities became independent states when the colonial powers withdrew. France, one of the mainland's leading colonial powers, divided its Southeast Asian empire into five units. Three of these units lay along the east coast: Tonkin in the north next to China, centered on the basin of the Red River; Cochin China in the south, with the Mekong Delta as its focus; and between these two, Annam. The other two French territories were Cambodia, facing the Gulf of Thailand, and Laos, landlocked in the interior. Out of these five French dependencies there emerged the three states of Indochina. The three east coast territories ultimately became one state, Vietnam; the other two (Cambodia and Laos) each achieved separate independence.

The British ruled two major entities in Southeast Asia (Burma and Malaya) in addition to a large part of northern Borneo and many small islands in the South China Sea. Burma was attached to Britain's Indian empire; from 1886 until 1937 it was governed from distant New Delhi. But when British India became independent in 1947 and split into several countries, Burma was not part of the grand design that created West and East Pakistan (the latter now Bangladesh), Ceylon (now Sri Lanka), and India. Instead, in 1948 Burma (now Myanmar) was given the status of a sovereign republic.

In Malaya, the British developed a complicated system of colonies and protectorates that eventually gave rise to the equally complex, far-flung Malaysian Federation. Included were the former Straits Settlements (Singapore was one of these colonies), the nine protectorates on the Malay Peninsula (former sultanates of the Muslim era), the British dependencies of Sarawak and Sabah on the island of Borneo, and numerous islands in the Strait of Malacca and the South China Sea. The original Federation of Malaysia was created in 1963 by the political unification of recently independent mainland Malaya, Singapore, and the former British dependencies on the largely Indonesian island of Borneo. Singapore, however, left the Federation in 1965 to become a sovereign city-state, and the remaining units were later restructured into peninsular Malaysia and, on Borneo, Sarawak and Sabah. Thus the term *Malaya* properly refers to the geographic area of the Malay Peninsula, including Singapore and other nearby islands; the term *Malaysia* identifies the politico-geographical entity of which Kuala Lumpur is the capital city.

The Hollanders took control of the "spice islands" through their Dutch East India Company, and the wealth that they extracted from what is today Indonesia brought the Netherlands its Golden Age. From the mid-seventeenth to the late-eighteenth century, the Dutch could develop their East Indies sphere of influence almost without challenge, for the British and French were preoccupied with the Indian subcontinent. By playing the princes of Indonesia's states against one another in the search for economic concessions and political influence, by placing the Chinese in positions of responsibility, and by imposing systems of forced labor in areas directly under its control, the Company had a ruinous effect on the Indonesian societies it subjugated. Java (Jawa), the most populous and productive island, became the focus of Dutch administration; from its capital at Batavia (now Jakarta), the Company extended its sphere of influence into Sumatra (Sumatera), Dutch Borneo

(Kalimantan), Celebes (Sulawesi), and the smaller islands of the East Indies. This was not accomplished overnight, and the struggle for territorial control was carried on long after the Dutch East India Company had yielded its administration to the Netherlands government. Dutch colonialism thus threw a girdle around Indonesia's more than 17,000 islands, paving the way for the creation of the realm's largest and most populous nation-state (nearly 220 million today).

In the colonial tutelage of Southeast Asia, the Philippines, long under Spanish domination, had a unique experience. As early as 1571, the islands north of Indonesia were under Spain's control (they were named for Spain's King

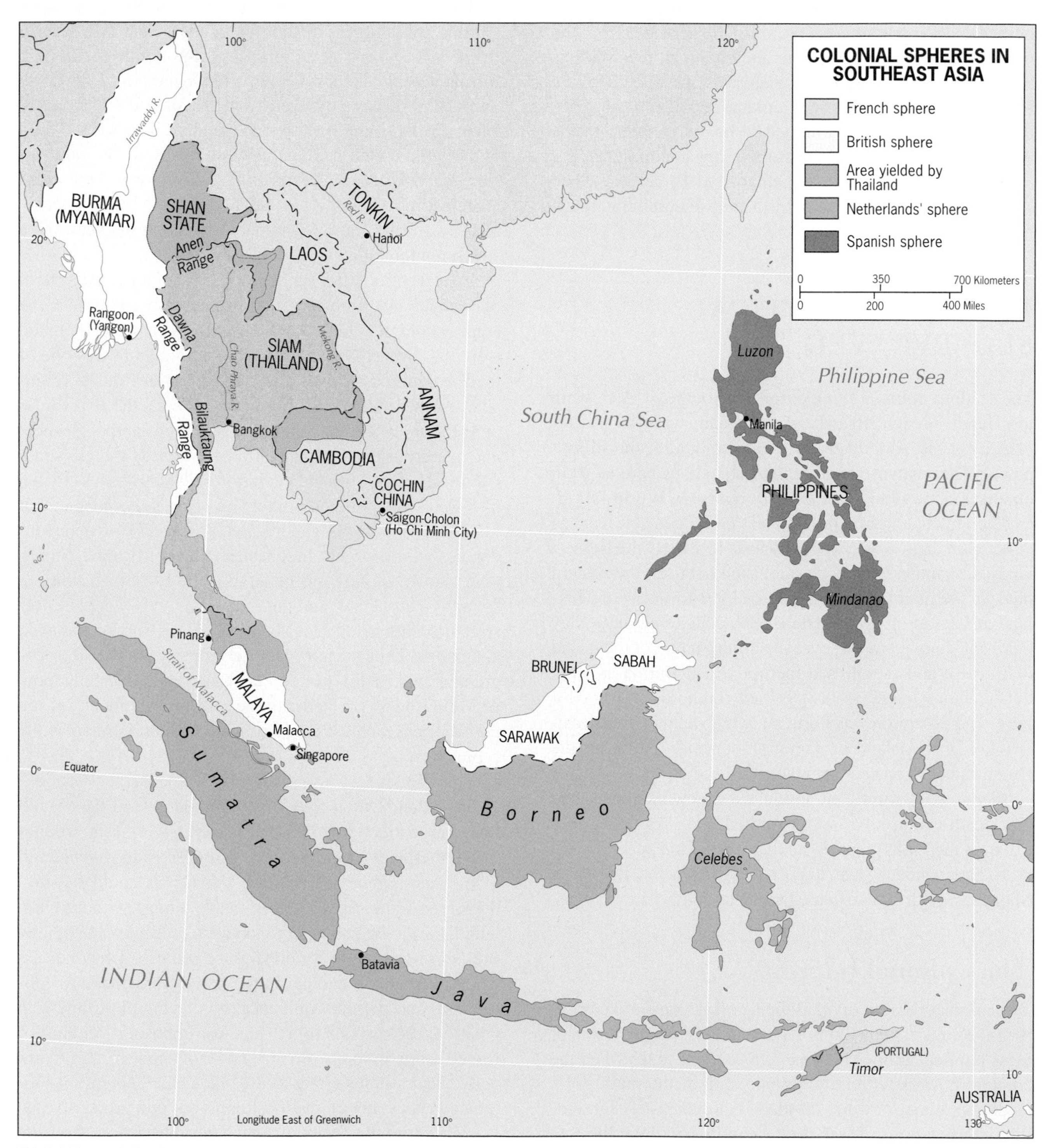

FIGURE 10-4

Philip II). Spanish rule began when Islam was reaching the southern Philippines via northern Borneo. The Spaniards spread their Roman Catholic faith with great zeal, and between them the soldiers and priests consolidated Hispanic dominance over the mostly Malay population. Manila, founded in 1571, became a profitable waystation on the route between southern China and western Mexico (Acapulco usually was the trans-Pacific destination for the galleons leaving Manila's port). There was much profit to be made, but the indigenous people shared little in it. Great landholdings were awarded to loyal Spanish civil servants and to men of the church. Oppression eventually yielded revolution, and Spain was confronted with a major uprising when the Spanish-American War broke out elsewhere in 1898.

As part of the settlement of that war, the United States replaced Spain in Manila. That was not the end of the revolution, however. The Filipinos now took up arms against their new foreign ruler, and not until 1905, after terrible losses of life, did American forces manage to "pacify" their new dominion. Subsequently, U.S. administration in the Philippines was more progressive than Spain's had been. In 1934, Congress passed the Philippine Independence Law, providing for a ten-year transition to sovereignty. But before independence could be arranged, World War II intervened. In 1941, Japan conquered the islands, temporarily ousting the Americans; U.S. forces returned in 1944 and, with strong Filipino support, defeated the Japanese in 1945. The agenda for independence was resumed, and in 1946 the sovereign Republic of the Philippines was proclaimed.

Today, all of Southeast Asia's states are independent, but centuries of colonial rule have left strong cultural imprints. In their urban landscapes, their education systems, their civil service, and countless other ways, this realm still carries the marks of its colonial past.

Cultural-Geographic Legacies

The French, who ruled and exploited a crucial quadrant of Southeast Asia, had a name for their empire: *Indochina*. That name would be appropriate for the rest of the realm as well because it suggests its two leading Asian influences for the past 2000 years. Periodic expansions of the Chinese Empire, notably along the eastern periphery, as well as the arrival of large numbers of Chinese settlers (see box titled "Overseas Chinese"), infused the region with cultural norms from the north. The Indians came from the west by way of the sea, as their trading ships plied the coasts, and settlers from India founded colonies on Southeast Asian shores in the Malay Peninsula, on the lower Mekong Plain, on Jawa and Bali, and on Borneo.

With the migrants from the Indian subcontinent came their faiths: first Hinduism and Buddhism, later Islam. The Muslim religion, promoted by the growing number of Arab traders who appeared on the scene, became the dominant religion in Indonesia (where nearly 90 percent of the population adheres to Islam today). But in Myanmar, Thailand, and Cambodia, Buddhism remained supreme, and in all three countries the overwhelming majority of the people are now adherents. In culturally diverse Malaysia, the Malays are Muslims (to be a Malay *is* to be a Muslim), and almost all Chinese are Buddhists; but most Malaysians of Indian ancestry remain Hindus. Although Southeast Asia has generated its own local cultural expressions, most of what remains in tangible form has resulted from the infusion of foreign elements. For instance, the main temple at Angkor Wat, constructed in Cambodia during the twelfth century, remains a monument to the Indian architecture of that time.

The *Indo* part of Indochina, then, refers to the cultural imprints from South Asia: the Hindu presence, the importance of Buddhism (which came to Southeast Asia via Sri Lanka [Ceylon] and its seafaring merchants), the influences of Indian architecture and art (especially sculpture), writing and literature, and social structures and patterns.

The *China* in the name Indochina signifies the role of the Chinese here. Chinese emperors coveted Southeast Asian lands, and China's power reached deep into the realm. Social and political upheavals in China sent millions of Sinicized people southward. Chinese traders, pilgrims, seafarers, fishermen, and others sailed from southeastern China to the coasts of Southeast Asia and established settlements there. Over time, those settlements attracted more Chinese emigrants, and Chinese influence in the realm grew (Fig. 10-5). Not surprisingly, relations between the Chinese settlers and the earlier inhabitants of Southeast Asia have at times been strained, even violent. The Chinese presence in Southeast Asia is long-term, but the invasion has continued into modern times. The economic power of Chinese minorities and their role in the political life of the realm have led to conflicts.

The Chinese initially profited from the arrival of the Europeans, who stimulated the growth of agriculture, trade, and industries; here the Chinese found opportunities they lacked at home. They harvested rubber, found jobs on the docks and in the mines, cleared the bush, and transported goods in their sampans. They brought useful skills with them, and as tailors, shoemakers, blacksmiths, and fishermen, they prospered. The Chinese were also astute in business; soon, they not only dominated the region's retail trade but also held prominent positions in banking, industry, and shipping. Thus they have always been far more important than their modest numbers in Southeast Asia would suggest. The Europeans used them for their own designs but found the Chinese to be stubborn competitors at times—so much so that eventually they tried to impose restrictions on Chinese immigration. When the United States took control of the Philippines, it also sought to stop the influx of Chinese into those islands.

Overseas Chinese

Southeastern China has been the source of Chinese emigrants to Southeast Asia for as long as 2000 years, but the largest movements have occurred over the past six centuries (Fig. 10-5). During the initially outward-looking Ming Dynasty (1368–1644) and early in the Manchu (Qing) Dynasty, small groups of Chinese emigrants, sometimes pushed by famine at home, sailed southward to establish small communities in Southeast Asian towns. During the early colonial period (1670–1870), growing commercial opportunities formed the pull factors that attracted greater numbers. But the largest exodus came during the late colonial period (1870–1940) when Southeast Asia's expanding job and trade opportunities attracted more than 20 million Chinese, many from Fujian and Guangdong Provinces. During the communist period, additional migrants illegally left China, many to or via Hong Kong.

Toward the end of their rule, colonial regimes began to restrict Chinese immigration, as did independent Thailand. The Chinese had become an influential class of middlemen and controlled ever more commerce, and colonial rulers as well as local peoples feared Chinese influence and "expansionism." During World War II and subsequent decolonization, Chinese were persecuted in Southeast Asia, especially in Malaya while under Japanese control, and in Indonesia in the 1960s when Chinese were targeted as communist sympathizers.

Not counting intermarriages, Southeast Asia today may contain as many as 30 million Chinese (more than half the world total of **Overseas Chinese**), and many have become very successful. When China began its new economic policy, Overseas Chinese invested heavily in the SEZs and Open Cities, playing a large role in stimulating the economic growth experienced there.

But success entails risks. Resentment against Chinese prosperity and business practices led to discriminatory legislation in Malaysia. When Indonesia's economy collapsed in 1998, police stood by as gangs looted, burned, murdered, and raped at will in Chinese commercial districts. Wealthier Chinese fled, exacerbating Indonesia's economic crisis.

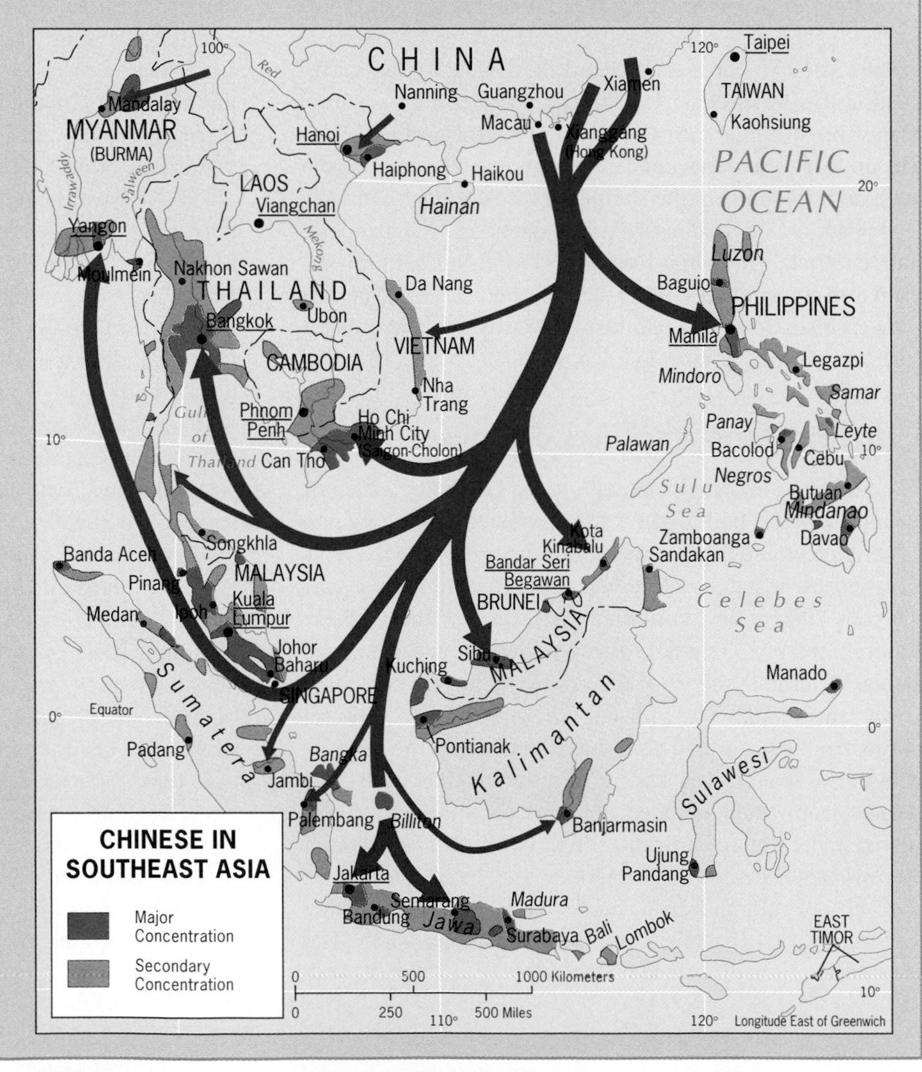

FIGURE 10-5

FROM THE FIELD NOTES

"Like most major Southeast Asian cities, Bangkok's urban area includes a large and prosperous Chinese sector. No less than 14 percent of Thailand's population of 63.1 million is of Chinese ancestry, and the great majority of Chinese live in the cities. In Thailand, this large non-Thai population is well integrated into local society, and intermarriage is common. Still, Bangkok's 'Chinatown' is a distinct and discrete part of the great city. There is no mistaking Chinatown's limits: Thai commercial signs change to Chinese, goods offered for sale also change (Chinatown contains a large cluster of shops selling gold, for example), and the urban atmosphere, from street markets to bookshops, is dominantly Chinese. This is a boisterous, noisy, energetic part of multicultural Bangkok, a vivid reminder of the Chinese commercial success in Southeast Asia."

When the European colonial powers withdrew and Southeast Asia's independent states emerged, Chinese population sectors ranged from nearly 50 percent of the total in Malaysia (in 1963) to barely over 1 percent in Myanmar. In Singapore, Chinese today constitute 76 percent of the population of 4.1 million; when Singapore seceded from Malaysia in 1965, the Chinese component in Malaysia was substantially reduced (it is about 25 percent today). In Indonesia, the percentage of Chinese in the total population is not high (no more than 3 percent), but the Indonesian population is so large that even this small percentage indicates a Chinese sector of more than 6 million. In 1998, this tiny minority reportedly controlled 70 percent of Indonesia's economy. In Thailand, on the other hand, many Chinese have married Thais, and the Chinese minority of about 14 percent has become a cornerstone of Thai society, dominant in trade and commerce.

In general, Southeast Asia's Chinese communities remained aloof and formed their own separate societies in the cities and towns. They kept their culture and language alive through social clubs, schools, and even residential suburbs that, in practice if not by law, were Chinese in character. At one time they were caught in the middle between the Europeans and Southeast Asians when the local people were hostile toward both white people and the Chinese. Since the withdrawal of the Europeans, however, the Chinese have become the main target of this antagonism, which remained strong because of the Chinese involvement in money lending, banking, and trade monopolies. Moreover, there is the specter of an imagined or real Chinese political imperialism along Southeast Asia's northern flanks.

The *China* in Indochina, therefore, represents a diversity of penetrations. Southern China was the source of most of the old invasions, and Chinese territorial consolidation provided the impetus for successive immigrants. Mongoloid racial features carried southward from East Asia mixed with the preexisting Malay stock to produce a transition from Chinese-like people in the northern mainland to darker-skinned Malay types in the distant Indonesian east. Although Indian cultural influences remained strong, people throughout Southeast Asia adopted Chinese modes of dress, plastic arts, types of houses and boats, and other cultural attributes. During the past century, and especially during the last half-century, renewed Chinese immigration brought skills and energies that propelled these minorities to comparative wealth and influence.

SOUTHEAST ASIA'S POLITICAL GEOGRAPHY

Southeast Asia's diverse and fractured natural landscapes, as well as varied cultural influences, have combined to create a complex political geography. As we noted earlier, political geography focuses on the spatial expressions of political behavior, and we were introduced to several aspects of it when we examined Western Europe's supranationalism, Eastern Europe's balkanization, Pakistan's irredentism toward Kashmiri Muslims, and the geopolitics of eastern Eurasia. Many political geographers have studied the fundamental causes of the cyclical rise and decline of states; their answers have ranged from environmental changes to biological forces. Friedrich Ratzel (1844–1904) conceptualized the state as a biological organism whose life, from birth through expansion and maturation to eventual senility and collapse, mirrors that of any living thing. Ratzel's **organic theory** of state development held that nations, being aggregates of human beings, would over the long term live and die as their citizens did.

4

Other political geographers sought to measure the strength of the forces that bind states (centripetal forces) and that divide them (centrifugal forces), and thus to assess a state's chances to survive separatism of the kind Canada and Sri Lanka are experiencing today, to name just two examples. Even in today's era of modern warfare, a stretch of water still affects the course of events. Taiwan would not be what it is without the 100 miles (160 km) of water that separates it from mainland China. Singapore's secession from (then) Malaya was facilitated because Singapore is an island; had it been attached to the Malay Peninsula, the centrifugal forces of separatism might not have prevailed. The role of physical geography in political events remains powerful.

In this section we pay particular attention to the boundaries and territories of the component states of Southeast Asia because this realm's political geography displays virtually every possible attribute of both. We focus first on the borders and then on the states' territorial shapes.

The Boundaries

Boundaries are sensitive parts of a state's anatomy: just as people are territorial about their individual properties, so nations and states are sensitive about their territories and borders. The saying that "good fences make good neighbors" certainly applies to states, but, as we know, the boundaries between states are not always "good fences."

5

Boundaries, in effect, are contracts between states. That contract takes the form of a treaty that contains the *definition* of the boundary in the form of elaborate description. Next, cartographers perform the *delimitation* of the treaty language, drawing the boundary on official, large-scale maps. And throughout human history, states have used those maps to build fences, walls, or other barriers in a process called *demarcation*.

Once established, we can classify boundaries geographically. Some are sinuous, conforming to rivers or mountain crests (*physiographic*) or coinciding with breaks or transitions in the cultural landscape (*anthropogeographic*). As any world political map shows, many boundaries are simply straight lines, delimited without reference to physical or cultural features. These *geometric* boundaries can produce problems when the cultural landscape changes where they exist.

In general, the boundaries of Southeast Asia were better defined than those of several other postcolonial areas of the world, notably Africa, the Arabian Peninsula, and Turkestan. The colonial powers that established the original treaties tried to define boundaries to lie in remote and/or sparsely peopled areas, for example, across interior Borneo. Nevertheless, certain Southeast Asian boundaries have produced problems, among them the geometric boundary between Irian Jaya, the portion of New Guinea ruled by Indonesia, and Papua New Guinea, the eastern component of the island.

Even on a small-scale map of the kind we use in this chapter, we can categorize the boundaries of this realm. A comparison between Figures 10-2 and 10-3 reveals that the boundary between Thailand and Myanmar over long segments is anthropogeographic, notably where the name *Karen*, the Myanmar minority, appears on Figure 10-3. Figure 10-1 shows that a large segment of the Vietnam–Laos boundary is physiographic-political, coinciding with the Annamese Cordillera (Mountains).

Boundaries also can be classified genetically, that is, as their evolution relates to the cultural landscapes they traverse. A leading political geographer, Richard Hartshorne (1899–1992), proposed a four-level *genetic boundary classification*. All four of these boundary types can be observed in Southeast Asia.

FROM THE FIELD NOTES

"I stood on the Laotian side of the great Mekong River which, during the dry season, did not look so great! On the opposite side was Thailand, and it was rather easy for people to cross here at this time of the year. But, the locals told me, it is quite another story in the wet season. Then the river inundates the rocks and banks you see here, it rushes past, and makes crossing difficult and even dangerous. The buildings where the canoes are docked are built on floats, and rise and fall with the seasons. The physiographic-political boundary between Thailand and Laos lies in the middle of the valley we see here."

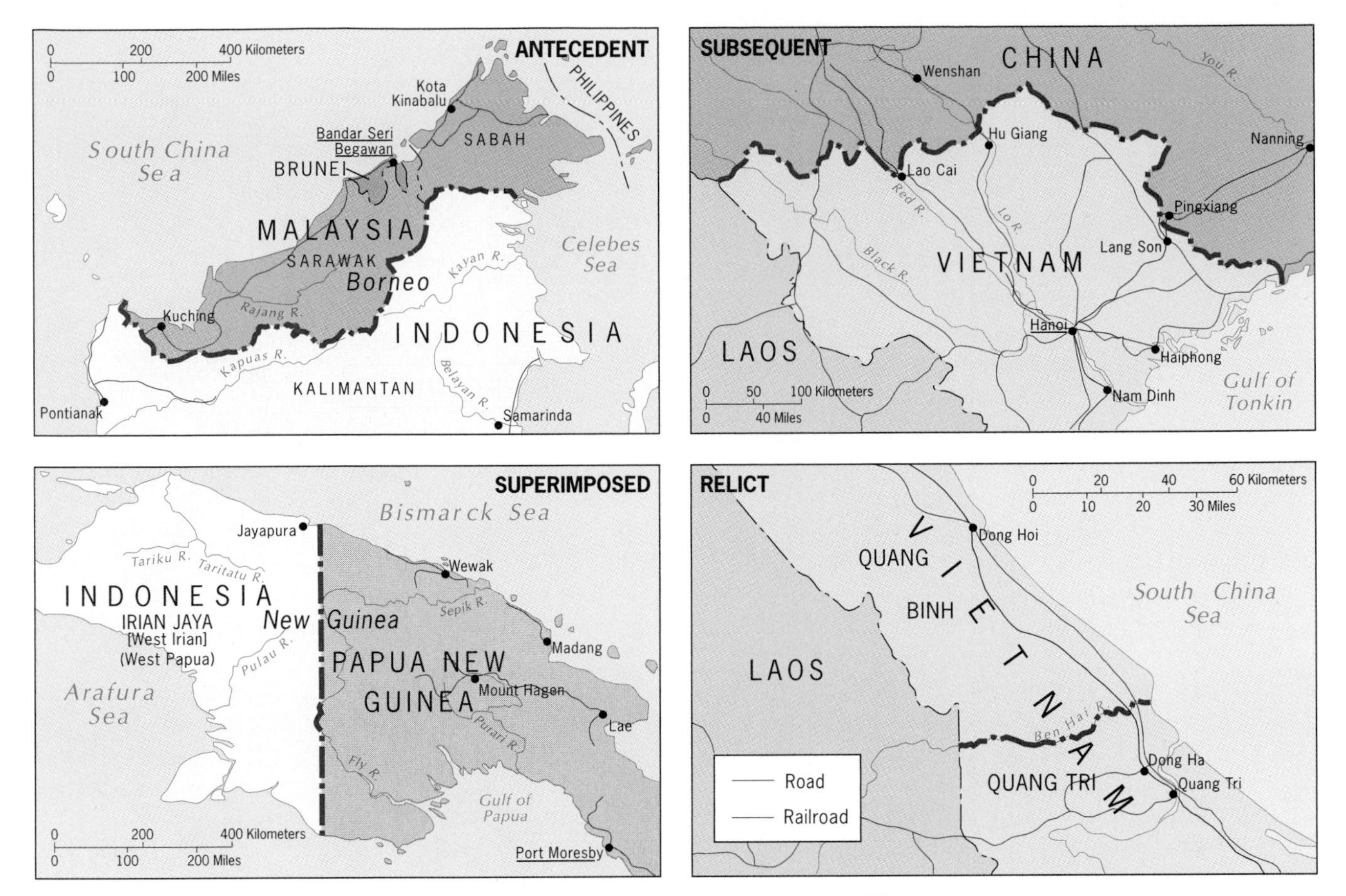

GENETIC POLITICAL BOUNDARY TYPES

FIGURE 10-6

Certain boundaries, Hartshorne reasoned, were defined and delimited before the present-day human landscape developed. In Figure 10-6 (upper-left map), the boundary between Malaysia and Indonesia on the island of Borneo is an example of the first boundary type, the **antecedent boundary**. Most of this border passes through sparsely inhabited tropical rainforest, and the break in settlement can even be detected on the small-scale world population map (Fig. I-9).

A second category of boundaries evolved as the cultural landscape of an area took shape, part of the ongoing process of accommodation. These **subsequent boundaries** are represented in Southeast Asia by the map in the upper right of Figure 10-6, which shows in some detail the border between Vietnam and China. This border is the result of a long process of adjustment and modification, the end of which may not yet have come.

The third category involves boundaries drawn forcibly across a unified or at least homogeneous cultural landscape. The colonial powers did this when they divided the island of New Guinea by delimiting a boundary in a nearly straight line (curved in only one place to accommodate a bend in the Fly River), as shown in the lower-left map of Figure 10-6. The **superimposed boundary** they delimited gave the Netherlands the western half of New Guinea. When Indonesia became independent in 1949, the Dutch did not yield their part of New Guinea, which is peopled mostly by ethnic Papuans, not Indonesians. In 1962 the Indonesians invaded the territory by force of arms, and in 1969 the United Nations recognized its authority there. This made the colonial, superimposed boundary the eastern border of Indonesia and had the effect of extending Indonesia from Southeast Asia into the Pacific Realm. Geographically, all of New Guinea forms part of the Pacific Realm.

The fourth genetic boundary type is the so-called **relict boundary**—a border that has ceased to function but whose imprints (and sometimes influence) are still evident in the cultural landscape. The boundary between the former North and South Vietnam (Fig. 10-6, lower-right map) is a classic example: once demarcated militarily, it has had relict status since 1976 following the reunification of Vietnam in the aftermath of the Indochina War (1964–1975).

Southeast Asia's boundaries have colonial origins, but they have continued to influence the course of events in postcolonial times. Take one instance: the physiographic boundary that separates the main island of Singapore from the rest of the Malay Peninsula, the Johor Strait (see Fig.

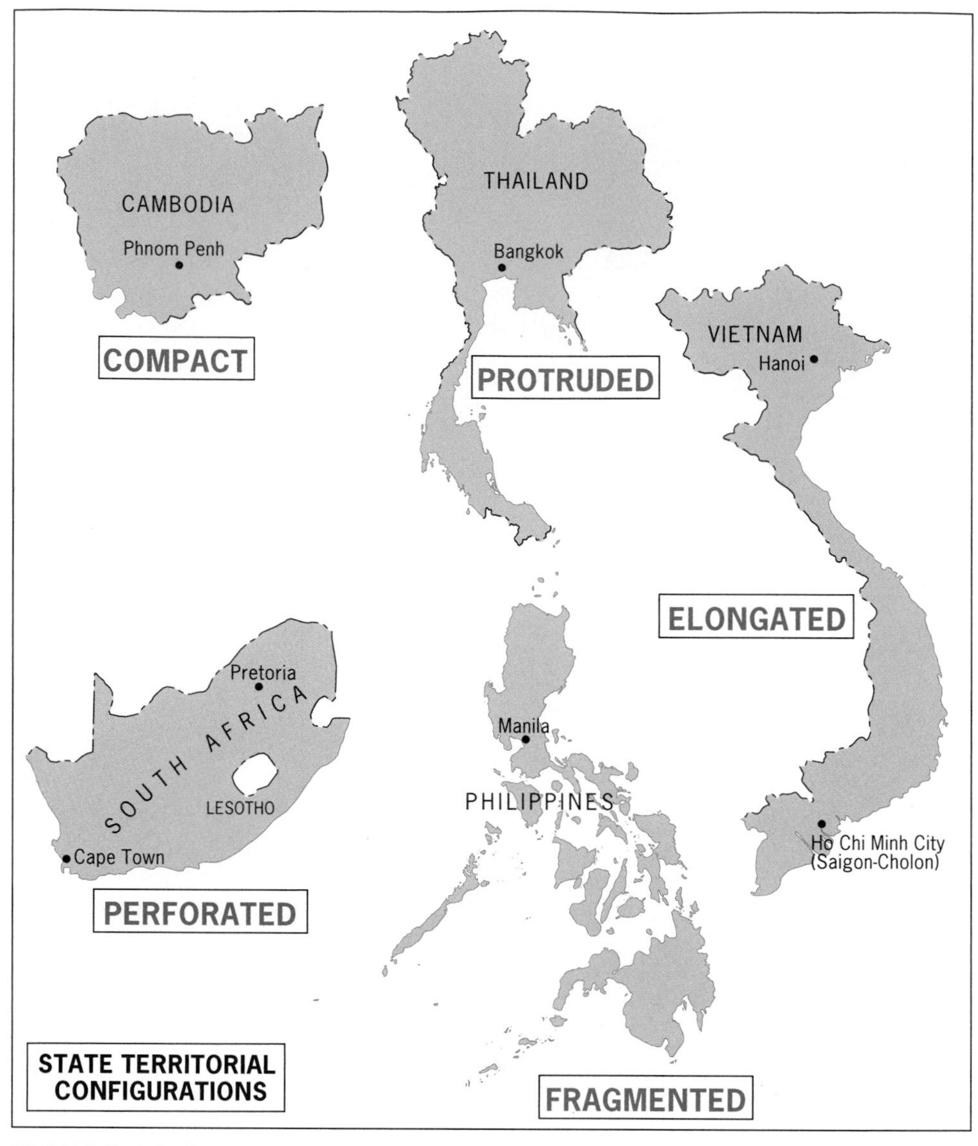

FIGURE 10-7

10-11). That physiographic-political boundary facilitated, perhaps crucially, Singapore's secession from the state of Malaysia. Without it, Malaysia might have been persuaded to stop the separation process; at the very least, territorial issues would have arisen to slow the sequence of events. As it was, no land boundary needed to be defined. The Johor Strait demarcated Singapore and left no question as to its limits.

State Territorial Morphology

Boundaries define and delimit states; they also create the mosaic of often interlocking territories that give individual countries their shape, also known as their *morphology*. The 10 **territorial morphology** of a state affects its condition, even its survival. Vietnam's extreme elongation has influenced its existence since time immemorial. As we will see, Indonesia has tried to redress its fragmentation into thousands of islands by promoting unity through the "transmigration" of Jawanese from the most populous island to many of the others.

Political geographers identify five dominant state territorial configurations, all of which we have encountered in our world regional survey but which we have not categorized until now. All but one of these shapes are represented in Southeast Asia, and Figure 10-7 provides the terminology and examples:

- **Compact states** have territories shaped somewhere between round and rectangular, without major indentations. This encloses a maximum amount of territory 11 within a minimum length of boundary. Southeast Asian example: Cambodia.

12 • **Protruded states** (sometimes called *extended*) have a substantial, usually compact territory from which extends a peninsular corridor that may be landlocked or coastal. Southeast Asian examples: Thailand, Myanmar.

13 • **Elongated states** (also called *attenuated*) have territorial dimensions in which the length is at least six times the average width, creating a state that lies astride environmental or cultural transitions. Southeast Asian example: Vietnam.

14 • **Fragmented states** consist of two or more territorial units separated by foreign territory or by water. Subtypes are mainland-mainland, mainland-island, and island-island. Southeast Asian examples: Malaysia, Indonesia, Philippines.

15 • **Perforated states** completely surround the territory of other states, so that they have a "hole" in them. No Southeast Asian example; the most illustrative current case is South Africa, perforated by Belgium-sized Lesotho.

In the discussion that follows, we will have frequent occasion to refer to this geographic property of Southeast Asia's states. For so comparatively small a realm with so few states, Southeast Asia displays a considerable variety of state morphologies. When we link these features to other geographic aspects (such as relative location), we obtain useful insights into the regional framework.

One point of caution: states' territorial morphologies do not determine their viability, cohesion, unity, or lack thereof; they can, however, influence these qualities. Cambodia's compactness has not ameliorated its divisive political geography, for example. But as we will find in the pages that follow, shape plays a key role in the still-unfolding political and economic geography of Southeast Asia.

REGIONS OF THE REALM

Southeast Asia's first-order regionalization must be based on its mainland-island fragmentation. But as we have noted there are physiographic, historical, and cultural reasons to include the Malaysian (southern) part of the Malay Peninsula in the insular region, as shown in Figure 10-2. Using the political framework as our grid, we see that the regions of Southeast Asia are constituted as follows:

Mainland Region Vietnam, Cambodia, Laos, Thailand, Myanmar (Burma)

Insular Region Malaysia, Singapore, Indonesia, East Timor, Brunei, Philippines

MAJOR CITIES OF THE REALM

City	Population* (in millions)
Bangkok, Thailand	7.8
Hanoi, Vietnam	4.0
Ho Chi Minh City (Saigon), Vietnam	8.3
Jakarta, Indonesia	12.2
Kuala Lumpur, Malaysia	1.5
Manila, Philippines	11.7
Phnom Penh, Cambodia	1.2
Singapore, Singapore	3.7
Viangchan, Laos	0.8
Yangon, Myanmar (Burma)	4.5

*Based on 2002 estimates.

Note, however, that the realm boundary excludes the Indonesian zone of New Guinea (Irian Jaya), which is part of the Pacific geographic realm.

MAINLAND SOUTHEAST ASIA

Five countries form the mainland region of Southeast Asia: two of them protruded, one compact, one elongated, and one landlocked. Two colonial powers, buffered by Thailand, shaped its modern historical geography. One religion, Buddhism, dominates cultural landscapes, but this is a multicultural, multi-ethnic region. Although one of the least urbanized regions in the world, it contains several major cities. And as Figure 10-2 shows, two countries (Vietnam and Myanmar) possess more than one core area each. We approach the region from the east.

ELONGATED VIETNAM

Consider this: when the Indochina War ended in 1975, Vietnam had less than half the population it has today. In 2002, Vietnam's population exceeded 80 million, about 60 percent of it under 21 years of age. For the great majority of Vietnamese, therefore, the terrible war of the 1960s and 1970s is history, not memory. What concerns most Vietnamese today is the need to overcome two decades of isolation, to reconnect to the world at large, and to join in the

FIGURE 10-8

economic boom on the Pacific Rim. In 1995, the United States moved to normalize relations with Hanoi, opening an embassy and authorizing interactions long prohibited in the aftermath of the conflict. Five years later, this was followed by a successful state visit by President Clinton.

Travel up the crowded road from the northern port of Haiphong to the capital, Hanoi, or sail up the river to the southern metropolis officially known as Ho Chi Minh City—but called *Saigon* by almost everyone there—and you are quickly reminded of the cultural effects of Viet-

Domino Theory

During the Indochina War (1964–1975), it was U.S. policy to contain communist expansion by supporting the efforts of the government of South Vietnam to defeat communist insurgents. Soon, the war engulfed North Vietnam as U.S. bombers attacked targets north of the border between North and South. In the later phases of the war, conflict spilled over into Laos and Cambodia. U.S. warplanes also used bases in Thailand. Like dominoes, one country after another fell to the ravages of the war or was threatened by it.

Some scholars warned that this domino effect could eventually affect not only Thailand but also Malaysia, Indonesia, and Burma (today Myanmar): the whole Southeast Asian realm, they predicted, could be destabilized. But, as we know, that did not happen. The war remained confined to Indochina. And the domino "theory" seemed invalid.

But is the theory totally without merit? Unfortunately, some political geographers to this day mistakenly define this idea in terms of communist activity. Communist insurgency, however, is only one way a country may be destabilized (Peru came close to collapse from it). But right-wing rebellion (Nicaragua's Contras), ethnic conflict (Bosnia), religious extremism (Algeria), and even economic and environmental causes can create havoc in a country. Properly defined, the **domino theory** holds that destabilization from any cause in one country can result in the collapse of order in a neighboring country, triggering a chain of events that can affect a series of contiguous states in turn.

A recent instance of the domino effect comes from Equatorial Africa, where the collapse of order in Rwanda spilled over into The Congo (then Zaïre) and affected Burundi, Congo, and Angola.

nam's elongation (Fig. 10-8). The French colonizers delimited Vietnam as a 1200-mile (2000-km) strip of land extending from the Chinese border in the north to the tip of the Mekong Delta in the south. Substantially smaller than California, this coastal belt was the domain of the Vietnamese (Fig. 10-3). The French recognized that Vietnam, whose average width is under 150 miles (240 km), was not a homogeneous colony, so they divided it into three units: (1) Tonkin, land of the Red River Delta and centered on Hanoi in the north; (2) Cochin China, region of the Mekong Delta and centered on Saigon in the south; and (3) Annam, focused on the ancient city of Hué, in the middle (Fig. 10-4). Today, the Vietnamese prefer to use *Bac Bo*, *Nam Bo*, and *Trung Bo* to designate these areas.

The Vietnamese (or Annamese, also Annamites, after their cultural heartland) speak the same language, although the northerners can easily be distinguished from southerners by their accent. As elsewhere in their colonial empire, the French made their language the *lingua franca* of Indochina, but their tenure was cut short by the Japanese, who invaded Vietnam in 1940. During the Japanese occupation, Vietnamese nationalism became a powerful force, and after the Japanese defeat in 1945, the French could not regain control. In 1954, the French suffered a disastrous final trouncing on the battlefield at Dien Bien Phu in the northwest and were ousted from the country.

But even after its forces routed the colonizers, Vietnam did not become a unified state. Separate regimes took control: a communist one in Hanoi and a noncommunist counterpart in Saigon. Vietnam's pronounced elongation had made things difficult for the French; now it played its role during the postcolonial period. Note, in Figure 10-8, that Vietnam is widest in the north and south, with a narrow "waist" in its middle zone. North and South Vietnam were worlds apart.

Many Americans still remember how the United States became involved in the inevitable conflict between communists and noncommunists in Vietnam during the 1960s and 1970s. At first, American military advisors were sent to Saigon to help the shaky regime there cope with communist insurgents. When the tide turned against the South, the United States committed military forces and equipment to halt the further spread of communism (see box titled "Domino Theory"); at one time, more than 500,000 U.S. soldiers were in Vietnam.

The conduct of the war, its mounting casualties, and its apparent futility created severe social tensions in the United States. If you had been a student during that period, your college experience would have been radically different from what it is today. Protest rallies, "teach-ins," anti-draft demonstrations, "sit-ins," marches, strikes, and even hostage-taking disrupted campus life. The Indochina War threatened the stability of American society. It drove an American president (Lyndon Johnson) from office in 1968 and destroyed the electoral chances of his vice president (Hubert Humphrey), who would not disavow the U.S. role in the war.

In 1975, the Saigon government fell and the United States was ousted—just a little over two decades after the French defeat at Dien Bien Phu. When the last helicopter left from the roof of the U.S. embassy in Saigon amid scenes of desperation and desertion, it marked the end of a sequence of events that closely conformed to the *insurgent state model* we introduced in the context of contemporary Colombia in Chapter 5 (pp. 265-266).

FROM THE FIELD NOTES
"After a four-hour, stop-and-go journey from the coast we arrived in Hanoi, Vietnam's capital. After experiencing Saigon's bustle and energy, we felt as though we were in a city where time had stood still. Here in the city center you could find little evidence of Vietnam's new economic era, no new, tall buildings as in Saigon, no significant signs of modernization. Hanoi's disadvantageous relative location, its weak surface links beyond its immediate hinterland, and the communist dogma of its administration combine to explain why Saigon is forging ahead while the capital, less than half Saigon's size, lags . . . As for the skyline of Saigon (Ho Chi Minh City), it continues to change as new commercial buildings steadily rise above the CBD, propelled by the forces shaping Pacific Rim-related modernization."

Vietnam Today

After the war's end, the Soviets, not the neighboring Chinese, became Vietnam's patrons. As Vietnam was finally unified in 1976 under a dogmatic communist regime, renewed conflict broke out, this time involving China (boundary issues in the north were the ostensible cause). But there was more to this situation. The Soviets exploited the historic distrust between Chinese and Vietnamese because Moscow was itself at odds with Beijing, and so life for Vietnam's Chinese minorities became much more difficult. Many Chinese (the number will never be known) joined in the tragic and disastrous exodus of boat people who sailed from Vietnam's coasts in small, often unseaworthy boats. Of the estimated 2 million of these refugees, more than half perished from storms, exposure, pirates, starvation, and sinkings. Most of those who survived were brought to the United States.

While accepting hundreds of thousands of Vietnamese refugees, the United States simultaneously placed an embargo on Vietnam, which isolated the communist country and stifled its economy. When the Soviet Union collapsed in 1991, Vietnam was in a desperate position—with one saving grace. With two major river deltas and plenty of fertile farmland, Vietnam can produce large harvests of rice, enough to feed its own population and export to foreign markets. Therefore, Vietnam could at least survive its boycott.

At the same time, Vietnam could do little to maintain, let alone improve, what was left of its infrastructure. In the late 1990s, makeshift one-lane bridges still slowed traffic on the congested road between the capital, Hanoi, and the port of Haiphong, a 60-mile (100-km) journey that takes four hours on a good day (Fig. 10-8). Dilapidated buildings, potholed roads, torn-up sidewalks, and malfunctioning utility systems and telephone lines bedevil the capital, where the first signs of modernization are just appearing. With 4 million residents, Hanoi anchors the northern (Tonkin) core area of Vietnam, the basin of the Red River (its agricultural hinterland). In the paddies, irrigation water is still raised by bucket. On the roads, goods move by human- or animal-drawn cart. As yet little has changed here—except people's expectations.

Elongated Vietnam's southern headquarters, Saigon-Cholon (officially named Ho Chi Minh City), is far ahead of the capital. More than 8 million people inhabit this urban agglomeration on the Saigon River, which contains about 10 percent of the country's entire population—and a far larger share of its best-educated and capable people. Saigon-Cholon (Cholon is the city's Chinatown, much diminished after the war but now reviving) is changing rapidly as high-rises built by foreign investors tower over the still-colonial townscape, modern hotels are opening, and a large Special Economic Zone has been laid out just downstream from the port. Unlike Hanoi, Saigon can be

Among the Realm's Great Cities . . .

Saigon

Officially, it is known as Ho Chi Minh City, renamed in 1975 after the ouster of American forces, but only bureaucrats and atlases call it that. To the people here, it is still Saigon, city on the Saigon River, served by, it seems, a thousand companies using the name. Saigon Taxi, Saigon Hotel, Saigon Pharmacy—there is hardly a city block without the name Saigon on some billboard.

The phenomenon is significant. Hanoi's northerners ordained the change, but this still is South Vietnam, and many southerners see the North as peopled by slow, dogmatic communists obstructing the hopes and plans for their vibrant city. In truth, Saigon may be at least twice as large as Hanoi, but like the capital it suffers from decades of deterioration, a broken-down infrastructure, and undependable public services. When, in the early 1990s, businesspeople could find no modern accommodations or working telephones, an Australian firm built an eight-story floating hotel with its own power plant, towed it up the Saigon River, and parked it on the downtown waterfront.

Therein lies Saigon's great advantage: large oceangoing vessels can reach the city in a three-hour sail up the meandering river. The Chinese developed its potential: in the 1960s, Chinese merchants controlled more than half of the South's exports and nearly all of its textile factories and foreign exchange. Saigon-Cholon (Cholon means "great market") was a Chinese success story, but after the war the Hanoi regime nationalized all Chinese-owned firms. Tens of thousands of middle-class Chinese left the city, its economy devastated.

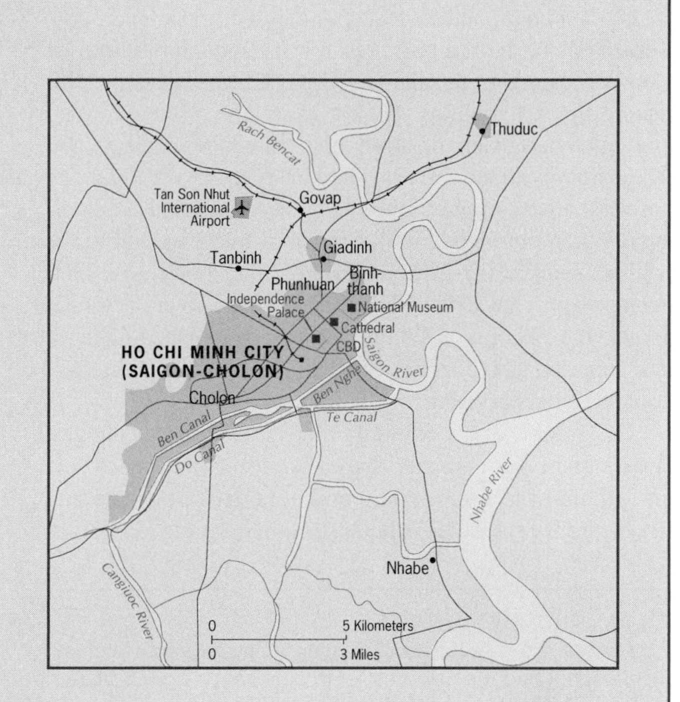

Today, Saigon (population: 8.3 million) is fast recovering. A modern skyline is rising above the colonial one. A large Special Economic Zone is being created south of the city. Factories are being built. The Pacific Rim era is beginning. Will the South rise again?

reached by oceangoing vessels, and it lies north of, rather than within, the Mekong Delta that is part of its hinterland. The city's streets are choked with bicycles, mopeds, handcarts, buses, taxis, and even a few private cars; consumer goods are everywhere, and the city throbs with commercial activity. Saigon, too, has serious infrastructural problems, but it is recovering from decades of neglect. One great need is a bridge to link the right bank of the Saigon River, on which downtown and most of Saigon lie, to the left bank, which remains merely a village.

A sense of distance and remoteness pervades north as well as south in Vietnam. To northerners, Saigon and the south are still symbols of the divisions that tore Vietnam apart during the war. To southerners, the bureaucrats from the north who run much of Saigon are a burden, obstructing the rush toward modernization. This sense of alienation is reinforced by the actual as well as the functional distance from Saigon to Hanoi. Imagine a country where the road trip from the capital to the largest city can take as long as a week!

By the opening of the century, Vietnam had not moved ahead economically as rapidly as had been expected, given its desirable Pacific Rim location, highly literate labor force, and earlier growth. Foreign investment has slowed, and plans to privatize state industries have been shelved. Like China, Vietnam remains a communist country, but unlike China it has not had the leadership of a Deng Xiaoping. Hanoi's avowed decision to mix communist ideology with market economics never resulted in far-reaching change: the emphasis remained on communist statism, not on economic reform. As a result, Vietnam is experiencing neither the turmoil seen in China nor the huge regional disparities arising there. During the mid-1990s, when Saigon prospered and Hanoi stagnated, it appeared that North and South were once again growing apart. Today, Saigon's exuberance is restrained by Hanoi's dogmatism—and with it, Vietnam's economic progress.

COMPACT CAMBODIA

Former French Indochina contained two additional entities—Cambodia and Laos. In Cambodia, the French possessed one of the greatest treasures of Hinduism: the city of Angkor, capital of the ancient Khmer Empire, and the temple complex known as Angkor Wat. The Khmer Empire prevailed here from the ninth to the fifteenth centuries, and King Suryavarman II built Angkor Wat to symbolize the universe in accordance with the precepts of Hindu cosmology. When the French took control of this area during the 1860s, Angkor lay in ruins, sacked long before by Vietnamese and Thai armies. The French created a protectorate, began the restoration of the shrines, established Cambodia's permanent boundaries, and restored the monarchy under their supervision.

Geographically, Cambodia enjoys several advantages, most notably its compact shape. Compact states enclose a maximum of territory within a minimum of boundary and are without peninsulas, islands, or other remote extensions of the national spatial framework. Cambodia had the further advantage of strong ethnic and cultural homogeneity: 90 percent of its 12.7 million inhabitants are Khmers, with the remainder equally divided between Vietnamese and Chinese. As Figure 10-8 shows, Southeast Asia's greatest river, the Mekong, enters Cambodia from Laos and crosses it from north to south, creating a great bend before flowing into southern Vietnam (see box titled "The Mighty Mekong"). Phnom Penh, the country's capital, lies on the river. The ancient capital of Angkor lies in the northwest, not far from the Tonle Sap, a major lake linked to, and filled by, the waters of the Mekong. Cambodia has a coastline on the Gulf of Thailand, but its core area lies in the interior; the Mekong is more important than the Gulf. Kompong Som, the port at the end of the railroad from Phnom Penh, has only limited facilities.

FROM THE FIELD NOTES

"Here near Siem Reap, along a small stream flowing into the Tonle Sap, I had the opportunity to observe one of Cambodia's many Mekong-Basin stilt villages. The water rises and falls with the seasons; it is comparatively dry now, and a pool of stagnant water is what is left of the river that will soon form here once again. In the past, stilt houses were built of traditional materials only, but now there are tin roofs and plastic sheeting. Conditions are not sanitary; there was refuse everywhere, waiting to be swept away by the rising waters. Outhouses such as those in the left foreground drained directly into the temporary pond, now mosquito-infested and severely polluted. I wondered how many infants would survive their first year here."

Cambodia's present social geography continues to suffer from the after-effects of the Indochina War. Neither its compactness nor its isolation could overcome the disadvantage of relative location near a conflict that was bound to spill across its borders (Fig. 10-8). Internal disharmony fueled by external interference led in 1970 to the ouster of the last king by the military; in 1975 that regime was itself overthrown by communist revolutionaries, the so-called Khmer Rouge. These new rulers embarked on a course of terror and destruction in order to reconstruct *Kampuchea* (as they called Cambodia) as a rural society. They drove townspeople into the countryside where they had no place to live or work, emptied hospitals and sent the sick and dying into the streets, outlawed religion and family, and in the process killed as many as 2 million Cambodians (out of a population of 8 million). In the late 1970s, Vietnam, having won its own war, invaded Cambodia to drive the Khmer Rouge away. But this action led to new terror, and a stream of refugees crossed the border into Thailand. Eventually, remnants of the Khmer Rouge managed to establish a base in the northwest of Cambodia, where their murderous leader, Pol Pot, who was never brought to account for his actions, committed suicide in 1998.

Cambodia's postwar trauma continues today. Once self-sufficient and able to feed others, it now must import food. Rice and beans are the subsistence crops, but the dislocation in the farmlands set production back severely, and continuing strife in the countryside has disrupted supply routes for years.

Cambodians hope that tourism, focused chiefly on the great temples at Angkor, will become a mainstay of the economy. But political stability is a prerequisite for this industry, and it has not yet returned to this ancient kingdom.

LANDLOCKED LAOS

North of Cambodia lies Southeast Asia's only landlocked country, Laos (Fig. 10-8). Interior and isolated, Laos changed

The Mighty Mekong

From its source among the snowy peaks of China's Qinghai and Xizang, the Mekong River rushes and flows some 2600 miles (4200 km) to its delta in southernmost Vietnam. This "Danube of Southeast Asia" crosses or borders five of the realm's countries (see photo, p. 504), supporting rice farmers and fishing people, forming a transport route where roads are few, and providing electricity from dams upstream. Tens of millions of people depend on the waters of the Mekong, from subsistence farmers in Laos to apartment dwellers in China. The Mekong Delta in southern Vietnam is one of the realm's most densely populated areas and produces enormous harvests of rice.

But problems loom. China has built two dams across the Lancang (as the Mekong is called there) to supply Yunnan Province with electricity. While such hydroelectric dams should not interfere with water flow, countries downstream worry that a severe dry spell in the interior would impel the Chinese to slow the river's flow to keep the reservoirs full. Cambodia is concerned over the future of the Tonle Sap, a large natural lake filled by the Mekong (Fig. 10-8). In Vietnam, farmers worry about salt water invading the paddies should the Mekong's level drop. And the Chinese may not be the only dam builders in the future: Thailand has expressed an interest in building a dam on the Thai-Laos border where it is defined by the Mekong.

In such situations, the upstream states have an advantage over those downstream. Several international organizations have been formed to coordinate development in the Mekong Basin, including the Mekong River Commission (MRC) founded half a century ago. China has offered to sell electricity from its dams to Thailand, Laos, and Myanmar. Coordinated efforts to reduce logging in the Mekong's drainage basin have had some effect. After consultations with the MRC, Australia built a bridge linking Laos and Thailand. There is even a plan to make the Mekong navigable from Yunnan to the coast, creating an alternative outlet for interior China.

Sail the Mekong today, however, and you are struck by the slowness of development along this artery. Wooden boats, thatch-roofed villages, and teeming paddies mark a river still crossed by antiquated ferries and flanked by few towns. Of modern infrastructure, one sees little. And yet the Mekong and its basin form the lifeline of mainland Southeast Asia's dominantly rural societies.

little during 60 years of French colonial administration (1893–1953). Then, along with other French-ruled areas, it became an independent state. Soon this well-entrenched, traditional kingdom fell victim to rivalries between traditionalists and the victorious communists, and the old order collapsed.

Laos has no fewer than five neighbors, one of which is the East Asian giant, China. The Mekong River forms a long stretch of its western boundary, and the important sensitive border with Vietnam to the east lies in mountainous terrain. With 5.5 million people (about half of them ethnic Lao, related to the Thai of Thailand), Laos lies surrounded by comparatively powerful states. The country has no railroads, just a few miles of paved roads, and very little industry; it is only 17 percent urbanized (the capital, Viangchan, lies on the Mekong and has an oil pipeline to Vietnam's coast). Laos remains the region's poorest and most vulnerable entity.

FROM THE FIELD NOTES
"Along a road in Xiangkhoang Province, in north-central Laos, I stopped to watch these people operate their gas-powered rice mill. I had some difficulty communicating with the people, but from the young boy I gathered that some of this equipment had been fashioned from U.S. war materiel, including the drum (I wondered what it might have contained) and the funnel, which was made from bomb casings."

PROTRUDED THAILAND

In virtually every way, Thailand is the leading state of the region under discussion. Its per-capita GNP during the economic boom of the 1990s was higher than that of Vietnam,

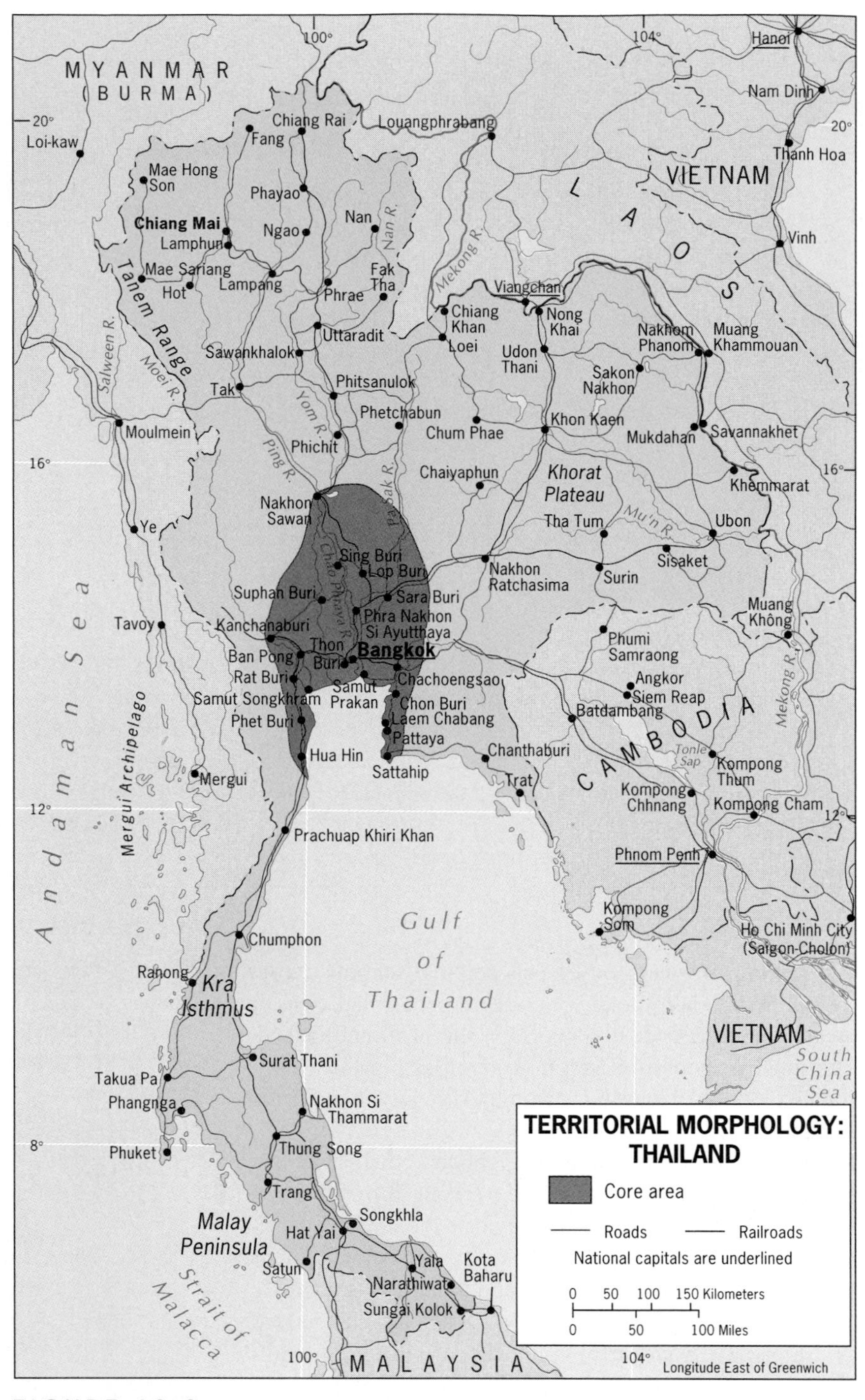

FIGURE 10-9

Cambodia, Laos, and Myanmar combined. Alone among its neighbors, Thailand was a strong participant in the Pacific Rim's economic development. Thailand's capital, Bangkok, is one of the two largest urban centers in the region and one of the world's most prominent primate cities. The country's population, 63.1 million in 2002, is growing at just about the slowest rate in the entire realm, almost equivalent to that of fully urbanized Singapore. Over the past few decades, only political instability and uncertainty have inhibited economic progress. Thailand is a constitutional monarchy; its moves toward stable democracy have been thwarted by graft, corruption, and, sometimes, violent confrontation.

Thailand is the textbook example of a protruded state. From a relatively compact heartland, in which lie the core area, capital, and major areas of productive capacity, a 600-mile (1000-km) corridor of land, in places less than 20 miles (32 km) wide, extends southward to the border with

Malaysia (Fig. 10-9). The boundary that defines this protrusion runs down the length of the Malay Peninsula to the Kra Isthmus, where neighboring Myanmar peters out and Thailand fronts the Andaman Sea (an arm of the Indian Ocean) as well as the Gulf of Thailand. In the entire country, no place lies farther from the capital than the southern end of this tenuous protrusion.

As Figure 10-2 shows, Thailand occupies the heart of the mainland region of Southeast Asia. While Thailand has no Red, Mekong, or Irrawaddy Delta, its central lowland is watered by a set of streams that flow off the northern highlands and the Khorat Plateau in the east. One of these streams, the Chao Phraya, is the Rhine of Thailand. From the head of the Gulf of Thailand to Nakhon Sawan, this river is a highway of traffic. Barge trains loaded with rice head for the coast, ferry boats head upstream, freighters transport tin and tungsten (of which Thailand is among the world's largest producers). Bangkok sprawls on both sides of the lower Chao Phraya, here flanked by skyscrapers, pagodas, factories, boatsheds, ferry landings, luxury hotels, and modest dwellings in crowded confusion. On the right bank of the Chao Phraya, Bangkok's west side, lie the city's remaining *klong* (canal) neighborhoods, where waterways and boats form the transport system. Bangkok still is known as the Venice of Asia, although many *klongs* have been filled in and paved over to serve as roadways.

During the Pacific Rim boom of the 1990s, Thailand's economy grew rapidly. Relative location, natural environment, and social conditions all contributed to this growth. Situated at the head of the Gulf of Thailand, the country's core area opens to the South China Sea and hence the Pacific Ocean. The Gulf itself has long been a rich fishing ground and more recently a source of oil (the major refinery lies in the port area of Laem Chabang, an industrial zone just to the north of the tourist resort of Pattaya). But the key to Thailand's economic explosion was the Thai workforce, laboring under often dreadful conditions for low wages to produce goods sold on foreign markets at cheap prices. This labor force attracted major foreign investment that, in turn, stimulated service industries.

For a time, Thailand became a Pacific Rim economic tiger, producing goods ranging from plastic toys to "Jap-

Among the Realm's Great Cities . . .

Bangkok

Bangkok (7.8 million), mainland Southeast Asia's second largest city, is a massive sprawling metropolis on the banks of the Chao Phraya River, an urban agglomeration without a center, an aggregation of neighborhoods ranging from the immaculate to the impoverished. In this city of great distances and high tropical temperatures, getting around is often difficult because roadways are choked with traffic. A (diminishing) network of waterways affords the easiest way to travel, and life focuses on the busiest waterway of all, the Chao Phraya. Ferries and water taxis carry tens of thousands of commuters and shoppers across and along this bustling artery, flanked by a growing number of high-rise office buildings, luxury hotels, and ultramodern condominiums. Many of these modern structures reflect the Thais' fondness for domes, columns, and small-paned windows, creating a skyline unique in Asia.

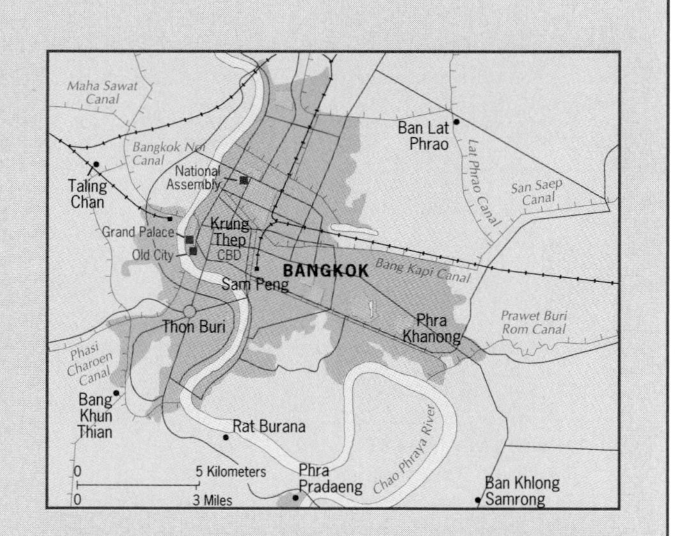

Gold is Buddhist Thailand's symbol, adorning religious and nonreligious architecture alike. From a high vantage point in the city, you can see hundreds of golden spires, pagodas, and façades rising above the townscape. The Grand Palace, where royal, religious, and public buildings are crowded inside a white, crenellated wall more than a mile long, is a gold-laminated city within a city embellished by ornate gateways, monsters, dragons, and statuary. Across the mall in front of the Grand Palace lie government buildings sometimes targeted by rioters in Bangkok's volatile political atmosphere. Not far away are Chinatown and myriad markets; Bangkok has a throbbing commercial life.

Decentralized, dispersed Bangkok has major environmental problems, ranging from sinking ground (resulting in part from excessive water pumping) to some of the world's worst air pollution. Yet industries, port facilities, and residential areas continue to expand as the city's hinterland grows and prospers with Pacific Rim investment. Bangkok is the pulse of Southeast Asia's mainland region.

anese" cars. But while the workers toiled, government and financial mismanagement planted seeds of ruin. In 1997, the country's currency, the *baht*, began a slide in value that undermined the economy and led to numerous bank failures. Exports and worker productivity declined, real estate values plunged, and Thailand became the first casualty in a series of economic setbacks that would eventually affect the entire western Pacific Rim.

Although Thailand and the other Pacific Rim economies suffered severe downturns, we should remember that the boom period had yielded many infrastructural improvements not only in the core area but also in distant areas such as the north, centered on Chiang Mai. Thailand, like other Pacific Rim economies, has improved highways, ports, power systems, airports, and housing. While a period of poor maintenance followed the events of 1997, Thailand's fundamental assets remained to serve the country well when the world economy revived.

FROM THE FIELD NOTES

"Thailand exhibits one of the world's most distinctive cultural landscapes, both in its urban and its rural areas. Graceful pagodas, stupas, and spirit houses of Buddhism adorn city and countryside alike. Gold-layered spires rise above, and beautify cities and towns that would otherwise be drab and featureless. The architecture of Buddhism has diffused into public architecture, so that many secular buildings, from businesses to skyscrapers, are embellished by something approaching a national style. The grandest expression of the form undoubtedly is a magnificent assemblage of structures within the walls of the Grand Palace in the capital, Bangkok, of which a small sample is shown here. Climb onto the roof of any tall building in the city's center, however, and the urban scene will display hundreds of such graceful structures, interspersed with the modern and the traditional townscape."

Thailand's natural environments and cultural landscapes continue to attract millions of visitors annually. The magnificent architecture of Buddhism (the country's old name, *Siam*, was a Chinese word meaning golden-yellow), swathed in gold and shining in the tropical sun, is without equal, as the photo on this page underscores. The warm coasts and beaches have long made Thailand a favorite destination for tourists from Germany to Japan. One of the world's most famous resorts is Phuket, on the distant southwest coast near the end of Thailand's Malay Peninsula protrusion.

Tourism is Thailand's leading source of foreign revenues, but there is a dark side to the industry. The Thai people's relaxed attitude toward sex has made Thai tourism in large part a sex industry, candidly advertised in such foreign markets as Japan where "sex tours" attract planeloads of male participants. When the AIDS pandemic reached Thailand in the 1980s it changed no habits, and by the mid-1990s the country was suffering one of the world's worst outbreaks of the disease. At the turn of the century, more than 1 million of Thailand's population of 60-plus million were infected.

Thailand's territorial morphology creates both problems and opportunities. Problems include the effective integration of the national territory through surface communications, which must cover long distances to reach remote locales; the influx of refugees from neighboring countries dislocated by internal conflict; and control over contraband (opium) drug production and trade in the interior "Golden Triangle" where Myanmar, Thailand, and Laos meet. The government's efforts to speed the development of rural areas, notably the northeast, through megaprojects such as large dams have disrupted local life, damaged ecologies, and deepened the split between city and countryside. On the other hand, Thailand's lengthy Kra Isthmus creates economic opportunities where southernmost Thailand lies close to Malaysia and Indonesia, raising hopes for a three-country development scheme in that area. Given political stability, Thailand's future as a regional leader appears secure.

EXTENDED MYANMAR

Myanmar (formerly Burma) shares with Thailand its protruded morphology but little else. Long languishing under one of the world's most corrupt military regimes, Myanmar is one of the planet's poorest countries where, it seems, time has stood still for centuries. Whereas air, railroad, and highway routes serve Thailand's peninsular protrusion, there is not even an all-weather track to the southern end of Myanmar's. Whereas busy shipping lanes converge on Thailand's coastal core area, Myanmar's Irrawaddy Delta is more reminiscent of Bangladesh.

Myanmar's territorial morphology is complicated by a shift in the Burmese core area that took place during colonial times. Prior to the colonial period, the focus of embryonic Burma lay in the so-called dry zone between the Arakan Mountains and the Shan Plateau. The urban focus of the state was Mandalay, which had a central situation and relative proximity to the non-Burmese highlands all around (Fig. 10-1). Then the British developed the agricultural potential of the Irrawaddy Delta, and Rangoon (now called Yangon) became the hub of the colony. The Irrawaddy waterway links the old and the new core areas, but the center of gravity has shifted to the south.

The political geography of Myanmar constitutes a particularly good example of the role and effect of state morphology on internal state structure. Not only is Myanmar a protruded state; in addition, its core area is surrounded on the west, north, and east by a horseshoe of great mountains—where many of the country's 11 minority peoples had their homelands before the British occupation. The colonial boundaries had the effect of incorporating these peoples into the domain of the Burman people, who constitute about two-thirds of the population (50.9 million today). When the British departed and Burma became independent in 1948, the less numerous peoples traded one master for another. In 1976, nine of these indigenous peoples formed a union to demand the right to self-determination in their homelands.

As Figure 10-3 shows, the peripheral peoples of Myanmar occupy a significant part of the state. The Shan of the northeast and far north, who are related to the neighboring Thai, account for about 9 percent of the population, or 4.6 million. The Karen (7 percent, 3.6 million) live in the neck of Myanmar's protrusion and have proclaimed that they wish to create an autonomous territory within a federal Myanmar. Although the powerful military have dealt these aspirations a series of setbacks, centrifugal forces continue to bedevil the central regime. Its response has been to exert power by all available means rather than to accommodate these forces, even to the point of stifling political discourse (let alone opposition) among the Burman themselves.

This is all the more tragic because Myanmar has economic potential far superior to that of Bangladesh, its neighbor to the west, and to other countries in its own region. It can feed itself, and it could export significant quantities of rice. Varied soils and diverse environments can yield a range of other crops. Tin and other metals have been found in the highlands. Today, Myanmar is exploiting the forests of the interior at an alarming rate; teak has risen to first place among exports by value—among legitimate exports, that is. In its corner of the "Golden Triangle," Myanmar also is the world's largest illicit producer of opium poppies and harvests other illegal crops for the drug trade as well. For a time, the regime in Yangon tried to control the trade, but its constant conflict with peoples outside the core area diverted its attentions and energies from that campaign.

For decades, the repressive policies of a brutal military regime have kept Myanmar in turmoil and all but halted its development. Infrastructure here is in truly dismal condition (in 1993 there was one working telephone in Myanmar capable of making and receiving overseas calls in Yangon, the capital). While more than 80 percent of the adult population is literate, there are very few jobs to employ their talents. Around the turn of the twenty-first century signs of change were in evidence, including the release of some political dissidents and the beginnings of tourism. But a change of heart by the regime will not soon undo the effects of policies that have pushed impoverished and exhausted Myanmar into the ranks of the world's poorest economies.

INSULAR SOUTHEAST ASIA

On the peninsulas and islands of Southeast Asia's southern and eastern periphery lie six of the realm's 11 states (Fig. 10-2). Few regions in the world contain so diverse a set of countries. Malaysia, the former British colony, consists of two major areas separated by hundreds of miles of South China Sea. The realm's southernmost state, Indonesia, sprawls across thousands of islands from Sumatera in the west to New Guinea in the east. North of the Indonesian archipelago lies the Philippines, a nation that once was a U.S. colony. These are three of the most severely fragmented states on Earth, and each has faced the challenges that such politico-spatial division brings. This insular region of Southeast Asia also contains two small but important sovereign entities: a city-state and a sultanate. The city-state is Singapore, once a part of Malaysia (and one instance in which internal centrifugal forces were too great to be overcome). The sultanate is Brunei, an oil-rich Muslim territory on the island of Borneo that seems transplanted from the Persian Gulf. In addition, a third small entity, East Timor, is in the process of achieving statehood. Few parts of the world are more varied or interesting geographically.

MAINLAND-ISLAND MALAYSIA

The state of Malaysia represents one of the three types of fragmented states discussed earlier: the mainland-island type, in which one part of the national territory lies on a continent and the other on an island. Malaysia is a colonial political artifice that combines two quite disparate components into a single state: the southern end of the Malay Peninsula and the northern part of the island of Borneo. Respectively, these are known as West Malaysia and East

Malaysia (Fig. 10-2). The name *Malaysia* came into use in 1963, when the original Federation of Malaya, on the Malay Peninsula, was expanded to incorporate the areas of Sarawak and Sabah in Borneo. When the name Malaya is used, it refers to the peninsular part of the Federation, whereas Malaysia refers to the total entity.

The Malays of the peninsula, traditionally a rural people, displaced older aboriginal communities there and today make up about 58 percent of the country's population of 24.2 million. They possess a strong cultural identity expressed in adherence to the Muslim faith, a common language, and a sense of territoriality that arises from their perceived Malayan origins and their collective view of Chinese, Indian, European, and other foreign intruders.

The Chinese came to the Malay Peninsula and to northern Borneo in substantial numbers during the colonial period, and today they constitute about one-fourth of Malaysia's population (they are the largest single group in Sarawak). During World War II, when the Japanese ruled Malaya, the Chinese were ruthlessly persecuted; many were driven into

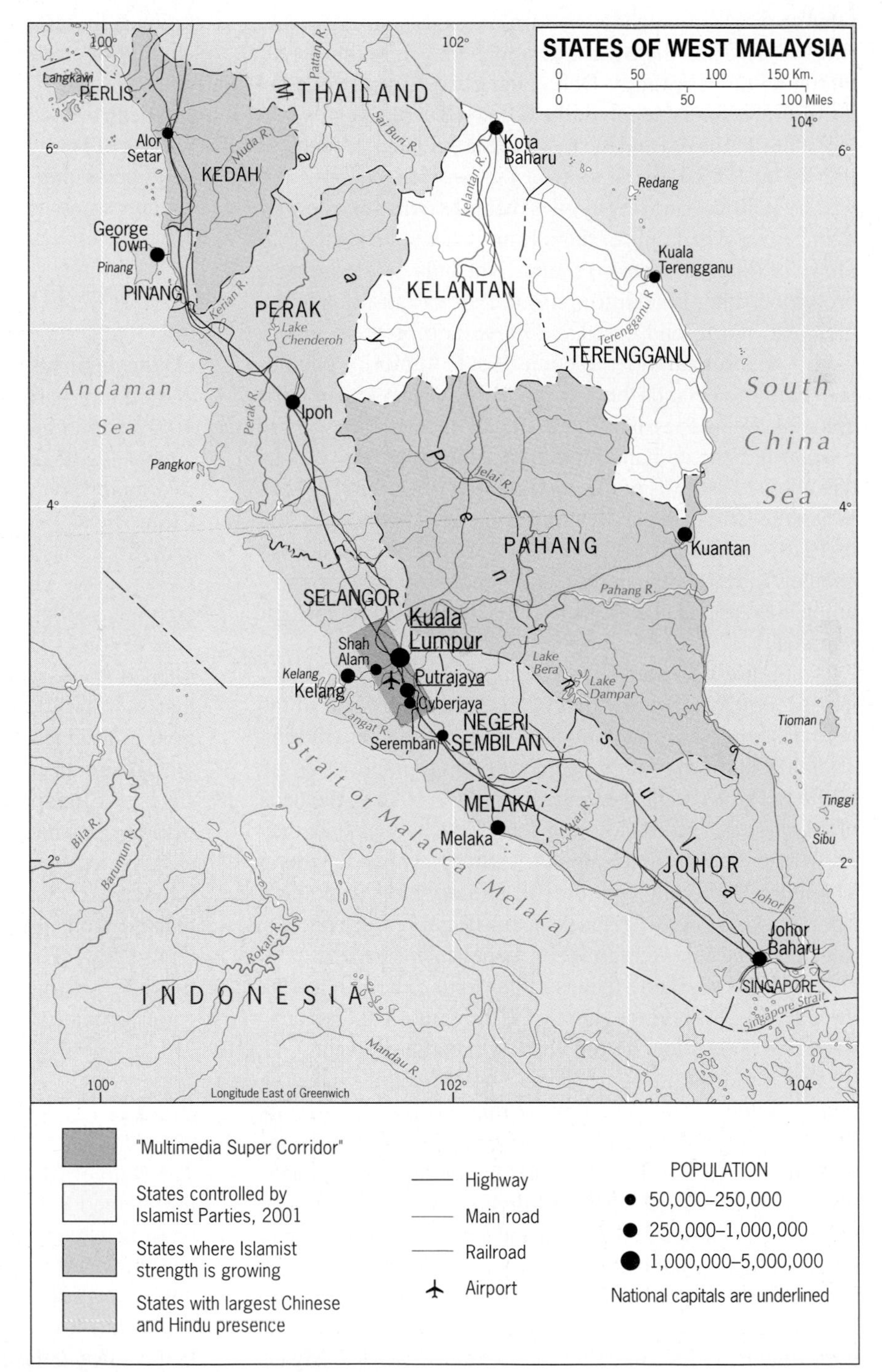

FIGURE 10-10

the forested interior where they founded a communist-inspired resistance movement that long destabilized the area. Today, the commercially active, urbanized Chinese sector still is victimized by discrimination, now by the dominant Malays.

Hindu South Asians were in this area long before the Europeans, and for that matter before the Arabs and Islam arrived on these shores. Today they still form a substantial minority of over 7 percent of the population, clustered, like the Chinese, on the western side of the peninsula (Fig. 10-3).

The populous peninsular part of Malaysia remains the country's dominant sector with 11 of its 13 States and nearly 80 percent of its population. Here the Malay-dominated government has strictly controlled economic and social policies while pushing the country's modernization. During the Asian economic boom of the 1990s, Malaysia's planners embraced the notion of symbols: the (then) capital, Kuala Lumpur, was endowed with the world's tallest building; a space-age airport outpaced Malaysia's needs; a high-tech administrative capital was built at Putrajaya, and a nearby development was called Cyberjaya—all part of a so-called *Multimedia Super Corridor* to anchor Malaysia's core area (Fig. 10-10).

The chief architect of this program was Malaysia's autocratic leader, Dr. Mahathir Mohamad. Not only did he have the support of a majority of the peninsula's Malays, but also his economic policies garnered the endorsement of most of Malaysia's ethnic Chinese. But Malaysia's headlong rush to modernize, coupled with some personal issues involving Mahathir and his rivals, caused a backlash among more conservative Muslims in the north, where an Islamic fundamentalist party made headway. By 2001, two northern States, tin-producing Kelantan and energy-rich but socially poor Terengganu, had Islamist State governments, and the fundamentalists' appeal in neighboring States was growing (Fig. 10-10).

The Malay Peninsula's primacy began long ago, during colonial times, when the British created a substantial economy based on rubber plantations, palm-oil extraction, and mining (tin, bauxite, copper, iron). The Strait of Malacca (Melaka) became one of the world's busiest and most strategic waterways, and Singapore, at the southern end of it, a prized possession (Singapore seceded from the Malaysian Federation in 1965).

Greater Malaysia survived decolonization, the loss of Singapore, and ethnic troubles; after 1980, its economy took off and became a major force on the Pacific Rim. Foreign manufacturers, attracted by the skills and modest wages of the Malaysian workforce, established hundreds of factories here (see box titled "Pinang: A Future Singapore?"). While the old colonial exports still rank high on the list, electronics have made their appearance in a big way. One benefit of the merger with East Malaysia was that sector's production of valuable timber and petroleum. Kuala Lumpur began to resemble Singapore, becoming a forest of high-rises. And Malaysia's GNP rose to third place in the realm, after all-urban Singapore and oil-rich Brunei, and well ahead of Thailand.

FROM THE FIELD NOTES

"Multinational, cosmopolitan Port Kelang, Malaysia made for an interesting field day. One busy side street had a row of restaurants with Chinese, Italian, Malaysian, and Vietnamese menus. Rugby was being played on one field; cricket on another. When I was here in 1995 the area was booming, but in 1999 there were signs of trouble resulting from the general economic downturn of the late 1990s. Having seen several sights of the kind shown in this photograph, I stopped to ask the lone supervisor here why there seemed to be no activity in this large project; building appeared to have come to a stop. 'This company built a whole lot of townhouse estates,' he said. 'People were upwardly-mobile and buying them as fast as we could build them. Then the crisis came and for many the money ran out.' In truth, Malaysia weathered the storm better than other Southeast Asian countries, but even here the impact was noticeable on the landscape."

Territorially, **East Malaysia** makes up 60 percent of the country, but Sabah and Sarawak contain only about 20 percent of Malaysia's population. In terms of development in the East, Sabah, also the more populous, is ahead of remote Sarawak (Fig. 10-6, upper-left map). Kota Kinabalu, Sabah's capital, is the most modern city in all of Borneo (not just the Malaysian part), with a substantial port handling timber, a large tourist-industry infrastructure including a modern airport, and natural gas reserves in production to the north.

The ethnic complexity of Sarawak and Sabah (East Malaysia) defies summation. Ethnic Chinese are the single largest group in Sarawak, but each State counts more than two dozen ethnic identities. As a result, politics in East Malaysia are fractious and often combative, leading to repeated intervention by the federal government. That government, based as it is in West Malaysia, sometimes is accused by Easterners of "colonial" exploitation and infringement. The sprawling country's greatest challenge continues to lie in its ethnic and cultural heterogeneity, heightened by its territorial fragmentation.

Pinang: A Future Singapore?

In the northwestern corner of Malaysia's Malay Peninsula, opposite the northern end of Indonesian Sumatera, where the Strait of Malacca (Melaka) opens into the Indian Ocean, lies an island called Pinang (Fig. 10-10). A thriving outpost during the British colonial period, Pinang (capital: George Town), like Singapore, was one of the so-called Straits Settlements.

Today, Singapore no longer is a part of Malaysia, but Pinang's importance is growing rapidly. Now connected to the mainland by one of Asia's longest bridges (9 miles [14 km]), the 100-square-mile island has a population of over 1 million. Chinese outnumber Malays by 60 to 32 percent, and George Town, the old colonial port, has become the focus of an expanding high-technology manufacturing complex for the international computer industry. The 65-story Tun Abdul Razak Complex, housing government offices, businesses, department stores, and recreational facilities, is just one landmark in an ultramodern skyline. From the top you can see modern industrial parks with neon-lit signs proclaiming the presence of such companies as Intel, SONY, Philips, Motorola, and Hitachi.

The rapid development of Pinang is a calculated strategy by the Malaysian government. The emphasis is on building a highly skilled labor force that is locally nurtured in technical schools and training facilities financed by the multinational corporations themselves. Malaysia makes infrastructure investments ranging from the new toll bridge to a major airport. George Town's colonial and multicultural townscape is attracting a growing tourist trade, and new hotels and resorts are enlarging the service industry.

Malaysia hopes that Pinang will become the leading corner of a major growth triangle (a concept that is gaining popularity along the Pacific Rim), together with Indonesian Sumatera to the west and southern Thailand to the north. Meanwhile, Pinang is following in Singapore's economic footsteps. But Kuala Lumpur's politicians surely hope that Chinese-dominated Pinang will not follow Chinese-dominated Singapore's political course

SINGAPORE

In 1965, a fateful event occurred in Southeast Asia: Singapore, crown jewel of British colonialism in this realm, seceded from the Malaysian Federation and became a sovereign state, albeit a ministate (Fig. 10-11). With its magnificent relative location, its human resources, and firm government, Singapore then overcame the limitations of space and the absence of raw materials to become one of the economic tigers on the Pacific Rim.

With a mere 240 square miles (600 sq km) of territory, space is at a premium in Singapore, and this is a constant worry for the government. Singapore's only local spatial advantage over Hong Kong is that its small territory is less fragmented (there are just a few small islands in addition to the compact main island). With a population of 4.1 million and an expanding economy, Singapore must develop space-conserving, high-tech industries. In this effort it has succeeded, and the urban area's high-rise buildings reflect the city-state's prosperity. You will find no slums in Singapore; the dilapidated streets in the old Chinatown were left merely to conserve a fragment of the city's past. Singapore mostly impresses with its newness and its modernity, and with the order of daily life.

As the map shows, Singapore lies where the Strait of Malacca (Melaka), leading westward to India, opens into the South China Sea and the waters of Indonesia (Fig. 10-11 inset). The port developed as part of the British Southeast Asian empire, and when independence came in 1963, it was made a part of the Malaysian Federation. However, two years later Singapore seceded from Malaysia and became a genuine city-state on the Pacific Rim. No reunification, merger, or annexation looms here.

Benefiting from its relative location, the old port of Singapore had become one of the world's busiest (by numbers of ships served) even before independence. It thrived as an ***entrepôt*** between the Malay Peninsula, Southeast Asia, Japan, and other emerging economic powers on the Pacific Rim and beyond. Crude oil from Southeast Asia still is unloaded and refined at Singapore, then shipped to Asian destinations. Raw rubber from the adjacent Peninsula and from Indonesia's island of Sumatera is shipped to Japan, the United States, China, and other countries. Timber from Malaysia, rice, spices, and other foodstuffs are processed and forwarded via Singapore. In return, automobiles, machinery, and equipment are imported into Southeast Asia through Singapore.

17

But that is the old pattern. Singapore's leaders want to redirect the city-state's economy and move toward high-tech industries for the future. In Singapore the government tightly controls business as well as other aspects of life. (Some newspapers and magazines have been banned for criticizing the regime, and there are even fines for such things as eating on the subway and failing to flush a public toilet.) Its overall success after secession has tended to keep the critics quiet: while GNP per capita from 1965 to 2000 multiplied by a factor of more than 15—to over U.S. $30,000—that of neighboring Malaysia reached just U.S.

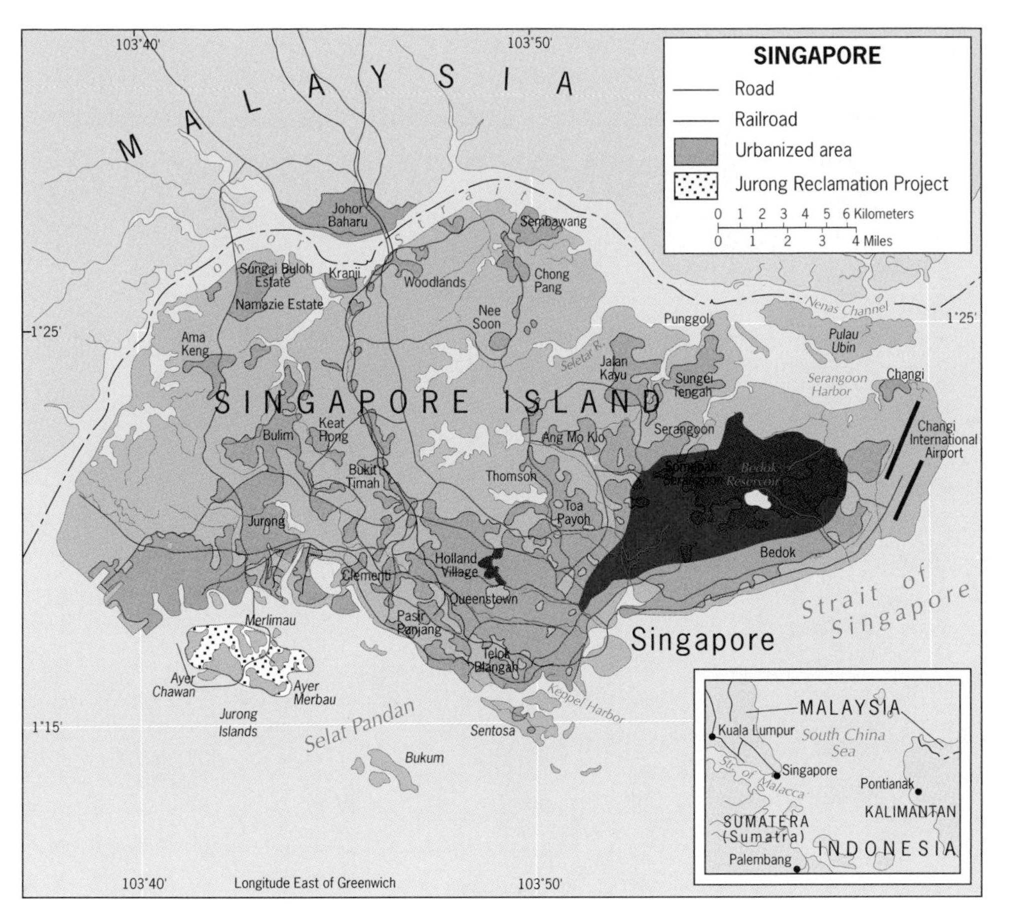

FIGURE 10-11

$3670. Among other things, Singapore became (and for many years remained) the world's largest producer of disk drives for small computers.

As multinational corporations settled in, Singapore's political stability on the volatile Pacific Rim was another advantage. Nonetheless, a problem emerged: unlike Taiwan, where brainpower is plentiful, Singapore's technical capacities were limited. As recently as 1995, only 8 percent of the population was university educated. So while the government reduced the incentives for low-technology manufacturers to come here and wages rose, the high-tech industries found themselves thinking about moving to nearby Malaysia and Thailand. That slowed Singapore's boom and challenged the government to find ways to stimulate a revival.

To accomplish this revival, Singapore has moved in several directions. First, it will focus on three growth areas: information technology, automation, and biotechnology. Second, there are notions of a "Growth Triangle" involving Singapore's developing neighbors, Malaysia and Indonesia; those two countries would supply the raw materials and cheap labor, and Singapore the capital and technical know-

FROM THE FIELD NOTES

"My first visit to Singapore was by air, and I got to know the central city, where I was based, rather well. But the second was by freighter from Hong Kong, and this afforded quite a different perspective. Here was one reason for Singapore's success: I have simply never seen a neater, cleaner, better organized, or more modern port facility. Even piles of loose items were carefully stacked. Loading and offloading went quietly and orderly. Singapore is one of the world's leading *entrepôts*, where goods are brought in, stored, and transshipped. Nobody does it better, and Singapore carefully guards its reputation for dependability, on-schedule loading, and lack of corruption."

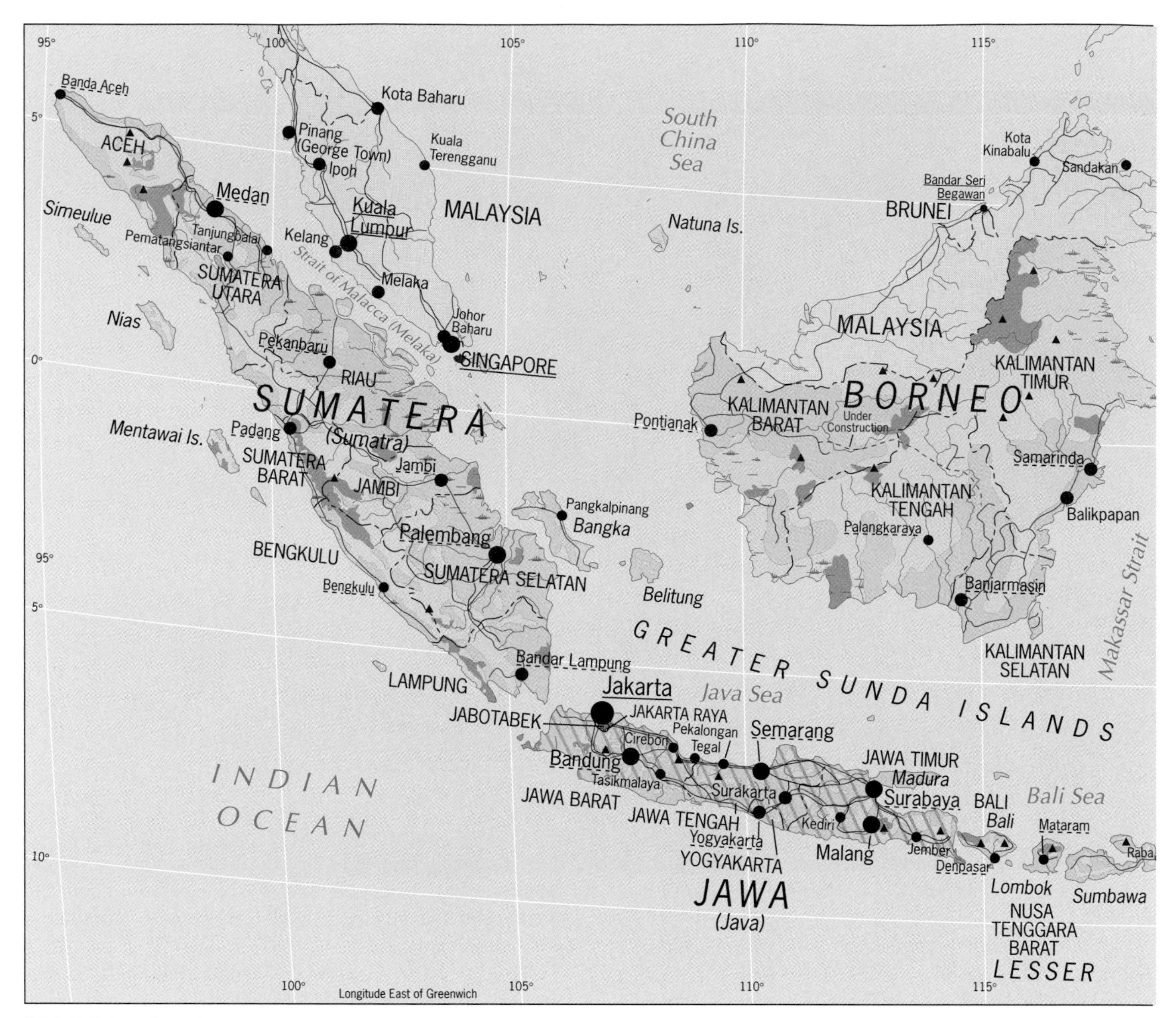

FIGURE 10-12

how. Third, Singapore opened its doors to capitalists of Chinese ancestry who left Hong Kong when China took it over and who wanted to relocate their enterprises here. Singapore's population is 76 percent Chinese, 15 percent Malay, and 7 percent South Asian. The government is Chinese-dominated, and its policies have served to sustain Chinese control. Indeed, Singapore's combination of authoritarianism and economic success often is cited in China itself as proving that communism and market economics can coexist.

Singapore continues to build on its success. It has constructed a second bridge across the Johor Strait to Malaysia to accommodate the spiraling truck traffic between the two countries. An active land reclamation program already has added over 4 percent more territory to the crowded island; a recent scheme connected the main island to the Jurong group of islands in the southwest (Fig. 10-11). All seven Jurong Islands, in turn, were glued together by dikes, their intervening waters drained to create space for a giant petrochemical complex. Key to Singapore's progress has been its comparative lack of corruption, attracting many companies to establish themselves here as a base from which to do business in notoriously corrupt countries elsewhere in the realm. Still, Singapore is the smallest of the Pacific Rim's economic tigers, and it faces growing competition near and far. Over the long term its relative location, which yielded its early wealth, may turn out to be its most enduring guarantee of survival—if not prosperity.

INDONESIA'S ARCHIPELAGO

The very term **archipelago** denotes fragmentation, and Indonesia is the world's most expansive archipelagic state. Spread across more than 17,000 islands, Indonesia's 219

18

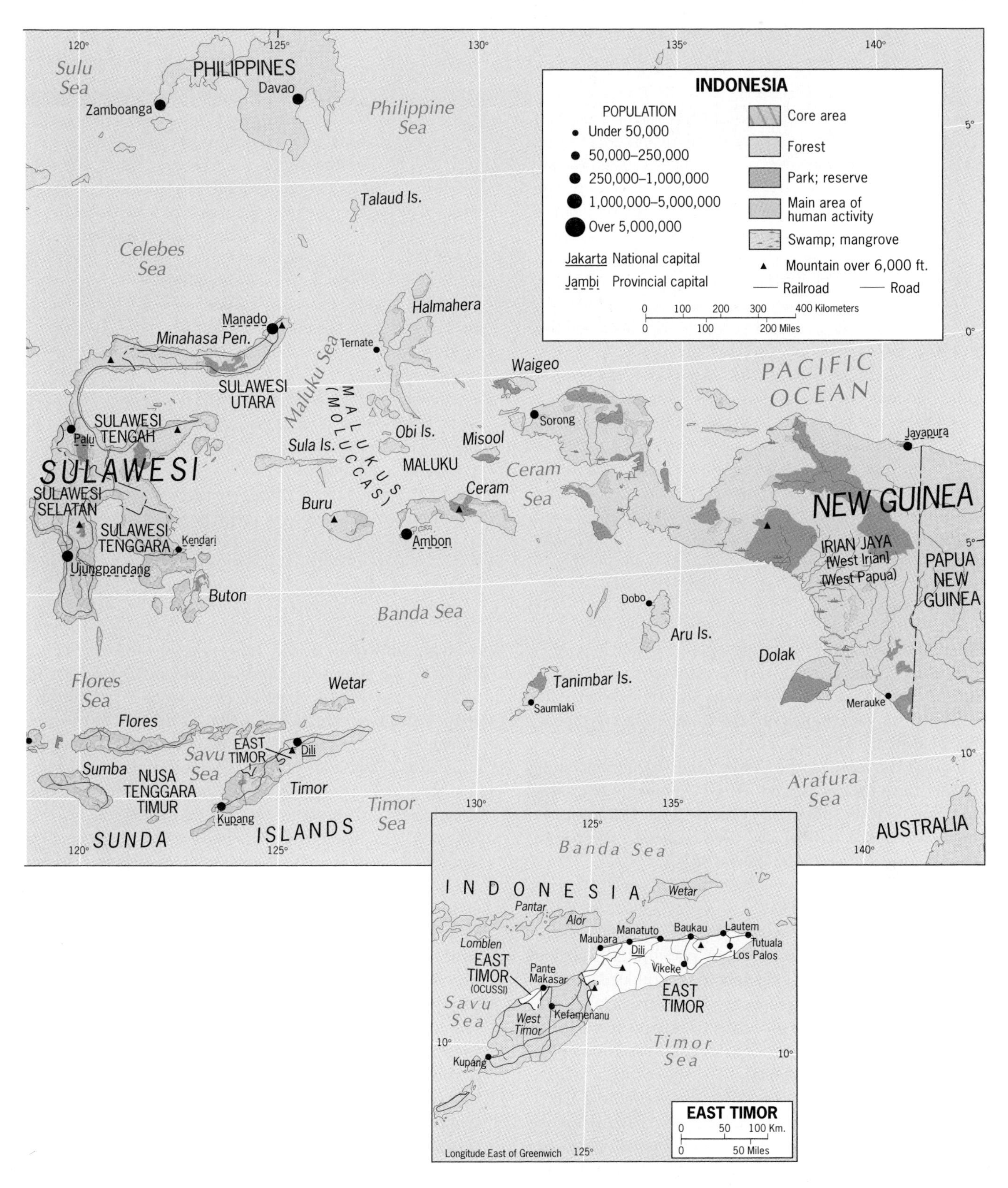

million people live separated and clustered—separated by water and clustered on islands large and small.

The map of Indonesia requires some attention (Fig. 10-12). Five large islands dominate the archipelago territorially, one of which, New Guinea in the east, is not part of the Indonesian culture sphere, although its western half is under Indonesian control. The other four major islands are collectively known as the Greater Sunda Islands: *Jawa*

Rich and Broken Brunei

Brunei is an anomaly in Southeast Asia—an oil-exporting Islamic sultanate far from the Persian Gulf. Located on the north coast of Borneo, sandwiched between Malaysian Sarawak and Sabah (Fig. 10-6, upper-left map), the Brunei sultanate is a former British-protected remnant of a much larger Islamic kingdom that once controlled all of Borneo and areas beyond. Brunei achieved full independence in 1984. With a mere 2225 square miles (5700 sq km)—slightly larger than Delaware—and only 345,000 people, Brunei is dwarfed by the other political entities of Southeast Asia (Fig. 10-1). But the discovery of oil in 1929 (and natural gas in 1965) heralded a new age for this remote territory.

Today, Brunei is one of the largest oil producers in the British Commonwealth, and recent offshore discoveries suggest that production will increase. As a result, the population is growing rapidly through immigration (64 percent of Brunei's residents are Malay, 20 percent Chinese), and the sultanate enjoys one of the highest standards of living in Southeast Asia (gross national product per capita today is estimated to exceed U.S. $25,000). Most of the people live near the oilfields in the western corner of the country and in the capital in the east—Bandar Seri Begawan. Evidence of a development boom can be seen in modern apartment houses, shopping centers, and hotels—a sharp contrast to many other towns on Borneo. There are some marked internal contrasts as well: Brunei's interior remains an area of subsistence agriculture and rural isolation, virtually untouched by the modernization of the burgeoning coastal zone.

Brunei's territory is not only small; it is fragmented. A sliver of Malaysian land separates the larger west from the east (where the capital lies), and Brunei Bay provides a water link between the two parts. It has been proposed that Brunei purchase this separating corridor from the Malaysians, but no action has yet been taken.

(Java), smallest but by far the most populous and important; *Sumatera* (Sumatra) in the west, directly across the Strait of Malacca from Malaysia; *Kalimantan,* the Indonesian sector of large, compact, minicontinent Borneo; and wishbone-shaped, distended *Sulawesi* (Celebes) to the east. Extending eastward from Jawa are the Lesser Sunda Islands, including Bali and, near the eastern end, Timor. Another important island chain within Indonesia is the Maluku (Molucca) Islands, between Sulawesi and New Guinea. The central water body of Indonesia is the Java Sea.

As the map shows, Indonesia does not control all of the archipelago. In addition to East Malaysia's territory on Borneo, there also is Brunei (see box titled "Rich and Broken Brunei"). And on the eastern part of the island of Timor, a long struggle against Indonesian rule led, in 1999, to a referendum on independence. The overwhelming majority of East Timorese voted in favor of sovereignty, and since then East Timor has been moving toward statehood under United Nations auspices.

Indonesia is a Dutch colonial creation, and the Dutch chose Jawa as their colonial headquarters, making Batavia (now Jakarta) their capital. Today, **Jawa** remains the core of Indonesia. With about 130 million inhabitants, Jawa is one of the world's most densely peopled places and one of the most agriculturally productive. Jawa also is the most highly urbanized part of a country in which more than 60 percent of the people still live on the land, and the Pacific Rim boom of the 1990s had strong impact here. The city of Jakarta, on the northwestern coast, became the heart of a larger conurbation now known as *Jabotabek*, consisting of the capital as well as Bogor, Tangerang, and Bekasi. During the 1990s the population of this megalopolis grew from 15 to 20 million, and it is predicted to reach 30 million by 2010. Already, Jabotabek is home to over 10 percent of Indonesia's entire population and 25 percent of its urban population. Thousands of factories, their owners taking advantage of low prevailing wages, were built in this area, straining its infrastructure and overburdening the port of Jakarta. On an average day, hundreds of ships lie at anchor, awaiting docking space to offload raw materials and take on finished products.

Indonesia's experience with rapid economic growth was tempered by the failures of its government. Suharto, the same president who had ruled the country for 25 years, and who had enriched himself, his family, and numerous friends and cronies, was reelected in 1997 for another five-year term. His authoritarian rule, credited by some for creating the stability that made Indonesia's economic boom possible, tolerated no free opposition or open election campaigns. But when Indonesia's economy was battered by the slump that had earlier struck Thailand and Malaysia, political turmoil forced Suharto to step down. During the rioting that accompanied this transition, the Jawanese, as they had done in the past, turned on the Chinese in their midst.

Always in Indonesia, Jawa is where the power lies. As a cultural group, the Jawanese constitute about 60 percent of the country's population. Sumatera, much larger than Jawa, has only about one-third as many people (45 million). Sulawesi has some 17 million, and Kalimantan, Indonesia's huge territory on Borneo, just 14 million. As for Irian Jaya in the far east, which is larger than California, its population is barely more than 2 million.

Diversity in Unity

Indonesia's survival as a unified state is as remarkable as India's and Nigeria's. With more than 300 discrete ethnic clusters, over 250 languages, and just about every religion practiced on Earth (although Islam dominates), actual and potential centrifugal forces are powerful here. Wide waters and high mountains perpetuate cultural distinctions and differences. Indonesia's national motto is *bhinneka tunggal ika*: diversity in unity.

What Indonesia has achieved is etched against the country's continuing cultural complexity. There are dozens of distinct aboriginal cultures; virtually every coastal community has its own roots and traditions. And the majority, the rice-growing Indonesians, include not only the numerous Jawanese—who are Muslims largely in name only and have their own cultural identity—but also the Sundanese (who constitute 14 percent of Indonesia's population), the Madurese (8 percent), and others. Perhaps the best impression of the cultural mosaic comes from the string of islands that extends eastward from Jawa to Timor (Fig. 10-12). The rice-growers of Bali adhere to a modified version of Hinduism, giving the island a unique cultural atmosphere; the population of Lombok is mainly Muslim, with some Balinese Hinduism; Sumbawa is a Muslim community; Flores is mostly Roman Catholic. In western Timor, Protestant groups dominate; in the east, where the Portuguese ruled, Roman Catholicism prevails. Nevertheless, Indonesia nominally is the world's largest Muslim country: overall, 88 percent of the people adhere to Islam, and in the cities the silver domes of neighborhood mosques rise above the townscape. But Islam is not (perhaps not yet) the issue it is in Malaysia, where observance generally is stricter and where minorities fear Islamization as Malay power grows.

Among the Realm's Great Cities . . .

Jakarta

Jakarta, capital of Indonesia, sometimes is called the Calcutta of Southeast Asia. Stand on the elevated highway linking the port to the city center and see the villages built on top of garbage dumps by scavengers using what they can find in the refuse, and the metaphor seems to fit. There is poverty here unlike that in any other city in the realm.

But there are other sides to Jakarta. Indonesia's economic progress has made its mark here, and the evidence is everywhere. Television antennas and satellite dishes rise like a forest from rusted, corrugated-iron rooftops. Cars (almost all, it seems, late-model), mopeds, and bicycles clog the streets, day and night. A meticulously manicured part of the city center contains a cluster of high-rise hotels, office buildings, and apartments. Billboards advertise planned communities on well-located, freshly cleared land.

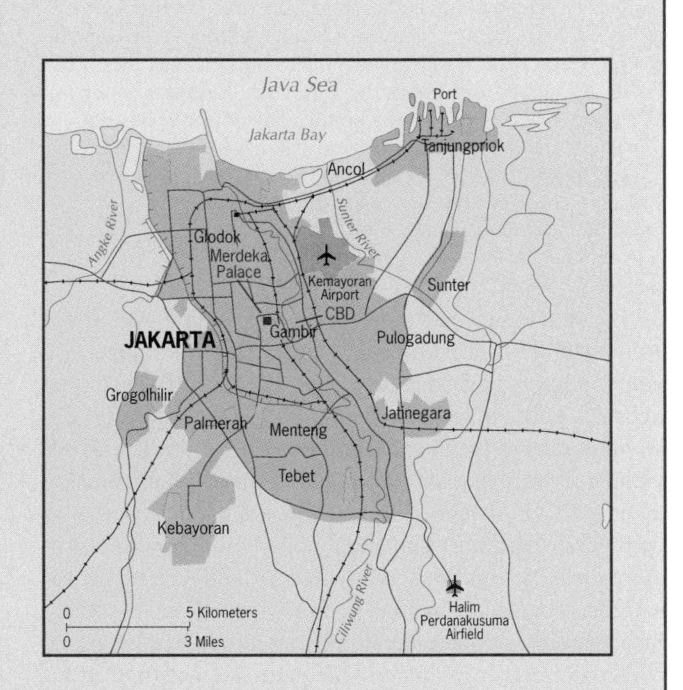

Jakarta's population is a cross-section of Indonesia's, and the silver domes of Islam rise above the cityscape alongside Christian churches and Hindu temples. The city always was cosmopolitan, beginning as a conglomerate of villages at the mouth of the Ciliwung River under Islamic rule, becoming a Portuguese stronghold and later the capital of the Dutch East Indies (under the name Batavia). Advantageously situated on the northwest coast of Jawa, Indonesia's most populous island, Jakarta bursts at the seams with growth. Sail into the port, and hundreds of vessels, carrying flags from Russia to Argentina, await berths. Travel to the outskirts, and huge shantytowns are being expanded by a constant stream of new arrivals. So vast is the human agglomeration—nobody knows how many millions have descended on this place (the official figure is 12.2 million)—that the majority live without adequate (or any) amenities.

A seemingly permanent dome of metallic-gray air, heavy with pollutants, rests over the city. The social price of economic progress is high, but Jakartans are willing to pay it.

FROM THE FIELD NOTES

"Getting to Ambon was no easy matter, and my boat trip was enlivened by an undersea earthquake that churned up the waters and shook up all aboard. The next morning our first view of the town of Ambon, provincial capital of Maluku, showed a center dominated by a large mosque with a modern minaret and a large white dome to the right of the main street leading to the port. My host from the local university told me that, as on many of the islands of the Malukus, about half of the people were Christians, not Muslims, and that some large churches were situated in the outskirts. 'But the relationships between Muslims and Christians are worsening,' he said. It had to do with the arrival of Jawanese, some of whom were 'agitators' and Islamic fundamentalists, stirring up religious passions. 'Have you heard of the Taliban?' he asked. 'Well, we have similar so-called religious students here, educated in Islamic schools that teach extremism.' . . . We drove from the town into the countryside to the west, toward Wakisihu and past the old Dutch Fort Rotterdam. 'Let me show you something,' he said as we turned up a dirt road. At its end, in the middle of a field, sat a large single structure called University of Islam, distinguished by a pair of enormous stairways. 'Can you call a single building like this where all they teach is Islam a university?' he asked. 'What they teach here is Islamic fundamentalism and intolerance.' He was prophetic. Weeks later, religious conflict broke out, the mosque in town was burned along with Christian churches; his own university lay in ruins."

The economic and political events of the late 1990s combined to foster Islamic fundamentalism in Indonesia, a potentially divisive development. Anger and frustration over the political system, rising prices, and growing joblessness during the economic downturn of the late 1990s yielded a ready market for a back-to-Islamic-basics movement. Already, Islamic fundamentalism had a strong foothold in northernmost Sumatera, notably in the Special Autonomous District of Aceh, where an anti-Jakarta movement has been accompanied by armed rebellion, and where Sharia law was proclaimed in 2000 with Jakarta's consent. Now other parts of Indonesia are getting a taste of it. If Indonesia's moderate and tolerant Islam were to give way to a more rigorous version, this part of the Pacific Rim would face a different future.

Transmigration and the Outer Islands

As noted earlier, Indonesia's population of 219 million makes it the world's fourth most populous country, but Jawa, we also noted, contains approximately 60 percent of it. With about 130 million people on an island the size of Louisiana, the population pressure is enormous here. Moreover, Indonesia's annual rate of population growth remains at 1.6 percent, resulting in a doubling time of 44 years. To deal with this problem, and at the same time to strengthen the core's power over outlying areas, the Indonesian government has long pursued a policy known as **transmigration**, inducing Jawanese (and Madurese from the adjacent island of Madura) to relocate to other islands. Several million Jawanese have moved to locales as distant as the Malukus, Sulawesi, and Sumatera; many Madurese were resettled in Kalimantan.

In the government's view, the transmigration policy would serve to counter centrifugal forces arising from Indonesia's geography, would spread the official language (Bahasa Indonesia, a modified form of Malay), and would bring otherwise unexploited land into production. It has done so, but it also has negative effects. In Kalimantan, the Muslim Madurese found themselves facing armed Dayaks, indigenous people who did not want to give up their land, and thousands were killed in what amounted to a regional civil war. Elsewhere, Jawanese often are blamed for introducing methods and practices of administration and business that run counter to local traditions. In Irian Jaya, where Indonesia is in effect a colonial power, aboriginal people have attacked Indonesian newcomers. Only the future will tell whether Jakarta's transmigration policy will ultimately succeed.

The Major Islands

Indonesia is an important state in a crucial geographic location. Each of its four major territorial components, in addition to Jawa, is geographically distinct.

Sumatera, Indonesia's westernmost major island, lies across the narrow Strait of Malacca from West Malaysia. Tropical rainforest still covers enough of it to sustain refuges for the orangutan population, but logging and human population growth are making ever greater inroads. Sharing Malaya's equatorial environments, Sumatera in colonial times became a base for rubber and palm-oil plantations. Its high relief and fertile volcanic soils make possible the cultivation of a wide range of crops, from vegetables in the highlands, tea and coffee on the slopes, to subsistence root crops in the lowlands. Sumatera and its neighboring small islands contain large coal reserves, petroleum and natural gas, tin, bauxite (for aluminum), and even gold and silver. As the map suggests, intra-island surface communications are still weak, although the north-south Sumatera Highway was completed in the 1980s. The highest population density is around the northern city of Medan; another cluster lies in the middle zone of the island; and a third, augmented by a large contingent of transmigrated Jawanese, focuses on Palembang and Bandar Lampung in the south. The northern part of the island has long been the most fractious. While the Batak people (see Fig. 10-3) adjusted to colonization and Westernization, and made Medan one of Indonesia's Pacific Rim boom towns, the Aceh fought the Dutch in a 30-year war that ended only in the twentieth century. In the 1990s, while Indonesia, Malaysia, and Thailand talked of an economic growth triangle including northern Sumatera, the Aceh mounted another rebellion—this time against Jakarta. Having just agreed to allow a referendum on independence to take place in East Timor, the now-democratic government under President Abdurrahman Wahid appeared ready, at the president's urging, to permit a vote in Aceh—but this idea was quickly dismissed. Then, even as the Aceh problem simmered, the House of Representatives of the central Sumateran Province of Riau approved a "declaration of independence." And the House of East Kalimantan proposed that the province be renamed the "Federal State of East Kalimantan." Devolutionary forces are at work throughout Indonesia.

Kalimantan is the Indonesian part of the minicontinent of Borneo, a slab of the Earth's crystalline crust whose backbone of mountains is of erosional, not volcanic, origin. Compact and massive—at over 290,000 square miles (750,000 sq km) it is larger than Texas—Borneo has a deep, densely forested interior that is a last refuge of some 35,000 orangutans. Elephants, rhinoceroses, and tigers still survive even as the loggers constrict their habitat. Numerous other species of fauna—insects, reptiles, mammals, birds—inhabit the rich mountain and lowland forests. So much of Borneo's Pleistocene heritage has survived principally because of its comparatively small human population. The Indonesian part, Kalimantan, constitutes 28 percent of the national territory, but, with 14 million people, it contains just 6 percent of Indonesia's population. Figure 10-12 shows that the main areas of human activity lie in the west, southeast, and east. All towns of any size, including Pontianak in the west and Balikpapan in the east, are on or near the coast; the rivers still form important routes into the interior. Parks and reserves to protect flora and fauna cover but a small portion of the vast territory. Not only the flora and fauna, but also the human population has ancient roots. Indigenous peoples, including some of the Dayak clans, continue their slash-and-burn subsistence in the interior; others have become sedentary farmers. But Indonesia's transmigration policy has brought about half a million Jawanese and Madurese to Kalimantan, and they are laying out farms and digging drainage canals, bringing drastic change to rural areas. Of Kalimantan's four provinces, only the southernmost exports significant raw materials, including oil from offshore reserves, coal, iron ore, and some gold and diamonds; but in the national picture, Indonesian Borneo remains little developed.

Sulawesi is an island that looks like a set of intersecting mountain ranges rising from the sea. Its northern end is a 500-mile (800-km) chain of extinct, dormant, and active volcanoes known as the Minahasa Peninsula, its future northern extension toward the Philippine Sea already marked by a ribbon of seafloor volcanoes just rising above the surface. So rugged is the relief on this island that you cannot take surface transportation from Palu to Ujungpandang. The ethnic mosaic is quite complex, with seven major groups inhabiting more or less isolated parts of the island as well as a growing number of transmigrated Jawanese. The two most populous and developed parts of Sulawesi are the southwestern peninsula centered on Ujungpandang and the eastern end of the Minahasa Peninsula, where Manado is the urban focus. Subsistence farming occupies most of the population of 17 million, but logging and wood products, some mining, and fishing augment the cash economy. Cultural landscapes are varied; in Minahasa, Christian churches remain numerous and form reminders of the colonial period, when relations between this province and the Dutch were favorable. During the liberation period Minahasans sided with the Dutch (as many of their neighbors in the Malukus did) and called their peninsula Holland's "twelfth province." It was an unfulfilled wish, souring relations between Jakarta and Manado for years afterward. But when President Suharto was forced to resign in 1998, his

FROM THE FIELD NOTES
"I drove from Manado on the Minahasa Peninsula in northeastern Sulawesi to see the ecological crisis at Lake Tondano, where a fast-growing water hyacinth is clogging the water and endangering the local fishing industry. On the way, in the town of Tomolon, I noticed this side street lined with prefabricated stilt houses in various stages of completion. These, I was told, were not primarily for local sale. They were assembled from wood taken from the forests of Sulawesi's northern peninsula, then taken apart again and shipped from Manado to Japan. 'It's a very profitable business for us,' the foreman told me. 'The wood is nearby, the labor is cheap, and the market in Japan is insatiable. We sell as many as we can build, and we haven't even begun to try marketing these houses in Taiwan or China.' At least, I thought, this wood was being converted into a finished product, unlike the mounds of logs and planks I had seen piled up in the ports of Borneo awaiting shipment to East Asia."

transitional successor, Habibie, was a son of Minahasa, and the past was forgotten.

Irian Jaya, like East Timor, fell to Indonesia well after the Dutch colonial era had ended, but unlike East Timor, the United Nations approved Indonesia's takeover in 1969. The Bahasa Indonesian name for New Guinea is *Irian*, and until 2000 this province was called Irian Jaya—Guinea West. The eastern half of New Guinea, across that classic superimposed, geometric boundary (Fig. 10-12), is an independent Papuan state (Papua New Guinea). In mid-2000, the president of Indonesia indicated that he would agree to the local demand to rename Irian Jaya *West Papua*, but the Indonesian parliament refused to approve it. In early 2001, the matter remained unresolved.

As an Indonesian province, Irian Jaya/West Papua has about 22 percent of the country's land but just 1 percent of its population, including over 200,000 ethnic Indonesians, most from Jawa and many of them concentrated in the provincial capital with the non-Papuan name of Jayapura. Forested, high-relief, glacier-peaked West Papua, lying as it does in a different geographic realm, is a world apart from teeming Jawa. It has what is reputedly the world's richest gold mine and second-largest open-pit copper mine; its forests not only harbor the indigenous peoples but also sustain a major logging industry.

Sanctioned as Indonesia's rule in West Papua may be, local opposition has continued for decades. The *Organisasi Papua Merdeka* (Free Papua Movement, or FPM) and occasional rebel attacks remind the Jakarta government of its role here. In 1999, during the turmoil in the other provinces, the FPM held rallies in Jayapura to promote its cause, and in 2000 a congress was held where independence was demanded and a Papuan flag, the Morning Star, displayed (see photo, p. 529). Later that year, Indonesian troops killed demonstrators in Merauke on the province's south coast. All this is deepening Papuan opposition to Jakarta's rule, and the time may come when Indonesia will rue its acquisition of West Papua, potential source of the kinds of centrifugal forces an archipelagic state must constantly confront.

EAST TIMOR

The easternmost of the lesser Sunda Islands is Timor, the eastern part of which was a Portuguese colony, overrun by Indonesia in 1975 and annexed in 1976, which became the scene of a bitter struggle for independence. When the people of East Timor in 1999 were allowed to express their views on independence, this Connecticut-sized territory with fewer than 800,000 inhabitants took the first steps toward nationhood.

Indonesia's armed forces reacted violently, adding hundreds more to the tens of thousands who had already died in the struggle against Jakarta. The country's modest infrastructure was devastated, and its people were dislocated. Foreign intervention, led by Australia (once Indonesia's sole supporter in its claim to East Timor), established some order. Reconstruction and social reorganization, under UN auspices, have been in progress since.

In Indonesia, many viewed the "loss" of East Timor as the first step in the disintegration of the state, to be followed by secession in Aceh, the Malukus, and West Papua. But in truth, East Timor was a special case, and if devolution were indeed to fragment Indonesia, East Timor was not the prototype. Unlike West Papua, the Malukus, or Aceh, Indonesia's control over East Timor defied UN rules. This was not merely an internal matter.

As the map (inset, Fig. 10-12) shows, East Timor's independence creates some politico-geographical complications. The political entity consists of a main territory, where the capital named Dili is located, and a small exclave on the north coast of (Indonesian) West Timor called Ocussi

This kind of scene strikes fear into the heart of the Indonesian government: a march by independence-minded Papuans carrying their flag and make-believe guns in the capital of what Indonesia still calls Irian Jaya (West Irian), Jayapura. This photograph was taken shortly after a meeting of 2000 representatives of Papuan groups in Jayapura, a meeting Indonesian President Wahid had intended (but later declined) to attend. More recently, reports of clashes and casualties suggest that Indonesia faces a growing problem in its Papuan territory.

(spellings vary, and it is sometimes mapped as Ocussi-Ambeno or Ambeno Province). Although there is a road from the exclave via Kefamenanu to the main territory, relations between East Timor and Indonesia may not make this link feasible, so that the two parts of the new state may have to be connected by boat traffic. A related implication of East Timor's (and Ocussi's) separation from Indonesia is that the territorial-sea boundaries (a concept discussed on p. 552) and other maritime jurisdictions have been changed radically. In the Timor Sea, where energy resources have been found, Indonesia and Australia are no longer the sole international neighbors: East Timor now adjoins Australia there. A reconfiguration of the maritime boundaries suggests that East Timor could claim a large share (perhaps as much as 85 percent) of the oil and gas reserves lying between it and Australia in the so-called Timor Gap.

Meanwhile, East Timor in 2001 was deciding on its principal language, its form of government, and other building blocks of a new state. Given the circumstances of its birth, East Timor will for some time rank with the poorest of the realm's countries.

FRAGMENTED PHILIPPINES

North of Indonesia, across the South China Sea from Vietnam, and south of Taiwan lies an archipelago of more than 7000 islands (only about 460 of them larger than one square mile in area) inhabited by 83.9 million people. The inhabited islands of the Philippines can be viewed as three groups: (1) Luzon, largest of all, and Mindoro in the north, (2) the Visayan group in the center, and (3) Mindanao, second largest, in the south (Fig. 10-13). Southwest of Mindanao lies a small group of islands, the Sulu Archipelago, nearest to Indonesia, where a Muslim-based insurgency has kept the area in turmoil.

Few of the generalizations we have been able to make for Southeast Asia could apply in the Philippines without qualification. The country's location relative to the mainstream of change in this part of the world has had much to do with this situation. The islands, inhabited by peoples of Malay ancestry with Indonesian strains, shared with much of the rest of Southeast Asia an early period of Hindu cultural influence, which was strongest in the south and southwest and diminished northward. Next came a Chinese invasion, felt more strongly on the largest island of Luzon in the northern part of the Philippine archipelago. Islam's arrival was delayed somewhat by the position of the Philippines well to the east of the mainland and to the north of the Indonesian islands. The few southern Muslim beachheads were soon overwhelmed by the Spanish invasion during the sixteenth century. Today the Philippines, adjacent to the world's largest Muslim state (Indonesia), is 83 percent Roman Catholic, 9 percent Protestant, and only 5 percent Muslim.

Out of the Philippines melting pot, where Mongoloid-Malay, Arab, Chinese, Japanese, Spanish, and American elements have met and mixed, has emerged the distinctive Filipino culture. It is not a homogeneous or a unified culture, but in Southeast Asia it is in many ways unique. One example of its absorptive qualities is demonstrated by the way the Chinese infusion has been accommodated: although the "pure" Chinese minority numbers less than 2 percent of the population (far lower than in most Southeast Asian countries), a much larger portion of the Philippine population carries a decidedly Chinese ethnic imprint. What has happened is that the Chinese have intermarried, producing a sort of Chinese-mestizo element that constitutes more than 10 percent of the total population. In another cultural sphere, the country's ethnic mixture and variety are paralleled by its great linguistic diversity. Nearly 90 Malay languages, major and minor, are spoken by the 84 million people of the Philippines; only about 1 percent still use Spanish. At independence in 1946, the largest of the Malay languages, Tagalog, or Pilipino, became the country's official language, and the educational system promotes its general use. English is learned as a subsidiary language and remains the chief *lingua franca*; an English-Tagalog hybrid ("Taglish") is increasingly heard today, cutting across all levels of society. The widespread use of English in the Philippines, of course, results from a half-century of American rule and influence, beginning in 1898 when the islands were ceded to the United States

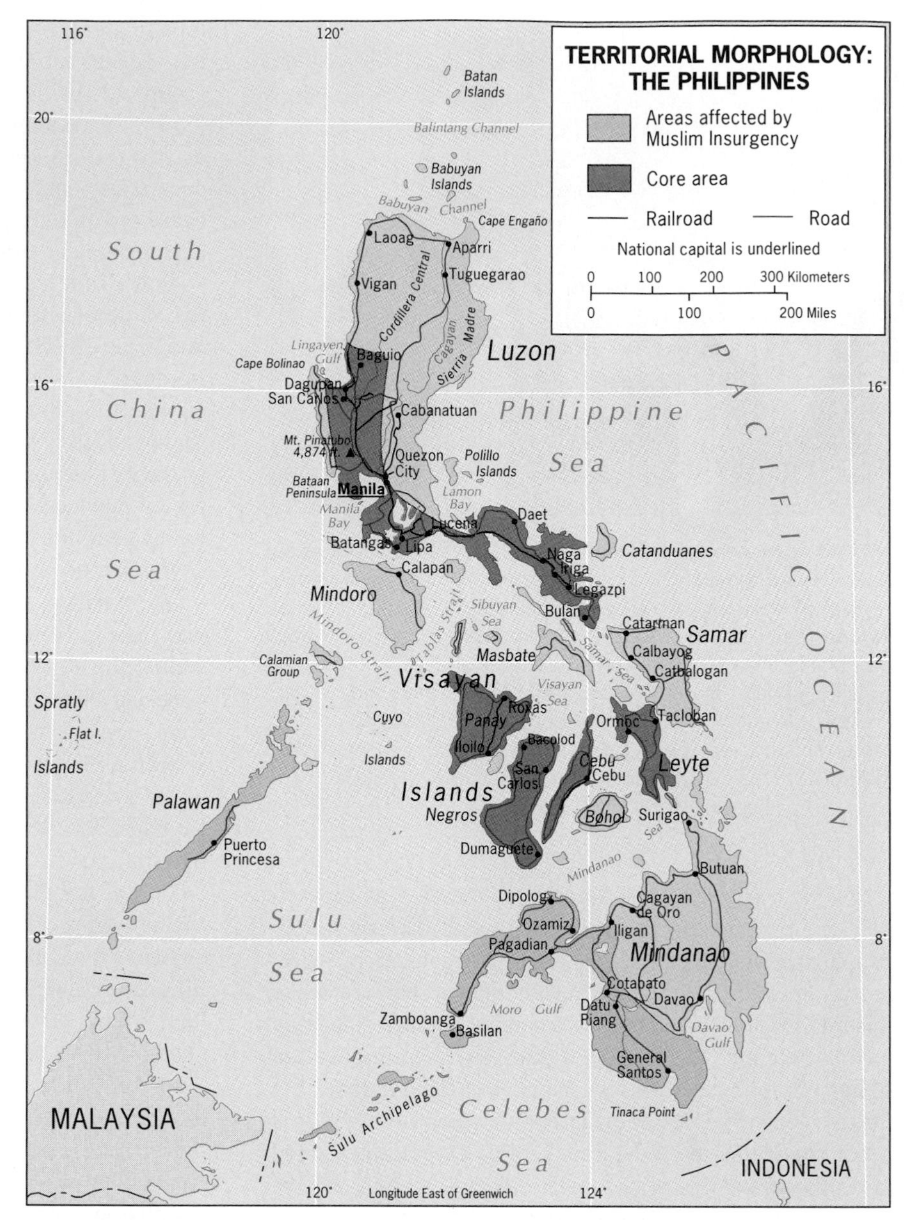

FIGURE 10-13

by Spain under the terms of the treaty that followed the Spanish-American War. The United States took over a country in open revolt against its former colonial master and proceeded to destroy the Filipino independence struggle, now directed against the new foreign rulers. It is a measure of the subsequent success of U.S. administration in the Philippines that this was the only dependency in Southeast Asia that sided against the Japanese during World War II in favor of the colonial power. The U.S. rule had its good and bad features, but the Americans did initiate reforms that were long overdue, and they were already in the process of negotiating a future independence for the Philippines when the war intervened in 1941.

The Philippines' population, concentrated where the good farmlands lie in the plains, is densest in three general areas (Fig. I-9): (1) the northwestern and south-central part of Luzon, (2) the southeastern extension of Luzon, and (3) the islands of the Visayan Sea between Luzon and Mindanao. Luzon is the site of the capital, Manila-Quezon City (11.7 million, nearly one-seventh of the entire national population), a major metropolis facing the South China Sea. Alluvial as well as volcanic soils, together with ample moisture in this tropical environment, produce self-sufficiency in rice and other staples and make the Philippines a net exporter of farm products despite a high population growth rate of 2.2 percent.

The Philippines seems to get little mention in discussions of developments on the Pacific Rim, and yet it would seem to be well positioned to share in the Pacific Rim's economic growth. Governmental mismanagement and political instability have slowed the country's participation, but during the 1990s the situation improved. Despite a series of jarring events—the ouster of U.S. military bases, the damaging eruption of a volcano near the capital, the violence of Muslim insurgents, and a dispute over the nearby Spratly Islands in the South China Sea—the Philippines made substantial economic progress during the decade. Its electronics and textile industries (mostly in the Manila hinterland) expanded continuously, and more foreign investment arrived. The centrally positioned Visayan island of Cebu experienced particularly rapid growth, based on its central location, good port, expanded airport, large and literate labor force, and a cadre of managers experienced in the service industries. A free-trade zone attracted foreign companies from the United States and Japan, and now products from toys to semiconductors flow from Cebu's manufacturers to world markets. But agriculture continues to dominate the Philippines' economy, unemployment remains high, further land reform is badly needed, and social restructuring (reducing the controlling influence over national affairs by a comparatively small group of families) must occur. However, progress is being made. The country now is a lower-middle-income economy, and given a longer period of stability and success in reducing the population growth rate, it will rise to the next level and finally take its place among Pacific Rim growth poles.

In recent years, the population issue has divided this dominantly Roman Catholic society, with the government promoting family planning and the clergy opposing it. But behind this debate lies another of the Philippines' assets: in a realm of mostly undemocratic regimes, the Philippines has come out of its period of authoritarian rule a rejuvenated, if not yet robust, democracy.

Among the Realm's Great Cities . . .

Manila

Manila, capital of the Philippines, was founded by the Spanish invaders of Luzon more than four centuries ago. The colonists made a good choice in terms of site and situation. Manila sprawls at the mouth of the Pasig River where it enters one of Asia's finest natural harbors. To the north, east, and south a crescent of mountains encircles the city, which lies just 600 miles (975 km) across the South China Sea from Hong Kong.

Manila, named after a flowering shrub in the local marshlands, is bisected by the Pasig, which is bridged in numerous places. The old walled city, Intramuros, lies to the south. Despite heavy bombing during World War II, some of the colonial heritage survives in the form of churches, monasteries, and convents. St. Augustine Church, completed in 1599, is one of the city's landmarks.

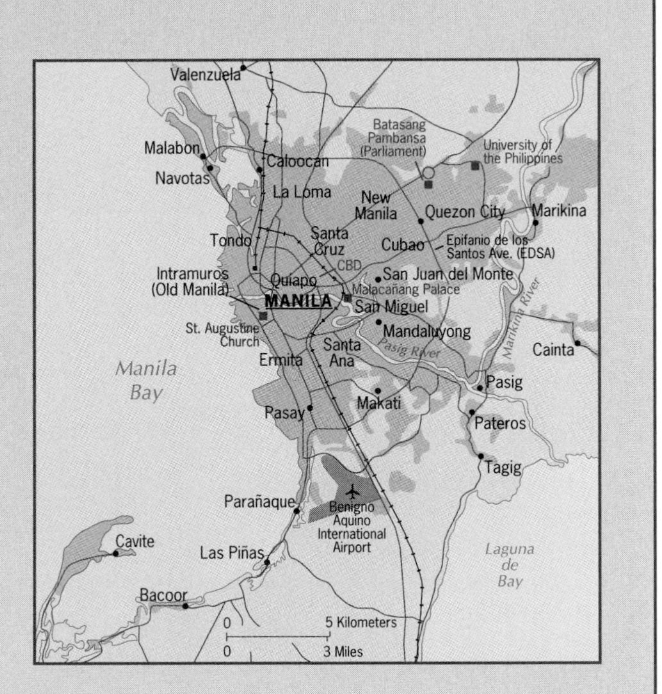

The CBD of Manila lies on the north side of the Pasig River. Although Manila has a well-defined commercial center with several avenues of luxury shops and modern buildings, the skyline does not reflect the high level of energy and activity common in Pacific Rim cities on the opposite side of the South China Sea. Neither is Manila a city of notable architectural achievements. Wide, long, and straight avenues flanked by palm, banyan, and acacia trees give it a look not unlike parts of San Juan, Puerto Rico.

In 1948, a newly built city immediately to the northeast of Manila was inaugurated as the *de jure* capital of the Philippines and called Quezon City. The new facilities were eventually to house all government offices, but many functions of the national government never made the move. In the meantime, Manila's growth overtook Quezon City's, so that it became part of the Greater Manila metropolis (which today is home to 11.7 million). Although the proclamation of Quezon City as the Philippines' official capital was never rescinded, Manila remains the *de facto* capital of the country today.

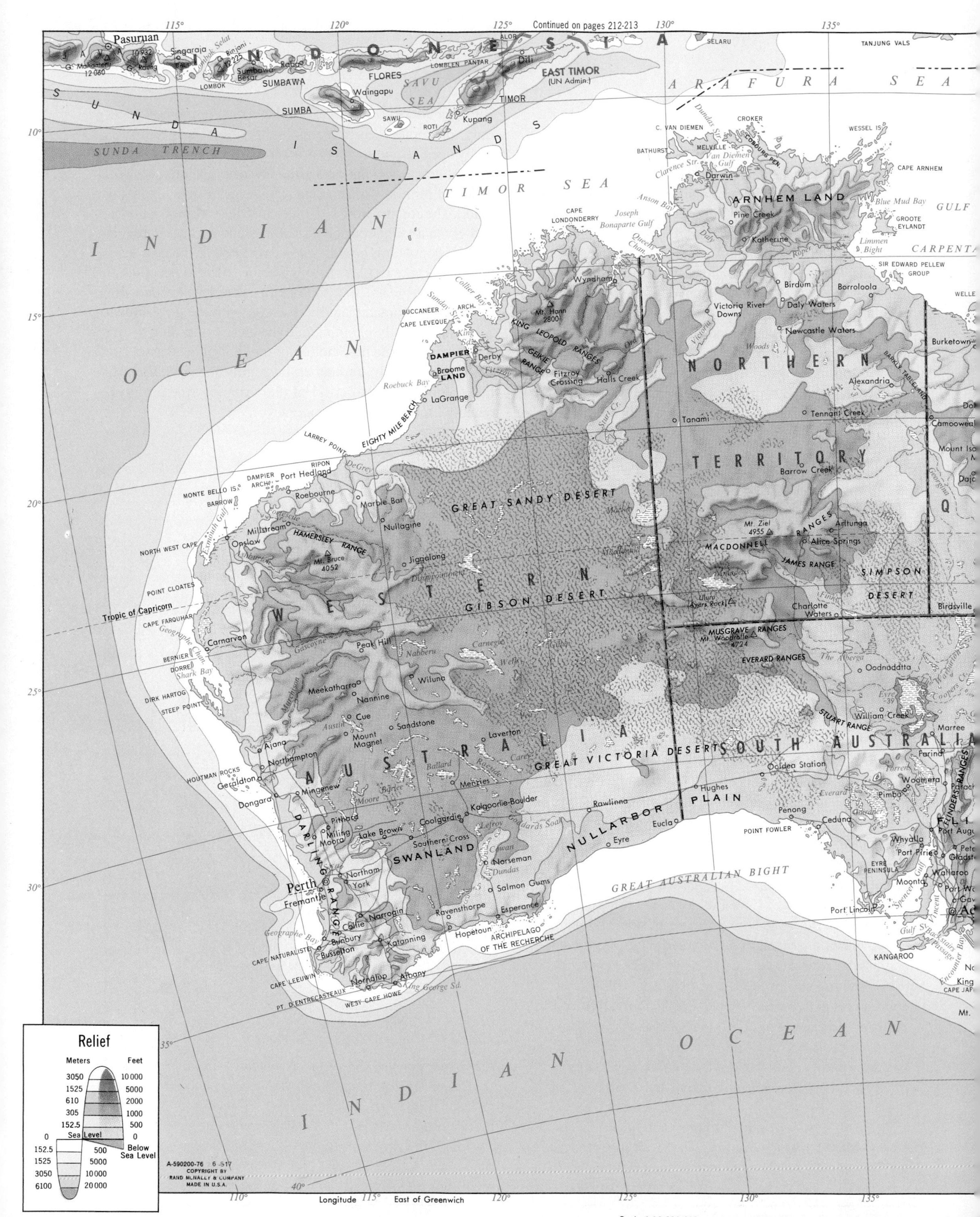

FIGURE 11-1

Scale 1:16 000 000; one inch to 250 miles. Lambert's Azimuthal, Equal Area Project

Elevations and depressions are given in feet

chapter 11

The Austral Realm

CONCEPTS, IDEAS, AND TERMS

1. Austral
2. Southern Ocean
3. Subtropical Convergence
4. West Wind Drift
5. Biogeography
6. Wallace's Line
7. Aboriginal population
8. Outback
9. Federation
10. Unitary state
11. Import-substitution industries
12. Aboriginal land issue
13. Immigration policies
14. Environmental degradation
15. Peripheral development

REGIONS

- AUSTRALIA
  - CORE AREA
  - OUTBACK
- NEW ZEALAND

Copyright © Rand McNally, 2000

DEFINING THE REALM

The Austral Realm is geographically unique. It is the only geographic realm that lies entirely in the Southern Hemisphere. It is also the only realm that has no land link of any kind to a neighboring realm, and is thus completely surrounded by ocean and sea. It is second only to the Pacific as the world's least populous realm. Appropriately, its name refers to its location (**Austral** means south)—a location far from the sources of its dominant cultural heritage but close to its newfound economic partners on the western Pacific Rim.

1

Two countries constitute this Austral Realm: Australia, in every way the dominant one, and New Zealand, physiographically more varied than its giant partner. Between them lies the Tasman Sea. To the west lies the Indian Ocean, to the east the Pacific, and to the south the frigid Southern Ocean (see box titled "Is There a Southern Ocean"?).

This southern realm is at a crossroads. On the doorstep of populous Asia, its Anglo-European legacies are now infused by other cultural strains. Polynesian Maori in New Zealand and Aboriginal communities in Australia are demanding better terms of life. Pacific Rim markets are buying huge quantities of raw materials. Japanese and other Asian tourists fill hotels and resorts. Queensland's tropical Gold Coast resembles Honolulu's Waikiki. The streets of Sydney and Melbourne display a multicultural panorama unimagined just two generations ago. All these changes have stirred political debate. Issues ranging from immigration quotas to indigenous land rights dominate, exposing social fault lines (city versus Outback in Australia, North and South in New Zealand). Aborigines and Maori were here first, and the Europeans came next. Now Asia looms in Australia's doorway.

◆ Major Geographic Qualities of the Austral Realm

1. Australia and New Zealand constitute a geographic realm by virtue of territorial dimension, relative location, and dominant cultural landscape.
2. Despite their inclusion in a single geographic realm, Australia and New Zealand differ physiographically. Australia has a vast, dry, low-relief interior; New Zealand is mountainous.
3. Australia and New Zealand are marked by peripheral development—Australia because of its aridity, New Zealand because of its topography.
4. The populations of Australia and New Zealand are not only peripherally distributed but also highly clustered in urban centers.
5. The realm's human geography is changing—in Australia because of Aboriginal activism and Asian immigration, and in New Zealand because of Maori activism and Pacific-islander immigration.
6. The economic geography of Australia and New Zealand is dominated by the export of livestock products (and in Australia also by wheat production and mining).
7. Australia and New Zealand are being integrated into the economic framework of the western Pacific Rim, principally as suppliers of raw materials.

LAND AND ENVIRONMENT

Physiographic contrasts between massive, compact Australia and elongated, fragmented New Zealand are related to their locations with respect to the Earth's tectonic plates (consult Fig. I-4). Australia, with some of the geologically most ancient rocks on the planet, lies at the center of its own plate, the Australian Plate. New Zealand, younger and less stable, lies at the convulsive convergence of the Australian and Pacific Plates. Earthquakes are rare in Australia and volcanic eruptions are unknown; New Zealand has plenty of both. This locational contrast is also reflected by differences in relief (Fig. 11-1). Australia's highest relief occurs in what Australians call the Great Dividing Range, the mountains that line the east coast from the Cape York Peninsula to southern Victoria, with an outlier in Tasmania. The highest point along these old, now eroding mountains is Mount Kosciusko, 7310 feet (2228 m) tall. In New Zealand, entire ranges are higher than this, and Mount Cook reaches 12,315 feet (3764 m).

West of Australia's Great Dividing Range, the physical landscape generally has low relief with some local exceptions such as the Macdonnell Ranges near the center; plateaus and plains dominate (Fig. 11-2). The Great Artesian Basin is a key physiographic region, providing underground water sources in what would otherwise be desert country, and crossed in the south by the continent's major river system, the Murray-Darling. The area mapped as *Shield* in Figure 11-2 contains Australia's oldest rocks and much of its mineral wealth.

Figure I-8 reveals the effects of latitudinal location and interior isolation on Australia's climatology. In this respect, Australia is far more varied than New Zealand, its climates ranging from tropical in the far north, where rainforests flourish, to Mediterranean in parts of the south. The interior is dominated by desert and steppe conditions, the steppes providing the grasslands that sustain tens of millions of livestock. Only in the east does Australia have an area of humid temperate climate, and here lies most of the country's economic core area. New Zealand, by contrast, is totally under the influence of the Southern and Pacific Oceans, creating moderate, moist conditions, temperate in the north and colder in the south.

Biogeography

One of this realm's defining characteristics is its wildlife. Australia is the land of kangaroos and koalas, wallabies and wombats, possums and platypuses. These and numerous other *marsupials* (animals whose young are born very early in their development and then carried in an abdominal pouch) owe their survival to Australia's early isolation dur-

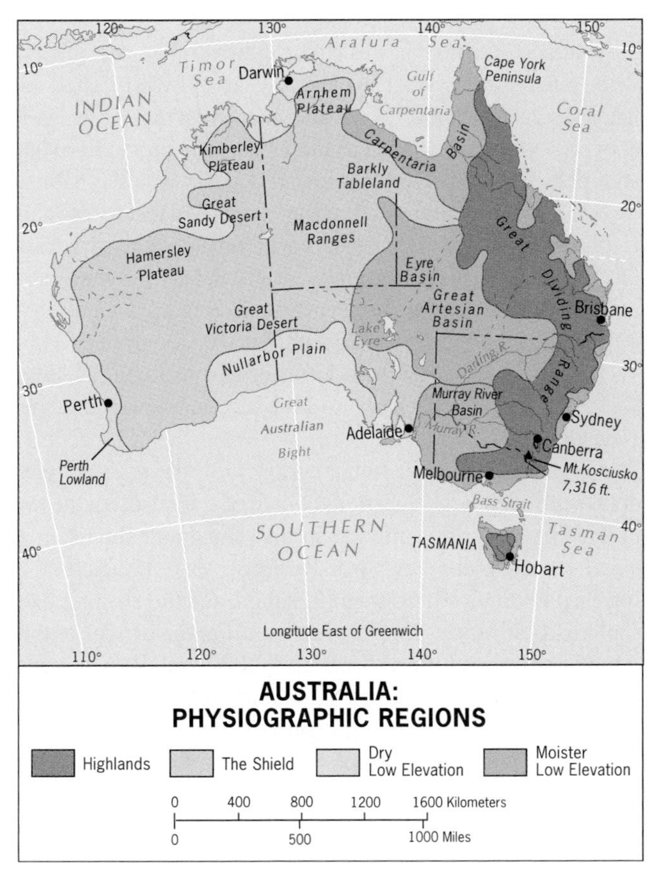

FIGURE 11-2

Is There a Southern Ocean?

Here is an interesting geographic experiment: try to find the Southern Ocean on commonly used maps of the world—commonly used, that is, here in the Northern Hemisphere. You are unlikely to find it, even on maps and globes published by the National Geographic Society, Rand McNally, and other major mapmaking organizations.

Now look in the library for atlases and maps published in Southern Hemisphere countries, for example, New Zealand and Australia. There the Southern Ocean is a prominent geographic feature. And it is not just a matter of opinion: to people whose countries adjoin it, the Southern Ocean is a significant factor in daily life. Its persistent westerly winds influenced the course of history. It is a crucial weather maker. Beneath its vast and frigid surface lie resources yet unknown. Its waters contain biota that are crucial to the global web of marine life.

How is it that the primary issue involving the **Southern Ocean** is whether it exists at all? The all-too-obvious answer is that this ocean, unlike the others on our planet, is not neatly bounded by continental coasts. Of course, it borders Antarctica to the south, but its northern limits are not visible on a relief map. So mapmakers simply extend the Atlantic, Pacific, and Indian Oceans all the way to Antarctic shores, ignoring the very existence of an ocean as large as the Indian.

For us geographers, it is a good exercise to turn the globe upside down now and then. After all, the usual orientation is quite arbitrary. Modern mapmaking started in the Northern Hemisphere, and the cartographers put their hemisphere on top and the other at the bottom. That is now the norm, and it can distort our view of the world. In bookstores in the Southern Hemisphere, you sometimes see tongue-in-cheek maps showing Australia and Argentina at the top, and Europe and Canada at the bottom. But this matter has a serious side. A reverse view of the globe shows us how vast the ocean encircling Antarctica is. The Southern Ocean may be remote, but its existence is real.

Where do the northward limits of the Southern Ocean lie? This ocean is bounded not by land but by a marine transition called the **Subtropical Convergence**. Here the cold, extremely dense waters of the Southern Ocean meet the warmer waters of the Atlantic, Pacific, and Indian Oceans. It is quite sharply defined by changes in temperature, chemistry, salinity, and marine fauna. Flying over it, you can actually observe it in the changing colors of the water: the Antarctic side is a deep gray, the northern side a greenish blue.

Although the Subtropical Convergence moves seasonally, its position does not vary far from latitude 40° South, which also is the approximate northern limit of Antarctic icebergs. Defined this way, the great Southern Ocean is a huge body of water that moves clockwise (from west to east) around Antarctica, which is why we also call it the **West Wind Drift**.

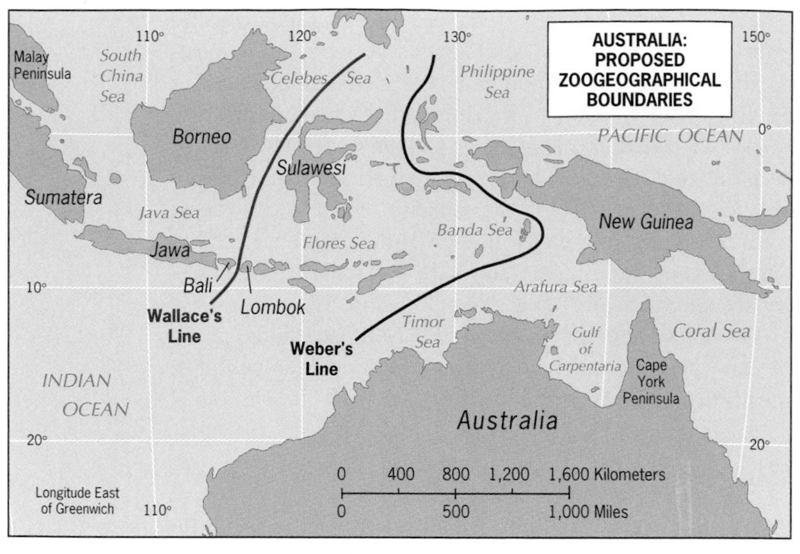

FIGURE 11-3

ing the breakup of Gondwana (see Fig. 7-3). Before more advanced mammals could enter Australia and replace the marsupials, as happened in other parts of the world, the landmass was separated from Antarctica and India, and today it contains the world's largest assemblage of marsupial fauna.

Australia's vegetation also has distinctive qualities, notably the hundreds of species of eucalyptus trees native to this geographic realm. Many other plants form part of Australia's unique flora, some with unusual adaptation to the high temperatures and low humidity that characterize much of the continent.

The study of fauna and flora in spatial perspective combines the disciplines of biology and geography in a field **5** known as **biogeography**, and Australia is a giant laboratory for biogeographers. In the Introduction (pp. 12-16), we noted that several of the world's climatic zones are named after the vegetation that marks them: tropical savanna, steppe, tundra. When climate, soil, vegetation, and animal life reach a long-term, stable adjustment, vegetation forms the most visible element of this ecosystem.

Biogeographers are especially interested in the distribution of plant and animal species, and in the relationships between plant and animal communities and their natural environments. (The study of plant life is called *phytogeography*; the study of animal life is called *zoogeography*.) These scholars seek to explain the distributions the map reveals. In 1876 one of the founders of biogeography, Alfred Russel Wallace, published a book entitled *The Geographical Distribution of Animals* in which he fired the first shot in a long debate: where does the zoogeographic boundary of Australia's fauna lie? Wallace's fieldwork in the area revealed that Australian forms exist not only in Australia itself but also in New Guinea and in some islands to the west. So Wallace proposed that the faunal boundary should lie between Borneo and Sulawesi, and just east of Bali (Fig. 11-3).

Wallace's Line soon was challenged by other researchers, **6** who found species Wallace had missed and who visited islands Wallace had not. There was no question that Australia's zoogeographic realm ended somewhere in the Indonesian archipelago, but where? Western Indonesia was the habitat of non-marsupial animals such as tigers, rhinoceroses, and elephants, as well as primates; New Guinea clearly was part of the realm of the marsupials. How far had the more advanced mammals progressed eastward along the island stepping stones toward New Guinea? The zoogeographer Max Weber found evidence that led him to postulate his own *Weber's Line*, which, as Figure 11-3 shows, lay very close to New Guinea.

Not all research in zoogeography or phytogeography deals with such large questions. Much of it focuses on the relationships between particular species and their habitats, that is, the environment they normally occupy and of which they form a part. Such environments change, and the changes can spell disaster for the species. In Australia, the arrival of the **Aboriginal population** (between 50,000 and 60,000 years **7** ago) had limited effect on the habitats of the extant fauna. But the invasion of the European colonizers and the introduction of their livestock led to the destruction of habitats and the extinction of many native species. Whether in East Africa or in Western Australia, the key to the conservation of what remains of natural flora and fauna lies in the knowledge embodied by the field of biogeography.

REGIONS OF THE REALM

Australia is the dominant component of the Austral Realm, a continent-scale country in a size category that also includes China, Canada, the United States, and Brazil. For two reasons, however, Australia has fewer regional divisions than do the aforementioned countries: Australia's relatively uncomplicated physiography and its diminutive human numbers. Our discussion, therefore, uses the core-periphery concept as a basis for investigating Australia and focuses on New Zealand as a region by itself.

AUSTRALIA

On January 1, 2001, Australia celebrated its 100th birthday as a state, the Commonwealth of Australia, recognizing (still) the British monarch as the head of state and entering its second century with a strong economy, stable political framework, high standard of living for most of its people, and favorable prospects ahead. Positioned on the Pacific Rim, ten times the size of Texas, well endowed with farmlands and vast pastures, major rivers, ample underground water, minerals, and energy resources, served by good natural harbors, and populated by nearly 20 million mostly-well-educated people, Australia is one of the most fortunate countries on Earth.

Not everyone in Australia shares adequately in all this good fortune, however, and the less advantaged made their voices heard during the celebrations. The country's indigenous (Aboriginal) population, though a small minority today of about 300,000, remains disproportionately disadvantaged in almost every way, from lower life expectancies to higher unemployment than average, from lower high school graduation rates to higher imprisonment ratios. But the nation is now embarked on a campaign to address these ills, with actions ranging from public demonstrations supporting reconciliation to official expressions of regret for past mistreatment and from enhanced social services to favorable court decisions over Aboriginal land claims.

When Australia was born as a federal state, its per-capita GNP, as reckoned by economic geographers, was the highest in the world. Australia's bounty fueled a huge flow of exports to Europe, and Australians prospered. That golden age could not last forever, and eventually the country's share of world trade declined. Still, Australia today ranks among the top 15 countries in the world in terms of GNP, and for the vast majority of Australians life is comfortable.

This is not to suggest that Australia has escaped every downturn in the global economy of which it is a part or that there are no concerns at all for the future. During the Pacific Rim boom of the 1980s and early 1990s, some locals wryly called Australia an NDC, a Newly Declining Country, a seller of raw materials, not finished ones; a purveyor of livestock, meat, and wheat on variable world markets; a society sinking deeply into debt. When the Pacific Rim economies stalled in the late 1990s, there were dire predictions about the impact on Australia. But Australia's economy proved to be remarkably adaptable. When its Asian markets dried up, Western markets soon took up the slack. The Australian economy weathered the Pacific Rim crisis better than expected.

The statistics bear witness to Australia's current good fortune. In terms of the indicators of development discussed in Chapter 9, Australia is far ahead of all its western Pacific Rim competitors except Japan (which has dropped in the rankings because of its own economic stagnation). As Australians celebrated their first century, they were, on average, earning far more than Thais, Malaysians, or Koreans. In terms of consumption of energy per person, the number of automobiles and miles of roads, levels of health, and literacy, Australia had all the properties of a developed country. Australian cities, where 85 percent of all Australians live, are not encircled by crowded shantytowns. Nor is the Australian countryside inhabited by a poverty-stricken peasantry.

MAJOR CITIES OF THE REALM

City	Population* (in millions)
Adelaide, Australia	1.1
Auckland, New Zealand	1.2
Brisbane, Australia	1.7
Canberra, Australia	0.3
Melbourne, Australia	3.3
Perth, Australia	1.4
Sydney, Australia	3.7
Wellington, New Zealand	0.4

*Based on 2002 estimates.

Distance

Australians often talk about distance. One of their leading historians, Geoffrey Blainey, labeled it a "tyranny"—an imposed remoteness from without and a divisive part of life within. Even today, Australia is far from nearly everywhere on Earth. A jet flight from Los Angeles to Sydney takes 14 hours nonstop and is correspondingly expensive. Freighters carrying products to European markets take ten days to two weeks to get there. Inside Australia, distances also are of continental proportions, and Australians pay the price—literally. Until some upstart private airlines started a price

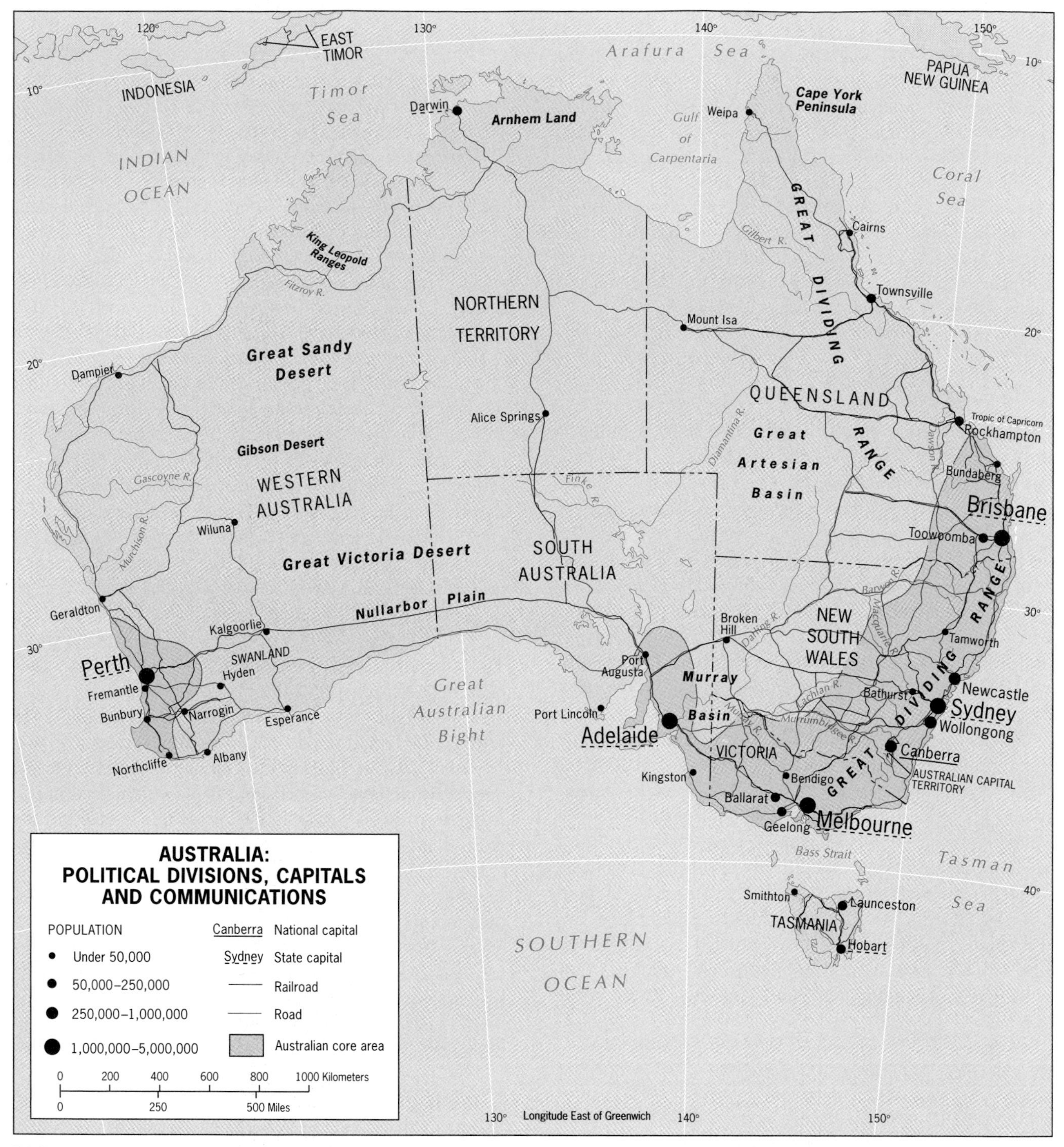

FIGURE 11-4

war, Australians paid more per mile for their domestic flights than air passengers anywhere else in the world.

But distance also was an ally, permitting Australians to ignore the obvious. Australia was a British progeny, a European outpost. Once you had arrived as an immigrant from Britain or Ireland, there were a wide range of environments, magnificent scenery, vast open spaces, and seemingly limitless opportunities. When the Japanese Empire expanded, Australia's remoteness saved the day. When immigration became an issue, Australia in its comfortable isolation could adopt an all-white admission policy that was not officially terminated until 1976. When boat people by the hundreds of thousands fled Vietnam in the aftermath of the Indochina War, almost none reached Australian shores.

Today Australia is changing and rapidly so. Immigration policy now focuses on the would-be immigrants' qualifications, skills, financial status, age, and facility in the English language. With regard to skills, high-technology specialists, financial experts, and medical personnel are especially welcome. Relatives of earlier immigrants, as well as a quota of genuine asylum-seekers, also are readily admitted. In recent years, total immigration has been limited to about 80,000 annually, but the number may have to increase if only to keep the country's population growing. According to recent data, Australia's natural rate of increase now barely hovers above zero.

Already, Australia's changed immigration policies have dramatically altered cultural landscapes, especially in the urban areas (see photo, p. 546). The country is fast becoming a truly multicultural society; in Sydney, for example, one in seven residents is of Asian ancestry, a ratio that will rise to one in five by 2010. During the 1990s, when for a time Japan was Australia's leading trade partner, Australian schools began to teach Japanese to tens of thousands of children. The Austral Realm is clearly in transition.

Core and Periphery

Australia is a large landmass, but its population is heavily concentrated in a core area that lies in the east and southeast, most of which faces the Pacific Ocean (here named the Tasman Sea between Australia and New Zealand). As Figure 11-4 shows, this crescent-like Australian heartland extends from north of the city of Brisbane to the vicinity of Adelaide and includes the largest city, Sydney, the capital, Canberra, and the second largest city, Melbourne. A secondary core area has developed in the far southwest, centered on Perth and its outport, Fremantle. Beyond lies the vast periphery, which the Australians call the **Outback**. 8

To better understand the evolution of this spatial arrangement, it helps to refer again to the map of world climates (Fig. I-8). Environmentally, Australia's most favored strips face the Pacific and Southern Oceans, and they are not large. We can describe the country as a coastal rimland with cities, towns, farms, and forested slopes giving way to the vast, arid, interior Outback. On the western flanks of the Great Dividing Range lie the extensive grassland pastures that catapulted Australia into its first commercial age—and on which still graze one of the largest sheep herds on Earth (over 160 million sheep, producing more than one-fifth of all the wool sold in the world). Where it is moister, to the north and east, cattle by the millions graze on ranchlands. This is frontier Australia, over which livestock have ranged for nearly two centuries.

Aboriginal Australians reached this landmass as long as 50,000 to 60,000 years ago, crossed the Bass Strait into Tasmania, and had developed a patchwork of indigenous cultures when Captain Arthur Phillip sailed into what is today Sydney Harbor (1788) to establish the beginnings of modern Australia. The Europeanization of Australia doomed the continent's Aboriginal societies. The first to suffer were those situated in the path of British settlement on the coasts, where penal colonies and free towns were founded. Distance protected the Aboriginal communities of the northern interior longer than elsewhere; in Tasmania, the indigenous Australians were exterminated in just decades after having lived there for perhaps 45,000 years.

Eventually, the major coastal settlements became the centers of seven different colonies, each with its own hinterland; by 1861, Australia was delimited by its now-familiar pattern of straight-line boundaries (Fig. 11-4). Sydney was the focus for New South Wales; Melbourne, Sydney's rival, anchored Victoria. Adelaide was the heart of South Australia, and Perth lay at the core of Western Australia. Bris-

FROM THE FIELD NOTES

"My most vivid memory from my first visit to Alice Springs is spotting vineyards and a winery in this parched, desert environment as the plane approached the airport. I asked a taxi driver to take me there, and got a lesson in economic geography. Drip irrigation from an underground water supply made viticulture possible; the tourist industry made it profitable. None of this, however, is evident from the view seen here: a spur of the Macdonnell Ranges overlooks a town of bare essentials under the hot sun of the Australian desert. What Alice Springs has is centrality: it is the largest settlement in a vast area Australians often call 'the centre.' Not far from the midpoint on the nearly 2000-mile (3200-km) Stuart Highway from Darwin on the Northern Territory's north coast to Adelaide on the Southern Ocean, Alice Springs also is the northern terminus of the Central Australian Railway, seen in the middle distance. The shipping of cattle and minerals is a major industry here. You need a sense of humor to live here, and the locals have it: the town actually lies on a river, the intermittent Todd River. An annual boat race is held, and in the absence of water the racers carry their boats along the dry river bed. No exploration of Alice Springs would be complete without a visit to the base of the Royal Flying Doctor Service, which brings medical help to outlying villages and homesteads."

bane was the nucleus of Queensland, and Hobart was the seat of government in Tasmania. The largest clusters of surviving Aboriginal people were in the so-called Northern Territory, with Darwin, on Australia's tropical north coast, its colonial city. Notwithstanding their shared cultural heritage, the Australian colonies were at odds not only with London over colonial policies but also with each other over economic and political issues. The building of an Australian nation during the late nineteenth century was a slow and difficult process.

A Federal State

On January 1, 1901, following years of difficult negotiations, the Australia we know today finally emerged: the Commonwealth of Australia, consisting of six States and two Federal Territories (Table 11-1; Fig. 11-4). The special status of Federal Territory was assigned to the Northern Territory to protect the interests of the Aboriginal population there. A second Federal Territory was delimited in southern New South Wales for the establishment of a new federal capital (Canberra). Both Sydney and Melbourne had fiercely competed for this honor, but it was decided that a completely new seat of government, in a specially designated area, was the best solution. In 1927, the government buildings were ready and the administration moved to Canberra, a city planned from scratch to serve a nation built from a group of quarrelsome colonies.

Today Canberra, Australia's only large city that is not situated on the coast, has a population of 342,000, less than one-tenth of Sydney's. As a growth pole, Canberra has been no great success. But as a symbol of Australian federalism, it has been a triumph. Australia's unification was made possible through the adoption of an idea with ancient Greek and Roman roots, one also familiar to Americans and Canadians: the notion of association. The term to describe it comes from the Latin *foederis*; in practice, it means alliance and coexistence, a union of consensus and common interest—a **federation**. It stands in contrast to the idea that states should be centralized, or unitary. For this, too, the ancient Romans had a term: *unitas*, meaning unity. Most European countries are **unitary states**, including the United Kingdom of Great Britain and Northern Ireland. Although the majority of Australians came from that tradition (a kingdom, no less), they managed to overcome their differences and establish a Commonwealth that was, in effect, a federation of States with different viewpoints, economies, and objectives, separated by vast distances along the rim of an island continent. And yet, the experiment succeeded.

9

10

An Urban Culture

During this century of federal association, the Australians developed an urban culture. Despite those vast open spaces and romantic notions of frontier and Outback, 85 percent of all Australians live in cities and towns. On the map, Australia's areal functional organization is similar to Japan's: large cities lie along the coast, the centers of manufacturing complexes as well as the foci of agricultural areas. Contributing to this situation in Japan was mountainous topography; in Australia, it was an arid interior. There, however, the similarity ends. Australia's territory is 20 times larger than Japan's, and Japan's population is nearly seven times that of Australia's. Japan's port cities are built to receive raw materials and to export finished products. Australia's cities forward minerals and farm products from the Outback to foreign markets and import manufactures from overseas. Distances in Australia are much greater, and spatial interaction (which tends to decrease with increasing distance) is less. In comparatively small, tightly organized Japan, you can travel from one end of the country to the other along highways, through tunnels, and over bridges with utmost speed and efficiency. In Australia, the overland trip from Sydney to Perth, or from Darwin to Adelaide, is time-consuming and slow. Nothing in Australia compares to Japan's high-speed bullet trains.

Table 11-1
STATES AND TERRITORIES OF FEDERAL AUSTRALIA, 2002

State	Area (1000 sq mi)	Estimated Population (millions)	Capital	Estimated Population (millions)
New South Wales	309.5	6.6	Sydney	3.7
Queensland	666.9	3.5	Brisbane	1.7
South Australia	379.9	1.5	Adelaide	1.1
Tasmania	26.2	0.5	Hobart	0.2
Victoria	87.9	4.9	Melbourne	3.3
Western Australia	975.1	1.9	Perth	1.4
Territory				
Australian Capital Territory	0.9	0.3	Canberra	0.3
Northern Territory	519.8	0.2	Darwin	0.1

Among the Realm's Great Cities . . .

Sydney

Slightly over two centuries ago, Sydney was founded by Captain Arthur Phillip as a British outpost on one of the world's most magnificent natural harbors. The free town and penal colony that struggled to survive evolved into Australia's largest city. Today Sydney (3.7 million) is home to nearly one-fifth of the country's entire population. An early start, the safe harbor, fertile nearby farmlands, and productive pastures in its hinterland combined to propel Sydney's growth. Later, as road and railroad links made Sydney the focus of Australia's growing core area, industrial development and political power augmented its primacy.

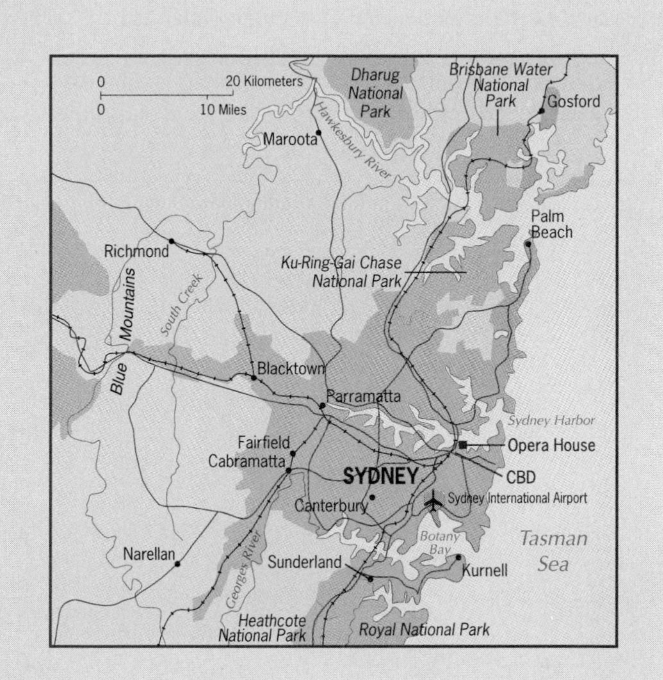

With its incomparable setting and mild, sunny climate, its many city beaches, and its easy reach to the cool Blue Mountains of the Great Dividing Range, Sydney is one of the world's most liveable cities. Good public transportation (including an extensive cross-harbor ferry system from the doorstep of the waterfront CBD), fine cultural facilities headed by the multi-theatre Opera House complex, and many public parks and other recreational facilities make Sydney attractive to visitors as well. A healthy tourist trade, much of it from Japan and other Asian countries, bolsters the city's economy. Sydney's hosting of the 2000 Olympic Games was further testimony to its rising visibility.

Increasingly, Sydney also is a multicultural city. Its small Aboriginal sector is being overwhelmed by the arrival of large numbers of Asian immigrants. The Sydney suburb of Cabramatta symbolizes the impact: more than half of its 80,000 residents were born elsewhere, mostly in Vietnam. Unemployment is high, drug use is a problem, and crime and gang violence persist. Yet, despite the deviant behavior of a small minority, tens of thousands of Asian immigrants have established themselves in some profession. As the photo and caption on p. 546 reveal, today things are definitely improving in this dynamic community.

These developments underscore Sydney's coming of age. The end of Australia's isolation has brought Asia across the country's threshold, and again the leading metropolis will show the way.

For all its vastness and youth, Australia nonetheless developed a remarkable cultural identity, a sameness of urban and rural landscapes that persists from one end of the continent to the other. Sydney, often called the New York of Australia, lies on a spectacular estuarine site, its compact, high-rise central business district overlooking a port bustling with ferry and freighter traffic. Sydney is a vast, sprawling metropolis with multiple outlying centers studding its far-flung suburbs; brash modernity and reserved British ways blend here. Melbourne, sometimes regarded as the Boston of Australia, prides itself on its more interesting architecture and more cultured ways. Brisbane, the capital of Queensland, which also anchors Australia's Gold Coast and adjoins the Great Barrier Reef, is the Miami of Australia; unlike Miami, however, its residents can find nearby relief from the summer heat in the mountains of its immediate hinterland (as well as at its beaches). Perth, Australia's San Diego, is one of the world's most isolated cities, separated from its nearest Australian neighbor by two-thirds of a continent and from Southeast Asia and Africa by thousands of miles of ocean.

And yet, each of these cities—as well as the capitals of South Australia (Adelaide), Tasmania (Hobart), and, to a lesser extent, the Northern Territory (Darwin)—exhibits an Australian character of unmistakable quality. Life is orderly and unhurried. Streets are clean, slums are few, graffiti rarely seen. By American and even European standards, violent crime (though rising) is uncommon. Standards of public transportation, city schools, and health-care provision are high. Spacious parks, pleasing waterfronts, and plentiful sunshine make Australia's urban life more acceptable than that almost anywhere else in the world. Critics of Australia's way of life say that this very pleasant state of affairs has persuaded Australians that hard work is not really necessary. Some experience in the commercial centers of the major cities contradicts that assertion: the pace of life is quickening. The country's cultural geography evolved as that of a European outpost, prosperous and secure in its isolation. Now Australia must reinvent itself as a major link in an Australo-Asian chain, a Pacific partner in a transformed regional economic geography.

Economic Geography

From the very beginning, however, goods imported from Britain (and later from the United States) were expensive,

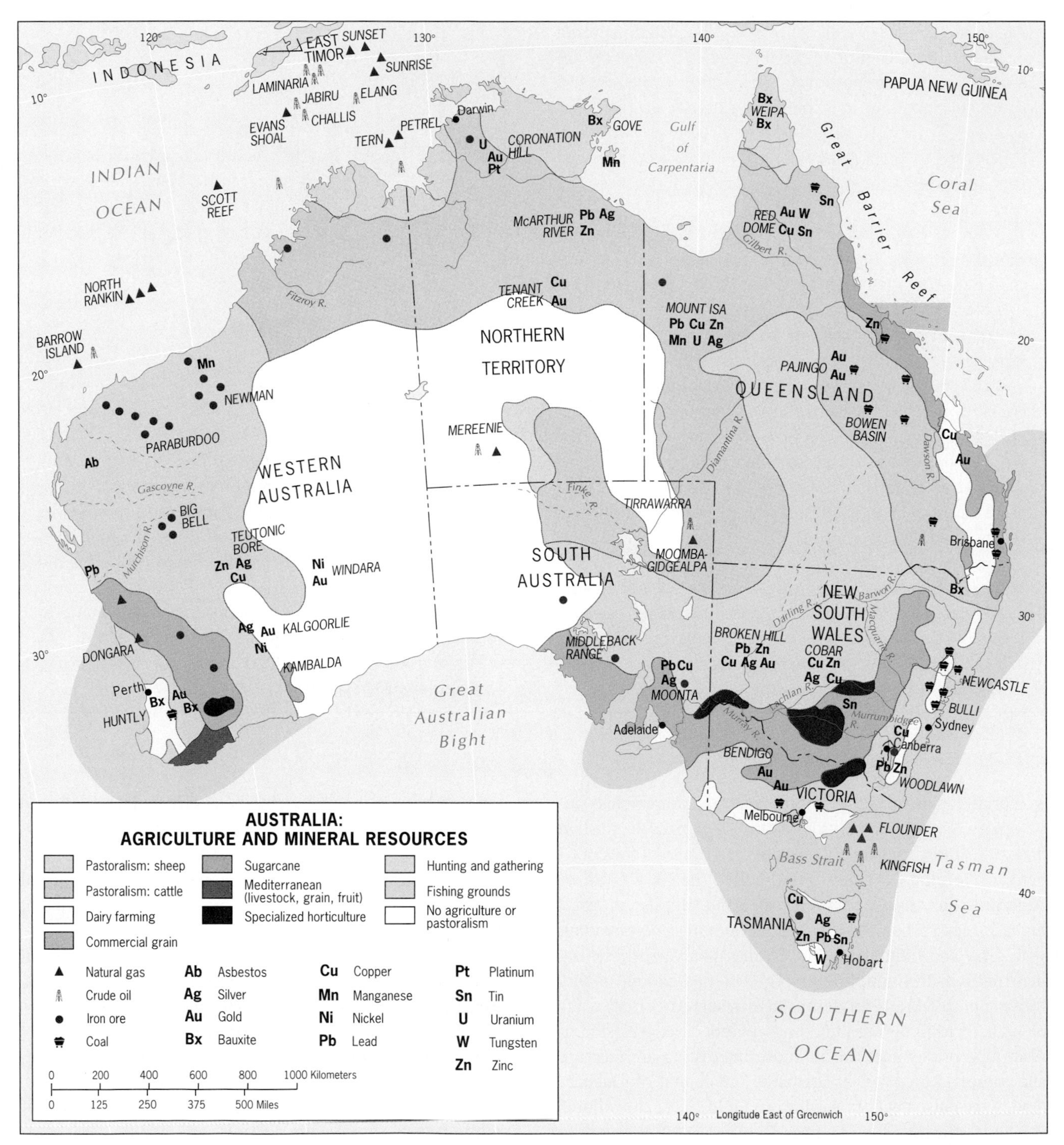

FIGURE 11-5

largely because of transport costs. This encouraged local entrepreneurs to set up their own industries in and near the developing cities—industries economic geographers call **import-substitution industries**.

When the prices of foreign goods became lower because transportation was more efficient and therefore cheaper, local businesses demanded protection from the colonial governments, and high tariffs were erected against imported goods. Local products now could continue to be made inefficiently because their market was guaranteed. If Japan could not afford this, how could Australia? We can see the answer on the map (Fig. 11-5). Even before federation in 1901, all the colonies could export valuable minerals whose earnings shored up those inefficient, uncompetitive local industries. By the time the colonies unified, pastoral industries were also contributing income. So the miners and the farmers paid for those imports Australians could not produce themselves, plus the products made in the cities. No wonder the cities grew: here were secure manufacturing jobs, jobs in state-run service enterprises, and jobs in the growing government bureaucracy. When we noted earlier that Australians once had the highest per-capita GNP in the world, this was achieved in the mines and on the farms, not in the cities.

But the good times had to come to an end. The prices of farm products fluctuated, and international market competition increased. The cost of mining, transporting, and shipping ores and minerals also rose. Australians like to drive (there are more road miles per person in Australia than in the United States), and expensive petroleum imports were needed. Meanwhile, the government-protected industries had been further fortified by strong labor unions. Not surprisingly, as the economy declined, the national debt rose, inflation grew, and unemployment crept upward. In 1999, only 22 percent by value of Australian exports were the kind of high-tech goods that have made East Asia's economic tigers so successful. That was more than double the 1985 figure but still far below what the country needed to produce.

Agricultural Abundance

And yet, Australia has material assets of which other countries on the Pacific Rim can only dream. In agriculture, sheep-raising was the earliest commercial venture, but it was the technology of refrigeration that brought world markets within reach of Australian beef producers. Wool, meat, and wheat have long been the country's big three income earners; Figure 11-5 displays the vast pastures in the east, north, and west that constitute the ranges of Australia's huge herds. The zone of commercial grain farming forms a broad crescent extending from northeastern New South Wales through Victoria into South Australia, and covers much of the hinterland of Perth. Keep in mind the scale of this map: Australia is only slightly smaller than the 48 contiguous States of the United States! Commercial grain farming in Australia is big business. As the climatic map would suggest, sugarcane grows along most of the warm, humid coastal strip of Queensland, and Mediterranean crops (including grapes for Australia's wines) cluster in the hinterlands of Adelaide and Perth. Mixed horticulture concentrates in the basin of the Murray River system, including rice, grapes, and citrus fruits, all under irrigation. And, as elsewhere in the world, dairying has developed near the large urban areas. With its considerable range of environments, Australia yields a diversity of crops.

Mineral Wealth

Australia's mineral resources, as Figure 11-5 shows, also are diverse. Major gold discoveries in Victoria and New South Wales produced a ten-year gold rush starting in 1851 and ushered in a new economic era. By the middle of that decade, Australia was producing 40 percent of the world's gold. Subsequently, the search for more gold led to the discoveries of other minerals. New finds are still being made today, and even oil and natural gas have been found both inland and offshore (see the symbols in Fig. 11-5 in the Bass Strait between Tasmania and the mainland, and off the northwestern coast of Western Australia). Coal is mined at many locations, notably in the east near Sydney and Brisbane but also in Western Australia and even in Tasmania; before coal prices fell, this was a valuable export. Major deposits of metallic and nonmetallic minerals abound—from the complex at Broken Hill and the mix of minerals at Mount Isa to the huge nickel deposits at Kalgoorlie and Kambalda, the copper of Tasmania, the tungsten and bauxite of northern Queensland, and the asbestos of Western Australia. A glance at the map reveals the wide distribution of iron ore (the red dots), and for this raw material as for many others, Japan has been Australia's best customer in recent years. In the late 1990s, Japan was buying more than one-third of all Australian mineral exports.

Manufacturing's Limits

Australian manufacturing, as we noted earlier, remains oriented to domestic markets. One cannot expect to find Australian automobiles, electronic equipment, or cameras challenging the Pacific Rim's economic tigers for a place on world markets—not yet, at any rate. Australian manufacturing is diversified, producing some machinery and equipment made of locally produced steel as well as textiles, chemicals, paper, and many other items. These industries cluster in and near the major urban areas where the markets are. The domestic market in Australia is not large, but despite declining real incomes, it remains relatively affluent. This makes it attractive to foreign producers, and Australia's shops are full of high-priced goods from Japan, South Korea, Taiwan, and Hong Kong. Indeed, despite its long-term protectionist practices, Australia still does not produce many goods that could be manufactured at home. Overall, the economy continues to display symptoms of a still-developing rather than a fully developed country.

Australia's Future

A new age is dawning in Australia, a country that not long ago (1988) celebrated the bicentennial of its first European settlement—two centuries as a European outpost facing the Pacific Ocean. Now, with Australia well into its third century, its European bonds are weakening and its Asian ties are strengthening.

On the face of it, Australia and its northern neighbors on the Pacific Rim would seem to exhibit a geographic complementarity: Japan and the economic tigers need Australia's excess food, metals, and minerals, and Australia needs the cheap manufactures Asia produces. But it is not that simple. Australia still has tariff barriers against imported goods, and the Asian countries on the Pacific Rim maintain import barriers against processed foods and minerals, thus discouraging Australia from refining its exports and earning more from them. Asia's manufacturers, furthermore, are more interested in the potential of large markets (such as Indonesia, with 219 million people) than Australia, and its investors like the low wages of East and Southeast Asia. In any case, Australia would find it difficult to open its economy and lower its protective tariffs without reciprocation from its Asian trading partners.

Australia's allies in the global core also have not been invariably helpful. U.S. government subsidies to American wheat farmers have enabled U.S. farmers to sell their wheat at lower prices than Australian farmers can afford to sell theirs. In the late 1990s, farm subsidies in the United States were under review—not because of long-term Australian appeals, but because of domestic political considerations. This underscores the risks inherent in an economy that depends substantially on revenues from agricultural staples.

FROM THE FIELD NOTES
"While in Darwin I had to visit a government office, and waiting in line with me was this Aboriginal girl, her brother, and her father. They lived about two hours from the city, she said (which I gathered was the length of their bus ride), and they had to come here from time to time for filing forms and visiting the doctor. When I asked whether they enjoyed coming to Darwin, all three shook their heads vigorously. 'It is not a friendly place,' she said. 'But we must come here.' In fact I had been surprised at the comparatively small number of Aboriginal people I had seen in Darwin, far fewer than in Alice Springs. Certainly laid-back Alice Springs is a very different place from larger, busier, and more modern Darwin."

Aboriginal Issues

As the twenty-first century begins, Australia faces other challenges. The Aboriginal population of 300,000 (including many of mixed ancestry) has been gaining influence in the country's affairs. In the 1980s, Aboriginal leaders began a campaign to obstruct exploration and mining on ancestral and sacred lands. Until 1992, Australians had taken it for granted that Aborigines had no rights to land ownership; the continent had been open and undemarcated, and there was no evidence of prior title. In that year, however, the Australian High Court ruled in favor of an Aborigine, Eddie Mabo, and his co-inhabitants of the Murray Islands in the Torres Strait. The court ruled that Mabo and his community owned customary title to their land. The ruling implied that Aborigines elsewhere, too, could claim title to traditional land.

When the federal government passed the law that would codify this far-reaching and astounding ruling, it protected Australia's pastoral leases (the tracts of land leased to ranchers of cattle and sheep by the Commonwealth) by excluding them—but not the holdings of mining companies. This led to an immediate Aboriginal victory at Coronation Hill in the Northern Territory (Fig. 11-5) where the administration prohibited the development of platinum, gold, and palladium deposits because they were deemed to lie on an Aboriginal community's sacred land.

Under the Native Title Bill, passed by the Australian Senate in 1993, Aborigines who had long lived on "vacant" land could petition the courts for title of that land: those who could prove that their land was taken away were entitled to compensation. And when a mining company's lease expired, Aborigines could reclaim the land. The result was a crisis of confidence in the mining industry and a sharp drop in investment in exploration, resource development, and infrastructure.

Soon the Aborigines tested the government's move to exclude the pastoral leases, and again they won in the High Court. In 1996, ruling on a case brought before it by the

Wik Aboriginal community in Queensland, the court decided that pastoral leases and native title could coexist, thereby voiding the government's exclusion of these leases from the Native Title Bill. This decision implied that vast areas (potentially as much as 78 percent of Australia) could be subject to Aboriginal claims (Fig. 11-6). Now not only miners and ranchers had reason to be concerned: activist Aboriginal groups announced that they would claim ownership to parcels of land in the heart of Sydney and other major cities.

Again the federal government tried to mitigate the impact of the High Court's ruling, this time by passing legislation that would abolish native title and thus preclude claims on pastoral leases where such title would "interfere" with the rights and work of the leaseholding pastoralist. Aborigines would be allowed to enter such land to hold ceremonies and gather food in their traditional ways, but they must not in any way obstruct pastoral activities by taking cultivated crops, moving livestock away, cutting fences, or opening gates. Moreover, all Aboriginal claims on offshore water and marine resources were voided. Aboriginal leaders responded by threatening to inundate the court system with numerous title claims that would cost those affected, and the government, hundreds of millions of dollars.

12 The **Aboriginal land issue** is essentially (though not exclusively) an Outback issue, notably in Western Australia, South Australia, and Queensland States. It is having a major impact on Australia's political geography; views on the matter in the large cities tend to differ sharply from those in the interior. It has had a polarizing effect on politics and society, strengthening political parties and movements opposed to making further "concessions" to Aborigines. In Queensland, the One Nation Party in 1998 got 25 percent of the vote on such a platform, forcing the ruling conservative party (the Liberals) that controlled the State government to make further concessions to its right wing, including a reduction in welfare awards to Aborigines. Polls indicate that most Australians nevertheless oppose the kind of extremism the anti-Aborigine (and anti-immigration) parties represent. But the national government must navigate a difficult course toward reconciliation.

It should be noted that only a few Aboriginal communities still pursue their original life of hunting, gathering, and fishing in the remotest corners of the Northern Territory, northern Queensland, and northernmost Western Australia. The remainder are scattered across the continent, subsisting on reserves the government set aside for them (much of the Northern Territory is so designated), working on cattle stations, or performing mostly menial jobs in cities and towns. In comparatively affluent Australia, the Aboriginal peoples suffer from poverty, disease, inadequate education, and even malnutrition. The Aboriginal rights question will be a major one in Australia's future.

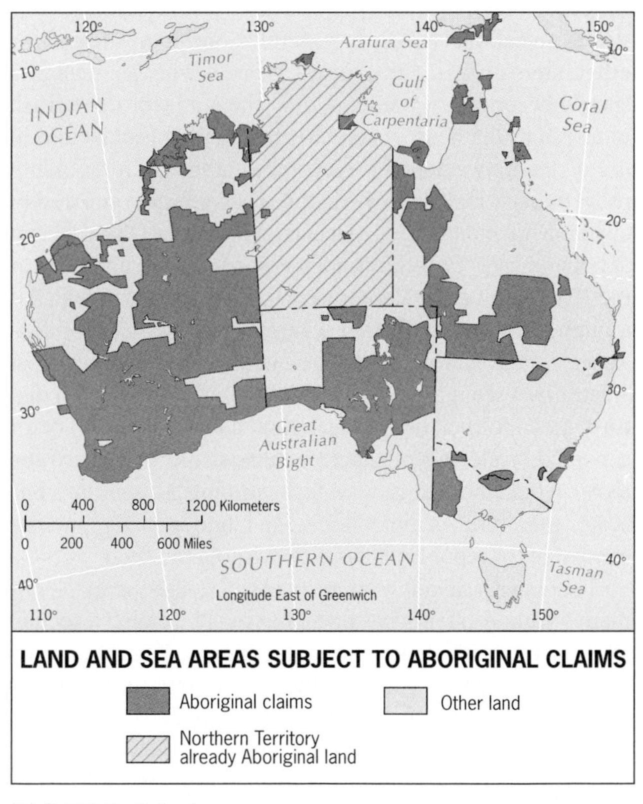

FIGURE 11-6

Immigration Issues

The population question has preoccupied Australia as long as Australia has existed—and even longer. Fifty years ago, when Australia had less than half the population it has today, 95 percent of the people were of European ancestry, and more than three-quarters of them came from the British Isles. Eugenic (race-specific) **immigration policies** maintained this situation until the 1970s. 13 Today, the picture is dramatically different: of 19.4 million Australians, only about one-third have British-Irish origins, and Asian immigrants outnumber both European immigrants and the natural increase each year. During the early 1990s, nearly 150,000 legal immigrants arrived in Australia annually, most from Hong Kong, Vietnam, China, the Philippines, India, and Sri Lanka. Immigration quotas have since been reduced, most recently to 80,000, but Asian immigrants continue to outnumber those from Western sources. Immigration from Britain and Ireland varies from 20,000 to 25,000 annually, and between 5000 and 10,000 New Zealanders move to Australia each year.

The immigration issue roils Australian society. In the mid-1990s, when hundreds of "boat people" landed on Australian shores as illegal immigrants, their arrival and subsequent treatment led to a wave of soul-searching. In 1998, when immigrant workers were among strike breakers at a major dockyard, labor unions denounced their role. Restrictions on foreign ownership of Australian real estate re-

The name Cabramatta conjures up varied reactions among Australians. During the 1950s and 1960s, many immigrants from southern Europe settled in this western suburb of Sydney attracted by affordable housing. During the 1970s and 1980s, Southeast Asians arrived in large numbers, and during this period Cabramatta, in the eyes of many, became synonymous with gang violence and drug dealing. More recently, however, Cabramatta's ethnic diversity has come to be viewed in a more favorable light, and is seen as the "multicultural capital of Australia," a tourist attraction and proof of Australia's capacity to accommodate non-Europeans. Meanwhile, Cabramatta has been spruced up with Oriental motifs of various kinds. This 'Freedom Gate' in the Vietnamese community is flanked by a Ming horse and a replica of a Forbidden City lion—all reflecting better times for an old transit point for immigrants and refugees.

flect Australian fears of what extremist groups call the "Asianization" of the country. At the same time, the success of many Asian settlers in Australia evinces the opportunities still available in this open and free society.

The Asian immigration issue is essentially an urban one: most Asian immigrants have settled in the cities and towns. Sydney has by far the largest contingent, and many ethnic districts have become part of its urban-cultural mosaic. In the 1970s Sydney, always Australia's most multicultural city, had Italian, Greek, and Yugoslavian neighborhoods. Today it also has Vietnamese, Laotian, Indian, and Chinese districts (see photo, above). Assimilation continues along a bumpy road, and the economy is the barometer: when the economy lags, social problems tend to worsen. Under any circumstances, however, as the ethnic complexion of Australia changes, multiculturalism will remain a challenge for the federation in the twenty-first century.

Environmental Degradation

Another growing issue involves the environment and conservation. Australia's wealth was derived not only from its ores and minerals and from soils and pastures: its ecology, too, paid a heavy price.

Great stands of magnificent forest were destroyed. In Western Australia, centuries-old trees were simply "ringed" and left to die, so that the sun could penetrate through their leafless crowns to nurture the grass below. Then the sheep could be driven into these new pastures. In Tasmania, where Australia's native eucalyptus tree reaches its greatest dimensions (comparable to North American redwood stands), tens of thousands of acres of this irreplaceable treasure have been lost to chain saws and pulp mills. Many of Australia's unique marsupial species have been destroyed, and many more are endangered or threatened. "Never have so few people wreaked so much havoc on the ecology of so large an area in so short a time," observed a geographer in Australia recently; but awareness of this **environmental degradation** 14 is growing. In Tasmania, the "Green" environmentalist political party has become a force in State affairs, and its activism has slowed deforestation, dam-building, and other "development" projects. Still, many Australians fear the environmentalist movement as an obstacle to economic growth at a time when the economy needs stimulation. This, too, is an issue for the future.

Status and Role

In recent years, two matters with political overtones have stirred up national debate in Australia, one domestic and the other international. The domestic issue is whether Australia should become a republic, ending the status of the British monarch as the head of state, or continue the status quo in the Commonwealth. In November 1999 a referendum proved that a majority of Australians were not prepared to abandon the monarchy—at least not in favor of "a president appointed by a two-thirds majority of the membership of the Commonwealth Parliament," as the ballot put it. Although polls had indicated a republican victory, the language on the ballot probably changed the outcome; most Australians seem to favor a republic, but more of them distrust their politicians. If the ballot had said "in favor of a popularly-elected president" chances are that Australia today would be on its way to becoming a republic. The issue, therefore, is not buried, and undoubtedly another campaign for republic status will be mounted.

The international matter is more consequential. For many years, Australia has had what may be called a special relationship with its neighbor, Indonesia. Australia needs Indonesia's help in curbing illegal seaborne immigration; Australia's and Indonesia's EEZ's (see p. 553) meet in the Timor Sea, where maritime boundaries allocate oil and natural gas reserves. It was thus convenient for Australia to go against international (UN) opinion and recognize Indonesia's 1976 annexation of East Timor. Canberra could deal directly with Jakarta on matters arising from this area. Although it could have made a critical difference, Australia gave neither recognition nor support to the rebel movement that subsequently fought for independence in East Timor.

Although successive Australian governments continued this policy, many Australians were uneasy about it, as revealed by editorials, press commentary, and letters to the media. When the East Timorese campaign for independence succeeded in 1999 and Indonesian troops began an orgy of destruction and murder, Australia took the lead by sending a peacekeeping force and spearheading the United Nations effort to stabilize the situation. But the entire issue led to a national debate over Australia's international role and its record on East Timor and Indonesia. Had Australia been too cooperative with a rapacious regime in Jakarta? Should Australia have acted as it did toward the rebels in East Timor? What lessons should Australia take from this experience to apply to other problem areas (for example, West Papua and Bougainville)?

Territorial dimensions, relative location, and raw-material wealth have helped determine Australia's place in the world and, more specifically, on the Pacific Rim. Australia's population, still under 20 million in the first decade of the new century, is smaller than Malaysia's and barely larger than that of the Caribbean island of Hispaniola. But Australia's importance in the international community far exceeds its human numbers.

FROM THE FIELD NOTES

"The drive from Christchurch to Arthur's Pass on the South Island of New Zealand was a lesson in physiography and biogeography. Here, on the east side of the Southern Alps, you leave the Canterbury Plain and its agriculture and climb into the rugged topography of the glacier-cut, snowcapped mountains. A last pasture lies on a patch of flatland in the foreground; in the background is the unmistakable wall of a U-shaped valley sculpted by ice. Natural vegetation ranges from pines to ferns, becoming even more luxuriant as you approach the moister western side of the island."

NEW ZEALAND

Fifteen hundred miles east-southeast of Australia, in the Pacific Ocean across the Tasman Sea, lies New Zealand. In an earlier age, New Zealand would have been part of the Pacific geographic realm because its population was Maori, a people with Polynesian roots. But New Zealand, like Australia, was invaded and occupied by Europeans. Today, its population of 3.9 million is almost 80 percent European, and the Maori form a minority of less than 400,000, with many of mixed Euro-Polynesian ancestry.

New Zealand consists of two large mountainous islands and many scattered smaller islands (Fig. 11-7). The two large islands, with the South Island somewhat larger than the North Island, look diminutive in the great Pacific Ocean, but together they are larger than Britain. In contrast to Australia, the two main islands are mainly mountainous or hilly, with several peaks rising far higher than any on the Australian landmass. The South Island has a spectacular snowcapped range appropriately called the Southern Alps, with numerous peaks reaching beyond 10,000 feet (3300 m). The smaller North Island has proportionately more land under low relief, but it also has an area of central highlands along whose lower slopes lie the pastures of New Zealand's chief dairying district. Hence, while Australia's land lies relatively low in elevation and exhibits much low relief, New Zealand's is on the average high and is dominated by rugged relief.

Human Spatial Organization

Thus the most promising areas for habitation are the lower-lying slopes and lowland fringes on both islands. On the North Island, the largest urban area, Auckland, occupies a comparatively low-lying peninsula. On the South Island, the largest lowland is the agricultural Canterbury Plain, centered on Christchurch. What makes these lower areas so attractive, apart from their availability as cropland, is their magnificent pastures. The range of soils and pasture plants allows both summer and winter grazing. Moreover, the Canterbury Plain, the chief farming region, also produces a wide variety of vegetables, cereals, and fruits. About half of all New Zealand is pasture land, and much of the farming provides fodder for the pastoral industry. Sixty million sheep and eight million cattle dominate these livestock-raising activities, with wool, meat, and dairy products providing nearly two-thirds of the islands' export revenues.

Despite their contrasts in size, shape, physiography, and history, New Zealand and Australia have much in common.

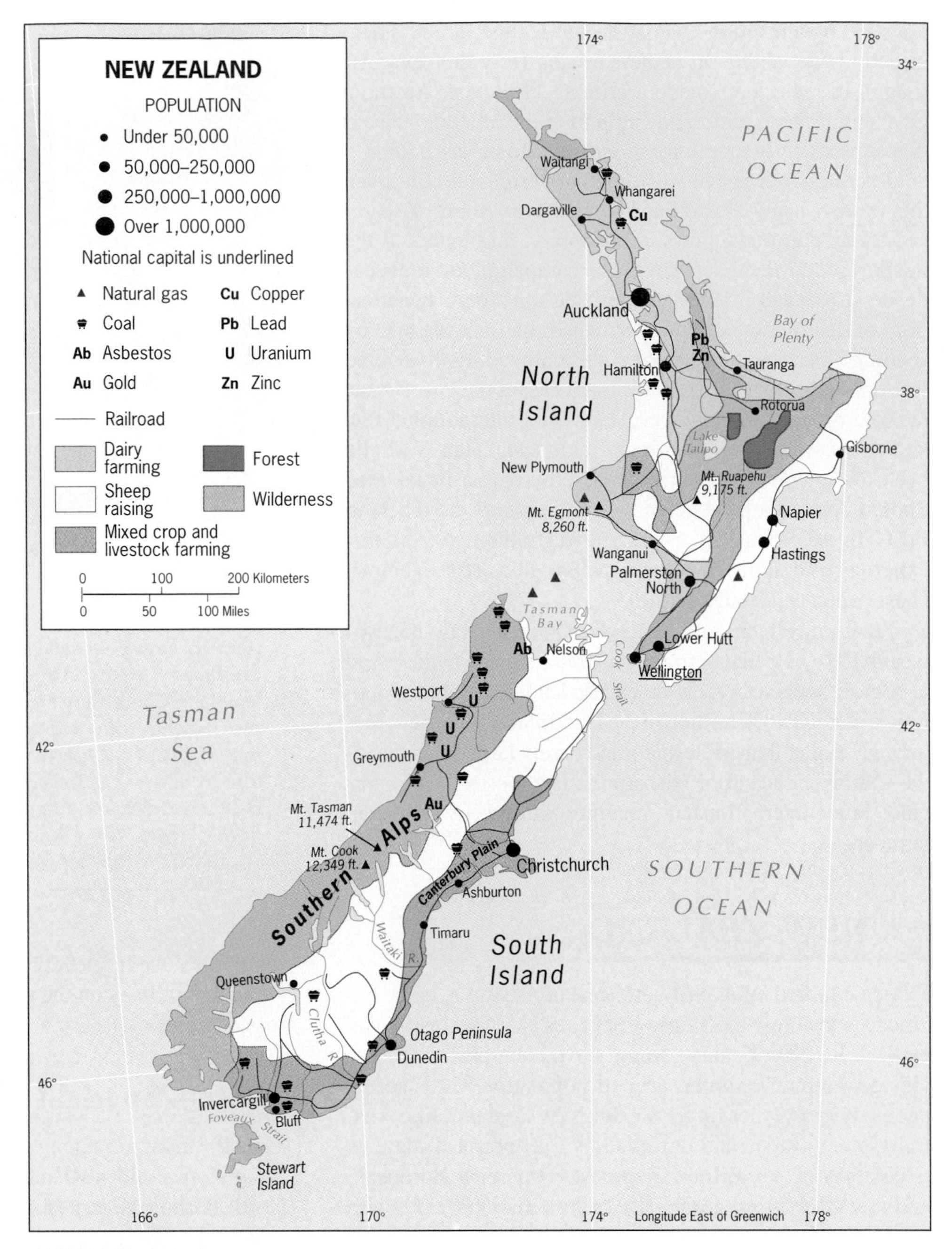

FIGURE 11-7

Apart from their joint British heritage, they share a sizeable pastoral economy, a small local market, the problem of great distances to world markets, and a desire to stimulate (through protection) domestic manufacturing. The high degree of urbanization in New Zealand (85 percent of the total population) again resembles Australia: substantial employment in city-based industries, mostly the processing and packing of livestock and farm products, and government jobs.

More remote even than Australia, New Zealand also has been affected by Pacific Rim development, though less so than its giant neighbor. Still, by the mid-1990s, Japan and other Pacific Rim economies were buying more of New Zealand's exports (mainly agricultural products) than either Australia or the United States. Britain today takes only about 6 percent of New Zealand's exports. Australia and the United States still send New Zealand most of its imports (about 40 percent combined), but the Pacific Rim's contribution has been rising. As Figure I-11 and Table I-2 show, New Zealand is a high-income economy, but it is not in the upper tier of the world's richer economies.

Spatially, New Zealand shares with Australia its pattern of **peripheral development** (Fig. I-9), imposed not by desert but by high rugged mountains. The country's major cities—Auckland and the capital of Wellington (together with its satellite of Hutt) on the North Island, and Christ-

15

church and Dunedin on the South Island—are all located on the coast, and the entire railway and road system is peripheral in its configuration (Fig. 11-7). This is more pronounced on the South Island than in the north because the Southern Alps are New Zealand's most formidable barrier to surface communications.

The Maori Factor and New Zealand's Future

One of New Zealand's historic sites is the wooden Treaty House at Waitangi, in the northern peninsula of the North Island. There, following a series of brutal conflicts, the British and the Maori signed the Treaty of Waitangi in 1840. The treaty granted the British sovereignty over New Zealand, but it also guaranteed Maori rights over tribal lands. February 6, the date of the signing of the treaty, is a public holiday in New Zealand called Waitangi Day. A ceremony at the Treaty House, attended by the prime minister and by Maori leaders, is a highlight of the day's events. Visitors to the ceremony in 1995 were in for a shock, however. Maoris disrupted the proceedings; the prime minister was jostled by members of the crowd; and an attempt was made to burn the Treaty House. For the first time in more than half a century, the ceremony had to be abandoned.

What precipitated this show of anger? In fact, it was only the latest in a series of events signaling a rise in Maori ethnic consciousness. Maori leaders insist that the terms of the Waitangi Treaty (partly abrogated in 1862) be enforced, that large tracts of land in urban as well as rural areas, amounting to more than half of the national territory, be awarded to their rightful Maori owners, and that Maori fishing rights be paramount over offshore waters.

Judicial rulings during the 1990s supported the Maori position, and Maori successes in one arena have led to new claims in others. Today, the Maori question has become the leading national issue in New Zealand, and Maori activism has risen sharply. In this context, the events of February 6, 1995 were no surprise.

The Maori appear to have reached New Zealand during the tenth century AD, and by the time the European colonists arrived they had had an enormous impact on the islands' ecosystems, especially on the North Island where most Maori lived. One of the Maori complaints is the persistently slow pace of integration of the Maori minority of under 400,000 into modern New Zealand society. In fact, the Maori have been joined by other Pacific islanders, so that Auckland today not only is New Zealand's largest city; it also may be the largest urban concentration of Polynesians anywhere, with over 100,000 Maori as well as Samoans, Cook Islanders, Tongans, and others making up one-sixth of the metropolitan population. Their neighborhoods give Auckland an ethnic patchwork that evinces the non-integration about which the Maori complain.

The Maori issue has other geographic overtones. Nearly all Maori still live on the North Island (where three-quarters of all New Zealanders reside), but Maori leaders are making huge claims, based on historic primacy, over much of the South Island. Remote and comparatively conservative, South Islanders have reacted to these claims much as their counterparts in Western Australia have to the Mabo and Wik decisions on Aboriginal rights.

The Waitangi Treaty was intended as the framework for a partnership between the British and the Maoris. If the current disputes can lead to a reconsideration and fulfillment of the treaty's terms in the context of modern New Zealand's human geography, then the country's future could be shaped by a harmonious, mainly bicultural society. If this effort fails, the specter of serious ethnic polarization threatens against a demographic background that projects the rapidly growing Polynesian population of New Zealand will double to 25 percent of the national total by 2012.

FROM THE FIELD NOTES
"It was Sunday morning in Christchurch on New Zealand's South Island, and the city center was quiet. As I walked along Linwood Street I heard a familiar sound, but on an unfamiliar instrument: the Bach sonata for unaccompanied violin in G minor—played magnificently on a guitar. I followed the sound to the artist, a Maori musician of such technical and interpretive capacity that there was something new in every phrase, every line, every tempo. I was his only listener; there were a few coins in his open guitar case. Shouldn't he be playing before thousands, in schools, maybe abroad? No, he said, he was happy here, he did alright. A world-class talent, a street musician playing Bach on a Christchurch side street, where tourists from around the world were his main source of income. Talk about globalization."

Dominant cultural heritage and prevailing cultural landscape form two criteria on which the delimitation of the Austral Realm is based. But in both Australia and New Zealand, the cultural mosaic is changing, and the convergence with neighboring realms is proceeding.

ARCTIC OCEAN
NOVOSIBIRSKIYE O-VA
Laptev Sea
East Siberian Sea
VRANGELYA I.
Beaufort Sea
ELLESMERE ISLAND
VICTORIA ISLAND
BAFFIN ISLAND
BROOKS RANGE
ALASKA (U.S.A.)
ALASKA RANGE
Arctic Circle
Lena
VERKHOYANSKIYE KHREBET
KORYAKSKIY KHREBET
RUSSIA
ST. LAWRENCE I.
Bering Str.
Anchorage
Gulf of Alaska
KODIAK ISLAND
ROCKY MOUNTAINS
CANADA
Hudson Bay
Sea of Okhotsk
P-OV KAMCHATKA
BERING SEA
ALEUTIAN ISLANDS
QUEEN CHARLOTTE ISLANDS
SAKHALIN I.
Petropavlovsk-Kamchatskiy
Vancouver
Victoria
Seattle
Portland
MONTRÉAL
TORONTO
CHICAGO
NEW YORK
UNITED STATES
St. Louis
Washington, D.C.
CHINA
HARBIN
SHENYANG
Vladivostok
KURILE ISLANDS
HOKKAIDO
NORTH KOREA
PYONGYANG
SOUTH KOREA
SEOUL
JAPAN
HONSHU
TOKYO
YOKOHAMA
KYUSHU
OSAKA
KITAKYUSHU
EAST CHINA SEA
RYUKYUS
San Francisco
SIERRA NEVADA
LOS ANGELES
SAN DIEGO
Missouri
Snake
Colorado
Mississippi
Rio Grande
New Orleans
Gulf of Mexico
Miami
INTERNATIONAL DATE LINE
MIDWAY IS. (U.S.A.)
BONIN IS. (JAPAN)
PACIFIC OCEAN
SIERRA MADRE OCCIDENTAL
HAVANA
BAHAMAS
CUBA
TAIPEI
TAIWAN
Tropic of Cancer
HAWAIIAN IS. (U.S.A.)
Honolulu
MEXICO
MEXICO CITY
JAMAICA
BELIZE
Belmopan
HONDURAS
GUATEMALA
GUATEMALA CITY
San Salvador
EL SALVADOR
Managua
NICARAGUA
Tegucigalpa
San José
COSTA RICA
PANAMA
BOGOTÁ
PHILIPPINE SEA
MANILA
PHILIPPINES
GUAM (U.S.A.)
NORTHERN MARIANA ISLANDS
WAKE I. (U.S.A.)
MARSHALL ISLANDS
Majuro
CAROLINE IS.
PALAU IS.
PALAU
FEDERATED STATES OF MICRONESIA
CELEBES SEA
HOWLAND I. (U.S.A.)
BAKER I. (U.S.A.)
KIRITIMATI
Equator
GALÁPAGOS IS. (Ecuador)
QUITO
ECUADOR
INDONESIA
NEW GUINEA
BISMARCK ARCH.
NAURU
KIRIBATI
JAVA SEA
PAPUA NEW GUINEA
Port Moresby
SOLOMON ISLANDS
Honiara
Funafuti
TUVALU
SAMOA
Apia
TOKELAU (N.Z.)
AMERICAN SAMOA
MARQUESAS IS.
ANDES
PERU
LIMA
ARAFURA SEA
EAST TIMOR
TIMOR SEA
CORAL SEA
VANUATU
NEW HEBRIDES
Port-Vila
WALLIS AND FUTUNA (Fr.)
FIJI
Suva
COOK ISLANDS (N.Z.)
SOCIETY IS.
TAHITI
ILES TUAMOTU
NEW CALEDONIA (Fr.)
LOYALTY IS.
TONGA
NIUE (N.Z.)
FRENCH POLYNESIA
Tropic of Capricorn
AUSTRALIA
GREAT DIVIDING RANGE
Brisbane
PITCAIRN (U.K.)
RAPA NUI (EASTER I.) (Chile)
Adelaide
Canberra
SYDNEY
Murray
NEW ZEALAND
NORTH ISLAND
Auckland
TASMAN SEA
MELBOURNE
PACIFIC OCEAN
SANTIAGO
CHILE
SOUTHERN OCEAN
TASMANIA
Hobart
SOUTH ISLAND
Wellington
Christchurch
Dunedin
Strait of Magellan
SOUTHERN OCEAN
Antarctic Circle
ANTARCTICA
Relief
Meters
Feet
3050
10 000
1525
5000
610
2000
305
1000
152.5
500
0
Sea Level
0
152.5
500
1525
5000
3050
10 000
6100
20 000
0 500 100 2000 Miles
0 1000 2000 Kilometers
Scale 1:87,500,000; one inch to 1,381 miles
Apinus II projection
Elevations and depressions are given in feet

FIGURE 12-1

chapter 12

The Pacific Realm

CONCEPTS, IDEAS, AND TERMS

1. **Marine geography**
2. **Territorial sea**
3. **High seas**
4. **Continental shelf**
5. **Exclusive Economic Zone (EEZ)**
6. **Maritime boundary**
7. **Median-line boundary**
8. **High-island cultures**
9. **Low-island cultures**
10. **Antarctic Treaty**

REGIONS

- MELANESIA
- MICRONESIA
- POLYNESIA

DEFINING THE REALM

Between the Americas to the east and the western Pacific Rim to the west lies the vast Pacific Ocean, larger than all the world's land areas combined. In this greatest of all oceans lie tens of thousands of islands, some large (New Guinea is by far the largest), most small (many are uninhabited). Together, the land area of these islands is a mere 376,000 square miles (974,000 sq km), about the size of Texas plus New Mexico, and over 90 percent of this lies in New Guinea.*

The Pacific geographic realm—land and water—covers nearly an entire hemisphere of this world, the one commonly called the Sea Hemisphere (Fig. 12-1). This Sea Hemisphere meets the Russian and North American realms in the far north and merges into the Southern Ocean in the south. Despite the preponderance of water, this fragmented, culturally complex realm does possess regional identities. It includes the Hawaiian Islands, Tahiti, Tonga, and Samoa—fabled names in a world apart.

In terms of modern cultural and political geography, Indonesia and the Philippines are not part of the Pacific Realm, although Indonesia's political system reaches into it; nor are Australia and New Zealand part of it. Before the European invasion and colonization, Australia would have been included because of its Aboriginal population and New Zealand because of its Maori population's Polynesian affinities. But the Europeanization of their countries has engulfed black Australians and Maori New Zealanders, and the regional geography of Australia and New Zealand today is decidedly not Pacific. In New Guinea, on the other hand, Pacific peoples remain numerically and culturally the dominant element.

Politico-geographically, the Pacific Realm is delimited into rectangular units marking the maritime boundaries of groups of islands (Fig. 12-1). The Pacific islands were colonized by the French, British, and Americans; an indigenous Polynesian kingdom in the Hawaiian Islands was annexed by the United States and is now the fiftieth State. Still, today, the map is an assemblage of independent and colonial territories. Paris controls New Caledonia and French Polynesia. The United States administers Guam and American Samoa, the Line Islands, Wake Island, Midway Islands, and several smaller islands; the United States also has special relationships with other territories, former dependencies that are now nominally independent. The British, through New Zealand, have responsibility for the

*The figures in Table I-2 do not match these totals because only the political entity of Papua New Guinea is listed, not the Indonesian part of the island (Irian Jaya [West Papua]). Here, as Figure 10-2 showed, the political and the realm boundaries do not coincide.

◆ Major Geographic Qualities of the Pacific Realm

1. The Pacific Realm's total area is the largest of all geographic realms. Its land area, however, is the smallest, as is its population.
2. The island of New Guinea, with 7.2 million people, alone contains over 80 percent of the Pacific Realm's population.
3. The Pacific Realm, with its wide expanses of water and numerous islands, has been strongly affected by United Nations Law of the Sea provisions regarding states' rights over economic assets in their adjacent waters.
4. The highly fragmented Pacific Realm consists of three regions: Melanesia (including New Guinea), Micronesia, and Polynesia.
5. Melanesia forms the link between Papuan and Melanesian cultures in the Pacific.
6. The Pacific Realm's islands and cultures may be divided into volcanic *high-island* cultures and coral-based *low-island* cultures.
7. In Micronesia, U.S. influence has been particularly strong and continues to affect local societies.
8. In Polynesia, local cultures are nearly everywhere severely strained by external influences. In Hawai'i, as in New Zealand, indigenous culture has been largely submerged by Westernization.
9. Indigenous Polynesian culture continues to exhibit a remarkable consistency and uniformity throughout the Polynesian region, its enormous dimensions and dispersal notwithstanding.

Pitcairn group of islands, and New Zealand administers and supports the Cook, Tokelau, and Niue Islands. Easter Island, the storied speck of land in the southeastern Pacific, is part of Chile. Indonesia rules West Papua.

Other island groups have become independent states. The largest are Fiji, once a British dependency, the Solomon Islands (also formerly British), and Vanuatu (until 1980 ruled jointly by France and Britain). Also on the current map, however, are such microstates as Tuvalu, Kiribati, Nauru, and Palau. Foreign aid is crucial to the survival of most of these countries. Tuvalu, for example, has a total area of about 10 square miles, a population of some 10,000, and a per-capita GNP of about $600, derived from fishing, copra sales, and some tourism. But what really keeps Tuvalu going is an international trust fund set up by Australia, New Zealand, the United Kingdom, Japan, and South Korea. Annual grants from that fund, as well as money sent back to families by workers who have left for New Zealand and elsewhere, allow Tuvalu to survive.

THE PACIFIC REALM AND ITS MARINE GEOGRAPHY

Certain land areas may not be part of the Pacific Realm (we cited the Philippines and New Zealand), but the Pacific Ocean extends from the shores of North and South America to mainland East and Southeast Asia and from the Bering Sea to the Subtropical Convergence. This means that several seas, including the Sea of Japan (East Sea), the East China Sea, and the South China Sea, are part of the Pacific Ocean. As we will see later, this relationship matters. Pacific coastal countries, large and small, mainland and island, compete for jurisdiction over the waters that bound them.

The Pacific Realm and its ocean, therefore, form an ideal place to focus on **marine geography**. This field encompasses a variety of approaches to the study of oceans and seas; some marine geographers focus on the biogeography of coral reefs, others on the geomorphology of beaches, still others on the movement of currents and drifts. A particularly interesting branch of marine geography has to do with the definition and delimitation of political boundaries at sea. Here geography meets political science and maritime law. 1

Littoral (coastal) states do not end where atlas maps suggest they do. States have claimed various forms of jurisdiction over coastal waters for centuries, closing off bays and estuaries and ordering foreign fishing fleets to stay away from nearby fishing grounds. Thus arose the notion of the **territorial sea**, where all the rights of a coastal state would prevail. Beyond lay the **high seas**, free, open, and unfettered by national interests. 2 3

It was in the interest of colonizing, mercantile states to keep territorial seas narrow and high seas wide, thus interfering as little as possible with their commercial fleets. In the seventeenth and eighteenth centuries the territorial sea was 3, 4, or at most 6 nautical miles wide, and the colonizing powers claimed the same widths for their colonies (1 nautical mile = 1.15 statute miles [1.85 km]).

In the twentieth century these constraints weakened. States without trading fleets saw no reason to limit their territorial seas. States with nearby fishing grounds traditionally exploited by their own fleets wanted to keep the increasing number of foreign trawlers away. States with 4 shallow **continental shelves**, offshore continuations of coastal plains, wished to control the resources on and below the seafloor, made more accessible by improving technology. States disagreed on the methods by which offshore boundaries, whatever their width, should be defined. Early efforts by the League of Nations in the 1920s to resolve these issues had only partial success, mainly in the technical area of boundary delimitation.

In 1945, the United States helped precipitate what has become known as the "scramble for the oceans." President Harry S Truman issued a proclamation that claimed U.S. jurisdiction and control over all the resources "in and on" the continental shelf down to its margin, around 100 fathoms (600 feet) deep. In some areas, the shallow continental shelf of the United States extends more than 200 miles offshore, and Washington did not want foreign countries drilling for oil just beyond the 3-mile territorial sea.

Few observers foresaw the impact the Truman Proclamation would have, not only on U.S. waters but on the oceans everywhere, including the Pacific. It set off a rush of other claims. In 1952, a group of South American countries, some with little continental shelf to claim, issued the Declaration of Santiago, claiming exclusive fishing rights up to a distance of 200 nautical miles (230 statute miles) off their coasts. Meanwhile, as part of the Cold War competition, the Soviet Union urged its allies to claim a 12-mile territorial sea.

Now the United Nations intervened, and a series of UNCLOS (United Nations Conference on the Law of the Sea) meetings began. These meetings addressed issues ranging from the closure of bays to the width and delimitation of the territorial sea, and after three decades of negotiations they achieved a convention that changed the political and economic geography of the oceans forever. Among its key provisions were the authorization of a 12-mile territorial sea for all countries and the establishment of a 200-mile (230 statute-mile) **Exclusive Economic Zone (EEZ)** over 5 which a coastal state would have total economic rights. Resources in and under this EEZ (fish, oil, minerals) belong to the coastal state, which could either exploit them or lease, sell, or share them as it saw fit.

These provisions had a far-reaching impact on the world's oceans and seas (Fig. 12-2), and especially so on the Pacific. Unlike the Atlantic Ocean, the Pacific is studded with islands large and small, and a microstate consisting of one small island suddenly acquired an EEZ covering 166,000 square nautical miles. European colonial powers still holding minor Pacific possessions (notably France) saw their maritime jurisdictions vastly expanded. Small low-income archipelagos could now bargain with large, rich fishing nations over fishing rights in their EEZs. And for all the UNCLOS Convention's provisions for the "right of innocent passage" of shipping through EEZs and via narrow straits, the world's high seas have obviously been diminished.

The extension of the territorial sea to 12 nautical miles and the EEZ to an additional 188 nautical miles (both are measured from the coastline) created new **maritime** 6 **boundary** problems. Waters less than 24 miles wide sepa-

FROM THE FIELD NOTES

"Having arrived late at night, I did not have a perspective yet of the setting of Pago Pago—but even before I got up I knew what the leading industry here had to be. An intense, overpowering smell of fish filled the air. Sure enough, daylight revealed a huge tuna processing plant right across the water, with fishing boats arriving and departing continuously. American Samoa has a huge Pacific-Ocean EEZ, one of the richest fishing grounds in the whole realm, and seafood exports rank ahead of tourism as the leading source of external income for the territory, an industry that is mostly American-owned. As the photograph shows, American Samoa is a group of 'high-islands' of volcanic origin, with considerable relief and environmental variety. Tuvalu, the country I had just left, was the complete opposite—a 'low-island' territory whose EEZ has been leased to Japanese fishing fleets, which do their processing on board and employ no locals."

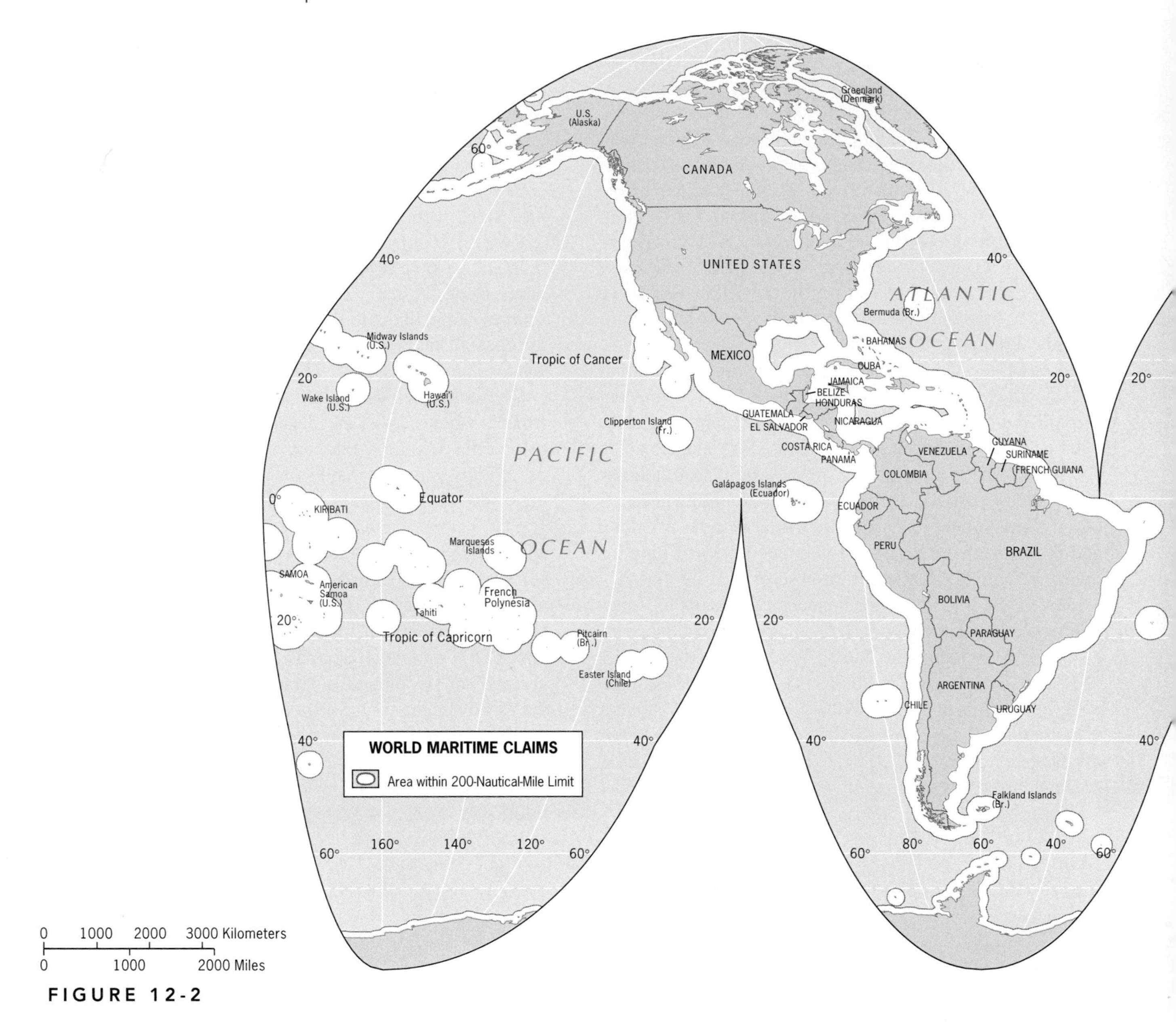

FIGURE 12-2

rate many countries all over the world, so that **median lines**, equidistant from opposite shores, have been delimited to establish their territorial seas. And even more countries lie closer than 400 nautical miles apart, requiring further maritime-boundary delimitation to determine their EEZs. In such maritime regions as the North Sea, the Caribbean Sea, and the Japan, East China, and South China Seas, a maze of maritime boundaries emerged, some of them subject to dispute. Political changes on land can lead to significant modifications at sea. East Timor's independence, and its new median-line EEZ boundary across the Timor Sea, greatly diminishes Australia's share of resources there.

The UNCLOS provisions also created opportunities for some states to expand their spheres of influence. Wider territorial-sea and EEZ allocations raised the stakes: claiming an island now entailed potential control over a huge maritime area. In Chapters 9 and 10, we referred to the large number of island disputes off mainland East Asia: between Japan and Russia, Japan and South Korea, Japan and China, China and Vietnam, and China and the Philippines. Ownership of many islands there is uncertain, and small specks of island territory have become large stakes in the scramble for the oceans. In the case of the Spratly Islands (Fig. 10-2), six countries claim ownership, including both Taiwan and China. China's island claims in the South China Sea sup-

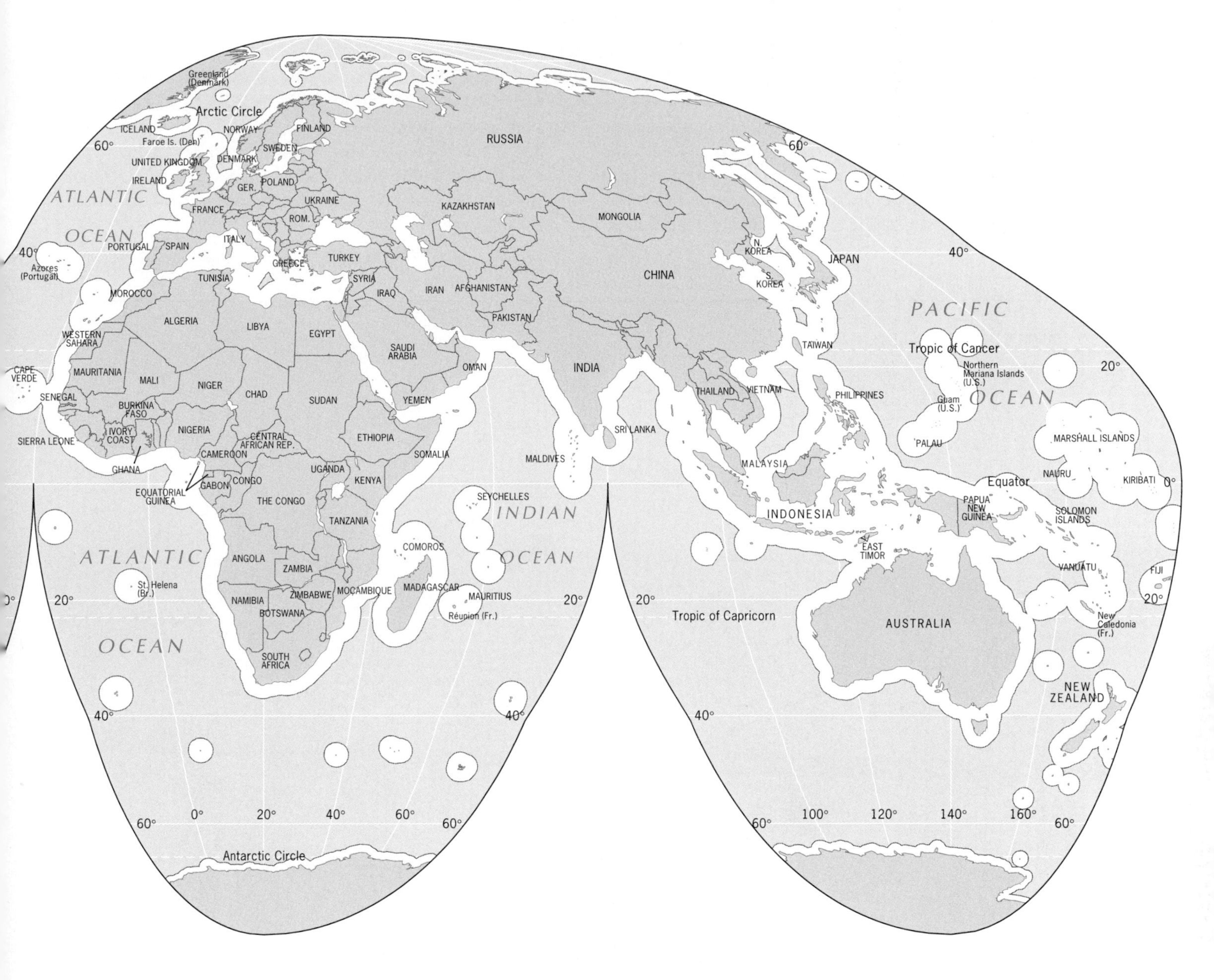

port Beijing's contention that this body of water is part and parcel of the Chinese state—a position that worries other states with coasts facing it.

Figure 12-2 reveals what EEZ regulations have meant to Pacific Realm countries such as Tuvalu, Kiribati, and Fiji. Those rectangular, geometrically bounded spaces shown on Figure 12-1 are now superseded by the nearly circular EEZs that surround the clusters of islands in this vast ocean space. Japan, Taiwan, and other fishing nations have purchased fishing rights in these EEZs from the island governments, although violations of EEZ rights do occur. Recently, Vanuatu and the Philippines were at odds over unauthorized Filipino fishing in Vanuatu's EEZ.

The process of boundary delimitation continues. In an earlier edition of this book, we included a map of the South China Sea and nearby waters off East and Southeast Asia, showing the median-line boundaries delimited according to UNCLOS specifications. But that map changed when coastal states engaged in bilateral and multilateral negotiations—and argued over island ownership with major boundary implications. The Pacific (and world) map of maritime boundaries remains a work in progress.

REGIONS OF THE REALM

In this realm as a whole, tourism may well be the largest overall revenue earner. Sail across the Pacific Ocean, and one spectacular vista follows another. Dormant and extinct volcanoes, sculpted by erosion into basalt spires draped by luxuriant tropical vegetation and encircled by reefs and lagoons, tower over azure waters. Low atolls with nearly snow-white beaches, crowned by stands of palm trees, seem to float in the water. Pacific islanders, where foreign influences have not overtaken them, appear to take life with enviable ease.

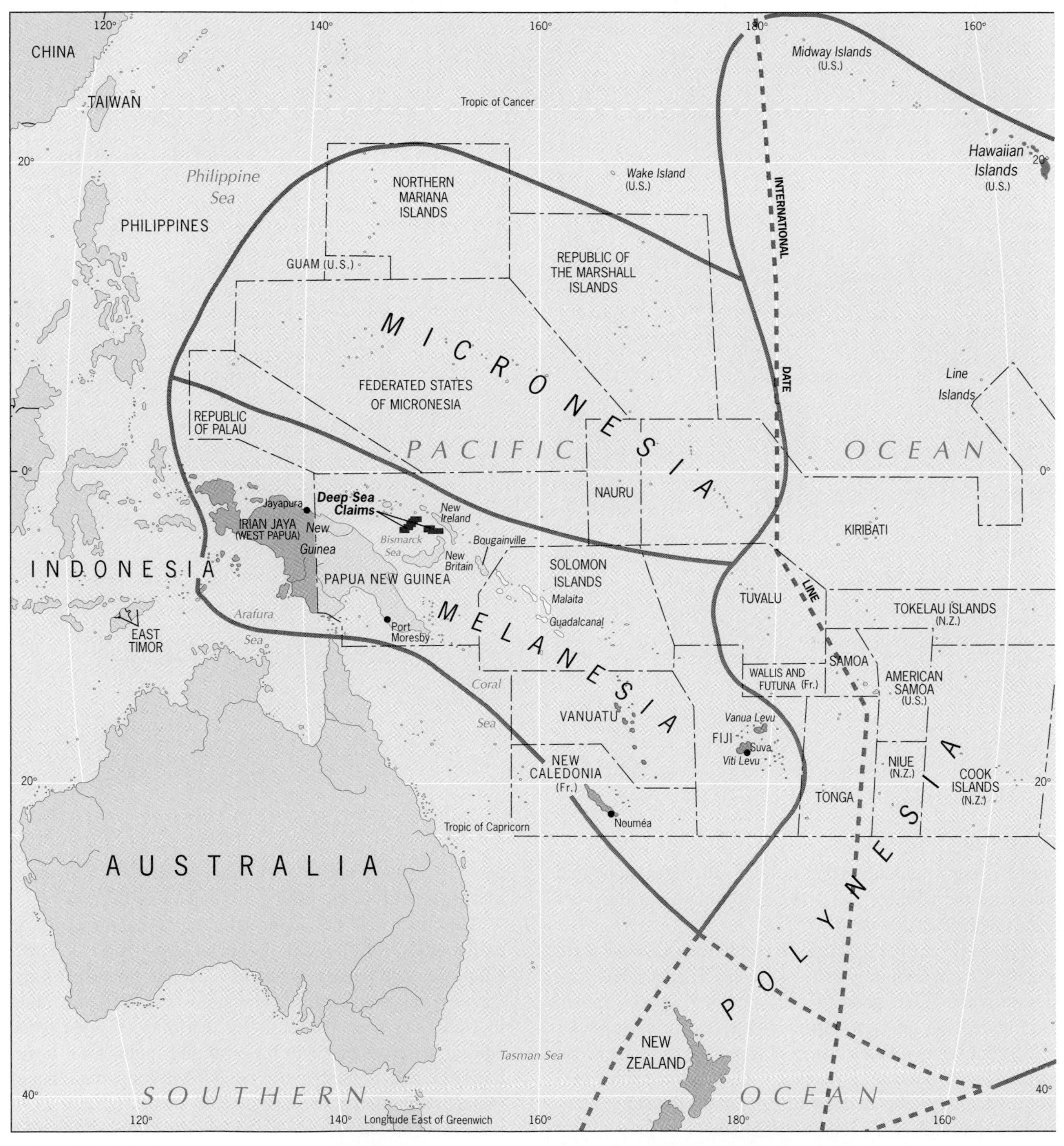

FIGURE 12-3

More serious investigations reveal that such Pacific sameness is more apparent than real. Even the Pacific Realm, with its long sailing traditions, its still-diffusing populations, and its historic migrations, has a persistent regional framework. Figure 12-3 outlines the three regions that constitute the Pacific Realm:

Melanesia Irian Jaya/West Papua (Indonesia), Papua New Guinea, Solomon Islands, Vanuatu, New Caledonia (France), Fiji

MAJOR CITIES OF THE REALM

City	Population* (in millions)
Honolulu, Hawai'i (U.S.)	1.1
Nouméa, New Caledonia	0.2
Port Moresby, Papua New Guinea	0.4
Suva, Fiji	0.3

*Based on 2002 estimates.

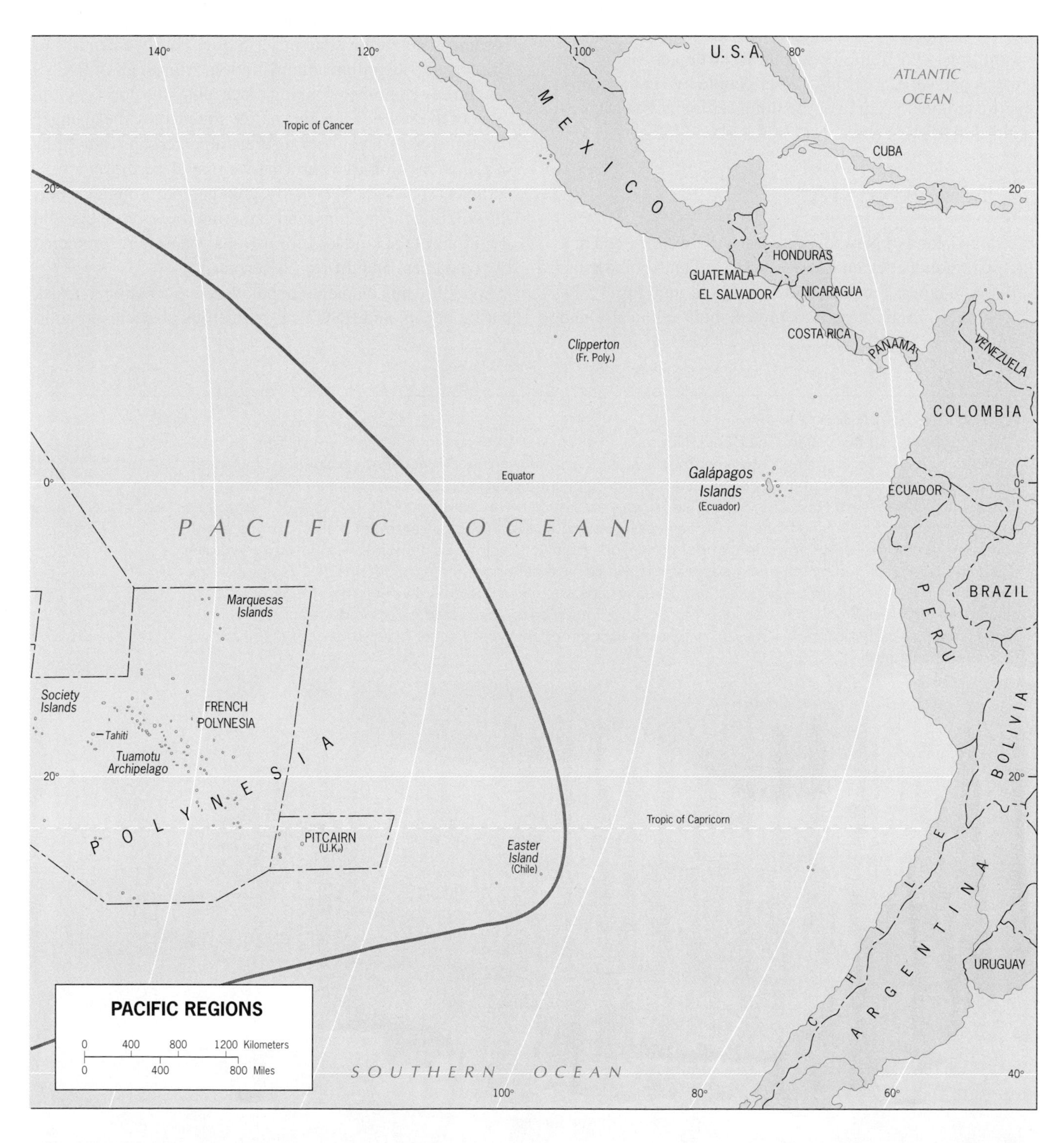

Micronesia Palau, Federated States of Micronesia, Northern Mariana Islands, Republic of the Marshall Islands, Nauru, western Kiribati, Guam (U.S.)

Polynesia Hawaiian Islands (U.S.), Samoa, American Samoa, Tuvalu, Tonga, eastern Kiribati, Cook and other New Zealand-administered islands, French Polynesia

Ethnic, linguistic, and physiographic criteria are among the bases for this regionalization of the Pacific Realm, but we should not lose sight of the dimensions. Not only is the land area small, but also the population of this entire realm (including Indonesia's West Papua) in 2002 was only about 9.7 million—7.5 million without West Papua—about the same as one very large city. Fewer people live in this realm than in another vast area of far-flung settlements, the oases of North Africa's Sahara.

MELANESIA

The large island of New Guinea lies at the western end of a Pacific region that extends eastward to Fiji and includes the Solomon Islands, Vanuatu, and New Caledonia (Fig. 12-3). The human mosaic here is complex, both ethnically and culturally. Most of the 7.2 million people of New Guinea (including the Indonesian part, West Papua [Irian Jaya], and the independent state of Papua New Guinea) are Papuans, and a large minority is Melanesian. Altogether there are as many as 700 communities speaking different languages; the Papuans are most numerous in the densely forested highland interior and in the south, while the Melanesians inhabit the north and east. The region as a whole has more than 6 million inhabitants, making this the most populous Pacific region by far.

With 5 million people today, Papua New Guinea (PNG) became a sovereign state in 1975 after nearly a century of British and Australian administration. Almost all of PNG's limited development is taking place along the coasts, while most of the interior remains hardly touched by the changes that transformed neighboring Australia. Perhaps four-fifths of the population lives in what we may describe as a self-sufficient subsistence economy, growing root crops and hunting wildlife, raising pigs, and gathering forest products. Old traditions of the kind lost in Australia persist here, protected by remoteness and the rugged terrain.

Welding this disparate population into a nation is a task hardly begun, and PNG faces numerous obstacles in addi-

FROM THE FIELD NOTES

"Arriving in the capital of New Caledonia, Nouméa, in 1996, was an experience reminiscent of French Africa 40 years earlier. The French tricolor was much in evidence, as were uniformed French soldiers. European French residents occupied hillside villas overlooking palm-lined beaches, giving the place a Mediterranean cultural landscape. And New Caledonia, like Africa, is a source of valuable minerals. It is one of the world's largest nickel producers, and from this vantage point you could see the huge treatment plants, complete with concentrate ready to be shipped (left, under conveyor). What you cannot see here is how southern New Caledonia has been ravaged by the mining operations, which have denuded whole mountainsides. Working in the mines and in this facility are the local Kanaks, Melanesians who make up about 44 percent of the population of about 230,000. Violent clashes between Kanaks and French have obstructed government efforts to change New Caledonia's political status in such a way as to accommodate pressures for independence as well as continued French administration."

tion to its cultural complexity. Not only are hundreds of languages in use, but more than half the population is illiterate. English, the official language, is used by the educated minority but is of little use beyond the coastal zone and its towns. The capital, Port Moresby, has about 350,000 inhabitants, reflecting the low level of urbanization (15 percent) in this low-income economy.

Yet Papua New Guinea is not without economic opportunities, and as Figure I-11 shows, it now ranks in the lower-middle-income group of states (in the 1980s it was one of the world's poorest). Oil was discovered in the 1980s, and by the late 1990s crude oil was PNG's largest export by value. Gold now ranks second, followed by copper, silver, timber, and several agricultural products including coffee and cocoa, reflecting the country's resource and environmental diversity. Pacific Rim developments have affected even PNG: most exports go to the nearest neighbor, Australia, but Japan ranks close behind.

Examine Figure 12-3, and you will see why Papua New Guinea has been beset by centrifugal forces since independence. PNG's maritime boundary cuts across an archipelago called the Solomon Islands, incorporating the largest of the Solomons, Bougainville, into the Papuan state. Almost immediately after independence, a separatist insurgency arose on Bougainville, supported by irredentist elements in the Solomon Islands. The rebels captured and closed Bougainville's large copper mine, cutting PNG's export revenues. A costly insurgency throughout the mid-1990s was not resolved until 1997 and then only tentatively following a political scandal in PNG involving the government's hiring of white mercenaries to defeat the rebels. At the outset of the twenty-first century there was tenuous stability and copper production had resumed, but the fundamental problem remained: the incorporation of a Melanesian island into a dominantly Papuan state.

This destructive episode highlights the vulnerability of Papua New Guinea in the postcolonial era. Indeed, the Bougainville issue is a direct legacy of colonialism. At independence in 1975 there was little enthusiasm in Bougainville for inclusion in PNG, but the Australians, who had administered both territories, insisted on the merger. Their reason was economic: revenues from the copper mine would supply much-needed cash from foreign sources to PNG's new government. But it was a forced marriage, leading to decades of strife costing more than the mine could ever repay.

As Figures 12-2 and 12-3 show, Papua New Guinea has a large EEZ, especially to the northeast where the large islands of New Britain and New Ireland encircle the Bismarck Sea. This is an area of tectonic-plate contact (see Fig. I-4), and such contact is associated with seafloor volcanic activity. Under certain circumstances, volcanic vents and upwelling, superheated waters combine to deposit rich lodes of dissolved minerals. When volcanic activity decreases, it leaves behind mounds of copper, silver, gold, zinc, and other minerals in concentrations far greater than anything found on land. Now mining technology is making their exploitation possible, and in 1997 Papua New Guinea issued the first mining certification to an Australian company to exploit the *chimneys* (vent-related stacks of ore) deep below the surface of the sea (Fig. 12-3). While the economic rewards will be high—the eastern claim has as much as 15 percent copper, 19 percent iron, 26 percent zinc, 7 percent silver—scientists fear that this kind of deep-sea mining will spread worldwide and damage the unique ecosystems associated with the still-hot vents. But under UNCLOS regulations, states have the rights to all economic assets in the EEZ, and for many the riches of the seafloor will prove irresistible.

Turning eastward, it is a measure of Melanesia's cultural fragmentation that as many as 120 languages are spoken in the approximately 1000 islands that make up the Solomon Islands (about 80 of these islands support almost all the people, numbering about 440,000). The islands' natural separation has helped limit ethnic conflict, but World War II bequeathed the Solomons a problem. When American forces took Guadalcanal from the Japanese, they brought in thousands of people from the neighboring island of Malaita (Fig. 12-3). After the war, most of these Malaitans stayed on and came to dominate life on Guadalcanal, controlling commerce and buying up land. Resentment among the local people grew, and in 1999 it exploded into violence. Attacks on Malaitans caused some 10,000 to flee the capital, Honiara; many sailed for Malaita. Then in June 2000, a group of Malaitans still on Guadalcanal took the country's prime minister hostage in a bid to extract concessions. Once again, foreigners created the state and relocated the people, leaving the locals to deal with the consequences after decolonization.

New Caledonia, still under French rule, is in a very different situation. Only about 45 percent of the population of about 230,000 are Melanesian; 37 percent are of French ancestry, many of them descended from the inhabitants of the penal colony France established here in the nineteenth century. Nickel mines, based on reserves that rank among the world's largest, dominate New Caledonia's export economy (see photo, p. 558). The mining industry attracted additional French settlers, and social problems arose. Most of the French population lives in or near the capital city of Nouméa, steeped in French cultural landscapes, in the southwestern quadrant of the island. The miners, mainly Melanesians, live in dormitory-style buildings near the huge refinery in the capital; those not directly involved farm for subsistence in the island's middle and northern areas. Melanesian demands for an end to colonial rule have led to violence, and the two communities are still in the process of coming to terms.

On its eastern margins, Melanesia includes one of the Pacific Realm's most interesting countries, Fiji. On two larger and over 100 smaller islands live more than 800,000 Fijians, of whom 51 percent are Melanesians and 44 percent

FROM THE FIELD NOTES

"Back in Suva, Fiji more than 20 years after I had done a study of its CBD, I noted that comparatively little had changed, although somehow the city seemed more orderly and prosperous than it was in 1978. At the Central Market I watched a Fijian woman bargain with the seller over a batch of taro, the staple starch-provider in local diets. I asked her how she would serve it. 'Well, it's like your potato,' she said. 'I can make a porridge, I can cut it up and put it in a stew, and I can even fry pieces of it and make them look like the French fries they give you at the McDonald's down the street.' Next I asked the seller where his taro came from. 'It grows over all the islands,' he said. 'Sometimes, when there's too much rain, it may rot—but this year the harvest is very good.' . . . Next I walked into the crowded Indian part of the CBD. On a side street I got a reminder of the geographic concept of agglomeration. Here was a colonial-period building, once a hotel, that had been converted into a business center. I counted 15 enterprises, ranging from a shoestore to a photographer and including a shop where rubber stamps were made and another selling diverse tobacco products. Of course (this being the Indian sector of downtown Suva) tailors outnumbered all other establishments."

South Asians, the latter brought to Fiji from India during the British colonial occupation to work on the sugar plantations. When Fiji achieved independence in 1970, the native Fijians owned most of the land and held political control, while the Indians were concentrated in the towns (chiefly Suva, the capital) and dominated commercial life. It was a recipe for trouble, and it was not long in coming when, in a later election, the politically active Indians outvoted the Fijians for seats in the parliament. A coup by the Fijian military was followed by a revision of the constitution, which awarded a majority of seats to ethnic Fijians. Before long, however, the Fijian majority splintered, but a coalition government for some time proved the constitution workable—until 2000. In May 1999, Fiji's first prime minister of Indian ancestry had taken office, angering some ethnic Fijians to the point of staging another coup. The prime minister and members of the government were taken hostage at the parliament building in Suva, and the perpetrators demanded that Fijians would henceforth govern the country.

This action, and Fiji's inability to counter it, had a devastating impact on the country. Foreign trading partners stopped buying Fijian products. The tourist industry suffered severely. And in the end, although the coup leaders were ousted and arrested, they secured a deal that gave ethnic Fijians the control they had sought. Fiji's future thus remains clouded.

Melanesia, the most populous region in the Pacific Realm, also is bedeviled by centrifugal forces of many kinds. No two countries present the same form of multiculturalism; each has its own challenges to confront, and some of these challenges spill over into neighboring (or more distant foreign) islands.

MICRONESIA

North of Melanesia and east of the Philippines lie the islands that constitute the region known as Micronesia (Fig. 12-3). The name (*micro* means small) refers to the size of the islands: the 2000-plus islands of Micronesia are not only tiny (many of them no larger than 1 square mile), but they are also much lower-lying, on an average, than those 8 of Melanesia. Some are volcanic islands (**high islands**, as the people call them), but they are outnumbered by islands 9 composed of coral, the **low islands** that barely lie above sea level. Guam, with 210 square miles (550 sq km), is Micronesia's largest island, and no island elevation anywhere in Micronesia reaches 3300 feet (1000 m).

The high-island/low-island dichotomy is useful not only in Micronesia, but also throughout the realm. Both the physiographies of these islands and the economies they support differ in crucial ways. High islands wrest substantial moisture from the ocean air; they tend to be well watered and have good volcanic soils. As a result, agricultural products show some diversity, and life is reasonably secure. Populations tend to be larger on these high islands than on the low islands, where drought is the rule and fishing and the coconut palm are the mainstays of life. Small communities cluster on the low islands, and over time, many of these have died out. The major migrations, which sent fleets to populate islands from Hawai'i to New Zealand, tended to originate in the high islands.

Until the mid-1980s, Micronesia was largely a United States Trust Territory (the last of the post–World War II trusteeships supervised by the United Nations), but that status has now changed. As Figure 12-3 shows, today Micronesia is divided into countries bearing the names of independent states. The Marshall Islands, where the United States tested nuclear weapons (giving prominence to the name *Bikini*), now is a republic in "free association" with the United States, having the same status as the Federated States of Micronesia and (since 1994) Palau. The Northern Mariana Islands are a commonwealth "in political union" with the United States. In effect, the United States provides billions of dollars in assistance to these countries, in return for which they commit themselves to avoid foreign policy actions that are contrary to U.S. interests. There are other conditions: Palau, for example, granted the United States rights to existing military bases for 50 years following independence.

Also part of Micronesia are the U.S. territory of Guam, where independence is not in sight and where U.S. military installations and tourism provide the bulk of income, and the remarkable Republic of Nauru. With a population of just 10,000 and 8 square miles of land, this microstate got rich by selling its phosphate deposits to Australia and New Zealand, where they are used as fertilizer. Per-capita incomes rose to U.S. $10,000, making Nauru one of the Pacific's high-income societies. But the phosphate deposits ran out, and an island scraped bare faced an economic crisis.

In this region of tiny islands, most people subsist on farming or fishing, and virtually all the countries need infusions of foreign aid to survive. The natural economic complementarity between the high-island farming cultures and the low-island fishing communities all too often is negated by distance, spatial as well as cultural. Life here may seem idyllic to the casual visitor, but for the Micronesians it often is a daily challenge.

POLYNESIA

To the east of Micronesia and Melanesia lies the heart of the Pacific, enclosed by a great triangle stretching from the Hawaiian Islands to Chile's Easter Island to New Zealand. This is Polynesia (Fig. 12-3), a region of numerous islands (*poly* means many), ranging from volcanic mountains rising above the Pacific's waters (Mauna Kea on Hawai'i reaches nearly 13,800 feet [over 4200 m]), clothed by luxuriant tropical forests and drenched by well over 100 inches of rainfall each year, to low coral atolls where a few palm trees form the only vegetation and where drought is a persistent problem. The Polynesians have somewhat lighter-colored skin and wavier hair than do the other peoples of the Pacific Realm; they are often also described as having an excellent physique. Anthropologists differentiate between these original Polynesians and a second group, the Neo-Hawaiians, who are a blend of Polynesian, European, and Asian ancestries. In the U.S. State of Hawai'i—actually an archipelago of more than 130 islands—Polynesian culture has been both Europeanized and Orientalized.

Its vastness and the diversity of its natural environments notwithstanding, Polynesia clearly constitutes a geographic region within the Pacific Realm. Polynesian culture, though spatially fragmented, exhibits a remarkable consistency and uniformity from one island to the next, from one end of this widely dispersed region to the other. This consistency is particularly expressed in vocabularies, technologies, housing, and art forms. The Polynesians are uniquely adapted to their maritime environment, and long before European sailing ships began to arrive in their waters, Polynesian seafarers had learned to navigate their wide expanses of ocean in huge double canoes as long as 150 feet (45 m). They traveled hundreds of miles to favorite fishing zones and engaged in inter-island barter trade, using maps constructed from bamboo sticks and cowrie shells and navigating by the stars. However, modern descriptions of a Pacific Polynesian paradise of emerald seas, lush landscapes, and gentle people distort harsh realities. Polynesian society was forced to get used to much loss of life at sea when storms claimed their boats; families were ripped apart by accident as well as migration; hunger and starvation afflicted the inhabitants of

FROM THE FIELD NOTES
"Space is at a premium in Tuvalu, a country of about 10 square miles (26 sq km) on 9 small islands forming a 360-mile (575-km) chain. I had the unusual opportunity to go ashore on the largest of these low islands, Funafuti Atoll, where about 4500 of the total population of about 11,000 live. The first building in the capital I saw was marked *Department of Fisheries*, and for good reason. Tuvalu may be territorially small, but it has a huge Exclusive Economic Zone, profitably leased to major fishing operations. Two other impressions: the obvious problems people here have with waste disposal (there is little room for dumps, and plastics are making this a crisis); and reminders of World War II. Rusting hulks of war materiel still lie where they were abandoned, a jarring sight in this isolated Pacific outpost."

smaller islands; and the island communities were often embroiled in violent conflicts and cruel retributions.

The political geography of Polynesia is complex. In 1959, the Hawaiian Islands became the fiftieth State to join the United States. The State's population is now 1.2 million, with over 80 percent living on the island of Oahu. There, the superimposition of cultures is symbolized by the panorama of Honolulu's skyscrapers against the famous extinct volcano at nearby Diamond Head. The Kingdom of Tonga became an independent country in 1970 after seven decades as a British protectorate; the British-administered Ellice Islands were renamed Tuvalu, and along with the Gilbert Islands to the north (now renamed Kiribati), they received independence from Britain in 1978. Other islands continued under French control (including the Marquesas Islands and Tahiti), under New Zealand's administration (Rarotonga), and under British, U.S., and Chilean flags.

In the process of politico-geographical fragmentation, Polynesian culture has suffered severe blows. Land developers, hotel builders, and tourist dollars have set Tahiti on a course along which Hawai'i has already traveled far. The Americanization of eastern Samoa has created a new society different from the old. Polynesia has lost much of its ancient cultural consistency; today, the region is a patchwork of new and old—the new often bleak and barren, with the old under intensifying pressure.

The countries and cultures of the Pacific Realm lie in an ocean on whose rim a great drama of economic and political transformation will play itself out during the twenty-first century. Already, the realm's own former margins—in Hawai'i in the north and in New Zealand in the south—have been so recast by foreign intervention that little remains of the kingdoms and cultures that once prevailed. Now the Pacific world faces changes far greater even than those brought here by European colonizers. Once upon a time the shores and waters of the Mediterranean Sea formed an arena of regional transformation that changed the world. Then it was the Atlantic, avenue of the Industrial Revolution and stage of fateful war. Now it seems to be the turn of the Pacific as the world's largest country and next superpower (China) faces the richest and most powerful (the United States). Giants will jostle for advantage in the Pacific; how will the weak societies of the Pacific Realm fare?

A FINAL CAVEAT: PACIFIC AND ANTARCTIC

South of the Pacific Realm lies Antarctica and its encircling Southern Ocean. The combined area of these two geographic expanses constitutes 40 percent of the entire planet—two-fifths of the Earth's surface containing one one-thousandth of the world's population.

Is Antarctica a geographic realm? In physiographic terms, yes, but not on the basis of the criteria we use in this book. Antarctica is a continent, nearly twice as large as Australia, but virtually all of it is covered by a dome-shaped icesheet nearly 2 miles (3.2 km) thick near its center. The continent often is referred to as the "white desert" because, despite all of its ice and snow, annual precipitation is low (less than 6 inches [15 cm] per year). Temperatures are frigid, with winds so strong that Antarctica also is called the "home of the blizzard." For all its size, no functional regions have developed here, no towns, no transport networks except the supply lines of research stations. And Antarctica still is a frontier, even a scientific frontier still slowly giving up its secrets. Underneath all that ice lie some 70 lakes of which Lake Vostok is the largest with over 5400 square miles (14,000 sq km). It may be as much as 2000 feet (600 m) deep. No one has yet seen a sample of its water.

Like virtually all frontiers, Antarctica has always attracted pioneers and explorers. Whale and seal hunters destroyed huge populations of Southern Ocean fauna during the eighteenth and nineteenth centuries, and explorers planted the flags of their countries on Antarctic shores. Between 1895 and 1914, the quest for the South Pole became an international obsession; Roald Amundsen, the Norwegian, reached it first in 1911. All this led to national claims in Antarctica during the interwar period (1918–1939).

The geographic effect was the partitioning of Antarctica into pie-shaped sectors centered on the South Pole (Fig. 12-4). In the least frigid area of the continent, the Antarctic Peninsula, British, Argentinean, and Chilean claims overlapped—and still do. One sector, Marie Byrd Land (shown in white on the map), was never formally claimed by any country.

Why should states be interested in territorial claims in so remote and difficult an area? Both land and sea contain raw materials that may some day become crucial: proteins in the waters, and fuels and minerals beneath the ice. Antarctica (5.5 million sq mi/14.2 million sq km) is almost twice as large as Australia, and the Southern Ocean (see box, p. 535) is nearly as large as the North and South Atlantic. However distant actual exploitation may be, countries want to keep their stakes here.

But the claimant states (those with territorial claims) recognize the need for cooperation. During the late 1950s, they joined in the International Geophysical Year (IGY) that launched major research programs and established a number of permanent research stations throughout the continent. This spirit of cooperation led to the 1961 signing of the **Antarctic Treaty**, which ensures continued scientific collaboration, prohibits military activities, protects the environment, and holds national claims in abeyance. In 1991, when the treaty was extended under the terms of the Wellington Agreement, concerns were raised that it does not do enough to control future resource exploitation.

10

In an age of growing national self-interest and increasing raw-material consumption, the possibility exists that Antarctica and its offshore waters may yet become an arena for international rivalry. Until now, its remoteness and its forbidding environments have saved it from that fate. The entire world benefits from this because evidence is mounting that Antarctica plays a critical role in the global environmental system, so that human modifications may have global (and unpredictable) consequences.

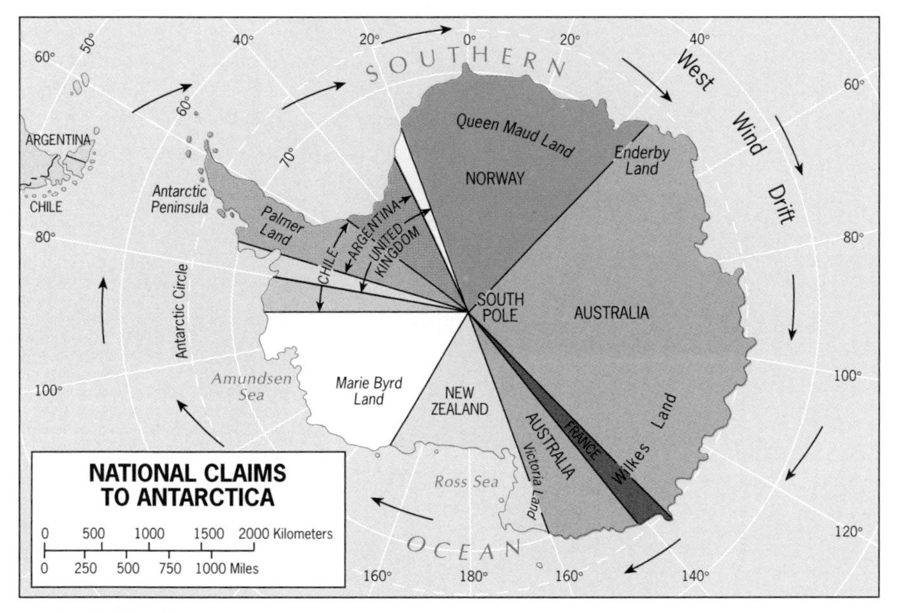

FIGURE 12-4

appendix A

Map Reading and Interpretation

As can be seen throughout this book, maps are tools that are very useful in gaining an understanding of patterns in geographic space. In fact, they constitute an important visual or *graphic communication* medium whereby encoded spatial messages are transmitted from the cartographer (mapmaker) to the map reader. Of course, this shorthand is necessary because the real world is so complex that a great deal of geographic information must be compressed into the small confines of maps that can fit onto the pages of this book. At the same time, cartographers must carefully choose which information to include; these decisions force them to omit many things in order to prevent cluttering a map with less relevant information. For instance, Figure B in this appendix shows several city blocks in central London but avoids mapping individual buildings because they would interfere with the main information being presented—the spatial distribution of cholera deaths.

MAP READING

Deciphering the coded messages contained in the maps of this book—map reading—is not difficult, and becomes quite easy with a little experience in using this "language" of geography. The need to miniaturize portions of the world on small maps is discussed in the section on map scale (pp. 7–8), and two additional contrasting examples are provided in Figures A and B. Orientation, or direction, on maps can

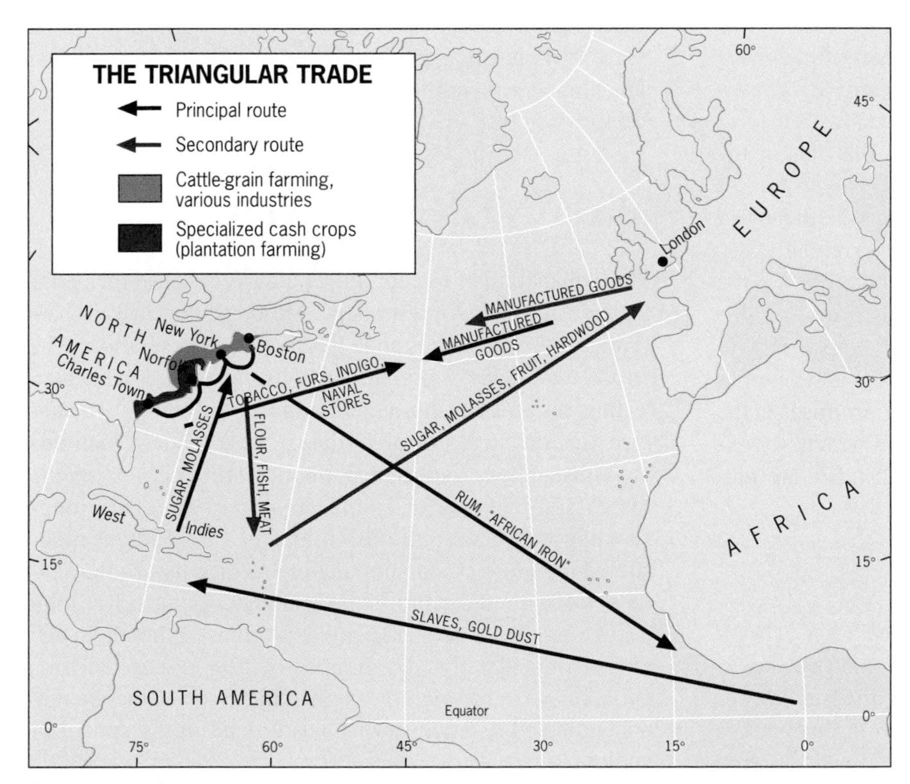

FIGURE A

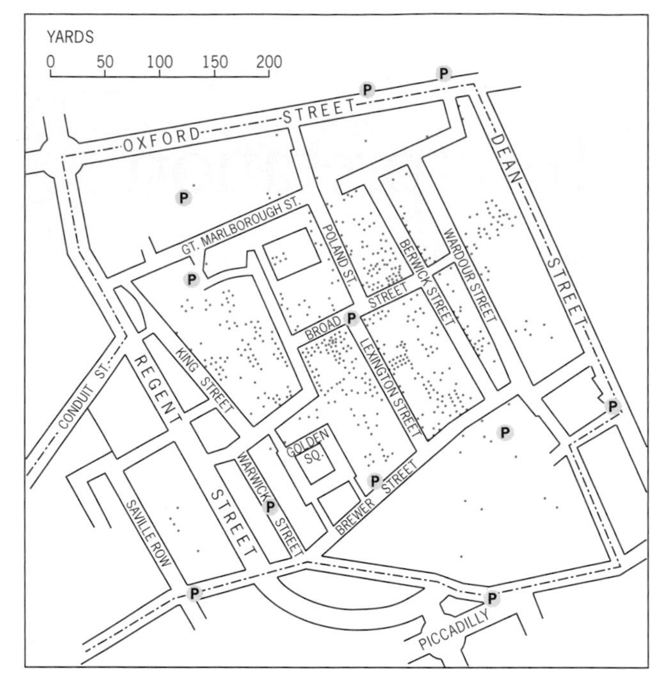

FIGURE B

usually be discerned by reference to the geographic grid of latitude and longitude. *Latitude* is measured from 0° to 90° north and south of the equator (parallels of latitude are always drawn in an east-west direction), with the equator being 0° and the North and South Poles being 90°N and 90°S, respectively. Meridians of *longitude* (always drawn north-south) are measured 180° east and west of the *prime meridian* (0°), which passes through the Greenwich Observatory next to London; the 180th meridian, for the most part, serves as the *international date line* that lies in the middle of the Pacific Ocean (see Fig. 12-1). Inspection of Figure A shows that north is not automatically at the top of a map; instead, the direction of north curves along every meridian, with all such lines of longitude converging at the North Pole. The many minor directional distortions in this map are unavoidable: it is geometrically impossible to transfer the grid of a three-dimensional sphere (globe) onto a two-dimensional flat map. Therefore, compromises in the form of *map projections* must be devised in which, for example, properties such as areal size and distance are preserved but directional constancy is sacrificed.

MAP SYMBOLS

Once the background mechanics of scale and orientation are understood, the main task of decoding the map's content can proceed. The content of most maps in this book is organized within the framework of point, line, and area symbols, which are made especially clear through the use of color. These symbols are usually identified in the map's *legend*, as in Figure A. Occasionally, the map designer omits the legend but must tell the reader verbally in a caption or within the text what the map is about. Figure B, for instance, is a map of cholera deaths in the London neighborhood of Soho during the outbreak of 1854, with each **P** symbol representing a municipal water pump and each green dot the location of a cholera fatality. *Point symbols* are shown as dots on the map and can tell us two things: the location of each phenomenon and, sometimes, its quantity. The cities of New York and London in Figure A and the dot pattern of cholera fatalities in Figure B (with each green dot symbolizing one death) are examples.

Line symbols connect places between which some sort of movement or flow is occurring. The "triangular trade" among Britain and its seventeenth-century Atlantic colonies (Fig. A) is a good example: each leg of these trading routes is clearly mapped, the goods moving along are identified, and the more heavily traveled principal routes are differentiated from the secondary ones. *Area symbols* are used to classify two-dimensional spaces and thus provide the cartographic basis for regionalization schemes (as can be seen in Fig. I-8 on pp. 14–15). Such classifications can be developed at many levels of generalization. In Figure A, blue and beige areas broadly offset land from ocean surfaces; more specifically, red and green area symbols along the eastern coast of North America delimit a pair of regions that specialized in different types of commercial agriculture. Area symbols may also be used to communicate quantitative information: for example, the light-tan-colored zones in Figure I-7 (p. 12) delimit semiarid areas within which annual precipitation averages from 12 to 20 inches (30 to 50 cm).

MAP INTERPRETATION

The explanation of cartographic patterns is one of the geographer's most important tasks. Although that task is performed for you throughout this book, readers should be aware that today's practitioners use many sophisticated techniques and cutting-edge computer technology to analyze vast quantities of areal data. These modern methods notwithstanding, geographic inquiry still focuses on the search for meaningful *spatial relationships*. This long-standing concern of the discipline is classically demonstrated in Figure B. By showing on his map that cholera fatalities clustered around municipal water pumps, Dr. John Snow was able to persuade city authorities to shut them off; almost immediately, the number of new disease victims dwindled to zero, thereby confirming Snow's theory that contaminated drinking water was crucial in the spread of cholera.

appendix B

Opportunities in Geography

The chapters of this book give an idea of the wide range of topics and interests that geographers pursue, particularly in the many discussions of concepts (regional and otherwise) and frequent references to the content of geography's systematic fields. There are specializations within each of those topical fields as well as in regional studies—but an introductory book such as this lacks enough space to discuss all of them. This appendix, therefore, is designed to help you, should you decide to major or minor in geography and/or to consider it as a career option.

AREAS OF SPECIALIZATION

As in all disciplines, areas of concentration or specialization change over time. In North American geography early in the twentieth century, there was a period when most geographers were physical geographers, and the natural landscape was the main object of geographic analysis. Then the pendulum swung toward human (cultural) geography, and students everywhere focused on the imprints of human activity on the surface of the Earth. Still later the analysis of spatial organization became a major concern. In the meantime, geography's attraction for some students lay in technical areas: in cartography, in aerial and satellite remote sensing, in computer-assisted spatial data analysis, and, most recently, in geographic information systems (GIS).

All this meant that geography posed (and continues to pose) a challenge to its professionals. New developments require that we keep up to date, but we must also continue to build on established foundations.

Regional Geography

One of these established foundations, of course, is regional geography, which encompasses a large group of specializations. Some geographers specialize in the theory of regions: how they should be defined, how they are structured, and how their internal components work. This leads in the direction of *regional science*, and some geographers have preferred to call themselves regional scientists. But make no mistake: regional science is regional geography.

Another, older approach to regional geography involves specialization in an area of the world ranging in size from a geographic realm to a single region or even a State or part of a State. There was a time when regional geographers, because of the interdisciplinary nature of their knowledge, were sought after by government agencies. Courses in regional geography abounded in universities' geography departments; regional geographers played key roles in international studies and research programs. But then the drive to make geography a more rigorous science and to search for universal (rather than regional) truths contributed to a decline in regional geography. The results were not long in coming, and in recent years you have probably seen the issue of "geographic illiteracy" discussed in newspapers and magazines. Now the pendulum is swinging back again, and regional geography and regional specialization are reviving. This is a propitious time to consider regional geography as a professional field.

Your Personal Interests

Geography, as we have pointed out, is united by several bonds, of which regional geography is but one. Regional geography exemplifies the spatial view that all geographers hold; the spatial approach to study and research binds physical and human geographers, regionalists, and topical specialists. Another unifying theme is an abiding interest in the relationships between human societies and natural environments. We have referred to that topic frequently in this book; as an area of specialization it has gone through difficult times. Perhaps more than anything, geography remains a field of *synthesis*, of understanding interrelationships.

Geography also is a field science, using "field" in another context. In the past, almost all major geography departments required a student's participation in a "field camp" as part of a master's degree program; thus many undergraduate programs included field experience. It was one of those bonding practices in which students and faculty with diverse interests met, worked together, and learned from one another. Today, few field camps of this sort are offered, but that does not change what geography is all about. If you see an opportunity for field experience with professional geographers—even just a one-day reconnaissance—take it. But realize this: a few days in the field with geographic instruction may hook you for life.

Some geographers, in fact, are far better field-data gatherers than analysts or writers. In this respect they are not alone: this also happens in archeology, geology, and biology (among other field disciplines). This does not mean that these field workers do not contribute significantly to knowledge. Often on a research team some of the members are better in the field, and others excel in subsequent analysis. From all points of view, however, fieldwork is important.

Geography, then, is practiced in the field and in the office, in physical and human contexts, in generality and detail. Small wonder that so many areas of specialization have developed! If you check the undergraduate catalogue of your college or university, you will see some of these specializations listed as semester-length courses. But no geography department, no matter how large, could offer them all.

How does an area of specialization develop, and how can one become a part of it? The way in which geographic specializations have developed tells us much about the entire discipline. Some major areas, now old and established, began as research and theory building by one scholar and his or her students. These graduate students dispersed to the faculties of other universities and began teaching what they had learned. Thus, for example, did Carl Sauer's cultural geography spread from the University of California, Berkeley during the middle decades of the twentieth century.

It is one of the joys of geography that the basics and methods, once learned, are applicable to so many features of the human and physical world. Geographers have specialized in areas as disparate as shopping-center location and glacier movement, tourism and coastal erosion, real estate and wildlife, retirement communities and sports. Many of these specializations began with the interests and energies of a single scholar. Some thrived and grew into major geographic pursuits; others remained one-person shows, but with potential. When you discuss your own interests with a faculty advisor, you may refer to a university where you would like to do graduate work. "Oh yes," the answer may be, "Professor X is in their geography department, working on just that." Or perhaps your advisor will suggest another university where a member of the faculty is known to be working on the topic in which you are interested. That is the time to write a letter or e-mail of inquiry. What is the professor working on now? Are graduate students involved? Are research funds available? What are the career prospects after graduation?

The Association of American Geographers, or **AAG** (1710 16th Street, N.W., Washington, D.C. 20009–3198 [**www.aag.org**]), recognizes no less than 54 so-called Specialty Groups. In academic year 2001–2002, the Specialty Group roster included the following branches of geography:

Africa
Aging and the Aged
Applied Geography
Asian Geography
Bible
Biogeography
Canadian Studies
Cartography
China
Climate
Coastal and Marine Geography
Contemporary Agriculture and Rural Land Use
Cryosphere (Earth's Ice System)
Cultural Ecology
Cultural Geography
Disability
Economic Geography
Energy and Environment
Environmental Perception and Behavioral Geography
Ethnic Geography
European
Geographic Information Systems (GIS)
Geographic Perspectives on Women
Geography Education
Geography of Religions and Belief Systems
Geomorphology (Landform Analysis)
Hazards
Historical Geography
History of Geography
Human Dimensions of Global Change
Human Rights
Indigenous Peoples
Latin American
Medical Geography
Microcomputers
Middle East
Military Geography
Mountain Geography
Political Geography
Population
Qualitative Research
Recreation, Tourism, and Sport
Regional Development and Planning
Remote Sensing

Rural Development
Russian, Central Eurasian, and East European
Sexuality and Space
Socialist Geography
Spatial Analysis and Modeling
Transportation Geography
Urban Geography
Values, Ethics, and Justice
Water Resources
Worldwide Web

If you contact the AAG at the Internet address given above, click on the "Specialty Groups" icon in the left-hand box on the home page and then scroll down the alphabetical listing to obtain the e-mail address of the current chairperson of any group. We encourage you to contact the professional geographer who chairs the group(s) you are interested in. These elected leaders are ready, enthusiastic, and willing to provide you with the information you are looking for.

All this may seem far in the future. Still, the time to start planning for graduate school is now. Applications for admission and financial assistance must be made shortly after the *beginning* of your senior year. That makes your junior year a year of decision.

AN UNDERGRADUATE PROGRAM

The most important concern for any geography major or minor is basic education and training in the field. An undergraduate curriculum contains all or several of the following courses (titles may vary):

1. Physical Geography of the Global Environment (weather and climate, natural landscapes, landforms, soils, moisture and surface water flows, elementary biogeography)
2. Introduction to Human Geography (basic principles of cultural and economic geography)
3. World Regional Geography (world geographic realms)

These beginning courses are followed by more specialized courses, including both methodological and substantive ones:

4. Geographic Information Technology
5. Cartographic Theory and Techniques
6. Introduction to Quantitative Methods of Analysis
7. Introduction to Geographic Information Systems (GIS)
8. Analysis of Remotely Sensed Data
9. Cultural Geography
10. Political Geography
11. Urban Geography
12. Economic Geography
13. Historical Geography
14. Geomorphology
15. Geography of United States-Canada, Europe, and/or other world geographic realms

You can see how Physical Geography of the Global Environment would be followed by the more specialized analysis of landforms covered in Geomorphology, and how Human Geography now divides into such areas as intermediate and/or advanced cultural and economic geo-graphy. As you progress, the focus becomes even more specialized. Thus Economic Geography may be followed by:

16. Industrial Geography
17. Transportation Geography
18. Geography of Development

At the same time, regional concentrations may come into sharper focus:

19. Geography of Western Europe (or other major regions)

In these more advanced courses, you will use the technical knowledge acquired from courses numbered **4** through **8** (and perhaps others). Now you can avail yourself of the opportunity to develop these skills further. Many departments offer such courses as these:

20. Advanced Quantitative Methods
21. Computer Cartography
22. Intermediate GIS
23. Advanced GIS
24. Advanced Satellite Imagery Interpretation

From this list (which represents only part of a comprehensive curriculum), it is evident that you cannot, even in four years of undergraduate study, register for all courses. The geography major in many universities requires a minimum of only 30 (semester-hour) credits—just 10 courses, fewer than half those listed here. This is another reason to begin thinking about specialization at an early stage.

Because of the number and variety of possible geography courses, most departments require that their majors complete a core program that includes courses in substantive areas as well as theory and methods. That core program is important, and you should not be tempted to put off these courses until your last semesters. What you learn in the core program will make what follows (or should follow) much more meaningful.

You should also be aware of the flexibility of many undergraduate programs, something that can be especially important to geographers. Imagine that you are majoring in geography and develop an interest in Southeast Asia. But the regional specialization of the geographers in your department may be focused somewhere else—say, Middle and South America. However, courses on Southeast Asia are indeed offered by other departments, such as anthropology, history, or political science. If you are going to be a regional specialist, those courses will be very useful and should be part of your curriculum—but you are not able to receive geography credit for them. After successfully completing those courses, however, you may be able to register for an independent study or reading course in the Geography of Southeast Asia if a faculty member is willing to guide you. Always discuss such matters with the undergraduate advisor or chairperson of your geography department.

LOOKING AHEAD

By now, as a geography major, you will be thinking of the future—either in terms of graduate school or a salaried job. In this connection, if there is one important lesson to keep in mind, it is to *plan ahead* (a redundancy for emphasis!). The choice of graduate school is one of the most important you will make in your life. The professional preparation you acquire as a graduate student will affect your competitiveness in the job market for years to come.

Choosing a Graduate School

Your choice of a graduate school hinges on several factors, and the geography program it offers is one of them. Possibly, you are constrained by residency factors, and your choice involves the schools of only one State. Your undergraduate record and grade point average affect the options. As a geographer, you may have strong feelings in favor of or against particular parts of the country. And although some schools may offer you financial support, others may not.

Certainly the programs and specializations of the prospective graduate department are extremely important. If you have settled on your own area of interest, it is wise to find a department that offers opportunities in that direction. If you have yet to decide, it is best to select a large department with several options. Some students are so impressed by the work and writings of a particular geographer that they go to his or her university solely to learn from and work with that scholar. In every case, information and preparation are crucial. Many a prospective graduate student has arrived on campus eager to begin work with a favorite professor, only to find that the professor is away on a sabbatical leave!

Fortunately, information can be acquired with little difficulty. One of the most useful publications of the AAG is its *Guide to Programs in Geography in the United States and Canada*, published each fall. A copy of this annual directory should be available in the office of your geography department, but if you plan to enter graduate school, a personal copy would be an asset. Not only does the *Guide* describe the programs, requirements, financial aid, and other aspects of geography departments in North America, but it also lists all faculty members and their current research and teaching specializations. Moreover, it also contains a complete listing of the Association's approximately 7,000 members and their specializations. And Internet addresses are now provided as well for accessing the Web sites of geography departments, which they increasingly utilize to supplement their entries in the *Guide* and provide even more detailed information about their programs. Although you will find the *Guide* to be very useful in your decision-making, we now also recommend that you explore the Web sites of departments you wish to consider.

You may discover one particular department that stands out as the most interesting and most appropriate for you. But do not limit yourself to one school. Consult the *Guide*'s 25-page table (entitled *Program Specialties*) that links 35 geographic specializations to every department listed, and then explore each appropriate match-up. After careful investigation, it is best to rank a half-dozen schools (or more), write or e-mail all of them for admission application forms, and apply to several. Multiple applications are a bit costly, but the investment is worth it.

Assistantships and Scholarships

One reason to apply to several universities has to do with financial (and other) support for which you may be eligible. If you have a reasonably well-rounded undergraduate program behind you and a good record of achievement, you are eligible to become a teaching assistant (TA) in a graduate department. Such assistantships usually offer full or partial tuition plus a monthly stipend (during the nine-month academic year). Conditions vary, but this position may make it possible for you to attend a university that would otherwise be out of reach. A tuition waiver alone can be well over $10,000 annually. Application for an assistantship is made directly to the department; you should write or e-mail the department contact listed in the latest AAG *Guide*, who will either respond directly to you or forward your inquiry to the departmental committee that evaluates applications.

What does a TA do? The responsibilities vary, but often teaching assistants are expected to lead discussion sections (of a larger class taught by a professor), laboratories, or other classes. They prepare and grade examinations and help undergraduates deal with problems arising from their courses. This is an excellent way to determine your own ability and interest in a teaching career.

In some instances, especially in larger geography departments, research assistantships are available. When a member of the faculty is awarded a large grant (e.g., by the National Science Foundation) for a research project, that grant may make possible the appointment of one or more research assistants (RAs). These individuals perform tasks generated by the project and are rewarded by a modest salary (usually comparable to that of a TA). Normally, RAs do not receive tuition waivers, but sometimes the department and the graduate school can arrange a waiver to make the research assistantship more attractive to better students. Usually, RAs are chosen from among graduate students already on campus who have proven their interest and ability. Sometimes, however, an incoming student is appointed. Always ask about opportunities.

Geography students also are among those eligible for many scholarships and fellowships offered by universities and off-campus organizations. When you write your introductory letter, be sure to inquire about all forms of financial aid.

JOBS FOR GEOGRAPHERS

Upon completing your bachelor's degree, you may decide to take a job rather than going on to graduate school. Again, this is a decision best made early in your junior year, for two main reasons: (1) so that you can tailor your curriculum for a vocational objective, and (2) so that you can start searching for a job well before graduation.

Internships

A very good way to enter the job market—and to become familiar with the working environment—is by taking an internship in an agency, office, or firm. Many organizations find it useful to have interns. Internships help organizations train beginning professionals and give companies an opportunity to observe the performance of trainees. Many an intern has ultimately been employed by his or her organization. Some employers have even suggested what courses the intern should take in the next academic year to improve future performance. For example, an urban or regional planning agency that employs an intern might suggest that the intern add a relevant GIS course or urban-planning course to his or her program of study.

Some organizations will appoint interns on a continuing basis, say, two afternoons a week around the year; others make full-time, summer-only appointments available. Occasionally, an internship can be linked to a departmental curriculum, yielding academic credit as well as vocational experience. Your department undergraduate advisor or the chairperson will be the best source of assistance.

One of the most interesting internship programs is offered by the National Geographic Society in Washington, D.C. Every year, the Society invites three groups of about eight interns each to work with the permanent staff of its various departments. Application forms are available in every geography department in the United States, and competition is strong. The application itself is a useful exercise, as it tells you what the Society (and other organizations) look for in your qualifications.

Professional Opportunities

A term you will sometimes see in connection with jobs is *applied geography*. This, one supposes, distinguishes geography teaching from practical geography. In fact, however, all professional geography in education, business, government, and elsewhere is "applied." In the past a large majority of geography graduates became teachers—in elementary and high schools, colleges, and universities. More recently, geographers have entered other arenas in increasing numbers. In part, this is related to the decline in geographic education in schools, but it also reflects the growing recognition of geographic skills by employers in business and government.

Nevertheless, what you as a geographer can contribute to a business is not yet as clear to the managers of many companies as it should be. The anybody-can-do-geography attitude is a form of ignorance you will undoubtedly confront. (This is so even in precollegiate education, where geography was submerged in social studies—and often taught by teachers who had never taken a course in the discipline!)

So, once employed, you may have to prove not only yourself but also the usefulness of the skills and capacities you bring to your job. There is a positive side to this. Many employers, once dubious about hiring a geographer, learn how geography can contribute—and become enthusiastic users of geographic talent. Where one geographer is employed, whether in a travel firm, publishing house, or planning office, you will soon find more.

Geographers are employed in business, government, and education. Planning is a profession that employs many geographers. Employment in business has grown in recent years (see the following section). Governments at national, State, and local levels have always been major employers of geographers. And in education, where geography was long in decline, the demand for geography teachers will grow again.

For more detailed information about employment, you should contact the AAG (at the address given earlier) for the inexpensive booklet entitled *Careers in Geography*; or check with your geography department, whose office is very likely to have a copy.

Business and Geography

With their education in global and international affairs, their knowledge of specialized areas of interest to business, and their training in cartography, GIS, methods of quantitative analysis, and writing, geographers with a bachelor's degree are attractive business-employment prospects. Some undergraduate students have already chosen the type of business they will enter; for them, there are departments that offer concentrations in their areas of interest. Several geography departments, for instance, offer a curriculum that concentrates on tourism and travel—a business in which geographic skills are especially useful.

These days, many companies want graduates with a strong knowledge of international affairs and fluency in at least one foreign language in addition to their skills in other areas. The business world is quite different from academia, and the transition is not always easy. Your employer will want to use your abilities to enhance profits. You may at first be placed in a job where your geographic skills are not immediately applicable, and it will be up to you to look for opportunities to do so.

One of our students was in such a situation some years ago. She was one of more than a dozen new employees doing what was essentially clerical work. (Some companies will use this type of work to determine a new employee's punctuality, work habits, adaptability, and productivity.) One day she heard that the company was considering establishing a branch to sell its product in East Africa. The student had a regional interest in East Africa as an undergraduate and had even taken a year of Swahili-language training. On her own time she wrote a carefully documented memorandum to the company's president and vice presidents, describing factors that should be taken into consideration in the projected expansion. That report evinced her regional skills, locational insights, and knowledge of the local market and transport problems, along with the probable cultural reaction to the company's product, the country's political circumstances, and (last but significantly) this employee's ability to present such issues effectively. She supported her report with good maps and several illustrations. Soon she received a special assignment to participate in the planning process, and her rise in the company's ranks had begun. She had seized the opportunity and demonstrated the utility of her geographic skills.

Geography graduates have established themselves in businesses of all kinds: banking, international trade, manufacturing, retailing, and many more. Should you join a large firm, you may be pleasantly surprised by the number of other geographers who hold jobs there—not under the title of geographer but under countless other titles ranging from analyst and cartographer to market researcher and program manager. These are positions for which the appointees have competed with other graduates, including business graduates. As we noted earlier, once an employer sees the assets a geographer can bring, the role of geographers in the company is assured.

Government and Geography

Government has long been a major employer of geographers at the national and State as well as local level. *Careers in Geography* (the AAG booklet) estimates that at least 2,500 geographers are working for governments, about half of them for the federal government. In the U.S. State Department, for instance, there is an Office of the Geographer staffed by professional geographers. Other agencies where geographers are employed include the Defense Mapping Agency, the Bureau of the Census, the U.S. Geological Survey, the Central Intelligence Agency, and the Army Corps of Engineers. Still other employers are the Library of Congress, the National Science Foundation, and the Smithsonian Institution. Many other branches of government also have positions for which geographers are eligible.

Opportunities also exist at State and local levels. Several States now have their own Office of the State Geographer; all States have agencies engaged in planning, resource analysis, environmental protection, and transportation policy-making. All these agencies need geographers who have skills in cartography, remote sensing, spatial database analysis, and the operation of geographic information systems.

Securing a position in government requires early action. If you want a job with the federal government, start at the beginning of your senior year. Every State capital and many other large cities have a Federal Job Information Center (FJIC). (The Washington office is at 1900 E Street, N.W., Washington, D.C. 20415.) You may request information about a particular agency and its job opportunities; it is appropriate to write or e-mail directly to the personnel office of the agency or agencies in which you are interested. You may also write to the Office of Personnel Management (OPM), Washington, D.C. 20415, which has offices in most large cities.

Planning and Geography

Planning has become one of geography's allied professions. The planning process is a complex one comprising people trained in many fields. Geographers, with their cartographic, locational, regional, and analytical skills, are sought after by planning agencies. Many an undergraduate student gets that first professional opportunity as an intern in a planning office.

Planning is done by many agencies and offices at levels ranging from the federal to the municipal. Cities have planning offices, as do regional authorities. Working in a planning office can be a very rewarding experience because it involves the solving of social and economic problems, the conservation and protection of the environment, the weighing of diverse and often conflicting arguments and viewpoints, and much interaction with workers trained in other fields. Planning is a superb learning experience.

A career in planning can be much enhanced by a background in geography, but you will have to adjust your undergraduate curriculum to include courses in such areas as public administration, public finance, and other related fields. Thus a career in planning itself requires early planning on your part. At many universities, the geography department is closely associated with the planning department, and your faculty advisor can inform you about course requirements. But if you have your eye on a particular office or agency, you should also request information from its director about desired and required skills.

Planning is by no means a monopoly of government. Government-related organizations such as the Agency for International Development (AID), the World Bank, and the International Monetary Fund (IMF) have planning offices, as do nongovernmental organizations (NGOs) such as banks, airline companies, industrial firms, multinational corporations, and research institutes. Private-sector opportunities for planners have been expanding, and you may wish to explore them. An important organization is the planners' equivalent of the geographers' AAG: the American Planning Association, 1776 Massachusetts Avenue, N.W., Washington, D.C. 20006. The AAG's *Careers in Geography* provides additional information on this expanding field and where you might pursue its study.

Teaching Geography

If you are presently a freshman or sophomore, your graduation may coincide with the end of a long decline in the geography-teaching profession. Just a few decades ago, teaching geography in elementary or high school was the goal of thousands of undergraduate geography majors. But then began the merging of geography into the hybrid field called "social studies," and prospective teachers no longer needed to have any training in geography in many States. In Florida, for instance, teachers were formerly required to take courses in regional geography and conservation (as taught in geography departments), but in the 1970s those requirements were dropped. Education planners fell victim to the myth that the teaching of geography does not require any training. What was left of geography often was taught by teachers whose own fields were history, civics, or even basketball coaching!

If you read the daily newspapers, you have seen reports of the predictable results. As mentioned earlier, "geographic illiteracy" has become a common complaint (often made by the same education planners who pushed geography into the social studies program and eliminated teacher-education requirements). Now States are returning to the education requirement so that teachers will learn some geography. And geography is returning to elementary and high school curricula—assisted in particular by the National Geographic Society, which supports State-level Geographic Alliances of educators and academic geographers. The need for geography teachers will soon be on the upswing again.

So this may be a good time to consider teaching geography as a career. You should do some research, however, because States vary in their progressiveness in this arena. You should also visit the School of Education in your college or university and ask questions about instructional opportunities in geography. In addition, contact not only the AAG but also the *National Council for Geographic Education (NCGE)* at 16A Leonard Hall, Indiana University of Pennsylvania, Indiana, PA 15705–1087 (**www.ncge.org**). Your State geographic society or Geographic Alliance also may be helpful; ask your department advisor or chairperson for details.

SOME FINAL THOUGHTS

The opportunities in geography are many, but often they are not as obvious as those in other fields. You will find that good and timely preparation produces results and, frequently, unexpected rewards.

The discussion in this appendix has provided a comprehensive answer to the oft-asked question, "What can one do with geography?" If you wish to explore this question further, along with the AAG's *Careers in Geography* we recommend *On Becoming a Professional Geographer*, a book edited by Martin S. Kenzer (Merrill/Macmillan, 1989).

We wish you every success in all your future endeavors. And, speaking for the entire community of professional geographers, we would be delighted to have you join our ranks should you choose a geography-related career.

appendix C

Pronunciation Guide

(Asterisks [*] denote "o" is pronounced as in "book".)
Abacha (ah-BAH-chah)
Abadan (ahba-DAHN)
Abakan (ahb-uh-KAHN)
Abdullah (ahb-DOO-lah)
Abidjan (abb-ih-JAHN)
Abkhazia (ahb-KAHZ-zee-uh)
Aboriginal (abb-uh-RIDGE-uh-null)
Abu Dhabi (ah-boo-DAH-bee)
Abuja (uh-BOO-juh)
Abyssinia (abb-ih-SINNY-uh)
Acacia (uh-KAY-shuh)
Acadia (uh-KAY-dee-uh)
Acapulco (ah-kah-POOL-koh)
Accra (uh-KRAH)
Aceh (ah-CHEH)
Acre (AH-kray)
Acropolis (uh-CROP-uh-liss)
Adamawa (add-uh-MAH-wuh)
Addis Ababa (adda-SAB-uh-buh)
Adelaide (ADDLE-ade)
Adan (AH-dahn/AYD'N)
Adirondack (add-ih-RONN-dak)
Adriatic (ay-dree-ATTIK)
Adygeya (ah-duh-GAY-uh)
Aegean (uh-JEE-un)
Afghanistan (aff-GHANN-uh-stan)
Afrikaans (uff-rih-KAHNZ)
Afrikaner (uff-rih-KAHN-nuh)
Agios Dimitrios (AH-jee-ohss deh-MEE-tree-ohss)
Agra (AHG-ruh)
Agri (AH-gree)
Agurchand (ah-ghoor-SHAHND)
Ahaggar (uh-HAH-gahr)
Ahmadabad (AH-muh-duh-bahd)
Ainu (EYE-noo)
Ajaria (uh-JAR-ree-uh)
Akan (AH-kahn)
Akashi Kaikyo (AH-kah-shee KYE-kyoh)
Akihito (ah-kee-HEE-toh)
Alagoas (ah-LAH-goh-ahss)
Al Andalus (ahl ahnda-LOOSE)
Alawite (AH-lah-wyte)
Alcazar (AHL-kah-zahr)
Aleppo (uh-LEP-poh)
Alevis (ah-LEE-veez)
Alexandria (ah-leg-ZAN-dree-uh)
Algae (AL-jee)
Algeria (al-JEERY-uh)
Algiers (al-JEERZ)
Al Hudaydah (ahl-hoh-DAY-duh)
Ali (ah-LEE)
Allah (AHL-ah)
Allahabad (ALLA-huh-bahd)
Allemanni (alla-MAH-nee)
Alluvial (uh-LOO-vee-ull)
Alma-Ata (ahl-muh-uh-TAH)
Al Madinah (ahl-mah-DEENA)
Almaty (ahl-mah-TUH)
Al-Quds (ahl-KOOTSS)
Al-Salam (ahl-sah-LAHM)
Altaic (al-TAY-ik)
Altay (AL-tye)
Altaya (al-TYE-uh)
Altiplano (ahl-tee-PLAH-noh)
Alpaca (al-PAK-uh)
Altun (ahl-TOON)*
Al-Wahhab (ahl-wah-HAHB)
Amador (AMMA-dor)
Amazonas (ahma-ZOAN-ahss)
Ambarawa (am-bah-RAH-wah)
Ambeno (AHM-bay-noh)
Ambon (ahm-BON)
Amerind (AMMER-rind)
Amhara (am-HAH-ruh)
Amharic (am-HAH-rik)
Amin, Idi (uh-MEEN, iddy)
Amman (uh-MAHN)
Amoy (ah-MOY)
Amritsar (um-RIT-sahr)
Amu-Darya (uh-moo-DAHR-yuh)
Amundsen, Roald (AH-moon-sun, ROH-ahl)
Amur (uh-MOOR)
Anatolia (anna-TOH-lee-uh)
Ancona (ahng-KOH-nuh)
Andalusia (ahn-duh-loo-SEE-uh)
Andaman (ANN-duh-mun)
Andes (ANN-deez)
Andhra Pradesh (ahn-druh pruh-DESH)
Angara (ahng-guh-RAH)
Angkor [Wat] (ANG-kor [WOT])
Angola (ang-GOH-luh)
Anhui (ahn-HWAY)
Ankara (ANG-kuh-ruh)
Annam (uh-NAHM)
Annam[ese]/[ite] (anna-[MEEZE]/[MITE])
Anshan (ahn-SHAHN)
Antalya (ahn-tahl-YAH)
Antananarivo (ahn-TAH-nah-nah-REEV)
Antecedent (an-tee-SEE-dunt)
Antigua (an-TEE-gwuh)
Antilles (an-TILL-eeze)
Antimony (ANN-tih-moh-nee)
Antioquia (ahn-tee-OH-kee-ah)
Antofagasta (untoh-fah-GAHSS-tah)
Antwerp (ANN-twerp)
Anuradhapura (UH-NOO-RAH-thah-poor-uh)
Anyang (ahn-YAHNG)
Aozou (OO-zoo)
Apapa (uh-PAHP-uh)
Apartheid (APART-hate)
Appennines (APP-uh-neinz)
Apure (ah-POOR-ray)
Aqaba (AH-kuh-buh)
Aqmola (ahk-moh-LAH)
Aquifer (ACK-kwuh-fer)
Arakan (ah-ruh-KAHN)
Aral (ARREL)
Aramco (uh-RAM-koh)
Arauca (ah-RAU-kah)
Arc de Triomphe (ark duh tree-AWMFF)
Archangel (ARK-ain-jull)
Archipelago (ark-uh-PELL-uh-goh)
Ardennes (ar-DEN)

Argentine (AR-jen-tyne)
Arguedas, Alcides (ahr-GWAY-dahss, ahl-SEE-duss)
Arica (ah-REE-kah)
Aridisol (uh-RIDDY-sol)
Arkhangelsk (ahr-KAHN-gyil-sk)
Armenia (ar-MEENY-uh)
Armuelles (ahr-MWAY-yace)
Artesian (ahr-TEE-zhun)
Aruba (uh-ROO-buh)
Arunachal Pradesh (AHRA-NAHTCH-ull pruh-DESH)
Aryan (AHR-yun)
Aryeetey-Attoh (ahr-YEE-tee ATTOH)
Asante (ah-SAHN-tay)
Ashanti (ah-SHAHN-tee)
Ashgabat (AHSH-gah-baht)
Ashkelon (AHSH-kih-lonn)
Aśoka (uh-SHOH-kuh)
Assad, Bashar (uh-SAHD, buh-SHAR)
Assad, Hafez al- (uh-SAHD, huh-FEZ ahl)
Assam (uh-SAHM)
Assyrian (uh-SEERY-un)
Astana (ah-STAH-nah)
Astrakhan (ASTRA-kahn)
Asturias (uh-STOOR-ree-uss)
Asunción (ah-soohn-see-OAN)
Aswan (as-SWAHN)
Atacama (ah-tah-KAH-mah)
Atatürk (ATTA-tyoork)
Atheism (AYTHEE-izm)
Athena (uh-THEENA)
Atoll (AY-toal)
Attenuated (uh-TEN-yoo-ay-ted)
Atyrau (ah-tah-RAU)
Auckland (AWK-lund)
Augelli (aw-JELLY)
Austral (AW-strull)
Australopithecene (aw-strallo-PITH-uh-seen)
Autocracy (aw-TOCK-ruh-see)
Avenida de 9 Julio (ah-veh-NEEDA day noo-AY-vay HOO-lee-oh)
Avenida Nicolas de Pierola (ah-veh-NEEDA NEEKO-lahss day pyair-ROH-lah)
Avignon (ah-veen-YAW)
Axum (AHKSS-oom)
Ayatollah (eye-uh-TOH-luh)
Azerbaijan (ah-zer-bye-JAHN)
Azeri (ah-ZAREY)
Bab el Mandeb (bab ull MAN-dub)
Babylon (BABBA-lon)
Bac Bo (BAHK-BOH)
Baden-Württemberg (BAHDEN VEERT-um-bairk)
Baganda (bah-GAHN-duh)
Bahasa (bah-HAH-sah)
Bahia (bah-EE-yah)
Bahrain (bah-RAIN)
Baht (BAHT)
Baja (BAH-hah)
Bajau (bah-YAU)
Bakassi (bah-KAH-see)
Baki (bah-KEE)
Baku (bah-KOO)
Bali [nese] (BAH-lee [NEEZ])
Balikpapan (bah-LIK-pah-pahn)
Balkan (BAWL-kun)
Balkhan (bahl-KAHN)
Balsas (BAHL-suss)
Baluchistan (buh-loo-chih-STAHN)
Banda (BAHN-duh)
Bandar Lampung (BAHN-dahr lahm-PUNG)
Bandar Seri Begawan (BUN-dahr SERRY buh-GAH-wun)
Bangalore (BANG-guh-loar)
Bangladesh [i] (bang-gluh-DESH [ee])
Banja Luka (BUN-yuh LOO-kuh)
Bantu (ban-TOO)
Banyamulenge (bahn-yah-moo-LENG-geh)
Baotou (bao-TOH)
Barbados (bar-BAY-dohss)
Barcelona (bar-suh-LOH-nuh)
Barents (BARRENS)
Barquisimeto (bar-key-suh-MAY-toh)
Barranquilla (bah-rahn-KEE-yah)
Barrio [*s*] (BAHR-ree-oh[ss])
Basalt (buh-SAWLT)
Bashkortostan (bahsh-KORT-uh-stahn)
Basilan (bah-see-LAHN)
Basilicata (bah-ZEE-lee-kah-tah)
Basque (BASK)
Basra (BAHZ-ruh)
Batak (buh-TAHK)
Batavia (buh-TAY-vee-uh)
Batista, Fulgencio (bah-TEESTA, fool-hen-see-oh)
Batumi (bah-TOO-mee)
Bauxite (BAWKS-site)
Bavaria (buh-VAIRY-uh)
Baykal [iya] (bye-KAHL [ee-ah]))
Bayou (BYE-yoo)
Bayovar (bye-YOH-vahr)
Beaumont (BOH-mont)
Bedouin (BEH-doo-in)
Beihai (bay-HYE)
Beijing (bay-ZHING)
Beira (BAY-ruh)
Beirut (bay-ROOT)
Bekasi (beh-KAH-see)
Belarus (bella-ROOSE)
Belarussia (bella-RUSH-uh)
Belém (bay-LEM)
Belgian (BEL-jun)
Belize (beh-LEEZE)
Belmopan (bell-moh-PAN)
Belo Horizonte (BAY-loh haw-ruh-ZONN-tee)
Bengal [i] (beng-GAHL [ee])
Benghazi (ben-GAH-zee)
Benguela (beng-GWELLA)
Beni (BAY-nee)
Benin (beh-NEEN)
Benue (BANE-way)
Berber (BERR-berr)
Berimbau (beh-RIM-bau)
Bering (BERRING)
Bern (BAIRN)
Betsimisaraka (bye-tsee-mee-SAH-rah-kah)
Beyoglu (bay-uh-GLOO)
Bhutan (boo-TAHN)
Bhutia (BOOH-tee-uh)
Biafra (bee-AH-fruh)
Bihar (bih-HAHR)
Bihe (bee-HAY)
Bilbao (bil-BAU)
Bilharzia (bill-HARZEE-uh)
Bioko (bee-OH-koh)
Biota (bye-OH-tuh)
Bishkek (BISH-kek)
Bitumen (bye-TOO-men)
Boeing (BOH-ing)
Boer (BOOR)
Bogor (BOH-goar)
Bogotá (boh-goh-TAH)
Bohai (bwoh-HYE)
Bohemia (boh-HEE-mee-uh)
Bolívar, Simón (boh-LEE-vahr, see-MOAN)
Bolshevik (BOAL-shuh-vick)
Bom Bahia (bom-bah-EE-yah)
Bombay (bom-BAY)
Bonaire (bun-AIR)
Bora Bora (boar-ruh-BOAR-ruh)
Boran (boar-RAN)
Bordeaux (boar-DOH)
Borneo (BOAR-nee-oh)
Bosnia-Herzegovina (BOZ-nee-uh hert-suh-goh-VEE-nuh)
Bosporus (BAHSS-puh-russ)
Botha (BAW-tuh)
Botswana (bah-TSWAHN-uh)
Bouchard, Lucien (boo-SHAR, looss-YAN)
Bougainville (BOO-gun-vil)
Boulangerie (boo-LAHN-zheh-ree)
Boulevard St. Laurent (boo-leh-VAHR san-law-RAH)
Bourgogne (boor-GOAN-yuh)
Bové, José (boh-VAY, zhoh-ZAY)
Brahmaputra (brahm-uh-POOH-truh)
Brandenburg (BRAHN-den-boorg)
Brasília (bruh-ZEAL-yuh)
Bratislava (BRUDDIS-lahva)
Bratsk (BRAHTSK)

Brazzaville (BRAHZ-uh-veel)
Brcko (BIRCH-koh)
Bremen (BRAY-mun)
Brisbane (BRIZZ-bun)
Brno (BURR-noh)
Brunei (BROO-nye)
Brunswick (BROON-svik)
Bucharest (BOO-kuh-rest)
Budapest (BOODA-pest)
Buddha (BOOD-uh)
Buddhism (BOOD-izm)
Buddhist (BOOD-ist)
Buenos Aires (BWAY-nohss EYE-race)
Buganda (boo-GAHN-duh)
Bulawayo (boo-luh-WAY-oh)
Bulgaria (bahl-GHAIR-ree-uh)
Bund (BUND)
Bureya (buh-RAY-yuh)
Burkina Faso (ber-keena FAHSSO)
Burmans (BURR-munz)
Burundi (buh-ROON-dee)
Buryat [iya] (boor-YAHT [ee-uh])
Buthelezi, Mangosuthu (boo-teh-LAY-zee, mungo-SOO-too)
Butte (BYOOT)
Byzantine (BIZZ-un-teen)
Cabimas (kah-BEE-mahss)
Cabinda (kuh-BIN-duh)
Cabora Bassa (kuh-boar-rah BAHSSA)
Cabramatta (kabbra-MADDA)
Cacao (kuh-KAY-oh/kuh-KAU)
Cadre (KAH-dray)
Cairo (KYE-roh)
Calabria (kuh-LAH-bree-uh)
Calais (kah-LAY)
Calcutta (kal-KUTT-uh)
Caldas (KAHL-dahss)
Calgary (KAL-guh-ree)
Cali (KAH-lee)
Caliph (KAY-liff)
Caliphate (KALLA-fate)
Callao (kah-YAH-oh)
Cambodia (kam-BOH-dee-uh)
Campeche (kahm-PAY-chee)
Canberra (KAN-burruh)
Cantabria (kahn-TAH-bree-uh)
Canterbury (KAN-ter-berry)
Canton (kan-TONN)
Capoeira (kah-poh-AY-ruh)
Caprivi (kuh-PREE-vee)
Caquetá (kah-kway-TAH)
Caracas (kah-RAH-kuss)
Carara (kuh-RAHR-ruh)
Caribbean (kuh-RIB-ee-un/karra-BEE-un)
CARICOM (CARRY-komm)
Carioca (kah-ree-OH-kah)
Carpathians (kar-PAY-thee-unz)
Cartagena (karta-HAY-nuh)
Casamance (KAH-zah-mahnss)
Casanare (kah-sah-NAH-ray)
Caspian (KASS-spee-un)
Caste (CAST)
Castile (kuh-STEEL)
Castile-La Mancha (kuh-STEEL luh-MAHN-chuh)
Castile-Leon (kuh-STEEL lay-OAN)
Catalan (katta-LAHN)
Catalonia (katta-LOH-nee-uh)
Cauca (KOW-kah)
Caucasian (kaw-KAY-zhun)
Caucasus (KAW-kuh-zuss)
Cay (KEE)
Cayambe (kah-YAHM-bee)
Ceará (see-ah-RAH)
Ceausescu, Nicolae (chow-SHESS-koo, NICK-oh-lye)
Cebu (seh-BOO)
Celebes (SELL-uh-beeze)
Celt [ic] (KELT [ick])
Cenozoic (senno-ZOH-ik)
Centavo (sen-TAH-voh)
Central (sen-TRAHL)
Centrifugal (sen-TRIFFA-gull)
Centripetal (sen-TRIPPA-tull)
Centro-Oeste (SENTRO oh-ESS-tee)
Cerrado (seh-RAH-doh)
Cerro de Pasco (serro day PAH-skoh)
Ceuta (see-YOO-tah)
Ceyhan (jay-HAHN)
Ceylon (seh-LONN)
Chaco (CHAH-koh)
Chaebol (JAY-BOAL)
Champagne (shahm-PAHN-yuh)
Champs Elysées (SHAWZ elly-ZAY)
Chang Jiang (chang jee-AHNG)
Ch'angan (CHAHNG-GAHN)
Changchun (CHAHNG-CHOON)*
Chao Phraya (CHOW pruh-yah)
Chaparé (chah-pah-RAY)
Charisma (kuh-RIZZ-muh)
Charleroi (SHARL-rwah)
Chateau Frontenac (shat-TOH FRAWN-teh-nahk)
Chavez (SHAH-vez)
Chechen (CHEH-chen)
Chechenya-Ingushetiya (cheh-CHEN-yuh in-goo-SHETTY-uh)
Chechnya (CHETCH-nee-uh)
Chelyabinsk (chel-YAH-bunsk)
Chengdu (chung-DOO)
Chennai (cheh-NYE)
Cherkessk (cher-KESK)
Chernobyl (see Chornobyl)
Chhattisgarh (CHADDISS-gahr)
Chiang Kai-shek (jee-AHNG kye-SHECK)
Chiang Mai (chee-AHNG mye)
Chiapas (chee-AHP-uss)
Chihuahua (chuh-WAH-wah)
Chile (CHILLI/CHEE-lay)
Chiloé (chee-luh-WAY)
Chilung [Jilong] (JEE-LOONG)
Chisinau (kee-shih-NAU)
Chittagong (CHITT-uh-gahng)
Cholla (JOH-luh)
Cholon (choh-LONN)
Cholula (choh-LOO-lah)
Chongqing (chong-CHING)
Chordata (kor-DATTA)
Chornobyl (CHAIR-noh-beel)
Chota Nagpur (choat-uh-NAHG-poor)
Chubu (CHOO-BOO)
Chukotka (chuh-KAHT-kuh)
Chukotskiy (chuh-KAHT-skee)
Chungyang (joong-YAHNG)
Chuquicamata (choo-kee-kah-MAH-tah)
Chuvash [iya] (choo-VAHSH [ee-uh])
Ciliwung (SILL-uh-wong)
Ciudad Alta (see-yoo-DAHD AHL-tuh)
Ciudad Baixa (see-yoo-DAHD bah-ZHEE-uh)
Ciudad del Este (see-yoo-DAHD del-ESS-stay)
Ciudad Guayana (see-yoo-DAHD gwuh-YAHNA)
Ciudad Juárez (see-you-DAHD WAH-rez)
Ciudades perdidas (see-you-DAH-dayss pair-DEE-duss)
Cliburn (KLYE-bern)
Coahuila (koh-uh-WEE-luh)
Coatzacoalcos (koh-aht-sah-koh-AHL-kohss)
Cochabamba (koh-chah-BUM-bah)
Cochin (KOH-chin)
Cocos (KOH-kuss)
Colombo (kuh-LUM-boh)
Colón (kuh-LOAN)
Comodoro Rivadavia (comma-DORE-oh ree-vah-DAH-vee-ah)
CONAIE (koh-NYE)
Conakry (koh-NAK-ree)
Confucius (kun-FEW-shuss)
Coniferous (kuh-NIFF-uh-russ)
Conquistador (koan-KEE-stuh-doar)
Constantinople (kon-stant-uh-NOPLE)
Conurbation (konner-BAY-shun)
Copenhagen (koh-pen-HAHGEN)
Copra (KOH-pruh)
Corcovado (koar-koh-VAH-doh)
Cordillera (kor-dee-YERRA)
Cordillera del Litoral (kor-dee-YERRA dale lee-toh-RAHL)
Córdoba (KORR-doh-buh)
Coromandel (kor-uh-MANDLE)
Corsica (KORR-sih-kuh)
Cortés, Hernán (kor-TAYSS, air-NAHN)
Costa del Sol (koh-stuh del SOAL)
Costa Rica (koh-stuh-REE-kuh)
Côte D'Ivoire (KOAT deev-WAH)
Coup d'état (koo-day-TAH)

Coveñas (koh-VAIN-yahss)
Creole [s] (KREE-oal[z])
Crete (KREET)
Crimea (cry-MEE-uh)
Crimean (cry-MEE-un)
Croat (KROH-aht/KROH-at)
Croatia (kroh-AY-shuh)
Cruz (KROOZ)
Cuiabá (koo-yuh-BAH)
Curaçao (koor-uh-SAU)
Cusiana-Cupiagua (koo-see-AHNA, koopee-YAH-gwah)
Cuyo (KOO-yoh)
Cuzco (KOO-skoh)
Cyclades (SIK-luh-deeze)
Cyprus (SYE-pruss)
Cyrenaica (sear-uh-NAY-ih-kuh)
Cyrillic (suh-RILL-ick)
Czar (ZAHR)
Czarina (zah-REE-nuh)
Czech (CHECK)
Czechoslovakia (check-uh-sloh-VAH-kee-uh)
Dagestan (dag-uh-STAHN)
Dahomey (dah-HOH-mee)
Dakar (duh-KAHR)
Dalai Lama (dah-lye LAHMA)
Dalian (dah-lee-ENN)
Dalit (DAH-lit)
Dalmatia (dal-MAY-shuh)
Damascus (duh-MASK-uss)
Damietta (dam-ee-ETTA)
Danube (DAN-yoob)
Daoism (DAU-ism)
Daqing (dah-CHING)
Dardanelles (dahr-duh-NELZ)
Dar es Salaam (dahr ess suh-LAHM)
Darien (dar-YEN)
Dayak (DYE-ack)
De Gaulle, Charles (duh-GAWL, SHARL)
Deccan (DECKEN)
Delhi (DELLY)
Delphi (DELL-fye)
Deltaic (dell-TAY-ick)
Deng Xiaoping (DUNG shau-PING)
Département (day-part-MAW)
Devolution (dee-voh-LOOH-shun)
Dhahran (dah-RAHN)
Dhaka (DAHK-uh)
Dhow (DAU)
Diaspora (dee-ASP-uh-ruh)
Dichotomy (dye-KOTT-uh-mee)
Dien Bien Phu (d'yen-b'yen-FOOH)
Dijon (dee-ZHAW)
Dili (DIH-lee)
Dinaric (dih-NAHR-rick)
Diyarbakir (dih-yahr-buh-KEER)
Djezira-al-Maghreb (juh-ZEER-uh ahl-mahg-GRAHB)
Djibouti (juh-BOODY)
Djouf (JOOF)
Dnieper (duh-NYEPPER)
Dniester (duh-NYESS-truh)
Dnipropetrovsk (duh-nep-roh-puh-TRAWFSSK)
Dodoma (DOH-duh-mah)
Dominica (duh-MIN-ih-kuh)
Donbas (DAHN-bass)
Donets (duh-NETTS)
Donetsk (duh-NETTSK)
Dostoyevsky (doss-stoy-YEFF-skee)
Douala (doo-AHLA)
Drakensberg (DRAHK-unz-berg)
Dravidian (druh-VIDDY-un)
Dresden (DREZ-den)
Druzba (DROOZE-bah)
Druze (DROOZE)
Dubai (doo-BYE)
Duchy (DUTCH-ee)
Dunedin (duh-NEED-nn)
Durban (DER-bun)
Dushanbe (doo-SHAHM-buh)
Dvina (duh-vee-NAH)
Dzibilchaltun (zeeb-eel-chahl-TOON)
Dzungaria (joong-GAH-ree-uh)
Ebola (ee-BOH-luh)
Ecuador (ECK-wah-dor)
Ecumene (ECK-yoo-meen)
Edinburgh (EDDIN-burruh)
Edo (EDD-oh)
Edgecumbe (EDGE-kum)
Eelam (EE-lum)
Eilat (AY-laht)
Eire (AIR)
Ejido[s] (eh-HEE-doh [ss])
El Aqsa (el AHK-suh)
El Cojo (ell-KOH-hoh)
El Niño (ell-NEEN-yoh)
El Paso (ell-PASSO)
El Puente (el-PWEN-tay)
Elam (EE-lum)
Elbe (ELB)
Elburz (el-BOORZ)
Ellesmere (ELZ-mear)
Ellinikon (eh-LINNY-konn)
Emirate (EMMA-rate)
Endemism (en-DEM-izm)
Endoreic (en-doh-RAY-ick)
England (ING-glund)
Enkare (en-KAH-ray)
Entebbe (en-TEBBA)
Entre Rios (en-truh-REE-ohss)
Entrepôt (AHNTRA-poh)
Eritrea (erra-TRAY-uh)
Erlitou (air-lee-TOH)
Esmeraldas (ezz-may-RAHL-dahss)
Estonia (eh-STOH-nee-uh)
Ethiopia (eeth-ee-OH-pea-uh)
Etorofu (etta-ROH-foo)
Eucalyptus (yoo-kuh-LIP-tuss)
Eunuch (YOO-nuck)
Euphrates (yoo-FRATE-eeze)
Eurasia (yoo-RAY-zhuh)
Evens (eh-VENSS)
Extremadura (ess-truh-muh-DOORA)
Façade (fuh-SAHD)
Faisalabad (fye-SAHL-ah-bahd)
Falkland [s] (FAWK-lund[z])
Farakka (fah-RAH-kuh)
Farghona (fahr-GOH-nuh)
Faro del Comercio (FAH-roh del koh-MAIR-see-oh)
Fauna (FAW-nuh)
Favela [s] (fah-VAY-lah[ss])
Fazenda (fah-ZENN-duh)
Fellaheen (fella-HEEN)
Fergana (fahr-guh-NAH)
Ferronorte (feh-roh-NOR-tay)
Fezzan (fuh-ZANN)
Fiji (FEE-jee)
Fijian (fuh-JEE-un)
Filipinos (filla-PEA-noze)
Finno-Ugric (finno-YOO-grick)
Fjord (FYORD)
Flores (FLAW-rihss)
Florianópolis (flaw-ree-ah-NOH-poh-leese)
Foederis (feh-DARE-iss)
Fonseca (fahn-SAY-kuh)
Formosa (for-MOH-suh)
Fortaleza (for-tuh-LAY-zuh)
Fox, Vicente (FOX, vee-SEN-tay)
Francophone (FRANK-uh-foan)
Frankfurt (FRUNK-foort)
Fría (FREE-uh)
Fuji (FOODGY)
Fujian (foo-jee-ENN)
Fukuoka (foo-kuh-WOH-kuh)
Fulani (foo-LAH-nee)
Funafuti Atoll (foo-nah-FOO-tee AY-toal)
Fungi (FUN-jye)
Fushun (foo-SHUN)*
Futa Jallon (food-uh juh-LOAN)
Fuzhou (foo-ZHOH)
Gaoxiong (see Kaohsiung)
Gabon (gah-BAW)
Gadhafi, Muammar (guh-DAHFI, MOO-uh-mar)
Gaelic (GALE-ick)
Gaillard (gil-YARD)
Galicia (guh-LEE-see-uh)
Galla (GAH-lah)
Gambia (GAM-bee-uh)
Gandhi (GONDY)
Ganga (GUNG-guh)
Ganges/Gangetic (GAN-jeez/gan-JETTICK)
Gansu (gahn-SOO)
Gao (GAU)
Gaoliang (gow-lee-AHNG)
Garagum (gah-rah-GOOM)

Garonne (guh-RON)
Gasohol (GAS-uh-hoal)
Gatún (guh-TOON)
Gauteng (GAU-teng)
Gavan (guh-VAHN)
Gaza (GAH-zuh)
Gdansk (guh-DAHNSK)
Gdynia (guh-DINNY-uh)
Geiger (GHYE-gherr)
Geneva (jeh-NEE-vuh)
Genoa (JENNO-uh)
Georges (ZHOR-zh)
Georgia (GEORGE-uh)
Gerardi, Juan José (heh-RAHR-dee, HWAHN hoh-ZAY)
Gezira (juh-ZEER-uh)
Ghana (GAH-nuh)
Ghanaian (gah-NAY-un)
Ghats (GAHTSS)
Gibraltar (jih-BRAWL-tuh)
Gilgit (GILL-gutt)
Ginza (GHIN-zuh)
Giralda (hih-RAHL-duh)
Giza (GHEE-zuh)
Glasgow (GLASS-skau)
Glasnost (GLUZZ-nost)
Goa (GOH-uh)
Gobi (GOH-bee)
Godthab (GAWT-hawb)
Goiás (goy-AHSS)
Golan (goh-LAHN)
Gondwana (gond-WON-uh)
Gorbachev, Mikhail (GOR-buh-choff, meek-HYLE)
Gorkiy (GORE-kee)
Gorno-Altay (gore-noh-AL-tye)
Gorno-Badakhshan (gore-noh-bah-dahk-SHAHN)
Gorod (guh-RAHD)
Göteborg (GOAT-uh-borg)
Gouda (GOO-dah)
Grande Arche (GRAWND-ARSH)
Grande Carajás (GRUNN-dee kuh-ruh-ZHUSS)
Greenwich (GREN-itch)
Grenada (gruh-NAY-duh)
Groznyy (GRAWZ-nee)
Guadalajara (gwah-duh-luh-HAHR-uh)
Guadalcanal (gwaddle-kuh-NAL)
Guadalquivir (gwahddle-kee-VEER)
Guadeloupe (GWAH-duh-loop)
Guajira (gwah-HEAR-ah)
Guam (GWAHM)
Guanabara (gwah-nah-BAR-uh)
Guangdong (gwahng-DUNG)
Guangxi Zhuang (gwahng-shee JWAHNG)
Guangzhou (gwahng-JOH)
Guaraní (gwah-rah-NEE)
Guatemala (gwut-uh-MAH-lah)
Guayaquil (gwye-ah-KEEL)
Guayas (GWYE-ahss)
Guelph (GWELF)
Guerrero (geh-RARE-roh)
Guiana [s] (ghee-AH-nah[z])
Guilin (gway-LIN)
Guinea (GHINNY)
Guinea-Bissau (ghinny-bih-SAU)
Guizhou (gway-JOH)
Gujarat (goo-juh-RAHT)
Gulag (GHOO-lahg)
Guri (GOOR-ree)
Gurkha (GHOOR-kuh)
Guryev (GHOOR-yeff)
Guyana (guy-AHNA)
Habibie (huh-BEE-bee)
Habomai (HAH-boh-mye)
Hacienda (ah-see-EN-duh)
Hägerstrand, Torsten (HAYGER-strand, TOR-stun)
Hague (HAIG)
Haifa (HYE-fah)
Haikou (HYE-KOH)
Hainan (HYE-NAHN)
Haiphong (hye-FONG)
Haiti (HATE-ee)
Hamburg (HAHM-boorg)
Han (HAHN)
Hang Bai (HAHNG-bye)
Hangzhou (hahng-JOH)
Hankou (hahn-KOH)
Hanoi (han-NOY)
Hanseatic (han-see-ATTIK)
Hanyang (hahn-YAHNG)
Harappa (huh-RAP-uh)
Harare (huh-RAH-ray)
Harbin (HAR-bin)
Harer (HAH-rahr)
Harijan (hah-ree-JAHN)
Hartshorne (HARTSS-horn)
Haryana (hah-ree-AHNA)
Hashish (hah-SHEESH)
Hausa (HOW-sah)
Hawai'i (huh-WAH-ee)
Hazara (huh-ZAHR-ah)
Hebei (huh-BAY)
Hegemony (heh-JEH-muh-nee)
Heian (HAY-ahn)
Heilongjiang (hay-long-jee-AHNG)
Hejira (heh-JEER-ruh)
Helsinki (hel-SINKEE)
Henan (heh-NAHN)
Herodotus (heh-RODDA-tuss)
Hesse (HESS)
Hexi (huh-SHEE)
Hezbollah (HIZZ-boh-luh)
Hidrovia (ee-DROH-vee-uh)
Hierarchical (hire-ARK-uh-kull)
Himachal Pradesh (huh-MAHTCH-ull pruh-DESH)
Himalayas (him-AHL-yuzz/himma-LAY-uzz)
Hindu (HIN-doo)
Hindu Kush (hin-doo KOOSH)*
Hindustan (hin-doo-STAHN)
Hindustani (hin-doo-STAHN-nee)
Hippocrates (hih-POCK-ruh-teez)
Hiroshima (hirra-SHEE-muh/huh-ROH-shuh-muh)
Hispaniola (iss-pahn-YOH-luh)
Ho Chi Minh (hoh-chee-MINN)
Hohhot (huh-HOO-tuh)
Hokkaido (hoh-KYE-doh)
Holistic (hoh-LISS-tick)
Holocene (HOLLO-seen)
Hominid (HOM-ih-nid)
Homo sapiens (hoh-moh SAY-pea-enz)
Homogeneity (hoh-moh-juh-NAY-uh-tee)
Honduras (hon-DURE-russ)
Hongkou (hong-KOH)
Honiara (hoh-nee-AHR-ah)
Honolulu (honn-uh-LOO-loo)
Honshu (HONN-shoo)
Hooghly (HOO-glee)
Hormuz (hoar-MOOZE)
Hotan (HOH-TAN)
Houghton (HOH-t'n)
Houphouët-Boigny, Félix (oo-FWAY bwah-NYEE, fay-LEEKS)
Hsinchu [Xinzhu] (shin-JOO)
Huai (HWYE)
Huallaga (wah-YAH-gah)
Huancayo (wahn-KYE-oh)
Huang [He] (HWAHNG [huh])
Huangpu (hwahng-POO)
Hubei (hoo-BAY)
Hué (HWAY)
Humboldt (HUMM-bolt)
Humus (HYOO-muss)
Hunan (hoo-NAHN)
Husayn (hoo-SINE)
Hussein, Saddam (hoo-SAIN, suh-DAHM)
Hutu (HOO-too)
Hwange (WAHNG-ghee)
Hyderabad (HIDE-uh-ruh-bahd)
Hyundai (HUN-dye)
Ibadan (ee-BAHD'N)
Iberia (eye-BEERY-uh)
Ibn Saud (ib'n sah-OOD)
Ibo [land] (EE-boh [land])
Ife (EE-fay)
Iguaçu (EE-gwah-soo)
Iki (EE-kee)
Ikoyi (ee-KOY-yee)
Île de France (EEL duh-FRAWSS)
Île de la Cité (EEL duh-la-see-TAY)
Illyrian (ih-LEER-ree-un)
Ilmen (yill-MEN)
Ilyich (ILL-yitch)
Imam (ih-MAHM)
Inchon (int-CHON)

Indonesia (indo-NEE-zhuh)
Ingush (in-GOOSH)
Ingushetiya (in-goo-SHETTY-uh)
Inkatha (in-KAH-tah)
Inle (IN-lay)
Interdigitate (intuh-DID-juh-tate)
Intramuros (in-trah-MOOR-rohss)
Inuit (IN-yoo-it)
Iquitos (ih-KEE-tohss)
Iran (ih-RAN/ih-RAHN)
Iranian (ih-RAIN-ee-un)
Iraq (ih-RAK/ih-RAHK)
Iraqi (ih-RAKKY)
Irian Jaya (IH-ree-ahn JYE-uh)
Irkutsk (ear-KOOTSK)
Irrawaddy (ih-ruh-WODDY)
Irredentism (irruh-DEN-tism)
Irtysh (ear-TISH)
Isfahan (iz-fuh-HAHN)
Iskenderun (iz-ken-duh-ROON)
Islam (iss-LAHM)
Islamabad (iss-LAHM-uh-bahd)
Islington (IZZ-ling-tunn)
Ismaili (izz-MYE-lee)
Isohyet (EYE-so-hyatt)
Israel (IZ-rail)
Istanbul (iss-tum-BOOL)
Isthmus/isthmian (ISS-muss/ ISS-mee-un)
Itaipu (ee-TYE-pooh)
Ituri (ih-TOORY)
Ivanovo (ee-VAH-nuh-voh)
Ivoirian (ih-VWAH-ree-un)
Izhevsk (EE-zheffsk)
Jabal (JAB-ull)
Jabotabek (juh-BOH-tah-bek)
Jaffna (JAHF-nuh)
Jains (JYE-nz)
Jakarta (juh-KAHR-tuh)
Jakota (juh-KOH-tuh)
Jalisco (huh-LISS-koh)
Jamaica (juh-MAKE-uh)
Jammu (JUH-mooh)
Jamshedpur (JAHM-shed-poor)
Java (JAH-vuh)
Jawa (JAH-vuh)
Jawanese (jah-vuh-NEEZE)
Jayapura (jye-uh-POOR-ruh)
Jazirah (juh-ZEER-uh)
Jebel Ali (JEH-bel ah-LEE)
Jerusalem (juh-ROO-suh-lum)
Jesuit (JEH-zoo-it)
Jharkhand (JAHR-kahnd)
Jiang Zemin (jee-AHNG zuh-MIN)
Jiangsu (jee-ahng-SOO)
Jiangxi (jee-ahng-SHEE)
Jilin (jee-LIN)
Jilong (jee-LUNG)
Jinmen Dao (JIN-MEN-dau)
Joao, Dom (ZHWOW, dom)
Johannesburg (joh-HANNIS-berg)
Johor (juh-HOAR)
Jomon (JOH-mon)
Jubail (joo-BILE)
Judaic (joo-DAY-ick)
Judaism (JOODY-ism)
Junggar (JOONG-gahr)
Junta (HOON-tah)
Jurong (juh-RONG)
Jutland (JUT-lund)
Kabardino-Balkar [iya] (kabber-DEE-noh bawl-KAR [ree-uh])
Kabila, Laurent (kuh-BEE-luh law-RAH)
Kabul (KAH-bull)
Kachins (kuh-CHINZ)
Kaduna (kah-DOO-nah)
Kaifeng (kye-FUNG)
Kai Tak (KYE TAK)
Kakadu (KAH-kuh-doo)
Kalaallit Nunaat (kuh-LAHT-lit noo-NAT)
Kalahari (kalla-HAH-ree)
Kalenjin (kuh-LEN-jin)
Kalgoorlie (kal-GHOOR-lee)
Kalimantan (kalla-MAN-tan)
Kaliningrad (kuh-LEEN-in-grahd)
Kalmyk (KAL-mik)
Kalmykiya (kal-MIK-ee-uh)
Kamba (KAHM-bah)
Kambalda (kahm-BAHL-duh)
Kamchatka (kum-CHAHT-kuh)
Kamehameha (kah-MAY-hah-may-hah)
Kampala (kahm-PAH-luh)
Kampuchea (kahm-pooh-CHEE-uh)
Kanaks (KAH-nahkss)
Kanarese (KAHN-uh-reece)
Kannada (KAHN-uh-duh)
Kano (KAH-noh)
Kansai (KAHN-SYE)
Kanto (KAN-toh)
Kaohsiung [Gaoxiong] (GAU-see-OOHNG)
Kara (KAHR-ruh)
Kara Kum (kahr-ruh KOOM)
Karachay (kah-ruh-CHYE)
Karachayevo-Cherkessiya (kahra-CHAH-yeh-vuh cheer-KESS-ee-uh)
Karachi (kuh-RAH-chee)
Karafuto (kahra-FOO-toh)
Karaganda (karra-gun-DAH)
Karakalpak (karra-kal-PAK)
Karakoram (kahra-KOR-rum)
Karamay (kah-RAH-may)
Kareliya (kuh-REE-lee-uh)
Karens (kuh-RENZ)
Kariba (kuh-REE-buh)
Karnataka (kahr-NAHT-uh-kuh)
Kashagan (KAH-shah-gahn)
Kashgar (KAHSH-gahr)
Kashi (KAH-shee)
Kashmir (KASH-meer)
Katanga (kuh-TAHNG-guh)
Kathmandu (kat-man-DOOH)
Katowice (kah-toh-VEE-tsuh)
Kattegat (KAT-ih-gat)
Kauai (KAU-eye)
Kawasaki (kah-wah-SAH-kee)
Kazakh (KUZZ-uck)
Kazakhstan (KUZZ-uck-STAHN/KUZZ-uck-stahn)
Kazan (kuh-ZAHN)
Kefamenanu (kuh-fahm-uh-NAH-noo)
Kelantan (keh-LAHN-tahn)
Kelang (kuh-LAHNG)
Kemal, Mustafa (keh-MAHL, moo-stah-FAH)
Kenya (KEN-yuh)
Kerala (KEH-ruh-luh)
Keratsinion (keh-rut-SINNY-onn)
Kerinci (kuh-REEN-chee)
Khabarovsk (kuh-BAHR-uffsk)
Khakassiya (kuh-KAHSS-ee-uh)
Khalistan (kahl-ee-STAHN)
Khan, Genghis (KAHN, JING-guss)
Khanate (KAHN-ate)
Khartoum (kar-TOOM)
Khatami (kah-TAH-mee)
Khiva (KEE-vuh)
Khmer [Rouge] (kuh-MAIR [ROOZH])
Khoi (KHOY)
Khoikhoi (KHOY-khoy)
Khoisan (khoy-SAHN)
Kholmsk (KAWLMSK)
Khomeini (hoh-MAY-nee)
Khorat (koh-RAHT)
Khurasan (koor-uh-SAHN)
Khyber (KYE-burr)
Khulna (KOOL-nah)
Kiev (Kyyiv) (KEE-yeff)
Kievan (kee-EVAN)
Kigali (kih-GAH-lee)
Kikuyu (kee-KOO-yoo)
Kikwit (KIK-wit)
Kilimanjaro (kil-uh-mun-JAH-roh)
Kilinochchi (kih-luh-NOCK-chee)
Kilwa (KEEL-wah)
Kinabalu (kin-ah-BAH-loo)
Kindu (KIN-doo)
Kinki (kin-KEE)
Kinmen (kin-MEN)
Kinshasa (kin-SHAH-suh)
Kiosk (KEE-osk)
Kirghiz (keer-GEEZE)
Kiribati (KIH-ruh-bahss)
Kirkuk (keer-KOOK)*
Kiruna (kih-ROONA)
Kisangani (kee-sahn-GAH-nee)
Kitakyushu (kee-TAH-KYOO-shoo)
Kivu (KEE-voo)
Klaipeda (KLYE-puh-duh)
Klyuchevskaya (klee-ooh-CHEFF-skuh-yuh)

Koala (kuh-WAH-luh)
Kobe (KOH-bay)
Kolkata (kol-KUTTA)
Kolkhoz (KOLL-koze)
Kolyma (koh-LEE-mah)
Komi (KOH-mee)
Kompong Som (kahm-pong SAWM)
Kompong Thom (kahm-pong TAWM)
Komsomolsk (komm-suh-MAWLSK)
Kongfuzi (kung-FOODZEE)
Kongzi (KUNG-dzee)
Königsberg (KAY-nix-bairk)
Konkan (KAHNG-kun)
Köppen (KER-pun)
Koran (kaw-RAHN)
Korea (kuh-REE-uh)
Korla (KOOR-LAH)
Koryakiya (kor-YAH-kee-uh)
Koryakskaya (kor-YAHK-skuh-yuh)
Kosciusko (kuh-SHOO-skoh)
Kosovar (KAW-suh-vahr)
Kosovo (KAW-suh-voh)
Kostunica, Vojislav (koss-too-NEE-chah, VOY-slahv))
Kota Kinabalu (KOH-tuh kin-ah-BAH-loo)
Kourou (koo-ROO)
Koutoubia (koo-TOO-bee-yuh)
Kowloon (kau-LOON)
Kra Isthmus (KRAH ISS-muss)
Krakow (KRAH-koov)
Krasnodar (KRASS-nuh-dahr)
Krasnovodsk (kruzz-noh-VAUGHTSK)
Krasnoyarsk (krass-nuh-YARSK)
Kribi (KREE-bee)
Krivoy Rog (krih-voy-ROAG)
Kronstadt (KROAN-shtaht)
Kruger (KROO-guh)
Krugersdorp (KROO-guzz-dorp)
Kryvyy Rih (kree-VEE-REE)
Kuala Lumpur (KWAHL-uh LOOM-poor)*
Kublai Khan (koob-lye KAHN)
Kufra (KOO-fruh)
Kumartuli (koo-MAR-too-lee)*
Kunashiri (koo-NAH-shuh-ree)*
Kunlun (KOON-LOON)*
Kunming (koon-MING)*
Kura (KOOR-uh)
Kurd [istan] (KERD [uh-stahn])
Kure (KOOH-ray)
Kurile (KYOOR-reel)
Kuroshio (koo-roh-SHEE-oh)
Kush (KOOSH)
Kushites (KOO-sheits)
Kuwait (koo-WAIT)
Kuybyshev (KWEE-buh-sheff)
Kuzbas (kooz-BASS)
Kuznetsk (kooz-NETSK)
Kwangju (GWONG-JOO)
Kwazulu (kwah-ZOO-loo)
Kyongju (GYOONG-JOO)
Kyongsang (GYOONG-SAHN)
Kyoto (kee-YOH-toh)
Kyrgyz (KEER-geeze)
Kyrgyzstan (KEER-geeze-stahn)
Kyushu (kee-YOO-shoo)
Kyyiv [See Kiev]
La Coruña (lah-kor-ROON-yah)
La Défense (lah-day-FAWSS)
La Paz (lah-PAHZ)
Ladakh (luh-DAHK)
Ladino (luh-DEE-noh)
Laem Chabang (lay-EMM chuh-BAHNG)
Lafaiete (lah-fuh-YAY-tuh)
Lagos (LAY-gohss)
Lahore (luh-HOAR)
Lamanai (LAH-mah-nay)
Lancang (LAHN-ZAHNG)
Land (LAHNT)
Länder (LEN-derr)
Lantau (LAHN-DAU)
Lanzhou (lahn-JOH)
Lao (LAU)
Laos (LAUSS)
Laotian (lay-OH-shun)
Latvia (LATT-vee-uh)
Lautoka (lau-TOH-kuh)
Laval (lah-VAHL)
La Violencia (lah vee-oh-LENN-see-uh)
Legume (LEG-gyoom)
Le Havre (luh-HAHV)
Leipzig (LYPE-sik)
Leith (LEETH)
Lemur (LEE-mer)
Lena (LAY-nuh)
Lenin (LENNIN)
León (lay-OAN)
Lesotho (leh-SOO-too)
Levant (luh-VAHNT)
Lhasa (LAH-suh)
Li (LEE)
Lianyungang (lee-en-yoong-GAHNG)
Liao (lee-AU)
Liaodong (lee-au-DUNG)
Liaoning (lee-au-NING)
Liberia (lye-BEERY-uh)
Libreville (LEE-bruh-veel)
Lichens (LYE-kenz)
Liechtenstein (LIK-ten-shtine)
Liège (lee-EZH)
Lima (LEE-muh)
Limpopo (lim-POH-poh)
Lingua franca (LEENG-gwuh FRUNK-uh)
Linguae francae (LEENG-gwee FRUNK-kee)
Lingala (ling-GAH-lah)
Litani (lih-TAH-nee)
Lithuania (lith-oo-AINY-uh)
Littoral (LIT-oh-rull)
Livingstonia (lih-ving-STOH-nee-uh)
Ljubljana (lee-oo-blee-AHNA)
Llama (LAH-muh)
Llano [*s*] (YAH-noh [ss])
Loess (LERSS)
Loihi (loh-EE-hee)
Loire (luh-WAHR)
Lombardy (LOM-bar-dee)
Lombok (LAHM-bahk)
Louvre (LOOV)
Lozi (LOH-zee)
Lualaba (loo-uh-LAH-buh)
Luanda (loo-AN-duh)
Lubumbashi (loo-boom-BAH-shee)
Lucknow (LUCK-nau)
Luhya (LOO-yuh)
Luleå (LOO-lee-oh)
Lund (LOOND)
Luo (LOO-oh)
Luxembourg (LUX-em-borg)
Luxor (LUK-soar)
Luzon (loo-ZAHN)
Lyon (lee-AW)
Maale (MAH-lay)
Maas (MAHSS)
Maasai (muh-SYE)
Maastricht (mah-STRICT)
Mabo (MAY-boh)
Macau (muh-KAU)
Macedonia (massa-DOH-nee-uh)
Machu Picchu (MAH-choo PEEK-choo)
Mackinder, Halford (muh-KIN-der, HAL-ferd)
Mactan (mahk-TAHN)
Madagascar (madda-GAS-kuh)
Madhya Pradesh (mahd-yuh-pruh-DESH)
Madras (muh-DRAHSS)
Madrid (muh-DRID)
Madura (muh-DOORA)
Madurai (mahd-uh-RYE)
Madurese (muh-dooh-REECE)
Magaz (muh-GAHZ)
Magdalena (mahg-dah-LAY-nah)
Magellan (muh-JELL-un)
Maghreb (mahg-GRAHB)
Magyar (MAG-yahr)
Mahakam (MAH-hah-kahm)
Maharajah (mah-hah-RAH-juh)
Maharashtra (mah-huh-RAH-shtra)
Mahathir Mohamad (MAH-hah-theer moh-HAH-mud)
Maize (MAYZ)
Makhachkala (muh-kahtch-kuh-LAH)
Makkah (MEK-ah)
Makung (mah-GOONG)
Malabar (MAL-uh-bahr)
Malabo (muh-LAH-boh)
Malacca (muh-LAH-kuh)

Málaga (MAHL-uh-guh)
Malagasy (malla-GASSY)
Malaita(ns) (mah-LAY-tah [nz])
Malawi (muh-LAH-wee)
Malay (muh-LAY)
Malaya (muh-LAY-uh)
Malayalam (mal-uh-YAH-lum)
Malaysia (muh-LAY-zhuh)
Maldives (MAWL-deevz)
Mali (MAH-lee)
Malmö (MAHL-meh)
Maluku(s) (mah-LOO-koo[z])
Manado (muh-NAH-doh)
Managua (mah-NAH-gwuh)
Manaus (muh-NAUSS)
Manchu (man-CHOO)
Manchukuo (mahn-JOH-kwoh)
Manchuria (man-CHOORY-uh)
Mandalay (man-duh-LAY)
Mandarin (MAN-duh-rin)
Mandela (man-DELLA)
Manipur (man-uh-POOR)
Manitoba (manna-TOH-buh)
Manzanillo (mun-zuh-NEE-yoh)
Mao Zedong (MAU zee-DUNG)
Maori (MAH-aw-ree/MAU-ree)
Maputo (mah-POOH-toh)
Maquiladora (mah-kee-luh-DORR-uh)
Mara (MAHRA)
Maracaibo (mah-rah-KYE-boh)
Marajó (mah-rah-ZHOH)
Mariana (marry-ANNA)
Mariinsk (muh-ryee-YEENSK)
Maritsa (muh-REET-suh)
Mariy [-el] (MAH-ree [el])
Marmagao (marma-GAU)
Marmara (MAH-muh-ruh)
Marquesas (mahr-KAY-suzz)
Marrakech (mahr-uh-KESH)
Marseille (mar-SAY)
Marsupial (mar-SOOPY-ull)
Martinique (mahr-tih-NEEK)
Massif (mass-SEEF)
Matadi (muh-TAH-dee)
Matao (mah-TAU)
Mato Grosso (mutt-uh-GROH-soh)
Mato Grosso do Sul (mutt-uh-GROH-soh duh-SOOL)
Matsu (mah-TSOO)
Matsu Tao (mah-TSOO DAU)
Mauna Kea (mau-nuh-KAY-uh)
Mauritania (maw-ruh-TAY-nee-uh)
Mauritius (maw-REE-shuss)
Mauryan (MAW-ree-un)
Maya (MYE-uh)
Mayan (MYE-un)
Mazar-e-Sharif (mah-ZAHR-ee-shah-REEF)
Mbeki, Thabo (mm-BEH-kee, TAH-boh)
Mecca (MEK-ah)
Mecsek (MAH-chek)
Medan (MAY-dahn)
Medellín (meh-deh-YEEN)
Medina (muh-DEENA)
Megalopolis (meh-guh-LOPP-uh-liss)
Meghalaya (may-guh-LAY-uh)
Meghna (MAIG-nuh)
Meiji (may-EE-jee)
Mekong (MAY-kong)
Melaka (see Malacca)
Melanesia (mella-NEE-zhuh)
Melbourne (MEL-bun)
Melilla (meh-LEE-yuh)
Mendoza (men-DOH-zah)
Mengkabong (MENG-kah-bong)
Menshevik (MEN-shuh-vick)
Merauke (muh-RAU-kuh)
Mercosul (mair-koh-SOOL)
Mercosur (mair-koh-SOOR)
Mérida (MAY-ree-dah)
Meridian (meh-RIDDY-un)
Merina (meh-REE-nuh)
Meroe (MEH-roh-ay)
Meru (MAY-roo)
Mesa (MAY-suh)
Mesabi (meh-SAH-bee)
Meseta (meh-SAY-tuh)
Meshketian (mesh-KETTY-un)
Mesoamerica (MEZZOH-america)
Mesopotamia (messo-puh-TAY-mee-uh)
Mesquita (meh-SKEE-tuh)
Mestizo (meh-STEE-zoh)
Meuse (MERZZ)
Mezzogiorno (met-soh-JORR-noh)
Miao (m'YOW)
Michoacán (mee-chuh-wah-KAHN)
Micronesia (mye-kroh-NEE-zhuh)
Milan (mih-LAHN)
Millau (mee-YOH)
Milosevic, Slobodan (mih-LAW-suh-vitch, SLOH-boh-dahn)
Minahasa (MEE-nah-hah-sah)
Minaret (MINNA-ret)
Minas Gerais (MEE-nuss zhuh-RICE)
Mindanao (min-duh-NAU)
Minh, Ho Chi (MINN, hoh chee)
Minifundia (minny-FOON-dee-uh)
Mirador(es) (meera-DOAR [ayss])
Miraflores (meera-FLAW-rayss)
Miskito (mih-SKEE-toh)
Mitú (mee-TOO)
Mius (mee-OOS)
Mizoram (mih-ZOR-rum)
Moçambique (moh-sum-BEEK)
Mogadishu (moo-gah-dee-SHOH)
Mohenjo Daro (moh-hen-joh-DAHRO)
Mojave (moh-HAH-vee)
Moldavia (moal-DAY-vee-uh)
Moldova (moal-DOH-vuh)
Moluccans (muh-LUCK-unz)
Molybdenum (muh-LIB-dun-um)
Mombasa (mahm-BAHSSA)
Mongol (MUNG-goal)
Mongolia (mung-GOH-lee-uh)
Monogamy (muh-NOG-ah-mee)
Montagnards (MON-tun-yardz)
Montaña (mon-TAHN-yah)
Montenegro (mon-teh-NEE-groh)
Monterrey (mon-teh-RAY)
Montevideo (moan-tay-vee-DAY-oh)
Montpellier (maw-pell-YAY)
Montreal (mun-tree-AWL)
Montserrat (mont-seh-RAHTT)
Moped (MOH-ped)
Moravia (more-RAY-vee-uh)
Mordoviya (mor-DOH-vee-uh)
Moscow (MAW-skau)
Moselle (moh-ZELL)
Mosque (MOSK)
Mount Isa (mount EYE-suh)
Mpumalanga (mm-pooma-LAHNG-guh)
Muara Muntai (moo-WAH-rah MOON-tye)
Mugabe (moo-GAH-bay)
Muglad (moo-GLAHD)
Muhajir (MOO-hah-jeer)
Muhammad (moo-HAH-mid)
Mujahideen (moo-jah-heh-DEEN)
Mulatto (moo-LAH-toh)
Mullah (MOOL-ah)*
Multan (mool-TAHN)
Mulukas (MAH-loo-kooz)
Mumbai (MOOM-bye)
Munich (MYOO-nik)
Murmansk (moor-MAHNTSK)
Murray (MUH-ree)
Musandam (muh-SAND-um)
Muscat (MUH-skaht)
Muscovy (muh-SKOH-vee)
Muslim (MUZZ-lim)
Mutrah (MUTT-truh)
Muuga (MOO-guh)
Myanmar (mee-ahn-MAH)
Mymensingh (mye-men-SING)
Nagasaki (nah-guh-SAHKEE)
Naga [land] (NAHGA [-land])
Nagorno-Karabakh (nuh-GORE-noh KAH-ruh-bahk)
Nagoya (nuh-GOYA)
Nagpur (NAHG-poor)
Nairobi (nye-ROH-bee)
Nakhodka (nuh-KAUGHT-kuh)
Nakhon Sawan (NUH-KAWN suh-WOON)
Nam Bo (nahm-BOH)
Namib (nah-MEEB)
Namibia (nuh-MIBBY-uh)
Nanjing (nahn-ZHING)
Naoroji (nau-ROH-jee)
Nara (NAHRA)

Nassau (NASS-saw)
Nasser, Gamal Abdel (NASS-er, guh-MAHL AB-dul)
Natal (nuh-TAHL)
Nauru (nah-OO-roo)
Nautical (NAW-dih-kull)
Naxcivan (nah-kee-chuh-VAHN)
Nazi (NAH-tsee)
Nazran (nahz-RAHN)
Ndebele (en-duh-BEH-leh)
Ndola (en-DOH-luh)
Negev (NEH-ghev)
Negro (NAY-groh)
Nehru, Jawaharlal (NAY-roo, juh-WAH-hur-lahl)
Nei Mongol (nay-MUNG-goal)
Nepal (nuh-PAHL)
Neva (NAY-vuh)
Nevis (NEE-vuss)
Nevsky Prospekt (NEFF-skee PROSS-spekt)
New Caledonia (noo-kalla-DOAN-yuh)
New Guinea (noo-GHINNY)
Newfoundland (NYOO-fun-lund)
Ngodi (eng-GOH-dee)
Ngorongoro (eng-gore-ong-GORE-roh)
Nha Trang (nah-TRAHNG)
Niamey (NEE-ah-may)
Nicaragua (nick-uh-RAH-gwuh)
Niger [Country] (nee-ZHAIR)
Niger [River] (NYE-jer)
Niger-Kordofanian (NYE-jer kor-doh-FAN-ee-un)
Nigeria (nye-JEERY-uh)
Nilotic (nye-LODDIK)
Ningbo (ning-BWOH)
Ningxia Hui (NING-shee-AH HWAY)
Nistru (NEE-stroo)
Niue (nee-OOH-ay)
Nizhniy Novgorod (NIZH-nee NAHV-guh-rahd)
Nkrumah, Kwame (en-KROO-muh, KWAH-mee)
Nobi (NOH-bee)
Nord-Pas de Calais (NORD pah-duh-kah-LAY)
Noriega (noar-ree-AY-gah)
Norilsk (nuh-REELSK)
Norte (NOR-tay)
Notre Dame (NOH-truh DAHM)
Nouméa (noo-MAY-uh)
Nouveau Quebec (noo-VOH kwih-BECK)
Nova Scotia (nova-SKOH-shuh)
Novgorod (NAHV-guh-rahd)
Novokuznetsk (noh-voh-kooz-NETSK)
Novorossiysk (noh-voh-ruh-SEESK)
Novosibirsk (noh-voh-suh-BEERSK)
Nubia (NOO-bee-uh)
Nueva Mendez (noo-AY-vuh MEN-dez)
Nuevo León (noo-AY-voh lay-OAN)
Nunavut (NOON-uh-voot)
Nungambakkam (noon-GAHM-bah-KAHM)
Nuuk (NEWK)
Nyika (nye-YEE-kuh)
Nyasaland (nye-ASSA-land)
Oahu (uh-WAH-hoo)
Oaxaca (wuh-HAH-kuh)
Obasanjo (oh-BAH-sahn-joh)
Oblast (OB-blast)
Obsidian (ob-SIDDY-un)
Occidental (oak-see-den-TAHL)
Ocho Rios (oh-choh REE-ohss)
Ochre (OH-ker)
Ocussi-Ambeno (oh-KOOH-see AHM-bay-noh)
Oder-Neisse (OH-der NEISS)
Odesa (oh-DESSA)
Ogaden (oh-gah-DEN)
Ogoni (oh-GOH-nee)
Oguz (uh-GOOZ)
Ohmae, Kenichi (OH-may, keh-NEE-chee)
Oirot (AW-ih-rut)
Oise (WAHZ)
Okavango (oh-kuh-VAHNG-goh)
Okhotsk (oh-KAHTSK)
Okrug (AW-krook)
Olduvai (OLE-duh-way)
Oman (oh-MAHN)
Omdurman (om-duhr-MAHN)
Omote Nippon (oh-MOH-tay NIP-on)
Oran (aw-RAHN)
Orangutan (aw-RANG-gyoo-tan)
Ordos (ORD-uss)
Ordzhonikidze (or-johnny-KIDD-zuh)
Øresund (ERR-uh-sun)
Oriental (orry-en-TAHL)
Oriente (orry-EN-tay)
Orinoco (orry-NOH-koh)
Orissa (aw-RISSA)
Oromo (AW-ruh-moh)
Orontes (aw-RAHN-teeze)
Osaka (oh-SAH-kuh)
Öskemen (ERSKEE-min)
Oslo (OZ-loh)
Oslofjord (OZ-loh-fyord)
Osman (oz-MAHN)
Ossetia (oh-SEE-shuh)
Ossetians (oh-SEE-shunz)
Ossies (OH-seez)
Ostrava (AW-struh-vuh)
Otavalo (oh-tah-VAH-loh)
Ottawa (OTTA-wuh)
Oweinat (oh-WEE-naht)
Oxisol (OXY-soll)
Oyo (OH-joh)
Padania (pah-DAHN-yah)
Pago Pago (PAH-goh PAH-goh)
Pagoda (puh-GOH-duh)
Pahlavi, Shah Muhammad Reza (puh-LAH-vee, shah mooh-HAH-mid RAY-zuh)
Pakhtuns (puck-TOONZ)
Pakistan (PAH-kih-stahn)
Palau (puh-LAU)
Pale (PAH-lay)
Palembang (pah-LEM-bahng)
Palenque (puh-LENG-kay)
Palestine (PAL-uh-stine)
Palestinian (pal-uh-STINNY-un)
Palk (PAWK)
Palo Alto (PALLOH AL-toh)
Palu (pah-LOO)
Pamir (pah-MEER)
Pamirs (pah-MEERZ)
Pampa (PAHM-pah)
Panache (pah-NAHSH)
Panaji (pah-NAH-jee)
Pangaea (pan-GAY-uh)
Papeete (pahp-ee-ATE-tee)
Papua New Guinea (pahp-OO-uh noo-GHINNY)
Papua [ns] (pahp-OO-uh [-unz])
Papuan (pahp-OO-un)
Pará (puh-RAH)
Paraguay (PAHRA-gwye)
Paraíba (pah-rah-EE-buh)
Paramillos (pah-rah-MEE-yohss)
Paraná (pah-rah-NAH)
Paria (PAHR-yah)
Pariah (puh-RYE-uh)
Parliament (PAR-luh-ment)
Pashtuns [see Pushtuns]
Pasig (PAH-sig)
Paseo del Norte (pah-SAY-oh del NOR-tay)
Patagonia (patta-GOH-nee-uh)
Pathan (puh-TAHN)
Pattaya (puh-TYE-uh)
Paulista [*s*] (pow-LEASH-tah [ss])
Pechora (peh-CHORE-ruh)
Pedro Miguel (PAY-droh mee-GWELL)
Peking (pea-KING)
Peloponnesus (pelloh-puh-NEEZE-uss)
Pelourinho (peh-loo-REEN-yoh)
Penghu (pung-HOO)
Peón (pay-OAN)
Peones (pay-OH-nayss)
Perestroika (perra-STROY-kuh)
Perm (PAIRM)
Permyakiya (pairm-YAH-kee-uh)
Pernambuco (pair-nahm-BOO-koh)
Peron, Juan (puh-ROAN, WAHN)
Persepolis (per-SEPP-uh-luss)
Persia (PER-zhuh)
Peshawar (puh-SHAH-wahr)
Peso (PAY-soh)
Petén (peh-TEN)
Petrograd (PETTRO-grahd)

Petronas (peh-TROH-nuss)
Petropavlovsk-Kamchatskiy (pit-ruh-PAHV-lufsk kahm-CHAHT-skee)
Pettah (PET-uh)
Phaleron (fuh-LEER-un)
Pharaonic (fair-ray-ONNICK)
Philippines (FILL-uh-peenz)
Phnom Penh (puh-NOM PEN)
Phoenicians (fuh-NEE-shunz)
Phu Bia (POO-bee-uh)
Phuket (POO-KETT)
Physiography (fizzy-OGG-ruh-fee)
Phytogeography (FYE-toh-jee-OG-gruh-fee)
Piazza (pea-AH-tsuh)
Piedmont (PEED-mont)
Pilipino (pill-uh-PEA-noh)
Pinang (puh-NANG)
Pinatubo (pin-uh-TOO-boh)
Pinheiros (PEEN-yay-ross)
Pinyin (pin-YIN)
Piraeus (puh-RAY-uss)
Pisac (pea-SAHK)
Pitcairn (PITT-kairn)
Pizarro, Francisco (pea-SAHRO, frahn-SEECE-koh)
Place de la Concorde (PLAHSS duh lah kon-KORD)
Placer [ing] (PLASS-uh [ring])
Plata, Rio de la (PLAH-tah, REE-oh day-lah)
Platypus (PLATT-uh-pus)
Plaza Azoguejo (PLA-sah ah-soh-GAY-hoh)
Plaza de Armas (PLAH-sah day AR-mahss)
Plaza de Mayo (PLAH-sah day MYE-oh)
Plaza de San Martin (PLAH-sah day sahn mahr-TEEN)
Plaza Mayor (PLAH-sah mye-YOAR)
Pleistocene (PLY-stoh-seen)
Plymouth (PLIH-muth)
Pointe-Noire (pwahnt-nuh-WAHR)
Politburo (POLLIT-byoor-roh)
Polonoroeste (POLLOH-nuh-roh-ESS-tee)
Polygamy (puh-LIG-uh-mee)
Polynesia (polla-NEE-zhuh)
Polytheism (polly-THEE-izm)
Pontianak (pahn-tee-AH-nahk)
Popocatépetl (poh-poh-kah-TEH-peh-til)
Port Moresby (port MORZ-bee)
Port Said (port-sah-EED)
Port-au-Prince (por-toh-PRANSS)
Porteños (por-TAIN-yohss)
Pôrto Alegre (POR-too uh-LEG-ruh)
Pôrto Velho (POR-too VELL-yoo)
Posavina (poh-suh-VEENA)
Potosí (poh-toh-SEE)
Povolzhye (puh-VOLL-zhuh)
Prague (PRAHG)
Prayagraj (PRAY-ag-rahdge)
Pretoria (prih-TOR-ree-uh)
Primorskiy (pree-MOHR-skee)
Príncipe (PREEN-see-pea)
Progeny (PRAH-juh-nee)
Prokofiev (pruh-KOFFEY-ev)
Proselytism (PRAH-sell-eh-tizm)
Provence (pro-VAHNSS)
Prudhoe (PROO-doh)
Prut (PRROOT)
Pudong (poo-DUNG)
Puebla (poo-EBB-luh)
Puerto Barrios (pwair-toh bah-REE-uss)
Puerto Caldera (pwair-toh kahl-DERRA)
Puerto Rico (pwair-toh REE-koh)
Puget (PYOO-jet)
Punjab (pun-JAHB)
Punjabi (poon-JAH-bee)
Punta Arenas (POON-tah ah-RAY-nahss)
Punta del Este (POON-tah del ESS-tay)
Pusan (BOO-SAHN)
Pushtun (PAH-shtoon)
Putin, Vladimir (POO-t'n, VLAH-duh-meer))
Putrajaya (poo-truh-JYE-UH)
Pyatigorsk (pyih-tyih-GORSK)
Pygmy (PIG-mee)
Pyongyang (pea-AWHNG-yahng)
Pyrenees (PEER-uh-neez)
Qaidam (CHYE-DAHM)
Qanat (KAH-naht)
Qaraghandy (kah-rah-GONDY)
Qasim (kah-SEEM)
Qatar (KOTTER)
Qattarah (kuh-TAR-ruh)
Qiang (chee-AHNG)
Qin (CHIN)
Qing (CHING)
Qingdao (ching-DAU)
Qinghai (ching-HYE)
Qingyi (ching-YEE)
Qinhuangdao (chin-hwahng-DAU)
Qiqihar (chee-CHEE-har)
Qoraqalpog (kora-kal-PAHG)
Quaternary (kwuh-TER-nuh-ree)
Quebec (kwih-BECK)
Quebecer (kwih-BECK-er)
Quebecois (kay-beh-KWAH)
Quebracho (kay-BROTCH-oh)
Quechua (KAYTCH-wah)
Quemoy (keh-MOY)
Quetzal (kay-DZAHL)
Quezon (KAY-soan)
Quinary (KWYE-nuh-ree)
Quito (KEE-toh)
Quran (kor-RAHN)
Raisina (rye-SEENA)
Raj (RAHDGE)
Rajang (rah-JAHNG)
Rajasthan (RAH-juh-stahn)
Rajin-Sombong (RAH-jin SAWM-bong)
Rajiv (ruh-ZHEEV)
Ramadan (rahma-DAHN)
Randstad (RUND-stud)
Rangoon (rang-GOON)
Rangpur (RHANG-poor)
Rarotonga (rarra-TAHNG-guh)
Rashid (ruh-SHEED)
Ratzel, Friedrich (RAHT-sull, FREED-rish)
Ravenstein (RAVVEN-steen)
Rawalpindi (rah-wull-PIN-dee)
Recife (ruh-SEE-fuh)
Reg (REGG)
Réunion (ray-yoon-YAW)
Reykjavik (RAKE-yah-veek)
Rhaeto-Romansch (RAY-shoh roh-MAHN-ssh)
Rhine (RYNE)
Rhodesia (roh-DEE-zhuh)
Rhône-Alpes (ROAN AHLP)
Rhône-Saône (ROAN say-OAN)
Riga (REEGA)
Rio Branco (REE-oh BRUNG-koh)
Rio de Janeiro (REE-oh day zhah-NAIR-roh)
Rio Grande do Norte (REE-oh GRUN-dee duh NORTAH)
Rio Grande do Sul (REE-oh GRUN-dee duh SOOL)
Rioja (ree-OH-hah)
Riverine (RIVER-reen)
Riyadh (ree-AHD)
Roma (ROH-muh)
Romania (roh-MAIN-yuh)
Rondônia (roh-DOAN-yuh)
Roraima (raw-RYE-muh)
Rostow (ROSS-stoff)
Rouen (roo-AW)
Rub al Khali (roob ahl KAH-lee)
Rue Notre Dame (ROO noh-truh DAHM)
Rue Saint-Louis (ROO sah-loo-WEE)
Ruhr (ROOR)
Ruijin (rway-JEEN)
Rumailah (roo-MYE-luh)
Rus (ROOSE)
Ruwenzori (roo-when-ZOHR-ree)
Rwanda (roo-AHN-duh)
Ryukyu (ree-YOO-kyoo)
Saarland (ZAHR-lunt)
Saba (SAY-buh/SAH-buh)
Sabah (SAHB-ah)
Sabinas (sah-BEE-nuss)
Saddhu (SAH-dooh)
Sahara (suh-HARRA)
Sahel (suh-HELL)

Saigon (sye-GAHN)
Saint-Denis (san-deh-NEE)
Sakartvelos (sah-KART-vuh-lohss)
Sakha (SAH-kuh)
Sakhalin (SOCK-uh-leen)
Salina Cruz (sah-LEE-nuh CROOZ)
Salinas (sah-LEE-nuss)
Salvador (SULL-vuh-dor)
Salween (SAHL-wain)
Samara (suh-MAH-ruh)
Samarinda (sah-mah-RIN-dah)
Samarqand (sah-mahr-KAHND)
Samoa (suh-MOH-uh)
Samoan (suh-MOH-un)
Samsung (SAHM-SOONG)
San (SAHN)
San Cristóbal (sahn kree-STOH-bahl)
San Joaquin (san-wah-KEEN)
San José (sahn hoh-ZAY)
San Juan (sahn HWAHN)
San Pedro Sula (sahn pay-droh-SOO-luh)
San Rafael (sahn rah-fye-ELL)
San'a (suh-NAH)
Sandinista (sahn-dee-NEE-stuh)
Santa Catarina (SUN-tuh kuh-tuh-REE-nuh)
Santiago (sahn-tee-AH-goh)
Santos (SUNT-uss)
Sanxia (sahn-SHAH)
São Francisco (sau frahn-SEECE-koh)
São Luis (sau loo-EECE)
São Paulo (sau PAU-loh)
São Tomé (sau TOH-may)
Sarajevo (sahra-YAY-voh)
Saratov (suh-RAHT-uff)
Sarawak (suh-RAH-wahk)
Sardinia (sahr-DINNY-uh)
Sarmatian (sahr-MAY-shee-un)
Saskatchewan (suss-KATCH-uh-wunn)
Saudi Arabia (SAU-dee uh-RAY-bee-uh)
Sauer (SAU-er)
Savoy (suh-VOY)
Saxony (SAX-uh-nee)
Scania (SKAIN-yuh)
Schengen (SHENG-gen)
Schistosomiasis (shistoh-soh-MYE-uh-siss)
Scythian (SITH-ee-un)
Sedentary (SEDDEN-terry)
Seikan (say-KAHN)
Seine (SENN)
Semarang (seh-MAHR-rahng)
Semeru (suh-MAY-roo)
Semitic (seh-MITTICK)
Sen-en (sen-NENN)
Sendero Luminoso (sen-DARE-oh loo-mee-NOH-soh)
Senegal (sen-ih-GAWL)
Seoul (SOAL)
Sepoy (SEE-poy)
Sepulchre (SEP-ul-kurr)
Serbia (SER-bee-uh)
Serer (seh-RAIR)
Sergipe (SAIR-zhee-pay)
Serov (SAIR-roff)
Serra dos Carajás (SERRA doo kuh-ruh-ZHUSS)
Sertão (sair-TOWNG)
Sesame (SESS-uh-mee)
Seto (SET-oh)
Sevastopol (seh-VASS-toh-pawl)
Seville (suh-VILL)
Seychelles (say-SHELLZ)
Sha Mian (shah mee-AHN)
Shaanxi (shahn-SHEE)
Shaba (SHAH-bah)
Shan (SHAHN)
Shandong (shahn-DUNG)
Shanghai (shang-HYE)
Shang-Yin (SHANG-YIN)
Shantou (SHAHN-TOH)
Shanxi (shahn-SHEE)
Shari (SHAH-ree)
Sharia (SHAH-ree-uh)
Shatt al Arab (shot ahl uh-RAHB)
Sheikdom (SHAKE-dum)
Shenyang (shun-YAHNG)
Shenzhen (shun-ZHEN)
Shevardnadze, Eduard (sheh-vart-NAHD-zeh, ED-wahrd)
Shi (SHERR)
Shia (SHEE-uh)
Shi'a (SHEE-uh)
Shihezi (shee-HAY-zee)
Shi'ism (SHEE-izm)
Shi'ite (SHEE-ite)
Shikoku (shick-KOH-koo)
Shikotan (shee-koh-TAHN)
Shimonoseki (shim-uh-noh-SECKEE)
Shiraz (shih-RAHZ)
Shirioshi (shih-ree-OH-shee)
Shogun (SHOH-goon)
Shona (SHOH-nuh)
Shostakovich (shosta-KOH-vitch)
Shwedogon (SHWAY-duh-gonn)
Siam (sye-AMM)
Siberia (sye-BEERY-uh)
Sichuan (zeh-CHWAHN)
Sicily (SISS-uh-lee)
Siddhartha (sid-DAHR-tuh)
Sidra (SIDD-ruh)
Siem Reap (SERM-REEP)
Sierra Madre Occidental (see-ERRA MAH-dray oak-see-den-TAHL)
Sierra Madre Oriental (see-ERRA MAH-dray aw-ree-en-TAHL)
Sikh (SEEK)
Sikhism (SEEK-izm)
Siking (see-KING)
Sikkim (SICK-um)
Silesia (sye-LEE-zhuh)
Sinai (SYE-nye)
Singapore (SING-uh-poar)
Sinhala (sin-HAHLA)
Sinhalese (sin-hah-LEEZE)
Sinicization (sine-ih-sye-ZAY-shun)
Sinicized (SYE-nuh-sized)
Sisal (SYE-sull)
Sitka (SIT-kuh)
Sivas (SEE-vahss)
Skagerrak (SKAG-uh-rak)
Skopje (SKAWP-yay)
Slamet (SLAH-met)
Slav (SLAHV)
Slavic (SLAH-vick)
Slovakia (sloh-VAH-kee-uh)
Slovenia (sloh-VEE-nee-uh)
Sofala (soh-FAH-luh)
Sofia (SOH-fee-uh)
Somali (suh-MAH-lee)
Somalia (suh-MAHL-yuh)
Song Qiang (SAWNG chee-AHNG)
Songhai (SAWNG-hye)
Songhua (SAWNG-hwah)
Sonora (suh-NORA)
Sotho (SOO-too)
Soufrière (soo-free-AIR)
Sovkhoz (SOV-koze)
Soweto (suh-WETTO)
Spatial (SPAY-shull)
Spratly (SPRAT-lee)
Spykman (SPIKE-mun)
Sri Lanka (sree-LAHNG-kuh)
Srinagar (srih-NUG-arr)
Srpska (SERP-skuh)
St. Eustatius (saint yoo-STAY-shuss)
St. Louis (sah-loo-EE)
St. Lucia (saint LOO-shuh)
St. Maarten (sint MAHRT-un)
Stalin (STAH-lin)
Stamboul (STAHM-bool)
Stanovoy (stuh-nah-VOY)
Steppe (STEP)
Stockholm (STOCK-hoam)
Strasbourg (STRAHSS-boorg)
Stupa (STOO-puh)
Stuttgart (SHTOOT-gart)*
Sudan (soo-DAN)
Sudanese (soo-duh-NEEZE)
Suharto (soo-HAHR-toh)
Sudd (SOOD)*
Sudeten (soo-DAYTEN)
Suez (SOO-ez)
Sui (SWAY)
Sulawesi (soo-luh-WAY-see)
Suleyman (SOO-lay-mahn)
Sultan (SULL-tun)
Sultanate (SULL-tuh-nut)
Sulu (SOO-loo)
Sumatera (suh-MAH-tuh-ruh)
Sumatra (suh-MAH-truh)

Sumbawa (soom-BAH-wuh)
Sumer (SOO-mer)
Sun Yat-sen (SOON yaht-SENN)
Sunda [nese] (SOON-duh [NEEZ])
Sung (SOONG)
Sunni (SOO-nee)
Supsa (SOOP-suh)
Suriname (soor-uh-NAHM-uh)
Suryavarman (soory-AHVA-mun)
Susten (SOOS-ten)
Sutlej (SUTT-ledge)
Suva (SOO-vuh)
Swahili (swah-HEE-lee)
Swazi [land] (SWAH-zee [land])
Syktyvkar (sik-tiff-KAR)
Syr-Darya (seer-DAHR-yuh)
Syria (SEARY-uh)
Szczecin (SHCHE-tseen)
Tabasco (tuh-BAH-skoh)
Tagalog (tuh-GAH-log)
Tahiti (tuh-HEET-tee)
Taiga (TYE-guh)
Taipa (TYE-pah)
Taipei [Taibei] (tye-BAY)
Taiwan (tye-WAHN)
Tajik (TUDGE-ick)
Tajikistan (tah-JEEK-ih-stahn)
Takla Makan (tahk-luh-muh-KAHN)
Taliban (TALLA-ban)
Tallinn (TALLEN)
Tamaulipas (tah-mau-LEEPUS)
Tamil [Nadu] (TAMMLE [NAH-doo])
Tampere (TAHM-puh-ray)
Tampico (tam-PEEK-oh)
Tana (TAHNA)
Tang (TAHNG)
Tanganyika (tan-gun-YEEKA)
Tangerang (tahn-guh-RAHNG)
Tangshan (tahng-SHAHN)
Tannu Tuva (tan-ooh-TOO-vuh)
Tanzania (tan-zuh-NEE-uh)
Tao (DAU)
Tarim (TAH-REEM)
Tashkent (Toshkent) (tahsh-KENT)
Tasman (TAZZ-mun)
Tasmania (tazz-MAY-nee-uh)
Tatar (TAHT-uh)
Tatarstan (TAHT-uh-STAHN)
Tatra (TAHT-truh)
Taxco (TAHSS-koh)
Tayshet (tye-SHET)
Tbilisi (tuh-BILL-uh-see)
Tchaikovsky (chye-KOFF-skee)
Teatro Colón (tay-AH-troh koh-LOAN)
Tecnópolis (tek-NOH-poh-leece)
Teda (TAY-duh)
Tegucigalpa (tuh-goose-ih-GAHL-puh)
Tehran (tay-uh-RAHN)
Tehuantepec (tuh-WHAHN-tuh-pek)
Tel Aviv-Jaffa (tella-VEEVE-JOFF-uh)
Telugu (TELLOO-goo)
Tema (TAY-muh)
Templada (tem-PLAH-dah)
Tengiz (TEN-ghiz)
Tenochtitlán (tay-noh-chit-LAHN)
Teotihuacán (tay-uh-tee-wah-KAHN)
Terai (teh-RYE)
Terek (TEH-rek)
Terengganu (teh-reng-GAH-noo)
Tertiary (TER-shuh-ree)
Tete (TATE-uh)
Thailand (TYE-land)
Thames (TEMZ)
Thamesmead (TEMZ-meed)
Thebes (THEEBZ)
Theocracy (thee-OCK-ruh-see)
Thessaloniki (thess-uh-luh-NEE-kee)
Thimphu (thim-POOH)
Thünian (TOO-nee-un)
Tian Shan (TYAHN SHAHN)
Tiananmen (TYAHN-un-men)
Tianjin (tyahn-JEEN)
Tiber (TYE-ber)
Tibet (tuh-BETT)
Tierra caliente (tee-ERRA kahl-YEN-tay)
Tierra del Fuego (tee-ERRA dale FWAY-goh)
Tierra fría (tee-ERRA FREE-uh)
Tierra helada (tee-ERRA ay-LAH-dah)
Tierra nevada (tee-ERRA neh-VAH-dah)
Tierra templada (tee-ERRA tem-PLAH-dah)
Tigrean (tih-GRAY-un)
Tigris (TYE-gruss)
Tijuana (tee-WHAHN-uh)
Tikhvin (TIK-vun)
Tilbury (TILL-buh-ree)
Timbuktu (tim-buck-TOO)
Timor (TEE-moar)
Tirane (tih-RAHN-uh)
Titicaca (tiddy-KAH-kuh)
Tlingit (TLING-git)
Tobago (tuh-BAY-goh)
Tocantins (toke-un-TEENS)
Tokaido (toh-KYE-doh)
Tokelau (TOH-kuh-lau)
Tokyo (TOH-kee-oh)
Toledo (toh-LAY-doh)
Tolstoy (TAWL-stoy)
Tolyatti (tawl-YAH-tee)
Tomolon (TOM-oh-loan)
Tondano (tonn-DAH-noh)
Tonga (TAHNG-guh)
Tongan (TONG-gun)
Tonkin (TAHN-KIN)
Tonle Sap (tahn-lay SAP)
Toponym (TOH-poh-nim)
Toponymy (toh-PONN-uh-mee)
Tordesillas (tor-day-SEE-yahss)
Torii (taw-REE)
Torres (TOAR-russ)
Torrijos (tor-REE-hohss)
Toshka (TOSH-kah)
Toshkent (tahsh-KENT)
Toyama (toh-YAH-muh)
Train à grande vitesse (TRAN ah-grawnd-vee-TESS)
Trajan (TRAY-junn)
Transcaucasia (tranz-kaw-KAY-zhuh)
Transdniestra (tranz-duh-NYESS-truh)
Transhumance (tranz-hyoo-MANSS)
Transvaal (TRUNZ-vahl)
Trias Monge, José (TREE-ahss MON-hay, hoh-ZAY)
Tripoli (TRIPPA-lee)
Tripolitania (trip-olla-TANEY-yuh)
Tripura (TRIP-uh-ruh)
Trondheim (TRAHN-hame)
Trung Bo (TROONG BOH)
Trypanosomiasis (try-pan-noh-soh-MYE-uh-sis)
Tselinograd (seh-LEENO-grahd)
Tsetse (TSETT-see)
Tsugaru (tsoo-GAH-roo)
Tsukuba (tsoo-KOOB-uh)
Tsumeb (SOO-meb)
Tsunami (tsoo-NAH-mee)
Tsushima (tsoo-SHEE-muh)
Tswana (TSWAHN-uh)
Tuareg (TWAH-regg)
Tubarão (too-buh-RAWNG)
Tuchueh (too-CHOO-way)
Tucson (TOO-sonn)
Tucumán (too-koo-MAHN)
Tucuruí (too-koo-roo-EE)
Tufo (TOO-foh)
Tukiu (too-KYOO)
Tula (TOO-luh)
Tun Abdul Razak (toon ahb-DOOL rah-ZAHK)
Tunis (TOO-niss)
Tunisia (too-NEE-zhuh)
Turfan (TER-fan)
Turin (TOOR-rin)
Turkana (ter-KANNA)
Turkestan (TER-kuh-stahn)
Türkmenbashi (tyoork-men-BAH-shee)
Turkmenistan (terk-MEN-uh-stahn)
Turku (TOOR-koo)
Tuscany (TUSS-kuh-nee)
Tutsi (TOOTSIE)*
Tuva (TOO-vuh)
Tuvalu (too-VAHL-oo)
Tuzla (TOOZ-lah)
Twa (TOO-wah)
Tyrol (tih-ROLL)
Tyumen (tyoo-MEN)
Tyva (TOO-vuh)
Tyvinian (too-VINNY-un)
Ubangi (oo-BANG-ghee)
Ubundu (oo-BOON-doo)*

Udaipur (uh-DYE-poor)
Udmurt [iya] (ood-MOORT [ee-uh])
Ufa (oo-FAH)
Uganda (yoo-GAHN-duh/yoo-GANDA)
Uhuru (oo-HOO-roo)
Ujungpandang (oo-JUNG PAHN-dahng)
Ukraine (yoo-CRANE)
Ulaanbaatar (oo-lahn-BAH-tor)
Ulsan (OOL-SAHN)
Uluru (ooh-LOO-roo)
Ulyanov (ool-YAH-noff)
UNCLOS (UNN-klohss)
Ungava (ung-GAH-vuh)
Ura Nippon (OOH-ruh NIP-on)
Urabá (oo-rah-BAH)
Ural (YOOR-ull)
Ural-Altaic (YOOR-ull-al-TAY-ick)
Urdu (OOR-doo)
Uruguay (OO-rah-gwye)
Ürümqi (oo-ROOM-chee)
Usinsk (oo-SINSK)
Ussuri (ooh-SOOR-ree)
Ust-Kamenogorsk
Utrecht (YOO-trekt)
Uttar Pradesh (ootar-pruh-DESH)
Uttaranchal (OO-tahr-rahn-CHARL)
Uyghur (WEE-ghoor)
Uzbek (OOZE-beck)
Uzbekistan (ooze-BECK-ih-stahn)
Uzen (oo-ZEN)
Vaal (VAHL)
Valdez (val-DEEZE)
Valencia (vuh-LENN-see-uh)
Valle Central (VAH-yay sen-TRAHL)
Valparaíso (vahl-pah-rah-EE-so)
Vanino (VAH-nih-noh)
Vanni (VAH-nee)
Vanuatu (vahn-uh-WAH-too)
Varanasi (vuh-RAHN-uh-see)
Varangian (vuh-RANGE-ee-un)
Veld (VELT)
Veneto (VENN-uh-toh)
Venezuela (veh-neh-SWAY-lah)
Ventspils (VENT-spilz)
Veracruz (verra-CROOZE)
Verde (VER-dee)
Vereeniging (fuh-REEN-ih-king)
Versailles (vair-SYE)
Viangchan (vyung-CHAHN)
Vienna (vee-ENNA)
Vietnam (vee-et-NAHM)
Vietnamese (vee-et-nuh-MEEZE)
Villahermosa (vee-yuh-air-MOH-suh)
Vilnius (VILL-nee-uss)
Vindhya (VIN-dyuh)
Visayan (vuh-SYE-un)
Vistula (VIST-yulluh)
Vladikavkaz (vlad-uh-kuff-KAHZ)
Vladivostok (vlad-uh-vuh-STAHK)
Vojvodina (VOY-vuh-deena)
Volgograd (VOLL-guh-grahd)
Vologda (VAW-lug-duh)
Volta Redonda (vahl-tuh rih-DONN-duh)
Von Thünen, Johann Heinrich (fon-TOO-nun, YOH-hahn HINE-rish)
Voortrekker (FOR-trecker)
Wabenzi (wah-BENZ-zee)
Wahhabism (wah-HAH-b'izm)
Wahid, Abdurrahman (wah-HEED, ahb-doo-RAH-mun)
Waikiki (wye-kuh-KEE)
Waitangi (WYE-tonggy)
Wakhan (wah-KAHN)
Wakisihu (wah-kee-SEE-hoo)
Walachia (wuh-LAH-kee-yuh)
Wallaby (WALLA-bee)
Wallonia (wah-LOANY-uh)
Walloon (wah-LOON)
Walvis (WAHL-vuss)
Wasatch (WAW-satch)
Weber (VAY-buh)
Wegener (VAY-ghenner)
Wei (WAY)
Wenzhou (whunn-JOH)
Weser (VAY-zuh)
Westermann (VESS-tair-munn)
Westphalia (west-FAIL-yuh)
Wildebeest (WIL-duh-beest)
Willamette (wuh-LAMM-ut)
Windhoek (VINT-hook)
Winnipeg (WIN-uh-peg)
Witbank (WHIT-bank)
Witwatersrand (WITT-waw-terz-rand)
Wolof (WOH-loff)
Wroclaw (VROH-tswahf)
Wuchang (woo-CHAHNG)
Wuhan (woo-HAHN)
Wuzhou (woo-JOH)
Xenophobia (zee-nuh-FOH-bee-uh)
Xhosa (SHAW-suh)
Xi (SHEE)
Xi Jiang (SHEE jee-AHNG)
Xia (SHAH)
Xiamen (shah-MEN)
Xian (shee-AHN)
Xianggang (see-AHNG-gahng)
Xiangkhoang (SHEE-AHNG-kwahng)
Xiaolangdi (SH-YAU-lahng-dee)
Xinjiang (shin-jee-AHNG)
Xizang (sheedz-AHNG)
Yacyretá (yah-see-ray-TAH)
Yakut [sk] (yuh-KOOT [sk])
Yakutiya (yuh-KOOTY-uh)
Yamalo-Nenetskiy (yuh-MAH-luh-nuh-NET-skee)
Yamoussoukro (yahm-uh-SOO-kroh)
Yamuna (YAH-muh-nuh)
Yanan (yen-AHN)
Yanbu (YAN-boo)
Yangon (yahn-KOH)
Yangpu (YAHNG-poo)
Yangzi [Yangtze] (YANG-dzee)
Yanomami (yah-noh-MAH-mee)
Yantai (yahn-TYE)
Yaohan (yau-HAHN)
Yaoundé (yown-DAY)
Yaroslavl (yar-uh-SLAHV-ull)
Yawata (yuh-WAH-tuh)
Yayoi (yah-YOH-ee)
Yeates (YATES)
Yekaterinburg (yeh-KAHTA-rin-berg)
Yeltsin, Boris (YELT-sinn, BAW-reese)
Yemen (YEMMEN)
Yemeni (YEH-meh-nee)
Yenisey (yen-uh-SAY)
Yerba maté (YAIR-bah mah-TAY)
Yerevan (yair-uh-VAHN)
Yerushalayim (yeh-roo-shuh-LYE-im)
Yevreyskaya (yev-RAY-sky-yah)
Yibin (EE-BIN)
Yichang (yee-CHAHNG)
Yi-Luo (YEE-loo-oh)
Yogyakarta (yah-gyuh-KAR-tuh)
Yokohama (yoh-kuh-HAH-muh)
Yonne (YAHN)
Yoruba (YAH-rooba)
Yuan (YOO-ahn)
Yucatán (yoo-kuh-TAHN)
Yugoslavia (yoo-goh-SLAH-vee-uh)
Yulara (yoo-LAH-rah)
Yumen (YOO-mun)
Yungas (YOONG-gahss)
Yunnan (yoon-NAHN)
Zagreb (ZAH-grebb)
Zagros (ZAH-gruss)
Zaïre (zah-EAR)
Zambezi (zam-BEEZY)
Zambia (ZAM-bee-uh)
Zamboanga (zahm-boh-AHNG-guh)
Zamfara (zahm-FAHR-rah)
Zanzibar (ZANN-zih-bar)
Zapata, Emiliano (zah-PAH-tah, eh-mee-lee-AH-noh)
Zapatista (zah-pah-TEESTA)
Zedillo, Ernesto (zeh-DEE-yoh, air-NES-toh)
Zeeland (ZAY-lund)
Zhanjiang (JAHN-jee-AHNG)
Zhejiang (JEJ-ee-AHNG)
Zhou (JOH)
Zhou Enlai (JOH en-lye)
Zhuhai (joo-HYE)
Zimbabwe (zim-BAHB-way)
Zionism (ZYE-un-izm)
Zoogeography (ZOH-oh-jee-OG-gruh-fee)
Zoroaster (zorro-AST-tuh)
Zoroastrian (zorro-ASTREE-un)
Zuider Zee (ZYDER ZEE)
Zulu (ZOO-loo)
Zürich (ZOOR-ick)

References and Further Readings

(Bullets [•] denote basic introductory works)

INTRODUCTION

Christopher, Anthony J. *The Atlas of States: Global Change, 1900–2000* (Chichester, UK: John Wiley & Sons, 1999).

Clark, Gordon L. et al., eds. *The Oxford Handbook of Economic Geography* (New York: Oxford University Press, 2000).

Claval, Paul, ed. *Introduction to Regional Geography* (Malden, Mass.: Blackwell, 1998).

Corbridge, Stuart E., ed. *Development Studies: A Reader* (London: Edward Arnold, 1995).

• de Blij, H. J. & Murphy, Alexander B. *Human Geography: Culture, Society, and Space* (New York: John Wiley & Sons, 7th rev. ed., 2002).

• de Blij, H. J. & Muller, Peter O. *Physical Geography of the Global Environment* (New York: John Wiley & Sons, 2nd rev. ed., updated version, 1998).

Douglas, Ian, Huggett, Richard, & Robinson, Mike, eds. *Companion Encyclopedia of Geography: The Environment and Mankind* (London & New York: Routledge, 1996).

Fenneman, Nevin M. "The Circumference of Geography," *Annals of the Association of American Geographers*, 9 (1919): 3–11.

• *Goode's World Atlas* (Skokie, Ill.: Rand McNally, 20th rev. ed., 2000).

Grove, Jean M. *The Little Ice Age* (London & New York: Methuen, 1988). Quotation taken from pp. 1–2.

Hanson, Susan, ed. *Ten Geographic Ideas That Changed the World* (New Brunswick, N.J.: Rutgers University Press, 1997).

Holt-Jensen, Arild. *Geography: History and Concepts–A Student's Guide* (Thousand Oaks, Cal.: Sage, 3rd rev. ed., trans. Brian Fullerton, 1999).

Jackson, John Brinckerhoff. *Discovering the Vernacular Landscape* (New Haven: Yale University Press, 1984).

Johnston, Ron J. et al., eds. *The Dictionary of Human Geography* (Malden, Mass.: Blackwell, 4th rev. ed., 2000).

Jordan, Terry G. "The Concept and Method," in Lich, Glen E., ed., *Regional Studies: The Interplay of Land and People* (College Station, TX: Texas A & M Press, 1992), pp. 8–24.

Livingstone, David. *The Geographical Tradition* (Cambridge, Mass.: Blackwell, 1992).

• Muehrcke, Phillip C. & Muehrcke, Juliana C. *Map Use: Reading-Analysis-Interpretation* (Madison, Wisc.: JP Publications, 3rd rev. ed., 1992).

National Research Council. *Rediscovering Geography: New Relevance for Science and Society* (Washington, D.C.: National Academy Press, 1997).

• Pattison, William D. "The Four Traditions of Geography," *Journal of Geography*, 63 (1964): 211–216.

Sauer, Carl Ortwin. "Cultural Geography," *Encyclopedia of the Social Sciences*, Vol. 6 (New York: Macmillan, 1931), pp. 621–623.

Sheppard, Eric & Barnes, Trevor, eds. *A Companion to Economic Geography* (Malden, Mass.: Blackwell, 2000).

Wegener, Alfred. *The Origin of Continents and Oceans* (New York: Dover, reprint of the 1915 original, trans. John Biram, 1966).

• Wheeler, James O., Muller, Peter O., Thrall, Grant I., & Fik, Timothy J. *Economic Geography* (New York: John Wiley & Sons, 3rd rev. ed., 1998).

Whittlesey, Derwent S., et al. "The Regional Concept and the Regional Method," in James, Preston E. & Jones, Clarence F., eds., *American Geography: Inventory and Prospect* (Syracuse, N.Y.: Syracuse University Press, 1954), pp. 19–68.

CHAPTER 1

Baldersheim, Harald & Stahlberg, Krister, eds. *Nordic Region-Building in a European Perspective* (Brookfield, Vt.: Ashgate, 1999).

• Berentsen, William H., ed. *Contemporary Europe: A Geographic Analysis* (New York: John Wiley & Sons [7th rev. ed. of George W. Hoffman, *Europe in the 1990s: A Geographic Analysis*], 1997).

Blacksell, Mark & Williams, Allan M., eds. *The European Challenge: Geography and Development in the European Community* (New York: Oxford University Press, 1994).

Burtenshaw, David, et al. *The European City: A Western Perspective* (New York: John Wiley & Sons, 1991).

Butlin, Robin A. & Dodgshon, R. A. *An Historical Geography of Europe* (New York: Oxford University Press, 1999).

Carter, Frank W. & Turnock, David. *Environmental Problems in Eastern Europe* (London & New York: Routledge, 2nd rev. ed., 1997).

Champion, Tony, et al. *The New Regional Map of Europe* (Amsterdam & New York: Elsevier, 1996).

Chisholm, Michael. *Rural Settlement and Land Use: An Essay in Location* (London: Hutchinson University Library, 3rd rev. ed., 1979).

• Clout, Hugh D., et al. *Western Europe: Geographical Perspectives* (New York: Wiley/Longman, 3rd rev. ed., 1994).

• Cole, John P. & Cole, Francis. *A Geography of the European Union* (London & New York: Routledge, 2nd rev. ed., 1997).

Delamaide, Darrell. *The New Superregions of Europe* (New York: Dutton, 1994).

• Diem, Aubrey. *Western Europe: A Geographical Analysis* (New York: John Wiley & Sons, 1979).

Embleton, Clifford, ed. *Geomorphology of Europe* (New York: Wiley-Interscience, 1984).

Emerson, Michael. *Redrawing the Map of Europe* (New York: St. Martin's Press, 1998).

Fernandez-Armesto, Felipe, ed. *The Times Guide to the Peoples of Europe* (Boulder, Colo.: Westview Press, 1995).

Glebe, Günther & O'Loughlin, John, eds. *Foreign Minorities in Continental European Cities* (Wiesbaden, West Germany: Franz Steiner Verlag, 1987).

• Gottmann, Jean. *A Geography of Europe* (New York: Holt, Rinehart & Winston, 4th rev. ed., 1969).

Graham, Brian, ed. *Modern Europe: Place, Culture, and Identity* (New York: Oxford University Press, 1998).

Guttman, Robert J., ed. *Europe in the New Century: Visions of an Emerging Superpower* (Boulder, Colo.: Lynne Rienner, 2001).

Haggett, Peter. *Locational Analysis in Human Geography* (London: Edward Arnold, 1965). Definition from p. 19.

Hall, Derek & Danta, Darrick, eds. *Reconstructing the Balkans: A Geography of the New Southeast Europe* (New York: John Wiley & Sons, 1996).

Hall, Ray & White, Paul, eds. *Europe's Population: Towards the Next Century* (London: UCL Press/Taylor & Francis, 1995).

"Happy Family? A Survey of the Nordic Countries," *The Economist*, January 23, 1999, special insert, 16 pp.

Heffernan, Michael. *The Meaning of Europe: Geography and Geopolitics* (New York: Oxford University Press, 1998).

Heffernan, Michael. *Twentieth-Century Europe: A Political Geography* (New York: John Wiley & Sons, 1997).

Hoffman, Eva. *Exit into History: A Journey Through the New Eastern Europe* (New York: Viking Press, 1993).

Hoggart, Keith, et al. *Rural Europe: Identity and Change* (New York: John Wiley & Sons, 1995).

Hudson, Ray & Williams, Allan M., eds. *Divided Europe: Society and Territory* (Thousand Oaks, Cal.: Sage, 1998).

Hupchick, Dennis P. & Cox, Harold E., eds. *A Concise Historical Atlas of Eastern Europe* (New York: St. Martin's Press, 1996).

Jefferson, Mark. "The Law of the Primate City," *Geographical Review*, 29 (1939): 226–232. Quotation taken from p. 226.

Johnson, Lonnie R. *Central Europe: Enemies, Neighbors, Friends* (New York: Oxford University Press, 1996).

Jones, Philip N. & Wild, Trevor. "Regional and Local Variations in the Emerging Economic Landscape of the New German Länder," *Applied Geography*, 17 (1997): 283–299.

Jönsson, Christer, Tägil, Sven, & Törnqvist, Gunnar. *Organizing European Space* (Thousand Oaks, Calif.: Sage, 2000).

• Jordan-Bychkov, Terry G. & Bychkova Jordan, Bella. *The European Culture Area: A Systematic Geography* (Lanham, Md.: Rowman & Littlefield, 4th rev. ed., 2001).

Keating, Michael. *Nations Against the State: The New Politics of Nationalism in Quebec, Catalonia, and Scotland* (New York: Macmillan, 1996).

Keating, Michael. *The New Regionalism in Western Europe: Territorial Restructuring and Political Change* (Northampton, Mass.: Edward Elgar, 1998).

King, Russel. *Mass Migration in Europe: The Legacy and the Future* (New York: John Wiley & Sons, 1995).

King, Russel, et al. *The Mediterranean: Environment and Society* (New York: John Wiley & Sons, 1997).

Kucera, Tomas, et al., eds. *New Demographic Faces of Europe* (New York: Springer Verlag, 2000)

Kürti, Laszlo & Langman, Juliet, eds. *Beyond Borders: Remaking Cultural Identities in the New East and Central Europe* (Boulder, Colo.: Westview Press, 1997).

Livi-Bacci, Massimo. *The Population of Europe* (Malden, Mass.: Blackwell, trans. Carl Ipsen, 2000).

Matvejevic, Predag. *The Mediterranean: A Cultural Landscape* (Berkeley, Cal.: University of California Press, trans. Michael H. Heim, 1999).

• McDonald, James R. *The European Scene: A Geographic Perspective* (Upper Saddle River, N.J.: Prentice-Hall, 1997).

Mortimer, Edward & Fine, Robert, eds. *People, Nation and State: The Meaning of Ethnicity and Nationalism* (London: I.B. Tauris, 1999).

Murphy, Alexander B. "Emerging Regional Linkages Within the European Community: Challenging the Dominance of the State," *Tijdschrift voor Economische en Sociale Geografie*, 84 (1993): 103–118.

Murphy, Alexander B. "Rethinking the Concept of European Identity," in Herb, Guntram H. & Kaplan, David H., eds., *Nested Identities: Nationalism, Territory, and Scale* (Lanham, Md.: Rowman & Littlefield, 1999), pp. 53–73.

Newhouse, John. *Europe Adrift* (New York: Pantheon, 1997).

O'Dowd, Liam & Wilson, Thomas M., eds. *Borders, Nations and States: Frontiers of Sovereignty in the New Europe* (Brookfield, Vt.: Ashgate, 1996).

Ohmae, Kenichi. *The End of the Nation-State: The Rise of Regional Economies* (New York: Free Press, 1996).

Ohmae, Kenichi. "The Rise of the Region State," *Foreign Affairs*, Spring 1993, 78–87.

Petrakos, George, ed. *Integration and Transition in Europe: Economic Geography of Interaction* (London & New York: Routledge, 2000).

Pinder, David, ed. *The New Europe: Economy, Society and Environment* (Chichester, UK & New York: John Wiley & Sons, 1998).

Pond, Elizabeth. *The Rebirth of Europe* (Washington, DC: Brookings Institution Press, 1999).

"Poverty in Eastern Europe: The Land That Time Forgot," *The Economist*, September 23, 2000, 27–30.

Rhodes, Martin, ed. *Regions and the New Europe: Patterns in Core and Periphery Development* (Manchester, UK: Manchester University Press, 1996).

• Shaw, Denis J. B., ed. *The Post-Soviet Republics: A Systematic Geography* (New York: John Wiley & Sons/Longman, 1995).

Sommers, Lawrence M. "Cities of Western Europe," in Brunn, Stanley D. & Williams, Jack F., eds., *Cities of the World: World Regional Urban Development* (New York: Harper & Row, 1983), pp. 84–121.

Specter, Michael. "Population Implosion Worries a Graying Europe," *New York Times*, July 10, 1998, A1, A6.

Townsend, Alan R. *Making a Living in Europe: Human Geographies of Economic Change* (London & New York: Routledge, 1997).

Turnock, David, ed. *East Central Europe and the Former Soviet Union: Environment and Society* (London: Arnold, 2001).

Unwin, Tim, ed. *A European Geography* (Harlow, UK & New York: Longman, 1998).

Wheeler, James O., Muller, Peter O., Thrall, Grant I., & Fik, Timothy J. *Economic Geography* (New York: John Wiley & Sons, 3rd rev. ed., 1998), Chapter 13.

Williams, Allan M. *The European Community* (Cambridge, Mass.: Blackwell, 2nd rev. ed., 1994).

Williams, Allan M. *The West European Economy: A Geography of Post-War Development* (Savage, Md.: Rowman & Littlefield, 1988).

Wintle, Michael, ed. *Culture and Identity in Europe: Perceptions of Divergence and Unity in Past and Present* (Brookfield, Vt.: Ashgate, 1996).

CHAPTER 2

"A Caspian Gamble: A Survey of Central Asia," *The Economist*, February 7, 1998, special insert, 18 pp.

• Bater, James H. *Russia and the Post-Soviet Scene: A Geographical Perspective* (New York: John Wiley & Sons, 1996).

Brawer, Moshe. *Atlas of Russia and the Independent Republics* (New York: Simon & Schuster, 1994).

• Brown, Archie, et al., eds. *The Cambridge Encyclopedia of Russia and the Former Soviet Union* (New York: Cambridge University Press, 1994).

Bychkova Jordan, Bella & Jordan-Bychkov, Terry G. *Siberian Village: Land and Life in the Sakha Republic* (Minneapolis: University of Minnesota Press, 2001).

Chew, Allen F. *Atlas of Russian History: Eleven Centuries of Changing Borders* (New Haven, Conn.: Yale University Press, 1967).

Chinn, Jeff & Kaiser, Robert. *Russians as the New Minority: Ethnicity and Nationalism in the Soviet Successor States* (Boulder, Colo.: Westview Press, 1996).

Dawisha, Karen & Parrott, Bruce. *Russia and the New States of Eurasia: The Politics of Upheaval* (New York: Cambridge University Press, 1994).

Demko, George J., et al., eds. *Population Under Duress: The Geodemography of Post-Soviet Russia* (Boulder, Colo.: Westview Press, 1998).

Dmitrieva, O. *Regional Development: The U.S.S.R. and After* (London: University College London Press, 1996).

Dobbs, Michael. *Down With Big Brother: The Fall of the Soviet Empire* (New York: Alfred A. Knopf, 1996).

Duncan, W. Raymond & Holman, G. Paul, Jr., eds. *Ethnic Nationalism and Regional Conflict: The Former Soviet Union and Yugoslavia* (Boulder, Colo.: Westview Press, 1994).

Dunlop, John B. *The Rise of Russia and the Fall of the Soviet Empire* (Princeton, N.J.: Princeton University Press, 1993).

Forsyth, James. *A History of the Peoples of Siberia: Russia's North Asian Colony, 1581-1990* (New York: Cambridge University Press, 1992).

Gachechiladze, Revaz. *The New Georgia: Space, Society, Politics* (College Station, Tex.: Texas A&M University Press, 1995).

Goltz, Thomas. *Azerbaijan Diary* (Armonk, N.Y.: M. E. Sharpe, 1998).

Hanson, Philip & Bradshaw, Michael, eds. *Regional Change in Russia* (Northampton, Mass.: Edward Elgar, 2000).

Harris, Chauncy D. "A Geographic Analysis of Non-Russian Minorities in Russia and Its Ethnic Homelands," *Post-Soviet Geography*, 34 (1993): 543–597.

Horensma, Pier. *The Soviet Arctic* (London & New York: Routledge, 1991).

Hosking, Geoffrey. *Russia: People and Empire* (Cambridge, Mass: Harvard University Press, 1997).

Howe, G. Melvyn. *The Soviet Union: A Geographical Study* (London & New York: Longman, 2nd rev. ed., 1986).

Hunter, Shireen T. *Transcaucasia in Transition: Nation-Building and Conflict* (Boulder, Colo.: Westview Press, 1994).

Huttenbach, Henry. *The Caucasus: A Region in Crisis* (Boulder, Colo.: Westview Press, 1996).

Ioffe, Gregory & Nefedova, Tatyana. *Continuity and Change in Rural Russia* (Boulder, Colo.: Westview Press, 1997).

Kaiser, Robert J. *The Geography of Nationalism in Russia and the U.S.S.R.* (Princeton, N.J.: Princeton University Press, 1994).

Kramer, Mark. *Travels With a Hungry Bear: A Journey to the Russian Heartland* (Boston: Houghton Mifflin, 1996).

Kraus, Michael & Liebowitz, Ronald D., eds. *Russia and Eastern Europe After Communism: The Search for New Political, Economic, and Security Systems* (Boulder, Colo.: Westview Press, 1995).

Lincoln, W. Bruce. *The Conquest of a Continent: Siberia and the Russians* (New York: Random House, 1994).

Lloyd, John. "The Russian Devolution," *The New York Times Magazine*, August 15, 1999, 34–41, 52, 61, 64.

• Lydolph, Paul E. *Geography of the U.S.S.R.* (Elkhart Lake, Wisc.: Misty Valley Publishing, 1990).

Lydolph, Paul E. *Climates of the Soviet Union* (Amsterdam: Elsevier, 1977).

Mackinder, Halford J. *Democratic Ideals and Reality* (New York: Holt, 1919).

Mastyugina, Tatiana & Perepelkin, Lev. *An Ethnic History of Russia: Pre-Revolutionary Times to the Present* (Westport, Conn.: Greenwood Press, 1996).

Mote, Victor L. *Siberia: Worlds Apart* (Boulder, Colo.: Westview Press, 1998).

Nijman, Jan. *The Geopolitics of Power and Conflict: Superpowers in the International System, 1945-1992* (New York: John Wiley & Sons/Belhaven, 1993).

Peterson, D. J. *Troubled Lands: The Legacy of Soviet Environmental Destruction* (Boulder, Colo.: Westview Press, 1993).

Pryde, Philip R., ed. *Environmental Resources and Constraints in the Former Soviet Republics* (Boulder, Colo.: Westview Press, 1995).

Remnick, David. *Lenin's Tomb: The Last Days of the Soviet Empire* (New York: Random House, 1993).

Remnick, David. *Resurrection: The Struggle to Build a New Russia* (New York: Random House, 1997).

Rodgers, Allan, ed. *The Soviet Far East: Geographical Perspectives on Development* (London & New York: Routledge, 1990).

• Shaw, Denis J. B. *Russia in the Modern World: A New Geography* (Malden, Mass.: Blackwell, 1999).

• Shaw, Denis J. B., ed. *The Post-Soviet Republics: A Systematic Geography* (New York: John Wiley & Sons/ Longman, 1995).

Shlapentokh, Vladimir, et al. *From Submission to Rebellion: The Provinces Versus the Center in Russia* (Boulder, Colo.: Westview Press, 1997).

Smith, Graham. *The Post Soviet States: Mapping the Politics of Transition* (London: Arnold, 1999).

Stewart, John Massey, ed. *The Soviet Environment: Problems, Policies and Politics* (New York: Cambridge University Press, 1992).

• Symons, Leslie, ed. *The Soviet Union: A Systematic Geography* (London & New York: Routledge, 2nd rev. ed., 1990).

"The Caucasus: Where Worlds Collide," *The Economist*, August 19, 2000, 17–19.

"The Rape of Siberia: The Tortured Land," *Time*, September, 4, 1995, pp. 42–53.

Thompson, John M. *Russia and the Soviet Union: An Historical Introduction from the Kievan State to the Present* (Boulder, Colo.: Westview Press, 3rd rev. ed., 1994).

Thubron, Colin. *Among the Russians* (New York: HarperCollins, 2000).

Thubron, Colin. *In Siberia* (New York: HarperCollins, 2000).

Tomikel, John & Henderson, Bonnie. *Russia and the Near Abroad* (Corry, Penn.: Allegheny Press, 1997).

Turnock, David, ed. *East Central Europe and the Former Soviet Union: Environment and Society* (London: Arnold, 2001).

Valencia, Mark J., ed. *The Russian Far East in Transition: Opportunities for Regional Economic Cooperation* (Boulder, Colo.: Westview Press, 1995).

Wixman, Ronald. *The Peoples of the U.S.S.R.: An Ethnographic Handbook* (Armonk, N.Y.: M. E. Sharpe, 1984).

CHAPTER 3

"A Survey of Canada," *The Economist*, July 24, 1999, special insert, 18 pp.

"A Survey of Silicon Valley," *The Economist*, special supplement, March 29, 1997, 20 pp.

Adams, John S. "Residential Structure of Midwestern Cities," *Annals of the Association of American Geographers*, 60 (1970): 37–62. Model diagram adapted from p. 56.

Allen, James P. & Turner, Eugene J. *We the People: An Atlas of America's Ethnic Diversity* (New York: Macmillan, 1987).

Ashbaugh, James G., ed. *The Pacific Northwest: Geographical Perspectives* (Dubuque, Iowa: Kendall-Hunt, 1994).

Atlas of North America: Space Age Portrait of a Continent (Washington, D.C.: National Geographic Society, 1985).

• Atwood, Wallace W. *The Physiographic Provinces of North America* (New York: Ginn, 1940).

Berry, Brian J. L. "The Decline of the Aging Metropolis: Cultural Bases and Social Process," in Sternlieb, George & Hughes, James W., eds., *Post-Industrial America: Metropolitan Decline and Inter-Regional Job Shifts* (New Brunswick, N.J.: Center for Urban Policy Research, Rutgers University, 1975), pp. 175–185.

• Birdsall, Stephen S., Florin, John W., & Price, Margo L. *Regional Landscapes of the United States and Canada* (New York: John Wiley & Sons, 5th rev. ed., 1999).

• Boal, Frederick W. & Royle, Stephen A., eds. *North America: A Geographical Mosaic* (London: Arnold, 1999).

• Bone, Robert M. *The Regional Geography of Canada* (Don Mills, Ontario: Oxford University Press Canada, 2000).

Bone, Robert M. *The Geography of the Canadian North: Issues and Challenges* (Toronto: Oxford University Press, 1992). Quotation taken from p. 12.

Borchert, John R. "American Metropolitan Evolution," *Geographical Review*, 57 (1967): 301–332.

Borchert, John R. "Futures of American Cities," in Hart, John Fraser, ed., *Our Changing Cities* (Baltimore, Md.: Johns Hopkins University Press, 1991), pp. 218–250.

Brinkley, Joel. "Information Superhighway is Just Outside the Beltway," *New York Times*, October 12, 1999, A1, A21.

Britton, John N. H., ed. *Canada and the Global Economy: The Geography of Structural and Technological Change* (Montreal: McGill-Queen's University Press, 1996).

Castells, Manuel & Hall, Peter. *Technopoles of the World: The Making of Twenty-First-Century Industrial Complexes* (London & New York: Routledge, 1994).

Dean, William G., et al., eds. *The Concise Historical Atlas of Canada* (Toronto: University of Toronto Press, 1998).

French, Hugh M. & Slaymaker, Olav H., eds. *Canada's Cold Environments* (Montreal: McGill-Queen's University Press, 1993).

Garreau, Joel. *Edge City: Life on the New Frontier* (New York: Doubleday, 1991).

Garreau, Joel. *The Nine Nations of North America* (Boston: Houghton Mifflin, 1981).

Gastil, Raymond D. *Cultural Regions of the United States* (Seattle: University of Washington Press, 1975).

Gates, Henry Louis, Jr. & Appiah, Kwame A., eds. *Africana: Encyclopedia of the African and African- American Experience* (New York: Basic Books, 1999).

• Getis, Arthur & Getis, Judith, eds. *The United States and Canada: The Land and the People* (Dubuque, Iowa: Wm. C. Brown, 1995).

Gober, Patricia. "Americans on the Move," *Population Bulletin*, 48 (November 1993): 1–40.

Gottmann, Jean. *Megalopolis: The Urbanized Northeastern Seaboard of the United States* (New York: Twentieth Century Fund, 1961).

Hall, John & Lindholm, Charles. *Is America Breaking Apart?* (Princeton, N.J.: Princeton University Press, 1999).

Harris, R. Cole. "Regionalism and the Canadian Archipelago," in McCann, Lawrence D., ed., *Heartland and Hinterland: A Geography of Canada* (Scarborough, Ont.: Prentice-Hall Canada, 1982), pp. 458–484.

Hart, John Fraser. *The Land That Feeds Us* (New York: W. W. Norton, 1991).

Hartshorn, Truman A. *Interpreting the City: An Urban Geography* (New York: John Wiley & Sons, 2nd rev. ed., 1992).

Hartshorn, Truman A. "The Changed South, 1947–1997," *Southeastern Geographer*, 37 (November 1997): 122–139.

Harvey, Thomas. "The Changing Face of the Pacific Northwest," *Journal of the West*, 37 (July 1998): 22–32.

Historical Atlas of the United States (Washington, D.C.: National Geographic Society, centennial ed., 1988).

Homberger, Eric. *The Penguin Historical Atlas of North America* (New York: Viking Penguin, 1996).

• Hunt, Charles B. *Natural Regions of the United States and Canada* (San Francisco: W. H. Freeman, 2nd rev. ed., 1974).

Janelle, Donald G., ed. *Geographical Snapshots of North America* (New York: Guilford Press, 1992).

Kaplan, Robert D. *An Empire Wilderness: Travels Into America's Future* (New York: Random House, 1998).

Lemco, Jonathan. *Turmoil in the Peaceable Kingdom: The Quebec Sovereignty Movement and Its Implications for Canada and the United States* (Toronto: University of Toronto Press, 1994).

Magocsi, Paul Robert, ed. *Encyclopedia of Canada's Peoples* (Toronto: University of Toronto Press, 1998).

• McCann, Lawrence D. & Gunn, Angus, eds. *Heartland and Hinterland: A Regional Geography of Canada* (Scarborough, Ont.: Prentice Hall Canada, 3 rev. ed., 1998).

McIlwraith, Thomas F. & Muller, Edward K., eds. *North America: The Historical Geography of a Changing Continent* (Lanham, Md.: Rowman & Littlefield, 2nd rev. ed., 2001).

McKee, Jesse O., ed. *Ethnicity in Contemporary America: A Geographical Appraisal* (Lanham, Md.: Rowman & Littlefield, 2nd rev. ed., 2000).

• McKnight, Tom L. *Regional Geography of the United States and Canada* (Upper Saddle River, N.J.: Prentice-Hall, 3rd rev. ed., 2001).

Meinig, Donald W. *The Shaping of America: A Geographical Perspective on 500 Years of History; Vol. 1: Atlantic America, 1492–1800* (New Haven, Conn.: Yale University Press, 1986).

Meinig, Donald W. *The Shaping of America: A Geographical Perspective on 500 Years of History; Vol. 2: Continental America, 1800–1867* (New Haven, Conn.: Yale University Press, 1993).

Meinig, Donald W. *The Shaping of America: A Geographical Perspective on 500 Years of History. Vol. 3: Transcontinental America, 1850–1915* (New Haven, Conn.: Yale University Press, 1998).

Muller, Peter O. *Contemporary Suburban America* (Englewood Cliffs, N.J.: Prentice-Hall, 1981).

Ohmae, Kenichi. *The End of the Nation-State: The Rise of Regional Economies* (New York: Free Press, 1996).

Orme, Antony, ed. *The Physical Geography of North America* (New York: Oxford University Press, 2001).

Pacione, Michael. *Urban Geography: A Global Perspective* (London & New York: Routledge, 2001).

Pandit, Kavita & Withers, Suzanne Davies, eds. *Migration and Restructuring in the United States: A Geographic Perspective* (Lanham, Md.: Rowman & Littlefield, 1999).

Pollard, Kelvin M. & O'Hare, William P. "America's Racial and Ethnic Minorities," *Population Bulletin*, 54 (September 1999): 1–48.

• Rooney, John F., Jr., et al., eds. *This Remarkable Continent: An Atlas of United States and Canadian Society and Cultures* (College Station, Tex.: Texas A&M University Press, 1982).

Saul, John Ralston. *Reflections of a Siamese Twin: Canada at the End of the Twentieth Century* (Toronto: Viking, 1997).

Selby, William A. *Rediscovering the Golden State: California Geography* (New York: John Wiley & Sons, 2000).

Shelley, Fred M., et al. *Political Geography of the United States* (New York: Guilford, 1996).

"The New Map of High-Tech: From Billville to Silicon Alley, The 13 Hottest Regions in America," *Wall Street Journal*, November 23, 1999, B1, B12.

Vance, James E., Jr. *This Scene of Man: The Role and Structure of the City in the Geography of Western Civilization* (New York: Harper's College Press, 1977). Urban realms model discussed on pp. 411–416.

Ward, David. *Cities and Immigrants: A Geography of Change in Nineteenth-Century America* (New York: Oxford University Press, 1971).

• Warkentin, John. *Canada: A Regional Geography* (Scarborough, Ont.: Prentice Hall Canada, 1997).

Wheeler, James O., Muller, Peter O., Thrall, Grant I., & Fik, Timothy J. *Economic Geography* (New York: John Wiley & Sons, 3rd rev. ed., 1998).

Yeates, Maurice H. *The North American City* (New York: Harper & Row, 4th rev. ed., 1990). Canadian urban evolution model discussed on pp. 60–67.

Yeates, Maurice H. *Main Street: Windsor to Quebec City* (Toronto: Macmillan of Canada, 1975).

Zelinsky, Wilbur. *Exploring the Beloved Country: Geographic Forays into American Society and Culture* (Iowa City, Iowa: University of Iowa Press, 1995).

• Zelinsky, Wilbur. *The Cultural Geography of the United States: A Revised Edition* (Englewood Cliffs, N.J.: Prentice-Hall, 2nd rev. ed, 1992).

Zelinsky, Wilbur. *The Enigma of Ethnicity: Another American Dilemma* (Iowa City, Iowa: University of Iowa Press, 2001).

CHAPTER 4

"A Silicon Republic . . . Lessons From [Costa Rica's] High-Tech Frontier," *Newsweek*, August 28, 2000, 42–44.

"After the Revolution: A Survey of Mexico," *The Economist*, October 28, 2000, special supplement, 16 pp.

Arreola, Daniel D. & Curtis, James R. *The Mexican Border Cities: Landscape Anatomy and Place Personality* (Tucson: University of Arizona Press, 1993).

Augelli, John P. "The Rimland-Mainland Concept of Culture Areas in Middle America," *Annals of the Association of American Geographers*, 52 (1962): 119–129.

Barker, David & McGregor, Duncan F. M., eds. *Environment and Development in the Caribbean: Geographical Perspectives* (Kingston, Jamaica: University of the West Indies Press, 1995).

Barton, Jonathan R. *A Political Geography of Latin America* (London & New York: Routledge, 1997).

• Blakemore, Harold & Smith, Clifford T., eds. *Latin America: Geographical Perspectives* (London & New York: Methuen, 2nd rev. ed., 1983).

• Blouet, Brian W. & Blouet, Olwyn M. *Latin America and the Caribbean: A Systematic and Regional Survey* (New York: John Wiley & Sons, 4th rev. ed., 2002).

Boswell, Thomas D. & Conway, Dennis. *The Caribbean Islands: Endless Geographical Diversity* (New Brunswick, N.J.: Rutgers University Press, 1992).

Butler, Edgar W., et al. *Mexico and Mexico City in the World Economy* (Boulder, Colo.: Westview, 2001).

• Clawson, David L. *Latin America and the Caribbean: Lands and Peoples* (Dubuque, Iowa: WCB/McGraw-Hill, 2nd rev. ed., 2000).

Clarke, Colin. *Class, Ethnicity, and Community in Southern Mexico: Oaxaca's Peasantries* (New York: Oxford University Press, 2000).

Collier, Simon, et al., eds. *The Cambridge Encyclopedia of Latin America and the Caribbean* (New York: Cambridge University Press, 2nd rev. ed., 1992).

D'Agostino, Thomas J. & Hillman, Richard S., eds. *Understanding the Contemporary Caribbean* (Boulder, Colo.: Lynne Rienner, 2001).

Davidson, William V. & Parsons, James J., eds. *Historical Geography of Latin America* (Baton Rouge, La.: Louisiana State University Press, 1980).

Elbow, Gary S. "Regional Cooperation in the Caribbean: The Association of Caribbean States," *Journal of Geography*, 96 (January/February 1997): 13–22.

Galloway, Jock H. "The Lesser Antilles," in Donald G. Janelle, ed., *Geographical Snapshots of North America* (New York: Guilford, 1992), pp. 16–19.

Gilbert, Alan. *Latin America* (London & New York: Routledge, 1990).

Gilbert, Alan, ed. *The Mega-City in Latin America* (New York & Tokyo: United Nations University Press, 1996).

Greenfield, Gerald M., ed. *Latin American Urbanization: Historical Profiles of Major Cities* (Westport, Conn.: Greenwood Press, 1994).

Griffin, Ernst C. & Ford, Larry R. "Cities of Latin America," in Brunn, Stanley D. & Williams, Jack F., eds., *Cities of the World: World Regional Urban Development* (New York: HarperCollins, 2nd rev. ed., 1993), pp. 224–265.

Grossman, Lawrence S. *The Political Ecology of Bananas: Contract Farming, Peasants, and Agrarian Change in the Eastern Caribbean* (Chapel Hill: University of North Carolina Press, 1998).

Gwynne, Robert N. & Kay, Cristobal. *Latin America Transformed: Globalization and Modernity* (New York: Oxford University Press, 1999).

Herzog, Lawrence A. *From Aztec to High-Tech: Architecture and Landscape Across the Mexico-United States Border* (Baltimore: Johns Hopkins University Press, 1999).

• James, Preston E. & Minkel, Clarence W. *Latin America* (New York: John Wiley & Sons, 5th rev. ed., 1986), Chaps. 2–16.

Klak, Thomas, ed. *Globalization and Neoliberalism: The Caribbean Context* (Lanham, Md: Rowman & Littlefield, 1997).

Kopinak, Kathryn. *Desert Capitalism: Maquiladoras in North America's Western Industrial Corridor* (Tucson: University of Arizona Press, 1996).

Lowenthal, David. *West Indian Societies* (New York: Oxford University Press, 1972).

MacLachlan, Ian & Aguilar, Adrian G. "Maquiladora Myths: Locational and Structural Change in Mexico's Export Manufacturing Industry," *The Professional Geographer*, 50 (August 1998): 315–323.

MacPherson, John. *Caribbean Lands* (London & New York: Longman, 4th rev. ed., 1980).

McCullough, David G. *The Path Between the Seas: The Creation of the Panama Canal, 1870-1914* (New York: Simon & Schuster, 1977).

"Mexico's New Frontier (Nuevo León and Baja California)," *The Economist*, February 8, 1997, 41–42.

Momsen, Janet, ed. *Women and Change in the Caribbean* (Bloomington, Ind.: Indiana University Press, 1994).

Myers, Norman. *The Primary Source: Tropical Forests and Our Future* (New York: W. W. Norton, 1984).

Paige, Jeffery M. *Coffee and Power: Revolution and the Rise of Democracy in Central America* (Cambridge, Mass.: Harvard University Press, 1997).

Pick, James B. & Butler, Edgar W. *Mexico Megacity* (Boulder, Colo.: Westview Press, 1997).

Portes, Alejandro & Grosfoguel, Ramón. "Caribbean Diasporas: Migration and Ethnic Communities," *Annals of the American Academy of Political and Social Science*, 513 (1994): 48–69.

Potter, Robert B. *The Urban Caribbean in an Era of Global Change* (Brookfield, Vt.: Ashgate, 2000).

• Preston, David A., ed. *Latin American Development: Geographical Perspectives* (London & New York: Longman, 2nd rev. ed., 1995).

• Richardson, Bonham C. *The Caribbean in the Wider World, 1492–1992: A Regional Geography* (New York: Cambridge University Press, 1992).

Rohter, Larry. "A Tiger in a Sea of Pussy Cats: Trinidad and Tobago Bid Goodbye to Oil, Hello to Gas," *New York Times*, September 4, 1998, C1-C2.

Rohter, Larry. "Asia Moves in on the Big Ditch," *New York Times*, December 19, 1999, WK 3.

Rohter, Larry. "Now Ruined Economies Afflict Central America," *New York Times*, November 13, 1998, A10.

Sargent, Charles S., Jr. "The Latin American City," in Blouet, Brian W. & Blouet, Olwyn M., eds., *Latin America and the Caribbean: A Systematic and Regional Survey* (New York: John Wiley & Sons, 3rd rev. ed., 1997), pp. 139–180. Diagram adapted from p. 173.

Sealey, Neil. *Caribbean World: A Complete Geography* (New York: Cambridge University Press, 1992).

The Border Economy (Dallas, Tex.: Federal Reserve Bank of Dallas, 2001).

"The Mexicans Treated as Aliens in Their Own Country: The Right Not to be Hispanic," *The Economist*, March 7, 1998, 88–89.

Thompson, Ginger. "Chasing Mexico's Dream Into Squalor: Misery on the Border," *New York Times*, February 11, 2001, 1, 6.

Trias Monge, José. *Puerto Rico: The Trials of the Oldest Colony in the World* (New Haven, Conn.: Yale University Press, 1997).

Ward, Peter M. *Mexico City* (Boston: G. K. Hall, 1990).

Watts, David. *The West Indies: Patterns of Development, Culture and Environmental Change Since 1492* (New York: Cambridge University Press, 1987).

West, Robert C. *Sonora: Its Geographical Personality* (Austin, Tex.: University of Texas Press, 1993).

• West, Robert C., Augelli, John P., et al. *Middle America: Its Lands and Peoples* (Englewood Cliffs, N.J.: Prentice-Hall, 3rd rev. ed., 1989).

Winn, Peter. *Americas: The Changing Face of Latin America and The Caribbean* (Berkeley, Cal.: University of California Press, updated ed., 1999).

CHAPTER 5

Augelli, John P. "The Controversial Image of Latin America: A Geographer's View," *Journal of Geography*, 62 (1963): 103–112. Quotation from p. 111.

Barton, Jonathan R. *A Political Geography of Latin America* (London & New York: Routledge, 1997).

Becker, Bertha K. & Egler, Claudio A. G. *Brazil: A New Regional Power in the World-Economy: A Regional Geography* (New York: Cambridge University Press, 1992).

• Blouet, Brian W. & Blouet, Olwyn M. *Latin America and the Caribbean: A Systematic and Regional Survey* (New York: John Wiley & Sons, 4th rev. ed., 2002).

Box, Ben. "Latin America's New Transportation Links," *1999 Britannica Book of the Year* (Chicago: Encyclopedia Britannica, 1999), pp. 412–413.

Brawer, Moshe. *Atlas of South America* (New York: Simon & Schuster, 1991).

"Brazil: The Northeast—Politics, Water and Poverty," *The Economist*, August 29, 1998, 36–38.

• Bromley, Rosemary D. F. & Bromley, Ray. *South American Development: A Geographical Introduction* (New York: Cambridge University Press, 2nd rev. ed., 1988).

Brooke, James. "The New South Americans: Friends and Partners," *New York Times*, April 8, 1994, A-3.

Browder, John D. & Godfrey, Brian J. *Rainforest Cities: Urbanization, Globalization, and Development of the Amazon Basin* (New York: Columbia University Press, 1997).

Brunn, Stanley D., Williams, Jack F., & Ziegler, Donald J., eds. *Cities of the World: Regional Urban Development* (Lanham, Md.: Rowman & Littlefield, 2002).

Caviedes, César N. *The Southern Cone: Realities of the Authoritarian State* (Totowa, N.J.: Rowman & Allanheld, 1984).

• Caviedes, César N. & Knapp, Gregory. *South America* (Englewood Cliffs, N.J.: Prentice Hall, 1995).

• Clawson, David L. *Latin America and the Caribbean: Lands and Peoples* (Dubuque, Iowa: WCB/McGraw-Hill, 2nd rev. ed., 2000).

Clawson, Patrick L. & Lee, Rensselaer W. *The Andean Cocaine Industry* (New York: St. Martin's Press, 1996).

Collier, Simon, et al., eds. *The Cambridge Encyclopedia of Latin America and the Caribbean* (New York: Cambridge University Press, 2nd rev. ed., 1992).

Crow, John A. *The Epic of Latin America* (Berkeley, Cal.: University of California Press, 4th rev. ed., 1992).

Denevan, William M. *Cultivated Landscapes of Native Amazonia and the Andes: Triumph Over the Soil* (New York: Oxford University Press, 2001).

Eakin, Marshall C. *Brazil: The Once and Future Country* (New York: St. Martin's Press, 1997).

Ford, Larry R. "A New and Improved Model of Latin American City Structure," *Geographical Review*, 86 (July 1996): 437–440.

Gade, Daniel W. *Nature and Culture in the Andes* (Madison, Wisc.: University of Wisconsin Press, 1999).

• Gilbert, Alan. *Latin America* (London & New York: Routledge, 1990).

Gilbert, Alan, ed. *The Mega-City in Latin America* (New York & Tokyo: United Nations University Press, 1996).

Godfrey, Brian J. "Brazil," *Focus*, Summer 1999, 28 pp.

Godfrey, Brian J. "Revisiting Rio de Janeiro and São Paulo," *Geographical Review*, 89 (January 1999): 94–121.

Goulding, Michael, et al. *Floods of Fortune: Ecology and Economy Along the Amazon* (New York: Columbia University Press, 1996).

Greenfield, Gerald M., ed. *Latin American Urbanization: Historical Profiles of Major Cities* (Westport, Conn.: Greenwood Press, 1994).

Griffin, Ernst C. & Ford, Larry R. "Cities of Latin America," in Brunn, Stanley D. & Williams, Jack F., eds. *Cities of the World: World Regional Urban Development* (New York: HarperCollins, 2nd rev. ed., 1993), pp. 224–265.

Griffin, Ernst C. & Ford, Larry R. "A Model of Latin American City Structure," *Geographical Review*, 70 (1980): 397–422. Model diagram adapted from p. 406.

Gugler, Josef, ed., *The Urban Transformation of the Developing World* (New York: Oxford University Press, 1996).

Gwynne, Robert N. & Kay, Cristobal. *Latin America Transformed: Globalization and Modernity* (New York: Oxford University Press, 1999).

Hansis, Randall. *The Latin Americans: Understanding Their Legacy* (New York: McGraw-Hill, 1997).

• James, Preston E. & Minkel, Clarence W. *Latin America* (New York: John Wiley & Sons, 5th rev. ed., 1986), Chaps. 17–37.

Keeling, David J. *Contemporary Argentina: A Geographical Perspective* (Boulder, Colo.: Westview Press, 1996).

Kelly, Philip. *Checkerboards and Shatterbelts: The Geopolitics of South America* (Austin: University of Texas Press, 1997).

LaFranchi, H. 'The Fertile Pampas Region Is On Its Way to Becoming a World Leader in Agro Business," *Christian Science Monitor*, November 26, 1997, 10–11.

Levine, Robert M. & Crocitti, John J., eds. *The Brazil Reader* (Durham, N.C.: Duke University Press, 1999).

McColl, Robert W. 'The Insurgent State: Territorial Bases of Revolution," *Annals of the Association of American Geographers*, 59 (1969): 613–631.

• Morris, Arthur S. *South America* (Totowa, N.J.: Barnes & Noble, 3rd rev. ed., 1987).

Myers, Norman. *The Primary Source: Tropical Forests and Our Future* (New York: W. W. Norton, 1984).

Pacione, Michael. *Urban Geography: A Global Perspective* (London & New York: Routledge, 2001).

Page, Joseph A. *The Brazilians* (Reading, Mass.: Addison-Wesley, 1995).

Potter, Robert B. *Third World Urbanization: Contemporary Issues in Geography* (New York: Oxford University Press, 1991).

• Preston, David A., ed. *Latin American Development: Geographical Perspectives* (London & New York: Longman, 2nd rev. ed., 1995).

Radcliffe, Sarah & Westwood, Sallie. *Remaking the Nation: Place, Identity and Politics in Latin America* (London & New York: Routledge, 1998).

Ramos, Alcida Rita. "South America's Indigenous Peoples," *2000 Britannica Book of the Year* (Chicago: Encyclopedia Britannica, 2000), pp. 520–521.

Ribeiro, Darcy. *The Brazilian People: The Formation and Meaning of Brazil* (Gainesville, Fla.: University Press of Florida, trans. Gregory Rabassa, 2000).

Rohter, Larry. "In Latin America, the Strongman Stirs in His Grave," *New York Times*, December 20, 1998, WK-4.

Rohter, Larry. "Weave of Drugs and Strife in Colombia," *New York Times*, April 21, 2000, A1, A10, All.

St. John, Ronald B. *The Bolivia-Chile-Peru Dispute in the Atacama Desert* (Durham, UK: International Boundaries Research Unit, University of Durham, Vol. 1, Boundary and Territory Briefing No. 6, 1994).

St. John, Ronald B. *The Boundary Between Ecuador and Peru* (Durham, UK: International Boundaries Research Unit, University of Durham, Vol. 1, Boundary and Territory Briefing No. 4, 1994).

Smith, Nigel J. H. *The Amazon River Forest: A Natural History of Plants, Animals, and People* (New York: Oxford University Press, 1999).

Smith, Nigel J. H., et al. *Amazonia: Resiliency and Dynamism of the Land and Its People* (New York & Tokyo: United Nations University Press, 1995).

Sponsel, Leslie E., ed. *Indigenous Peoples and the Future of Amazonia* (Tucson, Ariz.: University of Arizona Press, 1995).

Sternberg, Rolf. "Brazilian Amazonia: A Metamorphosis in Progress," *Revista Geografica*, 125 (Enero-Junio 1999): 5–47.

Tenenbaum, Barbara, ed. *Encyclopedia of Latin American History and Culture* (New York: Scribners, 5 vols., 1996).

"The Andean Coca Wars: A Crop That Refuses to Die," *The Economist*, March 4, 2000, 23–25.

Wilkie, Richard W. *Latin American Population and Urbanization Analysis: Maps and Statistics, 1950–1982* (Westwood, Calif.: UCLA Latin American Center, 1984).

Winn, Peter. *Americas: The Changing Face of Latin America and The Caribbean* (Berkeley, Cal.: University of California Press, updated ed., 1999).

World Bank. *World Resources, 1996–1997: The Urban Environment* (New York: Oxford University Press, 1996).

CHAPTER 6

Amery, Hussein A. & Wolf, Aaron T. *Water in the Middle East: A Geography of Peace* (Austin, Tex.: University of Texas Press, 2000).

Amirahmadi, Hooshang & El-Shakhs, Salah S., eds. *Urban Development in the Muslim World* (New Brunswick, N.J.: Rutgers University, Center for Urban Policy Research, 1993).

• Anderson, Ewan W. *The Middle East: Geography and Geopolitics* (London & New York: Routledge, 2000).

Azarya, Victor. *Nomads and the State in Africa: The Political Roots of Marginality* (Brookfield, Vt.: Ashgate, 1996).

• Beaumont, Peter, et al. *The Middle East: A Geographical Study* (New York: John Wiley & Sons/Halsted, 2nd rev. ed., 1988).

Blake, Gerald H., et al. *The Cambridge Atlas of the Middle East and North Africa* (New York: Cambridge University Press, 1988).

Bonine, Michael E., ed. *Population, Poverty, and Politics in Middle East Cities* (Gainesville, Fla.: University Press of Florida, 1997).

Chapman, Graham P. & Baker, Kathleen M., eds. *The Changing Geography of Africa and the Middle East* (London & New York: Routledge, 1992).

Chinn, Jeff & Kaiser, Robert J. *Russians as the New Minority: Ethnicity and Nationalism in the Soviet Successor States* (Boulder, Colo.: Westview Press, 1996).

Cloudsley-Thompson, J. L., ed. *Sahara Desert* (Oxford, UK: Pergamon Press, 1984).

Cohen, Saul B. "Middle East Geopolitical Transformation: The Disappearance of a Shatter Belt," *Journal of Geography*, 91 (January/February 1992): 2–10.

• Cressey, George B. *Crossroads: Land and Life in Southwest Asia* (Philadelphia: J. B. Lippincott, 1960).

Dowry, Alan. *The Jewish State: A Century Later* (Berkeley, Cal.: University of California Press, 1998).

• Drysdale, Alasdair & Blake, Gerald H. *The Middle East and North Africa: A Political Geography* (New York: Oxford University Press, 1985).

Ebel, Robert & Menon, Rajan, eds., *Energy and Conflict in Central Asia and the Caucasus* (Lanham, Md.: Rowman & Littlefield, 2000).

Engelmann, Kurt E. & Pavlakovic, Vjeran, eds. *Rural Development in Eurasia and the Middle East: Land Reform, Demo-*

graphic Change, and Environmental Constraints (Seattle: University of Washington Press, 2001).

Findlay, Allan M. *The Arab World* (London & New York: Routledge, 1994).

• Fisher, William B. *The Middle East: A Physical, Social and Regional Geography* (London & New York: Methuen, 7th rev. ed., 1978).

Freeman-Grenville, G. S. P. *The Historical Atlas of the Middle East* (New York: Simon & Schuster, 1993).

Fromkin, David. *A Peace to End All Peace: Creating the Modern Middle East, 1914–1922* (New York: Henry Holt, 1989).

Fuller, Graham E. & Francke, Rend Rahim. *The Arab Shi'a: The Forgotten Muslims* (New York: St. Martin's Press, 2000).

Gilbert, Martin. *Atlas of the Arab-Israeli Conflict* (New York: Oxford University Press, 1994).

Gilbert, Martin. *Israel* (New York: William Morrow, 1998).

Gleason, Gregory. *The Central Asian States: Discovering Independence* (Boulder, Colo.: Westview Press, 1997).

Goldscheider, Calvin. *Israel's Changing Society: Population, Ethnicity, and Development* (Boulder, Colo.: Westview Press, 1996).

• Gould, Peter R. *Spatial Diffusion* (Washington, D.C.: Association of American Geographers, Commission on College Geography, Resource Paper No. 4, 1969).

Gradus, Yehuda & Lipshitz, Gabi, eds. *The Mosaic of Israeli Geography* (Beersheva, Israel: Ben Gurion University Press, 1996).

Gurdon, Charles G., ed. *The Horn of Africa* (New York: St. Martin's Press, 1994).

Heathcote, Ronald L. *The Arid Lands: Their Use and Abuse* (London & New York: Longman, 1983).

• Held, Colbert C. *Middle East Patterns: Places, Peoples, and Politics* (Boulder, Colo.: Westview Press, 3rd ed., 2001).

Hiro, Dilip. *Dictionary of the Middle East* (New York: St. Martin's Press, 1996).

Hourani, Albert H. *A History of the Arab Peoples* (Cambridge, Mass.: Belknap/Harvard University Press, 1991).

Joffé, George, ed. *North Africa: Nation, State and Region* (London & New York: Routledge, 1993).

Kemp, Geoffrey & Harkavy, Robert E. *Strategic Geography and the Changing Middle East* (Washington, D.C.: Brookings Institution Press, 1997).

Kliot, Nurit. *Water Resources and Conflict in the Middle East* (London & New York: Routledge, 1994).

Lamb, David. *The Arabs: Journeys Beyond the Mirage* (New York: Random House, 1987).

Lemarchand, Philippe, ed. *The Arab World, The Gulf, and the Middle East: An Atlas* (Boulder, Colo.: Westview Press, 1992).

Lewis, Robert A., ed. *Geographic Perspectives on Soviet Central Asia* (London & New York: Routledge, 1992).

• Longrigg, Stephen H. *The Middle East: A Social Geography* (Chicago: Aldine, 2nd rev. ed., 1970).

Mansfield, Peter. *A History of the Middle East* (New York: Penguin Putnam, rev. ed., 1996).

Matvejevic, Predag. *The Mediterranean: A Cultural Landscape* (Berkeley, Cal.: University of California Press, trans. Michael H. Heim, 1999).

Meiselas, Susan. *Kurdistan: In the Shadow of History* (New York: Random House, 1998).

Morris, Benny. *Righteous Victims: A History of the Zionist-Arab Conflict, 1881–1999* (New York: Knopf, 1999).

Mostyn, Trevor, ed. *The Cambridge Encyclopedia of the Middle East and North Africa* (New York: Cambridge University Press, 1988).

Naipaul, V. S. *Beyond Belief: Islamic Excursions Among the Converted Peoples* (New York: Random House, 1998).

Newman, David. *The Dynamics of Territorial Change: A Political Geography of the Arab-Israeli Conflict* (Boulder, Colo.: Westview Press, 1998).

Park, Chris. *Sacred Worlds: An Introduction to Geography and Religion* (London & New York: Routledge, 1994).

Prescott, J. R.V. *Political Frontiers and Boundaries* (Winchester, Mass.: Allen & Unwin, 1987).

Pryde, Philip R., ed. *Environmental Resources and Constraints in the Former Soviet Republics* (Boulder, Colo.: Westview Press, 1995).

Rahman, Mushtaqur, ed. *Muslim World: Geography and Development* (Lanham, Md.: University Press of America, 1987).

Rashid, Ahmed. *Taliban: Militant Islam, Oil, and Fundamentalism in Central Asia* (New Haven: Yale University Press, 2000).

Robinson, Francis, ed. *Atlas of the Islamic World Since 1500* (New York: Facts on File, 1982).

Robinson, Francis, ed. *The Cambridge Illustrated History of the Islamic World* (New York: Cambridge University Press, 1996).

Sluglett, Peter & Farouk-Sluglett, Marion, eds. *The Times Guide to the Middle East: The Arab World and Its Neighbors, New Edition* (Boulder, Colo.: Westview Press, 1996).

Smart, Ninian. *The World's Religions* (New York: Cambridge University Press, 2nd rev. ed., 1998).

Soffer, Arnon. *Rivers of Fire: The Conflict Over Water in the Middle East* (Lanham, Md.: Rowman & Littlefield, trans. Murray Rosovsky & Nina Copaken, 1999).

Swearingen, Will D. & Bencherifa, Abdellatif, eds. *The North African Environment at Risk* (Boulder, Colo.: Westview Press, 1996).

Waines, David. *An Introduction to Islam* (New York: Cambridge University Press, 1995).

Wright, Robin. *The Last Great Revolution: Turmoil and Transformation in Iran* (New York: Knopf, 2000).

Zoubir, Yahia H., ed. *North Africa in Transition: State, Society, and Economic Transformation in the 1990s* (Gainesville, Fla.: University Press of Florida, 1999).

CHAPTER 7

Adams, William M. & Mortimore, Michael J. *Working the Sahel: Environment and Society in Northern Nigeria* (London & New York: Routledge, 1999).

Adams, William M., Goudie, Andrew S., & Orme, Antony R., eds. *The Physical Geography of Africa* (New York: Oxford University Press, 1996).

"Africa for Africans: A Survey of Sub-Saharan Africa," *The Economist*, September 7, 1996, special insert, 18 pp.

• Aryeetey-Attoh, Samuel, ed. *Geography of Sub-Saharan Africa* (Upper Saddle River, N.J.: Prentice- Hall, 1997).

Benneh, George, et al., eds. *Sustaining the Future: Economic, Social, and Environmental Change in Sub-Saharan Africa* (New York & Tokyo: United Nations University Press, 1996).

• Best, Alan C. G. & de Blij, H. J. *African Survey* (New York: John Wiley & Sons, 1977).

Binns, Tony. *Tropical Africa* (London & New York: Routledge, 1994).

Binns, Tony, ed. *People and Environment in Africa* (New York: John Wiley & Sons, 1995).

Black, Richard & Robinson, Vaughan, eds. *Geography and Refugees: Patterns and Processes of Change* (New York: John Wiley & Sons/Belhaven, 1993).

• Bohannan, Paul & Curtin, Philip D. *Africa and Africans* (Prospect Heights, Ill.: Waveland Press, 4th rev. ed., 1996).

Chapman, Graham P. & Baker, Kathleen M., eds. *The Changing Geography of Africa and the Middle East* (London & New York: Routledge, 1992).

Christopher, Anthony J. *Colonial Africa: An Historical Geography* (Totowa, N.J.: Barnes & Noble, 1984).

Christopher, Anthony J. *The Atlas of Changing South Africa* (London & New York: Routledge, 2 rev. ed., 2001).

Curtin, Philip D. *The Atlantic Slave Trade* (Madison, Wisc.: University of Wisconsin Press, 1969).

Curtin, Philip D., et al. *African History: From Earliest Times to Independence* (London & New York: Longman, 2nd rev. ed., 1995).

Curtis, Sarah & Taket, Ann. *Health and Societies: Changing Perspectives* (London & New York: Edward Arnold, 1996).

Davidson, Basil. *The Black Man's Burden: Africa and the Curse of the Nation-State* (New York: Times Books/Random House, 1992).

• de Blij, H. J. "Africa's Geomosaic Under Stress," *Journal of Geography*, 90 (January/February, 1991): 2–9.

Ezzell, Carol. "Care for a Dying Continent," *Scientific American*, May 2000, 96–105.

Fardon, Richard & Furniss, Graham, eds. *African Languages, Development, and the State* (London & New York: Routledge, 1994).

Foster, Harold D. *Health, Disease and the Environment* (New York: John Wiley & Sons, 1992).

Freeman-Grenville, G. S. P. *The New Atlas of African History* (New York: Simon & Schuster, 1991).

Gates, Henry Louis, Jr. & Appiah, Kwame A., eds. *Africana: Encyclopedia of the African and African- American Experience* (New York: Basic Books, 1999).

Gesler, Wilbert M. *The Cultural Geography of Health Care* (Pittsburgh: University of Pittsburgh Press, 1991).

Goliber, Thomas J. "Population and Reproductive Health in Sub-Saharan Africa," *Population Bulletin*, 52 (December 1997): 1–44.

Gooneratne, Wilbert & Obudho, Robert A., eds. *Contemporary Issues in Regional Development Policy: Perspectives from Eastern and Southern Africa* (Brookfield, Vt.: Ashgate, 1997).

• Gould, Peter R. *The Slow Plague: A Geography of the AIDS Pandemic* (Cambridge, Mass.: Blackwell, 1993).

Griffith, Daniel A. & Newman, James L., eds. *Eliminating Hunger in Africa: Technical and Human Perspectives* (Syracuse, N.Y.: Maxwell School, Syracuse University, Foreign and Comparative Studies/African Series 45, 1994).

Griffiths, Ieuan L. L. *The Atlas of African Affairs* (London & New York: Routledge, 2nd rev. ed., 1994).

• Grove, Alfred T. *The Changing Geography of Africa* (New York: Oxford University Press, 2nd rev. ed., 1994).

Harrison Church, Ronald J. *West Africa: A Study of the Environment and Man's Use of It* (London: Longman, 8th rev. ed., 1980).

Huke, Robert E. "The Green Revolution," *Journal of Geography*, 84 (1985): 248–254.

Jalloh, Alusine & Maizlish, Stephen E., eds. *The African Diaspora* (College Station, Tex.: Texas A&M University Press, 1996).

Kearns, Robin A. & Gesler, Wilbert M., eds. *Putting Health Into Place: Landscape, Identity, and Well-Being* (Syracuse, N.Y.: Syracuse University Press, 1998).

Khapoya, Vincent B. *The African Experience* (Englewood Cliffs, N.J.: Prentice-Hall, 1992).

Knight, C. Gregory & Newman, James L., eds. *Contemporary Africa: Geography and Change* (Englewood Cliffs, N.J.: Prentice-Hall, 1976).

Lamb, David. *The Africans* (New York: Random House, 1983).

Learmonth, Andrew T. A. *Disease Ecology: An Introduction* (New York: Blackwell, 1988).

Lemon, Anthony, ed. *Geography of Change in South Africa* (New York: John Wiley & Sons/Longman, 1995).

Lewis, Laurence A. & Berry, Leonard. *African Environments and Resources* (Winchester, Mass.: Unwin Hyman, 1988).

Maier, Karl. *Into the House of the Ancestors: Inside the New Africa* (New York: John Wiley & Sons, 1997).

Martin, Esmond Bradley & de Blij, H. J., eds. *African Perspectives: An Exchange of Essays on the Economic Geography of Nine African States* (London & New York: Methuen, 1981).

Marx, Anthony W. *Making Race and Nation: A Comparison of the United States, South Africa and Brazil* (New York: Cambridge University Press, 1998).

McEvedy, Colin. *The Penguin Atlas of African History* (New York: Penguin Putnam, rev. ed., 2000).

Meade, Melinda S. & Earickson, Robert J. *Medical Geography* (New York, Guilford, 2nd rev. ed., 2000).

Mehretu, Assefa. *Regional Disparity in Sub-Saharan Africa: Structural Readjustment of Uneven Development* (Boulder, Colo.: Westview Press, 1989).

Middleton, John, ed. *Encyclopedia of Sub-Saharan Africa* (New York: Simon & Schuster, 4 vols., 1994).

Moon, Graham & Jones, Kelvyn. *Health, Disease and Society: An Introduction to Medical Geography* (New York: Routledge & Kegan Paul, 1988).

Mortimore, Michael. *Adapting to Drought: Farmers, Famines and Desertification in West Africa* (New York: Cambridge University Press, 1989).

• Mountjoy, Alan & Hilling, David. *Africa: Geography and Development* (Totowa, N.J.: Barnes & Noble, 1987).

Murdock, George P. *Africa: Its Peoples and Their Culture History* (New York: McGraw-Hill, 1959).

Newman, James L. *The Peopling of Africa: A Geographic Interpretation* (New Haven, Conn.: Yale University Press, 1995).

O'Connor, Anthony M. *Poverty in Africa: A Geographical Approach* (New York: Columbia University Press, 1991).

Olson, James S. *The Peoples of Africa: An Ethnohistorical Dictionary* (Westport, Conn.: Greenwood Press, 1996).

Pakenham, Thomas. *The Scramble for Africa: The White Man's Conquest of the Dark Continent from 1876 to 1912* (New York: Random House, 1991).

Press, Robert M. *The New Africa: Dispatches From a Changing Continent* (Gainesville, Fla.: University Press of Florida, 1999).

Pritchard, J. M. *Landform and Landscape in Africa* (London: Edward Arnold, 1979).

Rakodi, Carole, ed. *The Urban Challenge in Africa: Growth and Management of Its Large Cities* (New York & Tokyo: United Nations University Press, 1996).

Reader, John. *Africa: A Biography of a Continent* (New York: Knopf, 1998).

Saff, Grant R. *Changing Cape Town: Urban Dynamics, Policy and Planning During the Political Transition in South Africa* (Lanham, Md.: University Press of America, 1998).

Senior, Michael & Okunrotifa, P. *A Regional Geography of Africa* (London & New York: Longman, 1983).

Siddle, David & Swindell, Ken. *Rural Change in Tropical Africa* (Cambridge, Mass.: Blackwell, 1994).

• Stock, Robert. *Africa South of the Sahara: A Geographical Interpretation* (New York: Guilford Press, 1995).

Stren, Richard & White, Rodney, eds. *African Cities in Crisis: Managing Rapid Urban Growth* (Boulder, Colo.: Westview Press, 1989).

Turner, Bill L. II, Hyden, Goran, & Kates, Robert W., eds. *Population Growth and Agricultural Change in Africa* (Gainesville, Fla.: University Press of Florida, 1993).

Waldmeir, Patti. *Anatomy of a Miracle: The End of Apartheid and the Birth of the New South Africa* (New York: W. W. Norton, 1997).

Wesseling, H. L. *Divide and Rule: The Partition of Africa, 1880-1914* (Westport, Conn.: Praeger, trans. Arnold J. Pomerans, 1996).

CHAPTER 8

Baxter, Craig. *Bangladesh: From a Nation to a State* (Boulder, Colo.: Westview Press, 1997).

Chapman, Graham P. *The Geopolitics of South Asia: From Early Empires to India, Pakistan and Bangladesh* (Brookfield, Vt.: Ashgate, 2000).

Chapman, Graham P., Dutt, Ashok K., & Bradnock, Robert W., eds. *Urban Growth and Development in Asia. Volume I: Making the Cities; Volume II: Living in the Cities* (Brookfield, Vt.: Ashgate, 1999).

Corbridge, Stuart E., guest ed. "India: 1947–1997," *Environment and Planning A*, Vol. 27, No. 12, 1997.

Corbridge, Stuart E. & Harriss, John. *Reinventing India: Liberalization, Hindu Nationalism and Popular Democracy* (Malden, Mass.: Blackwell/Polity, 2000).

Crossette, Barbara. *So Close to Heaven: The Vanishing Buddhist Kingdoms of the Himalayas* (New York: Knopf, 1995).

Crossette, Barbara. *India: Facing the Twenty-First Century* (Bloomington, Ind.: Indiana University Press, 1993).

Deshpande, C. D. *India: A Regional Interpretation* (New Delhi: Northern Book Centre/Indian Council of Social Science Research, 1992).

Dumont, Louis. *Homo Hierarchus: The Caste System and Its Implications* (Chicago: University of Chicago Press, 1970).

Dutt, Ashok K. "Cities of South Asia," in Brunn, Stanley D. & Williams, Jack F., eds., *Cities of the World: World Regional Urban Development* (New York: HarperCollins, 2nd rev. ed., 1993), pp. 351–387.

Dutt, Ashok K. & Geib, Margaret. *An Atlas of South Asia* (Boulder, Colo.: Westview Press, 1987).

Er-Rashid, Haroun. *Geography of Bangladesh* (Boulder, Colo.: Westview Press, 1977).

• Farmer, B. H. *An Introduction to South Asia* (London & New York: Routledge, 2nd rev. ed, 1993).

Gupta, Akhil. *Postcolonial Developments: Agriculture in the Making of Modern India* (Durham, N.C.: Duke University Press, 1998).

Haque, C. Emdad. *Hazards in a Fickle Environment: Bangladesh* (Hingham, Mass.: Kluwer, 1998).

Hasan, Mushirul. *Legacy of a Divided Nation: India's Muslims from Independence to Ayodhya* (Boulder, Colo.: Westview Press, 1997).

Huke, Robert E. "The Green Revolution," *Journal of Geography*, 84 (1985): 248–254.

Huyler, Stephen P. *Meeting God: Elements of Hindu Devotion* (New Haven, Conn.: Yale University Press, 1999).

Isaac, Kalpana. "Sri Lanka's Ethnic Divide," *Current History*, 95 (April 1996): 177–181.

Ives, Jack D. & Messerli, Bruno. *Himalayan Dilemma: Reconciling Development and Conservation* (London & New York: Routledge, 1989).

Jaffrelot, Christopher. *The Hindu Nationalist Movement in India* (New York: Columbia University Press, 1997).

Johnson, Basil L. C. *Bangladesh* (Totowa, N.J.: Barnes & Noble, 2nd rev. ed., 1982).

Johnson, Basil L. C. *India: Resources and Development* (Totowa, N.J.: Barnes & Noble, 1979).

Karan, Pradyumna P. & Ishii, Hiroshi. *Nepal: Development and Change in a Landlocked Himalayan Kingdom* (Tokyo: Institute for the Studies of Languages and Cultures of Asia and Africa, Tokyo University of Foreign Studies, 1994).

Karan, Pradyumna P. & Ishii, Hiroshi, eds. *Nepal: A Himalayan Kingdom in Transition* (New York & Tokyo: United Nations University Press, 1996).

Khilnani, Sunil. *The Idea of India* (New York: Farrar, Straus and Giroux, 1997).

Kosinski, Leszek A. & Elahi, K. Maudood, eds. *Population Redistribution and Development in South Asia* (Hingham, Mass.: D. Reidel, 1985).

Lall, Arthur S. *The Emergence of Modern India* (New York: Columbia University Press, 1981).

Lipner, Julius J. *Hinduism* (London & New York: Routledge, 1994).

Lukacs, John R., ed. *The People of South Asia: The Biological Anthropology of India, Pakistan, and Nepal* (New York: Plenum Press, 1984).

• McFalls, Joseph A., Jr. "Population: A Lively Introduction," *Population Bulletin*, 53 (September 1998): 1–48.

McGowan, William. *Only Man Is Vile: The Tragedy of Sri Lanka* (New York: Farrar, Straus & Giroux, 1992).

Mehta, Gita. *Snakes and Ladders: Glimpses of Modern India* (Garden City, N.Y.: Anchor Books, 1997).

Mitra, Subrata K., ed. *Subnational Movements in South Asia* (Boulder, Colo.: Westview Press, 1994).

Murphey, Rhoads. *A History of Asia* (New York: HarperCollins, 1992).

• Muthiah, S., ed. *An Atlas of India* (New York: Oxford University Press, 1992).

Newman, James L. & Matzke, Gordon E. *Population: Patterns, Dynamics, and Prospects* (Englewood Cliffs, N.J.: Prentice-Hall, 1984).

Noble, Allen G. & Dutt, Ashok K., eds. *India: Cultural Patterns and Processes* (Boulder, Colo.: Westview Press, 1982).

Paz, Octavio. *In Light of India* (Orlando, Fla.: Harcourt Brace, trans. Eliot Weinberger, 1997).

Roberts, Paul W. *Empire of the Soul: Some Journeys in India* (New York: Riverhead, 1999).

Robinson, Francis, ed. *The Cambridge Encyclopedia of India, Pakistan, Bangladesh, Sri Lanka, Nepal, Bhutan, and the Maldives* (New York: Cambridge University Press, 1989).

Samad, Yunas. *A Nation in Turmoil: Nationalism and Ethnicity in Pakistan* (Thousand Oaks, Cal.: Sage, 1995).

Sanderson, Warren C. & Tan, Jee-Peng. *Population in Asia* (Brookfield, Vt.: Ashgate, 1996).

Schwartzberg, Joseph E. *A Historical Atlas of South Asia: 2nd Impression, With Additional Material* (New York: Oxford University Press, 1992).

Shurmer-Smith, Pamela. *India: Globalization and Change* (London: Arnold, 2000).

Sopher, David E., ed. *An Exploration of India: Geographical Perspectives on Society and Culture* (Ithaca, N.Y.: Cornell University Press, 1980).

• Spate, Oskar H. K. & Learmonth, Andrew T. A. *India and Pakistan: A General and Regional Geography* (London: Methuen, 2 vols., 1971).

• Spencer, Joseph E. & Thomas, William L., Jr. *Asia, East by South: A Cultural Geography* (New York: John Wiley & Sons, 2nd rev. ed., 1971).

Srinivas, Smriti. *Landscapes of Urban Memory: The Sacred and the Civic in India's High-Tech City* (Minneapolis: University of Minnesota Press, 2001).

Thapar, Valmik. *Land of the Tiger: A Natural History of the Indian Subcontinent* (Berkeley, Cal.: University of California Press, 1998).

Tharoor, Shashi. *India: From Midnight to the Millennium* (New York: Arcade Publishing, 1997).

"Time to Let Go: A Survey of India," *The Economist*, February 22, 1997, special insert, 26 pp.

• Weightman, Barbara A. *Dragons and Tigers: Geography of South, East, and Southeast Asia* (New York: John Wiley & Sons, 2002).

Wirsing, Robert G. *India, Pakistan, and the Kashmir Dispute: On Regional Conflict and Its Resolution* (New York: St. Martin's Press, 1994).

Wirsing, Robert G. *War or Peace on the Line of Control? The India-Pakistan Dispute Over Kashmir Turns Fifty* (Durham, UK: International Boundaries Research Unit, University of Durham, Vol. 2, Boundary and Territory Briefing No. 5, 1998).

• Wolpert, Stanley. *India* (Berkeley, Cal.: University of California Press, updated ed., 1999).

Zachariah, K. C. & Rajan, S. Irudaya, eds. *Kerala's Demographic Transition* (Thousand Oaks, Cal.: Sage, 1998).

Zurick, David & Karan, Pradyumna P. *Himalaya: Life on the Edge of the World* (Baltimore: Johns Hopkins University Press, 1999).

CHAPTER 9

Atlas of Population, Environment, and Sustainable Development of China (Beijing & New York: Science Press, 2000).

Barnett, A. Doak. *China's Far West: Four Decades of Change* (Boulder, Colo.: Westview Press, 1993).

Benewick, Robert & Donald, Stephanie. *The State of China Atlas* (New York: Penguin Putnam, 1999).

• Burks, Ardath W. *Japan: A Postindustrial Power* (Boulder, Colo.: Westview Press, 3rd rev. ed., 1991).

Cannon, Terry & Jenkins, Alan, eds. *The Geography of Contemporary China: The Impact of Deng Xiaoping's Decade* (London & New York: Routledge, 1990).

Cartier, Carolyn. *Globalizing South China* (Malden, Mass.: Blackwell, 2000).

Chapman, Graham P. & Baker, Kathleen M., eds. *The Changing Geography of Asia* (London & New York: Routledge, 1992).

Chapman, Graham P., Dutt, Ashok K., & Bradnock, Robert W., eds. *Urban Growth and Development in Asia. Volume I: Making the Cities; Volume II: Living in the Cities* (Brookfield, Vt.: Ashgate, 1999).

Chinoy, Mike. *China Live: Two Decades in the Heart of the Dragon* (Atlanta: Turner Publishing, 1997).

Cook, Ian G., et al., eds. *Fragmented Asia: Regional Integration and National Disintegration in Pacific Asia* (Brookfield, Vt.: Ashgate, 1996).

Cressey, George B. *Asia's Lands and Peoples: A Geography of One-Third of the Earth and Two-Thirds of Its People* (New York: McGraw-Hill, 3rd rev. ed., 1963).

Cumings, Bruce. *Korea's Place in the Sun: A Modern History* (New York: W. W. Norton, 1997).

Cybriwsky, Roman A. *Tokyo: The Changing Profile of an Urban Giant* (Boston: G. K. Hall, 1991).

Dirlik, Arif, ed. *What Is In a Rim? Critical Perspectives on the Pacific Region Idea* (Lanham, Md.: Rowman & Littlefield, 2nd rev. ed., 1998).

Drakakis-Smith, David. *Pacific Asia* (London & New York: Routledge, 1992).

Dwyer, Denis J., ed. *China: The Next Decades* (New York: John Wiley & Sons/Longman, 1993).

Eccleston, Bernard, Dawson, Michael, & McNamara, Deborah, eds. *The Asia-Pacific Profile* (London & New York: Routledge, 1998).

Ginsburg, Norton S., ed. *The Pattern of Asia* (Englewood Cliffs, N.J.: Prentice-Hall, 1958).

"Go West, Young Han: Plans to Develop China's Western Provinces are About More than Economics," *The Economist*, December 23, 2000, 45–46.

Harrell, Stevan, ed. *Cultural Encounters on China's Ethnic Frontiers* (Seattle: University of Washington Press, 1995).

Harris, Chauncy D. "The Urban and Industrial Transformation of Japan," *graphical Review*, 72 (January 1982): 50–89.

Hillel, Daniel. "Lash of the Dragon: China's Yellow River Remains Untamed," *Natural History*, August 1991, 28–37.

Hoare, James & Pares, Susan. *Korea: An Introduction* (London & New York: Routledge, 1988).

• Hodder, Rupert. *The West Pacific Rim: An Introduction* (New York: John Wiley & Sons, 1992).

Hoh, Erling. "The Long River's Journey Ends: China's Three Gorges Dam Will Soon Transform the Yangtze," *Natural History*, July 1996, 28–38.

Hook, Brian, ed. *The Cambridge Encyclopedia of China* (New York: Cambridge University Press, 2nd rev. ed., 1994).

Hsieh, Chiao-min & Hsieh, Jean Kan. *China: A Provincial Atlas* (New York: Macmillan, 1992).

Hsieh, Chiao-Min & Lu, Max. *Changing China: A graphical Appraisal* (Boulder, Colo.: Westview Press, 1998).

Hsu, Francis L. K. & Serrie, Hendrick. *The Overseas Chinese: Ethnicity in National Context* (Lanham, Md.: University Press of America, 1998).

Kam Wing Chan. *Cities With Invisible Walls: Reinterpreting Urbanization in Post-1949 China* (New York: Oxford University Press, 1994).

Karan, Pradyumna P. & Stapleton, Kristin, eds. *The Japanese City* (Lexington, Ky.: University Press of Kentucky, 1997).

Knapp, Ronald G., ed. *Chinese Landscapes: The Village as Place* (Honolulu: University of Hawai'i Press, 1992).

• Kornhauser, David H. *Japan: graphical Background to Urban-Industrial Development* (London & New York: Longman, 2nd rev. ed., 1982).

Kristof, Nicholas D. & WuDunn, Sheryl. *Thunder From the East: Portrait of a Rising Asia* (New York: Knopf, 2000).

• Leeming, Frank. *The Changing graphy of China* (Cambridge, Mass.: Blackwell, 1993).

Le Heron, Richard & Park, Sam Ock, eds. *The Asian Pacific Rim and Globalization: Enterprise, Governance, and Territoriality* (Brookfield, Vt.: Ashgate, 1995).

Lin, rge C. S. *Red Capitalism in South China: Growth and Development of the Pearl River Delta* (Vancouver: University of British Columbia Press, 1997).

Linge, G. J. R. & Forbes, D. K., eds. *China's Spatial Economy* (New York: Oxford University Press, 1990).

Lo, Chor-Pang. *Hong Kong* (New York: John Wiley & Sons/Belhaven, 1992).

Lo, Fu-chen & Yeung, Yue-man, eds. *Emerging World Cities in Pacific Asia* (New York & Tokyo: United Nations University Press, 1996).

MacDonald, Donald. *A graphy of Modern Japan* (Ashford, UK: Paul Norbury, 1985).

Mackerras, Colin. *China's Minorities: Integration and Modernization in the Twentieth Century* (New York: Oxford University Press, 1994).

McKay, John & van Grunsven, Leo, eds. *Regional Change in Industrializing Asia: Economic Restructuring and Local Response* (Brookfield, Vt.: Ashgate, 1998).

Murphey, Rhoads. *A History of Asia* (New York: HarperCollins, 1992).

• Noh, Toshio & Kimura, John C., eds. *Japan: A Regional Geography of an Island Nation* (Tokyo: Teikoku-Shoin, 1985).

Ohmae, Kenichi. *The End of the Nation-State: The Rise of Regional Economies* (New York: Free Press, 1996).

Olds, Kris, et al., eds. *Globalization and the Asia Pacific: Contested Territories* (London & New York: Routledge, 2000).

Pannell, Clifton W. "China's Urban Transition," *Journal of Geography*, 94 (May/June 1995): 394–403.

Pannell, Clifton W., ed. *East Asia: Geographical and Historical Approaches to Foreign Area Studies* (Dubuque, Iowa: Kendall/ Hunt, 1983).

• Pannell, Clifton W. & Ma, Laurence J. C. *China: The Geography of Development and Modernization* (New York: Halsted Press/V. H. Winston, 1983).

• Pannell, Clifton W. & Veeck, Gregory. *East Asia* (Lanham, Md.: Rowman & Littlefield, 1999).

Porter, Jonathan. *Macau: The Imaginary City* (Boulder, Colo.: Westview Press, 1996).

Potter, Robert B., et al. *Geographies of Development* (Harlow, UK & New York: Longman, 1999).

Preston, Peter W. *Pacific Asia in the Global System: An Introduction* (Malden, Mass.: Blackwell, 1998).

Qing, Dai, Thibodeau, John G., & Williams, Philip, eds. *The River Dragon Has Come!: The Three Gorges Dam and the Fate of China's Yangtze River and Its People* (Armonk, NY: M. E. Sharpe, trans. M. Yi, 1997).

Reischauer, Edwin O. *The Japanese Today: Change and Continuity* (Cambridge, Mass.: Belknap/Harvard University Press, 1988).

Rozman, Gilbert, ed. *Japan and Russia: The Tortuous Path to Normalization, 1949–1999* (New York: St. Martin's Press, 2000).

Rumley, Dennis, et al., eds. *Global Geopolitical Change and the Asia-Pacific: A Regional Perspective* (Brookfield, Vt.: Ashgate, 1996).

Salisbury, Harrison E. *The Long March: The Untold Story* (New York: Harper & Row, 1985).

Shen, Jianfa. *Internal Migration and Regional Population Dynamics in China* (Amsterdam & New York: Elsevier, 1996).

Si-ming Li & Wing-shing Tang, eds. *China's Regions, Polity, and Economy: A Study of Spatial Transformation in the Post-Reform Era* (Hong Kong: The Chinese University Press, 2000).

Sit, Victor F. S. *Beijing: The Nature and Planning of a Chinese Capital City* (New York: John Wiley & Sons, 1995).

Smil, Vaclav. *The Bad Earth: Environmental Degradation in China* (Armonk, N.Y.: M. E. Sharpe, 1984).

• Smith, Christopher J. *China in the Post-Utopian Age* (Boulder, Colo.: Westview Press, 2000).

Smith, Patrick. *Japan: A Reinterpretation* (New York: Pantheon, 1997).

Spence, Jonathan D. *The Chan's Great Continent: China in Western Minds* (New York: W. W. Norton, 1998).

• Spencer, Joseph E. & Thomas, William L., Jr. *Asia, East by South: A Cultural Geography* (New York: John Wiley & Sons, 2nd rev. ed., 1971).

Starr, John Bryan. *Understanding China: A Guide to China's Economy, History, and Political Structure* (New York: Hill & Wang, 1997).

Terrill, Ross. *China in Our Time: The Epic Saga of the People's Republic from the Communist Victory to Tiananmen Square and Beyond* (New York: Simon & Schuster, 1992).

Theroux, Paul. "Going to See the Dragon: A Journey Through South China, Land of Capitalist Miracles, Where Yesterday's Rice Paddy Becomes Tomorrow's Metropolis, and a Thousand Factories Bloom," *Harper's Magazine*, October 1993, 33–56.

Trewartha, Glenn T. *Japan: A Geography* (Madison, Wisc.: University of Wisconsin Press, 2nd rev. ed., 1965).

van Kemenade, Willem. *China, Hong Kong, Taiwan Inc.: The Dynamics of a New Empire* (New York: Knopf, 1997). Quotation taken from p. 184.

Veeck, Gregory, ed. *The Uneven Landscape: Geographic Studies in Post-Reform China* (Baton Rouge, La.: Geoscience Publications, 1991).

Vogel, Ezra F. *The Four Little Dragons: The Spread of Industrialization in East Asia* (Cambridge, Mass.: Harvard University Press, 1991).

Wei, Yehua Dennis. *Regional Development in China: States, Globalization, and Inequality* (London & New York: Routledge, 2000).

• Weightman, Barbara A. *Dragons and Tigers: Geography of South, East, and Southeast Asia* (New York: John Wiley & Sons, 2002).

Weiping Wu. *Pioneering Economic Reform in China's Special Economic Zones: The Promotion of Foreign Investment and Technology Transfer in Shenzhen* (Brookfield, Vt.: Ashgate, 1999).

Witherick, M. & Carr, M. *The Changing Face of Japan* (London: Hodder & Stoughton, 1993).

Wittwer, Sylvan, et al. *Feeding a Billion: Frontiers of Chinese Agriculture* (East Lansing, Mich.: Michigan State University Press, 1987).

Yeung, Yue-man & Chu, David K.Y., eds. *Guangdong: Survey of a Province Undergoing Rapid Change* (Hong Kong: Chinese University Press, 1994).

Yeung, Yue-man & Hu, Xu-Wei, eds. *China's Coastal Cities: Catalysts for Modernization* (Honolulu: University of Hawai'i Press, 1992).

Yeung, Yue-man & Yun-wing, Sung, eds. *Shanghai: Transformation and Modernization Under China's Open Policy* (Hong Kong: Chinese University Press, 1996).

• Zhao Songqiao. *Geography of China: Environment, Resources, Population, and Development* (New York: John Wiley & Sons, 1994).

Zhao Songqiao. *Physical Geography of China* (New York & Beijing: John Wiley & Sons/Science Press, 1986).

CHAPTER 10

Broek, Jan O. M. "Diversity and Unity in Southeast Asia," *Geographical Review*, 34 (1944): 175–195.

Brookfield, Harold C., et al. *In Place of the Forest: Environmental and Socio-Economic Transformation in Borneo and the Eastern Malay Peninsula* (New York: United Nations University Press, 1995).

Burling, Robbins. *Hill Farms and Padi Fields: Life in Mainland Southeast Asia* (Englewood Cliffs, N.J.: Prentice-Hall, 1965).

Chapman, Graham P., Dutt, Ashok K., & Bradnock, Robert W., eds. *Urban Growth and Development in Asia. Volume I: Making the Cities; Volume II: Living in the Cities* (Brookfield, Vt.: Ashgate, 1999).

Cleary, Mark & Eaton, Peter. *Tradition and Reform: Land Tenure and Rural Development in South-East Asia* (New York: Oxford University Press, 1997).

Cox, Christopher R. *Chasing the Dragon: Into the Heart of the Golden Triangle* (New York: Holt/Marion Wood, 1996).

Daws, Gavan & Fujita, Marty. *Archipelago: The Islands of Indonesia* (Berkeley, Calif.: University of California Press, 1999).

• Dixon, Chris. *South East Asia in the World Economy* (New York: Cambridge University Press, 1991).

Dixon, Chris & Drakakis-Smith, David, eds. *Uneven Development in Southeast Asia* (Brookfield, Vt.: Ashgate, 1998).

Dwyer, Denis J., ed. *South East Asian Development* (New York: John Wiley & Sons/Longman, 1990).

Eccleston, Bernard, Dawson, Michael, & McNamara, Deborah, eds. *The Asia-Pacific Profile* (London & New York: Routledge, 1998).

• Fisher, Charles A. *Southeast Asia: A Social, Economic and Political Geography* (New York: E. P. Dutton, 2nd rev. ed., 1966).

• Fryer, Donald W. *Emerging South-East Asia: A Study in Growth and Stagnation* (New York: John Wiley & Sons, 2nd rev. ed., 1979).

Ginsburg, Norton S., et al., eds. *The Extended Metropolis: Settlement Transition in Asia* (Honolulu: University of Hawai'i Press, 1991).

• Glassner, Martin I. & de Blij, H. J. *Systematic Political Geography* (New York: John Wiley & Sons, 4th rev. ed., 1989).

"Good Fences: Borders are Arbitrary Abstractions, Economic Impediments and Surprisingly Ineradicable," *The Economist*, December 19, 1998, 19–22.

Hartshorne, Richard. "Suggestions on the Terminology of Political Boundaries," *Annals of the Association of American Geographers*, 26 (1936): 56–57.

Hill, Ronald D., ed. *South-East Asia: A Systematic Geography* (New York: Oxford University Press, 1979).

Hodder, Rupert. *The West Pacific Rim: An Introduction* (New York: John Wiley & Sons, 1992).

Hsu, Francis L. K. & Serrie, Hendrick. *The Overseas Chinese: Ethnicity in National Context* (Lanham, Md.: University Press of America, 1998).

Karnow, Stanley. *In Our Image: America's Empire in the Philippines* (New York: Random House, 1989).

Kim, T. J., et al., eds. *Spatial Development in Indonesia: Review and Prospects* (Aldershot, UK: Avebury, 1992).

Kristof, Nicholas D. & WuDunn, Sheryl. *Thunder From the East: Portrait of a Rising Asia* (New York: Knopf, 2000).

Leinbach, Thomas R. & Sien, Chia Lin, eds. *South-East Asian Transport* (New York: Oxford University Press, 1989).

Leinbach, Thomas R. & Ulack, Richard. "Cities of Southeast Asia,' in Brunn, Stanley D. & Williams, Jack F., eds., *Cities of the World: World Regional Urban Development* (New York: HarperCollins, 2nd rev. ed., 1993), pp. 389–429.

• Leinbach, Thomas R. & Ulack, Richard. *Southeast Asia: Diversity and Development* (Upper Saddle River, N.J.: Prentice Hall, 2001).

McGee, Terence G. *The Southeast Asian City: A Social Geography* (New York: Praeger, 1967).

McKay, John & van Grunsven, Leo, eds. *Regional Change in Industrializing Asia: Economic Restructuring and Local Response* (Brookfield, Vt.: Ashgate, 1998).

Mulder, Neils. *Inside Southeast Asia: Religion, Everyday Life, and Cultural Change* (Seattle: University of Washington Press, 2001).

Murray, Geoffrey & Perera, Audrey. *Singapore: The Global City-State* (New York: St. Martin's Press, 1996).

Neher, Clark D. & Marlay, Ross. *Democracy and Development in Southeast Asia: The Winds of Change* (Boulder, Colo.: Westview Press, 1996).

Parnwell, Michael J. G. & Bryant, Raymond L., eds. *Environmental Change in South-East Asia: People, Politics and Sustainable Development* (London & New York: Routledge, 1996).

• Prescott, J. R. V. *Political Frontiers and Boundaries* (Winchester, Mass.: Allen & Unwin, 1987).

Preston, Peter W. *Pacific Asia in the Global System: An Introduction* (Malden, Mass.: Blackwell, 1998).

Reynolds, Craig J. "A New Look at Old Southeast Asia," *Journal of Asian Studies*, 54 (1995): 419–446.

Rigg, Jonathan. *Southeast Asia—A Region in Transition: A Thematic Human Geography of the ASEAN Region* (New York: HarperCollins Academic, 1991).

• Rigg, Jonathan. *Southeast Asia: The Human Landscape of Modernization and Development* (London & New York: Routledge, 1997).

SarDesai, D. R. *Southeast Asia: Past and Present* (Boulder, Colo.: Westview Press, 4th rev. ed., 1997).

Schmidt, Johannes D., Hersh, Jacques, & Fold, Niels, eds. *Social Change in Southeast Asia* (London & New York: Longman, 1997).

• Spencer, Joseph E. & Thomas, William L., Jr. *Asia, East by South: A Cultural Geography* (New York: John Wiley & Sons, 2nd rev. ed., 1971).

Taylor, John G. *East Timor: The Price of Freedom* (London: Zed Books, 2000).

"The Faltering Firefighter: A Survey of Indonesia," *The Economist*, July 8, 2000, special insert, 16 pp.

Ulack, Richard & Pauer, Gyula. *Atlas of Southeast Asia* (New York: Macmillan, 1988).

• Weightman, Barbara A. *Dragons and Tigers: Geography of South, East, and Southeast Asia* (New York: John Wiley & Sons, 2002).

Wing-Kai Chiu, Stephen, et al. *City States in the Global Economy: Industrial Restructuring in Hong Kong and Singapore* (Boulder, Colo.: Westview Press, 1996).

Wurfel, David & Burton, Bruce, eds. *Southeast Asia in the New World Order* (New York: St. Martin's Press, 1996).

CHAPTER 11

"A Survey of Australia," *The Economist*, September 9, 2000, special insert, 16 pp.

Bambrick, Susan, ed. *The Cambridge Encyclopedia of Australia* (New York: Cambridge University Press, 1994).

• Barrett, Rees D. & Ford, Roslyn A. *Patterns in the Human Geography of Australia* (South Melbourne: Macmillan of Australia, 1987).

Beadle, N. C. W. *The Vegetation of Australia* (New York: Cambridge University Press, 1981).

Britton, Steve, et al., eds. *Changing Places in New Zealand: A Geography of Restructuring* (Dunedin, N.Z.: Allied Press, 1992).

Camm, J. C. R. & McQuilton, J., eds. *Australia: An Historical Atlas* (Broadway, N.S.W.: Fairfax, Syme & Weldon Associates, 1987).

Connell, John, ed. *Sydney: The Emergence of a World City* (New York: Oxford University Press, 2000).

Courtenay, Percy P. *Northern Australia: Patterns and Problems of Tropical Development in an Advanced Country* (London & New York: Longman, 1983).

Cumberland, Kenneth B. & Whitelaw, James S. *New Zealand* (Chicago: Aldine, 1970).

"Environment and Development in Australia," *Australian Geographer*, 19 (May 1988): 3–220.

Fitzpatrick, Judith M., ed. *Endangered Peoples of Oceania: Struggles to Survive and Thrive* (Westport, Conn.: Greenwood Press, 2000).

Gentilli, Joseph. *Australian Climate Patterns* (Melbourne: Thomas Nelson Press, 1982).

Head, Lesley. *Second Nature: The History and Implications of Australia as Aboriginal Landscape* (Syracuse, N.Y.: Syracuse University Press, 2000).

• Heathcote, Ronald L. *Australia* (New York: John Wiley & Sons/Longman, 2nd rev. ed., 1994).

Heathcote, Ronald L., ed. *The Australian Experience: Essays in Australian Land Settlement and Resource Management* (Melbourne: Longman Cheshire, 1988).

Heathcote, Ronald L. & Mabbutt, J. A. *Land, Water and People: Essays in Australian Resource Management* (Winchester, Mass.: Allen & Unwin, 1988).

Hofmeister, Burkhard. *Australia and Its Urban Centers* (Berlin: Gebrüder Borntraeger, 1988).

Holland, P. G. & Johnston, W. B., eds. *Southern Approaches: Geography in New Zealand* (Christchurch, N.Z.: New Zealand Geographical Society, 1987).

Hughes, Robert. *The Fatal Shore* (New York: Knopf, 1987).

Hughes, Robert. "The Real Australia," *Time*, September 11, 2000, 98–111.

Inglis, C., et al., eds. *Asians in Australia: The Dynamics of Migration and Settlement* (Sydney: Allen & Unwin, 1992).

• Jeans, Dennis N., ed. *Australia: A Geography. Volume 1: The Natural Environment* (Sydney: Sydney University Press, 1986); *Volume 2: Space and Society* (Sydney: Sydney University Press, 1987).

Lines, William J. *Taming the Great South Land: A History of the Conquest of Nature in Australia* (Berkeley, Cal.: University of California Press, 1992).

MacDonald, Glen M. *Biogeography* (New York: John Wiley & Sons, 2002).

McKnight, Tom L. *Australia's Corner of the World* (Englewood Cliffs, N.J.: Prentice-Hall, 1970).

• McKnight, Tom L. *Oceania: The Geography of Australia, New Zealand, and the Pacific Islands* (Englewood Cliffs, N.J.: Prentice Hall, 1995).

Meinig, Donald W. *On the Margins of the Good Earth: The South Australian Wheat Frontier, 1869–1884* (Chicago: Rand McNally, 1962).

• Morton, Harry & Johnston, Carol M. *The Farthest Corner: New Zealand, A Twice Discovered Land* (Honolulu: University of Hawai'i Press, 1989).

Parfit, Michael. "Australia—A Harsh Awakening" (with foldout map supplement "Australia Under Siege"), *National Geographic*, July 2000, 2–31.

Powell, Joseph M. *An Historical Geography of Modern Australia: The Restive Fringe* (New York: Cambridge University Press, 1988).

Powell, Joseph M. "Revisiting the Australian Experience: Transmillennial Conjurings," *Geographical Review*, 90 (January 2000): 1–17.

Rich, David C. *The Industrial Geography of Australia* (Sydney: Methuen, 1987).

• Robinson, Guy M., Loughran, Robert J., & Tranter, Paul J. *Australia and New Zealand: Economy, Society and Environment* (New York: Oxford University Press, 2000).

Rumley, Dennis. *The Geopolitics of Australia's Regional Relations* (Brookfield, Vt.: Ashgate, 1999).

Serventy, J. *Landforms of Australia* (New York: American Elsevier, 1968).

Smith, L. M. *The Aboriginal Population of Australia* (Canberra: Australian National University Press, 1980).

• Spate, Oskar H. K. *Australia* (New York: Praeger, 1968).

Walmsley, D. J. & Sorenson, A. D. *Contemporary Australia: Explorations in Economy, Society and Geography* (Melbourne: Longman Cheshire, 2nd rev. ed., 1993).

CHAPTER 12

• Bier, James A., cartographer. *Reference Map of Oceania: The Pacific Islands of Micronesia, Polynesia, Melanesia* (Honolulu: University of Hawai'i Press, 1995).

Blake, Gerald H., ed. *Maritime Boundaries: World Boundaries, Vol. 5* (London & New York: Routledge, 1994).

Blake, Gerald H., ed. *Maritime Boundaries and Ocean Resources* (Totowa, N.J.: Rowman & Littlefield, 1988).

Brookfield, Harold C. "The Pacific Realm," in Marvin W. Mikesell, ed., *Geographers Abroad: Essays on the Problems and Prospects of Research in Foreign Areas* (Chicago: University of Chicago, Department of Geography, Research Paper No. 152, 1973), pp. 70–93.

Brookfield, Harold C., ed. *The Pacific in Transition: Geographical Perspectives on Adaptation and Change* (New York: St. Martin's Press, 1973).

Brookfield, Harold C. & Hart, Doreen. *Melanesia: A Geographical Interpretation of an Island World* (New York: Barnes & Noble, 1971).

Bunge, Frederica M. & Cooke, Melinda W., eds. *Oceania: A Regional Study* (Washington, D.C.: U.S. Government Printing Office, 1984).

Carter, John, ed. *Pacific Islands Yearbook* (Sydney: Pacific Publications, annual).

• Chaturvedi, Sanjay. *The Polar Regions: A Political Geography* (New York: John Wiley & Sons, 1997).

Cole, George M. *Water Boundaries* (New York: John Wiley & Sons, 1997).

Connell, John. *Papua New Guinea: The Struggle for Development* (London & New York: Routledge, 1997).

Couper, Alastair D., ed. *Development and Social Change in the Pacific Islands* (London & New York: Routledge, 1989).

Crossley, L. *Explore Antarctica* (New York: Cambridge University Press, 1995).

Damas, David. *Bountiful Island: A Study of Land Tenure on a Micronesian Atoll* (Waterloo, Ont., Canada: Wilfrid Laurier University Press, 1996).

de Blij, H. J. "A Regional Geography of Antarctica and the Southern Ocean," *University of Miami Law Review*, 33 (1978): 299–314.

Dodds, Klaus J. *Geopolitics of Antarctica: Views From the Southern Oceanic Rim* (New York: John Wiley & Sons, 1997).

Fitzpatrick, Judith M., ed. *Endangered Peoples of Oceania: Struggles to Survive and Thrive* (Westport, Conn.: Greenwood Press, 2000).

Freeman, Otis W., ed. *Geography of the Pacific* (New York: John Wiley & Sons, 1951).

Friis, Herman R., ed. *The Pacific Basin: A History of Its Geographical Exploration* (New York: American Geographical Society, Special Publication No. 38, 1967).

• Glassner, Martin I. *Neptune's Domain: A Political Geography of the Sea* (Winchester, Mass.: Unwin Hyman, 1990).

Grossman, Lawrence S. *Peasants, Subsistence Ecology, and Development in the Highlands of Papua New Guinea* (Princeton, N.J.: Princeton University Press, 1984).

Hanlon, David & White, Geoffrey M., eds. *Voyaging Through the Contemporary Pacific* (Lanham, Md.: Rowman & Littlefield, 2000).

Howard, A., ed. *Polynesia: Readings on a Culture Area* (Scranton, Penna.: Chandler, 1971).

Howlett, Diana. *Papua New Guinea: Geography and Change* (Melbourne, Australia: Thomas Nelson, 1973).

Johnston, Douglas M. & Saunders, Philip M. *Ocean Boundary-Making: Regional Issues and Developments* (London & New York: Routledge, 1988).

Karolle, Bruce G. *Atlas of Micronesia* (Honolulu: Bess Press, 2nd rev. ed., 1995).

Kirch, Patrick V. *On the Road of Winds: An Archaeological History of the Pacific Islands Before European Contact* (Berkeley, Cal.: University of California Press, 2000).

Kirch, Patrick V. *The Wet and the Dry: Irrigation and Agricultural Intensification in Polynesia* (Chicago: University of Chicago Press, 1994).

Kissling, Christopher C., ed. *Transport and Communications for Pacific Microstates: Issues in Organization and Management* (Suva, Fiji: University of the South Pacific, Institute of Pacific Studies, 1984).

Kluge, P. F. *The Edge of Paradise: America in Micronesia* (New York: Random House, 1991).

Lea, John & Connell, John. *Urbanization in the Pacific* (London & New York: Routledge, 2001).

Leibowitz, Arnold H. *Embattled Island: Palau's Struggle for Independence* (Westport, Conn.: Praeger, 1996).

Lockhart, Douglas G., et al., eds. *The Development Process in Small Island States* (London & New York: Routledge, 1993).

• McEvedy, Colin. *The Penguin Historical Atlas of the Pacific* (New York: Penguin Putnam, 1999).

• McKnight, Tom L. *Oceania: The Geography of Australia, New Zealand, and the Pacific Islands* (Englewood Cliffs, N.J.: Prentice Hall, 1995).

Mitchell, Andrew. *The Fragile South Pacific: An Ecological Odyssey* (Austin: University of Texas Press, 1991).

"Mobility and Identity in the Island Pacific," special issue, *Pacific Viewpoint*, 26, No. 1 (1985).

Morgan, Joseph R. *Hawai'i: A Unique Geography* (Honolulu: Bess Press, 1996).

Nunn, Patrick. *Oceanic Islands* (Cambridge, Mass.: Blackwell, 1994).

Oliver, Douglas L. *Native Cultures of the Pacific Islands* (Honolulu: University of Hawai'i Press, 1989).

• Peake, Martin. *Pacific People and Society* (New York: Cambridge University Press, 1992).

Prescott, J. R.V. *The Maritime Political Boundaries of the World* (London & New York: Methuen, 1986).

Robillard, Albert B., ed. *Social Change in the Pacific Islands* (London & New York: Kegan Paul International, 1991).

Sager, Robert J. "The Pacific Islands: A New Geography," *Focus*, Summer 1988, pp. 10–14.

Spate, Oskar H. K. *The Pacific Since Magellan, Vol. II: Monopolists and Freebooters* (Minneapolis: University of Minnesota Press, 1983).

Spate, Oskar H. K. *The Pacific Since Magellan, Vol. III: Paradise Found and Lost* (Minneapolis: University of Minnesota Press, 1989).

Spate, Oskar H. K. *The Spanish Lake: A History of the Pacific Since Magellan, Vol. I* (Beckenham, UK: Croom Helm, 1979).

Sugden, David E. *Arctic and Antarctic: A Modern Geographical Synthesis* (Totowa, N.J.: Barnes & Noble, 1982).

Theroux, Paul. *The Happy Isles of Oceania: Paddling the Pacific* (New York: G. P. Putnam's Sons, 1992).

Ward, R. Gerard, ed. *Man in the Pacific Islands: Essays on Geographical Change in the Pacific* (New York: Oxford University Press, 1972).

Woodcock, Deborah W., ed. *Hawai'i: New Geographies* (Manoa, Hawai'i: University of Hawai'i, Department of Geography, 1999).

Zurick, David N. "Preserving Paradise," *Geographical Review*, 85 (April 1995): 157–172.

Glossary

Aboriginal population See **indigenous peoples**.*

Absolute location The position or place of a certain item on the surface of the Earth as expressed in degrees, minutes, and seconds of **latitude**, 0° to 90° north or south of the equator, and **longitude**, 0° to 180° east or west of the *prime meridian* passing through Greenwich, England (a suburb of London).

Accessibility The degree of ease with which it is possible to reach a certain location from other locations. *Inaccessibility* is the opposite of this concept.

Acculturation Cultural modification resulting from intercultural borrowing. In **cultural geography**, the term refers to the change that occurs in the culture of **indigenous peoples** when contact is made with a society that is technologically superior.

Agglomeration Process involving the clustering or concentrating of people or activities.

Agrarian Relating to the use of land in rural communities, or to **agricultural** societies in general.

Agriculture The purposeful tending of crops and livestock in order to produce food and fiber.

Alluvial Refers to the mud, silt, and sand (collectively *alluvium*) deposited by rivers and streams. *Alluvial plains* adjoin many larger rivers; they consist of these renewable deposits that are laid down during floods, creating fertile and productive soils. Alluvial **deltas** mark the mouths of rivers such as the Nile and the Ganges.

Altiplano High-elevation plateau, basin, or valley between even higher mountain ranges, especially in the Andes of South America.

Altitudinal zonation Vertical regions defined by physical-environmental zones at various elevations (see Fig. 4-13), particularly in the highlands of South and Middle America. See ***tierra caliente***, ***tierra templada***, ***tierra fría***, ***tierra helada***, and ***tierra nevada***.

American Manufacturing Belt North America's near-rectangular Core Region, whose corners are Boston, Milwaukee, St. Louis, and Baltimore. Dominated the industrial geography of the U.S. and Canada during the industrial age; still a formidable economic powerhouse that remains the realm's geographic heart.

Antecedent boundary A political boundary that existed before the **cultural landscape** emerged and stayed in place while people moved in to occupy the surrounding area.

Anthracite coal Hardest and highest carbon-content coal, and therefore of the highest quality.

* Words in boldface type are defined elsewhere in this Glossary.

Apartheid Literally, *apartness*. The Afrikaans term for South Africa's pre-1994 policies of racial separation, a system that produced highly segregated socio-geographical patterns.

Aquaculture The use of a river segment or an artificial pond for the raising and harvesting of food products, including fish, shellfish, and even seaweed (particularly in Japan).

Aquifer An underground reservoir of water contained within a porous, water-bearing rock layer.

Arable Land fit for cultivation by one farming method or another. See **physiologic density**.

Archipelago A set of islands grouped closely together, usually elongated into a *chain*.

Area A term that refers to a part of the Earth's surface with less specificity than **region**. For example, *urban area* alludes generally to a place where urban development has occurred, whereas *urban region* requires certain specific criteria upon which such a designation is based (e.g., the spatial extent of commuting or the built townscape).

Areal functional organization A geographic principle for understanding the evolution of regional organization, whose five interrelated tenets are applied to the spatial development of Japan on p. 479.

Areal interdependence A term related to **functional specialization**. When one area produces certain goods or has certain raw materials and another area has a different set of raw materials and produces different goods, their needs may be *complementary*; by exchanging raw materials and products, they can satisfy each other's requirements.

Arithmetic density A country's population, expressed as an average per unit area, without regard for its **distribution** or the limits of **arable** land—see also **physiologic density**.

Aryan From the Sanskrit *Arya* ("noble"), a name applied to an ancient people who spoke an Indo-European language and who moved into northern India from the northwest.

Atmosphere The Earth's envelope of gases that rests on the oceans and land surface and penetrates open spaces within soils. This layer of nitrogen (78 percent), oxygen (21 percent), and traces of other gases is densest at the Earth's surface and thins with altitude.

Autocratic A government that holds absolute power, often ruled by one person or a small group of persons who control the country by despotic means.

Balkanization The fragmentation of a **region** into smaller, often hostile political units.

Barrio Term meaning "neighborhood" in Spanish. Usually refers to an urban community in a Middle or South American city; also applied to low-income, inner-city concentrations of Hispanics in such western U.S. cities as Los Angeles.
Bauxite Aluminum ore; usually deposited at shallow depths in the wet tropics.
Biogeography The study of *flora* (plant life) and *fauna* (animal life) in spatial perspective.
Birth rate The *crude birth rate* is expressed as the annual number of births per 1000 individuals within a given population.
Bituminous coal Softer coal of lesser quality than **anthracite**, but of higher grade than **lignite**. When heated and converted to coking coal or *coke*, it is used to make steel.
Break-of-bulk point A location along a transport route where goods must be transferred from one carrier to another. In a port, the cargoes of oceangoing ships are unloaded and put on trains, trucks, or perhaps smaller river boats for inland distribution.
Buffer state See **buffer zone**.
Buffer zone A set of countries separating ideological or political adversaries. In southern Asia, Afghanistan, Nepal, and Bhutan were parts of a buffer zone between British and Russian-Chinese imperial spheres. Thailand was a *buffer state* between British and French colonial domains in mainland Southeast Asia.
Caliente See **tierra caliente**.
Cartogram A specially transformed map not based on traditional representations of **scale** or area.
Cartography The art and science of making maps, including data compilation, layout, and design. Also concerned with the interpretation of mapped patterns.
Caste system The strict **social stratification** and segregation of people—specifically in India's Hindu society—on the basis of ancestry and occupation.
Cay A low-lying small island usually composed of coral and sand. Pronounced *kee*, and often spelled "key."
Central business district (CBD) The downtown heart of a central city, the CBD is marked by high land values, a concentration of business and commerce, and the clustering of the tallest buildings.
Centrality The strength of an urban center in its capacity to attract producers and consumers to its facilities; a city's "reach" into the surrounding region.
Centrifugal forces A term employed to designate forces that tend to divide a country—such as internal religious, linguistic, ethnic, or ideological differences.
Centripetal forces Forces that unite and bind a country together—such as a strong national culture, shared ideological objectives, and a common faith.
Chaebol Giant corporation controlling numerous companies and benefitting from government connections and favors, dominant in South Korea's **economic geography**; key to the country's **development** into an **economic tiger**, but more recently a barrier to free-market growth.
Charismatic Personal qualities of certain leaders that enable them to capture and hold the popular imagination, to secure the allegiance and even the devotion of the masses. Gandhi, Mao Zedong, and Franklin D. Roosevelt were good examples in the 20th century.
China Proper The eastern and northeastern portions of China that contain most of the country's huge population; mapped in Figure 9-4.
Choke point A narrowing of an international waterway to a distance of less than 24 miles (38 km), necessitating the drawing of a **median line (maritime) boundary**. These are almost always strategic locations where, presumably, a blockading naval force could "choke" off the waterway. Examples are the Hormuz Strait between Oman and Iran at the entrance to the Persian Gulf, and the Strait of Malacca between Malaysia and Indonesia.
City-state An independent political entity consisting of a single city with (and sometimes without) an immediate **hinterland**. The ancient city-states of Greece have their modern equivalent in Singapore.
Climate The long-term conditions (over at least 30 years) of aggregate **weather** over a region, summarized by averages and measures of variability; a synthesis of the succession of weather events we have learned to expect at any given location.
Climate change theory An alternative to the **hydraulic civilization theory**; holds that changing **climate** (rather than a monopoly over **irrigation** methods) could have provided certain cities in the ancient **Fertile Crescent** with advantages over others.
Climatology The geographic study of **climates**. Includes not only the classification of climates and the analysis of their regional distribution, but also broader environmental questions that concern climate change, interrelationships with soil and vegetation, and human-climate interaction.
Coal See **anthracite coal**, **bituminous coal**, and **lignite**.
Collectivization The reorganization of a country's **agriculture** under communism that involves the expropriation of private holdings and their incorporation into relatively large-scale units, which are farmed and administered cooperatively by those who live there.
Colonialism See **imperialism**.
Commercial agriculture For-profit **agriculture**.
Common market A **free-trade area** that not only has created a **customs union** (a set of common tariffs on all imports from outside the area) but also has eliminated restrictions on the movement of capital, labor, and enterprise among its member countries.
Compact state A politico-geographical term to describe a **state** that possesses a roughly circular, oval, or rectangular territory in which the distance from the geometric center to any point on the boundary exhibits little variance. Poland and Cambodia are examples of this shape category.
Complementarity Exists when two regions, through an exchange of raw materials and/or finished products, can specifically satisfy each other's demands.
Confucianism A philosophy of ethics, education, and public service based on the writings of Confucius (*Kongfuzi*); traditionally regarded as one of the cornerstones of Chinese **culture**.
Coniferous forest A forest of cone-bearing, needleleaf evergreen trees with straight trunks and short branches, including spruce, fir, and pine.
Contagious diffusion The distance-controlled spreading of an idea, innovation, or some other item through a local population by contact from person to person—analogous to the communication of a contagious illness.
Conterminous United States The 48 **contiguous** or adjacent States that occupy the southern half of the North American realm. Alaska is not contiguous to these States because western Canada lies in between; neither is Hawai'i, separated from the mainland by over 2000 miles of ocean.
Contiguous Adjoining; adjacent.

Continental drift The slow movement of continents controlled by the **processes** associated with **plate tectonics**.

Continental shelf Beyond the coastlines of many landmasses, the ocean floor declines very gently until the depth of about 660 feet (200 m). Beyond the 660-foot line the sea bottom usually drops off sharply, along the *continental slope*, toward the much deeper mid-oceanic basin. The submerged continental margin is called the continental shelf, and it extends from the shoreline to the upper edge of the continental slope.

Continentality The variation of the continental effect on air temperatures in the interior portions of the world's landmasses. The greater the distance from the moderating influence of an ocean, the greater the extreme in summer and winter temperatures. Continental interiors also tend to be dry when the distance from oceanic moisture sources becomes considerable.

Conurbation General term used to identify a large multi-metropolitan complex formed by the coalescence of two or more major **urban areas**. The Atlantic Seaboard **Megalopolis**, extending along the northeastern U.S. coast from southern Maine to Virginia, is a classic example.

Copra The dried-out, fleshy interior of a coconut that is used to produce coconut oil.

Cordillera Mountain chain consisting of sets of parallel ranges, especially the Andes in northwestern South America.

Core See **core area**; **core-periphery relationships**.

Core area In geography, a term with several connotations. *Core* refers to the center, heart, or focus. The core area of a **nation-state** is constituted by the national heartland, the largest population cluster, the most productive region, and the part of the country with the greatest **centrality** and **accessibility**—probably containing the capital city as well.

Core-periphery relationships The contrasting spatial characteristics of, and linkages between, the *have* (core) and *have-not* (periphery) components of a national or regional system.

Corridor In general, refers to a spatial entity in which human activity is organized in a linear manner, as along a major transport route or in a valley confined by highlands. More specifically, the politico-geographical term for a land extension that connects an otherwise **landlocked state** to the sea.

Cultural diffusion The **process** of spreading and adoption of a cultural element, from its place of origin across a wider area.

Cultural ecology The multiple interactions and relationships between a **culture** and its **natural environment**.

Cultural environment See **cultural ecology**.

Cultural geography The wide-ranging and comprehensive field of geography that studies spatial aspects of human **cultures**.

Cultural landscape The forms and artifacts sequentially placed on the **natural landscape** by the activities of various human occupants. By this progressive imprinting of the human presence, the physical (natural) landscape is modified into the cultural landscape, forming an interacting unity between the two.

Cultural pluralism See **plural(istic) society**.

Cultural revival The regeneration of a long-dormant **culture** through internal renewal and external infusion.

Culture The sum total of the knowledge, attitudes, and habitual behavior patterns shared and transmitted by the members of a society. This is anthropologist Ralph Linton's definition; hundreds of others exist.

Culture area See **culture region**.

Culture hearth Heartland, source area, innovation center; place of origin of a major **culture**.

Culture region A distinct, culturally discrete spatial unit; a **region** within which certain cultural norms prevail.

Customs union A **free-trade area** in which member countries set common tariff rates on imports from outside the area.

Death rate The *crude death rate* is expressed as the annual number of deaths per 1000 individuals within a given population.

Deciduous A deciduous tree loses its leaves at the beginning of winter or the onset of the dry season.

Definition In political geography, the written legal description (in a treaty-like document) of a boundary between two countries or territories—see **delimitation**.

Deforestation The clearing and destruction of forests (especially tropical rainforests) to make way for expanding settlement frontiers and the exploitation of new economic opportunities.

Deglomeration Deconcentration.

Delimitation In political geography, the translation of the written terms of a boundary treaty (the **definition**) into an official cartographic representation (map).

Delta **Alluvial** lowland at the mouth of a river, formed when the river deposits its alluvial load on reaching the sea. Often triangular in shape, hence the use of the Greek letter whose symbol is Δ.

Demarcation In political geography, the actual placing of a political boundary on the **cultural landscape** by means of barriers, fences, walls, or other markers.

Demographic transition model Multi-stage model, based on Western Europe's experience, of changes in population growth exhibited by countries undergoing industrialization. High **birth rates** and **death rates** are followed by plunging death rates, producing a huge net population gain; this is followed by the convergence of birth and death rates at a low overall level. See Figure 8-10.

Demography The interdisciplinary study of population—especially **birth rates** and **death rates**, growth patterns, longevity, **migration**, and related characteristics.

Desert An arid area supporting sparse vegetation, receiving less than 10 inches (25 cm) of precipitation per year. Usually exhibits extremes of heat and cold because the moderating influence of moisture is absent.

Desertification The **process** of **desert** expansion into neighboring **steppelands** as a result of human degradation of fragile semiarid environments.

Development The economic, social, and institutional growth of national **states**.

Devolution The **process** whereby regions within a **state** demand and gain political strength and growing autonomy at the expense of the central government.

Dhows Wooden boats with characteristic triangular sails, plying the seas between Arabian and East African coasts.

Dialect Regional or local variation in the use of a major language, such as the distinctive accents of many residents of the U.S. South or New England.

Diffusion The spatial spreading or dissemination of a **culture** element (such as a technological innovation) or some other phenomenon (e.g., a disease outbreak). For the various channels of outward geographic spread from a source area, see **contagious**, **expansion**, **hierarchical**, and **relocation diffusion**.

Distance decay The various degenerative effects of distance on human spatial structures and interactions.
Diurnal Daily.
Divided capital In political geography, a country whose central administrative functions are carried on in more than one city is said to have divided capitals. The Netherlands and South Africa are examples.
Domestication The transformation of a wild animal or wild plant into a domesticated animal or a cultivated crop to gain control over food production. A necessary evolutionary step in the development of humankind: the invention of **agriculture**.
Domino theory The belief that political destabilization in one **state** can result in the collapse of order in a neighboring state, triggering a chain of events that, in turn, can affect a series of **contiguous** states.
Double cropping The planting, cultivation, and harvesting of two crops successively within a single year on the same plot of farmland.
Doubling time The time required for a population to double in size.
Dry canal An overland rail and/or road **corridor** across an **isthmus** dedicated to performing the transit functions of a canalized waterway. Best adapted to the movement of containerized cargo, there must be a port at each end to handle the necessary **break-of-bulk** unloading and reloading.
Ecology The study of the many interrelationships between all forms of life and the natural environments in which they have evolved and continue to develop. The study of *ecosystems* focuses on the interactions between specific organisms and their environments. See also **cultural ecology**.
Economic geography The field of geography that focuses on the diverse ways in which people earn a living, and how the goods and services they produce are expressed and organized spatially.
Economic tiger One of the burgeoning beehive countries of the western **Pacific Rim**. Following Japan's route since 1945, these countries have experienced significant modernization, industrialization, and Western-style economic growth since 1980. Three leading economic tigers today are South Korea, Taiwan, and Singapore. Term is increasingly used more generally to describe any fast-developing economy (e.g., Ireland's new reputation as the *Celtic Tiger*).
Economies of scale The savings that accrue from large-scale production wherein the unit cost of manufacturing decreases as the level of operation enlarges. Supermarkets operate on this principle and are able to charge lower prices than small grocery stores.
Ecosystem See **ecology**.
Ecumene The habitable portions of the Earth's surface where permanent human settlements have arisen.
Elite A small but influential upper-echelon social class whose power and privilege give it control over a country's political, economic, and cultural life.
El Niño-Southern Oscillation (ENSO) A periodic, large-scale, abnormal warming of the sea surface in the low latitudes of the eastern Pacific Ocean that has global implications, disturbing normal **weather** patterns in many parts of the world, especially South America.
Elongated state A **state** whose territory is decidedly long and narrow in that its length is at least six times greater than its average width. Chile and Vietnam are two classic examples.
Emigrant A person **migrating** away from a country or area; an out-migrant.
Empirical Relating to the real world, as opposed to theoretical abstraction.
Enclave A piece of territory that is surrounded by another political unit of which it is not a part.
Endemism Refers to a disease in a host population that affects many people in a kind of equilibrium without causing rapid and widespread deaths.
Entrepôt A place, usually a port city, where goods are imported, stored, and transshipped; a **break-of-bulk point**.
Environmental degradation The accumulated human abuse of a region's **natural landscape** that, among other things, can involve air and water pollution, threats to plant and animal ecosystems, misuse of **natural resources**, and generally upsetting the balance between people and their habitat.
Epidemic A local or regional outbreak of a disease.
Escarpment A cliff or very steep slope; frequently marks the edge of a plateau.
Estuary The widening mouth of a river as it reaches the sea; land subsidence or a rise in sea level has overcome the tendency to form a **delta**.
Ethnic cleansing The slaughter and/or forced removal of one **ethnic** group from its homes and lands by another, more powerful ethnic group bent on taking that territory.
Ethnicity The combination of a people's **culture** (traditions, customs, language, and religion) and racial ancestry.
European Union (EU) **Supranational** organization constituted by 15 European countries to further their common economic interests. The "4-5-6" method is the best way to learn the names of the current members: the 4 giants (Germany, France, Italy, the United Kingdom); the 5 inner countries, all neighbors of Germany (Denmark, the Netherlands, Belgium, Luxembourg, Austria); and the 6 outer countries (Finland, Sweden, Ireland, Portugal, Spain, and Greece).
European state model A **state** consisting of a legally defined territory inhabited by a population governed from a capital city by a representative government.
Exclave A bounded (non-island) piece of territory that is part of a particular **state** but lies separated from it by the territory of another state. Alaska is an exclave of the United Sates.
Exclusive Economic Zone (EEZ) An oceanic zone extending up to 200 **nautical miles** from a shoreline, within which the coastal **state** can control fishing, mineral exploration, and additional activities by all other countries.
Expansion diffusion The spreading of an innovation or an idea through a fixed population in such a way that the number of those adopting grows continuously larger, resulting in an expanding area of dissemination.
Extraterritoriality Politico-geographical concept suggesting that the property of one **state** lying within the boundaries of another actually forms an extension of the first state.
Favela Shantytown on the outskirts or even well within an urban area in Brazil.
Fazenda Coffee plantation in Brazil.
Federal state A political framework wherein a central government represents the various subnational entities within a **nation-state** where they have common interests—defense, foreign affairs, and the like—yet allows these various entities to retain

their own identities and to have their own laws, policies, and customs in certain spheres.

Federation See **federal state**.

Fertile Crescent Crescent-shaped zone of productive lands extending from near the southeastern Mediterranean coast through Lebanon and Syria to the **alluvial** lowlands of Mesopotamia (in Iraq). Once more fertile than today, this is one of the world's great source areas of **agricultural** and other innovations.

Fjord Narrow, steep-sided, elongated, and inundated coastal valley deepened by glacier ice that has since melted away, leaving the sea to penetrate.

Floodplain Low-lying area adjacent to a mature river, often covered by **alluvial** deposits and subject to the river's floods.

Forced migration Human **migration** flows in which the movers have no choice but to relocate.

Formal region A type of **region** marked by a certain degree of homogeneity in one or more phenomena; also called *uniform region* or *homogeneous region*.

Forward capital Capital city positioned in actually or potentially contested territory, usually near an international border; it confirms the **state's** determination to maintain its presence in the region in contention.

Fossil fuels The energy resources of **coal**, natural gas, and petroleum (oil), so named collectively because they were formed by the geologic compression and transformation of tiny plant and animal organisms.

Four Motors of Europe Rhône-Alpes (France), Baden-Württemberg (Germany), Catalonia (Spain), and Lombardy (Italy). Each is a high-technology-driven region marked by exceptional industrial vitality and economic success not only within Europe but on the global scene as well.

Fragmented state A **state** whose territory consists of several separated parts, not a **contiguous** whole. The individual parts may be isolated from each other by the land area of other states or by international waters. The United States and Indonesia are examples.

Francophone French-speaking. Quebec constitutes the heart of Francophone Canada.

Free-trade area A form of economic integration, usually consisting of two or more **states**, in which members agree to remove tariffs on trade among themselves. Usually accompanied by a **customs union** that establishes common tariffs on imports from outside the trade area, and sometimes by a **common market** that also removes internal restrictions on the movement of capital, labor, and enterprise.

Fría See **tierra fría**.

Frontier Zone of advance penetration, usually of contention; an area not yet fully integrated into a national **state**.

FTAA (Free Trade Area of the Americas) The ultimate goal of **supranational** economic integration in North, Middle, and South America: the creation of a single-market trading bloc (perhaps as soon as 2005) that would involve every country in the Western Hemisphere between the Arctic shore of Canada and Cape Horn at the southern tip of Chile.

Functional region A **region** marked less by its sameness than its dynamic internal structure; because it usually focuses on a central node, also called *nodal region* or *focal region*.

Functional specialization The production of particular goods or services as a dominant activity in a particular location.

Fundamentalism See **religious fundamentalism**.

Gentrification The upgrading of an older residential area through private reinvestment, usually in the downtown area of a central city. Frequently this involves the displacement of established lower-income residents, who cannot afford the heightened costs of living, and conflicts are not uncommon as such neighborhood change takes place.

Geographic realm The basic spatial unit in our world regionalization scheme. Each realm is defined in terms of a synthesis of its total human geography—a composite of its leading cultural, economic, historical, political, and appropriate environmental features.

Geography of development The field of geography concerned with spatial aspects and regional expressions of **development**.

Geometric boundaries Political boundaries **defined** and **delimited** (and occasionally **demarcated**) as straight lines or arcs.

Geomorphology The geographic study of the configuration of the Earth's solid surface—the world's landscapes and their constituent landforms.

Ghetto An intraurban region marked by a particular **ethnic** character. Often an inner-city poverty zone, such as the black ghetto in U.S. central cities. Ghetto residents are involuntarily segregated from other income and racial groups.

Glaciation See **Pleistocene Epoch**.

Globalization The gradual reduction of regional contrasts at the world scale, resulting from increasing international cultural, economic, and political exchanges.

Green Revolution The successful recent development of higher-yield, fast-growing varieties of rice and other cereals in certain developing countries.

Gross national product (GNP) The total value of all goods and services produced in a country during a given year.

Growth pole An urban center with certain attributes that, if augmented by a measure of investment support, will stimulate regional economic development in its **hinterland**.

Hacienda Literally, a large estate in a Spanish-speaking country. Sometimes equated with the **plantation**, but there are important differences between these two types of agricultural enterprise (see pp. 207–208).

Heartland theory The hypothesis, proposed by British geographer Halford Mackinder during the early twentieth century, that any political power based in the heart of Eurasia could gain sufficient strength to eventually dominate the world. Further, since Eastern Europe controlled access to the Eurasian interior, its ruler would command the vast "heartland" to the east.

Hegemony The political dominance of a country (or even a region) by another country. The former Soviet Union's postwar grip on Eastern Europe, which lasted from 1945 to 1990, was a classic example.

Helada See **tierra helada**.

Hierarchical diffusion A form of **diffusion** in which an idea or innovation spreads by trickling down from larger to smaller adoption units. An urban **hierarchy** is usually involved, encouraging the leapfrogging of innovations over wide areas, with geographic distance a less important influence.

Hierarchy An order or gradation of phenomena, with each level or rank subordinate to the one above it and superior to the one below. The levels in a national urban hierarchy are constituted by hamlets, villages, towns, cities, and (frequently) the **primate city**.

High island Volcanic islands of the Pacific Realm that are high enough in elevation to wrest substantial moisture from the tropical ocean air (see **orographic precipitation**). They tend to be well watered, their volcanic soils enable productive agriculture, and they support larger populations than **low islands**—which possess none of these advantages and must rely on fishing and the coconut palm for survival.

High sea Areas of the oceans away from land, beyond national jurisdiction, open and free for all to use.

Highveld A term used in Southern Africa to identify the high, grass-covered plateau that dominates much of the region. The lowest-lying areas in South Africa are called *lowveld*; areas that lie at intermediate elevations are the *middleveld*.

Hinterland Literally, "country behind," a term that applies to a surrounding area served by an urban center. That center is the focus of goods and services produced for its hinterland and is its dominant urban influence as well. In the case of a port city, the hinterland also includes the inland area whose trade flows through that port.

Historical inertia A term from manufacturing geography that refers to the need to continue using the factories, machinery, and equipment of heavy industries for their full, multiple-decade lifetimes to cover major initial investments—even though these facilities may be increasingly obsolete.

Holocene The current *interglaciation* epoch (the warm period of glacial contraction between the glacial expansions of an **ice age**); extends from 10,000 years ago to the present. Also known as the *Recent Epoch*.

Humus Dark-colored upper layer of a soil that consists of decomposed and decaying organic matter such as leaves and branches, nutrient-rich and giving the soil a high fertility.

Hydraulic civilization theory The theory that cities able to control **irrigated** farming over large **hinterlands** held political power over other cities. Particularly applies to early Asian civilizations based in such river valleys as the Chang (Yangzi), the Indus, and those of Mesopotamia.

Hydrologic cycle The system of exchange involving water in its various forms as it continually circulates among the **atmosphere**, the oceans, and above and below the land surface.

Ice age A stretch of geologic time during which the Earth's average atmospheric temperature is lowered; causes the equatorward expansion of continental icesheets in the higher latitudes and the growth of mountain glaciers in and around the highlands of the lower latitudes.

Iconography The identity of a region as expressed through its cherished symbols; its particular **cultural landscape** and personality.

Immigrant A person **migrating** into a particular country or area; an in-migrant.

Imperialism The drive toward the creation and expansion of a colonial empire and, once established, its perpetuation.

Import-substitution industries The industries local entrepreneurs establish to serve populations of remote areas when transport costs from distant sources make these goods too expensive to import.

Inaccessibility See **accessibility**.

Indentured workers Contract laborers who sell their services for a stipulated period of time.

Indigenous peoples Native or *aboriginal* peoples; often used to designate the inhabitants of areas that were conquered and subsequently colonized by the **imperial** powers of Europe.

Industrial Revolution The term applied to the social and economic changes in agriculture, commerce, and especially manufacturing and urbanization that resulted from technological innovations and specialization in late-eighteenth-century Europe.

Informal sector Dominated by unlicensed sellers of homemade goods and services, the primitive form of capitalism found in many developing countries that takes place beyond the control of government.

Infrastructure The foundations of a society: urban centers, transport networks, communications, energy distribution systems, farms, factories, mines, and such facilities as schools, hospitals, postal services, and police and armed forces.

Insular Having the qualities and properties of an island. Real islands are not alone in possessing such properties of **isolation**: an **oasis** in the middle of a **desert** also has qualities of insularity.

Insurgent state Territorial embodiment of a successful guerrilla movement. The establishment by anti-government insurgents of a territorial base in which they exercise full control; thus a state within a **state**.

Intercropping The planting of several types of crops in the same field; commonly used by **shifting cultivators**.

Interglaciation See **Pleistocene Epoch**.

Intermontane Literally, between mountains. The location can bestow certain qualities of natural protection or **isolation** to a community.

Internal migration **Migration** flow within a country, such as ongoing westward and southward movements toward the **Sunbelt** in the United States.

International migration **Migration** flow involving movement across an international boundary.

Intervening opportunity In trade or **migration** flows, the presence of a nearer opportunity that greatly diminishes the attractiveness of sites farther away.

Irredentism A policy of cultural extension and potential political expansion by a **state** aimed at a community of its nationals living in a neighboring state.

Irrigation The artificial watering of croplands.

Isohyet A line connecting points of equal rainfall total.

Isolation The condition of being geographically cut off or far removed from mainstreams of thought and action. It also denotes a lack of receptivity to outside influences, caused at least partially by poor **accessibility**.

Isotherm A line connecting points of equal temperature.

Isthmus A **land bridge**; a comparatively narrow link between larger bodies of land. Central America forms such a link between Mexico and South America.

Jakota Triangle The easternmost region of the East Asian realm, consisting of *Ja*pan, (South) *Ko*rea, and *Ta*iwan.

Juxtaposition Contrasting places in close proximity to one another.

Land alienation One society or culture group taking land from another. In Subsaharan Africa, for example, European **colonialists** took land from **indigenous** Africans and put it to new uses.

Land bridge A narrow **isthmian** link between two large landmasses. They are temporary features—at least in terms of geologic time—subject to appearance and disappearance as the land or sea level rises and falls.

Land reform The spatial reorganization of **agriculture** through the allocation of farmland (often expropriated from landlords) to **peasants** and tenants who never owned land.

Land tenure The way people own, occupy, and use land.

Landlocked An interior **state** surrounded by land. Without coasts, such a country is disadvantaged in terms of **accessibility** to international trade routes, and in the scramble for possession of areas of the **continental shelf** and control of the **exclusive economic zone** beyond.

Latitude Lines of latitude are **parallels** that are aligned east-west across the globe, from 0° latitude at the equator to 90° North and South latitude at the poles.

Leached soil Infertile, reddish-appearing, tropical soil whose surface consists of oxides of iron and aluminum; all other soil nutrients have been dissolved and transported downward into the subsoil by percolating water associated with heavy rainfall.

Leeward The protected or downwind side of a **topographic** barrier with respect to the winds that flow across it.

Lignite Low-grade, brown-colored variety of coal.

Lingua franca A "common language" prevalent in a given area; a second language that can be spoken and understood by many peoples, although they speak other languages at home.

Littoral Coastal or coastland.

Llanos The interspersed **savanna** grasslands and scrub woodlands of the Orinoco River's wide basin that covers much of interior Colombia and Venezuela.

Location theory A logical attempt to explain the locational pattern of an economic activity and the manner in which its producing areas are interrelated. The agricultural location theory that underlies the **von Thünen model** is a leading example.

Loess Deposit of very fine silt or dust that is laid down after having been windborne for a considerable distance. Notable for its fertility under **irrigation** and its ability to stand in steep vertical walls.

Longitude Angular distance (0° to 180°) east or west as measured from the *prime meridian* (0°) that passes through the Greenwich Observatory in suburban London, England. For much of its length across the mid-Pacific Ocean, the 180th meridian functions as the *international date line*.

Low island Low-lying coral islands of the Pacific Realm that—unlike **high islands**—cannot wrest sufficient moisture from the tropical ocean air to avoid chronic drought. Thus productive agriculture is impossible and their modest populations must rely on fishing and the coconut palm for survival.

Maghreb The region occupying the northwestern corner of Africa, consisting of Morocco, Algeria, and Tunisia.

Main Street Canada's dominant **conurbation** that is home to nearly two-thirds of the country's inhabitants; extends southwestward from Quebec City in the mid-St. Lawrence Valley to Windsor on the Detroit River.

Mainland-Rimland framework Twofold regionalization of the Middle American realm based on its modern cultural history. The Euro-Amerindian *Mainland*, stretching from Mexico to Panama (minus the Caribbean coastal strip), was a self-sufficient zone dominated by **hacienda land tenure**. The Euro-African *Rimland*, consisting of that Caribbean coastal zone plus all of the Caribbean islands to the east, was the zone of the **plantation** that heavily relied on trade with Europe.

Maquiladora The term given to modern industrial plants in Mexico's U.S. border zone. These foreign-owned factories assemble imported components and/or raw materials, and then export finished manufactures, mainly to the United States. Import duties are disappearing under **NAFTA**, bringing jobs to Mexico and the advantages of low wage rates to the foreign entrepreneurs.

Marchland An area or **frontier** of uncertain boundaries that is subject to various national claims and an unstable political history. Refers specifically to the movement of various armies across such zones.

Marine geography The geographic study of oceans and seas. Its practitioners investigate both the physical (e.g., coral-reef **biogeography**, ocean-**atmosphere** interactions, coastal **geomorphology**) and human aspects (e.g., **maritime boundary-making**, fisheries, beachside development) of oceanic environments.

Maritime boundary An international boundary that lies in the ocean. Like all boundaries, it is a vertical plane, extending from the seafloor to the upper limit of the air space in the atmosphere above the water.

Median line boundary An international **maritime boundary** drawn where the width of a sea is less than 400 **nautical miles**. Because the **states** on either side of that sea claim **exclusive economic zones** of 200 nautical miles, it is necessary to reduce those claims to a (median) distance equidistant from each shoreline. **Delimitation** on the map almost always appears as a set of straight-line segments that reflect the configurations of the coastlines involved.

Medical geography The study of health and disease within a geographic context and from a spatial perspective. Among other things, this field of geography examines the sources, **diffusion** routes, and distributions of diseases.

Megacity Informal term referring to the world's most heavily populated cities; in this book, the term refers to a **metropolis** containing a population of greater than 5 million.

Megalopolis When spelled with a lower-case *m*, a synonym for **conurbation**, one of the large coalescing supercities forming in diverse parts of the world. When capitalized, refers specifically to the multi-metropolitan corridor that extends along the northeastern U.S. seaboard from north of Boston to south of Washington, D.C. (Fig. 3-9).

Mercantilism Protectionist policy of European **states** during the sixteenth to the eighteenth centuries that promoted a state's economic position in the contest with rival powers. The acquisition of gold and silver and maintaining a favorable trade balance (more exports than imports) were central to the policy.

Meridian Line of **longitude**, aligned north-south across the globe, that together with **parallels** of **latitude** forms the global grid system. All meridians converge at both poles and are at their maximum distances from each other at the equator.

Mestizo Derived from the Latin word for *mixed*, refers to a person of mixed white and Amerindian ancestry.

Metropolis Urban **agglomeration** consisting of a (central) city and its suburban ring. See **urban (metropolitan) area**.

Metropolitan area See **urban (metropolitan) area**.

Migration A change in residence intended to be permanent. See also **forced**, **internal**, **international**, and **voluntary migration**.

Migratory movement Human relocation movement from a source to a destination without a return journey, as opposed to cyclical movement (see **nomadism**).

Model An idealized representation of reality built to demonstrate its most important properties. A **spatial** model focuses on a geographical dimension of the real world, such as the **von Thünen model** that explains agricultural location patterns in a commercial economy.

Monsoon Refers to the seasonal reversal of wind and moisture flows in certain parts of the subtropics and lower-middle latitudes.

The *dry monsoon* occurs during the cool season when dry offshore winds prevail. The *wet monsoon* occurs in the hot summer months, which produce onshore winds that bring large amounts of rainfall. The air-pressure differential over land and sea is the triggering mechanism, with windflows always moving from areas of relatively higher pressure toward areas of relatively lower pressure. Monsoons make their greatest regional impact in the coastal and near-coastal zones of South Asia, Southeast Asia, and East Asia.

Mosaic culture The emerging cultural-geographic framework of the United States, dominated by the fragmentation of specialized social groups into homogeneous communities of interest marked not only by income, race, and ethnicity but also by age, occupational status, and lifestyle. The result is an increasingly heterogeneous socio-spatial complex, which resembles an intricate mosaic composed of myriad uniform—but separate—tiles.

Mulatto A person of mixed African (black) and European (white) ancestry.

Multinationals Internationally active corporations that can strongly influence the economic and political affairs of many countries they operate in.

Muslim An adherent of the Islamic faith.

Muslim Front A term used by certain scholars for the African Transition Zone of northern Africa, which is primarily regarded as a still-expanding frontier of Islam that affects countries from Guinea in the west to the African Horn in the east (see Fig. 6-12).

NAFTA (North American Free Trade Agreement) The **free-trade area** launched in 1994 involving the United States, Canada, and Mexico.

Nation Legally a term encompassing all the citizens of a **state**, it also has other connotations. Most definitions now tend to refer to a group of tightly-knit people possessing bonds of language, **ethnicity**, religion, and other shared **cultural** attributes. Such homogeneity actually prevails within very few states.

Nation-state A country whose population possesses a substantial degree of **cultural** homogeneity and unity. The ideal form to which most **nations** and **states** aspire—a political unit wherein the territorial state coincides with the area settled by a certain national group or people.

NATO (North Atlantic Treaty Organization) Established in 1950 at the height of the Cold War as a U.S.-led **supranational** defense pact to shield postwar Europe against the Soviet military threat. NATO is now in transition, expanding its membership while modifying its objectives in the post-Soviet era.

Natural hazard A natural event that endangers human life and/or the contents of a **cultural landscape**.

Natural increase rate Population growth measured as the excess of live births over deaths per 1000 individuals per year. Natural increase of a population does not reflect either **emigrant** or **immigrant** movements.

Natural landscape The array of landforms that constitutes the Earth's surface (mountains, hills, plains, and plateaus) and the physical features that mark them (such as water bodies, soils, and vegetation). Each **geographic realm** has its distinctive combination of natural landscapes.

Natural resource Any valued element of (or means to an end using) the environment; includes minerals, water, vegetation, and soil.

Nautical mile By international agreement, the nautical mile—the standard measure at sea—is 6076.12 feet in length, equivalent to approximately 1.15 statute miles (1.85 km).

Neocolonialism The term used by developing countries to underscore that the entrenched colonial system of international exchange and capital flow has not changed in the postcolonial era—thereby perpetuating the huge economic advantages of the developed world.

Network (transport) The entire regional **system** of transportation connections and nodes through which movement can occur.

Nevada See **tierra nevada**.

Nomadism Cyclical movement among a definite set of places. Nomadic peoples mostly are **pastoralists**.

Nucleation Cluster; **agglomeration**.

Oasis An area, small or large, where the supply of water (from an **aquifer** or a major river such as the Nile) permits the transformation of the adjacent **desert** into productive cropland.

Occidental Western. Also see ***Oriental***.

Offshore banking Term referring to financial havens for foreign companies and individuals, who channel their earnings to accounts in such a country (usually an "offshore" island-state) to avoid paying taxes in their home countries.

OPEC (Organization of Petroleum Exporting Countries) The international oil *cartel* or syndicate formed by a number of producing countries to promote their common economic interests through the formulation of joint pricing policies and the limitation of market options for consumers. The 11 member-states (as of mid-2001) are: Algeria, Indonesia, Iran, Iraq, Kuwait, Libya, Nigeria, Qatar, Saudi Arabia, United Arab Emirates (UAE), and Venezuela.

Organic theory Friedrich Ratzel's theory of **state** development that conceptualized the state as a biological organism whose life—from birth through maturation to eventual senility and collapse—mirrors that of any living thing.

Oriental The root of the word *oriental* is from the Latin for *rise*. Thus it has to do with the direction in which one sees the sun "rise"—the east; *oriental* therefore means Eastern. ***Occidental*** originates from the Latin for *fall*, or the "setting" of the sun in the west; *occidental* therefore means Western.

Orographic precipitation Mountain-induced precipitation, especially where air masses are forced to cross **topographic** barriers. Downwind areas beyond such a mountain range experience the relative dryness known as the **rain shadow effect**.

Outback The name given by Australians to the vast, peripheral, sparsely-settled interior of their country.

Outer city The non-central-city portion of the American **metropolis**; no longer "sub" to the "urb," this outer ring was transformed into a full-fledged city during the late twentieth century.

Pacific Rim A far-flung group of countries and parts of countries (extending clockwise on the map from New Zealand to Chile) sharing the following criteria: they face the Pacific Ocean; they evince relatively high levels of economic development, industrialization, and urbanization; their imports and exports mainly move across Pacific waters.

Pacific Ring of Fire Zone of crustal instability along tectonic **plate** boundaries, marked by earthquakes and volcanic activity, that ring the Pacific Ocean basin.

Paddies (paddyfields) Ricefields.

Pandemic An outbreak of a disease that spreads worldwide.

Pangaea A vast, singular landmass consisting of most of the areas of the present-day continents. This supercontinent began to break up more than 200 million years ago when still-ongoing

plate divergence and **continental drift** became dominant processes (see Fig. 7-3).

Parallel An east-west line of **latitude** that is intersected at right angles by **meridians** of **longitude**.

Pastoralism A form of **agricultural** activity that involves the raising of livestock.

Peasants In a **stratified** society, peasants are the lowest class of people who depend on **agriculture** for a living. But they often own no land at all and must survive as tenants or day workers.

Peninsula A comparatively narrow, finger-like stretch of land extending from the main landmass into the sea. Florida and Korea are examples.

Peon (*peone*) Term used in Middle and South America to identify people who often live in serfdom to a wealthy landowner; landless **peasants** in continuous indebtedness.

Per capita Capita means *individual*. Income, production, or some other measure is often given per individual.

Perforated state A **state** whose territory completely surrounds that of another state. South Africa, which encloses Lesotho and is perforated by it, is a classic example.

Periodic market Village market that opens every third day or at some other regular interval. Part of a regional network of similar markets in a preindustrial, rural setting where goods are brought to market on foot and barter remains a major mode of exchange.

Periphery See **core-periphery relationships**.

Permafrost Permanently frozen water in the near-surface soil and bedrock of cold environments, producing the effect of completely frozen ground. Surface can thaw during brief warm season.

Physical geography The study of the geography of the physical (natural) world. Its subfields include **climatology**, **geomorphology**, **biogeography**, soil geography, **marine geography**, and water **resources**.

Physical landscape Synonym for **natural landscape**.

Physiographic political boundaries Political boundaries that coincide with prominent physical features in the **natural landscape**—such as rivers or the crest ridges of mountain ranges.

Physiographic region (province) A **region** within which there prevails substantial **natural-landscape** homogeneity, expressed by a certain degree of uniformity in surface **relief**, **climate**, vegetation, and soils.

Physiography Literally means *landscape description*, but commonly refers to the total **physical geography** of a place; includes all of the natural features on the Earth's surface, including landforms, **climate**, soils, vegetation, and water bodies.

Physiologic density The number of people per unit area of **arable** land.

Pilgrimage A journey to a place of great religious significance by an individual or by a group of people (such as a pilgrimage to Mecca for **Muslims**).

Plantation A large estate owned by an individual, family, or corporation and organized to produce a cash crop. Almost all plantations were established within the tropics; in recent decades, many have been divided into smaller holdings or reorganized as cooperatives.

Plate tectonics Plates are bonded portions of the Earth's mantle and crust, averaging 60 miles (100 km) in thickness. More than a dozen such plates exist (see Fig. I-4), most of continental proportions, and they are in motion. Where they meet one slides under the other, crumpling the surface crust and producing significant volcanic and earthquake activity; a major mountain-building force.

Pleistocene Epoch Recent period of geologic time that spans the rise of humankind, beginning about 2 million years ago. Marked by *glaciations* (repeated advances of continental icesheets) and milder *interglaciations* (icesheet contractions). Although the last 10,000 years are known as the **Holocene** Epoch, Pleistocene-like conditions seem to be continuing and we are most probably now living through another Pleistocene interglaciation; thus the glaciers likely will return.

Plural(istic) society A society in which two or more population groups, each practicing its own **culture**, live adjacent to one another without mixing inside a single **state**.

Polder Land reclaimed from the sea adjacent to the shore of the Netherlands by constructing dikes and then pumping out the water trapped behind them.

Political geography The study of the interaction of geographical area and political **process**; the spatial analysis of political phenomena and processes.

Pollution The release of a substance, through human activity, which chemically, physically, or biologically alters the air or water it is discharged into. Such a discharge negatively impacts the environment, with possible harmful effects on living organisms—including humans.

Population density The number of people per unit area. Also see **arithmetic density** and **physiologic density** measures.

Population distribution The way people have arranged themselves in geographic space. One of human geography's most essential expressions because it represents the sum total of the adjustments that a population has made to its natural, cultural, and economic environments.

Population explosion The rapid growth of the world's human population during the past century, attended by ever-shorter **doubling times** and accelerating *rates* of increase.

Population geography The field of geography that focuses on the spatial aspects of **demography** and the influences of demographic change on particular places.

Population implosion The opposite of the **population explosion**, refers to the declining populations of many European countries and Russia in which the **death rate** exceeds the **birth rate** and **immigration** rate.

Population movement See **migration**; **migratory movement**.

Population projection The future population total that demographers forecast for a particular country. For example, in Table I-2 such projections are given for all the world's countries for 2010.

Population (age-sex) structure Graphic representation (*profile*) of a population according to age and gender.

Postindustrial economy Emerging economy, in the United States and a handful of other highly advanced countries, as traditional industry is increasingly eclipsed by a higher-technology productive complex dominated by services, information-related, and managerial activities.

Primary economic activity Activities engaged in the direct extraction of **natural resources** from the environment such as mining, fishing, lumbering, and especially **agriculture**.

Primate city A country's largest city—ranking atop the urban **hierarchy**—most expressive of the national culture and usually (but not always) the capital city as well.

Process Causal force that shapes a spatial pattern as it unfolds over time.

Protruded state Territorial shape of a **state** that exhibits a narrow, elongated land extension (or *protrusion*) leading away from the main body of territory. Thailand is a leading example.

Push-pull concept The idea that **migration** flows are simultaneously stimulated by conditions in the source area, which tend to drive people away, and by the perceived attractiveness of the destination.

Qanat In **desert** zones, particularly in Iran and western China, an underground tunnel built to carry **irrigation** water by gravity flow from nearby mountains (where **orographic precipitation** occurs) to the arid flatlands below.

Quaternary economic activity Activities engaged in the collection, processing, and manipulation of *information*.

Quinary economic activity Managerial or control-function activity associated with decision-making in large organizations.

Rain shadow effect The relative dryness in areas downwind of mountain ranges caused by **orographic precipitation**, wherein moist air masses are forced to deposit most of their water content as they cross the highlands.

Realm See **geographic realm**.

Region A commonly used term and a geographic concept of central importance. An **area** on the Earth's surface marked by specific criteria.

Regional boundary In theory, the line that circumscribes a **region**. But razor-sharp lines are seldom encountered, even in nature (e.g., a coastline constantly changes depending upon the tide). In the **cultural landscape**, not only are regional boundaries rarely self-evident, but when they are ascertained by geographers they most often turn out to be **transitional** borderlands.

Regional complementarity See **complementarity**.

Regional disparity The spatial unevenness in standard of living that occurs within a country, whose "average," overall income statistics invariably mask the differences that exist between the extremes of the wealthy **core** and the poorer **periphery**.

Regional state A "natural economic zone" that defies political boundaries, and is shaped by the global economy of which it is a part; its leaders deal directly with foreign partners and negotiate the best terms they can with the national governments under which they operate.

Relative location The regional position or **situation** of a place relative to the position of other places. Distance, **accessibility**, and connectivity affect relative location.

Relict boundary A political boundary that has ceased to function, but the imprint of which can still be detected on the **cultural landscape**.

Relief Vertical difference between the highest and lowest elevations within a particular area.

Religious fundamentalism Religious movement whose objectives are to return to the foundations of that faith and to influence **state** policy. **Muslims** refer to this movement as *religious revivalism*.

Relocation diffusion Sequential **diffusion process** in which the items being diffused are transmitted by their carrier agents as they relocate to new areas. The most common form of relocation diffusion involves the spreading of innovations by a **migrating** population.

Revivalism (religious) See **religious fundamentalism**.

Rift valley The trough or trench that forms when a strip of the Earth's crust sinks between two parallel faults (surface fractures).

Sahel Semiarid **steppeland** zone extending across most of Africa between the southern margins of the arid Sahara and the moister tropical **savanna** and forest zone to the south. Chronic drought, **desertification**, and overgrazing have contributed to severe famines in this area since 1970.

Savanna Tropical grassland containing widely spaced trees; also the name given to the tropical wet-and-dry climate (*Aw*).

Scale Representation of a real-world phenomenon at a certain level of reduction or generalization. In **cartography**, the ratio of map distance to ground distance; indicated on a map as a bar graph, representative fraction, and/or verbal statement. *Macroscale* refers to a large area of national proportions; *microscale* refers to a local area no bigger than a county.

Scale economies See **economies of scale**.

Secondary economic activity Activities that process raw materials and transform them into finished industrial products. The *manufacturing* sector.

Sedentary Permanently attached to a particular area; a population fixed in its location. The opposite of **nomadic**.

Separate development The spatial expression of South Africa's "grand" **apartheid** scheme, whereby nonwhite groups were required to settle in segregated "homelands." The policy was dismantled when white-minority rule collapsed in the early 1990s.

Sequent occupance The notion that successive societies leave their cultural imprints on a place, each contributing to the cumulative **cultural landscape**.

Shantytown Unplanned slum development on the margins of cities in developing countries, dominated by crude dwellings and shelters mostly made of scrap wood, iron, and even pieces of cardboard.

Sharecropping Relationship between a large landowner and farmers on the land whereby the farmers pay rent for the land they farm by giving the landlord a share of the annual harvest.

Sharia The criminal code based in Islamic law that prescribes corporal punishment, amputations, stonings, and lashing for both major and minor offenses. Its occurrence today is associated with the spread of **religious fundamentalism** in Muslim societies.

Shatter belt Region caught between stronger, colliding external cultural-political forces, under persistent stress, and often fragmented by aggressive rivals. Eastern Europe and Southeast Asia are classic examples.

Shifting agriculture Cultivation of crops in recently cut and burned tropical-forest clearings, soon to be abandoned in favor of newly cleared nearby forest land. Also known as *slash-and-burn agriculture*.

Sinicization Giving a Chinese cultural imprint; Chinese **acculturation**.

Site The internal locational attributes of an urban center, including its local spatial organization and physical setting.

Situation The external locational attributes of an urban center; its **relative location** or regional position with reference to other non-local places.

Social stratification See **stratification (social)**.

Spatial Pertaining to space on the Earth's surface. Synonym for *geographic(al)*.

Spatial diffusion See **diffusion**.

Spatial interaction See **complementarity**, **transferability**, and **intervening opportunity**.

Spatial model See **model**.

Spatial process See **process**.

Spatial system The components and interactions of a **functional region**, which is defined by the areal extent of those interactions. Also see **system**.

Special Economic Zone (SEZ) Manufacturing and export center within China, created in the 1980s to attract foreign investment and technology transfers. Six SEZs—all located on southern China's Pacific coast—currently operate: Shenzhen, adjacent to Hong Kong; Zhuhai; Shantou; Xiamen; Hainan Island, in the far south; and still-building Pudong, across the river from Shanghai.

Squatter settlement See **shantytown**.

State A politically organized territory that is administered by a sovereign government and is recognized by a significant portion of the international community. A state must also contain a permanent resident population, an organized economy, and a functioning internal circulation system.

State capitalism Government-controlled corporations competing under free market conditions, usually in a tightly regimented society. South Korea is a leading example. Also see **chaebol**.

State planning Involves highly centralized control of the national planning process, a hallmark of communist economic systems. Soviet central planners mainly pursued a grand political design in assigning production to particular places; their frequent disregard of **economic geography** contributed to the eventual collapse of the U.S.S.R.

State territorial morphology A **state's** geographical shape, which can have a decisive impact on its spatial cohesion and political viability. A **compact** shape is most desirable; among the less efficient shapes are those exhibited by **elongated**, **fragmented**, **perforated**, and **protruded** states.

Stateless nation A **national** group that aspires to become a **nation-state** but lacks the territorial means to do so; the Palestinians and Kurds of Southwest Asia are classic examples.

Steppe Semiarid grassland; short-grass prairie. Also the name given to the semiarid climate type (*BS*).

Stratification (social) In a layered or stratified society, the population is divided into a **hierarchy** of social classes. In an industrialized society, the working class is at the lower end; **elites** that possess capital and control the means of production are at the upper level. In the traditional **caste system** of Hindu India, the "untouchables" form the lowest class or caste, whereas the still-wealthy remnants of the princely class are at the top.

Subduction In **plate tectonics**, the **process** that occurs when an oceanic plate converges head-on with a plate carrying a continental landmass at its leading edge. The lighter continental plate overrides the denser oceanic plate and pushes it downward.

Subsequent boundary A political boundary that developed contemporaneously with the evolution of the major elements of the **cultural landscape** through which it passes.

Subsistence Existing on the minimum necessities to sustain life; spending most of one's time in pursuit of survival.

Subtropical Convergence A narrow marine **transition zone**, girdling the globe at approximately latitude 40°S, that marks the equatorward limit of the frigid Southern Ocean and the poleward limits of the warmer Atlantic, Pacific, and Indian Oceans to the north.

Suburban downtown In the United States (and increasingly in other advantaged countries), a significant concentration of major urban activities around a highly **accessible** suburban location, including retailing, light industry, and a variety of leading corporate and commercial operations. The largest are now coequal to the American central city's **central business district (CBD)**.

Sunbelt The popular name given to the southern tier of the United States, which is anchored by the mega-States of California, Texas, and Florida. Its warmer climate, superior recreational opportunities, and other amenities have been attracting large numbers of relocating people and activities since the 1960s; broader definitions of the Sunbelt also include much of the western U.S., particularly Colorado and the coastal Pacific Northwest.

Superimposed boundary A political boundary emplaced by powerful outsiders on a developed human landscape. Usually ignores pre-existing cultural-spatial patterns, such as the border that still divides North and South Korea.

Supranational A venture involving three or more **states**—political, economic, and/or cultural cooperation to promote shared objectives. The **European Union** is one such organization.

System Any group of objects or institutions and their mutual interactions. Geography treats systems that are expressed **spatially**, such as in **functional regions**.

Systematic geography Topical geography: **cultural**, **political**, **economic geography**, and the like.

Taiga The subarctic, mostly **coniferous** snowforest that blankets northern Russia and Canada south of the **tundra** that lines the Arctic shore.

Takeoff Economic concept to identify a stage in a country's **development** when conditions are set for a domestic Industrial Revolution.

Taxonomy A **system** of scientific classification.

Technopole A planned techno-industrial complex (such as California's Silicon Valley) that innovates, promotes, and manufactures the products of the **postindustrial** informational economy.

Tectonics See **plate tectonics**.

Templada See **tierra templada**.

Terracing The transformation of a hillside or mountain slope into a step-like sequence of horizontal fields for intensive cultivation (as shown on the book's cover).

Territoriality A country's or more local community's sense of property and attachment toward its territory, as expressed by its determination to keep it inviolable and strongly defended.

Territorial sea Zone of seawater adjacent to a country's coast, held to be part of the national territory and treated as a segment of the sovereign **state**.

Tertiary economic activity Activities that engage in *services*—such as transportation, banking, retailing, education, and routine office-based jobs.

Tierra caliente The lowest of the **altitudinal zones** into which the human settlement of Middle and South America is classified according to elevation. The *caliente* is the hot humid coastal plain and adjacent slopes up to 2,500 feet (750 m) above sea level. The natural vegetation is the dense and luxuriant tropical rainforest; the crops include sugar and bananas in the lower areas, and coffee, tobacco, and corn along the higher slopes.

Tierra fría The cold, high-lying **altitudinal zone** of settlement in Andean South America, extending from about 6,000 feet (1,800 m) in elevation up to nearly 12,000 feet (3,600 m). **Coniferous** trees stand here; upward they change into scrub and grassland. There are also important pastures within the *fría*, and wheat can be cultivated.

Tierra helada In Andean South America, the highest-lying habitable **altitudinal zone**—ca. 12,000 to 15,000 feet (3,600 to 4,500 m)—between the tree line (upper limit of the ***tierra fría***) and the snow line (lower limit of the ***tierra nevada***). Too cold and barren to support anything but the grazing of sheep and other hardy livestock.

Tierra nevada The highest and coldest **altitudinal zone** in Andean South America (lying above 15,000 feet [4,500 m]), an uninhabitable environment of permanent snow and ice that extends upward to the Andes' highest peaks of more than 20,000 feet (6,000 m).

Tierra templada The intermediate **altitudinal zone** of settlement in Middle and South America, lying between 2,500 feet (750 m) and 6,000 feet (1,800 m) in elevation. This is the "temperate" zone, with moderate temperatures compared to the ***tierra caliente*** below. Crops include coffee, tobacco, corn, and some wheat.

Topography The surface configuration of any segment of **natural landscape**.

Toponym Place name.

Transculturation Cultural borrowing and two-way exchanges that occur when different **cultures** of approximately equal complexity and technological level come into close contact.

Transferability The capacity to move a good from one place to another at a bearable cost; the ease with which a commodity may be transported.

Transhumance Seasonal movement of people and their livestock in search of pastures. Movement may be vertical (into highlands during the summer and back to lower elevations in winter) or horizontal, in pursuit of seasonal rainfall.

Transition zone An area of **spatial** change where the peripheries of two adjacent **realms** or **regions** join; marked by a gradual shift (rather than a sharp break) in the characteristics that distinguish these neighboring geographic entities from one another.

Transmigration The policy of the Indonesian government to induce residents of the overcrowded, **core-area** island of Jawa to move to the country's other islands.

Treaty ports **Extraterritorial enclaves** in China's coastal cities, established by European colonial invaders under unequal treaties enforced by gunboat diplomacy.

Tropical deforestation See **deforestation**.

Tropical savanna See **savanna**.

Tsunami A seismic (earthquake-generated) sea wave that can attain gigantic proportions and cause coastal devastation.

Tundra The treeless plain that lies along the Arctic shore in northernmost Russia and Canada, whose vegetation consists of mosses, lichens, and certain hardy grasses.

Turkestan Northeasternmost region of the North Africa/Southwest Asia realm. Known as Soviet Central Asia before 1992, its five (dominantly Islamic) former S.S.R.'s have become the independent countries of Kazakhstan, Uzbekistan, Turkmenistan, Kyrgyzstan, and Tajikistan. Today Turkestan has expanded to include a sixth state, Afghanistan.

Unitary state A **nation-state** that has a centralized government and administration that exercises power equally over all parts of the state.

Urbanization A term with several connotations. The proportion of a country's population living in urban places is its level of urbanization. The **process** of urbanization involves the movement to, and the clustering of, people in towns and cities—a major force in every geographic realm today. Another kind of urbanization occurs when an expanding city absorbs rural countryside and transforms it into suburbs; in the case of cities in disadvantaged countries, this also generates peripheral **shantytowns**.

Urban (metropolitan) area The entire built-up, non-rural area and its population, including the most recently constructed suburban appendages. Provides a better picture of the dimensions and population of such an area than the delimited municipality (central city) that forms its heart.

Urban realms model A spatial generalization of the contemporary large American city. It is shown to be a widely dispersed, multi-centered **metropolis** consisting of increasingly independent zones or *realms*, each focused on its own **suburban downtown**; the only exception is the shrunken central realm, which is focused on the **central business district** (see Figs. 3-11 and 3-12).

Veld See **highveld**.

Voluntary migration Population movement in which people relocate in response to perceived opportunity, not because they are forced to migrate.

Von Thünen's Isolated State model Explains the location of agricultural activities in a commercial economy. A **process** of spatial competition allocates various farming activities into concentric rings around a central market city, with profit-earning capability the determining force in how far a crop locates from the market. The original (1826) Isolated State model now applies to the continental scale (see Fig. 1-6).

Weather The immediate and short-term conditions of the **atmosphere** that impinge on daily human activities.

Wet monsoon See **monsoon**.

Windward The exposed, upwind side of a **topographic** barrier that faces the winds that flow across it.

World geographic realm See **geographic realm**.

List of Maps and Figures

INTRODUCTION

CHAPTER 1

CHAPTER 2

CHAPTER 3

CHAPTER 4

CHAPTER 5

CHAPTER 6

CHAPTER 7

CHAPTER 8

CHAPTER 9

CHAPTER 10

CHAPTER 11

CHAPTER 12

APPENDIX A

Photo Credits

Introduction
Pages 3, 10, 13, 17, 31, 32: H. J. de Blij. **Page 6:** Craig Aurness/Westlight/Corbis Images.

Chapter 1
Page 47: H. J. de Blij. **Page 48:** Mike Mazzaschi/Stock, Boston. **Page 57:** Chris Kapolka/Stone. **Page 59:** Yann Arthus-Bertrand/Photo Researchers. **Page 65:** H. J. de Blij. **Page 72 (left):** H. J. de Blij. **Page 72 (right):** Barbara A. Weightman. **Pages 76, 81, 83:** H. J. de Blij. **Page 85:** Joseph Polleros/Regina Maria Anzenberger. **Pages 91 & 97:** H. J. de Blij. **Page 101:** Sovfoto/Eastfoto. **Page 104:** AP/Wide World Photos.

Chapter 2
Pages 109 & 114: H. J. de Blij. **Page 117:** George W. Moore. **Page 119:** Novosti/Liaison Agency, Inc. **Page 122:** Sovfoto/Eastfoto. **Page 127:** AFP/Corbis. **Page 131:** Joe Traver/Liaison Agency, Inc. **Page 142:** Itar Tass/Sovfoto/Eastfoto. **Page 143:** Novosti/Liaison Agency, Inc. **Page 145:** George W. Moore. **Page 147:** Mark Newman/Photo Researchers.

Chapter 3
Page 151 (left): Robert W. Cameron/Liaison Agency, Inc. **Page 151 (right):** Daniel Wray/The Image Works. **Page 152:** David Hiser/Stone. **Page 163:** © Uniphoto, Inc. **Page 165:** Renato Rotolo/Liaison Agency, Inc. **Page 173:** © Eastcott-Momatiuk/The Image Works. **Page 183:** SUPERSTOCK. **Page 187:** Courtesy of Port of New Orleans. Photo by Henry Clay. **Page 190:** Mike Yamashita/Woodfin Camp & Associates. **Page 193:** Kathleen Campbell/Liaison Agency, Inc.

Chapter 4
Page 201 (left): Reuters/Onorio Montas/Archive Photos. **Page 201 (right):** AP/Wide World Photos. **Page 202:** H. J. de Blij. **Page 208 (top):** © J.P. Courau/DDB Stock Photos. **Page 208 (bottom):** Jean-Gerard Sidaner/Photo Researchers. **Page 209:** H. J. de Blij. **Page 214:** Wesley Bocxe-Upon/Photo Researchers. **Page 221:** © Alex McLean/Landslides. **Page 224:** Trygve Bolstad/Panos Pictures. **Page 229:** Robert Harbison/1999 *The Christian Science Monitor*. **Page 232:** © Peter Poulides. **Page 234:** H. J. de Blij.

Chapter 5
Page 240: Courtesy Philip L. Keating, Indiana University. **Pages 242 & 247:** H. J. de Blij. **Page 250:** Courtesy Ryan Flahive. **Page 252:** H. J. de Blij. **Page 256:** Delfim Martins/D. Donne Bryant Stock Photography. **Page 260:** NRSC LTD/Photo Researchers. **Page 263:** Yann-Arthus-Bertrand/Corbis Images. **Page 270:** Courtesy Philip L. Keating, Indiana University. **Pages 274 & 275:** H. J. de Blij.

Chapter 6
Page 285: Courtesy Office of National Du Tourisme Tunisien. **Pages 288, 294, 301, 305, 306, 316, 319:** H. J. de Blij. **Page 329:** Courtesy of Chevron Corporation. **Page 330:** Courtesy of Worldsat International.

Chapter 7
Page 344: © Jose Azel/Aurora. **Page 346:** H. J. de Blij. **Page 354:** M & E Bernheim/Woodfin Camp & Associates. **Page 363:** Yann Arthus-Bertrand/Liaison Agency, Inc. **Pages 367, 370, 375, 381:** H. J. de Blij.

Chapter 8
Page 391: H. J. de Blij. **Page 392:** Corbis-Bettmann. **Page 398:** Barbara A. Weightman. **Page 400:** H. J. de Blij. **Page 403:** Nickelsberg/Liaison Agency, Inc. **Pages 406, 409, 412:** H. J. de Blij. **Page 413:** Barbara A. Weightman. **Page 415:** Steve McCurry/Magnum Photos, Inc. **Page 416:** Mark Henley/Impact Photos. **Pages 423 & 425:** H. J. de Blij.

Chapter 9
Pages 431, 441, 452: H. J. de Blij. **Page 455:** Dennis Cox/China Stock Photo Library. **Pages 458 & 461:** Barbara A. Weightman. **Page 466 (top):** Chris Stowers/Panos Pictures. **Page 466 (bottom):** AP/Wide World Photos. **Page 478:** H. J. de Blij. **Page 480:** Yann Arthus-Bertrand/Photo Researchers. **Pages 484 & 489:** H. J. de Blij.

Chapter 10
Pages 497 & 503: H. J. de Blij. **Page 504:** Barbara A. Weightman. **Page 510:** H. J. de Blij. **Pages 512 & 513:** Barbara A. Weightman. **Pages 516, 519, 521, 526, 528:** H. J. de Blij. **Page 529:** Kyodo News Service.

Chapter 11
Pages 539, 544, 547, 549: H. J. de Blij. **Page 546:** Trip/Eric Smith.

Chapter 12
Pages 553, 558, 560, 562: H. J. de Blij.

Index